Plant Virus Disease Control

Edited by

A. Hadidi
Agricultural Research Service
U.S. Department of Agriculture
Beltsville, Maryland

R. K. Khetarpal
National Bureau of Plant Genetic Resources
Ministry of Agriculture
New Delhi, India

H. Koganezawa
Shikoku National Agricultural Experiment Station
Ministry of Agriculture, Forestry, and Fisheries
Zentsuji, Kagawa, Japan

APS PRESS
The American Phytopathological Society
St. Paul, Minnesota

Cover photographs by L. Giunchedi, H.-T. Hsu, P. A. Burnett, and V. D. Damsteegt

This book has been reproduced directly from computer-generated copy submitted in final form to APS Press by A. Hadidi, senior editor of the volume. No editing or proofreading has been done by the Press.

Reference in this publication to a trademark, proprietary product, or company name by personnel of the U.S. Department of Agriculture or anyone else is intended for explicit description only and does not imply approval or recommendation to the exclusion of others that may be suitable.

Library of Congress Catalog Card Number: 98-70322
International Standard Book Number: 0-89054-191-4

Printed in the United States of America on acid-free paper

The American Phytopathological Society
3340 Pilot Knob Road
St. Paul, Minnesota 55121-2097, USA

CONTRIBUTORS

V. K. Baranwal
National Center for Integrated Pest
   Management
IARI Campus
New Delhi, India

M. Barba
Per La Patologia Vegetale
Via C. G. Bertero 22
00156 Rome, Italy
Istituto Sperimentale

P. A. Burnett
Lethbridge Research Center
P. O. Box 3000, Lethbridge
Alberta T1J 4B1, Canada

T. Candresse
INRA, Station de Pathologie Végétale
Centre de Recherches de Bordeaux, B. P. 81
33883 Villenave d'Ornon, Cedex
France

B. A. Chalhoub
INRA, Unité d'Amélioration des Plantes
78026 - Versailles
France

J. Cho
Department of Plant Pathology
University of Hawaii, Maui Branch
Kula, HI 96790, USA

A. Crescenzi
Dipartimento di Biologia Difesa e
Biotecnologie Agroforestali
Università degli Studi delle Basilicata
Via N. Sauro, 85
85100 Potenza, Italy

D. M. Custer
Department of Plant Pathology
University of Hawaii, Maui Branch
Kula, HI 96790, USA

J. L. Dale
Center for Molecular Biotechnology
School of Life Science
Queensland University of Technology
Brisbane, Queensland, Australia

V. D. Damsteegt
Foreign Disease-Weed Science
   Research Unit
Agricultural Research Service
U.S. Department of Agriculture
Frederick, MD 21702, USA

P. Das
Department of Biological Sciences
Wayne State University
Detroit, MI 48202, USA

A. K. Dhar
Agriculture and Agri-Food Canada
Research Center
P. O. Box 20280
Fredricton, New Brunswick E3B 427
Canada

M. Diekmann
International Plant Genetic
Resources Institute (IPGRI)
Via delle Sette Chiese 142
00145 Rome, Italy

S. Dinant
INRA, Unité de Biologie Cellulaire
78026 - Versailles, France

C. Duby
INA-PG, Départment Mathématiques
   et Informatique
16 rue Claude Bernard
75231 - Paris, France

G. Faccioli
Istituto di Patologia Vegetale
Università di Bologna
40126 Bologna, Italy

D. Fargette
LPRC, ORSTOM/CIRAD
34032 Montpellier, France

J. A. Foster
National Plant Germplasm Quarantine Center
Plant Protection and Quarantine
Animal and Plant Health Inspection Service
U.S. Department of Agriculture
Beltsville, MD 20705, USA

A. Franz
Federal Biological Research Center
    for Agriculture and Forestry
Institute for Biochemistry
    and Plant Virology
Messeweg 11/12
38104 Braunschweig, Germany

R. S. S. Fraser
Horticulture Research International
Worthing Road, Littlehampton
West Sussex BN17 6LP, U. K.

E. A. Frison
INIBAP
95 Rue Pierre Flourens
Parc Euromédecine
Bât G. - Montpellier 34000, France

D. Gallitelli
Dipartimento Protezione
Delle Piante Dalle Malattie
Università Degli Studi Di Bari
Via Amendola, 165/A
70126 Bari, Italy

I. D. Garg
Division of Plant Pathology
Central Potato Research Institute
Shimla 171001, India

S. M. Garnsey
U.S. Horticultural Research Laboratory
Agricultural Research Service
U.S. Department of Agriculture
2120 Camden Road
Orlando, FL 32803, USA

C. Gonsalves
Department of Plant Pathology
Cornell University
Geneva, NY 14456, USA

D. Gonsalves
Department of Plant Pathology
Cornell University
Geneva, NY 14456, USA

T. R. Gottwald
U.S. Horticultural Research Laboratory
Agricultural Research Service
U.S. Department of Agriculture
2120 Camden Road
Orlando, FL 32803, USA

A. Hadidi
National Germplasm Resources
    Laboratory
Agricultural Research Service
U.S. Department of Agriculture
Beltsville, MD 20705, USA

J. Hammond
Floral and Nursery Plants
    Research Unit
U.S. National Arboretum
Agricultural Research Service
U.S. Department of Agriculture
Beltsville, MD 20705, USA

R. W. Hammond
Molecular Plant Pathology Laboratory
Agricultural Research Service
U.S. Department of Agriculture
Beltsville, MD 20705, USA

R. M. Harding
Center for Molecular Biotechnology
School of Life Science
Queensland University of Technology
Brisbane, Queensland, Australia

V. Hari
Department of Biological Sciences
Wayne State University
Detroit, MI 48202, USA

W. Hartmann
Institute of Fruit, Vegetable
    and Viticulture
University of Hohenheim
D-70599 Hohenheim, Germany

W. E. Howell
Department of Plant Pathology
Washington State University
Irrigated Agriculture Research
    and Extension Center
Prosser, WA 99350-9687, USA

E. C. K. Igwegbe
Department of Biology
Medgar Evers College
CUNY
Brooklyn, NY 11225, USA

M. Jacquemond
INRA, Station de Pathologie Végétale
84143 Montfavet Cedex, France

R. K. Jain
Division of Mycology and Plant Pathology
Indian Agricultural Research Institute
New Delhi 110012, India

U. Jayasinghe
Centro Internacional De La Papa
Apartado 1558, Lima
Peru

W. Jelkmann
Federal Biological Research Center
  for Agriculture and Forestry
Institute for Plant Protection in Fruit Crops
Schwabenheimer Str. 101
D-69221 Dossenheim, Germany

A. T. Jones
Dpartment of Virology
Scottish Crop Research Institute
Invergowrie, Dundee DD2 5DA
Scotland, U. K.

W. Kaniewski
Monsanto Company
St. Louis, MO 63198, USA

J. M. Kaper
Molecular Plant Pathology Laboratory
Agricultural Research Service
U.S. Department of Agriculture
Beltsville, MD 20705, USA

L. Katul
Federal Biological Research Center
  for Agriculture and Forestry
Institute for Biochemistry
  and Plant Virology
Messeweg 11/12
38104 Braunschweig, Germany

H. Kegler
Institute of Phytopathology
D-06449 Aschersleben, Germany

R. K. Khetarpal
National Bureau of Plant Genetic Resources
Ministry of Agriculture
New Delhi - 110012 India

S. M. P. Khurana
Division of Plant Pathology
Central Potato Research Institute
Shimla 171001, India

R. Kisimoto
Entomology Laboratory
Faculty of Bioresources
Mie University
Tsu, Mie 514, Japan

H. Koganezawa
Shikoku National Agricultural
  Experiment Station
Ministry of Agriculture, Forestry
  and Fisheries
1-3-1, Senyu, Zentsuji
Kagawa 765, Japan

G. Krczal
Staatliche Lehr- und Forschungsanstalt
  für Landwirtschaft, Weinbau
  und Gartenbau
Projekgruppe Grüne Gentechnik
Breitenweg 71, D-67435 Neustadt a. d. W.
Germany

C. Lawson
Monsanto Company
St. Louis, MO 63198, USA

H. Lecoq
INRA, Station de Pathologie Végétale
Domaine Saint Maurice, BP94
84143 - Montfavet, France

M. Madkour
Agricultural Genetic Engineering
  Research Institute (AGERI)
Agricultural Research Center
9 Gamma Street
Giza 12619, Egypt

B. Maisonneuve
INRA, Unité d'Amélioration des Plantes
78026 - Versailles, France

K. M. Makkouk
Virology Laboratory - Germplasm
  Program
The International Center for Agriculture
Research in the Dry Areas (ICARDA)
P. O. Box 5466
Aleppo, Syria

F. Marani
Istituto di Patologia Vegetale
Università di Bologna
40126 Bologna, Italy

G. P. Martelli
Dipartimento Protezione
Delle Piante Dalle Malattie
Università Degli Studi Di Bari
Via Amendola, 165/A
70126 Bari, Italy

R. R. Martin
Horticulture Crops Research Laboratory
Northwest Center for Small Fruit Research
USDA-ARS
3420 N. W. Orchard Avenue
Corvallis, OR 97330, USA

R. F. L. Mau
Department of Entomology
University of Hawaii at Manoa
Honolulu, HI 96822, USA

Y. Maury
INRA, Unité de Pathologie Végétale
78026 - Versailles, France

D. P. Maxwell
Department of Plant Pathology
University of Wisconsin
Madison, WI 53706-1598, USA

G. I. Mink
Department of Plant Pathology
Washington State University
Irrigated Agriculture Research
  and Extension Center
Prosser, WA 99350-9687, USA

F. J. Morales
Virology Research Unit
Centro Internacional de Agricultura Tropical
AA 6713, Cali, Colombia

M. K. Nakhla
Department of Plant Pathology
University of Wisconsin
Madison, WI 53706-1598, USA

L. Nemchinov
National Germplasm Resources Laboratory
Agricultural Research Service
U.S. Department of Agriculture
Beltsville, MD 20705, USA

G. W. Otim - Nape
Namulonge Agricultural and Animal
  Production Research Institute
Kampala, Uganda

S-Z. Pang
Department of Plant Pathology
Cornell University
Geneva, NY 14456, USA

P. Piazzolla
Dipartimento di Biologia Difesa
  e Biotecnologie Agroforestali
Università degli Studi delle Basilicata
Via N. Sauro, 85
85100 Potenza
Italy

R. T. Plumb
Crop and Disease Management
  Department
IACR - Rothamsted
Harpenden, Hertfordshire AL5 2JQ
United Kingdom

G. P. Rao
Sugarcane Research Station
Kunřaghat, Gorakhpur, India

D. V. R. Reddy
Crop Protection Division
ICRISAT Asia Center, Patancheru
Andhra Pradesh 502 324, India

C. N. Roistacher
Department of Plant Pathology
University of California
Riverside, CA 92521-0122, USA

L. F. Salazar
Centro Internacional De La Papa
Apartado 1558
Lima, Peru

M. K. Satapathy
Post Graduate Department of Life Science
Regional College of Education
Bhubaneswar 751007, India

H. L. Shands
Agricultural Research Service
U.S. Department of Agriculture
Washington, D. C. 20250, USA

K. R. Sigvald
Swedish University of Agricultural
  Sciences
Research Information Center
P. O. Box 7044
75007 Uppsala, Sweden

R. P. Singh
Agriculture and Agri-Food Canada
Research Center
P. O. Box 20280
Fredricton, New Brunswick E3B 427
Canada

S. A. Slack
Department of Plant Pathology
Cornell University
Ithaca, NY 14853, USA

S. Spiegel
Department of Virology
The Volcani Center
Bet Dagan 50250, Israel

S. Srivastava
Laboratory of Plant Virology
Department of Botany
Lucknow University
Lucknow 226007, India

A. K. Stoner
National Germplasm Resources Laboratory
Agricultural Research Service
U.S. Department of Agriculture
Beltsville, MD 20705, USA

M. Tabler
Foundation for Research and Technology -
   Hellas
Institute of Molecular Biology and Biotechnology
P. O. Box 1527
GR-711 10 Heraklion, Crete, Greece

M. Tepfer
INRA, Laboratoire de Biologie Cellulaire
78026 Versailles Cédex, France

J. M. Thresh
Natural Resources Institute
Chatham, Kent ME4 4TB
U. K.

P. Tien
Department of Molecular Virology and
   Biotechnology
Institute of Microbiology
Chinese Academy of Sciences
Beijing, P. R. China 100080

M. Tsagris
Department of Biology
University of Crete
GR-711 10 Heraklion, Crete, Greece

A. Varma
Division of Mycology and Plant Pathology
Indian Agricultural Research Institute
New Delhi 110012, India

G. M. Timmerman - Vaughan
New Zealand Institute for Crop
   and Food Research Ltd.
Private Bag 4704
Christchurch, New Zealand

T. Verderevskaya
Horticultural Research Institute
14 Costiujeni Street
Kishinev 277072, Moldova

H. N. Verma
Laboratory of Plant Virology
Department of Botany
Lucknow University
Lucknow 226 007, India

H. J. Vetten
Federal Biological Research Center
   for Agriculture and Forestry
Institute for Biochemistry and Plant Virology
Messeweg 11/12
38104 Braunschweig, Germany

J. A. Walsh
Plant Pathology and Weed Science Department
Horticulture Research International
Wellesbourne, Warwick, CV35 9EF, U. K.

B. Walter
INRA, Laboratoire de Pathologie Végétale
Station de Recherche Vigne et Vin
68021 Colmar, France

R. Wample
Department of Plant Pathology
Washington State University
Irrigated Agriculture Research and Extension
   Center
Prosser, WA 99350-9687, USA

M. Wang
Department of Plant Pathology
University of Hawaii at Manoa
Honolulu, HI 96822, USA

H. E. Waterworth
National Germplasm Resources Laboratory
Agricultural Research Service
U.S. Department of Agriculture
Beltsville, MD 20705, USA

J. Watterson
Petoseed Research Center
37437 State Hwy 16
Woodland, CA 95695, USA

Y. Yamada
Entomology Laboratory
Faculty of Bioresources
Mie University
Tsu, Mie 514, Japan

R. K. Yokomi
U.S. Horticultural Research Laboratory
USDA-ARS
2120 Camden Road
Orlando, FL 32803, USA

Y. Yie
Department of Molecular Virology and Biotechnology
Institute of Microbiology
Chinese Academy of Sciences
Beijing, P. R. China 100080

CONTENTS

PRESENT STATUS OF CONTROLLING ECONOMICALLY IMPORTANT PLANT VIRUSES

FOREWORD

It is generally accepted that viral crop diseases rank second only to those caused by fungi in terms of economic importance. However, unlike fungal diseases where chemical methods aimed at prevention of infection have been quite successful, control of plant virus diseases has been much more problematic. Here control strategies have consisted mainly of approaches aimed at reducing or eliminating existing sources of infection within or outside the target crop, prevention of virus transmission, or minimizing the effects of infection on the economic usefulness of the plant. Cross protection, where the presence of a mild virus strain introduced via preinoculation of a plant prevents or minimizes the effects of its subsequent infection by a severe strain, has been selectively applied with some virus diseases. While breeding of resistant or tolerant cultivars via classical genetic means has proven to be one of the more effective means of controlling plant virus infections, in those cases where attempts at finding new natural sources of resistance have remained unsuccessful solutions are still wanting. It is here that more recently developed strategies that utilize transgenic plants containing virus resistance genes (often derived from the viral genome itself), introduced via recombinant DNA techniques, show great promise.

In actual practice, where do we now stand with plant viral disease control? This book, which in terms of successes in the field represents essentially the status *quo ante* 1997, suggests that in general there has been a consolidation and relative stabilization of tried-and-true virus control strategies. For instance, it shows that rigorous implementation of phytosanitary strategies has been effective in restoring large parts of the banana industry in Australia from near-devastation by banana bunchy top disease. For the immediate future such a strategy, aided by greatly improved methods for virus diagnosis and detection, and by technologies for the *in vitro* propagation of banana that insure the availability of large quantities of reliable planting materials, still constitutes the backbone of an economically viable banana industry in that part of the world. At the present time no cultivars resistant to banana bunchy top virus have as yet been identified. Prospects for the longer term future therefore seem to largely depend upon the promise held by alternative control methods such as development of transgenic banana containing virus resistance genes.

In the example just cited, the future development of transgenic virus resistance in the target crop seems a realistic prospect because of recent progress in understanding the molecular structure of the banana bunchy top viral genome and in the development of a transformation and regeneration system for banana. As our knowledge concerning the molecular structure of plant viruses is progressing steadily, expectations for future control on a practical scale of many other economically important viruses, using transgenic resistance strategies, can be similarly realistic. However, it should also be realized that many of the contributions in this book come from parts of the world where economic circumstances are such that on the one hand damages from viral crop disease can be absorbed with relative ease, while on the other hand the financial resources to do research and to develop virus control strategies based on the latest progress in biotechnology are also relatively abundant. Where but in the developed world can one practice not only the exploratory research for viral disease control, but also afford the additional luxury to spend considerable resources on currently fashionable risk assessment research. Although introduction of molecular genetically-based virus control technologies certainly pose a theoretical risk, the development of practical methods that are entirely risk-free will probably remain an unattainable utopia. Thus it is not too surprising that few if any of these technologies has as yet found large-scale application in crop protection. Furthermore, in the future setbacks will undoubtedly be encountered following attempts at large scale-field application. Such setbacks could range from the problems caused by viral mutations that can overcome the engineered resistance to the challenges posed by natural variability that could exist among virulent virus strains in different localities, each essentially requiring tailor-made control strategies. Repeated reengineering of plants would then be necessary to adapt to such variability, or broad spectrum resistance genes would have to be found and engineered into the target crop.

One control technology currently enjoying broad practical application in some parts of the world is viral satellite-based. It follows principles akin to those that may prevail widely in nature, where perhaps the ecological balance at the virus level in certain cases is maintained via viral satellites. Because of their ability to outcompete the replication of viruses, upon which they depend, satellites essentially exercise a form of molecular parasitism, and in the process they might prevent uncontrolled virus spread. The deliberate introduction of viral satellites into plants to protect them from virus attack could be considered an extension of this principle and, since they attenuate the viral disease symptoms, its practical application in the field for crop protection a form of biological control. As of the mid 1980s in China thousands of hectares of economically important crops have been protected from the disastrous effects of cucumber mosaic virus using a satellite variant which attenuates the viral symptoms. In Japan CMV satellite technology has enjoyed limited commercial introduction since the mid 1990s for the protection of processing- as well as consumption tomatoes. As with cross protection, in practice the satellite is introduced via preinoculation of crop plants at the seedling stage. Judging from the published reports, the method has more than proven its effectiveness and has not produced any adverse effects. More recently, cucumber mosaic virus satellite genes have also been engineered into select crop species (sometimes in combination with coat protein genes) and their expression levels have proven sufficient to afford the transformed plant lines protection against virus attack in the field. Here it should be noted however that the preinoculation method offers the clear advantage of being much easier adaptable to the natural virus variation that occurs with changing localities, and that perhaps for that reason it has been the more widely applied technology thus far. In spite of its successes, and its close resemblance to natural processes, the large-scale application of satellite-based plant protection technologies in the field has also not been immune to questions relating to its possible ecological risks however. It will be interesting to see whether in the future our preoccupation with risk assessment and with fail-safe methods in the development and application of new technologies for crop protection, will suceed in also negatively impacting these highly pragmatic and successful biological control efforts in the field.

J.M. Kaper

PREFACE

For over a century, plant viruses have been known to reduce the yield and quality of horticultural, ornamental, field and vegetable crops. At present, about one thousand plant virus diseases have been recognized worldwide. For the past several decades, information on various strategies to control these diseases have been accumulated. Some viruses may be controlled by using virus-free plant material, or by incorporating virus-resistant genes in desired cultivars. However, control strategies of most plant virus diseases, especially the vector-borne ones, may involve management of the disease itself: i.e., by taking steps to exclude or minimize the virus inoculum from a geographical area, and/or by incorporating resistance in plants. In general, these strategies have met with limited success due to the spread of viruses in new areas, lack of effective certification and/or quarantine programs, lack of a source of resistance in many commercial cultivars, and/or frequent breakdowns of resistance by emergence of virulent strains of the virus. The recent advent of genetically engineered resistance, often derived from the viral genome itself, has shown great promise as one of the major strategies for controlling plant viruses. For this reason, more plant virologists than ever have been recently involved in developing and producing genetically engineered virus-resistant plants.

The scientific literature concerning conventional and molecular aspects of plant virus disease control is amassing at a tremendous pace, but unfortunately this literature is widely scattered. We have attempted in this book to compile diverse information on different strategies of controlling plant virus diseases and to present it in a comprehensive manner. Thus, this book provides a single source of valuable information on: economic losses of crops due to virus infection, key strategies of virus control which include conventional control procedures as well as biotechnological and novel methods, and the present status of controlling economically important and major plant viruses. We have made special efforts to present the recent revolutionary advances made in molecular virology and their relevance in obtaining transgenic plants resistant to plant viruses. In addition, the following topics pertinent to the central theme of the book have also been included: advancements made in plant virus and viroid detection techniques, management of plant viroids, and elimination of plant viruses by thermotherapy and by tissue culture. This book also covers plant quarantine and certification programs. The latter include control of viruses affecting seeds, potato, deciduous fruit tress, grapevines, ornamental plants, strawberries, and citrus. This information will help anyone concerned with the safe movement of plant material across international boundaries or within a country. A number of chapters deal with a specific field, vegetable, or horticultural crop of economic importance. Each of these chapters summarizes the current state of knowledge concerning a virus disease of the crop in question. Among the crops covered are corn, wheat, barley, oats, rice, sugarcane, bean, faba bean, peanut, potato, tomato, cucumber, citrus, stone fruits, banana, and cassava. These chapters highlight how the different control strategies have been exploited for the successful control of viruses and the challenge involved in their control.

The Senior Editor would like to express his gratitude to Phillip Gaush for his invaluable expertise in preparing this book for publication; Jimmie Mowder and his staff for their unlimited support and the use of their computer facilities; and Joyce Mason and Nancy Ryan for typing manuscripts.

We hope that this book will serve as an excellent practical resource that will lead to the development and application of novel technologies for plant virus disease control.

A. Hadidi
R. K. Khetarpal
H. Koganezawa

INTRODUCTION

Viruses and viroids are known to cause considerable losses in crop yield and quality of plants and plant products. Because of the insidious nature of these pathogens it is very likely that they are responsible for far greater losses than is generally recognized. Not only do diseases caused by these pathogens result in significant economic losses, they can have a serious social impact in regions or countries, i.e., rice tungro in Southeast Asia, African cassava mosaic in Africa, plum pox in Europe and the Near East, cucumber mosaic in Italy, faba bean necrotic yellows in the Near East, potato spindle tuber in China, and coconut cadang-cadang in the Philippines are among the most notable.

Since these pathogens generally cannot be controlled by physical or chemical means like most fungi and bacteria, strategies for control must involve management of the diseases by exclusion or minimizing the organism in a geographical region and/or when feasible by incorporating resistance into economically important plants. A critical first step in controlling these pathogens is being able to accurately detect their presence in plant tissue.

This book is a comprehensive source of information devoted entirely to the strategies of controlling plant diseases caused by viruses and viroids. The first 32 chapters present a wealth of basic and applied information on these causative agents including the current state-of-the-art for their detection, their control through elimination or host plant and vector resistance, and disease free certification programs. The last 18 chapters describe the current status of controlling specific economically important plant viruses. In total, the book's contents create a framework for understanding and analyzing plant diseases caused by these pathogens and devising effective control measures.

The many plant virologists from all over the world who have contributed to this valuable book are to be complimented.

H.L. Shands
A.K. Stoner

CHAPTER 1

Economic Losses Due to Plant Viruses

H.E. Waterworth and A. Hadidi

No one will dispute the statement that viruses are responsible for considerable economic losses in crop productivity on a world wide basis. This in spite of years of research and education to deal with plant viruses. But without definition of the term 'considerable' the statement does not tell us much. Furthermore, data on magnitude losses by viruses is usually limited in scope and often contradictory. By most crude comparisons with other categories of pathogens, especially the fungi, losses to virus infections are considerably less. From a comprehensive survey of diseases of crops grown in North Carolina, Main and Gurtz estimate that viruses were responsible for 20% of all losses during 1988 (60). It is not uncommon to see in almost any list of diseases of a particular crop, far fewer recognized virus diseases than those caused by fungi. Furthermore, it is not uncommon to read crop loss estimate data given by specific named fungal disease while virus losses are lumped together as "Virus Diseases", "All Other", or "Miscellaneous" diseases (5,103). One can even surmise the relative less importance of virus disease losses compared with those incited by fungi, by the amount of financial resources invested in disease control research, by the experience of diagnosticians in plant disease clinics, by counting publications in plant pathology journals, or by almost any other comparative criteria.

Yet viruses almost certainly are responsible for far greater economic losses than is recognized because the nature in which losses occur is so very different than that attributed to bacteria or fungi. Losses are often more insidious, frequently less conspicuous and therefore go unnoticed and untreated. Many viruses are latent in some of their host species yet have adverse

affects. Latent infections usually result in plants that grow more slowly, attain a smaller size, produce less fruit, alter product composition, reduce life span of perennial crops and act in various other ways that go unnoticed. Only when comparative experiments are conducted with virus free and infected plants do we become aware that losses are occurring. It is conceivable that we sustain greater losses due to chronic and widespread latent infections by hundreds of viruses in many different crop genera for the reasons described below, than by those that incite an obvious disease.

This is especially true in perennial crops and those that are vegetatively propagated. Therefore, the term virus "infections" is more appropriate than the term virus "diseases" in order to consider all losses associated with viruses. The latter term was also used by Agrios in his review of virus losses (4).

For these reasons there have been very few comprehensive surveys undertaken with loss data by many different viruses on many crops within a country or major geographic area. Data presented in some of the few comprehensive surveys are estimates, based on extrapolations or educated guesses (5). There are several obvious reasons for not having accurate loss data which will probably remain the foreseeable future. Among them are variations in losses by a particular virus in a particular crop from year to year, different degrees of loss among regions within the same year, differences in loss assessment methodologies, viral losses are often complicated with other loss factors, variations in definitions of the term 'loss', and some components of loss are virtually unmeasurable. Economic losses occur in the form of cultural measures taken to avoid, reduce, or control yield and quality losses. Except for the skilled, virus induced symptoms can be confused with disorders due to pesticide toxicities, air pollution, genetic abnormalities, combined infections with fungi, and mineral excesses or deficiencies. Add to this the resources needed to obtain wide scale accurate loss data caused by several hundred viruses in all of the major crops of a country or region and one can understand why reliable and comprehensive virus loss data is not available.

Reviews on Plant Disease Losses

Many have reviewed the issue of losses due to plant diseases in general, covering various aspects of the subject. Entire books are devoted to the subject. One published by the FAO describes loss assessment methods (18). A book edited by Teng (92) includes chapters on the rationale and concepts of crop loss assessments, measurement of disease and pathogens, modeling of crop growth and yield for loss assessment, and disease progress curves as predictors for loss equations. Other chapters deal with methods of field data collection, generating the database for disease loss modeling, methods to generate levels of epidemics in loss experiments, quantifying the relationship between disease intensity and yield loss, and on use of principal components analysis and cluster analysis in crop loss assessment.

Others have discussed the nature of crop losses (59), identification and assessment of losses (94), management of crop losses (57), crop destruction, and classification of crop losses (58). James reviewed assessments of plant diseases and losses (44) and Nutter et al. (73) described terms and concepts for crop loss and provided a glossary of defined terms. Estimates of percent loss to most of the important diseases of major US crops prior to 1965, including viruses, is the subject of a book by the U. S. Department of Agriculture (5). Main and Gurtz (60) provided estimates of losses to major crop diseases in North Carolina from 1981 through 1988 to include percent loss, dollar value, acre equivalents, control costs, etc.

A few reviews focus on losses due to plant viruses in general. Agrios (4) discussed types of losses, relationships of losses to mode of virus transmission, provided examples of losses by viruses, and economic costs of virus management. Bos (10) reviewed experimental methods to determine losses, estimation of potential and actual losses from virus diseases, and prediction of crop losses, while Gibbs and Harrison discussed ways of preventing crop losses due to viruses (32).

There are a number of virus loss reviews that are more focused; some deal with many viruses of a specific crop, and some on a specific classification/group of viruses on many crops. For example, Nemeth (72) provides a comprehensive review on losses due to viruses in pome and stone fruit crops, Jellis and Boulton (45) and van der Zaag (99) reviewed virus losses in potato production, Brakke (11) described losses to viruses in cereal crops. Duffus (26) reviewed losses associated with the 'yellowing' viruses, while Frazier et al.(28) discuss economic importance of virus diseases of small fruits.

Finally, other reviews focus on losses due to a specific virus. Among these are barley yellow dwarf virus (34), tomato spotted wilt virus (19), potato leafroll virus (45), cassava mosaic virus (95), and rice tungro virus (39).

In this book, several chapters have dealt with losses due to a specific virus or viroid. The reader is referred to these chapters for crop loss information.

Losses from most virus diseases are difficult to measure unless crops are appreciably damaged. The prevalence of a virus or the extent of its attack are not reliable guides to the amount of damage caused. Attempts have been made to establish the relationships between final yield loss and certain indicators such as incidence, severity and duration of disease, or between combination of these, but the relationships are valid only under strictly defined conditions. Farmers and government agencies are faced with the questions of how damaging viruses actually are and how yield reductions can be assessed on a farm, district, region or country, or how losses can be predicted. Loss information is essential in order to determine economic thresholds for control measures and to enable administrators to assign research priorities.

SCOPE

Two categories of loss information, experimental and naturally occurring, are prevalent in the literature. There are literally thousands of reports that mention losses to virus diseases. Some are estimates, perhaps even "guesstimates" of losses involving one or many viruses under natural conditions. Data in other reports is based upon thorough detailed surveys, in some cases involving models and/or extrapolation. In both cases the data is usually limited to a specific field, state, or region and often for a specific year or time period.

There is another large body of virus 'loss'

Table 1. Some types of direct and indirect damage associated with plant virus infections[a]

Reduction in Growth

    -yield reduction (including symptomless infection)
    -crop failure

Reduction in Vigor

    -increased sensitivity to frost and drought
    -increased predisposition of pathogens and pests

Reduction in quality or market value

    -defects of visual attraction: size, shape, color
    -reduced keeping quality
    -reduced consumer appeal: grading, taste, texture, composition
    -reduced fitness for propagation

Costs of attempting to maintain crop health

    -cultural hygiene on farm including vector control
    -to produce virus free propagation materials
    -checking propagules and commodities on export/import (quarantine programs)
    -eradication programs
    -breeding for resistance
    -research, extension and education

---

[a] From Bos (10)

literature based upon experimentation. In these instances some of the plants are inoculated in greenhouses or field plots while others are not, so that differences on growth, yield, and product quality can be directly measured and losses determined. Obviously it is not possible nor important to include all of this information into a single review. Rather, an attempt has been made to represent a broad spectrum of examples of losses among several crops, various time periods, author opinions, geographic areas, and descriptions of "epidemics".

NATURE OF ECONOMIC LOSSES

In addition to the obvious detrimental effects, reduced yields and visual product quality, virus infections often do not induce noticeable disease. Many viruses exert their effects on plants in a variety of more subtle ways some of which have been attributed to other factors (Table 1). Discussion in this section will focus on some of the unusual effects that viruses have on plants or their products. Usually these effects can be measured only by controlled experiments. The nature in which some of these losses occur is almost impossible to attach an economic value to.

Loss of plant vigor is one of the more common forms in which viruses cause losses. Without nearby virus free plants for comparison the detrimental effects of infection may go unnoticed. For example, crinkle virus reduced strawberry plant vigor (64), as did bushy dwarf virus reduce height and diameter of red raspberry canes (24). Necrotic ringspot and prune dwarf viruses caused decline in vigor in stone fruit trees (66,72), and vein yellows virus in pear trees retarded tree growth (78). Latent viruses reduced growth and number of stool bed apple trees (14,15) and apple mosaic virus significantly reduced tree growth and number of laterals produced (79). Apple varieties grafted onto virus infected M.9 rootstocks appeared normal but were 50% smaller in trials than those on virus free stocks after one year (76), or new trees were appreciably delayed in producing their first crop (13). Beet yellows virus reduced leaf size of sugarbeet plants (42), and barley yellow dwarf virus caused reduced growth of cereal plants (11). From these and many other examples that could be cited, the detrimental affects of latent virus infection becomes readily apparent only in controlled experiments.

Virus infection also reduces the productive life of crops which may be blamed on other causes. This happens in annual crops such as potatoes infected with leafroll (99), but the effect is greater in perennial crops such as orchards, certain tropical crops, hayfields, and pastures where life span is reduced by several years. Leaf curl virus reduces life span of raspberry beds (86), and latent viruses in pome, stone and citrus fruit crops render orchards prematurely unproductive usually without obvious virus like disease symptoms (53,72,78). Tomato ringspot virus in peach trees (88), cherry leafroll virus in walnut trees (68), tristeza virus in citrus (53), and fanleaf and leafroll viruses in grapevines (35) are examples of viruses that significantly reduced the productive life of perennial crops.

There are numerous reports on the adverse effects that viruses have on vegetative propagation of fruit, nut, woody ornamental, and tropical crops. Grafting and budding failures were often simply explained as 'incompatibility'. It is now known that viruses latent in the rootstocks or scion variety were responsible in many cases (16). Apples, pears and quince, and most of the stone fruit were especially affected by this problem (72). Apple mosaic virus decreased budtake by 2% to 20% among several cultivars tested in trials (79), necrotic ringspot reduced cherry budtake as much as 96% among various combinations of cultivars and rootstocks (72). Latent virus infections also have adverse effects in stoolbeds. Campbell (14) reported the number of stools produced was reduced by 3% to 40% among four of the Merton Malling rootstock series. Fanleaf virus reduced the ability to produce rooted cuttings and success in grafting grapevines (100).

Viruses are responsible for a wide range of economic losses associated with seed production. Hundreds of viruses infect seed and many are transmitted to the resulting seedlings (67). Infected seed may appear normal depending on the virus and host species. Losses also occur when seed is aborted, discolored, cracked or reduced in size, weight, or viability. Neergaard (71) described the many ways in which viruses

affects seeds of various crops. For example, germination of wheat seed is reduced by barley yellow dwarf virus (34), as is apple seed from infected trees (72), and lettuce seed infected with mosaic virus (101). Only 8% of plum seed from Krikon stem necrosis virus infected trees germinated. Wheat striate mosaic virus was responsible for lighter kernels of wheat (6). Seed is smaller when wheat and barley are infected with BYDV(31) or when soybeans are infected with mosaic virus (71). Latent viruses in apple, cherry or peach trees reduced seed production (72) and in raspberry reduced seed yield by reducing flowering or aborting pollen (24).

Viruses also reduce yield in a number of unusual ways by affecting growth habit of their host crops. Viruses in strawberry reduced runner production (64), pox virus in plums caused 40% to 100% of fruit to drop before maturity among several cultivars (72), as did mosaic virus in apple trees (78). This virus also was responsible for less branching in young apple trees (79). Leafroll virus reduced the number of stems in potato plants (45,99) while paracrinkle virus produced tubers of less uniform size (32). Cocksfoot plants with streak virus grow taller but with fewer tillers than virus free plants, begin growing earlier in the spring and shade the more tillered adjacent plants. Therefore, infected plants compete more successfully with virus free adjacent plants which results in yield loss in hay crops (17).

The adverse effects of viruses in combination with other kinds of pathogens can be more than simply additive. There are many reports where latent virus infections exacerbate production problems caused by fungal pathogens, insects or weather conditions (10). For example, pea enation mosaic virus reduced yield of broad bean seed by 6%, its aphid vector alone reduced yield by 50%, but both reduced yield by 93% (10). Barley yellow dwarf virus enhanced mildew in oats, ergot disease in wheat (11) and, with take all fungus, had a negative effect on wheat yields (90) Wheat streak mosaic infection was associated with an increase in *Helminthosporium sativum* in wheat (3). Certain viruses in potato plants caused greater susceptibility to *Verticillium dahliae*, late blight (45), and blackleg diseases (77). *Alternaria* was more severe on sugar beets infected with beet western yellows virus (54). Apple trees with latent viruses were more susceptible to collar rot disease (15,72) as were Malling 9 rootstocks more severely infected by cytospora canker (7).

Virus infections can increase plant susceptibility to low temperatures. Leafroll virus in grapevines made them more susceptible to spring frosts (35), and wheat with BYMV did not survive the winter of 1977 in parts of the midwest (21). When growing under heat or drought stress, fanleaf infected vines were killed (100). Pome and stone fruit crops are more susceptible to frost damage when trees are infected with viruses (72).

Viruses are also responsible for some unusual effects on crops such as greater nutrient requirements of fruit trees, and reduced coloration in grapes which adversely affect fresh market sales and red wine color (35,72,95). Sugar content can also be reduced by viruses (95). Necrotic ringspot virus in peach trees (72), and beet western yellows virus each reduced sugar content of the harvest (91). Leafroll virus in grapevines reduced sugar content which in turn affected raisin production to the extent that some crops were not harvested. Certain viruses in cereal crops reduced milling qualities (6,34). Other examples could be given on some of the unusual effects that viruses have on plants or the quality of their harvest that affect economics of production.

SOME UNSCIENTIFIC WAYS IN WHICH ECONOMIC LOSSES DUE TO VIRUSES HAVE BEEN REPORTED

Technology or resources or both have not been available to obtain accurate virus crop infection loss data whether it be by large geographic areas, by many viruses that affect a particular crop, or by a single virus that affects many crops. This is especially true regarding tropical crops and those grown in some of the third world nations. As a result, the literature abounds with opinions regarding magnitude of losses, various interpretations of what constitutes an epidemic, and predictions regarding losses under vaguely described conditions. So this section is devoted to describing how destructive selected virus diseases have been, and with some virus/crop combinations, could occur again. Adjectives and quotations are given because in most cases hard data are not available.

Curly top of sugar beet and swollen shoot of cacao have had "particularly disastrous effects" according to Bos (10). "The former nearly ruined the sugar beet industry in the western United States towards the end of the 1920s, the latter killed millions of cacao trees in West Africa and caused numerous trees (which rose to 15 million a year) to be eradicated...." (10). In his review on

yellows viruses, Duffus stated that "virus diseases with special affinities to the phloem (the yellowing diseases)..... results in tremendous economic losses" (26). "Virus yellows on sugar beets reached epidemic proportions in the 1950s and was the most damaging sugar beet disease in the United States". It "maintained epidemic proportions in California from its discovery in 1951 until 1968" and was "one of the most serious diseases of any crop in northern Europe" (26). With regard to other members of the yellows group Duffus states that potato leaf roll virus "ranked with late blight and verticillium wilt as a cause of losses on potatoes, and that soybean dwarf virus is the "cause of serious losses in Japan". "Losses due to viruses exceed those caused by insects and nematodes" in potatoes in developing countries according to Page (74).

Thurston (97) described 19 important plant diseases which if established in additional parts of the world, would be especially destructive. Among these are four virus diseases: cassava mosaic, streak disease of maize, hoja blanca of rice, and swollen shoot of cacao. He indicated that "losses in Ghana have been devastating" regarding swollen shoot disease and that hoja blanca disease of rice caused "many fields in Columbia (to be) completely lost" in 1958. Thurston also stated that streak disease "on maize causes serious crop losses" and that cassava mosaic disease "seriously affects the food supply of millions of Africans". Thresh (96) considers cassava mosaic to be "the most damaging vector-borne pathogen of any African crop in a recent economic assessment".

Tristeza virus disease of citrus "has been one of the most destructive diseases of citrus worldwide" according to Lee and Rocha-Pena (53) and is "an important economic problem" according to Lister and Bar-Joseph (55). Tomato bushy stunt virus is an "economic threat to tomato crops grown for export" in Morocco and that "extraordinary outbreaks" occurred in 1978-1979 (61). Tomato spotted wilt virus "seriously affects production of food and ornamental crops worldwide" according to Cho et al. (19). Rice yellow mottle "is of considerable economic importance in W. Kenya...."(83). Rockow (81) states that luteoviruses "are of great economic importance" and that "the supply of virtually all grains throughout the world is affected by barley yellow dwarf virus". NEPO viruses "can have devastating effect on crops in areas where they occur, (that they) "kill plants in large areas of the crop or make it worthless" and can "result in total loss of crop in some fields (69). Among the rhabdoviruses Francki *et al.* (29) stated that maize mosaic "causes severe losses of corn in Venezuela", that beet leaf curl virus "has consistently caused serious crop losses and that lettuce necrotic yellows virus "causes a serious disease of lettuce crops in South Australia" (29).

Of other virus diseases, Abbott stated that sugarcane mosaic caused "enormous economic losses in the 1890s in practically every cane growing country" (2). Plum pox virus is "the most dangerous of the virus diseases of stone fruits" (72) and that it is "an extremely dangerous disease of plum trees" (84). Cotton leaf crumple virus "has become epidemic" since 1980 in Arizona (50). Diseases of rice caused by tungro virus, grassy stunt and ragged stunt viruses "are major diseases occurring in the Asian tropics" (93). Cucumber mosaic virus is responsible for "many important crop diseases" in several areas of the world (49). In Japan tomato "production areas have been shifted to escape virus infection. A large number of farmers have stopped cultivating tomatoes, and in some areas tomato production has been terminated" (82). In China the virus "has been a factor in limiting tobacco production for years" (82). Further examples of virus disease epidemics are described by Klinkowsk (51) and others. So much for "hard data" on economic losses due to virus infections.

EXAMPLES OF NATURAL LOSSES AND LOSS OF POTENTIAL YIELD

It is with some reluctance that this section is included for many reasons. Although there are hundreds of reports that include loss values, usually expressed as percent rather than in monetary terms, the data in many reports is of limited value. Too often the only data available is decades old meaning that some is of little value because, for example resistant varieties are now grown. In some cases the data is for a single year, or for an affected field or "area", a particular outbreak, and in others, the procedures employed to arrive at the loss values are not given. Definition of the term "loss", if provided at all, varies among reports. Is it loss of potential yield, losses to growers, to processors or loss to consumers?

In spite of the foregoing comments an attempt has been made to present examples of loss information taken at random from many reports. This data represents several decades, countries or

regions, time frames, and various data collection methods. In instances where the data were complex, in that it included different values from different fields in the same year, or varied from year to year within a field, they have been simplified to fit table format. **Usually only the largest of a range of loss values is shown.** In some reports the information is given only as losses "up to X%" without the lower values. **For these reasons the information in Table 2 is probably biased.** An attempt has been made to show some of the parameters on which the loss data is based. Some of the data is 30 to 50 years old and probably does not reflect present day losses. But if for no other reason, may reflect progress made in controlling some virus diseases.

Among the more thorough surveys of natural losses caused by virus infections are those that were conducted annually in North Carolina and data published in a series of reports (60). Most of the agronomic, vegetable, woody ornamental, fruits and nut, floral and bedding, turfgrass and forest tree crops were surveyed from 1981 through 1988 by persons with knowledge of the disease problems on the respective crops. For 1988 the results are provided in 111 tables and 27 graphs. The results are ranked in several formats: for all diseases by crop, (e.g. corn), value of loss by disease name (soybean mosaic) and pathogen category (virus), and by type of plant damage (decay). Data is also ranked by value of the loss by disease(e.g. tobacco mosaic, early blight), percent loss by pathogen category during 1988 and variations in losses by all viruses during the 8 year period of the survey. Among seven categories, viruses ranked fourth in causing losses following fungi, abiotic causes (frost, hail, drought, air pollution, toxic chemicals) and nematodes. Lesser losses were attributed to bacteria, physiological disorders, and undetermined causes. Viruses were responsible for 24% as much loss as fungi during the eight year period. Crop losses by symptom category ranged from 20.59% for "viruses" with value loss of $53.2 million to 0.01% for red crown rot and value of $34,721. One could reasonably assume that if such surveys were made in other states or countries similar types of loss data would be obtained. Figure 1 shows a comparison of % of total loss caused by fungi, nematodes and viruses from 1981 through 1988.

Much of the published loss data is in fact potential yield not realized. It is usually based on yields whether or not obvious disease was present.

Table 2. Examples of natural losses attributed to virus infections

Crop	Virus	Yield Reduction	Reported Parameters	Reference
abaca	mosaic	60%	when 80% plants infected in regions of Philippines	80
apple	mosaic	30% -40%	in infected trees w/symptoms	78
	mosaic	30%	in mature trees	84
banana	bunchy top	5000 Ac	abandoned in NSW before 1927	97
broadbean	pea enation	50%	of crop in affected regions	41
cacao	swollen shoot	50%	of mature trees on 250K acres in Ghana before 1954	97
cassava	mosaic	11%	of crop in Africa before 1956	97
	mosaic	5% -69%	among ecological zones in Africa before 1970	36
cherry	little cherry	25%-60%	reduced fruit size in region of B.C. 1947 -1962	72
	'viruses'	30%	of mature trees in English orchard	84
	yellows	50%	of infected trees before 1953	23
cotton	leaf crumple	80%	in some fields in Arizona before 1985	50
grapevine	leafroll	5%	in California until 1970	35
	leafroll	8%	in USA during 1950s	26
lettuce	tomato spotted wilt	50% -90%	in Oahu, HI in 1940s	19
	beet yellow stunt	50% -85%	in affected fields before 1972	26
	necrotic yellows	50%	of crop in south Australia before 1963	89
melons	zuccini yellow mosaic & watermelon mosaic	40% -50%	in desert valleys of southern CA during 1984	70
orange	tristeza	10K trees	in 3 provinces in Spain before 1988	53
pangola grass	stunting	50% -60%	on 2800 Ac in Suriname & Guyana before 1965	97
peach	mosaic	200K trees	in southwest USA before 1953	23
peanut	groundnut rosette	560,000 tons	in Nigeria in 1975	1
plum	pox	83%	in regions of former Czechoslovakia before 1965	72
potato	'mild mosaic viruses'	18%	in Maine with secondary PVX infections before 1951	45
	'viruses'	2%	of crop in UK in 1980	102
	PVX	30 bls/Ac	in field w/50% plants infected	9
	leafroll	3%	in USA in the 1950s	26
rice	tungro	1%	in Malaysia 1981-1984	93
	tungro	100%	in Indonesia 1969-1971	93
	hoja blanca	50%	entire crop Venezuela 1956	97
soybean	'viruses'	0.76%	Southern USA average for 1974-1994	103
	'viruses'	0.19%	Southern USA average for 1984-1994	103
spinach	beet western	2%	in USA in the 1950s	26
sugarbeet	yellows	6%	in USA in the 1950s	26
	yellows	5%	in England 1970 -1973	40
	yellow vein	9%	in Great Britain 1946-1962	26
	yellow vein	61%	in Sweden before 1949	26
	yellow vein	22%	in central California before 1960	26

Table 2. (Continued)

tomato	spotted wilt	75% -1C0%	in Oahu, Hawaii in 1940s	19
	cucumber mosaic	30% -6C%	in PRC before 1990	98
walnut	cherry leafroll	10%	of California crop in recent years	personal communication
wheat	barley yellow dwarf (BYDV)	10bu/Ae	in Judith Valley Montana in 1980	31
	streak mosaic	7%	of crop in Kansas 1947	65
	streak mosaic	18%	of crop in Alberta, Canada in 1963	6
	streak mosaic	1.8%	of crop in Kansas, 1976 -1983	11
	barley yellow dwarf	7%	south Manitoba, Canada in 1978	34
	barley yellow dwarf	25% & 60%	of crop in New Zeland in the 1960s & crop in Chile in 1979	11
	soil borne wheat mosaic	20%	of crop in areas of 13 fields in NB, 1972 -1974	75
	soil borne wheat mosaic	2.9%	of crop in Kansas, 1976 -1983	11
	spindle streak	40%	of crop in areas of fields in Ontario before 1970	11

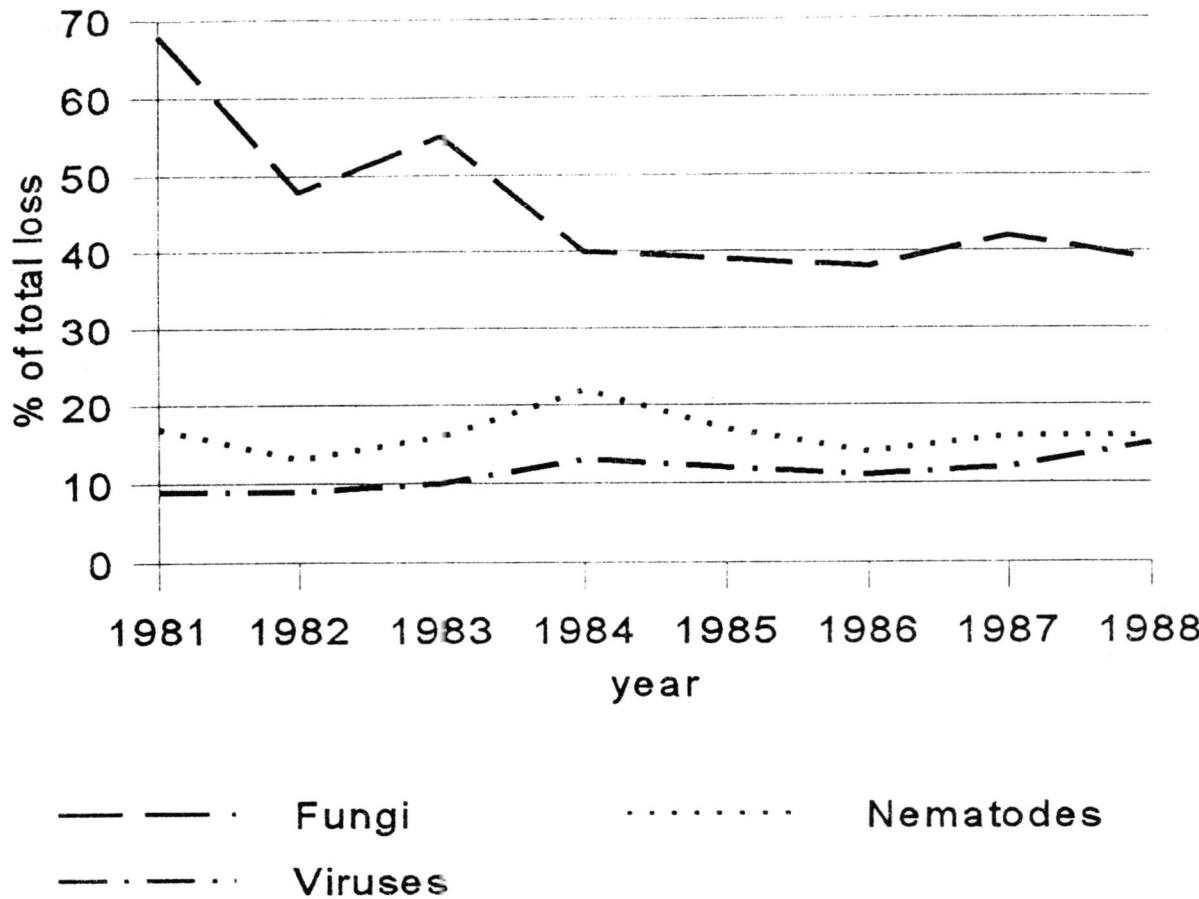

Figure 1. Percent loss among all crops caused by fungi, nematodes and viruses from 1981 to 1988 in North Carolina. Remaining losses were due to abiotic factors, bacteria, physiological or undetermined causes (5% to 37%). (From Main and Gurtz , 60)

Table 3. Examples of yield increases among crops when virus was not present in greenhouse or field trials

Crop	Virus	Yield Increase	Parameters	Reference
apple	3 latents	27%	of MM 104 rootstocks produced	14
	russet ring	60%	of the number of boxes graded "Extra Fancy"	87
barley	barley yellow dwarf	45% & 65%	in 2 row & 6 row vars in Montana	31
	bean yellow mosaic	85%	ditto	48
	cucumber mosaic	87%	ditto	48
	pea leafroll	94%	ditto	48
cotton	leaf crumple	27%	no. bolls from plants inoculated young & set in field	12
grape	'viruses'	300%	cv Sauvignon in 1975 New Zeland trial, 1995	95
lettuce	beet western yellows	44%	seed yields in greenhouse trials	
	beet western yellows	31%	in cv "Little Gem" in England	101
	mosaic	85%	ditto	101
	mixed infection	46% -84%	ditto	101
pepper	cucumber mosaic	11 -56%	among 3 sites, 5 year trials	98
potato	M	10%	in cv 'King Edward' in England	45
	S & X	25%	marketable yield, cv 'Netted Gem'	45
	leafroll	65% -92%	plots planted with infected tubers	38
	leafroll & Y	29% -83%	repeated planting infected tubers	45
	S & X	11% -38%	in 2 cvs in OR, CA,& B.C	104
	S & X	5 -15%	where all plants secondarily infected	99
rice	2 tungro viruses	12 -90%	among 7 vars inoculated 1 week after soaking	93
	2 tungro viruses	20 & 90%	among 9 vars infected with one or two viruses	39
	ragged stunt	22 -100%	depend on inoculation time & among 6 cvs	93
strawberry	veinbanding & latent 'C'	88%	by 3rd yr after inoculation	8
	3 viruses	19%	one yr after infection	62
sweet potato	virus complex	25 -35%	among 4 cvs in 1978 trials	20
tomato	cucumber mosaic	19% -52%	among 6 sites in 1987 trial	98
wheat	barley yellow dwarf	40% -90%	among plants of susceptible vars inoculated early	11
	barley yellow dwarf	29% -43%	when cv 'Nesma' was inoculated at stages 1-2 vs stages 3-4	27
	soil borne wheat mosaic	0% -80%	among 53 cvs, 1952 trial	11

Table 4. Some useful methods to prevent or reduce detrimental effects of virus infections

Objective	Methods	Example crops/ where employed	Reference
Exclude virus from states/countries	Quarantines & inspections	worldwide, on many genera	49
Exclude virus from planting site	Plant virus free plants	vegetatively propagated crops	53,72,74,78,85,95
	Sow virus free seed	lettuce,beans,peas,barley,tomatoes	63,67,101
Prevent infection	Grow cvs resistan to virus or vectors	potato,wheat,rice,lettuce,maize	11,52,63,74,97,101, 102
	Cross protection with mild strains	citrus	37,53
	Protect plants with competing satellites	CMV in tobacco,pepper,tomato	82,98
	Grow transgenic cvs	potato,citrus,vegetables	19,33,53,56,98
Reduce sources of inoculum	Eliminate crop reuse	cotton,sugarbeet,potato,papaya	12,25,52,54,63,85
	Eliminate alternate hosts & weeds	maize,wheat	22,52
	Rouge infected plants	cassava,fruit trees	52
	Wholesale eradication programs	swollen shoot-Ghana	43
		citrus tristeza-Israel	43,53
		banana -Australia	97
	Sanitation	citrus	52,53
Avoid virus by cultural methods	Alter planting date	cereal viruses	11,31
	Alter plant spacing	sugarbeet, peanut,vegetables	1,46,52,63
	Isolate field	vegetables	52,63
Reduce spread of viruses	Control insect, mite & nematode vectors	sugarbeet/yellows	41,63
		vegetables/TSWV	19
	Reduce cultivation	vegetables/TSWV	19
	Use physical barriers	tobacco/TSWV	52
Reduce impact of infection	Grow tolerant cvs	wheat,cocoa,	22,97
		citrus, peanut	52,53
	Selected rootstocks	citrus	53

Some of the key parameters are also given.

SOME STRATEGIES TO REDUCE ECONOMIC LOSSES

Fungicides when applied to crop plants protect them or reduce invasion by fungal pathogens and is a widely practiced means to manage diseases. Unfortunately, there is no such direct method yet to control diseases or latent infections caused by viruses. Most of the useful procedures are designed to reduce sources of infection within or from outside the crop, to limit spread by vectors including persons, and to minimize the effect of infection on yield (Table 4). Usually available procedures do not offer permanent solutions to a virus disease problem in a particular area. Control is often a continuing process in which organization of control procedures, care by growers and cooperation among them is necessary year by year. The primary exception is where a source of resistance to a virus has been found and incorporated into an agriculturally acceptable variety.

Strategies to reduce losses by a broad spectrum of plant viruses is the subject of many thorough reviews. Among them are book chapters by Agrios (4), Gibbs and Harrison (32), Garrett (30) and Matthews (63). Hull and Davies outline nonconventional approaches to control plant virus diseases (43) while Gibbs *et al.* describe future virus control strategies (33). Finally, a series of Compendia, each on a specific crop or group of related crops and published by the American Phytopathology Society provide information on control strategies for each of many viruses.

REFERENCES

1. A'Brook, J. 1964. The effect of planting date and spacing on the incidence of groundnut rosette disease and of the vector *Aphis craccivore*, Koch at Mokawa, Northern Nigeria. Ann. Appl. Biol. 54: 199-208.

2. Abbott, E.V. 1953. Sugarcane and its diseases. Pages 526-535 in: Plant Diseases, Yrbk. of Agric. for 1953. US Dept Agric, Agric Res. Serv.

3. Adlakha, K.L., and Raychaudhuri, S.P. 1975. Interaction between *Hilminthosporium sativum* and wheat mosaic streak virus. Zeit. Pflanz. Pflanzenschutz. 82: 201-206.

4. Agrios, G.N. 1990. Economic considerations. Pages 1-22 in: Plant Viruses, Vol. II, Pathology. C.L. Mandahar, ed. CRC Press.

5. Annonymous. 1965. Losses in Agriculture. US Dept Agric. Agr Res. Service Hdbk No 291. 120 Pages.

6. Atkinson, T.G., and Grant, M. N. 1967. An evaluation of streak mosaic losses in winter wheat. Phytopathology 57: 188-192.

7. Baumann, G. 1980. Investigations on virus-indexed and virus-free plants of different clones of M.9 apple rootstock. Acta. Hort. 114: 171-184.

8. Bolton, A.T. 1974. Effects of three virus diseases and their combinations on fruit yield of strawberries. Can. J. Plant Sci. 54: 271-275.

9. Bonde, R. 1954. Potato X virus causes large loss; better seed is answer. Maine Farm Res. 2 (2): 10-12.

10. Bos, L. 1982. Crops losses caused by viruses. Crop Protection 1: 263-282.

11. Brakke, M.K. 1987. Virus diseases of wheat. Agronomy 13: 585-624.

12. Butler, G.D. Jr., Brown, J.K., and Henneberry, T.J. 1986. Effect of cotton seedling infection by cotton leaf crumple virus on subsequent growth and yield. J. Econ. Ent. 79:208-211.

13. Campbell, A.I. 1963. The effect of some latent virus infections on the growth and cropping of apples. J. Hort. Sci. 38: 15-19.

14. Campbell, A.I. 1965. The effect of some apple viruses on clonal rootstock production. Zast. bilja No 85-88: 261-265.

15. Campbell, A.I. 1969. The effect of some apple viruses on the susceptibility of two clonal rootstocks to collar rot caused by *Phytophthora cactorum*. J. Hort. Sci. 44: 69-73.

16. Campbell, A.I. 1980. The effects of viruses on the growth, yield and quality of three apple cultivars on healthy and infected clones of four rootstocks. Acta. Hortic. 114: 185-191.

17. Catherall, P.L., and Griffiths, E. 1966. Influence of cocksfoot streak virus on the growth of cocksfoot swards. Ann. Appl. Biol. 57: 149-155.

18. Chiarappa, L., ed. 1971. Crop loss assessment methods. FAO Manual on the Evaluation and Prevention of Losses by Pests, Disease, and Weeds. Alden Press, Oxford.

19. Cho, J.J., Mau, R.F.L., German, T.., Hartmann, R. W., and Yudin, L. S. 1989. A multidisciplinary approach to management of tomato spotted wilt virus in Hawaii. Plant Dis. 73: 375-383.

20. Chung, M.L., Hsu, Y.H., Chen, M.J., and Chiu, R.J. 1986. Virus diseases of sweet potato in Taiwan. Pages 84-90 in: Plant Virus Diseases of Hort. Crops. Taipei.

21. Cisar, G., Brown, C.M., and Jedlinski, H. 1982. Effect of fall or spring infection and sources of tolerance of barley yellow dwarf of winter wheat. Crop Sci. 22: 474-478.

22. Clement, D.L., Lister, R.M., and Foster, J.E. 1986. ELISA based studies on the ecology and epidemiology of barley yellows dwarf virus in Indiana. Phytopathology 76: 86-92.

23. Cochran, L.C., and Reeves, E.L. 1953. Virus diseases of stone fruits. Pages 714-721 in: Plant Diseases, Yrbk. of Agriculture for 1953. US Dept Agric.

24. Converse, R.H. 1985. Latent viruses: harmful or harmless? HortScience 20: 845-848.

25. deBokx, J.A., ed. 1972. Viruses of Potatoes and Seed-Potato Production. Pudoc Press, Wageningen.

26. Duffus, J.E. 1977. Aphids, viruses, and the yellow plague. Pages 361-383 in: Aphids as Virus Vectors. K. F. Harris and K. Maramorosch, eds. Academic Press.

27. El-Yamani, M., and Hill, J.H. 1990. Identification and importance of barley yellow dwarf virus in

Morocco. Plant Dis. 74: 291-294.

28.   Fraizer, N.W., and Mellor, F.C. 1970 Strawberry crinkle. Pages 18-23 in: Virus Diseases of Small Fruits and Grapevines. N. W. Fraizer, ed. Univ. of Calif. Press, Berkeley.

29.   Francki, R.I.B., Kitajima, E.W., and Peters, D. 1981. Rhabdoviruses. Pages 455-489 in Plant Virus Infections Comparative Diagnosis. E. Kurstak, ed. Elsevier/North Holland Press.

30.   Garrett, R.G. 1986. Prologue: A basis for control. Pages 1-9 in: Plant Virus Epidemics, Monitoring, Modelling and Predicting Outbreaks. G D. McLean, R.G. Garrett, and W.G. Ruesink, eds. Academic Press.

31.   Geske, S., and Riesselman, J. 1988. Earley yellow dwarf and wheat streak mosaic in small grains. Montguide. MT State Univ. Ext. Service Publ. 8802. 3 Pages.

32.   Gibbs, A., and Harrison, B. 1976 Ways of preventing crop losses. Pages 219-221 in: Plant Virology, The Principles. Halsted Press.

33.   Gibbs, A., Skotnicki, A., and Skotnick, M. 1986. Plant virus control strategies: Future prospects. Pages 513-523 in: Plant Virus Epidemics, Monitoring, Modelling and Predicting Outbreaks. G. D. McLean, R.G. Garrett, and W G. Ruesink, eds. Academic Press.

34.   Gill, C.C. 1980. Assessment of losses on spring wheat naturally infected with barley yellow dwarf virus. Plant Dis. 64: 197-203.

35.   Goheen, A.C. 1970. Grape leafroll. Pages 209-212 in: Virus Diseases of Small Fruits and Grapevines. N. W. Fraizer, ed. Univ. of Calif. Press, Berkeley.

36.   Hahn, S.K., Ikotun, T., Theberge R.L., and Swennen, R. 1989. Major economic diseases of cassava, plantain and cooking/starchy bananas in Africa. Trop. Agr. Res. Series No. 22: 106-112.

37.   Hamilton, R.I. 1980. Defenses riggered by previous invaders: Viruses. Pages 279-303 in: Plant Disease, An Advanced Treatise. J.G. Horsfall and E.B. Cowling, eds. Academic Press, N Y.

38.   Harper, F.R., Nelson, G.A., and Pittman, U.J. 1975. Relationship between leaf roll symptoms and yield in Netted Gem potato. Phytopathology 65: 1242-1244.

39.   Hasanuddin, A., and Hibino, H. 1989 Grain yield reduction, growth retardation, and virus concentration in rice plants infected with tungo-associated viruses. Trop. Agr. Res. Series No. 22: 56-73.

40.   Heathcote, G.D. 1978. Review of losses caused by virus yellows in English sugar beet crops and the cost of partial control with insecticides. Plant Path 27: 12-17.

41.   Heathcote, G.D, and Gibbs, A.J. 962. Plant Pathol. 11: 69-73.

42.   Hull, R. 1968. The effect of infection with beet yellows virus on the growth of sugarbeet. Am. Soc. Sugarbeet Techol. 15: 192-199.

43.   Hull, R. and Davies, J.W. 1992. Approaches to nonconventional control of plant virus diseases. Critical Revs in Plt Sci. 11: 17-33.

44.   James, W.C. 1974. Assessment of plant diseases and losses. Annu. Rev. Phytopath. 12: 27-48.

45.   Jellis, G.J., and Boulton, R.E. 1984. Damage and loss caused by potato diseases. Pages 255-266 in: Plant Disease: Infection, Damage and Loss. R.K.S. Wood and G.J. Jellis, eds. Blackwell Press.

46.   Johnstone, G.R., Koen, T.B., and Conley, H.L.

47.   1982. Incidence of yellows in sugar beet as affected by variation in plant density and arrangement. Bull. of Entomol. Res. 72: 289-294.

47.   Kahn, R.P., Waterworth, H.E., Gillaspie, A.G., and Foster, J.A. 1979. Detection of viruses or virus-like agents in vegetatively propagated plant importations under quarantine in the United States, 1968-1978. Plant Dis. Rptr. 63: 775-779.

48.   Kaiser, W.J., and Danesh, D. 1971. Biology of four viruses affecting *Cicer arietinum* in Iran. Phytopathology 61: 372-375.

49.   Kaper, J. M. 1993. Satellite-mediated symptom modulation: an emerging technology for the biological control of viral crop disease. Microb Releases 2: 1-9.

50.   Kingdon, L. 1985. Cotton leaf crumple causes yield loss. Ariz Farmer-Stockman 64 (April): 41.

51.   Klinkowski, M. 1960. Die wirtschaftliche bedeutung pflanzlicher virosen. Angewandte Botank 34: 165-178.

52.   Kranz, J., Schmutterer, H. and Koch, W. 1977. Pages 3-42 in: Diseases Pests and Weeds in Tropical Crops. Wiley Press, NY

53.   Lee, R.F., and Rocha-Pena. 1992. Citrus tristeza virus. Pages 226-249 in: Plant Diseases of International Importance, Vol III. J. Kumar, H.S. Chaube, U S. Singh, and A.N. Mukhopadhyay, eds. Prentice Hall, NJ.

54.   Lewellen, R.T., and Skoyen, I.O. 1984. Beet western yellows can cause heavy losses in sugarbeet. Calif. Agric 38 (Jan-Feb): 4-5

55.   Lister, R.M., and Bar-Joseph, M. 1981. Closteroviruses. Pages 809-844 in: Plant Virus Infections Comparative Diagnosis. E. Kurstak, ed. Elsevier/North Holland Press.

56.   Love, J.M., and Tauer, L.W. 1988. Biotechnology and the economics of reducing viral disease losses in U. S. potato and tomato production. Appl. Agric. Res. 3: 187-194.

57.   MacKenzie, D.R. 1983. Toward the management of crop losses. Pages 82-92 in: Challenging problems in plant health. T. Kommedahl and P. H. Williams, eds. Am. Phytopath. Soc. Press., St. Paul.

58.   Main, C.E. 1977. Crop destruction--The raison d'etre of plant pathology. Pages 55-78 in: Plant Disease An Advanced Treatise, Vol I. J.G. Horsfall and E.B. Cowling. eds. Academic Press.

59.   Main, C.E. 1983. Nature of crop losses: An overview. Pages 61-68 in: Challenging Problems in Plant Health. T. Kommedahl and P. Williams, eds. Am. Phytopathol. Soc. Press, St. Paul.

60.   Main, C.E., and Gurtz, S.K. 1989. 1988. Estimates of crop losses in North Carolina due to plants diseases and nematodes. N.C. State Univ. Spl. Publ. No. 8. 209 Pages.

61.   Martelli, G.P. 1981. Tombusviruses. Pages 61-90 in: Plant Virus Infections Comparative Diagnosis. E. Kurstak, ed. Elsevier/North Holland Press.

62.   Martin, L.W., and Converse, R.H. 1977. Influence of recent and chronic virus infections on strawberry growth and yield. Phytopathology 67: 573-575.

63.   Matthews, R.E.F. 1991. Economic importance and control. Pages 591-634 in: Plant Virology, 3rd Ed. Academic Press.

64.   McGrew, J.R. 1970. Strawberry Latent C. Pages 16-18 in: Virus Diseases of Small Fruits and Grapevines. Univ of Calif Press, Berkeley.

65.   McKinney, H.H. 1953. Virus diseases of cereal crops. Pages 350-360 in: Plant Diseases, Yrbk. of

Agric. for 1953. USDA

66.   Millikan, D.F. 1955. The influence of infection by ring spot virus upon the growth of one year old Montmorency nursery trees. Phytopathology 45: 565-566.

67.   Mink, G.I. 1993. Pollen and seed transmitted viruses and viroids. Annu. Rev. Phytopathol. 31: 375-402.

68.   Mircetich, S.M., and Rowhani, A. 1984. The relationship of cherry leafroll virus and blackline disease of English walnut trees. Phytopathology 74: 423-428.

69.   Murant, A.F. 1981. Nepoviruses. Pages 197-238 in: Plant Virus Infections Comparative Diagnosis. E. Kurstak, ed. Elsevier/North Holland Press.

70.   Nameth, S.T., Laemmlen, F.F., and Dodds, J.A. 1985. Viruses cause heavy melon losses in desert valleys. Calif. Agric. July/Aug: 28-29.

71.   Neergaard, P. 1977. Seed borne viruses. Pages 71-117 in: Seed Pathology, Vol I. P. Neergaard, ed. Halsted Press.

72.   Nemeth, M. 1986. Virus, mycoplasma and rickettsia diseases of fruit trees. Martinus Nijhoff Publ. 841 Pages.

73.   Nutter, F.W., Teng, P.S., and Royer, M.H. 1993. Terms and concepts for yield, crop loss, and disease thresholds. Plant Dis. 77: 211-215.

74.   Page, O.T. 1994. Potato disease research in developing countries. Canad. J. of Plant Pathol. 16: 135-138.

75.   Palmer, L.T., and Brakke, M.K. 1975. yield reduction in winter wheat infected with soilborne wheat mosaic virus. Plant Dis. Rptr. 59: 469-471.

76.   Parry, M.S. 1980. Evidence of clonal variation and latent virus effect on the vigour of Cox's Orange Pippin apple trees on M.9 rootstocks. J. Hort. Sci. 55: 439-440.

77.   Perombelon, M.C.M. 1972. The extent and survival of contamination of potato stocks in Scotland by *Erwinia carotovora* var. *caratovora* and *E. carotovara* var. *atroseptica*. Ann. Appl. Biol. 71: 111-117.

78.   Posnette, A.F. 1989. Losses caused by virus and viruslike diseases. Pages 3-6 in: Virus and Virus-like Diseases of Pome Fruits and Simulating Noninfectious Disorders. Wash. State Univ. Coop Ext. Publ. SP 0003.

79.   Rebandel, Z., Zawadzka, B., and Wierszyllowski, J. 1979. Effect of apple mosaic virus on bud take and growth of trees in the nursery. Fruit Sci. Repts. VI (1): 9-17.

80.   Reyes, T.T. 1989. Selected economically important diseases of some major crops in the Philippines. Trop. Agr. Res. Series 22: 11-20.

81.   Rockow, W. F., and Duffus, J. E. 1981. Luteoviruses and yellows diseases. Pages 147-170 in: Plant Virus Infections Comparative Diagnosis. E. Kurstak, ed. Elsevier/North Holland Press.

82.   Sayama, H., Sato, T., Kominato, M., Natsuaki, T., and Kaper, J.M. 1993. Field testing of a satellite-containing attenuated strain of cucumber mosaic virus for tomato protection in Japan. Phytopathology 83: 405-410.

83.   Sehgal, O.P. 1981. Southern bean mosaic virus group. Pages 91-121 in: Plant Virus Infections Comparative Diagnosis. E. Kurstak, ed. Elsevier/North Holland Press.

84.   Self, B. 1980. ELMA plant material --A Review. Plantsman 1: 213-223.

85.   Shepard, J.F., and Claflin, L.E. 1975. Critical analyses of the principles of seed potato certification. Annu. Rev. Phytopathl. 13: 271-293.

86.   Stace-Smith, R., and Converse, R. 1970. Raspberry leaf curl. Pages 120-122 in: Virus Diseases of Small Fruits and Grapevines. N.W. Fraizer, ed. Univ of Calif Press, Berkeley.

87.   Starcher, D.B. 1960. The economic importance of russet ring on golden delicious apples. Proc. of 51st Annu. Meet. Wash. State Hort. Assoc. 55: 120-121.

88.   Stouffer, R.F., and Lewis, R.H. 1969. The present status of stem pitting in Pennsylvania. Plant Dis. Rptr. 53: 429-434.

89.   Stubbs, L.L., and Grogan, R.C. Title Kurstak Page 488. Aust. J. Agr. Res. 14: 439-459.

90.   Sward, R.J., and Kollmorgen, J.F. 1986. The separate and combined effects of barley yellow dwarf virus and take-all fungus (*Gaeumannomyces graminis* var. *tritici*) on the growth and yield of wheat. Aust. J. Agric. Res. 37: 11-22.

91.   Tamaki, G., Fox, I., Butt, B.A., and Richards, A.W. 1978. Relationships among aphids, virus yellows, and sugarbeet yields in the Pacific northwest. J. of Econ. Entom. 71: 654-656.

92.   Teng, P.S., ed. 1987. Crop loss assessment and pest management. Am. Phytopathol. Soc. Press, St. Paul. 270 Pages.

93.   Teng, P.S., Hibino, H., and Leung, H. 1989 Yield loss due to rice virus diseases in Asian tropics. Trop. Agr. Res. Series No. 22: 93-100.

94.   Teng, P.S., and Oshima, R.J. 1983. Identification and assessment of losses. Pages 69-81 in: Challenging Problems in Plant Health. T. Kommedahl and P.H. Williams, eds. Am. Phytopathol. Soc. Press, St. Paul.

95.   Thomas, W. 1976. The impact of virus diseases. Wine Review (2): 21-31.

96.   Thresh, J.M., Fargette, D., and Otim-Nape, G.W. 1994. Effects of African cassava mosaic geminivirus on the yield of cassava. Trop. Sci. 34: 26-42.

97.   Thurston, D.H. 1973. Threatening plant diseases. Annu. Rev. of Phytopathol. 11: 27-52.

98.   Tien, P., and Gusui, W. 1990. Satellite RNA for the biocontrol of plant disease. Adv. in Virus Res. 39: 321-339.

99.   van der Zaag, D.E. 1987. Yield reduction in relation to virus infection. Pages 146-1?? in: Viruses of Potato. J.A. deBok and van der Want, eds.

100.   Vuittenez, A. 1970. Fanleaf of grapevine. Pages 217-228 in: Virus Diseases of Small Fruits and Grapevine. N. W. Fraizer, Ed. Univ of Calif Press, Berkeley.

101.   Walkey, D.G.A., and Payne. C.J. 1990. The reaction of two lettuce cultivars to mixed infection by beet western yellows virus, lettuce mosaic virus and cucumber mosaic virus. Plant Pathol. 39: 156-160.

102.   Wastie, R.L., and Solomon, R.M. 1988. The contribution and value of resistant cultivars to disease control. Pages 103-112 in: Control of Plant Diseases: Costs and Benefits. B.C. Clifford and E. Lester, eds. Blackwell Sci. Publications.

103.   Wrather, J.A., Chambers, A.Y., Fox, J.A., Moore, W.F., and Sciumbato, G.L. 1995. Soybean disease loss estimates for the Southern United States, 1974 to 1994. Plant Dis. 79: 1076-1079.

104.   Wright, N.S. 1970. Combined effects of potato viruses X and S on yield of Netted Gem and White Rose potatoes. Am. Potato J. 47: 475-478.

CHAPTER 2

Breeding for Resistance to Plant Viruses

R.K. Khetarpal, B. Maisonneuve, Y. Maury, B. Chalhoub, S. Dinant, H. Lecoq and A. Varma

The goal of plant breeding is to contribute to a qualitative and quantitative improvement in crop production. The "green revolution" and the phenomenal increase in food production during recent years owe a great deal to the development of disease resistant cultivars of food crops. In the agrarian-based developing countries, breeding for resistance to diseases has played a significant role as crop losses directly imbalance the socio-economic status of the country. Among plant pathogens, viruses are known to cause significant losses to most of the major crops around the world (see chapter by Waterworth and Hadidi, 1). Therefore, extensive work has been done on breeding for resistance to plant viruses (36,39,40,121). The use of resistant varieties of plants, if available, is an effective strategy for minimizing the losses caused by viral diseases. This is largely due to intrinsic properties of the plant viruses which do not permit their control by simple physical and chemical methods.

The major advantage of breeding for resistance to viruses is that once a resistant cultivar is developed no specific action is required by farmers to achieve control. Also, host resistance to viruses is selective and environmentally sound as compared to the use of pesticides applied for controlling certain vectors of plant viruses such as insects and fungi. Though the initial cost of producing a new resistant variety may be high, in the long term if the resistance lasts, it turns out to be economical.

Breeding for resistance to viruses or to any pathogen is most frequently used in annual field and vegetable crops which are generally grown over large areas. In case of ornamentals and fruit crops owing to their different modes of reproduction and multiplication, breeding for resistance is more difficult and thus is not frequently undertaken. The development of high quality cultivars with a good level of disease resistance is a challenge for the breeders. The primary aim, therefore, is to assemble the sources of genotypic resistance which can be used for developing desired cultivars. This is accomplished by a large scale exploration and exploitation of the plant genetic resources. The main sources of genotypic resistance to pests and pathogens are most often the centers of origin as well as the area of diversification of cultivated plants. In these locations plants have been exposed to the selective pressure of pests and pathogens over the years and thus have developed resistance (78). To breed for resistance to viruses, generally the plant breeders screen for resistance, identify the sources of resistance and study the inheritance of resistance. Further studies on specificity and on the mode of action of the resistance gene are often undertaken by virologists.

The recent advances made in molecular biology and plant transformation and regeneration technologies have widened the scope of conferring resistance to viruses in plants. It is now possible to engineer resistance to viral pathogens in plants using transgenes from a wide range of organisms (see related chapters in this book, 53,156). The

---

Standard acronyms of the virus names are used in the text: BCMV- bean common mosaic potyvirus; BCTV- beet curly top geminivirus; BSMV- barley stripe mosaic hordeivirus; BWYV- beet western yellows luteovirus; BYDV- barley yellow dwarf luteovirus; BYMV- bean yellow mosaic potyvirus; CYVV- clover yellow vein potyvirus; LMV- lettuce mosaic potyvirus; MSV- maize streak geminivirus; PLRV- potato leaf roll luteovirus; PMV- potato M carlavirus; PSbMV- pea seedborne mosaic potyvirus; PVX- potato X potexvirus; PVY- potato Y potyvirus; PeMoV- peanut mottle potyvirus; RBDV- raspberry bushy dwarf ilarvirus; RBSDV- rice black-streaked dwarf fijivirus; RRSV- raspberry ringspot nepovirus; SMV- soybean mosaic potyvirus; TMV- tobacco mosaic tobamovirus; ToMV- tomato mosaic tobamovirus; TuMV- turnip mosaic potyvirus; WMV2- watermelon mosaic potyvirus-2.

strategy which involves the introduction of transgenes into a crop in a single step is called *horizontal gene transfer* (HGT) whereas conventional breeding involving incorporation of resistance genes from parents to offsprings by crossings has been termed *vertical gene transfer* (VGT) (108).

The present chapter deals with steps involved in developing breeding programs for resistance to viruses, lists the various viral resistance genes known so far and highlights the advances made in molecular biology that is broadening the possibilities of developing resistant varieties in the future. The different kinds of resistance, the biochemistry of resistance mechanisms, and the different molecular strategies of conferring resistance to viruses have been dealt with in detail in subsequent chapters in this book.

SCREENING FOR RESISTANCE

Any strategy of breeding for virus resistance requires a good knowledge of the virus and its different strains. A correct identification of viral strains is necessary for precisely determining the number of genes governing resistance in a genitor (112). The choice of viral strain(s) to be used for inoculation is important because resistance is pathotype specific such as in case of tomato to TMV (100), pea to PSbMV (2), lettuce to LMV (104) and beans to BCMV (29,132). The inoculum should preferably represent the maximum possible variability of the virus. A collaboration of the breeder with a virologist is thus necessary for effectively screening the material.

It is essential to ensure that the inoculum produces, at high frequencies, typical symptoms on susceptible hosts. The plants to be tested should generally be young and preferably at uniform stage of development. Regarding the mode of inoculation, mechanical sap inoculation is the most convenient method of infecting plants on a large scale. The inoculation may be done manually or by using an inoculation gun. If the virus is not sap-transmitted, insect vectors have to be used for inoculation purposes as in case of luteoviruses. In such cases the viruliferous insects need to be reared on infected plants. In case of vegetatively propagated crops such as raspberry, graft inoculations are made (55). The plants after inoculation need to be strictly protected from other viruses in order to avoid confusions, especially for crops like grapevine where symptoms appear after

few months or years. Appearance of symptoms often forms the basis of screening. It is advisable to monitor presence of virus in symptomless plants with sensitive serological or molecular detection techniques. There may be cases where inoculation response is highly variable in the plant population: from complete resistance to partial resistance with different grades of symptom intensities in-between. In such cases the range is measured by a scoring system often denoted by a scale as in case of corn inoculated with MSV (31) or cabbage inoculated with TuMV (148).

Sometimes screening for resistance is also carried out under natural infection conditions as has been done for white clovers to different viruses (22), garlic germplasm to garlic mosaic disease (69), etc. This is only possible if the disease recurs at the same area on a particular crop every year owing to the presence of vectors and of reservoirs of virus nearby and that there is no risk of mixed infections. Otherwise the resistant genotypes can be overlooked, unless serological or molecular tests are conducted to ensure the presence or absence of the target virus. Field screening of rice for resistance to RBSDV was carried out by growing host plants of the vectors of the virus, to increase vector populations, close to the rice field (56). The screening done under natural infection conditions for 2-3 years takes into account the field resistance but does not ensure the testing against different strains of the virus. The multilocational screening for resistance helps in exposing the genotypes to diverse geographical isolates of the virus.

In case of seed transmitted viruses of the dicots, an initial screening of genetic resources may be done by enzyme-linked immunosorbent assay (ELISA) testing the testae in the laboratory. This is based on the fact that during multiplication of the germplasm in the field, the infection normally becomes generalized on the susceptible lines by vector dissemination of the virus from a plant infected through seed. Since testae are the maternal tissues, the plant infection is generally accompanied by the infection of the testae. The initial screening of testae, without involving a lot of labor and time, permitted elimination of 85% of pea accessions susceptible to PSbMV from field testing (68).

SOURCES OF RESISTANCE AND BREEDING PROGRAMS

In a resistance breeding program the first

CHAPTER 2

step is to find a source of resistance to the particular virus. Sources of resistance to many of the viruses have been extensively reviewed (60,113). The breeder screens a collection of genotypes of cultivated species, existing or old varieties and land races. If no resistance is found then the related species are screened. If useful source of resistance is identified in cultivated species or closely related and sexually compatible species, the breeder crosses it with a cultivar having desirable agronomical traits. The breeding strategy for a crop is chosen keeping in view the biology of its reproduction (i.e. self-pollinated or self-incompatible), type of cultivar (i.e. F₁ hybrid, homozygous line or vegetative clone) and inheritance of the resistance (i.e. monogenic, oligogenic or polygenic; dominant or recessive). In case resistance sources are identified only in related wild species that are difficult or impossible to be crossed with cultivated species, techniques of interspecific crosses are adopted such as *in vitro* culture of immature embryo that was used to introduce resistance to BWYV from *Lactuca virosa* to *L. sativa* (83,84), and protoplast fusion (20). Fastidious numerous backcrosses are often necessary to obtain a fertile resistant line with agronomical characters of cultivated species. The final objective is to combine the resistance with good agronomical traits.

RESISTANCE GENES

Inheritance

The knowledge of inheritance of resistance is required in proper exploitation of available genetic resources and in developing strategies for deploying resistance genes. The studies on inheritance of host resistance has involved the classical crossing of resistance and susceptible parents and determination of the reactions of F1, F2 and backcrosses to infection with virus. A perusal of the available literature revealed that genetic analysis of resistance has been carried out for 179 host-virus combinations (Table 1) which are listed in the Appendix.

It was found that in 78% of the host-virus combinations reported, resistance is monogenic and the rest are either oligo- or polygenic. A total of 51% resistance cases are under the control of dominant genes and 35% are governed by recessive genes (Table 1). However, complex situations have also been observed. There are 12% cases where resistance may just not be completely

dominant or recessive and instead it may be gene-dosage dependent i.e. it may show incomplete dominance. Fraser (38) postulated that resistance may appear to be fully dominant or recessive when relationships are judged on the basis of symptom severity, but can show gene dosage dependence when virus multiplication is measured. Thus many genes reported as dominant or recessive may turn out to be incompletely dominant if specifically worked out. Also environmental effects on expression of resistance may prevent the fit of a simple gene model (33). The mode of inheritance of resistance in different species or germplasm of a crop may vary against the same virus (Appendix) or even against the different strains of the same virus such as of SMV in soybean (136,160). If resistance is governed by more than one gene, the different genes may not necessarily be inherited in the same manner e.g. in soybean accession PI1486355 out of the two genes governing resistance, one is dominant and the other is incompletely dominant (82).

In some cases effectiveness of resistance gene(s) for a particular virus was shown to be affected by minor (or modifier) genes as for corn resistant to MSV (31). Indirect effects of host genetic background on the effectiveness of a particular gene e.g. *mo₁* gene in lettuce conferring resistance to LMV (149) and *Yd2* gene in barley that confers resistance to BYDV (18,19,59) have also been reported.

There are also few interesting examples where polygenic resistance to a virus in a particular plant species has been accumulated from originally susceptible plants. Cisar *et al.* (21) bred wheat resistant to BYDV from susceptible oat germplasm. It is believed that such a resistance is the result of accumulation of minor genetic factors (minor genes) present in each parent.

The analysis of the *gene for gene relationship* has been reported for a limited number of host-virus combinations viz, *Capsicum*/TMV, tomato/TMV, potato/PVX and *Phaseolus vulgaris*/BCMV (38). Such few cases occur, as compared to the host-fungal interactions, presumably because of the fact that in most cases only one allele of resistance is known and in fact there may be more examples.

Cluster of Genes

In several plant species, resistance to different pathogens were shown to be controlled by tightly linked or allelic genes. This clustering

Table 1. Summary of number of virus resistance genes reported

Resistance gene	Monogenic	Oligo- or polygenic
Dominant	81	10
Recessive	43	20
Incompletely dominant	15	6
(Nature unknown)	–	4
Total number of resistance genes	139	40

of resistance genes initially described for fungal diseases (116) is now reported for some viral diseases. In case of *Pisum sativum* resistance to lentil strain of PSbMV, BYMV, WMV-2, CYVV, and the NL-8 strain of BCMV were proposed to be controlled by tightly linked recessive genes clustered on chromosome 2 (111,112). Similar relationships were also suggested for resistance in *Phaseolus vulgaris* to several potyviruses (34,73) and in some cultivars of *Cucurbitaceae* species to several viruses (109). In case of bread wheat it was found that the *Bdv1* gene conferring tolerance to BYDV and *Lr34* and *Yr18* genes conferring adult plant resistance to rust fungi are either very closely linked or under pleiotropic genetic control (129). In lettuce the *Tu* gene conferring resistance to TuMV is tightly linked to several *Dm* genes that confer resistance to the downy mildew fungus (162). It is now left to be seen if the clustering of resistance genes is common in plants.

Based on genetic experiments it can not be clearly shown as how these resistance genes are clustered. Some hypothesis suggest that clusters of resistance genes might have arisen by duplication of an ancestral gene, followed by the divergence of the copies which were specifically exploited by different variants of the virus. In this context it is interesting to note that on the coat-protein-based dendrogram of potyviruses (122) PSbMV was found to be closer to those potyviruses against which the resistance is governed by the recessive genes clustered on the same chromosome of pea (K. Masmoudi and Y. Maury, unpublished).

In other cases, specificities to viral pathotypes were shown to be allelic variation of the same locus. This is for example true for the *Tm-2* gene in tomato to ToMV (48), the *mo₁* gene in lettuce to LMV (104) or the *Yd2* gene in barley to BYDV (19). The recent advances made in molecular characterization of resistance genes would clarify whether it is a case of clustering or allelic variability of resistance genes.

Mode of Action

Based on the site of action of the functional products, the resistance genes may operate through resistance to vectors, resistance to transmission by vectors or seeds, resistance to symptom development (tolerance), resistance to virus multiplication, resistance to intercellular or systemic movement, hypersensitivity and immunity to virus infection (36). From the mechanism point of view, resistance to virus may be constitutive or induced. *Constitutive resistance* involves the inhibition of essential viral function e.g. the product of *Tm1* gene of tomato would confer resistance to TMV by inhibiting the function of the virus-encoded replicase (89). *Induced resistance*, which is more common, is by the elicitation of a defense reaction in the host plant following the recognition of the product of a part of the viral genome named as "avirulence gene" and may be associated to a hypersensitive reaction (local necrotic lesions) at the site of infection e.g. viral coat protein acts as an elicitor for the *N'* mediated resistance in tobacco against avirulent strains of TMV (24). However, the induced resistance does not always involve the formation of necrotic lesions. The coat protein of avirulent strains of PVX induce resistance in potato lines having the *Rx* gene and on which no symptoms appear concomitantly (71). This induced resistance is effective against a large spectrum of pathogens (71) which is not the case in constitutive resistance.

From a practical point of view the resistance of plants to vectors appears to be an attractive mode of controlling plant viruses. For a plant breeder, screening progeny plants for vector resistance seems to be relatively easy because vector resistance is mostly inherited as a single major gene. Also in case of resistance to virus vectors, there is little chance for the virus to mutate into virulent strains due to lack of direct interaction between the resistance gene product and the virus (36). However, possible appearance of new biotypes of vector due to selection pressure imposed by cultivating resistant cultivars may pose serious problems as was reported with rice brown planthoppers on rice (131). Also breeding for resistance to one vector can result in an increased susceptibility to colonization by another (3) and vector resistance may sometimes not prove durable on field release because of variation within the vector population as was observed for leafhoppers (25). Therefore it seems that strategy of vector resistance could play a significant role in virus control if used in combination with virus resistance.

In case of seed transmitted viruses, seeds can be the primary source of inoculum. The blocking of the annual reappearance of viral diseases through breeding for non-seed-transmissibility could be an effective control strategy. In case of barley, resistance to seed transmission of BSMV was observed in a cultivar (Modjo13212) and through breeding this gene was incorporated in a high yielding susceptible cultivar 'Vantage'. Resistance to seed transmission of PeMoV in two lines of peanut (157), of PSbMV in a cultivar of pea (150), of SMV in 19 lines of soybean (44) and of LMV in lettuce cultivars containing the mol^1 and mol^2 gene (8,104,123) has also been recorded.

DURABILITY AND BREAKDOWN OF RESISTANCE

Durable resistance is the one that has remained effective while cultivars possessing it have been widely cultivated in an environment favoring the disease (57). The reason why some resistances are durable is unknown. The term durability of resistance is useful for retrospective judgement and not for looking forward as it does not imply that resistance will remain permanently effective. Durable resistance can be complete or quantitative and therefore cannot be defined in epidemiological terms. It focuses attention on a particular attribute of resistance that is of practical importance in future breeding programs.

The viruses, owing to the small size of their genome, evolve rapidly and may come up with new genetic variants that may overcome the resistance. It is important to mention that the frequent use of the expression *broken down* for a resistance that has been overcome, is misleading. As proposed by Walkey (147), it is more appropriate to use the term *breaking down* instead of *broken down*, because the cultivar still maintains the resistance to the strains of the virus for which it was tested, but the cultivar does not possess enough resistance against the new variant (virulent strain) of the virus. It is a difficult task to generalize on the overall durability of different viral resistance genes. This is because from published literature it is often not known whether the resistance genes have been tested against different existing isolates of the virus or not. Most of the resistance genes have been overcome by virulent isolates (38,145). This means that the variability of the virus genome is so high that generally there exists a variant which is selected under the selective pressure of the resistance gene. It was also noted that virulent isolates have been more common against the dominant and gene dosage dependent alleles than against the recessive ones (38).

Nevertheless, breeding for resistance to viruses has been successful in certain cases where the resistance has lasted for a considerable period (Table 2). The examples of durable resistance include both monogenic and polygenic resistance whereas theoretically polygenic resistance should be more durable. Even some of the durable genes have been reported to be overcome by virulent isolates under controlled conditions and few in the field e.g. a TMV isolate from *Capsicum* was found to overcome the *N* gene of TMV resistance in tobacco after 50 years of its durability (23). But since these isolates have not become prevalent, it should not be discouraging to the breeding programs. Besides, the resistance genes that have been overcome can always be useful in other parts of the world or can still be deployed in combination with other genes with additive effects.

It was reported that some virulent isolates lack competitive fitness which would allow them to become prevalent in the field e.g. resistant breaking isolates of RRSV have not become

Table 2. Examples of durable resistance to plant viruses

Resistance gene	Virus	Host
$Tm-2^2$	TMV	tomato
Ry	PVY	potato
N	TMV	tobacco
Bu	RBDV	raspberry
Tu	TuMV	lettuce
Rpv	PeMoV	peanut
I	BCMV	snap & dry beans
Nx,Nb	PVX	potato
$mo1^1, mo1l^2$	LMV	lettuce
(Polygenic)	BWYV	sugarbeet
(Polygenic)	BCTV	sugarbeet

prevalent because their rate of seed transmission is lower (49). Also isolates of TMV capable of overcoming $Tm-2^2$ gene in tomato were found to have a reduced level of multiplication on resistant plants (38). In case of *Phaseolus vulgaris* resistance conferred by the dominant I gene against BCMV has been effective in dry bean and snap bean for about 40 years (159) and the virulent strains though reported have not become prevalent in the field. Resistance to BCTV in sugarbeet and to TuMV in lettuce has been noted since more than 50 and 20 years, respectively (30). In Britain potato cultivars such as Epicure and King Edward have shown field resistance against PVX for more than 50 years (121). Similarly durability of resistance in potato to other viruses such as PVY, PMV and PLRV have also been reported (138). The reason for durability of virus resistance in these crops is not well known but is assumed to be due to the involvement of more than one type of resistance mechanism each being inherited independently.

PROBLEMS AND PROSPECTS

Resistance breeding has been often approached too simplistically. The epidemiological parameters such as survival of the virus in alternate hosts or in vectors, vector-crop relationship, etc. are seldom analyzed. Therefore, the use of resistant cultivars in few cases has even turned an insignificant disease into a significant one. The serious epidemics of tungro disease on the introduced high yielding cultivars of rice in

India and the Philippines, and the maize dwarf mosaic disease on new American hybrid corn in Italy illustrates the problems that can occur (17). There may be hidden viruses which may become more important on the resistant cultivars as these viruses have never been taken into consideration in the breeding stations.

The overall process of incorporation of resistance by crossings and repeated back crossings takes a number of years and even then it is not possible to predict in advance that a newly constructed genome in a cultivar will possess durable resistance. Because of this uncertainity it is required to maintain genetic diversity between cultivars for their resistance characters in order to eventually exploit them in a way that can reduce the major epidemics due to evolution of virulent strains of the viruses and emergence of new viruses.

Above all, resistance to a virus is only one of the many characters which have to be taken into account by the breeders. The difficulties lie mainly in combining sufficient resistance with the other agronomical characters required to develop an acceptable variety. In some high yielding tobacco cultivars, resistance to TMV was found to be accompanied by a poorer quality in the product (58). A cultivar carrying resistance to a given virus may be susceptible to other viruses or pathogens in the field. The bean genotypes possessing monogenic dominant resistance to BCMV revealed severe disease symptoms when infected by comoviruses in South America (92). Similarly lettuce genotype carrying gene for resistance to downy mildew fungus in California was severely attacked by TuMV (161). The incorporation of genetic resistance to multiple pathogens as can be exemplified by incorporation of resistance to TMV, cyst nematode, root knot nematodes and wild fire bacteria from *Nicotiana repanda* into *N. tabacum* (47), would of course be an ideal solution. But such examples are few.

There are relatively few sources of natural resistance genes for viruses of major crops and most natural resistance genes have been overcome by virulent viral isolates. It may take only a few nucleotide/amino acid substitutions to change a virus to a virulent form as has been reported for TMV (89) and PVX (117). Thus the continuous search for new resistance sources remains a challenge for the breeders. In certain cases no resistance could be found for particular crops and viruses e.g. for lettuce no resistance to BWYV was found in a large number of cultivars and breeding lines (151); many horticultural crops have no source of resistance to a number of viruses that are known to affect them (37). Also, often resistance genes are only identified in wild relatives such as the resistance to BWYV found in *Lactuca virosa* (83). Most of the resistance sources from wild species cannot be used for conventional breeding purposes due to barriers of sexual or somatic hybridiztion.

The recent advances in molecular biology has significant role in facilitating different steps of breeding itself and so also in incorporating novel resistance genes. The development of molecular markers in breeding for resistance would shorten the selection process. Two generations of molecular markers have been used: restriction fragment length polymorphism (RFLP) as codominant markers and polymerase chain reaction (PCR) based markers that are often dominant. The recent development of PCR offers several types of markers relatively easy to use: random amplified polymorphic DNA (RAPD) (155), Sequence characterized amplified regions (SCAR) (99) or Amplified Fragment Length Polymorphism (AFLP) (146). In a segregating population, comparison of DNA bulks (bulk segregant analysis) of two plants differing in disease resistance is a very quick method to identify some molecular markers without using the near isogenic lines (90). The analysis of individual plants of the segregating population with the polymorphic markers between both the bulks give an estimation of the genetic distance between the markers and the resistance gene. Identification of two codominant markers flanking the resistance gene offers the possibilities

to use them for assisting in the selection process. Besides, the use of molecular markers for identifying gene clusters would facilitate breeding for resistance to several viruses as cultivars bred for resistance to a particular virus would also carry factor for resistance to other viruses. Above all, in case of natural resistance genes which are recently being cloned, the identification of genes on the basis of conserved amino acid regions would enable plant breeders to monitor resistance gene segregation using appropriate DNA probes instead of testing progeny for resistance and susceptibility (134).

The HGT seems to be promising in overcoming certain inherent constraints encountered in the conventional breeding for resistance to viruses. It is now possible to introduce desired genes into plants not only from different species or genera or families but also from viruses and other microorganisms by using genetic transformation techniques. Transformation is now becoming possible for most of the cultivated plants including cereals which of late were considered as recalcitrant species (74). Genetic engineering for viruses include three kinds of transgenes (64) involving different strategies i.e. (i) *pathogen-derived transgenes* which include resistance mediated by viral genes such as coat-protein, replicase, movement protein, polyprotein protease, satellite RNAs, sense and antisense RNA. (ii) *plant-derived transgenes* which include plant genes encoding for PR proteins, antiviral proteins and natural host resistance genes, and (iii) *non plant, non pathogen derived transgenes*: which involve resistance mediated by antibodies, mammalian oligoadenylate synthetase enzyme (induced by interferons), etc.

The HGT is now being standardized for a number of host-virus system. In some cases the process of HGT, however, takes as long as VGT because the technique of transformation needs to be standardized and a large number of transformants need to be tested for identifying the one which can serve as a genitor for introducing the required gene into a cultivar. The level of resistance provided by HGT may be similar to that of VGT. In both cases the level of resistance can be low or high. The technical ability to isolate and transfer genes has surpassed the level of understanding of their biochemical properties required for their rational manipulation (35). The possible limitation could be that in the coming years unexpected effects may appear such as gene silencing (86) which can eventually alter the efficiency of protection. Besides, the trained manpower and the financial input required for launching a program on HGT seems to be a limiting factor for the developing countries.

The few constraints of HGT should not reflect a myopic view on its vast potential. It is a very promising strategy for imparting resistance to crops that are propagated vegetatively. Above all the molecular biologists can provide breeders with a wide range of transgenes from which the most appropriate individuals for the breeding program can be screened. Newer sources of antiviral transgenes with potentials of imparting resistance in plants continue to be discovered and there is no limit to designing new resistance genes (53). Most crop species are infected with more than one virus. Fortunately, recent reports of transgenes with broad spectrum resistance, effective against a range of viruses (13,27,93,133) are widening the potential of such kind of resistances.

Among the molecular strategies the plant-derived transgene i.e. the use of natural resistance gene should be promising. Molecular cloning of natural resistance genes which enable plants to resist a diverse range of pathogens has revealed that proteins encoded by these genes have several features in common (134). This suggest that plants may have evolved common signal transduction mechanisms to a wide range of unrelated pathogens. Once the molecular signals involved in recognition of the pathogens and in the expression of resistance are properly characterized, more effective and broad spectrum strategies of controlling plant diseases could be envisaged.

By expanding and diversifying the resistance gene-pool, HGT can significantly contribute to pyramiding of resistance genes (66) that, in fact, is important in breeding for durable understanding the mechanism of action of natural resistance. It appears that the time is not far when resistance genes for recognizing pathogen elicitors would give clues for understanding the mechanism of durability of resistance and thus lead to novel method of disease control. It is not possible to speculate the time when HGT would be completely independent to create new resistant varieties. But, it can be safely assumed that in the near future HGT would join hands with VGT to accelerate the development of virus resistant varieties and would make possible most of the things that were considered as impossible in traditional breeding for resistance to viruses.

To sum up, for an effective viral disease control, the breeding for resistance (HGT or VGT) would always have an important role to play. Resistance is known to occur at different stages of the viral infection, therefore, accumulation of different mechanisms of resistance within a cultivar should increase the protection level (77). The comparative study of virus spread in resistant and susceptible cultivars is necessary to properly evaluate the field effectiveness of resistances, particularly when these are partial, strain-specific or composite i.e. combining different resistance mechanisms in the same genotype (76) However, when partial resistances are not sufficient to give high level of protection complementary strategy of preventing the viruses to become important in a specific agro-ecosystem should be adopted (17,75). This would help in reducing the infection pressure on the crops and so also the probability of the emergence of resistance-breaking strains of the virus or vector. The overall solution would thus come from a close collaboration of the breeders, plant pathologists and molecular biologists.

APPENDIX

Different Kinds of Genetic Resistance Reported So Far in Various Host-virus Combinations [a,b]

[a] Attempts were made to present the list as complete as possible, but in no way it may be considered as thoroughly complete.
[b] Reference numbers in the brackets refer to the review articles in which the mode of inheritance of resistance for the given host-virus system has been listed.

Monogenic dominant resistance

Virus	Host	Reference
Barley yellow mosaic potyvirus	*Hordeum vulgare*	139
Bean common mosaic potyvirus	*Phaseolus vulgaris*	(113)
Bean yellow mosaic potyvirus	*Phaseolus vulgaris*	(113)
	Vicia faba	(124)
Beet necrotic yellow vein furovirus	*Beta maritima*	152
	Beta vulgaris	79
Beet western yellows luteovirus	*Lactuca virosa*	83
Blackeye cowpea mosaic potyvirus	*Phaseolus vulgaris*	(113)
	Vigna unguiculata	(113)
Broadbean wilt fabavirus	*Phaseolus vulgaris*	110
Cherry leaf roll nepovirus	Juglans hybrids	28
Cowpea aphidborne mosaic potyvirus	*Phaseolus vulgaris*	(113)
Cowpea chlorotic mottle bromovirus	*Glycine max*	12
	Vigna unguiculata	(36)
Cucumber green mottle mosaic tobamovirus	*Cucumis anguira*	26
Cucumber mosaic cucumovirus	*Cucumis melo*	105
	Lactuca saligna	(36)
	Vigna unguiculata	(36)
Cowpea mosaic comovirus	*Vigna unguiculata*	107
Maize dwarf mosaic potyvirus	*Zea mays*	(113)
Mungbean yellow mosaic geminivirus	*Vigna mungo*	(124)
	Vigna radiata	(124)
Okra yellow vein mosaic virus	*Abelmoschus manihot*	140
Papaya ring spot potyvirus	*Cucumis melo*	(113)
	Cucumis metuliferus	(113)
	Cucumis sativus	(113)
Passionfruit woodiness potyvirus	*Phaseolus vulgaris*	(113)
Pea enation mosaic enamovirus	*Pisum sativum*	125

Monogenic dominant resistance (cont.)

Virus	Host	Reference
Peanut mottle potyvirus	*Glycine max*	(124)
Peru tomato potyvirus	*Lycopersicon esculentum*	(113)
	Lycopersicon hirsutum	(113)
Potato A potyvirus	*Solanum chacoense*	(113)
	Solanum stoloniferum	(113)
	Solanum tuberosum	(113)
Potato leafroll luteovirus	*Solanum chacoense*	16
Potato S carlavirus	*Solanum andigena*	4
	Solanum tuberosum	61
	X *Solanum tuberosum* ssp. *andigena*	
Potato X potexvirus	*Capsicum pendulum*	(36)
	Solanum acaule	120
	Solanum tuberosum ssp. *andigena*	153
	Solanum tuberosum ssp. *tuberosum*	(36)
Potato Y potyvirus	*Nicotiana binavidesi*	(113)
	Solanum hougasii	(113)
	Solanum microdontum	(113)
	Solanum stoloniferum	(113)
	Solanum tuberosum	(113)
	Nicotiana tabacum	67
Raspberry bushy dwarf (?)ilarvirus	*Rubus idaeus*	(36)
Rice black-streaked dwarf fijivirus	*Oryza sativa*	56
Rice hoja blanca tenuivirus	*Oryza sativa*	11
Rice stripe tenuivirus	*Oryza sativa*	81
Soil borne wheat mosaic furovirus	*Triticum sativum*	91
Southernbean mosaic sobemovirus	*Phaseolus vulgaris*	(36)
	Vigna unguiculata	52
Soybean mosaic potyvirus	*Glycine max*	(113)
Sugarcane mosaic potyvirus	*Saccharum spontaneum*	(113)
	Sorghum bicolor	(113)
	Zea mays	101
Tobacco etch potyvirus	*Capsicum frutescens*	(113)
Tobacco mosaic tobamovirus	*Capsicum annuum*	(36)
	Nicotiana glutinosa	(36)
	Nicotiana sylvestris	(36)
	Lycopersicon esculentum	(36)
Tobacco ringspot nepovirus	*Vigna unguiculata*	(36)
Tomato chlorotic spot tospovirus	*Lycopersicon esculentum*	14

Monogenic dominant resistance (cont.)

Virus	Host	Reference
Tomato spotted wilt tospovirus	*Capsicum chinense*	15
	Lycopersicon esculentum	(36)
Tomato yellow leaf curl geminivirus	*Lycopersicon esculentum*	(36)
Turnip mosaic potyvirus	*Brassica napus* ssp. *oliefera*	(113)
	Brassica napus ssp. *rapifera*	(113)
	Brassica napus group *Napobrassica*	127
	Cichorium intybus X *C. endiva*	115
	Lactuca sativa	(113)
Watermelon mosais potyvirus-2	*Cucumis melo*	41
	Cucumis sativus	(113)
	Cucurbita moschata	42
	Phaseolus vulgaris	(113)
Wheat streak mosaic potyvirus	*Zea mays*	(113)
Zucchini yellow mosaic potyvirus	*Cucumis melo*	(113)
	Cucurbita ecuadorensis	(113)
	Cucurbita moschata	(113)

Monogenic recessive resistance

Virus	Host	Reference
Alfalfa mosaic alfamovirus	*Medicago sativa*	(36)
Barley yellow dwarf luteovirus	*Hordeum vulgare*	137
Barley mild mosaic baymovirus	*Hordeum vulgare*	97
Barley yellow mosaic potyvirus	*Hordeum vulgare*	65
Bean common mosaic potyvirus	*Phaseolus vulgaris*	(113)
	Pisum sativum	(113)
Bean leafroll luteovirus	*Pisum sativum*	5
Bean yellow mosaic potyvirus	*Pisum sativum*	(113)
	Vicia faba	(113)
	Vigna mungo	98
Beet western yellow luteovirus	*Lactuca sativa*	103
Bidens mottle potyvirus	*Lactuca sativa*	(113)
Blackeye cowpea mosaic potyvirus	*Vigna unguiculata*	(113)
Clover yellow vein potyvirus	*Phaseolus vulgaris*	(113)

Monogenic recessive resistance (cont.)

Virus	Host	Reference
Clover yellow vein potyvirus (cont.)	*Pisum sativum*	(113)
Cucumber mosaic cucumovirus	*Capsicum annuum*	128
	Cucumis sativus	87
Lettuce mosaic potyvirus	*Lactuca sativa*	(113)
Mungbean yellow mosaic geminivirus	*Vigna radiata*	(124)
Passionfruit woodiness potyvirus	*Pisum sativum*	(113)
Pea seedborne mosaic potyvirus	*Lens culinaris*	(113)
	Pisum sativum	(113)
Pepper mottle potyvirus	*Capsicum annuum*	(113)
Peru tomato potyvirus	*Lycopersicon esculentum*	(113)
Potato S carlavirus	*Solanum tuberosum*	62
Potato Y potyvirus	*Capsicum chinense*	(113)
	Lycopersicon hirsutum	141
	Nicotiana tabacum	(113)
Raspberry bushy dwarf (?)ilarvirus	*Rubus idaeus*	70
Rice tungro spherical machlovirus	*Oryza sativa*	126
Tobacco etch potyvirus	*Capsicum annuum*	(113)
	Capsicum frutescens	(113)
	Lycopersicon esculentum	(113)
	Nicotiana tabacum	135
Tobacco mosaic tobamovirus	*Phaseolus vulgaris*	(36)
	Capsicum annuum	128
Tomato spotted wilt tospovirus	*Lycopersicon esculentum*	(36)
Tomato yellow leaf curl geminivirus	*Lycopersicon cheesmani*	50
Turnip mosaic potyvirus	*Mathiola incana*	(113)
White lupin mosaic potyvirus	*Pisum sativum*	114
Zucchini yellow fleck potyvirus	*Cucumis sativus*	43
Zucchini yellow mosaic potyvirus	*Citrullus lanatus*	(113)
	Cucumis sativus	51

Monogenic incompletely dominant resistance

Virus	Host	Reference
Barley yellow dwarf luteovirus	*Hordeum vulgare*	119
	Triticum aestivum	130
Beet mosaic potyvirus	*Beta vulgaris*	(113)
Cucumber mosaic cucumovirus	*Nicotiana* sp.	85
Muskmelon yellows virus disease	*Cucumis melo*	32
Peanut mottle potyvirus	*Glycine max*	(113)
Potato Y potyvirus	*Capsicum annuum*	(113)
Rice stripe tenuivirus	*Oryza sativa*	142
Soybean mosaic virus	*Glycine max*	(113)
	Phaseolus vulgaris	(113)
Tobaco mosaic tobamovirus	*Capsicum chinense*	(36)
	Lycopersicon esculentum	(36)
Tobacco vein mottling potyvirus	*Nicotiana tabacum*	(36)
Tomato spotted wilt tospovirus	*Lactuca sativa*	95
Tomato yellow leaf curl geminivirus	*Lycopersicon pimpinellifolium*	6

Oligogenic or polygenic dominant resistance

Virus	Host[c]	Reference
Alfalfa mosaic alfamovirus	*Phaseolus vulgaris*[2]	(36)
Arabis mosaic nepovirus	*Rubus idaeus*[2]	55
Maize dwarf mosaic potyvirus	*Zea mays*[2-3]	(113)
Potato Y potyvirus	*Solanum tuberosum*[d]	(36)
Raspberry ringspot nepovirus	*Rubus idaeus*[2]	55
Rice stripe tenuivirus	*Oryza sativa*[2]	81
Tomato blackring nepovirus	*Rubus idaeus*[2]	55
Tomato yellow leaf curl geminivirus	*Lycopersicon hirsutum* F. *glabratum*[2]	7
Turnip mosaic potyvirus	*Brassica campestris* ssp. *pekinensis*[2]	(113)
Wheat spindle streak mosaic potyvirus	*Triticum aestivum*[2]	144

[c]
[d]   Numbers given as superscripts represent the number of genes governing oligogenic resistance and the symbol. Represents polygenic resistance.

Oligogenic or polygenic recessive resistance

Virus	Host[c]	Reference
Barley mild mosaic bymovirus	*Hordeum vulgare*[2]	96
Bean golden mosaic geminivirus	*Phaseolus vulgaris*[2]	(124)
Bean yellow mosaic potyvirus	*Phaseolus coccineus*[2-3]	
	Phaseolus vulgaris[3]	54
	Vigna radiata[2]	98
	Vigna radiata X *V. mungo*[2]	98
	Vigna radiata X *V. subglobata*[2]	98
Clover yellow vein potyvirus	*Phaseolus vulgaris*[2]	(113)
Cowpea chlorotic mottle bromovirus	*Glycine max*[2]	45
Cucumber green mottle mosaic tobamovirus	*Cucumis melo*[d]	118
Cucumber mosaic cucumovirus	*Cucurbita pepo*[2]	102
	Cucumis melo[3]	63
Groundnut rosette luteovirus	*Arachis hypogea*[2]	94
Mungbean yellow mosaic geminivirus	*Vigna mungo*[2]	46
	Vigna radiata[2]	(124)
Tobacco etch potyvirus	*Nicotiana tabacum*[2-3]	(113)
Tobacco mosaic tobamovirus	*Capsicum annuum*[2]	(36)
Tomato spotted wilt tospovirus	*Lycopersicon esculentum*[d]	72
Turnip mosaic potyvirus	*Brassica campestris* ssp. *pekinensis*[2]	158
	Brassica oleracea ssp. *gemmifera*[4]	(113)

c
d   Numbers given as superscripts represent the number of genes governing oligogenic resistance and the symbol.
Represents polygenic resistance.

Oligogenic or polygenic incompletely dominant resistance

Virus	Host[c]	Reference
Barley stripe mosaic hordeivirus	*Hordeum vulgare*[2]	(36)
Barley yellow dwarf luteovirus	*Hordeum vulgare*[4]	(36)
Cucumber mosaic cucumovirus	*Capsicum annuum*[d]	106
	Cucumis melo[3]	(36)
	Nicotiana tabacum[2]	143
Rice grass mosaic potyvirus	*Lolium perenne*[2]	154

c
d   Numbers given as superscripts represent the number of genes governing oligogenic resistance and the symbol.
Represents polygenic resistance.

Polygenic resistance

Virus	Host	Reference
Beet curly top geminivirus	*Beta vulgaris*[d]	88
Beet western yellows luteovirus	*Beta vulgaris*[d]	80
Maize streak geminivirus	*Zea mays*[d]	31
Potato leafroll luteovirus	*Solanum tuberosum*[d]	9,10

[d] Represents polygenic resistance.

REFERENCES

1. Agrios, G.N. 1990. Economic Considerations. Pages 1-22 in : Plant Viruses, Vol. II, Pathology. C.L. Mandahar, ed. CRC Press, Boca Raton, FL, U.S.A.

2. Alconero, R., Provvidenti, R., and Gonsalves, D. 1986. Three pea seed-borne mosaic virus pathotype from pea and lentil germplasm. Plant Dis. 70 : 783-786.

3. Arnold, M.H., Innes, N.L., and Brown, S.J. 1976. Resistance Breeding. Pages 175-179 in : Agricultural Research for Development, M.H. Arnold, ed., Cambridge University Press.

4. Baerecke, M.L. 1967. Tests of Sacc and Saco-crosses for resistance to potato virus S. Eur. Potato J. 10: 206-219.

5. Baggett, J.R., and Hampton, R.O. 1991. Inheritance of viral bean leaf roll tolerance in peas. J. Am. Soc. Hort. Sci. 116 : 728-731.

6. Banerjee, M.K., and Kalloo. 1987. Sources and inheritance of resistance to leaf curl virus in *Lycopersicon*. Theor. Appl. Genet. 73: 707- 710.

7. Banerjee, M.K., and Kalloo. 1987. Inheritance of resistance to tomato leaf curl virus in *Lycopersicon hirsutum* f. *glabratum*. Euphytica 36 : 581-584.

8. Bannerot, H., Boulidard, L., Marrou, J., and Duteil, M. 1969. Etude de l'hérédité de la tolérance au virus de la mosaique de la laitue chez la variété *gallega de invierno*. Ann. de Phytopathol. 1 : 219-226.

9. Barker, H. 1987. Multiple components of the resistance of potatoes to potato leaf roll virus. Ann. Appl. Biol. 111 : 641-648.

10. Barker, H. 1989. Analysis of polygenically controlled resistance in potato to potato leaf roll virus. Pages 55-61 in: Proc. IV International Plant-Virus Epidemiology Workshop, Montpellier, France.

11. Beachell, H.M., and Jennings, P.R. 1961. Mode of inheritance of hoja blanca resistance in rice. Page : 11 in : Proc. of Rice Technical Working Group, Lafayette (1960).

12. Bijaisoradat, M., and Kuhn C.W. 1985. Nature of resistance in soybean to cowpea chlorotic mottle virus. Phytopathology 75 : 351-355.

13. Blaise, F., Kusiak, C., Dinant, S., Astier-Manifacier, S., and Albouy, J. 1995. Etude de la résistance de tabacs transgéniques exprimant la capside du virus de la laitue (LMV) vis-^-vis du virus Y de la pomme de terre (PVY). Ann du Tabac. 27. : 87-94.

14. Boiteux, L.S., and Giordano, L. De.B. 1993. Genetic basis of resistance against two Tospovirus species in tomato (*Lycopersicon esculentum*). Euphytica 71 : 151-154.

15. Boiteux, L.S., and Avila, A.C.De. 1994. Inheritance of a resistance specific to tomato spotted wilt tospovirus in *Capsicum chinense* 'PI 159236'. Euphytica 75: 139-142.

16. Brown, C.R., and Thomas, P.E. 1994. Resistance to potato leafroll virus derived from *Solanum chacoense*: characterisation and inheritance. Euphytica 74 : 51-57.

17. Buddenhagen, I.W. 1983. Crop improvement in relation to virus diseases and their epidemiology. Pages 25-37 in : Plant Virus Epidemiology: The Spread and Control of Insect-borne Viruses. R.T. Plumb and J.M. Thresh, eds. Blackwell Scientific Publications, Oxford, U.K.

18. Catherall, P.L., and Hayes, J.D. 1966. Assessment of varietal reaction and breeding for resistance to the yellow dwarf virus in barley. Euphytica 15 : 39-51.

19. Chalhoub, B.A., Sarrafe, A., and Lapierre, H. 1995. Partial resistance in the barley (*Hordeum vulgare* L.) cultivar 'CHIKURINE IBARAKI 1' to two PAV-like isolates of barley yellow dwarf virus: allelic variability of the Yd2 gene locus. Plant Breeding 114 : 303-307..

20. Chupeau, M.C., Maisonneuve, B., Bellec, Y., and Chupeau, Y. 1994. A *Lactuca* universal hybridizer, and its use in creation of fertile interspecific somatic hybrids. Mol. Gen. Genet. 245 : 139-145.

21. Cisar,G., Brown, C.M., and Jadlinski, H. 1982. Diallele analyses for tolerance in winter wheat to the barley yellow dwarf virus. Crop Sci. 22 : 328-333.

22. Cope, W.A., Walker, S.K., and Lucas, L.T. 1978. Evaluation of selected white clover clones for resistance to viruses in the field. Plant Dis. Reptr. 62 : 267-270.

23. Csillery, G., Tobias, I., and Rusko, J. 1983. A new pepper strain of tomato mosaic virus. Acta Phytopathol. Acad. Sci. Hung. 18: 195-200.

24. Culver, J.N., and Dawson, W.O. 1991. Tobacco mosaic virus elicitor coat protein genes produce a

hypersensitive phenotype in transgenic *Nicotiana sylvestris* plants. Mol. Plant-Microbe Interact. 4 : 458-463.

25. Dahal, G.H., Hibino, H., Cabunagan, R.C., Tiongco, E.R., Flores, Z.M., and Aguiero, V.M. 1990. Changes in cultivar reaction to tungro due to changes in virulence of the leafhopper vector. Phytopathology 80:659-665.

26. den Nijs, A.P.M. 1982. Inheritance of resistance tocucumber green mottle mosaic virus (Cgm) in *Cucumis anguira* L. Rep. Cucurb. Genet. Coop. No. 5 : 57-59.

27. Dinant, S., Blaise, F., Kusiak, C., Astier-Manifacier, S., and Josette, A. 1993. Heterologous resistance to potato virus Y in transgenic tobacco plants expressing the coat protein gene of lettuce mosaic potyvirus. Phytopathology 83 : 818-824.

28. Dosba, F., and Germain, E. 1993. Behaviour of the progeny of Juglans interspecific hybrids towards cherry leaf roll virus (CLRV). Acta Hortic. No 311 : 68-72.

29. Drijfhout, E. 1978. Genetic interaction between *Phaseolus vulgaris* and bean common mosaic virus with implications for strain identification and breeding for resistance. Agric. Res. Rep. Wageningen 872 : 1-98.

30. Duffus, J.E. 1987. Durability of resistance. Ciba Foundation Symposium 133 : 196-199.

31. Efron, Y., Kim, S.K., Fajemisin, J.M., Mareck, J.H., Tang, C.Y., Dabrowski, Z.T., Rossel, H.W., Thottappilly, G., and Buddenhagen, I.W. 1989. Breeding for resistance to maize streak virus: a multidisciplinary team approach. Plant Breeding 103: 1-36.

32. Esteva, J, and Nuez, F. 1992. Tolerance to a whitefly-transmitted virus causing muskmelon yellows disease in spain. Theor. Appl. Genet. 84 : 693-697.

33. Fischer, D.B., and Rufty, R.C. 1993. Inheritance of partial resistance to tobacco etch virus and tobacco vein mottling virus in burley tobacco cultivar Sota 6505. Plant Dis. 77 : 662-666.

34. Fisher, M.L., and Kyle, M.M. 1994. Inheritance of resistance to potyviruses in *Phaseolus vulgaris* L. : cosegregation of phenotypically similar dominant responses to nine potyviruses. Theor. Appl. Genet. 89 : 818-823.

35. Fraley, R.T. 1989. Genetic engineering for crop improvement. Pages 395-407 in : Plant Biotechnology, Shain-dow Kung, and C.J. Arntzen, eds., Butterworths, London.

36. Fraser, R.S.S. 1986. Genes for resistance to plant viruses. CRC Critical Reviews in Plant Sciences : 3 (3) : 257-294.

37. Fraser, R.S.S. 1988. Virus recognition and pathogenicity: implications for resistance mechanism and breeding. Pesticide Sci. 23 : 267-275.

38. Fraser, R.S.S. 1990. The genetics of resistance to plant viruses. Annu. Rev. Phytopathol. 28 : 179-200.

39. Fraser, R.S.S. 1992. The genetics of plant-virus interactions: implications for plant breeding. Euphytica 63: 175-185.

40. Gallais, A., and Bannerot, H. 1992. Amélioration des èspece végétale cultivées, 768p, INRA, Paris.

41. Gilbert, R.Z., Kyle, M.M., Munger, H.M., and Gray, S.M. 1994. Inheritance of resistance to watermelon mosaic virus in *Cucumis melo* L. HortScience. 29 : 107-110.

42. Gilbert-Albertini, F., Lecoq, H., Pitrat, M., and Nicolet, J.L. 1993. Resistance of *Cucurbita moschata* to watermelon mosaic virus type-2 and its genetic relation to resistance to zucchini yellow mosaic virus. Euphytica 69 : 231-237.

43. Gilbert-Albertini, F., Pitrat, M., and Lecoq, H. 1995. Inheritance of resistance to zucchini yellow fleck virus in *Cucumis sativus* L. HortScience 30 : 336-337.

44. Goodman, R.M., Bowers, G.R., and Paschal, E.H. 1979. Identification of soybean germplasm lines and cultivars with low incidence of soybean mosaic virus transmission through seeds. Crop Sci. 19 : 264-267.

45. Goodrick, B.J., Kuhn, C.W., and Boerma, H.R. 1991. Inheritance of non necrotic resistance to cowpea chlorotic virus in soybean. J. Heredity 82 : 512-514.

46. Govindraj, P., and Subramanian, M. 1991. Resistance study on yellow mosaic virus in blackgram. Ind. Phytopath. 44 : 219-221.

47. Gwynn, G.R., Reilly, K.R., Komm, J.J., Burk, L.G., and Reed, S.M. 1986. Genetic resistance to tobacco mosaic virus, cyst nematodes, root knot nematodes, and wildfire from *Nicotiana repanda* incorporated into *N. tabacum*. Plant Dis. 70 : 958-962.

48. Hall, T.J. 1980. Resistance at the *Tm-2* locus in the tomato to tomato mosaic virus. Euphytica 29 : 189-197.

49. Hanada, K., and Harrison, B.D. 1977. Effects of virus genotype and temperature on seed transmission of nepoviruses. Ann. Appl. Biol. 85 : 79-92.

50. Hassan, A.A., Mazayd, H.M., Moustafa, S.E., Nassar, S., Nakhla, M., and Sims, W.L. 1984. Inheritance of resistance to tomato yellow leaf curl virus derived from *L. cheesmannii* and *L. hirsutum.*. HortScience 19 : 574-575.

51. Hayja, Z.A., and Al-Shahwan, I.M. 1991. Inheritance of resistance to zucchini yellow mosaic virus in cucumber. Zeitsch. Pflanzenkr. Pflanzensch. 98 : 301-304.

52. Hobbs, H.A., Kuhn, C.W., Papa, K.E., and Brantley, B.B. 1987. Inheritance of non necrotic resistance to southern bean mosaic virus in cowpea. Phytopathology 77 : 1624-1629.

53. Hull, R. and Davies, J.W. 1992. Approaches to non-conventional control of plant virus diseases. Crit. Rev. Pl. Sci. 11 : 17-33.

54. Ito, H., and Banba, H. 1990. Evaluation of germplasm collections and inheritance of resistance to necrotic strain of bean yellow mosaic virus (BYMV-N) in French bean (*Phaseolus vulgaris* L.). Bull. Hokkaido Prefect. Agric. Expt. Stn. 61 : 7-12.

55. Jennings, D.L. 1964. Studies on the inheritance in the Red Raspberry of immunities from three nematode-borne viruses. Genetica 35 : 152-164.

56. Jin, T.D., Paek, R.K., and Chang, K.Y. 1987. Studies on the screening methods of rice black-streaked dwarf virus disease for breeding. Res. Rep. of the Rural Develop. Admins., Plant Environ., Mycology and Farm Products Utilization (Korea Republic) 29 : 226-241.

57. Johnson, R. 1981. Genetic background of durable resistance. Pages 5-26 in : Durable Resistance in Crops. F. Lamberti, J.M. Waller, and N.A. Vander Graff, eds. Plenum Press, New York and London.

58. Johnson, C.S., and Main, C.E. 1983. Yield/quality tradeoffs of tobacco mosaic virus-resistant cultivars

in relation to disease management. Plant Dis. 67 :
886-890.

59.   Jones, A.T., and Catherall, P.L. 1977. The
relationship between growth rate and the expression
of tolerance to barley yellow dwarf virus in barley.
Ann. Appl. Biol. 65 : 137-145.

60.   Kalloo. 1988. Disease resistance in vegetable crops.
Pages 1-66 in: Vegetable Breeding, Vol. II. CRC
Press, Inc., Boca Raton, FL, U.S.A

61.   Kamenik, J., and Kamenikova, L. 1992. Genetics of
the hypersensitive resistance of potatoes to potato S
carlavirus, PVS. Genetika a Slechteni 28 : 53-63.

62.   Kamenik, J., and Kamenikova, L. 1992. Genetics of
extreme resistance to potato virus S in potatoes.
Genetika a Slechteni 28 : 129-141.

63.   Karchi, Z., Cohen, S., and Govers, A. 1975.
Inheritance of resistance to cucumber mosaic virus
in melon. Pytopathology 65 : 479-481.

64.   Kavanagh, T.A., and Spillane, C. 1995. Strategies
for engineering virus resistance in transgenic plants.
Euphytica : 85-149-158.

65.   Kawada, N. 1991. Resistant cultivars and genetic
ancestry of the resistant genes to barley yellow
mosaic virus in barley (Hordeum vu'gare L.). Bull.
Kyushu Natl. Agric. Expt. Stn. 27 : 65-79.

66.   Kelly, J.D., Afanador, L., and Haley, S.D. 1995.
Pyramiding genes for resistance to bean common
mosaic virus. Euphytica 82 : 207-212.

67.   Keum, W.S., Chung, Y.H., Jung, S.H., Choi, S.J.,
and Lee, S.C. 1991. Inheritance of resistance to
potato virus Y vein necrosis strain of N. africana. J.
Kor. Soc. Tob. Sci. 13 : 48-51.

68.   Khetarpal, R.K., Maury, Y., Cousin, R., Burghofer,
A., and Varma, A. 1990. Studies on resistance of
pea to pea seed-borne mosaic virus and new
pathotypes. Ann. Appl. Biol. 116 : 297-304.

69.   Khetarpal, R.K., Kumar, C.A., Ram Nath, Thomas,
T.A., and Venkataswaran, K. 1991. Garlic mosaic
virus disease: occurrence and screening for
tolerance. Indian J. Pl. Genet. Resources 4 : 84-87.

70.   Knight, V.H., and Barbara D.J. 1981. Suceptibility
of red raspberry varieties to raspberry bushy dwarf
virus and its genetic control. Euphytica 30 : 803-
811.

71.   Kohm, B.A., Goulden, M.G., Gilbert, J.E.,
Kavanagh, T.A., and Baulcombe, D.C. 1993. A
potato virus X resistance gene mediates an induced
nonspecific resistance in protoplasts. The Plant Cell
5 : 913-920.

72.   Kumar, N., and Irulappan, I. 1992. Inheritance of
resistance to spotted wilt virus in tomato
(Lycopersicon esculentum Mill.). J. Gen. Breed. 46
: 113-117.

73.   Kyle, M.M., and Provvidenti, R. 1993. Inheritance
of resistance to potyviruses in Phaseolus vulgaris.
L. II Linkage relations and utility of a dominant
gene for lethal systemic necrosis to soybean mosaic
virus. Theor. App. Gen. 86 : 189-196

74.   Law, C.N. 1995. Genetic manipulation in plant
breeding - prospects and limitations. Euphytica 85
: 1-12

75.   Lecoq, H., and Pitrat, M. 1983. Field experiments
on the integrated control of aphid-borne viruses in
muskmelon. Pages 177-184 in : Plant Virus
Epidemiology - The Spread and Control of Insect-
borne Viruses. R.T. Plumb and J.M. Thresh, eds.
Blackwell Scientific Publications, Oxford, U.K.

76.   Lecoq, H., and Pitrat, M. 1989. Effects of resistance
on the epidemiology of virus diseases of cucurbits.

in : Proc. of Cucurbitaceae 89 - Evaluation and
Enhancement of Cucurbit Germplasm, Charleston,
SC, 185 Pages.

77.   Lecoq, H., Pochard, E., Pitrat, M., Laterrot, H., and
Marchoux, G. 1982. Identification et exploitation de
résistance aux virus chez les plantes mara"chères.
Cryptog. Mycol. 3 : 333-345.

78.   Leppik, E.E. 1970. Gene centers of plants as sources
of disease resistance. Annu. Rev. Phytopathol. 8 :
323-344.

79.   Lewellen, R.T. 1991. Registration of rhizomania -
resistant germplasm of Beta vulgaris. Crop Sci. 31
: 244-245.

80.   Lewellen, R.T., and Skoyen, I.O. 1984. Beet western
yellows can cause heavy losses in sugarbeet. Calif.
Agric. 38 : 4-5.

81.   Lu, J.A., Wan, C.Z., Zou, F.M., and Fan, H.L. 1992.
Inheritance of resistance to rice stripe virus (RSV)
in Yunnan Province, China. Intern. Rice Res.
Newsletter 17 : 7-8.

82.   Ma, G., Chen, P., Buss, G.R., and Tolin, S.A. 1995.
Genetic characteristics of two genes for resistance
to soybean mosaic virus in PI486355 soybean.
Theor. Appl. Genet. 91 : 907-914.

83.   Maisonneuve, B., Chovelon, V., and Lot, H. 1991.
Inheritance of resistance to beet western yellows
virus in Lactuca virosa L. HortScience 26 : 1543-
1545.

84.   Maisonneuve, B., Chupeau, M.C., Bellec, Y. and
Chupeau, Y. 1995. Sexual and somatic hybridization
in the genus Lactuca. Euphytica 85 : 281-285.

85.   Masuta, C., Suzuki, M., Kuwata, S., Takanami, Y.,
and Koiwai, A. 1993. Yellow mosaic symptoms
induced by Y satellite RNA of cucumber mosaic
virus is regulated by a single incompletely dominant
gene in wild Nicotiana species. Phytopathology 83
: 411-413.

86.   Matzke, M.A., and Matzke, J.M. 1995. How and
why do plants inactivate homologous (trans)genes?
Plant Physiol. 107 : 679-685.

87.   Maule, A.J., Boulton, M.I., and Wood, K.R. 1980.
Resistance of cucumber protoplasts to cucumber
mosaic virus: a comparative study. J. Gen. Virol. 51
: 271-279.

88.   McFarlane, J.S. 1969. Breeding for resistance to
curly top. J. Inst. Sugar Beet Res. 4 : 73-83.

89.   Meshi, T., Motoyoshi, F., Adachi, A., Watanabe, Y.,
Takamatsu, N., and Okada, Y. 1988. Two
concomitant base substitutions in the putative
replicase gene of tobacco mosaic virus confer the
ability to overcome the effects of a tomato
resistance gene, Tm-1. EMBO J. 7 : 1575-1581.

90.   Michelmore, R.W., Paran, I., and Kesseli, R.V.
1991. Identification of markers linked to disease-
resistance genes by bulk segregant analysis : A rapid
method to detect markers in specific genomic
regions by using segregating populations. Proc. Natl.
Acad. Sci., U.S.A. 88 : 9828-9832.

91.   Modawi, R.S., Heyne, E.G., Brunetta, D., and
Willis, W.G. 1982. Genetic studies of field reaction
to wheat soil borne mosaic virus. Plant Dis. 66 :
1183-1184.

92.   Morales,F.J., and Castano, M. 1992. Increased
disease severity induced by some comoviruses in
bean genotypes possessing monogenic dominant
resistance to bean common mosaic potyvirus. Plant
Dis. 76 : 570-573.

93.   Namba, S., Ling, K., Gonzalves, C., Slightom, J.L.,
and Gonzalves, D. 1992. Protection of transgenic

plants expressing the coat protein gene of watermelon mosaic virus II or zucchini yellow mosaic virus against potyviruses. Phytopathology 82 : 940-946.

94. Olorunju, P.E., Kuhn, C.W., Demski, J.W., Misari, S.M., and Ansa, O.A. 1992. Inheritance of resistance in peanut to mixed infections of groundnut rosette virus (GRV) and groundnut rosette assistor virus and a single infection of GRV. Plant Dis. 76 : 95-100.

95. O'Malley, P.J., and Hartman, R.W. 1989. Resistance to tomato spotted wilt virus in lettuce. HortScience 24 : 360-362.

96. Ordon, F., and Friedt, W. 1991. Genetic analysis of exotic barley germplasms for resistance to soil-borne mosaic viruses. Barley Genetics Newsletter 21 : 69-72.

97. Ordon, F., and Friedt, W. 1993. Mode of inheritance and genetic diversity of BaMMV resistance of exotic barley germplasms carrying genes different from ym4. Theor. Appl. Genet. 86 : 229-233.

98. Pal, S.S., Dhaliwal, H.S., and Bains, S.S. 1991. Inheritance of resistance to yellow mosaic virus in some Vigna species. Plant Breeding 106 : 168-171.

99. Paran, I., and Michelmore, R.W. 1993. Development of reliable PCR-based markers linked to downy mildew resistance genes in lettuce. Theor. Appl Genet. 85:985-993.

100. Pelham, J. 1972. Strain-genotype interaction of tobacco mosaic virus in tomato. Ann. Appl. Biol. 71 : 219-228.

101. Perseley, D.M., Martin, I.F., and Greber, G.S. 1981. The resistance of maize inbred lines to sugarcane mosaic virus in Australia. Aust. J. Agric. Res. 32 : 741-748.

102. Pink, D.A.C. 1987. Genetic control of resistance to cucumber mosaic virus in Cucurbita pepo. Ann. Appl. Biol. 111 : 425-432.

103. Pink, D.A.C., Walkey, D.G.A., and McClement, S.J. 1991. Genetics of resistance to beet western yellows virus in lettuce. Plant Pathology 40 : 542-545.

104. Pink, D.A.C., Lot, H., and Johnson, R. 1992. Novel pathotypes of lettuce mosaic virus - breakdown of a durable resistance? Euphytica 63 : 169-174.

105. Pitrat, M., and Lecoq, H. 1980. Inheritance of resistance to cucumber mosaic virus transmission by Aphis gossypii in Cucumis melo. Phytopathology 70 : 958-961.

106. Pochard, E. 1977. Methods for the study of partial resistance to cucumber mosaic virus in Capsicum. Pages 93-104 in : Proc. of the 3rd Congress of Eucarpia on Genetics and Selection of Pepper, INRA, Avignon-Montfavet, France.

107. Ponz, F. Russel, M.L., Rowhani, A., and Bruening, G. 1988. A cowpea line has distinct genes for resistance to tobacco ringspot virus and cowpea mosaic virus. Phytopathology 78 : 1124-1128.

108. Prins, T.W., and Zadoks, J.C. 1994. Horizontal gene transfer in plants, a biohazard? Outcome of literature review. Euphytica 76: 133-138.

109. Provvidenti, R. 1986. Viral diseases of cucurbits and sources of resistance. Pages 1-16 in : Food and Fertilizer Technology Center. Taipei, Taiwan, Tech. Bull. No. 93.

110. Provvidenti, R. 1988. Inheritance of resistance to broadbean wilt virus in bean. HortScience 23 : 895-896.

111. Provvidenti, R. 1990. Inheritance of resistance to pea mosaic virus in Pisum sativum. J. Heredity 81 : 143-145.

112. Provvidenti, R., and Alconero, R. 1988. Inheritance of resistance to a lentil strain of pea seed borne mosaic virus in Pisum sativum. J. Heredity 79 : 45-47.

113. Provvidenti, R., and Hampton, R.O. 1992. Sources of resistance to viruses in the potyviridae. Arch. Virol. (Suppl. 5) : 189-211.

114. Provvidenti, R., and Hampton, R.O. 1993. Inheritance of resistance to white lupin mosaic virus in common pea. HortScience 28 : 836-837.

115. Provvidenti, R., Robinson, R.W., and Shail, J.W. 1979. Chicory : A valuable source of resistance to turnip mosaic virus for endive and escarole. J. Am. Soc. Hort. Sci. 104 : 726-728.

116. Pryor, T., and Ellis, J. 1993. The genetic complexity of fungal resistance genes in plants. Adv Plant. Pathol. 10 : 281-305

117. Querci, M., Baulcombe, D.C., Goldbach, R.W., and Salazar, L.F. 1995. Analysis of the resistance breaking determinants of potato virus X (PVX) strain HB on different potato genotypes expressing extreme resistance to PVX. Phytopathology 85 : 1003-1010.

118. Rajamony, L., More, T.A., and Seshadri, V.S. 1990. Inheritance of resistance to cucumber green mottle mosaic virus in muskmelon (Cucumis melo L.). Euphytica 47 : 93-97.

119. Rasmusson, D.C., and Schaller, 1959. The inheritance of resistance in barley to the yellow dwarf virus. Agronomy J. 51 : 661-664.

120. Ross, H. 1954. Die vererbung der "Immunitat" gegen das X-virus in tetraploidem Solanum acaule. Caryologia Suppl. 6 : 1128-1132.

121. Russell, G.E. 1978. Plant Breeding for Pest and Disease Resistance, 485p, Butterworths, London.

122. Rybicki, E.P., and Shukla, D.D. 1992. Coat-protein phylogeny and systematics of potyviruses. Arch. Virol. (Suppl. 5) : 139-170.

123. Ryder, E.J. 1970. Inheritance of resistance to common lettuce mosaic. J. Am. Soc. Hortic. Sci. 95 : 378-389.

124. Schmidt, H.E. 1992. Bean mosaics. Pages 40-73 in : Plant Diseases of International Importance, Vol II. H.S Chaube, U.S. Singh, A.N. Mukhopadhayay, and J. Kumar, eds. Prentice Hall, New Jersey, U.S.A.

125. Schroeder, W.T., and Barton, D.W. 1958. The nature and inheritance of resistance to the pea enation mosaic virus in garden pea. Pisum sativum L. Phytopathology 48 : 628-632.

126. Shahjahan, M., Imbe, T., Jalani, B., Zarki, A.H., and Othman, O. 1991. Inheritance of resistance to rice tungro spherical virus in rice (Oryza sativa L.). Pages 247-254 in : In Rice Genetics II - Proc. of the Second Int. Rice Gen. Symp., 14-18 May, 1990.

127. Shattuck, V.I., and Stobbs L.W. 1987. Evaluation of rutabaga cultivars for turnip mosaic virus resistance and inheritance of resistance. HortScience 22 : 935-937.

128. Singh. J., and Thakur, M.R. 1977. Genetics of resistance to tobacco mosaic virus, cucumber mosaic virus and leaf curl virus in hot pepper (Capsicum annuum L.). Capsicum 77 : 119-126.

129. Singh, R.P. 1993. Genetic association of gene Bdv1 for tolerance to barley yellow dwarf virus with genes Lr34 and Yr18 for adult plant resistance to rusts in bread wheat. Plant Dis. 77 : 1103-1106.

130. Singh, R.P., Burnett, P.A., Albarran, M., and Rajaram, S. 1993. Bdv1 : a gene for tolerance to

barley yellow dwarf virus in bread wheats. Crop Sci. 33 : 231-234.

131.   Sogawa, K. 1982. The rice brown planthopper: feeding, physiology and host plant interactions. Annu. Rev. Entomol. 27 : 49-73.

132.   Spence, N.J., and Walkey, D..G.A. 1995. Variation for pathogenicity among isolates of bean common mosaic virus in Africa and a reinterpretation of the genetic relationship between cultivars of *Phaseolus vulgaris* and pathotypes of BCMV. Plant Pathol. 44 : 527-546.

133.   Stark, D.M., and Beachy, R.N. 1989. Protection against potyvirus infection in transgenic plants: Evidence for broad spectrum resistance. Bio/Technology 7 : 1257-1262.

134.   Staskawicz, B.J., Ausubel, F.M., Baker, B.J., Ellis, G.E., and Jones, J.D.G. 1995. Molecular genetics of plant disease resistance. Science 268 : 661-667.

135.   Stover, R.H. 1951. Association in tobacco of the severe symptom response to etch virus and the white burley character. Phytopathology 41 : 125-1126.

136.   Sun, Z.Q., Liu, Y.Z., Sun, D.M., and Lao, L. 1990. Inheritance of resistance to soybean mosaic virus strains 1,2 and 3 in soybean (*Glycine max*). Oil Crops of China No. 2 : 20-24.

137.   Suneson, C.A. 1955. Breeding for resistance to barley yellow dwarf virus in barley. Agronomy J. 47 : 283.

138.   Swiezynski, K.M., Chrzanowska, M., Dziewonska, M.A., and Was, M. 1993. Durability of resistance to viruses in the potato. Page 350 in : Durability of Disease Resistance. T. Jacobs and J.E. Parlevliet, eds. Kluwer Academic Publishers, Dordrecht, Netherlands.

139.   Takahashi, R., Hayashi, R., Inouye, T., Moriya, I., and Hirao, C. 1973. Studies on resistance to yellow mosaic disease in barley I. Tests for varietal reactions andgenetic analysis of resistance to the disease. Ber Ohara Inst. 16 : 1-17.

140.   Thakur, M.R. 1976. Inheritance of resistance to yellow vein mosaic in a cross of okra species, *Abelmoschus esculentus* x *A. manihot* sub sp. *manihot*. SABRAO J. 8 : 69-71.

141.   Thomas, J.E., and Mc Grath, D.J. 1988. Inheritance of resistance to potato virus Y in tomato. Aust. J. Agric. Res. 39 : 475-479.

142.   Toriyama, K. 1972. Breeding for resistance to major rice diseases in Japan. Pages 253-281 in : Rice Breeding. IRRI, Los Banos, Philippines.

143.   Troutman, J.L., and Fulton, R.W. 1958 Resistance in tobacco to cucumber mosaic virus. Virology 6 : 303-316.

144.   van Koevering, M., Haufler, Z., Fulbright, D.W., Isleib, T.G., and Everson, E.H. 1987. Heritability of resistance in winter wheat to wheat spindle streak mosaic virus. Phytopathology 77 : 742-744.

145.   Varma, A., Khetarpal, R.K., and More, T.A. 1992. Genetical resistance in plants to viruses. Pages 127-128 in : Proc. of Golden Jubilee Symposium on Genetic Research and Education: Current Trends and the Next Fifty Years, Vol I, New Delhi, India.

146.   Vos, P., Hogers, R., Bleeker, M., Reijans, M., Lee, Theo van de, Hornes, M., Frijters, A., Pot, J., Peleman, J., Kuiper, M., and Zabeau M. 1995.

AFLP: a new technique for DNA fingerprinting. Nucleic Acids Res. 23 : 4407-4414.

147.   Walkey, D.G.A. 1991. Applied Plant Virology, 338p, Chapman and Hall, London, New York and Melbourne.

148.   Walkey, D.G.A., and Neely, H.A. 1980. Resistance in white cabbage to necrosis caused by turnip and cauliflower mosaic viruses and pepper-spot. J. Agric. Sci. 95 : 703-713.

149.   Walkey, D.G.A., Ward, C.M., and Phelps, K. 1985. Studies on lettuce mosaic virus resistance in commercial lettuce cultivars. Plant Pathology 34 : 545-551.

150.   Wang, D., Maule, A.J., Biddle, A.J., Woods, R.D., and Cockbain, A.J. 1993. The susceptibility of pea cultivars to pea seedborne mosaic virus infection and virus seed transmission in the U.K. Plant Pathol. 42 : 42-47.

151.   Watts, L.E. 1975. The response of various breeding lines of lettuce to beet western yellows virus. Ann. Appl. Biol. 81 : 393-397.

152.   Whitney, E.D. 1989. Identification, distribution, and testing for resistance to rhizomania in *Beta maritima*. Plant Dis. 73 : 287-290.

153.   Wiersema, H.T. 1961. Methods and means used in breeding potatoes with extreme resistance to viruses X and Y. Pages 30-36 in : Proc. 4th Conf. Potato Virus Dis. H. Veenman and Zonen, Wageningen, The Netherlands.

154.   Wilkins, P.W. 1987. Transfer of polygenic resistance to ryegrass mosaic virus from perennial to italian ryegrass by backcrossing. Ann. Appl. Biol. 111 : 409-413.

155.   Williams, J.G.K., Kubelik, A.R., Livak, K.J., Rafalski, J.A., and Tingey, S.V. 1990. DNA polymorphisms amplified by arbitrary primers are useful as genetic markers. Nucleic Acids Res. 18 : 6531-6536.

156.   Wilson, T.M.A. 1993. Strategies to protect crop plants against viruses : Pathogen-derived resistance blooms. Proc. Natl. Acad. Sci., U.S.A. 90 : 3134-3141.

157.   Wyne, J.C., and Beute, M.K. 1991. Breeding for disease resistance in peanut (*Arachis hypogea* L.). Annu. Rev. Phytopathol. 29 : 279-303.

158.   Yoon, J.Y., Green, S.K., and Opena, R.T. 1993. Inheritance of resistance to turnip mosaic virus in Chinese cabbage. Euphytica 69 : 103-108.

159.   Zaumeyer, W.J., and Meiners, J.P. 1975. Disease resistance in beans. Annu. Rev. Phytopathology 13 : 313-334.

160.   Zhang, Y.D., Gai, J.Y., and Ma, Y.H. 1989. Inheritance of resistance to two local soybean mosaic virus strains S_A and S_C in soybeans. Acta Agron. Sinica 15 : 213-220.

161.   Zink, F.W., and Duffus, J.E. 1969. Relationship of turnip mosaic virus suceptibility and downy mildew (*Bremia lactucae*) resistance in lettuce. J. Am. Soc. Hort. Sci. 94 : 403-407.

162.   Zink, F.W., and Duffus, J.E. 1973. Inheritance and linkage of turnip mosaic virus and downy mildew (*Bremia lactucae*) reaction in *Lactuca serriola*. J. Am. Soc. Hort. Sci. 98 : 49-51.

CHAPTER 3

Control of Plant Virus Diseases by Cross Protection

H. Lecoq

Virus diseases are prevalent in cultivated plants in highly intensive agriculture systems as well as under more traditional farming conditions. Limiting the economic impact of plant viruses is particularly difficult to achieve because of the systemic and non-curable nature of virus infections in the field, the lack of resistant or tolerant cultivars of good agronomic quality, and the diversity of viruses and their natural vectors. Keeping this in mind, any control method for an economically important virus should be considered for application. Among these methods, cross protection is particularly attractive since it uses viruses themselves to combat viral diseases.

Cross protection was discovered by McKinney in 1929 (16), when he observed that a tobacco plants systemically infected by a "green" strain of tobacco mosaic virus (TMV) were protected from infection by another strain of this virus inducing yellow mosaic symptoms. This phenomenon was subsequently generalized to apply to many other plant viruses except for a limited number of viruses (including geminiviruses such as beet curly top virus) (15). Cross protection was widely used to establish relationships among virus strains. It was also of potential interest for protecting plants against viruses in the field (15).

Cross protection has not only been observed to occur between viruses, it was also demonstrated to occur between viroids (21) or plant virus satellites (10). Several comprehensive reviews have been published recently on cross protection, and on the mechanisms which might be involved in this interaction (4-7,22,23,25). Therefore, we will focus here on the applied aspects of implementing this technique for a few well known virus/host combinations.

CROSS PROTECTION: PRINCIPLES, PROCEDURES AND POSSIBLE RISKS

General Comments

The terminology commonly used refers to the virus strain which induces cross protection, as the 'protecting strain' and to the strain which is used to evaluate cross protection efficiency as the 'challenging strain'. In laboratory experiments challenging strains are often chosen because they induce either severe or easily recognizable (i.e. aucuba, yellow) symptoms. Challenging strains can be inoculated mechanically, by grafting or using their natural vectors at different times after the inoculation of the protecting strains. Interestingly, in some cases there is evidence of some replication of the challenging virus strain, but it fails to produce its typical symptoms (7).

When field application is considered, the protecting strain is generally an isolate which induces mild symptoms and does not affect the marketable yield of the crop. However, when losses from the disease is great, some yield reduction induced by the protecting strain might be economically tolerable. In the literature, protecting strains used for controlling severe strains are often referred to as 'mild', 'attenuated', 'hypovirulent' or 'avirulent' strains (4,5,9,22). In this article hereafter, protecting strains will be referred to as 'mild' strains or isolates. Cross protection using mild strains has been used in experiments against viruses belonging to different groups (Table 1), infecting vegetable crops or fruit trees most frequently, probably because it is on these hosts that viruses have the most deleterious effects on the cash value of the crop.

An 'ideal' mild isolate to be used in the

field for cross protection should have the following characteristics:

i)   It should induce milder symptoms than isolates commonly encountered in the fields, and should not alter the marketable yield and the quality of the crop. It should also be mild in all its cultivated hosts including those which are not targets for the cross protection.

ii) It should be fully systemic and invade all host tissue. Indeed, cross protection effectiveness depends on the presence of the mild strain in all tissues to be protected from severe strains inoculation.

iii) It should be genetically stable, and not evolve towards more severe forms at times.

iv) It should not be easily disseminated by vectors, in order to limit any non-intentional spread to other crops or fields.

v)   It should provide a protection towards the widest possible range of severe challenging isolates.

vi) The protective inoculum should be easy to produce, check for purity, store, and provide to farmers. A simple inoculation procedure should be designed so that it does not require expensive equipment or specific training in order to be applied in the fields.

Selection of Mild Strains

Mild strains may be obtained by various methods (Table 1). Some were obtained as

Table 1 . Viruses for which mild strains have been used for cross protection in the fields (on an experimental [E] or a commercial [C] basis), means by which mild strains have been obtained, and crops which have been cross protected

Virus	Origin of mild strains	Crops (experimental [E] or commercial [C] field tests)
Badnavirus		
cocoa swollen shoot	field isolates (8)	cocoa [E]
Closterovirus		
citrus tristeza	field isolates (3,18)	citrus [C]
Nepovirus		
arabis mosaic	field isolates (9)	grapevine [E]
Potyvirus		
papaya ringspot	mutagenesis (31)	papaya [C], squash [E]
zucchini yellow mosaic	greenhouse variants (12)	squash [C], cucumber [C]
soybean mosaic	cold treatment (11)	soybean [E]
Tobamovirus		
tomato mosaic	mutagenesis (24) heat treatment (22) field isolates (4)	tomato [C]

naturally occurring variants - plants (most often trees) with mild symptoms may be observed in the fields while most of the other plants show severe symptoms. Frequently mild virus subcultures may be isolated from such plants. Other mild strains have been obtained in the laboratory, either from single local lesion isolations from samples with severe symptoms, or from plants inoculated by severe isolates, but which developed spontaneously axillary branches with mild symptoms (12). Heat or cold treatments may also yield mild isolates (11,22), and mild variants were obtained in the laboratory, after mutagenesis treatment (generally with nitrous acid) followed by single local lesion selection (24,31).

At this stage of mild strain isolation, special care should be taken to ensure that the protecting isolate is not contaminated with a severe isolate or another virus. Various cloning and control proceedures are, therefore, necessary to ensure the purity and safety of the mild isolate. These include single local lesion transfers, differential host analysis, serological assays and examination with the electron microscope.

After a mild isolate has been selected, different steps should be followed in order to evaluate its' potential for practical applications. A preliminary evaluation should be performed in the laboratory or in a protected environment in order to evaluate the effect of the mild strain on symptom expression as swell as marketable yield production. This should also provide information about the protective effect of the mild isolate against a range of challenging virus isolates from various geographic origins under a variety of inoculation mechanisms (artificial or vector) and environmental conditions.

Subsequently, experiments should be conducted under field conditions in areas where natural epidemics caused by severe isolates occur. Local regulations governing biological control experiments should be obeyed while maintaining strict scientific supervision. These tests may reveal unsuspected problems, such as possible synergetic effects with unrelated viruses or reversion towards a more severe form of the virus. Field evaluation of mild strains will be achieved only when large scale plots are used at commercial farms in different ecological situations. These trials will confirm the effectiveness of the technique and provide a first hand evaluation of the economic feasibility of cross protection.

Mild Strain Production and Application

Mild strain increase should be conducted in a highly protected environment under very strict phytosanitary supervision in order to eliminate risks of contamination by undesirable viruses, bacteria or fungi. Certified seeds and sterilized potting soils for plant production should be used and disinfection of insect-proof greenhouses or growing cabinet structures as well as regular pesticide treatments should be performed. Random serological tests should also be performed in order to detect any possible contamination. Mild strain production on a large scale may be conducted by research institutes, agricultural technical centers as well as private companies.

It is essential to provide farmers (or farm advisers) with an easy and an efficient inoculation technique for the mild strain. Various techniques have been tested. Mechanical inoculation (with hair brushes, cotton swabs, sponge pads or forceps (5, 19)) is labor intensive and time consuming; it may also favor the non-intentional spread of other mechanically transmissible severe viruses. Different grafting techniques have proved to be very efficient to inoculate mild strains to woody plants (3,17). Another alternative is the use of spray guns (with adapted air pressures and nozzle sizes) (4,19,32); it has been shown to be efficient in inoculating relatively stable viruses (tobamoviruses and potyviruses) to seedlings and, for some crops, to adult plants.

Another parameter important for implementing cross protection is the 'safety' period after mild strain inoculation. This period may be defined as the time necessary for the mild strain to invade its host before providing a full protection. This may depend upon the host, the virus and the environmental conditions.

Mode of Action and Risk Assessment

Although cross protection has been used in practice in a number of crops and has been studied with an academic approach in many host/virus combinations, little is known about its' mechanism of operation. One theory involves precursor or receptor site exhaustion, while another involves the synthesis of specific inhibitors of related viruses by the cross protected plants (25,27). Two other hypotheses have been more recently developed. One hypothesis suggests that the nucleic acid of the challenging strain might be sequestered by the coat protein of the protecting strain, preventing subsequent translation and replication of the challenging virus RNA (25,27). The other involves hybridization between (+) and (-) RNA strands of protecting and challenging strains preventing replication and/or translation of the challenging strain (23). The latter theory is particularly attractive since it takes into account the cross protection phenomenon observed with viroids or satellite RNAs. However, different mechanisms may be involved in different virus/host combinations.

More than any other control method, cross protection has been associated with potential hazards (5,7,15). Several limitations or possible risks associated with this control method are as follows:

i)   An incomplete protection may occasionally be observed; indeed some apparent "breakdowns" of cross protection have been reported, especially in senescent plants (5,12,20).

ii)   The protecting strain may possibly spread to other hosts in which it may have more severe effects (5).

iii) Amplified disease symptoms caused by a synergism with other viruses might spread readily in the cross protected crop (14,20).

iv)   Heteroencapsidation or heteroassistance in mixed infection with another virus may modify virus transmission, specificity or efficiency (1,12).

v)   genetic recombination between the protecting strain and other virus(es) in mixed infection may occur (5).

vi)   The protecting virus may mutate into a more severe form that would cause a destructive disease.

Despite these anticipated difficulties, the method has been and is effective with a number of virus diseases and crops. Nevertheless, due to the constraints linked to its' implementation in the field, cross protection should be considered only for viruses which present a real threat to a crop and for which no other alternative control method is presently available. It should be considered only if farmers will gain significant economic benefits from its' use without taking risks with the health of their crops.

Under these conditions, the use of cross protection should not introduce additional specific risks in comparison to those risks which may arise by letting a severe strain spread naturally in the fields without appropriate control measure.

EXAMPLES OF FIELD APPLICATIONS

Tobacco (or Tomato) Mosaic Tobamovirus

Tobacco (or Tomato) mosaic tobamovirus (ToMV) induces very severe diseases in various Solanaceous crops including tomato. It provokes severe mosaic on tomato leaves, flower abortion and necrotic symptoms on the fruits (internal browning) which render the crop unmarketable. It is seed transmitted in tomatoes and readily spread by mechanical cultivation (2).

Mild strains were obtained either from naturally occurring strains or by culturing a severe strain in tissue held above 34°C (5). The mild strain MII-16, which was extensively used in Western Europe was obtained by nitrous acid mutagenesis followed by single local lesion selection (24). Cross protection against ToMV in tomatoes has been widely used in Europe, Canada, New Zealand and Japan (2,4,20,22). Average yield increase of early heated crops associated with the use of cross protection was estimated to reach 15% in the Netherlands (2).

In France, a subculture of MII-16 was used commercially from 1972 to 1983 (14,19). Preliminary experiments and pilot tests demonstrated that it did not reduce the yield but only slightly delayed fruit setting. It provided very efficient protection against ToMV severe, however, the risk of synergism with other viruses (and particularly with cucumber mosaic virus, common in southern France) limited the use of ToMV cross protection (14).

The inoculum (purified virus preparation) to be used by farmers was prepared by an agricultural technical center (19,20) under very strict control measures. Every purified preparation was carefully analyzed by checking its' specific infectivity (estimated by counting the number of local lesions produced on tobacco cv Xanthi), the absence of severe strain contaminants (by inoculating a set of tomato plants), and its' cross protective potential (by challenging cross protected plants with an aucuba strain of ToMV) (19).

Two inoculation procedures were proposed to farmers; one used forceps dipped in the inoculum solution, while the other used a spray gun (19). This latter method was adopted preferentially as it helped to limit the risk of mechanically transmiting any severe virus strain which might be present in the seedling population. After inoculation with the mild strain, tomato leaves developed yellow spots which served as a good indicator to the farmer that the innoculation succeeded.

Besides the real success of ToMV cross protection scheme (more than 13 million tomato plants were cross protected in France in 1981, 5 years after the first release of ToMV resistant cultivars), some drawbacks were noticed:

i) In some cases, generally associated with stress situations for the plants, more pronounced symptoms were observed. However, it is not known whether the symptoms were signs of cross protection "breakdowns" or whether they were only more severe responses of the plants to the mild strain under certain environmental conditions.

ii) Farmers tended to neglect classical prophylactic measures when using cross protection. In particular, they sometimes failed to use disinfected seed lots, leading to accidental spread of seed borne severe strains.

iii) The mild strain belongs to ToMV pathotype 1. Therefore, it induces necrotic reactions on plants possessing the $Tm2_2$ gene in the heterozygous form and some problems were noticed on resistant plants when they were grown next to susceptible cross protected crops.

Nevertheless, cross protection proved to be a very efficient and popular method to control ToMV in tomato in France and many other countries. Only the development of resistant cultivars of good agronomic quality has progressively reduced its' use. This may be summarized by Broadbent's comment on the mild strain MII-16 : 'Apart from the breeding of resistant cultivars, its development has done more than any other factor to decrease crop losses from TMV in tomato' (2).

Zucchini Yellow Mosaic Potyvirus

Zucchini yellow mosaic virus (ZYMV) is now regarded as one of the major pathogens of cucurbits in most regions of the world. ZYMV is very efficiently transmitted by several aphid species in a non-persistent manner. In zucchini squash, symptoms are particularly pronounced and include mosaic and distortions of the leaves and fruit. Infected plants generally cease to produce marketable fruit 1 or 2 weeks after inoculation which leads to significant economic losses, particularly when epidemics start early after planting. In melons, cucumbers and watermelons, symptoms are also very severe (12).

ZYMV presents a high potential for variability, and variants differing in symptomatology, host range, virulence towards resistance genes, and aphid transmissibility have been reported (13). Control measures by cultural practices, including the use of plastic mulches or oil sprays may provide a temporary protection for the crops, but this is generally not sufficient to prevent significant economic losses. Resistance to ZYMV has not been identified so far in *Cucurbita pepo* gene banks. The only available resistance to ZYMV has been reported in the related species *C. moschata* or resistance may be obtained through genetic engineering using the coat protein mediated resistance strategy. Until these resistance mechanisms are available in commercial cultivars, cross protection appears to be an interesting alternative approach to ZYMV control in zucchini squash (12,28,29).

A mild variant (ZYMV-WK) was obtained from a melon axillary branch showing very attenuated symptoms although it was produced on a plant infected by a severe isolate (12). Extracts from this branch were mechanically inoculated to a melon plant which subsequently exhibited mild symptoms. ZYMV-WK produces very mild symptoms both in cucurbits (melon, cucumber, watermelon or squash) and non-cucurbits (*Chenopodium amaranticolor*), and it is poorly aphid transmissible (12). ZYMV-WK only slightly affected the production of zucchini squash and, as mentioned for ToMV in tomato, it slightly delays fruit setting rather than decreasing the production potential of individual plants.

In the field, under severe ZYMV epidemic conditions, ZYMV-WK provided a very effective protection in temperate, Mediterranean and tropical environments (12,28,29). No synergetic effect was observed with any of the several other viruses commonly infecting cucurbits (12,29). Consequently, ZYMV-WK mild isolate may be used in protected crops as well as in crops grown in the open.

In laboratory experiments, ZYMV-WK was observed to be efficient in protecting plants against severe ZYMV isolates from various origins (Mediterranean Basin, northern Europe, Africa, North and Central America, etc.). However, ZYMV-WK was not efficient in controlling severe isolates originating from Reunion, Mauritius and Madagascar, which are also serologically distantly related to the type strain of ZYMV (unpublished data). Therefore, it seems advisable to carry out preliminary cross protection experiments in controlled environmental conditions with local ZYMV isolates before contemplating its use in the field.

Two inoculation methods were used: one used an inoculation bottle and the other used a spray-gun. Both were equally effective; the spray-gun proved to be more rapid while using more inoculum per plant (crude extracts of mild strain infected plants) (G. Ginoux and E. Lecoq, unpublished data). ZYMV-WK conferred a full protection 14 days after inoculation, although some level of protection was observed 7-8 days after inoculation (28).

In the later stage of crop development a few plants may occasionally develop more severe symptoms on leaves and moderate symptoms on fruits (12). It is not yet known whether the apparent partial breakdown of the cross protection in these plants is due to high inoculum pressure, physiological changes in the plants or the evolution of the mild isolate.

The spread of cucumber mosaic comovirus and watermelon mosaic potyvirus type 2 (WMV2) was very similar in control and cross-protected plots suggesting that an infection by ZYMV-WK did not specifically interfere with the spread of these two aphid transmitted viruses (12). Although ZYMV-WK is poorly aphid transmissible, it was transmitted, to a significant extent, from plants grown in the field and co-infected with WMV2. In fact, it was shown that WMV2 provided a functional helper component to assist transmission of the mild isolate (12). ZYMV-WK does not appear to be seed transmitted (12, Lecoq, unpublished data); this may be very useful to plant breeders, who may not satisfactorily achieve crosses or seed increase, when ZYMV epidemics are prevalent.

ZYMV-WK proved to be efficient in protecting zucchini squash against severe ZYMV infections under field conditions. Preliminary observations indicate that this isolate is also very efficient in protecting watermelon and cucumbers.

Papaya Ringspot Potyirus

Papaya ringspot virus (PRSV) is an aphid borne potyvirus which causes important economical losses to papaya, an important tropical fruit crop. Symptoms on papaya include plant stunting, a severe mosaic on leaves, spotting on fruits which reduce their quality or render them unmarketable. PRSV also causes a severe disease in cucurbits (6,30,32).

Two mild variants were obtained by mutagenesis with nitrous acid treatment of crude virus preparations and after selection from 663 single local lesion isolates (31). These mild strains proved to be very effective both in laboratory tests and in pilot field tests in Hawaii. Protection was observed in Taiwan while cross protection was generalized in some regions where more than a million papaya seedlings were cross protected in 1986 (32). However, severe symptoms were noticed on some cross protected plants, particularly under severe inoculum pressure. The PRSV mild isolates were derived from a Hawaiian severe strain : they may not completely protect against the Taiwan isolates. A search for mild variants derived from a Taiwanese severe strain has yielded promising new mild isolates (32). Recently, a study was done to compare the specificity of the protection conferred by classical cross protection using a mild variant of PRSV and by transgenic papaya expressing the coat protein gene of the same isolate (26). In both cases, a high level of protection was observed with severe PRSV isolates from Hawaii. In contrast, a range of reactions varying from delayed and attenuated symptoms to severe symptoms was observed with a series of other isolates from different geographical origins (26).

Citrus Tristeza Closterovirus

Citrus tristeza virus is prevalent in many regions where citrus is grown. This closterovirus is transmitted by aphids in a semi-persistent mode. It causes very important losses, particularly in certain scion/rootstock combinations where it may cause reduction in fruit size and yield, stem-pitting, decline and the premature death of the plant. Mild isolates were obtained in Brazil, Australia, South Africa and Reunion Island from trees noticed as showing no or very discrete symptoms in areas where all other trees were severely infected (17,18).

A unique aspect of this system is that the

protecting virus may be inoculated by grafting scions originating from mild strain-infected mother plants, which limits the problems of inoculum production and storage. Another peculiarity is that it appears to be necessary to select a mild variant for each scion/rootstock combination. For instance, in Brazil, of 45 selected field isolates tested for their 'mildness' 3 were found to be satisfactory for Pera orange, 2 others for Gallego lime and another for Ruby Red grapefruit. The best isolates for Pera orange were not convenient for the other crops and consequently, specific mild isolates have to be used for each citrus type.

Cross protection has been widely used with great success in Brazil; more than 8 million Pera orange trees were cross protected in 1980 (17) and more than 50 million trees in 1987 (27). Cross protection was also a success in other countries such as South Africa. This disease was probably ideal for the development of cross protection since it is often endemic and impossible to eradicate, it causes major crop losses by spreading rapidly and, most importantly, because there is no alternative control method.

Cocoa Swollen Shoot Badnavirus

Cocoa swollen shoot virus is a very severe disease of cocoa in Western Africa. It is transmitted by mealybugs and induces mosaic or vein banding symptoms on the leaves, typical swellings of shoots and premature death of the tree (8).

Attempts to control the disease by cross protection have been conducted, with some success, in Ghana for many years. One major barrier to the implementation of this method was that cross protection was incompatible with the eradication strategy which was first chosen to control the disease in this country (8). Now that eradication seems to have failed, cross protection to control CSSV in cocoa is reconsidered for regions where the virus is endemic.

Progress recently achieved in grafting or rooting cocoa will probably contribute to the development of this control method.

CONCLUDING REMARKS

Cross protection has proved to be very effective in controlling several major viral disease problems when no alternative control method was available. It was successfully applied to a variety of crops which were cultivated under a variety of conditions (annual and perennial crops cultivated in modern greenhouses or in the field using traditional tropical farming methods), against a variety of viruses including tobamo-, poty-, clostero- and badnaviruses.

The versatility of this method makes it easy to use in a timely manner as new problems arise, and it can be applied easily to a range of different cultivars. Cross protection should be integrated into the crop management system as the use of specific cultural practices along with cross protection may enhance its' field efficiency and counteract slight yield losses which are occasionally observed after mild strain inoculation.

REFERENCES

1. Bourdin, D., and Lecoq, H. 1991. Evidence that heteroencapsidation between two potyviruses is involved in aphid transmission of a non transmissible isolate from mixed infections. Phytopathology 81:1459-1464.
2. Broadbent, L. 1976. Epidemiology and control of tomato mosaic virus. Annu. Rev. Phytopathol. 14:75-96.
3. Costa, A.S., and Muller, G.W. 1980. Tristeza control by cross protection : a U.S. Brazilian cooperative success. Plant Dis. 64:538-541.
4. Fletcher, J.T. 1978. The use of avirulent strains to protect plants against the effects of virulent strains. Ann. Appl. Biol. 89:110-114.
5. Fulton, R.W. 1986. Practices and precautions in the use of cross protection for plant virus disease control. Annu. Rev. Phytopathol. 24:67-81.
6. Gonsalves, D., and Garnsey, S.M. 1989. Cross-protection techniques for control of plant virus diseases in the Tropics. Plant Dis. 73:592-597.
7. Hamilton, R.I. 1985. Using plant viruses for disease control. HortScience 20:848-852.
8. Hughes, J. d'A., and Ollenu, A.A. 1994. Mild strain protection of cocoa in Ghana against cocoa swollen shoot virus - a review. Plant Pathol. 43:442-457.
9. Huss, B., Walter, B., and Fuchs, M. 1989. Cross-protection between arabis mosaic virus and grapevine fanleaf virus in Chenopodium quinoa. Ann. Appl. Biol. 114:45-60.
10. Jacquemond, M., and Tepfer, M. 1995. Satellite RNA-mediated resistance to plant viruses : are the ecological risks well assessed? In this book.
11. Kosaka, Y., and Fukunushi, T. 1993. Attenuated isolates of soybean mosaic virus derived at low temperature. Plant Dis. 77:882-886.
12. Lecoq, H., Lemaire, J.M., and Wipf-Scheibel, C. 1991. Control of zucchini yellow mosaic virus in squash by cross protection. Plant Dis. 75:208-211.
13. Lecoq, H., and Purcifull, D.E. 1992. Biological variability of potyviruses, an example: zucchini yellow mosaic virus. Arch. Virol. Suppl.5:229-234.
14. Marrou, J., and Migliori, A. 1972. La prémunition, une nouvelle méthode de protection des cultures contre le virus de la mosaïque du tabac. P.H.M.-

Revue Horticole 124:27-31.

15.   Mathews, R.E.F. 1991. Plant Virology, third edition. Academic Press Inc, San Diego. 835 pages.

16.   McKinney, H.H. 1929. Mosaic diseases in the Canary Islands, West Africa and Gibraltar. J. Agric. Res. 39: 557-578.

17.   Muller, G.W., 1980. Use of mild strains of citrus tristeza virus (CTV) to reestablish commercial production of 'Pera' sweet orange in Sao Paulo, Brazil. Proc. Flo. State Hort. Soc. 93:62-64.

18.   Muller, G.W., and Costa, A.S. 1987. Search for outstanding plants in tristeza infected citrus orchards : the best approach to control the disease by preimmunization. Phytophylactica 19:197-198.

19.   Navatel, J.C. 1977. Lutte contre la mosaïque du tabac sur tomate : prémunition et variétés résistantes. CTIFL Documents n°53, 109-119.

20.   Navatel, J.C., Trapateau, M., and Marchoux, G. 1983. La prémunition : méthode de protection contre les virus. Bilan de la lutte contre la mosaïque du tabac. Quelques perspectives nouvelles. C. R. Col. ACTA 'Faune et flore auxilliaire en agriculture', Paris, May 1983, 36-42.

21.   Nibblett, C.L., Dickson, E., Fernow, K., Horst, R.K., and Zaitlin, M. 1978. Cross protection among four viroids. Virology 91:198-203.

22.   Oshima, N. 1975. The control of tomato mosaic virus disease with attenuated virus of tomato strain of TMV. Rev. Plant Prot. Res. 8:126-135.

23.   Palukaitis, P., and Zaitlin, M. 1984. A model to explain the 'cross protection' phenomenon shown by plant viruses and viroids. Pages 420-429 in: Plant-microbe interactions: molecular and genetics perspectives. T. Kosuge, and E.W. Nester, eds. MacMillan, New York.

24.   Rast, A.T.B. 1972. MII-16, an artificial symptomless mutant of tobacco mosaic virus for seedling inoculation of tomato crops. Neth. J. Plant Pathol. 78:110-112.

25.   Sherwood, J.L. 1987. Mechanisms of cross-protection between plant virus strains. Pages 136-150 in : Plant resistance to viruses, D. Evered and S. Harnett, eds. John Wiley and Sons, Chichester, New York.

26.   Tennant, P.F., Gonsalves, C., Ling, K.S., Fitch, M., Manshardt, R., Slightom, J.L., and Gonsalves, D. 1994. Differential protection against papaya ringspot virus isolates in coat protein gene transgenic papaya and classically cross-protected papaya. Phytopathology 84:1359-1366.

27.   Urban, L.A., Sherwood, J.L., Rezende, J.A.M., and Melcher, U. 1990. Examination of mechanisms of cross protection with non-transgenic plants. Pages 415-426 in : Recognition and response in plant-virus interactions. R.S.S. Fraser, ed. Springer-Verlag, Berlin.

28.   Walkey, D.G.A., Lecoq, H., Collier, R., and Dobson, S. 1992. Studies on the control of zucchini yellow mosaic virus in courgettes by mild strain protection. Plant Pathol. 4:762-771.

29.   Wang, H.L., Gonsalves, D., Provvidenti, R., and Lecoq, H. 1991. Effectiveness of cross-protection by a mild strain of zucchini yellow mosaic virus in cucumber, melon and squash. Plant Dis. 75:203-207.

30.   Wang, H.L., Yeh S.D., Chiu R.J., and Gonsalves D. 1987. Effectiveness of cross protection by mild mutants of papaya ringspot virus for control of ringspot disease of papaya in Taiwan. Plant Dis. 71:491-497.

31.   Yeh, S.D., and Gonsalves, D. 1984. Evaluation of induced mutants of papaya ringspot virus for control by cross protection. Phytopathology 74:1086-1091.

32.   Yeh, S.D., Gonsalves, D., Wang, H.L., Namba, R., and Chiu, R.J. 1988. Control of papaya ringspot virus by cross protection. Plant Dis. 72:375-380.

CHAPTER 4

Control of Virus Infection in Crops Through Breeding Plants for Vector Resistance

A.T. Jones

Apart from immunity or strong resistance to plant viruses, virus control in crops is largely dependent on planting virus-free material in 'clean' areas and controlling virus inoculum entering the crop and/or spreading within it. As the spread of most plant viruses is dependent on a vector, traditionally virus movement into and within crops has been based on the control of virus vectors, usually by pesticide applications and/or cultural methods. The effectiveness of these approaches, either alone or together, varies enormously and is dependent on several interacting factors, including the vector species, vector numbers, mode of virus transmission, virus and vector host range and, in some instances, environmental conditions. However, even where satisfactory control has been achieved using pesticides, it is now widely accepted that such inputs, even in high-unit-output developed countries, are unsustainable economically (because of escalating costs in the production and application of pesticides and the development of new ones), environmentally (in terms of killing non-target organisms and contamination of the food chain, water and environment), and in some instances, in efficiency (in terms of missing target organisms). Against this background, there is increasing pressure to develop methods of virus control that are in keeping with a sustainable approach to agricultural production; breeding plants for virus vector resistance is one such approach (51).

Initial studies in this area utilized crop plants bred specifically for resistance to major plant pests but that were known also to be virus vectors. Whilst such material has been found

useful in some situations (38,49), the plant characteristics required to deter or prevent virus acquisition and/or transmission are often very different from those that may provide satisfactory control of vectors as pests. For example, plants bred for tolerance to pests can be spectacularly effective in decreasing losses due to pest attack even though the colonization and multiplication of pests is not deterred entirely. However, such plant characteristics are unlikely to be very useful in decreasing virus incidence, especially of non-persistent viruses. Nevertheless, the use of some of these pest-resistant crop plants for controlling virus infection has indicated the potential of vector resistance for controlling plant viruses and has stimulated the investigation of this approach for a wider range of plant/virus combinations. Thus, in 1976, Kennedy (57) cited only eight instances of an association between resistance to aerial vectors and a decrease in virus incidence or infection and some of these reports were only passing comments or casual observations by authors. However, in an extensive review of this subject area in 1987, Jones (51) cited more than 20 examples of different crop/virus combinations where vector resistance was found to decrease virus incidence in the field, several additional examples where such a decrease was reported in laboratory studies, and evidence of more than 15 important crop/virus combinations where vector resistance was being actively used or studied as an approach to virus control. Since that review, further crop/virus combinations have been added to the lists. There is therefore considerable interest in the potential to control viruses by this means and increasing evidence for its effectiveness against viruses transmitted by most of the main categories of aerial plant virus vectors and some indications that this approach might also be successful for viruses transmitted by nematode and fungal vectors (51). This chapter does not seek to cover all of these examples but seeks rather to identify some of the important mechanisms involved in vector resistance, because an increase in our detailed understanding of these mechanisms is likely to produce new approaches and techniques for the further improvement of vector resistance for virus control. However, this approach to virus control, as with many others, is not without difficulties and possible hazards, and therefore some of these will also be highlighted.

SOME PRELIMINARY CONSIDERATIONS

The relations between virus, vector and host plant involve many interacting factors. Changes in any one of these factors may have pronounced influences on infection of plants with viruses and on virus spread between plants and between crops. In addition, many of these interactions are influenced by environmental and cultural conditions. Some of the more important interacting factors are : for vector/host plant interactions, the host range of the vector, its mobility and the type, and the level and stability of vector resistance in the host plant; for virus/vector interactions, the vector relations of the virus (non-persistent, semi-persistent, persistent circulative, persistent propagative), and the role of any helper virus in transmission; for virus/host plant interactions, the host range of the virus and the infectibility and susceptibility of the plant to virus and the concentration and location of virus particles in the plant and their availability to the vector. In a few instances, virus infection may decrease the vector resistance of the host plant. For example, the yellowing of leaves, caused by infection with some viruses, makes them more attractive to aphids (9), and infection of black currant with the agent of reversion disease causes loss of pubescence in the buds, making them more prone to infestation by vector mites (108). In addition, virus infection of some vectors may influence their feeding behavior (21). Despite these examples, unless stated otherwise, it will be assumed that vector-resistant and vector-susceptible plants are equally prone to infection with virus and attention will focus on the vector/host plant interactions.

The analysis of vector/host plant interactions, as might be expected, has been best studied for aerial vectors, particularly aphids, planthoppers and leafhoppers. Nevertheless, many of the principal features of these interactions that affect virus infection and that are given below may also apply to other vector species including nematodes (33,34,92) and fungi (72,87). In 1951, Painter (81) identified three kinds of resistance in plants to insects : *non-preference* (an adverse effect on insect behavior), *antibiosis* (an adverse effect on insect growth, reproduction and survival) and *tolerance* (ability of a plant to support insect colonization without adverse effects on its own growth and development). These continue to receive, wide acceptance. However, the term non-preference refers to the behavioral response of the insect to the plant and might be better replaced by the term *antixenosis* to describe the properties of the plant responsible for non-preference (86). Furthermore, tolerance, is not strictly an expression of resistance to an insect but resistance to the effects on the plant of insect colonization and feeding. As indicated earlier, although tolerance is an important character from the point of view of breeding for pest resistance, it is likely to have little or no effect on resistance to virus transmission. Indeed, in some circumstances it is likely to increase vector numbers and virus infection (57) and will therefore not be considered further. Areas of overlap occur in the expression of antixenosis and antibiosis and they may be difficult to distinguish when both forms of resistance occur in the same plant. However, antixenosis is most often expressed by interfering with the location of the host plant by the vector and the initial identification and establishment of suitable feeding sites on the plant; whereas antibiosis, although sometimes influencing the establishment of feeding sites, is usually expressed in relation to the growth and development of the vector and its progeny. Each of these expressions of resistance may lead prematurely to vector dispersal and thereby influence secondary virus spread (58). However, it is important to realize that the resistance terms used are only descriptions of the expression of resistance and not of its cause. It is useful therefore to identify some of the factors contributing to expressions of resistance to vectors in plants following the natural sequence of events in the acquisition and/or transmission of plant viruses by aerial vectors arriving in a crop.

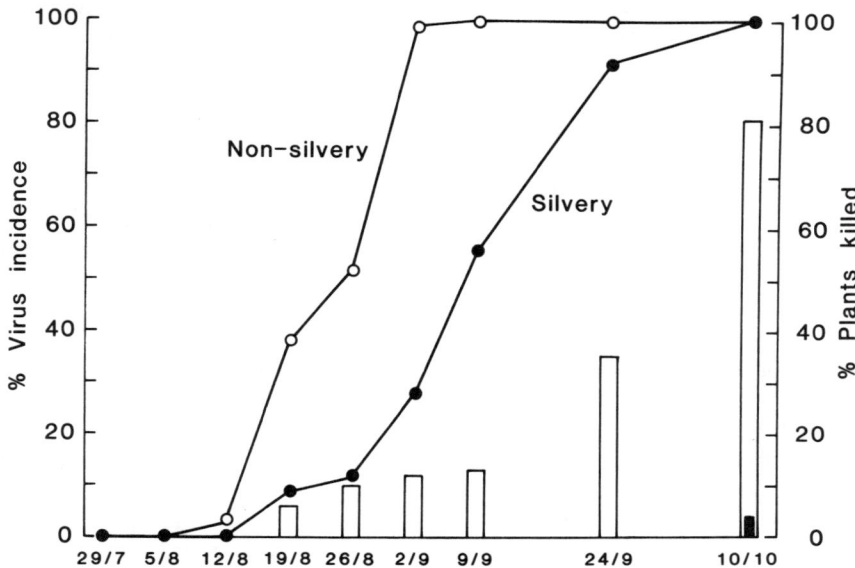

Figure 1. Effect of a silvery leaf character in *Cucurbita pepo* on the incidence (%) of non-persistently transmitted viruses (lines) and on the percentage of plants killed by virus infection (histograms) in 1981. The delay in virus infection of silvery plants was sufficient to prevent death in all except a few plants in early October. Drawn from data by Davis and Shifriss (24).

SOME MECHANISMS INVOLVED IN VIRUS VECTOR RESISTANCE

Interference with Location of Host Plants by Vectors

The many factors influencing host recognition by virus vectors is a rapidly expanding field of study but remains incompletely understood. In some instances, host finding may occur simply by a random encounter, but the well documented responses of insects to color and other physical and chemical stimuli indicate that active mechanisms are involved in host recognition. This is supported by the finding of a greater abundance of secondary rhinaria on antenna of alate aphids compared to apterous ones, and that some of these rhinaria respond to plant volatiles (84). Changes in the occurrence or strength of plant signals to which vectors respond may deter alighting responses or feeding altogether, thereby preventing virus inoculation or acquisition. The significant influence of plant color on vector alighting and

virus transmission in crops was demonstrated by Shifriss (99) and Davis and Shifriss (24). They found that, compared to similar plants of a normal green color, cucurbits with silvery leaves reflected much higher than normal amounts of UV and blue light, and in the field were infested with fewer aphids and had a lower incidence of infection with non-persistent aphid-borne viruses in the early stages of an epidemic (Figure 1; 24,99). Although in these experiments all plants of both kinds were ultimately infected with virus, the delay in the epidemic in the silvery plants was sufficient to prevent plant death in all except a few plants at harvest and consequently they were able to produce a saleable crop (Figure 1).

In wheat, one probable character for resistance to the grain aphid, *(Sitobion avenae)*, is a dark glossy green color of the upper leaves (70). This non-glaucous character is associated with a decrease in diketones and hydroxydiketones when compared with the bluish-green waxy bloom of *S. avenae* susceptible genotypes. It is known that diketones in normal leaf surface wax strongly absorb UV light and this may be the cause of the noticeable differences in color between the glaucous and on non-glaucous types, which in turn may influence the alighting response of aphids. In field experiments, infestation by *S. avenae* was 75% less on non-glaucous than glaucous wheat (2). Decreases in aphid numbers on bloom-less or non-glaucous plants have also been reported for *Schizaphis graminum* on sorghum (103) and for *Brevicoryne brassicae* and *Aleurodes brassicae* on brassicas (29,106). This character is relatively easy to select in segregating populations One of the components of field-resistance of some groundnut cultivars to bud necrosis disease, caused by tomato spotted wilt virus transmitted by thrips, is believed to be the dark green color of its leaves which, in comparison to the lighter colored leaves of field-susceptible cultivars, possibly interferes with the alighting response of the thrips vector, thereby decreasing virus incidence (3). Müller (77) reported that brown colored lettuce cultivars were colonized by aphids to a much lesser extent and were less frequently infected with lettuce mosaic virus, than green cultivars. Similarly, Ellic and Cole (29) found that red colored cabbage and Brussels sprout plants were more resistant to colonization by aphids than green plants.

The effects of crop canopy on the alighting response of insects is also well known (1,36) and, another possible component of field-resistant ground nuts to bud necrosis disease, is its spreading growth habit which, in contrast to the upright habit of many field-susceptible cultivars, rapidly covers the ground surface. This spreading growth habit would minimize more rapidly the checkerboard effect of the crop and UV reflection and in so doing probably interferes with the alighting response of the thrips vector (3).

Interference with the Initial Settling and Feeding Behavior of Vectors

Having located a host, chemical and physical stimuli from the plant surface are involved in the identification and establishment of suitable feeding sites. If the resistance is sufficiently strong to prevent probing and feeding altogether, no acquisition or inoculation of virus will occur but, if the vector is required to probe the plant before it is deterred from feeding, inoculation of non-persistent viruses is likely to occur. However, a delay in making the first probe, if sufficiently long, may exceed the time for which

non-persistent viruses are retained by vectors, so lessening the probability of virus transmission. If the vector is sufficiently deterred from feeding after several brief exploratory probes, it may leave the crop. Where this happens, the spread of persistent and semi-persistent viruses might be expected to decrease, because little transmission is likely to occur in these brief probes and because any secondary spread would be minimized. In some specific instances, vectors may be immobilized by sticky plant exudates that may, or may not, prevent virus inoculation but will decrease possible secondary spread.

Plants show a range of physical and chemical attributes that can interfere with this initial settling phase before feeding begins.

Trichomes

Dunn (26) has pointed out that plant trichomes are a form of plant resistance that is common to insects with biting, piercing or sucking mouthparts and Webster (111) has listed more than 30 insect pests in which resistance is determined by genes that control pubescence, or are closely linked to these genes. For example, this character is associated with resistance to aphids in cereals (89), thrips in cassava (11), leafhoppers in potato (109) and beetles in cereals (96). Sometimes this character may provide resistance in a single genotype to a whole range of insects as in the glandular hairs of tomato that are effective against mites, whiteflies and aphids (74), and pubescence in wheat that is effective against aphids, beetles and hessian fly (89). The effectiveness of hairs in resistance to insects may depend on whether hairs are hooked, long, secrete toxic or sticky substances, are on the angle of their emergence from the leaf surface, and on their density on the plant surface (35,109).

The mode of action of plant trichomes in influencing vector feeding and movement can be through - physical interference that decreases the accessibility of vectors to feeding sites on the plant; chemical interference through the release of viscous or sticky exudates that hamper vector movement and probing; the release of toxic chemicals that immobilize or kill vectors or inhibit probing. Sometimes, the interference caused by viscous or toxic chemicals are difficult to distinguish from one another or may occur together on the same plant and it is therefore convenient to consider them together but noting these distinctions where these are known.

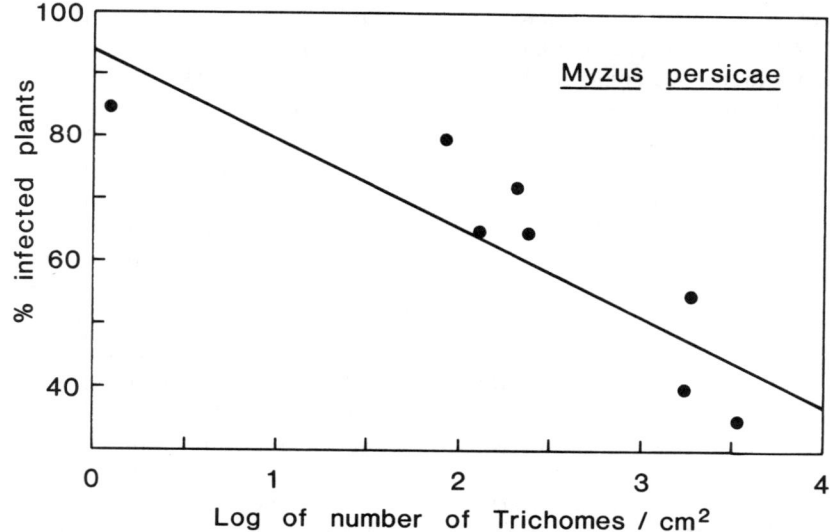

Figure 2. The relationship between log of trichome number/cm^2 and percent infection with soybean mosaic potyvirus by *Myzus persicae* in isolines of cv. Clark soybean differing in trichome density. Single viruliferous *M. persicae* were allowed 1 min acquisition access periods to isoline test plants kept in a glasshouse. Data from Gunasinghe *et al.* (35).

Physical Interference by Trichomes

In laboratory studies in the USA with soybean isolines differing in the extent of leaf pubescence, decreases in the transmission efficiency of soybean mosaic virus (SMV) by *M. persicae*, *Aphis citricola* and *Rhopalosiphum maidis* was correlated with increased density of leaf trichomes (Figure 2). In the field, spread of SMV in soybean isolines was negatively correlated with the density of leaf hairs and field epidemics of SMV were delayed (35).

Chemical Interference by Trichome Exudates

Although some plant hairs interfere with insect movement and probing solely in a physical way, many plant species have hairs associated with glands that release sticky exudates. The stickiness of these exudates and probably the chemicals they contain can interfere with vector movement and feeding on the plant. Some of the best studied glandular trichomes are probably those of *Solanum berthaultii*. Leaves of this species have two types of glandular trichome; type A are short with a four-lobed gland at its apex, type B are longer with an ovoid gland at its tip. The exudates from type A glands harden after rupture whilst those from type B remain viscous (31). The presence of both kinds of trichome and their density on the

leaf surface are important factors in the effectiveness of this resistance mechanism (109). Aphids are trapped in the sticky exudates of type B hairs and their feeding is impaired.

Using an electronic recording system, Lapointe and Tingey (66) showed that exudates from type B hairs had a major influence on aphid feeding behavior. In contrast to leaves without them, aphids probed leaves with type B trichomes less promptly, made briefer probes and fed for shorter total times. The delay in making the first probe was up to 20-30 min and these workers postulated that vector efficiency might be decreased through exceeding the retention period of non-persistent viruses such as PVY. However, in the field, plants of some lines were readily infected with PVY but much less so with the persistently transmitted virus, potato leafroll (PLRV; 88). This resistance to infection with PLRV is probably caused in part by the impaired feeding behavior and by the fact that the aphids and their progeny are immobilized on the plant and unable to disperse, and so decreasing secondary spread.

The stickiness of exudates from leaf hairs of *Lycopersicon pennellii*, and derivatives from this species determine the effectiveness of resistance to the whitefly vector *B. tabaci* (15). The degree of resistance is also influenced by the amount of exudate produced which is itself influenced by environmental conditions (14).

In some accessions of *L. hirsutum* f. *glabratum*, Berlinger *et al.* (14) found that the trichomes released 2-tridecanone, a compound toxic to whiteflies, and there was some evidence that the concentration of this compound in different accessions was correlated with their whitefly resistance.

Other Anatomical or Morphological Features

Apart from trichomes, plants may have other anatomical or morphological features that influence the settling of vectors. Chamberlain and Evans (20) found far fewer numbers of the mite, *Aculoides dubius* on ryegrass genotypes with shallow leaf ridges than on those with deep ridges. Although it is uncertain whether these mites, in addition to the known vector, *Abacarus hystrix*, transmit ryegrass mosaic virus (RMV), these authors suggest that incorporation of shallow ridging in leaves of ryegrass cultivars might be a means of restricting infection with RMV by decreasing mite numbers.

Matthieu *et al.* (73) observed that in leek cultivars, the incidence of infection with leek chlorotic streak virus decreased with increasing depth of epicuticular wax on the leaves. Presumably, the wax layer decreased the efficiency of virus transmission by the aphid vectors. In very recent work in India on pigeonpea sterility mosaic, the agent of which is transmitted by the eriophyid mite *Aceria cajani*, lines of pigeonpea showing field resistance to infection and to the mite vector were found to have a much thicker cuticle and cell wall (>50-100%) than field susceptible lines (85). Furthermore, this increased thickness was usually greater than the observed stylet length of the mite vector. It seems likely therefore that the resistance of these plants to the vector mite and the disease agent it transmits, is due to the inability of the vector to penetrate their epidermal cells, thereby preventing satisfactory feeding and transmission of the agent (85).

Surface Waxes and Plant Volatiles

Plants contain a wide range of chemical compounds on, or near, the leaf surface that act either as attractants or deterrents to different arthropods. Often the ratio of these different compounds is critical in determining the response, or strength of response to target organisms. For some plant/vector combinations the repellent effect of some compounds can be very pronounced. One

such form of strong resistance occurs in red raspberry cultivars to the important virus vector aphid, *Amphorophora idaei* (50). The strongest resistance is controlled by single dominant genes (56,63) and less effective resistance probably by several minor genes (46,63). Birch and Jones (16) found that when *A. idaei* were placed on detached leaves of resistant cultivars floating in water, most aphids walked off the leaves and drowned in the water within 18-24 h. Field experiments showed that, under very high inoculum pressure from viruliferous *A. idaei*, the spread of the four viruses transmitted by this aphid (50) was much less rapid in genotypes containing either of the two forms of resistance to *A. idaei* when compared to *A. idaei*-susceptible genotypes (Figure 3; 48,49). Later studies of commercial crops in the UK over more than 20 years have shown that in large crops (0.5-1 ha), the incidence of infection with *A. idaei*-transmitted viruses is usually <5% 8-15 yrs after planting and most such infections occur at the periphery of the crop and especially where they adjoin crops of *A. idaei*-susceptible cultivars (49, 50, 52; A. T. Jones, unpublished). The cause(s) of this strong deterrent to settling, probing and colonization by *A. idaei*, seemed to be located on, or in, the wax layer on the leaf surface because removal of this layer, by treatment with a mixture of acetone and cellulose acetate, greatly decreased the level of resistance to *A. idaei* in resistant cultivars (53); behavioral studies of the aphids on these plants also supported this conclusion (17, 53). GC-MS analysis of dichloromethane extracts of the leaf surface of progeny plants segregating for resistance to *A. idaei*, found three classes of dominant compounds (esters, tocopherols and amyrins) and discriminant analysis of the peak areas of 13 such compounds indicated that resistance in the progeny seedlings could be estimated with 96% success based on these compounds (90).

In laboratory studies, the application to plants of some plant fatty acids, particularly dodecanoic acids, has been shown to decrease aphid settling on leaves and the transmission of persistent and semi-persistent viruses (32, 97). However, the application of such materials to plants in the field did not decrease virus spread (43).

By contrast, it is known that most plant species produce chemical compounds that play a major role in attracting and inducing feeding by insects. Many such compounds have been isolated and several may occur in a single plant species. For example, in relation to the attraction of the

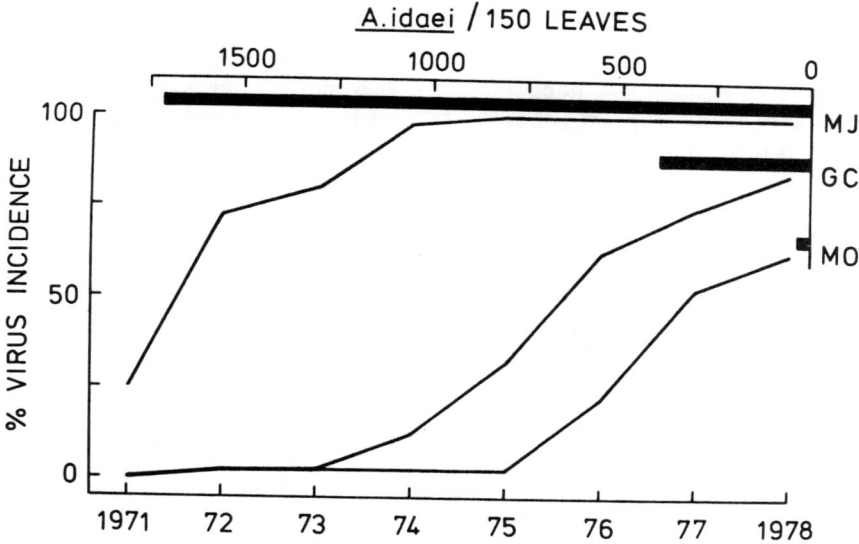

Figure 3. Effect of resistance to *A. idaei* in raspberry on decreasing the incidence (%) of viruses transmitted by this aphid vector in small field trials in Scotland during 1971-1978. Histograms indicate the relative numbers of *A. idaei*/150 leaves during peak aphid populations. Cultivars and their resistance to *A. idaei* are: MJ - Malling Jewel (no resistance), GC - Glen Clova (minor gene resistance), MO - Malling Orion (major gene resistance). Data from Jones (48,49).

brown planthopper *(Nilaparvata lugens)* to rice, more than 30 compounds with interactive effects have been found in steam distillates of plants. The planthopper is attracted only weakly or not at all by any one of these compounds but attractiveness is generated when the compounds are mixed together in proportions resembling those occurring in plants (102). Conceivably, therefore, the relative concentrations of specific chemical compounds in plants, or the absence of some compounds, may determine the relative attractiveness of plants and/or their ability to induce a feeding response and thereby influencing whether virus inoculation occurs.

Interference with Sustained Feeding Behavior of Vectors

When resistance in plants is due solely to antibiosis, the fact that vectors are likely to remain on such plants for lengthy periods (often days), would seemingly restrict the usefulness of this form of resistance in preventing virus transmission. The most likely benefit might be a decrease in secondary spread through decreasing vector populations and possibly vector activity, but this would probably only affect the spread of persistently transmitted viruses (30, 57). However, detailed studies on insect feeding behavior have shown that, in several instances, antibiosis seems

to operate by preventing phloem-finding or phloem ingestion and such effects may influence the likelihood of virus acquisition and/or inoculation, both of persistent and semi-persistent viruses.

Dreyer and Campbell (25) reported that the resistance of a sorghum genotype to *Schizaphis graminum* seemed to be caused by increased methylation of the middle lamella pectin, so hindering aphid penetration of the host plant tissue and restricting access to the phloem. This interpretation was supported by the fact that enzyme preparations, from an aphid biotype capable of overcoming this host plant resistance were found to depolymerize pectin from the resistant sorghum twice as rapidly as enzyme preparations from normal biotypes. The increased depolymerization activity was related to the increased amount of pectin methylesterase activity. No data are available on the influence of this form of resistance on virus transmission. However, for most persistent and semi-persistent viruses that must enter the phloem to infect plants, this mechanism of vector resistance could decrease the likelihood of both inoculation and acquisition of such viruses.

Oya and Sato (80) found that the leafhopper *Nephotettix cincticeps* excreted as much honeydew on *N. cincticeps*-resistant as on - susceptible rice cultivars, indicating that it fed to similar extents on these two kinds of plant.

However, whilst insects on susceptible plants fed from the phloem as expected, those on resistant plants were found to feed mainly from the xylem. Although the stylets of a few leafhoppers terminated in the phloem of resistant plants, these insects seemed unable to feed to any extent from these cells. Similar data were obtained for the leafhopper, *N. virescens,* on rice (7,60). By contrast, Auclair and Baldos (6) found that the planthopper, *Sogatella furcifera* located the phloem in both *S. furcifera*-resistant and susceptible rice plants. However, ingestion of plant contents on resistant plants, as measured by the volume of honeydew excreted, was 267-fold less than on susceptible plants. Because the rice viruses transmitted by these leafhopper species and the planthopper vector *S. oryzicola* probably must enter the phloem to infect plants, this altered feeding behavior on resistant plants explains not only their increased mortality and decreased fecundity, but probably also accounts for the decreased incidence in such plants of viruses transmitted by these vectors (40, 47). Possibly, the reported changes in the feeding behavior of some aphid species on aphid-resistant plants may also explain the resistance of such plants to virus infection (37, 75).

Studies in plant chemistry have identified many varied chemical compounds with antifeedant activity to a range of plant pests. Such compounds include, phenolic acids, isothiocynates and compounds from catabolism of glucosinolates, hydroxamic acids, diterpenoid acids and drimanes (32,82,84,93,105). More recently, lectins have been associated with large decreases in passive phloem uptake by aphids (22). Whilst the presence of these or other compounds may go some way to explain the deterrent effect on vector feeding and thereby the transmission of some plant viruses, particularly those transmitted in a persistent manner, as indicated below, it does not explain all such instances.

Specific Interference with Vector Transmission of Virus

The specific resistance in some muskmelon lines to the inoculation by *Aphis gossypii* of several non-persistent viruses, is associated with resistance to this aphid vector (67-69,94). Kennedy, *et al.* (59) and Kennedy and Kishaba (58) found that this resistance was due to antixenosis and to altered feeding behavior but not to any translocated chemical in the plant. Although

significantly more aphid stylets terminated in the phloem of *A. gossypii*-resistant plants than of susceptible ones, electronic monitoring of aphid feeding showed that ingestion from phloem sieve cells on resistant plants was less frequent and of shorter duration than on susceptible ones. Some resistance factor(s) therefore seem to interfere with ingestion. Although this probably explains the mechanism of resistance to *A. gossypii*, it is difficult to see how this could account for the resistance to inoculation by *A. gossypii* of several non-persistent viruses (68,69). Earlier reports describe resistance to virus transmission by aphids in other crops but aphid resistance is not recorded in these studies. Thus, Sylvester and Simons (104) found that *M. persicae* readily acquired turnip mosaic virus from, but rarely transmitted it to, *Brassica chinensis* and Simons and Moss (101) showed a similar situation with *M. persicae*, PVY and a pepper *(Capsicum annuum)* cultivar. These phenomena highlight our relative ignorance of the specific factors involved in the processes of acquisition and inoculation of viruses by vectors and may indicate that these processes differ in detail if not in principle.

POSSIBLE ADDITIONAL BENEFITS FROM USING VECTOR RESISTANT MATERIAL

In addition to decreasing the incidence or rate of spread of infection with specific viruses, cultivation of vector-resistant material is likely to offer other benefits. Other factors being equal, the cultivation of such material should provide a substantial financial benefit in the control of some viruses through the decreased use of pesticides to control the vector. In some situations this saving can be considerable. For example, Jennings and Pineda (47) reported that in some regions of Colombia, as many as 15 insecticide applications were applied to individual rice crops to control the planthopper *S. oryzicola,* the vector of rice hoja blanca virus, and that such applications achieved only moderate control. With the introduction of vector-resistant cultivars, few or no insecticides were necessary and the cultivation of such material was at no extra cost to the grower. In field trials in the Philippines, Heinrichs *et al.* (42) made a cost analysis of the effectiveness of insecticide applications to control *N. virescens* and the tungro virus complex it transmits in rice cultivars differing in resistance to this vector. They found that insecticide applications were neither economic

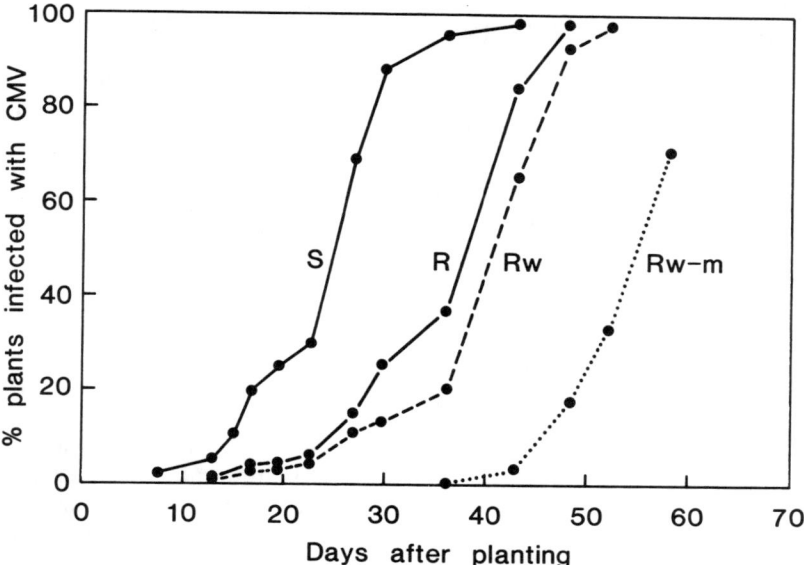

Figure 4. Progress of epidemics of cucumber mosaic virus (CMV) in experimental plots of a *Cucumis melo* line (R) containing resistance to some CMV strains and the main aphid vector, *A. gossypii* in southern France. It shows the increased control of infection using integrated control measures. w, plots weeded; w-m plots weeded and mulched; S, a breeding line with no resistance to CMV or *A. gossypii*. From Lecoq and Pitrat (67).

nor effective for tungro disease control in cultivars with little or no resistance to *N. virescens,* whereas the low incidence of tungro disease in pesticide-treated cultivars with strong *N. virescens*-resistance was also obtained in the absence of these chemicals.

In some situations there may be no suitable alternative to breeding vector-resistant plants for effective virus control, either because of (a) difficulties in getting pesticides to reach the target organism, for example, once the crop canopy has closed, (b) the need for multiple pesticide applications, for example for whiteflies or, (c) serious problems caused by the development of pesticide-resistant biotypes of vector organisms (39).

Even when plant resistance to arthropods is only moderate or incomplete, the performance of pesticides when applied to such crops can be greatly enhanced when compared with non-resistant crops (19). This enhanced effectiveness on such resistant plant material is due not only to the lower numbers of the target organisms, but also to the greater sensitivity of the surviving organisms to the chemical and this is probably caused by their decreased body weight and generally lower biological activity (41).

One serious disadvantage of most pesticides is that non-target organisms are often killed or seriously affected by them, including the predators and pathogens of the target organism themselves. One of the principal causes of the resurgence of the planthopper *N. cincticeps* on rice in Japan is attributed to the removal of natural enemies killed by insecticides used to control the planthopper (61). A consequence of lower numbers of applications and/or concentrations of pesticides on vector-resistant plants therefore is to increase opportunities for biological control through natural enemies and parasites (12,55). Even with *S. berthaultii,* whose glandular trichomes provide a non-specific form of arthropod resistance, moderate levels of this form of resistance were compatible with biological control (78,79).

Finally, slight vector resistance, which on its own may not be sufficient to provide satisfactory virus control may, in combination with other control measures, offer significant improvements in virus control. For example, in combination with slight virus resistance (76, 94) and with cultural methods of mulching and weed control (Figure 4; 67), epidemics of non-persistent viruses were significantly delayed.

POSSIBLE PROBLEMS FROM THE PRODUCTION AND USE OF VECTOR-RESISTANT CULTIVARS

In some situations a virus vector may be a pest in its own right and choosing a suitable form of resistance to it may provide a bonus in protection from virus infection. Indeed, in the past, that is what has most often occurred. In many other instances, however, virus vectors pose little problem as pests. In such situations vector resistance may not be the best or easiest method of virus control, especially if other approaches, such as plant immunity or resistance to virus, or cultural methods, can provide adequate control. For viruses that are spread by several vector species or, with crops that are infected with several viruses having different vectors, breeding for effective resistance to all of these vectors may be impractical unless the resistance mechanism is non-specific, such as that provided by plant trichomes. Nevertheless, resistance to several vector species has been incorporated into single cultivars of crop plants (82,113), some resistance genes provide resistance to a range of different pests (82), and resistance genes to some pests are sometimes closely linked to genes to different pests (54).

An important consideration for the breeder is the ease of screening progeny for vector resistance. Most examples of strong resistance to vectors are governed by single major genes, making the selection of resistant progeny plants in a breeding program relatively straightforward. Polygenic resistance, which is often a less pronounced but more durable form of resistance, may be more difficult to detect, and some of the more subtle resistance mechanisms, such as resistance to vector transmission, may entail laborious tests of several kinds to identify resistant progeny. Because the scale of screening is often limited by time, space and cost, some of the less obvious forms of resistance will commonly be overlooked. For example, whilst polygenic resistance to the main aphid vector of four important viruses of red raspberry in the UK is very effective in decreasing infection with such viruses in commercial crops (virus incidence rarely exceeds 1-5% after 10 years cultivation) but, in the very small experimental plots commonly used for seedling assessment by breeders, the incidence of infection was 75% (49,52, A.T. Jones, unpublished). Some useful forms of resistance are therefore likely to be missed by screening progeny

seedlings in small plots commonly used by plant breeders. Care is also needed to assess whether results observed in greenhouse and laboratory studies are maintained in the field, as environmental conditions are known to affect the expression of some resistance mechanisms. For example, it has been shown that the efficacy of some forms of insect resistance can be greatly influenced by, temperature through affecting the viscosity of trichome exudates (110), light either increasing or decreasing resistance (28; A.T. Jones and A.N.E. Birch, unpublished) and leaf age and position through decreasing the epicuticular wax on older leaves (13).

In addition to the difficulties encountered in producing vector-resistant cultivars, some potential hazards exist in the widespread cultivation of such cultivars. In a free choice situation there is the potential danger of the heavy colonization by vectors of non-resistant cultivars or other crop species with the consequent increased incidence of viruses in these unprotected crops (57). However, whereas this may sometimes happen in small plots of breeders' selections and remains a potential hazard in commercial situations, there is little evidence that this has occurred in crops to any significant extent (112).

A potentially more serious problem is that posed by the emergence of new vector biotypes induced by the selection pressure imposed by cultivating vector-resistant cultivars. Several dramatic examples of major epidemics caused by such biotypes are recorded among insects feeding on rice (102). However, there are other examples where vector resistance has been durable for many decades e.g. resistance to the raspberry aphid, *A. agathonica* in North America (23). In general, resistance-breaking biotypes occur most commonly in monophagous vector species and where plant resistance is determined by single genes, and they seem less common in polyphagous vectors (presumably because the selection pressure is lessened due to the occurrence of alternative hosts) and where resistance is controlled by several genes and is non-specific e.g. plant hairs. However, the development of new vector biotypes in response to the selection pressure of growing resistant cultivars need not necessarily be more troublesome than the wild type, and may be less so due to poor biological fitness. For example in the UK, biotypes 2 and 4 of the large raspberry aphid, *A. idaei,* are reported to have a high overwintering mortality and seem less fitted for survival than biotypes 1 and 3 (18, 62). Although little work has been done to study the vector efficiency of different biotype vectors of viruses, Saksena, *et al.* (95) found differences in efficiency of transmission of barley yellow dwarf virus by biotypes of *R. maidis,* and large differences have been observed in the ability to transmit virus by clones and biotypes of other vector aphids (91, 100, 107).

Finally, where resistance is specific to one vector species, breeding for resistance to it can increase susceptibility or sensitivity to colonization by another. Examples of this occur in cotton where the introduction of jassid-resistant cultivars resulted in greatly increased susceptibility to whiteflies and aphids and their associated viruses, especially in India and Sudan (4). A similar situation occurred in wheat in North America where pubescent cultivars were developed to provide resistance to the cereal leaf beetle and hessian fly and which caused infestations of the vector mite, *Eriophyes tulipae,* that were attracted by the leaf hairs and resulted in a higher incidence of wheat streak mosaic virus (38).

Breeders need therefore to be aware of these possible hazards in cultivating vector-resistant cultivars and to take precautions to assess their likely occurrence before the introduction of such material to widespread commercial use.

CONCLUSIONS

The wide range of variation that exists in plants for resistance to virus vectors and the opportunity it presents for virus control is only now beginning to be studied in depth to any significant extent. Many of these recent studies have shown the very complex and often subtle interactions that occur between virus vectors and their plant hosts. Whilst such studies often reveal our comparative ignorance about many of these interactions, they also provide insights into possible new approaches for both pest and virus control. The identification of some of the mechanisms that underlie certain forms of vector resistance in plants indicate that many preconceived ideas on the limited value of such forms of resistance for virus control need to be reappraised. For example, near immunity to vectors, was commonly regarded as the only effective mechanism for the control of non-persistent aphid-borne viruses, if not all viruses (30,57,64) but, in some circumstances, this need not necessarily be a pre-requisite for good control of these viruses. Other studies have shown the potential of even slight vector resistance,

especially when combined with some forms of virus resistance (35,49,68,69,76). The use of such partially resistant plant material in integrated control programs involving chemical, cultural and/or biological control methods, may increase further the effectiveness obtained in the control of virus incidence (Fig. 4; 67). For many crops therefore, and sometimes in existing commercially acceptable material, there is variation of several kinds to be exploited.

However, not all forms of vector resistance are equally effective in controlling virus incidence and some forms may actually increase it (5,8,57). Furthermore, many problems exist in the detection and exploitation of some of the mechanisms of resistance that promise to be useful. In particular, improved screening methods are needed to select some of the less obvious mechanisms of vector resistance and field scale testing methods are needed for evaluating the benefits of some forms of resistance that are not apparent in small scale tests. In addition, the influence of environmental factors on the expression of some forms of vector resistance requires further study.

New techniques may offer solutions to some of these problems. For example, biochemical techniques for the rapid detection of specific secondary plant metabolites which are closely correlated with pest resistance in some plants, could give considerable assistance to the plant breeder in selecting resistant plants and seedling progeny. Furthermore, at the basic level, they may aid research on the precise mechanism(s) involved in resistance. Advanced genetical techniques enable the transfer of chromosome pieces between species and even genera (65,71). In more recent years, powerful and precise technologies have become available to isolate and introduce specific 'foreign' pest-resistance genes into plants by genetic engineering. The success of such technologies for pest control has already been demonstrated not only using genes derived from plants, such as the trypsin inhibitor gene from cowpea (44) and a lectin gene from snowdrop (45), but also those from other distinct organisms such as *Bacillus thuringiensis* (10). However, as with many resistance genes that occur naturally within a plant species or genus, not all transgenes will be effective in preventing the transmission of certain viruses, and the widespread cultivation of plants containing such genes is likely to exert selection pressure on the vector population for resistance-breaking biotypes. However, the development of such biotypes might be minimized and virus control improved through using gene promoters that express these transgenes only in specific plant tissues, such as phloem (98) or, only under specific circumstances, such as wounding (83). In these ways the potential usefulness of such transgenes is likely to be prolonged.

As the new millennium approaches, the current and prospective research on vector resistance for the control of plant viruses, outlined briefly in this chapter, suggests that many new opportunities exist using conventional genetics and genetic engineering to achieve further significant progress in this area of science.

REFERENCES

1. A'Brook, J. 1964. The effect of planting date and spacing on the incidence of groundnut rosette disease and of the vector, *Aphis craccivora* Koch, at Mokwa, Northern Nigeria. Ann. Appl. Biol. 54: 199-208.

2. Acreman, T.M. 1984. The contribution of resistance to cereal aphid control. Brighton Crop Prot. Conf. 1984 - Pests Dis. Vol. 1: 31-36.

3. Amin, P. W. 1985. Apparent resistance of groundnut cultivar Robut 33-1 to bud necrosis disease. *Plant Dis.* 69, 718-719.

4. Arnold, M. H., Innes, N. L., and Brown, S. J. 1976. Resistance breeding. Pages 175-195 in: Agricultural Research for Development. M. H. Arnold, ed. Cambridge University Press.

5. Atiri, G.I., Ekpo, E.J.A., and Thottappilly, G. 1984. The effect of aphid-resistance in cowpea on infestation and development of *Aphis craccivora* and the transmission of cowpea aphid-borne mosaic virus. Ann. Appl. Biol. 104: 339-345.

6. Auclair, J.L., and Baldos, E. 1982. Feeding by the whitebacked planthopper, *Sogatella farcifera*, within susceptible and resistant rice varieties. Entomol. Exp. Appl. 32: 200-203.

7. Auclair, J.L., Baldos, E., and Heinrichs, E.A. 1982. Biochemical evidence for the feeding sites of the leafhopper *Nephotettix virescens* within susceptible and resistant rice plants. Insect Sci. Its Appl. 3: 29-34.

8. Baerecke, M.L. 1958. Blattrollresistenz-zuchtung. Pages 97-106 in: Handbuch der Pflanzenzuchtung 3. H. Kappert and W. Rudorf, eds. Parey, Berlin.

9. Baker, P.F. 1960. Aphid behaviour on healthy and on yellows-virus-infected sugar beet. Ann. Appl. Biol. 48: 384-391.

10. Barton, K. A., Whiteley, H. R., and Yang, N. S. 1987. *Bacillus thuringiensis* delta-endotoxin expressed in transgenic *Nicotiana tabacum* provides resistance to lepidopteran insects. Pl. Physiol. 85: 1103-1109.

11. Bellotti, A., and van Schoonhoven, A. 1978. Mite and insect pests of cassava. Annu. Rev. Entomol. 23: 39-67.

12. Bergman, J.M., and Tingey, W.M. 1979. Aspects of interaction between plant genotypes and biological control. *Bull. Entomol. Soc. Amer.* 25: 275-279.

13. Bergman, D.K., Dillwith, J.W., Zarrabi, A.A., and Berberet, R.C. 1991. Epicuticular lipids of alfalfa leaves relative to position on the stem and their correlation with aphid (Homoptera: Aphididae) distributions. Environ. Entomol. 20: 470-476.

14. Berlinger, M.J., Dahan, R., and Shevach-Urkin, E. 1983. Breeding for resistance to whiteflies in tomatoes - in relation to integrated pest control in greenhouses. Bull. SROP 6: 172-176.

15. Berlinger, M.J., Dahan, R., and Shevach-Urkin, E. 1984. Resistance to the tobacco whitefly, *Bemisia tabaci*, in tomato and related species: a quick screening method. Bull. SROP 7: 39-40.

16. Birch, A.N.E., and Jones, A.T. 1988. Levels and components of resistance to *Amphorophora idaei* in raspberry cultivars containing different resistance genes. Ann. Appl. Biol. 113: 567-578.

17. Birch, A.N.E., Jones, A.T., Woodford, J.A.T., and Jones, D.R. 1991. Behavioural responses of the virus vector aphid, *Amphorophora idaei*, to resistance in red raspberry. Proc. Int. Symp., Aphid-Plant Interactions : Populations to Molecules. Oklahoma State University, August 1990. 265 Pages.

18. Briggs, J.B. 1965. The distribution, abundance and genetic relationships of four strains of the rubus aphid *(Amphorophora rubi* (Kalt.)) in relation to raspberry breeding. J. Hortic. Sci. 40: 109-117.

19. Campbell, W.V., and Wynne, J.C. 1985. Influence of the insect-resistant peanut cultivar Nc6 on performance of soil insecticides. J. Econ. Entomol. 78: 113-116.

20. Chamberlain, J.A., and Evans, P.E. 1979. Relationship between depth of leaf ridging and numbers of a possible mite vector. Rep. - Welsh Plant Breed. Sta. (Aberystwyth, Wales), 1978. 197 Pages.

21. Chang, V.C.S., and Ota, A.K. 1978. Feeding activities of *Perkinsiella* leafhoppers and Fiji disease resistance of sugarcane. J. Econ. Entomol. 71: 297-300.

22. Cole, R.A. 1994. Isolation of a chitin-binding lectin, with insecticidal activity in chemically-defined synthetic diets, from two wild brassica species with resistance to cabbage aphid *Brevicoryne brassicae*. Entomol. Exp. Appl. 72: 181-187.

23. Daubeny, H.A. 1980. Red raspberry cultivar development in British Columbia with special reference to pest response and germplasm exploitation. Acta Hortic. 112: 59-67.

24. Davis, R.F., and Shifriss, O. 1983. Natural virus infection in silvery and non-silvery lines of *Cucurbita pepo*. Plant Dis. 67: 397-380.

25. Dreyer, D.L., and Campbell, B.C. 1984. Association of the degree of methylation of intercellular pectin with plant resistance to aphids and with induction of aphid biotypes. Experientia 40: 224-226.

26. Dunn, J.A. 1978. Resistance to some insect pests in crop plants. Appl. Biol. III: 43-85.

27. Dwivedi, S.L., Nigam, S.N., Reddy, D.V.R., Reddy, A.S., and Rango Rao, G.V. 1995. Progress in breeding groundnut varieties resistant to peanut bud necrosis virus and its vector. Pages 35-40 in: Recent Studies on Peanut Bud Necrosis Disease: Proc. of a Meeting on 20 March 1995, Icrisat Asia Centre, India. A.A.M. Buiel, J.E. Parlevliet and J.M. Lenné, eds.

28. Elden, T.C., and Kenworthy, W.J. 1995. Physiological responses of an insect-resistant soybean line to light and nutrient stress. J. Econ. Entomol. 88: 430-436.

29.   Ellis, R., and Cole, R.A. 1995. Sources of resistance to cabbage aphid in brassicas. Pages 29-30 in: Rep. - Hort. Res. Int. (Wellesbourne, Engl.) 1993-1994.

30.   Gibson, R.W., and Plumb, R.T. 1977. Breeding plants for resistance to aphid infestation. Pages 473-500 in: Aphids as Virus Vectors. K.F. Harris and K. Maramorosch, eds. Academic Press.

31.   Gibson, R.W., and Turner, R.H. 1977. Insect trapping hairs on potato plants. Proc. Natl. Acad. Sci., (USA). 22: 272-277.

32.   Gibson, R.W., Rice, A.D., Pickett, J.A., Smith, M.C., and Sawicki, R.M. 1982. The effect of the repellents dodecanoic acid and polygodial on the acquisition of non-, semi- and persistent plant viruses by the aphid *Myzus persicae*. Ann. Appl. Biol. 100: 55-59.

33.   Giebel, J. 1974. Biochemical mechanisms of plant resistance to nematodes: a review. J. Nematol. 6: 175-184.

34.   Gommers, F.J. 1981. Biochemical interactions between nematodes and plants and their relevance to control. Helminthol. Abstr. Ser. B 50: 9-24.

35.   Gunasinghe, U.B., Irwin, M.E., and Kampmeier, G.E. 1988. Soybean leaf pubescence affects aphid vector transmission and field spread of soybean mosaic virus. Ann. App. Biol. 112: 259-272.

36.   Halbert, S.E., and Irwin, M.E. 1981. Effect of soybean canopy closure on landing rates of aphids with implications for restricting spread of soybean mosaic virus. Ann. Appl. Biol. 98: 15-19.

37.   Haniotakis, G.E., and Lange, W.H. 1974. Beet yellows virus resistance in sugar beets: mechanism of resistance. J. Econ. Entomol. 67: 25-28.

38.   Harvey, T.L., and Martin, T.J. 1980. Effects of wheat pubescence on infestations of wheat curl mite and incidence of wheat streak mosaic. J. Econ. Entomol. 73: 225-227.

39.   Heinrichs, E.A. 1979. Control of leafhopper and planthopper vectors of rice viruses. Pages 529-560 in: Leafhopper Vectors and Plant Disease Agents. K. Maramorosch and K.F. Harris, eds. Academic Press, New York.

40.   Heinrichs, E.A., and Rapusas, H.R. 1983. Levels of resistance to the whitebacked planthopper, *Sogatella furcifera* (Homoptera: Delphacidae), in rice varieties with different resistance genes. Environ. Entomol. 12: 1793-1797.

41.   Heinrichs, E.A., Fabellar, L.T., Basilio, R.P., Wen, T-C., and Medrano, F. 1984. Susceptibility of rice planthoppers *Nilaparvata lugens* and *Sogatella furcifera* (Homoptera: Delphacidae) to insecticides as influenced by level of resistance in the host plant. Environ. Entomol. 13: 455-458.

42.   Heinrichs, E.A., Rapusas, H.R., Aquino, G.B., and Palis, F. 1986. Integration of host plant resistance and insecticides in the control of *Nephotettix virescens* (Homoptera: Cicadellidae), a vector of rice tungro virus. J. Econ. Entomol. 79: 437443.

43.   Herrbach, E. 1987. Effect of dodecanoic acid on the colonisation of sugar beet by aphids and the secondary spread of virus yellows. Ann. Appl. Biol. 111: 477-482.

44.   Hilder, V.A., Gatehouse, A.M.R., Sheerman, S.E., Baker, R.F., and Boulter, D. 1987. A novel mechanism of insect resistance engineered into tobacco. Nature (London) 330: 160-163.

45.   Hilder, V.A., Powell, K.S., Gatehouse, A.M.R., Gatehouse, J.A., Gatehouse, L.N., Shi, Y., Hamilton, W.D.O., Merryweather, A., Newell, C.A.,

Timans, J.C., Peumans, W.J., van Damme, E., and Boulter, D. 1995. Expression of snowdrop lectin in transgenic tobacco plants results in added protection against aphids. Transgenic Res. 4: 18-25.

46.   Jennings, D.L. 1963. Preliminary studies on breeding raspberries for resistance to mosaic disease. *Hortic. Res.* 2: 82-96.

47.   Jennings, P.R., and Pineda, A.T. 1970. Effect of resistant rice plants on multiplication of the planthopper, *Sogatodes oryzicola* (Muir). Crop Sci. 10: 689-690.

48.   Jones, A.T. 1976. The effect of resistance to *Amphorophora rubi* in raspberry (*Rubus idaeus*) on the spread of aphid-borne viruses. Ann. Appl. Biol. 82: 503-510.

49.   Jones, A.T. 1979. Further studies on the effect of resistance to *Amphorophora idaei* in raspberry (*Rubus idaeus*) on the spread of aphid-borne viruses. Ann. Appl. Biol. 92: 119-123.

50.   Jones, A.T. 1986. Advances in the study, detection and control of viruses and virus diseases of *Rubus*, with particular reference to the United Kingdom. Crop Res. 26: 127-171.

51.   Jones, A.T. 1987. Control of virus infection in crop plants through vector resistance : a review of achievements, prospects and problems. Ann. Appl. Biol. 111: 745-772.

52.   Jones, A.T., and Murant, A.F. 1975. Etiology of a mosaic disease of 'Glen Clova' red raspberry. Hortic. Res. 14:89-95.

53.   Jones, A.T., Birch, A.N.E., Griffiths, D.W., Robertson, G.W., McNicol, J., and Hall, J. 1991. Characterisation of resistance in raspberry to the virus vector aphid, *Amphorophora idaei*. Proc. Int. Symp., Aphid-Plant Interactions : Populations to Molecules, Oklahoma State University, August 1990. 302 Pages.

54.   Kaloshian, I., Lange, W.H., and Williamson, V.M. 1995. An aphid-resistance locus is tightly linked to the nematode-resistance gene, *Mi*, in tomato. Proc. Natl. Acad. Sci. USA 92: 622-625.

55.   Kartohardjono, A., and Heinrichs, E.A. 1984. Populations of the brown planthopper, *Nilaparvata lugens* (Stal.) (Homoptera: Delphacidae), and its predators on rice varieties with different levels of resistance. Environ. Entomol. 13: 359-365.

56.   Keep, E., and Knight, R.L. 1967. A new gene from *Rubus occidentalis* L. for resistance to strains 1, 2, and 3 of the Rubus aphid *Amphorophora rubi* Kalt. Euphytica 16: 209-214.

57.   Kennedy, G.G. 1976. Host plant resistance and the spread of plant viruses. Environ. Entomol. 5: 827-832.

58.   Kennedy, G.G., and Kishaba, A.N. 1977. Response of alate melon aphids to resistant and susceptible muskmelon lines. J. Econ. Entomol. 70: 407-410.

59.   Kennedy, G.G., McLean, D.L., and Kinsey, M.G. 1978. Probing behavior of *Aphis gossypii* on resistant and susceptible muskmelon. J. Econ. Entomol. 71: 13-16.

60.   Khan, Z.R., and Saxena, R.C. 1984. Electronically recorded waveforms associated with feeding behavior of green leafhopper (GLH). Int. Rice Res. Newsletter 9: 8-9.

61.   Kiritani, K. 1979. Pest management in rice. Annu. Rev. Entomol. 24: 279-312.

62.   Knight, R.L., and Alston, F.H. 1974. Pest resistance in fruit breeding. Pages 73-83 in: Biology of Pest and Disease Control. D.P. Jones and M.E. Salomon,

eds. Oxford, Blackwell Scientific Publications.

63. Knight, R.L., Briggs, J.B., and Keep, E. 1960. Genetics of resistance to *Amphorophora rubi* (Kalt.) in the raspberry. II. The genes A2-A7 from the American variety, Chief. Genet. Res. 1: 319-331.

64. Knight, R.L., Keep, E., and Briggs, J.B. 1959. Genetics of resistance to *Amphorophora rubi* (Kalt.) in the raspberry. I. The gene Al from Baumforth A. J. Genet. 56: 261-280.

65. Knight, R.L., Keep, E., Briggs, J.B., and Parker, J.H. 1974. Transference of resistance to blackcurrant gall mite, *Cecidophyopsis ribis*, from gooseberry to blackcurrant. Ann. Appl. Biol. 76:123-130.

66. Lapointe, S.L., and Tingey, W.M. 1984. Feeding response of the green peach aphid (Homoptera: Aphididae) to potato glandular trichomes. J. Econ. Entomol. 77: 386-389.

67. Lecoq, H., and Pitrat, M. 1983. Field experiments on the integrated control of aphid-borne viruses in muskmelons. Pages 169-176 in: Plant Virus Epidemiology. R.T. Plumb and J. M. Thresh, eds. Blackwell Scientific Publications.

68. Lecoq, H., Cohen, S., Pitrat, M., and Labonne, G. 1979. Resistance to cucumber mosaic virus transmission by aphids in *Cucurbita melo*. Phytopathology 69: 1223-1225.

69. Lecoq, H., Labonne, G., and Pitrat, M. 1980. Specificity of resistance to virus transmission by aphids in *Cucumis melo*. Ann. Phytopathol. 12: 139-144.

70. Lowe, H.J.B., Murphy, G.J.P., and Parker, M.L. 1985. Non-glaucousness, a probable aphid-resistance character of wheat. Ann. Appl. Biol. 106: 555-560.

71. Martin, T.J., Harvey, T.L., Bender, C.G., and Seifers, D.L. 1984. Control of wheat streak mosaic virus with vector resistance in wheat. Phytopathology 74: 963-964.

72. Matta, A. 1982. Mechanisms in non-host resistance. Pages 119-141 in: Active Defence Mechanisms in Plants. R.K.S. Wood, ed. Polonium Press, New York and London.

73. Matthieu, J.L., Mourns, M., Ceustermans, N., Bent, M., and Verhoyen, M. 1984. Incidence du revêtment cireux épicuticulaire sur la sensibilitié ce différentes variétés de poireaux au virus de la stiure chlorotique du poireau *(Cilium proem* L.). Meded. Rijksfac. Landbouwwet., Gent. 49: 433-441.

74. Maxwell, F.G., Jenkins, J.N., and Parrott, W.L. 1972. Resistance of plants to insects. Adv. Agron. 24: 187-265.

75. McMurtry, J.A., and Stanford, E.H. 1960. Observations of feeding habits of the spotted alfalfa aphid on resistant and susceptible alfalfa plants. J. Econ. Entomol. 53: 714-717.

76. Moyer, J.W., Kennedy, G.G., and Romanow, L.R. 1985. Resistance to watermelon mosaic virus. II. Multiplication in *Cucumis melo*. Phytopathology 75: 201-205.

77. Müller, H.J. 1964. Uber die Auflagdichte von Aphidon auf farbige salatpfianzen. Entomol. Exp. Appl. 7: 85-104.

78. Obrycki, J.J., and Tauber, M.J. 1984. Natural enemy activity on glandular pubescent potato plants in the greenhouse: an unreliable predictor of effects in the field. Environ. Entomol. 13: 679-683

79. Obrycki, J.J., Tauber, M.J., and Tingey, W.M. 1983. Predator and parasitoid interaction with aphid-resistant potatoes to reduce aphid densities: a two-year field study. J. Econ. Entomol. 76: 456-462.

80. Oya, S., and Sato, A. 1981. Differences in feeding habits of the green rice leafhopper, *Nephotettix cincticeps* Ubler (Homoptera: Deltocephalidae), on resistant and susceptible rice varieties. Appl. Entomol. Zool. 16: 451-457.

81. Painter, R.H 1951. Insect Resistance in Crop Plants. New York, Macmillan. 520 Pages.

82. Panda, N., and Khush, G.S. 1995. Host Plant Resistance to Insects. CAB Int., Wallingford, Engl. 431 Pages.

83. Pena-Cortes, H., Sanchez-Serrano, J., Rocha-Sosa, M., and Willmitzer, L. 1988. Systemic induction of proteinase-inhibitor-II gene expression in potato plants by wounding. Planta 174: 84-89.

84. Pickett, J.A., Wadhams, L.J., Woodcock, C.M., and Hardie, J. 1992. The chemical ecology of aphids. Annu. Rev. Entomol. 37: 67-90.

85. Reddy, M.V., Sheila, V.K., Murthy, A.K., and Padma, N. 1996. Mechanism of resistance to pigeonpea sterility mosaic pathogen. Int. J. Trop. Plant Pathol. (in press).

86. Renwick, J.A.A. 1983. Nonpreference mechanisms: plant characteristics influencing aphid behaviour. Pages 199-213 in: Plant Resistance to Insects. P.A. Hedin, ed. American Chemical Society, Washington.

87. Ride, J.P. 1985. Non-host resistance to fungi. Pages 29-61 in: Mechanisms of Resistance to Plant Diseases. R.S.S. Fraser, ed. Martinus Nijhoff/Dr W. Junk Publications, Dordrecht.

88. Rizvi, S.A.H., and Raman, K.V. 1983. Effect of glandular trichomes on the spread of potato virus Y (PVY) and potato leaf roll virus (PLRV) in the field. Pages 162-163 in: Proc. Int. Cong. Res. for the Potato in the year 2000. W. J. Hooker, ed.

89. Roberts, J.J., and Foster, J.E. 1983. Effect of leaf pubescence in wheat on the bird cherry oat aphid (Homoptera: Aphidae). J. Econ. Entomol. 76: 1320-1322.

90. Robertson, G.W., Griffiths, D.W., Birch, A.N.E., Jones, A.T., McNicol, J.W., and Hall, J.E. 1991. Further evidence that resistance in raspberry to the virus vector aphid, *Amphorophora idaei*, is related to the chemical composition of the leaf surface. Ann. Appl. Biol. 119: 443-449.

91. Rochow, W.F. 1960. Specialization among greenbugs in the transmission of barley yellow dwarf virus. Phytopathology 50: 881-884.

92. Rohde, R.A. 1972. Expression of resistance in plants to nematodes. *Annu. Rev. Phytopathol.* 10: 233-252.

93. Rose, A.F., Jones, K.C., Haddon, W.F., and Dreyer, D.L. 1981. Grindelane diterpenoid acids from *Grindelia humilis:* feeding deterrency of diterpene acids towards aphids. Phytochemistry 20: 2249-2253.

94. Romanow, L. R., Moyer, J.W., and Kennedy, G.G. 1986. Alteration of efficiencies of acquisition and inoculation of watermelon mosaic virus 2 by plant resistance to the virus and to an aphid vector. Phytopathology 76: 1276-1281.

95. Saksena, K.N., Singh, S.R., and Sill, W.H. 1964. Transmission of barley yellow-dwarf virus by four biotypes of the corn leaf aphid, *Rhopalosiphum maidis*. J. Econ. Entomol. 57: 569-571.

96. Schillinger, J.A., and Gallun, R.L. 1968. Leaf pubescence of wheat as a deterrent to the cereal leaf beetle, *Oulema melanopus* (L.). Ann. Entomol. Soc. Amer. 61: 900-906.

97.   Sherwood, M.H., Greenway, A.R., and Griffiths, D.C. 1981. Responses of *Myzus persicae* (Sulzer) (Hemiptera: Aphididae) to plants treated with fatty acids. Bull. Entomol. Res. 71: 133-136.

98.   Shi, Y., Wang, M.B., Hilder, V.A., Gatehouse, A.M.R., Boulter, D., Powell, K.S., and Gatehouse, J.A. 1994. Phloem-specific expression of GUS and GNA directed by RSs1 promoter in transgenic tobacco plants. L. Expt. Bot. 45: 623-631.

99.   Shifriss, O. 1981. Do *Cucurbita* plants with silvery leaves escape virus infection? Origin and characteristics of NJ260. Cucurbit Genet. Cooperative Rep. 4: 4243.

100.   Simons, J.N. 1966. Effects of temperature and length of acquisition feeding time on transmission of nonpersistent viruses by aphids. J. Econ. Entomol. 59: 1056-1062.

101.   Simons, J.N., and Moss, L.M. 1963. The mechanism of resistance to potato virus Y infection in *Capsicum annuum* var. Italian El. Phytopathology 53: 684-691.

102.   Sogawa, K. 1982. The rice brown planthopper: feeding, physiology and host plant interactions. Annu. Rev. Entomol. 27: 49-73.

103.   Starks, K.J., and Weibel, D.E. 1981. Resistance in bloomless and sparse-bloom sorghum to greenbugs. Environ. Entomol. 10: 963-965.

104.   Sylvester, E.S., and Simons, J.N. 1951. Relation of plant species inoculated to efficiency of aphids in the transmission of *Brassica nigra* virus. Phytopathology 41: 908-910.

105.   Thackray, D.J., Morse, S., and Leech, C. 1988. The role of hydroxamic acids in wheat and maize in conferring resistance to aphids. Aspects Appl. Biol. 17: 225-227.

106.   Thompson, K.F. 1963. Resistance to the cabbage aphid (*Brevicoryne brassicae*) in Brassica plants. Nature (London) 198: 209-210.

107.   Thottappilly, G., Tsai, J.H., and Bath, J.. 1972. Differential aphid transmission of two bean yellow mosaic virus strains and comparative transmission by biotypes and stages of the pea aphid. Annu. Entomol. Soc. Am. 65: 912-915.

108.   Thresh, J.M. 1970. Reversion of blackcurrant. Pages 82-84 in: Virus Diseases of Small Fruits and Grapevines. N.W. Frazier, ed. University of California Press, Berkeley.

109.   Tingey, W.M., and Laubengayer, J.E. 1981. Defence against the green peach aphid and potato leafhopper by glandular trichomes of *Solanum berthaultii*. J. Econ. Entomol. 74: 721-725.

110.   Walters, D.S., Harman, J., Craig, R., and Mumma, R.O. 1991. Effect of temperature on glandular trichome exudate composition and pest resistance in geraniums. Entomol. Exp. Appl. 60: 61-69.

111.   Webster, J.A. 1975. Association of plant hairs and insect resistance: an annotated bibliography. USDA Miscell. Publ. No. 1297 18 Pages.

112.   Webster, J.A., and Smith, D.H. 1984. Effect of resistant wheat (*Triticum aestivum* L. em Thell) on cereal leaf beetle (*Oulema melanopus* (L.)) populations and damage to adjacent oats (*Avena sativa* L.). Prot. Ecol. 6: 251-255.

113.   Wood, E.A., Sebesta, E.E., Webster, J.A., and Porter, D.R. 1995. Resistance to wheat curl mite (Acari: Eriophyidae) in greenbug-resistant 'Gaucho' triticale and 'Gaucho' x wheat crosses. J. Econ. Entomol. 88: 1032-1036.

CHAPTER 5

Biochemistry of Resistance to Plant Viruses

R.S.S. Fraser

There are many ways to control diseases caused by plant viruses. However, unlike control of fungal pathogens, one method which is not available in practice for viral pathogens is the use of chemical pesticides. Numerous chemicals have been reported to control virus multiplication and the formation of disease symptoms (7), but none has been adopted for practical use in crop production systems. Many of these chemicals have phytotoxic effects. They are also potentially damaging to animals, humans and the environment. Given the increasing stringency of regulations for the registration of chemical pesticides in many countries, it is unlikely that the future will see any significant introduction of chemical pesticides for viruses either.

The single case where chemical treatment of virus infections can have a practical value in disease management, is in the preparation of healthy propagation material from totally infected lines of crops such as garlic. The chemicals are applied to meristem tip or callus cultures in combination with plant hormone treatments, and often heat treatments, leading to virus free lines (52).

Virus diseases may also be avoided or controlled by planting certified virus-free material of vegetatively propagated crops such as potatoes or perennial trees, by testing seed lots for the absence of seed transmitted viruses, by using cultural practices designed to avoid transmission of mechanically transmitted viruses, and by seeking to prevent transmission of vector transmitted viruses by controlling the vector or its access to host plants. All of these methods can make a valuable contribution to practical disease control in specific favorable situations but they do not offer the whole answer for all situations and are sometimes of little practical benefit. Overall, the greatest opportunities for meaningful control of virus diseases are through host plant resistance.

This chapter deals with the biochemistry of resistance mechanisms, and the underlying genetic control.

TYPES OF RESISTANCE

Plant viruses show a very wide diversity of particle size, shape and structure, form and genetic structure of their genomes, and in mechanisms by which the genome is expressed and replicated (29). They cause diverse forms of pathogenesis on their equally diverse host plants (16). It would, therefore, be expected that plants would have evolved a variety of mechanisms of resistance to counter different types of viral attack. Before considering examples and mechanisms it will be useful to consider overall types of resistance mechanism from a theoretical stand point. We can envisage three main types of mechanism operating at different levels of host population complexity.

At the level of the plant species, if the whole species is immune to a particular virus it is said to display non-host resistance to that virus. Thus, for example, bean common mosaic virus (BCMV) is restricted to *Phaseolus vulgaris* and a few other leguminous hosts. All other species tested appear incapable of supporting multiplication or symptom development by BCMV (5).

In species which are normally susceptible to a particular virus, some cultivars, or breeding lines, may display heritable resistance to that virus. Sometimes this resistance has been introduced by crossing with resistant individuals of related wild species. This cultivar resistance is the form used by the plant breeder in practical crop protection and in many cases the genetic basis has been determined. However, it should be noted that resistance breaking (virulent) strains of the virus, with the ability to overcome particular resistance genes, have evolved in many cases.

Acquired resistance (sometimes called induced resistance) occurs in susceptible individuals of the species and is induced by a prior infection, or by a chemical or cultural treatment. Generally this type of resistance is not heritable. However, it has recently become possible to create a special form of this type of resistance by genetic transformation of susceptible plants with DNA copies of portions of the viral genome or associated nucleic acids. Under control of a suitable promoter to ensure expression of active RNA or protein moieties, these transgenes can confer resistance to infection by the whole virus. If they are stably integrated into the host chromosome they will be heritable and, in effect, a lasting form equivalent to cultivar resistance will have been created.

TARGETS OF THE RESISTANCE MECHANISMS

The pathogenic cycle of any plant virus is complex. It involves replication of virus particles causing the primary infection in each initially-infected cell, secondary cycles of replication resulting from spread of the virus from cell to cell and over long distances through the plant, and spread of the progeny particles by passive means or through vector transmissions to initiate fresh cycles of infection in new plants. These "macro-economic" cycles of replication and spread each involve "micro-economic" molecular processes such as uncoating of the virus genome, expression of virus genes, replication of the genome, synthesis of virus proteins, assembly of progeny particles, molecular interaction with host components involved in replication and protein synthesis, molecular interaction with host components involved in cell to cell and long distance spread and molecular interaction with components of the possible vectors. Any of these macro- or micro-economic components might form the target for a host resistance mechanism. The potential targets are shown diagrammatically in Figure 1 (19). This scheme also covers the three broad classes of resistance introduced above.

The resistance mechanisms in Figure 1 operating at different target sites in the viral replicative cycle, may be classified into two types. Positive mechanisms are where the resistant plant contains some property which actively or directly inhibits some phase of the viral replicative cycle. Examples could include anything which reduces

the plants attractiveness to a vector of the virus: numerous well established types include hairiness, secretion by the plant of insect repelling pheromones, and antifeedants (28,42). Inhibitors of viral replicase activity, inhibitors of symptom formation, or inhibitors of cell-to-cell spread of the virus, would also fall into this category. It should be noted that the inhibitory mechanism in the positive model could be constitutive (ie., present in the plant prior to infection). However, the resistance mechanism could also be induced as a result of an initially successful infection. An example is the commonly occurring necrotic local-lesion response to viruses in which a small number of cells become infected around each primary inoculation point, but further spread of the virus is halted, generally within two to three days of the initial infection event. The implication is that early in infection there is a recognition event between plant and virus components which induces a resistance response via a signal transduction pathway. These mechanisms are considered in more detail in later sections.

The alternative form of resistance mechanism is the negative one, where resistant plants lack some component required by the virus to complete its replicative cycle. Examples could include a suitable host coded subunit of the functional virus replicase, or a host component to recognize the virus movement protein and allow modification of plasmodesmata to permit cell to cell spread. Such negative mechanisms should be constitutive and, if subjected to genetic analysis, should turn out to be controlled by recessive alleles.

Non-host Resistance

What precisely is non-host resistance and how does it work? These are amongst the most intriguing questions in plant virology today. The answers are probably complex and it is likely that several mechanisms may be involved. Recent studies using protoplast systems and sequencing of virus genomes are beginning to illuminate the mechanisms in more detail.

The question of the mechanism of non-host immunity is the mirror image of an equally intriguing question - what controls the host range of a particular plant virus? Why do some virus have very wide host ranges, while others appear able to infect only a very small number of closely related hosts? How is it that a given pair of viruses may be able to infect some hosts in

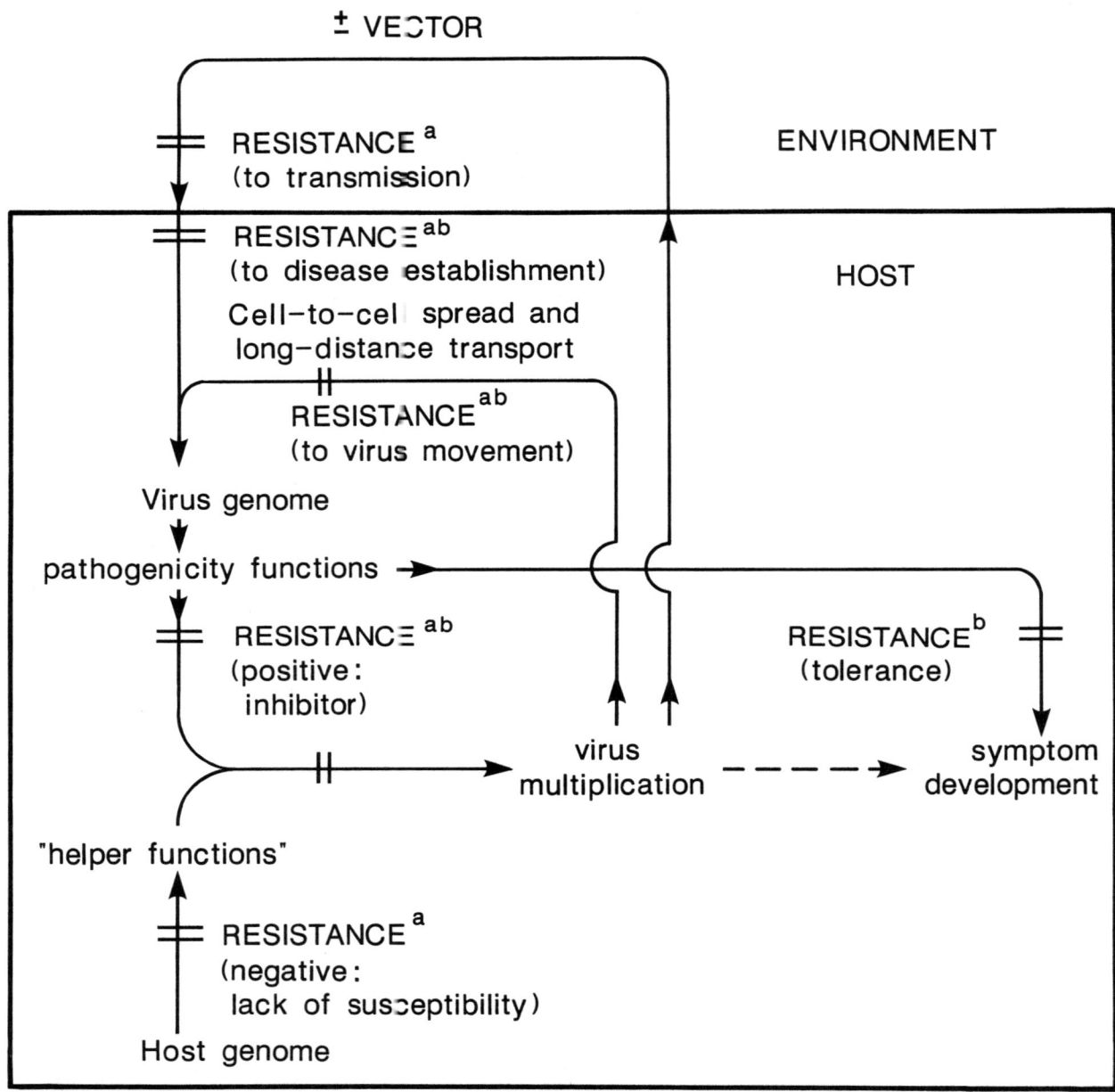

Figure 1. The replicative cycle of a plant virus showing various possible targets of cultivar resistance mechanisms.
[a] Targets which might also be involved in the termination of host range and non-host immunity.
[b] Targets that might also be involved in various types of acquired resistance mechanism. Modified from Fraser 1990 (20).

common, but each virus may also be able to infect a separate range of other hosts inaccessible to the other virus (a phenomenon known as host range congruence). In a series of classical papers, Bald and Tinsley (2-4) postulated that viruses may produce various "pathogenicity factors" which are required for successful multiplication and spread. Plants produce various "susceptibility factors" which can be used by viruses to assist multiplication and spread. Only when there is functional compatibility between the appropriate virus and plant coded factors would successful plant pathogenesis be established. Given a suitable number of pathogenicity and susceptibility factors in each plant-virus interaction, and given a suitable level of variation in pathogenicity and susceptibility factors in all plants and viruses, it is clear that this model should be able to explain the great diversity of response in plant-virus interactions.

With the molecular genetic analysis of genome expression and function now well advanced for many groups of plant viruses (29), a good deal is known about the viral pathogenicity factors and their functions, and it would certainly appear that a sufficient number of factors with sufficient variation in each, exists to satisfy the Bald and Tinsley model. The determinants of pathogenicity of barley stripe mosaic virus (BSMV) to oat have been mapped to the α RNA of the viral genome (56). There were six amino acid differences between BSMV isolates, pathogenic and non-pathogenic, from oats. Interestingly, the pathogenicity determinants mapped in an open reading frame with a sequence suggesting a methyl transferase/RNA helicase activity, but the non-pathogenic isolate appeared to be defective in cell-to-cell movement in oat, as it could replicate in oat protoplasts (59).

Where knowledge is more deficient is in understanding the diversity and variability in host susceptibility factors, Indeed, there is disappointingly little knowledge of how host-coded factors are specifically involved in virus replication and spread. The existing examples involve host-coded components of viral replicases, but more information is required (33,49).

Many host range studies have merely involved mechanical inoculation of target species with the virus or attempts at vector infection. Failure of the species to develop visible symptoms or gross evidence of virus multiplication, was taken as an indicator of non-host status. However, closer examination in some plant-virus combinations revealed the phenomenon of

subliminal infection where the virus appears able to multiply in a very small number of cells, presumably the results of the initial inoculation, but fails to spread to further cells in the leaf (9,53).This mechanism was supported by reports that virus multiplication could be established by inoculation of protoplasts of some plants previously thought to be non-hosts for the viruses involved (25). Experiments involving detection of subliminal infection and successful infection of protoplasts from ostensibly non-host species, are much more time consuming than "inoculate and watch" tests of non-host immunity. Inevitably, then, very few host-virus combinations have been examined in the greater level of analytical detail. It is possible that the cells of more plant species are potentially hosts for more plant viruses than previously thought. It is also possible that failure of cell-to-cell spread after the initiation of a point infection is an important mechanism in many cases of apparent non-host resistance (55). The difference between a true and complete non-host immunity, and one which operates by preventing cell-to-cell spread after allowing a minute amount of multiplication, can be quantified in semantic terms as precisely one cell.

An alternative mechanism proposed by Holmes (35) is that non-host plants contain a large number of individual components inhibitory of virus formation, and that these operate in an additive manner to give complete suppression of infection. It may be that plants have evolved gradual incremental resistance mechanisms which have led to viruses hosted during their evolutionary history being totally excluded from the species. Unfortunately there is no direct experimental evidence for this "pyramiding" approach to the exclusion of viruses. It is, however, intriguing that particular plant groups, such as the Pteridophytes and Gymnosperms appear to be remarkably free of virus infections. Alternatively, perhaps plant virologists have not looked hard enough.

Cultivar Resistance

Genetics of Plant Virus Interactions

The first studies of mechanisms of resistance to plant viruses involved determination of the genetics of inheritance of the resistances as adjuncts to their use in breeding programs. It is now more than half a century since the demonstration by Holmes (34) that the hypersensitive resistance to tobacco mosaic virus (TMV) in certain tobacco lines was caused by a single dominant gene N. Since then a large number of crop-virus combinations have been examined, Reviews by Fraser (18,20,21) summarize findings for 87 randomly chosen examples of crop species resistant to particular viruses. These reviews also consider various conceptual models of plant virus interactions at the genetic level.

Most of the examples in the survey involved resistance determined at a single genetic locus. In some crop species examined there were several sources of resistance, each capable of operating independently. Thus, in tomato, resistance to tomato mosaic virus (ToMV, a close relative to TMV) is controlled by genes at two loci, $Tm-1$ and $Tm-2$. Two alleles exist at the $Tm-2$ locus - $Tm-2$ itself and $Tm-2^2$ (32). In contrast, there are a few well documented cases where genes at different loci show epistatic effects. Thus, in resistance to BCMV in *Phaseolus vulgaris* the gene $bc-u$ has little effect on its own, but is required for full expression of resistance conditioned by genes at each of three other loci designated $bc-1$, $bc-2$, and $bc-3$ (12).

In the survey a slight majority of the resistance alleles examined appeared to be dominant over susceptibility. About one quarter showed evidence of gene-dosage dependent effects, i.e. resistance was more effective in homozygous resistant plants than in heterozygotes. Somewhat less than one quarter of the sample showed resistance as recessive to susceptibility,

The range of dominance/recessiveness behavior of alleles for resistance to plant viruses is interesting and might suggest a variety of mechanisms. In previous reviews (20,21). I have suggested that the dominant and gene-dosage dependent resistances may be associated with positive (inhibitor-producing) mechanisms, while the recessive forms might be associated with negative (lack of susceptibility) mechanisms. It was suggested that such mechanisms might be very difficult for the virus to overcome by mutation and should, therefore, prove very durable and effective against all isolates of the virus. It is interesting to compare this with resistance to fungal pathogens in plants, where recessiveness is very rare. However, one well characterized example, the *mlo* genes for resistance to powdery mildew in barley, have also been extremely durable and are effective against all isolates of the fungus (37).

In the plant/virus survey referred to earlier, more than half of the examples of resistance genes had been overcome by virulent (resistance breaking) isolates of the virus involved. Many of the other genes did not appear to have been adequately tested by exposure to a large number of isolates of the virus. There are very few examples of resistance genes which have proved completely robust over a long period of exposure to their virus in practical crop protection situations in different areas, as well as in the virologist's laboratory. Notable examples include $bc-3$ for BCMV resistance in *Phaseolus vulgaris*, and Ry for PVY resistance in potato (36). The N gene in tobacco remained robust for 50 years despite being used intensively in experimental virology and notably being challenged by numerous experiments involving viral mutagenesis. However, it has now been reported that a pepper isolate of TMV, Ob, will overcome this resistance (10). The $Tm-2^2$ resistance in tomato to ToMV has been outstandingly durable in commercial tomato cultivars. Rare resistance breaking isolates have been found (15) but these are unable to establish themselves fully on resistant crops and so far have posed no threat to the practical value of the resistance (21).

Mechanisms of Resistance

Broadly speaking, we still do not have a deep understanding of how most mechanisms of resistance to viruses in plants operate, although there are a few notable exceptions. Paradoxically, some of the most valuable information on resistance mechanisms is coming from studies of the virus component of the interaction. The ability to clone and sequence DNA copies of viral genomes and to construct infectious artificial recombinants between strains with different patterns of interaction with resistance genes has allowed identification of the viral function which interacts with the resistance gene product. Molecular genetic mapping of the host genome has also provided RFLP and RAPD markers which are sufficiently closely linked to known resistance genes (31,58) to make their isolation and sequencing likely in the very near future. So far, this approach has been used to isolate the first plant gene for fungus resistance (41) and virus resistance genes should not be far behind. The recent identification of resistance and susceptible responses to a virus in the model plant species *Arabidopsis thaliana* (11) with its small and intensively mapped genome, may offer a fast route to this objective.

Broadly speaking, the different types of genetic control of plant resistance to viruses seem to be associated with different types of mechanism (20). Dominant resistance is strongly associated with mechanisms which localize virus around the site of initial infection. The strongest cases of this are exemplified by *Tm-2* and *Tm-2²* in tomato where inoculating virus is confined to a single infected cell in resistant plants (46). The implication is that the resistance operates by preventing the virus from modifying plasmodesmata to allow cell-to-cell spread, or by otherwise blocking the transport process. It is interesting that resistance-breaking isolates overcoming both alleles have been shown to be altered in the 30kDa viral movement protein which controls cell-to-cell spread (6,44). Artificial recombinants between the resistance breaking and avirulent strains have conclusively shown that the resistance breaking property is located in the 30kDa protein.

A very common form of virus-localizing resistance is the hypersensitive response where the virus may spread to inoculate a few hundred or thousand cells from the point of infection, but further spread is then inhibited and a necrotic local lesion forms. Generally the evidence suggests that the necrosis and cell death is not the primary block to virus movement, but may be a secondary defense reaction (17). Virus particles can be detected outside the necrotic area in some cases, but are clearly unable to spread further. Perhaps some form of modification or gating of the plasmodesmata has become operative in these cases.

The determinant of virulence/avirulence has been shown to be located in the viral coat protein gene for at least two examples of hypersensitive defense reactions. These are the *N'* gene for TMV resistance in tobacco and the *Nx* gene for potato virus X (PVX) resistance in potato (48,51). Presumably the different forms of coat protein are involved in recognition events with host components which may trigger a cascade of events including eventual virus localization and the necrotic response. The coat protein gene and its product need not be directly involved in the ultimate expression of the defense mechanism.

For the *N* gene for TMV resistance in tobacco the determinant of virulence/avirulence was shown to be located in the gene for the viral replicase (47). It is interesting that *N* and *N'* appear to recognize quite different viral functions as determinants of virulence/avirulence. It has been suggested that *N* and *N'* are allelic, but the evidence for allelism is far from complete (13).

Further interesting observations on the

potato *Rx* gene came from David Baulcombe's group. Using synthesized recombinants between PVX strains inducing different responses, they showed that the determinant of the hypersensitive response is a threonine residue at position 121 of the viral coat protein. This feature induced hypersensitivity not only in potato carrying *Rx*, but also in *Gomphrena globosa*. This suggests that there are homologous components in potato and *Gomphrena* which recognize threonine 121 in the PVX coat protein and activate the resistance response (30). Further experiments showed that the resistance induced by PVX in *Rx* plants was also effective against the unrelated virus CMV (38).

In *N* gene tobacco, Loebenstein and colleagues have produced evidence for an inhibitor of virus replication (IVR) which may be involved in the resistance response. This protein does inhibit the multiplication of several plant viruses in test systems (26) and there is good circumstantial evidence in support of its association with *N* gene activity (27). However, its mode of action remains to be established and it is not clear whether the inhibition of TMV multiplication is sufficient to explain the complete cessation of virus spread in *N* gene tobacco. In an alternative model, Moser *et al.* (45) have shown that the 30kDa movement protein of TMV in the cell wall fraction decreased sharply in amount as the virus became localized and necrosis began.

Resistance mechanisms involving gene-dosage dependent alleles seem to be broadly associated with mechanisms which permit some spread of the virus through the plant, but which tend to reduce multiplication. Thus, the *Tm-1* gene in tomato permits full systemic spread of ToMV but inhibits its multiplication generally by about 60% in heterozygous plants and up to 95 % in homozygotes (23). Interestingly, TMV strains which overcome the *Tm-1* gene have an alteration in the viral replicase gene and again artificial recombination has shown that this is responsible for resistance breaking behavior (43). The suggestion in this case is that the product of the resistance gene is an inhibitor of viral replicase activity. Clearly this would reduce virus multiplication but would not necessarily interfere with systemic spread. Clearly, also, homozygous plants with twice the level of expression of inhibitor would be expected to be more resistant than heterozygotes. Presumably, the replicase of resistance breaking isolates is not recognized or inhibited by the *Tm-1* gene product.

Little seems to be known about

mechanisms of resistance associated with recessive genes, although any associated with negative mechanisms involving the lack of susceptibility factors would, by definition, be difficult to study. These mechanisms should tend to be associated with complete immunity types of response and there does seem to be a little evidence from studies of recessive mechanisms that this may occur. However, more work is clearly needed.

Pokeweed antiviral protein (PAP) reduces the infectivity of a number of plant viruses when co-inoculated to leaves of susceptible species (8). PAP is a member of a larger class of ribosome-inactivating proteins. These toxins operate at very low concentrations to cause complete ribosome inactivation in cells by site-specific removal of a single adenine base from the larger ribosomal RNA, thus interfering with elongation factor binding (14). PAP is a cell wall protein. A model for its antiviral mechanism was that when damage occurred during virus infection or vector transmission, the PAP selectively entered infected cells, disrupted protein synthesis, and caused local suicide with a restriction of virus multiplication. A problem with this model was that pokeweed ribosomes appeared unaffected by PAP. However, recent experiments with ribosomes prepared from protoplasts of pokeweed (hence not exposed to PAP during homogenization of leaf material) have clearly shown that pokeweed ribosomes are indeed inactivated by PAP. The local suicide model of virus resistance in plants which accumulate RIPs in their cell walls, therefore seems to be validated. Genes for RIPs (including PAP) have been cloned (39) and will be transferred to crop plants for tests of antiviral activity. In this case, however, the human and animal side effects will need careful evaluation.

Acquired Resistance

This covers a diversity of mechanisms. Some provide very effective control of virus infections; others have been much more difficult to substantiate as meaningful resistance mechanisms against viruses although they may well have effects against fungal and bacterial invaders.

Systemic Acquired Resistance and the Pathogenesis of Related Proteins

When plants are inoculated by viruses to which they respond by formation of necrotic local lesions, they often respond to a second inoculation at a later date by forming lesions which are smaller or less numerous than those formed on a previously uninoculated control plant (50). This form of apparent resistance is associated with the accumulation of the so-called pathogenesis related (PR) proteins which tend to share common properties (54). Thus, one group in tobacco is soluble in low pH buffers, resistant to digestion by trypsin, and located in the apoplast. Salicylic acid appears to be the endogenous systemic signal responsible for the induction of PR proteins and the apparent resistance (57). An antiviral function was long argued for PR proteins (1) but there appears to be little direct evidence for this. Transgenic plants expressing cloned PR protein genes do not appear to have enhanced virus resistance (40). Many PR proteins have now been found to have functions such as chitinase and glucanase activities, which could involve them in resistance to fungi and bacterial pathogens (24). Moreover, it has been questioned whether the smaller lesions formed on plants showing acquired resistance are actually associated with any reduction of virus multiplication (22).

REFERENCES

1. Antoniw, J.F., and White, R.F.. 1986. Changes with time in the distribution of virus and PR protein around single local lesions of TMV-infected tobacco. Plant Mol. Biol. 6:145-150.
2. Bald, J.G., and Tinsley, T.W. 1967. A quasi-genetic model for plant virus host ranges. I. Group reactions within taxonomic boundaries. Virology 31:616-624.
3. Bald, J.G., and Tinsley, T.W. 1967. A quasi-genetic model for plantvirushost ranges. II. Differentiation between host ranges. Virology 32:321-327.
4. Bald, J.G., and Tinsley, T.W. 1967. A quasi-genetic model for plant virus host ranges. III. Congruence and relatedness. Virology 32:328-336.
5. Bos, L. 1971. Bean common mosaic virus. in: CMI/AAB Descriptions of Plant Viruses, Vol. 73.
6. Calder, V.L., and Palukaitis, P. 1992. Nucleotide sequence analysis of the movement genes of resistance breaking strains of tomato mosaic virus, J. Gen. Virol. 73:165-168.
7. Cassells, A.C. 1983. Chemical control of virus diseases of plants. Pages 119-122 in: Progress in Medicinal Chemistry, Vol. 20. G.P. Ellis and G.B. West, eds. Elsevier, Amsterdam.
8. Chen, Z.C., White, R.F., Antoniw, J.F., and Lin, Q. 1991. Effect of pokeweed antiviral protein (PAP) on the infection of plant viruses. Plant Pathol. 40:612-620.
9. Cheo, P.C., and Garard, J.S. 1971. Differences in virusirus-replicating capacity among species inoculated with tobacco mosaic virus. Phytopathology 61:1010-1012.
10. Csillery, G., Tobias, I., and Rusko, J.. 1983. A new pepper strain of tomato mosaic virus. Acta

Phytopathol. Acad. Sci. Hungaricae. 18:195-200.

11. Dempsey, D.A., Wobbe, K.K., and Klessig, D.F. 1993. Resistance and susceptible responses of *Arabidopsis thaliana* to turnip crinkle virus. Phytopathology 83:1021-1024.

12. Drijfhout, E. 1978. Genetic interaction between *Phaseolus vulgaris* and bean common mosaic virus with implications for strain identification and breeding for resistance. Agric. Res. Reports, Wageningen 872:1-98.

13. Dunigan, D.D., Golemboski, D.B., and Zaitlin, M. 1987. Analysis of the *N* gene of *Nicotiana*. Pages 120-129 in: Plant Resistance to Viruses, Ciba Foundation Symposium 133. John Wiley & Sons, Chichester, UK.

14. Endo, Y., Mitsui, K., Motizuki, M., and Tsurugi, K. 1987. The mechanism of action of ricin and related toxin lectins on eukaryotic ribosomes. The site and the characteristics of the modification in 28S ribosomal RNA caused by the toxins. J. Biol. Chem. 262:5908-5912.

15. Fletcher, J.T. 1992. Disease resistance in protected crops and mushrooms. Euphytica 63:33-49.

16. Fraser, R.S.S. 1971. Plant viruses as agents to modify the plant phenotype for good or evil. Pages 1-24 in: Genetic Engineering with Plant Viruses. T.M.A. Wilson and J.W. Davies, eds. CRC Press, Boca Raton, 1992, 1.

17. Fraser, R.S.S. 1985. Mechanisms of genetically controlled resistance and virulence: virus diseases. Pages 143-196 in: Mechanisms of Resistance to Plant Diseases. R.S.S. Fraser, ed. Martinus Nijhoff/Dr W Junk, Dordrecht.

18. Fraser, R.S.S. 1986. Genes for resistance to plant viruses. CRC Crit. Rev. Plant Sci. 3:257-294.

19. Fraser, R.S.S. 1990. The genetics of plant-virus interactions: mechanisms controlling host range, resistance and virusirulence in recognition and response. Pages 71-91 in: Plant-Virus Interactions. R.S.S. Fraser, ed. Springer-Verlag, Heidelberg, Germany.

20. Fraser, R.S.S. 1990. The genetics of resistance to plant viruses. Annu. Rev. Phytopathol. 28:179-200.

21. Fraser, R.S.S. 1992. The genetics of plant-virus interactions: implications for plant breeding. Euphytica 63:175-185.

22. Fraser, R.S.S., and Clay, C.M. 1983. Pathogenesis-related proteins and acquired systemic resistance: causal relationship or separate effects? Neth. J. Plant Pathol. 89:283-292.

23. Fraser, R.S.S., and Loughlin, S.A.R. 1980. Resistance to tobacco mosaic virus in tomato: effects of the *Tm-1* gene on virus multiplication. J. Gen. Virol. 48:87-96.

24. Fritig, B., Kauffmann, S., Dumas, B., Geoffroy, P., Kopp, M., and Legrand, M. 1987. Mechanism of thehypersensitivity reaction of plants. Pages 92-102 in: Plant Resistance to Viruses. Ciba Foundation Symposium 133. D. Evered and S. Harnett, eds. John Wiley & Sons, Chichester, UK.

25. Furusawa, I., and Oknno, T. 1978. Infection with BMV of protoplasts derived from five plant species. J. Gen. Virol. 40:489-491.

26. Gera, A., Loebenstein, G., Salomon, R., and Frank, A. 1990. An inhibitor of virus replication (IVR) from protoplasts of a hypersensitive tobacco cultivar infected with tobacco mosaic virus, is associated with a 23K protein species. Phytopathology 80:78-81.

27. Gera, A., Tam, Y., Teviruserovirussky, E., and Loebenstein, G. 1993. Enhanced tobacco mosaic virus production and suppressed synthesis of the inhibitor of virus replication in protoplasts and plants of local lesion responding cultivars exposed to 35°C. Physiol. Molec. Plant Pathol. 43:299-306.

28. Gibson, R.W., Pickett, J.A., Dawson, G.W., Rice, A.D., and Stribley, M.F. 1984. Effects of aphid alarm pheromone derivativiruses and related compounds on non- and semi-persistent plant virus transmission by *Myzus persicae*. Ann. Appl. Biol. 104:203-209.

29. Goldbach, R., Eggen, R., de Jager, C., van Kammen, A., van Lent, J., Rezelm an, G., and Wellink, J. 1990. Genetic organization, evolution and expression of plant viral RNA genomes. Pages 147-162 in: Recognition and Response in Plant-Virus Interactions. R.S.S. Fraser, ed. Springer-Verlag, Heidelberg, Germany.

30. Goulden, M.G., and Baulcombe, D.C.. 1993. Functionally homologous host components recognize potato virus X in *Gomphrena globosa* and potato. Plant Cell 5:921-930.

31. Haley, S.D., Afanador, L., and Kelly, J.D. 1994. Identification and application of a random amplified polymorphic DNA marker for the I gene (potyvirus resistance) in common bean. Phytopathology 84:157-160.

32. Hall, T.J. 1980. Resistance at the *Tm-2* locus in the tomato to tomato mosaic virus. Euphytica 29:189-197.

33. Hayes, R.J., and Buck, K.W. 1990. Complete replication of a eukaryotic virus RNA *in vitro* by a purified RNA-dependent RNA polymerase. Cell 63:363-368.

34. Holmes, F.O. 1938. Inheritance of resistance to tobacco mosaic virus in tobacco. Phytopathology 28:553-561.

35. Holmes, F.O. 1955. Additive resistances to specific viral diseases in plants. Ann. Appl. Biol. 42:129-139.

36. Jellis,G. J. 1992. Multiple resistance to diseases and pests in potatoes. Euphytica 63:51-58.

37. Jøgensen, J.H. 1992. Discovery, characterization and exploitation of Mlo powdery mildew resistance in barley. Euphytica 63:141-152.

38. Köhm, B.A., Goulden, M.G., Gilbert, J.E., Kavanagh, T.A., and Baulcombe, D,C. 1993. A potato virus X resistance gene mediates an induced, non-specific resistance in protoplasts. Plant Cell 5:913-920.

39. Lin, Q., Chen, S.C., Antoniw, J.F., and White, R.F. 1991. Isolation and characterization of a cDNA clone encoding the antiviral protein from *Phytolacca americana*. Plant Mol. Biol. 17:609-614.

40. Linthorst, H.J.M., Meuwissen, R.L.J., Kauffmann, S., and Bol, J.F. 1989. Constitutive expression of pathogenesis-related proteins PR- 1, GRP, and PR-S in tobacco has no effect on virus infection. Plant Cell 1:285-291.

41. Martin, G.B., Brommonschenkel, S.H., Chunwongse, J., Frary, A., Ganal, M.W., Spivey, R., Wu, T., Earle, E.D., and Tanksley, S.D. 1993. Map-based cloning of a protein kinase gene conferring disease resistance in tomato. Science 262:1432-1436.

42. Martin, T.J., Harvey, T.L., Bender, C.G., and Seifers, D.L. 1984. Control of wheat streak mosaic virus with vector resistance in wheat.

Phytopathology 74:963-964.

43. Meshi, T., Motoyoshi, F., Adachi, A., Watanabe, Y., Takamatsu, N., and Okada, Y. 1988. Two concomitant base substitutions in the putative replicase genes of tobacco mosaic virus confer the ability to overcome the effects of a tomato resistance gene, *Tm-1*. EMBO J. 7:1575-1582.

44. Meshi, T., Motoyoshi, F., Maeda, T., Yoshiwoka, S., Watanabe, H., and Okada, Y. 1989. Mutations in the tobacco mosaic virus 30-kD protein gene overcome *Tm2* resistance in tomato. Plant Cell 1:515-522.

45. Moser, O., Gagey, M.J., Godefroy-Colburn, T., Stussi-Garaud, C., Ellwart-Tschurtz, M., Nitschko, H., and Mundry, K.W. 1988. The fate of the transport protein of tobacco mosaic virus in systemic and hypersensitive tobacco hosts. J. Gen. Virol. 69:1367-1378.

46. Nishiguchi, M., and Motoyoshi, F. 1987. Resistance mechanisms of tobacco mosaic virus strains in tomato and tobacco. Pages 38-46 in: Plant Resistance to Viruses, Ciba Foundation Symposium 133. D. Evered and S. Harnett, eds. John Wiley & Sons, Chichester, UK.

47. Padgett, H.S., and Beachy, R.N. 1993. Analysis of a tobacco mosaic virus strain capable of overcoming *N* gene-mediated resistance. Plant Cell 5:577-586.

48. Pfitzner, U.M., and Pfitzner, A.J. 1992. Expression of a viral avirulence gene in transgenic plants is sufficient to induce the hypersensitive defense reaction. Molec. Plant-Microbe Interact. 6:318-321.

49. Quadt, R., Kao, C.C., Browning, K.S., Hershberger, R.P., and Anlquist, P. 1993. Characterisation of a host protein associated with brome mosaic virus RNA dependent RNA polymerase. Proc.Nat. Acad. Sci. USA 90:14981502.

50. Ross, A.F. 1961. Systemic resistance induced by localized virus infections in plants. Virology 14:340-358.

51. Santa Cruz, S., and Baulcombe, D.C. 1993. Molecular analysis of potato virus X isolates in relation to the potato hypersensitivity gene *Nx*. Molec. Plant-Microbe Interact. 6:707-714.

52. Stace-Smith, R. 1990. Page 295-320 in: Tissue Culture in Plant Viruses, Vol. II. C.L. Mandahar, ed. CRC Press, Boca Raton, Florida.

53. Sulzinski, M.A., and Zaitlin, M. 1982. Tobacco mosaic virus replication in resistant and susceptible plants: in some resistant species virus is confined to a small number of initially infected cells. Virology 121:12-19.

54. Van Loon, L.C. 1983. The induction of pathogenesis-related proteins by pathogens and specific chemicals. Neth. J. Plant Pathol. 88:265-273.

55. Van Loon, L.C. 1987. Disease induction by plant viruses. Adv. Virus Res. 33:205-256.

56. Weiland, J.J., and Edwards, M.C. 1994. Evidence that the -a gene of barley stripe mosaic virus encodes determinants of pathogenicity to oat (*Avena sativa*). Virology 201:116-126.

57. Yalpani, N., Shulaev, V., and Raskin, I. 1993. Endogenous salicylic acid levels correlate with accumulation of pathogenesis-related proteins and virus resistance in tobacco. Phytopathology 83:702-708.

58. Yu, Y.G., Saghai Maroof, M.A., Buss, G.R., Maughan, P.J., and Tolin, S.A. 1994. RFLP and microsatellite mapping of a gene for soybean mosaic virus resistance. Phytopathology 84:60-64.

59. Zheng, Y., and Edwards, M.C. 1990. Expression of resistance to barley stripe mosaic virus in barley and oat protoplasts. J. Gen. Virol. 71:1865-1868.

CHAPTER   6

Coat Protein and Replicase-mediated Resistance to Plant Viruses

W. Kaniewski and C. Lawson

Genetic resistance to virus infection and disease in cultivated plants is the preferred approach to preventing losses due to viruses. However, for many virus diseases, effective resistance genes from plants have not been identified and there is a serious need for new sources of virus resistance in many crops. Although traditional breeding methods for virus resistance genes have delivered new resistant cultivars, the process is labor intensive, time consuming, and often undesirable traits must be selected out in order for a new variety to be commercially acceptable. Successful genetic improvement of crop plants by methods developed through biotechnology has resulted in introducing new genes for virus resistance into existing commercial crop varieties. These new genes are derived from the viruses themselves in a concept referred to as Pathogen Derived Resistance (PDR) (74). The first demonstration of virus resistance in transgenic plants utilized the tobacco mosaic virus (TMV) coat protein (CP) gene expressed in tobacco plants (64). This resistance demonstration was referred to as coat protein-mediated resistance (CP-MR). Since that time the CPs from almost all plant virus groups have been expressed as transgenes in a variety of crops and model test plants. In many examples, very high levels of resistance to virus infection have been demonstrated in these transgenic plants.

A range of virus resistance has often been observed in transgenic plants expressing viral genes. At one extreme, immunity to infection by homologous and closely related viruses has been claimed. Moderate resistance is often observed as some plants escaping infection, reduced titer of virus in infected plants, and suppressed symptoms. Low resistance usually is observed as a delay in symptoms and titer development by days or weeks, but eventually all challenged plants become virus infected and show symptoms.

In addition to CP genes, other genetic elements from plant viruses are also being tested in transgenic plants for the ability to confer resistance to virus infection or disease (39,93). In particular, the expression of replicase (RNA dependent RNA polymerase) native genes and modified versions have been shown to confer resistance to virus infection. The modified proteins are thought to be defective in functions critical for the normal replication or spread of the virus in the plant. Both, the expression of native viral replicase genes and modified replicase genes have resulted in transgenic plants demonstrating high levels of resistance to virus infection or disease. The CP-MR and Replicase-MR are the two PDR strategies most likely to be commercially available in the near future, especially in vegetable and grain crops. These highly effective strategies offer a unique opportunity to combine biotechnology approaches and classical breeding derived genes to achieve highly durable resistances and broad efficacy. The following discussion focuses on CP-MR and Replicase-MR and some considerations necessary to bring plant products from these technologies into commercialization.

GENETIC MODIFICATION OF PLANTS FOR VIRUS RESISTANCE

Genetic engineering for virus resistance should be considered as a logical option in all cases where substantial losses occur in cultivated crops and natural resistance cannot be found or is difficult to incorporate into leading varieties by breeding.

Initially, losses caused by a virus disease should be characterized and quantified. Preliminary epidemiological studies should be used to determine the aggressiveness and prevalence of strains within the targeted growing area. In the case of wide strain diversity in targeted areas, the most common and aggressive field isolate should be selected for cloning. In cases where more than one strain causes substantial losses, genes from multiple strains can be selected for transformation (40,81).

Whenever possible, major crop varieties should be considered for transformation to allow for immediate use of transgenic plants. In cases where target varieties cannot be transformed, other varieties can be used to introduce genes of interest, followed by breeding into commercial varieties.

Viral genes should be selected based on analogy with previous performance experience. In most cases CP would be considered first because there are many examples in which CP has conferred high resistance to viruses from various groups (25). The coat protein gene is generally well characterized and therefore easily cloned. Following expression in plants, the coat protein can be detected using readily available antisera generated to virions. In the case of luteoviruses, CP has conferred only weak to moderate resistance (37,41,88), so the use of other genes such as replicase should be considered. For virus groups where high resistance has not yet been achieved, basic research must to be done to select the best resistance genes or modifications of commonly used genes.

The development of transgenic plants resistant to viral diseases requires an effective expression cassette for the expression of viral transgenes. The expression cassette consists of a transcriptional promoter and terminator which control the production of the viral gene messenger RNA in transgenic plants. Generally, constitutive promoters that result in high levels of gene expression have achieved the greatest success. The cauliflower mosaic virus (CaMV) 35S promoter, in a single or double copy version (enhanced 35S) is used most frequently to express viral transgenes (42), although several other constitutive promoters have also been used successfully. In some cases, high levels of expression are required to achieve virus resistance. For example, expression of TMV CP in tobacco under the control of the nopaline synthase (NOS) promoter, which is known to express poorly in plants, did not result in measurable resistance in any of 36 transgenic lines generated, while expression of the same gene by the 35S promoter resulted in a high frequency of resistant tobacco

and tomato lines (W. Kaniewski, unpublished). The use of the pal2 promoter which is tissue specific and expresses genes at lower levels than the 35S promoter was not as effective in conferring resistance to TMV (71).

In situations where a virus is restricted to certain cells or tissues, the use of tissue specific promoters may be considered. For example, a phloem specific promoter, glutamine synthase (GS), was used to express the CP of potato leafroll virus (PLRV), a virus restricted to phloem tissue in potato. This construct, however, did not yield any plant lines that were significantly more resistant than plants transformed with a construct utilizing the 35S promoter (W. Kaniewski, unpublished). Tissue specific promoters such as pal2 (epidermis, xylem) and rolC (phloem, leaf hairs) driving TMV CP were marginally or not effective in producing resistant plants, respectively (71). Therefore the use of tissue specific promoters may not improve resistance when compared to strong constitutive promoters that are also expressing within these same tissues.

Gene expression from a particular construct may vary considerably between plant species and therefore the expression cassette should be optimized for each specific crop. The optimization of gene expression for enhanced resistance involves many additional genetic elements. These include the effects of translational enhancing mRNA leaders, the presence of 3' untranslated regions, and transcriptional terminators. The addition of appropriate introns and/or modifications of coding sequences for increased translation can also substantially improve transgene expression. These types of experiments, supported by many examples, were recently reviewed by Hanley-Bowdoin and Hemenway (28).

Despite all the information available on the successful use of PDR, there is no guarantee that a specific construct will result in a high frequency of highly resistant plants. Therefore, to select the best construct for commercial resistance, a crop should be transformed with several constructs. The use of double gene constructs containing either the same gene driven by different promoters or two different resistance genes may provide broader or more durable resistance.

In addition to the gene expression cassette, transformation vectors typically contain marker genes used to select or screen for transformants. There is no evidence that these genes influence resistance.

Plant Transformation

Direct gene transfer, using methods developed by plant biotechnology, can now be applied by plant breeders to broaden the gene pool of crop species. Plant transformation with viral genes is becoming a common tool to generate virus resistant plants (7). Introducing new genes into plants can now be performed on most crops by either an *Agrobacterium*-mediated method or via direct DNA delivery (particle gun) into plant tissues that subsequently differentiate into whole plants (31).

Agrobacterium-mediated transformation has been the method of choice for most dicotyledonous plants and particle gun delivery for most monocotyledonous plants. Recently, an efficient *Agrobacterium*-mediated transformation system has been developed for rice (30). Transformation and regeneration protocols are being continually improved for existing transformation-competent crops and new methods developed for recalcitrant crops.

Transformation events are generally differentiated from nontransformed tissue by the utilization of selectable or scorable markers. These are delivered concomitantly with the gene for the trait of interest into the plant of choice. The most utilized selectable markers are those that detoxify aminoglycoside antibiotics and various herbicides (31).

Scorable markers are sometimes used to aid in the identification of transgenic plants. The gene for β-glucuronidase (GUS) has been often included in gene vectors as a marker to determine if a regenerated plant has been successfully transformed (32). However, this gene is not necessary in the expression vector if assays are available for the viral gene product or the selectable marker gene product. The assay of activity or presence of selectable markers are often useful to confirm transformation. The most commonly used selectable marker is neomycin phosphotransferase II (NPT II) which can be easily detected using commercially available ELISA kits. Assay of the plant tissue on media containing the antibiotic is also often used and is generally reliable. Although, the presence of a scorable or selectable marker does not guarantee expression from the viral gene cassette.

Plants transformed with viral genes are usually assayed for viral protein expression to confirm the successful transformation and integrity of the expression cassette driving these genes.

Direct detection confirms viral gene expression and allows for the estimation of the protein expression levels.

Gene Expression Analysis

Transgenic plants can be assayed for presence of the transgene by southern blot analysis or PCR, and expression of transgene RNA by RT-PCR or northern blot analysis. Western blot analysis is done to detect the expression of the viral gene and to visually determine that the gene product is of the expected size.

The most useful assay is direct enzyme linked immunosorbent assay (ELISA), which confirms protein expression and can accurately quantitate levels of the viral transgene product. Successful development of an ELISA for detection of viral protein expression in plants, depends on the availability of high quality antisera and antibody conjugated to one of several enzyme detection systems. Good antisera are usually obtainable for most plant viruses and the ELISA can be successfully developed for CP expression. However, the assay can be impacted by several parameters. The expression levels of the viral protein must be high enough to be reliably detected by the assay. The ELISAs for most CP assays have sensitivities of 0.1 - 1.0 ng/sample. Extractability for the detection and quantitation of the protein in an ELISA compatible buffer can also be a concern. Viral proteins may complex with membrane fractions of the plant cell making recovery difficult without the use of ionic detergents.

Double antibody sandwich ELISA (DAS-ELISA) may fail to detect CP transgene expression if the CP gene is from a potyviruses. The dominating N-terminal epitope can make some polyclonal antibody preparations behave more like a monoclonal antibody. A DAS-ELISA, using the same coating antibody and conjugated antibody may work well for intact virus, but the sensitivity of such an assay for CP monomers can be 1000 fold lower than for the respective virions. This problem can be partially solved by using a different secondary antibody, such as PTY-1 monoclonal antibody generated to the core protein portion of potyvirus CP (34) and to detect its presence in the assay using anti-mouse antibody enzyme conjugate. This system has a sensitivity of about 2 ng/sample and so is useful for the detection of CP expression in the transgenic plants with the higher expression levels (W. Kaniewski, unpublished).

Detection of protein expression of some viral transgenes has not been achieved, yet the transgenic plants are highly resistant to virus infection. This has been especially true for the replicase genes and may have to do with the translatability, stability of the protein or extractability of the protein from the plant tissue (23). Possibly, only minute concentrations of the protein are necessary for virus resistance.

Strategies for Selection of Virus Resistant Plants for Vegetatively Propagated Crops and for True Seeded Crops

Single transformation events are always precursors of individual transgenic lines. Such individual plants confirmed as transgene expressors need to be propagated for resistance experiments. It is generally not recommended to challenge newly regenerated transformed plants with viruses. Epigenetic effects of tissue culture hormones can affect the susceptibility of the plants. The plants are highly stressed and this can also increase their susceptibility to virus infection. Moreover results from resistance experiments with a single plant can be very misleading if the plant is not immune to virus infection.

Vegetatively propagated plants can be increased by tissue culture micropropagation or by rooting nodal cuttings in potting media. Large numbers of clonally propagated plants of each transgenic line can be evaluated for transgene expression, plant type and virus resistance in the growth chamber or greenhouse. Micropropagated plants have also been used successfully for large scale field resistance experiments (35,80). These are satisfactory methods useful for early screening of lines for resistance, but the final selection of commercially acceptable plant lines will need to be with the type of planting material normally used for the crop.

A different strategy is applied for plants propagated through seeds. Several levels of expression and resistance analysis are required. Usually seeds from R_0 transformants are planted and progeny (R_1) assayed for gene segregation by ELISA. A 3:1 gene segregation ratio identifies a single copy insertion which facilitates the selection of homozygous lines and is useful for breeding into other varieties. Lines which segregate aberrantly are discarded. If the ELISA is quantitative, then 4-6 of the highest R1 expressors

are selfed and saved for seed production. The remaining expressing R_1 plants can be challenged with virus for evaluation of resistance. Lines which don't demonstrate virus resistance are discarded. The resistant lines for which plants were saved from the highest R_1 expressors are further evaluated in the R_2 generation. These progeny are tested by ELISA to determine if they are homozygous for the gene of interest. The use of seed from homozygous resistant lines is recommended for field testing for resistance and plant type evaluation.

Regeneration of plants from tissue culture often produces off-type transgenic plant lines (20), some of which may show altered susceptibilities to virus infection due to physical malformations or demonstrate genetically recessive traits which give virus resistance (82). This observation should be of particular concern when concluding that transgenic plants from a given gene construction are resistant to virus infection from R_0 and R_1 plant analysis. It is usually necessary to confirm that the transgene is in the segregating progeny and correlates with the resistance phenotype. In vegetatively propagated crops, regeneration from tissue culture can result in the selection of lines which show virus resistance phenotype. This trait in these crops may be a valuable observation, although the resistance is not necessary because of transgene expression (65).

Criteria for Commercially Useful Lines

Prior to commercial use of virus resistant transgenic plants, selected lines must be thoroughly examined for a few years in multilocation field tests (36). Plants should maintain all the important features of the original cultivar used for transformation. Growth, vigor, and yield can not be affected (33). Genetic stability of the resistance in selected lines must be documented in multigeneration tests. Lines with single gene insertions are preferred because of simplicity of eventual future crossings. These are generally the same considerations taken for resistance genes developed through classical breeding methods.

Only lines resistant to the majority of naturally occurring severe virus isolates have the potential for commercial value. The level of virus resistance, durability of resistance in natural field locations, and good plant type are the major criterions in selecting lines for practical use. The selection and evaluation processes should follow general field test evaluation protocols used in breeding for virus resistance. Field immunity to virus infections would obviously be preferable, but any substantial increased resistance over that presently available in current varieties could be useful. Long term durability of the engineered resistance is another concern. As with the evaluation of any new disease resistance genes, the selection of strains or isolates that can overcome the resistance is a concern and evaluation of this potential should be tested as extensively as possible early in the research program. To date, there have been no reports of the selection of new resistance breaking strains from any of the field tests of transgenic plants resistant to virus disease. Starting with highly resistant lines and a resistance management strategy, the risk of selecting resistance breaking isolates can be reduced. The possibility of combining alternative strategies (5) or the expression of multiple CPs from two diverse virus strains can make the resistance broader and more durable (66,81).

Regulatory Issues

The potential benefits of transgenic virus resistance include increased yield, reduced pesticide use to control vectors of viruses, improved crop quality, and increased benefits of multiple virus resistance traits. This technology offers a high agronomic benefit with low risk. Indeed, in the USA, Asgrow Co. has received USDA regulatory approval for transgenic squash resistant to cucumber mosaic virus (CMV), zucchini yellow mosaic virus (ZYMV) and watermelon mosaic virus II (WMVII). This is the first commercially available plant product engineered for virus resistance (24, 83).

It is likely that most of the viral genes used to produce PDR in crop plants will be exempted from Governmental regulation as more information about these genes is generated.

Some questions remain about possible agronomic risks of recombination of viruses with the viral transgene mRNA and the potential impact of the large scale application of this technology. During recent open discussions sponsored by USA Governmental agencies on the subject of recombination of viruses with a viral transgene, it was concluded, that the probability of more virulent virus strains arising from recombination with viral transgenes in plants is vanishingly small. Many reasons for this conclusion were

discussed. These include, that only a portion of the virus genome is in the genetically modified plants and the probability of virus recombination with these transcripts is likely to be reduced when compared to that of intact viruses in naturally occurring mixed infections, where recombination could occur between any of the viral genes. A virus genome may be size limited and any stable recombination should result in a genome of similar size to the parent, this also limits the extent of heterologous recombination. In addition to the physical constraints to recombination, recombination must result in a viable competitive virus which is distinguishable from the parents in factors such as viability, stability, fitness, competition and virulence, vector transmission or pathogenicity. Highly virus resistant crops will reduce the amount of the target virus pathogen in the crop and the overall occurrence of plants infected with multiple viruses. Therefore, crop plants genetically improved for virus resistance will likely provide an environment less conducive to virus evolution by recombination. However, investigation of the recombination potential of plant viruses with plant produced viral mRNA from transgenes is being pursued by academic and industrial laboratories to determine if novel viral pathogens are likely to arise by recombination with plant viral transgene mRNA transcripts.

COAT PROTEIN-MEDIATED RESISTANCE

Many CP genes have been shown to be highly effective in preventing or reducing infection and disease from the homologous and closely related viruses (24), although, the frequency of obtaining plants with high resistance can vary widely. Many virus CP genes expressed in transgenic plants deliver only a small percentage of highly resistant lines, with the majority of lines showing moderate levels of resistance or susceptible responses. Other CP genes have produced highly resistant plants with frequencies approaching 100% (W. Kaniewski, unpublished). It is interesting, that the same gene expressed in one plant species can generate highly resistant plants and in other species only weakly resistant or susceptible lines are observed. For certain virus groups (gemini, luteovirus), CP constructs have not produced plants with the same high levels of resistance as obtained for many other plant-viral transgene combinations.

CP-MR Efficacy

Homologous CP-MR has been demonstrated with viral CP genes from most virus groups. It is quite possible that it can be achieved with all plant RNA viruses and some of the DNA viruses. The type and level of resistance observed can vary widely and the ability to confer agronomically useful resistance must be demonstrated under field conditions for each application.

Of considerable importance is the necessity to determine the ability of CP-MR to provide resistance to strains and field isolates for each target virus and crop. There is strong experimental evidence that for many viruses, CP-MR is effective against strains of the virus. Resistance to diverse virus strains or isolates has been reported for TMV (58), tomato mosaic virus (ToMV) (73), CMV (55,67), alfalfa mosaic virus (AlMV) (46,84), potato virus X (PVX) (29), potato virus Y (PVY) (62), PLRV (37), tomato spotted wilt virus (TSWV) (18) and others.

There are also some examples of good homologous resistance, but poor resistance to distant strains. CMV-C CP gave good resistance to strains from subgroup I, but poor resistance to WL strain from subgroup II (67). Tobacco expressing tobacco rattle virus (TRV) CP is resistant to the homologous TCM strain, but not to the PLB strain with only 39% protein sequence homology (90). Papaya ringspot virus (PRSV) CP in papaya provides resistance to some strains, but not to others (24). TSWV CP gave resistance to strains from the same subgroup, but not from the other (61). CP resistance to selected viruses within a virus group was reported for tobamoviruses (57), potyviruses (56,76), carlaviruses (51) and tobraviruses (90). The tobamovirus group experiments show a need for at least 60% amino acid homology of the CP to observe resistance. The only report of CP-MR across virus groups was by Anderson et al. (1), which was not well documented. These results could not be repeated in our laboratory.

The resistance from CP expression can be generally fairly broad against members of the same virus group. However, the level of resistance may range from apparent immunity to a delay in symptom development, and broad resistance does not necessarily occur in every case.

Correlation of CP Expression with Resistance

Transgene CPs of different viruses are expressed using a variety of promoters and expression of certain CP genes vary widely between plant lines and even more between different transformed plant species. Some CP genes always express at high levels while others express at very low levels. Moreover, expression of transgenes can vary with the age of the plant, position of tissue sample collection within the plant, efficiency of extraction, and type of assay performed. Much of the published CP transgene expression data is not accurate enough to compare transgenic lines and it is even more difficult when trying to compare expression results between laboratories.

Results of CP-MR for some virus groups strongly indicates that there is a positive correlation between CP expression and resistance. For examples, refer to the following publications: TMV (58,64), CMV (17), AlMV (84,89), and PVX (29). Low expression of CMV CP in tomato produced plants almost completely susceptible to infection, however when tomatoes were transformed with an improved construct resulting in higher CP expression, a high level of resistance to infection was observed at a high frequency (W. Kaniewski, unpublished). However, within the virus groups where there is usually a positive correlation between CP expression and resistance, there are often high expressing plant lines which exhibit low or no resistance. This could be explained by patchy expression of the CPs by the promoter which then allows the challenging virus to get established and eventually overcome the CP-MR, although this is not well documented. Our experiments with ToMV (73), AlMV (84), CMV (17) and PVX (29) indicate that a certain level of expression is necessary for a high level of resistance. Lines expressing above this threshold can be highly resistant, but some lines are only moderately resistant and others are susceptible. The observed plant response is most probably dependent on the promoter used for CP gene expression, but specific properties of the virus and differential expression due to the location of the gene insertion in the plant genome may also be important. In one example, a CMV-CP gene provides tobacco plants with extreme resistance to CMV infection (17), however, the same gene construct in tomato did not deliver any highly resistant plant line. In this case, an out of frame translational start codon upstream of the native CMV-CP start codon reduced protein translation levels from the expression cassette. The expression in tomato was greatly reduced compared to the tobacco lines.

However, many published examples of CP-MR for potyviruses have generally not seen a direct correlation between transgene expression levels and virus resistance (43,45,76). In our early experiments, a noninfectible PVX/PVY potato line expressed low levels of PVY CP, but comparable levels of mRNA. Similar results were observed by Stark and Beachy (76) with soybean mosaic virus (SMV) CP. Extremely low expression or perhaps non-expression of TEV CP resulted in very high resistance (44). In general, the level of potyvirus CP expression in transgenic plants is not a predictor of the resistance. For tospoviruses there is also no correlation between CP transgene expression and resistance to the virus (86). However, it is possible that high expression of the TSWV CP gene confers low to moderate resistance to other tospoviruses, but no detectable CP expression and low production of CP transcripts confers a high level of resistance very specific to the homologous virus (22).

CP-MR Mechanism

The possible mechanisms of CP mediated resistance have been discussed extensively in review publications (25,28,70,77). Strong support for prevention of uncoating mechanism theory were evidences that RNA inoculation overcame the resistance. This was reported for TMV (60,69), AlMV (46), and TSV (90). For AlMV (84) as well as for PVX (29), potato virus S (PVS) (51), arabis mosaic virus (ArMV) (8) and grapevine chrome mosaic virus (GCMV) (9) resistance was effective also against viral RNA. This suggests that besides interference of virus uncoating, some other steps in virus replication or spread are affected. Clark *et al.* (16), constructed TMV coat protein mutants that were defective in virion assembly, but were still able to confer resistance. This suggests that interaction of the CP with the virus RNA to affect translation or interaction with a host component is responsible for the resistance.

Beachy (7) suggested CP involvement in cell to cell and long distance spread of the virus. Evidence for CP involvement in virus spread has been demonstrated for TMV (94) and for bean yellow mosaic virus (BYMV) (27).

None of the proposed theories explain all of the transgenic plant-virus resistance observations. Specifically lack of CP expression correlating to resistance does not fit to any of the suggested theories. Another example of different mechanisms was observed in Russet Burbank potato transformed with double CP gene construct which expresses PVX and PVY CP (43). This potato was mechanically not infectible by both viruses, but after graft inoculation with a double infected wild type scion, PVX was able to spread throughout the plant and could be easily detected in the transgenic plant tissues while PVY was never detected in the transgenic portion of the grafted plant (W. Kaniewski, unpublished). This indicates 2 different mechanisms of interactions of CP with these infecting viruses. Even for the same plant-CP combination clear differences can be found. Tobacco expressing AlMV CP was resistant to virus, but easily infectible by RNA or even RNA without RNA 3 (89), but another tobacco expressing AlMV-CP was resistant to virus and RNA inoculation (84).

For potyviruses, there may be CP and RNA mechanisms which can confer resistance in a transgenic plant. This may involve a resistance mechanism mediated by a plant response to degrade translation defective transcripts of tobacco etch potyvirus (TEV) (44). Recently, it was discovered that the initiation of translation was critical for conferring virus resistance in plants engineered to express modified CP genes from peanut stripe potyvirus (PStV) (15).

All these observations suggest, that CP resistance cannot be explained by one mechanism. Different mechanisms could be involved for different plant-virus combinations, and multiple mechanisms are probably involved even for individual transgenic plants. Separate mechanisms could be speculated for truncated or mutated coat proteins.

Resistance to Vector Transmission

CP-MR is very effective for most virus groups when the transgenic plants are challenged by mechanical inoculation of the virus. In nature, arthropods or other vectors are often the most significant vehicle by which viruses spread in agricultural fields. When possible, evaluation of resistance to natural vector transmission should be included early in the plant selection scheme. In most applications of CP-MR, effective resistance was observed for both mechanical and vector transmission. Potatoes which express PVX and

PVY CP (43) were resistant to aphid transmitted PVY. Resistance to PLRV can be only evaluated using aphid transmission and CP expressing potatoes were resistant (37,41,88).

Tobacco, tomato and cucumber expressing CMV CP are highly resistant to aphid transmission (24,67,81) in growth chambers and field experiments. TSWV CP plants are resistant to thrips transmission (22). The CP gene of potato mop-top virus (PMTV) expressed in *N. benthamiana* conferred resistance to infection by mechanical and fungal inoculation (68).

The only known example where plants were resistant to mechanical, but not to vector transmission is tobacco expressing TRV CP and nematode transmission of this virus (63).

Field Testing

The first field test of transgenic plants was for the confirmation of tomato resistance to TMV (58). Since then field tests have been conducted with many different transgenic plants resistant to various viruses. The practical value of economically important crops expressing CP-MR can be demonstrated only by effective field testing (37). The criteria for effective field resistance to viral disease can vary significantly from growth chamber to the field depending on the crop and the virus. The field test designs should also reflect the level of resistance that is observed and/or desired (36).

Agronomic performance and preservation of cultivar identity are other important criteria which are evaluated by field testing for virus resistant, transgenic plants prior to commercialization (33).

REPLICASE-MEDIATED RESISTANCE

There are fewer examples of plants transformed with replicase genes than with CP genes, however this technology is the most advanced behind CP-MR and the current results indicate that a high level of resistance can be produced. Among the replicase-MR examples are genes which have various deletions, sequence modifications, truncations and full-length replicase genes (13).

The replicase-MR was first demonstrated in tobacco plants containing a truncated gene of TMV (23). Since then, replicase-MR has been observed for PVX (6,10), PVY (4), AlMV (11),

cymbidium ringspot virus (CyRSV) (72), CMV (2), pea early browning virus (PEBV) (49), PLRV (38,80), cowpea mosaic virus (CPMV) (75) and TRV (3).

Replicase-MR Efficacy

This strategy is the second most advanced application of the use of viral-derived genes for plant virus resistance. In most published applications of replicase-MR, either the plants are extremely resistant to homologous virus challenge or not resistant (10,23). Levels of intermediate resistance were reported for CMV replicase expressing tobacco (2). Some replicase genes did not deliver plants with any enhanced resistance to disease symptoms (53,79,92).

Initial studies with truncated TMV replicase gene (23) showed extreme specificity. Transgenic tobacco plants were completely resistant to the homologous TMV strain, but susceptible to closely related strains or isolates of TMV and ToMV (W. Kaniewski, unpublished). Tobacco plants expressing a defective CMV replicase gene were resistant to several, but not all strains from the homologous subgroup and susceptible to all strains tested from the heterologous subgroup (95). Tobacco expressing PVX full length replicase gene was highly resistant to homologous strain, but susceptible to distantly related South American strain (10). *N. benthamiana* transformed with CPMV replicase gene was highly resistant to homologous virus and closely related strains, but not to other comoviruses, like cowpea severe mosaic virus (CPSMV) (75). *N. benthamiana* plants transformed with the truncated PEBV replicase gene were resistant to various isolates, but not resistant to some other tobraviruses (49). Potatoes expressing full length PVY or PLRV replicase genes are highly resistant to all strains and isolates so far tested (W. Kaniewski, unpublished). Several mechanisms may be contributing to the observed virus resistance which is reflected by the various replicase gene modifications expressed in transgenic plants. For gemini virus resistance, the expression of a truncated DNA polymerase gene (C1 of TYLCV) may by a dominant negative mechanism reduce infectivity and delay virus titer increase in transgenic *N. benthamiana* (59). In general, with increasing numbers of examples of transgenic plants expressing various viral replicase genes, it can be expected that different types and levels of resistance will be observed. The preliminary observations suggest that unmodified full-length replicase genes provide broader, less specific resistance than the modified defective replicase genes.

Correlation of Replicase Expression with Resistance

There are examples where replicase protein expression is key to the resistance phenotype (12,49) and examples where the resistance could be RNA mediated (18,52,75). The expression of the replicase protein has been detected for CMV (14) and CyRSV (48), and the DNA polymerase for TYLCV (59). However, the detection of replicase gene expression is often difficult. Usually it is necessary to produce replicase protein in *E. coli* to prepare specific antisera. However, even high titer antisera which can detect low amounts of purified antigen may not be effective in detecting replicase expression in transgenic plants. Also, antisera generated to synthetic peptides of replicase epitopes may not always be of adequate sensitivity. Therefore, there has not been enough data generated to correlate replicase protein expression with resistance. However, Noris *et al.* (59) can detect the truncated C1 gene in transgenic plants and correlate it's presence with inhibition of virus infection.

Replicase-MR Mechanism

The plant RNA virus replicase is used in combination with host factors to make copies of the viral genome and control some aspects of virus gene expression. The replicase enzyme has RNA polymerase activity, binds RNA and nucleotide triphosphates, and a helicase domain(s) may also be recognized as a function of the replicase for many plant viruses (26). There are many possible mechanisms in which the expression of replicase genes in plants could interfere with normal virus replication. Three recent reviews have covered the previous examples of replicase-MR and discussed possible mechanisms (6,13,93). There are no clear demonstrations of the mechanisms or of how various coding sequence modifications function to confer virus resistance when these genes are expressed in transgenic plants. Szybalski (78) had previously speculated that a trans-dominant-lethal viral gene, especially replicase genes could protect the host (also plants) from viral infection. He demonstrated the feasibility of this approach using mutated replicase

genes of bacterial virus.

Currently, the predominant demonstrations are mutations in domains controlling critical functions of the replicase protein which could result in a "dominant negative" interaction. This defective gene product is thought to compete with the native replicase during plant virus infections, possibly at several levels during virus replication. The catalytic domain of the replicase has most often been targeted for modifications (11,47). Unfortunately, in many examples some critical gene constructs or experiments were not conducted, making it difficult to conclude what activities, or lack thereof, are responsible for the resistance phenotype. Longstaff (47) created mutations which inactivated an infectious PVX clone and some of these mutations conferred resistance when expressed in transgenic plants. However, the full-length unmodified replicase gene was not included in the test. The modifications of the GDD domain may play a different role in conferring resistant phenotypes than originally postulated. Subsequently, the finding that the full-length replicase gene of PVX conferred resistance and all constructs modified and unmodified were effective has suggested an RNA-mediated mechanism for PVX (54). This is also the suggested mechanism for replicase-MR to CPMV (75). However, this finding is not universal for all plant RNA viruses. The full-length replicase gene of PVY, but not the deletion of the GDD domain conferred resistance to virus infection (4). The full-length sense replicase gene of PLRV and a truncated sense open reading frame, but not a misframe (translation not in the replicase reading frame) full-length gene conferred resistance to this virus (38). A translatable gene for TMV 54 kD, but not an untranslatable gene was necessary for resistance (12). Braun and Hemenway (10) found that the full-length and the 5' portion of the PVX replicase gene conferred resistance, but the 3' portion which contained the GDD domain was not effective.

It appears that the replicase-MR observed for TMV is a dominant negative effect of the expression of a truncation mutant of TMV replicase (13). The use of the full-length gene (183 kD) or 126 kD sequence did not produce any plant lines which demonstrated resistance to TMV infection (19,23), but an insertion in the helicase domain resulted in resistant lines (19). These gene constructs would suggest two distinct activities, an enzyme fragment which may interfere with the assembly of a larger replication complex, and a dysfunctional enzyme which would interfere with the activity of the replication complex.

The expression of full-length functional replicase genes appears to confer a high level of resistance to an increasing number of viruses (4,10,38,72,75). The tobamovirus group may be the exception to this finding, since only a truncated replicase gene or defective gene was effective for this virus group (12,19,23).

The CMV replicase gene with an internal deletion confers a high level of resistance (2). The native full-length gene was not tested, therefore it is not possible to conclude which activity is contributing to virus resistance in this example. In another example, mutant AlMV replicase gene expression in transgenic plants conferred resistance to virus infection (11). In two previous studies with unmodified genes, it was not reported if infectious virus could be recovered from the inoculated transgenic plants (79,92). It was noted that RNA 1 and RNA 2 decreased with time in these plants suggesting resistance to virus multiplication. If adequately tested, full-length unmodified replicase genes from multipartite viruses may be found to be effective in conferring some level of resistance (75). Although plants maybe susceptible to virus disease symptoms, reducing virus multiplication could be effective for preventing virus spread in the field.

Viral replicase genes are effective in conferring resistance to infection by both RNA and virions (2,10,47). There appear to be multiple mechanisms involved which depend on the virus, plant, and type of gene applied. Whatever the mechanism, the discovery that expression of various replicase sequences can confer a very high level of resistance is extremely useful for engineering plants for virus resistance.

Resistance to Vector Transmission

There is very limited data about vector virus transmission to replicase expressing plants. Our data shows a high level of resistance to aphid transmission in PLRV replicase expressing plants (38). Aphid transmission of two CMV strains were tested and resistance was reported (95). In general, it has been demonstrated that plants highly resistant to mechanical inoculation are also resistant to vector transmission.

Field Testing

There are few published reports of field testing of transgenic plants expressing replicase

genes for virus resistance. We have field tested extensively, (1992-95) potatoes expressing unmodified full length PLRV replicase gene (38,79). A relatively high percentage of lines showed field immunity to PLRV infection. No virus or virus disease symptoms were present by any analysis. Plants grown from tubers in subsequent seasons from PLRV challenged plants were also healthy. The highly resistant lines were tested in a natural exposure test where the plants were exposed to any naturally occurring PLRV isolates. Resistance breaking isolates were not found in the transgenic resistant lines suggesting a broad based resistance to the PLRV field isolates at this test site. This indicates that at least for PLRV, expression of replicase genes can lead to virus resistant plants of commercial value.

SUMMARY

The discovery that the expression of some viral RNAs and proteins in plants confers resistance to the virus from which it was derived, is an example of the practical application of biotechnology to solve agronomic problems. Like all genes used for disease resistance in agronomic crops, effective management will be necessary to maintain the efficacy and durability of the trait. These plants with 'novel' genes for virus resistance obtained by biotechnology, may allow for the construction of more robust resistance systems (21).

Virus resistance in plants achieved by the expression of viral coat protein genes and replicase genes are alternatives to the chemical and cultural methods currently being applied to control virus diseases. The high level of efficacy observed for the CP-MR and replicase-MR to plant virus infection indicates that these technologies will be effective new tools for protecting crop losses due to virus disease.

REFERENCES

1.   Anderson, E.J., Stark, D.M., Nelson, R.S.,Powell, P.S., Tumer, N.E., and Beachy, R.N. 1989. Transgenic plants that express the coat protein genes of tobacco mosaic virus or alfalfa mosaic virus interfere with disease development of some nonrelated viruses. Phytopathology 79:1284-1290.

2.   Anderson, J.M., Palukaitis, P., and Zaitlin, M. 1992. A defective replicase gene induces resistance to cucumber mosaic virus in transgenic tobacco plants. Proc. Natl. Acad. Sci. USA 89:8759-8763.

3.   Angenent, G.C., van den Ouweland, J.M.W., and Bol, J.F. 1990. Susceptibility to virus infection of transgenic tobacco plants expressing structural and

nonstrucural genes of tobacco rattle virus. Virology 175:191-198.

4. Audy, P., Palukaitis, P., Slack, S., and Zaitlin, M. 1994. Replicase-mediated resistance to potato virus Y in transgenic tobacco plants. Mol. Plant-Microbe Interact. 7:15-22

5. Barker, H., Webster, K.D., Jolly, C.A., Reavy, B., Kumar, A., and Mayo, M.A. 1994. Enhancement of resistance to potato leafroll virus multiplication in potato by combining the effects of host genes and transgenes. Mol. Plant-Microbe Interact. 7:528-530.

6. Baulcombe, D. 1994. Replicase-mediated resistance: A novel type of virus resistance in transgenic plants. Trends in Microbiology 2:60-63.

7   Beachy, R.N. 1993. Introduction: Transgenic resistance to plant viruses. Sem. Virol. 4:327-328.

8. Bertioli, D. J., Cooper, J. I., Edwards, M. L., and Hawes, W. S. 1992. Arabis mosaic nepovirus coat protein in transgenic tobacco lessens disease severity and virus replication. Ann. Appl. Biol. 120:47-54.

9   Brault, V., Candresse, T., le Gall, O., Delbos, R.P., Lanneau, M., and Dunez, J. 1993. Genetically engineered resistance against grapevine chrome mosaic nepovirus. Plant Mol. Biol. 21:89-97.

10. Braun, C.J., and Hemenway, C. L. 1992. Expression of amino-terminal portions or full-length viral replicase genes in transgenic plants confers resistance to potato virus X infection. Plant Cell 4:735-744

11. Brederode, F.T., Taschner, P.E.M., Posthumus, E., and Bol, J.F. 1995. Replicase mediated resistance to alfalfa mosaic virus. Virology 207:467-474.

12 Carr, J. P., Marsh, L.E., Lonmonossoff, G.P., Sekiya, M.E., and Zaitlin, M. 1992. Resistance to tobacco mosaic virus induced by the 54-kDa gene sequence requires expression of the 54-kDa protein. Mol. Plant-Microbe Interact. 5:397-404

13 Carr, J.P., and Zaitlin, M. 1993. Replicase-mediated resistance. Seminars in Virology 4: 339-347.

14. Carr, J.P., Gal-On, A., Palukaitis, P., and Zaitlin, M. 1994. Replicase-mediated resistance to cucumber mosaic virus in transgenic plants involves suppression of both virus replication in the inoculated leaves and long-distance movement. Virology 199:439-447.

15. Cassidy, B.G., and Nelson, R.S. 1995. Differences in protection phenotypes in tobacco plants expressing coat protein genes from peanut stripe potyvirus with or without an engineered ATG. Mol. Plant-Microbe Interact. 8:357-365.

16. Clark, W.G., Fitchen, J.H., and Beachy, R.N. 1995. Studies of coat protein-mediated resistance to TMV I. The pm2 assembly defective mutant confers resistance to TMV. Virology 208:485-491.

17. Cuozzo, M., O 'Connell, K., Kaniewski, W.K., Fang, R-X., Chua, N-H., and Tumer, N. 1988. Viral protection in transgenic tobacco plants expressing the cucumber mosaic virus coat protein or its antisense RNA. Bio/Technology 6: 549-557.

18. de Haan, P., Gielen, J.JL., Prins, M., Wijkamp, I.G., Van Schepen, A., Peters, D., Van Grinsven, M.Q.J.M., and Goldbach, R. 1992. Characterization of RNA-mediated resistance to tomato spotted wilt virus in transgenic tobacco plants. Bio/Technology 10:1133-1137.

19. Donson, J., Dearney, C.M., Turpen, T.H., Khan, I.A., Kurath, G., Turpen, A.M., Jones, G.E., Dawson, W.O., and Lewandowski, D.J. 1993. Broad resistance to tobamoviruses is mediated by a modified tobacco mosaic virus replicase transgene. Mol. Plant-Microbe Interact. 6:635-642.

20. Evans, D.A., Sharp, W.R., and Medina-Filho, H.P. 1984. Somaclonal and gametoclonal variation. Amer. J. Bot. 71:759-774.

21. Fraser, R.S.S. 1988. Virus recognition and pathogenicity: Implications for resistance mechanisms and breeding. Pestic. Sci. 23:267-275.

22. Goldbach, R., and de Haan, P. 1993. Prospects of engineered forms of resistance against tomato spotted wilt virus. Sem. Virol. 4:381-387.

23. Golemboski, D.B., Lomonossoff, G.P., and Zaitlin, M. 1990. Plants transformed with a tobacco mosaic virus nonstructural gene sequence are resistant to the virus. Proc. Natl. Acad. Sci. USA 87:6311-6315.

24. Gonsalves, D., and Slightom, J.L. 1993. Coat protein-mediated protection: analysis of transgenic plants for resistance in a variety of crops. Sem. Virol. 4:397-406.

25. Grumet, R. 1994. Development of virus resistant plants via genetic engineering. Pages 47-79 in: Plant Breeding Reviews, vol. 12. J. Janick, ed. John Wiley & sons. New York.

26. Habili, N., and Symons, R. H. 1989. Evolutionary relationship between luteoviruses and other RNA plant viruses based on sequence motifs in their putative RNA polymerases and nucleic acid helicases. Nucleic Acids Res. 17:9543-9555.

27. Hammond, J.K., and Kamo, K.K. 1993. Transgenic coat protein and antisense RNA resistance to bean yellow mosaic potyvirus. Acta Hort. 336:171-181.

28. Hanley-Bowdoin, L. and Hemenway, C. 1992. Expression of plant viral genes in transgenic plants. Pages 251-296 in: Genetic engineering with plant viruses. T. Michael, A. Wilson, and J.W. Davies, eds. CRC Press, Boca Raton, Florida.

29. Hemenway, C., Fang, R-X., Kaniewski, W., Chua, N-H., and Tumer, N. 1988. Analysis of the mechanism of protection in transgenic plants expressing the potato virus X coat protein or its antisense RNA. EMBO J. 7: 1273-1280.

30. Hiei, Y., Ohta, S., Komari, T., and Kumashiro, T. 1994. Efficient transformation of rice (Oryza sativa L.) mediated by Agrobacterium and sequence analysis of the boundaries of the T-DNA. Plant J. 6:271-282.

31. Hinchee, M.A., Corbin, D.R., Armstrong, C. L., Fry, J.E., Sato, S.S., DeBoer, D.L., Peterson, W. L., Armstrong, T.A., Connor-Ward, D.V., Layton, J.G., and Horsch, R.B. 1994. Plant Transformation. Pages 231-270 in: Plant Cell and Tissue Culture. I.K. Vasil and T.A. Thorpe, eds. Kluwer Acad. Publ. Netherlands.

32. Jefferson, R.A., Kavanagh, T.A., and Beven, M.W. 1987. GUS fusions: β-glucuronidase as a sensitive and versatile gene fusion maker in higher plants. EMBO J. 6:3901-3907.

33. Jongedijk D., Huisman M.J., and Cornelissen B.J.C. 1993. Agronomic performance and field resistance of genetically modified, virus-resistant potato plants. Sem. Virol. 4:407-416.

34. Jordan R. 1992. Potyviruses, monoclonal antibodies, and antigenic sites. Pages 81-95 in. Potyvirus taxonomy, Archives of Virology 5. O.W. Barnett Jr., ed. Springer-Verlag Wien, New York.

35.   Kaniewski, W., Lawson, C., Sammons, B., Haley, L., Hart, J., Delannay, X., and Tumer N. 1990. Field resistance of transgenic Russet Burbank potato to effects of infection by potato virus X and potato virus Y. Bio/Technology 8:750-754.

36.   Kaniewski, W.K., and Thomas, P.E. 1993. Field testing of virus resistant transgenic plants. Sem. Virol. 4:389-396.

37.   Kaniewski, W., Lawson, C., Thomas, P. 1993. Agronomically useful resistance in transgenic Russet Burbank potato containing a PLRV CP gene. Abstract. IX International Congress of Virology, Glasgow, Scotland.

38.   Kaniewski, W., Lawson, C., Loveless, J., Thomas, P., Mowry, T., Reed, G., Mitsky, T., Zalewski, J., and Muskopf, Y. 1994. Expression of potato leafroll virus (PLRV) replicase genes in Russet Burbank potatoes provide immunity to FLRV. Pages 289-292 in: Proceed. rd EFPP Conference, M. Manka, ed. J. Phytopathol.

39.   Kaniewski, W.K., and Lawson, E.C. 1994. Biotechnology strategies for virus resistance in plants. Pages 147-154 in: Proceed. 3rd EFPP Conference, M. Manka, ed. J. Phytopathol.

40.   Kaniewski, W.K., Mitsky, T and Loveless, J. 1994. Highly resistant transgenic tomatoes expressing cucumber mosaic virus coat protein genes. Pages 293-296 in: Proceed. 3rd EFPP Conference, M. Manka ed. J. Phytopathol.

41.   Kawchuk, L., Martin R.R., and McPherson J. 1990. Sense and antisense RNA-mediated resistance to potato leafroll virus in Russet Burbank potato plants. Mol. Plant-Microbe Interact. 3: 301-307.

42.   Kay, R., Chan, A., Daly, M., and McPherson, J. 1987. Duplication of CaMV 35S promoter sequences creates a strong enhancer for plant genes. Science 236:1299-1302.

43.   Lawson, C., Kaniewski, W., Haley, L., Rozman, R., Newell, C., Sanders, P., and Tumer, N. 1990. Engineering resistance to mixed virus infection in a commercial potato cultivar: resistance to potato virus X and potato virus Y in transgenic Russet Burbank. Bio/Technology 8:127-134.

44.   Lindbo, J. A., and Dougherty, W. G. 1992. Untranslatable transcripts of the tobacco etch virus coat protein gene sequence can interfere with tobacco etch replication in transgenic plants and protoplasts. Virology 189:725-733.

45.   Ling, K., Namba, S., Gonsalves, C., Slightom, J.L., and Gonsalves, D. 1991. Protection against detrimental effects of potyvirus infection in transgenic tobacco plants expressing the papaya ringspot virus coat protein gene. Bio/Technology. 9:752-758.

46.   Loesch-Fries, L.S., Merlo, D., Sinner, T., Burhop, L., Hill, K., Krahn, K., Jarvis, N., Nelson, S., and Halk, E. 1987. Expression of alfalfa mosaic virus RNA 4 in transgenic plants confers virus resistance. EMBO J. 6:1845-1851.

47.   Longstaff, M., Brigneti, G., Boccard, F., Chaoman, S., and Baulcombe, D. 1993. Extreme resistance to potato virus X infection in plants expressing a modified component of the putative viral replicase. EMBO J. 12:379-386.

48.   Lupo, R., Rubino, L., and Russo M. 1994. Immunodetection of the 33K/92K polymerase protein of cymbidium ringspot virus-infected and in transgenic plant tissue extracts. Arch. Virol. 138:135-143.

49.   MacFarlane, S.A., and Davies, J.W. 1992. Plants transformed with a region of the 201-kilodalton replicase gene from pea early browning virus RNA1 are resistant to virus infection. Proc. Natl. Acad. Sci. USA 89:5829-5833.

50.   MacKenzie D.J., and Tremaine J.H. 1990. Transgenic Nicotiana debneyii expressing viral coat protein are resistant to potato virus S infection. J. Gen. Virol. 71:2167-2170.

51.   MacKenzie, D.J., Tremaine, J.H., and McPherson, J. 1991. Genetically engineered resistance to potato virus S in potato cultivar Russet Burbank. Mol. Plant-Microbe Interact. 4:95-102.

52   Marsh, L.E., Pogue, G.P., Connell, J.P., and Hall, T.C. 1991. Artificial defective interfering RNAs derived from brome mosaic virus. J. Gen.Virol. 72:1787-1792.

53.   Mori, M., Mise, K., Okuno, T., and Furusawa, I. 1992. Expression of brome mosaic virus-encoded replicase genes in transgenic tobacco plants. J. Gen. Virol. 73:169-172.

54.   Mueller, E., Gilbert, J., Davenport, G. Brigneti, G., and Baulcombe, D.C. 1995. Homology-dependent resistance: transgenic virus resistance in plants related to homology-dependent gene silencing. Plant J. 7:1001-1013.

55.   Namba, S., Ling, K., Gonsalves, C.,Gonsalves, D. , and Slightom, J. L. 1991. Expression of the gene encoding the coat protein of cucumber mosaic virus (CMV) strain WL appears to provide protection to tobacco plants against infection by several different CMV strains. Gene 107:181-188.

56.   Namba, S., Ling, K., Gonsalves, C., Slightom, J.L., and Gonsalves, D. 1992. Protection of transgenic plants expressing the coat protein gene of watermelon mosaic virus II or zucchini yellow mosaic virus against six potyviruses. Phytopathology 82:940-946.

57.   Nejidat, A., and Beachy, R.N. 1990. Transgenic tobacco plants expressing a coat protein gene of tobacco mosaic virus are resistant to some other tobamoviruses. Mol. Plant-Microbe Interact. 3:247-251.

58.   Nelson, R.S., McCormick, S.M., Delanney, X., Dube, P., Layton, J., Anderson, E.J., Kaniewska, M., Proksch, R.K., Horsch, R.B., Rogers, S.G., Fraley, R.T., and Beachy, R.N. 1988. Virus tolerance, plant growth, and field performance of transgenic tomato plants expressing coat protein from tobacco mosaic virus. Bio/Technology. 6:403-409.

59.   Noris,E., Accotto, G.P., Tavazza, R. Brunetti, A., Crespi, S., and Tavazza, M. 1996. Resistance to Tomato Yellow Leaf Curl Geminivirus in Nicotiana benthamiana plants transformed with a truncated viral C1 gene. Virology 224:130-138.

60.   Osbourn, J. K., Watts, J. W., Beachy, R. N., and Wilson, T. M. A. 1989. Evidence that nucleocapsid disassembly and a later step in virus replication are inhibited in transgenic tobacco protoplasts expressing TMV coat protein. Virology 172:370-373.

61.   Pang, S-Z., Nagpala, P., Wang, M., Slightom, J.L. and Gonsalves, D. 1992. Resistance to heterologous isolates of tomato spotted wilt virus in transgenic tobacco expressing its nucleocapsid protein gene. Phytopathology 82:1223-1229.

62.   Perlak, F., Kaniewski, W., Lawson, C., Vincent, M., and Feldman, J. 1994. Genetically improved

potatoes: Their potential role in integrated pest management. Pages 451-454 in: Proceed. 3rd EFPP Conference, M. Manka, ed. J. Phytopath.

63   Ploeg, A.T., Mathis, A., Bol, J.F., Brown, D.J.F., and Robinson, D.J. 1993. Susceptibility of transgenic tobacco plants expressing tobacco rattle virus coat protein to nematode-transmitted and mechanically inoculated tobacco rattle virus. J. Gen. Virol. 74:2709-2715.

64.   Powell-Abel, P., Nelson, R.S., De, B., Hoffmann, N., Rogers, S.G., Fraley, R.T., and Beachy, R.N. 1986. Delay of disease development in transgenic plants that express the tobacco mosaic virus coat protein gene. Science 232:738-743.

65.   Presting, G.G., Smith, O.P., and Brown, C.R. 1995. Resistance to potato leafroll virus in potato plants transformed with the coat protein gene or with vector control constructs. Phytopathology 85:436-442.

66.   Prins, M., de Hann, P., Luyten, R., van Veller, M., van Grinsven, M.Q.J.M., and Goldbach, R. 1995. Broad resistance to tospoviruses in transgenic tobacco plants expressing three tospoviral nucleoprotein gene sequences. Mol. Plant- Microbe Interact. 8:85-91.

67.   Quemada, H. D., Gonsalves, D., and Slightom, J.L. 1991. Expression of coat protein gene from cucumber mosaic virus strain C in tobacco: Protection against infections by CMV strains transmitted mechanically or by aphids. Phytopathology 81:794-802.

68.   Reavy, B., Arif, M., Kashiwazaki, S., Webster, K.D., and Barker, H. 1995. Immunity to potato mop-top virus in Nicotiana benthamiana plant expressing the coat protein gene is effective against fungal inoculation of the virus. Mol. Plant- Microbe Interact. 8:286-291.

69.   Register, J. and Beachy, R.N. 1988. Resistance of TMV in transgenic plants results from interference with an early event in infection. Virology 166:524-532.

70.   Reimann-Philipp, U., and Beachy, R. N. 1993. The mechanism(s) of coat protein-mediated resistance against tobacco mosaic virus. Sem. Virol. 4:349-356.

71.   Reimann-Philipp, U., and Beachy, R.N. 1993. Coat protein-mediated resistance in transgenic tobacco expressing the tobacco mosaic virus coat protein from tissue-specific promoters. Mol. Plant-Microbe Interact. 6:323-330.

72.   Rubino, L., Lupo, R., and Russo, M. 1993. Resistance to cymbidium ringspot tombusvirus infection in transgenic Nicotiana benthamiana plants expressing a full-length viral replicase gene. Mol. Plant-Microbe Interact. 6:729-734.

73.   Sanders, P. R., Sammons, B., Kaniewski, W., Haley, L., Layton, J., Lavallee, B.J., Delannay, X., and Turner, N. E. 1992. Field resistance of transgenic tomatoes expressing the tobacco mosaic virus or tomato mosaic virus coat protein genes. Phytopathology 82:683-690.

74.   Sanford, J. C., and Johnston, S. A. 1985. The concept of parasite-derived resistance deriving resistance genes from the parasite's own genome. J. Theor. Biol. 113:395-405.

75.   Sijen, T., Wellenk, J., Hendriks, J., Verner, J., and van Kammen, A. 1995. Replication of cowpea mosaic virus RNA1 or RNA2 is specifically blocked in transgenic Nicotiana benthamiana plants

expressing the full-length replicase or movement protein genes. Mol. Plant-Microbe Interact. 8:340-347.

76.   Stark, D.M., and Beachy, R.N. 1989. Protection against potyvirus infection in transgenic plants: evidence for broad spectrum resistance. Bio/Technology 7:1257-1262.

77.   Sturtevant, A.P. and Beachy, R.N. 1993. Virus resistance in transgenic plants: coat protein mediated resistance. Pages 93-114 in: Transgenic plants. A. Hiatt, ed. Marcel Dekker Inc., New York.

78.   Szybalski, W. 1991. Protection of plants against viral diseases by cloned viral genes and anti-genes. Gene 107:177-179.

79.   Taschner, P.E.M., van der Kuyl, A.C., Neeleman, L., and Bol, J.F. 1991. Replication of an incomplete alfalfa mosaic virus genome in plants transformed with viral replicase genes. Virology 181:445-450.

80.   Thomas, P., Kaniewski, W., Reed, G. and Lawson, C. 1994. Transgenic resistance to potato leafroll virus in Russet Burbank potatoes. Pages 551-554 in: Proceed. 3rd EFPP Conference, M. Manka, ed. J. Phytopathol.

81.   Tomassoli, L., Kaniewski, W., Ilardi, V., Barba, M. 1995. Produzione di piante di pomodoro geneticamente modificate per la resistenza al virus del mosaico del cetriolo. L 'Informatore Agrario 4:1-8.

82.   Toyoda, H., Chatani, K., Matsuda, Y., and Ochi, S. 1989. Multiplication of tobacco mosaic virus in tobacco callus tissues and in vitro selection for viral disease resistance. Plant Cell Reports 8:433-436.

83.   Tricoli, D.M., Carney, K.J., Russell, P.F., McMaster, J.R., Groff, D.W., Hadden, K.C., Himmel, P.T., Hubbard, J.R., Boeshore, M.L., Reynolds,J.F. Quemada, H.D., 1995. Field evaluation of transgenic squash containing single or multiple virus coat protein gene constructs for resistance to cucumber mosaic virus, watermelon mosaic virus 2, and/or zucchini yellow mosaic virus. Bio/Technology 13:1458-1465.

84.   Tumer, N., Hemenway, C., O 'Connell, K., Cuozzo, M., Fang, R-X ., Kaniewski, W. and Chua, N-H. 1987. Expression of coat protein genes in transgenic plants confers protection against alfalfa mosaic virus, cucumber mosaic virus and potato virus X. Pages 351-356 in: Plant Molecular Biology. D. von Wettstein and N-H. Chua, ed. Plenum Publishing Corporation.

85.   Tumer, N.E., Kaniewski, W.K., Haley, L., Gehrke, L., Lodge, J.K., and Sanders P. 1991. The second amino acid of alfalfa mosaic virus coat protein is critical for coat protein-mediated protection. Proc. Natl. Acad. Sci. USA 88:2331-2335

86.   Vaira, A.M., Semeria, L., Crespi, S., Lisa, V., Allavena, A., and Accotto, G. P. 1995. Resistance to tospoviruses in Nicotiana benthamiana transformed with the N gene of tomato spotted wilt virus: correlation between transgene expression and protection in primary transformants. Mol. Plant-Microbe Interact. 8:66-73.

87.   van der Vlugt, R.A.A., and Goldbach, R.W. 1993. Tobacco plants transformed with the potato virus YN coat protein gene are protected against different PVY isolates and against aphid-mediated infection. Trans. Res. 2:109-114.

88.   van der Wilk, F., Willink, D.P.L., Huisman, M.J., Huttinga, H., and Goldbach, R. 1991. Expression

of the potato leafroll luteovirus coat protein gene in transgenic potato plants inhibits viral infection. Plant Mol. Biol. 17:431-439.

89.   van Dun, C.M.P., Bol, J.F., and van Vloten-Doting, L. 1987. Expression of alfalfa mosaic virus and tobacco rattle virus coat protein genes in transgenic tobacco plants. Virology 159:299-305.

90.   van Dun, C.M.P., and Bol, J.B. 1988. Transgenic tobacco plants accumulating tobacco rattle virus coat protein resist infection with tobacco rattle virus and pea early browning virus. Virology 167:649-652.

91.   van Dun, C.M.P., Overduin, L., van Vloten-Doting, L., and Bol, J.F. 1988. Transgenic tobacco expressing tobacco streak virus or mutated alfalfa mosaic virus coat protein does not cross protect against alfalfa mosaic virus infection. Virology 164:383:389.

92.   van Dun, C.M.P., van Vloten-Doting, L., and Bol, J.F. 1988. Expression of alfalfa mosaic virus cDNA 1 and 2 in transgenic tobacco plants . Virology 163:572-578.

93.   Wilson, T.M.A. 1993. Strategies to protect crop plants against viruses: Pathogen derived resistance blossoms. Proc. Natl. Acad. Sci. USA 90:3134-3141.

94.   Wisniewski, L.A., Powell, P. A., Nelson, R.S., and Beachy, R.N. 1992. Local and systemic spread of tobacco mosaic virus in transgenic tobacco. Plant Cell 2:559-567.

95.   Zaitlin, M., Anderson, J.M., Perry, K.L., Zhang, L., and Palukaitis, P. 1994. Specificity of replicase-mediated resistance to cucumber mosaic virus. Virology 201:200-205.

CHAPTER 7

Antisense RNA and Ribozyme-mediated Resistance to Plant Viruses

M. Tabler, M. Tsagris, and J. Hammond

A broad spectrum of biotechnological strategies has been developed and tested for generating plants that are protected against viruses. Plant transformation with DNA sequences derived from the genome of the viral pathogen has proven to be an efficient strategy in conferring viral resistance, in line with the prediction of Sanford and Johnston (94) that pathogen-derived gene expression may interfere with the replication of the pathogen itself if gene products are present in either dysfunctional form, in excess, or at the wrong developmental stage. The most prominent and most widely used technique of pathogen-derived resistance is the coat protein (CP) mediated protection (also called "CP-mediated resistance") in which a sense RNA encoding the CP is expressed. The expression of antisense RNA can also be considered as a pathogen-derived protection strategy. For updated reviews on the current achievements in controlling plant virus diseases by pathogen-derived resistance in transgenic plants and other protection strategies for crop plants, the reader is referred to Scholthof et al. (99), Wilson (132), Lomonossoff (67), and Hammond (36), plus other chapters in this volume. Recently other, non pathogen-derived protection approaches have been successfully tested, like the expression of the mammalian 2'-5' oligoadenylate (2-5A) synthetase which results in simultaneous protection against different RNA viruses (122). In this chapter we review various aspects related to the antisense and ribozyme technologies as tools for viral protection in plants, and outline some future perspectives.

ANTISENSE TECHNOLOGY IN PLANTS

The antisense technology is based on the capability of complementary nucleic acids to form double-stranded helices. Any single-stranded nucleic acid present in a living cell is therefore susceptible to bind via Watson-Crick base pairing to a nucleic acid of complementary (antisense) polarity. Specific binding of the "antisense" nucleic acid results in interference with the biological function of the "sense" nucleic acid, either by arresting the sense nucleic acid or by induced degradation of the double-stranded complex. In principle, both the sense and the antisense nucleic acid can be either DNA or RNA. Endogenous expression of antisense RNA in transgenic plants has been extremely successful for suppression of a broad spectrum of endogenous genes in plants (for review, see 73,93, 116,126,127). Since the antisense gene specifically affects the expression of a given target gene, it can be used to create phenotypic, dominant negative mutants in which the undesired gene is suppressed or ablated. Some typical examples for successful gene suppression by antisense RNA in plants will be briefly summarized.

The first example was the suppression of the chalcone synthase (CHS) of petunia (128). CHS is the key enzyme for the biosynthesis of flavonoids which are involved in several biological processes, including flower pigmentation. Antisense plants showed various alterations in flower pigmentation, ranging from no visible effect to complete loss of pigmentation. So far the most spectacular achievements have been made by suppressing several genes that are involved in tomato fruit ripening. Two enzymes that are involved in cell wall degradation have been downregulated, polygalacturonidase (PG) (102,106,107) and pectin methylesterase (PME) (121). It is anticipated that the inhibition of cell wall degrading enzymes will inhibit or retard fruit softening and therefore improve the shelf-life of the fruits. Such tomatoes have been tested under field conditions (58) and commercialized. Detailed reviews on fruit ripening and its manipulation by antisense RNA are available (22,34,100).

Two enzymes involved in ethylene synthesis have been suppressed, the key enzyme of ethylene biosynthesis, ACC synthase (78), and ACC oxidase or ethylene forming enzyme (EFE) (35). Inhibition of the ACC synthase prevents the respiratory burst during fruit ripening, and mature tomatoes stay green in air, neither softening nor developing aroma for at least 120 days. Fruit ripening can be induced by treatment with either propylene or ethylene, which overcomes the antisense phenotype.

Another example where crops can be "designed", is the seed-specific antisense-mediated inhibition of the stearoyl-ACP desaturase in Brassica (56). After inhibition of the key enzyme that is responsible for desaturation of fatty acids, the fatty acid composition in rape seed was dramatically changed, now containing stearate levels of up to 40% which represents a much higher portion of saturated fatty acids.

Müller-Röber et al. (74) intended to analyze whether ADP-glucose is the sole precursor for starch biosynthesis in potato tubers. Using antisense RNA, they suppressed the ADP glucose pyrophospholyase (AGPase) which generates ADP-glucose from glucose-1 phosphate and ATP. In transgenic plants in which the AGPase gene had been effectively suppressed, the tubers contained almost no starch, indicating that starch biosynthesis was completely blocked. Instead of starch, these tubers contained about 30% sucrose and 8% glucose in their dry mass. Even the small subunit of the ribulose-,1,5-bisphosphate carboxylase-oxygenase (RUBISCO) could be successfully inhibited (92).

ANTISENSE RNA AGAINST PLANT VIRUSES

In view of the efficiency in down-regulating gene expression by mRNA-directed antisense RNAs, it could be expected that replication of plant viruses would also be successfully suppressed by antisense RNAs directed against the pathogenic viral RNA. However, viral RNA targets are different from mRNA in several respects and following sections will discuss the situation separately for RNA viruses, DNA viruses, and viroids. We must also discuss the differences between antisense RNA and sense RNA in the induction of RNA-mediated resistance.

The general low efficiency of antisense RNA as an antiviral agent against RNA viruses contrasts with the effectiveness of mRNA-directed antisense RNA in gene suppression. One reason that may account for this difference is that antisense RNA and the viral target RNA are synthesized in different cellular compartments. Whereas the stably transformed antisense gene is transcribed in the nucleus, the viral target RNA is synthesized by the viral replicase, typically in the cytoplasm. Presumably, only a small fraction of the nuclear antisense RNA is actually transported in intact form to the site of virus replication. Duplex formation is, however, a bi-molecular reaction, and requires that the antisense RNA is present in at least stoichiometric amounts at the site of virus replication to be effective.

Since the local cytoplasmic concentration of the viral RNA target may be high but the concentration of intact antisense RNA is relatively low, it is understandable that protection has often been observed only at low concentrations of inoculum. As a small amount of genomic (-) strand RNA is sufficient to act as template for production of a much larger amount of virion (+) sense RNA, interference in the production of the initial genomic minus strand appears to be the critical step. It must be remembered, however, that the amount of viral RNA infecting a new cell (either as a result of vectored transmission, cell-to-cell, or long-distance transport within a plant) will be limited, and it should be possible to have stochiometric amounts of antisense RNA present. Indeed, recovery from established infections has been reported (37), suggesting that lower amounts of inoculum reach successive cells, increasing the ratio of antisense RNA: genomic RNA. Hammond and Kamo (37) also showed that plant lines expressing higher levels of antisense RNA were either better protected against initial infection, or recovered sooner and more completely than lines with lower levels of expression.

By contrast, if mRNA is the target, both the sense and the antisense RNA are transcribed in the nucleus. Consequently, antisense RNA can exert its inhibitory effect in the nucleus, as well as after export from it and thus has several chances to bind to the target RNA. In line with this it is now becoming clear that all the steps in the informational flow from DNA to protein, including RNA transcription, RNA processing, RNA transportation from nucleus to cytoplasm and, eventually various steps of translation (initiation, ribosome-binding, elongation termination) are susceptible to antisense RNA interference (see 25,43,76).

Besides cellular compartmentation, there is a second major difference between viral RNA and endogenous mRNA: the mRNA has no natural antisense counterparts. This statement may not be completely correct since there is some indication that, as in prokaryotes, a few eukaryotic RNAs also exist in both polarities (for review see 55,59,75), but these cases can be considered as exceptions to the rule. RNA viruses, however, replicate via a duplex RNA, no matter whether the RNA genome is single-stranded of (+) or (-) polarity, or even double-stranded. In other words, the occurrence of antisense RNA is an integral part of the replication cycle. Because of this it is also hard to distinguish between sense and antisense RNA. In the case of plus strand RNA viruses, which represent about 75% of all plant viruses known to date (71), the genomic (+) strand (or subgenomic RNAs) serves as mRNA, whereas the antigenomic (-) strand acts as template for the synthesis of the (+) strand. Both processes are required for virus protection and must be somehow coordinated to avoid collision between the translating ribosome and the viral replicase.

Since the two processes proceed simultaneously, it is rather philosophical to discriminate between sense and antisense RNA. For example, the sense RNA [of (+) polarity] can be considered as an antisense RNA for the (-) strand template. In fact, artificially introduced RNAs of either (-) or (+) polarity could in principle interfere with virus replication, provided they are able to bind to the viral RNA in the process of translation and replication, respectively.

In this context it is of interest that an increasing number of examples indicate that "CP-mediated" resistance can be the result of diverse mechanisms and that protein synthesis is actually not required for conferring resistance by "CP-mediated protection" to some RNA viruses. Examples have been described where protection against potato leafroll virus (PLRV) could be observed in the absence of detectable levels of CP (130). Similar results have been reported for PVY (40). Even expression of untranslatable RNA transcripts, not necessarily derived from the CP gene, resulted in virus protection for PVY (27,129) or TEV (64). This type of protection is therefore called "RNA-mediated resistance". One could speculate that the observed protection is actually due to interference of two complementary

RNAs. Since the viral (-) RNA is present in much lower concentrations than the (+) RNA, lower concentrations of the complementary RNA would be required to cause interference. Therefore, the (+) RNA could make a better "antisense" RNA than the antigenomic RNA. In fact Lindbo and Dougherty (64) speculated that untranslatable or truncated (65) sense RNA provides better protection because, unlike the translatable transcript, it is not associated with ribosomes and is therefore better accessible for an RNA:RNA interaction with the antigenomic RNA strand.

RNA-mediated inhibition in transgenic tobacco has also been reported for tomato spotted wilt virus (TSWV) which is a negative-strand RNA virus (17). The smallest of the three genomic RNAs of TSWV, the S RNA, has an ambisense strategy of gene expression. The genomic RNA S encodes the 52.4 kD nonstructural NS gene whereas its complementary RNA encodes the viral nucleoprotein N of 28.8 kD. Expression of the N gene RNA transcript, which is of antisense polarity versus the genomic RNA, resulted in RNA-mediated resistance. Again, translation of the N gene was not required for protection, indicating that protection was due to RNA duplex formation. In line with this, protection was observed against viruses that are closely related in the N gene, but not against strains which show a higher degree of sequence variability. However, Pang et al. (79,80) have shown that resistance to closely-related tospoviruses was conferred by an RNA-mediated mechanism, but that expression of the N gene protein was required for resistance to more distantly-related tospoviruses. A similar difference has been suggested with potyviruses (J. Hammond, submitted). Virus protection could be also achieved by expressing the 54 kD open reading frame of TMV (32). The corresponding protein has never been detected during TMV infection, so that it also seemed conceivable here that resistance was due to RNA-mediated protection. However, by using an 54 kD protein sequence which was either deficient in the AUG initiation codon or had a frameshift mutation (10), it was demonstrated that the resistance to TMV actually requires the expression of the (undetectable) protein. If the protection mechanism were RNA-mediated, the small changes in the RNA chain should not alter the resistance.

Additional work from Dougherty's group (21,33,66,108,109,113) suggests that plants have a mechanism for sequence-specific removal of RNAs that are present in excess. According to this model, plants may either have this mechanism induced by a high level of transcription of the transgene alone, leading to extreme resistance to infection by viruses closely related to the transgene; or the mechanism may be induced by the combined level of the transgene and replicating viral RNA, requiring establishment of an infection followed by recovery (21). In the resistant state, such plants exhibit high levels of transgene transcription and low levels of steady state transgene transcript accumulation (21,108,113). Goodwin *et al.* (33) showed that such RNA-mediated resistance was also dependent to some extent on the transgene copy number; three or more copies were necessary to establish the highly resistant state in which plants did not become infected, while one or two copies of the transgene were sufficient for inducible resistance (infection followed by recovery). Goodwin *et al.* (33) also provided evidence to suggest that specific RNA sequences within the transgene transcript are targeted for cleavage, which might explain why RNA-mediated resistance is limited to closely-related isolates. Transgene copy number was not the only factor, as some plant lines with high transgene copy numbers were fully susceptible to infection; in these lines very low levels of transgene transcription were detected, and it is possible that the high transgene number led to gene-silencing (33).

There is no inherent reason why an antisense RNA should not induce the RNA surveillance system proposed by Dougherty *et al.* (21), provided the threshold level of transgene transcript necessary to activate the surveillance system is reached; it is, however, less likely that the combination of antisense transgene transcript and viral minus sense together would exceed the threshold level if the transgene transcript alone does not, as far less viral (-) strand than (+) strand is produced. We shall now return to discussion of viral resistance conferred by antisense RNA as it is usually understood, RNA complementary to the viral genomic strand.

RNA Viruses

Most of the early experiments in which antisense RNA was tested as an antiviral agent for plant RNA viruses were actually designed as internal negative controls of CP-mediated protection. The CP gene was introduced into plants in sense and antisense orientation, resulting in the expression of CP mRNA or antisense RNA, respectively. This was tested for the potato virus X (PVX) (44) and the cucumber mosaic virus (CMV) (14,89). When the transgenic antisense host plants were challenged with the virus, protection was only achieved at low inoculum concentrations. At high concentrations the protection was overcome, unlike the case where the CP mRNA and CP was expressed. Similar results have been obtained with tobacco mosaic virus (TMV), although the little resistance observed was abolished when the TMV 3' non-coding region was deleted from the transgene construct, even in a line expressing higher levels of the antisense RNA than lines with the 3' sequence (86). A better degree of inhibition could be observed by expression of antisense RNA directed against potato leafroll virus (PLRV) in potato (54,130). This virus, however, accumulates slowly, does not reach high titers and therefore may not be representative. Nelson *et al.* (77) used two short (51nt) antisense RNAs, one directed against the Ω segment of the 5' untranslated leader, and a second that overlapped the initiation codon and 5' portion of the 126K and 183K genes of TMV, to produce transgenic tobacco plants. Some plants expressing the Ω antisense RNA remained symptomless until senescence; the partial resistance was shown to be due to inhibition of replication, as levels of genomic RNA were reduced 10 to 20-fold, compared to 6 to 7-fold reduction in coat protein accumulation (produced from a 3'-coterminal subgenomic RNA). There was up to 50-fold reduction of infectious virions (77).

LeClerc and AbouHaidar (62) showed that plants expressing antisense RNA from potato aucuba mosaic potexvirus (PAMV) were highly resistant to 0.1μg/ml of PAMV, and that many plants did not show symptoms even when virus titers were as high as in non-transgenic controls. Seven of 14 plants of one line escaped infection with 1μg/ml of PAMV, whereas most plants expressing PAMV CP were infected at this level of inoculum (62).

Further cases where attenuation could be obtained by expressing antisense RNA have been described (16,26,27,37,49,50,51,64,65). Hammond and Kamo (37) reported that one line expressing antisense RNA to bean yellow mosaic potyvirus (BYMV) was effectively immune to BYMV (up to at least 100μg/ml), and other lines recovered from BYMV infection. As might be expected from an RNA-mediated reaction, resistance conferred by antisense RNA is generally virus-specific (37,108). However, it has been reported that antisense RNA also had an inhibitory effect on related viruses (see also section on DNA viruses). Fang and

Grumet (26) challenged tobacco plants expressing an CP antisense gene of zucchini yellow mosaic virus (ZYMV) with two further potyviruses: potato virus Y (PVY) and the tobacco etch virus (TEV). In both cases the antisense plants showed a delay of symptom expression versus the untransformed control. If an unrelated virus (TMV) was used, no difference could be observed. There are at least two plausible explanations of these results. One is that specific target sequences (33; and see above) within the ZYMV CP gene are conserved between ZYMV, PVY and TEV; while the overall nucleotide homology is 50-60%, there are localized regions of high homology (J. Hammond, unpublished). Another is that the ZYMV antisense construct was a reversed CP construct that employed the 5' untranslated region of TEV as a translational enhancer; the potyvirus 5' untranslated region contains a subsequence that is highly conserved among different viruses, and may play a role in initiation of virion assembly, as a polymerase recognition sequence, or for VPg binding (133). As such, a highly conserved sequence with a role of such significance in the potyvirus life cycle could well interfere directly as an antisense RNA, or act as a specific target sequence to initiate RNA degradation by a sequence-specific nuclease activity (21,33).

Yepes et al. (134) reported resistance to tomato ringspot nepovirus from an antisense CP construct lacking the viral 3' sequences. Resistance of different lines varied from a slight delay in symptom appearance to complete resistance to mechanical inoculation. Nine of 93 R0 lines produced no visible symptoms, and no virus was detected in upper, non-inoculated leaves. Three of the nine lines were still fully resistant in the R2 generation, while 30-60% of inoculated plants of the remaining six lines remained asymptomatic following inoculation. In this case it is notable that the antisense transcripts were not detectable by Northern analysis in some lines showing complete resistance, and that only low levels of transcripts were detected in any resistant line (134).

Prins et al. (87) suggested that resistance to tospoviruses in plants expressing N-gene RNA might be based on an antisense inhibition of virus multiplication as a result of a direct RNA:RNA interaction, although sense inhibition of the N-gene would be equally possible; it had earlier been shown that an untranslatable N-gene RNA was equally as effective as the translatable version (17). The low levels of N-gene transcript and N-protein produced (87) suggest that this, and the resistance to tomato ringspot virus (134), may be

cases of "RNA-mediated" rather than "antisense-mediated" resistance.

Lindbo and Dougherty (30) suggested that RNA:protein interactions, such as occur between ribozymes and translatable RNAs, or between viral replicases and the 5' and 3' untranslated regions or internal subgenomic RNA promoters of viral genomes, would interfere with RNA:RNA interactions; they therefore suggested that such important regulatory regions would not be effective targets for antisense RNA resistance. The effective construct of Hammond and Kamo (37), however, includes the BYMV 3' non-coding region, while Huntley and Hall (49-51) observed that effective reduction in BMV replication was conferred by an antisense RNA to a subgenomic promoter region. These results suggest that protein:RNA interactions do not effectively interfere in the presumed RNA:RNA binding necessary for antisense RNA to be effective.

There are now sufficient examples of significant resistance from antisense RNA to viruses of several taxonomic groupings to suggest that effective resistance will be found in many cases if enough transgenic lines are examined. The results of Hammond and Kamo (37), and Yepes et al. (134) suggest that perhaps 10% of lines confer useful levels of resistance.

DNA Viruses

DNA viruses, unlike RNA viruses, have a nuclear phase during which the viral DNA is transcribed into RNA. This resembles the situation of mRNA and the antisense technology should be expected to work much better here. Indeed it was found that expression of antisense RNA directed against the AL1 gene of tomato golden mosaic virus (TGMV), a geminivirus, efficiently inhibited viral replication (15). Moreover, the same plants expressing the TGMV antisense gene showed considerable protection against beet curly top virus (BTCV), whereas no protection was observed with another geminivirus, the African cassava mosaic virus (ACMV) (2). Both viruses have the same overall similarity (about 64%) with the region of TGMV which was expressed as an antisense gene in transgenic tobacco. However, BCTV, unlike ACMV, contains local domains with similarity levels of up to 82%. This demonstrated that no absolute sequence complementarity is required for suppression of related viruses, but the similarity must exceed a certain threshold level to allow duplex formation. Similarly, van der Krol et *al.* (127) observed that the chsA antisense RNA also suppressed the chsJ genes, which share 86% sequence similarity. Moreover, inhibition also worked in the heterologous system, when the petunia chs antisense gene was expressed in tobacco (128). The corresponding tobacco gene shares about 80% sequence similarity (127).

Since CP-mediated protection has so far shown little success for DNA viruses, the antisense approach is one alternative for this group of viruses. Only one report (60) notes CP-mediated resistance, to tomato yellow leafcurl virus. A defective interfering DNA (30,31,110), and a replication associated protein (48) have also been shown to confer virus-specific resistance against geminiviruses. To our knowledge the antisense approach has not been tested for pararetroviruses, which include economically important viruses such as the rice tungro bacilliform virus (RTBV) (39,88).

Viroids

Viroids are a group of subviral plant pathogens which could be another potential target for antisense RNA. Viroids consist of only a covalently closed circular RNA, of 246-375 nucleotides depending on the viroid species (reviewed in 18). Unlike viruses, viroids are never encapsidated nor are they translated into peptides. Viroids are replicated by RNA polymerase II in a rolling-circle-like mechanism in the nucleus of the infected cell (98). Whereas their cellular localization makes them a good target for endogenously expressed antisense RNA, their unique structure, which has a high degree of secondary structure and self-complementarity in each of the (+) and the (-) RNA (90) may prevent efficient duplex formation. Moreover, as in the case of RNA viruses, (+) and (-) RNAs are part of the replication cycle, so the discrimination between sense and antisense RNA is again rather philosophical, especially since neither strand serves as a mRNA template.

Matousek and co-workers incubated natural PSTVd (+) RNA with either complementary DNA oligonucleotides (68) or *in vitro* synthesized RNA of (-) polarity (69) prior to inoculation, which resulted in reduced or inhibited infectivity. In further work they tested transgenic plants expressing either of two short (18nt) RNAs complementary to PSTVd (+) RNA in the upper central conserved region (CCR), or a longer RNA complementary to the viroid (-) RNA in the left half of the viroid rod structure (70). The CCR was targeted because of its role in viroid processing and structural transitions; the second antisense RNA, complementary to the viroid (-) strand, was selected because there is a much lower level of (-) strand intermediates in infected cells than circular (+) strand progeny (70). It was not possible to detect the short RNAs in transgenic plants, but when plants were challenged a delay in PSTVd replication was observed; however the levels of viroid ultimately reached those of the controls. In plants that expressed RNA complementary to the viroid (-) strand, PSTVd levels were also reduced compared to the controls, and remained so for longer than in the plants with the short RNAs complementary to the (+) strand (70). They suggest that antisense inhibition could be completely overcome if some threshold level of PSTVd was achieved in cells (70).

Atkins *et al.* (1) also targetted both (+) and (-) strands, of citrus exocortis viroid (CEVd), but using nearly full length antisense RNAs directed against each strand, and produced in transgenic plants. The antisense RNA targeted against CEVd (-) strand resulted in a moderate reduction in viroid accumulation in inoculated plants (1). Plants expressing antisense RNA against the viroid (+) strand actually produced more CEVd RNA (three- to eight-fold the level in control plants at 34 days after infection), an unexpected result (1). The greater effectiveness of antisense RNA against the viroid (-) strand may reflect the lower amounts of (-) strand produced in a normal viroid infection, or may result from competiotion between the transgene transcript and the viroid (+) strand for host factors, analagous to a defective interfering RNA (1).

IMPROVEMENTS IN ANTISENSE TECHNOLOGY FOR PLANT VIRUSES

The available data indicate that the antisense approach for inhibiting virus replication in plants has had variable successful with both DNA and RNA viruses. In contrast, expression of (+) sense RNA has resulted in several cases of successful resistance, which might be due to RNA-RNA interaction or to a sequence-specific nuclease activity. One could argue that there is no need for antisense RNA technology if virus protection can be achieved by other biotechnological approaches in plants. However, it might be helpful to eventually combine several protection strategies in order to maximize resistance, and reducing the possibility that a single protection mechanism could be overcome by a resistance-breaking strain. Last but not least, it is of course a scientific challenge to make the antisense approach work.

The following section will therefore focus on some principles of duplex RNA formation and will outline what could be done to improve the antisense strategy for virus protection, no matter whether genomic or antigenomic RNA is expressed.

Optimizing the Binding of Antisense RNA

One of the general questions in antisense technology that has been recently discussed by Nellen and Lichtenstein (76) is "what makes an RNA anti-sense-itive?". As an extension to this one could also ask: what makes a complementary RNA an effective antisense RNA?

The process of double-strand formation will release energy (the longer the RNA, the more free energy) and one might therefore expect that longer RNAs are more effective as antisense RNA. However, it seems that there is no common rule with regard to length of antisense RNA. Antisense RNAs derived from the entire cDNA of an mRNA have been effective in gene suppression. However partial cDNA antisense genes, complementary to the 5' or 3' region worked equally well (for review, 116) and the smallest effective antisense RNA comprised only 41 nucleotides (9).

More important than the absolute amount of energy potentially freed is the activation energy for duplex formation. Assuming formation of a fully double-stranded structure, the secondary structure of each sense and antisense RNA has to be overcome. In other words, duplex formation is controlled kinetically and not by the absolute amount of energy that can be released. The complex relationship between RNA structure and speed of duplex formation is best illustrated in the case of naturally occurring prokaryotic antisense RNAs. Many of these antisense RNAs are relatively short (about 70 nucleotides) and have a typical stem-loop structure, (reviewed in 104). For example, the copy number of the *E. coli* plasmid R1 is under the control of the antisense RNA called copA. This system has been studied in great detail by Nordström and co-workers. Sense and antisense RNA are encoded from the same DNA, but are transcribed in opposite directions. Point

mutations therefore do not change anything in the complementarity of the two RNAs. Nevertheless, when a single nucleotide is changed in sense and antisense RNA, in a region that forms a loop structure (in both the sense and the antisense RNA), the binding rate constant can drop by a factor of three to four, though the RNAs are still perfectly complementary (82). In line with this, the plasmid copy number will rise by the same factor. It is obvious that the binding rate constant is influenced by the secondary structure of sense and antisense RNA. Each will influence the "kissing structures" that are involved in the initial contacts of two RNAs which may lead to the formation of a duplex RNA (compare 25).

Unquestionably, the endogenous sense RNA cannot be altered. The target region and length of the antisense RNA can, however, be varied. This will influence the secondary structure of the antisense RNA, and hence the binding kinetics so that it can be "optimized" for efficient binding to its corresponding target RNA. The complex relationship between length of the antisense RNA and its binding rate has recently been convincingly demonstrated by Rittner *et al.* (91). Using a fixed 5' terminus of the antisense RNA directed against human immunodeficiency virus type 1 (HIV-1), they varied the 3' terminus and thus the length of the RNA and selected for antisense RNAs that were able to bind rapidly. The binding rate constants for RNAs of defined length showed an oscillating type of correlation depending on the length of the RNA and the resulting secondary structure. Antisense RNAs which differ in just a few nucleotides at the 3' terminus may have binding rate constants varying by two log units, which could be correlated with changes in the secondary structure of the antisense RNA. Moreover, the data on the *in vitro* binding rates correlated with the effectiveness in inhibiting HIV-1 replication when tested *in vivo*. These results clearly underline the importance of testing the antisense RNA *in vitro* before using it for suppression of a target RNA *in vivo*, which may allow selection of much more potent antisense RNAs.

Before optimizing the actual length of the antisense RNA the target RNA can also be scanned by computer-aided search for effective RNA target sequences. By setting a window from 50 to 400 nucleotides, Sczakiel and co-workers (101) calculated the energy profile for the genomic and antigenomic HIV RNA which reflects the local folding potential. It was observed that domains having significant energy minima (high ∆G values) correlated with those target regions that had been most effective in inhibition of HIV replication.

Extension of the Half-life of Antisense RNA

Naturally occurring antisense RNAs in prokaryotic organisms are extremely stable, presumably due to their characteristic stem-loop structure, which protects them from exonucleolytic degradation. It has been reported that the RNA-OUT which controls the insertion sequence IS 10 is stable for up to 70 minutes *in vivo*, which is up to three bacterial generations (104). By adding such protective stem-loop structures to the termini of the antisense RNA degradation by exonucleases may be minimized. Such an approach has recently been taken by Heinrich *et al.* (42). An alternative would be to couple the antisense RNA to a tRNA and to express it from a promoter specific for the DNA dependent RNA polymerase III as recently reported by Bourque and Folk (3). Another approach for protecting the antisense RNA is to include a polyadenylation signal into the antisense gene.

Cytoplasmic Expression of Virus-directed Antisense RNA

Protective structures, as discussed above will help to increase cytoplasmic concentrations of antisense RNA. An alternative and much more efficient way to obtain high levels of antisense RNA would be to express it form an episomal replication system as opposed to a stably transformed chromosomal gene. Several examples exist where plant RNA viruses have been used as vector systems (for review see 53). For example, Joshi *et al.* (52), have replaced the open reading frame b of RNA ß of barley stripe mosaic virus with the firefly luciferase gene, which was expressed upon transfection into tobacco and maize protoplasts. Donson *et al.*, have used a TMV-based vector to produce the bacterial protein neomycin phosphotransferase (NPTII) (20)., and Chapman *et al.* (12) have expressed ß-glucuronidase (GUS) in a vector based on potato virus X (PVX). Similar results have been obtained by Dolja *et al.* who inserted GUS into the polyprotein of tobacco etch virus (TEV) (19). Thus it is conceivable that antisense RNAs may also be expressed with the help of such a virus-based vector system. Indeed, Broglio *et al.* (4) have reported that a virus-based vector system

can be used to test the efficacy of constructs before producing transgenic plants.

RIBOZYMES

RNA enzymes (ribozymes) are derived from naturally occurring self-cleaving RNAs (for reviews see 11,24). Several classes of catalytic RNAs are known. Since they can be used to specifically bind and cleave a particular RNA target they can be considered as an extension and improvement of the antisense technology. Like antisense RNAs, ribozymes bind to their target, but unlike antisense RNAs they also induce specific cleavage of the complementary RNA. In line with this, ribozymes have also been described as antisense RNAs with a "warhead" (132).

One class of self-cleaving RNAs contains a so-called hammerhead structure (28,29) and is found amongst some subviral plant pathogens which include certain satellite RNAs that can accompany some RNA viruses and two viroids[1] which do not require a helper virus for replication, unlike satellite RNAs (reviewed in 5,6,114,-115,117).

The catalytic domain of hammerhead RNAs has been utilized by Haseloff and Gerlach to create RNA enzymes (ribozymes) which irreversibly cleave the substrate RNA in trans (38). The ribozyme consists of two flanking arms which are complementary to the target RNA, resembling an antisense RNA. However, in addition it carries the catalytic hammerhead domain between these flanking regions. The attraction of the design of Haseloff and Gerlach is that only minimal sequence requirements must be fulfilled for selection of a cleavable site. Only a three nucleotide motif, for example GUC, needs to

---

[1]Because it is often stated that the hammerhead domain is found "in plant viroids", it should be stressed that the majority of viroids neither contain a hammerhead structure, nor undergo self-cleavage (75). Such a structure has only been found in avocado sunblotch viroid (ASBVd) (68,69) and in peach latent mosaic viroid (PLMVd) (76). By contrast, the potato spindle tuber viroid (PSTVd) has a so-called central conserved region (CCR) which is involved in enzyme-catalyzed processing (77) and a domain related to it is found in all non-hammerhead-type viroids, which represent the vast majority of viroids discovered to date. Possibly the processing enzyme (cleavage as well as ligation) for these viroids is an RNase of the T1-type (78-80). The occurrence of a hammerhead domain in viroids should therefore be considered an exception to the rule.

be present in the target RNA. Effective *in vitro* cleavage has also been obtained adjacent to GUA,GUU, CUC, and UUC trinucleotides (83,103). A target-specific ribozyme RNA is then engineered which contains the catalytic domain and antisense sequences specific for the sequence context of the selected three nucleotide motif within the target RNA.

The antisense regions facilitate binding of the ribozyme to its target and correct positioning of the catalytic domain with respect to the cleavable motif in the target RNA. The length of the sequences that flank the catalytic domain influence the specificity and stability of binding. However, specificity of binding does not necessarily increase with the length of the flanking sequences (46).

A number of examples have been described where the ribozyme approach has been used for gene suppression *in vivo* (7,13,42,57,63, 95-97,105). A hammerhead ribozyme directed against mRNA of the neomycin phosphotransferase gene (npt), has been tested by Steinecke *et al.* in plant cells (112). For this purpose, DNAs encoding the ribozyme and the target RNA, respectively, were co-transfected in tobacco protoplasts. Although usually hard to detect, the authors could visualize the cleaved substrate RNA, indicating that the ribozyme approach indeed worked *in vivo*. Cleavage in tobacco protoplasts has also been reported for chloramphenicol acetyl transferase (CAT) mRNA (83,85). However, it is fair to say that GUS-directed ribozymes, which were catalytically active *in vivo*, did not result in suppression of GUS activity when expressed in protoplasts of *Arabidopsis thaliana* (72).

In principle, one ribozyme molecule could cleave in truly catalytic manner more than one substrate molecule. Such a mode of reaction would require three phases: (i) binding of the ribozyme to the target, (ii) actual cleavage of the target RNA and (iii) dissociation of the ribozyme RNA. In the case of the RNase P ribozyme it seems that the dissociation step is the rate-limiting step (120). For accomplishing catalytic turnover there is a conflict of interest with regard to the length of the flanking sequences: they should be long enough to allow efficient binding, but they should be short enough to facilitate dissociation of the ribozyme from its cleaved target RNA. However, sequences that are too short may not be able to compete with the secondary structure of the target RNA so that no efficient binding is achieved *in vivo*, which might have been the case when GUS-directed ribozymes failed to suppress GUS activity (72).

Furthermore, relatively small molecules might not be sufficiently stable *in vivo*, unless they are chemically modified (40,41,81). Unlike potential pharmaceutical applications of chemically modified ribozymes, their use in plants may be limited to protoplast systems. An alternative to ribozymes with catalytic turnover is the use of long antisense RNAs which carry a catalytic hammerhead domain (e.g.16,97). Such catalytic antisense RNAs, which can include several catalytic domains in one single molecule, will only cleave the target RNA stoichiometrically, because the duplex RNA is too stable to allow dissociation. It should be stressed however, that *in vivo* the duplex RNAs may become unwound by proteins, so that catalytic turnover is conceivable (compare 76).

For convenient generation of catalytic antisense RNAs, we have developed a technique that allows the insertion of a ribozyme cassette into certain restriction sites of a cDNA (118). The resulting recombinant DNA construct allows the synthesis of antisense RNA carrying the catalytic domain of the hammerhead ribozymes. About 25 different restriction recognition sequences can be used, which overlap with a cleavable motif for the hammerhead ribozymes. For example, each SalI-site (G/TCGAC) contains the GUC cleavage motif in the transcribed RNA against which a ribozyme can be directed. After SalI digestion of the cDNA and removal of the protruding ends, a SalI-specific ribozyme DNA cassette is inserted, containing the catalytic ribozyme domain plus three of the four nucleotides that had been removed by trimming. Transcription delivers the desired antisense-ribozyme. For further details on "selectable ribozyme cassettes" and on the restriction sites that can be used for insertion of such cassettes, see the report of Tabler and Tsagris (118).

So far, we have not tested this approach of modified, catalytic antisense RNAs for gene suppression in plants. However, in collaboration with the laboratory of G. Sczakiel, catalytic domains were incorporated into an HIV-1 antisense RNA. In transfection experiments it was shown by Homann *et al.* that the catalytic antisense RNA was about four to seven fold more effective in inhibiting the HIV-1 replication than the corresponding antisense control (47). Furthermore, a catalytic RNA which had lost its cleavage activity due to a one nucleotide deletion, only had a inhibitory effect comparable to the antisense RNA. This indicates that the cleavage activity of the catalytic antisense RNAs could improve the antisense effect.

Ribozymes with long hybridizing arms and multiple catalytic domains have been used in a number of studies (16), and shown more effective than short-armed ribozymes when directed against CAT mRNA in either animal or plant cells (8,84), although the effect may have been due to increased stabilization of the ribozyme or its antisense RNA carrier than to the increased opportunity for hybridization to the target RNA (16).

There are a few reports of attempts to use ribozymes against plant viral RNAs, either *in vitro* or *in vivo*. Intermolecular processing of PLRV coat protein and polymerase genes (51), and of the 5' region of TMV RNA (16), has been observed *in vitro*, but with low efficiency at temperatures comparable to the *in planta* situation. At 25° C, only 36% of the substrate was cleaved after 6 h incubation with a six-fold molar excess of RNA containing three hammerhead ribozymes, whereas 82% of substrate was cleaved after 1 h at 50° C; there were also differences in the efficiency of cleavage of the three ribozyme sites (16). Edington and Nelson (23) noted some reduction in TMV replication in protoplasts of transgenic plants expressing a ribozyme, but Wilson (132) suggests that very high levels of ribozyme to target RNA may be necessary, possibly requiring use of a mild virus strain to amplify a ribozyme (targeted to a different virus) as a sub-genomic RNA. In transgenic plants a multiple ribozyme construct against TMV was found to be no more effective than the antisense RNA in which the ribozymes were embedded for targetting; these results suggest that addition of the ribozyme domains to the antisense RNA sequences did not affect the ability of the antisense RNA to confer resistance to TMV in any way (16). The effect of the ribozymes was additionally analyzed by creating a TMV mutant that would not be cleaved by the ribozymes, but no differential resistance was observed, indicating that the ribozymes did not contribute to resistance *in planta* (16). Atkins *et al.* (1) compared resistance to CEVd in transgenic plants expressing either antisense RNA or antisense RNA with three hammerhead ribozyme domains embedded; separate constructs directed against both CEVd (+) and (-) strands were tested. Although the ribozyme constructs were shown to cleave the viroid RNAs *in vitro*, the ribozyme constructs actually had reduced effects when compared to the antisense RNA alone when tested *in planta*, suggesting that the ribozyme sequences were ineffective under physiological

conditions, and that the observed effect was due entirely to the antisense RNA (1). Wegener *et al.* (131) reported reduction of NPT II activity in transgenic plants, but this also may have been due to antisense RNA activity rather than to the activity of the ribozyme.

It was also possible to attenuate the expression of the white gene which is responsible for the red eye pigmentation in *Drosophila melanogaster*. Transgenic flies expressing a white-directed catalytic antisense RNA showed a clear cut reduction in eye pigmentation when tested in a genetic background that had reduced levels of white expression a priori (42). A ribozyme targetted to the *fushi tarazu* gene of *Drosophila* also yielded the expected mutant phenotype in transgenic flies (135).

It will have to be verified whether the modification of antisense RNA by incorporation of a catalytic ribozyme domain also results in improvement of the antisense technology in plants, so that catalytic antisense RNA can be used as efficient antiviral agent. Results obtained to date are not encouraging, and may reflect some inherent differences between plants and animals; such differences may be overcome through greater understanding of the factors affecting ribozyme efficiency. Since ribozymes are modified antisense RNAs any improvement as outlined for antisense RNA will be also beneficial for ribozyme technology.

REFERENCES

1.   Atkins, D., Young, M., Uzzell, S., Kelly, L., Fillatti, J., and Gerlach, W.L. 1995. The expression of antisense and ribozyme genes targeting citrus exocortis viroid in transgenic plants. J. Gen. Virol. 76:1781-1790.

2.   Bejarano, E.R. and Lichtenstein, C.P. 1994. Expression of TGMV antisense RNA in transgenic tobacco inhibits replication of BCTV but not ACMV geminiviruses. Plant Mol. Biol. 24:241-248.

3.   Bourque, J.E. and Folk, W.R. 1992. Suppression of gene expression in plant cells utilizing antisense sequences transcribed by RNA polymerase III. Plant Mol. Biol. 19:641-647.

4.   Broglio, E.P., Abouzid, A.A., Hiebert, E., and Powell, C.A. 1995. A virus-based vector system for evaluating the usefulness of geminivirus gene constructs in conferring pathogen-derived resistance. Abstr. Phytopathology 85:1142.

5.   Bruening, G. 1989. Compilation of self-cleaving sequences from plant virus satellite RNAs and other sources. Methods Enzymol. 180:546-558.

6.   Bruening, G., Buzayan, J.M., Hampel, A. and Gerlach, W.L. 1988. Replication of small satellite RNAs and viroids: possible participation of non-enzymatic reactions. Pages 127-145 in: RNA Genetics Vol II: Retroviruses, Viroids and RNA

Recombination. E. Domingo, J.J. Holland, and P. Alquist, eds. CRC Press, Boca Raton, FL .

7. Cameron, F.H. and Jennings, P.A. 1989. Specific gene supression by engineered ribozymes in monkey cells. Proc. Natl. Acad. Sci. USA 86:9139-9143.

8. Cameron, F.H., and Jennings, P.A. 1994. Multiple domains in a ribozyme construct confer increased suppressive activity in monkey cells. Antisense Res. Develop. 4:87-94.

9. Cannon, M., Platz, J., O'Leary, M., Sookdeo, C., Cannon, F. 1990. Organ-specific modulation of gene expression in transgenic plants using antisense RNA. Plant Mol. Biol. 15:39-47.

10. Carr, J.P., Marsh, L.E., Lomonossoff, G.P., Sekiya, M.E. and Zaitlin, M. 1992. Resistance to tobacco mosaic virus induced by the 54-kDa gene sequence requires expression of the 54-kDa protein. Mol. Plant Microbe Interact. 5:397-404.

11. Cech, T.R. 1987. The chemistry of self-cleaving RNA and RNA enzymes. Science 231:1532-1539.

12. Chapman, S., Kavanagh, T. and Baulcombe, D. 1992. Potato virus X as a vector for gene expression in plants. Plant J. 2:549-557.

13. Cotten, M. and Birnstiel, M.L. 1989. Ribozyme mediated destruction of RNA in vivo. EMBO J. 8:3861-3866.

14. Cuozzo, M., O'Connell, K.M., Kaniewski, W., Fang, R.-X., Chua. N.-H., and Tumer, N.E. 1988. Viral protection in transgenic tobacco plants expressing the cucumber mosaic virus coat protein or its antisense RNA. Bio/Technology 6:549-557.

15. Day, A.G., Bejarano, E.R., Buck, K.W., Burrell, M., and Lichtenstein, C.P. 1991. Expression of an antisense viral gene in transgenic tobacco confers resistance to the DNA virus tomato golden mosaic virus. Proc. Natl. Acad. Sci. 88:6721-6725.

16. De Feyter, R., Young, M., Schroeder, K., Dennis, E.S., and Gerlach, W. 1996. A ribozyme gene and an antisense gene are equally effective in conferring resistance to tobacco mosaic virus on transgenic tobacco. Mol. Gen. Genet. 250:329-338.

17. De Haan, P., Gielen, J.J., Prins, M., Wijkamp, I.G., van Schepen, A., Peters, D., van Grinsven, M.Q. and Goldbach, R. 1992. Characterization of RNA-mediated resistance to tomato spotted wilt virus in transgenic tobacco plants. Bio/Technology 10:1133-1137.

18. Diener, T.O. 1987. The viroids. Plenum Press, New York, NY.

19. Dolja, V.V., McBride, H.J., and Carrington, J.C. 1992. Tagging of plant potyvirus replication and movement by insertion of ß-glucuronidase into the viral polyprotein. Proc. Natl. Acad. Sci. USA 89:10208-10212.

20. Donson, J., Kearney, C.M., Hilf, M.E., and Dawson, W.O. 1991. Systemic expression of a bacterial gene by a tobacco mosaic virus-based vector, Proc. Natl. Acad. Sci. USA 88:7204-7208.

21. Dougherty, W.G., Lindbo, J.A., Smith, H.A., Parks, T.D., Swaney, S., and Proebsting, W.M. 1994. RNA-mediated virus-resistance in transgenic plants: Exploitation of a cellular pathway possibly involved in RNA degradation. Mol. Plant-Microbe Interact. 7:544-552.

22. Eckes, P. 1992. Inhibition of fruit ripening by antisense-RNA technology. Angew. Chem. Int. Ed. Engl. 31:175-177.

23. Edington, B.V., and Nelson, R.S., 1992. Utilization of ribozymes in plants. Plant viral resistance. Pages 209-221 in: Gene Regulation: Biology of Antisense RNA and DNA . R.P. Ericksen, and J.G. Izant, eds. Raven Press, New York, NY.

24. Edgington, S.M. 1992. Ribozymes: stop making sense. Bio/Technology 10:256-262.

25. Eguchi, Y., Taleo, I., and Tomizuwa, J. 1991. Antisense RNA. Annu. Rev. Biochem. 60:631-652.

26. Fang, G., and Grumet, R. 1993. Genetic engineering of potyvirus resistance using constructs derived from the zucchini yellow mosaic virus coat protein gene. Mol. Plant Microbe Interact. 6:358-367.

27. Farinelli, L., and Malnoë, P. 1993. Coat protein gene-mediated resistance to potato virus Y in tobacco: Examination of the resistance mechanisms - Is the transgenic coat protein required for protection? Mol. Plant Microbe Interact. 6:284-292.

28. Forster, A.C., and Symons, R.H. 1987. Self-cleavage of plus and minus RNAs of a virusoid and a structural model for the active sites. Cell 49:211-220.

29. Forster, A.C., and Symons, R.H. 1987. Self-cleavage of virusoid RNA is performed by the proposed 55-nucleotide active site. Cell 50:9-16.

30. Frischmuth, T., and Stanley, J. 1991. African cassava mosaic virus DI DNA interferes with the replication of both genomic components. Virology 183:539-544.

31. Frischmuth, T., and Stanley, J. 1994. Beet curly top virus symptom amelioration in Nicotiana benthamiana transformed with a naturally occurring viral subgenomic DNA. Virology 200:826-830.

32. Golemboski, D.B., Lomonossoff, G.P., and Zaitlin, M. 1990. Plants transformed with tobacco mosaic virus nonstructural gene sequence are resistant to the virus. Proc. Natl. Acad. Sci. USA 87:6311-6315.

33. Goodwin, J., Chapman, K., Swaney, S., Parks, T.D., Wernsman, E.A., and Dougherty, W.G. 1996. Genetic and biochemical dissection of transgenic RNA-mediated virus resistance. Plant Cell 8:95-105.

34. Gray, J., Picton, S., Shabbeer, J., Schuch, W., and Grierson, D. 1992. Molecular biology of fruit ripening and its manipulation with antisense genes. Plant Mol. Biol. 19:69-87.

35. Hamilton, A.J., Lycett, G.M., and Grierson, D. 1990. Antisense gene that inhibits synthesis of the hormone ethylene in transgenic plants. Nature 346:284-287.

36. Hammond, J. 1996. Biotechnology and resistance. Acta Hortic. 432:246-256.

37. Hammond, J., and Kamo, K.K. 1995. Effective resistance to potyvirus infection conferred by expression of antisense RNA in transgenic plants. Mol. Plant-Microbe Interact. 8:674-682.

38. Haseloff, J., and Gerlach, W.L. 1988. Simple RNA enzymes with new and highly specific endoribonuclease activities. Nature 334:585-591.

39. Hay, J.M., Jones, M.C., Blakesbrough, M., Dasgupta, I., Davies, J.W., and Hull, R. 1991. Analysis of the sequence of an infectious clone of rice tungro bacilliform virus, a plant pararetrovirus. Nucleic Acids Res. 19:2615-2621.

40. Heidenreich, O., and Eckstein, F. 1992. Hammerhead ribozyme-mediated cleavage of the long terminal repeat RNA of human immunodeficiency virus type 1. J. Biol. Chem. 267:1904-1909.

41. Heidenreich, O., Pieken, W., and Eckstein, F. 1993.

Chemically modified RNA: approaches and applications. FASEB J. 7:90-96.

42. Heinrich, J.-C., Tabler, M., and Louis, C. 1993. Attenuation of white gene expression in transgenic Drosophila melanogaster: Possible Role of a catalytic antisense RNA. Dev. Genet. 14:258-265.

43. Hélène, C., and Toulmé, J.-J. 1990. Specific regulation of gene expression by antisense, sense and antigene nucleic acids. Biochim. Biophys. Acta 1049:99-125.

44. Hemenway, C., Fang, R.-X., Kaniewski, W.K., Chua, N.-H. and Tumer, N.E. 1988. Analysis of the mechanism of protection in transgenic plants expressing the potato virus X coat protein or its antisense RNA. EMBO J. 7:1273-1280.

45. Hernández, C. and Flores, R. 1992. Plus and minus RNAs of peach latent mosaic viroid self-cleave in vitro via hammerhead structures. Proc. Natl. Acad. Sci. USA 89:3711-3715.

46. Herschlag, D. 1991. Implications of ribozyme kinetics for targeting the cleavage of specific RNA molecules in vivo: more isn't always better. Proc. Natl. Acad. Sci. USA 88:6921-6925.

47. Homann, M., Tzortzakaki, S., Rittner, K., Sczakiel, G. and Tabler, M. 1993. Incorporation of the catalytic domain of a hammerhead ribozyme into antisense RNA enhances its inhibitory effect on the replication of human immunodeficiency virus type 1. Nucleic Acids Res. 21:2809-2814

48. Hong, Y., and Stanley, J. 1996. Virus resistance in Nicotiana benthamiana conferred by African cassava mosaic virus replication-associated protein (AC1) transgene. Mol. Plant-Microbe Interact. 9:219-225.

49. Huntley, C.C., and Hall, T.C. 1993. Minus-sense transcripts of brome mosaic virus RNA-3 intercistronic region interfere with viral replication. Virology 192:290-297.

50. Huntley, C.C., and Hall, T.C. 1993. Interference with brome mosaic virus replication by targeting the minus strand promoter. J. Gen. Virol. 74:2445-2452.

51. Huntley, C.C., and Hall, T.C. 1995. Interference with brome mosaic virus replication in transgenic rice. Mol. Plant-Microbe Interact. 9:164-170.

52. Joshi, R.L., Joshi, V. and Ow, D.W. 1990. BSMV genome mediated expression of a foreign gene in dicot and monocot plant cells. EMBO J. 9:2663-2669.

53. Joshi, R.L., and Joshi, V. 1991. Strategies for expression of foreign genes in plants - potential use of engineered viruses. FEBS Lett. 281:1-8.

54. Kawchuk, L.M., Martin, R.R. and McPherson, J. 1991. Sense and antisense RNA-mediated resistance to potato leafroll virus in Russet Burbank potato plants. Mol. Plant Microbe Interact. 4:247-253.

55. Kimelman, D. 1992. Regulation of eukaryotic gene expression by natural antisense transcripts. Pages 1-10 in: Gene regulation: biology of antisense RNA and DNA. R.P. Erickson and J G. Izant, eds. Raven Press Ltd., New York, NY.

56. Knutzon, D.S., Thompson, G.A., Radke, S.E., Johnson, W.B., Knauf, V.C. and Kridl, J.C. 1992. Modification of Brassica seed oil by antisense expression of a stearoyl-acyl carrier protein desaturase gene. Proc. Natl. Acad. Sci. USA 89:2624-2628.

57. Koizumi, M., Kamiya, H. and Ohtsuka, E. 1992. Ribozymes to inhibit transformation of NIH3T3 cells by the activated c-Ha-ras gene. Gene 117:179-184.

58. Kramer, M., Sanders, R.A., Sheehy, R.E., Melis, M., Kuehn, M.,and Hiatt, W.R. 1990. Field evaluation of tomatoes with reduced polygalacturonase by antisense RNA. Pages 347-355 in: Horticultural Biotechnology, Plant Biology Vol. 11. A.B. Bennett and S.D. O'Neill, eds. Alan R. Liss, New York, NY.

59. Krystal, G.W. 1992. Regulation of eukaryotic gene expression by naturally occurring antisense RNA. Pages 11-20 in: Gene regulation: biology of antisense RNA and DNA. R.P. Erickson and J.G. Izant, eds. Raven Press Ltd., New York, NY.

60. Kunik, T., Salomon, R., Zamir, D., Navot, N., Zeidan, M., Michelson, I., Gafni, Y., and Czosnek, H. 1994. Transgenic tomato plants expressing the tomato yellow leaf curl virus capsid protein are resistant to the virus. Bio/Technology 12:500-504.

61. Lamb, J.W., and Hay, R.T. 1990. Ribozymes that cleave potato leafroll virus RNA within the coat protein and polymerase genes. J. Gen. Virol. 71:2257-2264.

62. LeClerc, D., and AbouHaidar, M.G. 1995. Transgenic tobacco plants expressing a truncated form of the PAMV capsid protein (CP) gene show CP-mediated resistance to potato aucuba mosaic virus. Mol. Plant Microbe Interact. 8:58-65.

63. L'Huillier, P.J., Davies, S.R. and Bellamy, A.R. 1992. Cytoplasmic delivery of ribozymes leads to efficient reduction in α-lactalbumin mRNA levels in C127I mouse cells. EMBO J. 11:4422-4418.

64. Lindbo, J.A. and Dougherty, W.G. 1992. Untranslatable transcripts of the tobacco etch virus coat protein gene sequence can interfere with tobacco etch virus replication in transgenic plants and protoplasts. Virology 189:725-733.

65. Lindbo, J.A. and Dougherty, W.G. 1992. Pathogen-derived resistance to a potyvirus: Immune and resistant phenotypes in transgenic tobacco expressing altered forms of a potyvirus coat protein nucleotide sequence. Mol. Plant Microbe Interact. 5:144-153.

66. Lindbo, J.A., Silva-Rosales, L., Proebsting, W.M., and Dougherty, W.G. 1993. Induction of a highly specific antiviral state in transgenic plants: Implications for regulation of gene expression and virus resistance. Plant Cell 5:1749-1759.

67. Lomonossoff, G.P. 1995. Pathogen-derived resistance to plant viruses. Annu. Rev. Phytopathol. 33:323-343.

68. Matoušek, J., Trněná, L., Arnold, L., Dědič, P., Rakousky, S., Steger, G. and Riesner, D. 1993. Inhibition of potato spindle tuber viroid (PSTVd) infectivity with DNA oligonucleotides. Biochimie 75:63-69.

69. Matoušek, J., Trněná, L., Rakousky, S., and Riesner, D. 1994a. Inhibition of potato spindle tuber viroid (PSTVd) infectivity with antisense RNA transcripts. J. Phytopathol. 140:10-24.

70. Matoušek, J., Schröder, A.R.W., Trněná, L., Reimers, R., Baumstark, T., Dědič, P., Vlasák, J., Becker, I., Kreutzaler, F., Fladung, M., and Riesner, D. 1994. Inhibition of viroid infection by antisense RNA expression in transgenic plants. Biol. Chem. Hoppe-Seyler 375:765-777.

71. Matthews, R.E. 1991. Plant Virology, 3rd ed., Academic Press, San Diego, CA.

72. Mazzolini, L., Axelos, M., Lescure, N. and Yot, P.

1992. Assaying synthetic ribozymes in plants: high level expression of a functional hammerhead structure fails to inhibit target gene activity in transiently transformed protoplasts. Plant Mol. Biol. 20:715-731.

73. Mol, J.N., van der Krol, A.R., van Tunen, A.J., van Blokland, R., de Lange, P. and Stuitje, A.R. 1990. Regulation of plant gene expression by antisense RNA. FEBS Lett. 268:427-430.

74. Müller-Röber, B., Sonnewald, U. and Willmitzer, L. 1992. Inhibition of the ADP-glucose pyrophosphorylase in transgenic potatoes leads to sugar-storing tubers and influences tuber formation and expression of tuber storage protein genes. EMBO J. 11:1229-1236.

75. Murray, J.A. and Crockett, N. 1992. Antisense techniques: an overview. Pages 1-49 in: Antisense RNA and DNA. J.A. Murray, ed. Wiley- Liss, New York, USA.

76. Nellen, W. and Lichtenstein, C. 1993. What makes an mRNA anti-sensitive? Trends Biochem. Sci. 18:419-423.

77. Nelson, A., Roth, D.A., and Johnson, J.D. 1993. Tobacco mosaic virus infection of transgenic *Nicotiana tabacum* plants is inhibited by antisense constructs directed at the 5' region of viral RNA. Gene 127:227-232.

78. Oeller, P.W., Min-Wong, L., Taylor, L.P., Pike, D.A. and Theologis, A. 1991. Reversible inhibition of tomato fruit senescence by antisense RNA. Science 254:437-439.

79 Pang, S.-Z., Nagpala, P., Wang, M., Slightom, J.L., and Gonsalves, D. 1992. Resistance to heterologous isolates of tomato spotted wilt virus in transgenic tobacco expressing its nucleocapsid protein gene. Phytopathology 82:1223-1229.

80. Pang, S.-Z., Slightom, J.L., and Gonsalves, D. 1993. Different mechanisms protect transgenic tobacco against tomato spotted wilt and impatiens necrotic spot *Tospoviruses*. Bio/Technology 11:819-824.

81. Paoella, G., Sproat, B.S. and Lamond, A.I. 1992. Nuclease resistant ribozymes with high catalytic activity. EMBO J. 11:1913-1919.

82. Persson, C., Wagner, E.G., and Nordström, K. 1988. Control of replication of plasmid R1: kinetics of *in vitro* interaction between the antisense RNA, CopA, and its target, CopT. EMBO J. 7:3279-3288.

83. Perriman, R., Delves, A., and Gerlach, W. 1992. Extended target-site specificity for a hammerhead ribozyme. Gene 113:157-163.

84. Perriman, R., Graf, L., and Gerlach, W. 1993. A ribozyme that enhances gene suppression in tobacco protoplasts. Antisense Res. Develop. 3:253-263.

85. Perriman, R., Breuning, G., Dennis, E.S., and Peacock, W.J. 1995. Effective ribozyme delivery in plant cells. Proc. Natl. Acad. Sci. USA 92:6175-6179.

86. Powell, P.A., Stark, D.M., Sanders, P.R. and Beachy, R.N. 1989. Protection against tobacco mosaic virus in transgenic plants that express tobacco mosaic virus antisense RNA. Proc. Natl. Acad. Sci. USA 86:6949-6952.

87. Prins, M., de Haan, P., Luyten, R., van Veller, M., van Grinsven, M.Q.J.M., and Goldbach, R. 1995. Broad resistance to tospoviruses in transgenic tobacco plants expressing three tospoviral nucleoprotein gene sequences. Mol. Plant-Microbe Interact. 8:85-91.

88. Qu, R., Bhattacharyya, M., Laco, G.S., de Kochko, A., Subba Rao, B.L., Kaniewska, M.B., Elmer, J.S., Rochester, D.E., Smith, C.E. and Beachy, R.N. 1991. Characterization of the genome of rice tungro bacilliform virus: comparison with commelina yellow mottle virus and caulimoviruses. Virology 185:354-364.

89. Rezian, M.A., Scene, K.G. and Ellis, J.G. 1988. Antisense RNAs of cucumber mosaic virus in transgenic plants assessed for control of the virus. Plant Mol. Biol. 11:463-471.

90. Riesner, D. 1990. Structure of viroids and their replication intermediates. Are thermodynamic domains also functional domains? Semin. Virol. 1:83-99.

91. Rittner, K., Burmester, C. and Sczakiel, G. 1993. *In vitro* selection of fast-hybridizing and effective antisense RNAs directed against the human immunodeficiency virus type 1. Nucleic Acids Res. 21:1381-1387.

92. Rodermel, S.R., Abbott, M.S. and Bogorad, L. 1988. Nuclear-organelle interactions: nuclear antisense gene inhibits ribulose bisphosphate carboxylase enzyme levels in transformed tobacco plants. Cell 55:673-681.

93. Rothstein, S.J. and Lagrimini, L.M. 1989. Silencing gene expression in plants. Oxford Surveys of Plant Molecular and Cell Biology 6:221-246.

94. Sanford, J.C. and Johnston, S.A. 1985. The concept of parasite-derived resistance - deriving resistance genes from the parasite's own genome. J. theor. Biol. 113:395-405.

95. Sarver, N., Cantin, E.M., Chang, P.S., Zaia, J.A., Ladne, P.A., Stephens, D.A. and Rossi, J.J. 1990. Ribozymes as potential anti-HIV-1 therapeutic agents. Science 247:1222-1225.

96. Saxena, S.K. and Ackerman, E.J. 1990. Ribozymes correctly cleave a model substrate and endogenous RNA *in vivo*. J. Biol. Chem. 265:17106-17109.

97. Scanlon, K.J., Jiao, L., Funato, T., Wang, W., Tone, T., Rossi, J.J. and Kashani-Sabet, M. 1991. Ribozyme-mediated cleavage of c-fos mRNA reduces gene expression of DNA synthesis enzymes and metallothionein. Proc. Natl. Acad. Sci. USA 88:10591-10595.

98. Schindler, I., and Mühlbach, H.-P. 1992. Involvement of nuclear DNA-dependent RNA polymerases in potato spindle tuber viroid replication: a reevaluation. Plant Science 84:221-229.

99. Scholthof, K.-G., Scholthof, H.B. and Jackson, A.O. 1993. Control of plant virus diseases by pathogen-derived resistance in transgenic plants. Plant Physiol. 102:7-12.

100. Schuch, W., Bird, C.R., Ray, J., Smith, C.J., Watson, C.F., Morris, P.C., Gray, J.E., Arnold, C., Seymour, G.B., Tucker, G.A. and Grierson, D. 1989. Control and manipulation of gene expression during tomato fruit ripening. Plant Mol. Biol. 13:303-311.

101. Sczakiel, G., Homann, M. and Rittner, K. 1993. Computer-aided search for effective antisense RNA target sequences of the human immunodeficiency virus type 1. Antisense Res. Develop. 3:45-52.

102. Sheehy, R.E., Kramer, M. and Hiatt, W.R. 1988. Reduction of polygalacturonase activity in tomato fruit by antisense RNA. Proc. Natl. Acad. Sci. USA 85:8805-8809.

103. Shimayama, T., Nishikawa, S., and Taira, K. 1995.

Generality of the NUX rule: Kinetic analysis of the results of systematic mutations in the trinucleotide at the cleavage site of hammerhead ribozymes. Biochemistry 34:3649-3654.

104. Simons, R.W. and Kleckner, N. 1988. Biological regulation by antisense RNA in prokaryotes. Annu. Rev. Genet. 22:567-600.

105. Sioud, M. and Drlica, K. 1991. Prevention of human immunodeficiency virus type 1 integrase expression in Escherichia coli by a ribozyme. Proc. Natl. Acad. Sci. USA 88:7303-7307.

106. Smith, C.J., Watson, C.F., Ray, J. Bird, C.R., Morris, P.C., Schuch, W. and Grierson, D. 1988. Antisense RNA inhibition of polygalacturonase gene expression in transgenic tomatoes. Nature 334:724-726.

107. Smith, C.J., Watson, C.F., Morris, P.C., Bird, C.R., Seymour, G.B., Gray, J.E., Arnold, C., Tucker, A., Schuch, W., Harding, S. and Grierson, D. 1990. Inheritance and effect on ripening of antisense polygalacturonase genes in transgenic tomatoes. Plant Mol. Biol. 14:369-379.

108. Smith, H.A., Swaney, S.L., Parks, T.D., Wernsman, E.A., and Dougherty, W.G. 1994. Transgenic plant resistance mediated by untranslatable RNAs: Expression, regulation, and fate of nonessential RNAs. Plant Cell 6:1441-1453.

109. Smith, H.A., Powers, H., Swaney, S., Brown, C., and Dougherty, W.G. 1995. Transgenic potato virus Y resistance in potato: Evidence for an RNA-mediated cellular response. Phytopathology 85:864-870.

110. Stanley, J., Frischmuth, T., and Elwood, S. 1990. Defective viral RNA ameliorates symptoms of geminivirus infection in transgenic plants. Proc. Natl. Acad. Sci. USA 87:6291-6295.

111. Steger, G., Baumstark, T., Mörchen, M., Tabler, M., Tsagris, M., Sänger, H.L. and Riesner, D. 1992. Structural requirements for viroid processing by RNaseT1. J. Mol. Biol. 227:719-737.

112. Steinecke, P., Herget, T. and Schreier, P.H. 1992. Expression of a chimeric ribozyme gene results in endonucleolytic cleavage of target mRNA and a concomitant reduction of gene expression in vivo. EMBO J. 11:1525-1530.

113. Swaney, S., Powers, H., Goodwin, J., Silva-Rosales, L., and Dougherty, W.G. 1995. RNA-mediated resistance with non-structural genes from the tobacco etch virus genome. Mol. Plant-Microbe Interact. 8:1001-1011.

114. Symons, R.H. 1989. Self-cleavage of RNA in the replication of small pathogens of plants and animals. Trends Biochem. Sci. 14:445-450.

115. Symons, R.H. 1990. The fascination of low molecular weight pathogenic RNAs. Sem. Virology 1:75-81.

116. Tabler, M. 1993. Antisense RNA in plants: a tool for analysis and suppression of gene function. Pages 237-258 in: Morphogenesis in Plants: Molecular Approaches, K.A. Roubelakis-Angelakis, and K. Tran Thanh Van, eds. NATO ASI Series A253. Plenum, New York, NY.

117. Tabler, M. and Tsagris, M. 1990. Viroid replication mechanisms. Pages 185-205 in: Recognition and response in plant-virus interactions. R.S. Fraser, ed. NATO ASI Series H41. Plenum, New York, NY.

118. Tabler, M. and Tsagris, M. 1991. Catalytic antisense RNAs produced by incorporating ribozyme

119. Tabler, M., Tzortzakaki, S. and Tsagris, M. 1992. Processing of longer-than-unit-length potato spindle tuber viroid RNAs into infectious monomeric circular molecules by a G-specific endoribonuclease. Virology 190:746-753.

120. Talljö, A. and Kirsebom, L.A. 1993. Product release is a rate-limiting step during cleavage by the catalytic RNA subunit of Escherichia coli RNase P. Nucleic Acids Res. 21:51-57.

121. Tieman, D.M., Harriman, R.W., Ramamohan, G. and Handa, A.K. 1992. An antisense pectin methylesterase gene alters pectin chemistry and soluble solids in tomato fruit. Plant Cell 4:667-679.

122. Truve, E., Aaspôllu, A., Honkanen, J., Puska, R., Mehto, M., Hassi, A., Teeri, T.H., Kelve, M., Seppänen, P. and Saarma, M. 1993. Transgenic potato plants expressing mammalian 2'-5' oligoadenylate synthetase are protected from potato virus X infection under the field conditions. Bio/Technology 11:1048-1052.

123. Tsagris, M., Tabler, M. and Sänger, H.L. 1987. Oligomeric potato spindle tuber viroid (PSTV) RNA does not process autocatalytically under conditions where other RNAs do. Virology 157:227-231.

124. Tsagris, M., Tabler, M., Mühlbach, H.-P. and Sänger, H.L. 1987. Linear oligomeric potato spindle tuber viroid (PSTV) RNAs are accurately processed in vitro to the monomeric circular viroid proper when incubated with a nuclear extract from healthy potato cells. EMBO J. 8:2173-2183.

125. Tsagris, M., Tabler, M. and Sänger, H.L. 1991. Ribonuclease T1 generates circular RNA molecules from viroid-specific RNA transcripts by cleavage and intramolecular ligation. Nucleic Acids Res. 19:1605-1612.

126. Van der Krol, A.R., Mol, J.N. and Stuitje, A.R. 1988. Modulation of eukaryotic gene expression by complementary RNA or DNA sequences. BioTechniques 6:958-976.

127. Van der Krol, A.R., Mol, J.N. and Stuitje, A.R. 1988. Antisense genes in plants: an overview. Gene 72:45-50.

128. Van der Krol, A.R., Lenting, P.E., Veenstra, J., van der Meer, I.M., Koes, R.E., Gerats, A.G., Mol, J.N. and Stuitje, A.R. 1988. An anti-sense chalcone synthase gene in transgenic plants inhibits flower pigmentation. Nature 333:866-869.

129. Van der Vlugt, R.R, Ruiter, R.K. and Goldbach, R. 1992. Evidence for sense RNA-mediated protection to PVY^N in tobacco plants transformed with the viral coat protein cistron. Plant Mol. Biol. 20:631-639.

130. Van der Wilk, F., Posthums-Lutke Willink, D., Huisman, M.J., Huttinga, H. and Goldbach, R. 1991. Expression of the potato leafroll luteovirus coat protein gene in transgenic plants inhibits viral infection. Plant Mol. Biol. 17:431-439.

131. Wegener, D., Steinecke, P., Herget, T., Petereit, I., Phillip, C., and Schreier, P. 1994. Expression of a reporter gene is reduced by a ribozyme in transgenic plants. Mol. Gen. Genet. 245:465-470.

132. Wilson, T.M. 1993. Strategies to protect crop plants against viruses: Pathogen-derived resistance blossoms. Proc. Natl. Acad. Sci. USA 90:3134-3141.

133. Yeh, S-D., Jan, F-J., Chiang, C.H., Doong, T-J., Chen, M-C., Chung, P-H., and Bau, H-J. 1992. Complete nucleotide sequence and genetic

organization of papaya ringspot virus RNA. J. Gen. Virol. 73:2531-2541.

134. Yepes, L.M., Fuchs, M., Slightom, J.L., and Gonsalves, D. 1996. Sense and antisense coat protein gene constructs confer high levels of resistance to tomato ringspot nepovirus in transgenic *Nicotiana* species. Phytopathology 86:417-424.

135. Zhao. J.J., and Pick, L. 1993. Generating loss-of-function phenotypes of the *fushi tarazu* gene with a targetted ribozyme in *Drosophila*. Nature 365:448-451.

CHAPTER 8

Satellite RNA-mediated Resistance to Plant Viruses: Are The Ecological Risks Well Assessed?

M. Jacquemond and M. Tepfer

The term satellite was proposed first by Kassanis in 1962 (49), to describe a small virus serologically unrelated to, but dependent on tobacco necrosis virus (TNV) for replication. After the description by Schneider in 1969 (85) of a small RNA that not only depends on the tobacco ringspot virus (TRSV) genome for its replication, but also is encapsidated by TRSV coat protein, the term satellite has been generally applied to small RNAs showing the following principal features: small non-infectious RNAs dependent on a helper virus for replication, showing essentially no sequence homology with the helper genome. The satellite group includes the satellite viruses, which are encapsidated in capsid protein encoded by the satellite genome, and the satellite RNAs, which are encapsidated in capsid protein encoded by the helper virus. Since these first discoveries, satellite viruses and RNAs have been found associated with many viruses, differing by their morphology and their mechanisms of genomic expression. Two major classes of satellite RNAs have been described; large satellite RNAs with an open reading frame whose translation is necessary for their replication, and small satellite RNAs, including those associated with cucumber mosaic virus (CMV), whose biological properties do not require their translation. Molecular features of satellite RNAs and their role in symptomatology have been reviewed recently by Collmer and Howell and Roossinck et al. (5,79).

Satellite RNAs are replicated by using the viral RNA-specific replication complex established early during virus infection. Moreover, it has been demonstrated in the case of CMV that the viral RNA-dependent RNA polymerase purified from infected plants can synthesize in vitro full-length

molecules of the satellite RNA of this virus (27, 103, 104). Synthesis of both the helper genome and satellite RNA may have different consequences: i) the two can replicate simultaneously without any observable interference; ii) presence and synthesis of a satellite RNA can modulate the symptoms induced by the helper virus, leading to either an attenuation or an exacerbation of symptom gravity. Satellite RNAs that modify symptom expression are of course of greater interest, not only for fundamental research, but also as natural agents for controlling certain viral diseases. The satellite RNAs associated with CMV fall into the symptom modulating group, and considerable attention has been drawn to this model during recent years for two main reasons : i) CMV affects many cultivated crops, and resistant varieties are available for only a few of them, ii) CMV satellite RNA is considered as a natural parasite of the virus, leading most often to symptom attenuation.

Symptom modulation mediated by CMV satellite RNA has been extensively studied by many laboratories. For most of the hosts investigated so far, the presence of a satellite RNA in a CMV strain correlates with a major reduction in the synthesis of genomic RNAs, and as a consequence, with a striking attenuation of symptom gravity (35,44,70,99). However, a few hosts, particularly tomato (Lycopersicon esculentum) and certain wild Lycopersicon species, may develop considerably aggravated symptoms, depending on the satellite RNA strain. In tomato, CMV without satellite RNA generally induces symptoms of mosaic and/or stunting and/or fernleaf, all of which are strongly attenuated by classic satellite RNAs. Certain other satellite RNAs, however, lead to either increased mosaic symptoms, (leaf yellowing or whitening) or to lethal necrosis (necrogenic satellite RNA) (23,36,47,48,73). Kaper has recently discussed the principal parameters, including the satellite RNA sequence, the helper strain and the host, which are the major elements of the complex relationship that determines the type and the gravity of the symptoms developed by infected tomato plants (41).

This story would be of limited interest if CMV was not endemic in many countries, if lethal necrosis was not the cause of dramatic losses in tomato crops, and if methods based on cross-protection had not been described for controlling CMV in tomato since 1982 (32). The situation is thus very complex: on one hand, satellite RNA can be used as a natural agent capable of limiting the impact of CMV infection, while on the other hand, satellite RNA constitutes a serious danger for some hosts. Two approaches have been developed in order to exploit the beneficial properties of satellite RNA: i) cross-protection using a CMV strain bearing a satellite RNA that is non-necrogenic on tomato; ii) use of transgenic plants expressing a satellite RNA gene. Both approaches have been applied in the open field in China (94), and cross-protection has also been field-tested in other countries including Japan, the U.S. and Italy (19,66,84). Release of a CMV strain, although attenuated, or release of transgenic plants expressing a viral gene in nature, raises important questions about safety. Will these previous open field trials, or experimental data obtained by different groups worldwide, allow us to estimate the risks associated with the use of satellite RNA for controlling CMV? Does the ability of some satellite RNA variants to induce tomato lethal necrosis constitute a too important disadvantage for the use of non-necrogenic ones? Presentation of results necessary for an in-depth discussion of these points will be the principal purpose of this review. In order to strengthen certain points, we will present here certain of our previously unpublished results. For clarity, the methods used in these experiments will be described in a separate section at the end of the text.

NATURAL OCCURRENCE OF SATELLITE RNA IN CMV ISOLATES

Epidemiological Surveys

Although there are many reports dealing with CMV occurrence in numerous species, presence of a satellite RNA has not often been evaluated. Attempts to detect a satellite RNA have been done essentially by groups already involved in satellite RNA studies, or when its presence was the cause of a severe epidemic, for instance of tomato lethal necrosis.

The first outbreaks of tomato necrosis that can be attributed to CMV satellite RNA seem to have occurred in France during the seventies. Although this characteristic syndrome was rapidly shown to be associated with CMV (57,75), the direct involvement of a satellite RNA was only demonstrated some years later (48). Epidemiological surveys carried out during 1974 to 1978 at Montfavet have shown the independent nature of how the two disease phenotypes (mosaic/fernleaf or necrosis) develop in naturally infected tomato plants (76). Whereas fernleaf disease was broadly distributed in fields, as a result of transmission by winged aphids, necrosis had a patchy distribution, which corresponded to secondary infections by apterous aphids of plants adjacent to a necrotic one. Moreover, it appeared that severe epidemics of necrosis were observed essentially when aphid populations were particularly dense, which under the climatic conditions of southeastern France corresponds to the spring growing season. Finally, previous infection by a severe isolate (as shown by pronounced fernleaf or stunting) neither protected against nor stimulated the further development of necrosis (76).

These experiments were pursued during 1980-1984, and focused on the presence of satellite RNA in CMV isolated either from tomato plants or from wild species growing in proximity to the experimental tomato fields. These studies were based only on biological tests, since our aim was to characterize the biological properties of any satellite RNAs present. These results have not been published previously, because most of the isolates from tomato and nearly all of the isolates from wild species had to be amplified first in tobacco before analysis. Formally speaking, since tobacco strongly stimulates satellite RNA replication, detection of a satellite RNA after even one passage through this host does not guarantee that the satellite RNA population detected represents accurately that of the original isolate. Nevertheless, we have thought it appropriate to present these results here, since they illustrate for France a situation very similar to that described in other countries (8,18,21,22,38,50,74). Presence of a non-necrogenic satellite RNA was based upon

Table 1. Occurrence of CMV and satellite RNAs in weeds in proximity to tomato crops

	1981	1982	1983	Total
Number of Samples	478	692	604	1774
Number of CMV isolates[a]	68	81	57	206
Presence of a necrogenic satellite RNA[b]	2	3	1	6
Presence of non-necrogenic satellite RNA[c]	2	11	1	14
% infected tomato plants	12.0	33.3	22.2	
% necrotic plants	1.4	32.0	4.7	

[a] Tissues were ground in 0.03M disodium phosphate containing 0.2% DIECA and used for inoculating a tobacco, a melon and two tomatoes.
[b] Number of isolates inducing necrosis on tomato.
[c] Number of isolates protecting infected tomato plants against super-infection with a necrogenic strain.

the ability of the isolate to protect against necrosis induced by super infection with a standard necrogenic inoculum. Tomato plants were sampled in 1983 and 1984. No CMV outbreak occurred in 1983; among the 200 samples analyzed, only 13 were CMV-positive, and none contained a satellite RNA. In 1984, a severe outbreak of necrosis occurred, since 75 % of the plants were necrotic. Necrogenic CMV was recovered from 53 of the 60 necrotic plants analyzed. Moreover, CMV was recovered from 77 samples among the 86 harvested on plants developing either fernleaf symptoms or no symptoms at all. Thirteen of these proved to be necrogenic on tomato, and the plants from which they had been isolated later developed necrosis. Only nine isolates were able to protect tomato against necrosis, suggesting that they bore a non-necrogenic satellite RNA. Its presence was confirmed by analysis on 2.4 % polyacrylamide gels of total viral RNAs purified from infected (but not super-infected) tomato plant. Wild species, known to be natural hosts of CMV (77), were also checked monthly during the growing season, each year from 1981 to 1983. CMV was recovered in 12 % of the plants, but very few isolates contained a satellite RNA (necrogenic or non-necrogenic), even after a first passage through tobacco (Table 1). All isolates collected in 1983 were amplified two more times on tobacco in order to evaluate their ability to produce a satellite RNA. Increasing the number of passages through tobacco indeed increased the number of isolates that had amplified a satellite RNA of both types (Table 2).

In Italy, tomato necrosis was not reported in 1982, and attempts to detect a satellite RNA in six natural CMV isolates failed (74). Nevertheless, after one to four passages through tobacco, a satellite RNA was found to be present. Severe necrogenic outbreaks developed in 1988, and a satellite RNA was found in isolates arising only from necrotic plants, which was then further characterized (8,18,43). The following year, before tomatoes were planted, a survey of weedy species growing in a previously affected area was carried out. CMV was recovered from 5 of the 12 species indexed, and a satellite RNA was detected only in *Amaranthus*, which could thus be considered to be a source of CMV containing satellite RNA (19). Apart from necrosis, two other syndromes were observed in tomato plants in different fields: classical shoestring-fernleaf and a previously undescribed fruit necrosis. The latter disease was characterized by internal browning of fruits, and was found to be associated with CMV in plants showing almost no other symptoms. Although no satellite RNA was detected in preliminary experiments (18), a non-necrogenic satellite RNA was potentially associated with this syndrome and has been sequenced (8,9). Finally, a fourth separate disease was described, and named tomato top stunting; the satellite RNA detected in these plants is thought to be responsible for this syndrome (25).

A similar situation has been described in Spain, where surveys in vegetable crops during 1979 to 1982 showed that lethal necrosis was not present, and satellite RNA was not detected in the 34 CMV isolates examined (21). However, as in the cases described above, a satellite RNA emerged after serial passage of two isolates, but depending on the host and the experiment, it was

Table 2. Emergence of a satellite RNA following serial passages of natural isolates through tobacco

Number of passages through tobacco	Total number of isolates	Emergence of a necrogenic satellite RNA	Emergence of a non-necrogenic satellite RNA
1	57	1	1
2	55	3	4
3	48	5	5

either a necrogenic or a non-necrogenic one (22). Necrogenic disease was first detected in a localized region in Spain in 1986, and has spread dramatically since that time. Jorda *et al.* (38) have described different syndromes that could be considered to be similar to those observed in Italy: i) lethal necrosis on plants and necrotic symptoms on fruits, ii) internal browning of fruits, iii) typical fernleaf, iv) curling and stunting, which they named curl-stunt syndrome, and which, like the Italian tomato top stunting, is characterized by severe internode shortening. As could be expected, a satellite RNA was present in all isolates arising from necrotic plants, but also in 30 % of the isolates from plants developing fernleaf symptoms at the time of sampling, and in samples from plants showing the curl-stunt syndrome. Analysis by electrophoresis under semi-denaturing conditions of certain isolates corresponding to these different syndromes showed that satellite RNAs having different mobilities could be present, either alone or in association. Studies of the biological properties of individual purified satellite RNAs allowed characterization of three types: type 1 does not modify symptoms induced by the helper virus, type 2 provokes curl-stunt and type 3 typical lethal necrosis. Curiously, no satellite RNA with ameliorative function was detected. Finally, internal browning of fruit was clearly associated with CMV, but not with satellite RNA (38).

Tomato lethal necrosis has not been reported from the US, but a satellite RNA has been associated with another characteristic disease of tomato plants, white leaf syndrome (23). This disease was observed as early as 1970 in New York state. In a recent epidemiological survey in vegetable crops in New York state and Bermuda, CMV was shown to be frequent on peppers and cucurbits, but quite rare in tomatoes (9 % of the samples were CMV-positive in ELISA tests) (50).

Among the 106 isolates collection, only two contained a satellite RNA. They had been recovered from a pepper and a tomato in two different counties of New York state. Moreover, the tomato sample had been harvested in a farm where a severe outbreak of white leaf disease had previously occurred. Eleven isolates were multiplied in tobacco, but none of them contained a satellite RNA, even after three passages (50).

Finally, although no surveys have been published, it should also be mentioned that CMV outbreaks (including necrosis) have caused severe damage in Japan since 1970, where certain production areas have had to be shifted in order to escape CMV epidemics (84).

Tomato is not the only host that has been frequently found to be infected with a CMV isolate containing satellite RNA. Such isolates are frequent in *Nicotiana glauca* in Israel and Spain, but are not associated with a particular symptomatology (38,80). Three types of small molecules that could be distinguished by electrophoretic analysis were found either alone or in association with Israeli CMV strains. Two of them are undoubtedly CMV satellite RNAs, since they specifically hybridize with a satellite RNA probe. The third seems to have another unknown origin (80). Finally, satellite RNAs occur naturally in CMV isolates from tobacco in Japan. One of these Japanese satellite RNAs, found in the Y strain, has two particular features: it induces brilliant yellow mosaic symptoms in tobacco (91), and it is longer than other satellite RNAs, bearing a large insertion in the central part of the molecule (30).

Considered together, these results lead to three main conclusions:

i) Satellite RNA is not as common in CMV isolates as one might have thought, considering the severity of the outbreaks that can occur in tomato fields, except precisely in cases where these outbreaks develop. When an outbreak occurs, satellite RNA is detected not only in tomato, but also in neighboring weeds. Tobacco and certain other *Nicotiana* species are perhaps exceptions to this general rule, since satellite RNA seems to be common in these hosts, independently of a characteristic outbreak. However, this last remark may be premature, since no epidemiological surveys equivalent to those carried out for tomato have been done for *Nicotiana* species.

ii) Several natural variants can be distinguished on the basis of the modulation of the symptoms induced on tomato: necrogenic satellite RNA (France, Japan, Italy, Spain), non-necrogenic satellite RNA possessing ameliorative functions (France), non-necrogenic satellite RNA inducing no modification of the symptomatology (Spain), non-necrogenic satellite RNA inducing unusual curling and stunting (Italy and Spain, although it is perhaps premature to consider that it is the same syndrome that has been described in these two countries), non-necrogenic satellite RNA potentially associated with internal browning (Italy but not Spain), white leaf satellite RNA (US). Moreover, more than one variant can be detected in a single isolate, a situation probably due to mixed or successive infections. It should be noted that at least part of the heterogeneity of satellite RNA populations described in many CMV strains currently studied in different laboratories may be due to forced and successive amplification of these strains through different hosts, which could also be at the origin of emergence of new variants.

iii) Although satellite RNA is not often detectable in CMV isolated from naturally infected plants in the field, it is however often amplified during successive mechanical passages in different hosts, and particularly in tobacco. Of course, experimental laboratory conditions are artificial, and can apply selection pressure that could favor the amplification of an extremely minor component that might not be amplified in nature. In any case, these observations are the starting point of an important debate concerning the usual state of CMV satellite RNAs in nature. One possibility is that the newly amplified satellite RNAs have arisen *de novo*, due to some as yet unidentified modification of the interactions between the virus and host plants (particularly solanaceous hosts). Indeed, at this time the question of the origin of CMV satellite RNAs is completely unresolved. The second possibility is that the satellite RNA is a broadly disseminated natural component of CMV, but which would generally be present in

infinitesimal amounts. In this case, the unresolved question is, what factors normally down-regulate satellite RNA replication, but which are modified under experimental conditions? At least one factor that could play a role in maintaining satellite RNAs at a low level in nature would be counter selection for transmission by insects (see below).

Natural Transmission of Satellite RNA

Natural transmission of the satellite RNA by aphids, simultaneously with the helper virus, was first indirectly shown by Mossop and Francki in 1977 (69), who were studying the genetic determinants of aphid transmissibility of CMV. Since the satellite nature of the "fifth component" of CMV was not yet established, its presence was rather a disadvantage for this study, since its possible role in aphid transmission could not be ruled out.

Results of experimental transmission by aphids of a necrogenic strain were reported by one of us in 1982 (31). It was shown that the probability of transmission of the satellite RNA was dependent on the host species used as the source plant. When this host was melon, which replicates satellite RNA to a level lower than solanaceous plants, some of the tomato test plants developed only severe fernleaf symptoms characteristic of the helper virus without satellite RNA. This suggests that the necrogenic satellite RNA had not been transmitted to these plants, or if so, only at a very low level. In contrast, when the source plants were either tobacco or tomato, both of which accumulate satellite RNA to high levels, the satellite RNA was always transmitted with the helper virus and all the tomato test plants developed necrosis. However, the efficiency of virus transmission from these sources was considerably reduced. This last result is crucial from an epidemiological point of view: satellite RNA not only decreases the replication of its helper virus, but also reduces the efficiency of some host plants as a source for transmission by aphids. Thus, plants in which the satellite RNA is poorly replicated (such as melon) are good source plants for the virus, but from which the satellite RNA is not always transmitted; in contrast, plants that amplify satellite RNA to high levels (tomato or tobacco) are poor source plants for the virus, but when virus is transmitted, the prevalence of satellite RNA is high. The presence of satellite RNA can lead to a kind of double suicide (of the satellite and its helper virus)!

Transmission of satellite RNA by contact has been also demonstrated (71). When plants (tomato or tobacco) were first infected with a satellite-RNA-free CMV strain and super-infected with a satellite-RNA-containing one by several means (mechanical inoculation, viruliferous aphids, use of a contaminated razor blade or rubbing a leaf with a leaf detached from a diseased plant), they all later accumulated a satellite RNA. Since the relationship between satellite RNA and helper virus has essentially no specificity, transmission by contact should be kept in mind, since it could have important implications in the field.

CROSS-PROTECTION WITH A CMV STRAIN THAT CONTAINS SATELLITE RNA

Viral Cross-protection and Interference Between Satellite RNAs

Cross-protection has been described as early as 1933 as "a protective inoculation against a plant virus" (82). It corresponds to inoculation with a mild strain of a virus in order to prevent infection by more virulent ones. Many studies have since shown that this method can be employed for any virus for which sufficient strains are available. Cross-protection does not always imply an inhibition of the replication of the super-infecting (challenge) strain. In the case of CMV, different approaches have been developed to check for the presence of the challenge strain in the protected plants. From electrophoretic analysis of virions or double stranded viral RNAs, under conditions where two CMV strains are easily differentiated, it has been shown that in protected tomato, tobacco or squash, the challenge strain could be recovered from the infected leaves, but not from the upper ones (13, 14). On the other hand, a study involving the use of a monoclonal antibody specific for the challenge strain showed that this strain was clearly detected in upper leaves of both protected tobacco and pepper plants, although its accumulation was decreased when compared to plants that had not been preinoculated (105).

Different groups have constructed mild CMV strains or characterized naturally occurring ones (19,32,66,84,92,94,95,105,109). CMV satellite RNA may contribute to cross-protection at two levels: i) its addition to a more severe strain is a means of obtaining a mild strain that conserves its properties following serial passages in an adapted host plant; ii) inoculation of tomato plants with a strain containing a non-necrogenic satellite RNA protects them against challenge infection by a necrogenic strain (32). Thus, this type of attenuated strain may be used for traditional (general) cross-protection, and also for specific protection against tomato necrosis, although the target molecules within the challenge strain are the genomic RNAs in the first case, with, in addition, the necrogenic satellite RNA in the second one. Moreover, mild strains constructed by addition of satellite RNA have proved to be particularly efficient for protecting against virulent satellite-RNA-free strains in different hosts (32,66,92,105,109).

Apart from cross-protection, interference between necrogenic and non-necrogenic satellite RNAs is also known to occur when they are brought in the same inoculum. The biological effect specific to the necrogenic satellite RNA dominates in tomato, even if present in a 200-fold dilution compared to the non-necrogenic molecule. More precisely, all tomato plants died from necrosis when infected with a CMV inoculum containing 1 µg/ml of the non-necrogenic R satellite RNA and 5×10^{-3} µg/ml of the necrogenic I17N satellite RNA (36). When lower amounts of necrogenic satellite were added, the percentage of necrotic plants was reduced until all of them develop the attenuated symptoms characteristic of the non-necrogenic satellite RNA. Similar results have been more recently obtained by studying the non-necrogenic S and the necrogenic D satellite RNAs (90).

Molecular Analysis of Progeny Satellite RNA

Because of the preferential expression of the necrogenic satellite RNA, analysis of the satellite molecules amplified is crucial. Electrophoresis through 9 % polyacrylamide gels under semi-denaturing conditions allows differentiation of certain satellite RNA sequence variants (47,101). Using this method, Smith et al. (90) have detected the presence of the two satellite RNAs in plants infected with an inoculum containing two types of molecules, even when the necrogenic D satellite RNA was in 1000-fold dilution compared to the non-necrogenic S satellite RNA. Similarly, they showed that, depending on its concentration in the challenge inoculum, D satellite RNA could also be detected in total nucleic acid extracts from plants where the two satellite RNAs had been brought at different times (cross-protection experiments). Unfortunately, it is not clear in this work if each sample represented tissues from a single plant or from a group of several ones, and what symptoms they showed at the time of sampling. In a previous study, the same group had used Northern assays after electrophoresis of total nucleic acids through a 6 % polyacrylamide gel under semi-denaturing conditions, which allowed the separation of the double-stranded forms of the two satellite RNAs (66). They showed that the double-stranded RNA of the necrogenic D satellite was detected in challenged tomato plants which remained asymptomatic.

The polyacrylamide gel techniques described above do not allow discrimination between the extremely closely related satellite RNAs that we have studied. For this reason, we have developed an RT-PCR procedure for sequence analysis of the satellite RNA progeny in tomato plants during cross-protection experiments involving different non-necrogenic satellite RNAs and the necrogenic I17N satellite RNA whose sequence has been already published (34). The non-necrogenic variants were the R satellite RNA, of which the first reported sequence has recently been revised (34,97), and variants obtained by exchange of parts of R and I17N satellite RNAs, which will be presented further in a later section. The satellite RNAs in the inocula were produced by *in vitro* transcription from the corresponding cloned satellite cDNAs (Figure 2 presented later), and added at a concentration of 1 µg/ml to viral helper RNA (10 µg/ml) prepared from the satellite-RNA-free strain I17F. Challenge infection was done two weeks after the first infection, with viral RNAs from the I17N strain bearing the necrogenic I17N satellite RNA, at a concentration of 50 µg/ml. Three weeks after this challenge infection, the plants in which a non-necrogenic satellite RNA had been amplified following the first inoculation were completely protected and did not develop any necrotic symptoms (presented later in Table 6), while plants inoculated first with I17F viral RNA alone all died if challenge infected, or always developed severe fern leaf symptoms if not. Viral RNAs were purified from individual protected plants; at least one plant has been analyzed for each inoculum where a non-necrogenic transcript had been added. As could be expected, since these purifications were done late in infection, and since a satellite RNA was present, virus yields were very low, and viral RNAs

consisted mainly of satellite RNA molecules. Sequencing of all satellite RNA progeny (corresponding to 6 individual plants) did not reveal the presence of I17N molecules.

The purified viral RNAs were also used as inoculum for two young tomato plants at a concentration of 20 µg/ml (a higher than usual concentration was needed because of the large amount of satellite RNA in these preparations). Following infection, these plants developed attenuated fern leaf symptoms, typical of the presence of non-necrogenic satellite RNA. In some cases, weak symptoms of necrosis (local necrotic lesions) could be observed on a few leaves late in infection (more than one month after inoculation), which never became lethal. Similar results were obtained when crude extracts from protected plants were used as inoculum. Taken together, and bearing in mind the preferential expression in tomato of a necrogenic satellite RNA when present in the same inoculum as a non-necrogenic one, these results show that following super-infection the necrogenic satellite RNA may be replicated in the protected plants, but only at a very low level. Detection of the minority population is thus possible only by using the biological test, which ensures sufficient amplification of the necrogenic satellite RNA to observe induction of very mild necrotic symptoms, due to necrogenic molecules that had not been detected by RT-PCR sequencing.

There is thus an apparent discrepancy between the results of the two groups. We can only tentatively propose that this could be due to differences in experimental conditions and in the analysis carried out. Particularly, it should be mentioned that Montasser *et al.* (66) and Smith *et al.* (90) have used natural satellite RNA preparations in their work, while we have used RNA transcripts produced *in vitro* from sequenced cDNAs, and that the helper strains and satellite RNA variants differed as well. Efficiency of a helper genome (either protective or used as a challenge inoculum) to initiate infection and to replicate, or efficiency of a given strain to accumulate a particular satellite RNA, could be crucial in determining the possibility of synthesis of either the challenge genome (as illustrated above for conventional cross-protection experiments) or the challenge satellite RNA. Moreover, the delay between the two infections was also different, which is also crucial, since it determines amounts of non-necrogenic satellite RNA and genomic RNAs in the leaves that are super-infected. If one supposes that super-infection is done on a leaf where protective viral RNAs are present in small amounts (too short or too long delay between the two inoculations, older lower leaves and not upper ones...), the challenge genomic RNAs can initiate infection and thus permit the multiplication of the necrogenic satellite RNA, although this infection can remain local, and will not necessarily provoke appearance of necrotic symptoms. Finally, we have analyzed by RT-PCR viral RNAs extracted from purified viral particles very late after inoculation, while the other authors have analyzed total nucleic acids by Northern experiments. Nevertheless, evaluation of the biological properties of the viral RNAs purified from efficiently protected plants allows us to suggest that, under our experimental conditions, accumulation of the necrogenic satellite RNA remains at a low level or, at least, very few molecules are encapsidated in these plants. These data are important in assessing potential risks, because it raises the question of whether or not such plants would be sources of a necrogenic RNA that could be transmitted by aphids. Although this has not been tested, the main features of aphid-mediated transmission of satellite RNA, which were presented above, would suggest that transmission of a super-infecting necrogenic variant in a population of non-necrogenic satellite RNA would be far from certain to occur, since the necrogenic molecule would represent such a small proportion of the satellite RNA population.

Testing Cross-protection Under Natural Growing Conditions

Greenhouse Experiments

Greenhouse experiments are necessary for evaluating the effect of a protecting inoculum on plant growth, fruit quality and yield before testing it in field cross-protection experiments. This avoids not only a voluntary release of CMV in the field, but also uncontrolled super-infection by other strains or other viruses, which would introduce a bias into the observations. To our knowledge, there is only one published report of greenhouse testing of mild CMV strains, which was carried out in cross-protection experiments against severe satellite-RNA-free strains (66). This work demonstrated very clearly the effect of cross-protection conferred by a mild CMV strain against a severe one, as judged by fruit quality and quantity, as well as by overall plant development (height and weight of shoot system). Moreover, it showed than the protection is observed sooner and

is more efficient if a non-necrogenic satellite RNA is added to the protecting mild strain.

A similar assay was carried out in Montfavet during the spring and summer of 1983 in order to evaluate cross-protection against lethal necrosis (M. Daniel and M. Jacquemond, unpublished). The three strains mentioned above were used (the satellite-RNA-free I17F, as well as R and I17N, which contain a non-necrogenic and a necrogenic satellite RNA, respectively). The experimental protocol is presented in the materials and methods section. The results presented in Table 3 clearly illustrate the protective effects of the non-necrogenic satellite RNA against a necrogenic inoculum, since complete protection was conferred by the I17F strain to which R satellite RNA had been added. Nevertheless, these results also show that the level of protection is dependent on the helper strain, since less effective protection was obtained with the R strain itself. With the latter strain, only half of the twenty plants tested were efficiently protected, while three died rapidly due to necrosis, and the seven others later developed partial, though non-lethal, necrotic symptoms. According to the different criteria taken into account in grouping CMV strains (10,58,72), I17F belongs to group I, while R belongs to group II. The latter strain is less efficiently replicated in tomato plants, and does not persist when the ambient temperature is high. The presence and the biological properties of the virus were checked by biological test (inoculation of young tomato plants with a crude leaf extract) six weeks after the challenge infection. Non-necrogenic CMV was recovered from all four of the plants analyzed that had been first inoculated with I17F strain to which R satellite RNA had been added. In plants first infected with the R strain (including its natural satellite RNA), and then challenge infected, necrogenic CMV was recovered from the plants developing partial necrosis. However, the biological test alone did not make it possible to determine if it was the R strain which had amplified the necrogenic I17N satellite RNA or the I17N strain itself. Finally, non-necrogenic CMV was recovered from seven protected plants among compared to healthy controls. This reduction the ten analyzed, which is consistent with the low amount of R strain that replicated during the hot summer period. As was the case for tomato plants infected with the CMV strain16 studied by Montasser et al. (66), I17F had a clear negative effect on the plants, which always developed severe fernleaf symptoms by the end of the

Table 3. Effects of preinoculation and of satellite-RNA-mediated protection on growth and fruit yields of tomato plants

Treatment[a]	% necrotic plants[b]	Mean total weight of fruit per plant (Kg)[c]	Mean weight of leaves per plant at end of expeiment (g)[c]	Mean weight of stem per plant at end of experiment (g)[c]
-/-	0	4.32o	609.5s	393.7x
I17F/-	0	2.36p	240.0t	198.0y
I17F/I17N	100	0	0	0
I17F+R satRNA/-	0	3.35q	368.4u	276.9z
I17F+R satRNA/I17N	0	3.25q	309.7v	267.3z
R/-	0	3.82r	499.6w	375.2x
R/I17N	50[d]	PP[d]: 3.78r UPP[d]: 2.51p	PP[d]: 584.0s UPP[d]: 403.4u	PP[d]: 407.6x UPP[d]: 360.1x

[a] Each treatment corresponds to 20 tomato plants cv. Melody GH 31 (this cultivar is resistant to *Verticillium*, *Fusarium* and TMV).
[b] Symptoms observed at the end of the culture; according to initial inoculum the symptoms were severe fernleaf (I17F), attenuated fernleaf (I17F+R sat-RNA), no apparent symptoms (R).
[c] Values followed by the same letter do not differ significantly at the 5% level according to Duncan's multiple range test.
[d] Half of the twenty plants tested were efficiently protected (PP), while 3 died soon from necrosis, and the 7 others (UPP) developed partial, though non-lethal, necrotic symptoms.

experiment. Although R satellite RNA attenuates the gravity of the disease induced by I17F, fruit yields were still significantly lower corresponds mainly to a decrease of fruit size, rather than to a decrease in their number (Figure 1).

Open Field Experiments

Experiments in the open field have been carried out in four countries, with clearly different aims. In the US, assays in simulated epidemic conditions have been done in order to evaluate the efficiency of cross-protection against tomato lethal necrosis, even if this disease has not been reported in any state of that country. This efficiency was clearly proven (66). It is interesting to note that the satellite-RNA-containing S strain used in this work, which belongs to the same group as R strain, conferred a high level of protection, either against the necrogenic D strain or the satellite-RNA-free 16 strain. Very few healthy controls showed CMV symptoms, and the measurements of plant height or fruit yield showed that the S strain does not significantly affect plants.

In Italy, lethal necrosis has caused considerable losses in tomato crops during successive years (19). Cross-protection has thus

been developed as practically the only means of limiting severity of the disease. The S strain proved to be efficient under natural epidemic conditions, where 40 % of the untreated plants developed necrosis. As a consequence, fruit yield was doubled in the protected plants.

Japan is also a country where CMV cross-protection has been applied recently on a large scale, because CMV has caused increasing losses in several areas, leading to a halt in tomato cultivation in some of them (84). Once again, protection was efficient and preinoculated plants produced about 30 % more healthy fruits than untreated ones. As is the case for I17F strain combined with R satellite RNA, the protective strain itself caused a significant reduction in fruit yield, and thus would only be suitable for use in areas where particularly serious CMV outbreaks occur regularly (84).

China is undoubtedly the country where cross-protection has been the most extensively applied, since it has been used there to protect several different crops against CMV, but not tomato against lethal necrosis, since this disease has not yet been reported in China. Construction of mild strains by adding non-necrogenic satellite RNAs to local satellite-RNA-free isolates, and

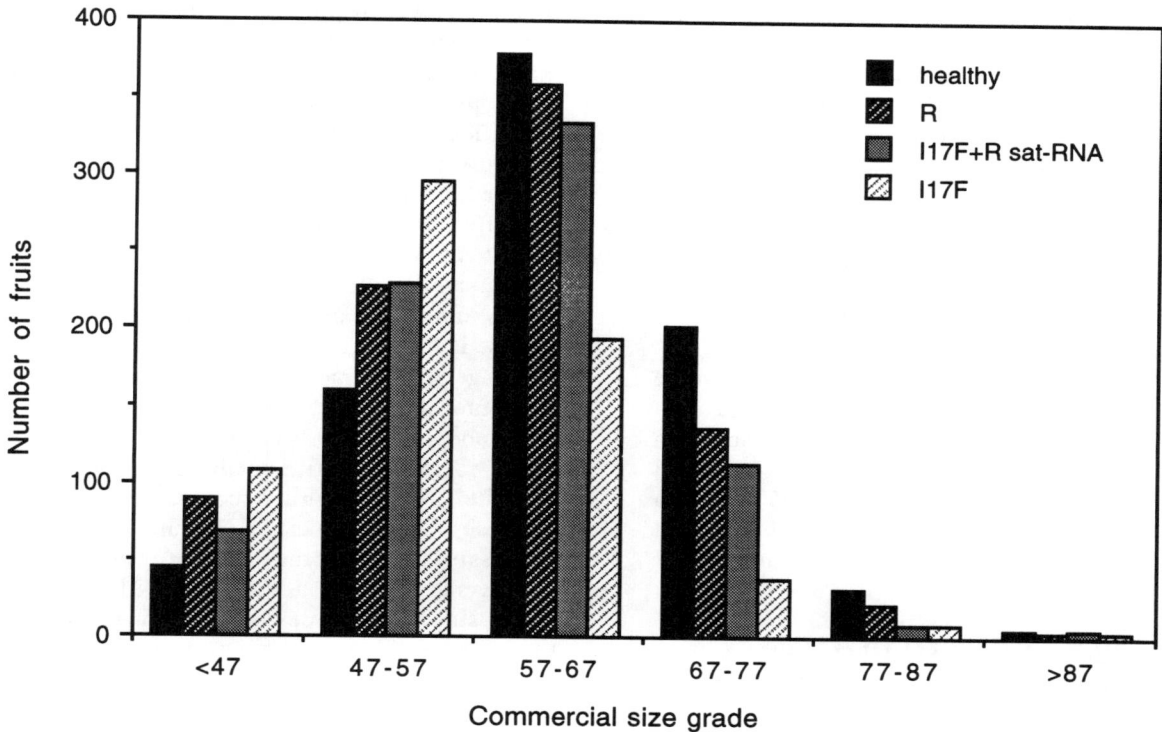

Figure 1. Number of fruits per commercial size grade produced by non-infected controls and by plants infected with the three CMV inocula tested in the cross-protection experiments under greenhouse conditions

verification of their innocuity on several species have been described by Tien *et al.* (95). Such cross-protecting inocula were then used during 8 years in many crops, in particular sweet pepper(3,000 ha) and tomato, but also hot pepper, cucumber and tobacco (94). Three main conclusions were drawn by the authors from their assays: I) an important increase in fruit yields was obtained in treated crops; ii) inoculation with the attenuated CMV strain activates defense mechanisms in the plants, which also became resistant to fungi; iii) no modifications of the natural populations of viruses were observed during this period and, in particular, tomato lethal necrosis was not observed in the treated areas.

Safety Precautions for Cross-protection

Emergence of a variant with increased pathogenicity during the necessary large-scale bulk production of the protective inoculum, and damage to another species which could be naturally infected in the open field, are the two major elements of risk associated with large-scale use of cross-protection.

In order to try to avoid the presence of several variants or the emergence of a satellite RNA from the helper strain itself, the protective inoculum has been prepared in three cases by mixing the helper genome and the satellite RNA purified by gel electrophoresis (19,66,94,95). Moreover, the Chinese group has also cloned the inoculum by isolation from a single local lesion obtained on a hypersensitive host (94,95). Such precautions are necessary but not sufficient, since it has been demonstrated that after serial passages through either tomato or tobacco, a new variant has emerged in the protective inocula used in Japan and China, respectively (84,93). This illustrates the necessity of regular verification of the protective inoculum by one of the methods available, either polyacrylamide gel electrophoresis under semi-denaturing conditions or, as has been proposed by Tien *et al.* (93), by temperature-gradient gel electrophoresis, since this method can also allow detection of new variants. Of course, an even better method would be to use a satellite RNA prepared from an *in vitro* transcript or, ideally, an entirely artificial inoculum in which the CMV genomic RNAs are also produced by transcription *in vitro*.

In order to assess the second risk, the protective inoculum has been tested on a large number of hosts susceptible to the virus. Except for squash, the only cucurbitaceous host which is known to stimulate the replication of certain satellite RNA variants to a level similar to solanaceous plants, the protective inoculum has remained symptomless or has caused a clear attenuation of the gravity of the disease (84,94,95).

Finally, possible interferences between the protective inoculum and other viruses that can naturally infect tomato plants have also been studied. No synergistic effects have been observed either with TMV or PVY (66,84,94, 95), or with PVX (66,94,95). Moreover, an antagonistic effect was described in the case of potato spindle tuber viroid (PSTV) (66).

TRANSGENIC PLANTS EXPRESSING A SATELLITE RNA GENE

Inoculation Tests in the Laboratory

The first transgenic plants expressing a CMV satellite RNA gene were obtained in 1986 (3). Two genes bearing greater than unit-length satellite cDNA (1.3 or 2.3 copies) were independently introduced into the tobacco genome. Following infection with a satellite-RNA-free strain, unit-length satellite RNA was amplified to a high level from the endogenous precursor transcript, and was encapsidated in viral particles (3). Moreover, these plants developed the expected attenuated symptoms (26). Replication of CMV genomic RNAs and yield of the virus were greatly reduced in these plants. A particularly interesting finding was that when the plants were infected with tomato aspermy virus (TAV) a related cucumovirus that is able to support CMV satellite RNA replication, amplification of the satellite RNA and attenuation of symptoms were also observed, but without decrease in either the virus titer or genomic RNA replication.

Similar tolerance to CMV was also observed in tobacco plants expressing a monomeric copy of satellite RNA (33). Moreover, this work showed that tolerance was expressed whatever the means of infection of the transgenic

plants (mechanical inoculation or aphid transmission). More recently, we have shown that a similar level of protection can be conferred by genes coding for either a (+) or a (-) satellite RNA, since plants expressing either type of gene developed attenuated symptoms. Nevertheless, amplification of the satellite RNA in plants transformed with the (-) sense genes seemed to be delayed, resulting in lower accumulation of satellite RNA and a lesser decrease in amplification of genomic RNAs (97). If one wishes to use satellite RNA genes in solanaceous plants, which normally amplify satellite RNA to high levels, it might be worth considering using a (-) sense gene as a means of reducing the satellite RNA titer in infected plants, which would thus be less good sources for satellite RNA transmission.

Having confirmed the effectiveness of a satellite RNA gene in tobacco plants (60,94), the Chinese and Japanese groups have also successfully transformed tomato with a non-necrogenic satellite RNA gene (81,94). Plants expressing the gene developed faint symptoms of the disease following infection with a satellite-RNA-free strain. Moreover, timing of fruit set and fruit yield per plant were the same as in healthy control plants (81).

More recently, the S satellite RNA, which has been used for cross-protection experiments in the US and Italy, has also been successfully transferred into the tomato genome (64). Following infection with the helper genome, the plants did not develop the chlorosis and shoestring leaf symptoms typical of the helper strain, thus confirming the high level of tolerance conferred by the satellite RNA gene.

A higher tolerance has been conferred to transgenic tobacco plants by expressing simultaneously a satellite RNA gene and a gene coding for the viral capsid protein (CP) (108). Two expression cassettes, consisting of one or the other viral gene under the control of the CaMV 35S promoter and polyadenylation signals, were transferred to the tobacco genome. Both genes were expressed in transformants, and the CP level in plants expressing the two genes was similar to that in plants expressing the CP gene alone. In this particular case, the individual genes conferred only partial protection, but when both genes were present, the resistance was about twice that conferred by each viral gene independently. Such results clearly demonstrate that combining resistance genes that are thought to act at different steps of the viral infection greatly improves the

Table 4. Symptoms developed by tomato plants expressing a necrogenic satellite RNA and infected by the satellite-RNA-free strain I17F with or without R non-necrogenic satellite RNA added

Inoculum (μg/ml) and plants[a]	Number of plants	Severe fernleaf	Attenuated fernleaf	Necrosis
I17F viral RNA (10)				
Control plants	19	19	0	0
Line RCN4-1	17	4	0	13
Line RCN4-10	19	12	0	7
I17F RNA (10)+R satRNA (5×10^{-4})				
Control plants	20	1	19	0
Line RCN4-1	19	1	8	10
Line RCN4-10	19	1	14	4
I17F RNA (10)+R satRNA (1)				
Control plants	19	0	0	19
Line RCN4-1	18	0	6	12
Line RCN4-10	19	0	17	2

[a] Tomato (cv Red Cherry) were infected when 15 days old. The transformed RCN4-1 plant has been described by Tousch et al. (96).

resistance level of the transformed plants (108).

Transgenic Plants as a Tool for Studying Disease Modulation by Satellite RNA

Apart for their potential for use in plant protection, transgenic plants can also be considered as a tool in fundamental studies, for instance those aimed at determining the mechanisms of symptom modulation by satellite RNAs. It has been shown that tobacco plants expressing either a monomer or a dimer of a satellite RNA responsible for bright yellowing did not display any yellowing symptoms before being infected with the helper virus (60). In a similar way, transgenic tomato plants expressing a necrogenic satellite RNA were morphologically indistinguishable from non-transformed plants before inoculation (65,96). As expected, however, they died from necrosis once infected with the helper virus. To conclude from these data that satellite RNA cannot by itself induce these symptoms would be premature. The low amount of RNA transcript synthesized by the plant, and the presence of non-viral sequences flanking the satellite RNA could also account for its ineffectiveness.

What would happen if transgenic tomato plants expressing a satellite RNA of one type were infected with an inoculum containing a satellite RNA of another type? This question is particularly pertinent in the case of tomatoes expressing a non-necrogenic satellite RNA. Unfortunately, no data are available from the groups possessing such plants. However, we have studied the inverse combination. The F1 progeny from two tomato lines expressing a necrogenic satellite RNA gene at different levels, one of which was described in Tousch et al. (96), were checked for the presence of the transcript by dot blot hybridization. In both cases, the transcript was detected in 75% of the plants, corresponding to one locus of insertion of the gene. They were infected with I17F viral RNAs with or without R satellite RNA. Results presented in Table 4 show that the level of transcript synthesized in one line was high enough to ensure amplification of the necrogenic satellite RNA in all the transformed plants (76 % developed necrosis). For the other line, only 37 % of the plants developed necrosis, which corresponds to half of the transformed plants, suggesting that the concentration of the transcript was too low to ensure optimal amplification of satellite RNA.

When the non-necrogenic R satellite RNA was included in the inoculum, it could compete for replication with the endogenous necrogenic RNA transcript, leading to a situation similar to what was obtained by infecting tomato plants with an inoculum containing the two satellite RNAs (36). Thus, depending on the expression level of the necrogenic gene and on the concentration of the non-necrogenic satellite RNA in the inoculum, the tomato plants developed symptoms characteristic of either one or the other of the two satellite RNAs. Nevertheless, even at a high concentration in the inoculum, the non-necrogenic satellite RNA could not completely overcome the necrogenic one synthesized by the plant. From these results, and knowing the dominance of expression of a necrogenic satellite RNA in mixed infections, one can express doubts about the effectiveness of resistance towards a necrogenic satellite RNA of tomato plants expressing a non-necrogenic one.

Transgenic Plants in the Open Field

There are two reports, both from China, dealing with release of plants expressing a satellite RNA gene in the open field (94,107). According to the first assay carried out in 1990 , either tobacco or tomato plants expressing the satellite RNA gene were less affected by CMV than non-transformed controls following mechanical inoculation of the virus, although the decrease of the gravity of the symptoms was less pronounced in tobacco. As a consequence, leaf yield from transformed and control tobacco plants were not significantly different, whereas fruit yield was about 50% higher in transgenic tomato plants than in control ones. Nevertheless, the beneficial effects of preinoculation with a mild CMV strain, such as earlier flowering and maturation, or induction of resistance towards fungi, was not observed in transgenic plants (94). In the second study, larger areas planted with transgenic tomato plants were observed under natural infection conditions. The tolerance was less pronounced, and as a consequence, increase of fruit yield in transformants was much lower (7 to 11%) (107). In the same paper, Yie and Tien also reported that more encouraging results have been obtained during a preliminary field test of tomatoes expressing both satellite RNA and capsid genes.

POTENTIAL ECOLOGICAL RISKS ASSOCIATED WITH THE USE OF SATELLITE RNA IN THE FIELD

Risk associated with satellite-RNA-mediated protection, either by cross-protection or by use of transgenic plants, has two components:

i) Genetic drift: Since viral replicases have a high rate of misincorporation and lack proof-reading activity, viral RNAs generally have a high mutation rate, and thus occur as heterogeneous populations, within which different sub-classes will be more or less predominant, depending on the physical and biological environment (15). Considering the extremely high level of amplification of CMV satellite RNAs in certain infected plants, one could predict that mutation-driven variation within satellite RNA populations would be particularly rapid. Since the sequence differences between satellite RNAs that are either ameliorative or necrogenic on tomato can be extremely slight (see below), evaluation of the frequency of mutation in satellite RNAs, determination of the conditions favorable to the selection of different variants and, finally, better knowledge of satellite RNA evolution in natural isolates are crucial.

ii) Natural dissemination of satellite RNA by aphids: Pre-inoculation of plants with an inoculum containing satellite RNA, or natural infection of transgenic plants, both lead to the presence of large amounts of satellite RNA, which makes these plants potential sources of CMV and satellite RNA for aphid-mediated transmission. Although satellite RNA is not always transmitted, in part because its presence can reduce the efficiency of a source plant, there is a clear potential for transmission of newly arisen variants to other hosts, where they could have deleterious effects.

Sequence Data

The sequence of at least 42 satellite RNA variants has been reported by several different groups (Table 5). The origins of these variants and the sequencing results obtained require a few general comments:

- None of these variants were directly sequenced from viral RNAs purified from naturally infected plants. However, in three cases (PG, Tfn, TTS) the satellite RNA had been observed in the natural isolate, and sequenced after a single passage through the original host. These sequences

Table 5. Sizes and biological properties on tomato plants of satellite RNA variants

Name	Length (nt)	Biological properties on tomato	Reference
D	335	necrogenic	46, 51, 53, 78
1	334	ameliorative	7
Q	336	ameliorative	24
Y	368	necrogenic	30
	369	necrogenic	11, 61 (revision)
	369	necrogenic/ameliorative[a]	106
S	339	ameliorative	2
Yn	334	necrogenic	42
G	334	ameliorative	20
B1	337	chlorosis	20, 53
B2	340	ameliorative	20
B3	337	ameliorative/chlorosis[b]	20
WL1	336	ameliorative	20, 53
WL2	342	chlorosis/whitening[b]	20
WL3	340	ameliorative	86
E	339	ameliorative	29
OY2	386	ameliorative	29
I17N	335	necrogenic	34, 37
R	334	ameliorative	34, 97 (revision)
Sq10	335	necrogenic	46
Ch20	335	necrogenic	46
X2nT3	334	necrogenic	46
X7	334	necrogenic	46
X12	336	necrogenic	46
X15	336	necrogenic	46
X2c	337	non-necrogenic	46
Ra	338	ameliorative	11
S19	370	ameliorative	55 (revision), 63
T73	335	ameliorative	63
K8	339	ameliorative	17
F	338	ameliorative	28
PG	334	necrogenic	43
T	334	necrogenic/aggravated stunting[b]	45
T43	384	necrogenic	59
T62	384	necrogenic	59
iX	334	necrogenic	68
As	382	?	4
28-19	388	?	4
O(n2)	382	?	4
Tfn	390	non-necrogenic	9
TTS	339	top stunting	25
J876	387	necrogenic	98
D27	391	non-necrogenic	98
Ko2	368	ameliorative	84

[a] Depending on tomato varieties and environmental conditions.
[b] Depending on helper strains.

can thus be considered to represent natural isolates of satellite RNAs (9,25,43).

- One variant (Q) was absent from the parental CMV isolate, and then arose later, after successive amplifications (24). We have observed a similar situation for the D strain, which yielded the first sequenced satellite RNA (78). This seems also to have been the case for the In-CMV strain, which needed five and eight passages in squash and *Chenopodium* respectively, in order to accumulate a satellite RNA (46). Emergence of satellite RNA in strains studied for many years, and in some cases, before the discovery of the "fifth component", is not surprising, considering the stability and high biological activity of the molecule and the forced multiplication of the strains in greenhouses.

- A single CMV strain can accumulate different variants depending on the host and the number of passages. This feature has been exploited by Kaper *et al.* (46) and Palukaitis (73), with the aim of selecting variants possessing different biological properties. A third unintentional example is given by the R strain. This strain had been kept in our laboratory as dried infected leaves of either tobacco or cucumber plants. Whereas the former material yielded the ameliorative R satellite RNA following multiplication in tomato plants (34), the latter yielded the ameliorative Ra molecule (11). A similar situation is described for D.7 satellite RNA, but in this case there were also differences in the biological properties of the satellite RNA amplified (necrogenic versus non-necrogenic) (98).

- An intermediate situation is illustrated by the progressive selection of a new variant which can finally overcome the initially predominant molecule (39). The two populations were then clearly detected simultaneously, either because they had different electrophoretic properties, or different biological properties, or both (2,39,42,45,84,98).

- Heterogeneity of RNA populations has been clearly illustrated. It was first observed by Richards *et al.* (78) by direct sequencing of different preparations of the same satellite RNA. Later, sequencing different cDNA clones obtained from the same RNA preparation, or from the same strain in different laboratories, has shown numerous variations (11,34 37,46,51,-53,78,84,98,106). Nevertheless, when tested for, such variations have not usually been correlated with any modification of biological properties (34,51).

- Based upon their sizes, satellite RNAs can be divided into two groups: i) the small satellite RNAs ranging in length from 334 to 342 nucleotides, which includes the majority (30 of 42) of the strains described worldwide; ii) larger satellite RNAs, a group composed of several Japanese isolates approximately 370 nucleotides in length, and still larger (380 to 405 nucleotides) isolates from several countries (Japan, Italy, Korea) that were described more recently (4,9,59).

- Based on sequence homology, 25 satellite RNAs have been segregated into three phylogenetic groups, I, IIa and IIb, which are not correlated with geographical and host origins (16). All group I RNAs are non-necrogenic, while all but two group IIa and IIb RNAs (R and OY2) are necrogenic. The authors of this study also concluded that the secondary structure of the satellite RNAs is an important functional constraint on their evolution. The results obtained agree well with the secondary structure model proposed earlier for Q satellite RNA (24), and further extended to seven other molecules (20,30).

- The sequences of necrogenic satellite RNAs are highly conserved, those of non-necrogenic ones (i.e. all others, whatever the symptoms developed) are more variable (46). One exception is the French R satellite RNA, which, despite its ameliorative function, more closely resembles a necrogenic satellite RNA than a non-necrogenic one (34,97).

Genetic Determinants of Satellite-RNA-associated Symptomatology

Definition of a Necrogenic Domain Within the Satellite RNA

Primary Structure Analysis

In order to clarify the important question of how a satellite RNA can provoke the necrotic syndrome on tomato, recombination and mutagenesis of cDNAs, followed by analysis of the biological properties of the corresponding transcripts synthesized *in vitro*, have been carried out by different groups. Initial experiments have shown that the necrogenic domain is located towards the 3' end of the molecule (11,52). Alignment of the various sequences in this region delimited a region of approximately fifteen residues where the necrogenic sequences are highly conserved, whereas non-necrogenic ones differ at one to 9 positions. Previous work dealing with necrogenic/non-necrogenic satellite RNA

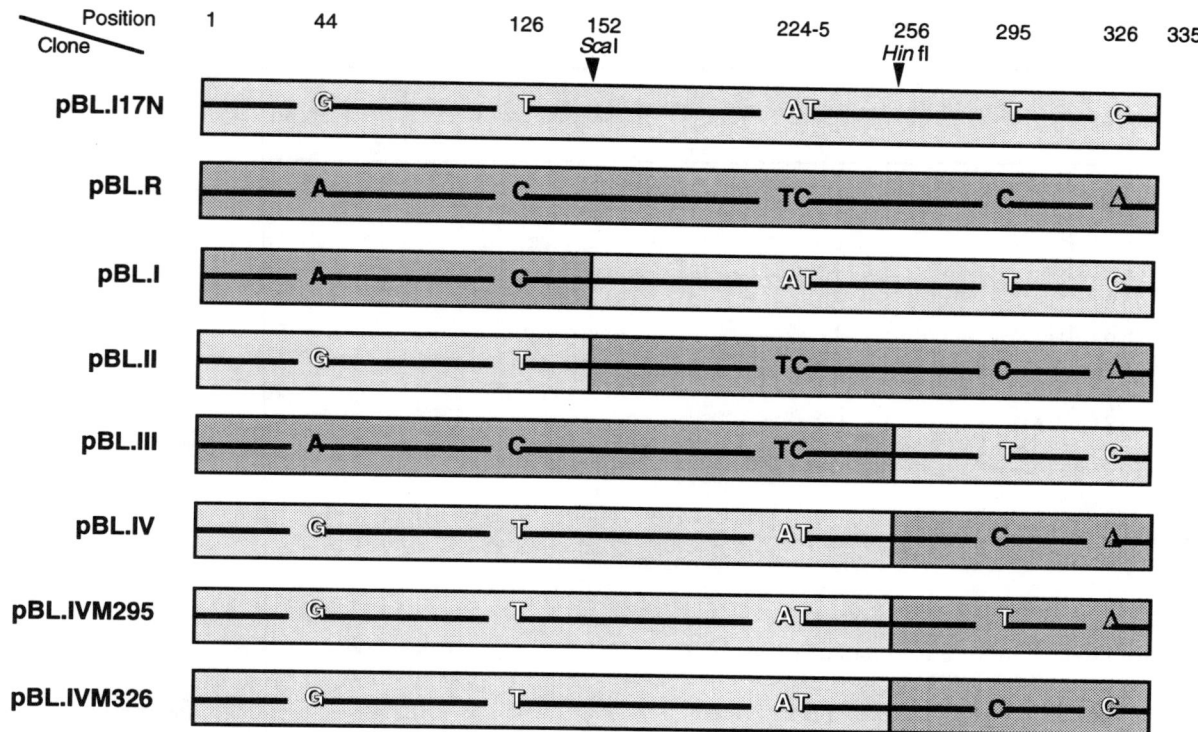

Figure 2. Description of recombinant and mutant cDNAs constructed from those corresponding to I17N and R satellite RNAs

pairs, belonging both to groups I and IIb, has made it possible to show that three positions (corresponding to nt 285, 290 and 292 of the satellite RNAs studied here) played a crucial role in the necrogenic response, in that a single change at one of these positions could render necrogenic an ameliorative RNA (12,63,87). Devic *et al.* (12) have also proposed that spacing between nucleotides corresponding to 289 and 290 can also play a role in determining necrogenicity. It has also been observed that all necrogenic satellite RNAs show a triplet deletion (between nts 290-291) compared to the majority of non-necrogenic ones. Nonetheless certain non-necrogenic satellite RNAs also have this triplet deletion (R, for instance), and also, Sleat and Palukaitis have found that introduction of the triplet deletion and the adjacent nucleotides into a non-necrogenic satellite RNA, reproducing the sequence of a necrogenic sequence, was not sufficient to render it necrogenic (87).

We have studied the necrogenic determinants in a pair of satellite RNAs of a different phylogenetic group (IIa) from those used in the studies described above. The two satellite RNAs I17N and R (necrogenic and non-

necrogenic, respectively) are identical at all the positions shown to determine necrogenicity in previous studies. Comparison of their sequences suggested that another position, close to the three previous ones, could be determinant for the necrogenicity of I17N satellite RNA. The following experiments were carried out to test this hypothesis. Figure 2 presents the two initial cDNA clones, pBL.I17N and pBL.R, as well as the successive recombinants and mutants we have constructed from them. These cDNAs vary at six positions: 5 substitutions at positions 44, 126, 224,225 and 295, and a one base deletion/insertion at position 326 (34,97). Recombination using the unique *Sca*I site, located at position 152 of the two initial cDNAs, yielded pBL.I and pBL.II, which possess the 3' end of I17N and R cDNAs, respectively. pBL.III and pBL.IV were obtained in a similar manner by using the unique *Hin*fI site at position 256. pBL.III and pBL.IV differ from pBL.R and pBL.I17N by only the last two variations at the 3' end. pBL.IV, which yielded a non-necrogenic transcript, was further used in mutagenesis experiments in order to try to restore necrogenicity. This yielded pBL.IVM295 and pBL.IVM326, which were identical to pBL.I17N

Table 6. Biological properties of RNA transcripts produced from pBL.I17N, pBL.R and from recombinants and mutants derived from them[a]

Clone	N° of infected plants	Biological property	% Expression[b]
PBL.I17N	80	necrogenic	47
pBL.R	80	ameliorative	48
pBL.I	90	necrogenic	45
pBL.II	60	ameliorative	41
pBL.III	120	necrogenic	56
pBL.IV	80	ameliorative	50
pBL.IVM295	80	necrogenic	49
pBL.IVM326	90	ameliorative	51

[a] The data presented are the sum of the results obtained during 6 different assays (10 or 20 plants were inoculated by clone and by assay).
[b] Percent plants having amplified a satellite RNA following inoculation.

at positions 295 and 326 respectively. Mutant clones were verified by DNA sequencing.

The T7 promoter of pBluescript KS(+) recombinants linearized at the EcoRI site produced transcripts approximately 650 nt long. They bore the satellite RNA sequence plus the following flanking sequences: 49 nt from pBluescript and a poly(C) tail 15 to 20 nt long at the 5' end, and at the 3' end poly(A) poly(G) tails 40 to 50 nt long plus 210 nt arising principally from the pAT153 cloning vector. Their biological activity was considerably reduced compared to native satellite RNA, which is correlated with the presence of flanking sequences. Nevertheless, when they were included in a CMV inoculum at a concentration of 1 µg/ml, a satellite RNA was amplified in half of the plants inoculated. The results presented in Table 6 correspond to the sum of several independent assays. No significant difference of efficiency could be observed between the various constructions. In each experiment, the control plants inoculated with I17F alone did not develop necrosis, showing the absence of a necrogenic satellite RNA in the inoculum, and if super-infected with I17N did develop necrosis, showing that the I17F RNA preparation was not contaminated with a non-necrogenic satellite RNA. As expected, RNA transcripts from the four recombinant cDNAs had the same biological properties (necrogenic or non-necrogenic) as the parent RNA from which their 3' region was derived. This confirmed that sequences necessary for necrosis are located within the last 79 nt of the

molecule. Analysis of the two mutants showed that pBL.IVM295 produced a necrogenic transcript, whereas pBL.IVM326 yielded an ameliorative one. This shows that the one base deletion/insertion at position 326 is not involved in the necrotic response, whereas the substitution at position 295 (C in ameliorative satellite RNAs and U in necrogenic ones), was determinant for necrogenicity.

The 3' half of progeny satellite RNA populations was sequenced, using individual preparations from one plant infected with each of transcripts R, I, II and IV, as well as five preparations corresponding to transcripts IVM295 and IVM326. In all cases, the sequence of progeny satellite RNA populations corresponded to those expected from the cDNA at the positions where differences were expected. The sequences obtained were also examined for unexpected mutations at other sites. In a total of 14 sequences obtained, 4 point mutations were observed occurring in a single progeny RNA population. In most cases, results were best interpreted as representing a mixed RNA population, with both the novel and the expected nt present. These could be interpreted either as the result of drift in the satellite RNA population, or could be due to misincorporation during PCR amplification.

As shown in Figure 3, the non-necrogenic R satellite RNA carries the potential necrogenic determinant described by others. Nevertheless it differs at a nearby position, nt 295, shown in our work to confer the ameliorative function. This

```
Y Sat-RNA(a)        CUAAGGCUUAUGCUAUGCUGAUCUCCGUGAAUGUCUAUACAUUCCUCUACAGGACCC       necrogenic
                                  ⇓                                                                   ⇓
                                  U                                                            ameliorative


Y Sat-RNA(b)        CUAAGGCUUAUGCUAUGCUGAUCUCCGUGAAUGUCUAUACAUUCCUCUACAGGACCC       necrogenic
                             ⇓     ⇓ ⇓                                                                 ⇓
                             A     G U                                                        ameliorative


W11 Sat-RNA(c)      CUUAGACUUAGGUUAUGCUGAUCUCCGUGAAUGUCUACACAUUCCUCUACAGGACCC       ameliorative
                             ⇓     ⇓ ⇓                                                                 ⇓
                             G     U C                                                         necrogenic


R Sat-RNA(d)        CUAAGGCUUAUGCUACGCUGAUCUCCGUGAAUGUCUA.UCAUUCCUC.ACAGGACCC       ameliorative
                                  ⇓                                                                   ⇓
                                  U                                                            necrogenic
```

Figure 3. Alignment of the 55 3'-terminal residues of the satellite RNA variants mutated from a necrogenic form towards a non-necrogenic one or *vice versa*. The arrows indicate the positions found to be determinant for necrogenicity.a, b and c correspond to references 63, 12 and 89, respectively. d: this work.

illustrates that the delimitation of a functional (e.g. necrogenic) domain by recombination and mutational analysis will be limited by the molecules taken into account. Also, beyond the small necrogenic domain defined in the region of nts 285-295, it is not unreasonable to expect that more distant interactions may play a role in necrogenicity. Secondary structure, and more generally the three dimensional structure of satellite RNAs, would be expected to play an important role in determining the physiological consequences of interaction with different hosts.

Roles of Secondary and Tertiary Structure

Convincing recent results do not support the idea that a putative polypeptide encoded by a satellite RNA could play any role in pathogenicity (6,37,63). Thus, as is the case of viroids, satellite RNAs probably act directly as RNA molecules. If so, various aspects of their spatial structure will be of utmost importance not only in the regulation of satellite RNA replication, but also in the determination of necrogenicity. That regions other than the necrogenic domain can modulate necrogenicity is illustrated by the results obtained with recombinants made from parts of Y satellite RNA. This RNA was first described as necrogenic (91), and used in mutagenesis experiments for the delimitation of its necrogenic domain by two groups (11,12,63). Then, in 1986 controversy about its necrogenicity arose (42). Additional

further data suggest that Y satellite RNA is indeed necrogenic, but expression of its necrogenicity can be dependent on the experimental conditions, as will be described below in further detail. The complexity of the situation is illustrated by the results obtained by the Japanese group itself. In 1988, Y satellite RNA was shown to be necrogenic when co-inoculated with the Y strain as helper but not with the O strain of the virus (62), but then in 1989 it was used with the O strain in the study of its necrogenic domain (63)!

Under conditions where Y satellite RNA does not induce lethal necrosis, recombinants between the 5' half of a necrogenic (D) or a non-necrogenic (S) molecule, and the 3' half of Y satellite RNA (including the necrogenic domain) appeared to provoke a partial necrosis from which the plants recovered. In contrast, the inverse combination (5' half of Y and necrogenic domain from D) proved to be unable to induce necrosis. These results suggested that more than the necrogenic domain was required for necrogenicity (40). Moreover, a single base mutation upstream to the necrogenic domain (nt 278) corresponding to the residue present in D satellite RNA, was further shown to enhance the necrogenicity of the previous DY recombinant (102). These authors proposed that loss of necrogenicity in Y and certain recombinant RNAs could be associated with the large insertion typical of Y in the 5' half of the molecule. This hypothesis was not supported by the study of another satellite RNA bearing a large insertion in the central part of the molecule and which was shown to be fully necrogenic (98). The authors proposed that the modified necrogenicity of Y satellite RNA would be due to differences in secondary structure. However, under their conditions, Y replicates to a much lower level than strongly necrogenic molecules (98). Since we have shown that when the archetype strongly necrogenic satellite RNA, I17N, is present in extremely low levels, this molecule can also induce partial, non-lethal necrosis (see above), the reduced replication of Y satellite could be a sufficient explanation of its lower necrogenicity in some cases. Thus we prefer to consider the Y satellite to be necrogenic, and that its' potential reduced necrogenicity is a secondary result due to environmental conditions, including host genotype.

The secondary structure of several satellite RNAs has been studied by RNase sensitivity studies. Examination of the numerous sequences available supports the model developed, since many complementary mutations in base-paired regions can be observed. Recently, Tousignant and Kaper (98) have used the FOLD/SQUIGGLES program of the GCG package to predict the secondary structure of several satellite RNAs. Unfortunately, this technique must be used with great precaution, since it has a rather low predictive value. In fact, this is clearly illustrated by the radically different structures they proposed for five satellite RNAs, none of which resembles the model based on experimental results. One has additional reason to doubt the validity of these structures, since one would expect that conservation of essential satellite RNA functions (replication, encapsidation) would require structural conservation, as indeed the experimental results suggest. However, Tousignant and Kaper suggest that a particular stem-loop structure would be present only in necrogenic satellite RNAs (98). When we used the same programs on I17N and R satellite RNAs, the stem-loop is predicted in both, which casts additional doubt on the validity of this proposed structure and on its potential role in necrogenicity.

Other Typical Syndromes Associated with a Satellite RNA

The molecular basis of two other typical syndromes associated with a satellite RNA have also been investigated: brilliant yellowing of tobacco induced by Y satellite RNA and chlorosis of either tobacco or tomato induced by certain B or WL variants (73,91). Domains responsible for these chlorotic symptoms are all located in the 5' half of the molecules (11,52,63). The tobacco yellowing domain of Y satellite RNA lies within the large insertion in this molecule. It was first proposed that a secondary structure between nts 125-170 could be correlated with tobacco yellowing (63). Recombinants and mutagenesis experiments have then localized the domain responsible for this syndrome to two residues downstream from that region. As is the case for tomato necrosis, these residues, although very close, differed depending on the satellite RNA pair considered (37,55).

Certain B and WL variants are able to induce chlorosis on either tobacco or tomato, but never on both (73). A single residue has been shown to determine the host specificity of this response, and a nearby one to determine the ability to induce chlorosis on tobacco (89). Interestingly, these residues lie within a 9 nt sequence segment that is similar to the tobacco yellowing domain previously determined for Y satellite RNA. There is at this time no evidence that yellowing and chlorosis result from the same mechanisms. Nevertheless, existence of a consensus sequence and characterization within it of different residues as determinant for symptom induction, strongly suggest that a functional residue may require a correct nucleotide context to be expressed.

Other Factors Involved in Pathogenicity

Accumulation of a satellite RNA and symptom induction result from a complex trilateral interaction between the helper genome, the satellite RNA and the host species. The viral genome plays a crucial role in both its ability to assist a satellite RNA, and determining the type of symptoms the infected plants will develop. Although the satellite-RNA-helper strain relationship is not highly specific, there is one CMV strain - CMV-Ix - which has proven to be unable to assist certain variants (44,45). The situation is more complicated when TAV is used as the helper, since three different situations can be observed: i) Some TAV strains cannot assist Y or Ix satellite RNAs (37,67). In the latter case, it has been shown that the satellite RNA was in fact synthesized in the inoculated leaves, but was not encapsidated, and did not spread from these leaves (67). ii) The satellite RNA is produced, an attenuation of the symptoms is observed, but the multiplication of the helper genome is not affected (3,67). iii) The symptoms characteristic of the satellite RNA are not observed, although it was efficiently replicated (37,67,88). Modulation of the symptoms by different helper strains is well documented for CMV and tomato as indicated in Table 5, and previously evoked for Y satellite RNA. Such specificity is more stringent for tobacco chlorosis, which occurs only when the helper strain belongs to group II (88).

The third partner, the host species, also determines symptom expression. The best examples are given by the bright yellowing which was observed on *N. tabacum*, *N. glutinosa* and *N. sylvestris*, but not on *N. clevelandii* (91), and by tomato lethal necrosis, which was expressed by 96 different tomato cultivars, but not by several accessions of different exotic *Lycopersicon* species, although satellite RNA was replicated in these plants, which was still able to induce necrosis on tomato test plants (100).

A fourth partner has been recently identified, in the case of the Y satellite RNA under conditions of lowered necrogenicity. New experiments, examining the role of temperature and other environmental factors have been stimulated by the controversy over the necrogenicity of this molecule. Previous work had already shown that depending on the tomato cultivar and the helper strain, necrogenicity of this RNA could be strongly reduced (62). More recently, it was shown that temperature could also modify the tomato response (40). Finally, when a similar assay was carried out in different laboratories, variations in the development and the gravity of the necrogenic disease suggested that other environmental factors, such as light or soil nutrients, could also be involved (106).

Satellite RNA Evolution

Selection of Satellite RNA Variants

Selection pressure on satellite RNA variants exerted by the host has already been mentioned (see sequence data). This is generally interpreted as favoring the emergence, within a population of different molecules, of ones with the greatest fitness. The effects of host selection pressure can be surprisingly subtle, since a variant selected on one host (tomato) has been shown to differ from the native Ix satellite RNA amplified on either tobacco or squash by only one residue, a U/C substitution (68). This variation was specific for tomato, since a new passage through squash led to the recovery of the original nucleotide. Since the parent satellite RNA was obtained by

purification from total viral RNA, either a point mutation or, more probably, presence of the two types of molecules at different levels in the population could explain this selection. This apparently accidental selection was observed in a single experiment out of three, and no modification of the biological properties of the new variant was observed that could explain its appearance. As a matter of fact, the U/C substitution at the same position has already been described in a cDNA clone of the necrogenic D satellite RNA, although its appearance was not correlated with amplification within different hosts (51). These results are a further illustration of the molecular plasticity of the satellite RNA, and are reason for concern, since we know that certain single nucleotide changes can drastically modify symptom expression. Nevertheless, though mutational drift will occur, the selection pressure is not the same for all nucleotides, since maintenance of essential functional domains will clearly constrain genetic drift.

Mutation of Satellite RNAs upon Successive Multiplications

A rigorous study of this type requires the use of a pure preparation of satellite RNA which can only be obtained by transcription *in vitro*. Kurath and Palukaitis described such a study in 1990 (54). RNA transcripts from two cDNA clones of D satellite RNA, corresponding to the major component of the initial RNA population (prototype) and to a minor one (variant), were serially passaged through different hosts (three *Nicotiana* species, tomato, squash and *Gomphrena globosa*). A highly variable region in D satellite RNA had already been reported around position 225 (46,53,78). Heterogeneity appeared only in that region and after only a few passages of the prototype RNA, whatever the host species, though heterogeneity was more acute in tobacco (54). No heterogeneity was observed in the progeny obtained from the variant RNA, even after 10 passages through tobacco, although as noted by the authors, the method used would not make it possible to detect micro heterogeneity. This suggested that the variant RNA remained as a minor component because it had no selective advantage, although it was perfectly functional. Moreover, it did not seem to respond to selection pressure in the same way as the prototype RNA (54).

Evolution of Satellite RNA in Natural Populations

Satellite RNA evolution in natural CMV isolates can be studied according to two main approaches with clearly distinct aims. The first approach consists of a systematic enumeration of the variants present in natural isolates, and of characterization of their principal structural and biological features. This labor intensive type of study is the only way to obtain an overall view of the evolution of the population. The second more limited approach consists in careful monitoring of the biological properties of CMV isolates regarding the presence of satellite RNA, particularly in areas known to be either satellite-RNA-free or not, and where satellite-RNA-mediated protection has been developed.

Using the first approach, the Spanish group has obtained striking results in a three years study of an epidemic of tomato necrosis in eastern Spain (1). During this period, satellite RNA became increasingly prevalent in CMV isolated from tomato. However, heterogeneity of the satellite RNA was already high during the first year, suggesting that the satellite RNAs were already highly diversified in other plant species previous to the first year of the epidemic of necrosis.

The second approach is also of considerable interest, because it more directly addresses questions pertinent to estimation of the real risks associated with the use of satellite-RNA-mediated protection. According to Yie and Tien (107), no necrogenic satellite RNA or other harmful variant has been detected during a large survey, despite ten years of cross-protection practiced in China. In Italy, a survey within weed species has also been carried out in an area where cross-protection had been applied the previous year and in two nearby fields. In that case also, no modification of the viral population was observed (8).

CONCLUDING REMARKS

Satellite RNAs of CMV have been extensively studied, particularly during the last few years. Although there were only three groups in three countries working on CMV satellite RNA at the end of the seventies, at least 10 groups in 7 countries are now carrying out such studies. Increasing interest for the model was naturally due to the modulation of symptoms this small RNA can provoke in tomato or tobacco. It was thus thought to be of particular interest, either for the study of the molecular basis of pathogenicity, or for developing new control strategies against CMV.

From the pathogenicity studies, it has become evident that the genetic determinants of pathogenicity were not as simple as was initially thought, particularly since the results obtained were dependent on the satellite RNA pair considered. Progressively, the idea that rather than primary sequence differences, it is the modifications of the spatial structure of the satellite RNA that is determinant, has been adopted by the various groups. In any case, it may be premature to define a particular secondary structure responsible for necrosis, since at this time there is in fact no evidence that the (+) strand satellite RNA is directly responsible for necrosis, rather than the (-) strand or a double-stranded form. In addition, the roles of the helper strain and the host have also been clearly demonstrated. So, from an initially simple model - small molecule, thus rather easy to clone and sequence, with an unusual biological activity, making it possible to initiate, without major difficulties, the replication from an RNA transcript, inducing very typical symptoms on some hosts - current thinking has evolved towards a very complex multilateral relationship.

Constitutive expression of a satellite RNA gene in either transformed tobacco or tomato plants has been successfully obtained by several groups. In each case, the plants have developed an interesting tolerance behavior towards the virus. Although European groups have not yet tested these plants in open field, this has been done in China. It is evident that the absence of legislation concerning the release of genetically modified organisms in China has favored their testing in natural growing conditions. Although we have been among the first to describe the two satellite RNA-mediated protection strategies, we have nevertheless never proposed their use in the open field, for several different reasons.

Cross-protection seems to us a clearly dangerous method under the conditions where it is currently applied. The protective inocula still consist of natural populations of either genomic RNAs or satellite RNA. Moreover, the helper strain is an aphid-transmissible one. All these elements lead to evident dangers at different levels:

- Selection of one satellite RNA variant from a natural population, even if present as a minor component, has been demonstrated in many cases under laboratory conditions. It has to be noted that all the studies describing the selection of a variant within a population have been carried out by mechanical inoculation. Is the selection pressure the same when the virus is aphid-transmitted?

- A satellite-RNA-free strain can be amplified in plants infected by viruliferous aphids having acquired the virus on a host infected with a satellite-RNA-containing one. Thus, a more aggressive strain can arise from an attenuated one because of the loss of the satellite RNA. Finally, protection of one crop can lead to a severe outbreak in a nearby one.

- Either selection of a variant, or genetic drift within a population, can potentially have dramatic effects for a particular host. Such effects are not predictable, but description of variants with new biological properties in Italy and Spain (tomato curl-stunt) is a serious danger signal.

- Once a satellite RNA is present in an area, a large number of variants can be detected in naturally infected plants. Description in Spain of 62 variants from 42 natural CMV isolates from a total of 57 examined is somewhat preoccupying, since it clearly shows a high level of genetic divergence of the molecule.

In certain cases, it can also be argued that intentional use of a satellite RNA would be less deleterious than a natural epidemic of necrosis. The use of cross-protection or transgenic plants, in areas where tomato crops have been severely affected for several successive years (Italy or Japan), can then be defended. This idea is supported by the fact that cross-protection has apparently not modified the viral population in the area concerned.

The major risk concern is centered on the potential of a beneficial satellite RNA to acquire deleterious functions by genetic drift or by recombination. Considering that new deleterious disease syndromes are still being associated with satellite RNAs, it is difficult to envisage disarming the molecule for all possible negative effects. Nonetheless, transgenic plants have a clear advantage over cross-protection, since in the former case the satellite RNA is only synthesized if the plant is infected, and in that case from an RNA of defined sequence. Since in at least one study in the laboratory it was shown that a particular satellite RNA variant was stable through several passages, it is possible that, even in the field, genetic drift of a particular well-adapted molecule would be minimal.

The second factor of risk is of course that a potentially deleterious molecule could be transmitted to new host plants. It has often been proposed to design a satellite RNA that would not be encapsidated, and would thus be non-transmissible. The inability of certain natural cucumovirus strains to encapsidate a particular satellite RNA is evidence that this might be possible. Another, though less effective way to potentially reduce transmission, would be to reduce the level of expression of the satellite RNA gene to the strict minimum required for protection, for instance by using a (-) strand gene.

One can propose two primary goals for future study, if the considerable potential benefits of satellite RNAs are to be used in transgenic plants on a large scale. It is clear that we need to know more about how satellite RNAs function, and in particular about how they can cause deleterious effects on certain hosts. In any case, the truly key problem is to develop non-transmissible satellite RNAs, since only in this case will we have any assurance of preventing transmission of satellite RNAs to new hosts, with perhaps at this time unknown deleterious effects.

APPENDIX

Virus Strains

The satellite-RNA-free I17F, and the necrogenic satellite-RNA-containing I17N strains have been obtained during the same cloning, by serial passages of a single local lesion through *Vigna sinensis*, of the same isolate harvested on a naturally infected necrotic tomato plant in 1975. The R strain, which bears the non-necrogenic R satellite RNA, has been isolated from *Ranunculus asiaticus* (36). These strains are maintained on tomato plants.

Virus and Viral RNA Purification

Virus was purified according to Lot *et al.* (56) approximately two weeks after infection. Viral RNA was extracted from the virions by the phenol-sodium dodecyl sulfate method and analyzed on either 2.4% or 5% polyacrylamide gels. Before loading, RNAs were partially denatured by heating at 65°C for 3 min and quick cooling. This treatment interrupts hydrogen bonds between molecules, and enhances the proportion of small molecules (subgenomic and satellite RNA) relative to larger ones (genomic RNAs). Satellite RNA was purified by sedimentation of viral RNA through a sucrose density gradient.

Cross-protection Experiments Under Greenhouse Conditions

Plants and Experimental Design

Tomato plants of cv. Melody CH 31 were used. This cultivar is resistant to *Verticillium*, *Fusarium* and TMV. Seedlings were transplanted in March, when 15 days old, and then infected. The experimental plants were divided into 4 blocks according to 2 gradients: a temperature gradient due to the cooling system of the greenhouse; a gradient corresponding to the position of the assay within the greenhouse. Each block consisted of 40 plants in 4 rows. Three first inoculum were used: viral RNA from the I17F strain (20 µg/ml) with or without R satellite RNA (0.05 µg/ml) and viral RNA from the R strain (40 µg/ml). Half of the infected and control plants were super-infected with viral RNA from the I17N strain (40 µg/ml) 2 weeks later; all the infected plants developed the expected symptoms at that time. This has led to 8 different treatments which were applied once in each block.

Additional nutrients were supplied weekly. Insecticide treatments were also applied weekly. Moreover, plants were treated against *Botrytis cinerea* from the middle of July since an epidemic started then. Fruits were collected twice weekly from the end of June until the end of the experiment at the beginning of August.

Presence of CMV and Satellite RNAs

50 days after the challenge infection, the presence of both CMV and satellite RNA was evaluated by biological test. Tissues were ground in 0.03M disodium phosphate containing 0.2% DIECA, and used as inoculum for a tobacco, a melon and two tomato plants. If necessary, one of the two tomatoes was challenge-infected 2 weeks later with I17N strain, in order to detect the non-necrogenic R satellite RNA. Among the control plants, which were 21 days old when infected with I17N strain, only 5 have developed necrotic symptoms; the virus was not detected in young leaves of the 15 others. Results concerning these plants are not presented, because they are not significantly different from those obtained for healthy plants.

Recombination and Mutagenesis Experiments

Satellite RNA cDNA Clones

All usual molecular biology techniques, such as restriction endonuclease digestions, ligations, transformation of *Escherichia coli*, were as described by Sambrook *et al* (83). The cDNAs synthesized from the necrogenic I17N and the non-necrogenic R satellite RNAs have already been described (34). The *Bam*HI inserts of the pAT153 recombinant plasmids, each bearing the cDNA sequence plus 188 bp from the vector, were cloned into pBluescript KS(+) in such a way that (+) RNA transcripts will be produced using the T7 promotor of the vector. Recombinant cDNAs shown in Figure 2 were constructed by exchange of parts of the two initial cDNAs at the unique *Sca*I or *Hin*fI sites. Mutagenesis of pBL.IV, at positions 295 or 326, was carried out using the site-directed mutagenesis kit from Boehringer Mannheim under conditions specified by the manufacturer. Briefly, the *Bam*HI insert was cloned into a M13mp9 phage carrying amber mutations in essential genes. The resulting single stranded DNA, produced in an amber suppressor host, was hybridized with the RF DNA of a M13mp9 plasmid without these mutations. A 20-mer oligonucleotide complementary to the cDNA, except for a single base mismatch, was hybridized to the previous form. After filling-in and ligation, the phage carrying the mutated DNA could be selected by growing in a suppressor-negative strain. Success of mutagenesis has been verified by DNA sequencing of the *Sca*I-*Bam*HI fragment bearing the 3' end of the (+) strand.

In Vitro Transcription

Recombinant plasmids purified on a CsCl gradient were linearized at the *Eco*RI site. RNA transcription was carried out in a 50 µl reaction volume containing 40 mM Tris, pH 8.0, 8 mM magnesium chloride, 25 mM sodium chloride, 10 mM DTT, 0.4 mM of each rNTP, 1 µg of linearized DNA, 20 units of T7 RNA polymerase, for 1h at 37°C. This yielded at least 1 µg of RNA transcript. Concentrations were estimated on 1.2% agarose gels using purified satellite RNA as standards. RNA transcripts were used without further treatment.

Bioassay of the Transcripts

The satellite-RNA-free I17F strain was used as helper. Inocula contained 10 µg/ml of viral RNA plus 1 µg/ml of transcript. In each experiment 10 or 20 young tomato plants (cv. Monalbo) were inoculated at the first leaf stage. Plants infected with a non-necrogenic inoculum, and half the control plants inoculated with the helper RNA alone, were super-infected 2 weeks later with viral RNA from the I17N strain (which carries the necrogenic I17N satellite RNA) (50 µg/ml).

Sequencing Progeny Satellite RNA

Satellite cDNA was synthesized using the following protocol: 5 µg of purified total viral RNA and 150 picomoles of a 9-mer oligonucleotide complementary to the 3' end of all our satellite RNAs (5'GGGTCCTGT3') were heated 3 min at 80°C and quick cooled. cDNA was synthesized for 1 h at 37°C in a 50 µl reaction volume containing 40 units of RNasin, 50 mM Tris, pH 8.3, 75 mM KCl, 3 mM MgCl$_2$, 10mM DTT, 1 mM of each dNTP, 500 units of Mu-MLV reverse transcriptase. Nucleic acids were extracted once with phenol, once with phenol/chloroform, then ethanol precipitated. The nucleic acid pellet was dissolved in water and treated with RNase A for 30 min at 37°C. Using one tenth of the cDNA synthesized, PCR amplification was carried out with the following oligonucleotides: 5'TCAGTACTACACTCTCA3', corresponding to residues 148 to 164 of the satellite RNAs and including the ScaI site (Figure 2), and 5'GTAAGTATGGGTCCTGT3' corresponding to the oligonucleotide used for cDNA synthesis plus 8 nucleotides added to enhance amplification. PCR conditions were: 2 cycles of 1 min at 94°C, 1 min at 40°C, 1 min at 72°C, and 33 cycles of 1 min at 94°C, 1 min at 45°C, 1.5 min at 72°C. The amplified fragment was purified following electrophoresis on a 5% polyacrylamide gel. Sequencing of the DNA was performed using the TaqDyeDeoxy™ Terminator cycle sequencing kit from Applied Biosystems, under the following conditions: 95°C for 5 min and 24 cycles of 1 min at 95°C, 15 sec at 45°C, 3 min at 60°C. The reaction products were extracted with phenol/chloroform once and ethanol precipitated, then analyzed on an automatic sequencer from the same manufacturer.

REFERENCES

1. Aranda, M.A., Fraile, A., and Garcia-Arenal, F. 1993. Genetic variability and evolution of the satellite RNA of cucumber mosaic virus during natural epidemics. J. of Virol. 67: 5896-5901.

2. Avila-Rincon, M.J., Collmer, C.W., and Kaper, J.M. 1986. In vitro translation of cucumoviral satellites. I. Purification and nucleotide sequence of cucumber mosaic virus-associated RNA 5 from cucumber mosaic virus strain S. Virology 152: 446-454.

3. Baulcombe, D.C., Saunders, G.R., Bevan, M.W., Mayo, M.A., and Harrison, B.D. 1986. Expression of biologically active viral satellite RNA from the nuclear genome of transformed plants. Nature 321: 446-449.

4. Choi, J.K., Sano, T., Uyeda, I., and Shikata, E. 1992. Complementary DNA cloning and nucleotide sequence of three satellite RNAs associated with cucumber mosaic virus. Korean J. Plant Pathol. 6: 491-496.

5. Collmer, C.W., and Howell, S.H. 1992. Role of satellite RNA in the expression of symptoms caused by plant viruses. Annu. Rev. Phytopathol. 30: 419-442.

6. Collmer, C.W., and Kaper, J.M. 1988. Site-directed mutagenesis of potential protein-coding regions in expressible cloned cDNAs of cucumber mosaic viral satellites. Virology 163: 293-298.

7. Collmer, C.W., Tousignant, M.E., and Kaper, J.M. 1983. Cucumber mosaic virus-associated RNA 5. X. The complete nucleotide sequence of a CARNA 5 incapable of inducing tomato necrosis. Virology 127: 230-234.

8. Crescenzi, A., Barbarossa, L., Gallitelli, D., and Martelli, G.P. 1993. Cucumber mosaic cucumovirus populations in Italy under natural epidemic conditions and after a satellite-mediated protection test. Plant Dis. 77: 28-33.

9. Crescenzi, A., Grieco, F., and Gallitelli, D. 1992. Nucleotide sequence of a satellite RNA of a strain of cucumber mosaic virus associated with a tomato fruit necrosis. Nucleic Acids Res. 20: 2886.

10. Devergne, J.C., and Cardin, L. 1973. Contribution à l'étude du virus de la mosaïque du concombre (CMV). IV. Essai de classification de plusieurs isolats sur la base de leur structure antigénique. Ann. Phytopathol. 5: 409-430.

11. Devic, M., Jaegle, M., and Baulcombe, D. 1989. Symptom production on tobacco and tomato is determined by two distinct domains of the satellite RNA of cucumber mosaic virus (strain Y). J. Gen. Virol. 70: 2765-2774.

12. Devic, M., Jaegle, M., and Baulcombe, D. 1990. Cucumber mosaic virus satellite RNA (strain Y): Analysis of sequences which affect systemic necrosis on tomato. J. Gen. Virol. 71: 1443-1449.

13. Dodds, J.A. 1982. Cross-protection and interference between electrophoretically distinct strains of cucumber mosaic virus in tomato. Virology 118: 235-240.

14. Dodds, J.A., Lee, S.Q., and Tiffany, M. 1985. Cross protection between strains of cucumber mosaic virus: Effect of host and type of inoculum on accumulation of virions and double-stranded RNA of the challenge strain. Virology 144: 301-309.

15. Domingo, E., and Holland, J.J. 1988. High error rates, population equilibrium, and evolution of RNA

replication systems. Pages 3-36 in: RNA Genetics, Volume III. E. Domingo, J.J. Hol and and P. Alquhist, eds. CRC Press, Inc, Florida USA.

16. Fraile, A., and Garcia-Arenal, F. 1991. Secondary structure as a constraint on the evolution of a plant viral satellite RNA. J. Mol. Biol. 221: 1065-1069.

17. Fraile, A., Moriones, E., and Garcia-Arenal, F. 1990. Characterization of a satellite RNA associated with strain K8 of cucumber mosaic virus. Nucleic Acids Res. 18: 4593.

18. Gallitelli, D., Di Franco, A., Vovlas, C., and Kaper, J.M. 1988. Infezioni miste del virus de mosaico del cetriolo (CMV) e di potyvirus in col ore ortive di Puglia e Basilitica. Inf. Fitopatol. 38: 57-64.

19. Gallitelli, D., Vovlas, C., Martelli, G., Montasser, M.S., Tousignant, M.E., and Kaper J.M. 1991. Satellite-mediated protection of tomato against cucumber mosaic virus: II. Field test under natural epidemic conditions in southern Italy. Plant Dis. 75: 93-95.

20. Garcia-Arenal, F., Zaitlin, M., and Palukaitis, P. 1987. Nucleotide sequence analysis of six satellite RNAs of cucumber mosaic virus: Primary sequence and secondary structure alterations do not correlate with differences in pathogenicity. Virology 158: 339-347.

21. Garcia-Luque, I., Diaz-Ruiz, J.R., Rubio-Huertos, M., and Kaper, J.M. 1983. Cucumovirus survey in Spanish economically important crops. Pyhtopath. Medit. 22: 127-132.

22. Garcia-Luque, I., Kaper, J.M., Diaz-Ruiz, J.R., and Rubio-Huertos, M. 1984. Emergence and characterization of satellite RNAs associated with Spanish cucumber mosaic virus isolates. J. Gen. Virol. 65: 539-547.

23. Gonsalves, D., Provvidenti, R., and Edwards, M.C. 1982. Tomato white leaf: The relation with an apparent satellite RNA and cucumber mosaic virus. Phytopathology 72: 1533-1538.

24. Gordon, K.H.J., and Symons, R.H. 1983. Satellite RNA of cucumber mosaic virus forms a secondary structure with partial 3'-terminal homology to genomal RNAs. Nucleic Acids Res. c: 947-960.

25. Grieco, F., Cillo, F., Barbarossa, L., and Gallitelli, D. 1992. Nucleotide sequence of a cucumber mosaic virus satellite RNA associated with a tomato top stunting. Nucleic Acids Res. 24: 6733.

26. Harrison, B.D., Mayo, M.A., and Baulcombe, D.C. 1987. Virus resistance in transgenic plants that express cucumber mosaic virus satellite RNA. Nature 328: 799-802.

27. Hayes, R.J., Tousch, D., Jacquemond, M., Pereira, V.C., Buck, K.W., and Tepfer, M. 1992. Complete replication of a satellite RNA in vitre by a purified RNA dependent-RNA polymerase. J.Gen. Virol. 73: 1597-1600.

28. Hidaka, S., Hanada, K., and Ishikawa, K. 1990. In vitro messenger properties of a satellite RNA of cucumber mosaic virus. J. Gen. Virol. 71: 439-442.

29. Hidaka, S., Hanada, K., Ishikawa, K., and Miura, K. 1988. Complete nucleotide sequence of two new satellite RNAs associated with cucumber mosaic virus. Virology 164: 326-333.

30. Hidaka, S., Ishikawa, K., Takanami Y., Kubo, S., and Miura, K. 1984. Complete nucleotide sequence of RNA 5 from cucumber mosaic virus (strain Y). FEBS Lett. 174: 38-42.

31. Jacquemond, M. 1982. L'ARN satellite du virus de la mosaïque du concombre. IV. Transmission expérimentale de la maladie nécrotique de la tomate par pucerons. Agronomie 2: 641-646.

32. Jacquemond, M. 1982. Phénomènes d'interférence entre les deux types d'ARN satellite du virus de la mosaïque du concombre. Protection des tomates vis à vis de la nécrose létale. C. R. Acad. Sci. Paris 294: 991-994.

33. Jacquemond, M., Amselem, J., and Tepfer, M. 1988. A gene coding for a monomeric form of cucumber mosaic virus satellite RNA confers tolerance to CMV. Molecular Plant-Microbe Interactions 1: 311-316.

34. Jacquemond, M., and Lauquin, G.J.M. 1988. The cDNA of cucumber mosaic virus-associated satellite RNA has in vivo biological properties. Biochem. Biophys. Res. Comm. 151: 388-395.

35. Jacquemond, M., and Leroux, J.P. 1982. L'ARN satellite du virus de la mosaïque du concombre. II. Etude de la relation virus-ARN satellite chez divers hôtes. Agronomie 2: 55-62.

36. Jacquemond, M., and Lot, H. 1981. L'ARN satellite du virus de la mosaïque du concombre. I. Comparaison de l'aptitude à induire la nécrose létale de la tomate d'ARN satellites isolés de plusieurs souches du virus. Agronomie 1: 927-932.

37. Jaegle, M., Devic, M., Longstaff, M., and Baulcombe, D. 1990. Cucumber mosaic virus satellite RNA (Y strain): Analysis of sequences which affect yellow mosaic symptoms on tobacco. J. Gen. Virol. 71: 1905-1912.

38. Jorda, C., Alfaro, A., Aranda, M.A., Moriones, E., and Garcia-Arenal, F. 1992. Epidemic of cucumber mosaic virus plus satellite RNA in tomatoes in eastern Spain. Plant Dis. 76: 363-366.

39. Kaper, J.M. 1983. Perspective on CARNA 5, cucumber mosaic virus-dependent replicating RNAs capable of modifying disease expression. Pl. Mol. Biol. Rep. 1: 49-54.

40. Kaper, J.M. 1992. Satellite-induced viral symptom modulation in plants: A case of nested parasitic nucleic acids competing for genetic expression. Res. Virol. 143: 5-10.

41. Kaper, J.M. 1993. Satellite-mediated symptom modulation: An emerging technology for the biological control of viral crop disease. Microb. Releases 2: 1-9.

42. Kaper, J.M., Duriat A.S., and Tousignant M.E. 1986. The 368-nucleotide satellite of cucumber mosaic virus strain Y from Japan does not cause lethal necrosis in tomato. J. Gen. Virol. 67: 2241-2246.

43. Kaper, J.M., Galitelli, D., and Tousignant, M.E. 1990. Identification of a 334-ribonucleotide viral satellite as principal ætiological agent in a tomato necrosis epidemic. Res. Virol. 141: 81-95.

44. Kaper, J.M., and Tousignant, M.E. 1977. Cucumber mosaic virus-associated RNA-5. I. Role of host plants and helper strains in determining amount associated with virions. Virology 80: 186-195.

45. Kaper, J.M., Tousignant, M.E., and Gelekta, L.M. 1990. Cucumber mosaic virus-associated RNA 5. XII. Symptom-modulating effect is codetermined by the helper virus satellite replication support function. Res. Virol. 141: 487-503.

46. Kaper, J.M., Tousignant, M.E., and Steen, M.T. 1988. Cucumber mosaic virus-associated RNA 5. XI. Comparison of 14 CARNA 5 variants relates ability to induce tomato necrosis to a conserved nucleotide sequence. Virology 163: 284-292.

47. Kaper, J.M., Tousignant, M.E., and Thompson, S.M. 1981. Cucumber mosaic virus-associated RNA-5. VIII. Identification and partial characterization of a CARNA-5 incapable of inducing tomato necrosis. Virology 114: 526-533.

48. Kaper, J.M., and Waterworth, H.E. 1977. Cucumber mosaic virus-associated RNA-5 causal agent for tomato necrosis. Science 196: 429-431.

49. Kassanis, B. 1962. Properties and behaviour of a virus depending for its multiplication on another. J. Gen. Microbiol. 27: 477-488.

50. Kearney, C.M., Zitter, T.A., and Gonsalves, D. 1990. A field survey for serogroups and the satellite RNA of cucumber mosaic virus. Phytopathology 80: 1238-1243.

51. Kurath, P., and Palukaitis, P. 1987. Biological activity of T7 transcripts of a prototype clone and a sequence variant clone of a satellite RNA of cucumber mosaic virus. Virology 159: 199-208.

52. Kurath, G., and Palukaitis, P. 1989. Satellite RNAs of cucumber mosaic virus: Recombinants constructed in vitro reveal independent functional domains for chlororis and necrosis in tomato. Molecular Plant-Microbe Interactions 2: 91-96.

53. Kurath, G., and Palukaitis, P. 1989. RNA sequence heterogeneity in natural populations of three satellite RNAs of cucumber mosaic virus. Virology 173: 231-240.

54. Kurath, G., and Palukaitis, P. 1990. Serial passage of infectious transcripts of a cucumber mosaic virus satellite RNA results in sequence heterogeneity. Virology 176: 8-15.

55. Kuwata, S., Masuta, C., and Takanami, Y. 1991. Reciprocal phenotype alterations between two satellite RNAs of cucumber mosaic virus. J. Gen. Virol. 72: 2385-2389.

56. Lot, H., Marrou, J., Quiot, J.B., and Esvan, C. 1972. Contribution à l'étude du virus de la mosaïque du concombre. I. Méthode de purification rapide du virus. Ann. Phytopathol. 4: 25-38.

57. Marrou, J., and Duteil, M. 1974. La nécrose de la tomate. I. Reproduction des symptômes de la maladie par inoculation mécanique de plusieurs souches du virus de la mosaïque du concombre. Ann. Phytopathol. 6: 155-172.

58. Marrou, J., Quiot, J.B., Marchoux, G., and Duteil, M. 1975. Caractérisation par la symptomatologie de quatorze souches du virus de la mosaïque du concombre et de deux autres cucumovirus: Tentative de classification. Med. Fac. Landbouww. Rijks Univ. Ghent. 40: 107-121.

59. Masuta, C., Hayashi, Y., Wang, W.Q., and Takanami, Y. 1990. Comparison of four satellite RNA isolates of cucumber mosaic virus. Ann. Phytopath. Soc. Japan 56: 207-212.

60. Masuta, C., Komari, T., and Takanami, Y. 1989. Expression of cucumber mosaic virus satellite RNA from cDNA copies in transgenic tobacco plants. Ann. Phytopath. Soc. Japan 55: 49-55.

61. Masuta, C., Kuwata, S., and Takanami, Y. 1988. Effects of extra 5' non-viral bases on the infectivity of transcripts from a cDNA clone of satellite RNA (strain Y) of cucumber mosaic virus. J. Biochem. 104: 841-846.

62. Masuta, C., Kuwata, S., and Takanami, Y. 1988. Disease modulation on several plants by cucumber mosaic virus satellite RNA (Y strain). Ann. Phytopathol. Soc. Japan 54: 332-336

63. Masuta, C., and Takanami, Y. 1989. Determination

64. McGarvey, P.B., and Kaper, J.M. 1993. Transgenic plants for conferring virus tolerance: Satellite approach. Pages 277-296 in: Transgenic Plants, Volume I. S.D. Kung and R. Wu, eds, Academic Press, Inc., New York, USA.

65. McGarvey, P.B., Kaper, J.M., Avila-Rincon, M.J., Pena, L., and Diaz-Ruiz, J.R. 1990. Transformed tomato plants express a satellite RNA of cucumber mosaic virus and produce lethal necrosis upon infection with viral RNA. Biochem. Biophys. Res. Comm. 170: 548-555.

66. Montasser, M.S., Tousignant, M.E., and Kaper, J.M. 1991. Satellite-mediated protection of tomato against cucumber mosaic virus: I. Greenhouse experiments and simulated epidemic conditions in the field. Plant Dis. 75: 86-92.

67. Moriones, E., Diaz, I., Rodriguez-Cerezo, E., Fraile, A., and Garcia-Arenal, F. 1992. Differential interactions among strains of tomato aspermy virus and satellite RNAs of cucumber mosaic virus. Virology 186: 475-480.

68. Moriones, E., Fraile, A., and Garcia-Arenal, F. 1991 Host-associated selection of sequence variants from a satellite RNA of cucumber mosaic virus. Virology 184: 465-468.

69. Mossop, D.W., and Francki, R.I.B. 1977. Association of RNA 3 with aphid transmission of cucumber mosaic virus. Virology 81: 177-181.

70. Mossop, D.W., and Francki, R.I.B. 1979. Comparative studies on two satellite RNAs of cucumber mosaic virus. Virology 95: 395-404.

71. Ohki, S.T., Tanaka, H., and Inouye, T. 1989. Cucumber mosaic virus satellite RNA transmissible to plants infected with a different isolate of CMV. Ann. Phytopathol. Soc. Japan 55: 69-71.

72. Owen, J., and Palukaitis, P. 1988. Characterization of cucumber mosaic virus. I. Molecular heterogeneity maping of RNA3 in eight CMV strains. Virology 166: 495-502.

73. Palukaitis, P. 1988. Pathogenicity regulation by satellite RNAs of cucumber mosaic virus: minor nucleotide sequence changes alter host response. Molecular Plant-Microbe Interactions 1: 175-181.

74. Piazzolla, P., Gallitelli, D., and Savino, V. 1982. Appearance of satellite RNA (CARNA 5) in six cucumber mosaic virus isolates from the open field. Phytopath. Medit. 21: 32-34.

75. Putz, C., Kuszala, J., Kuszala, M., and Spindler, C. 1974. Variation du pouvoir pathogène des isolats du virus de la mosaïque du concombre associée à la nécrose de la tomate. Ann. Phytopathol. 6: 139-154.

76. Quiot, J.B., Leroux, J.P., Labonne, G., and Renoust, M. 1979. Epidémiologie de la maladie filiforme et de la nécrose de la tomate provoquées par le virus de la mosaïque du concombre dans le sud-est de la France. Ann. Phytopathol. 11: 393-408.

77. Quiot, J.B., Marchoux, G., Douine, L., and Vigouroux, A. 1979. Ecologie et épidémiologie du virus de la mosaïque du concombre dans le sud-est de la France. V. Rôle des espèces spontanées dans la conservation du virus. Ann. Phytopathol. 11: 325-348.

78. Richards, K.E., Jonard, G., Jacquemond, M., and Lot, H. 1978. Nucleotide sequence of cucumber mosaic virus-associated RNA 5. Virology 89: 395-408.

79. Roossinck, M.J., Sleat, D., and Palukaitis, P. 1992. Satellite RNAs of plant viruses: Structures and biological effects. Microbiological Reviews 56: 265-279.

80. Rosner, A., Bar-Joseph, M., Moscovitz, M., and Meravech, M. 1983. Diagnosis of specific viral RNA sequences in plant extracts by hybridization with a polynucleotide kinase-mediated, 32P-labeled, double-stranded RNA probe. Phytopathology 73: 699-702.

81. Saito, Y., Komari, T., Masuta, C., Hayashi, Y., Kumashiro, T., and Takanami, Y. 1992. Cucumber mosaic virus-tolerant transgenic tomato plants expressing a satellite RNA. Theor. Appl. Genet. 83: 679-683.

82. Salaman, R.N. 1933. Protective inoculation against a plant virus. Nature 131: 468.

83. Sambrook, J., Fristch, E.F., and Maniatis, T. 1989. Molecular Cloning: A Laboratory Manual, Second Edition. Cold Spring Harbor University Press, Cold Spring Harbor, USA.

84. Sayama, H., Sato, T., Kominato, M., Natsuaki, T., and Kaper, J.M. 1993. Field testing of a satellite-containing attenuated strain of cucumber mosaic virus for tomato protection in Japan. Phytopathology 83: 405-410.

85. Schneider, I.R. 1969. Satellite-like particle of tobacco ringspot virus that resembles tobacco ringspot virus. Science 166: 1627-1629.

86. Sleat, D.E. 1990. Nucleotide sequence of a new satellite RNA of cucumber mosaic virus. Nucleic Acids Res. 18: 3416.

87. Sleat, D.E., and Palukaitis, P. 1990. Site-directed mutagenesis of a plant viral satellite RNA changes its phenotype from ameliorative to necrogenic. Proc. Natl. Acad. Sci. 87: 2946-2950.

88. Sleat, D.E., and Palukaitis, P. 1990. Induction of tobacco chlorosis by certain cucumber mosaic virus satellite RNAs is specific to subgroup II helper strains. Virology 176: 292-295.

89. Sleat, D.E., and Palukaitis, P. 1992. A single nucleotide change within a plant virus satellite RNA alters the host specificity of disease induction. The Plant Journal 2: 43-49.

90. Smith, C.R., Tousignant, M.E., Geletka, L.M., and Kaper, J.M. 1992. Competition between cucumber mosaic virus satellite RNAs in tomato seedlings and protoplasts: A model for satellite-mediated control of tomato necrosis. Plant Dis. 76: 1270-1274.

91. Takanami, Y. 1981. A striking change in symptoms on cucumber mosaic virus-infected tobacco plants induced by a satellite RNA. Virology 109: 120-126.

92. Tien, P., and Chang, X.H. 1983. Control of two seed-borne virus diseases in China by the use of protective inoculation. Seed Sci. & Technol. 11: 969-972.

93. Tien, P., Steger, G., Rosenbaum, V., Kaper, J., and Riesner, D. 1987. Double-stranded cucumovirus associated RNA 5: Experimental analysis of necrogenic and non-necrogenic variants by temperature-gradient gel electrophoresis. Nucleic Acids Res. 15: 5069-5083.

94. Tien, P., and Wu, G. 1991. Satellite RNA for the biocontrol of plant disease. Adv. Virus Res. 39: 321-339.

95. Tien, P., Zhang, X., Qiu, B., Qin, B., and Wu, G.

96. 1987. Satellite RNA for the control of plant diseases caused by cucumber mosaic virus. Ann. appl. Biol. 111: 143-152.

96. Tousch, D., Jacquemond, M., and Tepfer, M. 1990. Transgenic tomato plants expressing a cucumber mosaic virus (CMV) satellite RNA gene: Inoculation with CMV induces lethal necrosis. C. R. Acad. Sci. Paris 331: 377-384.

97. Tousch, D., Jacquemond, M., and Tepfer, M. 1994. Replication of cucumber mosaic virus (CMV) satellite RNA from (-) sense transcripts produced either in vitro or in transgenic plants. J. Gen. Virol. 75: 1009-1014.

98. Tousignant, M.E., and Kaper, J.M. 1993. Cucumber mosaic virus-associated RNA 5. XIII. Opposite necrogenicities in tomato of variants with large 5' half insertion/deletion regions. Res. Virol. 144: 349-360.

99. Waterworth, H.E., Kaper, J.M., and Tousignant, M.E. 1979. CARNA-5, the small cucumber mosaic virus-dependent replicating RNA, regulates disease expression. Science 204: 845-847.

100. White, J. L., and Kaper, J. M. 1987. Absence of lethal stem necrosis in select Lycopersicon spp. infected by cucumber mosaic virus strain D and its necrogenic satellite CARNA 5. Phytopathology 77: 808-811.

101. White, J. L., and Kaper, J. M. 1989. A simple method for detection of viral satellite RNAs in small plant tissue samples. J. Virol. Methods 23: 83-94.

102. Wu, G., and Kaper, J.M. 1992. Widely separated sequence elements within cucumber mosaic virus satellites contribute to their ability to induce lethal tomato necrosis. J. Gen. Virol. 73: 2805-2812.

103. Wu, G. S., Kaper, J.M., and Jaspars, E.M.J. 1991. Replication of cucumber mosaic virus satellite RNA in vitro by an RNA-dependent RNA polymerase from virus infected tobacco. FEBS Lett. 292: 213-216.

104. Wu, G., Kaper, J.M., and Kung, S.D. 1993. Replication of satellite RNA in vitro by homologous and heterologous cucumoviral RNA-dependent RNA polymerases. Biochimie 75: 749-755.

105. Wu, G., Kang, L., and Tien, P. 1989. The effect of satellite RNA on cross-protection among cucumber mosaic virus strain. Ann. Appl. Biol. 114: 489-496.

106. Wu, G., Kaper, J.M., Tousignant, M.E., Masuta, C., Kuwata, S., Takanami, Y., Pena, L., and Diaz-Ruiz, J.R. 1993. Tomato necrosis and the 369 nucleotide Y satellite of cucumber mosaic virus: Factors affecting satellite biological expression. J. Gen. Virol. 74: 161-168.

107. Yie, Y., and Tien, P. 1993. Plant virus satellite RNAs and their role in engineering resistance to virus diseases. Seminars in Virology 4: 363-368.

108. Yie, Y., Zhao, F., Zhao, S.Z., Liu, Y.Z., and Tien, P. 1992. High resistance to cucumber mosaic virus conferred by a satellite RNA and coat protein in transgenic commercial tobacco cultivar G-140. Molecular Plant-Microbe Interactions 5: 460-465.

109. Yoshida, K., Goto, T., and Iizuka, N. 1985. Attenuated isolates of cucumber mosaic virus produced by satellite RNA and cross-protection between attenuated isolates and virulent ones. Ann. Phytopathol. Soc. Japan 51: 238-242.

CHAPTER 9

Alternative Strategies for Engineering Virus Resistance in Plants

R.R. Martin

The use of parts of viral genomes to transform plants to provide disease resistance was first suggested by Hamilton (21). This concept was based on observations dating back to 1929 that mild strains of a virus could protect plants from related severe strains. Practical use of cross protection has been used extensively in tomato, citrus and papaya as a means of protecting plantings from infection by severe virus strains (35). With developments in molecular biology in the 1970's and an increased understanding of the mechanisms by which *Agrobacterium tumefaciens* transfers genes to plant genomes, the tools to provide protection against virus infection using a portion of viral genomes became a reality (8). In the decade since the first viral gene was transferred to plants other strategies for developing resistance have been explored including some that do not require any viral sequences.

Since the first demonstration that virus coat protein expressed in plants provides some level of resistance (39) this approach has been used to develop resistance to a large number of different viruses (6). However, there are reasons to look at alternatives to coat protein-mediated resistance (CPMR). First, it is likely that strains able to circumvent resistance will evolve, particularly as CPMR becomes widely used in crop protection. It can be expected that multiple protection strategies will be necessary to maximize resistance when transgenic crops are subjected to field conditions (42). Second, viruses are very efficient in terms of utilizing the host's genes and cellular synthesis systems. All viral genes tend to be essential and unique to pathogenic functions. This means that nearly any viral gene may be a suitable target for the genetic engineering of pathogen-derived resistance (41). Third, CPMR usually provides resistance to low levels of inoculum. In most cases this is sufficient, however, insect or nematode vectors may provide enough inoculum at the cellular level (not at the leaf or whole plant level) to overcome the resistance and establish an infection. Fourth, heterologous encapsidation as demonstrated for potyviruses may alter the ecology and epidemiology of some viruses (13,29). Fifth, with the possibility of recombination as demonstrated for DNA viruses (16) and for RNA plant viruses (20, 49) it may be undesirable to use viral genes as sources of resistance. Strategies that do not use any viral sequences for developing resistance avoid the potential risk associated with recombination. In cases where the coat protein is involved in transmission, recombination of the gene with a different virus could alter its epidemiology and ecology.

When one considers these points it seems imperative that we look at alternative strategies to enhance the usefulness of engineered resistance. Also, an alternative strategy may provide either a higher level of resistance or perhaps a more durable form of resistance. Another consideration is the risk associated with the use of various strategies for engineering resistance. This is an area that needs more research so that risk assessment is based on scientific data rather than unsupported assumptions.

The recent demonstration of recombination between a transgene and a disabled virus (20) raises many questions about the risk associated with use of viral genes to develop resistance. It would be desirable to see a study using a viral transgene not containing the 3' noncoding region, because the 3' noncoding region provides the polymerase a recognition site in the transgene which increases the chance of recombination if the template switching mechanism of RNA recombination is correct. It would also be desirable to challenge the plants with a competent virus since this is likely to be the situation in the field. One strategy to reduce the possibility of recombination is to use partial genes for developing resistance, for example, it may be possible to develop resistance with part of the coat protein that lacks the determinant(s) involved in transmission.

There are several alternative strategies for engineering virus resistance in plants. Some of these have been reviewed recently (35, 51). Of course one of the primary goals in the production of insect and pathogen resistant engineered crops is to reduce the use of pesticides. Billions of dollars are spent each year to protect crops from diseases and insects (1) and this technology should provide a means of reducing our dependence on chemicals for their control yet make it possible to maintain a relatively cheap and high quality food supply.

In this chapter alternative strategies for engineering resistance to plant viruses not covered in other chapters will be discussed. I will try to highlight the advantages and disadvantages of each strategy.

INDUCED HYPERSENSITIVE RESPONSE

Making use of a toxin gene under the control of a virus-specific promoter should cause cell death when that gene is expressed. The gene should only be expressed upon virus infection of primary cells and should be expressed early in infection before there is any cell-to-cell movement of the virus. There are a number of toxins that could be used, such as proteases, RNAses or ribosome-inhibiting proteins (51). The virus-specific sequences might be either the 3' or 5' noncoding regions, or a subgenomic promoter that likely contains polymerase recognition sequences. Use of a construct containing the 3' noncoding region followed by the toxin gene in the antisense orientation should be a good strategy for several reasons. To be effective, the polymerase must copy the viral genomic RNA very early during infection starting at the 3' end. The gene of interest should also be inserted in the antisense orientation, thus reducing the potential risk of nonspecific expression.

With potato leafroll virus (PLRV), Kawchuk et al. (26) reported that the coat protein gene together with a 192 base pair leader sequence in an antisense orientation provided the same level of resistance as that observed when the construct was used in the plus-sense. They speculated that this protection was due to the construct acting as an antisense that hybridized with viral RNA and inhibited replication. Miller and Young (37) suggested that it is possible this protection was due to the viral polymerase copying the RNA transcript from the antisense to give plus-sense RNA which then was translated to produce coat protein which provided the protection that was observed. If the latter hypothesis is correct, then using a toxin gene in place of the coat protein could provide a hypersensitive response.

It may be possible to fuse several subgenomic promoters or 3' noncoding regions together to develop multiple virus resistance from a single construct. In this manner, one could protect a crop against the two or three most serious virus diseases with a single construct. In North America in excess of 15 million dollars are spent annually to control the aphid-borne viruses PLRV and potato virus Y (PVY) in potato (1). Resistance to these two viruses would greatly reduce the amount of pesticides used in potato production.

This type of virus resistance must be very well regulated. If the expression of the toxin gene is delayed, then rather than having a hypersensitive response at the single cell level one could end up with the entire plant being killed. If only a few plants in a field become infected this would not be serious. However, if a large flight of viruliferous aphids moved into a crop, entire fields could be lost if the expression of the toxin gene was not tightly controlled.

INHIBITION OF AMBER CODONS

Many viruses make use of amber codons as part of their gene expression strategy. The genomes of tobamoviruses contain an amber codon in their polymerase genes which results in the synthesis of a 126 kD protein when it acts as a termination codon and of a 183 kD readthrough protein when the amber codon is inhibited. The luteoviruses make use of this strategy to produce a protein that consists of the coat protein plus the aphid transmission factor. When the amber codon acts as a stop codon it results in the production of the 22 - 24 kD coat protein, but when the codon is inhibited a 60 - 80 kD protein is produced. Electrophoresis of purified luteoviruses result in two protein bands in polyacrylamide gels, one about 23 kD and the other of about 75 kD. The high molecular weight band is recognized by antisera to the 23 kD protein, but antiserum to the portion of the 75 kD protein downstream from the amber codon only recognizes the 75 kD protein and not the 23 kD protein. Based on staining intensity it is estimated that the amber codon reads through less than 10% of the time, ie. the high molecular weight protein is about 10% as abundant as the 23 kD protein in purified particles (34,48). It is also possible that the amber codon reads through all the time and that some post-translational processing results in a predominance of the 23 kD protein.

Suppressor tRNAs inhibit amber codons so that they read through all the time (22). Use of suppressor tRNA genes should alter the balance of viral components during infection and could result in some level of resistance. If the 23 kD protein of luteoviruses does result from translational termination due to the amber codon, then suppressing the amber codon may alter the relative amounts of the two proteins. If the amber codon is inhibited and all or most of the coat protein produced is the 60 - 80 kD protein then virus assembly may be blocked due to steric hindrance. Partial suppression of an amber codon may provide some level of resistance. In the tobamoviruses, the presence of the amber codon suggests a possible role of the 126 kD polymerase during virus replication. If this codon is suppressed and always reads through to produce only 183 kD protein, then it is possible that viral replication would be affected.

This strategy offers the advantage that no viral sequences are used to develop the resistance genotype. This preempts the concerns about recombination. A second advantage is that a single construct could provide resistance to a number of viruses. A disadvantage is that the amber codon supressor may affect translation of some host genes.

INSECT TRANSMISSION FACTORS

Specific interactions between some viruses and their vectors enable the vector to transmit the virus from infected plants to other plants. This is true for various types of vectors; including insects, mites, fungi and nematodes (35). In many cases the mechanism of the specific interaction is unknown but some of the interactions have been studied in detail. The best understood interaction is that of the potyviruses, where an interaction between the coat protein, a virus-encoded helper component and aphid stylet are involved in virus transmission. Specific sites of recognition have also been defined between luteoviruses and their aphid vector.

The potyvirus aphid-transmission factor or "helper component" acts as a bridge between the virus and the aphid stylet. Some potyvirus isolates are not aphid-transmissible because, in some cases, they produce a a defective helper component. In other cases potyviruses are not aphid transmissible because of a mutation in a DAG motif near the N-terminus (3). Mutational analysis of the DAG sequence in tobacco vein mottling virus has shown that the third position of the motif must be a G whereas there is some though limited flexibility in the first two positions (4). One strategy to engineer resistance to potyviruses might be to transform plants with a defective helper component that would either bind to the aphid stylet but not to the virus particle or would bind to the virus particle but not to the stylet. A second strategy would be to transform plants with a coat protein construct mutated in the DAG region. In either case the defective proteins would have to compete with the functional virus-encoded proteins before inhibition of transmission was obtained. It is likely that either of these mutated proteins, even if expressed at very high levels, would not completely block transmission simply because a small percentage of wild type protein may result in efficient transmission.

Two different virus-vector interactions have been identified among luteoviruses. The first is at the hindgut and may be group-specific. In work reported by Gildow (18) involving five different BYDV and four different aphid vectors the different BYDV isolates were transported specifically from the hindgut to the haemocoel regardless of whether or not the aphid was a vector of that virus. At the same time, non-luteoviruses were not transported across the hindgut membrane but rather passed through the aphid. In all but one case the specificity was at the group level, ie. all BYDV strains moved across the hindgut membrane in all but one combination. This suggests that blocking of a single receptor might provide resistance to all luteoviruses. In the cereal crops where multiple luteoviruses infect a single crop this would be a very useful system for engineering resistance. Moreover, the same construct might be useful for engineering resistance to luteoviruses in a wide range of crops.

In addition to the nearly group-specific receptor at the hindgut there appears to be a virus-specific receptor at the basal lamella of the salivary gland (19). This is the point where the virus moves from the haemocoel into the salivary gland and is then available to be egested during feeding. This step is required for the transmission of luteoviruses. Luteovirus injected into the haemocoel of a nonvector aphid cannot be transmitted, whereas a vector aphid injected with luteovirus will become viruliferous. Blocking of the receptor at the basal lamella of the salivary gland should provide virus-specific resistance, which could be desirable in crops where only one luteovirus is known to cause disease, ie. potato.

MOVEMENT PROTEINS

One mechanism for virus resistance in plants is the inability of a virus to move from cell to cell. There are many examples in which isolated protoplasts are susceptible to infection by a specific virus while whole plants of the same species and cultivar are resistant (25). Proteins that facilitate movement of virus particles from cell-to-cell or long distance movement within a plant have been identified among the gene products of several plant viruses (14, 30, 36). In some cases these proteins have been localized at the plasmodesmata in plants transgenic for the movement protein (2, 14). It also has been shown that the size of the pores of the plasmodesmata increase during virus infection (52) presumably to facilitate virus movement. When the movement protein was mutated to delete amino acids 3, 4 and 5 from the N-terminus in an infectious clone of tobacco mosaic virus (TMV) the local and long distance movement of the virus was limited (15). Plasmodesmata of transgenic plants containing this dysfunctional movement protein were not modified to the same extent as those in plants containing wild-type movement protein (27). Also the dysfunctional movement protein limited local and systemic movement of TMV and several other tobamoviruses. Plants containing this mutant protein of TMV also conferred varying levels of resistance to a number of viruses that are not members of the tobamovirus group, including tobacco rattle tobravirus, peanut chlorotic streak caulimovirus, tobacco ringspot nepovirus and limited levels of resistance to cucumber mosaic and alfalfa mosaic viruses (9). However, when functional movement protein was used the symptoms development was accelerated and severity of symptoms increased.

Transgenic plants containing a mutant form of the 13 kD gene of the triple gene block from white clover mosaic virus (WClMV) were resistant to infection by strains of WClMV (7). These plants were also resistant to two potexviruses that were tested and potato virus S, a member of the carlavirus group, but not to tobacco mosaic virus. The carlaviruses also contain a triple gene block very similar to potexviruses.

The results to date suggests that mutant movement protein should be used rather than wild-type movement protein for engineering virus resistance. Also, the use of mutant movement protein to engineer virus resistance may be broad spectrum compared to other strategies. If this holds true it maybe possible to engineer resistance to several viruses in a crop with a single construct. The use of a mutant protein reduces the probability that the transgene could result in a functional new virus if it were to recombine with a virus infecting the transgenic plant.

PLANT ANTIBODIES

The expression of functional antibodies in plants was first reported in 1989 (23). In this case the light chain and heavy chain were expressed in separate plants and these plants were crossed to obtain plants that expressed both light and heavy chains. Many groups have since worked on expressing antibody genes in plants with very limited success. Recently, Van Engelen et al. (45) have been able to produce functional antibodies in plants by cloning light and heavy chain genes in the same T-DNA but each driven by a different promoter. In this way both genes were expressed in a coordinated manner and in the same cell types. This resulted in assembly of functional antibodies in the plant.

Single chain antibodies (scFv) consist of the variable domains of the heavy and light chains joined by a flexible peptide linker. These antibodies have the advantage of always being expressed in equimolar amounts and at the same time and site, avoiding problems with assembly. They are relatively small molecules (about 30 kD) and can be cloned using the polymerase chain reaction (PCR) making use of primers to conserved regions of the IgG genes (12).

Recently, scFv specific for the coat protein of artichoke mottle crinkle virus, a tombusvirus, have been expressed in tobacco plants (44). The presence of the scFv resulted in lower virus titers in systemically infected leaves and a delay in symptom expression.

There are some advantages to using scFv for engineering resistance to plant viruses. The obvious one is that there are no viral sequences being expressed in the transgenic plant. This negates the concerns about the risks associated with expressing viral genes in plants. The cloning strategy is the same for the production of any scFv. The scFv approach could be applied to any stage of the life cycle of a virus. It is likely that a scFv which interferes with early stages of the virus life cycle would be most effective. For example, antibodies that interact with genes that are expressed in low copy number during virus infection, ie. antibodies against the replicase for many viruses, should be more effective than antibodies directed against the coat protein. For viruses that express polyproteins that are processed by proteases to yield functional proteins, a scFv that binds to the cleavage site preventing access to the site by the protease should provide effective resistance. In this case, the scFv antibody targets an early stage of replication when only a few copies of the target would be present.

Anti-idiotypic antibodies which bind to specific domains that interact with host or vector receptors may serve to saturate the receptors and thus prevent virus from interacting with the receptor. Anti-idiotypic antibodies are antibodies made against the antigen-interacting end of an antibody and should have the same surface conformation as the antigen used to elicit the original antibody. Thus, an anti-idiotypic antibody that interacts with the "functional end" of an antibody which binds to a virus-specific domain in the viral capsid should have the same surface conformation as the viral capsid domain. Using antibodies to block virus receptors in a vector may or may not be successful depending on the number of receptors that need to be blocked. Also, the level of expression must be high enough that the expressed antibody outcompetes the virus particles for the receptors.

PROTEASE INHIBITORS

The Potyviridae, Nepoviruses and Comoviruses use a polyprotein strategy of gene expression. Thus, a single large polyprotein is produced from each RNA and then cleaved to yield the individual proteins required for virus replication. There is evidence that the

luteoviruses (47) and carlaviruses (28) also make use of proteolysis as a strategy to produce some of their functional proteins. Virus-encoded proteases are part of the polyprotein produced by these viruses. Thus, any inhibitor of the protease will likely prevent virus replication beyond production of the initial polyprotein. An antibody that binds to the cleavage site making it inaccessible to the protease, or one that binds to the protease making it unable to interact with the cleavage sites should prevent processing of the polyprotein. Since plants also have proteases, an antibody that binds to the protease cleavage site would be preferrable to one that binds to the protease. An antibody that binds to the protease could cross react with a plant protease and interfere with normal protein catabolism in the host plant.

Other inhibitors of the virus specific protease should also prevent processing of the polyprotein. Immunity in 'Arlington' cowpea to cowpea mosaic virus has been shown to result from an inhibition of the proteolytic processing of the polyprotein (38). A similar inhibition that prevents some insects from digesting their foods may be a useful strategy for controlling insect pests (40).

Plants transformed with the nuclear inclusion A (NIa) protease of two different potyviruses were resistant to subsequent infection by the respective viruses (32, 46). In the case of the NIa gene of tobacco vein mottle virus, plants challenged with other potyviruses were susceptible, suggesting that this form of induced resistance is virus-specific.

A protease inhibitor has the advantage that no viral sequences are required to develop resistance. A second advantage is that the inhibitor may provide resistance to some insect pests that make use of a protease similar to the viral protease. A possible disadvantage is the protease inhibitor may interfere with protein cycling in the plant.

IDENTIFICATION OF HOST RESISTANCE GENES

Host resistance to virus infection may be effective for long periods of time, e.g. resistance to raspberry bushy dwarf virus in the variety Willamette has lasted for over 40 years and in most areas where raspberries are grown this variety is still resistant (11) Resistance to virus vectors can also be long lasting, e.g. a single gene for resistance to the raspberry aphid,

Amphorophora agathonica, in North America has been effective at controlling aphid-borne viruses in raspberry for 50 years (10). With crops such as, potato, tomato, corn, lettuce and wheat and arabidopsis excellent genetic maps have been developed that will increase the likelihood of identifying resistance genes (43). In addition to phenotypic markers that have been identified in crop plants, RFLPs (17, 43) and RAFDs markers are being used to develop complete genetic maps for most of the major crops. The more detailed these genetic maps are the easier it will be to identify specific resistance genes.

The N gene that mediates resistance to TMV in tobacco is the first virus resistance gene that has been isolated from plant (50). The authors suggested that the N gene mediates rapid gene induction because of its similarity to well studied animal genes. Thus, this gene may initiate a cascade of reactions that result in an hypersensitive response. When the N gene was transformed into susceptible tobacco, the resulting plants were resistant to TMV. This gene apparently has homologies to other disease resistant genes isolated from plants. If these resistant genes represent a family of genes it may be possible to identify resistance genes from other plants that will provide resistance to a wide range of viruses and other plant pathogens.

Identification of proteins in plants that are specific for disease resistance may provide another means of identifying host resistance genes. Proteins correlated with resistance to barley yellow dwarf virus have been identified in barley genotypes carrying the Yd2 gene for resistance to this virus (24). Antibodies specific for this protein should provide a means of identifying this gene in a genetic library from resistant barley.

The narrow genetic base of many of the agricultural crops grown in the developed countries results in few plant disease resistant genes in commercial cultivars. Efforts should be made to broaden the genetic base with the goal of introducing sources of resistance to many different pathogens. Such efforts are underway in breeding programs for many crop and ornamental plants.

In many cases there are no known sources of resistance to a virus in a crop, e.g. potato leafroll resistance in potato, barley yellow dwarf resistance in wheat, strawberry mild yellow edge resistance in strawberry and blueberry shock ilarvirus in blueberry. However it may be possible to introduce resistance genes from related species via protoplast fusion as has been done for several disease resistance genes in potato (5).

There are obvious advantages to making use of plant resistance genes whenever possible. There is less public concern with genes introduced into a crop by conventional plant breeding than by genetic engineering. The transfer of a plant gene from one plant to another by genetic engineering is more readily accepted than is moving genes between the animal and plant kingdoms by the same process.

If possible it would be best to obtain the disease resistance by conventional breeding rather than by genetic engineering. Recently, in March, 1995, the environmental protection agency (EPA) in the United States proposed to regulate plants genetically engineered for pest resistance in the same manner as pesticides are regulated. Thus, a disease resistant plant obtained through genetic engineering would be regulated whereas, another plant with resistance to the same disease obtained by conventional breeding would not be regulated. It is likely that this type of proposal will not become law and if it does it will likely be reversed. However, it does point out the high level of public concern about the technology of genetic engineering. This technology is probably the most discussed and regulated of any technology in history. Resistance introduced from wild species by protoplast fusion which is then used to develop commercial cultivars by conventional breeding likely would not be regulated in the same manner, even though in this case hundreds of new genes are being introduced to the crop.

CONCLUSIONS

There are many different strategies that can be employed to develop virus-resistant plants. Engineering resistance to plant viruses, using a strategy that interferes with an early step in the virus life cycle should be more effective than interfering with a later step. One must also consider the risk associated with the various strategies. The plantibody approach has the advantage that no viral sequences are being incorporated into the plant. Strategies that make use of minimal viral sequences could be advantageous in that less than full genes would be present in the transgenic plant.

We should learn from experiences with other genetic systems such as single dominant genes used in the past for disease resistance, and the use of chemicals to control insects and diseases. There are many examples of disease resistance based on single dominant genes breaking down after a short period of use in the field. Moreover, many pathogens and insects have developed resistance to pesticides used in crop production. The traits we are dealing with are single dominant traits and we should expect that viruses will overcome the resistances that we are now trying to develop. We should continue to search for alternate strategies for developing virus resistance and successful strategies should be used wisely to maximize the length of time that they will be effective in the field. We should recognize that these forms of resistance are another tool for controlling plant diseases which should be incorporated into an overall integrated crop management scheme.

Genetic engineering has been given a lot of publicity and promotion and has also been very closely scrutinized. We must remember it is another tool to be used in our attempts to reduce pesticide use, decrease cost of crop production, increase yields and provide a safe and cheap food supply. This technology will not replace the need for plant breeding but rather complement the efforts of plant breeders. Finally, we must consider the demands of the consumer, for the very best plant (ie. highest level of virus resistance) will not be of much use if it is deemed unacceptable by consumers.

REFERENCES

1.   Agricultural Chemical Usage, 1992 Field Crops Summary, U.S. Dept. Agric., National Agric. Statistics Service, March 1993.
2.   Atkins, D., Hull, R., Wells, B., Roberts, K., Moore, P., and Beachy, R.N. 1991. The tobacco mosaic virus 30K movement protein in transgenic tobacco plants is localized to plasmodesmata. J. Gen. Virol. 72:209-211.
3.   Atreya, P.L., Atreya, C.D., and Pirone, T.P. 1991. Amino acid substitutions in the coat protein result in loss of insect transmissibility of a plant virus. Proc. Natl. Acad. Sci. USA 88:7887-7891.
4.   Atreya, P.L., Lopez-Moya, J.J., Chu, M., Atreya, C.D., and Pirone, T.P. 1995. Mutational analysis of the coat protein N-terminal amino acids involved in potyvirus transmission by aphids. J. Gen. Virol. 76:265-270.
5.   Austin, S., Ehlenfeldt, M.K., Baer, M.A., and Helgeson, J.P. 1986. Somatic hybrids produced by protoplast fusion between S. tuberosum and S. brevidens: phenptic variation under field conditions. Theor. Appl. Genet. 71:682-690.
6.   Beachy, R.N., Loesch-Fries, S., and Tumer, N.E. 1990. Coat protein-mediated resistance against virus infection. Annu. Rev. Phytopathol. 28:451-474.
7.   Beck, D.L., Van Dolleweerd, C.J., Lough, T.J., Balmori, E., Voot, D.M., Anderson, M.T., O'Brien, I.E.W., and Forster, R.L.S. 1994. Disruption of virus movement confers broad-spectrum resistance against systemic infection by plant viruses with a triple gene block. Proc. Natl. Acad. Sci. USA 91:10310-10314.
8.   Bevan, M.W., Mason, S.E., and Goelet, P. 1985. Expression of tobacco mosaic virus coat protein by a cauliflower mosaic virus promoter in plants transformed by Agrobacterium. EMBO J. 4:1921-1926.
9.   Cooper, B., Lapidot, M., Heick, J.A., Dodds, J.A., and Beachy, R.N. 1995. A defective movement protein of TMV in transgenic plants confers resistance to multiple viruses whereas the functional analog increases susceptibility. Virology 206:307-313.
10. Daubeny, H.A. 1986. The British Columbia raspberry breeding program since 1980. Acta Hortic. 183:47-57.
11. Daubeny, H.A., Lawrence, F.J., and McGregor, G.R. 1989. 'Willamette' red raspberry. Fruit Var. J. 43:46-48.
12. Davis, G.T., Bedzyk, W.D., Voss, E.W., and Jacobs, T.W. 1991. Single chain antibody (SCA) encoding genes: one-step construction and expression in eukaryotic cells. Bio/Technology 9: 165-169.
13. Farinelli, L., Malnoae, P., and Collet, G.F. 1992. Heterologous encapsidation of potato virus strain O (PVYO) with the transgenic coat protein of PVY strain N (PVYN) in Solanum tuberosum cv. Bintje. BioTechnology 10:1020-1025.
14. Fujiwara, T., Giesman-Cookmeyer, D., Ding, B., Lommel, S.A., and Lucas, W.J. 1993. Cell-to-cell trafficking of macromolecules through plasmodesmata potentiated by the red clover necrotic mosaic virus movement protein. Plant Cell 5:1783-1794.
15. Gafny, R., Lapidot, M., Berna, A., Holt, C.A., Deom, C.M., and Beachy, R.N. 1992. Effects of terminal deletion mutations on the function of the movement protein of tobacco mosaic virus. Virology 187:499-507.
16. Gal, S., Pisan, B., Hohn, T., Grimsley, N., and Hohn, G. 1991. Genomic homologous recombination in planta. EMBO J. 10:1571-1578.
17. Gebhardt, C. 1994. RFLP mapping in potato of qualitative and quantitative genetic loci conferring resistance to potato pathogens. Amer. Pot. J. 71:339-345
18. Gildow, F.E. 1993. Evidence for receptor-mediated endocytosis regulating luteovirus acquistion by aphids. Phytopathology 83:270-277.
19. Gildow, F.E., and Gray, S.M. 1993. The aphid salivary gland basal lamina as a selective barrier associated with vector- specific transmission of barley yellow dwarf luteoviruses. Phytopathology 83:1293-1302.
20. Greene, A.E., and Allison, R.F. 1994. Recombination between viral RNA and transgenic plant transcripts. Science 263:1423-1425.
21. Hamilton, R.I. 1980. Defenses triggered by previous invaders: viruses. Pages 279-303, in: Plant Disease: An Advanced Treatise, Vol. 5, J.G. Horsfall and E.B. Cowling, eds. Academic Press, New York.
22. Hatfield, D.L., Smith, D.W.E., Lee, B.J. Worland, P.J., and Oroszian S. 1990. Structure and function of suppressor tRNAs in higher eukaryotes. Critical Rev. Biochem. and Molec. Biol. 25:71-96.

23.   Hiatt, A., Cafferkey, R., and Bowd sh, K. 1989. Production of antibodies in transgenic plants. Nature 342:76-78.

24.   Holloway, P.J., and Heath, R. 1992. Identification of polypeptide markers of barley yellow dwarf virus resistance and susceptibility genes ir non-infected barley (*Hordeum vulgare*) plants.   Theor. Appl Genet. 85:346-352.

25.   Hull, R. 1989. The movement of viruses in plants. Annu. Rev. Phytopathol. 27:213-240.

26.   Kawchuk, L.M., Martin, R.R., and McPherson, J. 1991. Sense and antisense RNA-mediated resistance to potato leafroll virus in Russet Burbank potato plants. Mol. Plant-Microbe Interact. 4:247-253.

27.   Lapidot, M., Gafny, R., Ding, B., Wolf, S., Lucas, W.J., and Beachy, R.N. 1993. A dysfunctional movement protein of tobacco mosaic virus that partially modifies the plasmodesmata and limits virus spread in transgenic plants. Plant J. 4:959-970.

28.   Lawrence, D.M., Rozanov, M.N., and Hillman, B.I. 1995. Autocatalytic processing of the 223-kDa protein of blueberry scorch carlavirus by a papain-like proteinase. Virology 207:127-135.

29.   Lecoq, H., Ravelonandro, M., Wipf-Scheibel, C., Monsion, M., Raccah, B., and Durez, J. 1993. Aphid transmission of a non-aphic-transmissible strain of zucchini yellow mosaic potyvirus from transgenic plants expressing the capsid protein of plum pox potyvirus. Mol. Plant-Microbe Interact. 6:403-406

30.   Letho, K., Grantham, G.L., and Dawson, W.O. 1990. Insertion of sequences containing the coat protein subgenomic RNA promoter and leader in front of the tobacco mosaic virus 3CK ORF delays its expesssion and causes defective cell-to-cell movement. Virology 174:145-157.

31.   Logemann, J., Jach, G., Tommerup, H., Mundy, J., and Schell, J. 1992. Expression of a barley ribosome-inactivating protein leads to increased fungal protection in transgenic tobacco plants. BioTechnology 10:305-308.

32.   Maiti, I.B., Murphy, J.F., Shaw, J G., and Hunt, A.G. 1993. Plants that express a potyvirus proteinase gene are resistant to virus infection. Proc. Natl. Acad. Sci. USA 90:6110-6114.

33.   Martin, R.R. 1994. Genetic engineering of potatoes. Amer. Pot. J. 71:347-358.

34.   Martin, R.R., Keese, P.K., Young, M.J., Waterhouse, P.M., and Gerlach, W.L. 1990. Evolution and molecular biology of luteoviruses. Annu. Rev. Phytopathol. 28:341-63.

35.   Matthews, R.E.F. 1991. Plant Virology, 3rd ed. Academic Press, New York. 835 Pages

36.   Meshi, T., Watanabe, Y, Saito, T., Sugimoto, A., Maeda, T., and Okada, Y. 1987. Function of the 30 kD protein of tobacco mosaic virus: Involvement in cell-to-cell movement and dispensability for replication. EMBO J 6, 2557-2563.

37.   Miller, W.A., and Young M.J. 1995 Prospects for genetically engineered resistance to barley yellow dwarf viruses. Pages 345-369 in: Barley Yellow Dwarf:40 Years of Progress. C.J. D'Arcy and P.A. Burnett, eds. APS Press, St. Paul, Minnesota.

38.   Ponz, F., Glascock, C.B., and Breuning, G. 1988.

An inhibitor of polyprotein processing with the characteristics of a natural virus resistance factor. Mol. Plant- Microbe Interact. 1:25-31.

39.   Powell-Abel, P., Nelson, R.S., De, B., Hoffman, N., Rogers, S.G., Fraley, R.T., and Beachy, R.N. 1986. Delay of disease development in transgenic plants that express the tobacco mosaic virus coat protein gene. Science 232:738-743.

40.   Ryan, C.A. 1990. Protease inhibitors in plants: Genes for improving defenses against insects and pathogens. Annu. Rev. Phytopathol. 28:425-449.

41.   Sanford, J.C., and Johnston, S.A. 1985. The concept of parasite-derived resistance-Deriving resistance genes from the parasite's own genome. J. Theor. Biol. 113:395-405.

42.   Scholthof, K.G., Scholthof, H.B., and Jackson, A.O. 1993. Control of plant virus disease by pathogen-derived resistance in transgenic plants. Plant Physiol. 102:7-12.

43.   Tanksley, S.D., Young, N.D., Paterson, A.H., and Bonierbale, M.W. 1989. RFLP mapping in plant breeding: New tools for an old science. Bio/Technology 7:257-264.

44.   Taviadorakl, P., Benvenuto, E., Trinca, S., De Martinis, D., Cattaneo, A., and Galeffi, P. 1993. Transgenic plants expressing a functional single-chain Fv antibody are specifically protected from virus attack. Nature 366:469-472.

45.   van Engelen, F.A., Schouten, A., Molthoff, J.W., Roosien, J., Salinas, J., Dirkse, W.G., Schots, A., Bakker, J., Gommers, F.J., Jongsma, M.A., Bosh, D., and Stiekema, W.J. 1994. Coordinated expression of antibody subunit genes yields high levels of functional antibodies in roots of transgenic tobacco. Plant Molecular Biology 26:1701-1710.

46.   Vardi, E., Sela, I., Edelbaum, O., Livneh, O., Kuznetsova, L., and Stram, Y. 1993. Plants transformed with a cistron of a potato virus Y protease (NIa) are resistant to virus infection. Proc. Natl. Acad. Sci. USA 90:753-7517.

47.   Wang, J.Y., Chay, C., Gildow, F.E., and Gray, S.M. 1995. Readthrough protein associated with virions of barley yellow dwarf luteovirus and its potential role in regulating the efficiency of aphid transmission. Virology 206:954-962.

48.   Waterhouse, P.M., Martin, R.R., and Gerlach, W.L. 1989. BYDV-PAV virions contain readthrough protein. Phytopathology 79:1215.

49.   White, K.A., and Morris, T.J. 1994. Recombination between defective tombusvirus RNAs generates functional hybrid genomes. Proc. Natl. Acad. Sci. USA 91:3642-3644.

50.   Whitman, S., Dinesh-Kumar, S.P., Choi, D., Hehl, R., Corr, C., and Baker, B. 1994. The product of the tobacco mosaic virus resistnace gene N: Similarity to Toll and the Interleukin-1 receptor. Cell: 78:1101-1115.

51.   Wilson, T.M.A. 1993. Strategies to protect crop plants against viruses: Pathogen derived resistance blossoms. Proc. Natl. Acad. Sci. USA 90:3134-3141.

52.   Wolf, S., Deom, C.M., Beachy, R.N., and Lucas, W.J. 1989. Movement protein of tobacco mosaic virus modifies plasmodesmatal size exclusion limit. Science 246:377-379.

CHAPTER 10

Controlling Mosaic Virus Diseases Under Field Conditions Using Multiple Gene Strategies in Transgenic Plants

Y. Yie and P. Tien

Since it was first demonstrated that expression of viral coat protein in transgenic tobacco plants can mediate resistance against tobacco mosaic virus (TMV) infection in 1986 (1), almost every part of viral genes and virus-targeted sequences have been tested for their capacity to protect against virus infection (2-10,12,14-19,22). The pathogen-targeted transgenic resistance strategies involve antiviral ribozymes, antisense RNAs or antibody genes, pokeweed antiviral protein genes etc.; the tested viral pathogen-derived sequences include the viral coat protein (CP) gene, viral non-structural protein genes (e.g. viral replicase gene, movement protein gene, etc.) and satellite RNA (Sat-RNA) (for review, see reference 24). Of all viral genes tested, the CP genes mediated protection has been widely studied and has been proven to still be effective in obtaining resistance to viruses in transgenic plants. For example, over 20 viruses in at least 10 different taxonomic groups, in a wide variety of dicot species were tested between 1986 and 1993 (24). Today, the number of tested viruses increases rapidly.

However, there are some limitations if only one kind of pathogen-derived or virus-targeted gene is introduced into plants. First of all, most engineering resistance has been shown to be short of actual immunity to viral infection. Transgenic plants expressing viral CP only showed a delay of symptom development. After a period of time, the virus(es) overcame the protection and symptoms developed. Moreover, the extent of CP mediated protection can also be affected by the form and concentration of the virus(es) inoculated. When transgenic tobacco plants expressing TMV CP (1) or alfalfa mosaic virus (AlMV) CP (13,23)

were inoculated with viral nucleic acids or viruses at high concentration, their resistance reduced sharply or was lost completely. The above limitation is also true for Sat-RNA mediated resistance. The discovery of some Sat-RNAs with ability to attenuate symptoms of their helper viruses led to their use as a biological control agent (BCA) (21). In our laboratory, the protective effects of cucumber mosaic virus (CMV) Sat-RNA have been demonstrated in greenhouse and field tests. In 1986, Baulcombe et al. (3) first reported that the virus concentration accumulated in transgenic tobacco plants expressing the Sat-RNA of CMV was 5-20% of that accumulated in control plants and that the transgenic plants grown in the greenhouse showed symptom attenuation after challenge with CMV (3,8). The same results were obtained in our laboratory in 1988. However, field tests carried out in China from 1990 to 1992 showed that the resistance of the transgenic plants was not enough to protect tobacco and tomato plants from the damage caused by natural virus infection (20). It is suggested that Sat-RNA may confer some extent of resistance at late stages after virus infection. The levels of expression of Sat-RNA are rather low in transgenic plants. Second, the engineered resistance against different virus strains by one kind of gene is rather narrow. Transgenic tobacco plants expressing TMV RNA nucleotides 3472-4916 of the U1 strain (a putative 54kDa ORF corresponding to most of the 126 kDa UAG-codon read-through domain of the 183 kDa replicase) conferred high resistance only to inoculation with homologous or closely related strains of TMV. The fact is that the virus infection in fields is caused by not only different strains of a virus, but also by mixed infection of different viruses. Analysis of 393 defined field trials of transgenic plants (25 species) performed in 21 countries between 1986 and 1991, revealed that 50 of the trials involved "virus-resistance' traits (24). Field results have shown that transgenic plants with CP-mediated resistance may not behave as predicted from greenhouse/growth chamber experiments. Usually, they exhibited great susceptibility to virus challenge.

To overcome these limitations and produce high and broad resistant transgenic plants, we developed a multiple-gene approach based on the assumption that every distinct resistance gene could target against different stages and different sites of viral replication and contribute some resistance to plants. For example, the CP gene could interfere with uncapsidation of virions in early infection and spread of mature virions in later infection. As described above, Sat-RNAs could interfere with the virus replication at the late stage of infection. Therefore, transgenic plants expressing both CP and Sat-RNA would show higher resistance. The transgenic plants expressing multiple genes of two or more viruses would show wide resistance spectrum.

It has been shown in our laboratory that the transgenic commercial tobacco cultivar G-140 plants expressing both CMV CP and Sat-RNA had higher levels of resistance to CMV infection than those expressing only CP or Sat-RNA alone (25,27). In addition, to facilitate production of multigene transgeneic plants, a new procedure for rapid production of homozygous transgenic line was developed in our laboratory. Three homozygous transgenic lines expressing both CMV CP and Sat-RNA had been obtained using this procedure. Field tests showed that all three lines had high levels of resistance to virus infection under both natural field infection and mechanical inoculation conditions (26). Recently, transgenic tobacco plants expressing multiple resistance genes derived from CMV, TMV and PVY (potato virus Y) were obtained and tested under field conditions. The results showed that the transgenic plants had higher and more wide-range resistance to control mixed viral infections under natural conditions.

In this chapter, we will present the procedure for rapid production of homozygous transgenic plant lines and some results of our field tests.

PROCEDURE FOR RAPID PRODUCTION OF HOMOZYGOUS TRANSGENIC LINES

In our laboratory, an attempt was made to develop a new procedure for obtaining homozygous transgenic lines rapidly with the best expressers of introduced traits. The procedure usually used for plant transformation by Ti-plasmid can be successfully applied to obtain transgenic plants. However, it needs several generations to select homozygous transgenic lines, since transformation of foreign genes into plants genomes via *A. tumefaciens* is a parasexual process (11). This is a very laborious and time-consuming procedure, often involving years of work.

In our procedure (Figure 1), the foreign genes are indirectly introduced into haploid plants from anther culture. The transformed haploid plants are analyzed for stable expression of the transgenes and tested for the introduced traits. The best expressers were made diploid with colchicine. The seeds produced by the homozygous diploid plants were then used in field tests.

To speed up the screening processes for the best expressers, the transgenic haploid plants are tested under high pressure. For example, transgenic plants are challenged with a high concentration of virus. In general, the transgenic plants expressing resistance genes show different levels of resistance, which may be caused, among other factors, by differences in the copy number or in the integration site of the introduced genes. When a low inoculation concentration of virus is used to screen the resistance of transformed plants, those plants with weak resistance will also be selected and used in the next tests, which will subsequently bring much more additional selections. When the transgenic plants are challenged with high concentration of virus inoculum, most plants showing weak resistance do not pass the test.

Using this procedure, we had successfully selected homozygous transgenic lines of tobacco in only two years. The plants of these transgenic lines showed high and stable inheritable resistance to CMV infection under field conditions.

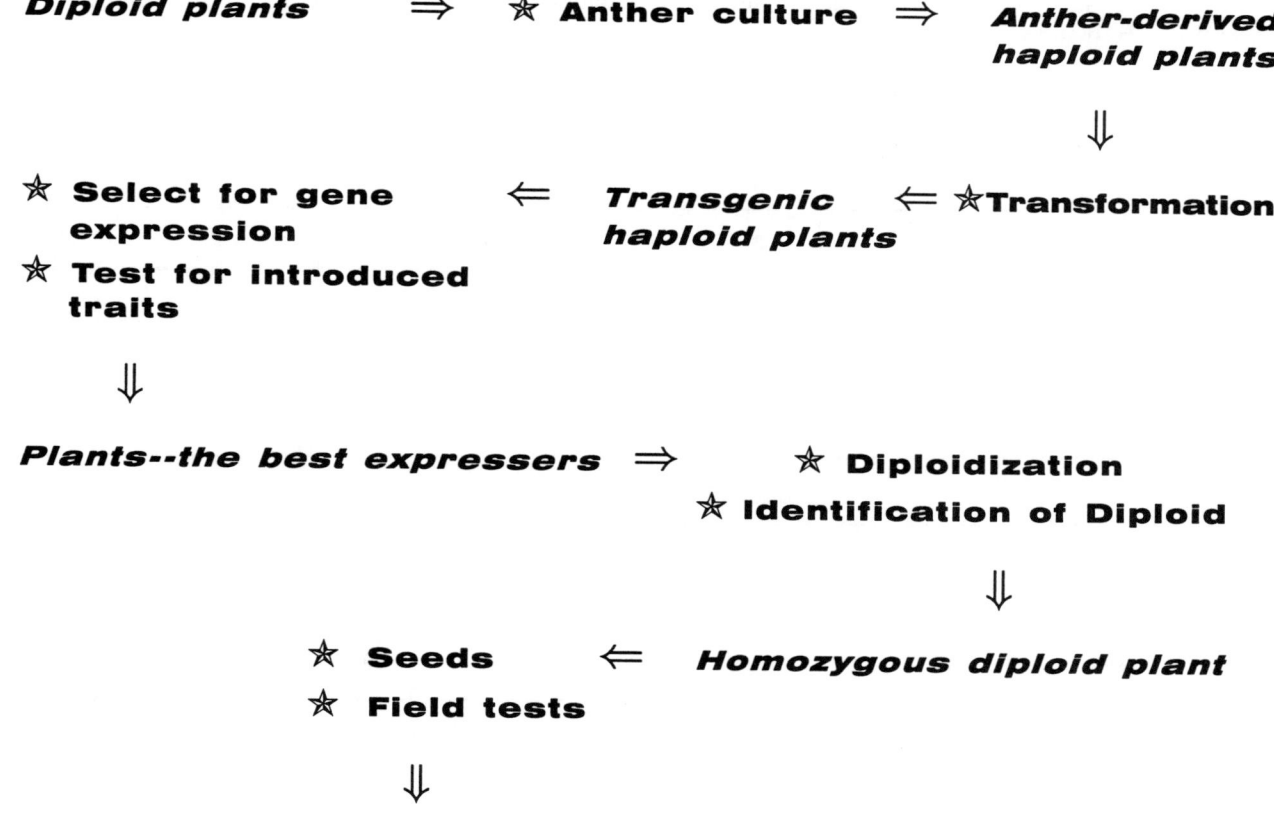

Figure 1. Procedure for fast production of homozygous transgenic lines

FIELD TESTS FOR HOMOZYGOUS TRANSGENIC TOBACCO LINES EXPRESSING BOTH CMV CP AND SAT-RNA

From 1990 to 1991, a chimerical gene expression vector pRCPI containing both CMV CP and Sat-RNA was constructed and transformed into commercial tobacco cultivar G-140 plants. Transformed plants were tested for both the expression of genes and their levels of resistance to CMV in the greenhouse. When challenged with 50 mg/ml of the B strain of CMV (containing no satellite RNA), the virus concentration in transgenic plants expressing only Sat-RNA was about 10-20% of that in non-transformed plants, while in CP+Sat-RNA transformed plants it was only 4-5%. By comparison of disease incidence and disease index among CP+Sat, CP alone or Sat-RNA alone transformed plants, the levels of resistance of plant expressing both CP and Sat-RNA were about twice as strong as that of plants expressing CP alone or Sat-RNA alone. Since the CP level expressed in CP+Sat transgenic plants is similar to that in CP alone transgenic plants, the two-fold increase in levels of resistance of CP+Sat-RNA transgenic plants was not due to the increase in level of CP expression. On the other hand, the level of Sat-RNA accumulation in CP+Sat-RNA transgenic plants was lower than that in the Sat-RNA alone transgenic plants. This may be due to the effect of the CP gene conferring resistance to an early stage in CMV infection, reducing the replication of CMV as Sat-RNA does (27).

In 1992, the new procedure described above was employed in rapid production of homozygous transgenic lines in our work. Leaf discs of haploid commercial cultivar NC89 tobacco plants from anther cultures were transformed with the chimerical vector pRCPI.

Table 1. Resistance of transgenic tobacco lines to CMV under natural infection and mechanical inoculation conditions (Hainan test)

NATURAL INFECTION[a]

transgenic genotype	Disease incidence of three replicates (%)				Significance[d]		Disease index[c] of three replicates (%)				Significance[d]	
	n_q	n_r	n_s	Avg	p<.05	p<.01	n_q	n_r	n_s	Avg	p<.05	p<.01
CP+Sat A	9.5	7.1	6.2	7.6	a	a	3.6	3.8	4.0	3.8	a	a
CP+Sat B	9.2	7.1	8.3	8.2	a	a	3.6	3.9	4.0	3.9	a	a
CP+Sat C	13.8	7.6	7.4	8.3	a	a	4.0	3.9	4.0	3.9	a	a
CP T	27.2	11.2	21.6	20.0	a	a	10.8	11.6	11.9	11.4	b	a
Controls	87.4	66.6	71.0	75.0	a	a	50.7	63.4	58.6	57.5	c	b

MECHANICAL INFECTION[b]

transgenic genotype	Disease incidence of three replicates (%)				Significance[d]		Disease index[c] of three replicates (%)				Significance[d]	
	n_q	n_r	n_s	Avg	p<.05	p<.01	n_q	n_r	n_s	Avg	p<.05	p<.01
CP+Sat B	10.6	15.0	14.3	13.3	a	a	7.6	7.9	9.1	8.3	a	a
CP+Sat A	27.8	30.4	20.4	26.2	b	ab	12.5	14.1	18.3	14.9	b	ab
CP+Sat C	29.2	34.7	20.4	28.1	b	ab	12.7	14.9	19.5	15.7	b	ab
CP T	79.4	82.1	66.5	76.0	c	c	40.3	41.3	50.0	44.5	c	b
Controls	100	100	100	100	d	d	100	100	100	100	d	c

[a] Data collected after transplanting for 90 days under natural infection. Disease incidence: t0.05 SD=9.98, t0.01=14.2; Disease index: t0.05 SD=6.18, t0.01 SD=9.29.

[b] Data collected after inoculation for 70 days under mechanical inoculation. Disease incidence: t0.05 SD=9.61, t0.01 SD=14.02; Disease index: t0.05 SD=7.00, t0.01 SD=10.67.

[c] Disease index was calculated from the formula: {[S(number of plants of each grade)×(disease grade)]/[(total number of plants)×(highest grade)]}×100 Disease grade was rated on a scale of 0-4, where 0=no symptoms and 4=most severe symptoms.

[d] Different symbols represent statistically significant difference at p<0.05 and p<0.01 level. For example, 'a' compared with 'a' means that there is no statistically significant difference between the two data points; an 'a' compared to a 'd' means that there is a statistically significant difference between the two data points.

One hundred and twelve kanamycin resistance transformed haploid plants were subjected to selection based on the expression of both genes. Eighty-nine transgenic haploid plants expressing both CP and Sat-RNA were selected and tested for their resistance to CMV. The transgenic haploid plants were transplanted into the greenhouse in April 1992 and mechanically inoculated with high concentration of CMV at 200 mg/ml 5 days after transplanting. Only five plants, showing no symptoms 30 days after challenge, were selected and diploidized with colichicine. Three homozygous diploid lines (A,B and C) with high level of resistance to CMV were obtained after only one generation. Seeds of the homozygous transgenic plants were recovered in fall 1992 by self-fertilization and were applied to further field tests.

The T1 plants from homozygous transgenic lines were subjected to field test in subtropical Hainan Province from October 1992 to March 1993. The field was 0.67 ha in area. Test materials involved five homozygous transgenic lines, three lines A, B and C expressing both CMV CP and Sat-RNA; one line expressing only CMV CP; the control NC89 plants, which were transformed with empty vector pROKII alone. Tobacco seedlings were grown on seedling beds in the greenhouse and then transplanted in the field at the four-leaf stage. No insecticide was applied in the field and insects, except aphids, were removed by hand if necessary. One portion of the test plants was left for natural viral infection, plants of the other portion was mechanically inoculated with tissue extracts of satellite-free CMV infected tobacco leaves. Symptom development was scored at 10-day intervals during the whole growth period.

The T2 plants from transgenic lines were subjected to field test in Henan Province from March to October 1993. The performance of this field test was the same as that in Hainan except

Table 2. Resistance of transgenic tobacco lines to CMV under natural infection and mechanical inoculation conditions (Henan test)

NATURAL INFECTION[a]

transgenic genotype	Disease incidence of three replicates (%)				Significance[d]		Disease index[c] of three replicates (%)				Significance[d]	
	n_q	n_r	n_s	Avg	p<.05	p<.01	n_q	n_r	n_s	Avg	p<.05	p<.01
CP+Sat A	5.2	4.0	5.5	4.9	a	a	2.5	2.6	3.3	2.8	a	a
CP+Sat B	7.2	4.2	6.0	5.8	a	a	3.6	3.3	4.3	3.7	a	a
CP+Sat C	4.6	10.0	4.5	6.5	a	a	3.2	4.0	3.0	3.4	a	a
CP T	13.0	16.3	15.5	14.9	b	ab	7.1	7.6	10.0	8.5	b	b
Controls	17.4	26.0	28.9	24.1	c	b	10.6	11.0	17.8	14.1	c	c

MECHANICAL INFECTION[b]

	n_q	n_r	n_s	Avg	p<.05	p<.01	n_q	n_r	n_s	Avg	p<.05	p<.01
CP+Sat B	10.0	18.8	19.2	16.0	a	a	6.3	5.7	5.7	5.9	a	a
CP+Sat A	11.4	15.1	20.0	15.5	a	a	7.1	6.4	6.5	6.7	a	a
CP+Sat C	11.4	12.0	38.1	20.5	a	a	6.1	8.3	10.7	8.7	a	a
CP T	52.2	34.0	57.7	48.0	b	b	34.2	28.5	30.1	30.9	b	b
Controls	100	100	100	100	c	c	89.2	79.3	82.3	83.7	c	c

[a] Data collected after transplanting for 90 days under natural infection conditions. Disease incidence: t0.05 SD=7.20, t0.01 SD=10.50; Disease index: t0.05 SD=3.9, t0.01 SD=5.7.

[b] Data collected after inoculation for 70 days under mechanical inoculation conditions. Disease incidence: t0.05 SD=17.9, t0.01 SD=26.10; Disease index: t0.05 SD=7.26, t0.01 SD=10.57.

[c] See table 1c

[d] Different symbols represent statistically significant difference at p<0.05 and p<0.01 level (see table 1d).

that the field was enlarged to 2 ha. In this test, agronomic characteristics besides the relative levels of resistance including plant height, stem girth, internode length, overall size of leaf, grade percentage of the cured leaf and yield were evaluated.

Identification of homozygous transgenic lines was performed on T1 plants and T2 plants. Random samples from T1 or T2 plants were analyzed for the stable expression of both CP and sat-RNA. All analyzed plants were positive for the expression of both genes, which supported that the plants were homozygous.

Analysis for the Relative Levels of Resistance of Transgenic Lines

One of the main purposes in the tests was to evaluate the high resistance stability to CMV infection The relative levels of resistance of transgenic lines were subsequently evaluated by comparing the values of disease incidence and disease index among all test plants.

In the field test carried out in Hainan, the test consisted of two parts: one for resistance analysis under natural aphid infection, in which the test field was situated in the area where CMV was widely prevalent, the other was for resistance analysis under mechanical inoculation conditions. Table 1 summarizes the values of disease incidence and disease index of the test materials under natural infection and mechanical infection. Under natural infection, 75% of the control plants (transformed with pROKII only) developed mosaic symptoms 90 days after transplanting. Some plants were severely symptomatic (disease index of 57.5%). In contrast, the transgenic lines A,B and C of NC89, which express both CMV CP and Sat-RNA, had much fewer plants with CMV symptoms with a disease incidence of 7.6-9.6% and a disease index of 3.8-4.0. For the transgenic line T, expressing only CMV CP, 20% of plants showed symptoms and the disease index was 11.4%, much less than the control NC89, but much higher than the homozygous diploid lines A,B and C. At the end of the test, the levels of symptoms in transgenic lines A, B and C had not increased, some even became lower. In the CP transgenic line and control NC89 plants, the symptoms were more severe. Under mechanical inoculation conditions, the transgenic lines A, B and C showed even better resistance to CMV infection than controls. All of the mechanically inoculated control NC89 plants were severely symptomatic 30 days after inoculation, with 100% disease incidence and 100% disease index. The transgenic lines A, B and C averaged markedly less infection (disease incidence=22.2%, disease index=13.0%). As observed under mechanical inoculation, the plants of transgenic line B showed the highest resistance: of 105 plants of line B, the disease incidence and disease index were 13.3% and 8.3% respectively. This trend was also observed under natural conditions. Symptom development in transgenic T plants was similar to that occurring under natural conditions, i.e. less than control NC89 and more severe than transgenic lines A, B and C.

Seeds obtained from T1 plants of transgenic lines A, B and C that did not show symptoms following inoculation with CMV in Hainan test were used for field tests in Henan. Under the natural infection conditions, 24% of the control NC89 plants were symptomatic 50 days after transplanting, and about 29% developed very severe mosaic and stunt symptoms (with the severity grade 4 as shown in Table 2). At that time, the percentage of symptomatic plants of transgenic lines A, B and C averaged 5.8% and none of them had the severity of grade 4. About 15% of the CP clone transgenic line plants showed symptoms. The results above point to the fact that the transgenic lines A, B and C showed even higher resistance to CMV infection compared with the field test conducted in Hainan.

The results from the test under mechanical inoculation conditions clearly showed the high resistance in transgenic lines A, B and C. Ten days after inoculation, only 6.8%, 4.5%, and 6.8% of the transgenic lines A, B and C, respectively, were symptomatic while in the control group NC89, 39% of plants showed symptoms. Subsequently, the infection rate of the control grew rapidly and the symptom expression became more severe. Fifty days after inoculation, all of the control plants were severely symptomatic.This is in contrast with 17.2% of transgenic plants lines A, B and C being mildly symptomatic. Disease incidence of transgenic line T plants, which expressed only CMV CP, was 48%, less than control plants and higher than transgenic lines A, B and C, showing the same response as mentioned above.

The data from the field test in Henan were consistent with those from the test in Hainan, which demonstrates that the plants of transgenic lines A, B and C have high and heritable resistance to CMV infection under field conditions.

Evaluation of Agronomic Characteristics and Yields of Transgenic Lines

To assess the potential of transgenic lines A, B and C in field application, a number of agronomic properties were investigated in the Henan field test. In general , there was no obvious difference between the transgenic lines and the non-transformed NC89 plants in the main agronomic characteristics such as plant shape, leaf shape, growth rates, etc., which suggested that the intrinsic cultivar characteristics had not been changed after transformation. However, when challenged with CMV, plants of transgenic lines A, B and C were superior to control NC89 plants, as is judged by these criteria (Tables 3, 4 and 5). The differences of the above traits were statistically significant between non-transformed NC89 and the transgenic lines A, B and C.

FIELD TEST FOR MULTIPLE-GENE RESISTANCE TO TWO VIRUSES OF TRANSGENIC TOBACCO COMMERCIAL VARIETY K326

As described above, three homozygous transgenic lines expressing both CMV CP and Sat-RNA showed high levels of resistance to CMV under field conditions. It is logical to suggest that if more than two genes of different types are transformed into one plant, stronger resistance will be achieved. Also, it is logical to suggest that if multiple genes from different viruses are transformed into one plant, wide-range and high resistance to mixed virus infections will be achieved.

Recently, a multiple-gene strategy of genetically engineered resistance against viruses was applied to control CMV, TMV and PVY mixed infections in our laboratory. Leaf discs of haploid plants from anther culture of a commercial tobacco cultivar K326 were co-transformed by vector pRCPI and vector pROTM containing 54K gene or (and) vector pROPY containing PVY CP

Table 3. Survey of agronomic properties of the tested transgenic lines in Henan

Transgenic lines	Plant height (cm)		Number of cured leaves per plant	Overall leaf size(cm)		Stem girth (cm)	Internode length (cm)
	no topping	topping		length	width		
CP+Sat A	165.0	112.0	23.4	66.6	33.8	9.8	3.90
CP+Sat B	170.0	111.0	23.8	61.4	31.6	9.1	3.98
CP+Sat C	173.0	111.0	23.1	63.4	32.6	9.5	3.88
CP T	150.0	94.5	22.1	58.0	30.0	8.7	3.70
Controls	148.5	92.0	21.5	57.4	28.7	8.6	3.70

Table 4. Comparison of yield of cured leaf of the transgenic lines under Henan field conditions

Transformed genotype	Transgenic lines	Yield of cured leaf of three replicates (Kg/mu[a])			Average	Significance[b]	
		I	II	III		$p<0.05$	$p<0.01$
CP+Sat	B(NC89)	195.5	198.0	202.1	198.5	a	a
	C(NC89)	193.4	189.6	190.9	190.6	b	b
	A(NC89)	186.4	181.6	188.4	185.5	c	c
CP	T(NC89)	182.9	175.6	182.1	180.2	d	cd
Controls	NC89	177.6	170.8	178.2	175.5	de	d

[a] 1mu=0.027ha.
[b] Different symbols represent statistically significant difference at $p<0.05$ and $p<0.01$ level. Here, t0.05 SD=5.02, t0.01 SD=7.31 (see Table 1d).

Table 5. Comparison of quality of tobacco leaf of the transgenic lines under Henan field conditions

Transformed genotype	Transgenic lines	Ratio of upper-grade cured leaf of three replicates (%)			Average	Significance[a]	
		I	II	III		$p<0.05$	$p<0.01$
CP+Sat	B(NC89)	36.8	34.8	38.6	36.7	a	a
	A(NC89)	36.1	32.5	34.1	34.2	ab	a
	C(NC89)	33.7	32.5	36.2	34.1	ab	a
CP	T(NC89)	34.6	33.6	31.0	33.1	bc	a
Controls	NC89	33.4	30.3	28.3	30.7	c	b

[a] Different symbols represent statistically significant difference at $p<0.05$ and $p<0.01$ level. Here, t0.05 SD=3.34, t0.01 SD=4.86 (see Table 1d).

Table 6. Comparison of disease incidence and disease index among transgenic lines under natural conditions

Transgenic genotype	Disease incidence at different days after transplanting (%)			Disease index[a] (%)
	30	50	75	
CMV CP-Sat+PVY CP	15.0	32.5	30.0	9.5
CMV CP-Sat-TMV 54K	14.5	27.0	24.0	10.5
CMV CP+Sat	18.0	49.0	47.5	14.4
PVY CP	18.5	58.5	62.0	24.6
TMV 54K	21.0	51.5	55.0	20.2
Control	24.0	78.5	94.0	67.5

[a] See Table 1c.

gene. Homozygous transgenic lines expressing CMV CP+CMV Sat-RNA+TMV 54K or CMV CP+CMV Sat-RNA+PVY CP were obtained and tested in field. The results of the two-year field tests showed that both disease incidence and disease index of the multiple-gene transgenic lines were much lower than two-gene or single-gene transgenic lines. The yield and quality of tobacco leaves were superior to controls. It is suggested that the multiple-gene strategy is an effective way for higher genetically engineered resistance to control mixed virus infection.

Rapid Production of Homozygous Transgenic Lines Expressing Multiple Genes

Plant expression vectors were constructed in our laboratory, including pRCPI , pROTM (containing TMV 54K gene) and pROPY. Vector pRCPI was co-transformed with pROTM or with pROPY via *Agrobacterium tumefaciens* into haploid plants. The kanamycin-resistance transformed haploid plants were analyzed for expression of the introduced genes. Fifty transgenic plants were found to express CMV CP+CMV Sat-RNA+PVY CP (termed CSP plants), 47 plants to express CMV CP+CMV Sat-RNA+TMV 54K (CST plants). Meanwhile, 30 plants to express both CMV CP and Sat-RNA (CS plants), 27 plants to express only PVY CP (PY plants), and 30 plants to express TMV 54K (TM plants) were selected as controls.

These transgenic plants were transplanted into the greenhouse and were mechanically inoculated with tissue extracts of mixed virus infected tobacco leaves. The transgenic plants showing no symptoms or only mild symptoms 30 days after inoculation were selected and diploidized with colchicine. Among these homozygous diploid plants, there were 4 plants expressing CMV CP+CMV Sat-RNA+PVY CP, 5 plants expressing CMV CP+CMV Sat-RNA+TMV 54K. The seeds produced by the homozygous diploid plants were then used in field tests.

Analysis for the Resistance to Mixed Viral Infections of the Homozygous Transgenic Lines in Field Tests

The Field Test of T1 Plants

To evaluate the levels of the resistance to mixed viral infections of the homozygous transgenic lines expressing multiple genes, the T1 plants of these lines were subjected to field test in Henan Province. The test materials involved fifteen individual transgenic lines, 5 CSP, 4 CST, 2 CS, 2 PY and 2 TM lines, as described above. Non-transformed K326 plants were set as control. The field was situated in the area where CMV, TMV and PVY were widely prevalent. Every test material was divided into two parts: one was tested under natural conditions and the other under mechanical infection. No insecticide was applied during the whole experiment. Both the symptom development and the agronomic characteristics were investigated at 10-day intervals during the

Table 7. Comparison of disease incidence and disease index among transgenic lines under mechanical infection

Transgenic genotype	Disease incidence at different days after transplanting (%)			Disease index[a] (%)
	15	30	45	
CMV CP-Sat+PVY CP	31.0	67.5	74.0	30.1
CMV CP-Sat-TMV 54K	22.5	62.0	62.0	28.3
CMV CP+Sat	47.0	80.5	84.0	42.4
PVY CP	68.5	100	100	90.4
TMV 54K	67.5	100	100	91.6
Control		97.0	100	100

[a] See table 1c

whole growth period.

Under natural infection conditions, there was little difference in disease incidence among all test plants, non-transformed and from homozygous transgenic lines, 30 days after transplanting. Later, the infection rate grew with significant difference, the control plants and those from CS, PY and TM lines became infected rapidly. Fifty days after transplanting, 49% of the CS plants became infected, much less than control K326 plants (disease incidence=78.5%), but much more than the CST or CSP plants (for disease incidence, CSP=32.5%, CST=27%) (see Table 6). After that time, the symptoms of the plants from CST, CSP and CS lines became milder and the disease incidence reduced as they grew, while the symptoms of those from PY and TM lines remained with few changes and insignificant increase in disease incidence. In contrast, symptoms of non-transformed K326 plants still developed quickly with great increase in disease incidence. Seventy-five days after transplanting, 94% of control K326 plants were symptomatic and most plants developed severe mosaic and stunted symptoms (disease index=67.5%). For the CSP plants expressing CMV CP+CMV Sat-RNA+PVY CP and CST plants expressing CMV CP+CMV Sat-RNA+TMV 54K, the levels of disease incidence and disease index were much lower than those of the other plants. Also, the plants from CS line, expressing both CMV CP and Sat-RNA, had lower levels of disease incidence and disease index than those from PY and TM lines expressing only one kind of genes. Significant differences in

disease incidence and disease index among the three-gene, two-gene and single-gene transgenic lines showed the difference in levels of resistance to mixed viral infections.

The results from the test under mechanical inoculation conditions were consistent with those from the test under natural infection conditions (Table 7). Fifteen days after inoculation, 97% of control K326 plants were symptomatic, while the infection rate of transgenic lines ranged from 22.5 to 67.5%. Obviously, CSP and CST lines were the best expressers, with only 31% and 22.5% of disease incidence, respectively. Afterwards, the infection rate of all transgenic lines increased rapidly with few changes in symptoms. Thirty days after inoculation, all control K326 plants and those from one-gene transgenic PY and TM lines showed symptoms, compared to 62 to 80.5% of infected plants in transgenic lines expressing two or three genes. Later, the symptom expression in plants from PY and TM lines became more and more severe, while the symptom expression in plants from CSP, CST and CS lines as well as the disease incidence remained with few changes throughout the whole test. Forty-five days after inoculation, all control K326 plants and a majority of the plants from PY and TM lines showed severe symptoms.

The Field Test of T2 Plants

For each transgenic genotype described above, only one transgenic line that showed both higher virus-resistance and better agronomic

Table 8. Resistance of transgenic tobacco lines to CMV under natural infection and mechanical inoculation conditions (Henan test)

NATURAL INFECTION[a]

transgenic genotype	Disease incidence of three replicates (%)				Significance[d]		Disease index[c] of three replicates (%)				Significance[d]	
	n_q	n_r	n_s	A~g	p<.05	p<.01	n_q	n_r	n_s	Avg	p<.05	p<.01
CMV CP-Sat +PVY CP	15.3	14.0	13.3	14.2	a	a	5.75	4.92	4.17	4.95	a	a
CMV CP-Sat -TMV 54K	15.0	15.7	16.3	15.7	a	a	5.92	6.08	6.75	6.75	a	a
CMV CP+Sat	26.3	26.7	27.0	27.7	b	b	10.92	12.08	11.08	11.36	b	b
PVY CP	65.3	70.7	60.3	65.4	c	c	26.83	29.17	26.17	27.39	c	c
TMV 54K	61.0	73.3	68.0	67.4	c	c	24.17	28.67	27.17	26.67	c	c
Controls	91.0	94.3	87.3	90.9	d	d	42.42	47.25	36.74	42.81	d	d

MECHANICAL INFECTION[b]

transgenic genotype												
CMV CP-Sat +PVY CP	27.5	32.5	30.0	30.0	a	a	13.75	15.63	12.50	13.96	a	a
CMV CP-Sat -TMV 54K	35.0	37.5	27.5	33.3	a	a	16.88	21.88	13.75	17.50	a	a
CMV CP+Sat	52.5	47.5	57.5	52.5	b	b	28.13	25.63	30.63	28.13	b	b
PVY CP	92.5	87.5	90.0	90.0	c	c	72.50	66.88	75.00	71.46	c	c
TMV 54K	85.0	87.5	90.0	89.2	c	c	68.75	75.00	79.38	74.38	c	c
Controls	100	100	100	100	d	c	99.38	100	100	100	d	d

[a] Data collected after transplanting for 90 days under natural infection. Disease incidence: t0.05 SD=6.28, t0.01=8.93; Disease index: t0.05 SD=3.63, t0.01 SD=5.16.
[b] Data collected after inoculation for 70 days under mechanical inoculation. Disease incidence: t0.05 SD=7.65, t0.01 SD=10.89; Disease index: t0.05 SD=6.59, t0.01 SD=9.38.
[c] See table 1c
[d] Different symbols represent statistically significant difference at p<0.05 and p<0.01 level (see Table 1d).

characteristics was selected from the T1 plants. It was subjected to further field testing in 1995. The performance of the field test was the same as that in the T1 field test. Every test line was tested under natural conditions and under mechanical infection conditions.

The multiple-gene transgenic plants showed highest levels of resistance to mixed viral infections under natural infection conditions or mechanical inoculation conditions (Table 8). Seventy-five days after transplanting, the percentage of symptomatic plants of transgenic CSP and CST lines averaged 14.2%, 15.7% respectively, while that of the control, transgenic PY and TM lines was higher, but with a very low disease index (4.95%). The percentage of symptomatic plants of transgenic CST line expressing CMV CP+CMV Sat-RNA+TMV 54K

averaged 15.7% with a very low disease index (6.75%). The results were the same in the field test of T1 plants, the transgenic CS line expressing both CMV CP and Sat-RNA had higher resistance than the single-gene transgenic PY and TM lines, but had lower resistance than the multiple-gene transgenic CSP and CST lines. At that time, only 27.7% of CS line plants were symptomatic, while the percentage of infected PY and TM lines plants averaged 65.4%, 67.4% respectively. As the plants grew, the control K326 plants, as well as PY and TM line plants, developed more and more severe symptoms. The results showed that PY and TM lines had lower resistance to mixed viral infections. When tested under mechanical inoculation with tissue extracts of mixed CMV, TMV and PVY infected tobacco leaves, the differences in levels of resistance to mixed viral

Table 9. Comparison of yield and quality of cured leaf among the transgenic lines

transformed genotype	Yield of cured leaf (Kg/mu[a])				Significance[b]		Ratio of uppergrade cured leaf (%)				Significance[c]	
	I	II	III	Avg	p<.05	p<.01	I	II	III	Avg	p<.05	p<.01
CMV CP-Sat +PVY CP	197.5	190.2	194.0	193.9	a	a	30.6	27.8	31.0	29.8	a	a
CMV CP-Sat -TMV 54K	194.3	187.2	180.1	186.9	b	a	32.3	25.5	27.6	18.5	a	a
CMV CP+Sat	188.6	187.4	181.0	185.7	b	b	28.2	25.5	26.1	26.6	b	b
PVY CP	181.9	183.7	178.8	181.5	bc	b	28.1	25.3	23.0	25.5	bc	b
TMV 54K	182.9	176.8	180.0	179.9	c	bc	28.4	22.6	23.3	24.8	c	b
Controls	186.5	170.6	172.8	173.3	d	c	26.4	24.3	25.9	25.5	c	b

[a] 1mu=0.027ha.

[b] Different symbols represent statistically significant difference at p<0.05 and p<0.01 level. Here, t0.05 SD=6.71,t0.01 SD=8.12 (see Table 1d).

[c] Different symbols represent statistically significant difference at p<0.05 and p<0.01 level. Here, t0.05 SD=2.71,t0.01 SD=3.86 (see Table 1d).

infections among these different transgenic lines became more obvious. Fifteen days after inoculation, all the control K326 plants and over 50% of the transgenic PY and TM line plants had developed symptoms. Moreover, the symptoms of the control K326 plants became more and more severe after that time and every one had severe mosaic and stunted symptoms 75 days after inoculation. As observed in test under natural conditions, the single-gene transgenic PY and TM lines showed much lower levels of resistance to mixed virus infection than multiple-gene transgenic CS, CSP and CST lines, indicating that the single-gene-mediated resistance might be overcome by high virus pressure, for example, by mixed virus infection.

The differences in levels of resistance among the transgenic lines were obviously shown by comparing their growth parameters (Table 9). Control K326 plants developed symptoms in the earlier stages of growth and the symptoms became more severe, their growth showed significant reduction.The same results were shown in single-gene transgenic PY and TM lines, which had less resistance. On the contrary, the multiple-gene transgenic CSP and CST lines displayed some superiority in growth due to their high levels of resistance in field. As a result, there were obviously superior in growth parameters such as yield and quality of the cured leaves, etc.

CONCLUSIONS

The data from the field tests carried out in China demonstrated that a multiple-gene strategy is a practical and effective way to obtain high levels of resistance to virus infection, especially to mixed viral infections. It is possible to use this strategy to obtain homozygous transgenic lines that can be used for practical plant production. The homozygous transgenic lines expressing both CMV CP and Sat-RNA showed higher levels of resistance than the transgenic T line that expressed only CMV CP under the natural conditions under which CMV was widely prevalent in test field. As a result, these transgenic lines showed no significant reduction in growth and had obvious superiority in growth parameters such as the yield and ratio of upper-grade cured leaf. The high levels of resistance to mixed viral infections were obtained in homozygous transgenic lines expressing multiple genes. The transgenic line, which expressed CMV CP+CMV Sat RNA+PVY CP, had much higher levels of resistance under the field infection condition under which the mixed CMV, TMV and PVY were widely prevalent. The same results were shown in the transgenic line expressing CMV CP+CMV Sat-RNA+TMV 54K. The single- gene transgenic lines, for example, that expressed only TMV 54K or only PVY CP, had weak resistance under high virus pressure and might be overcome.

As a strategy, the use of multiple-gene in

obtaining high resistance to mixed viral infections has certain advantages as reviewed previously, but it has some limitations. There are difficulties in obtaining multiple-gene transgenic plants. One way is to clone multiple genes into one plant expression vector and transform this multiple-gene vector into plants. Obviously, there is limitation in techniques. The other is to clone each gene into a vector and transform plants separately. Transgenic plants that are stable in inheriting and expressing multiple genes are then obtained by crossing. This is a very laborious and time-consuming procedure. Also, the multiple-gene transgenic plants are obtained by co-transformation as described above. However, the efficiency of co-transformation is rather low and the procedure for screening positive plants with multiple traits is laborious.

REFERENCES

1. Abel, P.P., Nelson, R.S., De, R., Hoffmann, N., Rogers, S.G., Fraley, R.T., and Beachy, R.N. 1986. Delay of disease development in transgenic plants that express the tobacco mosaic virus coat protein gene. Science 232:738-743.

2. Angenent, G.C., van den Ouweland, J.M.W., and Bol, J.F. 1990. Susceptibility to virus infection of transgenic tobacco plants expressing structural and non-structural genes of tobacco rattle virus. Virology 175:191-198.

3. Baulcombe, D.C., Saunders, G.R., Bevan, M.W., Mayo, M.A., and Harrison. B.D.1986. Expression of biologically active virus satellite RNA from the nuclear genome of transformed plants. Nature 321:446-449.

4. Cuozzo, M., O'Connell, K.M., Kaniewski, W., Fang, R.X., Chua, N.H., and Tumer, N.E. 1988. Viral protection in transgenic tobacco plants expressing the cucumber mosaic virus coat protein or its antisense RNA. Bio/Technology 6:549-557.

5. De Haan, P., Gielen, Jan, J.L., Prins M., Wijkamp, I.G., Schepen, A. van, Peters, D., Grinsven, M.Q.J.N. van, and Goldbach. R. 1992. Characterization of RNA-mediated resistance to tomato spotted wilt virus in transgenic tobacco plants. Bio/Technology 10:1133-1137.

6. Dun, C.M.P. van, and Bol, J.F. 1988. Transgenic tobacco plants accumulating tobacco rattle virus and pea early-browning virus coat protein resist infection with tobacco rattle virus. Virology 161:649-52.

7. Golemboski, D.B., Lomonossoff, G P., and Zaitlin, M. 1990. Plants transformed with a tobacco mosaic virus nonstructural gene sequence are resistant to the virus. Proc. Natl Acad. Sci. USA 87:6311-6315.

8. Harrison, B.D., Mayo, M.A., and Baulcombe, D.C. 1987. virus resistance in transgenic plants that express cucumber mosaic virus satellite RNA. Nature 328:799-802.

9. Haseloff, J., and Gerlach, W.L. 1988. Simple RNA enzymes with new and highly specific endoribonuclease activities. Nature 334:585-591.

10. Hemenway, C., Fang, R.X., Kaniewski, J.J., Chua, N.H., and Tumer, N.E. 1988. Analysis of the mechanism of protection in transgenic plants expressing the potato virus X coat protein or its antisense RNA. EMBO J. 7:1273-1280.

11. Hillman, J.R. 1992. Opportunities and problems in plant biotechnology - an overview. Pages   in: Opportunities and Problems in Plant Biotechnology. W. Powell and J.R. Hillman, eds. Royal Society of Edinburgh, Edinburgh, UK.

12. Lawson, C., Kaniewski, W., Haley, L., Rozman, R., Newell, C., Sanders, P., and Tumer, N.E.1990. Engineering resistance to mixed virus infection in a commercial potato cultivar: Resistance to potato virus X and potato virus Y in transgenic Russet Bubank. Bio/Technology 8:127-134.

13. Loesch-Fries, L.S., Merlo, D., Zinnen, T., Burhop, L., Hill, L., Krahn, K., Jarvis, N., Nelson, S., and Halk, E. 1987. Expression of alfalfa mosaic virus RNA4 in transgenic plants confers virus resistance. EMBO J. 6:1845-1851.

14. MacFarlane, S.A., and Davies, J.W. 1992. Plants transformed with a region of the 201-kilodalton replicase gene from pea early browning virus RNA1 are resistant to virus infection. Proc. Natl. Acad. Sci. USA 89:5829-5833.

15. MacKenzie, D.J., and Tremaine, J.H. 1990. Transgenic Nicotina debneyii expressing viral coat protein are resistance to potato virus S infection. J. Gen. Virol. 71:2167-2170.

16. Nejidat, A., and Beachy, R.N.1990. Transgenic tobacco plants expressing a coat protein gene of tobacco mosaic virus are resistant to some tobamo viruses. Mol. Plant-Microbe Interact. 3:247-251.

17. Nelson, R.S., McCormick, S.M., Delannay, X., Dube, P., Layton, J., Anderson, E.J., Kaniewska, M., Proksch, R.K., Horsch, R.B., Rogers, S.G., Fraley, R.T., and Beachy, R.N. 1988. Virus tolerance, plant growth, and field performance of transgenic tomato plants expressing coat protein from tobacco mosaic virus. Bio/Technology 6:403-409.

18. Rezaian, M.A., Skene, K.G.M., and Ellis, J.G. 1988. Antisense RNAs of cucumber mosaic virus in transgenic plants assessed for control of the virus. Plant Mol. Biol. 11:463-471.

19. Stark, D.M., and Beachy, R.N. 1989. Protection against potyvirus infection in transgenic plant: Evidence for broad spectrum resistance. Bio/Technology 7:1257-1262.

20. Tien, P., and Wu, G.S. 1991. Satellite RNA for the biocontrol of plant disease. Advances in Virus Research 39:321-339.

21. Tien, P., and Zhang, X.H. 1983. Control of two plant viruses by protection inoculation in China. Seed Sci. Technol. 11:969-972.

22. Tumer, N.E., O'Connell, K.M., Nelson, R.S., Sanders, P.R., Beachy, R.N., Fraley, R.T., and Shah, D.M. 1987. Expression of alfalfa mosaic virus coat protein gene confers cross protection in transgenic tobacco and tomato plants. EMBO J. 6:1181-1188.

23. van Dun, C.M.P., Bol, J.F., and Van Vloten-Doting, L. 1987. Expression of alfalfa mosaic virus and tobacco rattle virus coat protein genes in transgenic plants. Virology 159:299-311.

24. Wilson, T.M.A. 1993. Strategies to protect crop plants against viruses: pathogen-derived resistance blossoms. Proc. Natl Acad. Sci. USA 90.3132-3141.

25. Yie, Y., and Tien P. 1993. Plant virus satellite

RNAs and their role in engineering resistance to virus diseases. Seminars Virol. 4:363-368.

26. Yie, Y., Wu, Z.X., Wang, S.Y., Zhao, S.Z., Zhang, T.O., Yao, G.Y., and Tien, P. 1995. Rapid production and field testing of homozygous transgenic tobacco lines with virus resistance conferred by expression of satellite RNA and coat

protein of cucumber mosaic virus. Transgenic Research 4:256-263.

27. Yie, Y., Zhao F., Zhao, S.Z., Liu Y. L., and Tien P. 1992. High resistance to cucumber mosaic virus conferred by satellite RNA and coat protein in transgenic commercial tobacco cultivar G-140. Mol. Plant-Microbe Interact. 5:460-465.

CHAPTER 11

Releasing Genetically Engineered Virus Resistant Crops - Risk Assessment

G.M. Timmerman-Vaughan

The use of genetic engineering technology has become a powerful and successful strategy for producing virus resistant crop plants. In 1986, Powell-Abel et al. (65) first described the use of the tobacco mosaic virus (TMV) coat protein gene to produce virus resistant plants. Subsequently, similar experiments have been described using coat protein genes from a wide range of plant viruses, as discussed in review articles (45,80). Besides the use of coat protein genes, approaches involving the integration of other viral sequences into plant genomes have succeeded in producing virus resistant plants. These include the use of RNA replicase coding regions (12,34,54,55), mutant viral movement protein genes (8), untranslatable coat protein mRNA sequences (53), proteinase sequences (56) and transcripts from the nucleoprotein gene of an enveloped negative strand RNA virus (23). In general, these strategies have been more successful in producing virus resistance than the use of antisense RNA (18,41,66).

The common feature of these approaches is that the gene for resistance was derived from the pathogen's own genome, a concept that was proposed by Sanford and Johnston (75). The surprising array of viral genes able to produce observable resistance suggests that the development of plant viral disease can be disrupted by a number of different mechanisms. Understanding the precise mechanism(s) whereby viral gene products expressed from a transgene disrupts viral disease presents a special challenge, especially given our relative ignorance of the molecular events occurring during plant virus replication and spread.

The aim of this chapter is to consider the safety issues that may be posed by the widespread release of transgenic, virus-resistant crop plants.

Two broad issues will be considered: a) the application of transgenic technology to crop plants; and b) the use of plant virus derived resistance genes in crop plants.

GENERAL SAFETY ISSUES FOR TRANSGENIC CROP PLANTS

General Comments

A challenge in planning to release genetically modified crop plants from contained, small-scale trials into general agriculture is to identify whether these may pose risks to the environment, agriculture and human health. In most cases, genetically engineered crop plants contain a small number of well characterized genes which have been introduced deliberately. The resulting plants are generally indistinguishable from the parent organism, except they may express novel phenotypes resulting from the inserted genes. Since genetic manipulation methods have been used, the nature of the genetic change is readily characterized. It then becomes important to assess which, if any, of these novel crops have the potential to become pests. Recommendations for the release of genetically modified organisms, including crop plants, have been made in a number of published reports (59-62). Many governments have developed regulatory policies for assessing and managing the risks that may accompany the small scale field release of genetically modified organisms.

Many genetically engineered crop plants have a number of features in common. For example, the method used for DNA insertion into dicotyledonous plants is usually *Agrobacterium* Ti-plasmid mediated transformation. The result is the insertion of a piece of DNA into the plant genome. In addition, most transgenic plants contain and express a selectable marker gene which usually encodes resistance to an antibiotic or herbicide. Our current understanding of the safety of these common features of genetically engineered crop plants are discussed below.

Insertion of Novel DNA into Crop Plant Genomes - Risk Assessment

Use of Agrobacterium Ti-plasmid Mediated Transformation

Plants have been transformed by various physical means including microprojectile bombardment, micro-injection and electroporation, as well as through the use of a biological vector system. The most commonly used biological method is *Agrobacterium*-mediated transformation using Ti- or Ri-plasmids. This exploits the natural ability of the bacterium to transfer DNA from the bacterial cell into the plant nucleus, where it becomes stably integrated. The parent organism, either *A. tumefaciens* or *A. rhizogenes*, is a plant pathogen which causes crown gall disease in some plants. A central step in the development of crown gall disease is transfer of a portion of the Ti- or Ri-plasmid DNA, the T-DNA, into the plant nucleus. For plant genetic engineering purposes, the plasmid has usually been "disarmed" by removing the disease causing genes from the T-DNA. Historically, there has been concern about this method of plant transformation because it is based on the use of a known plant pathogen. However, general agreement has developed that there is negligible possibility of the disarmed vector system acting as a plant pest or mediating further transfer of DNA from the transformed plant (46,61). Huttner *et al.* (46) have made a complete compilation of the substantial arguments that the use of *A. tumefaciens* Ti-plasmid for the gene transfer process is safe and does not itself present any risk to the environment, agriculture or human health.

DNA Insertions into the Genome

A related question is whether any risk is associated with the presence of a DNA insertion in the plant genome. In most cases any genetic change is the result of a single insertion of a well characterized DNA fragment and the only change to the function of the organism or its genome results from expression of the gene(s) on the inserted DNA. Phenotypically, most plants regenerated from a transformation experiment resemble the parental line, and those plants selected for release are likely to be identical to the parental line but with the added desired phenotype(s). Unexpected phenotypes can be produced after a transformation experiment, however. Although DNA insertion could activate or inactivate host gene(s) in the immediate vicinity of the DNA insertion itself (pleiotropic effects), changes to plant phenotype are more likely to arise from "somaclonal variation", which produces uncharacterized changes that occur during tissue culture and regeneration (50). The tissue culture and regeneration processes have been shown to trigger genetic change. For example, regeneration of corn plants from tissue culture has been shown to activate transposable elements (63). In the vast majority of cases, pleiotropic effects of DNA insertion or somaclonal variation have deleterious effects on plant phenotype and robustness. Such plants are readily observed and eliminated. However, the possibility of unexpected phenotypes arising highlights the importance of examining plant phenotype carefully, both in the greenhouse and in the field.

Use of Selectable Markers - Risk Assessment

General Comments

The inclusion of selectable marker genes in the transferred DNA is common to most transgenic plants. These genes are usually antibiotic or herbicide resistance genes and they are included on the DNA fragment being transferred into plant cells so that a positive selection can be applied for the growth of transformed cells. The presence of these genes also helps confirm the transgenic nature of regenerated plants or subsequent progeny and will serve as ready tags to identify plants as transgenic.

NPTII Genes

The most widely used selectable marker in transgenic crop plants is the gene neomycin phosphotransferase ll (NPTII) which confers resistance to the neomycin/kanamycin group of antibiotics. The safety of the NPTII gene for use in crop plants was reviewed by Flavell *et al.* (29). In their article they examined the evidence that the NPTII gene is safe. The questions they addressed were: whether the NPTII gene product is toxic; whether eating the NPTII protein will affect antibiotic therapy; whether spread of the NPTII gene from plants to bacteria (if it should occur) will affect antibiotic therapy; and whether release of the NPTII gene from plants into the environment could cause damage. More recently, Fuchs *et al.* (32) tested the safety of the NPTII protein for human or animal consumption. They studied NPTII protein degradation in gastric fluids as well as the protein's acute toxicity in rodent feeding studies. These experiments supported the assertion that the NPTII gene is safe for widespread use.

Use of Other Selectable Marker Genes

In addition, other selectable marker genes besides NPTII may be used in crop plant genetic engineering. These include genes encoding proteins which inactivate antibiotics such as hygromycin B (23), bleomycin (42), methotrexate (38), and gentamicin (40); or herbicides such as phosphinothricin (21), sulfonyl-urea type herbicides (37) and the plant hormone analog 2,4 D (76). These selectable marker genes are less widely used than the NPTII gene, but are finding increased use. Before these other genes are considered to be safe to the environment and to human health, similar questions to those examined for the NPTII gene must be addressed.

Pollen Transfer of Herbicide Resistance Genes

Pollen transfer into wild relatives of crop plants is an obvious concern when herbicide-resistance genes are employed as selectable markers. Many crop plants are able to hybridize with wild relatives (36) and therefore the use of herbicide resistant transgenic crop plants could result in pollen transfer of herbicide-resistance genes to other plants, some of which may be considered weed species. In many cases, the use of herbicide resistance genes in transgenic plants will have minimal environmental impact; for example, in vegetatively propagated crops, crops that are harvested before flowering, and crops that do not hybridize with any weedy relatives. When considering the use of herbicide resistance genes in transgenic crop plants, factors to consider might include: the existence of weedy relatives which hybridize with the crop plant, whether these relatives are economically important weeds, and whether the herbicide inactivated by the inserted resistance gene product is used for control of wild relatives. A review discussing prediction and management of weed problems that might arise from use of transgenic plants was published (47).

Transferring Regulatory Emphasis to Plant Phenotype

A consensus of opinion has developed that the focus when evaluating and regulating genetically modified organisms should be the possible risks associated with plant phenotype(s) rather than the genetic techniques used. For example, this is a major conclusion of the comprehensive document prepared for the Ecological Society of America by Tiedje *et al.* (79). A body of evidence now exists to permit redirection of regulatory and environmental concern away from Ti-plasmid mediated transformation (46) and use of the NPTII selectable marker gene (29,32). Emphasis should now be placed on the novel combinations of properties produced by genetic engineering and whether these pose risks to the environment, agriculture and human health. Although novel plant phenotypes are produced by genetic engineering, they are also produced by more traditional activities such as plant breeding or by mutagenesis. In many cases, release of transgenic crop plants may not present new risks. In the discussion below, I consider the safety of releasing genetically engineered virus-resistant crop plants based on current knowledge. Our ability to estimate the risk associated with such releases in many cases is limited by our relative ignorance of the molecular nature of plant viral replication, evolution, and epidemiology.

USE OF PATHOGEN-DERIVED VIRUS RESISTANCE GENES - RISK ASSESSMENT

General Comments

The versatility of genetic manipulation technology for producing virus resistant crop plants has been demonstrated with the development of plants resistant to a wide array of plant viruses (45,80). This is a very promising application of genetic engineering technology and molecular biological research. Plant viral disease accounts for significant losses in crop yield and quality of produce and therefore novel methods for achieving virus resistance are likely to be warmly welcomed in agriculture. Introgression of naturally occurring virus resistance genes by plant breeding methods is a more traditional method for controlling plant viral disease (30,31). This approach, although proven and very powerful, is limited by the resistance genes available for any given crop species. Research indicates that genetic engineering of crop plants to express sequences such as coat protein or replicase genes may produce resistance to most if not all positive strand RNA viruses in any crop plant for which there is a transformation protocol (45,80). It is now essential to demonstrate whether widespread deployment of transgenic plants containing and expressing viral genes is safe for the environment, agriculture and human health. In this section the topics to be considered include: possible effects of these genes on viral pathogens, especially on their spread to non-host species and on viral evolution; possible effects of these genes on plant communities, especially the possibility of spread via pollen transfer; and possible effects on human health and food safety.

Heterologous Encapsidation, Vector Spread and Viral Epidemiology

In an editorial, De Zoeten (24) suggested that the expression in transgenic plants of plant viral coat protein molecules may change the epidemiology of plant viruses, increasing their spread to non-host species via a mechanism termed "heterologous encapsidation". Heterologous encapsidation is a term describing two similar phenomena: a) "transencapsidation", in which the nucleic acid of one plant virus is coated with capsid protein molecules of a second plant virus, and b) "phenotypic mixing" in which viral nucleic acid is coated with a mixture of two (or more) capsid proteins.

Concern about constitutive expression of plant viral coat protein genes in transgenic plants is due to the role of coat protein molecules in transmission by vectors, which was documented after the study of mixed infections. Most naturally occurring spread of plant viruses is mediated by the action of vectors, usually arthropods such as aphids. Three groups of plant viruses for which the virus-aphid interaction has been characterized are the luteoviruses (17,72), the potyviruses (64), and the cucumoviruses (33). Although the aphid characteristics specifying selectivity for transfer have not been identified, considerable progress has been made toward characterizing the viral components involved in aphid transmissibility and

determination of vector range. In luteoviruses such as barley yellow dwarf virus (17,72), aphid transmission specificity is determined by the viral coat protein. Similarly, in cucumber mosaic virus (CMV), a cucumovirus, aphid transmission is associated with strain related differences in the coat protein (33). In potyviruses there are two virus-encoded proteins which affect virus transmissibility, the coat protein (4,9,44,64) and the helper component (64).

Mixed virus infections are common m nature. In a minority of cases of mixed infections there are changes to the characteristics of viral transmission by aphids, such as when a non-transmissible virus becomes aphid transmissible in the presence of the second, co-infecting virus (73). This phenomenon is termed "dependent transmission". For example, in plants infected with two strains of barley yellow dwarf virus (RPV and MAV), the vector specificity is altered so that the MAV isolate, which is not usually transmissible by the aphid *Rhopalosiphum padi*, becomes transmissible by this vector (17,72). Both studies showed that the change in aphid transmissibility was due to heterologous encapsidation of the MAV genome with a combination of MAV and RPV capsid protein molecules. Similarly, a poorly aphid transmissible strain of CMV, CMV-6, became highly aphid transmissible either when acquired by aphids from mixed infections with other CMV strains or after reassembly with purified coat protein from a highly transmissible strain (33). There have been a number of studies of changes in potyviral trnnsmissibility from mixed infections (9,44,73). For example, Hobbs and Mclaughlin (44) examined the increased aphid transmissibility of been yellow mosaic virus (BYMV) strain "Scott" that occurred in mixed infections with pea mosaic virus. They produced evidence that both coat protein and helper components affected the aphid transmissibility of an isolate of BYMV-Scott. In a second study, Bourdin and Lecoq (9) showed that a non-aphid-transmissible isolate of zucchini yellow mosaic virus became aphid transmissible when present in mixed infections with papaya ringspot virus. The role of the coat protein was demonstrated by showing that heterologous encapsidation occurred. Furthermore, the helper component from zucchini yellow mosaic virus was shown to be fully functional and therefore was excluded as the reason for the increased transmissibility.

These studies, therefore, showed that heterologous encapsidation significantly altered the viral vector specificity and resulted in transmission by aphids that were not normally vectors. The relevance of these observations to transgenic plants expressing coat protein genes was demonstrated by Farinelli *et al.* (28) who observed heterologous encapsidation in transgenic plants expressing the PVY strain N coat protein and infected with PVY strain O. Other cases of mixed infection have been described including cases where the viruses involved were not from the same taxonomic groups (73). The mechanisms explaining these observations have not yet been characterized. Possible explanations include a specific interaction involving coat protein molecules. However, another possible explanation is that the concentration of the usually non-transmissible virus may be much greater in a mixed infection. Rochow (73) points out that virus concentration can play a role in the transmission of a virus by aphids. In a relevant paper, Vance (81) demonstrated that the concentration of potato virus X (a potexvirus) is much greater when in mixed infection with PVY (a potyvirus) than it is on its own.

These observations do not answer the question of whether heterologous encapsidation alters viral epidemiology so that "new" diseases arise. Since the viral genome is not altered, a "new" virus does not result from heterologous encapsidation. The heterologously encapsidated genome may be carried by a non-host aphid to a new plant species, but if this plant is not a host to the viral genome, the plant will not become diseased. Generally, non-susceptibility of a plant to a particular virus reflects a block to either replication or to cell-to-cell movement (20). Since the heterologously encapsidated viral genome is unchanged, it still lacks the sequences needed to establish an infection in a non-host plant. In this situation, heterologous encapsidation does not represent a risk since no further spread of the virus will occur.

However, Rochow (73) proposed a hypothetical scenario whereby heterologous encapsidation and changed vector specificity could result in creation of a "new" disease. A heterologously encapsidated virus particle may be picked up by a species of aphid that would normally not transmit the virus. This aphid may in turn feed on a susceptible plant that is not colonized by the usual aphid vectors for the plant virus. Therefore, the virus has been introduced to a new host because of the effect of heterologous encapsidation on aphid vector transmission. The result might be a new viral disease. Continued vector-mediated spread within the new host

population would depend on the existence of a suitable vector which feeds on the new host plant and can transmit the virus. Obviously this situation is not unique to transgenic plants and would be expected to occur when heterologously encapsidated virions are produced in double infected plants.

Clearly, complex interactions exist among plants, plant viruses and arthropod vectors, making the prediction of risk difficult. As discussed above, heterologous encapsidation is one way that plant viruses can overcome vector specificity. There are clearly situations involving transgenic plants where the possible environmental or epidemiological ramifications of heterologous encapsidation will be negligible, such as in cases where mixed infections are already common. In potato for example, mixed infections are very common involving viruses such as PVY, potato virus X, potato leafroll virus, potato virus A and potato virus S. Therefore, the use of transgenic potatoes containing coat protein genes from potato viruses will not introduce new combinations. Similar situations will exist for many crop plant species. Peas for example may be infected with alfalfa mosaic virus (AMV), soybean dwarf virus, CMV and pea seed-borne mosaic virus simultaneously, or tomatoes with tomato spotted wilt virus, and CMV or AMV. The situation might conceivably be different if the transgenic plant was engineered to contain and express a gene from a virus that does not normally infect that plant species. Therefore, the regulatory emphasis given to assessing the risk of changes to viral epidemiology should be appropriate to the novelty of the gene(s) involved. Less emphasis should be required where the proteins expressed in transgenic plants mimic those already found commonly in nature.

RNA Recombination and Viral Evolution

When considering the general release of transgenic virus resistant plants, a fundamental concern is our ignorance of the potential for new viruses to evolve as a result of RNA recombination involving the mRNA molecules expressed from a transgene containing plant viral sequences and infecting plant viral genomes. In the following section, the recent literature on plant viral recombination is reviewed and issues pertinent to the use of virus-derived sequences in transgenic crop plants are examined.

Although RNA recombination in a plant virus was first observed in 1986 (13), subsequent studies showed recombination in members of taxonomically diverse groups of positive strand RNA plant viruses. RNA recombination has been observed in laboratory experiments involving: the bromoviruses brome mosaic virus (BMV) (13,14,70,71) and cowpea chlorotic mottle virus (CCMV)(3); the tobamovirus tobacco mosaic virus (TMV) (7,19,51,69); and alfalfa mosaic virus (AMV) (82). Indirect evidence for the role of RNA recombination in the evolution of viral genomes has largely been obtained by comparative nucleotide sequence analysis. Sequence analysis has revealed natural RNA rearrangements in the bromoviruses (2), the tobraviruses (35), the nepovirus tomato ringspot virus (74), beet necrotic yellow vein virus (10), and barley stripe mosaic virus (26). RNA recombination produced the defective interfering (DI) RNAs found in three groups of plant viruses: the tombusviruses (43,48), the carmoviruses (52), and the potexviruses (84). Recombination has been reported involving satellite RNAs of the carmovirus turnip crinkle virus (16,85). In addition, sequence analysis has provided evidence for recombination between potato leafroll virus genomic RNA and a chloroplast exon sequence (57). The RNA genome of beet yellows virus, a closterovirus, appears to have evolved after multiple recombination events involving its own genome, cellular RNA, and RNA from other viruses (11). Clearly, recombination occurs widely among RNA plant viruses, can be observed within the modern timeframe, and has been a potent mechanism for viral evolution.

The mechanism of RNA recombination involving plant virus genomes is not known. Studies in animal viruses suggested two possible general mechanisms: copy choice (also called template switching) or breakage-and-religation (49). Among plant virologists the copy choice mechanism is generally favored and was reviewed by Bujarski and Nagy (15). The model for the copy choice mechanism is that the RNA-replicase complex, including the nascent transcript, dissociates from the template and then binds to and copies a new template, forming a recombinant transcript derived from two template RNAs. It is unclear how closely the two template molecules must interact. Many of the recombination events occur at or near regions of sequence homology or similarity. It is also speculated that RNA recombination may be favored in regions where the RNA template has extensive secondary structure, causing the RNA replicase complex to pause (49). The copy-choice mechanism could explain both homologous and non-homologous recombination, both of which are observed in plant viruses.

Recent research has also indicated that the success of recombinant plant viral RNAs is strongly influenced by selection pressures operational within the plant. In many experimentally induced cases of RNA recombination (2,13,14,69,70,71,82), the recombination events were observed because they restored wild type sequences and infectivity to mutant templates. In TMV, RNA recombination quickly deleted extra genes from genomes containing duplications of the coat protein gene (7), or the 30K protein gene (51) so that most of the viral progeny genomes were unit length.

Studies in plant as well as animal viruses have indicated that, in the presence of selection pressures, RNA recombinants are prevalent in non coding regions. The BMV recombinants resulted from exchanges within 3' non-coding regions (13,14,70,71), while the CCMV recombinations occurred in the intercistronic region between the 3a and coat protein genes (2). It was not possible to determine the exact site of recombination in AMV (82) because of the extensive sequence similarity (97%) between the molecules under investigation. Experiments conducted with animal RNA viruses also indicated the presence of recombination "hot spots". These generally occurred in intergenic regions, as discussed in a review by Lai (49). Unfortunately it has been difficult to distinguish whether apparent "hot spots" are preferred sites for recombination or are the accepted recombinations, due to selection pressures. Subsequently, however, the suggestion that apparent "hot spots" reflect selection processes has been supported by experimental evidence from an animal virus system. Banner and Lai (6) used a polymerase chain reaction (PCR) assay to detect recombination events, thereby avoiding the need for a selection step to detect recombinants. They found that RNA recombination occurred randomly.

The research summarized above clearly indicates that RNA recombination has been an important process shaping plant viral genomes for many groups of plant viruses. The next question, therefore, is whether the transcripts found in transgenic virus resistant crop plants will become included in the RNA recombination processes, resulting in the development of new viruses. Three relevant issues are examined in the discussion below.

Likelihood That mRNA and an Infecting Viral Genome Will Recombine

The possibility of recombination between plant viral genomic RNA and mRNA expressed from the viral sequences inserted into a genetically engineered plant must be considered. As mentioned above, nucleotide sequence data provide evidence of recombination between plant viral RNAs and cellular, transcribed sequences. Mayo and Jolly (57) found that the 5' end of the genome of a potato leafroll virus isolate contained a sequence nearly identical (109/119 bases) to an exon of a tobacco chloroplast open reading frame (ORF). In addition, the beet yellows virus genome was shown to contain sequences similar to heat shock proteins (1). The time frame over which these events occurred is unknown. The PLRV recombination appears to have occurred since PLRV diverged from other luteoviruses since the chloroplast ORF was only observed in the strain characterized by Mayo and Jolly (57). Although recombination between viral and cytoplasmic RNAs occurred, we have no way of estimating the likelihood of such an event occurring again. Furthermore, our ability to estimate the probability of such a recombination event occurring is limited by our ignorance of the mechanism(s) for homologous and non homologous recombination. Important factors affecting recombination frequencies between cellular and viral RNAs might include: the extent of sequence similarity between two RNA molecules; the cytoplasmic concentration of the two RNA molecules, and the presence of viral RNA replicase binding sites on the RNA substrates for recombination. Since homologous recombinants may be observed more often than non-homologous recombinants (58), genomic RNA from related viruses may have a better chance of recombining with transgene mRNA than unrelated viruses. With regard to the possible effect of RNA concentration, Knorr et al. (48) found that the recombination events that led to de novo DI RNA formation only occurred at high multiplicities of infection. The amount of mRNA expressed from incorporated virus-derived resistance genes, where characterized, has generally proven to be low compared with the levels occurring during virus infection (80). Therefore, low transcript concentrations may reduce the likelihood of recombinants occurring.

Another factor of unknown importance is the compartmentalization of RNA replication events in the cell. A common feature of plant and animal viral RNA replication systems is localization in membrane-associated compartments (22,39,67,68). There is no indication what access a cytoplasmic mRNA molecule might have to viral RNA replicase complexes.

An experiment using AMV tested whether recombination was detectable between a mutant plant viral genomic RNA and coat protein mRNA expressed from a transgene (25). Plants expressing the wild type AMV coat protein gene were inoculated with an RNA3 genome which contained a deletion in the coat protein gene. When the plants were examined there was no evidence for replication of these RNA3 molecules in either inoculated or systemic leaves, indicating that no RNA recombination occurred. The experiment also showed that the amount of coat protein expressed in the transgenic plants was inadequate to provide the replication function needed to complement the RNA3 mutation. In contrast, recombination involving AMV RNA3 was detected when plants were coinfected with mutant RNA3 genomes, as described above (82). There was a strong selection for recombinants in both experiments. By undergoing recombination, RNA3 molecules could have regained wild type function, permitting viral replication and spread.

Other experiments to test whether recombination is detectable between viral genomes and transgene mRNAs can be designed readily. It will be important for such experiments to be conducted to make accurate assessments of the chance that new RNA plant viruses could evolve via RNA recombination involving the transcripts from viral genes transformed into crop plants. Attempts to detect recombinants not subject to selection pressure could be made by PCR in experiments similar to those carried out by Banner and Lai (6). Alternately, attempts to select recombinants could be made based on the ability of sequences in the transgene message to rescue the infectivity of mutant viral genomes. In addition, systematic examination of the progeny viruses that have replicated in transgenic plants containing virus-derived resistance genes should produce valuable information about viral sequence changes that might occur in these plants. Sequence data obtained after inoculations with viruses related but not identical to the virus that provided the resistance gene might be particularly interesting.

Likelihood of Recombinants Surviving In Vivo Selection Pressures to Replicate and Spread Within the Host Plant

As discussed in a previous section, experimental evidence indicates that selection pressures may strongly limit the sequence combinations that survive and result in competitive plant viruses. Surviving recombinants often seem to result from exchanges outside of coding regions. Therefore, one strategy which may minimize the production by RNA recombination of successful viral genomes containing transgene sequences is to avoid the inclusion of intergenic, non-coding regions in the transcripts expressed from virus-derived resistance genes transformed into transgenic plants.

Likelihood of Truly Novel Combinations Being Produced via RNA Recombination

Because mixed infections are common (73), ample opportunities exist for RNA recombination to occur between viral genomes as long as the viruses are found in the same cells. The extent of the viral spread in mixed infections can range from "systemic infection" at one extreme, where viral replication occurs throughout the plant, through a "subliminal infection" where viral replication occurs only in the initially infected cell(s). Subliminal infections are common and may often go undetected (4,77).

This implies that many opportunities exist in nature for viable, successful recombinants to form between different viral genomes. Therefore, novel combinations produced from recombination between a transgene mRNA and viral genome may indeed be rare. In some situations, new combinations may be possible, such as deployment of virus-derived resistance genes so that they overcome geographic rather than biological boundaries.

Is Pollen Transfer of Resistance Genes to Related Weed Species a Threat?

Another possible concern is that these novel virus resistance genes could be transferred from genetically engineered crop species into closely related weed species by pollen transfer, resulting in a weed population with a competitive advantage because of its new virus resistance phenotype. Since some crop plants and wild relatives are able to hybridize (36), such an event is possible. The questions to ask are when or if such hybridizations pose an ecological or environmental threat.

There will be many instances when the prospect of pollen transfer of virus resistance genes from a transgenic crop plant poses no threat. For example, crop species may not hybridize with any wild relatives to produce viable progeny. Or, if hybridization occurs, the wild relative may already be resistant to the plant virus of interest. Therefore, for many instances where genetically engineered virus resistance genes would be used, all concerns about effects on wild relatives can be eliminated.

The question of pollen transfer of disease resistance genes into wild species is also relevant to new crop varieties produced by traditional plant breeding. The resistance genes employed are often derived by making wide crosses which include wild relatives of the crop plant being improved. This demonstrates that many disease resistance genes deployed via traditional plant breeding may function in a number of genetic backgrounds, including wild relatives of the same or closely related species. Furthermore, the genes introduced by plant breeders include genes for resistance to fungal and bacterial pathogens as well as to viruses. Plant breeding has a long history and the environmental and ecological risks associated with breeding for disease resistance have been accepted. From this perspective, it seems appropriate to ask whether pollen transfer from crop plants containing virus resistance genes produced using gene splicing methods poses any greater threat than deployment of new disease-resistant varieties produced by plant breeding methods.

Do Virus-derived Resistance Genes Pose Any Threat to Human Health?

The prospect that foods produced from genetically engineered plants may soon be marketed has prompted governments and other organizations to reexamine the assessment of the safety of foods produced by biotechnology. A FAO/WHO report provides a good review of important considerations (27). A conclusion reached is that modified plant foods produced by biotechnology (including new cultivars from traditional plant breeding) should be evaluated for safety by examining nutritional and toxicological properties as well as allergenic potential.

From a historical perspective, plant viral gene products expressed in transgenic plants appear to be ideally safe means of protecting crop plants from viral disease. Unless resistant varieties or rigorous control methods are used, plant viruses, and therefore their molecular constituents, are regularly found infecting crop plants. Systemically infected plant tissues contain much higher concentrations of viral gene products, such as coat protein molecules, than are expressed from the genes inserted into genetically engineered crop plants. A priori, therefore, toxicity of plant viral gene products themselves is not a concern. Further investigation of the safety of foods derived using these genes must be directed at possible pleiotropic effects that may have resulted from transformation or tissue culture procedures.

SUMMARY

Virus resistant crop plants have been produced by genetic engineering methodology using genes derived from viral pathogens. This approach is versatile and effective, and is likely to be deployed widely in agriculture. Therefore, assessments must be made of whether crop plants, genetically engineered to contain viral genes, pose risks to the environment, agriculture or human health.

In most cases, transgenic plants only differ from their parental variety by insertion and perhaps expression of a well characterized DNA fragment. Evaluation of hundreds of transgenic lines has shown that the process of Ti-plasmid mediated transformation is safe (46) and that the most commonly used selectable marker gene, NPT1I, is also safe (29,32). Examination of plant phenotype remains important (79), however, so that unwanted pleiotropic effects of tissue culture or DNA insertion can be detected. Genetic engineering research has illuminated the importance of such examination for crops produced by all biotechnological means, including traditional plant breeding methods, where large segments of uncharacterized DNA are introgressed by sexual crossing.

Future research must consider the effects of widespread deployment of virus resistant crop plants. In many cases, risk assessment will be simplified because the interactions between the virus, crop plant, other members of the plant community and arthropod vectors are well characterized. The method is so versatile, however, that it can also be applied to provide protection against less well characterized viral diseases. In these cases, risk assessment may not

be complete until the disease and its epidemiology are better characterized. Clearly our understanding of plant virus-host interactions and the effect of plant viruses on plant communities will increase dramatically as a result of the efforts made to introduce this means of viral disease control into agriculture.

ACKNOWLEDGEMENTS

Thanks to Nick Ashby, John Fletcher and Dave Beck for critical reading of the manuscript.

REFERENCES

1. Agranovsky, A.A., Boyko, V.P., Karasev, A.Y., Koonin, and Dolja, V.V. 1991. Putative 65 kDa protein of beet yellows closterovirus is a homologue of HSP70 heat shock proteins. J. Mol. Biol. 217:603-610.

2. Allison, R.F., Janda, M., and Ahlquist, P. 1989. Sequence of cowpea chlorotic mottle virus RNAs 2 and 3 and evidence of a recombination event during bromovirus evolution. Virology 172:321-330.

3. Allison, R., Thompson, C., and Ahlquist, P. 1990. Regeneration of a functional RNA virus genome by recombination between deletion mutants and requirement for cowpea chlorotic mottle virus 3a and coat genes for systemic infection. Proc. Natl. Acad. Sci. USA 87:1820-1824.

4. Atabekov, J.G., and Dorokhov, Y.L. 1984. Plant virus-specific transport function and resistance of plants to viruses. Adv. Virus Res. 29:313-364.

5. Atreya, P.L., Atreya, C.D., and Pirone, T.P. 1991. Amino acid substitutions in the coat protein results in loss of insect transmissibility of a plant virus. Proc. Natl. Acad. Sci. USA 88:187-191.

6. Banner, L.R., and Lai, M.M. 1991. Random nature of coronavirus RNA recombination in the absence of selection pressure. Virology 185:441-445.

7. Beck, D.L., and Dawson, W.O. 1990. Deletion of repeated sequences from tobacco mosaic virus mutants with two coat protein genes. Virology 177:462-469.

8. Beck, D.L., Van Dolleweerd, C.J., Lough, T.J., Balmori, E., Voot, D.M., Andersen, M.T., O'Brien, I.E.W., and Forster, R.L.S. 1994. Disruption of virus movement confers broad-spectrum resistance against systemic infection by plant viruses with a triple gene block. Proc. Natl. Acad. Sci. USA 91(22):10310-10314.

9. Bourdin, D., and Lecoq, H. 1991. Evidence that heteroencapsidation between two potyviruses is involved in aphid transmission of a non-aphid-transmissible isolate from mixed infections. Phytopathology 81:1459-1464.

10. Bouzoubaa, S., Niesbach-Klosgen, U., Jupin, I., Guilley, H., Richards, K., and Jonard, G. 1991. Shortened forms of beet necrotic yellow vein virus RNA-3 and -4: internal deletions and subgenomic RNA. J. Gen. Virol. 72:259-266.

11. Boyko, V.P., Karasev, A.V., Agranovsky, A.A., Koonin, E.V., and Dolja, V.V. 1992. Coat protein gene duplication in a filamentous RNA virus of

plants. Proc. Natl. Acad. Sci. USA 89:9156-9160.

12. Braun, C.J., and Hemenway, C.L. 1992. Expression of amino-terminal portions or full-length viral replicase genes in transgenic plants confers resistance to potato virus X infection. The Plant Cell 4:735-744.

13. Bujarski, J.J., and Kaesberg, P. 1986. Genetic recombination between RNA components of a multipartite plant virus. Nature 321:528-531.

14. Bujarski, J.J., and Dzianotti, A.M. 1991. Generation and analysis of non-homologous RNA-RNA recombinants in brome mosaic virus. sequence complementarities at crossover sites. J. Gen. Yirol. 65:4153-4159.

15. Bujarski, J.J., and Nagy, P.D. 1994. Genetic RNA-RNA recmbination in positive stranded RNA viruses of plants. Pages 1-24 in. Homologous recombination and gene silencing in plants. Kluwer Academic Publishers, The Netherlands.

16. Cascone, P.J., Carpenter, C.D., Li, X.H., and Simon, A.E. 1990. Recombination between satellite RNAs of turnip cnnkle virus. EMBO J. 9:1709-1715.

17. Creamer, R., and Falk, B.W. 1990. Direct detection of transcapsidated barley yellow dwarf luteoviruses in double infected plants. J. Gen. Virol. 71:211-217.

18. Cuozzo, M., O Connel, K.M., Kaniewski, W., Fang, R.X., Chua, N.H., and Tumer, N.E. 1988. Viral protection in transgenic tobacco plants expressing the cucumber mosaic virus coat protein or its anti sense RNA. Bio/Technology 6:549-557.

19. Dawson, W.O., Lewandowski, D.J., Hilf, M.E., Bubrick, P., Raffo, A.J., Shaw, J.J., Grantham, G.L., and Desjardins, P.R. 1989. A tobacco mosaic virus hybrid expresses and loses and added gene. Virology 172:285-292.

20. Dawson, W.O., and Hilf, M.E. 1992. Host range deterriunants of plant viruses. Annu. Rev. Pl. Physiol. and Pl. Molec. Biol. 43:527-555.

21. De Block, M., Botterman, J., Vandewiele, M., Dockx, J., Thoen, C., Gossele, Y., Rao Movva, N., Thompson, C., Van Montagu, M., and Leemans, J. 1987. Engineering herbicide resistance in plants by expression of a detoxifying enzyme. EMBO J. 6:2513-2518.

22. De Graaff, M., Coscoy, L., and Jaspars, E.M.J. 1993. Localization and biochemical characterization of alfalfa mosaic virus replication complexes. Virology 194:81-881.

23. De Haan, P., Gielen, J.J.L., Prins, M., Wijkamp, I.G., van Schepen, A., Peters, D., van Grinsven, M.Q.J.M., and Goldbach, R. 1992. Characterization of RNA-rdediated resistance to tomato spotted wilt virus in transgenic tobacco plants. Bio/Technology 10:1133-1137.

24. De Zoeten, G.A. 1991. Risk assessment: Do we let history repeat itself? Phytopathology 81:585-586.

25. Dore, J.M., van Dun, C.M.P., Pinck, L., and Bol, J.F. 1991. Alfalfa mosaic virus RNA3 mutants do not replicate in transgenic plants expressing RNA3-specific genes. Gen. Virol. 72:253-258.

26. Edwards, M.C., Petty, I.T.D., and Jackson, A.O. 1992. RNA rccombination in the genome of barley stripe mosaic virus. Virology 189:389-392.

27. FAO/WHO. 1991. Strategies for assessing the safety of foods produced by biotechnology. World Health Organization, Geneva.

28. Farinelli, L., Malnoe, P., and Collet, G.F. 1992. Heterologous encapsication of potato virus Y strain O (PYYO) with the transgenic coat protein of PVY strain N (PVYN) in Solanum tuberosum cv. Bintje. Bio/technology 10:1020-1025.

29. Flavell, R.B., Dart, E., Fuchs, R.L., and Fraley, R.T. 1992. Selectable marker genes: safe for plants? Bio/Technology 10:141-144.

30. Fraser, R.S.S. 1985. Genes for resistance to plant viruses. 1985 CRC Crit. Rev. in Pl. Sci. 3.257 294.

31. Fraser, R.S.S. 1990. The genetics of resistance to plant viruses. Annu. Rev. of Phytopathol. 28. 179-200.

32. Fuchs, R.L., Ream, J.E., Hammond, B.G., Naylor, Leimgruber, R.M., and Berberich S.A. 1993. Safety assessment of the neomycin phosphotransferase II (NPTII) protein. Bio/Technology 11:1543-1547.

33. Gera, A., Boebenstein, G., and Raccah, B. 1979. Protein coats of two strains of cucumber mosaic virus affect transmission by Aphis gossypii. Phytopathology 69.396-399.

34. Golemboski, D.B., Lomonosoff, G.P., and Zaitlin, M. 1990. Plants transformed with a tobacco mosaic virus nonstructural gene sequence are resistant to the virus. Proc. Natl. Acad. Sci. USA 87.6311-6315.

35. Goulden, M.G., Lomonosoff, G.P., Wood, K.R., and Davies, J.W. 1991. A model for generation of tobacco rattle virus (TRY) anomalous isolates. pea early browning virus RNA 2 acquires TRY sequences from both RNA-1 and RNA-2. J. Gen. Virol. 72:1751-1754.

36. Harlan, J.R. 1965. The possible role of weed races in the evolution of cultivated plants. Euphytica 14:173-176.

37. Haughn, G.W., Smith, J., Mazur, B., and Somerville, 1988. Transformation with a mutant Arabidopsis acetolactate synthase gene renders tobacco resistant to sulfonyl-urea herbicides. Mol. Gen. Genet. 211:266-271.

38. Hauptmann, R.M., Vasil, V., Ozias-Akins, P, Tabaeizadeh. Z., Rogers, S.G., Fraley, F.T., Horsch, R.B., and Yasil, I.K. 1988. Evaluation of selectable markers for obtaining stable transformants in the Gramineae. Plant. Physiol. 86:602~06.

39. Hayes, R.J., and Buck, K.W. 1990. Complete replication of a eukaryotic virus RNA in vitro by a purified RNA-dependent RNA polymerase. Cell 63:363-368.

40. Hayford, M.B., Medford, J.I. , Hoffman, N.L. , Rogers, and Klee, H.J. 1988. Development of a plant transformation selection system based on expression of genes encoding gentamicin acetyltransferases. Plant Physiol. 86:1216-1222.

41. Hemenway, C., Fang, R.X., Kaniewski, W.K., Chua, N.H., and Tumer, N.F. 1988. Analysis of the mechanism of protection in transgenic plants expressing the potato virus X coat protein or its antisense RNA. EMBO J. 7:1273-1280.

42. Hille, J., Yerheggen, F., Roelvink, P., Franssen, van Kammen, A., and Zabel, P. 1986. Bleomycin resistance: a new dominant selectable marker for plant cell transformation. Plant Molec. Biol. 7:171-176.

43. Hillman, B.I., Carrington, J.C., and Morris, T.J. 1987. A defective interfering RNA that contains a mosaic of a plant virus genome. Cell 51:427A33.

44. Hobbs, H.A., and McLaughlin, M.R. 1990. A non-aphid-transmissible isolate of bean yellow mosaic virus-Scott that is transmissible from mixed infections with pea mosaic virus-204-1.

Phytopathology 80:268-272.

45. Hull, R., and Davies, J.W. 1992. Approaches to nonconventional control of plant virus diseases. Crit. Rev. Pl. Sci. 11:17-33.

46. Huttner, S.L., Arntzen, C., Beachy, F., Bruening, Nester, E., Qualset, C., and Vidaver, A. 1992. Revising oversight of genetically modified plants. Biotechnology 10:967-971.

47. Keeler, K.H., and Turner, C.E. 1991. Management of transgenic plants in the environment. Pages 189-218 in: Risk Assessment in Genetic Engineenng. M. Levin and H. Straus, eds. McGraw-Hill, Inc., New York.

48. Knorr, D.A., Mullin, R.H., Hearne, P.Q., and Morris, T.J. 1991. *De novo* generation of defective interfering RNAs of tomato bushy stunt virus by high multiplicity passage. Virology. 193 202.

49. Lai, M.M.C. 1992. RNA recombination in animal and plant viruses. Microbiol. Rev. 56.61 79.

50. Larkin, P.J., and Scowcroft, W.R. 1981 Somaclonal variation a novel source of variability from cell cultures for plant improvement. Theor. Appl. Gen. 60:197-214.

51. Lehto, K., Grantham, G.L., and Dawson, W.O. 1990. Insertion of sequences containing the coat proteinsubgenomic RNA promoter and leader in front of the tobacco mosaic virus 30K ORF delays its expression and causes defective cell-to- cell movement. Virology 174:145-157.

52. Li, X., Heaton, L.A., Morris, T.J., and Simon, A.E. 1989. Turnip crinkle virus defective interfering RNAs intensify viral symptoms and are generated *de novo*. Proc. Natl. Acad. Sci. USA 86:9173-9177.

53. Lindbo, J.A., and Dougherty, W.G. 1992. Untranslatable transcripts of the tobacco etch virus coat protein gene sequence can interfere with tobacco etch virus replication in transgenic plants and protoplasts. Virology 189:725-733

54. Longstaff, M., Brigneti, G., Boccard, F., Chapman, and Baulcombe, D. 1993. Extreme resistance to potato virus X infection in plants expressing a modified component of the putative viral replicase. EMBO J. 12:379-386.

55. MacFarlane, S.A., and Davies, J.W. 1992. Plants transformed with a region of the 201-kilobase replicase gene from pea early browning virus RNA 1 are resistant to virus infection. Proc. Natl. Acad. Sci. USA 89:5829-5833.

56. Maiti, I.B., Murphy, J.F., Shaw, J.G., and Hunt, A.G. 1993. Plants that express a potyvirus proteinase gene are resistant to virus infection. Proc. Natl. Acad. Sci. USA 90:6110-6114.

57. Mayo, M.A., and Jolly, C.A. 1991. The 5'-terminal sequence of potato leafroll virus RNA: evidence for recombination between virus and host RNA. J. Gen. Viiol. 72:2591-2595.

58. Nagy, P.D., and Bujarsky, J.J. 1992. Genetic recombination in brome mosaic virus: effect of sequence and replication of RNA on accumulation of recombinants. J. Virol. 66:6824-6828.

59. National Academy of Sciences. 1987. Introduction of Recombinant DNA-engineered Organisms into the Environment: Key Issues. National Academy Press, Washington D.C.

60. National Research Council. 1989. Field Testing Genetically Modified Organisms: Framework for Decisions. National Academy Press, Washington D.C.

61. Organisation for Economic Cooperation and Development. 1990. Good Developmental Practices for Small Scale Field Research with Genetically Modified Plants and Microorganisms. OECD, Paris.

62. Organisation for Economic Cooperation and Development. 1992. Safety Considerations for Biotechnology. OECD, Paris.

63. Peschke, Y.M., Phillips, R.L., and Gengenbach, B.G. 1987. Discovery of transposable element activity among progeny of tissue culture derived maize plants. Science 238:804-807.

64. Pirone, T.P., and Thornbury, D. W. 1983. Role of virion and helper component in regulating aphid transmission of tobacco etch virus. Phytopathology 73.872 875.

65. Powell-Abel, P., Nelson, R.S., De, B., Hoffman, Rogers, S.G., Fraley, R.T., and Beachy, R.N. 1986. Delay of disease development in transgenic plants that express the tobacco mosaic virus coat protem gene. Science 232:738-743.

66. Powell-Abel, P., Stark, D.M., Sanders, P.R., and Beachy, R.N. 1989. Protection against tobacco mosaic virus in transgenic plants that express tobacco mosaic virus antisense RNA. Proc. Natl. Acad. Sci. USA 86:6949-6952.

67. Quadt, R. , and Jaspars, E.M.J. 1990. RNA polymerases of plus-strand RNA viruses of plants. Mol. Plant-Microbe Interact. 2: 219-223.

68. Quadt, R., Rosdorff, H.J.M., Hunt, T.W., and Jaspars, E.M.J. 1991. Analysis of the protein composition of alfalfa mosaic virus RNA-dependent RNA polymerase. Virology 182:309-315.

69. Raffo, A.J., and Dawson, W.O. 1991. Construction of tobacco mosaic virus subgenomic replicons that are replicated and spread systemically in tobacco. Virology 184:277-289.

70. Rao, A.L.N., and Hall, T.C. 1990. Requirement for a viral trans-acting factor encoded by brome mosaic virus RNA-2 provides strong selection *in vivo* for functional recombinants. J. Virol. 64:2437-2441.

71. Rao, A.L.N., Sullivan, B.P., and Hall, T.C. 1990. Use of *Chenopodium hybridum* facilitates isolation of brome mosaic virus RNA recombinants. J. Gen. Virol. 71:1403-1407.

72. Rochow, W.F. 1970 Barley yellow dwarf virus: phenotypic mixing and vector specificity. Science 167:875-878.

73. Rochow, W.F. 1972. The role of mixed infections in the transmission of plant viruses by aphids. Annu. Rev. of Pl. Pathol. 10:101-124.

74. Rott, M.E., Tremaine, J.H., and Rochon, D.M. 1991. Comparison of the 5' and 3' termini of tomato ringspot virus RNA1 and RNA2: evidence for RNA recombination. Virology 185:468-472.

75. Sanford, J.C., and Johnston, S.A. 1985. The concept of parasite-derived resistance - deriving resistance genes from the parasite's own genome. J. Theor. Biol. 113:395-405.

76. Streber, W.R., and Willmitzer, L. 1989. Transgenic tobacco plants expressing a bacterial detoxifying enzyme are resistant to 2,4-D. Bio/Technology 7:811-816.

77. Sulzinski, M.A.,and Zaitlin,M. 1982. Tobacco mosaic virus replication in resistant and susceptible plants: in some resistant species virus is confined to a small number of initially infected cells. Virology 121:12-19.

78. Thornbury, D.W., Hellmann, G.M., Rhoads, R.E., and Pirone T.P. 1985. Purification and characterization of potyvirus helper component.

Virology 144:260-267.

79. Tiedje, J.M., Colwell, R.K.,Grossman, Y.L., Hodson, R.E., Lenski, R.E., Mack, R.N., and Regal, P.J. 1989. The planned introduction of genetically engineered organisms: ecological considerations and recommendations. Ecology 70:298-315.

80. Timmerman, G.M. 1991. Genetic engineering for resistance to viruses. in: Advanced Methods in Plant Breeding. D. Murray, ed. CAB International, Oxford, U.K.

81. Vance, V.B. 1991. Replication of potato virus X RNA is altered in coinfections with potato virus Y. Virology 182:486-494.

82. Van der Kuyl, A.C., Neeleman, L., and Bol, J.F. 1991. Complementation and recombination between alfalfa mosaic virus RNA3 mutants in tobacco plants. Virology 183:731-738.

83. Waldron, C., Murphy, E.B., Roberts, J.L., Gustafson, Armour, S.L., Malcolm, S.K. 1985. Resistance
to hygromycin B. Plant Mol. Biol. 5:103-108.

84. White, K.A., Bancroft, J.B., and Mackie, G.A. 1991. Defective RNAs of clover yellow mosaic virus encode nonstructural/coat protein fusion products. Virology 183:479-486.

85. Zhang, C., Cascone, P.J., and Simon, A.E. 1991. Recombination between satellite and genomic RNAs of turnip crinkle virus. Virology 184:791-794.

CHAPTER 12

Antiviral Substances of Plant Origin

H.N. Verma, V.K. Baranwal and S. Srivastava

Plant viruses are a major cause of concern among agriculturists because of their economic impact on crops they infect. To reduce losses, scientists have explored several strategies to control virus infection. However, it is the exploitation of inherent resistance phenomenon and manipulation of inducible defence in plants which is receiving much attention by researchers. The common approach for introducing resistance against viruses in crops has been through conventional plant breeding. Limited success has been achieved through this method. Recent advances in molecular biology of resistance to virus infection have opened new approaches of making susceptible crops resistant against virus infection. These approaches include pathogen derived resistance to viruses (coat protein mediated resistance, movement protein mediated resistance, replicase and protease mediated resistance) and virus resistance through transgenic expression of antiviral proteins of non-viral origin (7).

The resistance induced by biotic and abiotic moieties has been termed as acquired /induced resistance. Acquired/induced resistance may be localized or systemic. The localized resistance has been reviewed by Loebenstein (38) and is manifested only in those areas where exogenous application has been made. Most of the antiviral plant extracts belong to this category. Such plant extracts manifest their activity in early stage of virus infection in plants (4). The practical utility of such virus inhibitors causing localized resistance has often been questioned. However, expression of pokeweed antiviral protein (PAP) from *Phytolacca* in transgenic tobacco and potato plants has created new possibilities for introducing resistance to a broad spectrum of plant viruses in susceptible crops (37).

The systemic acquired/induced resistance (SAR) by biotic or abiotic agents has now been recognized to play an important role in defense against plant viruses. Much of early work on this remained concentrated on cross protection. It is accomplished by direct interference of a mild strain of the same or related virus. The mechanism of cross protection has been dealt in detail by Ponz and Bruening (44) as well as in this book. It has been shown that it is mainly the coat protein of the virus which induces the resistance in cross protection (8) and such resistance has been designated as coat protein (CP) mediated resistance (54). The protection is relatively, specific and acts only against related viruses showing high level of homology in the amino acid sequence of their coat proteins. For example, in tobacco the expression of the coat protein of the soybean mosaic virus (SMV) results in protection against infection with potato virus Y (PVY) and tobacco etch virus (TEV) which are all members of the potyvirus group (52). The homology between the amino acid sequence of the coat proteins of these viruses is about 55%. However, tobacco plants expressing SMV coat protein gene do not show resistance to the unrelated tobacco mosaic virus (a member of the tobamovirus group).

The time taken for resistance development, induced by biotic and abiotic agents has been shown to vary. Ross (50,51) showed that the resistance developed in a tobacco variety after 7 days of infection with TMV in infected and uninfected leaves. Resistance acquired by tobacco after inoculation of lower leaves with *Thielaviopsis basicola* was detectable within 3 days, reached a maximum within 6-8 days and then declined (24). Similarly, systemic resistance induced in *Datura stramonium* by TMV could be detected in 4 days after introducing the inoculum and reached a maximum in 6 days. The localized resistance induced by chemicals like salicylic acid was observed after 10 days of its application (71).

The effect of polyacrylic acid (PA) was maximum when the injection of PA was made 2 to 3 days before inoculation of TMV to Xanthi-nc tobacco. In contrast, the resistance induced by botanicals is rather fast. Inhibition of the infection of sunnhemp rosette virus (SRV), a local isolate of TMV, was completely achieved in untreated upper leaves of *Cyamopsis tetragonoloba* whose basal leaves were treated with leaf extract of different species of *Clerodendrum* 24 hours before virus inoculation (68). Ostermann *et al.* observed that a very short period was required for inducing resistance by leaf extract from carnation plants (41).

The earliest demonstration of systemic resistance inducing ability by botanicals was made by Mckeen (39). He showed the inhibition of cucumber mosaic virus infection in untreated opposite primary leaf of cowpea whose other primary leaf was treated with extracts of pepper (*Capsicum frutescens*). He speculated that the pepper extract translocated from one primary leaf to the other leaf. It was only hypothetical as no experimental proof was given. Thereafter it was only in mid 1970s that the group working at Botany Department, Lucknow University, Lucknow came across brinjal plant whose leaf extract was able to induce systemic resistance in *Nicotiana glutinosa* against TMV and in *N. tabacum* against tobacco ringspot virus (60). Now a few more plants have been shown to induce much higher systemic induced resistance in different test hosts against different viruses. The response, however, varies in different hosts against different viruses. It appears that the botanicals inducing SAR adopt a different biochemical chain of events in inducing resistance against virus infection in plants compared to pathogen or chemical induced resistance. This chapter focuses mainly on the recent developments on botanical mediated plant virus resistance, because it is an area where there has been an explosion of research in the past decade (13).

BOTANICAL MEDIATED ANTIVIRAL PATHWAY/VIRUS RESISTANCE

Almost all of the well studied virus inhibiting substances from higher plants are now recognized as basic proteins. The proteins from *Mirabilis jalapa*, *Dianthus caryophyllus* and *Phytolacca americana* catalytically damage eukaryotic ribosomes and are hence called ribosome inactivating proteins or RIPs (5).

The RIPs have been shown to be N-glycosidases which remove a specific adenine base in a conserved loop of the 28S rRNA of eukaryotic organisms (15,16) or the 23S rRNA of prokaryotes (23). Such damaged ribosomes can no longer bind to the elongation factor-2 (18,49). RIPs, thus, damage ribosomes, arrest protein synthesis and cause cell death.

RIPs show antiviral activity against both animal and plant viruses (5) and can be classified into two types (56). Type 1: RIPs consist of a single polypeptide chain which is enzymatically active. These are scarcely toxic to animals and inhibit protein synthesis in cell free systems, but have little or no effect on whole cells. The three well known antiviral proteins (PAPs, Dianthins and MAP) contributing towards virus resistance belong to this category. Type 2: RIPs contain two types of polypeptide chains. Chain A is linked to chain B through a disulphide bond. Chain B binds the toxin to the cell surface and the A chain enzymatically inactivates the ribosomes (40). These are toxic to cells and inhibit protein synthesis in intact cells and in cell free systems.

Several similarities exist among the Type 1 RIPs. They are all basic proteins with a molecular weight in the range of 26-32 kDa. They show a high isoelectric point and are unusually stable, being resistant to denaturing agents and proteases. A majority of the RIPs are glycoproteins.

The proteins from *Boerhaavia diffusa* and *Clerodendrum aculeatum* indirectly prevent virus infection in plants. They themselves are not antiviral but provoke the plant system to produce a new protein in the treated plants which is the actual virus inhibitory protein. These proteins thus induce antiviral state in the plants through formation of *de novo* synthesized protein and perhaps are active in signalling the activation of defence mechanisms in susceptible hosts and hence have been called as systemic resistance inducers (SRIs).

CHARACTERISTICS OF ANTIVIRAL PROTEINS MEDIATING VIRAL DEFENCE IN PLANTS

Ribosome Inactivating Proteins

Pokeweed Antiviral Proteins (PAPs)

The antiviral protein present in the leaves

of *Phytolacca americana* (14) was purified to homogeniety and its molecular weght (Mr) determined as 29 kDa (28). This protein, called PAP, has a pI of 8.1 (30). The antiviral effect of PAP was most pronounced when it was coinoculated with the virus. PAP also inhibited virus infection when applied prior to virus challenge. Local lesion formation by TMV on *N. tcbacum* cv. Xanthi-nc was inhibited by nearly 70% even after 48 hours of treatment. The virus inhibitory effect of PAP increased with the decrease in the time lapse between treatment and challenge inoculation. PAP was less effective in preventing virus infection when applied a short time after virus inoculation. No inhibition was observed when PAP was applied 50 minutes after virus infection (11). PAP reduced infectivity of several mechanically transmitted RNA and DNA viruses when the purified virus or sap from virus infected plants was mixed with an equal volume of PAP solution and the mixture rubbed on the leaves of the local-lesion hosts (*N. glutinosa* /TMV; *Chenopodium quinoa* /CMV; *C. amaranticolor* /TMV, CMV, alfalfa mosaic virus, PVY; *Gomphrena globosa* /PVX) or systemic hosts (*Brassica campestris* /cauliflower mosaic virus; *N. benthamiana* / African cassava mosaic virus). PAP, thus, appears to be a general inhibitor of virus infection (11,53,57). PAP also shows antiviral activity against several animal viruses. It is toxic to cells infected with poliovirus (58) and influenza virus (57). It inhibits multiplication of herpes simplex virus type 1 (2) and human immunodeficiency virus (72). PAP did not show any detectable activity against two bacterial and six fungal pathogens (11). *P. americana* is now known to contain three other proteins with similar biological properties. PAP II is a basic protein with a Mr of 30 kDa and has a pI of 8.3 (29,30). PAP is synthesized at all stages of development and PAP II synthesis occurs with the ageing of plants (30). Tryptic maps suggest that little homology exists in the primary structures of PAP and PAP II and that the two proteins are possibly produced from different genes which are expressed at different times during the growth of the pokeweed plant (29).

PAP-S purified from the seeds of *P. americana* possessed a Mr of 31 kDa and it partially cross-reacted with anti-PAP (5). Amino acid sequence analysis of PAP and PAP-S clearly indicates that the two species of PAP are different. They vary considerably in the amino-terminal region (25). The complete amino acid sequence of PAP-S has been determined (35). The protein

consists of 261 amino acids and has an Mr of 29.167 kDa.

PAP, PAPII, PAP-S and PAP-R present in the roots of the pokeweed plant (9) are RIPs.

A cDNA clone for PAP was isolated and introduced into plants of *N. tabacum* cv Samsun, *N. benthamiana* and *Solanum tuberosum* cv Russet Burbank by transformation with *A. tumefaciens*. The transgenic plants that expressed PAP showed resistance to PVX, PVY and CMV after mechanical inoculation and aphid transmission (37).

Carnation Antiviral Proteins (DIANTHINS)

Sap from carnation leaves shows virus inhibitory activity (45,46,59). Dianthin 30 and Dianthin 32 were isolated from the leaves of *Dianthus caryophyllus* (55). Local lesion production by TMV on *N. glutinosa* was inhibited by 100% when the inhibitor was coinoculated with the virus (53). These are probably the same proteins purified by Ragetli and Weintraub (45). The Mr as determined by SDS -PAGE are 29.5 kDa and 31.7 kDa, respectively (55). Immunoelectrophoretical methods revealed that Dianthin 32 is distributed in the growing shoots and in the young and old leaves of *D. caryophyllus* and Dianthin 30 is distributed throughout the plant (48). The two are glycoproteins containing mannose and show a weak cross reaction. The nucleotide sequence of cDNA coding for Dianthin 30 has been determined (36). The carnation proteins are also inducers of systemic resistance (43).

Mirabilis Antiviral Protein (MAP)

The roots,leaves and stem of *Mirabilis jalapa* show high inhibitory activity against plant viruses. The *Mirabilis jalapa* leaf extract, when used as a foliar spray 24 hours prior to virus inoculation, suppressed the disease symptoms on a few systemic hosts (tomato/tomato yellow mottle virus; *Cucumis melo* var. momordica/CMV; *Cucumis sativus* /cucumber green mottle mosaic virus; tomato/tomato yellow mosaic virus; urd/yellow mosaic of urd) (63). A 50-60% reduction of the virus content in the treated plants was observed in the infectivity assays. *Mirabilis jalapa* extract was able to check the population of aphids and whiteflies and, thereby, control the natural spread of a few viruses on the systemic hosts (63).

The mirabilis antiviral protein (MAP)

isolated from the roots inhibits mechanical transmission of TMV, PVY, cucumber green mottle mosaic virus and turnip mosaic virus on local-lesion and systemic hosts and can induce systemic resistance of low order when applied to basal leaves of Xanthi-nc tobacco 24 hours prior to TMV inoculation of upper leaves (34). The purified protein consists of a single polypeptide without a sugar moiety and has a Mr of 24.2 kDa. It is a basic protein rich in lysine content, with a pI of 9.8 (34). The complete amino acid sequence of MAP has been determined. It consists of 250 amino acids and its Mr as determined from the amino acid sequence is 27,833 Da. The native MAP was resistant to protease digestion (20). MAP has two cysteine residues at positions 36 and 220 that are thought to make the disulfide bond (20). The presence of this bond lowers its enzymic activity and its elimination increases the ribosome inhibitory activity of MAP (21).

The E. coli transformants carrying a synthetic gene of MAP produced low amounts of MAP similar to the native MAP (20). The MAP cDNA consists of 1066 nucleotides and encodes for 278 amino acids. The extra 28 amino acid residues belong to the amino-terminal extra peptide which is comparable to the signal peptides of plant proteins (31). The product encoded by this cDNA is confirmed to be MAP precursor.

A genomic gene for MAP isolated by PCR revealed the presence of a short intron which is absent in the genes for other RIPs. It is suggested that the MAP genomic gene might be the least evolved among the known RIPs (32).

MAP produced by Mirabilis jalapa cells in suspension culture showed comparable biological activity with that of the roots and leaves and also reacted positively with anti-MAP (26). Several nutritional and hormonal factors also affect the formation of MAP by M. jalapa cells in suspension culture (27).

Infection of N. glutinosa leaves by TMV was inhibited by all RIPs tested (6). A positive correlation between the PAP concentration and the magnitude of inhibition of virus infection and inactivation of the ribosomal RNA was shown by Chen et al. (12). The mechanism by which the RIPs exert antiviral activity is, possibly, through inhibition of protein synthesis due to inactivation of the ribosomes of the infected cells. Ready et al. (47) suggested that PAP present in the cell wall matrix could enter the cells along with the virus, and once inside, it could inactivate ribosomes and block viral replication. Such a mechanism could be effective against almost all mechanically

transmitted viruses. PAPs antiviral function, in case of aphid transmitted viruses may be through a selective internalization of PAP into virus infected cells. RIPs have been shown to be more toxic to virus infected cells as compared to non-infected cells and they inhibit virus expression without inhibiting the host cell gene expression (10).

Systemic Resistance Inducers

Boerhaavia diffusa

The resistance inducer present in the roots of B. diffusa (BD) induces strong systemic resistance against several mechanically transmitted spherical and tubular viruses in local-lesion hosts (Datura metel, N. tabacum var. Ky-58 Burley, N. glutinosa/TMV; Cyamopsis tetragonoloba/SRV; Vigna sinensis/SRV, GMV; C. amaranticolor/TMV, SRV, GMV, TRSV; Spinacea oleracea/GMV) and systemic hosts (N. tabacum cv. NP-31/TMV; Crotolaria juncea/SRV; Gomphrena globosa/GMV; N. glutinosa/TRSV; V. radiatus, V. mungo/yellow mosaic of mung and urd; tomato/CMV, TMV) when applied 24 hours before virus inoculation or when tested after mixing with the inoculum (3,61,69). The resistance inducer is a glycoprotein (70-80% protein + 8-13% carbohydrate) with a molecular weight in the range of 16-20 kDa as determined by gel filtration chromatography (67). The SRI is a basic glycoprotein with a pI greater than 9.0 and gives a molecular weight of 30 kDa on SDS polyacrylamide gels (unpublished data). The resistance inducing protein is found to be extremely thermostable (61). Following treatment with the systemic resistance inducing protein, the hosts produce a virus inhibitory agent (VIA). The VIA shows characteristics of a protein and reduces infectivity of the viruses, both in vitro and in vivo (62). The VIA synthesized is neither host nor virus specific and is not synthesized in the presence of actinomycin D (62).

Clerodendrum aculeatum

Leaf extract from Clerodendrum aculeatum when sprayed prevents infection of mechanically and white fly transmitted viruses in several local-lesion hosts (C. amaranticolor /TMV, SRV, tomato yellow mosaic virus, GMV; D. stramonium, D. metel, N. glutinosa, N. tabacum var. Ky-58/TMV; C. tetragonoloba /SRV) and a systemic host (tobacco/tobacco leaf curl virus) (66,68). The resistance induced is of a systemic nature and can be reversed by the simultaneous application of actinomycin D (68). The inhibitor in crude sap was resistant to denaturation by organic solvents and was extremely thermostable (68). The purified inhibitor is a basic glycoprotein with a pI of 8.65. It has a Mr of 34 kDa and retains its biological activity upon incubation with pronase, trypsin, proteinase K and pepsin (H.N. Verma, unpublished).

The first systemic resistance inducer was, however, isolated from a fungus *Trichothecium roseum*. It was characterized as polysaccharide and was named T-poly (19). A polysaccharide from *Fomes fomentarius* (Basidiomycetes) has also been demonstrated to induce systemic resistance against TMV in Xanthi-nc tobacco (1). The other systemic resistance inducers from higher plants have not been thoroughly characterized (70). In case of plant fungal interaction only a specific region of glycoprotein moiety from *Phytophthora megasperma* f.sp. *glycinea* functioned as signal for production of coumarin phytoalexin by parsley cells and protoplasts. N-linked glycosidic moiety played no role (42). Thus more studies are required to define which part of systemic resistance inducing glycoprotein is active in signalling the activation of defence mechanism in host-virus interaction.

Mechanism of Systemic Induced Resistance by Botanicals

Systemic induced resistance against virus infection by botanicals has not been fully understood. The botanical resistance inducers themselves do not act on the virus directly. Verma and Awasthi (62) demonstrated the *de novo* synthesis of an antiviral substance(s) in untreated leaves of *N. glutinosa*, whose basal leaves were treated with root extract of *B. diffusa*. These induced substances inhibited almost completely tobacco mosaic virus in *N. glutinosa*, *Datura stramonium* and *D. metel* but inhibition of tobacco ring spot virus or gomphrena mosaic virus in *Chenopodium amaranticolor* was less pronounced. In another study Verma and Dwivedi (64) found that virus inhibiting agent (VIA) induced in *C. tetragonoloba* inhibited completely the infection of tobamoviruses in all the 7 hypersensitive hosts tested. Yet in another study by Khan and Verma (33), it was observed that VIA produced in *C. tetragonoloba* following treatment with extract of

Pseuderanthemum bicolor inhibited completely SRV, TMV, cucumber green mottle mosaic virus and PVX in their respective hypersensitive hosts viz., *C. tetragonoloba*, *D. stramonium*, *C. amaranticolor* and *G. globosa*. No attempt has been made about comparative studies for VIA production in different hosts by different resistance inducers. However, the VIA produced by *B. spectabilis* in *N. tabacum* cv. samsun NN, *N. glutinosa*, *D. stramonium* was less effective against TMV (64). Thus it seems induction of systemic resistance by botanicals is non-specific and is effective against a broad spectrum of viruses.

It has been demonstrated that the production of VIA is maximum after 24 hours of application of *B. diffusa* root extract in *N. glutinosa*. However, *B. spectabilis* induced maximum VIA activity after 48 hours of its application in *N. glutinosa*. Thus VIA production in a host is time specific but a general phenomenon. Properties of VIA produced in *C. tetragonoloba* after application of *B. spectabilis* leaf extract has been studied by Verma and Dwivedi (64). The VIA could be precipitated by ammonium sulphate and hydrolyzed by trypsin but not by ribonuclease indicating that it was proteinaceous rather than a nucleic acid. The VIA induced by *Pseuderanthemum bicolor* in *C. tetragonoloba* was also proteinaceous with a Mr of 15 kDa and contained 18 amino acids (33). The production of VIA was sensitive to actinomycin D (64). Thus actinomycin D sensitive mechanisms operated for the production of VIA.

It is reasonable to think that a mobile inducing signal may be produced in treated leaves after these botanical resistance inducers bind with the host plant surface. The mobile signal produces VIA in the entire plant system. It appears that the putative messenger becomes active soon after induction and starts producing VIA, reaches a maximal concentration and then starts to decline. The role of salicylic acid (SA) as endogenous signal in systemic acquired resistance has been studied (17) but role of SA in systemic resistance induced by botanical substances has not yet been studied.

Commonly associated with systemic acquired resistance induced by pathogens is the systemic synthesis of several families of serologically distinct low molecular weight pathogenesis related (PR) proteins. The localization and timing of some PR proteins suggested their possible involvement in acquired resistance against viruses. However, definite proof that the induction of PR proteins cause the acquired resistance has not been given. Plobner and Leiser (43) did not find the production of PR proteins during systemic resistance induced by carnation extract in Xanthi-nc tobacco plants. Absence of PR proteins during induction of SAR by botanicals suggests that another biochemical chain of reaction other than one operating in pathogen/chemical induced SAR might be operating during botanical induced systemic resistance against virus infection in plants.

The agricultural role of endogenous antiviral substances of plant origin has been reviewed by Verma *et al.* (70). Hansen (22), while reviewing antiviral chemicals for plant disease control, gave the following characteristics of ideal antiviral compound that will serve all purposes for virus disease management in crops:

1. Soluble in water or non phytotoxic solvents.
2. Effective against at least some agriculturally or horticulturally important viruses at non phytotoxic concentration.
3. Easily taken up by plants and distributed throughout the system.
4. Non toxic by itself and its catabolic forms to humans, plants and wildlife.

Botanical resistance inducers can be classified as ideal virus suppressing agents as they qualify all the characteristics of ideal antiviral compound. The resistance inducing proteins from *Boerhaavia diffusa* and *Clerodendrum aculeatum* can be applied directly by spraying on systemic hosts for management of some commonly occurring virus diseases under greenhouse conditions or field conditions in microplots. However, their effectiveness on larger scale in fields needs to be evaluated.

The induced antiviral state in *N. glutinosa* by SRI from *C. aculeatum* decreased considerably after 3 days (65). However, the resistance inducing ability of *C. aculeatum* SRI could be enhanced up to 6 days by priming it with certain proteinaceous additives (65). The activity is probably enhanced due to modification or enhanced stability of proteinaceous inducers by these additives. Habuka *et al.* (21) have demonstrated that by removing intramolecular disulphide bond by genetic engineering in *Mirabilis* antiviral protein (MAP), the activity could be enhanced by 22 times above the native MAP. Similar studies in case of proteinaceous systemic resistance inducers might lead to production of highly concentrated preparation of systemic resistance inducers. However, the expression of resistance inducers in transgenic plants may overcome the constraint of their large scale production. Thus, one unexploited approach to engineering virus resistance is the manipulation of inducible defence in plants. The production of systemic resistance by use of botanical genes will be effective against a broad spectrum of viruses and will not break down when plants will be exposed to high temperatures.

FUTURE STUDIES

Systemic resistance inducers (SRIs) and ribosome inactivating proteins (RIPs) of plant origin are very promising agents for preventing infection of diseases in several susceptible hosts. A high level of systemic resistance/tolerance could be induced following application of these antiviral proteins prior to infection by the viruses. Systemic resistance inducers are non-phytotoxic, promote plant growth, improve crop yield and quality and show a broad spectrum of protection. For large scale production of resistance inducing proteins to manage virus diseases in the field, modern techniques, like tissue culture and genetic engineering need to be explored. It would be a significant advancement, if the antiviral protein genes could be cloned and used in transgenic plants to obtain broad spectrum virus resistance, in addition to current approaches of inserting viral genes, coat protein, satellite RNA and antisense strategies to obtain resistant plants. A begining has already been made towards development of virus resistant crop plants with the insertion of the PAP gene in tobacco and potato plants.

REFERENCES

1. Akoi, M., Tan, M., Fukushima, A., Hieda, T., Kubo, S., Takabayashi, M., Ono, K., and Mikami, Y. 1993. Antiviral substances with systemic effects produced by basidiomycetes such as *Fomes fomentarious*. Biosci. Biotechnol. and Biochem. 57: 278-282.

2. Aron, G. M. and Irvin, J. D. 1980. Inhibition of herpes simplex virus multiplication by the pokeweed antiviral protein. Antimicrob. Agents Chemother. 17, 1032-1033.

3. Awasthi, L. P., Chaudhury, B., and Verma, H. N. 1984. Prevention of plant virus diseases by *Boerhaavia diffusa* inhibitor. Int. J. Trop. Pl. Dis. 2: 41-44.

4. Baranwal, V. K., and Verma H. N. 1992. Localized resistance against virus infection by leaf extract of *Celosia cristata*. Plant Pathol. 41: 633-638.

5. Barbieri, L., and Stirpe, F. 1982. Ribosome

inactivating proteins from plants: Properties and possible uses. Cancer Surveys 1: 489-520.

6. Barbieri, L., Battelli, M. G., and Stirpe, F. 1993. Ribosome-inactivating proteins from plants. Biochim. Biophys. Acta. 1154: 237-282.

7. Baulcombe, D. 1994. Novel strategies for engineering virus resistance in plants. Current Opinion In Biotechnology 5:117-124.

8. Beachy, R. N., Loesch-Fries, S., and Turner, N. E. 1990. Coat protein mediated resistance against virus infection. Annu. Rev. Phytopathol. 28 451-457.

9. Bolognesi, A., Barbieri, L., Abbondanza, A., Falasca, A. I., Carnicelli, D., Battell, M. G., and Stirpe, F. 1990. Purification and properties of new ribosome inactivating proteins with RNA N-glycosidase activity. Biochem. Biophys. Acta. 1087: 293-302.

10. Bonness, M. S., Ready, M. P., Irvin, J. D., and Mabry, T. J. 1994. Pokeweed antiviral protein inactivates pokeweed ribosomes; implications for the antiviral mechanisms. The Plant Journal 5: 173-183.

11. Chen, Z. C., White, R. F., Antoniw, J. F., and Lin, Q. 1991. Effect of pokeweed antiviral protein (PAP) on the infection of plant viruses . Plant Pathol. 40: 612-620.

12. Chen, Z. C., Antoniw, J. F., and White, R. F. 1993. A possible mechanism for the antiviral activity of pokeweed antiviral protein. Physiol. Mol. Pl. Pathol. 42: 249-258.

13. Chessin, M., DeBorde, D., and Zipf, A. 1995. Antiviral proteins in higher plants. CRC Press, Boca Raton, Florida.

14. Duggar, B. M., and Armstrong, J. K. 1925. The effect of treating the virus of tobacco mosaic with the juices of various plants. Annal. Mo. Bot. Gard. 12:359-366.

15. Endo, Y., and Tsurugi, K. 1987. RNA N-glycosidase activity of ricin A chain. Mechanism of action of the toxic lectin ricin on eukaryotic ribosomes. J. Biol. Chem. 262: 8128-8130.

16. Endo, Y., Mitsui, K., Motizuki, M., and Tsurugi, K. 1987. The mechanism of action of ricin and related toxic lectins on eukaryotic ribosomes. The site and the characteristics of the modification in 28S ribosomal RNA caused by the toxins. J. Biol. Chem. 262: 5908-5912.

17. Gaffney, T., Friedrich, L., Vernooij, B., Negrotto, D., Nye, G., Uknes, S., Ward, E., Kessmann, H., and Ryals, J. 1993. Requirement of salicylic acid for the induction of systemic acquired resistance. Science 261:754-756.

18. Gessner, S. L., and Irvin, J. D. 1980. Inhibition of elongation factor2-dependent translocation by the pokeweed antiviral protein and ricin. J. Biol. Chem. 255: 3251-3253.

19. Gupta, B. M., Chandra, K., Verma, H. N., and Verma, G. S. 1974. Induction of antiviral resistance in Nicotiana glutinosa plants by treatment with Trichothecium polysaccharide and its reversal by actinomycin D. J. Gen. Virol. 24: 211-213.

20. Habuka, N., Murakami, Y., Noma, M., Kudo, T., and Horikoshi, K. 1989. Amino acid sequence of Mirabilis antiviral protein, total synthesis of its gene and expression in Escherichia coli. J. Biol. Chem. 264: 6629-6637.

21. Habuka, N., Miyano, M., Kataoka, J., Tsuge, H., Ago, H., and Noma, M. 1991. Substantial increase of the inhibitory activity of Mirabilis antiviral protein by an elimination of the disulfide bond with genetic engineering. J. Biol. Chem. 266: 23558-23560.

22. Hansen, J. 1989. Antiviral chemicals for plant disease control. Crit. Rev. Plant Science. 8, 45-88.

23. Hartley, M. R., Legname, G., Osborn, R., Chen, Z., and Lord, J. M. 1991. Single-chain ribosome inactivating proteins from plants depurinate Escherichia coli 23S ribosomal RNA. FEBS Lett. 290: 65-68.

24. Hecht, E. I., and Bateman, D.F. 1964. Nonspecific acquired resistance to pathogens resulting from localized infections by Thielaviopsis basicola or viruses in tobacco leaves. Phytopathology 54: 523-530.

25. Houston, L. L., Ramkrishnan, S., and Hermodson, M. A. 1983. Seasonal variation in different forms of pokeweed antiviral protein, a potent inactivator of ribosomes. J. Biol Chem. 256: 9601-9604.

26. Ikeda, T., Takanami, Y., Imaizumi, S., Matsumoto, T., Mikami, Y., and Kubo, S. 1987. Formation of anti plant viral protein by Mirabilis jalapa L. cells in suspension culture. Plant Cell Report. 6: 216-218.

27. Ikeda, T., Niino, K., Kataoka, J., and Matsumoto, T. 1987. Effects of nutritional and hormonal factors on the formation of an anti plant viral protein by Mirabilis jalapa L. cells in suspension culture. Agric. Biol. Chem. 51: 3119-3124.

28. Irvin, J. D. 1975. Purification and partial characterization of the antiviral protein from Phytolacca americana which inhibits eukaryotic protein synthesis. Arch. Biochem. Biophys. 169: 522-528.

29. Irvin, J. D., Kelly, T., and Robertus, J. D. 1980. Purification and properties of a second antiviral protein from Phytolacca americana which inactivates eukaryotic ribosomes. Arch. Biochem. Biophys. 200: 418-425.

30. Irvin, J. D. 1983. Pokeweed antiviral protein. Pharmac. Ther. 21: 371-387.

31. Kataoka, J., Habuka, N., Furuno, M., Miyano, M., Takanami, Y., and Koiwai, A. 1991. DNA sequence of Mirabilis antiviral protein (MAP), a ribosome inactivating protein with an antiviral property, from Mirabilis jalapa L. and its expression in Escherichia coli. J. Biol. Chem. 266: 8426-8430.

32. Kataoka, J., Miyano, M., Habuka, N., Masuta, C., and Koiwai, A. 1993. A genomic gene for MAP, a ribosome inactivating protein from Mirabilis jalapa, contains an intron Nucleic Acids Res. 21: 1035.

33. Khan, M. M. A., and Verma, H. N. 1990. Partial characterization of an induced virus inhibitory protein associated with systemic resistance in Cyamopsis tetragonoloba (L.) Taub. plants. Ann. Appl. Biol. 117: 617-623.

34. Kubo, S., Ikeda, T., Imaizumi, S., Takanami, Y., and Mikami, Y. 1990. A potent plant virus inhibitor found in Mirabilis jalapa L. Ann. Phytopathol. Soc. Jpn. 56: 481-487.

35. Kung, S. S., Kimura, M., and Funatsu, G. 1990 The complete amino acid sequence of antiviral protein from the seeds of pokeweed (Phytolacca americana). Agric. Biol. Chem. 54: 3301-3318.

36. Legname, G., Bellosta, P., Gromo, G., Modena, D., Kun, J. N., Roberts, L. M., and Lord, J. M. 1991. Nucleotide sequence of cDNA coding for Dianthin 30, a ribosome inactivating protein from Dianthus caryophyllus. Biochem. Biophys. Acta 1090:

119-122.

37. Lodge, K. J., Kaniewski, W. K., and Tumer, N. E. 1993. Broad spectrum virus resistance in transgenic plants expressing pokeweed antiviral protein. Proc. Natl. Acad. Sci. USA. 90: 7089-7093.

38. Loebenstein, G. 1972. Localization and induced resistance in virus infected plants. Annu. Rev. Phytopathol. 10: 177-206.

39. McKeen, C.D. 1956. The inhibitory activity of extracts of *Capsicum frutescens* on plant virus infections. Canad. J. Bot. 34: 891-903.

40. Olsnes, S., and Pihl, A. 1982. Toxic lectins and related proteins. Pages 51-105: Molecular action of toxins and viruses. P. Cohen and S. Van Heyningen. eds. Elsevier Biomedical Press, Amsterdam.

41. Ostermann, W. D., Meyer, U., and Leiser, R. M. 1987. Induction of plant virus resistance, 2., leaf extracts from carnation plants (*Dianthus caryophyllus*) as inducer of resistance. Zen-trabl. Mikrobiol. 142: 229-238.

42. Parker, J. E., Schutte, W., Hahlbrock, K., and Scheel, D. 1991. An extracellular glycoprotein from *Phytophthora megasperma* f. sp.glycinea elicits phytoalexin synthesis in cultured parsley cells and protoplasts. Mol. plant-Microbe Interact. 4: 19-27.

43. Plobner, L., and Leiser, R-M. 1990. Induction of virus resistance by carnation protein. VIIIth International congress of virology, Berlin (Abstr.).

44. Ponz, F., and Bruening, G. 1986. Mechanism of resistance to plant viruses. Annu. Rev. Phytopathol. 24: 355-381.

45. Ragetli, H. W. J., and Weintraub, M. 1962. Purification and characteristics of a virus inhibitor from *Dianthus caryophyllus* i) Purification and activity. Virology 18: 232-240.

46. Ragetli, H.W.J., and Weintraub, M. 1962. Purification and characteristics of a virus inhibitor from *Dianthus caryophyllus* L. ii) Characterization and mode of action. Virology 18: 241-248.

47. Ready, M. P., Brown, D. T., and Robertus, J. D. 1986. Extracellular localization of pokeweed antiviral protein. Proc. Natl. Acad. Sci. USA. 83: 5053-5056.

48. Reisbig, R. R., and Bruland, O. 1983. Dianthin 30 and 32 from *Dianthus caryophyllus*: Two inhibitors of plant protein synthesis and their tissue distribution. Arch. Biochem. Biophys. 224: 700-706.

49. Rodes , T. L., and Irvin, J. D. 1981 Reversal of the inhibitory effects of the pokeweed antiviral protein upon protein synthesis. Biochim. Biophys. Acta. 652: 160-167.

50. Ross, A. F. 1961. Systemic acquired resistance induced to localized virus infections in plants. Virology 14: 329-339.

51. Ross, A. F. 1961. Systemic acquired resistance induced to localized virus infections in plants. Virology 14: 340-358.

52. Stark, D.M., and Beachy, R.N. 1989. Protection against potyvirus infection in transgenic plants. Evidence for broad spectrum resistance. Bio/Technology 7: 1257-1262.

53. Stevens, W. A., Spurdon, C., Onyon, L. J., and Stirpe, F. 1981. Effect of inhibitors of protein synthesis from plants on tobacco mosaic virus infection. Experientia 37: 257-259.

54. Stiekema, W.J., Visser, B., and Florack, D.E.A. 1993. Is durable resistance against virus and bacteria attainable via biotechnology?: Durability of disease resistance. Th. Jacobs and J. E. Parlevliet.

55. Stripe, F., Williams, D.G., Onyon, L.J., Legg, R.F., and Stevens, W. A. 1981. Dianthins, ribosome damaging proteins with antiviral properties from *Dianthus caryophyllus* L.(carnation). Biochem. J. 195: 399-405.

56. Stirpe, F., Barbieri, L., Battelli, M. G., Soria, M., and Douglas, A. 1992. Ribosome-inactivating proteins from plants: Present status and future prospects. Bio/technology 10: 405-412.

57. Tomlinson, J.A., Walker, V.M., Flewett, T.H., and Barclay, G. R. 1974. The inhibition of infection of cucumber mosaic virus and influenza virus by extracts from *Phytolacca americana*. J. Gen. Virol. 22: 225-232.

58. Ussery, M.A., Irvin, J.D., and Hardesty, B. 1977. Inhibition of polio virus replication by a plant antiviral peptide. Ann. N.Y. Acad. Sci. 284: 431-440.

59. Van Kammen, A., Norordam, D., and Thung, T.H. 1961. The mechanism of inhibition of infection with tobacco mosaic virus by an inhibitor from carnation sap. Virology 14: 100-108.

60. Verma, H.N., and Mukherjee, K. 1975. Brinjal leaf extract induced resistance to virus infection in plants. Indian J. Exptl. Biol. 13: 416-417.

61. Verma, H. N., and Awasthi, L. P. 1979. Antiviral activity of *Boerhaavia diffusa* root extract and the physical properties of the virus inhibitor. Can J. Bot. 57: 926-932.

62. Verma, H. N., and Awasthi, L.P. 1980. Occurrence of a highly antiviral agent in plants treated with *Boerhaavia diffusa* inhibitor. Can. J. Bot. 58: 2141-2144.

63. Verma, H.N., and Kumar, V. 1980. Prevention of plant virus diseases by *Mirabilis jalapa* leaf extract. New Botanist 7: 87-91.

64. Verma, H. N., and Dwivedi S. D. 1984. Properties of a virus inhibiting agent, isolated from plants following treatment with *Bougainvillea spectabilis* leaf extract. Physiol. Pl. Pathol. 25: 93-101.

65. Verma, H. N., and Varsha. 1994. Induction of durable resistance by primed *Clerodendrum aculeatum* leaf extract. Indian Phytopathol. 47: 19-22.

66. Verma, H. N., and Varsha. 1995. Prevention of natural occurrence of tobacco leaf curl disease by primed *Clerodendrum aculeatum* leaf extract. Pages 202-206 in: Detection of Plant Pathogens and their Management. J. P. Verma, A. Verma and D. Kumar. eds. Angkor Publishers (P) Ltd. New Delhi.

67. Verma, H. N., Awasthi, L. P., and Saxena, K. C. 1979. Isolation of the virus inhibitor from root extract of *Boerhaavia diffusa* inducing systemic resistance in plants. Can. J. Bot. 57: 1214-1217.

68. Verma, H. N., Chowdhury, B., and Rastogi, P. 1984. Antiviral activity in leaf extracts of different *Clerodendrum* L. species. Z. Pflanzenkr. Pflanzensch. 91: 34-41.

69. Verma, H. N., Rastogi, P., Prasad, V., and Srivastava, A. 1985. Possible control of natural virus infection on *Vigna radiatus* and *Vigna mungo* by plant extracts. Ind. J. Pl. Pathol. 3: 21-24.

70. Verma, H.N., Varsha, and Baranwal, V.K. 1995. Agricultural role of endogenous antiviral substances of plant origin. Pages 23-37 in: Antiviral Proteins in Higher Plants. M. Chessin, D. DeBorde and A. Zipf. eds. CRC Press, Boca Raton, Florida.

eds. Kluwer Academic Publishers, The Netherlands.

71.   White, R. F. 1979. Acetylsalicylic acid (aspirin) induces resistance to tobacco mosaic virus in tobacco. Virology 99: 410-412.

72.   Zarling, J.M., Moran, P.A., Haffer, O., Sias, J., Richman, D.D., Spina, C.A., Myers, D.A.,

Kuelbelbeck, V., Ledbetter, J. A., and Uckun, F. M. 1990. Inhibition of HIV replication by pokeweed antiviral protein targeted to CD4+ cells by monoclonal antibodies. Nature 347: 92-95.

Resistance to Plant Viruses - an Overview

J. Hammond

Two groups of chapters deal with resistance to, or control of, plant virus diseases. These can be separated into what can be viewed as natural resistances or approaches, and genetic engineering approaches, and will be discussed in that order.

Khetarpal *et al.* (Chapter 2) describe traditional resistance breeding (i.e. using resistance genes derived from other cultivars or related species). Where resistance has not been identified in other cultivars or breeding lines, the most likely source will be wild accessions or related species from the center of diversity of the crop in question. They discuss the mechanics of screening for resistance, and the range of virus isolates that should be tested to ensure adequate resistance to the spectrum of isolates prevalent in areas where the crop will be grown. They also note the necessity of screening the inoculum for the presence of other viruses that might interfere with visual assessment, and the desirability of monitoring symptomless plants for latent infection (although tolerance may prove to be the best available form of resistance in some instances).

The form of inheritance of the resistance - as a major or minor gene, dominant or recessive, monogenic or polygenic, all affect the breeding strategy, as do the ploidy level of the crop, self-fertility or incompatibility, and the necessity of embryo rescue or other *in vitro* techniques. The source of the gene (in a cultivar or a distantly related species) and these factors will all determine how many back-cross generations are required before the desired resistance can be introgressed into an agronomically adapted line. The majority of characterized virus resistance genes are monogenic and dominant (see also Fraser, Chapter 5), but the expression of resistance may be affected by gene dosage or environmental conditions. In some instances virus resistance may be linked to other resistance genes - or to susceptibility to another pathogen.

Resistance to viruses may be via resistance to vectors (covered in depth by Jones, Chapter 4); to transmission by vectors or through seed; to symptom development; to virus multiplication; to local or long-distance spread of the virus through the plant (including hypersensitivity); or through immunity. Resistance may be either constitutive, or induced. Immunity is constitutive, whereas a hypersensitive response requiring initial viral replication to elicit the reaction is an example of an induced resistance.

Vector resistance (Khetarpal *et al.*, Chapter 2; Jones, Chapter 4) may not be durable because of the plasticity of vector populations, or because factors that enhance resistance to one vector may favor another; resistance to seed transmission, on the other hand, is highly valued because of the immediate dramatic reduction in the inoculum source within the crop.

Resistance-breaking isolates (Khetarpal *et al.*, Chapter 2; Fraser, Chapter 5) are reported more frequently with dominant, single gene resistance than with recessive or polygenic resistance. Some resistance breaking strains are, however, less fit in other attributes necessary for spread or persistence, and may not come to predominate. It is desirable to deploy different resistance genes in different cultivars, or there may be catastrophic losses over a significant area when a resistance-breaking strain arises.

Molecular mapping techniques that will aid introgression of resistance genes into cultivars are briefly reviewed by Khetarpal *et al.* (Chapter 2). Such techniques will also aid in the isolation of naturally occurring resistance genes that can then be transferred directly into cultivars by genetic engineering. Overall the best and most durable results will be obtained when different resistance genes with effects on different stages of the viral life cycle (Fraser, Chapter 5) are pyramided into a single cultivar, as it is unlikely that a virus isolate will be able to overcome multiple resistances at once. Combinations of genetic engineering approaches (e.g. Yie and Tien, Chapter 10) with naturally occurring resistance genes increases the hurdles the virus must overcome.

Jones (Chapter 4) notes that vectored spread from *outside* the crop is often not affected by vector control *within* the crop, as transmission often occurs before the vector is killed by the pesticides applied. Chemical vector control frequently has the penalty of "collateral damage" to non-target organisms, and environmental degradation. Vector resistance has the dual attraction of control of the virus, and limiting damage by the vector itself, which may be quite considerable. Most of the chapter is directed towards aerial vectors (primarily aphids), but consideration is also given to fungal and nematode vectors. The different types of interaction that influence virus/vector/plant relationships are addressed; these may be bilateral or trilateral interactions, such as the increased attraction of aphids to the yellowed leaves that result from some virus infections.

Vector/plant relationships include *Antixenosis*, or non-preference due to physical or chemical properties of the plant that discourage feeding; *Antibiosis*, or adverse effects on growth, reproduction or survival of the vector; and *Tolerance*, the ability of the plant to grow being little affected by colonization. Of these, tolerance to the vector will cause little or no reduction in virus transmission - and indeed may increase it. Antixenosis and antibiosis may occur either separately or together, and either may in some cases cause increased vector movement with concomitant virus spread (especially in the case of stylet-borne viruses).

Vector resistance may include interference in the ability of the vector to locate the plant, such as differences in plant coloration between cultivars, or absence of volatile compounds that may attract the vectors. Differences in the rapidity of canopy coverage, affecting the balance of UV light reflectance between the plant and to soil, may also affect vector ability to locate the plant. Chemical or physical stimuli, including the presence or absence of pubescence, trichomes, or glandular hairs with sticky secretions may interfere in settling and feeding behavior, and influence susceptibility to persistently-transmitted more than stylet-borne viruses. Surface waxes can act as a physical barrier to penetration. Such traits have the major advantage of having no negative effects on non-target organisms, including natural predators of the vector, leaving increased opportunity for biological control of the vector to reduce populations further. If the vector by itself causes little or no damage, vector resistance alone is not the best strategy - in such cases direct resistance to the virus is likely to be far more effective, if it is available. Vector resistance can be drastically influenced by temperature, and caution must also be taken to ensure that the factors that increase resistance to one vector do

not increase susceptibility to another. The combination of vector resistance with other types of virus resistance is often more than additive, and, as noted above, will likely be more durable than either alone.

Fraser (Chapter 5) divides the mechanisms of virus resistance into *Non-host resistance*, *Heritable resistance*, and *Acquired resistance*. The targets of resistance mechanisms may be any stage of the viral lifecycle capable of interruption, to interfere in the completion of the lifecycle, or at least to limit the extent of replication or spread. These can be divided into "macro" elements that may be viewed broadly as epidemiological aspects, and "micro" elements that are more biochemical or molecular events. Positive resistance implies the presence of a mechanism that inhibits some phase of virus replication or spread, and includes physical barriers such as glandular hairs, or thick cuticles, or a hypersensitive response, and inhibitors of viral replication. Positive resistance is primarily dominant or dosage dependent (i.e. homozygous plants are more resistant than heterozygotes). Negative resistance implies the lack of a necessary component (e.g. a host factor) necessary for the virus to complete its life cycle; these might include a protein that interacts with the viral movement protein, or a compatible host-coded replicase subunit. Such resistance is typically recessive.

Non-host resistance is inability to support viral replication, and is effective against all isolates. Many cases may involve failure to move from cell to cell, and thus many "non-hosts" may be in fact subliminally infected; this has been demonstrated in some cases by the ability of the virus to replicate efficiently in protoplasts of "non-hosts". *Heritable resistance* is present in some but not all genotypes, and can be crossed into other genotypes, but may not be effective against all virus isolates. Heritable resistance appears to be dominant in a slight majority of characterized cases, but may also be dosage-dependent; it is much less commonly recessive. *Acquired resistance* (also called induced resistance) is gained as a result of exposure of susceptible plants to the virus, another pathogen, a chemical, or a cultural treatment, and is not heritable (except in as far as it can be induced in the progeny). Pathogen-derived resistance in transgenic plants may be regarded as a special case of acquired resistance that would be heritable, as expression of the viral gene that induces the resistance is heritable.

Fraser (Chapter 5) also notes some

characteristics of different types of resistance. Many resistance genes are eventually overcome by resistance-breaking strains. There are few examples of isolates remaining effective for long periods against multiple isolates; in such cases resistance-breaking strains tend to be less fit in other ways, and do not predominate (also noted by Khetarpal *et al.*, Chapter 2). Dominant resistance is strongly correlated with localization of initial infection - blockage of cell-cell movement at a very early stage of infection, and often with a hypersensitive response; in some cases this has been shown to be determined by the viral coat protein, although the tobacco *N* gene response is due to the viral replicase. In contrast to dominant resistance, gene dosage-dependent mechanisms generally permit viral spread, but limit the extent of replication, and often significantly reduce symptom severity. In some cases, if replication remains below a threshold level, no visible symptoms are produced. As there is a greater inhibition in homozygous plants, the resistance level is correspondingly greater. Recessive genes are harder to study, and much less has been determined; however, in the absence of a necessary factor (a negative mechanism), complete immunity is expected.

Verma *et al.* (Chapter 12) describe the use of antiviral substances of plant origin. In many cases these result in *localized acquired (or induced) resistance* only where exogenous application of the antiviral extract has been made - usually resulting in fewer local lesions, but not preventing systemic infection in hosts that are susceptible. In other cases *systemic acquired resistance* may result. *Systemic acquired resistance (SAR)* includes cross-protection induced by a mild strain of the same virus, or a closely related virus (subject of Chapter 3 by Lecoq); there is also some *SAR* against unrelated viruses in certain cases, induced either by localized infection by another virus, or by chemicals including salicylic acid and polyacrylic acid. Some botanical extracts induce *SAR* after local application, with shorter induction periods than salicylic acid or virus infection. Such botanical-induced *SAR* is not general, but limited to certain viruses in particular hosts - presumably through induction of limited cascades of biochemical events that are dependent upon the botanical extract applied, and upon the specific host.

Almost all botanical viral inhibitory substances that have been characterized are basic proteins; many are ribosome inactivating proteins (RIPs), and most are glycoproteins. RIPs are divided into two classes - single chain or two disulphide-linked chains. Both types are extremely stable proteins. Some non-RIP botanicals are not directly antiviral, but induce production of an inhibitory substance in treated plants; these botanicals have been termed systemic resistance inducers. The systemic resistance inducers have little or no virus or host specificity, and the active antiviral is probably a protein. Botanicals do not appear to induce pathogenesis-related (PR) proteins. There is potential for expression of botanicals in transgenic plants - and indeed pokeweed antiviral protein, a RIP, has been expressed in transgenic plants with some resistance to several viruses.

Lecoq (Chapter 3) has reviewed the requirements and effectiveness of cross-protection for obtaining resistance to severe virus isolates in regions where such isolates are endemic and no other resistance is effective. Cross-protection is applicable to most virus groups (although not for geminiviruses such as beet curlytop virus), and also for viroids and satellite RNAs. Several specific applications are discussed; potential mechanisms have been extensively reviewed elsewhere and are only cited here. It is noted that cross-protection does not exclude the challenge (severe) isolate, which often undergoes limited replication, but fails to produce the typical symptoms (note that limited replication in some naturally or transgenic resistant plants may also result in symptom suppression).

If the severe disease is prevalent, then some yield loss due to the protecting isolate may be acceptable, because expected losses due to severe isolate infection would be far worse. Otherwise the protecting isolate should ideally be mild (in both the host to be protected and in other crops which it is able to infect); fully systemic in all tissues of the crop to be protected; genetically stable; non-transmissible or poorly transmissible by vectors to limit unintended spread; protect against a wide spectrum of other isolates; and easy to produce, check for purity, store, and inoculate. Such isolates may be selected from among field isolates, or obtained in the lab by repeated passage or by mutation. Inoculation techniques suitable for reliable infection of large numbers of plants must be available - often for infection of seedlings prior to transplant to the field, by grafting in the case of woody perennials, or by field inoculation in some cases.

Limitations or hazards include incomplete protection in some instances; the potential for spread to unprotected crops which may suffer economic losses; synergism with other viruses able to infect the protected crop; heteroencapsidation; and either recombination or mutation to create more severe isolates. Despite these possible limitations, cross-protection has proven highly effective in a number of crops, but should be limited to situations where no practical alternatives are available and the virus is a real threat to productivity or quality of the crop. A number of examples are given of the application to greenhouse, field and orchard crops, under both traditional and modern management systems, and against viruses from various groups. The method is readily adapted to new virus isolates or crop cultivars (as compared to transgenic approaches, which are more labor-intensive and take longer to adapt), and can be included easily in integrated pest management systems.

From cross-protection induced by deliberate virus infection we move on to consider the genetically engineered equivalent, coat protein-mediated and replicase-mediated resistance, which are reviewed by Kaniewski and Lawson (Chapter 6). Genetic engineering offers an alternative where no effective resistance genes have been identified in a crop or its close relatives - or as a way of pyramiding natural and transgenic resistances to create more durable multi-faceted resistance.

In practice, a range of resistance phenotypes is observed with most transgene constructs, from slight delay in symptom production or severity in some lines to extreme resistance in other lines - at least to the homologous virus or closely related viruses or isolates. While other genes and types of construct are being examined, it is most likely that coat protein (CP) and replicase genes will be deployed widely in the near term. Results so far suggest that CP may confer somewhat broader, but less complete resistance than replicase genes, which appear to offer greater extremes of resistance (almost all or nothing), but with much greater limitation of the range of isolates against which resistance is effective. Where there are distinct serogroups of strains, both CP- and replicase-mediated resistance are less likely to confer broad spectrum resistance, although there may be significant differences in degree of heterologous strain resistance between lines expressing the same construct.

When setting out to establish CP- or replicase-mediated resistance, it is first necessary to know something of the spectrum of virus isolates that occurs in the area where the crop is to

be grown - and the most common and aggressive strain should be chosen for cloning of the protective construct. If multiple strains are present, then multiple constructs can be prepared and evaluated. The target for transformation will be either major cultivars, or advanced breeding lines that can be crossed into cultivars. The CP gene has proven fairly effective in most virus groups examined - although experience with potato leafroll luteovirus CP in potato has been disappointing; this may be partly due to the ploidy level of the genome, but results with replicase seem more promising.

Kaniewski and Lawson (Chapter 6) discuss the practicality of tissue-specific expression from appropriate promoters - but note that results so far support the use of strong constitutive promoters. Other characteristics that determine effectiveness of constructs include the efficiency of expression, which can be influenced by translational enhancers, 3' untranslated regions (which contribute to mRNA stability), transcriptional terminators, introns, and codon usage. Marker genes for efficient selection and identification of transgenic lines are also important; various types of assay have been used to determine whether the desired transgene is present, but PCR, Northern blotting and ELISA each have their place. Further selection is necessary for preliminary selection of resistant lines, followed by multiplication of promising lines, and ultimately field-testing under realistic conditions. Laboratory tests cannot always predict field performance, as gene expression may be modified by environmental conditions and the presence of other pathogens. Similarly, mechanical transmission may not reveal differences between lines that will be revealed by vectored transmission in the field. Examination of field performance is also necessry to confirm that agronomic and varietal traits have not been adversely affected during the transformation and regeneration processes. As already noted, pyramiding of different transgenes (and combination of transgenes with natural resistance) is likely to increase the durability of resistance; as noted by Fraser (Chapter 5), combining resistances to different stages of the viral life cycle is likely to yield the most lasting resistance.

Kaniewski and Lawson (Chapter 6) also address regulatory issues - most plant virus genes will likely be de-regulated (with the probable exception of functional movement proteins). A consensus appears to have been reached that it is unlikely that viruses will recombine with transgenes to create novel viruses of greater virulence than other prevalent strains - there is already great potential for recombination to occur within mixed infections. Little added risk is to be expected as the result of deployment of transgenic plants expressing viral genes; viral replication in plants deployed because of their resistance is likely to be reduced, producing and environment in which recombination may be less likely than in mixed infections.

In many cases there is little correlation between the CP expression level and the degree of protection - especially with potyviruses. This is due in part because some cases of "CP-mediated" resistance are in fact RNA-mediated - which is discussed by Tabler *et al.* (Chapter 7). Gene silencing is another mechanism that may contribute to resistance anticipated to function otherwise. The actual mechanisms of resistance that are apparently mediated by protein are not well understood, but do differ between virus groups. It is highly likely that in many cases there are multiple mechanisms induced by a single construct, such that resistance to the homologous isolate may be RNA-mediated, while resistance against more distantly related viruses or isolates may be mediated by the expressed protein. Copy number and position effects cause some of the variation between lines and the relative contribution of different aspects of resistance.

Replicase-mediated resistance is generally more limited to closely-related isolates than CP-mediated resistance, and resistance is usually either extreme or negligible, with few intermediate phenotypes. It is suggested that un-modified replicase genes confer broader specificity of resistance than modified defective replicase genes (dominant negative mutants). Various mechanisms for replicase-mediated resistance have been proposed, as for CP-transgenes. In some instances protein expression appears to be required, although other cases could be RNA-mediated.

Jacquemond and Tepfer (Chapter 8) provide an extensive review of cucumber mosaic virus satellite RNA, describing the properties of satellite RNAs that make them of value for controlling disease, and also the features that can cause more severe disease in certain host/virus isolate/satellite combinations. Experience in the field has so far shown that use of virus and satellite RNA to cross-protect crops is highly effective over large acreages, and that similar results may be obtained with transgenic plants expressing only the satellite RNA (see also Chapter 10, by Yie and Tien). However,

Jacquemond and Tepfer urge caution in the field utilization of either cross-protection or transgenic satellite approaches despite the apparent lack of evidence for any non-necrogenic satellite mutating into a necrogenic form.

Satellite RNAs are rarely found in field isolates, and only at very low frequencies in wild plants. Whether they are present in more isolates at levels below even the detection threshold of PCR is uncertain, but passage of some isolates in certain crop plants, and especially tobacco and tomato, seems to amplify satellite RNAs to a very high degree. Because of the inherent errors of RNA polymerase, the sequence amplified may not be representative of any satellite present in the original host.

Because satellite RNA reduces helper virus replication, there is generally less virus available for vectored transmission. The relative ability of a particular host plant to replicate the satellite RNA has a profound effect on the proportion of plants from which both virus and satellite can be transmitted. Cross-protection does not necessarily mean complete suppression of the challenge (severe) virus isolate - and in some cases the challenge isolate can be readily located in tissues of cross-protected plants that show only the mild symptoms of the protective inoculum. If the challenge virus contains a necrogenic satellite, this can often out-compete the protective satellite. For these reasons Jacquemond and Tepfer (Chapter 8) urge that protective isolates should be minimally aphid-transmissible, or preferably non aphid-transmissible, to limit vectored transfer of either protective non-necrogenic or necrogenic satellites to other crops. Field results have shown that in most plants protection by non-necrogenic satellites and supporting CMV isolates is sufficient to keep replication of necrogenic satellites to a minimum, and to reduce virus replication significantly. Indeed, such cross-protection has been the only possible way to continue growing tomatoes in some regions where losses to CMV combined with necrogenic satellites were catastrophic. Selection of an appropriate CMV isolate for maintenance of the protective non-necrogenic satellite is important for persistence of protection under varying field conditions; if lack of aphid transmissibility can be found in an appropriate CMV isolate, this would certainly be beneficial. Cross-protection has been reported to have an added benefit of inducing systemic acquired resistance to some fungal pathogens, and also an antagonistic effect on potato spindle tuber viroid replication.

Comparison of Chinese field tests of transgenic plants expressing satellite RNA with non-transgenic plants pre-inoculated with a mild CMV strain containing a non-necrogenic satellite leads to some interesting conclusions. The transgenic plants were less severely infected than non-transgenic controls, but had no benefits of the systemic acquired resistance to fungi observed in plants cross-protected with a mild CMV + satellite; nor did the transgenic plants show earlier flowering and maturation as the cross-protected plants did.

Jacquemond and Tepfer (Chapter 8) address the ecological risks of genetic drift in a satellite RNA population, the possibility of a necrogenic satellite evolving from a non-necrogenic one, and the risks of vectored satellite spread into crops in which either the original satellite or a sequence variant might have adverse effects. In this regard the transgenic approach has some advantages, because the satellite RNA is not replicated unless the plant is infected with CMV, and up to that point the satellite sequence is maintained in the transgene DNA. The sequences of necrogenic satellites are more highly conserved than those of non-necrogenic satellites - which suggests that selection of ameliorative non-necrogenic satellites with the least sequence similarity to necrogenic types would be preferable for either cross-protection or transgenic approaches. The molecular basis of necrogenicity is discussed, but while the necrogenic determinant(s) have been localized to the 3' portion of the satellite RNA, the minimal sequences necessary have not been satisfactorily defined. There are probably secondary or tertiary structures that modify the effect of the necrogenic domain.

Yie and Tien (Chapter 10) describe some of the unsatisfactory aspects of transgenic resistance using satellite RNA or CMV CP alone, and the approach taken in China to pyramid these and other transgenes (PVY CP or TMV 54K replicase genes) into single lines. To enable rapid production of lines homozygous for each transgene, they transformed haploid tobacco tissues and then used colchicine to obtain diploid plants after successful transformation. As expected, pyramiding genes led to higher resistance and lower disease incidence, with significant differences between plants with single genes and multiple genes. Plants expressing both CMV CP and satellite RNA were significantly more resistant than either alone (about doubling resistance), and adding resistance to PVY or TMV increased yield and quality further under conditions in which all three viruses were present. The transgenic plants were also found to be true to type under field conditions, with most plants being indistinguishable from non-transgenic controls apart from the disease resistance. As noted in Chapters 5 (Fraser) and 2 (Khetarpal et al.), pyramiding these genes led to greater resistance, and should also contribute to durability of the resistance.

Tabler et al. (Chapter 7) discuss antisense RNA and ribozymes, and some of the mechanisms that may influence RNA-mediated resistance of various kinds - and how RNA-mediated mechanisms may form a component of the resistance conferred by many constructs designed to express proteins. Indeed, it has been shown that in some instances resistance to closely related viruses or isolates may be conferred by RNA, in lines that transcribe RNA at a high level but have low steady state levels of transgene transcript; other lines transformed with the same construct may have lower levels of resistance to more distantly related viruses or isolates, that appears to be dependent upon the level of expressed protein. This type of dichotomy of mechanisms probably explains the lack of correlation between levels of CP expression and resistance with in several cases, including potyviruses.

Tabler et al. (Chapter 7) also discuss why the concept of "antisense" RNA has much less meaning in regard to plant viruses than in relation to regulation of nuclear genes. This is because an mRNA-sense transcript may act as an antisense RNA in interacting with viral (-) RNA - which is typically present in much lower amounts than the viral (+) strand, and hence makes a more effective target. Antisense RNA, untranslatable "sense" RNA and protein expression are compared, in addition to reviewing the development of antisense RNA and its use for regulation of nuclear genes. Whereas initial results with antisense RNA for control of viral replication were generally significantly less successful than for gene regulation, and more directly comparable, less effective than equivalent CP expression, recently there have been several instances of highly effective virus resistance conferred by expression of antisense RNA.

A similar discussion of utilization of ribozymes against targets in vitro and in vivo, and comparison of regulation of gene expression versus suppression of virus replication deals with results obtained so far - which are very limited compared to most other approaches - and the prospects for improvement. The theoretical basis for improvement of both ribozymes and antisense RNA are discussed, including factors that affect association and dissociation, targetting, and transcript stability versus ability to interact with the target.

Martin (Chapter 9) describes various alternative strategies for expressing either viral or non-viral genes to obtain virus resistance, plus the advantages and potential disadvantages of each. These include expression of a toxic gene product that is only induced by viral infection, leading to cell death (of the virus-infected cells) and an induced hypersensitive response. For this strategy to be effective, it is imperative that the toxin be induced prior to any cell-to-cell movement. It is potentially possible to create such a construct inducible by any of several viruses - but the toxin must be tightly regulated to prevent expression in the absence of viral infection, and to allow rapid and effective expression before virus spread occurs. Otherwise systemic necrosis of the plant might result, rather than virus resistance.

Suppression of amber read-through codons might interfere with virus regulation and interfere with the replication cycle. Typically read-through products comprise less than 10%, and sometimes significantly less, of the expression of the reading frame, so shifting the balance of amber-terminated to read-through product might initiate a feedback mechanism that would shut down replication. Such a mechanism might confer resistance to multiple viruses, with the added advantage of having no viral genes present for recombination. A potential disadvantage would be interference with translation of any cellular genes that rely on partial readthrough.

Defective insect transmission factors - CP or helper component - might interfere with vectored transmission - but would function by competition with the wild-type product produced during replication, not by exclusion of a viral function. Dysfunctional movement proteins (MPs) have been shown to reduce spread of not only the homologous virus, but also of viruses that have movement proteins that share sequence similarity. The TMV MP has been shown to reduce infection of several viruses with related MPs, and a defective triple gene block from white clover mosaic virus conferred resistance to other viruses with a triple gene block. This strategy has the advantage that potential recombinants would gain a dysfunctional gene, and would presumably not be viable.

Plantibodies - plant-expressed antibodies or single-chain antibodies - could be targetted at various stages in the viral life cycle - although the steps most likely to be effective are those with limiting amounts of viral gene product, and not anything directly related to CP, which is typically the easiest gene product against which to raise antibody. An advantage is that cloning is essentially the same for any antibody gene. A related possibility is the use of anti-idiotype antibodies to block receptors - although putative viral receptors may have other functions unrelated to virus infection.

Protease inhibitors to interfere with virus maturation could include antibodies to the protease active site or target site; the active site is more likely to be conserved between proteases of different viruses, and perhaps with insect pest proteases, but might also be shared with normal cellular proteases.

Host resistance genes might be expected to confer long-lasting resistance, due to their long co-existence with the virus or vector. Genome mapping will certainly lead to identification and isolation of more resistance genes. However, such genes may require interactions with other genes or gene products to confer the resistance phenotype, and may require specific chromosome contexts for maximal activity - clusters of resistance genes against various pathogens have been identified in several species. Host resistance genes have the probable advantage of reduced public resistance to expression as transgenes, compared to viral genes or genes derived from other phylla.

One system not mentioned here (apart from those covered in other chapters) is the 2-5A-synthase (1,2) derived from animal systems, and normally induced by cytokines, that has been shown to function in plants against several plant viruses. A potential disadvantage of this system was observed in double transgenic plants, expressing both the 2',5' oligoadenylate synthase and ribonuclease L, which were resistant to CMV but developed systemic necrosis when inoculated with PVY (1).

Overall, it is noted that single dominant genes from other sources transferred by traditional breeding methods have often broken down over relatively short time in field deployment, and we should not expect virus-derived resistance to be too different - although non-viral gene strategies may be expected to last for longer, and all types of resistance are likely to be more effective when pyramided.

Timmerman-Vaughn (Chapter 11) reviews potential safety issues raised by widespread use of virus-resistant transgenic crops. Do such crops pose risks to the sustainability of current agricultural practices, to the environment, or to human health? If there are any such risks, do the benefits of increased production outweigh the risks, especially in the face of ever-increasing populations and increasing losses of agricultural land to other uses or environmental degradation? The gene transfer processes have been demonstrated not to have any significant risks to the environment, agricultural practices, or human health, although there is still significant resistance to the use of transgenic crops from some quarters. Some of the opposition is based on the perception that agribusiness will increasingly displace small farmers, but this is a trend that is already firmly established and cannot fairly be attributed to genetic engineering; others oppose on principle the use of methods that circumvent nature, but plant breeders have been finding ways of circumventing natural barriers to gene transfer via the use of embryo rescue, wide hybrids and bridging crosses, among other techniques, for years. Without using the techniques that were available, the increase in the world food supply referred to as the Green Revolution would not have been possible.

Insertion of DNA into the genome may cause changes in plant phenotype, either altering gene regulation by insertional mutagenesis, or somaclonal variation as a result of the tissue culture necessary for transformation and regeneration. Such changes can also cause activation of dormant transposable elements. Most such phenotypic alterations are deleterious and would be screened out during selection of transgenic lines. It is unlikely that phenotypes escaping such selection would have adverse consequences beyond losses in yield or quality of the crop, which would probably lead to withdrawal of affected genotypes from the marketplace.

The selectable markers, - typically antibiotic or herbicide resistance genes - that are used to initially identify transgenic plants, probably have no adverse effects on human or animal health. Extensive studies of NPT II, the gene used for resistance to kanamycin, have supported this position. Herbicide resistance genes raise different concerns, and would be better not used in cases where the crop is able to hybridize with weedy wild relatives. One example where use of herbicide resistance would be very unwise is sorghum; sorghum crosses freely with Johnsongrass, which is already a difficult problem to control. Herbicide resistance in Johnsongrass would create problems of a much greater

magnitude. In other crops herbicide resistance is much less likely to "escape" to weedy plants, and could be used safely.

Pathogen-derived resistance carries certain risks - of recombination between pathogens and the transgene transcript, or complementation, or heteroencapsidation of plant viruses. Heterologous encapsidation has been demonstrated to occur in natural mixed infections, and results in the transmission of isolates that are not normally vector transmitted from a single infection. Such non-transmissible or poorly-transmissible isolates result from mutations in one or more viral genes involved in transmission - but the occurrence and persistence of such isolates in nature is presumably due to heteroencapsidation; mixed infections are commonly found in both crops and free-living plants.There is therefore little if any additional risk from heteroencapsidation in transgenic plants expressing viral coat protein, unless heteroencapsidated virus is enabled to establish infection in a host it would otherwise be unable to infect. This is extremely unlikely even if the transgenic plant expresses coat protein from a virus not able to infect that host, as the transgene coat protein will not be available in any non-transgenic plant to which the virus is transmitted. As many aphid virus vectors are polyphagous and also capable of transmitting multiple viruses, and make test feeding probes to determine the suitability of different plants as hosts, it is likely that viruses endemic to an area have already encountered all of the potential plant hosts.

Recombination between a virus and a transgene of viral origin raises somewhat different concerns. Again, it must be borne in mind that mixed infections of different viruses are common in both natural and agricultural situations. Where a virus is able to infect a transgenic host, however, the level of transgene transcript is usually significantly below levels of replicating viral RNA, probably reducing the possibility of recombination compared to a mixedly-infected cell.

Recombination between viral genomes has been found to occur largely in intragenic or non-coding regions, and where there is significant secondary structure, which may cause the replicase to pause. Design of transgenes to exclude such regions, where practical, will probably do much to reduce the possibility of successful recombination. In particular, absence of the viral 3' non-coding region would require a double recombination event to generate a functional virus. However, the 3' non-coding region may contribute to the virus resistance in some cases, and omission from the construct might result in lower resistance.

Transfer of resistance genes from crop plants to weedy relatives must also be considered. There are only limited studies of the effects of plant virus infection on free-living plant populations, with little evidence that viruses significantly influence such populations. In general endemic viruses have co-evolved with native plant populations, and there will be greater genetic variability in the native plants than in crops planted in the same area. Some of the native populations probably already carry some degree of resistance, or at least tolerance, to endemic viruses. Introgression of a transgene conferring virus resistance might reduce the reservoir of virus in native plants, with the potential for a reduction of transmission into the crop.

Overall these chapters give an understanding of the different aspects of plant resistance to viruses, and the approaches for utilization of available germplasm and of genetic engineering. As a practical matter, in the first instance it may not be necessary to know the actual mechanism of resistance, of whatever origin. However, given that effective resistance can be conferred by multiple types of natural resistance gene, and by various types of transgene construct, knowledge of the mechanisms of each may allow selection of the types and combinations that are least likely to be overcome by mutants of the virus in question. Reliance should not be placed on a single gene deployed in most or all cultivars grown in an area, as this may result in catastrophic losses if a resistance-breaking strain emerges. Combinations of compatible resistance mechanisms are likely to be more durable, while other combinations might actually encourage the evolution of resistance-breaking strains. Even if the most durable kinds of resistance against current viruses are pyramided together, we should remember that new diseases periodically arise, and efforts to obtain and deploy virus resistance will continue to be necessary.

REFERENCES

1. Ogawa, T., Hori, T., and Ishida, I. 1996. Virus-induced cell death in plants expressing the mammalian 2',5' oligoadenylate system. Nature Biotechnology 14:1566-1569.

2. Truve, E., Aaspõllu, A., Honkanen, K., Puska, R., Mehto, M., Hassi, A., Teeri, T.H., Kelve, M., Seppänen, P., and Saarma, M. 1993. Transgenic potato plants expressing mammalian 2'-5' oligoadenylate synthetase are protected from potato virus X infection under the field conditions. Bio/Technology 11:1048-1052.

CHAPTER 13

Forecasting Aphid-borne Virus Diseases

K.R. Sigvald

Plant viruses cause extensive damage to a variety of crops in Scandinavia many other countries. The economic losses from such virus damage vary greatly from year to year and region to region. Normally, in Scandinavia, virus diseases are less important than other pest and disease problems; nevertheless, during certain years plant viruses infecting cereals, potatoes and sugar beets, in particular, cause great damage which result in serious economic losses to the growers.

Aphids play an important role in the epidemics of both persistently and non-persistently transmitted viruses. Many of the non-persistently transmitted viruses are known to be vectored by aphids, and a number of species have been confirmed as vectors during the last few decades. However, only a small proportion of the world's aphid species have been tested to determine whether or not they can act as vectors.

Interactions among host plants, viruses, vectors and the environment are very complicated. Environment influences the crops, vectors and viruses. For example, temperature not only directly affects vector behavior but also influences virus multiplication and its' translocation in the plant.

Barley yellow dwarf virus (BYDV) is one of the most important aphid-borne virus diseases on cereals in Scandinavia, especially on spring barley and oats. It is of less importance on winter wheat. The incidence of BYDV varies greatly between regions and during different years, depending mainly on vector population and, to a lesser extent, on the availability of virus source. Fluctuations in BYDV incidence from year to year are mostly related to differences in population levels of the main virus vector, *Rhopalosiphum padi*. The grain aphid, *Sitobion avenae*, is less important. However, mild winters seem to favor the latter species, which appears to be a significant vector in some years especially in the southern regions of Scandinavia. The rose grain aphid, *Metopolophium dirhodum* is of only minor importance.

Beet yellows virus has occasionally caused substantial damage in southern regions of Scandinavia. During the 1940s, this virus was a major problem, probably owing to the cultural practices used at the time. In 1959, there was an epidemic of beet yellows virus in southern Sweden which resulted in huge economic losses. The main vector of the virus is *Myzus persicae*, however, other aphid species such as *Macrosiphum euphorbiae*, *Aphis fabae* and *Aulacorthum solani* are also of some importance.

In Scandinavia, the aphid-borne potato virus Y (PVY) is more frequent on potatoes than potato leaf roll virus (PLRV), which is also spread by aphids. This could be due to lower population of *M. persicae* (the main vector of PLRV) in potato fields and a higher population of the aphid species vectoring PVY. However, the degree of spread of PVY varies greatly between regions and different years. In the northern regions of Scandinavia the spread of PVY is minimal because vectors are uncommon. By contrast, in southern regions the spread of PVY has led to serious problems for seed potato growers during some years. Nevertheless, few fields are infected so severely to warrant rejection of the seed potatoes produced.

In this chapter the epidemiology of certain plant viruses will be discussed along with possible forecasting of aphid-borne virus incidence. Taking this into account, potato virus Y will be treated in detail.

FORECASTING NON-PERSISTENTLY TRANSMITTED APHID-BORNE VIRUSES

Many aphid species are important vectors

of non-persistently transmitted viruses. PVY and soybean-mosaic virus (SMV) are two good examples. The adverse effects of PVY on seed potato production have led to large economic losses in many countries. Most viruses transmitted by aphids in a non-persistent manner are believed to be acquired and transmitted within fields by the probing of aphids that move from plant to plant. Hille Ris Lambers (7) stated that this was the case for PVY. Both PVY and SMV are transmitted mainly by aphid species that do not feed on the crop that they inadvertantly infect with the virus (12,16).

After probing for 5-10 s, the aphids acquire the virus and are immediately able to transmit it to other plants. However, most aphids only remain viruliferous for about 30 min. Starvation prior to the probing of source plants greatly increases the proportion of aphids transmitting non-persistently transmitted viruses.

POTATO VIRUS Y

In Sweden and many other countries in northern Europe, PVY is one of the most important virus diseases of potato. Although the spread of PVY^O in Sweden is negligible during most years, occasional outbreaks of this virus have been recorded since 1970, especially in southern and central regions (19). In the northern regions of Sweden, which are important areas for seed potato production, the incidence of PVY^O has been very low, mostly because few vectors and few virus source plants are present in the fields (17).

Although PVY can be transmitted by aphid species that feed preferentially on potato, other species that do not colonize potato seem to be far more important, e.g. 'R. padi, Brachycaudus helichrysi, Acythosiphon pison and Phorodon humili (4,9,10,14,21,22). In Sweden, and in many other countries where PVY is a serious problem for seed potato producers, the relationship between aphid migration and the spread of PVY has been studied by exposing bait plants to vectors in the field (1,13,18,21). Winged aphids (alates) have also been collected and placed on test plants to determine whether or not they are viruliferous (2,3,6,22). There are great differences in virus-transmission efficiency between aphid species (5,8,14).

During the last decade there has been increasing interest in developing methods for PVY forecasting. The main variables used when forecasting the incidence of PVY include the number of alates and their vector efficiency, the time of aphid migration in relation to plant growth, and the availability of virus source (15,16). Simulation models have also been used to describe the epidemiology of non-persistently transmitted viruses (12,16). The role of aphid behavior in the epidemiology of PVY has recently been studied using simulation models (11). The following is a description of a simulation model designed especially for predicting conditions for PVY epidemiology in potato fields in Sweden.

Simulation Model for Potato Virus Y

Results from field experiments, laboratory studies, and the literature were used when developing a simulation model for PVY^O in Sweden. Data were also collected from a number of potato fields in southern and central Sweden from 1982 through 1985, in order to evaluate the model (16).

Data obtained in field experiments from 1975 through 1979 in southern Sweden (18) were used to calculate how long it takes before a potato plant inoculated with PVY can act as a virus source under field conditions. In these field experiments, the main aphid migration was found to take place during July, at about the same time as the potato plants became infected with PVY. About 3 weeks later the progeny tubers became infected (18). Based on these studies it was concluded that alates, not wingless aphids (apterae), were mainly responsible for the spread of PVY (19). The latent period was estimated from field experiments (1975-1979) in southern Sweden. The results of those experiments indicated that progeny tubers become infected about 3 weeks after the foliage has been inoculated by migrating aphids (16).

From 1974 to 1986, the flight activity of alates was monitored using yellow water traps. Aphids were collected three times a week, and eight aphid species were identified : A. pisum, Aphis fabae gr., Aphis nasturii + A. frangulae, Brevicoryne brassicae, M. dirhodum, M. persicae, R. padi and S. avenae. All other less common species were assigned to a category called "Other aphid species". From 1982 through 1985, data were collected from about 180 potato fields in southern Sweden. The types of data collected from each field were: cultivar, planting date, proportion of infected potato plants serving as a source of virus (field inspection), date of emergence, date of flowering, date of progeny tuber formation, date of removal of PVY-diseased potato plants, the number removed per ha, mineral oil application rate, and date of haulm destruction. From each field, samples of 300 progeny tubers were collected after harvesting and tested for PVY (greenhouse test).

Model Description

Results from these studies have provided the background needed to develop a dynamic simulation model for PVY^O in Sweden. The model describes relationships between important variables and parameters and gives a good overview of the epidemiology of PVY^O. The simulation model is designed to describe plant-disease-vector dynamics at the level of the individual potato field. Some of the most important variables are: alates as virus vectors, PVY^O diseased potato plants as virus sources, the susceptibility of the potato crop in relation to plant age, date of haulm destruction, and PVY^O-infected progeny tubers. The model output predicts the extent to which the proportion of progeny tubers infected with PVY^O will increase during late summer.

The epidemiology of PVY^O in a given field is described in the relational diagram, which shows the relationship between different variables and parameters (16) - the time step is one day. The present model differs from that presented earlier (16). The parameters and variables that have been added or changed in this model, in comparison with the earlier one (16), are described below.

Newly PVY^O-infected Plants

The number of newly PVY^O-infected potato plants depends on the rate variable (spread of PVY) and on the time that it takes for newly PVY-infected plants to become totally virus infected.

Totally PVY^O-infected Potato Plants

Progeny tubers of potato plants become PVY^O infected between 12 days (when young) and 24 days (when old) after inoculation : The younger the potato plants are at inoculation, the sooner the progeny tubers become infected. Thus, potato plants in the model cannot act as a virus source until they are totally infected. This state variable also includes the potato plants grown from PVY^O-infected seed potato.

Spread of PVY^O

This rate variable is proportional to the infection risk and is influenced by the degree of mature plant resistance, vector efficiency, cultivar-related susceptibility, mineral oil application rate, date of haulm destruction and risk for PVY^O infection estimated from virus sources in potato fields outside the studied field.

Latent Period

The latent period (i.e. the time elapsed before a newly PVY^O-infected potato plant can act as a virus source) is related to plant age. Viral multiplication and translocation probably occur more rapidly in young potato plants than in older ones (15). The latent period is set to 14 days for haulms during the first 5 weeks after emergence and is increased by 2 days each week thereafter. Its' maximum length is 26 days.

Cultivar Susceptibility, Mineral Oil Application Rate and Date of Haulm Destruction

Potato cultivars differ in susceptibility and are thus assigned different susceptibility factors: the greater the susceptibility, the higher the factor value, which ranges from 1.4 to 0.6. The use of mineral oil decreases the risk of PVY^O-infection in progeny tubers. If no mineral oil is applied a value 1.0 is used : the other values used are 0.7 for 1-3 sprayings, 0.6 for 4 sprayings, 0.5 for 5 sprayings, and 0.4 for 6 sprayings. Results from several field experiments show that the proportion of PVY^O-infected progeny tubers can be reduced by 40-60% when mineral oil is used against PVY^O. After the date of haulm destruction, it is assumed that flow between the state variables in the model ceases.

Removal of PVY-diseased Potato Plants and Risk for Virus Spread from Virus Sources Outside the Potato Field

The risk for spread of PVY^O can be decreased by removing PVY^O-diseased potato plants from the field before the main aphid flight. In the relational diagram, data on the number of PVY^O-diseased potato plants serving as a virus source and date of removal are included. This is an important factor because removal of a virus source before the main flight of aphids will have a great influence on the proportion of progeny tubers infected.

The main PVY^O source is within the

potato field. However, the fewer the PVY°-diseased plants within a given field, the greater becomes the importance of a virus source in other potato fields in the area. This factor is taken into account by basing it on the disease incidence in the region during the previous year. The variables of infection risk, mature plant resistance, vector efficiency, proportion of healthy potato plants, and proportion of progeny tubers infected with PVY° have been described earlier (16).

Results from Testing the Simulation Model

After writing the DYNAMO-code, the program was run several times using a range of aphid numbers, initial proportions of potato plants infected with PVY°, levels of mature plant resistance, length of latent period, dates of removal of PVY° source, dates of haulm destruction, etc., to compare the general behavior of the model with that of the real system.

Predicted proportions of progeny tubers infected with PVY° were compared with measured values obtained in field experiments (18). There was good agreement between the model output and data from field experiments in 1976 and 1977 (Figure 1 and 2). When aphid migration took place in July, the model output correctly predicted that the proportion of progeny tubers infected with PVY° would increase during August, about 3 weeks later (Figure 1). By changing the date of emergence it was also possible to compare the model output with results from field experiments (15) in which mature plant resistance had been studied.

Sensitivity Analyses

The importance of different factors within the PVY° model was assessed by varying different variables and parameters. The magnitude of the changes was restricted to keep values within the ranges determined in the field and laboratory studies, i.e. usually no more than a 50 percent increase or decrease in most variables. Most of the aphid data used were from the 1975-1977 field experiments. However, when manipulating efficiency factors, data from a number of other years and regions were also used.

Changing the proportion of PVY°-diseased plants serving as a virus source had a great influence on the proportion of potato tubers with PVY° : thus an increase in the former from 0.1% to 0.5% caused the latter to increase from 20% to 60%.

Aphid migration differs greatly between years and regions. Therefore data obtained on aphid migration from about 10 different regions and years were used in the sensitivity analyses in which vector efficiency values were varied (50% decrease or increase). Such changes had a great influence on model output for *R. padi*, *Aphis* sp. and "other aphid species", but only slightly affected the output for *M. persicae* (Table 1).

Latent period length, susceptibility of the potato crop, the extent to which PVY°-diseased potato plants were removed, mineral oil application rate, cultivar-associated susceptibility, and the risk of infection posed by virus sources in other potato fields were also varied in the sensitivity analyses.

Changing the proportion of potato plants removed (virus sources) had a great influence on the proportion of PVY°-infected progeny tubers if this change was made before the aphid flight (Figure 3). Results from these simulations were in good accordance with observations made in seed potato fields.

Comparison of Predicted Values with Field Estimates of the Proportion of Progeny Tubers Infected with PVY

Data from about 70 potato fields were used for parameter estimation and for simulations. When testing the validity of the model, however, we used data from 100 other potato fields, including the proportion of plants infected, date of emergence, cultivar, degree to which PVY°-diseased potato plants were removed, mineral oil application rate, and date of haulm destruction. There was a good correlation between predicted values and observed results $r^2=0.80$, $p<0.001$ (arcsin/transf.).

We also compared forecasted values with actual data from seed potato fields in different regions of Sweden. From each of the seed potato fields a sample of 300 progeny tubers was tested for PVY (greenhouse test). Infection rates in each potato field were assigned to classes, i.e. 0-2, 2-8, 8-15, 15-40 and above 40% PVY-infected progeny tubers and then compared with predicted values. The forecasted values were considered as correct if they were within the 95% confidence limit of class limits. Otherwise, the forecasted value was considered too high or too low. In general, there was a close correlation between forecasted values and results from tested samples (Table 2). During

1976

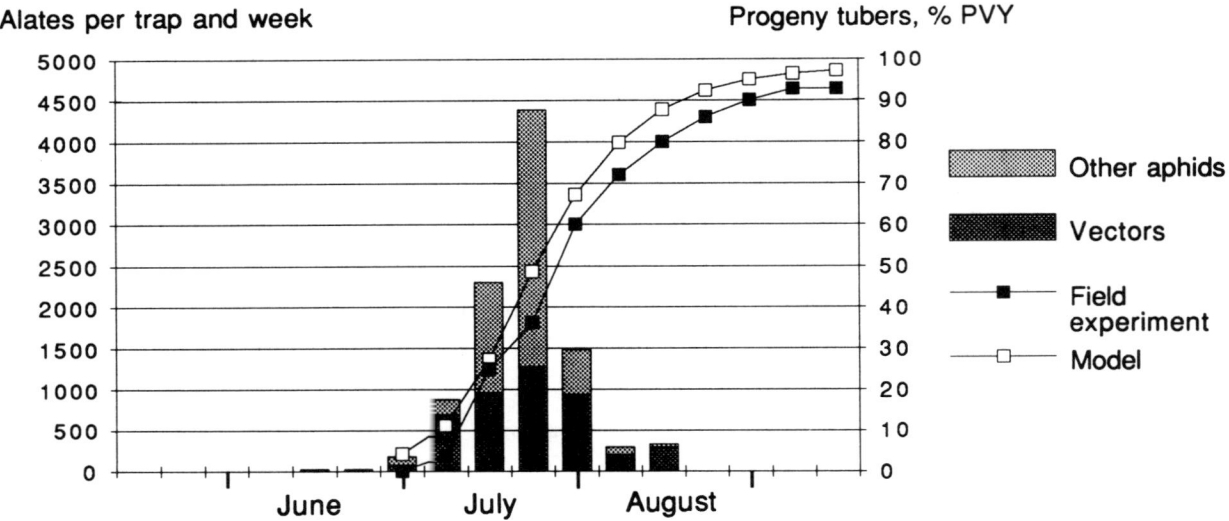

Figure 1. Relationship between model output and field measurements in 1976 in southern Sweden.

1977

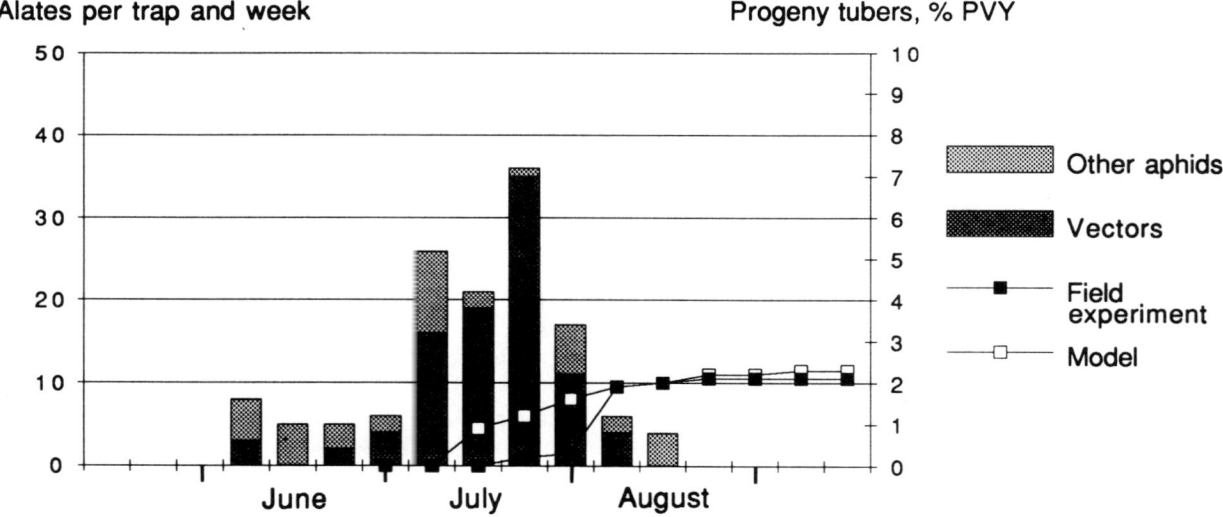

Figure 2. Relationship between model output and field measurements in 1977 in southern Sweden.

Table 1. Sensitivity analyses of the vector efficiency for the aphid species *R. padi*, *A. fabae* and *M. persicae* (The percentage PVY^O-infected progeny tubers [forecasted values] for 50% increase or decrease of the initial efficiency factors)

% PVY^O infected tubers	*R. padi*		*A. fabae*		*M. persicae*	
	-50%	+50%	-50%	+50%	-50%	+50%
55.4	34.1	72.9	49.3	60.8	53.9	56.8
10.5	9.0	12.1	10.3	10.7	10.5	10.6
8.2	7.3	9.3	8.1	8.4	8.1	8.3
1.3	1.1	1.5	1.3	1.3	1.2	1.3
0.6	0.6	0.6	0.6	0.6	0.6	0.6
0.9	0.8	1.0	0.8	1.0	0.9	0.9
22.8	16.1	30.2	21.2	24.4	22.6	23.1

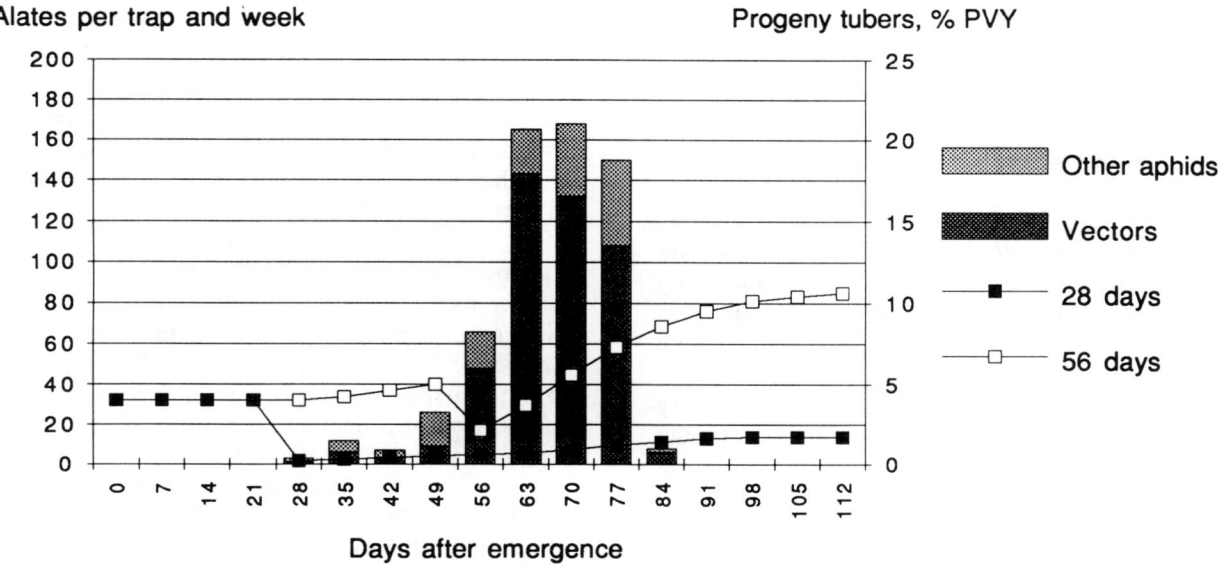

Figure 3. Simulation of the epidemiology of PVY-different dates of removal of PVY^O-diseased potato plants acting as virus sources.

Table 2. Relationship between forecasts of proportions of PVY^O-infected progeny tubers and measured values from tested samples

Year and province[a]	No. of fields	Proportion of fields, forecasted values (%)		
		Correct	High	Low
1982, G, R	35	74	9	17
1983, G, R	28	73	5	22
1984, G, R	26	96	0	4
1986, G	51	90	6	4
1986, R	126	89	0	11
1986, E	32	81	12	6
1986, AC	84	100	0	0
1987, G	50	98	0	2
1987, AC+BD	22	100	0	0
1987, R	136	99	0	1
1987, E	34	100	0	0
1987, M	66	85	0	15

[a] Each letter represents a different province

years with low vector numbers most of the tested seed potato fields fell into the lowest class, and forecasts were correct for more than 95% of the fields under such circumstances.

Discussion of Potato Virus Y Epidemiology

Model Behavior and Comparison with Field Data

The computer-based simulation model has become an essential tool for use in analyzing results from epidemiological studies. Simulations help us to better understand the relative importance of different variables and parameters, thereby providing a basis upon which to identify additional ways to improve the model. In the present study, there was a close correlation between the predicted proportion of progeny tubers infected and disease incidence estimates made in the 1975-1977 field experiments (18), indicating that the model predictions can be

accurate. Similarly, in 1976, when there was a very large aphid migration early in the summer, the model accurately predicted the proportion of progeny tubers infected with PVY^O in early September (measured incidence=93%, predicted incidence=96), Furthermore, in 1977, when vector populations were low, both predicted and measured values were ca 2% PVY-infected progeny tubers. Thus the model has great flexibility and provides accurate predictions under a variety of conditions.

Simulation models offer many advantages in virus epidemiology work. For example, they can be used to evaluate the relative importance of different variables at the field level. The influence of the proportion of PVY^O-diseased potato plants serving as virus sources on the proportion of progeny tubers infected with PVY^O can easily be demonstrated for any given level and pattern of vector intensity. Indeed, the simulations showed that the incidence of diseased plants had a great influence on the proportion of PVY-infected progeny tubers (16). The influence of latent period length can also be simulated. If very young potato plants are inoculated with PVY^O they can act as a virus source relatively early in the season, when many alates are still migrating, and thus increase the risk for PVY^O infection of progeny tubers. On the other hand, if alates migrate late in the season, the risk for PVY^O infection of progeny tubers will decrease dramatically because of greater mature plant resistance and a longer latent period. In this case as well, there was good agreement between field data and model output (15).

By taking the vector efficiency of a given aphid species into account along with the timing of its main flight period and the number of individuals migrating, the importance of the species as a PVY^O vector during a given year can be estimated. Thus, an aphid species with a high vector efficiency that migrates early in the season, when potato plants are very susceptible, may nevertheless be given a low vector-importance rating in the model if the aphids generally migrate in low numbers (as is the case for *M. persicae*) (17,18). By contrast *R. padi*, which is only a moderately efficient vector, can receive a high vector-importance rating during some years since large migrations can occur relatively early in the season in Sweden (Tables 3-5). Hence, by running such simulations we can get a better understanding of the relative importance of different aphid species as vectors of PVY, thereby helping us in developing priorities for future research.

The effects of various cultural practices on disease incidence can easily be simulated. Removal of PVY-diseased potato plants before the main flight of the primary vectors can greatly reduce the proportion of progeny tubers infected with PVY^O. Removal of diseased plants during or after the aphid flight period has little effect on the results, whereas removal of virus sources before the aphid flight had a great effect (Figure 3).

The results from simulated removals of PVY-diseased potato plants acting as virus sources are in accordance with field observations. The predicted proportion of PVY-infected progeny tubers for a potato field in central Sweden was 10% when running the model with data from field inspections. by contrast, values obtained in a greenhouse test were mcuh higher, i.e. 45% PVY-infected progeny tubers. This discrepancy can be ascribed mainly to the removal of PVY-diseased potato plants acting as a virus source during and after the main aphid flight period but before the field inspection. When running the model with data on the number of potato plants removed and date of removal, the predicted values were in accordance with the greenhouse test.

The date of haulm destruction also has a great influence on the proportion of PVY-infected progeny tubers. Early haulm destruction decreases the PVY infection risk for progeny tubers. Values obtained in the simulations agreed well with those obtained in field studies. In a number of potato fields sampled in southern Sweden in 1990 at different times of haulm destruction, the proportion of PVY-infected progeny tubers increased from 4% to 16% over a 20 day period. Similar results were obtained in 1992 in five potato fields sampled in central Sweden. Simulations can also predict the effect of planting sprouted seed potatoes early and applying mineral oil.

The good correlation between predicted infection-rates and the incidence of infection in samples of progeny tubers (r^2=0.80, $p<0.001$) in the two studied regions (R and G) suggests that the model should be useful in predicting the risk of PVY^O spread.

This study shows that the PVY^O simulation model can be used in forecasting the risk of virus spread in Sweden as well as in other countries. There are great differences in virus spread between different years and regions (17). During years when the incidence of PVY is low in Sweden, there is no need to test progeny tubers after the harvest. Thus, by using the simulation model to forecast PVY incidence, farmers would be able to skip post harvest testing during low-disease years, thereby reducing their operational costs. Similarly, disease incidence can be reduced by taking prophylactic measures at an early stage. However, such measures, e.g. mineral oil application, are too expensive to be used on a routine basis. Thus, if the risk for PVY spread could be predicted, such treatments could be used more selectively, thereby reducing costs.

The seed potato grower would also benefit by being able to predict the proportion of progeny tubers infected in late summer. If there is a great risk that the level of infection of the tuber yield will exceed the threshold set for seed potatoes, it may be more profitable to delay haulm destruction and market the potatoes for consumption or industrial use (starch or ethanol). During the last five years the forecasting method presented here has shown great promise when applied under practical conditions.

Table 3.   Number of alatae caught in suction traps in southern and northern Sweden in 1989

Aphid species	Southern (M)			Northern (AC)		
	May	June	July	May	June	July
Acyrthosiphom pisum	18	77	263	0	0	0
Aphis sp.	7	67	558	0	0	0
Brachycaudus helichrysi	104	400	983	0	0	0
Metopolophium dirhodum	10	158	664	0	0	0
Myzus persicae	0	0	0	0	0	0
Phorodon humuli	2	37	83	0	0	0
Rhopalosiphum padi	45	2405	6482	0	165	136
Other aphid speices	234	2161	3646	2	88	136

Table 4.   Number of alatae caught in suction traps in northern Sweden in 1987 and 1990

Aphid species	1987			1990		
	May	June	July	May	June	July
Acyrthosiphom pisum	0	0	0	0	0	3
Aphis sp.	0	1	11	0	3	1
Brachycaudus helichrysi	0	0	0	0	0	1
Metopolophium dirhodum	0	0	0	0	0	0
Myzus persicae	0	0	2	0	0	0
Rhopalosiphum padi	0	3	40	3	298	2226
Other aphid speices	0	2	228	2	137	286

SUCTION TRAPS IN FORECASTING PLANT VIRUSES

Aphid migration has been studied in Sweden for many years by using suction traps at 12 m above ground level (24). Between 1974 and 1984, only one or two suction traps were being used on a regular basis. In 1984, we began to increase the number of suction traps, and at present there are eight traps operating at various latitude in the country (Figure 4). The distance between the most southern and northern suction traps is approximately 1400 km.

Suction traps are used for various types of forecasting. The direct damage caused by different aphid species can be predicted by studying the spring migration of aphid species such as *R. padi*. The traps are also used to predict population levels of fruit fly in different regions of Sweden. Suction traps are also very useful in predicting the incidence of virus disease such as PVY, PLRV,

Table 5. Number of winged aphids caught in suction traps in southern (M) and central (C) Sweden in 1988

Aphid speicies	Southern (M)					
	May	June	July	August	September	October
Acyrthosiphom pisum	0	2	54	1	3	2
Aphis fabae gr	0	1	287	12	0	0
Aphis sp.	0	6	164	5	1	0
Brachycaudus helichyrysi	0	11	26	9	7	12
Brevixoryne brassicae	0	0	16	7	0	0
Metopolophium dirhodum	0	57	242	2	0	0
Myzus persicae	0	1	21	9	7	13
Rhopalosiphum padi	1	802	2618	300	283	854
Sitobion avenae	0	2	431	38	6	0
Other aphid speices	1	455	629	864	1391	820

Aphid speicies	Central (C)					
	May	June	July	August	September	October
Acyrthosiphom pisum	0	6	0	0	1	0
Aphis fabae gr	0	57	99	1	0	0
Aphis sp.	0	113	120	1	0	0
Brachycaudus helichrysi	0	34	88	8	0	0
Metopolophium dirhodum	0	35	0	0	0	0
Myzus persicae	0	5	2	2	0	0
Rhopalosiphum padi	335	6908	221	111	558	7
Sitobion avenae	0	7	25	16	71	0
Other aphid speices	337	2324	711	475	475	20

BYDV, and BYV.

Since 1984, relationships between suction trap catches and the spread of PVY have been studied in seven regions of Sweden. In each region, data from more than 500 potato fields have been collected concerning cultivar, proportion of PVY-diseased potato plants acting as a virus source (field inspection), and proportion of PVY-infected progeny tubers. Preliminary results show that there is a close correlation between suction trap catches and virus spread if the proportion of virus source and mature plant resistance are also considered. The relationship was rather weak when only the number of alates or the number of known PVY vectors was considered.

Suction trap catches varied greatly from year to year and region to region (Tables 3-5). *R. padi* was the most common aphid species caught in the traps in southern, central and northern parts of Sweden. In 1988, there were large migrations in

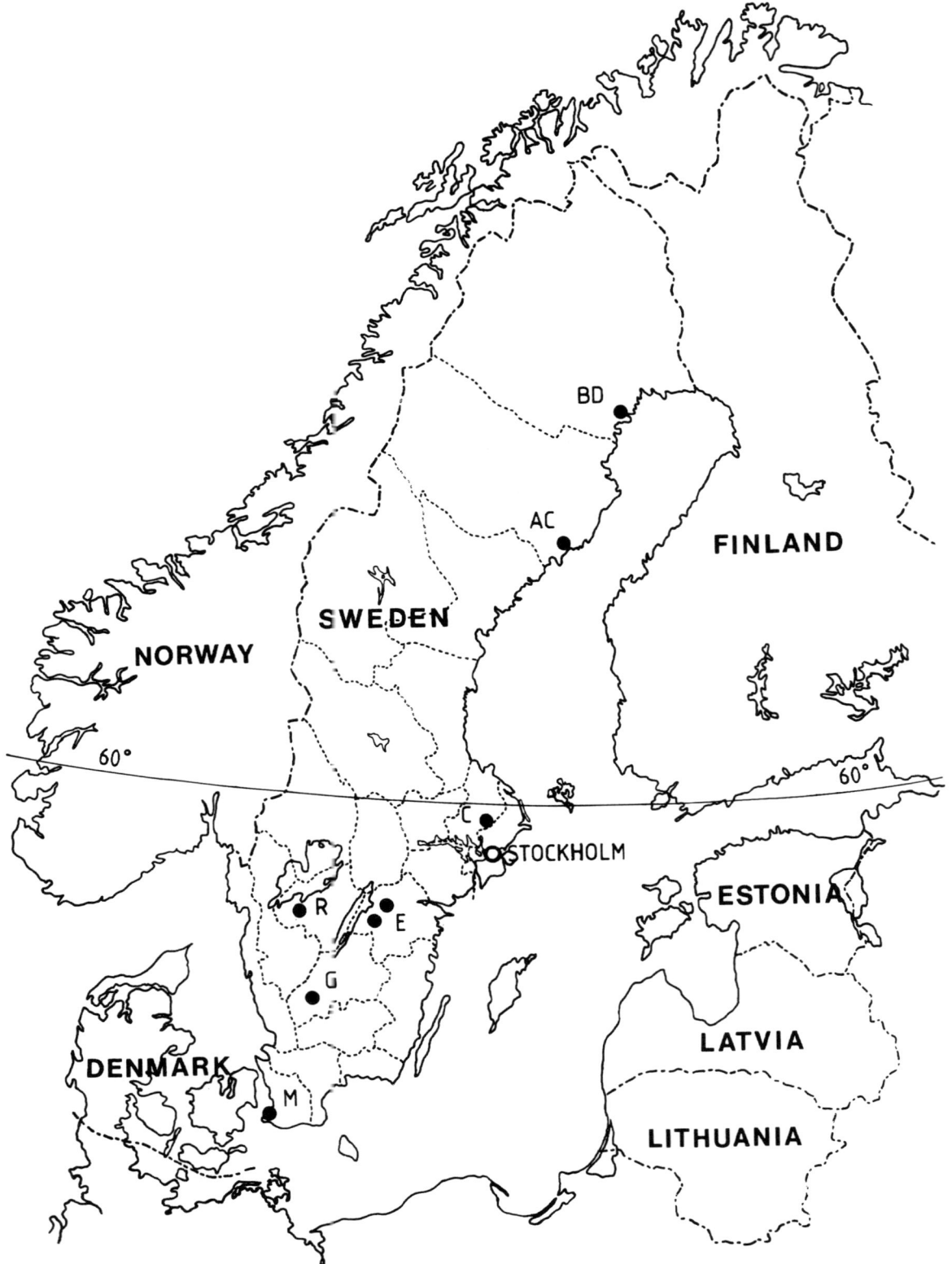

Figure 4. Location of suction traps in Sweden. ● = suction trap

both southern and central Sweden (Table 5). Differences from year to year and region to region in aphid migration have a great influence on the spread of several aphid-borne plant viruses. Cultural practices, sowing date and crop age-structure are also very important factors to consider when estimating risk for spread of plant viruses.

BARLEY YELLOW DWARF VIRUS

Relationships between virus, vectors, host plants, and the environment have been studied in many countries in order to forecast the incidence of barley yellow dwarf virus (BYDV). In many European countries, the virus causes great yield reductions, especially in winter wheat and winter barley. The spread and infectivity of aphid vectors of BYDV in the fall have been studied in England using suction traps (20).

During the last 10 years the incidence of this virus in spring cereals seems to have increased in several countries, possibly owing to changes in cultural practices. In Scandinavia, most damage caused by virus-transmitting aphids involves spring cereals. In 1973, 1978, 1980, 1985 and 1988 damaging outbreaks of BYDV occurred in a number of regions of Scandinavia. For example, in 1973 the damage was responsible for crop losses exceeding 50 percent in several oat fields in the Southern Highlands of Sweden.

The most important aphid vector of BYDV in Sweden is R. padi, although S. avenae can also be of importance in certain years. M. dirhodum is of little importance in Sweden. Results from suction trap catches show that R. padi is the dominant vector because it migrates earlier and in greater numbers than S. avenae and M. dirhodum (Figures 5-7).

In Sweden R. padi eggs overwinter on the winter host Prunus padus. That means that spring migrating R. padi are free from BYDV when leaving the winter host. After acquiring the virus in grassland or leys, R. padi can transmit it to spring-sown fields, e.g. oats and barley. Apterae are mainly responsible for further virus spread within the field.

In 1985 and 1988, the crop losses due to BYDV in oat fields occurred in eastern parts of central Sweden, including the Southern Highlands. The two main reasons for the great spread of BYDV in Sweden were an infestation by the main vector, R. padi, and late sowing of crops during those years.

R. padi numbers differ greatly from year to year and region to region in Sweden (Figures 5-7). The incidence of BYDV seems to be closely correlated with number of spring-migrating R. padi, but weather conditions during early summer are also important. Thus, high temperatures and drought favor aphid multiplication in spring cereals which can in turn enhance the spread of BYDV within the fields by apterous R. padi.

A comparison of the number of spring-migrating R. padi with disease incidence measurements during certain years indicated that this aphid species is responsible for most of the virus spread. However, large populations of this aphid need not be a threat if the virus source is infrequent in the region. Most virus spread occurs in regions with a high proportion of grassland and grass leys. However, during the last 5 to 10 years, exacerbated by a series of mild winters, the incidence of BYDV seems to have increased in winter wheat and winter barley.

Further studies are needed to improve our ability to predict the spread of BYDV. For example, we have to learn more about the relative importance of the virus source in different types of grassland, the behaviors of different aphid species, factors affecting the susceptibility of the crop, and the influence of various weather factors. The influence of predators also has to be considered when studying the intra-field spread of BYDV. In addition, we need to consider the risk of spread of BYDV and other persistently transmitted viruses from regions other than the ones that have been studied. Vectors can be carried by the wind 500 km or more. Suction traps could be a good tool when conducting such studies.

BEET YELLOWS VIRUS

In Sweden the invasion of virus transmitting aphids with southerly winds seems to be the most important factor causing outbreaks of BYV in southern parts of the country (23). Wiktelius showed that the number of days with frost during winter and the occurrence of southerly winds explained most of the variation in the spread of BYV (23).

The incidence of BYV in a single field varies depending on sowing date, plant density, insecticide use, and the frequency of the local virus source. Elimination of the local virus source and control treaments targeting aphids in early June can decrease virus incidence by 60-70%. The main vectors of BYV in Sweden are M.

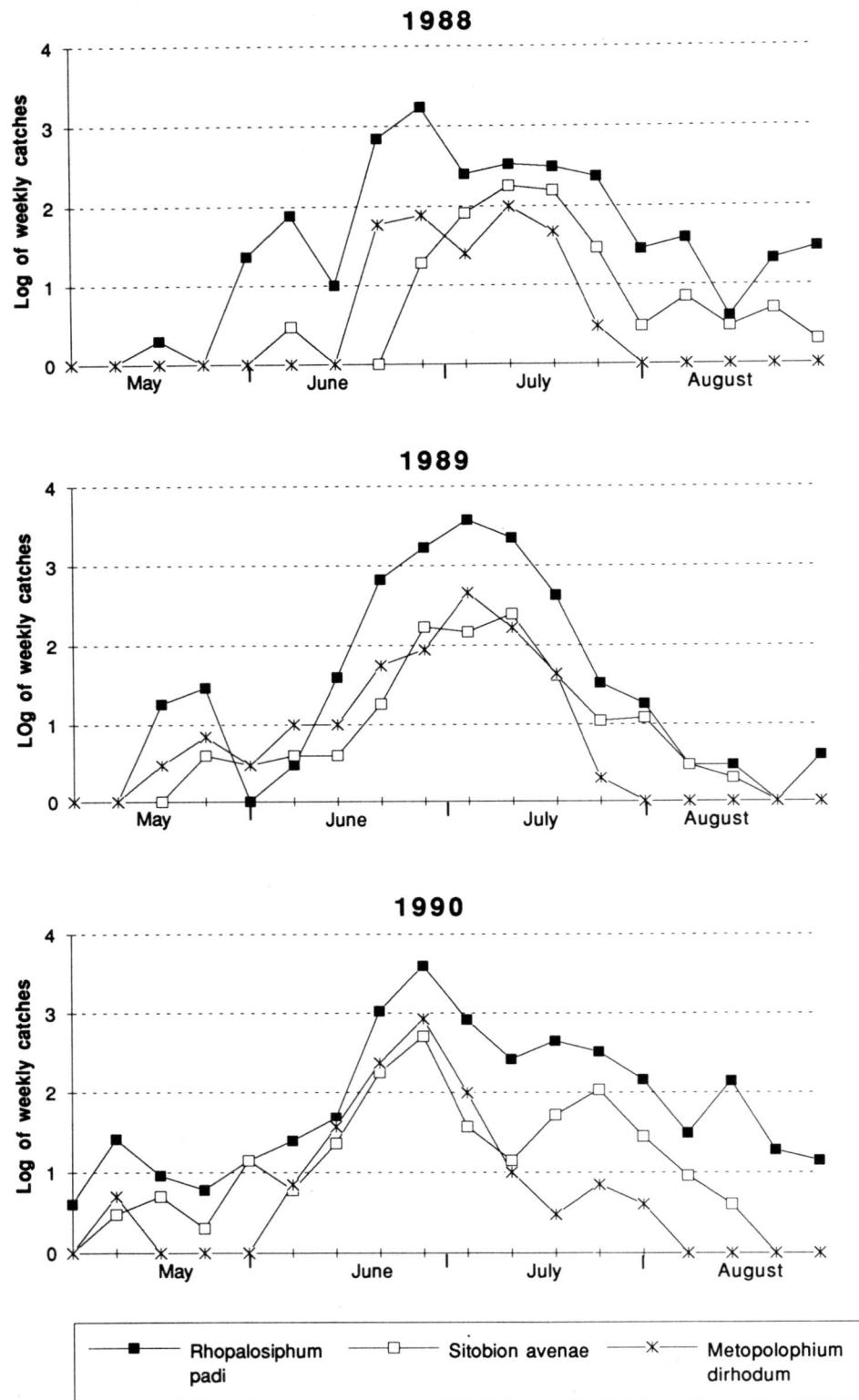

Figure 5. Weekly suction trap catches of aphids in southern Sweden (M) in 1988, 1989, and 1990.

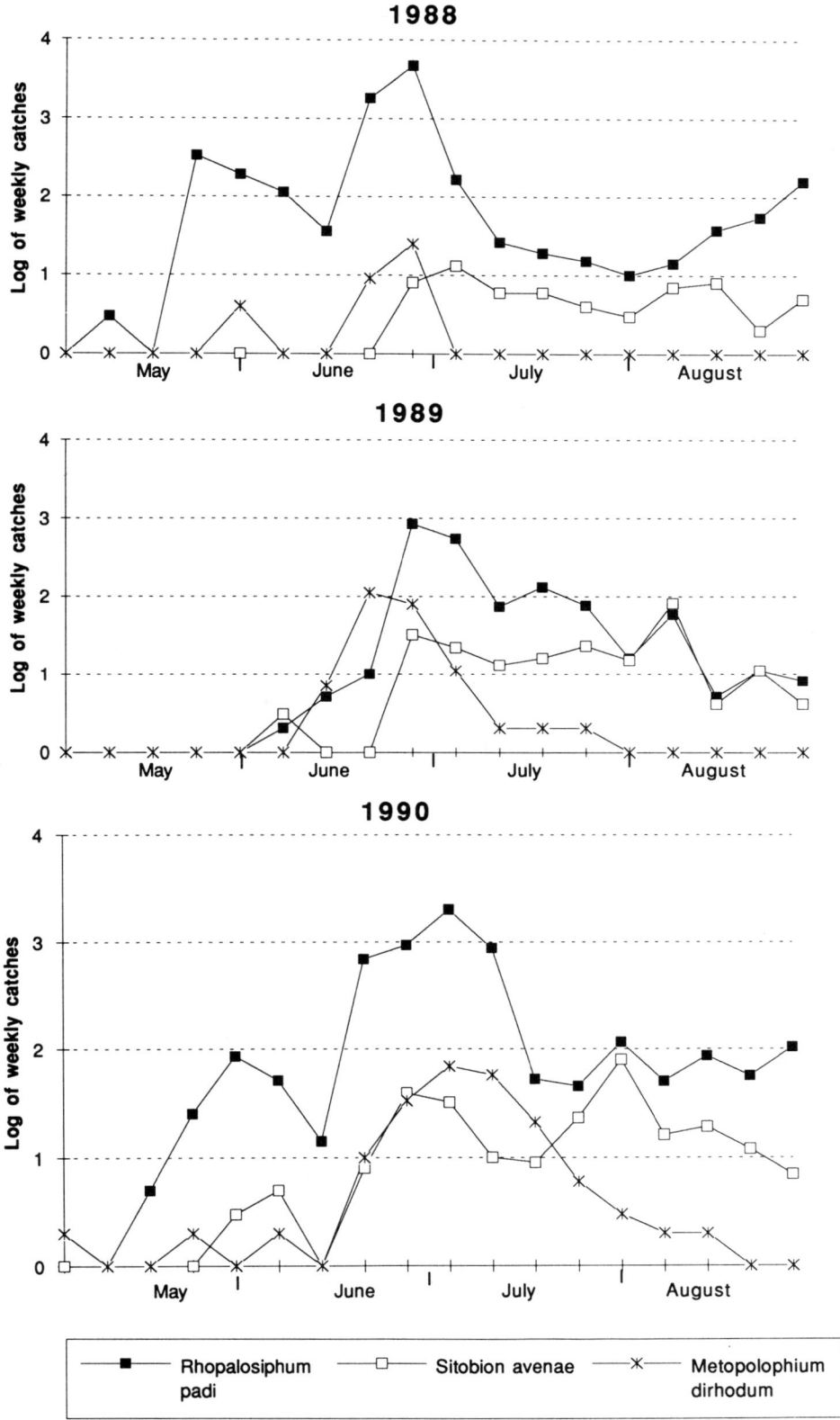

Figure 6. Weekly suction trap catches of aphids in central Sweden (C) in 1988, 1989, and 1990.

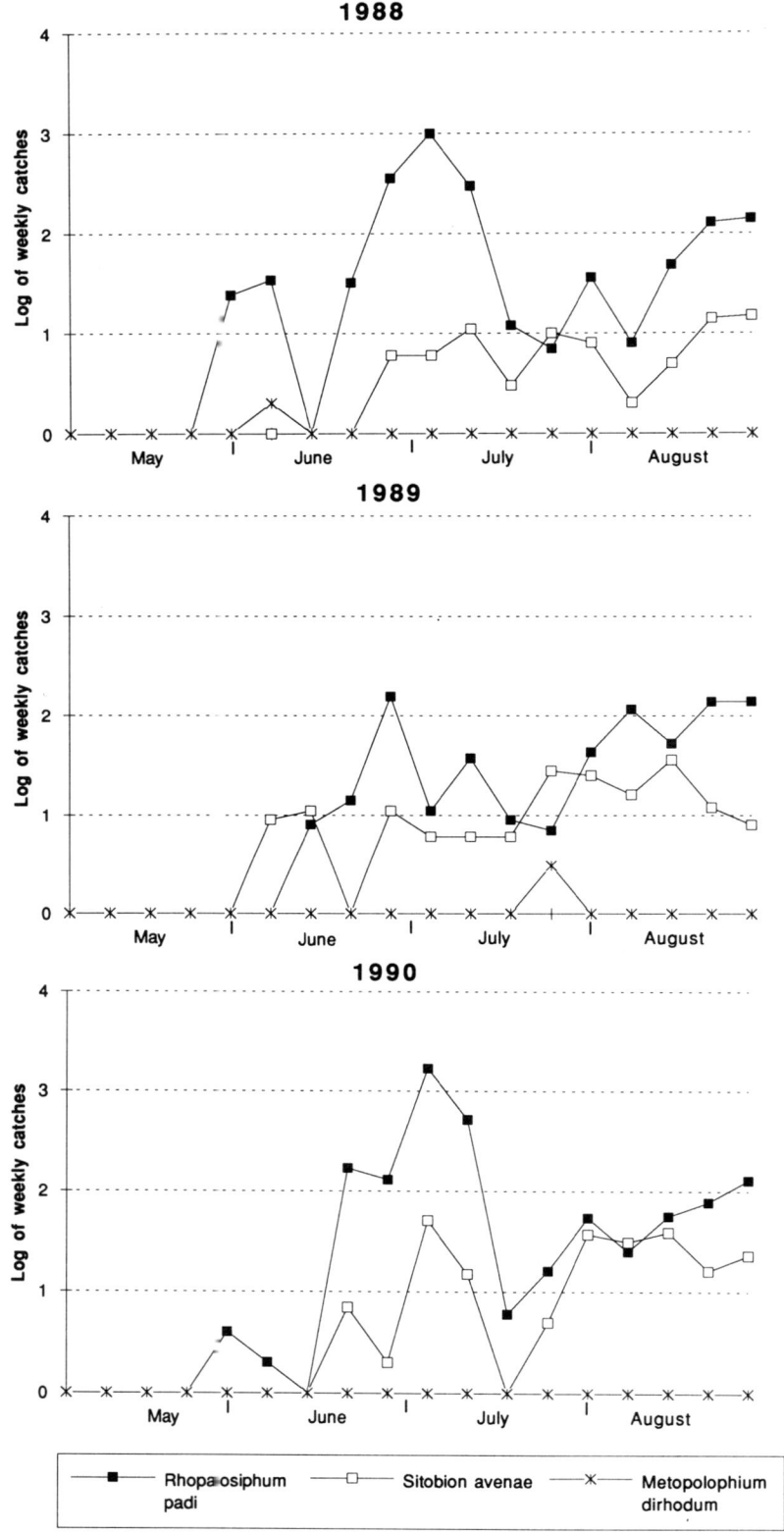

Figure 7. Weekly suction trap catches of aph ds in northern Sweden (AC) in 1988, 1989, and 1990.

persicae, M. euphorbias and *A. fabae.* Rather few of these species are caught in suction traps, which may explain the low incidence of BYV.

ACKNOWLEDGEMENTS

The studies on the epidemiology of potato virus Y have been supported by grants from the Association of Swedish Potato Growers. We thank Karin Nordin for valuable assistance with data collection and Lillian Holmer and Ulla Larsson for sorting insects and identifying aphid species. I am also grateful to the Swedish Seed Testing and Certification Institute for providing me with data on disease incidence in the field and in samples of progeny tubers. I also want to thank Felix AB, Estrell AB, IVK Potatis AB, Sterkelsen and Solanum for support. Studies on aphid migration using suction traps have been supported by the Swedish Agricultural Board.

REFERENCES

1. Bokx, J.A. de. 1979. Determination of infection pressure of potato virus Y^N with potato plants. Mededelingen van de Faculteit Landbouwetenschappen Rijksuniversiteit. Gent 44/2 : 653-656.

2. Bokx, J.A. de, and Piron, P.G.M. 1984. Aphid trapping in potato fields in the Netherlands in relation to transmission of PVY^N. Mededelingen van de Faculteit Landbouwwetenschappen Rijksuniversitetet Gent 49/2b : 443-452.

3. Bokx, J.A. de, and Piron, P.G.M. 1985. Aphid trapping in potato fields and transmission of potato virus Y^N. Mededelingen van de Faculteit Landbouwwetenschappen Rijksuniversitetet Gent 50/2b : 483-492.

4. Edwards. A.R. 1963. A non-colonizing aphid vector of potato virus diseases. Nature 200 : 1233-1234.

5. Harrington, R., and Gibson, R.W. 1989. Transmission of potato virus Y by aphids trapped in potato crops in southern England. Potato Res. 32 : 167-174.

6. Harrington, R., Katis, N., and Gibson, R.W. 1986. Field assessment of the relative importance of different aphid species in the transmission of potato virus Y. Potato Res. 29 : 76.

7. Hille Ris Lambers, D. 1972. Aphids : Their life cycles and their role as vectors. Page 36 in: Viruses of Potatoes and Seed Potato Production. J.A. de Bokx, ed. PUDOC, Wageningen.

8. Katis, N., and Gibson, R.W. 1985. Transmission of potato virus Y by cereal aphids. Potato Res. 28 : 65-70.

9. Kostiw, M. 1979. Transmission of potato virus Y by *Rhopalosiphum padi* L. Potato Res. 22 : 237-238.

10. Kostiw, M. 1980. Transmission of potato viruses by some aphid species. Tagungsberichte der Academi der Landwirtschafts wissenschaften der DDR Berlin 1984 339-244.

11. Nemecek, T., 1993. The Role of Aphid Behaviour in the Epidemiology of Potato Virus Y: A Simulation Study. Dissertation, Swiss Federal Institute of Technology, Zurich.

12. Ruesink, W.G., and Irwin, M.E. 1986. Soybean mosaic virus epidemiology: a model and some implications. Pages 295-313 in: Plant Virus Epidemics Monitoring. Modeling and Predicting Outbreaks. G.D. McLean, R.G. Farret, W.G. Ruesink, eds. Academic Press, Australia.

13. Ryden, K., Brishammar, S., and Sigvald, R. 1983. The infection pressure of potato virus Y^O and the occurrence of winged aphids in potato fields in Sweden. Potato Res. 26:229-235.

14. Sigvald, R. 1984. The relative efficiency of some aphid species as vectors of potato virus Y^O (PVY^O). Potato Res. 27: 285-290.

15. Sigvald, R. 1985. Mature plant resistance of potato plants against potato virus Y^O (PVY^O). Potato Res. 28:135-143.

16. Sigvald, R. 1986. Forecasting the incidence of potato virus Y^O. Pages 419-441 in: Plant Virus Epidemics - Monitoring, Modelling and Predicting Outbreaks. G.D. McLean, R.G. Farret, W.G. Ruesink, eds. Academic Press, Australia.

17. Sigvald, R. 1987. Aphid migration and the importance of some aphid species as vectors of potato virus Y^O (PVY^O) in Sweden. Potato Res. 30:267-283.

18. Sigvald, R. 1989. Relationship between aphid occurrence and spread of potato virus Y^O (PVY^O) in field experiments in southern Sweden. J. Appl. Ent. 108:34-43.

19. Sigvald, R. 1990. Aphids on potato foliage in Sweden and their importance as vectors of potato virus Y^O. Acta Agric. Scand. 40:53-58.

20. Tatchell, G.M., and Plumb, R.T. 1992. Spread and infectivity of aphid vectors of barley yellow dwarf virus in autumn in southern England, 1980-1990. Pflanzenschutz-Nachrichten Bayer 45:443-454.

21. van Hoof, H.A. 1977. Determination of the infection pressure of potato virus Y^N. Neth. J. Pl. Pathol. 83:123-127.

22. Van Hoof, H.A. 1980. Aphid vectors of potato virus Y. Neth. J. Pl. Pathol. 86:159-162.

23. Wiktelius, S. 1977. The importance of southerly winds and other weather data on the incidence of sugar beet yellowing viruses in southern Sweden. Swed. J. Agric. Res. 7:85-95.

24. Wiktelius, S. 1982. Flight phenology of cereal aphids and possibilities of using suction trap catches as an aid in forecasting outbreaks. Swedish J. Agric. Res. 12:9.

CHAPTER 14

Chemical Control of Insect and Nematode Vectors of Plant Viruses

M.K. Satapathy

The problem of increasing the world's food supply by increasing agricultural production to ensure the demands of an ever increasing agricultural production to ensure the demands of an ever increasing world population is an important issue today when global population exceeds 5 billion. Though striking increase in crop productivity has been achieved with development of short duration high yielding varieties, intensified cultivation development of irrigation system and increased application of fertilizer, the process has been impaired by losses due to various pests and diseases including some that were previously unknown or unimportant (84). According to some reports (40) crop losses due to pests, diseases and weeds stand at 47.1 percent for rice, 24.4 percent for wheat, 35.6 percent for corn, 54.0 percent for sugarcane and 33.9 percent for cotton.

Vector borne plant viruses which are extensively distributed and cause economically important diseases (50) form an important component of plant viruses. Though use of resistant varieties for management of vector borne plant virus diseases provide an ideal solution, development of resistant varieties is a long run process and often varieties break down after few years of intensive cultivation with arrival of virulent vector or the virus. Moreover in the absence of resistance source in the germplasm, the problem becomes more challenging. Hence attempts are being made to find out new chemicals and the methods of their use to minimize the disease spread through the control of the vectors.

VECTORS OF PLANT VIRUSES

Vectors of plant viruses belong to three major groups of phylum Arthropoda and Nematoda in the animal kingdom and division Mycota (Fungi in the plant kingdom). Among arthropod vectors, approximately 99 percent are the insects (29). The order Homoptera contains the largest and the most important insect vectors of plant viruses (29,50). Homopterous vectors include aphids (Aphidoidea: mainly Aphididae), mealybugs (Coccoidea: Pseudococcidae), Whiteflies (Aleyrodoidea: Aleyrodidae), Psyllids (Psylloidea: Psyllidae) in the sub order Sternorrhyncha; and leafhoppers (Cicadoidea: cicadellidal), spittle bugs (Cicadoidea: cercopidae), treehoppers (Cicadoidea: membracidae) and plant hoppers (Fulgoroidea: Delphacidae and cixiidae) in the sub order Auchenorrhyncha (28,30). A few insect vectors of plant viruses belong to other orders, such as true bugs (Hemiptera), the thrips (Thysanoptera), the beetles (Coleoptera) and the grasshoppers (Orthoptera). In other arthropod groups, only the mites (order Acarina-family Eriophyidae) belonging to the class Arachnida are important plant virus vectors.

Besides insects, few species of soil inhabiting ectoparasitic nematodes belonging to the phylum Nematoda represent an interesting group of plant virus vectors. All the nematoda vectors belong to the order Dorylaimida and include species from the genera, *Trichodorus, Paratrichodorus, Longidorus* and *Xiphinema* (82). Nematodes of the *Trichodorus* (about 20 species) and *Paratrichodorus* (about 15 species) of family Trichodoridae transmit tobraviruses exemplified by tobacco rattle and pea early browning viruses, while nematodes of the *genera Longidorus* (30 species) and *Xiphinema* (70 species) of family *Longidoridae* vector mainly nepoviruses such as tobacco ring spot, tomato ring spot and cherry leaf roll (31).

Few plant viruses are vectored by soil

inhabiting fungi *Olpidium* spp. belonging to the Chytridiales. It transmits tobacco necrosis and cucumber necrosis viruses. *Polymyxa* and *Spongospora* (Plasmodiophorales) which are obligate parasites and colonize readily only a limited number of plant species in the field soil, and many more species under experimental conditions (41) transmit a number of viruses. For example, the former transmits wheat mosaic virus, and beet necrotic yellow vein and the later transmits potato mop top virus.

Control of fungal vectors is discussed in another chapter while that of insect and nematode vectors is dealt with here.

TRANSMISSION AND VIRUS-VECTOR INTERACTION

Usually virus transmission is specific. Viruses which are vectored by one major taxonomic group are not transmitted by vectors from another group.

Insect borne plant viruses are divided into three major groups, non-persistent, semi-persistent, and persistent (79,89). The terms stylet borne and circulative are also often used (42) for non-persistent and persistent viruses, respectively.

Most of the aphid borne viruses are non-persistent in nature where the viruses are carried on or near the mouthparts of the insect and do not multiply in the insect (90). Non-persistent viruses have short acquisition and short inoculation feeding time whereby aphids acquire and transmit viruses within seconds or minutes (27,50), and the insects loose infectivity rapidly after leaving the infected plant.

Some aphid species and mostly leafhoppers, planthoppers, and tree hoppers transmit viruses in persistent or circulative manner. In persistent transmission, acquisition feeding time is longer and there is a latent period following acquisition feeding. Having acquired the viruses the insects retain the ability to transmit it for a few days. Sometimes it is able to transmit the virus for the remainder of its life. The virus is retained through the molt of the insect. Persistent viruses are of two categories; barely yellow dwarf which do not multiply within their vector, *Acyrthosiphon pisum* (56) and those which do such as sowthistle yellow vein transmitted by *Hyperomyzus lactucae* (81). Persistent viruses that multiply within their vectors are called propagative viruses and are sometimes transmitted through eggs (transovarial transmission) as seen

with sowthistle yellow vein virus in its vector *H. Latucae* (80) and rice dwarf virus in *Nephotettix cincticeps* (22). As compared to non-persistent viruses, persistent viruses usually show a high level of specificity in their vector relationship and transmission efficiency does not increase by fasting prior to acquisition feeding.

Semi-persistent viruses such as beet yellows (aphid borne); maize chlorotic dwarf, and rice tungro transmitted by *Graminella nigriforms* (55) and *N. virescens* (3,36), respectively, have transmission properties intermediate between persistent and non-persistent viruses. These viruses do not circulate within their vectors (29) but their vector retains the ability to transmit them for a few days. Though the vector acquires the virus after short acquisition feeding the transmission efficiency improves if the acquisition feeding time increases. Starving the insect before acquisition feeding does not improve the transmission efficiency.

Nematode vectors transmit viruses by feeding on roots of infected plants and then moving on to roots of healthy plants. Larvae of second to fourth instars and adults act as vectors (31). Virus is more readily acquired than inoculated. Although the process appear to take a few minutes, the frequency of transmission increases with increasing access period up to 48 h (6). There is no evidence of any latent period between acquisition and inoculation but the virus can persist in the nematodes for months or years (6,87). The viruses neither multiply nor pass through the nematode eggs.

CHEMICALS

Principle of Vector Control

Control of insects as vectors of plant viruses is a different and more difficult problem than controlling them as pests (3). To control insects as pests it requires only bringing down the population below the number at which their feeding directly does not harm a crop. But presence of relatively few insects which do not cause direct damage may cause substantial spread of the disease in the field as vectors of the virus. Threshold level of insects as vectors is different from that of pests (62). Aphids cause damage as pests mainly because of the large infestations of wingless individuals which are relatively immobile and unimportant as vectors of viruses compared with the winged forms that are responsible for starting the infestation. The problem of controlling aphids as pests is mainly the problem of preventing the wingless form from multiplying to damaging numbers whereas the problem of controlling virus diseases is to kill the winged forms before they infect healthy plants (13).

The success of the control of the insect transmitted virus diseases through chemicals (insecticides) varies with different viruses and mainly dependent upon the ability of the insecticides to prevent virus transmission through killing the insect vector. The nature of the virus-vector relationship plays an important role in the prevention ability of the insecticides. Generally it seems difficult to prevent the infection of aphid-borne non-persistent viruses and spread of the virus under field conditions by insecticides, as they are carried by transient aphids and can be acquired or transmitted within a minute (88) or less (78). On the contrary it is easier to prevent the spread of leafhopper transmitted phloem borne persistent viruses by insecticides since the vectors of these viruses take comparatively longer to acquire and transmit the virus (78). Besides the transmission property, the effectiveness of an insecticide also depends on the source of infection which is within the plot or outside the field or spread occurs mainly by virus being brought into the crop by already viruliferous insects. For preventing the spread of non-persistent viruses, an insecticide would have to kill incoming winged insects very quickly, whereas a slower acting insecticide could be expected to prevent the spread of persistent viruses, particularly spread from plant to plant within a crop (12). Moreover the insecticide should have long residual toxicity against insect vector so that minimum time it is applied in a cropping season to prevent insect infestation.

Insecticides in Vector Control

Earlier attempts on control of virus vectors using insecticides were not very promising due to absence of fast-acting insecticides and lack of adequate knowledge on virus vector relationship. Some of the earlier insecticides used were sulphur, nicotine, rotenone, pyrethrum, etc. Dusting of sulphur, nicotine and a mixture of copper carbonate, lead arsenate, sulphur and nicotine dusts at two-week intervals for two months on tomato plants failed to control tomato spotted wilt in Australia (12). These insecticides killed the vector visiting the crop immediately

after application but not those coming between two applications due to lack of residual toxicity. However, Magee et al. (47) using tartar emetic and sugar spray applied twice weekly reduced tomato spotted wilt from 27 percent in untreated plots to 9 percent in treated ones.

With the development of synthetic insecticides such as DDT, BHC, parathion, etc., they were further evaluated against different virus diseases. DDT was very effective in lettuce yellows control and spread of tomato spotted wilt within the field. Broadbent et al. (14) could reduce potato virus Y infected plants by spraying DDT emulsion, endrin and parathion in the fields. The virus was not introduced by incoming aphids and spread within the crop was also reduced. The same insecticides were more effective for potato leafroll, the aphid borne persistent virus. The effectiveness of DDT and parathion in controlling aphids and reducing potato leafroll has also been found by Gibson et al. (23) but a completely healthy crop could not be produced as aphids coming into the crop were already infective. Parathion, TEPP and BHC were very successful in controlling strawberry yellows transmitted by Capitophorus fragaefolii (11). Three parathion sprays at weekly intervals also controlled insect vector, and decreased infection of pea enation mosaic from 54 to 9.0 percent (19).

Major limitations with contact insecticides like DDT and BHC are that they lack in persistency and the new foliage that comes out between successive applications provide favorable sites for infestation by viruliferous vectors. This brought the development of systemic insecticides which have penetrating ability and long residual toxicity.

Application of systox to potatoes has been reported (44,69) effective in reducing potato leafroll. Broadbent et al. (14) found schradan, endrin, systox and mipapox at 100 gl/acre at 10-14 day interval to prevent potato leafroll spread.

Hull and Gates (37) followed by Hull (38) reported that sugar beet yellows spread by Myzus persicae was considerably reduced by spraying the crop with systemic insecticides. Spraying nurseries with E.605 (Organophosphorus) three times reduced the incidence of disease to 1/9 of the control plots. Heurer (35) found that spraying sugar beet with endothion, phosfomidon, demeton or demeton-O-methyl delayed the appearance of yellows and increased the yield by more than 35 percent. Soil application of insecticides were much more effective (33). Hoja blanca disease of rice transmitted by Sogatodes orizicola was reduced (48) from 23 to 4 percent by spraying with demeton-O-methyl, 3, 7 and 11 weeks after germination.

Singh *et al.* (75) reported that one application of carbofuran or disulfoton at 1.5 kg. a.i./ha reduced leaf curl disease of chilies through the control of its whitefly vector, *Bemisia tabaci*. Good control of the spread of tomato leaf curl disease (72,74) and yellow vein mosaic disease (YVMD) of okra (71) has been reported through control of their whitefly vector by use of insecticides. However, contradictory results have also been observed where insecticides reduced the population of whitefly but there was little effect on the incidence and spread of YVMD (18). Some insecticide treated plots had even higher disease than the untreated control plots as with aphid transmitted maize dwarf virus (70). This was possibly due to increased activity of the aphid vector because of application of sublethal dose insecticides (8).

Phorate, phosphomidon, carbaryl (57) and carbofuran (53,69,73) were very effective in the control of *N. virescens*, the vector of semipersistent and phloem borne rice tungro viruses. Besides carbofuran, a carbamate insecticide BP and isoprocarb were also very effective in controlling vector population and reducing disease spread (64). Diazinon was effective on cultivar Pankaj but not on Ratna. The differential behavior of insecticides is said to be due to varietal resistance besides cultivar - insecticide interaction.

Synthetic pyrethroids which are very effective at low doses and have less mammalian toxicity were very promising in insect vector control (5,21). For example, cypermethrin reduced rice tungro disease incidence to 3.0 percent in cultivar TN1 and 0.3 percent in cultivar Ratna, and very effectively controlled vector (both adult and nymphs) population and helped increase yields. No adults and nymphs of insect vector were observed in cypermethrin treated plots. Besides cypermethrin, other pyrethroids like decamethrin (2), deltamethrin (7) and alphacypermethrin (46) were also effective in reducing vector population and disease spread. Deltamethrin and permethrin were also reported to bring about an effective reduction in the transmission of some nonpersistent viruses like potato virus Y (24,25,59) but not bean yellow mosaic virus (61). An intermediate type of result was reported (60) with deltamethrin and lambda-cyhalothrin which had the greatest impact on secondary spread of cowpea mosaic virus and cucumber mosaic virus within cowpea plots and less on the initial introduction of winged aphids.

Methods of Application

In the last decade much progress has been made in insecticide application for control of insects. Foliar sprays have been commonly used but sprays and dusts are often ineffective because the residues are readily washed off the plants by frequent rains. Sometimes ineffective control by insecticides is attributed to its failure to penetrate the canopy. granular insecticides applied as paddy water application are absorbed into the plant and remain effective for long periods. However, microbial and chemical degradation of the insecticides needs frequent application for season long control. This led to alternate applications like seed treatment, root soak, root coat and root zone applications.

Treatment of rice seeds with carbofuran (F) was highly effective in reducing green leafhopper vector and tungro disease spread (52,67). Root soak treatment with cartap hydrochloride (43) and methyl demeton-S (49) were effective in the control of rice dwarf and tungro virus vectors, respectively. To improve the performance of insecticides used in root soak treatment, root coat method was developed in which a sticking agent is added permitting insecticides to stick to the root surface so that additional insecticide is absorbed by the plant for several days after planting. This method had been found promising for controlling rice tungro vectors. (39,62).

The success of root coat treatment of seedlings led to the concept of root zone application in which insecticides sealed in capsules or mudballs are pushed near the roots of seedlings a few days after transplanting. This method of application was proved very successful because the chemical being concentrated in the root zone, shielded from sunshine, heat and volatilization is slowly released and is less harmful to parasites and predators of rice plants. Carbofuran, isoprocarb and BPMC were very effective in control of green leafhopper vectors and spread of tungro virus disease in rice fields (58,65,66). Root zone application and soil incorporation were equally effective in reducing tungro disease incidence and vector control as compared to foliar spray, root soak and root coat treatment. A dose of 0.5 kg. a.i./ha of carbofuran applied at root zone was superior to 6.0, 5.0 and

9.5 kg. a.i./ha of the same insecticide used by broadcast, root soak and root coat methods, respectively (66). Besides rice, single application of granular insecticides like disulfoton, phorate and methocate in soil during planting kept aphids from colonizing the plants for a long period and controlled virus spread (15,16) in potatoes. Soil incorporation of disulfoton and phorate gave good control of aphids at least ten weeks after planting because of their slow release (17,76).

Mechanism of Action

Mechanism of action of insecticides in controlling virus vectors depends on their speed of action. Persistent insecticides that paralyze or kill almost immediately or repel and prevent insects from feeding are very effective in disease management (3,12,13). Inappropriate insecticides or insecticides with sublethal doses cause hyperactivity and bring dispersal or increased movement of viruliferous vectors leading to further disease spread. From studies on the control of tungro viruses, it was observed (3,64) that for successful transmission of tungro viruses, the viruliferous insect must feed at least for 10 min and an insecticide to be effective should be able to kill the insect very quickly (<10 min) before it transmits the viruses into the host tissues. Carbofuran, isoprocarb, BPMC and bendicarb kill the vector very fast and are antifeedant in nature. Cypermethrin is said to have repellant nature and quick knock down effect (2). Besides killing the viruliferous insects, effective insecticides also reduce probing frequency (43) and feeding activity (58) in the phloem which are essential for tungro virus transmission.

Oil

Oil sprays have been effective in controlling the spread of a number of non-persistent viruses and few persistent viruses (4,9,91). Though mechanism of action is not well understood, it is said (86) that oils alter the surface structure or charge on the stylets, thus limiting the ability to adsorb or release virus particles. Though oils are said to have no or very limited effect on vectors, Satapathy and Anjaneyulu (68) observed mineral oils like Bayol 50, Sandoz and Sunco 7N to prevent tungro virus transmission by killing the green leafhopper vector, similar to insecticides (20).

Nematicides

Nematode transmitted viruses persist for long periods in the soil and in their nematode hosts (51). Moreover, movement and dispersal of nematodes being generally slow, chemical treatment to kill the vectors is the most effective and practical method of control (45). However, the major limitations in nematicide treatment in controlling/killing nematodes is that they often occur at considerable depth beyond the reach of the nematicides and uniform application to field soil is an ideal process never achieved in practice.

The nematicides used for vector control fall into two groups, fumigant and nonfumigants. 1,3-D (1,3 dichloropropane), methyl bromide, EDB (Ethylene dibromide) and DBCP (1,2-dibromo-3-chloropropane) are the common fumigant where as aldicarb, oxamyl, plenamiphos and quintozone form nonfumigants.

Harrison *et al.* (32) found DDF (dichloropropane-dichloropropene) and methyl bromide to be effective nematicides in controlling arabis mosaic virus by killing 99 percent of the eelworm vector, *X. diversicaudatum* with pre-planting soil treatment. The chemical was effective up to 70 cm depth in the soil. Soil fumigation with quintozine (Pentachloronitrobenzene) reduced population of *L. elongatus* and gave full protection against tomato black ring and raspberry ringspot viruses (54,83). Aldicarb and phorate have been effective in the control of *Trichodorus* spp. transmitting tobacco rattle virus in potatoes (1,77). Another chemical, oxamyl, which is applied as granules or as a foliar spray has been shown to reduce the spread of nematode transmitted viruses by inhibiting their feeding and preventing the eelworm from acquiring the virus. This has been found effective for potato spraing disease in the field. Fungicide pentachloro nitrobenzene has proved effective in the control of *L. elongatus* transmitting raspberry ringspot virus (54,85).

CONCLUDING REMARKS

Though insecticides have commonly been used in the vector control, it is not only uneconomical but further aggravates the problem of undesirable residues, environmental pollution, pest resistance and harmful effects on non-target organisms including parasites and predators of the target pests. Often repeated insecticidal treatments in the field with high vector population fail to control diseases (34) caused by non-persistent and semipersistent viruses. Resurgence of other unwanted or unknown pests is also sometimes associated with insecticide application.

With increasing concern over environmental effect of continued reliance on toxic chemicals for pest control, greater attention is being devoted to the development of alternate, less toxic and less disruptive methods for pest control in general and vector borne viruses in particular. Use of host resistance, barrier crops, manipulation of cultural practices and use of botanicals and biological control methods (use of predators, parasites and pathogens of insect vector) are being used to reduce the vector population (26). Judicious use of selective pesticides together with intensive development and application of biological, cultural and other control measures popularly known as integrated management is considered the most reasonable and feasible alternative to chemical control of insect pests including plant virus vectors.

REFERENCES

1. Alphey, T.J.W., Corper, J.I., and Harrison, B.D. 1975. Systemetic nematicides for the control of trichodorid nematodes and of potato spraing disease caused by tobacco rattle virus. Plant Pathol. 24: 117-121.

2. Anjaneyulu, A., and Bhaktavatsalam, G. 1987. Evaluation of synthetic pyrethroid insecticides for tungro management. Trop. Pest. Mgmt. 33: 323-326.

3. Anjaneyulu, A., Satapathy, M.K., and Shukla, V.D. 1994. Rice Tungro. Oxford & IBH Publishing Co., New Delhi.

4. Asies, C.J. 1975. Control of the spread of tulip breaking virus in tulips with mineral oil sprays. Neth. J. Plant Pathol. 81: 67-70.

5. Atiri, G.I., and Ligan, D. 1986. Effect of pyrethroids (cypermethrin and deltamethrin) on the disease expression of cowpea aphid-borne mosaic virus. Agric. Ecosystems Environ. 15: 31-37.

6. Ayala, A., and Allen, M.W. 1968. Transmission of the california tobacco rattle virus (CTRV) by three species of the nematode genus *Trichodorus*. J .Agric. Univ. Puerto Rico. 52: 101-125.

7. Bhaktavatsalam, G., and Anjaneyulu, A. 1986. Evaluation of some new synthetic pyrethoids for tungro (RTV) disease and vector control. Int. Rice Res. Newsl. 11(5): 21.

8. Boiteau, G., Kings, R.R., and Levesque, D. 1985. Lethal and sublethal effects of aldicarb on two potato aphids (Homoptera Aphidae) : *Myzus persicae* (Sulzer) and *Macrosiphum euphorbiae* (Thomas). J.Econ. Entomol. 78 : 41-44.

9. Bradley, R.H.E. 1963. Some ways in which a paraffin oil impedes aphid transmission of potato virus Y. Can. J. Microbiol. 9: 369-380.

10. Bradley, R.H.E., and Ganong, R.Y. 1955. Evidence that potato virus is carried near the tip of the stylets of the aphid vector, *Myzus persicae* (Sulz.). Can. J.

Microbiol. 1:755-782.

11. Breakey, E.P., and Campbell. L. 1951. Reduction of strawberry yellows transmitted by *Capitophorus fragaefolii* through insecticides. Plant Dis. Reptr. 35: 63-69.

12. Broadbent, L. 1957. Insecticidal control of the spread of plant viruses. Annu. Rev. Entomol. 2: 339-354.

13. Broadbent, L. 1968. Disease control through vector control. Pages 593-630 in: Viruses, Vectors and Vegetation. K. Maramorosch, ed. Interscience Publishers, New York.

14. Broadbent, L., Burt, P.E., and Heathcote, G.D. 1956. The control of potato virus diseases by insecticides. Ann. Appl. Biol. 44: 256-274.

15. Broadbent, L., and Burt, P.E. 1962. Systemic insecticides and virus control in potatoes. Pages 81-85 in: Proc. Brit. Insect. Fungi. Conf., 1st, 1961.

16. Burt, P.E., Broadbent, L., and Heathcote, G.D. 1960. The use of soil insecticides to control potato aphids and virus disease. Ann. Appl. Biol. 48: 580-590.

17. Close, R.C. 1967. Granular insecticides for aphid control. Proc. NZ. Weed Pest Control Conf. 22: 222-226.

18. Dahal, G., Neupane, F.P., and Baral, D.R. 1992. Effect of planting and insecticides on the incidence and spread of yellow vein mosaic of okra in Nepal. Int. J. Trop. Pl. Dis. 10: 109-124.

19. Davis, A.C., McEwen, F.L., and Schroeder, W.T. 1961. Control of pea enation mosaic in peas with insecticides. J.Econ. Entomol. 54: 161-166.

20. De Ong, E.R. 1948. Chemistry and use of insecticides. Reinhold Publishing Corp. New York, USA.

21. Elliot, N., James, N.F., and Potter, C. 1978. The future of pyrethroids in insect control. Annu. Rev. Entomol. 23: 443-469.

22. Fukushi, T. 1933. Transmission of the virus through the eggs of an insect vector. Proc. Impr. Acad. (Tokyo) 9: 457-460.

23. Gibson, K.E., Landis, B.J., and Klostermeyer, E.C. 1951. Effect of aphid control on the spread of leafroll in potatoes. Am. Potato J. 28: 658-666.

24. Gibson R.W. 1983. The ability of different pyrethroids to control spread of potato viruses by aphids, Proc. 10th Int. Cong. Pl. Protect. 3: 1192.

25. Gibson, R.W., Rice, A.D., and Sawicki, R.M. 1982. Effects of the pyrethroid deltamethrin on the acquisition and inoculation of viruses by *Myzus persicae*. Ann. Appl. Biol. 100: 49-54.

26. Harpaz, I. 1982. Nonpesticidal control of vector borne diseases. Pages 1-22 in: Pathogens, Vectors and Plant Disease Approaches to Control. K.F. Harris, and K. Maramorosch, eds. Academic press, New York.

27. Harris, K.F. 1977. An ingestion - egestion hypothesis of noncirculative virus transmission. Pages 165-220 in: Aphids as virus vectors. K.F. Harris, and K.Maramorosch, eds. Academic Press, New York.

28. Harris, K.F. 1980. Aphids, Leafhoppers and Planthoppers Pages 1-13 in: Vectors of Plant Pathogens. K.F. Harris, and K. Marmorosch, eds. Academic Press, New York.

29. Harris, K.F. 1981. Arthropod and nematode vectors of plant viruses. Annu. Rev. Phytopathol. 19: 391-426.

30. Harris, K.F., and Maramorosch, K, 1980. Vectors

of Plant Pathogens. Academic Press, New York.

31. Harrison, B.D. 1977. Ecology and control of viruses with soil inhibiting vectors. Annu. Rev. Phytopathol. 15: 331-360.

32. Harrison, B.D., Peachey, J.E., and Winslow, R.D. 1963. The use of nematicides to control the spread of arabis mosaic virus by *Xiphinema diversicaudatum* (micol). Ann. Appl. Biol. 52: 243-255.

33. Heathcote, G.D. 1978. Review of losses caused by virus yellows in English sugar beet crops and the cost of partial control with insecticides. Plant Pathol. 27: 12-17.

34. Heinrichs, E.A. 1979. Control of leafhopper and plant hopper vectors of viruses. Pages 529-560 in: Leafhopper Vectors and Plant Disease Agents. K. Maramorosch, and K.F. Harris, eds. Academic Press, New York.

36. Hibino, H. 1989. Insect borne viruses of rice. Pages 209-241 in: Advances in Diseases and Vector Research. K.F. Harris ed. Springler and Verlag, New York.

35. Heuver, H. 1960. Bestrijding Van vergeling ziekte in Bieten. Land bouwvoorlichting 17. 81-87.

37. Hull, J., and Gates, L.F. 1953. Experiments on the control of beet yellows virus in sugarbeet seed crops by insecticidal sprays. Ann. Appl. Biol. 40: 60-78.

38. Hull, R. 1959. Sugar beet yellows in Great Britain. Plant Pathol. 8: 145.

39. IRRI, 1976. International Rice Research Institute. Annual Report for 1975, Los Banos, Philippines, 479 Pages.

40. Ishikura, H. 1991. Plant protection in developing countries. Where do we go from here. Pages 356-359 in: International Plant Protection: Focus on the Developing World. Proc. 11th Int. Cong. of Plant Protection, Vol. II. E.D. Magallona, ed.

41. Jones. R.A.C., and Harrison, B.D. 1972. Ecological studies on potato mop-top virus in Scotland. Ann. Appl. Biol. 71: 47-57.

42. Kennedy, J.S., Day, U.F., and Eastop, V.F. 1962. A conspectus of aphids as vectors of plant viruses. Commonw, Inst. Entomol. London.

43. Kono, Y., Nagarashi, D., and Sakai, M. 1975. Effect of cartap, chlordimeform and diazinon on the probing frequency of green rice leafhopper. Appl. Ent. Zool 10: 58-60.

44. Lammbers, H.R., Reestman, A.J., and Schepers, A. 1953. Insecticides against aphid vectors of potato. Neth. J. Agric. Sci. 1: 188-201.

45. Lamberti, F., and Basile, M. 1982. Chemical control of nematode vectors. Pages 57-69 in: Pathogens, Vectors and Plant Diseases: Approaches to control. K.F. Harris, and K.Maramorosch, eds. Academic Press, New York.

46. Macatula, R.F., Valencia, S.L., and Mochida, 0. 1987. Evaluation of 12 insecticides against green leafhopper for preventing rice tungro virus disease. IRRI Res. Pap. Ser. No.128. 9 Pages.

47. Magee, C.J., Morgan, W.L., and Johnston, A.N. 1942. Control of tomato spotted wilt with tartar emetic and sugar spray. Australian Inst. Agr. Sci. 8: 115-117.

48. Malaguti, G., and Angeles. 1957. Ensayos preliminares sorbe, eluse di insecticesen el control de los vectores dela "hosa blance" dal Arroz. Agron Trop (Maracar vensuela) 7: 161-163.

49. Mami, M., and Jeyaraj, S. 1976. Laboratory studies on the control of rice green leafhopper *Nephotettix*

virescens (Dist.) by seed and seeding dip treatments with systemic insecticides. Labder J. Sci. Tech. 13:14-18.

50. Maramorosch, K. 1963. Arthropod transmission of plant viruses. Annu. Rev. Entomol. 8: 369-414.

51. Martelli, G.P. 1978. Nematode borne viruses of grapevine, their epidemiology and control. Nematol. Mediterr. 6: 1-27.

52. Mitra, D.K. Raychaudhuri, S.P., Everett, T.R., Ghosh, A., and Niazi, F.R. 1970. Control of the rice green leafhopper with insecticidal seed treatment and pretransplant seedling soak. J. Econ. Entomol. 63:1958-1961.

53. Mochida, O., Valencia, S.L., and Basilio, R.P. 1986. Chemical control of green leafhoppers to prevent virus diseases, especially tungro disease, on susceptible/ intermediate rice cultivars in the tropics. Pages 195-208 in: Trop. Agric. Res. Ser. No. 19. Tropical Agriculture Research Centre, Tsukuba, Japan.

54. Murant, A.F., and Taylor. C.E. 1965. Treatment of soil with chemicals to prevent transmission of tomato black ring and raspberry ring spot viruses by *Longidorus elongatus* (deman). Ann. Appl. Biol. 55: 227-237.

55. Nault, L.R., Styer, W.E., Knoke, J.K., and Pitre, H.N. 1973. Semipersistent transmission of leafhopper-borne maize chlorotic dwarf virus. J. Econ. Entomol. 66: 1271-1273.

56. Paliwal, Y.C., and Sinha, R.C. 1970. On the mechanism of persistence and distribution of barley yellow dwarf virus in an aphid vector. Virology 42: 668-680.

57. Pathak, M.D., Vea, E., and John, V.T. 1967. Control of insect vectors to prevent virus infection of rice plants. J.Econ Entomol. 60: 218-224.

58. Rapusas, H.R., Heinrichs, E.A., Aquino, G.B., and Basilio, R.P. 1986. Root zone application of insecticides for control of the rice tungro virus vector. J. Pl. Prot. Trop. 3: 111-119.

59. Rice, A.D., Gibson, R.W., and Stribley, M.F. 1983. Effects of deltamethrin on walking, flight and potato virus Y-transmission by pyrethroid-resistant *Myzus persicae*. Ann. Appl. Biol. 102: 229-236.

60. Roberts, J.W.F., Hodgson, C.J., Jackai, L.E.N., Thottappily, G. and Singh, S.R. 1993. Interaction between two synthetic pyrethroids and the spread of two non-persistent viruses in cowpea. Ann. Appl. Biol. 122: 57-67.

61. Sassen, B. 1983. The effect of two pyrethroids on the feeding behavior of three aphid species and transmission of two different viruses. Pflanzenk. Pflanzenp. und Pf lanzenschutz. 90: 119-126.

62. Satapathy, M.K. 1983. Prevention and control of rice tungro virus disease. Ph.D. thesis, Utkal University, Bhubaneswar.

63. Satapathy, M.K., and Anjaneyulu, A. 1984. Cypermethrin, a synthetic pyrethroid in the control of rice tungro virus disease and its vector. Trop. Pest Mgmt. 30: 170-178.

64. Satapathy, M.K., and Anjaneyulu, A. 1986. Prevention of rice tungro virus disease and control of the vector with granular insecticides. Ann. Appl. Biol. 108: 503-510.

66. Satapathy, M.K., and Anjaneyulu, A. 1991. Rice tungro management by different methods of carbofuran application. Oryza 28: 377-387.

67. Satapathy, M.K., and Anjaneyulu, A. 1991. Carbofuran seed treatment and reduction of tungro

incidence in experimental rice nursery. Oryza 28: 469-472.

68. Satapathy, M.K., and Anjaneyulu, A. 1992. Prevention of rice tungro virus infection by oils. Indian Phytopathol. 45: 407-411.

69. Schepers, A., Restman, A.J., and Lambers, H.R.D. 1955. Some experiments with systox. Pages 75-83 in: Proc. 2nd Conf. Potato virus Dis., 1954.

70. Seifers, D.L., and Harvey, T.L. 1989. Effect of carbofuran on transmission of maize dwarf virus in sorghum mechanically and by the aphid Schizaphis graminum. Plant Dis. 73: 61-63.

71. Shastry, K.S.M., and Singh, S.J. 1973. Restriction of yellow vein mosaic virus spread of okra through the control of the vector whitefly. Indian J. Mycol. Pl. Path. 3:76-81.

72. Shastry, K.S.M., and Singh, S.J. 1974. Control of the spread of the tomato leaf curl virus by controlling the whitefly (Bemisia tabaci gen). Indian J. Hort. 31: 178-181.

73. Shukla, V.D., and Anjaneyulu, A. 1980. Evaluation of systematic insecticides for control of rice tungro. Plant Dis. 64: 790-792.

74. Singh, S.J., Shastry, K.S.M., and Shastry, K.S. 1973. Effect of oil sprays on the control of tomato leaf curl virus in the field. Indian J. Agric. Sci. 43: 669-672.

75. Singh, S.J. Shastry, K.S. and Shastry, K.S.M. 1979. Efficacy of different insecticides and oil in the control of leaf curl virus disease of chilies. J. Plant Dis. and Protect. 86: 253-256.

76. Smith, H.C., Close, R.C., and Rough, B.F.A. 1964. The efficiency of granular insecticides in controlling virus diseases of crop. Proc. NZ weed pest control. Conf. 17: 168-174.

77. Sykes, G.B. 1975. The effect of phorate, a wide spectrum pesticide, on the occurrence of spraing in twelve cultivars of potato. Plant Pathol. 24: 71-73.

78. Sylvester, E.S. 1949. Beet mosaic virus-green peach aphid relationships. Phytopathology 39: 417-424.

79. Sylvester, E.S. 1958. Aphid transmission of plant viruses. Proc. 10th Int. Congr. Entomol. 3: 195-204.

80. Sylvester, E.S. 1969. Virus transmission by aphids - a view point. Pages 159-173 in: Virus, Vectors and Vegetation. K. Maramorosch, ed. Interscience Publishers, New York.

81. Sylvester, E.S., and Richardson, J. 1970. Infection of Hyperomyzus iactuocae by sowthistle yellow vein virus, Virology 42: 1023-1042.

83. Taylor, C.E., and Murant, A.F. 1968. Chemical control of raspberry ringspot and tomato black ring viruses in strawberry. Plant Pathol. 17: 171-178.

84. Thresh, J.M. 1988. Rice viruses and the green revolution. Aspects of Appl. Biol. 17: 187-194.

85. Trudgill, D.L., and Alphey, T.J.W. 1976. Chemical control of the virus vector nematode Longidorus elongatus and Pratylenchus crenatus in raspberry plantation. Plant Pathol. 25: 15-20.

86. Vanderveken, J. 1977. Oils and other inhibitors of non-persistent virus transmission. Pages 435-454 in : Aphids as Virus Vectors. K.F. Harris and K. Maramorosch, eds. Academic Press, New York.

87. Van Hoof, H.A. 1970. Some observations on retention of tobacco rattle virus in nematodes. Neth J. Plant Pathol. 76: 329-330.

88. Watson, M.A. 1946. The transmission of beet mosaic and beet yellows viruses by aphids, a comparative study of a non-persistent virus and a persistent virus having host plants and vectors in common. Proc. Roy. Soc. London Ser. B. 133: 200-219.

89. Watson, M.A., and Roberts, F.M. 1939. A comparative study of the transmission of Hyocyamus virus 3, potato virus Y, and cucumber virus by vectors Myzus persicae (Sulz.), M. circumflexus (Buckton) and Microsiphum gei (Koch). Proc. Roy. Soc. London Ser. B 127: 543-576.

90. Watson, M.A., and Plumb, R.T. 1972. Transmission of plant pathogenic viruses by aphids. Annu. Rev. Entomol. 17: 425-452.

91. Zitter, T.A., and Simons, J.N. 1980. Management of viruses by alternation of vector efficiency and by cultural practices. Annu Rev. Phytopathol. 18: 289-310.

CHAPTER 15

Chemical Control of Fungal Vectors of Plant Viruses

J.A. Walsh

At least 30 plant viruses are known to have, or are suggested to have fungal vectors. Relative to those transmitted by arthropod vectors this may seem a small number, but, they include the causal agents of several important diseases of arable and horticultural crops. Recent practices, particularly the intensification of production, the switch to winter cereals, increased irrigation of some crops and the centralized propagation of transplanted crops have exacerbated disease problems and disseminated diseases more rapidly and widely. Several of the diseases are of local importance but those of barley, wheat, sugar-beet and lettuce cause major diseases in Europe, North America, Japan and China: yield losses of 30-40% have been reported in cereals, sugar-beet growing is uneconomic in parts of southern Europe, and entire lettuce crops have had to be ploughed in.

Although there have been many reviews covering different aspects of fungally-transmitted viruses (3,4,10,16,23,29,40), their control by chemical means has not previously been reviewed. Chemical control of zoosporic fungi was reviewed fourteen years ago (9). Previous reviewers (3,4,40) have classified the viruses into various groups based on the virus particle morphology, fungal vector species and mode of transmission (Table 1). As can be seen, only five species of fungi act as vectors, all are lower fungi and are obligate plant parasites infecting roots. Several workers have questioned the taxonomic affinity of these organisms with the true fungi and suggest affinities with the protists. As none of them can be cultured away from their hosts they are difficult to work with experimentally, much of the evidence that they transmit viruses is circumstantial (2) and Koch's postulates has not been fulfilled for many of the viruses listed in Table 1. These vectors all have thick-walled resting spores, some of which have been shown to survive for over 20 years (12).

The spores are quite resistant to many chemicals making their control and that of the viruses they transmit very difficult and their eradication in contaminated soils all but impossible. When they germinate they release motile spores (zoospores) which are propelled by one (Chytridiomycetes) or two (Plasmodiophoromycetes) flagella. The zoospore is the only stage of the life cycle that exists (in a non-quiescent manner), albeit very briefly, outside plant roots. It is the stage that is most vulnerable to chemical control. However, as the root system of one plant can release millions of zoospores, a very high percentage kill is necessary to achieve good disease control and avoid virus spread. Zoospores infect root cells and produce zoosporangia, which unlike resting spores cannot survive outside roots. These in turn release more zoospores which reinfect and continue this cycle, or produce resting spores which are released into soil when roots decay. *Songospora subterranea* is unusual because it causes diseases in its own right, whereas the other fungi are only considered to cause significant damage when they transmit a virus to their host. The form of *S. subterranea* that infects potatoes causes the disease known as powdery scab (24) and the form that infects watercress causes crook root disease (44).

Two distinct relationships between vector fungi and the viruses they transmit have been proposed (40). These relationships were originally termed non-persistent and persistent, but it has been argued that these terms are inappropriate (13). The two relationships are defined by whether the virus is acquired *in vivo* or *in vitro* and by whether the virus survives externally or internally to the resting spores in the absence of a living host plant. As these characters are correlated in all relationships described so far, only two groups were described: one in which the

Table 1. Fungally vectored viruses

Virus	Vector
Chytridiomycete vector (*Olpidium* spp.)	
<u>Isometric virus particles</u>	
Cucumber leafspot virus (CLSV)	*Olpidium radicale*
Cucumber necrosis virus (CNV)	*O. radicale*
Melon necrotic spot virus (MNSV)	*O. radicale*
Red clover necrotic mosaic virus (RCNMV)	*O. radicale*
Squash necrosis virus (SqNV)	*O. radicale*
Tobacco necrosis virus (TNV)	*O. brassicae*
Tobacco necrosis satellite virus (TNSV)	*O. brassicae*
<u>Rod-shaped particles</u>	
Lettuce big-vein virus (LBVV)	*O. brassicae*
Tobacco stunt virus (TSV)	*O. brassicae*
<u>Uncharacterized agents</u>	
Freesia leaf necrosis virus	*O. brassicae*
Lettuce ring necrosis virus	*O. brassicae*
Pepper yellow vein virus (PYVV)	*O. brassicae*
Plasmodiophorid vector	
<u>Rod-shaped particles (Furoviruses)</u>	
Beet necrotic yellow vein virus (BNYVV)	*Polymyxa betae*
Beet soil-borne virus (BSBV)	*P. betae*
Oat golden stripe virus (OGSV)	*P. graminis*
Peanut clump virus (PCV)	*P. graminis*
Potato mop top virus (PMTV)	*Spongospora subterranea f. sp. subterranea*
Rice stripe necrosis virus (RSNV)	*P. graminis*
Soil-borne wheat mosaic virus (SBWMV)	*P. graminis*
<u>Filamentous particles (Baymoviruses)</u>	
Barley mild mosaic virus (BaMMV)	*P. graminis*
Barley yellow mosaic virus (BaYMV)	*P. graminis*
Oat mosaic virus (OMV)	*P. graminis*
Rice necrosis mosaic virus (RNMV)	*P. graminis*
Wheat spindle streak mosaic virus (WSSMV)	*P. graminis*
Wheat yellow mosaic virus (WYMV)	*P. graminis*
<u>Isometric particles</u>	
Watercress yellow spot virus (WYSV)	*Spongospora subterranea f. sp. nasturtii*

viruses were acquired *in vitro* and survived externally and one in which viruses were acquired *in vivo* and survived internally. These relationships are broadly accepted with the qualification that *in vitro* acquisition is not restricted to the laboratory but has also been shown to occur naturally when a virus is absorbed by zoospores in soil water (10). Obviously the mode of acquisition and transmission of viruses by fungal vectors is likely to have important implications for chemical control.

Three major types of chemical control agents have been used depending upon the stage of the fungal life cycle targeted for control or the substrate in which the crop is grown: 1) Surfactants, heavy metal ions and to a lesser extent recognized fungicides have been used in nutrient or aquatic crop production systems; 2) various soil amendments and fungicides have been used to control fungal and hence virus spread in field crops; and 3) soil partial sterilants or disinfectants are used to eradicate or reduce contamination of the resting spore stages of the fungi in field and greenhouse environments as well as on planting material. Antiviral compounds have been tested for their ability to control some of the diseases but to date none are in commercial use.

This chapter reviews the use of chemicals for the control of viruses with fungal vectors under the categories outlined above.

CHEMICAL CONTROL OF VECTORS AND VIRUSES IN NUTRIENT OR AQUATIC GROWING SYSTEMS

The first record of chemical control of a fungal vector of a virus in this environment was in 1956 (43) when the use of zinc-fritted glass was described for the control of crook root (*Spongospora subterranea* f.sp. *nasturtii*) disease of watercress. Interestingly, at that time the crook root agent was not known to transmit any viruses. Zinc at a concentration of 2 ppm was shown to completely suppress infection of watercress roots *in vitro*. At that time the use of soluble zinc compounds to control the disease was not considered practical because of the large quantities of water passing through the beds in which the watercress was grown (up to 0.9×10^6 liters per ha per day). Consequently a zinc frit (zinc oxide dissolved in glass and milled to a fine powder) was tested and when applied at 34 g/m^2 gave good control of crook root. The zinc content of the frit

was found to be critical and needs to be at least 23% (w/w) zinc oxide. The treatment was effective for only one season and repeated annual applications sometimes caused iron deficiency symptoms in crops. The frit was therefore replaced by direct applications of zinc sulphate at a constant measured rate from gravity-controlled drip feed equipment into water flows measured using graduated weirs (47). During 1962-80, these methods lessened the incidence of crook root significantly when the amount of zinc in the outflow was maintained at the minimum level (0.05 - 0.1 μg Zn/ml) required to kill the zoospores (48). At some watercress farms, however, only partial control was obtained due largely to the intermittent use of zinc, problems in measurement of water flow or mixing, maintenance of the required strength of stock solutions of zinc sulphate and problems with dilution of zinc by natural springs rising in watercress beds. More recently some growers have switched to using zinc chloride as it is cheaper, zinc solutions are sprayed through fine nozzles into water courses to aid mixing and special slow-release zinc-containing pellets have been developed. As for most of the time crook root has been controlled by zinc, it was not known to act as a vector of virus, data on the effectiveness of zinc in controlling spread of watercress yellow spot virus (WYSV) has only recently been obtained. The author found in a survey of watercress growers in the UK that virus infection was present at all but two farms, one of these farms was the only site where zinc treatments were applied throughout the whole year (57) (most growers only apply zinc in the winter months when WYSV symptoms and crook root are most severe). Also, when monitored over a period of 27 months, the percentage infection of plants by WYSV and crook root in beds treated with zinc was generally lower than those in untreated beds (57) and when zinc treatments were integrated with strategically timed cultural practices even better control was achieved (55). The mode of action of zinc is thought to be through killing the zoospore stage (45). The environmental impact of zinc in rivers and streams downstream of watercress beds is causing concern in the UK and consequently alternative control strategies are being sought. Under a voluntary agreement between growers and the UK rivers regulatory authority, growers discharge zinc at 0.075 ppm into water courses, however, there is uncertainty about the future use of zinc for control of crook root and WYSV. It was also found that bicarbonate at a concentration of >500 ppm in spring water supplying watercress beds gave good control of crook root (46). However, the costs associated with maintaining the necessary bicarbonate content of water were uneconomic.

Many greenhouse-grown horticultural crops are now cultivated in various soil-less substrates irrigated with nutrient solution (hydroponic systems) which may or may not be recirculated. This is an ideal environment for many zoosporic fungi and most of the viruses transmitted by *Opidium brassicae* and *O. radicale* have been found causing damage in such environments. In the UK, lettuce big-vein virus (LBVV) started to cause major problems in lettuce grown by the re-circulated nutrient film technique in 1978 and on some holdings 100% infection of crops was observed. Laboratory experiments following these outbreaks identified chemicals controlling the zoospore stage of *O. brassicae* (49). These included copper (4 μg/ml), zinc (10 μg/ml), and the fungicide carbendazim (as Bavistin WP, a wettable powder containing 50% (w/w) carbendazim: BASF). Further tests showed that a "blank" formulation of Bavistin containing only surface active agents (surfactants) was equally effective in killing zoospores. This led to tests on a range of surfactants of which Agral, Centrimide, Deciquam 222, Ethylan CPX, Hyamine 1622, Manoxol/OT and sodium lauryl sulphate were found to be toxic to zoospores at concentrations of 1-10 μg/ml. A commercial trial was subsequently carried out with Agral (non-ionic liquid containing 90% (v/v) alkyl phenol ethylene oxide condensate: ICI Plant Protection Division) added at a concentration of 20 μg/ml at 4 day intervals to the circulating nutrient solution in a greenhouse containing c.140,000 lettuce plants, where the two preceding lettuce crops had shown 100% LBVV symptoms. After 6-7 wk, when the plants had reached maturity none of them showed LBVV symptoms and no *O. brassicae* was observed in their roots (50). In an adjacent area of lettuce irrigated with untreated nutrient solution >90% of plants had developed severe LBVV symptoms and their roots were infected by *O. brassicae*. Agral has also been shown to be lethal to zoospores of *O. radicale* and when added twice weekly at 20 μg/ml to the hydroponic nutrient solution supplying a cucumber crop gave good control of *O. radicale* and melon necrotic spot virus (MNSV) increasing cucumber yields 2-3 fold (51).

Despite these notable successes in the control of viruses in hydroponic systems by surfactants, doubt has been expressed about the validity of the original results with Bavistin and the reliability of the zoospore assay deployed (14). Later experiments showed that the related benzimidazole fungicide benomyl did not stop *O. brassicae* zoospore motility but did prevent infection and reproduction. It was found that the *Olpidium* thallus was killed after infection which was consistent with the fact that the active ingredient (methyl-2-benzimidazole carbamate) is absorbed by the plant and is effective there by interfering with fungal mitosis (19). Our results support this finding in that neither Bavistin WP nor DF formulations were very effective in killing *O. brassicae* zoospores *in vitro* but gave good LBVV symptom suppression in pot tests and field trials (56). It has also been found that various fungicides, for example metalaxyl, stopped zoospore motility *in vitro* but not infection or reproduction *in vivo* (14). Zinc has been shown to be only partially effective *in vivo* (53). No doubt some of these differences can be explained by the inactivation or adsorption of compounds in soil and it has also been shown that inactivated zoospores can still be infective. It is certainly the case that zoospore assays *in vitro* are no guarantee that any particular product will be effective or ineffective *in vivo*. Despite this caveat such assays successfully identified Agral as an effective chemical for control of viruses in hydroponic systems. In some NFT conditions however, Agral has been found to provide no control of LBVV symptoms (33) nor lettuce ring necrosis virus also transmitted by *O. brassicae* (53).

A number of fungicides have been tested for their ability to control viruses in hydroponic systems. Although carbendazim precipitates in solutions at pH values higher than pH 3.0 a slurry of Bavistin (0.06 g product/L) gave good control of LBVV symptoms and *O. brassicae* in a recirculated hydroponic system (49). Although not tested alone thiophanate-methyl as a carbendazim regenerating product (25 mg a.i./L) gave good control of lettuce ring necrosis symptoms and its vector *O. brassicae* when used in conjunction with zinc (>.05 mg/L) in a recirculated hydroponic system (53).

Current trends in the treatment of recirculated nutrient solutions in hydroponic growing systems include novel chemical treatments such as ozone as well as non-chemical physical treatments including pasteurization, filtration, ultraviolet light and slow sand filtration. To date only UV light appears to have been tested against fungal virus vectors and has been shown to reduce

Table 2. Agents that have been shown to successfully control vectors or viruses in nutrient or aquatic growing systems

Virus	Vector	Agent of control	Effective concentration	Reference
WYSV	S. subterranic	zinc	>0.05 ppm	55
-	S. subterranic	bicarbonate	>500 ppm	46
LBVV	O. brassicae	Agral	20 ppm	50
LBVV	O. brassicae	Bavistin	0.06 g/L	49
MNSV	O. radicale	Agral	20 ppm	51
LBVV and 'lettuce ring necrosis'	O. brassicae	thiophanate-methyl	25 mg a.i./L	53
		+ zinc	>0.5 mg/L	53
LBVV and 'lettuce ring necrosis'	O. brassicae	Ultraviolet light	dependant on flow rate	20

spread of LBVV and lettuce ring necrosis (20). The limitation of most of these treatments is that they will only prevent disease spread to plants upstream of infection foci, plants downstream are likely to be infected by zoospores before such zoospores are killed by the control treatment.

SOIL AMENDMENTS AND FUNGICIDES FOR CONTROL OF VIRUS OR VECTOR SPREAD IN FIELD CROPS

Following the earlier success of zinc in controlling S. subterranea in watercress beds, treatment of soil with zinc frit (35% (W/W) zinc oxide), zinc sulphate or zinc oxide for the control of potato mop-top virus (PMTV) was tested. The frit was phytotoxic because of its boron content and the results were variable. Although all compounds reduced spread of PMTV and S. subterranea they did not eliminate the fungus from the soil and the amount of zinc required to give control increased with the increase in clay content of the soil (17). No control of LBVV symptoms in field grown lettuce (30) and in pot experiments (J Walsh, unpublished, 36) has been obtained with zinc although it has been shown to inactivate

zoospores of the vector in vitro (49). Zinc sulphate was ineffective in controlling Polymyxa betae the vector of beet necrotic yellow vein virus (BNYVV) in greenhouse pot tests (8). Of the other metallic ions, copper in the form of copper oxychloride was found to inhibit P. betae in pot tests (8) but had no effect on rhizomania in field trials (7, 15). Similar results were obtained with cuprous oxide when tested against S. subterranea in pot and field trials (17). Copper sulphate has been shown to inactivate O. brassicae zoospores in vitro (49), but has not been tested in the field. Mercury, usually in the form of mercurous chloride (calomel) has been widely tested and again despite suppression of P. betae in pot tests (37) no benefits could be demonstrated in field experiments (5). Boron has been shown to inactivate zoospores of O. brassicae in vitro (49) but has not been tested in field experiments and can be phytotoxic.

Again following the success demonstrated in hydroponic systems, several surfactants have been tested for their ability to reduce the spread of viruses in soil. Neither Agral nor ammonium lauryl sulphate reduced infection of sugar beet by P. betae in soil at the concentrations tested (8,37).

Table 3. Soil amendments and fungicides shown to control virus or vector spread in field crops

Virus	Vector	Substance	Rate of use	Reference
PMTV	*S. subterranea*	zinc frit	1980 kg/ha	17
PMTV	*S. subterranea*	zinc sulphate	1320 kg/ha	17
PMTV	*S. subterranea*	zinc oxide	680-1360 kg/ha	17
PMTV	*S. subterranea*	sulphur	1180 kg/ha	17
WSSMV	*P. graminis*	nitrogen (urea, uric acid or ammonium nitrate)	700 kg/ha	39
WSSMV	*P. graminis*	turkey manure	70 tonnes/ha	39
WSSMV	*P. graminis*	mercuric chloride	1.25 g/L	39
WSSMV	*P. graminis*	captan	10 g/L	39
WSSMV	*P. graminis*	2, 4, 5-trichlorophenol	2.5 g/L	39
LBVV	*O. brassicae*	quintozene	78 kg/ha	31
TNV	*O. brassicae*	captan	1 g/kg soil	41
PMTV	*S. subterranea*	mercurous chloride	75 kg a.i./ha	17
-	*P. betae*	prothiocarb	1 mg/seed	25
	O. brassicae	prothiocarb	"	25
LBVV	*O. brassicae*	Bavistin	20 mg/plant	56
LBVV	*O. brassicae*	benomyl	2.5 mg/plant	14
LBVV	*O. brassicae*	thiabendazole	5 mg/plant	Walsh and McPherson, unpublished
BNYVV	*P. betae*	Fluazinam	10-50 ppm	52

Agral gave no suppression of LBVV symptoms in pot or field experiments despite its efficacy in hydroponic systems (J. Walsh and M. McPherson, unpublished). Despite these negative results, disease development of a Plasmodiophorid fungus not known to transmit a plant virus has been restricted in the field by Agral (0.1% a.i.) (22).

Various amendments have been tested for their ability to reduce the transmission of fungally vectored viruses. Decreasing the pH of infective soil to 5.0 by applying sulphur greatly decreased the infection of potatoes with PMTV and *S. subterranea* in field experiments (17). This treatment did not eliminate either disease and when the pH of treated soil was raised, the transmission of PMTV resumed. Soil pH was also shown to affect development of rhizomania of beet caused by BNYVV and again the application of sulphur to lower soil pH to 5.5 controlled the disease (1). Acid soils in West Africa are considered to be deleterious to *P. graminis* and peanut clump virus (PCV) (42). A disadvantage of such a low soil pH is an adverse effect on plant growth. Soil applications of chemical or organic fertilizers containing high levels of nitrogen were reported to reduce the incidence of wheat spindle streak mosaic virus (WSSMV) (39).

One of the first attempts to control a fungal virus vector by using a fungicide was in 1960 when quintozene at 50 ppm was shown to

Table 4.   Soil partial sterilants and disinfectants shown to eradicate or reduce vector inoculum.

Virus	Vector	Substance	Rate of use	Reference
		SOIL TREATMENTS		
LBVV	*O. brassicae*	formaldehyde	8% solution	26
LBVV	*O. brassicae*	chloropicrin	786 L/ha	31
LBVV	*O. brassicae*	PCNB	79 kg/ha	31
LBVV	*O. brassicae*	methyl bromide	224 kg/ha	14
LBVV	*O. brassicae*	methyl bromide	500 kg/ha	59
LBVV	*O. brassicae*	D-D	0.2 ml/L	6
TNV	*O. brassicae*	methyl bromide	976 kg/ha	41
Freesia leaf necrosis virus	*O. brassicae*	methyl bromide, steam	750 kg/ha	54
BNYVV	*P. betae*	various	various	8,37
SBWMV	*P. graminis*	chloropicrin	0.5ml/L soil	32
SBWMV	*P. graminis*	carbon disulphide	15m/L soil	32
SBWMV	*P. graminis*	D-D	0.3ml/L soil	32
SBWMV	*P. graminis*	ethyl alcohol	30%	32
SBWMV	*P. graminis*	formaldehyde	0.8% solution	32
WSSMV	*P. graminis*	D-D	2 ml/L	39
WSSMV	*P. graminis*	metham sodium	2 ml/L	39
WSSMV	*P. graminis*	methyl bromide	1.5 ml/L	39
WSSMV	*P. graminis*	ethyl alcohol	95%	39
WSSMV	*P. graminis*	formaldehyde	2%	39
PCV	*P. graminis*	dibromochloro-phenol	"nematicidal dose rates"	42
PCV	*P. graminis*	D-D	"	42
		PRE-PLANTING TREATMENTS		
PMTV	*S. subterranea*	various	various	27
LBVV	*O. brassicae*	mercuric chloride	0.05%	21
Freesia leaf necrosis virus	*O. brassicae*	formaldehyde	0.8%	54
LBVV	*O. brassicae*	Iodel F.D.	8.3%	Walsh, unpublished
LBVV	*O. brassicae*	Jet 5	2%	Walsh, unpublished
LBVV	*O. brassicae*	sodium hypochlorite	26%	Walsh, unpublished

give good control of *O. brassicae* on lettuce in sand (36). Further tests showed that this chemical also suppressed the incidence of LBVV in field crops (31). Captan controlled TNV also transmitted by *O. brassicae* (41). A range of fungicides were tested for their ability to control PMTV in the field, but only mercurous chloride showed promise (17). However, its use commercially was not considered practical because of its cost and toxicity to humans. Mercuric chloride, captan, and 2, 4, 5-trichlorophenol as soil drenches have been shown to give good control of WSSMV in laboratory experiments (39). Dibromochloropropane, carbofuran, and aldicarb have been shown to control Indian peanut clump virus (IPCV) at nematicidal dose rates (35). The first report of a systemic fungicide giving control was prothiocarb which was reported to give substantial control of *P. betae* and *O. brassicae* when incorporated into the coating of sugarbeet pills at the rate of 1 mg per seed (25). This was followed by various reports of Bavistin and benomyl, incorporated into peat blocks of transplants or drenched prior to transplanting, reducing the incidence of LBVV (14,50). Benomyl is degraded to methyl benzimidazole-2-yl carbamate (carbendazim) which is the active ingredient of Bavistin, so it is considered that their modes of action in soil applications are identical (14). Fosetyl-aluminum has exhibited activity against *P. betae* in pot tests (37) and reduced the incidence of LBVV in a polytunnel experiment (56). Thiabendazole has been shown to give some control of LBVV in pot tests and field trials but can be phytotoxic to lettuce (J. Walsh and M. McPherson, unpublished). Many of these fungicides often give variable results under field conditions.

Despite this limited success in the control of viruses transmitted by *O. brassicae* and the effect of a number of these fungicides on *P. betae* in pot experiments, until recently no fungicides had been shown to control rhizomania in the field. The relatively new fungicide 'Fluazinam' has been reported to give good control of rhizomania (52) and also showed promise in pot tests for control of LBVV (56). Despite these promising results it has still to be tested rigorously under commercial conditions.

SOIL PARTIAL STERILANTS OR DISINFECTANTS FOR ERADICATION OR REDUCTION OF VECTOR INOCULUM

The first report on the effect of soil sterilants on the incidence of a virus (LBVV) was as early as 1940 (26) but it was not until 22 years later that the disease was shown to be transmitted by *O. brassicae*. Since then many different soil partial sterilants have been tested for their ability to control many of the viruses listed in Table 1. Chloropicrin, pentachloronitrobenzene (PCNB)(31), methyl bromide (14,59), and 1, 3-dichloroporopene with 1, 2-dichloropropane (D-D)(6), have been shown to be partially or completely effective in controlling LBVV when injected into soil, although some findings were inconclusive or contradictory. Methyl bromide killed *O. brassicae* and prevented transmission of TNV to cucumbers and beans in greenhouses in New Zealand (41) and methyl bromide and steam was shown to control *O. brassicae* transmitting freesia leaf necrosis virus (54). Methyl bromide, metham sodium, chloropicrin, dazomet, dichloropropene and D-D have been shown to control rhizomania and their use for this purpose has been reviewed elsewhere (8,37). In California, Telone (dichloropropene) is used routinely to reduce the damage caused by this disease (38). Various treatments (listed in Table 4) have been found to control SBWMV (32), WSSMV (39) and PCV (42) all transmitted by *P. graminis*.

Despite these apparent successes, the expense of such control methods is often prohibitive for field application. In comparative tests, invariably methyl bromide proved to be the most effective, however it is also the most toxic and expensive. Many fumigants are also hazardous to the environment, methyl bromide is listed as an ozone-depleting substance and the Environmental Protection Agency of the USA has ruled that its production and consumption is to be eliminated by 2001. Problems have also been experienced with excessive bromine residues in plants, especially lettuce. Such treatments are usually only partially effective as they only kill resting spores of vectors in the upper layers of the soil and re-infestation from lower layers occurs during the growing season and treatments must be repeated each time a susceptible crop is grown.

Control measures aimed at killing fungal spores contaminating planting material or surfaces likely to come into contact with planting material can be very effective in eliminating or reducing virus spread. In such circumstances it is often possible to use some of the more toxic substances with less risk to the environment and the final produce than when attempting control in soil. As early as 1911 it was shown that chemical treatments of potato tubers infected by *S. subterranea* could reduce the amount of infection in the resulting crop (34). Since then many different chemicals have been tested for this purpose and found to decrease the introduction and spread of PMTV. This subject has been reviewed elsewhere (27). Dipping diseased lettuce plants in mercuric chloride solution prevented the development of LBVV symptoms (21). It has also been shown that rigorous cleaning and formaldehyde treatments of freesia corms and cormels to remove and kill the vector *O. brassicae* reduced dissemination of freesia leaf necrosis virus (54). The disinfectant Iodel F.D. was found to kill *O.brassicae* resting spores after treatment for 1h and was recommended for washing down propagation areas and machinery etc (48). Following the recognition that LBVV was being disseminated from centralized lettuce propagators a comprehensive study on the effect of disinfectants on the viability of *O. brassicae* resting spores was carried out (56). Of the 32 products tested only 7 killed all resting spores; when tested in the presence of soil and with a contact time of 20 minutes, only three of these (Jet 5, Iodel F.D., and sodium hypochlorite) were still effective.

MISCELLANEOUS ALTERNATIVE TREATMENTS

Alternative chemical approaches include direct control of the viruses. Ribavirin was shown to reduce the severity of LBVV symptoms in lettuce and the titer of virus in *O. brassicae* (11). The antiviral compound DHT (2, 4-dioxohexhydro - 1,3,5, -triazine) has been shown to suppress synthesis of BNYVV, but only at concentrations phytotoxic to sugar beet plants (38). The active ingredient of Bavistin (carbendazim) is known to prevent symptoms of an aphid-transmitted virus in lettuce and although it was claimed its mode of action in controlling LBVV was through inactivation of zoospores, the possibility that it may have a direct effect on symptom expression cannot be discounted.

Steam sterilization has been shown to be effective in disinfecting soil containing *O. brassicae* (54). Flash flame pasteurization of soil has been shown to reduce seedling infection by BNYVV (1), and solarization during the hot summer period in India for a minimum of 70 days reduced the incidence of PCV (35).

FUTURE RESEARCH AND TRENDS

There is a need to develop more benign or specific fungicides and soil partial sterilants for the control of viruses with fungal vectors. The fungicide Fluazinam has shown potential to control a number of viruses, but it remains to be seen whether this potential is realised under field conditions. A new fumigant, carbonyl sulphide, developed in Australia is reported to be environmentally acceptable and a possible replacement for methyl bromide. In the current agricultural economic environment where the costs of developing and registering new fungicides are so great, it is unlikely that very specific compounds will be developed for fungal vectors. Development of such products is especially difficult as many that gave good control in laboratory or pot experiments, have been found to be ineffective in the field. A relatively unexplored area of potential control is the use of biological control agents. *Trichoderma harzianum* has been shown to parasitise and degrade the resting spores of *P. betae* under laboratory conditions (18), but was not successful as a control agent in field trials (8). Similarly *Pseudomonas fluorescens* was ineffective in French trials (7). Various biological agents are currently being evaluated for potential to control LBVV (58). The longevity of resting spores of vector fungi indicates that they are remarkably resistant to microbial degradation. For biological control agents to be effective it is likely that the soil environment will have to be manipulated to promote their activity (or disadvantage the vector), or they will have to be used in conjunction with other agents as part of an integrated control strategy. Benefits from manipulating the soil environment with turkey manure (39) and straw (29) have already been observed. As chemical control agents are withdrawn, or banned and tolerance levels of their residues in crops are reduced, new control strategies will need to be evolved. The most important component of such strategies in the immediate future is likely to be natural plant resistance. In the longer term, transgenic plants with genetically engineered resistances to virus and vector are likely to become available. Transformed sugar-beet expressing the BNYVV coat protein gene specifically in the roots have already been produced (28). Hygiene and cultural practices are also likely to be important components, the former can be very effective in the greenhouse environment, particularly where artificial growing media are used. Such media can be pasteurized and other surfaces can be sterilized with steam or chemical agents. Exciting developments in the production of serological and molecular probes for the detection and quantification of vectors are taking place. Serological detection of *S. subterranea* resting spores in soil has been achieved and comparisons with baiting tests have shown that the antiserum can distinguish between lightly and heavily infested soils (Walsh, Merz and Harrison, unpublished). Such developments could help in planning crop rotations and lead to disease avoidance by choosing not to grow susceptible crops in heavily contaminated fields. The components of future integrated control strategies will be decided by their availability, suitability, effectiveness, conditions and regulations prevailing locally and of course cost. Although chemical means of control used alone or at reduced levels as part of an integrated strategy are likely to be with us for some time to come, current trends suggest their days are numbered.

ACKNOWLEDGMENTS

Research on viruses and their fungal vectors at HRI is funded by the Ministry of Agriculture, Fisheries and Food and aspects concerning chemical control by the Horticultural Development Council. I am very grateful for the excellent support provided by Judith Bambridge, Peni McKay and Janet Hughes for typing this document.

REFERENCES

1. Abe, H. 1987. Studies on the ecology and control of *Polymyxa betae* Keskin as a fungal vector of the causal virus (beet necrotic yellow vein virus) of rhizomania disease of sugar beet. Bull. Hokkaido Prefect. Agric. Exp. Stn 60: 99.
2. Adams, M. J. 1988. Evidence for virus transmission by plasmodiophorid vectors. Pages 203-211 in: Viruses with fungal vectors. J. I. Cooper and M. J. C. Asher eds. Association of Applied Biologists, Wellesbourne, U.K.
3. Adams, M. J. 1990. Epidemiology of fungally transmitted viruses. Soil use and Management 6: 184-188.
4. Adams, M. J. 1991. Transmission of plant viruses by fungi. Ann. appl. Biol. 118: 479-492.
5. Alghisi, P., and D'Ambra, V. 1966. Studies on rhizomania in sugar beet. Riv. Patol. Veg. Ser. IV2: 3-41.
6. Allen, M. W. 1948. Relation of soil fumigation, nematodes and inoculation technique to big-vein disease of lettuce. Phytopathology 38: 612-627.
7. Anon. 1979-1985. Compte Rendu des Travaux, Institut Technique Francais de la Betterave Industrielle, Paris.
8. Asher, M. J. C. 1988. Approaches to the control of fungal vectors of viruses with special reference to rhizomania. Pages 615-627 in: Proc. British Crop Protection Conf. Pest and Diseases -1988. BCPC, Surrey.
9. Bruin, G. C. A., and Edginton, L. V. 1983. The chemical control of diseases caused by zoosporic fungi. Pages 193-232 in: Zoosporic plant pathogens a modern perspective. S. T. Buczacki ed. Academic Press, London.
10. Campbell, R.N. 1979. Fungal vectors of plant viruses. Pages 8-24 in: Fungal viruses. H.P. Molitoris, M. Hollings and H.A. Wood, eds. Springer-Verlag, Berlin.
11. Campbell, R. N. 1980. Effects of benomyl and ribavirin on the lettuce big vein agent and its transmission. Phytopathology 70: 1190-1192.
12. Campbell, R. N. 1985. Longevity of *Olpidium brassicae* in air-dry soil and the persistence of the lettuce big-vein agent. Can. J. Bot. 63: 2288-2289
13. Campbell, R. N. 1993. Persistence: A vector relationship not applicable to fungal vectors. Phytopathology 83: 363-364.
14. Campbell, R. N., Greathead, A. S., and Westerlund, F. V. 1980. Big vein of lettuce: infection and methods of control. Phytopathology 70: 741-746.
15. Canova, A., Giunchedi, L., Credi, R., and Arlotti, D. 1975. Notes on studies and experiments with rhizomania in sugar beet. Atti del Convegno Giornate Bieticole Italiane, Barga di Lucca: 391-401.
16. Cooper, J. I., and Asher, M. J. C. eds. 1988. Viruses with fungal vectors. Developments in Applied Biology 2, Association of Applied Biologists, Wellesbourne, U.K.
17. Cooper, J. I., Jones, R. A. C., and Harrison, B. D. 1976. Field and glasshouse experiments on the control of potato mop-top virus. Ann. Appl. Biol. 83: 215-230.
18. D'Ambra, V., and Mutto, S. 1986. Parasitism of *Trichoderma harzianum* on cystospores of *Polymyxa betae*. J. Phytopathol. 115: 61-72.
19. Davidse, L. C. 1977. Mode of action, selectivity and mutagenicity of benzimidazole compounds. Neth. J. Plant Pathol. 83 (Suppl. 1): 135-144.
20. Gharbi, S., and Verhoyen, M. 1993. Sterilisation par irradiation UV des solutions nutritives en vue d'eviter les infections virales propagees par *O. brassicae* en culture hydroponique de laitue. Med. Fac. Landbouww. Rijksuniv. Gent 58/3a: 1113-1124.
21. Grogan, R. G., Zink, F. W., Hewitt, W. B., and Kimble, K. A. 1958. The association of *Olpidium* with the big-vein disease of lettuce. Phytopathology 48: 292-296.
22. Humpherson-Jones, F. M. 1993. Effect of surfactants and fungicides on clubroot (*Plasmodiophora brassicae*) of brassicas. Ann.

23. Hiruki, C., and Teakle, D. S. 1987 Soil-borne viruses of plants. Pages 177-215 in: Current Topics in Vector Research, Volume 2. K. F Harris ed. Springer-Verlag, New York.

24. Hims, M. J., and Preece, T. F. 1975. *Spongospora subterranea* f. sp. *subterranea* CMI Descriptions of Pathogenic Fungi and Bacteria. No. 477. Commonwealth Agricultural Bureaux. England.

25. Horak, I., and Schlosser, E. 1978 Effect of prothiocarb on *Polymyxa betae* and *Olpidium brassicae*. Med. Fac. Landbouww. Rijkuniv. Gent 43/2: 979-987.

26. Jagger, I. C. 1940. Brown blight of lettuce. Phytopathology 30: 53-64.

27. Jones, R. A. C. 1988. Epidemiology and control of potato mop-top virus. Pages 255-270 in: Viruses with fungal vectors. J. I. Cooper and M. J. C. Asher eds. Association of Applied Biologists, Wellesbourne.

28. Kallerhoff, J., Perez, P., Gerentes, D., Poncetta, C., Ben Tahar, S., and Perret, J. 1990 Sugar-beet transformation for rhizomania resistance. Page 91 in: Proc. of First Symposium of the International Working Group on Plant Viruses with Fungal Vectors. R. Koenig ed. Ulmer, Stuttgart.

29. Maraite, H. 1991. Transmission of viruses by soil fungi. in: Biotic interactions and soil-borne diseases. Developments in Agricultural and Managed-Forest Ecology 23. A. B. R. Beemster, G. J. Bollen, M. Gerlach, M. A. Ruissen, B. Schippers and A. Tempel. eds.

30. Marlatt, R. B. 1959. The effects of applications of micro-nutrients to Arizona field-grown lettuce. Plant Dis. Rep. 43: 1019-1022.

31. Marlatt, R. B., and McKittrick, R. T. 1963. Fungicidal control of big-vein in the irrigated lettuce crop. Phytopathology 53: 597-599.

32. McKinney, H. H., Paden, W. R., and Koehler, B. 1957. Studies on chemical control and overseasoning of, and natural inoculation with, the soil-borne virus of wheat and oats. Plant Dis. Rep. 41: 256-266.

33. Meunier, S., and Verhoyen, M. 1989. Inhibition de la transmission du big vein virus (maladie des grosses nervures) dans les cultures de laitue. Med. Fac. Landbouww. Rijksuniv. Gent 54/2b: 657-664.

34. Pethybridge, G. H. 1911. Investigations on potato diseases. J. Dep. Agric. and Tech. Instr. Ireland 11: 417-449.

35. Reddy, D. V. R., Nolt, B. L., Hobbs, H. A., Reddy, A. S., Rajeshwari, R., Rao, A. S., Reddy, D. D. R., and McDonald, D. 1988. Clump virus in India: isolates, host range, transmission and management. Pages 239-246 in: Viruses with fungal vectors. J. I. Cooper and M. J. C. Asher eds. Association of Applied Biologists, Wellesbourne.

36. Rich, S. 1960. Tetrachlor controls *Olpidium* on lettuce. Plant Dis. Rep. 44: 325-353.

37. Schaufele, W. R. 1987. Trials on the chemical control of *Polymyxa betae*. Proc. 50th Winter Congress of the International Institute for Sugar Beet Research, Brussels 2: 97-110.

38. Schlosser, E. 1988. Epidemiology and management of *Polymyxa betae* and beet necrotic yellow vein virus. Pages 281-292 in: Viruses with fungal vectors. J. I. Cooper and M. J. C. Asher eds. Association of Applied Biologists, Wellesbourne.

39. Slykhuis, J. T. 1970. Factors determining the

40. Teakle, D. S. 1983. Zoosporic fungi and viruses-double trouble. Pages 233-248 in: Zoosporic plant pathogens a modern perspective. S. T. Buczacki ed. Academic Press, London.

41. Thomas, W. 1973. Control of *Olpidium brassicae*, the vector of cucumber systemic necrosis and bean stripple streak virus diseases. New Zealand J. Exp. Agric. 1: 92-96.

42. Thouvenel, J.-C., Fauquet, C., Fargette, D., and Fishpool, L. D. C. 1988. Peanut clump virus in West Africa Pages 247-254 in: Viruses with fungal vectors. J. I. Cooper and M. J.C. Asher eds. Association of Applied Biologists, Wellesbourne.

43. Tomlinson, J. A. 1956. Control of watercress crook root disease by zinc-fritted glass. Nature 178: 1301-1302.

44. Tomlinson, J. A. 1958. Crook root of watercress. III The causal organism *Spongospora subterranea* (Wallr.) Lagerh. f. sp. *nasturtii* f. sp. nov. Trans. Br. Mycol. Soc. 41: 491-498.

45. Tomlinson, J. A. 1958. Crook root of watercress. II The control of the disease with zinc-fritted glass and the mechanism of its action. Ann. appl. Biol. 46: 608-621.

46. Tomlinson, J. A. 1958. Crook root of watercress. I Field assessment of the disease and the role of calcium bicarbonate. Ann. appl. Biol. 46: 593-607.

47. Tomlinson, J. A. 1960. Crook root of watercress. A review of research. N.A.A.S. q. Rev. 49: 13-19.

48. Tomlinson, J. A. 1988. Chemical control of *Spongospora* and *Olpidium* in hydroponic systems and soil. Pages 293-303 in: Viruses with fungal vectors. J. I. Cooper and M. J. C. Asher eds. Association of Applied Biologists, Wellesbourne.

49. Tomlinson, J. A., and Faithfull, E. M. 1979. Effects of fungicides and surfactants on the zoospores of *Olpidium brassicae*. Ann. appl. Biol. 93: 13-19.

50. Tomlinson, J. A., and Faithfull, E. M. 1980. Studies on the control of lettuce big-vein disease in recirculated nutrient solutions. Acta Hortic. 98: 325-332.

51. Tomlinson, J. A. and Thomas, B. J. 1986. Studies on melon necrotic spot virus disease of cucumber and on the control of the fungus vector (*Olpidium radicale*). Ann. appl. Biol. 108: 71-80.

52. Uchino, H., Kanzawa, K., and Tamada, T. 1993. Effect of Fluazinam on infection of sugar beets by *Polymyxa betae*, vector of beet necrotic yellow vein virus. Pages 153-156 in: Proc. Symposium of Intnl. Working Grp. on Plant Viruses with Fungal Vectors, 2nd. C. Hiruki, ed. American Society of Sugar Beet Technologists, Denver, Co.

53. Vanachter, A., Van Wambeke, E., and Van Assche, C. 1992. Influence of thiophanate-methyl and zinc on the development of lettuce ring necrosis disease and its vector *Olpidium brassicae* in NFT grown lettuce. Med. Fac. Landbouww. Rijksuniv. Gent 57/2a: 249-257.

54. Van Dorst, H. J. M., and Peters, D. 1988. Experiences with the freesia leaf necrosis agent and its presumed vector, *Olpidium brassicae*. Pages 315-322 in: Viruses with fungal vectors. J. I. Cooper and M. J. C. Asher eds. Association of Applied Biologists, Wellesbourne.

55. Walsh J. A. 1992. Resistant watercress. Grower

development of wheat spindle streak mosaic caused by a soil-borne virus in Ontario. Phytopathology 60: 319-331.

118 (6): 18-21.

56.   Walsh, J. 1992. Catching up with big vein. Grower 118 (23): 11-15.

57.   Walsh, J. A., and Clay, C. M. 1993. A summary of research on watercress yellow spot virus and its fungal vector *Spongospora subterranea* f. sp. *nasturtii*. Pages 111-114 in: Proc. of the Second Symposium of the International Working Group on Plant Viruses with Fungal Vectors. C. Hiruki, ed.

American Society of Sugar Beet Technologists, Denver, Co.

58.   Walsh, J., Clay, C., and McPherson, M. 1992. Detection and control of lettuce big-vein virus. Horticulture Research International Annual Report for 1991-92, Wellesbourne: 28-29.

59.   White, J. G. 1983. The use of methyl bromide and carbendazim for the control of lettuce big-vein disease. Plant Pathol. 32: 151-157.

CHAPTER 16

Exclusion of Plant Viruses

J. A. Foster and A. Hadidi

As humans, we are involved in an epic conflict. This struggle began in prehistoric times and will continue for the foreseeable future. Our enemies are aggressive and relentless, attacking every day everywhere on earth. Victory for us means survival and the continuing development of our civilization; defeat means starvation and chaos. Our enemies in this struggle are the pests that compete with us for our food supply (8,75).

This conflict intensified when humans first cultivated plants and raised animals as reliable sources of food (47). From time to time, our ancestors watched helplessly as insects devoured their plants, weeds overran their crops, and diseases of mysterious origin caused devastation. Over the centuries, we have learned about these pests by studying them and have recently applied this knowledge to develop control measures which minimize pest damage (84). In many cases, we have reached a tenuous equilibrium with our adversaries (7,47); some day even total victory over some of them may be within our capabilities.

Unfortunately, our enemies continue to receive reinforcements from overseas (7,8,47,75). These invaders often leave behind any agents that control them and find our crops especially susceptible. During our studies of these imported pests, we have been horrified to discover that most did not have the capability to travel here on their own but instead were brought inadvertently by humans (8,75). We have contributed significantly to our own misery. In order to save us from ourselves, we have developed pest control strategies, referred to collectively as exclusion (35,59,61), in an attempt to prevent, or at least delay, introduction of foreign pests into our agriculture.

None of these invading pests is more insidious than the plant viruses. They are small enough to avoid our watchful eyes, hide inside most parts of infected hosts, and only occasionally reveal their presence by causing symptoms. Only viroids and certain small prokaryotes are as elusive. Since the existence of viruses was first discovered around the turn of this century (72), virologists have learned a great deal about these pathogens and how they are disseminated. This knowledge has been fundamental to the development of procedures to exclude plant viruses. Therefore, a summary of our current understanding of virus dissemination is an appropriate beginning for a discussion of plant virus exclusion.

MOVEMENT OF PLANT VIRUSES TO A NEW LOCATION

Plant viruses can not control their dispersal. These viruses are not capable of self-propulsion, like insects, and they don't have a mechanism for discharging a propagule into wind or water, like certain weeds or fungi. Instead, plant viruses are spread passively, usually in or on vegetative plant parts, seeds, pollen, vectors, soil or water. Therefore, to be correct, these viruses do not move from place to place but are being moved over these distances in or on an appropriate carrier. A carrier in this context is any contaminated living organism or inert material that moves or is moved from one location to another.

In an effort to organize our knowledge concerning plant virus dissemination over large distances, we can postulate a general sequence of events which must occur to establish any virus in a new area. In the currently affected area, the virus must be accessible to a carrier which is capable of moving or being moved over large distances. During carrier movement, the virus must survive the journey. Upon arrival in the new location, this foreign pathogen must infect a plant host. Once the initial infection has occurred, spread and impact of the virus becomes dependent on its genetic capabilities, plant hosts, environmental conditions, vectors or other means of spread, and human activities in the new area. Thus, if we try to predict the likelihood of a particular virus being spread to any location, the risk of this undesirable event would be related to the probabilities that carriers would be contaminated and moved, virus would survive the journey, virus would infect a plant in the new location, and virus would become established and spread in the new area (15).

Virus Sources

The probability that a carrier will acquire or be infected or contaminated by a virus depends on virus ecology at that time and location (76,89,122). Virus incidence will increase in proportion to the quality, quantity, accessibility and distribution of local virus reservoirs; number, distribution and susceptibility of host plants; quantity, activity, transmission efficiency, and lack of specificity of capable vectors; compatibility of the environmental conditions; favorable impact of animal and human activities; and adaptability of the virus. Virus prevalence will be decreased by genetic limitations of the virus, scarcity or inaccessibility of local virus reservoirs, host rarity or resistance, scarcity or inefficiency of vectors, abundance of vector parasites and predators, unfavorable environmental conditions, or detrimental influences of humans and other animals. Virus ecology is a dynamic process which changes over time and varies in each location. However, after reviewing our knowledge of virus ecology, we can distinguish wild, cultivated and experimental virus sources as three basic ecosystems which seem to be fundamentally different but which are each capable, under proper conditions, of contributing contaminated carriers that are moved over large distances.

Wild Sources

Wild plants are prized by plant collectors who seek new species for research or new genes to improve the performance of commercial cultivars. Explorations have been and will continue to be carried out in remote areas of the world in search

of better germplasm (23,36,86,87). Plant explorers usually focus their efforts in locations where the desired species has evolved over centuries. These areas are referred to as gene centers or centers of diversity (23,36,48,55, 66,92,106). Unfortunately, a center of diversity for a given plant species is usually also where the greatest diversity of pests has evolved in parallel with the different host genotypes (18,48,55,66, 106). Therefore, wild plants that explorers collect in centers of diversity are also the most likely carriers of new and damaging pests (14,58,66). In their enthusiasm for the treasures they have gathered, plant collectors often carry or ship this germplasm, along with any pests present, throughout the world.

Our knowledge of virus ecology among wild plants is fragmentary at best. In the past, wild plants have usually been studied in the context of their impact on virus epidemiology in nearby cultivated plants (14,24,114). Unfortunately, there has been little or no monetary incentive to study viruses in native plants of no current economic value (55).

On the basis of the limited number of studies and small percentage of wild areas sampled, we can assume that many of the viruses in wild plants have not yet been discovered (14,55). Many of these unknown viruses may be similar to characterized viruses, but others may be distinct. During the last few decades, new viruses with unique characteristics have been periodically discovered. In view of the history of plant virology, we can't assume that all viruses of wild plants will fit neatly into our current concepts of virus taxonomy.

Even in the cases of viruses that have been detected in wild plants, we know very little about the incidence and dissemination of these viruses in these plants under natural conditions (14,24,55,114). We could speculate that each of these viruses will be spread in a manner similar to more thoroughly studied viruses with similar characteristics. However, our present knowledge of virus epidemiology has been gathered from research on monocultures of highly bred and selected crop species; consequently, the principles we have deduced from research on these commercial crops may not be entirely applicable to a more natural situation (14,15,55,114). At the present time, it would be impossible to calculate the probability that a collected wild plant part would be infected, a captured vector would be viruliferous, or a soil or water sample would be contaminated, even with a described virus known to occur in the area where any of these carriers was collected.

Cultivated Sources

Over many centuries, our ancestors recognized the potential of certain plant products for use as food, fiber, building materials or barter. Each of these valued plants was gradually improved, at first by crude selection of better plants and more recently by plant breeding. Consequently, each of the plant products we enjoy today is the result of a considerable investment of human resources over the centuries. When any of these assets has been threatened by pests, we have been willing to commit significant resources for research and application of control measures to minimize the impact of these pests (84). As a result of our previous commitments to the study of pests in cultivated crops, we now know more about these pests and their epidemiology in crop production areas than we know about any other pest-plant ecosystem.

On the basis of our present knowledge, we can distinguish two distinct scenarios for initiation of a virus epidemic. Viruses can be introduced into the production field by planting infected crop propagules, or growing crops can be inoculated by vectors, either present in the field or moving into the field from virus reservoirs outside the field (14,90,113,115,122).

In the first situation, humans are ultimately responsible for any damage resulting from virus introduction into a crop production area in the seeds or clonal propagules they use to plant that crop. Seeds infected with seed-borne viruses, such as bean common mosaic virus in bean seed or lettuce mosaic virus in lettuce seed (71,122), or infected clonal propagules, such as potato tubers with potato virus Y (101,122) or citrus budwood with citrus tristeza virus (5), are considered the primary sources for epidemics of these viruses. After probing or feeding on these infected seedlings or clones, vectors can then spread these viruses to healthy plants throughout the cultivated field. If vectors do not occur in the planting, damage should be confined to the initially infected plants.

The most effective strategy for controlling viruses introduced only in crop propagules is to ensure that all propagules planted in the crop production area are free of these viruses. If the virus does not occur in the emerging crop, nearby wild plants or migrating vectors, there will be no virus present for potential vectors to spread. The principle of controlling certain virus diseases by starting with virus-tested plants or their progeny is the justification for many certification programs now used throughout the world (9,13,44,45,65, 73,79,82,83,90,100). For the viruses introduced only in crop propagules, the probability that virus infection will occur in any plant produced in a proper certification program should approach zero.

Unfortunately, planting certified propagules will not prevent infection by endemic viruses which are spread to crop plants from other sources. The probability of infection in any plant at any given time may depend on the number, location and type of local virus reservoirs; incidence, activity, transmission efficiency and specificity of vectors, host susceptibility, environmental conditions, man's cultural practices, and the capabilities of the virus (76,89,122). With current control techniques, we can decrease, but not entirely prevent, virus inoculation of cultivated plants by vectors.

In recent years, spread of many viruses under a variety of conditions has been monitored and evaluated. Development of virus epidemics in time and space has been depicted by plotting disease progress curves (115) and gradients (113). Spatial patterns of infected plants have also been recorded at different times during epidemics and analyzed for their degree of randomness or clustering (42,68-70). Temporal and spatial aspects of virus spread can now be approximated by mathematical equations (3,68). Models have been proposed for estimating disease progress and impact on yield of viruses transmitted in seeds (95), viruses introduced in clonal propagules (101), and viruses carried into fields by vectors (76). By modifying certain parameters in these models, we may be able to simulate how viruses would be spread under different circumstances. With field studies and/or modeling, the critical factor or factors which determine whether or not an epidemic will occur can be identified; these factors may be different for different viruses and may change for a particular virus in different areas (117). In certain situations, we may be able to use models for some of the most thoroughly studied viruses to forecast an epidemic with reasonable accuracy (30,76). However, current models are not consistently accurate in all situations, so more refinements need to be made. At this point in time, not all viruses of quarantine significance have been studied sufficiently for development of a predictive model. In the future, models may help us assess the probability of exporting infected plant propagules.

During the study of virus epidemiology in cultivated plants, some data on the incidence of viruliferous aerial vectors (53,76,89,90) or spatial distribution of infectious soil-borne vectors (11,56,109) has been collected, but little effort has been made to predict the probability of collecting viruliferous vectors or contaminated soil or water during sampling of cultivated fields. With our current knowledge, the risks of exporting aerial or soil-borne carriers with plant material or as samples for study can not be accurately assessed for any location at any given time.

The risks of collecting plant material, vectors, soil or water for export from fields of cultivated crops may not be confined to those viruses which are economically important in that location. Viruses which are overlooked due to the impact of more damaging pests in these fields may be quite destructive after being moved to a new location. In any year, a new virus may be introduced into the cultivated field in new crop propagules (7,14,15,24,55,114). Citrus tristeza virus, grapevine fanleaf virus, many potato viruses, and seed-borne legume viruses are examples of viruses which were probably spread to new areas in infected plant propagules. On the other hand, in any year a new virus may be spread from the native flora to crops planted in a new location (7,14,15,18,24,55,114). Cacao swollen shoot virus, maize streak virus, rice yellow mottle virus, maize rough dwarf virus, and rice hoja blanca virus are all considered to be viruses of wild plants which were spread to crops grown in new areas. Even though the risk of exporting a new virus from a cultivated field may be less than from wild areas, this risk is still significant enough to warrant caution on the part of importing countries.

Experimental Sources

A significant percentage of the agricultural material exchanged between countries originates in research facilities. The curator of a plant collection or repository may ship germplasm (14,15,52,55,88); plant breeders may exchange advanced breeding lines, selections or cultivars (15,45,106); biotechnologists may export transgenic plants (92); bacteriologists, mycologists or nematologists may move pathogen cultures (33); virologists may be sources for virus indicator plants, virus cultures, or tested cultivars used as mother plants in certification programs; entomologists or acarologists may transport living arthropods (33); and agronomists, geologists, or pathologists may ship soil samples (109). Before export, these virus carriers may be maintained in field plots, greenhouses, screenhouses, laboratories, growth chambers or incubators. Since the carrier's history and maintenance conditions vary with each source, the risk that exported research material is infected must be assessed on a case-by-case basis.

Viruses present in carriers exported from a research facility may have been present in material originally brought into the facility (14,15,45-55,71,106). Any arriving arthropod, nematode, fungal, prokaryote, viral or viroid culture may be labeled as pure but instead may be contaminated with viruses. New soil or water samples may contain viruliferous vectors or virus particles. In an effort to preserve the genetic characteristics of introduced plants, propagators use buds, scionwood, cuttings or tissue culture to reproduce each parent plant asexually and, often unintentionally, perpetuate any viruses from the parent plant in the progeny. Seed-transmitted viruses are transferred from one generation to another in seeds by scientists unaware of infections in their plants. A biotechnologist may inadvertently select an infected cell line to transform. From any of these original sources, local vectors may then spread introduced viruses and increase the number of infected carriers in the facility. Perpetuation of viruses in research plant material is regrettable but understandable. Over many plant generations, plants with severe host-virus interactions have been rogued out as unthrifty, and symptomless host-virus combinations have been selected (18,20,114). Since the presence of a virus in any carrier may not be obvious, even to a virologist, the only possibility of identifying which plants, vectors, soil or water samples are contaminated is to test each new potential carrier after it arrives but before distribution from the research facility (15,45,52,88,106).

Even originally healthy carriers maintained by a careful scientist in a research facility can become infected or contaminated by viruses endemic to the area (106). Aerial or soil-borne vectors can transmit domestic viruses from local weeds or crops to susceptible research plants grown in field plots. Plants maintained in a greenhouse or screenhouse are less likely to become infected by local viruses. However, the degree of protection afforded by these facilities depends on the capacity of the structures to exclude viruliferous vectors and the effectiveness of phytosanitary procedures at preventing vectors from entering with humans, plant material or equipment used in the research. Maximum protection of plants destined for export is provided by tissue culture (52,58,60,86,88). If tissue cultures are established from virus-tested plants, this plant material could be distributed to requestors in other countries with little or no risk of contamination by local viruses. Certain research facilities, including many International Agriculture Research Centers, now distribute virus-tested plants (32).

Virus Dissemination

The means by which viruses are moved from their sources to new areas can be separated into natural and man-made pathways (59,61,106). Natural pathways would include the activities of flora, fauna and microbes as well as effects of meteorological and geological factors. Man is not involved in dissemination of viruses by these natural forces. On the other hand, man-made pathways include all of the methods by which humans intentionally or unintentionally move viruses or their carriers. By modifying human behavior, the probability that viruses will be moved from place to place by human activities can be reduced.

Natural Pathways

Certain viruses may be carried over large distances by their aerial vectors. There is evidence that aphids (110,116), leafhoppers (111,116), planthoppers (93,111,116), and possibly other arthropods (116) which are known to be virus vectors can migrate or be carried by air currents for hundreds of kilometers. Large animals could also carry contaminated soil on their bodies (10,49,56,108) or infected plant material in their digestive systems (63) to new areas, but evidence for virus dissemination by these methods under natural conditions is lacking. Most viruses transported large distances by aerial vectors are probably transmitted by insects in a semi-persistent or persistent manner (61,116,122). However, spread of nonpersistent viruses to distant locations may be possible (116,121). Experiments conducted under artificial conditions indicate that not all members of a vector population can acquire and transmit virus from an infected plant. Of those that acquire a persistent virus, most transmit virus only intermittently over the remainder of their lives. The impact of a long flight, and the

environmental conditions encountered during that flight, on transmission capability of a vector are not known. However, within a large vector population, only one individual needs to acquire virus, survive the journey, and transmit virus to a susceptible host in order to spread the virus to a new area. The discoveries of maize mosaic virus (17) and rice hoja blanca virus (116) in U.S. states along the Gulf of Mexico, rice grassy stunt and ragged stunt viruses in Japan (93,116), and beet yellows virus in Sweden and Finland (116) are considered examples of virus dissemination by migrating or wind-blown vectors.

Seed-transmitted viruses (13,71,80) may be carried in infected seeds over large distances (13,14,49). Lighter seeds can be blown to other areas by wind (13,14). Seeds may drop into waterways and be carried downstream to other areas (13). Animals could pick up infected seeds inadvertently on their bodies (14) or consume infected seeds or fruit containing infected seeds (13,14,63) and then days later deposit these seeds many kilometers from where they were acquired. Common sense would suggest that exposure to atmospheric conditions, water or an animal's digestive processes will significantly decrease the viability of seeds below that obtained from seeds recently harvested from an infected plant. Perhaps, these conditions would more severely impact weaker virus-infected seeds than more vigorous healthy seeds.

Pollen-transmitted viruses (71,80) can also be carried in or on pollen for large distances by vectors or wind (49). Wind dispersal of infected pollen is highly probable. However, this pollen will lose viability within a finite period, so the distance that a pollen-transmitted virus could be moved in infected pollen by wind may be limited (71). The probability that infected pollen would be carried to the flower or leaf of a distant virus host before expiring should be increased if the pollen is carried by vectors, such as bees, which seek the appropriate plant.

Some viruses can survive as naked particles outside a living cell for the time required for long distance dissemination. Certain viruses remain infectious in rivers, streams and drainage water (49,63); some of these and others occur in soil (63). There is no data to suggest that air-borne virus particles will remain viable long enough to be blown by wind to a distant location. However, infectious tobacco mosaic virus has been detected in aerosols from a field of diseased tobacco plants (4). On the other hand, virus-infected fungal resting spores could be blown a large distance with dust particles (10,49) or carried downstream in drainage water, waterways or floods (15,49). Viruliferous nematodes could also be spread by waterways or during floods (10).

Man-made Pathways

The carriers which pose the greatest risk to any agriculture are the plant parts imported by humans for vegetative propagation (15,52,75,81). These plant propagules may include the treasures discovered by ordinary travelers overseas, nursery stock imported for commercial sale, or germplasm collected by explorers for plant breeding or research. Regardless of how this plant material is handled in transit, transported clonal propagules will usually contain all of the viruses present in the parent plant from which the propagules were taken. As long as infected cells survive the journey and remain viable during propagation of the host in the new location, successful dissemination of the virus or viruses to the new area is a certainty.

In general, shipment of seeds poses less risk to the importing country's agriculture, because only a small percentage of the known viruses are reported to be seed-transmitted in a comparatively small number of plant species (71,106). When viruses are seed-transmitted, the percentage of viable infected seed produced from infected plants varies with the plant, virus and environment (13,71). The risk of disseminating viruses in seed is less when few parent plants are inoculated late in the growing season; however, the percentage of infected seed in a harvested seed lot increases with earliness (13,71) and greater incidence of infection for a given cultivar. In most cases, the probability that a seed-transmitted virus may occur in any particular seed may be low, but shipment of millions of these seeds will make virus transport and entry inevitable.

Even fewer viruses are known to be transmitted by pollen (71,80). With modern techniques of pollen storage, this rather fragile gamete can be transported large distances without a significant reduction in viability. In the case of tobacco ringspot virus in soybean pollen, fresh infected pollen was less viable than similar healthy pollen (120). However, the impact of storage and transport on viability of infected pollen upon arrival in a new location is not known.

With recent improvements in transportation and plant storage, perishable commodities imported for consumption or analysis are now routinely shipped between countries (8,35,75). In most cases, all of the viruses found in the parent plant will also be present in fleshy tissues of the fruit or vegetable we eat as food. Within a short period of time, the perishable commodity is either consumed or becomes unsuitable for vector probing or feeding. However, an infected foreign fruit or vegetable can arrive in a new area in good enough condition for accidental or unauthorized propagation, and seeds in imported fruits or grains may be inadvertently or deliberately sown to yield infected seedlings. Therefore, viruses are routinely imported with perishable commodities, but the effectiveness of these carriers in introducing viruses to a new area is determined by how each perishable material is handled after entry in the new location.

Exchange of pathogen cultures between scientists also poses a risk to the importing country's agriculture (33,35). Since viruses are obligate parasites, virus cultures are often shipped in host tissue. Consequently, other pests, including other viruses, may be present in or on host tissue containing the requested virus. In most cases, virologists are knowledgeable concerning the characteristics of viruses that they investigate and will take all reasonable precautions when packaging virus cultures to prevent virus escape during transport. However, scientists are individuals with their own perceptions and prejudices; some may not be as careful at times as they should be.

Vectors that may be carrying viruses are also shipped between countries (33,35). Entomologists, acarologists, nematologists and mycologists may import foreign organisms as cultures without realizing one or more viruses are also present. Pollinating insects may carry virus-infected pollen. Movement of apiaries has been associated with introduction of prunus necrotic ringspot virus into cherry orchards (78). Viruliferous vectors may also occur on imported plant material (33,35). Vectors may acquire viruses from plant material which they infest during shipment and after entry carry these viruses to nearby plants. Vectors that transmit persistent viruses may even be viruliferous with a virus which can't infect the host on which the vector is found during transit. Vectors may also be trapped in containers or vehicles moving between countries.

Soil or water containing viruses or viruliferous vectors can be transported from place to place on vehicles, equipment, or plant material (8,10,33,35,-49,56,108) or as samples for analysis

(35). If soil remains damp throughout the journey, viruliferous nematodes may survive. Both dried and wet soil can be contaminated with naked virus particles or fungi carrying viruses.

Virus Introduction

Upon arrival in a new area, most viruses must infect a growing plant to survive. The exceptions are those plant viruses which can be passed from one vector generation to the next through arthropod eggs or fungal resting spores (33). However, even these exceptional viruses must eventually infect a susceptible plant to have significant impact. The probability of this initial infection in a new environment depends on the method of introduction into the new area.

Vegetative propagation by humans is the most efficient method of introducing any viruses present in imported plant material (15,75). The propagator actually facilitates the survival of the virus by growing the infected host in an environment where the host, and therefore the virus, will thrive. Only in the unusual case when a virus is localized in an unselected portion of the imported plant material would vegetative propagation fail to introduce all of the viruses present in imported plant material.

The probability of introducing seed-transmitted viruses by humans planting infected seed is not a certainty but is considered high (13,15,71) because of the large volume of seed usually planted. Some of the infected seed may be weaker than healthy seed at harvest; consequently, this infected seed may not survive the journey or may not be vigorous enough to germinate. Thus, planting an infected seed lot may result in selecting those virus isolates which have a minimal debilitating impact on infected seed. Once infected seedlings germinated, they would most likely encounter a favorable environment for survival, because humans planting this seed chose the location with the expectation that seedlings would prosper there.

After arrival in a new area, hitchhiking or migrating vectors are also likely to choose a food host which is susceptible to inoculation with any viruses they may be carrying. The probability of introduction is greater for viruses which can infect a wide range of plants or the most prevalent crop in the new location (116). However, the time table for viruses carried nonpersistently by a vector is quite short, so in order for these viruses to be introduced, the viruliferous vector must probe a susceptible plant soon after arrival (116,121). Viruses which persist in their vectors may have days, even the vector's lifetime, of opportunities to be inoculated into a susceptible host. Therefore, the probability of nonpersistent viruses being introduced in distant locations is significantly less than that of semi-persistent or persistent viruses.

The risk of virus introduction into a new area by most other means is probably minimal, because virus inoculation into a susceptible host in a conducive environment would be a more random and unlikely event. Infected seeds which are blown, washed or carried by large animals to a new area may not find an environment suitable for germination before they perish or are eaten; consequently, the probability of virus introduction is less for seeds moved by natural forces than for seeds imported by humans (13). The chances of infected pollen landing on the pistil of a compatible species after a long journey by wind and then fertilizing the egg of that plant in preference to locally produced pollen seem to be remote. Infected pollen carried by an insect or human to distant flowers would seem to have a higher probability of fertilizing the appropriate species. However, in studies with nepoviruses, infected pollen was less viable than (120) and competed poorly with (67) virus-free pollen. If deposition of foreign pollen on a leaf or flower of a virus host is all that is necessary for local insects to inoculate a susceptible host with a foreign virus (80), the likelihood of virus introduction is higher. Soil carrying viruliferous nematodes, inoculative fungi, or naked virus particles must be deposited in the vicinity of a susceptible virus host for introduction to be probable, because individual virus particles or these vectors can not move a long distance to find the host necessary for virus survival (10,49,108). Perishable commodities are more likely to be consumed or destroyed than discarded or planted in a location where seeds from fruit or grain would germinate, or a portion of a vegetable would grow, to produce a virus-infected plant. If scientists who import foreign viruses are careful and handle these viruses in proper containment facilities, these pathogens are unlikely to spread to susceptible local plants. In conclusion, for the cases of lowest risk, the direction provided by humans or vectors would not be present to improve the probability of initial virus infection.

Virus Establishment and Spread

In the more spectacular cases, an exotic virus is introduced into an area with efficient and prevalent vectors, susceptible hosts, and a favorable environment (14,15). Virus is spread rapidly and causes damage soon after introduction. If an outbreak is discovered early, host removal may result in eradication (2,21,90,91), but later only containment or suppression may limit virus damage. Previous efforts to eradicate, contain or suppress banana bunchy top virus (21), banana mosaic virus (cucumber mosaic virus)(1), cacao swollen shoot virus (85), citrus tristeza virus (5,22), little cherry virus (103), plum pox virus (2,97), potato virus Y-N (102), sugarcane Fiji disease virus (26), and tomato spotted wilt virus in greenhouses (91) have been costly. When a virus has spread throughout its natural range within an area, exclusion efforts become pointless and should be abandoned.

In most cases, however, introduction of an exotic virus does not immediately result in obvious damage to the appropriate crop (5,15,90,106,107). The virus isolate, native vectors, host cultivars, or environmental conditions are not conducive to an epidemic at that time. If the virus doesn't infect a host or vector which carries virus from one crop season to the next, the virus may perish when initially infected hosts or vectors expire. The human population may never know that entry of an exotic virus occurred. In other cases, an exotic virus may survive in a biennial or perennial host or in vectors for years before changes in the virus, host, vectors, or environment result in virus spread and damage to plants desirable to humans (5,15,90,106,107). In fact, foreign viruses may be present in your area now, unbeknownst to you, and continue to survive until conditions facilitate their spread and discovery.

Even though virus introduction signifies the failure of exclusion efforts, events following entry can have a significant impact on the risk assessments which form the basis of future exclusion efforts. As we gain more knowledge about establishment and spread of viruses in new areas, we may find that viruses we now fear may not be significant and some of those that don't seem important now should deserve more attention.

GOAL OF EXCLUSION

On the basis of their own personal experiences with pests (8,75) as well as some knowledge of previous crop disasters (7,8,15,35,47,54,57,74,75,84), the public has recognized the limitations of individual pest control efforts and adopted a more comprehensive collective approach which involves applying pest management programs over a large area to benefit the general population (8,75). Every community, state, provincial or federal program must originate from a general consensus developed by citizens and their leaders in that location on the necessity of such efforts. As long as the majority of the population supports, or at least accepts, the general consensus for these efforts, citizens will be willing to make certain reasonable sacrifices and direct their leaders to enact and enforce laws committing resources to specific pest control problems. However, if activities derived from these laws are not perceived by the public as a reasonable commitment for a desirable result, the consensus for the program will dissipate, and the collective effort will fail.

In the past, each society through its leaders has committed only limited resources to pest exclusion efforts. Wealthy societies with valuable agricultural industries have usually made a greater commitment of personnel, facilities and equipment than poorer societies, even though the impact of a devastating pest is usually greater on less affluent people. However, no society is rich or committed enough to support an exclusion effort designed to prevent introduction of all pests that could occur on all imported cargo. In the real world, choices must be made to decide which pests are important enough to justify the expenditure of limited resources to exclude them (75).

Each society will usually try to protect its most valuable resources. People grow different food, fiber and ornamental plants in different parts of the world. Some of these plants may be critical for survival of the human population or the source of economic wealth gained from sale of plant products. Native or recently introduced plants may also be important for maintaining environmental stability. In every area, the types of plants growing in that area could be ranked in their order of overall importance (75). This ranking of plants will differ from area to area and change through time.

Once the most important plants in a particular area have been identified, records of natural pest occurrence on each type of plant can be collected and evaluated. Some pests will be widespread wherever the particular plant is grown, but other pests remain localized in their geographic distribution. Commitment of limited resources to exclude pests that are already widespread in or on the particular plant in the local area is very difficult to justify (34). Importers of plant material may consider such exclusion efforts an unnecessary burden and be tempted to smuggle plants into the area to circumvent the imposed restrictions on their activities. Trading partners may consider such exclusion efforts as a trade barrier and take retaliatory measures against commodities they import from this area. Exclusion efforts against pests that do not occur in the local area are much easier to justify to importers, trading partners and the general public (59).

The effort to exclude only foreign pests may be modified under special circumstances. In some cases, introduction of a new strain or race of a widespread pest may be more devastating than the impact of most of the exotic pests known to damage the same plant. Therefore, foreign strains or races of a domestic pest may be added to the list of targeted pests (58,59,74). Another exception to excluding only foreign pests would be domestic pests that are the target of eradication, suppression or containment programs (34,35,-59,61). A domestic pest which is confined to a portion of the area by such a program will not be controlled successfully if the same pest is allowed to enter through ports in unaffected parts of the area.

The commitment of resources to exclude a targeted pest must be motivated by a realistic expectation of success in preventing or delaying introduction of that pest. Exclusion efforts can be successful in modifying human behavior when the humans involved understand the consequences of their actions. However, humans have had very little success in the past trying to control wind, rain, migrating vectors, or any other natural phenomenon. Until we can manipulate nature, we will not be able to prevent pest introduction by natural pathways. At this time in history, it would be a waste of limited resources to try. Consequently, all foreign pests that have a high probability of spreading to the area by natural means should be eliminated from the list of targeted pests. Those pests remaining on our list would most likely be introduced by human activities that can be regulated (34,35,58,59,61, 86,88,106,107).

In addition to an expectation of success, exclusion efforts must have a perception of benefit which exceeds the cost of the effort (8,35,58,75). The pest excluded should have the capacity to cause significant damage of important plants under conditions present where the plants are grown. For plant viruses, the general consensus is that the most damaging viruses, such as plum pox virus or tomato yellow leaf curl virus, are important enough to justify exclusion efforts. On the other hand, other viruses, such as the cryptic viruses (12,80), cause little or no damage; consequently, little benefit would be derived from the cost of excluding them. However, most viruses fall between the scourge and the curiosity. Unfortunately, there is no general consensus on which of these viruses are worth an exclusion effort and which are not. Until a consensus can be reached on standards for pest risk assessment of different viruses, allocation of resources to exclude viruses will be the subject of prolonged debate. This debate can be quite unsettling to the general public and agricultural industries that are concerned about how their activities are regulated and how public funds are spent.

While the debate over risk continues, leaders of each society must make informed decisions every year on the expenditure of limited resources for exclusion efforts. Priorities must be set. In most cases, the pests that these leaders should target for exclusion should be those exotic pests that have the potential to damage important local crops and can be realistically excluded by regulating human activities (15,34,35,61,74,88). These leaders should obtain input for these decisions from knowledgeable scientists as well as importers, producers and sellers of plant material affected by targeted pests (35,74,75,88). Economics and politics, as well as biology, may influence their decisions (8,35,59,61). Consequently, since each country has different environmental conditions, an unique array of pests, a distinct list of important plants, and unduplicated economic and political concerns, the perception of the risk posed by particular viruses will not be the same in different countries. In certain situations, neighboring countries may be able to agree on the risk of some types of agricultural material.

The implementation of exclusion efforts, often referred to as pest risk management, is designed to minimize the probability that a targeted pest will be moved to a new area. In order to be successful, exclusion efforts must impede pest movement along each of the pest's potential pathways of entry (35). For example, a foreign nepovirus may be spread to a new area during movement of infected vegetative propagules, infected seeds or pollen, soil or water infested with viruliferous nematodes, and nematode or virus cultures (10,49,108); consequently, safeguards must be taken during importation of each of these potentially contaminated articles to prevent virus introduction. The probability of success is improved if two or more barriers to virus dissemination are implemented along each pathway, but only if each impediment affects the virus differently (34,35). For example, the risk of introducing a virus in infected plant parts is reduced by importation from countries where quarantine programs attempt to exclude infected propagules, survey and containment or suppression programs limit virus incidence, and certification programs provide plant material for export. The hope is that each independent safeguard will further decrease the probability that the undesirable virus will be introduced and become established in the domestic agriculture. If the probability of virus introduction can not be minimized to an acceptable level, because of inadequate knowledge or resources, potential virus carriers should not be imported (8).

EFFORTS TO EXCLUDE PLANT VIRUSES

At the present time, the options available to humans for excluding submicroscopic pathogens, including viruses, are quite limited in comparison to the procedures used to prevent entry of other pests. Unlike arthropods, weeds, fungi or nematodes, viruses can not be seen with the naked eye or even with assistance of a light microscope. In addition, viruses occur inside cells of the imported host or vector and therefore are not exposed for detection by visual observations like surface pests. Consequently, inspection for the actual pathogen is not practical for viruses. Use of a diagnostic symptom to indicate the presence of a virus would be an alternative if development of such a symptom in any plant part was reliable. Unfortunately, all strains of a virus do not consistently induce diagnostic symptoms in all hosts under a wide range of environmental conditions (8,13,15,20,52,58,88,107). In fact, strains which are symptomless or less stressful in the most prevalent plant genotypes eventually predominate over time by natural selection (14,15) or human choice (18,20,58,86,114), and therefore are most likely to be carried in exported plant material. Consequently, use of visual inspections, regardless of whether they are conducted on imported plant material at a port of entry or during

plant growth in post entry quarantine, will not be consistently effective in excluding all virus strains along any of their pathways of entry.

Application of treatments to potentially contaminated plant material has been effective at eliminating certain arthropods, weeds, nematodes and fungi, but useful eradication procedures for viruses in infected plants have not yet been developed. Heat, cold or chemical treatments have limited virus multiplication and spread in treated plants and if combined with tip or meristem culture can result in the production of uninfected plants (15,32,45,46,52,65, 73,79,82,86,88,90). Unfortunately, each of these resulting plants must be tested for virus, because even these treatments do not always result in virus elimination. At the present time, no viricide treatment is available that can be routinely applied to thousands of imported or exported plants or plant parts with the expectation that all treated plants will survive and none will carry viable virus.

For virus exclusion, therefore, our options are reduced to vector elimination, restrictions on use of potential virus carriers, and virus testing of plant propagules either before export or after entry. Since the first two options involve regulation of potential virus carriers, these exclusion measures will be discussed under this general topic. Virus testing and controlled propagation before export are the basis of foreign certification programs, and the same procedures upon entry involve quarantine. However, for the purposes of this discussion, we would like to expand the concept of virus exclusion beyond current regulatory, certification and quarantine procedures to include international cooperation, research, and public education. Even though these activities do not directly result in virus exclusion, they have potential for improving the probability of success for current and future exclusion efforts.

Regulation of Potential Virus Carriers Upon Entry

Implementation of exclusion efforts is authorized by the laws and regulations of most political entities (8,31,58,59,61,75). These statutes may impose prohibitions, grant exceptions to these prohibitions, require import permits, specify authorized ports of entry, require phytosanitary certificates or certificates of origin, stipulate inspections, authorize treatments, deny entry of contaminated articles, and prescribe safeguards after entry (58,59). In essence, these documents specify to some degree how the most important plant parts, products and accessories will be handled to maximize the probability that a wide array of quarantine pests will be excluded.

All quarantine laws and regulations target arthropods and plant pathogens which are considered a threat to the local agriculture or environment. In some cases, foreign pests have been reported to be vectors of plant viruses (33); therefore, efforts to exclude these vectors may also result in virus exclusion. However, certain foreign viruses are transmitted by vectors which are widely distributed throughout agricultural areas of the world. *Myzus persicae* (Sulz.) or *Bemisia tabaci* (Genn.) have a wider geographic distribution than most of the viruses each transmits. Therefore, justification for eliminating these widespread vectors from imported agricultural products may not be exclusion of the vector as a pest but exclusion of the pathogen that the vector may be carrying. Very few quarantine laws or regulations target widespread vectors of foreign viruses in order to block the vector pathway of entry for foreign viruses (33,35).

Most quarantine laws and regulations restrict importation of natural soil (31,33-35,57,59,81), because so many animal and plant pests can survive in this medium. Soil may arrive at a port of entry as a sample for use or analysis or as a contaminant of any agricultural product or general cargo. The laws or regulations may require sterilization of all soil discovered by regulatory officials. Sterilization would destroy any stages of virus vectors as well as any naked virus particles present in soil. In special circumstances, an exception to the prohibition on entry of soil may be granted, and untreated foreign soil may be sent under permit to an authorized laboratory for analysis or use after entry. However, the precautions for containment, use and disposal of that soil must effectively minimize the probability of escape by any soil-borne pests, including viruses and virus vectors, into the local environment.

Scientists may provide a pathway for plant virus entry when they import cultures for research (31,33,35,57,59,81). The escape of the gypsy moth (*Lymantria dispar* L.) from a U.S. lab (84), Africanized honey bee (*Apis mellifera* L.) from a Brazilian facility (62), and tobacco blue mold fungus (*Peronospora tabacina* D.B. Adam) from an English lab (41) are examples which demonstrate the plausibility of this pathway of pest entry. Many quarantine laws or regulations require a permit for importation of pest cultures (31,33,35,59) so that an unbiased review of the risks associated with each importation can be made. Approval for importation of virus cultures should be granted only after evaluation of the probability of virus escape and the benefits of the proposed research. Importation of viruses for laboratory research only involves minimal risk as long as human access to the pathogen is limited and disposal of infected plant material destroys the virus. However, inoculation of plants with an imported virus, especially if it involves a vector, is a much greater risk, and appropriate precautions and restrictions must be imposed on this research to minimize the probability of virus being spread into the local environment. In certain cases, regulatory officials may not approve this type of research, because the benefits would not justify the risk.

Quarantine laws and regulations also specify which perishable plant material can be imported for consumption, manufacture or analysis (31,35,57,59,81). The justification for regulating imported fruits and vegetables, cut flowers, or plant material for processing is the possibility that mobile pests, such as insects or fungi, may spread from these imported plant parts to susceptible domestic hosts. Viruses are rarely targeted by these regulatory activities, because the vectors necessary for virus spread are not attracted to senescing fruits, vegetables, flowers or foliage and because normal packaging and handling in warehouses, vehicles and stores minimize the probability of virus acquisition by vectors. In most cases, the risk that a foreign virus will be introduced by this pathway is quite small. However, certain imported vegetables, such as potato tubers or beet roots, and cut flowers can be propagated after importation, either deliberately or accidently (59). Therefore, these propagules represent a greater risk as a pathway for virus entry than other plant material in the perishable category. In addition, many shipments of fruits or grains contain seeds which may carry seed-transmitted viruses and which may upon germination produce virus-infected plants where they are discarded or deliberately planted. Very few quarantine laws or regulations prescribe additional precautions for propagatable perishable material (31). Plants and plant parts imported for propagation are usually the most strictly regulated of all imported articles, because this material will most likely be propagated near susceptible hosts in an environment conducive to pest establishment and spread (35,75,81). If the targeted virus or viruses have not been reported to be transmitted through seed to germinating seedlings, the importer may be encouraged to import true seed of the virus host instead of vegetative propagules. In the majority of cases when foreign plant propagules known to be hosts of targeted viruses are imported for propagation, quarantine laws or regulations require virus testing of this plant material before it is exposed to the local environment. This testing may be carried out in an approved foreign certification program before export or during quarantine after entry.

Foreign Certification Before Entry

The goal of plant certification programs is to provide local growers with healthy, high quality propagules for planting their crops (9,13,44,45,65,73,79,82,83,100). These programs range from individual efforts by enlightened growers or nurserymen to complicated national programs defined by written regulations and administered by government agencies (45). For control of viruses and other submicroscopic pathogens, these programs must produce healthy mother plants of each cultivar by therapy and pathogen testing, maintain these tested plants in isolated areas or facilities to protect them from reinfection, and manage production of certifiable propagules in a manner and under conditions which minimize the possibility of infection by endemic pests. Use of certified propagules for planting crops and implementation of standard control measures to reduce infection by endemic pests will minimize the final number of infected plants and delay the time of infection so that yield and quality of the agricultural product is usually significantly improved over that obtained from uncertified plants (79).

Growers in other countries may also purchase certified plants from sources with reputable certification programs. Certified foreign plants may be more desirable than local types, because they are unique cultivars or germplasm, less expensive, or free of damaging pathogens. If a pathogen of quarantine significance occurs in the exporting country, plants or propagating material from the local certification program may be the only plant material that can be imported from that country with minimal risk of introducing the pathogen. Regulatory officials of the importing country may consider importation of certified plants more desirable than importation and quarantine of uncertified plants, because exotic

pathogens are being eliminated from the plant material in the exporting country, instead of after entry in the quarantine facility of the importing country. Acceptance of plants from a foreign certification program may also relieve some of the burden on overworked quarantine personnel and crowded quarantine facilities in the importing country. These benefits of importing certified plants motivate nurserymen, growers, sellers, and regulatory officials to locate and evaluate certification programs in other countries.

Acceptance of plant material from foreign certification programs is not without risk. Plants or plant parts are being moved from an area with hazardous pests to an area free of those pests on the basis of a testing and propagation program designed and implemented by humans who are not any less fallible than you or I. If the certification program fails to eliminate one or more significant pathogens from certified plants before shipment, the exporter may only lose some overseas customers when the problem is discovered, but importers and their neighbors may incur significant losses, increased control costs, domestic quarantines, and restrictions on the sale of their products after an exotic pathogen from certified plants becomes established and spreads. Consequently, importing countries must carefully evaluate the risks of accepting foreign certified plant material.

Testing procedures used to demonstrate the absence of virus and virus-like pathogens in mother plants should be the most sensitive and reliable techniques that are available at the present time. For many well-characterized viruses, the enzyme-linked immunosorbent assay (ELISA) serology and nucleic acid detection tests presently in use are a significant improvement over tests that were commonly used to detect these viruses a few decades ago. However, for other viruses and virus-like pathogens, detection still involves grafting plant parts from each mother plant onto sensitive indicator plants (9,45,73,82,83). Since research has continued to improve our detection capabilities in the past, there is no reason to believe that our current techniques are the best that we can achieve. Decades from now, we may marvel at how crude our detection procedures were in the 1990s. Therefore, it is quite possible that our current detection procedures may not be adequate to detect certain pathogen strains that are irregularly distributed or present in very low concentrations in certain hosts. Consequently, despite the best efforts of competent scientists who test mother plants for a certification program, these plants could still be infected, and progeny of these infected plants could be labeled as certified and exported.

Spread of local endemic viruses into a certification planting is also a very real possibility. Tested plants and perhaps the next generation are usually confined to screenhouses, greenhouses, or isolated farms to prevent infection by endemic viruses. With the exception of greenhouse ornamentals, later propagation stages are usually planted outdoors. Most certification programs require certifiable fields to be located in areas where vectors of endemic viruses are absent or virus spread is minimal. Efforts to monitor virus spread in the vicinity of certifiable fields may or may not be required. As an alternative, virus dissemination may be assessed on the basis of the presence or absence of symptomatic plants in the field during a few visual inspections each growing season. However, visual inspection, even under the best conditions, only detects a certain percentage of the virus-infected plants (13,65,73, 82,100). Random testing may be conducted on mother plants and perhaps their first generation progeny (9,13,44,73,79,82,83,100) but is rarely done in later propagation stages of certification schemes. Plants with unusual symptoms may be tested (9,44,73,79,100) or just discarded (44,65,79,82,83,100). Vector control, soil tests or treatments, decontamination of equipment, and a certain minimal separation from the nearest potentially infected crop hosts may also be specified to decrease the probability of pathogen spread to certifiable plants. Isolation from weed hosts may or may not be required. Unfortunately, there is no consensus on how great the isolation distance from the nearest hosts must be to ensure that certifiable plants remain free of endemic viruses (90,113). The minimum distance would depend on local conditions and vary from year to year (82,90,113).

If a local virus is spread to plants in a certification program, the impact of that virus will be greater if the infected plants belong to an early propagation stage than to a late stage. One or two recently infected plants out of thousands or millions in the final propagation stage before export would have a relatively low probability of becoming sources of an epidemic in the importing country. Since certification programs involve sequential propagation of plants from tested mother plants, any plants infected just before sale will soon be flushed out of the system to make room for the next generation of healthy plants. However, if virus spread occurs into mother plants

or initial propagation stages, many plants in subsequent propagation steps will be contaminated, and local vectors may increase the percentage of infected plants with each generation. The probability of virus infection in certified plants is higher the greater the number of propagation stages in the certification scheme (9) and the longer plants in each stage are exposed to the outside environment. Certification schemes which allow plants to be certified for years at each of many levels are especially vulnerable. For the importing country, risk of virus infection is much lower if propagules taken directly from tested mother plants are imported and then propagated to commercial quantities than when thousands of certified plants which have been grown for years outdoors in the exporting country are imported.

In the past, some certification programs have had problems with dishonesty. The temptation to label uncertified stock as certified or ignore some certification standards is especially strong in years when the quantity of certified stock is inadequate to fill all the orders. The exporter fears he or she will lose future sales to customers who do not receive all of the plants they ordered. Perhaps, no one will notice the uncertified stock necessary to fill the order. However, if the presence of uncertified stock or other irregularities are discovered, either by inspection of the product or accompanying paperwork, testing of samples, or introduction of a damaging pest, the exporter may lose the privilege to sell his or her products to that customer as well as every other potential customer in the importing country. The certification system of the exporting country would also be suspect, if not discredited. Plant protection officials of the importing country should monitor shipments of certified plant material closely and test random samples as a check on the foreign exporters as well as a prudent safeguard to protect their domestic agriculture (86,107).

Despite the doubts and potential problems with accepting foreign certification, many successful agreements have been implemented between trading partners without the movement of targeted viruses (35,58,86). In the future, importing countries will probably require more stringent certification standards for more types of plant material to ensure the quality and freedom from pests necessary for improving their agriculture.

Quarantine After Entry

The purpose of quarantine is to isolate foreign plants from domestic plants and vectors immediately after importation and until standard detection procedures fail to demonstrate the presence of any damaging pests. In a few special cases, this isolation can be achieved by growing foreign plants in a field that is a safe distance from potential hosts or vectors of targeted pests (8,58,86,88,119). Such an isolated field could also be used for growing foreign plants known to carry only viruses which are transmissible solely by plant propagation. However, for most plant material imported for propagation, quarantine involves confinement to a quarantine facility. Processing plants through a quarantine program may require considerable resources, but the cost of quarantine is usually less than the losses caused year after year by quarantine pests after entry (104).

Quarantine facilities usually consist of an inspection area, greenhouses, screenhouses, and laboratories which are sealed to contain imported plants and any accompanying pests as well as to prevent access by domestic vectors to foreign plants during the quarantine period (8,39,60,-99,104,119). These facilities vary in size and design in different parts of the world (60). This variation may result from differences in local resources, crops being processed, targeted pests, number of accessions imported, and sophistication of the pathogen tests conducted. These facilities may serve one or more crop industries in a state or province, an entire nation, or a group of neighboring countries. In most cases, however, the general procedure for processing an imported plant in any of these quarantine facilities is similar. Therefore, in order to discuss the exclusion of viruses during importation of hosts through quarantine, processing of a hypothetical virus host will be followed from arrival to release in a quarantine facility.

When a shipment containing a virus host arrives at a quarantine facility, the package is usually taken to an inspection room where it is opened and the contents are inspected for any visible pests that may have accompanied the plant material (8,39,60,86,99,104). This room is designed so that any mobile pests, such as insects, or any air-borne pests, such as fungi, will not escape to the outside environment. Treatments may be required routinely or if certain pests are detected (8,39,60,86,99,104). Since imported plant material will be propagated in a different part of the facility, it is important to eliminate in the inspection room any potential vectors as well as any pests whose damage will interfere with symptom expression or pathogen testing later in the quarantine process.

The next step is propagation of imported plant material. Good horticultural skills are usually necessary to rescue plant propagules that have endured a long journey, possibly from the opposite hemisphere, and must then grow in an alien environment (86,104). However, standard horticultural procedures may have to be modified to avoid spreading viruses from one plant introduction to another. Imported plants must be spaced far enough apart to avoid contact (8,60,119). Decontamination of tools and worker's hands before working with a new accession may be necessary (39). All conditions which inhibit symptom expression or reduce virus titer must be avoided (60), even if these conditions are optimal for plant growth. The objective of virus exclusion by quarantine is to detect and eliminate viruses from imported plants, not to grow beautiful plant introductions.

After a plant introduction is established in the quarantine facility, each plant must be labeled in a distinctive manner. Each seedling, rooted cutting, sprouted tuber, or grafted rootstock derived from an imported genotype is usually handled as a distinct subclone of that accession (8,58,86). One or more of the typical subclones of each introduction will be subjected to pathogen testing, and the others will be held in reserve. If a virus or any other quarantine pest is detected in tested subclones, those in reserve will be checked in hope of finding one or more subclones free of the pathogen (86). With many viruses, discovery of a pathogen-free subclone propagated from an infected import is sometimes possible, because not all seedlings (15,45) or buds (79,86) from an infected mother plant are infected.

As foreign plants grow in quarantine, some may exhibit disease symptoms (39,86,99,104). These symptoms may be induced by one or more pests, different types of injury, host genetic mutations, or a combination of these factors. Therefore, detection of virus symptoms may be dependent on the competence of the observer, timing of the inspection, and absence of any other pest, genetic, environmental or culturally induced injuries of the foreign plants (34,107). These symptoms may be the first evidence of a targeted virus or an indication that a new virus is present in the import. If subsequent virus testing doesn't suggest a known pathogen as the incitant of observed symptoms, the responsible quarantine official must employ other techniques to determine whether an undescribed pathogen is present or the symptoms are caused by an abiotic or genetic factor.

Techniques used to detect viruses in any quarantined plant should be the most accurate and reliable methods known at that time for each of the quarantine pathogens of that plant genus. However, since virologists know more about some viruses than others, the detection techniques for the viruses of any given plant will vary in sophistication (35,86). The protocol for testing any given plant may involve any combination of graft transmission, mechanical transmission, serology, electron microscopy, immunosorbent electron microscopy, ds RNA detection, nucleic acid hybridization, or gel electrophoresis of polymerase chain reaction (PCR) products. Since the viruses of each plant genus are different, the combination of detection techniques as well as the specific procedures will be distinctive for each type of plant (86).

Each detection test must be performed as carefully and accurately as humanly possible. The credibility of the quarantine program depends on detecting pathogens when they are present but also on negative results when they are absent. Growth of introductions and indicator plants as well as environmental conditions must be optimal for biological tests, and reagents must be active and mixed in the appropriate manner for lab tests. Both negative and positive controls must be included in each group of tests as a check on the reagents, equipment, environment, indicator plants, and worker's performance (82,107). Workers must be properly trained to prevent mix-ups or contaminations. Proper attention to the details of labeling and record keeping will ensure that the data for each test is accurately registered and applicable to the appropriate plant introduction. Contamination must be avoided by adequate spacing of indicator plants and eliminating virus from worker's hands and tools, or using new tools, before testing the next introduction. An aggressive pest elimination program in greenhouses, screenhouses and controlled environment rooms will also help minimize the possibility of contamination (60,99,104). If possible, the plants to be tested, growing indicator plants, infected plants used as positive controls, healthy plants used as negative controls, and inoculated or exposed indicator plants should be maintained in separate locations within the quarantine facility. After use, all exposed plant material, soil, and equipment must be sterilized before removal from the facility (8,39,60,99,104). The domestic agriculture is threatened whenever plant propagules infected with a quarantine pest, or any other contaminated carriers, are released from a quarantine program, regardless of whether this error resulted from poor testing technique, sloppy data management, contamination, or inadequate disposal procedures.

Importation of certain genetically-engineered plants through quarantine may create additional complications. Since transgenic plants have exhibited increased resistance but not immunity to virus infection, these plants are potential carriers of foreign viruses. Serology procedures for detection of a targeted virus may give a positive result when plants engineered to produce coat protein of that virus are tested, regardless of whether infectious virus is present in the tested plant or not. The presence or absence of a virus detected by a molecular probe or by gel electrophoresis after PCR may not be discernible if the host plant is genetically engineered to produce a segment of nucleic acid which reacts to the molecular probe or which is amplified by PCR. In these situations, modified procedures or entirely different techniques must be used to test these transgenic plants for targeted viruses.

Virus detection procedures currently used during quarantine must be viewed in perspective as the best methods we have at this point in human history. Even our most sophisticated procedures being conducted by our most competent virologists have deficiencies. All infected plants may not always be detected, because mild or unusual strains are present or the virus occurs in low concentration or is too irregularly distributed in the plant (107). At the present time, the responsible pathologist can only minimize the number of misses by duplicating his or her tests or by using different techniques to detect the same virus in each imported plant. Therefore, release of infected plant material, even from the best quarantine program, is a possibility.

When a valuable plant introduction is found infected during one or more pathogen tests, therapy procedures may be implemented to obtain the genotype free of any detected pathogens (15,82,86). In some cases, virus may be so erratically distributed in the infected plant or within a seed lot that propagation of individual buds (79,86) or seedlings (15,45) will result in production of healthy plants. However, in most cases, application of heat in the form of hot air is used as a therapy treatment. In special cases, some promising results have been reported using chemotherapy (45,46). Unfortunately, none of these therapeutic procedures achieves elimination of the targeted virus or viruses every time. For many plants, therapy treatment suppresses virus titer below the detectable level but does not entirely eliminate virus from the shoot tip or meristem. Consequently, each plant which results from a therapy procedure must be tested for the virus or viruses previously detected in the treated plant after a time period sufficient for virus multiplication to detectable levels (45,82). This interval between treatment and testing may be months or years and will vary with the virus-host combination. There is no consensus among virologists concerning what is a safe time interval to wait before retesting so that all infected plants will be detected by virus testing.

After all reasonable efforts to detect quarantine pests in an imported plant have failed, the tested plant and its progeny can be released from quarantine and distributed to interested individuals, businesses, or repositories for propagation. In special cases, imported plants which carry only widespread domestic viruses could also be distributed if these viruses don't represent a threat to the domestic agriculture (86). Early release of partially tested plants for restricted use in a secure location is also possible (86). However, in all cases, all distributed propagules must originate from tested plants or their progeny (8,13,15,58,86). Release of an entire shipment after pathogen tests on only a sample may result in entry of virus-infected plant material.

International Cooperation

In the past, countries that import or export agricultural products have negotiated agreements with their trading partners to ensure quality and minimize pest risk of plant material exchanged in international trade. In special cases, quarantine in a third country is arranged so that pests contaminating exported propagules will be eliminated during plant growth or testing in the intermediate country before plants arrive in the importing country (8,15,32,58,60,86,88). As interest in international trade of agricultural products increases, exporting countries may have to develop more certification programs in order to sell their products to countries concerned with quarantine pests. Countries that are not discriminating about

the health of plant material they import will be sold uncertifiable plant products that will have a greater risk of carrying damaging pests.

Most regulatory officials recognize that pest exclusion is not just a national priority but a regional goal (74). If an exotic virus becomes established in a neighboring country, there is a high probability that this virus will eventually be spread across the border into your country despite the best efforts of humans to prevent virus dissemination. Consequently, regulatory officials of neighboring countries have formed regional plant protection organizations, such as the European and Mediterranean Plant Protection Organization (EPPO), to cooperate in their pest exclusion and control efforts (8,58,74,88). The goals of many of these regional organizations include development of standards for pest risk assessment and management, harmonization of entry requirements for member countries, and regional acceptance of certification procedures to promote trade of certified plant material both within the region and beyond. EPPO has published data sheets (27,74) and detection procedures (28) for quarantine pests and certification schemes for different floral and fruit crops (29). These regional plant protection organizations communicate and negotiate with other regional organizations or countries to promote safe international trade.

Many other organizations contribute to the exclusion effort by compiling and disseminating information on pests of quarantine significance. The best scientific journals for plant pathology regularly publish research articles on quarantine pests by authors from different countries. Summaries describing original research on these pests can be found in the Review of Plant Pathology and other more specialized abstracting journals. Databases which may include unpublished information are also being developed and improved (16,94). At regular intervals, meetings are sponsored by international organizations for exchange of information and research techniques on damaging pests (19). In the case of viruses, the International Congress of Virology meets every three years. Symposia focusing on submicroscopic pathogens of a particular crop are organized every few years by the International Organization of Citrus Virologists, International Council for the Study of Viruses and Virus Diseases of the Grapevine, International Society for Horticultural Science (ISHS) Working Group on Virus Diseases of Fruit Trees, ISHS Working Group on Virus Diseases of

Small Fruits, ISHS Working Group on Virus Diseases of Ornamental Plants, ISHS Working Group on Vegetable Viruses, International Working Group of Legume Virologists, and International Society of Sugarcane Technologists. Original research publications, reports and virus-tested germplasm are distributed by the International Agriculture Research Centers (8,15,88,118). In an effort to promote safe international exchange of plant material, the Food and Agriculture Organization and the International Plant Genetic Resources Institute have gathered together leading experts on the pests of a particular crop and encouraged them to reach agreement on recommendations for the production, pathogen testing, therapy, and distribution of germplasm for that crop. The guidelines published for various crops are an initial effort to reach an international consensus on how to move plant material of these crops safely between countries (15,32). All of these activities contribute to the knowledge of foreign pests and help alert scientists, regulatory officials, agricultural industries, and the public to the dangers of exchanging plant material, vectors and other pest carriers.

Research

For some diseases, their most fundamental characteristic is not yet known. The pathogen or pathogens responsible for causing each of these diseases have not been identified (34,35). In general, most of these diseases are reported in trees and woody shrubs or in crops grown in less developed countries. At the present time, the pathogens responsible for these diseases are detected by transmission to a sensitive indicator plant which reacts to infection by developing characteristic symptoms. Identification and characterization of these presently unknown causal agents will reveal whether quarantine pests are involved and suggest how we can successfully exclude them.

For known viruses, identification and characterization have led to development of serological or nucleic acid techniques (25,77,107) which are more rapid and reliable than detection procedures used in the past. The relatively recent improvements in serology, including ELISA and immunosorbent electron microscopy, and in viral nucleic acid detection, such as dot-blot hybridization or gel electrophoresis after PCR (43,50), are indications of what progress can be made in the future. However, techniques of the future must overcome the present detection problems caused by low virus titer, irregular virus distribution in the plant, and virus variability. The recent development of PCR provides a laboratory technique for amplifying a small portion of viral nucleic acid up to detectable levels as a means of solving the problem of low virus titer. Future research should focus on devising techniques to predict or identify which tissues of a plant are infected so some infected cells will be sampled from every infected plant tested. Each newly developed procedure must be tested against a wide range of virus strains to determine its usefulness for detecting all strains of that virus. Additional effort must also be committed to technology transfer so that any new procedure can be implemented beyond the research laboratory to resolve regulatory or epidemiological problems.

Once sensitive and reliable detection procedures have been developed, studies on virus epidemiology and damage can progress. As our knowledge improves, we may be able someday to predict long distance dissemination, establishment in a new location, subsequent spread, and eventual losses sufficiently to identify the real threats to any country's agriculture. At the present time, research to assess the probability of long distance virus movement by natural or man-made pathways needs more emphasis. Future studies should determine which viruses will spread in a new environment in a manner predicted by research on virus epidemiology in currently affected locations. For certain viruses, only relatively small refinements in previously published models may be necessary to predict virus epidemics in a new area. With other viruses, more research is necessary to identify the factors critical to virus spread in a new location. Research on loss assessment must also be emphasized more to translate virus incidence into quantifiable damage (40,54,112). More knowledge of the factors which determine plant damage may eventually help us predict the impact of any foreign virus after entry. Damage caused by or predicted for viruses must be expressed in a manner comparable to that of other pests and abiotic factors for that crop (54). Then regulatory officials, agricultural industries, the general public, and their leaders may be willing to commit the resources necessary to exclude those viruses that would rank among the most damaging pests of the most important crops.

Research in the fields of plant breeding and genetic engineering will also influence future virus exclusion efforts. Scientists involved in this research will continue to identify plant characteristics that they consider desirable; these traits will dictate the types and sources of imported germplasm. Agricultural industries may also introduce new commercial cultivars from other countries. These decisions will influence the risk of importing foreign pests. For the foreseeable future, scientists will be trying to develop new cultivars that are more resistant to virus infection or vector infestation, either by classical plant breeding (18,38,90,96,98) or genetic engineering (6,51,64). Widespread use of resistant genotypes may significantly alter the risk assessment for a virus. Planting resistant cultivars in the exporting country may lower the risk that infected material will be exported, and use of the same cultivars in the importing country may reduce the potential for spread and damage from a disaster to a mere nuisance. However, since viruses can mutate and overcome plant resistance (18,38,90,98), it would not be wise to let the momentary success of resistant cultivars deceive us into thinking that a potentially damaging virus could be introduced without future consequences.

Research on the physiology of virus infection (37,105) may eventually help the exclusion effort. At some time in the future, virologists may discover a metabolite or enzyme which consistently reacts to virus infection but not to infection or injury by other pests or abiotic factors. The test devised from this research would have to be simple, rapid and usable for a variety of plant tissues so that regulatory officials could test thousands of imported plant parts in a short time at a port of entry. Future research may also lead to the development of procedures for preventing virus multiplication or cell-to-cell spread. If scientists could develop a therapy treatment which is 100% effective in eliminating viruses without significantly damaging the treated plant, rapid and safe importation of virus hosts would be possible.

Public Education

Even though the general public through its politicians allocates resources for the pest exclusion effort in many countries, not all citizens accept and comply with all import regulations at all times. A small percentage of the population will always attempt to beat the system and smuggle in foreign plant material for personal gain (8,58,75,86,88). Importers may feel that pests targeted for exclusion are inconsequential or that the quarantine system doesn't work quickly enough

for their needs. Among some individuals and groups, quarantine has a negative image (58,61,88). In many cases, however, average citizens will import hazardous plant material out of ignorance of quarantine laws or unintentionally (75). Communication between potential importers and regulatory officials is vital to the exclusion effort (34,52,86). In many countries, considerable time has elapsed since the crisis which compelled passage of current quarantine laws and regulations; consequently, periodic efforts must be made to remind the public and appropriate agricultural industries of why pest exclusion is necessary (61,74,75). For foreign insects and weeds, the probability of convincing importers to comply with plant quarantine regulations is high, because damage caused by these pests is obvious and within the normal experience of anyone who grows plants (8,75). However, symptoms caused by viruses and other submicroscopic pathogens are often unnoticeable to the average citizen, and damage may occur in very subtle ways. Virologists have a more challenging educational responsibility.

PERSPECTIVE

The effort to exclude plant viruses has reached an awkward age. Unlike our ancestors, we now have some knowledge of the pathology and epidemiology of these viruses. Unfortunately, we do not know enough about their capabilities to predict with much certainty which viruses can be moved to new locations and cause damage of their hosts under different conditions. At the present time, we can not focus our exclusion efforts on the most potentially damaging foreign viruses but instead must spread our resources to target many more viruses with perceived damage potential in our environment.

When we move plant material and other potential virus carriers from place to place, we now realize that we could also be moving damaging viruses as well. However, our capacity to detect these foreign viruses has not developed sufficiently for us to distinguish virus-infected plants from uninfected plants by using a rapid test at the time of entry. Current detection procedures require days, weeks, months and sometimes years depending on the targeted virus. In addition, past virus research has not led to the development of a therapy procedure which is 100% effective; consequently, virus exclusion can not presently be achieved by exposing all imported plants to a general therapeutic treatment. Even though we have learned a great deal about viruses during the last century, we still must limit exporters to shipping only certified plants, impose restrictions on use of potential virus carriers, or subject importers to delays while virus testing is carried out in quarantine if virus exclusion is to be successful.

Unfortunately, we do not have unlimited time to improve our virus exclusion efforts. The resources committed to virus exclusion are being strained even now. As the speed and accessibility of air transportation has increased in recent decades, more and more businesses, scientists, and citizens have become involved with international exchange of plant material and other virus carriers. Unfortunately, the increased frequency of international exchanges as well as improved accessibility to more remote locations increases the risk of importing damaging pests. Those involved often view the long quarantine periods caused by current pathogen detection procedures as a hindrance to their activities. Hopefully, before the public's patience is exhausted, we will develop more accurate and timely procedures for virus exclusion. Progress in recent years suggests that we can meet this challenge.

REFERENCES

1. Adam, A. V. 1962. An effective program for the control of banana mosaic. Plant Dis. Reptr. 46: 366-370.
2. Adams, A. N. 1978. The incidence of plum pox virus in England and its control in orchards. Pages 213-219 in: Plant Disease Epidemiology. P. R. Scott and A. Bainbridge, eds. Blackwell Scientific, Oxford.
3. Allen, R. N. 1983. Spread of banana bunchy top and other plant virus diseases in time and space. Pages 51-59 in: Plant Virus Epidemiology. R. T. Plumb and J. M. Thresh, eds. Blackwell Scientific, Oxford.
4. Banttari, E. E., and Venette, J. R. 1980. Aerosol spread of plant viruses: potential role in disease outbreaks. Ann.New York Acad. Sci. 353: 167-173.
5. Bar-Joseph, M., Marcus, R., and Lee, R. F. 1989. The continuous challenge of citrus tristeza virus control. Annu. Rev. Phytopathol. 27: 291-316.
6. Beachy, R. N., ed. 1993. Transgenic resistance to plant viruses. Seminars in Virology 4 (6): 327-416.
7. Bennett, C. W. 1973. A consideration of some of the factors important in the growth of the science of plant pathology. Annu. Rev. Phytopathol. 11: 1-10.
8. Berg, G. H. 1991. Plant Quarantine, Theory and Practice. OIRSA, San Salvador.
9. Bernhard, R. 1986. Certification des arbres fruitiers. EPPO Bull. 16: 245-253.
10. Boag, B. 1986. Detection, survival and dispersal of soil vectors. Chapter 7 in: Plant Virus Epidemics:

Monitoring, Modelling and Predicting Outbreaks. G. D. McLean, R. G. Garrett and W. G. Ruesink, eds. Academic Press Australia, North Ryde.
11. Boag, B., and Brown, D. J. F. 1989. Field sampling for detecting virus transmitting nematodes and their associated viruses. Mededelingen van de Faculteit Landbouwwetenschapen Rjksuniversiteit Gent 54 (3b): 1125-1132.
12. Boccardo, G., Lisa, V., Luisoni, E., and Milne, R. G. 1987. Cryptic plant viruses. Ad. Virus Res. 32: 171-214.
13. Bos, L. 1977. Seed-borne viruses. Chapter 5 in: Plant Health and Quarantine in International Transfer of Genetic Resources. W. B. Hewitt and L. Chiarappa, eds. CRC Press, Inc., Cleveland.
14. Bos, L. 1981. Wild plants in the ecology of virus diseases. Chapter 1 in: Plant Diseases and Vectors: Ecology and Epidemiology. K.Maramorosch and K. F. Harris, eds. Academic Press, Inc., New York.
15. Bos, L. 1992. New plant virus problems in developing countries: a corollary of agricultural modernization. Ad. Virus Res. 41: 349-407.
16. Boswell, K. F., Dallwitz, M. J., Gibbs, A. J., and Watson, L. 1986. The VIDE (Virus Identification Data Exchange) project: a data bank for plant viruses. Rev. Pl. Path. 65:221-231.
17. Bradfute, O. E., and Tsai, J. H. 1983. Identification of maize mosaic virus in Florida. Plant Dis. 67: 1339-1342.
18. Buddenhagen, I. W. 1983. Crop improvement in relation to virus diseases and their epidemiology. Pages 25-37 in: Plant Virus Epidemiology. R. T. Plumb and J. M. Thresh, eds. Blackwell Scientific, Oxford.
19. Chiarappa, L. 1970. Phytopathological organizations of the world. Annu. Rev. Phytopathol. 8: 419-440.
20. Converse, R. H. 1985. Latent viruses: harmful or harmless? HortScience 20: 845-848.
21. Dale, J. L. 1987. Banana bunchy top: an economically important tropical plant virus disease. Ad. Virus Res. 33:301-325.
22. Dodds, J. A., and Gumpf, D. J. 1991. Citrus tristeza virus in central California. Citrograph 76(3): 4-11.
23. Dorofeyev, V. F., ed. 1992. Origin and Geography of Cultivated Plants by N. I. Vavilov. D. Love, translator. Cambridge University Press, Cambridge.
24. Duffus, J. E. 1971. Role of weeds in the incidence of virus diseases. Annu. Rev. Phytopathol. 9: 319-340.
25. Duncan, J. M., and Torrance, L. (eds.) 1992. Techniques for the Rapid Detection of Plant Pathogens. Blackwell Scientific, Oxford.
26. Egan, B. T., and Ryan, C. C. 1986. Predicting disease incidence and yield losses in sugarcane in a Fiji disease epidemic. Chapter 22 in: Plant Virus Epidemics: Monitoring, Modelling and Predicting Outbreaks. G. D. McLean, R. G. Garrett and W. G. Ruesink, eds. Academic Press Australia,North Ryde.
27. European and Mediterranean Plant Protection Organization. 1978-1989. Data sheets on quarantine organisms. EPPO Bull.8(2), 9(2), 10(1), 11(1), 12(1), 13(1), 14: 5-71, 16: 13-78,18: 497-567, 19: 667-756.
28. European and Mediterranean Plant Protection Organization. 1990-1994. Quarantine procedure No. 1-45. Phytosanitary procedure No. 46-58. EPPO

Bull. 20: 229-282, 21: 233-266, 22: 217-252, 23: 203-214, 24: 315-346.

29. European and Mediterranean Plant Protection Organization. 1991-1994. Certification scheme No. 1-11. EPPO Bull.21: 267-290, 22: 253-296, 23: 215-252, 24: 347-367, 857-889.

30. Ferriss, R. S., and Berger, P. H. 1993. A stochastic simulation model of epidemics of arthropod-vectored plant viruses. Phytopathology 83: 1269-1278.

31. Food and Agriculture Organization of the United Nations (FAO). 1990. FAO Digest of Plant Quarantine Regulations. FAO, Rome.

32. Food and Agriculture Organization of the United Nations (FAO) and International Plant Genetic Resources Institute (IPGRI). 1989 et seq. Technical Guidelines for the Safe Movement of Germplasm. FAO, IPGRI, Rome.

33. Foster, J. A. 1982. Plant quarantine problems in preventing the entry into the United States of vector-borne plant pathogens. Chapter 8 in: Pathogens, Vectors, and Plant Diseases: Approaches to Control. K. F. Harris and K. Maramorosch, eds. Academic Press, Inc., New York.

34. Foster, J. A. 1988. Regulatory actions to exclude pests during the international exchange ofplant germplasm. HortScience 23: 60-66.

35. Foster, J. A. 1991. Exclusion of plant pests by inspections, certifications and quarantines. Pages 311-338 in: CRC Handbook of Pest Management in Agriculture, Vol. 1, 2nd ed. D. Pimentel, ed. CRC Press, Inc.,Boca Raton.

36. Frankel, O. H., and Bennett, E., eds. 1970. Genetic Resources in Plants - Their Exploration and Conservation. Blackwell Scientific, Oxford.

37. Fraser, R. S. S. 1987. Biochemistry of Virus-Infected Plants. Research Studies Press, Ltd., Letchworth.

38. Fraser, R. S. S. 1990. The genetics of resistance to plant viruses. Annu. Rev. Phytopathol. 28: 179-200.

39. Gillaspie, A. G., and McKnew, C. C. 1975. Improved sugarcane quarantine facilities and procedures at Beltsville, Maryland. Sugar J. 38(3): 40-43.

40. Grainger, J. 1979. Scientific proportion and economic decisions for farmers. Annu. Rev. Phytopathol. 17: 223-252.

41. Granhall, I. 1981. Old and new problems in the field of plant quarantine. EPPO Bull. ll: 139-144.

42. Gray, S. M., Moyer, J. W., and Bloomfield, P. 1986. Two-dimensional distance class model for quantitative description of virus-infected plant distribution lattices. Phytopathology 76: 243-248.

43. Hadidi, A., Levy, L., and Podleckis, E. V. 1995. Polymerase chain reaction technology in plant pathology. Chapter 13 in: Molecular Methods in Plant Pathology. R. P. Singh and U. S. Singh, eds. CRC Press, Inc., Boca Raton.

44. Hamilton, R. I. 1983. Certification schemes against seed-borne viruses in leguminous hosts, present status and further areas for research and development. Seed Sci. Tech. 11: 1051-1062.

45. Hansen, A. J. 1985. An end to the dilemma - virus-freeall the way. HortScience 20: 852-859.

46. Hansen, A. J. 1988. Chemotherapy of plant virus infections. Chapter 21 in: Applied Virological Research Vol. 1. New Vaccines and Chemotherapy. E. Kurstak, R. G. Marusyk, F. A. Murphy, M. H. V. van Regenmortel, eds. Plenum, New York.

47. Harlan, J. R. 1976. Diseases as a factor in plant evolution. Ann. Rev. Phytopathol. 14: 31-51.

48. Harlan, J. R. 1981. Ecological settings for the emergence of agriculture. Pages 3-22 in: Pests, Pathogens and Vegetation. J. M. Thresh, ed. Pitman, London.

49. Harrison, B. D. 1977. Ecology and control of viruses with soil-inhabiting vectors. Ann. Rev. Phytopathol. 15: 331-360.

50. Henson, J. M., and French, R. 1993. The polymerase chain reaction and plant disease diagnosis. Ann. Rev. Phytopathol. 31: 81-109.

51. Hiatt, A., ed. 1993. Transgenic Plants: Fundamentals and Applications. Marcel Dekker, Inc., New York.

52. International Board for Plant Genetic Resources (IBPGR). 1988. IBPGR Advisory Committee on In Vitro Storage. Conservation and Movement of Vegetatively Propagated Germplasm: In Vitro Culture and Disease Aspects. IBPGR, FAO, Rome.

53. Irwin, M. E., and Goodman, R. M. 1981. Ecology and control of soybean mosaic virus. Chapter 6 in: Plant Diseases and Vectors: Ecology and Epidemiology. K. Maramorosch and K. F.Harris, eds. Academic Press, Inc., New York.

54. James, W. C., Teng, P. S., and Nutter, F. W. 1991. Estimated losses of crops from plant pathogens. Pages 15-51 in: CRC Handbook of Pest Management in Agriculture, Vol. 1, 2nd ed. D. Pimentel, ed. CRC Press, Inc., Boca Raton.

55. Jones, R. A. C. 1981. The ecology of viruses infecting wild and cultivated potatoes in the Andean region of South America. Pages 89-107 in: Pests, Pathogens and Vegetation. J. M. Thresh, ed. Pitman, London.

56. Jones, R. A. C., and Harrison, B. D. 1972. Ecological studies on potato mop-top virus in Scotland. Ann. Appl. Biol. 71: 47-57.

57. Joshi, N. C. 1989. Plant quarantine in India. Rev.Trop. Pl. Path. 6: 181-200.

58. Kahn, R. P. 1977. Plant quarantine: principles, methodology, and suggested approaches. Chapter 25 in: Plant Health and Quarantine in International Transfer of Genetic Resources. W. B. Hewitt and L. Chiarappa, eds. CRC Press, Inc., Cleveland.

59. Kahn, R. P. 1982. The host as a vector: exclusion as a control. Chapter 7 in: Pathogens, Vectors, and Plant Diseases: Approaches to Control. K. F. Harris and K.Maramorosch, eds. Academic Press, Inc., New York.

60. Kahn, R. P. 1983. A model plant quarantine station:principles, concepts, and requirements. Pages 303-333 in: Exotic Plant Quarantine Pests and Procedures for Introduction of Plant Materials. K. G. Singh, ed. Association of Southeast Asian Nations Plant Quarantine Centre and Training Institute, Serdang, Selangor, Malaysia.

61. Kahn, R. P. 1991. Exclusion as a plant disease control strategy. Ann. Rev. Phytopathol. 29: 219-246.

62. Kerr, W. E. 1956. Introducao de abelhas africanas no Brasil. Brasil Apicola 3: 211-213.

63. Koenig, R. 1986. Plant viruses in rivers and lakes. Ad. Virus Res. 31: 321-333.

64. Lal, R., and Lal, S. 1993. Genetic Engineering of Plants for Crop Improvement. CRC Press, Inc., Boca Raton.

65. Lawson, R. H. 1981. Controlling virus diseases in major international flower and bulb crops. Plant

Dis. 65: 780-786.

66. Leppik, E. E. 1970. Gene centers of plants as sources of disease resistance. Ann. Rev. Phytopathol. 8: 323-344.

67. Lister, R. M., and Murant, A. F. 1967. Seed-transmission of nematode-borne viruses. Ann. Appl. Biol. 59: 49-62.

68. Madden, L. V., and Campbell, C. L. 1986. Descriptions of virus disease epidemics in time and space. Chapter 13 in: Plant Virus Epidemics: Monitoring, Modelling and Predicting Outbreaks. G. D. McLean, R. G. Garrett and W. G. Ruesink, eds. Academic Press Australia, North Ryde.

69. Madden, L. V., Louie, R., and Knoke, J. K. 1987. Temporal and spatial analysis of maize dwarf mosaic epidemics. Phytopathology 77: 148-156.

70. Madden, L. V., Pirone, T. P., and Raccah, B. 1987. Analysis of spatial patterns of virus-diseased tobacco plants. Phytopathology 77: 1409-1417.

71. Mandahar, C. L. 1981. Virus transmission through seed and pollen. Chapter 8 in: Plant Diseases and Vectors: Ecology and Epidemiology. K. Maramorosch and K. F. Harris, eds. Academic Press, Inc., New York.

72. Markham, R. 1977. Landmarks in plant virology: genesis of concepts. Ann. Rev. Phytopathol. 15: 17-39.

73. Martelli, G. P., ed. 1992. Grapevine Viruses and Certification in EEC Countries: State of the Art. Istituto Agronomico Mediterraneo, Bari.

74. Mathys, G., and Baker, E. A. 1980. An appraisal of the effectiveness of quarantines. Ann. Rev. Phytopathol. 18:85-101.

75. McCubbin, W. A. 1954. The Plant Quarantine Problem. Ejnar Munksgaard, Copenhagen.

76. McLean, G. D., Garrett, R. G., and Ruesink, W. G. (eds.) 1986. Plant Virus Epidemics: Monitoring, Modelling and Predicting Outbreaks. Academic Press Australia, North Ryde.

77. Miller, S. A., and Martin, R. R. 1988. Molecular diagnosis of plant disease. Ann. Rev. Phytopathol. 26: 409-432.

78. Mink, G. I. 1983. The possible role of honeybees in long-distance spread of prunus necrotic ringspot virus from California into Washington sweet cherry orchards. Pages 85-91 in: Plant Virus Epidemiology. R. T. Plumb and J. M. Thresh, eds. Blackwell Scientific, Oxford.

79. Mink, G. I. 1991. Control of plant diseases using disease-free stocks. Pages 363-391 in: CRC Handbook of Pest Management in Agriculture, Vol. 1, 2nd ed. D. Pimentel, ed. CRC Press, Inc., Boca Raton.

80. Mink, G. I. 1993. Pollen-and seed-transmitted viruses and viroids. Ann. Rev. Phytopathol. 31: 375-402.

81. Morschel, J. R. 1979. Controlling the movement of plant diseases and pests into, out of, and within Australia. Pages 35-42 in: Plant Health, The Scientific Basis for Administrative Control of Plant Diseases and Pests. D. L. Ebbels and J. E. King, eds. Blackwell Scientific, Oxford.

82. Navarro, L. 1986. Citrus certification in Mediterranean countries. EPPO Bull. 16: 227-238.

83. Navatel, J. C., and Fournier, B. 1986. La production de plants de fraisiers certifies dans quelques pays europeens. EPPO Bull. 16: 369-373.

84. Ordish, G. 1976. The Constant Pest. Peter Davies, Ltd., London.

85. Owusu, G. K. 1983. The cocoa swollen shoot disease problem in Ghana. Pages 73-83 in: Plant Virus Epidemiology. R. T. Plumb and J. M. Thresh, eds. Blackwell Scientific, Oxford.

86. Parliman, B. J., and White, G. A. 1985. The plant introduction and quarantine system of the United States. Plant Breeding Rev. 3: 361-434.

87. Perdue, R. E., and Christenson, G. M. 1989. Plant exploration. Plant Breeding Rev. 7: 67-94.

88. Plucknett, D. L., and Smith, N. J. H. 1988. Plant quarantine and the international transfer of germplasm. CGIAR Study Paper No. 25. The World Bank, Washington.

89. Plumb, R. T., and Thresh, J. M. (eds.) 1983. Plant Virus Epidemiology. Blackwell Scientific, Oxford.

90. Quiot, J. B., Labonne, G., and Marrou, J. 1982. Controlling seed and insect-borne viruses. Chapter 6 in: Pathogens, Vectors, and Plant Diseases: Approaches to Control. K. F. Harris and K. Maramorosch, eds. Academic Press, Inc., New York.

91. Rautapaa, J. 1992. Eradication of *Frankliniella occidentalis* and tomato spotted wilt virus in Finland: a case study on costs and benefits. EPPO Bull. 22: 545-550.

92. Rissler, J., and Mellon, M. 1993. Perils Amidst The Promise, Ecological Risks of Transgenic Crops in a Global Market. Union of Concerned Scientists, Cambridge.

93. Rosenberg, L. J., and Magor, J. I. 1983. A technique for examining the long-distance spread of plant virus diseases transmitted by the brown planthopper, *Nilaparvata lugens* (Homoptera: Delphacidae), and other wind-borne insect vectors. Pages 229-238 in: Plant Virus Epidemiology. R. T. Plumb and J. M. Thresh, eds. Blackwell Scientific, Oxford.

94. Royer, M. H., and Dowler, W. M. 1988. A world plant pathogen database. Plant Dis. 72: 284-288.

95. Ruesink, W. G., and Irwin, M. E. 1986. Soybean mosaic virus epidemiology: a model and some implications. Chapter 14 in: Plant Virus Epidemics: Monitoring, Modelling and Predicting Outbreaks. G. D. McLean, R. G. Garrett and W. G. Ruesink, eds. Academic Press Australia, North Ryde.

96. Russell, G. E. 1978. Plant Breeding for Pest and Disease Resistance. Butterworth and Co., Ltd., London.

97. Savino, V., Digiaro, M., Martelli, G. P., and Di Terlizzi, B. 1992. Plum pox virus outbreaks in Apulia and Basilicata (southern Italy). Acta Hortic. 309: 125-128.

98. Shaner, G. 1991. Genetic resistance for control of plant disease. Pages 495-540 in: CRC Handbook of Pest Management in Agriculture, Vol. 1, 2nd ed. D. Pimentel, ed. CRC Press, Inc., Boca Raton.

99. Sheffield, F. M. L. 1958. Requirements of a post-entry quarantine station. FAO Plant Prot. Bull. 6: 149-152.

100. Shepard, J. F., and Claflin, L. E. 1975. Critical analyses of the principles of seed potato certification. Annu. Rev. Phytopathol. 13: 271-293.

101. Sigvald, R. 1986. Forecasting the incidence of potato virus Y-O. Chapter 21 in: Plant Virus Epidemics: Monitoring, Modelling and Predicting Outbreaks. G. D. McLean, R. G. Garrett and W. G. Ruesink, eds. Academic Press Australia, North Ryde.

102. Singh, R. P. 1992. Incidence of the tobacco veinal necrotic strain of potato virus Y (PVY-N) in Canada

in 1990 and 1991 and scientific basis for eradication of the disease. Can. Pl. Dis. Survey 72: 113-119.

103. Slykhuis, J. T., Yorston, J., Raine, J., McMullen, R. D., and Li, T. S. C. 1980. Current status of little cherry disease in British Columbia. Can. Pl. Dis. Survey 60: 37-42.

104. Smee, L. 1975. The post-entry quarantine of imported plant material in Australia. PANS 21: 168-174.

105. Sreenivasulu, P., Naidu, R. A., and Nayudu, M. V. 1989. Physiology of Virus Infected Plants. South Asian Publishers, New Delhi.

106. Stace-Smith, R. 1985. Role of plant breeders in dissemination of virus diseases. HortScience 20: 834-837.

107. Stace-Smith, R., and Martin, R. R. 1989. Plant quarantine diagnostic problems: viruses. Chapter 13 in: Plant Protection and Quarantine. Vol. II. Selected Pests and Pathogens of Quarantine Significance. R. P. Kahn, ed. CRC Press, Inc., Boca Raton.

108. Taylor, C. E. 1980. Nematodes. Chapter 16 in: Vectors of Plant Pathogens. K. F. Harris and K. Maramorosch, eds. Academic Press, Inc., New York.

109. Taylor, C. E., and Brown, D. J. F. 1976. The geographical distribution of *Xiphinema* and *Longidorus* nematodes in the British Isles and Ireland. Ann. Appl. Biol. 84: 383-402.

110. Taylor, L. R. 1986. The distribution of virus disease and the migrant vector aphid. Chapter 3 in: Plant Virus Epidemics: Monitoring, Modelling and Predicting Outbreaks. G. D. McLean, R. G. Garrett and W. G. Ruesink, eds. Academic Press Australia, North Ryde.

111. Taylor, R. A. J. 1985. Migratory behavior in the Auchenorrhyncha. Chapter 11 in: The Leafhoppers and Planthoppers. L. R. Nault and J. G. Rodriguez, eds. John Wiley & Sons, Inc., New York.

112. Teng, P. S. (ed.) 1987. Crop Loss Assessment and Pest Management. American Phytopathological Society Press, St. Paul.

113. Thresh, J.M. 1976. Gradients of plant virus diseases. Ann. Appl. Biol. 82: 381-406.

114. Thresh, J.M. 1982. Cropping practices and virus spread. Annu. Rev. Phytopathol. 20: 193-218.

115. Thresh, J.M. 1983. Progress curves of plant virus disease. Adv. Appl. Biol. 8: 1-85.

116. Thresh, J.M. 1983. The long-range dispersal of plant viruses by arthropod vectors. Phil. Trans. R. Soc. London B 302: 497-528.

117. Thresh, J.M. 1986. Plant virus disease forecasting. Chapter 18 in: Plant Virus Epidemics: Monitoring, Modelling and Predicting Outbreaks. G. D. McLean, R. G. Garrett and W. G. Ruesink, eds. Academic Press Australia, North Ryde.

118. Thurston, H.D. 1977. International crop development centers: a pathologist's perspective. Annu. Rev. Phytopathol. 15: 223-247.

119. Veenenbos, J. A. J., and Treur, A. 1984. Plant quarantine facilities in the Netherlands. EPPO Bull. 14: 389-391.

120. Yang, A. F., and Hamilton, R. I. 1974. The mechanism of seed transmission of tobacco ringspot virus in soybean. Virology 62: 26-37.

121. Zeyen, R. J., and Berger, P. H. 1990. Is the concept of short retention times for aphid-borne nonpersistent plantviruses sound? Phytopathology 80: 769-771.

122. Zitter, T. A. 1977. Epidemiology of aphid-borne viruses. Chapter 16 in: Aphids as Virus Vectors. K. F. Harris and K. Maramorosch, eds. Academic Press, Inc., New York.

CHAPTER 17

IPGRI's Role in Controlling Virus Diseases in Plant Germplasm

E.A. Frison and M. Diekmann

The International Plant Genetic Resources Institute (IPGRI) is one of the 16 International Agricultural Research Centers (IARCs) of the CGIAR (Consultative Group on International Agricultural Research). It was established in 1974 as the International Board for Plant Genetic Resources (IBPGR) and has its headquarters in Rome, Italy. IPGRI's mandate is to advance the conservation and use of genetic resources for the benefit of present and future generations.

IPGRI's activities in the field of germplasm health can be divided into three categories. The first activity relates to strategic research with the objective of developing techniques that will improve the safety and efficiency of germplasm movement and involves essentially biotechnology. The second, and perhaps most obvious, activity consists mainly of assembling and disseminating information relating to quarantine. In this area IPGRI closely collaborates with the Plant Protection Service of FAO (Food and Agriculture Organization of the United Nations). The third activity concerns the phytosanitary aspects of genebank management. In this area IPGRI contributes to improving the awareness of curators and other scientists working with germplasm of the deleterious effect of seedborne pathogens in the conservation of the genetic diversity of crop genepools and to the development of new concepts to overcome these problems. In all three fields, viruses play an important role.

STRATEGIC RESEARCH

Movement of germplasm can be jeopardized because of the lack of rapid and reliable indexing methods for diseases, especially viruses, of quarantine concern (9,10,12). The development of therapy techniques allows infected germplasm, especially vegetatively propagated material, to be cleaned up and moved safely. As a model, research on the development of in vitro therapy techniques for rosaceous crops has been supported by IPGRI (25,26).

Virus-detection techniques are often developed for diagnostic purposes and are therefore very specific in differentiating between strains of a given virus. This may, however, be a handicap for quarantine testing where it is essential to detect all possible strains. Broader spectrum tests can be developed at little extra expense if the researcher has 'quarantine indexing' in mind at the development stage. When 'probes' are developed, be they nucleic acid hybridization probes or monoclonal antibodies, a number of clones are obtained from which a selection can be made. Some will be very specific, while others will have a broader spectrum. Besides stimulating and catalyzing work in this area, IPGRI has undertaken research on the development of broad-spectrum tests for viruses (11,15).

There is also a need for rapid and simple virus-detection tests. Techniques requiring purification steps or sophisticated equipment can be satisfactorily applied in a well-equipped laboratory and when small numbers of samples are involved, but they are highly impractical for large-scale indexing. For example, ^{32}P-labelled nucleic acid probes are valuable tools for research purposes but their short shelf life makes their use difficult in remote areas or in a laboratory that does not have the ability to produce its own labelled probes. IPGRI is also involved in collaborative research in this area (e.g. 2-4,16, 17,27,28).

The combination of in vitro therapy and in vitro indexing results in the preferred mode of

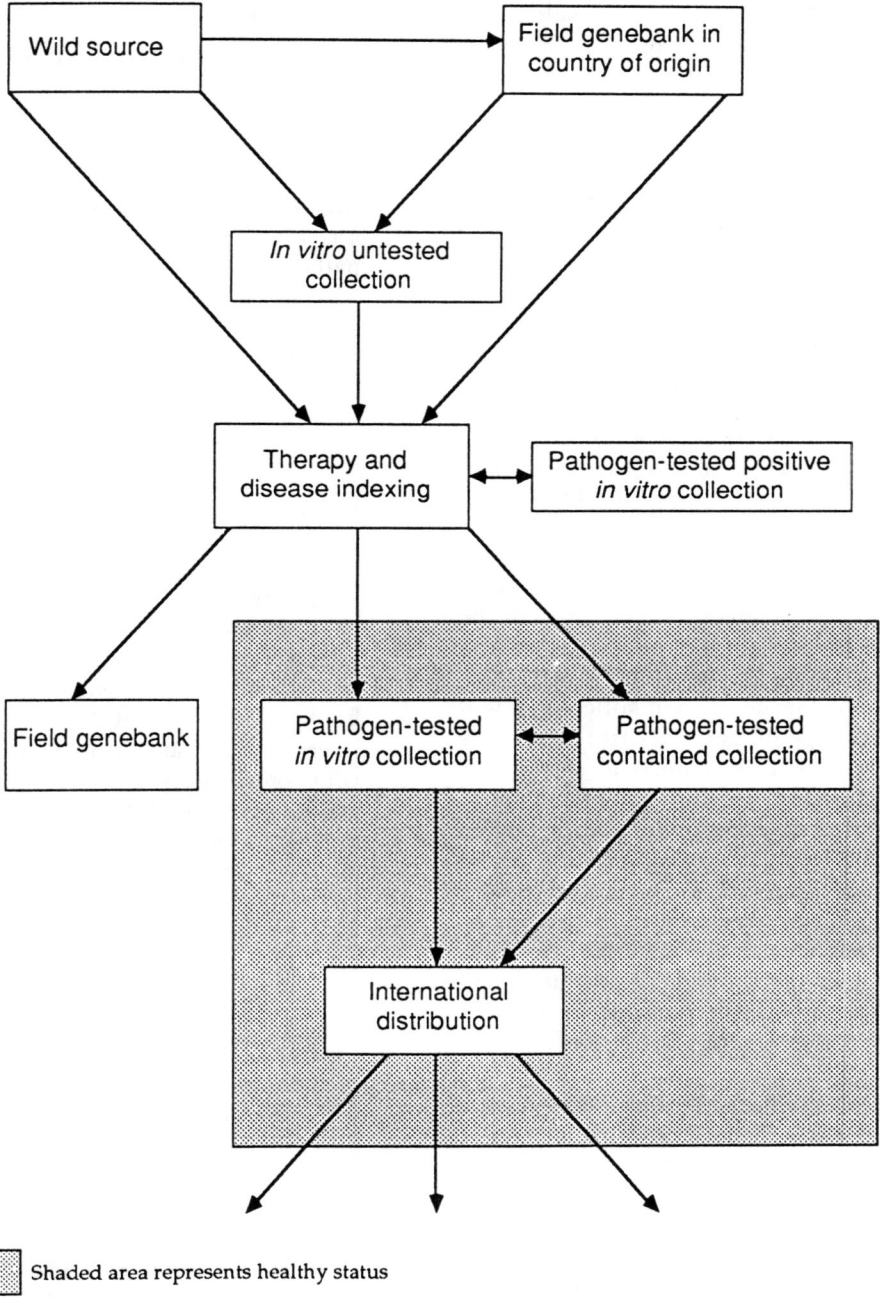

Shaded area represents healthy status

Figure 1. Flow diagram for germplasm of vegetatively propagated crops

germplasm movement in the form of tissue cultures, as it provides a contained system. This will further decrease risks of pathogen escape, as the plantlets will remain in tubes during the whole process. Strategic research in these areas is one of the priorities of IPGRI's research program (24).

A conceptual framework for assessing the factors associated with the movement of vegetatively propagated crops, which is based on the assumption that *in vitro* methods are available (as is the case for many crops), was developed at the 1987 IBPGR-sponsored meeting in Raleigh. The flow diagram (Figure 1) illustrates the major elements of this framework.

Ideally, wild material is 'cleaned' before being stored in an *in vitro* genebank. The cleaning operation generally includes therapy (meristem tip culture, possibly combined with thermotherapy) to free the material from pathogens and indexing to establish whether the material is indeed free of pathogens, especially viruses. Once the material is considered to be 'healthy' it can be multiplied *in vitro* for international movement.

In cases when material cannot be 'cleaned' before being stored, because adequate techniques are not yet available, it should be kept in a 'pathogen-untested *in vitro* collection'. If valuable material is tested and is found to contain one or more pathogens, it can be kept in a 'pathogen-tested-positive *in vitro* collection' Material in these two types of collections should not be released, but can be kept for conservation and research purposes under the supervision of a pathologist. Maintenance of such collections will allow the safe storage of material prior to therapy and indexing, and will accommodate anticipated improvements in indexing and therapy techniques. This will also provide opportunities for selection of resistant or tolerant genotypes, the rapid development of polycross nurseries under containment and preservation of pathogens. International exchange of material in such collections - under the supervision of quarantine authorities - has to be made possible in order to allow proper duplication of conservation collections.

Seeds play a major role in plant genetic resources activities. The vast majority of germplasm accessions are stored and moved in the form of seeds. The IARC germplasm collections, which form the largest international repository of genetic diversity of the major food crops of the world, consist of about 95% seed propagated crops; only 5% (e.g. potato, cassava, *Musa* spp.) are vegetatively propagated or kept and exchanged

in the form of tissue culture (5).

Because of the destructive nature of most seed-health testing methods, problems occur when accessions have only a small number of seeds. This is especially the case with wild relatives of crops, for which it is often difficult to collect large samples of seeds. For this reason, research to develop non-destructive seed-health tests has been commissioned by IPGRI (19,20).

INFORMATION NEEDS FOR QUARANTINE

Collecting, conservation and utilization of plant genetic resources are essential components of international crop-improvement programs. Inevitably, the movement of germplasm involves a risk of accidentally introducing plant pests and pathogens along with the host plant material. Viruses pose a particular risk because they are difficult to detect and identify. In order to minimize this risk, effective testing procedures are required to ensure that distributed material is free of pests and pathogens which are of quarantine concern.

The increasing volume of germplasm exchanged internationally, coupled with recent rapid advances in biotechnology, has created a need for crop-specific overviews of the existing knowledge on all aspects related to the phytosanitary safety of germplasm transfer. A first step was the Special Task Force convened by IBPGR in 1975. It resulted in the publication of a book entitled 'Plant Health and Quarantine in International Transfer of Genetic Resources' (18), with contributions from 31 specialists. One of the major recommendations of a subsequent meeting organized by IBPGR in Raleigh, N.C. in 1987 on *in vitro* culture and disease aspects of conservation and movement of vegetatively propagated crops was the development of crop-specific guidelines on indexing and therapy (21).

This has prompted FAO and IBPGR to launch a collaborative program for the safe and expeditious movement of germplasm (14). The aim of the joint FAO/IPGRI program is to generate a series of crop-specific technical guidelines that provide relevant information on disease indexing and other procedures that will help to ensure phytosanitary safety when germplasm is moved internationally.

The technical guidelines are produced by panels of specialized plant pathologists and virologists who meet to pool their knowledge. The

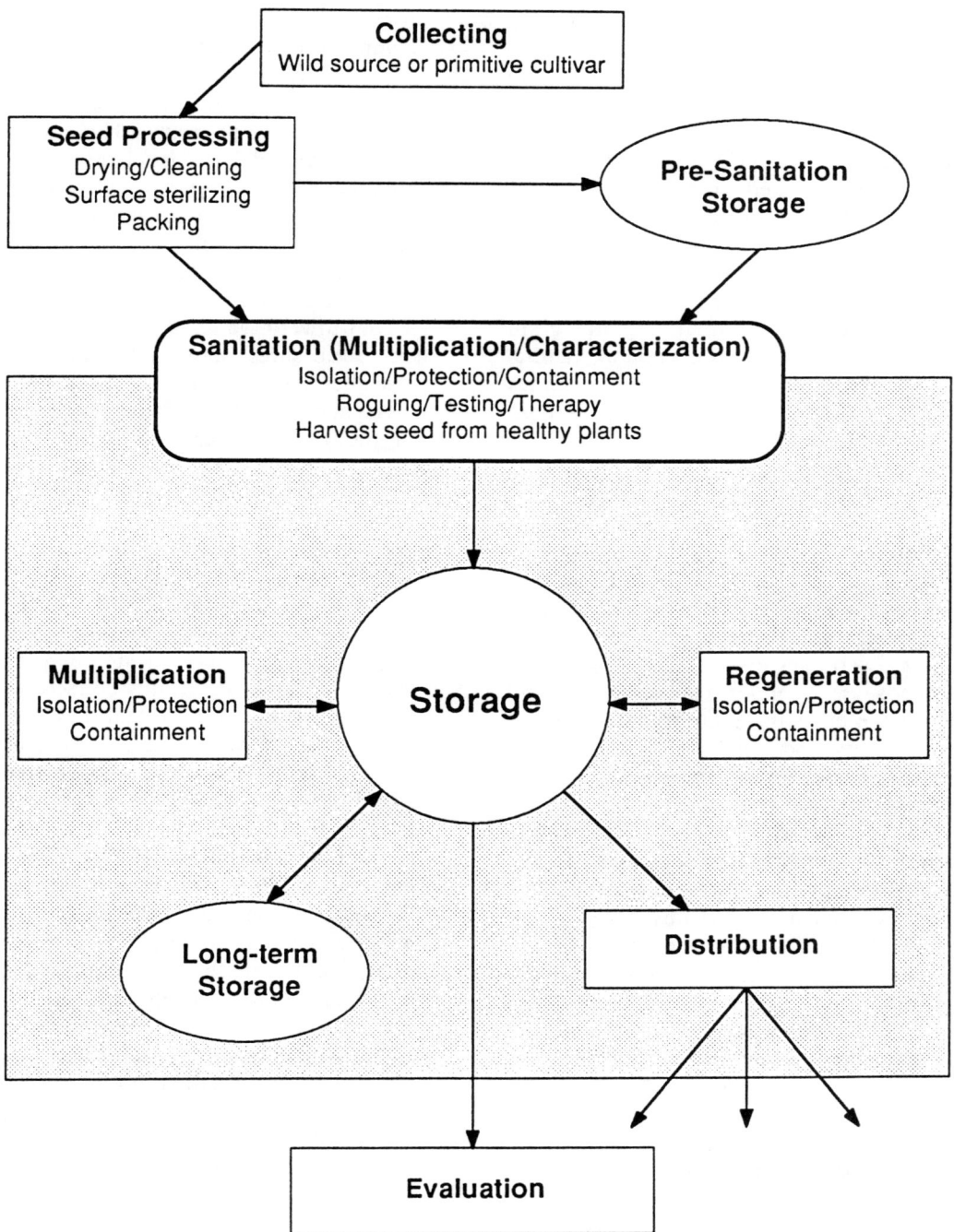

Figure 2. Phytosanitary aspects of genebank management

guidelines reflect the consensus of the specialists attending the meeting, based on the best scientific knowledge available at the time of the meeting. The guidelines are divided into two parts. The first part makes recommendations on how best to move germplasm of the crop concerned and lists institutions recovering and/or maintaining healthy germplasm. The second part covers the important pests and diseases of quarantine concern, giving a description of therapy and indexing methodologies.

The guidelines are widely distributed to plant protection services, genebanks and libraries of agricultural universities and research stations all over the world. So far priority has been given to vegetatively propagated crops for which the risks involved with the movement of germplasm are greater. Guidelines have been produced for cassava, *Citrus*, cocoa, coconut, edible aroids, grapevine, legumes, *Musa*, small fruits (*Fragaria, Rubus, Ribes* and *Vaccinium*), small grain temperate cereals, sugarcane, sweet potato, vanilla, yam, and stone fruits. In the near future guidelines will be published on *Allium* spp., *Eucalyptus* spp., potato and pome fruits.

GENEBANK MANAGEMENT

Phytosanitary aspects of genebank management have long been neglected and few genebanks have a plant pathologist among their staff. Poor phytosanitary status of seeds in a genebank has several negative effects. Besides the fact that seeds are vehicles for pathogens and create quarantine concern, microorganisms affect the vigor, viability and quality of seeds during storage, particularly under short-term storage conditions.

Pathogens can usually survive very well in seeds stored under conditions of long-term storage of germplasm (at low moisture content and low temperature). A summary is given by Agarwal and Sinclair (1). Besides its negative effect on viability, seed infection can seriously affect the characterization and evaluation of germplasm accessions, as some characters might be confused with disease symptoms not recognized as such. Disease symptoms can also mask interesting characters. It is therefore important that characterization and evaluation be carried out using uninfected seed (13).

A poor phytosanitary status of a collection can also lead to genetic erosion. Diseases may eliminate susceptible accessions with otherwise

interesting characters. However, roguing activities also can cause genetic erosion (2,23). Similar concerns are voiced about the application of pesticides, particularly herbicides, in germplasm multiplication. Phytotoxic effects may wipe out susceptible accessions, especially of wild relatives which are not normally subject to pesticide application. Removing a certain percentage of the seed quantity available through random sampling for testing may also change the genetic composition of the accession (8).

With good management practices from a phytosanitary point of view it is possible to overcome these problems. The diagram in Figure 2 shows a model for genebank management with an ideal flow of operations, from collecting to long-term storage and evaluation. While it is recognized that it will not be possible to apply this model immediately to all collections, mainly for financial and logistic reasons, it should be seen as a goal toward which all genebanks should strive. Some measures imply no or minimal additional costs but have a significant effect. For example the incidence of seedborne bean common mosaic virus in the USDA *Phaseolus* germplasm collection could be reduced by 66%, simply as a result of a winter greenhouse increase, when vector populations are low, as opposed to a summer greenhouse increase (22). The germplasm health activities at IARCs with substantial germplasm exchange are described by Diekmann (6,7).

The essential features of the model diagram are the sanitation step before storage and the maintenance of 'healthy collections'. The term 'healthy', although not ideal, has been used here because it is a relative and holistic concept. The terms 'disease free' or 'pathogen-free' would be too absolute and in most cases incorrect.

Ideally, all germplasm accessions should undergo a sanitation step before being stored. The methods involved will vary greatly according to the species concerned, the area of origin of the material and in the area where the sanitation is carried out, and the availability of therapy and indexing methods. If the sanitation can not be carried out immediately, the material should be stored with a temporary status until it can undergo sanitation. This is called 'pre-sanitation storage' in the diagram and includes also the case where a subsample of the original collection is put in long-term storage for safety reasons.

The logical moment to apply sanitation is when the accession is grown out for the first time, be it for increase (multiplication) or for characterization of the material. Germplasm should be grown under containment or isolation to avoid accidental escapes of pests and diseases and also to avoid contamination with locally present diseases. The accessions should be regularly inspected by a plant pathologist and infected plants should be rogued if feasible. Therapy methods should be applied to produce healthy material if the percentage of infection is too high and roguing would cause genetic erosion in material that can not be replaced or if its replacement would be too expensive. Appropriate protection methods, including pesticide treatments, should be applied when relevant. Seeds should be harvested only from healthy looking plants which have been tested for latent infections by viruses if necessary.

Once the material is 'healthy', it should be maintained under conditions allowing it to stay 'healthy'. This implies that appropriate measures have to be taken during storage and especially when material is grown out for regeneration and multiplication. In the diagram, the area in grey represents the 'healthy' status of germplasm.

LOOKING AHEAD

IPGRI's mandate to contribute to sustainable improvements in the productivity of agriculture and forestry is pursued by international cooperation. This is particularly essential in the area of germplasm health. By organizing workshops and conferences, in many cases together with other institutions, by publishing and distributing information material, and by commissioning research in all areas related to germplasm health, IPGRI hopes to minimize the risk of pathogen spread with germplasm exchange and to protect germplasm collections from pathogen infection. IPGRI would like to invite scientists interested in germplasm health to participate in a dialogue and in collaborative research with the aim of improving the health status of germplasm collections.

REFERENCES

1. Agarwal, V.K., and Sinclair, J.B. 1987. Principles of Seed Pathology. Vol. I and II. CRC Press, Boca Raton, FL, U.S.A.
2. Alconero, R., Weeden, N.F., Gonsalves, D., and Fox, D.T. 1985. Loss of genetic diversity in pea germplasm by the elimination of individuals infected by pea seedborne mosaic virus. Ann. Appl. Biol. 106:357-364.
3. Bateson, M.F., and Dale, J.L. 1995. Banana bract mosaic virus: characterization using potyvirus specific degenerate PCR. Arch. Virology 140:515-527.
4. Burns, T.M., Harding, R.M., and Dale, J.L. 1994. Evidence that banana bunchy top virus has a multiple component genome. Arch. Virology 137:371-380.
5. Cooper, D., Engels, J., and Frison, E. 1994. A multilateral system for plant genetic resources: imperatives, achievements and challenges. Issues in Genetic Resources No. 2, May 1994. International Plant Genetic Resources Institute, Rome, Italy.
6. Diekmann, M. 1988. Seed Health Testing and Treatment of Germplasm at the International Center for Agricultural Research in the Dry Areas (ICARDA). Seed Sci. & Technol. 16:405-416.
7. Diekmann, M. 1992. Seed Health Measures at International Agricultural Research Centers. Pages 367-376 in: Seed Pathology. Proceedings of the CTA Seminar held at Copenhagen, Denmark, June 20-25, 1988. S.B. Mathur, and J. Jorgensen, eds. CTA, Wageningen, The Netherlands.
8. Diekmann, M. 1993. Sampling germplasm for seed health testing. Pages 126-131 in: Proc. 1st ISTA Plant Disease Committee Symposium on Seed Health Testing, August 9 to 11, 1993 in Ottawa. J. W. Sheppard, ed. Agriculture Canada Central Seed Laboratory, Ottawa, Canada.
9. Frison, E.A. 1991. Plant health research at the International Board for Plant Genetic Resources (IBPGR). Pages 271-275 in: Crop Genetic Resources of Africa, Volume II: Proc. of an International Conference on Crop Genetic Resources of Africa, October 17-20, 1988, Ibadan, Nigeria. N.Q. Ng, P. Perrino, F. Attere and H. Zedan, eds. Int'l. Institute of Tropical Agriculture, Ibadan/ International Board for Plant Genetic Resources, Rome/ United Nations Environment Programme, Nairobi/ Consiglio Nazionale delle Ricerche, Rome.
10. Frison, E.A. 1992. Int'l. movement of plant genetic resources. Pages 361-365 in: Seed Pathology, Proceedings of the CTA Seminar held at Copenhagen, Denmark, June 20-25, 1988. S.B. Mathur, and J. Jorgensen, eds. CTA, Wageningen, The Netherlands.
11. Frison, E.A. 1993. Production of cross-reacting and heterospecific monoclonal antibodies for plant virus detection. Doctoral, Faculté des Sciences Agronomiques, Gembloux.
12. Frison, E.A. 1994. IBPGR concerns on seed-borne disease. Pages 32-40 in: Proc. Conference on the Potential of Biotechnology to Minimize Seed-Borne Diseases, September 28 - October 1, 1989, Puyallup, Washington. Research and Extension Center, Washington State University, Puyallup, Washington.
13. Frison, E.A., Bos, L., Hamilton, R.I., Mathur, S.B., and Taylor, J.D., eds. 1990. FAO/IBPGR Technical Guidelines for the Safe Movement of Legume Germplasm. Food and Agriculture Organization of the United Nations, Rome/International Board for Plant Genetic Resources, Rome.
14. Frison, E.A.,and Putter, C.A.J. 1988. FAO/IBPGR technical guidelines for the safe movement of germplasm. Page 279 in: Abstr. 5th International Congress of Plant Pathology, 20-27 August, 1988, Kyoto.
15. Frison, E.A., and Stace-Smith, R. 1992. Cross-reacting and heterospecific monoclonal antibodies produced against arabis mosaic nepovirus. J. Gen.

Virol. 73:2525-2530.

16.   Harding, R.M., Burns, T.M., and Dale, J.L. 1991. Virus-like particles associated with banana bunchy top disease contain small single-stranded DNA. J. Gen. Virology 72:225-230.

17.   Harding, R.M., Burns, T.M., Hafner, G., Dietzgen, R.G., and Dale, J.L. 1993. Nucleotide sequence of one component of the banana bunchy top virus genome contains a putative replicase gene. J. Gen. Virology 74:323-328.

18.   Hewitt, W.B., and Chiarappa, L., eds. 1977. Plant Health and Quarantine in International Transfer of Genetic Resources. CRC Press, Inc.

19.   Higley, P.M., McGee, D.C., and Burris, J.S. 1993. Development of methodology for non-destructive assay of bacteria, fungi and viruses in seeds of large-seeded field crops. Seed Sci. & Technol. 21:399-409.

20.   Higley, P.M., McGee, D.C., and Burris, J.S. 1994. Effects of non-destructive tissue extraction on the viability of corn, soybean and bean seeds. Seed Sci. & Technol. 22:245-252.

21.   IBPGR. 1988. IBPGR Advisory Committee on *In Vitro* Storage. Conservation and Movement of Vegetatively Propagated Germplasm: *In Vitro* Culture and Disease Aspects. International Board for Plant Genetic Resources, Rome.

22.   Klein, R.E., Wyatt, S.D. and Kaiser, W.J. 1989. Influence of propagation on incidence of seedborne bean common mosaic virus in the USDA *Phaseolus*

germplasm collection. Plant Dis. **73**:759-761.

23.   Recchio-Demmin, B.E., McFerson, J.R., and Kresovich, S. 1990. Genetic impact of the pea seedborne mosaic virus eradication program on the USA national *Pisum* collection. Pisum Newsletter 22:46-47.

24.   Spiegel, S., Frison, E.A., and Converse, R.H. 1993. Recent developments in therapy and virus-detection procedures for international movement of clonal plant germ plasm. Plant Dis. 77:1176-1180.

25.   Spiegel, S., and Martin, R.R. 1992. Detection of strawberry mild yellow-edge disease in micropropagated strawberry plantlets. Acta Hortic. 308:61-68.

26.   Stein, A., Spiegel, S., Faingersh, G., and Levy S. 1991. Responses of micropropagated peach cultivars to thermotherapy for the elimination of Prunus necrotic ringspot virus. Ann. Appl. Biol. 119:265-271.

27.   Winter, S., Purac, A., Legget, F., Frison, E.A., Rossel, H.W., and Hamilton, R.I. 1992. Partial characterization and molecular cloning of a closterovirus from sweet potato infected with the sweet potato virus disease complex from Nigeria. Phytopathology 82:869-875.

28.   Winter, S., Purac, A., and Hamilton, R.I. 1992. Development of a nucleic acid hybridization assay for detection of a closterovirus associated with the sweet potato virus disease complex. Abstr. Phytopathology 82:1144.

CHAPTER 18

Seed Certification for Viruses

Y. Maury, C. Duby and R.K. Khetarpal

Seed certification for a crop is comprised of legal norms to be qualified for ensuring genetic identity, physical purity, germinability and freedom from seed-transmitted pathogens and weeds. A number of pathogenic fungi, viruses, viroids and bacteria are known to be transmitted through seeds. Virus seed transmission may play a key role in the epidemiology of the disease, i.e. viruses having no overwintering host plants such as soybean mosaic potyvirus (32) or viruses having no known vector such as barley stripe mosaic hordeivirus (67). In these cases, epidemic outbreaks of the virus disease are completely dependent on the primary virus inoculum brought by seeds at the beginning of the growing season. Moreover, even low rates of seed transmission in conjunction with secondary spread by vectors can result in the introduction of viruses into new areas which may lead to the development of disease epidemics. This can be illustrated by the introduction of peanut stripe virus to the U.S. (12,71).

With the increasing global seed trade and exchange, such possible introduction of seed-transmitted viruses needs to be carefully controlled. Therefore certification for seed-transmitted viruses, as a short term approach, will help in keeping the disease under control.

The objective of developing certification programs for seed-transmitted viruses is to ensure that the level of infection of a seed lot destined for growers is acceptable in a given agricultural context or that a given seed sample to be exploited for developing a crop variety is free from virus(es). Moreover the seed material under exchange needs to be certified as virus-free in order to minimize the risk associated with the introduction of exotic viruses or virulent strains to an area where they have not been previously reported.

The aim of the present chapter is to review certain aspects of seed transmission of viruses concerning certification, to analyze the present status of quality control of seed for viruses, and to discuss different perspectives in the frame of certification programs.

SEED TRANSMISSION OF VIRUSES

Aspects of Seed Transmission Related to Certification

According to Stace-Smith and Hamilton (66) about 18 percent of the described plant viruses are seed-transmitted in one or more hosts.

In this chapter these acronyms for virus names have been used, viz, AlMV (alfalfa mosaic virus), BBSV (broadbean stain comovirus), BBTMV (broadbean true mosaic comovirus), BCMV (bean common mosaic potyvirus), BSMV (barley stripe mosaic hordeivirus), BYMV (bean yellow mosaic potyvirus), CABMV (cowpea aphid borne mosaic potyvirus), CMV (cucumber mosaic cucumovirus), CMoV (cowpea mottle carmovirus), CYVV (clover yellow vein potyvirus), LMV (lettuce mosaic potyvirus), LALV (Lucerne australian latent nepovirus), MDMV (maize dwarf mosaic potyvirus), PCV (peanut clump furovirus), PeMoV (peanut mottle potyvirus), PEMV (pea enation mosaic virus), PLRV (potato leaf roll luteovirus), PNRSV (prunus necrotic ringspot ilarvirus), PSbMV (pea seedborne mosaic potyvirus), PStV (peanut stripe potyvirus), PVA (potato A potyvirus), PVM (potato M carlavirus), PVS (potato S carlavirus), PVX (potato X potexvirus), PVY (potato Y potyvirus), SBMV (southern bean mosaic sobemovirus), SCMV (subterranean clover mottle sobemovirus), SMV (soybean mosaic potyvirus), SqMV (squash mosaic comovirus), TMV (tobacco mosaic tobamovirus), TSV (tobacco streak ilarvirus) and ULCV (urdbean leaf crinkle virus).

Table 1. Number of seed-transmitted virus in different virus groups[a]

Number of virus	Virus groups
1	alfamovirus, capillovirus, fabavirus, machlovirus, enamovirus, bymovirus, rymovirus, tombusvirus, tospovirus
2	bromovirus, furovirus, hordeivirus, tobravirus
3	carmovirus, potexvirus, tobamovirus, tymovirus
4	carlavirus, cucumovirus, rhabdovirus, sobemovirus
7	comovirus
8	ilarvirus
18	nepovirus
20	potyvirus

[a] Additionally, there are 2 ungrouped and 7 unknown viruses reported to be seed transmitted

There are, however, varying reports regarding the number of viruses that are seed-transmitted. In order to avoid confusion in the literature, Mink (59), by excluding the viruses that appeared only once in the literature and also by avoiding synonyms, reported the total number of seed-transmitted viruses to be 108 belonging to 25 taxonomic group (Table 1). However, all the seed-transmitted cryptoviruses listed in the 5th ICTV report (14) will not be discussed in this chapter as they do not appear to have any economic significance. It is also worth noting that, so far, seed transmission has never been demonstrated for phloem-limited viruses: all the seed-transmitted viruses are thus mechanically transmissible.

The recent status of knowledge on the mechanism of seed transmission has been reviewed by Johansen et al. (33). It is now an established fact that seed-transmitted viruses are carried within the embryo. In an infected plant, the virus infects a large number of seed testas, as it infects extensively the maternal parenchymatous tissues. However, virus in mature testas does not play any role in seed transmission due to the impossibility for the virus to move to the embryo after the resorption of the suspensor (69). Exceptionally, mature testas infected by certain stable viruses (e.g. TMV) may be a source of inoculum for the seedling population. In the unusual mode of transmission of this virus through tomato and pepper seeds, after elimination of surface contaminants by heat or chemical treatments (10), an increase of soil infectivity can be observed during the decomposition of infected testas.

All of the trials of heat treatment for eradicating seed-transmitted viruses have failed or lead to a loss in seed viability. This implies that virus-free seeds may be available either from virus-free mother plants, particularly for germplasm collections, or through selection of virus-free seed lots through quality control.

Technical Approach for Assessing Seed Transmission Rates

In a seed-testing station generally many seed lots need to be tested and low rates of infection have to be detected in large samples. As a consequence of increased sensitivity of virus detection techniques, the methodology for large-scale testing can now be much simplified.

Biological assays, i.e. "grow-out" and infectivity tests require a long time for standardization and are laborious and time-consuming especially when working with bulk samples. These tests can now be used as independent controls of the results obtained by serological assays.

The enzyme-linked immunosorbent assay (ELISA) offers greater sensitivity of detection than earlier techniques, enables a large number of samples to be analyzed per day, and has become a technique of choice for large scale testing of seed-transmitted viruses (57). Recently, a new variant of ELISA, the tissue-blot immunoassay (TBIA), was found to be very useful in testing for seed-transmitted viruses in germinated seeds (45,48). It should, however, be noted that during large scale utilization of a serological technique, there is an inherent risk of selecting variants which may not react with specific antibodies. That is why periodical control of the ELISA results by biological indexing would reveal the emergence of variants which escape ELISA detection.

There is no doubt that the recent development of the polymerase chain reaction (PCR) technique and its adoption for detecting viruses in seeds has resulted in an immense gain in sensitivity of detection. Recently, using reverse transcription - PCR, certain viruses have been detected in seeds such as PSbMV in pea seeds (43), CMV in lupin seeds (70), BCMV in bean seeds (64) and AMV, BYMV, CYVV, CMV and SCMV in germplasm of subterranean clover and medic seeds (5). The use of PCR for seed testing of viruses on a routine basis would require extraction of RNA from a large number of seeds which is laborious, time consuming and inconvenient. Moreover, nucleic acid extraction of samples cannot be exploited for simultaneous detection of seed transmitted bacteria and fungi because the progress in the detection of these two types of pathogens is facilitated by bio-PCR, which detects the pathogen after enrichment on culture media. In this case, immunocapture - PCR (IC-PCR), a variant of PCR which utilizes antibodies to trap viral particles without prior RNA extraction, was presumed to facilitate the routine use of PCR for routine seed testing. However, A lower level of sensitivity and lack of reliability of IC-PCR has recently been observed for group testing of pea seed infected with PSbMV (61). Thus, this technique needs to be adapted to seed testing before it can be routinely used.

The present discussion is based on the application of the ELISA technique for virus detection in seeds.

Correlation Between Detection of Virus in Infected Embryos and Seed Transmission

Analyses of the following systems: BSMV/barley (47), LMV/lettuce (18), PeMoV/peanut (7), PSbMV/pea (26,56), SMV/soybean (9,46,53), and SqMV/cucurbits (60), have shown a large variation of virus concentration among infected embryos. The virus has even been found in the axis and not in the cotyledons of about 20% of the infected embryos of soybean and pea (50,53) and also in a proportion of peanut (71) and bean embryos (42).

However, such a low concentration is positively detected by ELISA when extracting the whole soybean embryos at the 200 w/v dilution and 500 w/v for detecting SMV and PSbMV in pea embryos (50,53).

Good correlations have been found between the percentage of ELISA-positive embryos in a given seed lot and the percentage of infected seedlings raised from the same lot for SMV/ soybean seed (54), PeMoV/peanut seed (7) and PSbMV/pea seed (56). Johansen et al. (33) consider that infectivity assays often indicate an absence of intact virions in cotyledons. If the virus is inactivated in cotyledons, ELISA-positive embryos where the virus is present in the cotyledons and not in the axis would be false-positive embryos. This distribution of virus in embryos has been found very rarely for SMV/soybean (55) and more frequently for BlCMV/cowpea seed (19). However, the data presented in the latter paper do not demonstrate that the BlCMV from cowpea cotyledons is not seed-transmitted.

Avoidance of Interference from Non-embryonic Tissues During Extraction

Since the presence of virus in the seed coat does not relate to virus transmission from seeds to seedlings, whole-seed serological assays are not suitable for estimating rates of seed transmission (33).

While preparing samples for determining the transmission rate in routine conditions on large number of seeds it is, therefore, necessary that the viral antigen from non-embryonic tissues (i.e. the seed coat in general) not be simultaneously extracted in order to prevent false-positive reactions. This introduces potential complications as manual processing of large numbers of seeds is poorly compatible with the routine.

Surprisingly, a lack of viral antigen in testas has been reported for mature peanut seed transmitting PStV (11,68) and PeMoV (1), as well as for LMV/lettuce seed (14). This could also be due to a failure to extract the virus from testas as demonstrated in the case of SMV/soybean (55), or a failure to detect the virus as shown for PSbMV/pea (50). In the latter case it was observed that in testas of infected pea seeds, the PSbMV capsid is partially cleaved during the maturation process. Such partial cleavage of the N- and C-terminal regions of the capsid is well known for potyviruses (3,31). As the N-terminal region of the capsid protein is the most immunogenic, early bleeding antisera detected PSbMV specifically in the embryos but not in testas, thus, enabling the use of whole seed for testing without prior decortication.

In other virus/sed systems, testas have to be removed in order to avoid the false-positive reactions and to correctly determine the seed transmission rate of a seed lot. In the case of BSMV/barley, the embryo has to be separated not only from testas but also from endosperm to avoid the false-positive reactions while detecting the seed transmitted virus (23,47).

Determination of Seed Transmission Rate

If the percent seed transmission of a seed lot is high, individually excised embryos can be tested by ELISA. However, if the percent transmission is assumed to be low, a large number of embryos need to be tested utilizing the group testing method. Due to the variation in virus titer in different embryos, it is not possible to correlate, as previously suggested for BSMV (47), an ELISA absorbance value for a sample of embryos with the number of infected embryos in the sample. Therefore, the group-testing method adapted for determining seed transmission rates consists of dividing a representative sample of the seed lot to be analyzed into N groups of n seeds, each made at random. So, N tests are done instead of $N \times n$ done in case of single embryo tests. The percentage p of transmission can be estimated as a function of the number of ELISA negative groups (Y):

$$ p = 1 - \left(\frac{Y}{N} \right)^{\frac{1}{n}} $$

and the 1-α confidence interval of p is:

$$ 1 - \left(\frac{Y + \frac{t_\alpha^2}{2} + t_\alpha \sqrt{Y\left(1 - \frac{Y}{N}\right) + \frac{t_\alpha^2}{4}}}{N + t_\alpha^2} \right)^{\frac{1}{n}} < p < 1 - \left(\frac{Y + \frac{t_\alpha^2}{2} - t_\alpha \sqrt{Y\left(1 - \frac{Y}{N}\right) + \frac{t_\alpha^2}{4}}}{N + t_\alpha^2} \right)^{\frac{1}{n}} $$

where t_α is the value read from a table of the standard Normal variable such that $Pr(Z > t_\alpha) = \alpha/2$; this value is 1.96 for a 1-α=0.95 confidence level. The magnitude of the confidence interval determines the precision of the method (55).

The number of seeds to be tested per group has a limit value depending on the dilution limit of the embryos having the lowest titer of virus. Therefore, with a given antiserum, a workable group size limit is determined, coherent with a probability Ps1=1 of detecting one infected embryo in such a group of seeds (55-57).

SEED CERTIFICATION FOR VIRUSES

Seed lots subjected to certification are either bulk material for direct sowing in the field or a small sample of germplasm for research or of breeding lines for multi-locational trials. The approach in developing a seed certification program against viruses would therefore depend on the size of the seed lot, its inherent vulnerability to harbor viruses and the purpose for which it is to be used.

In the frame of international exchange, according to the General Agreement on Tariff and Trade (GATT), any phytosanitary measure should be based on *Pest Risk Analysis*. The pest risk analysis enables each country to define its *Appropriate Level of Protection* by fixing a *tolerance limit* which refers to the inoculum threshold that can be tolerated in an agricultural context. For determining the inoculum threshold, it is worthwhile to collect detailed epidemiological information on the seed transmission rate, the degree of susceptibility of various cultivars, the date and intensity of transmission by vectors in relation to climatic conditions, the resulting percentage of infected plants in the field and yield reductions. All the collected data can be integrated into a dynamic simulation model. A model, flexible enough to be useful over a wide range of environmental and agroclimatic conditions, has been developed for SMV/soybean (32,63). This model predicted that in geographical areas where vector intensity is low before flowering a tolerance limit of 1% could be acceptable; in areas where the vector intensity is high in the early growing season, the tolerance limit should be 0.01% (i.e. 1 seed infected out of 10,000). Furthermore, in Northeast China it was found that 0.01% tolerance is needed in areas where vector population is high, whereas a tolerance of 0.05% is enough to control the disease near the sea-shore where vector population is low (22).

A classical example of the use of different tolerance limits in different geographical regions is the certification of lettuce for LMV. Stringent seed certification measures have been adopted due to the high economic value of the crop and the ability of the virus to provoke an epidemic through a very low level of primary inoculum in seeds. In Europe, an inoculum threshold of 0.1% was considered tolerable (49). However, in California lettuce crops seed lots have to index 0 in 30,000 (21). When tested in 60 groups of 500 seeds each, the infection rate of such seed lots is less than 0.012% at the 95% level of probability. It should be noted that when lettuce is infected, the market value is immediately reduced because the infected leaves repel the buyers. For this reason, the disease has gained a lot of attention with regards to certification for many years.

Setting a Tolerance Limit

The *tolerance limit* refers to the certification standard. It is the maximum limit of infection in a seed lot that can be tolerated in an agricultural context. In the case of domestic trade of commercial seed lots, the certification program is of interest when the host range of the seed-transmitted virus is narrow and when no significant reservoir of the virus exists in the proximity of the crop. In this case, the seed-transmitted virus can play a key role as a primary inoculum.

Such tolerance levels are inappropriate in the case of germplasm accessions and breeding lines where a *zero tolerance* is advocated, mainly because the carry-over of infection by these materials can jeopardize the crop improvement program. Such zero tolerance should theoretically be applied also to the international exchange of commercial seed lots when a destructive virus exists in the exporting country i.e. the country where the seed lot is produced and/or when the same is not known to occur in the importing country. Since it is not feasible to achieve zero tolerance while certifying bulk seed lots, it is in the larger interest to assess the risk associated with the possible introduction of a viral disease not known to occur in the importing country (35).

Evaluation of a Bulk Seed Lot with Respect to a Tolerance Limit

Is the infection of a given seed lot above or under the tolerance limit? This is the question that has to be answered in the frame of certification. The certification approach is not the same as for determining the seed transmission rate.

A statistical approach was developed by Geng *et al.* (17) that involves a *non tolerable* level of *infection* (**Int**) and a lower *tolerable* level (**It**).

The procedure also includes the following two probabilities of errors:

α = probability of rejecting a seed lot with **It** level

β = probability of accepting a seed lot with **Int** level

Recently, Masmoudi *et al.* (51) extended this approach to the use of group analysis for detecting PSBMV in pea seeds. The procedure consists in testing **k** groups of N seeds and the decision rule is: *the seed lot is rejected if at least one group of N seeds is found infected.*

The number **k** of groups to be tested can be calculated from the double inequality as given below :

$$\frac{\text{Log } \beta}{\text{Log}\left[1-\text{Ps}_1\left(1-e^{-N\text{Int}}\right)\right]} \leq k \leq \frac{\text{Log }(1-\alpha)}{\text{Log}\left[e^{-N\text{It}}+N\text{Ite}^{-N\text{It}}\left(1-\text{Ps}_1\right)\right]}$$

in which **Ps₁** is the probability of detection of one infected embryo in such groups of seeds.

This double inequality can be solved for **k** only if the higher bound is greater than the lower bound, which is possible if the difference between **It** and **Int** is large enough, according to the values of α, β and **N**.

Taking into account **Ps₁**, it is possible to increase the group size and reduce the number of groups. **Ps₁** values have to be experimentally determined. As an example, **Ps₁** values of 1, 0.81, 0.77 and 0.72 have been experimentally determined for 30, 100, 200 and 500 pea seeds per group, respectively, when using an antiserum to PSbMV. In this system where inoculum thresholds of 0.1% to 0.5% are considered, the number of groups to be tested is very low (37 and 7 groups of 200 seeds, respectively)(51).

It is important to underline this significant simplification in regard to the number of ELISA samples to be examined and the ease with which a seed lot can thus be certified by this approach.

QUARANTINE

Avoiding the movement of viruses along

with the transport of the infected seed material is the first and foremost step in excluding the establishment of a viral disease n a given geographical area. This calls for an efficient quarantine system especially for the importing country. However, certification of imported bulk quantities as completely virus-free in quarantine is difficult because such testing would require a lot of space, material and manpower. In this sense, a certain degree of flexibility in approach is required and the major objective should be to prohibit release of the material found to be infected with a devastating virus or virulent strains unknown in the importing country. The philosophy and methodology of exclusion of plant viruses through quarantine measures have been dealt in detail elsewhere (8) as well as in this book.

Certification for Germplasm and Breeding Material

Plant genetic resources are the key elements in any crop improvement program. Germplasm which represents a rich variability in plant genotype is collected from diverse agroclimatic zones. Theoretically in those geographical regions where sources of resistance in wild genotypes are found there is a strong possibility of the emergence of virulent isolates of the virus because of selective inoculum pressure. In the case of seed-transmitted viruses there is also the possibility of perpetuating such pathotypes through infected seeds of susceptible germplasm lines. Therefore it is essential to certify that seeds of genetic resources and breeding lines at different stages of utilization and exchange are free from viruses (30).

Germplasm Evaluation and Trials

Germplasm materials are generally multiplied and evaluated in the field. During the course of multiplication seed-transmitted virus(es) serve as primary source(s) of inocula and in the presence of insect vectors are transmitted to neighboring susceptible lines. Hence the virus is finally spread to different lines of the collection in which it becomes seed-borne. This is why a very large percentage of the germplasm collections from different regions are found to be infected (Table 2). Since germplasm is collected from diverse regions and is often exchanged internationally, considerable viral strain variation can be expected in these collections.

Since the genetic resources are the basic material used for breeding purposes, they may harbor a primary source of inoculum for the breeding lines. Therefore, it is necessary to ensure that foundation seed and breeders seed are also free from viruses so that they will not be perpetuated in newly developed varieties (28). Hamilton (25) gathered information on certification programs adopted in temperate and tropical agricultural regions with respect to PSbMV, BCMV, PMV, PSV, CAbMV, CMV and SMV for controlling diseases at the level of germplasm screening, seed increase and in the breeding lines. He found that there was a significant gap between the technical knowledge available on certification and what was being done in practice. Nowadays, with the tremendous advances in detection techniques, certification for freedom from seed-transmitted viruses should be more readily attained.

Germplasm Exchange and Quarantine

The increase in exchange of seeds of germplasm accessions that are likely to be contaminated by viruses has necessitated the need to certify them to be free from viruses through stringent quarantine measures. It may be noted that the quarantine measures to be applied for genetic resources are not the same as for commercial seed lots. The *tolerance limits* in these cases need to be *zero*, keeping in mind the possibility of introducing unknown viruses or virulent strains into a country. SMV, PSbMV and PStV are the classical examples of economically significant, seed-transmitted viruses of soybean, pea and peanut, respectively, which are known to have spread to different countries through seeds of infected germplasm (12,20,27).

Seeds of germplasm material are usually exchanged in small quantities. As quarantine processing of only a portion of the seed lot does not indicate the status of the untested portion, it is necessary to plant seed of an imported accession in isolation. Plant expressing suspicious symptoms should be rogued and only seeds from plants shown to be virus-free by appropriate tests should be harvested and released to breeders for use in crop improvement programs (34,38).

By adopting such a procedure a number of important seed-transmitted viruses viz, CABMV, SBMV, SMV and many others were intercepted in accessions of legumes imported into Australia (34). Similarly SMV, CMoV, CABMV, BYMV, BBSV, BCMV and PSbMV were intercepted in seeds of legume germplasm and breeding lines imported from different countries into India (39,40). The importance of interception can be highlighted by the fact that the economically damaging virus of cowpea, CMoV, is still not known to occur in India where cowpea is an important source of protein for a predominantly vegetarian population. Moreover, other intercepted viruses are known to possess different virulent strains.

This "grow-out" procedure is likely to help in detecting those viruses which are present at concentrations below the serological detection limit or which remain in the latent form in the early growth stage of the plant. Moreover this is the only procedure which can reveal seed infection by an unknown or uncharacterized virus through symptom expression. Finally, the chance of rejecting an accession is almost zero as it is very rare to observe transmission of a virus from all seeds of a lot.

Germplasm Conservation

Conservation of virus-free seeds of germplasm collections requires the same attention as is now being given to the conservation of vegetatively propagated crops *in vitro* cultures. Conserving virus-infected seeds in gene banks would not be worthwhile in the long run as conserving seed lots with a high percentage of virus infection is wasteful considering the time, money and energy spent on gene banks. Production of virus free seeds for conservation can be achieved simply by identifying virus free plants from which harvested seeds are to be used for conservation. The more rapid alternative which consists in testing a representative sample of the seed lots for presence of viruses before being conserved would not be fool proof.

Seeds of germplasm stored in a gene bank must also be periodically multiplied in the field for either replenishing the depleting stock or to circumvent the problems of seed viability. Such multiplication needs to be protected from vectors for ensuring freedom from seed-transmitted viruses.

PERSPECTIVES

A quality control of seed for viruses can always contribute to the short-term management of the diseases for which seed transmission plays a key role in the epidemiology. However, in the past, presumably due to the unavailability of a simplified procedure, this approach has been effective only for very few seed transmitted viruses such as LMV and BSMV causing serious losses on lettuce and barley, respectively. Virologists are generally more involved in resistance programs, which are expected to be the most economical method of virus control. However, a perusal of the literature revealed that resistance genes have been reported against only 30 of the 108 seed transmitted viruses known (R.K. Khetarpal, unpublished). Hence diseases caused by economically important viruses for which resistance genes have yet to be found (Table 3) or for which no resistant cultivar is available in an agroclimatic zone, can be kept under check in the short term via a seed certification program.

Above all, even after the development of resistant cultivars, two arguments may call for a complementary use if a preventative method of virus control at the seed level: either the preference for culinary qualities of local susceptible cultivars or the breakdown of resistance by the emergence of virulent strains. The BCMV/beans system is an example of both. The disease was considered to be of minor importance in the U.S. after the development of resistant cultivars carrying the *I* gene (62). However, severe epidemics due to the necrotic strain NL3 have recently occurred. This virulent strain has low seed transmissibility in several cultivars but is highly seed transmitted in some of them (65) which could immediately justify their control at the seed level. Presumably, for both reasons, this virus has been recently included in the Certified Seed Scheme of the South African Dry Bean Producers Organization (G. Pietersen, Pretoria, personal communication).

In the case of vegetatively propagated crops, such as potato, ornamentals and fruit trees, certification against viruses is often rigorously followed. This is because if the mother plant is infected with any virus, most of the propagules originating from it would also be infected. Under conditions where vectors are active, large scale multiplication of certified material is worthwhile because absence of virus can be monitored in a few mother plants using highly sensitive techniques. Following multiplication of this nuclear stock in the field where it can be partially recontaminated, seed lots can be classified for sale on the basis of the level of their seed infection. In France, for certification of potatoes against viruses (PLRV, PVA, PVM, PVS, PVX and PVY) the seed lots of tubers are classified into four classes

Table 2. Some examples of germplasm collections infected with seed transmitted viruses

Virus	Germplasm	Country	Reference
BCMV	*Phaseolus vulgaris*	U.S.A.	29,41
	Vigna mungo	India	2
CABMV	*V. unguiculata*	Iran	36
PSbMV	*Pisum sativum*	France, India	37,58
		U.S.A.	27
		U.K.	52
		Canada	24
		New Zealand	16
	Lens culinaris	U.S.A.	29
PeMoV	*Arachis hypogaea*	India (ICRISAT)	7
SMV	*Glycine max*	U.S.A.	20
ULCV	*Vigna radiata*	India	6
	V. mungo	India	a

[a] D.B. Parakh and R.K. Khetarpal, unpublished

Table 3. Examples of some important seed-transmitted viruses for which resistant cultivars are not known

Virus	Host
BLCMV	*Vigna unguiculata*
BBTMV	*Vicia faba*
BBSV	*Vicia faba*
CABMV	*Vigna unguiculata*
LALV	*Medicago sativa*
PCV	*Arachis hypogea*
PNRSV	*Prunus*
SBMV	*Phaseolus vulgaris*
SqMV	Cucurbits
TSV	*Glycine max*
ULCV	*Vigna radiata, V. mungo*

viz., Super Elite, Elite, A and B based on infection levels of 0.25%, 0.33%, 1% and 3%, respectively, for plants in field-testing and 1%, 2%, 3%, and 10%, respectively, for tubers in the laboratory-testing (4).

In the case of seed-transmitted viruses, since relatively few of the seeds collected from a diseased plant are infected, certification procedures in the past were cumbersome. They were limited to viruses having a drastic effect on yield like LMV/lettuce and BSMV/barley. For viruses having a moderate effect on yield, a certification has been developed in China for SMV and Healthy seed is produced in specific virus free localities (J.Q. Guo, Harbin, personal communication). However, perhaps due to low plant to seed transmission linked to late infection of soybean, such a certification has not been considered to be worthwhile in the U.S.

Nowadays, a very small number of tests is enough to conclude that infection of a seed lot is above or under the tolerance limit (a test on 7 groups of 200 seeds in the case of PSbMV/pea seed with respect to a 0.5% tolerance limit). Thus, quality control is easily applicable at the seed level for a number of crops such as grain legumes (peanut , cowpea, bean, faba bean, etc.) Which play a vital role as the main source of protein in a number of countries. These important crops are indeed known to be infected with a number of seed transmitted viruses which might not necessarily have serious economic importance, yet this approach would lead to a gain in productivity.

In both private and public seed production sectors, there is a growing need for reliable quality control methodologies. The International Seed Testing Association, sponsored by the member governments, and the recently created International Seed Health Initiative, sponsored by the private vegetable seed companies in France, Israel, Japan, The Netherlands and the U.S., have now decided to work jointly in this direction.

Seed producers often classify their seed after quality control into different classes as is done for potato tubers. Such a classification of seed could help for defining the appropriate inoculum threshold in a given agricultural context without investing in an extensive program of generating epidemiological data. Information collected for a few years on the incidence of the disease could be related to the virus input according to the class of seed sown. Further surveys could reveal whether or not a readjustment of the tolerance limit value is needed.

As a first step, the classification of seed seems to be more practical and has fewer inconveniences than certification: it would inform the grower about the level of primary inoculum in different classes of seed lots of the cultivar selected according to agronomic considerations. It would also allow the grower to decide whether or not to purchase a given class of an infected seed lot if virus-free ones are not available or are too costly. Seed classification would finally lead to the use of commercial seed lots within a defined tolerance.

When an appropriate tolerance has been defined in a country, quality control performed by seed firms on a voluntary basis is not satisfactory. The experience in Europe with the system LMV/lettuce indicates that a legal decision pertaining to certification should be made by the respective governments and 'super quality control' be performed at random on samples of seed lots sold in these countries. Such a legal decision would be relevant only if the importation of seed into these areas is done according to the

appropriate tolerance.

In case of genetic resources, especially those under exchange, the certification norms vary from country to country based on the presence or absence of the disease in a region and also on the importance attached to the danger in movement of different viral strains along with the seed material. Technical guidelines jointly issued by FAO and IPGRI (formerly IBPGR) (15) and the check-list of seed-transmitted viruses (44) are useful references for ensuring safe movement of germplasm material.

Finally, the certification for seed-transmitted viruses would be of relevance only if simultaneously it is ensured that commercial seed lots, germplasm and breeding materials are also free from other destructive seed-transmitted pathogens such as fungi and bacteria. This might call for an integrated certification program for seed-transmitted pathogens, especially in the case of crops such as bean, pea, soybean, etc. which are often simultaneously infected with different pathogens. Also, The efforts being made to simplify the seed certification approach would be worthwhile only if the importance of seed transmitted viruses in seed trade and exchange is properly appreciated by both the public and private sectors.

ACKNOWLEDGMENTS

The authors are extremely grateful to Dr. R.I. Hamilton (Pacific Agriculture Research Center, Vancouver, Canada), Dr. K.M. Makkouk (ICARDA, Aleppo, Syria) and Dr. I.M. Smith (EPPO, Paris) for their constructive criticism of the manuscript.

REFERENCES

1. Adams, D.B., and Kuhn, C.W. 1977. Seed transmission of peanut mottle virus. Phytopathology 67:1126-1129.
2. Agarwal, V.K., Nene, Y.L., Beniwal, S.P.S., and Verma, H.S. 1979. Transmission of bean common mosaic virus through urdbean (*Phaseolus mungo*) seeds. Seed Sci. & Technol. 7:103-108.
3. Allison, R.F., Dougherty, W.G., Parks, T.D., Willis, L., Johnston, R.E., Kelly, M., and Armstrong, F.B. 1985. Biochemical analysis of the capsid protein gene and capsid protein of tobacco etch virus: N-terminal amino acids are located on the virion's surface. Virology 147:309-316.
4. Anonymous. 1995. Procédures techniques et administratives relatives au plant de Pomme de terre, mise à jour du 26 Septembre, 1995. Norms set by Groupement National Interprofessionnel des Semences, Service Officiel de Contrôle, Paris,

France.

5.   Bariana, H.S., Shannon, A.L., Chu, P.W.G., and Waterhouse, P.M. 1994. Detection of five seed borne legume viruses in one sensitive multiplex polymerase chain reaction test. Phytopathology 84:1201-1205.

6.   Beniwal, S.P.S., Chaubey, S.N., and Bharathan, N. 1980. Presence of urdbean leaf crinkle virus in seeds of mungbean germplasm. Indian Phytopath. 33:360-361.

7.   Bharathan, N., Reddy, D.V.R., Rajeshwari, R., Murthy, V.K., and Rao, V.R. 1984. Screening peanut germplasm lines by enzyme-linked immunosorbent assay for seed transmission of peanut mottle virus. Plant Dis. 68:757-758.

8.   Bos, L. 1992. New plant virus problems in developing countries: a corollary of agricultural modernization. Adv. Vir. Res. 41:349-407.

9.   Bossennec, J.M., and Maury, Y. 1978 Use of the ELISA technique for the detection of soybean mosaic virus in soybean seeds. Ann. Phytopathol. 10:263-268.

10.   Broadbent, L.H. 1963. The epidemiology of tomato mosaic III. Cleaning virus from hands and tools. Ann. Appl. Biol. 52:225-232.

11.   Demski, J.W., and Warwick, D. 1986. Direct test of peanut seed for the detection of peanut stripe virus. Peanut Sci. 13:38-40.

12.   Demski, J.W., Reddy, D.V.R., Sowell, G. Jr., and Bays, D. 1984. Peanut stripe virus - a new seedborne potyvirus from China infecting groundnut (Arachis hypogea). Ann. Appl. Biol. 105:495-501.

13.   Falk, B.W., and Purcifull, D.E. 1983. Development and application of an ELISA test to index lettuce seeds for lettuce mosaic virus in Florida. Plant Dis. 67:413-416.

14.   Francki, R.I.B., Fauquet, C.M., Knudson, D.L., and Brown, F. 1991. Classification and nomenclature of viruses. 450 Pages. Rep. Int. Comm. Taxon. Viruses, 5th Arch. Virol. Suppl. 2.

15.   Frison, E.A., Bos, L., Hamilton, R.I., Mathur, S.B., and Taylor, J.D., eds. 1990. FAO/IBPGR Technical Guidelines for the Safe Movement of Legume Germplasm. Food and Agricultural Organization of the United Nations, Rome/International Board of Plant Genetic Resources, Rome.

16.   Fry, P.R., and Young, B.R. 1980. Pea seed-borne mosaic virus in New Zealand. Aust. Plant Pathol. 9:10-11.

17.   Geng, S., Campbell, R.N., Carter, M., and Hills, F.J. 1983. Quality control programs for seed borne pathogens. Plant Dis. 67:236-242.

18.   Ghabrial, S.A., Li, Debao, and Shepherd, R.J. 1982. Radio-immunoassays of bean common mosaic virus for detection of lettuce mosaic virus in lettuce seed. Plant Dis. 66:1037-1040.

19.   Gillaspie, Jr., A.G., Hopkins, M.S., and Pinnow, D.L. 1993. Relationship of cowpea seed part infection and seed transmission of blackeye cowpea mosaic potyvirus in cowpea. Plant Dis. 77:875-877.

20.   Goodman, R.M., and Oard, J.H. 1980. Seed transmission and yield losses in tropical soybeans infected by soybean mosaic virus. Plant Dis. 64:913-914.

21.   Grogan, R.G. 1983. Lettuce mosaic virus control by use of virus indexed seeds. Seed Sci. & Technol. 11:1043-1049.

22.   Guo, J.Q. 1992. The critical levels of seed borne incidence for controlling the damage to soybean caused by soybean mosaic virus. J. Northeast Agric. College 23:220-225.

23.   Hamilton, R.I. 1965. An embryo test for detecting seed-borne barley stripe mosaic virus in barley. Phytopathology 55:798-799.

24.   Hamilton, R.I. 1977. Detecting pea seed-borne mosaic virus. Can. Agric. 22:15-17.

25.   Hamilton, R.I. 1983. Certification schemes against seedborne viruses in leguminous hosts, present status and future areas for research and development. Seed Sci. & Technol. 11:1051-1062.

26.   Hamilton, R.I., and Nichols, C. 1978. Serological methods for detection of pea seed borne mosaic virus in leaves and seeds of Pisum sativum. Phytopathology 68:539-543.

27.   Hampton, R.O., and Braverman, S.W. 1979. Occurrence of pea seed borne mosaic virus and new virus immune germplasm in the plant introduction collection of Pisum sativum. Plant Dis. Rep. 63:95-99.

28.   Hampton, R.O., Waterworth, H., Goodman, R.M., and Lee, R. 1982. Importance of seed borne viruses in crop germplasm. Plant Dis. 66:977-978.

29.   Hampton, R.O. 1983. Seedborne viruses in crop germplasm resources: disease dissemination risks and germplasm reclaimation technology. Seed Sci. & Technol. 11:535-554.

30.   Hampton, R.O., Kraft, J.M., and Muehlbaeur, F.J. 1993. Minimizing the threat of seed borne pathogens in crop germplasm: elimination of pea seed borne mosaic virus from the USDA-ARS germplasm collection of Pisum sativum. Plant Dis. 77:220-224.

31.   Hiebert, E., Tremaine, J.H., and Ronald, W.P. 1984. The effect of limited proteolysis on the amino acid composition of five potyviruses and on the serological reaction and peptide map of the tobacco etch capsid protein. Phytopathology 74:411-416.

32.   Irwin, M.E., and Goodman, R.M. 1981. Ecology and control of soybean mosaic virus. Pages 182-215 in: Plant Diseases and Vectors, Ecology and Epidemiology. K. Maramorosch and K.F. Harris, eds. Academic Press, New York.

33.   Johansen, E., Edwards, M.C., and Hampton, R.O. 1994. Seed transmission of viruses: current perspectives. Annu. Rev. Phytopathol. 32:363-386.

34.   Jones, D.R. 1987. Seedborne diseases and the international transfer of plant genetic resources: an Australian Perspective. Seed Sci. & Technol. 15:765-776.

35.   Kahn, R.P. 1979. A concept of pest-risk analysis. EPPO Bull. 9:119-130.

36.   Kaiser, W.J., and Mossahebi, G.H. 1975. Studies with cowpea aphid-borne mosaic virus and its effect on cowpea in Iran. F.A.O. Plant Prot. Bull. 23:33-39.

37.   Khetarpal, R.K., Maury, Y., Cousin, R., Burghofer, A., and Varma, A. 1990. Studies on resistance of pea to pea seed borne mosaic virus and new pathotypes. Ann. Appl. Biol. 116:297-304.

38.   Khetarpal, R.K., Ram Nath, and Parakh, D.B. 1991. Quarantine procedures for the introduction of virus-free exotic plant germplasm in India. Page 257 in: Proc. International Conference of Virology in the Tropics, December 2-6, 1991, Lucknow, India.

39.   Khetarpal, R.K., Singh, S., Parakh, D.B., and Ram Nath. 1993. Interception of seed transmitted

viruses in imported legumes. Pages 33-34 in: Indian Society of Plant Genetic Resources - Dialogue 1993 on Plant Genetic Resources: Developing National Policy, N.B.P.G.R., New Delhi.

40. Khetarpal, R.K., Parakh, D.B., Singh, S., Ram Nath, Jain, R.K., and Varma, A. 1994. Bean common mosaic virus detected by DAC-Indirect ELISA in exotic *Phaseolus vulgaris*. Indian J. Virol. 10:13-16.

41. Klein, R.E., Wyatt, S.D., and Kaiser, W.J. 1988. Incidence of bean common mosaic virus in USDA *Phaseolus* germplasm collection. Plant Dis. 72:301-302.

42. Klein, R.E., Wyatt, S.D., Kaiser, W.J., and Mink, G.I. 1992. Comparative immunoassays of bean common mosaic virus in individual bean (*Phaseolus vulgaris*) seed and bulked bean seed sample. Plant Dis. 76:57-59.

43. Kohnen, P.D., Dougherty, W.G., and Hampton, R.O. 1992. Detection of pea seedborne mosaic potyvirus by sequence specific enzymatic amplification. J. Virol. Methods 37:253-258.

44. Kumar, C.A., Khetarpal, R.K., Parakh, D.B., Singh, S., and Ram Nath. 1994. Check-list on seed transmitted viruses: leguminous hosts. 14 Pages. Technical Bulletin of National Bureau of Plant Genetic Resources, New Delhi, India.

45. Lin, N.S., Hsu, Y.H., and Hsu, H.T. 1990. Immunological detection of plant viruses and a mycoplasma-like organism by direct tissue blotting on nitrocellulose membranes. Phytopathology 80:824-828.

46. Lister, R.M. 1978. Application of enzyme-linked immunosorbent assay for detecting viruses in soybean seed and plants. Phytopathology 68:1393-1400.

47. Lister, R.M., Caroll, T.W., and Zaske, S.K. 1981. Sensitive serological detection of barley stripe mosaic virus in barley seed. Plant Dis. 65:809-814.

48. Makkouk, K.M., Hsu, H.T., and Kumari, S.G. 1993. Detection of three plant viruses by dot-blot and tissue-blot immunoassay using chemiluminescent and chromogenic substrates. J. Phytopathol. 139:97-102.

49. Marrou, J., and Messiaen, C.M. 1967. The *Chenopodium* test: a critical method for detecting seed transmission of lettuce mosaic virus. Proceedings of the International Seed Testing Association 32:49-57.

50. Masmoudi, K., Suhas, M., Khetarpal, R.K., and Maury, Y. 1994. Specific serological detection of the transmissible virus in pea seed infected by pea seed borne mosaic virus. Phytopathology 84:756-760.

51. Masmoudi, K., Duby, C., Suhas, M., Guo, J.Q., Guyot, L., Olivier, V., Taylor, J., and Maury, Y. 1994. Quality control of pea seed for pea seed borne mosaic virus. Seed Sci. & Technol. 22:407-414.

52. Matthews, P., Dawson, J.R.O., and Hills, G.H. 1981. Pea seed-borne mosaic virus (PSbMV). Pages 31-32, John Innes Seventy-first Annual Report (1980).

53. Maury, Y., Bossennec, J.M., Boudazin, G., and Duby, C. 1983. The potential of ELISA in testing soybean seed for soybean mosaic virus. Seed Sci. & Technol. 11:491-503.

54. Maury, Y., Bossennec, J.M., and Vetten, H.J. 1984. Pages 32-38, Report of 18th International Seminar on Seed Pathology, 9-16 July, 1984, Washington, USA.

55. Maury, Y., Duby, C., Bossennec, J.M., and Boudazin, G. 1985. Group analysis using ELISA : determination of the level of transmission of soybean mosaic virus in soybean seed. Agronomie 5:405-415.

56. Maury, Y., Bossennec, J.M., Boudazin, G., Hampton, R.O., Pietersen, G., and Maguire, J.D. 1987. Factors influencing ELISA evaluation of transmission of pea seed borne mosaic virus in infected pea seed: seed group size and seed decortication. Agronomie 7:225-230.

57. Maury, Y., and Khetarpal, R.K. 1989. Testing seeds for viruses using ELISA. Pages 31-49 in: Perspectives in Phytopathology. V.P. Agnihotri, N. Singh, H.S. Chaubey, U.S. Singh and T.S. Dwivedi, eds. Today and Tomorrow Printers and Publishers, New Delhi, India.

58. Maury, Y., and Khetarpal, R.K. 1992. Pea seedborne mosaic virus. Pages 74-92 in: Plant Diseases of International Importance, Vol. II. H.S. Chaube, U.S. Singh, A.N. Mukhopadhyay and J. Kumar, eds. Prentice Hall, Inc., New Jersey, U.S.A.

59. Mink, G.I. 1993. Pollen and seed transmitted viruses and viroids. Annu. Rev. Phytopathol. 31:375-402.

60. Nolan, P.A., and Campbell, R.N. 1984. Squash mosaic virus detection in individual seeds and seed lots of cucurbits by enzyme-linked immunosorbent assay. Plant Dis. 68:971-975.

61. Phan, T.T.H., Khetarpal, R.K., Lee, T.A.H., and Maury, Y. 1996. Comparison of Immunocapture-PCR and ELISA in quality control of pea seed borne mosaic virus. In: Seed Health Testing Towards the 21st Century, 2nd ISTA-PDC Symposium, August 5-8, Cambridge, U.K.

62. Provvidenti, R. 1990. Reaction of some leading bean cultivars to African and indigenous strains of bean common mosaic virus. Annual Report of the Bean Improvement Co-operative 33:167-168.

63. Ruesink, W.G., and Irwin, M.E. 1986. Soybean mosaic virus epidemiology: a model and some implications. Pages 295-313 in: Plant Virus Epidemics - Monitoring, Modelling and Predicting Outbreaks. G.D. Mc Lean, R.G. Garrett and W.G. Ruesink, eds. Academic Press, U.S.A.

64. Saiz, M., Castro, S. Blas, C. De, and Romero, J. 1994. Serotype - specific detection of bean common mosaic potyvirus in bean leaf and seed tissue by enzyme amplification. J. Virol. Methods 50:145-154.

65. Spence, N.J., and Walkey, D.G.A. 1994. Bean common mosaic virus and related viruses in Africa. NRI Bulletin 63, Chatham, U.K. Natural Resources Institute.

66. Stace-Smith, R., and Hamilton, R.I. 1988. Inoculum thresholds of seedborne pathogens: viruses. Phytopathology 78:875-880.

67. Timian, R.G. 1974. The range of symbiosis of barley and barley stripe mosaic virus. Phytopathology 64:342-345.

68. Xu, Z., Chen, K., Zhang, Z., and Chen, J. 1991. Seed transmission of peanut stripe virus in peanut. Plant Dis. 75:723-726.

69. Wang, D., and Maule, A.J. 1992. Early embryo invasion as a determinant in pea of the seed transmission of pea seed borne mosaic virus. J. Gen.

Virol. 73:1615-1620.

70.   Wylie, S., Wilson, C.R., Jones, R.A.C., and Jones, M.G.K. 1993. A polymerase chain reaction assay for cucumber mosaic virus in lupin seeds. Aust. J. Agric. Res. 44:41-51.

71.   Zettler, F.W., Elliott, M.S., Purcifull, D.E., Mink, G.I., Gorbet, D.W., and Knauft, D.A. 1993. Production of peanut seed free of peanut stripe and peanut mottle viruses in Florida. Plant Dis. 77:747-749.

CHAPTER 19

Control of Viruses Affecting Potatoes Through Seed Potato Certification Programs

S.A. Slack and R.P. Singh

Seed potato certification programs have been the fundamental management tool for the control of tuber-borne virus diseases throughout the 20th century. These programs have been instrumental in the identification of clonal seed stocks, which are free or relatively free of tuber-borne pathogens and in the identification and verification of cultivars. Further, they have put into place an exquisite monitoring system for detecting the onset of disease epidemics and the shift in pathogen populations which may affect disease incidence and severity. These programs were largely established to address the concern of "running out" or "degeneration" of clonal seed stocks. We now know that running out or degeneration problems were largely caused by virus infections. This chapter will address those viruses that have been significant to the potato industry and that are controlled through seed potato certification programs.

EARLY HISTORY AND PRINCIPLES

In the late 1800's it was known that potato cultivars would rapidly degenerate or run out after several years of propagation. By 1900 Dutch and German agriculturalists had found that symptoms of degeneration, such as leaf curling, rolling, crinkling and blotching, were transmissible from plant-to-plant and from generation-to-generation through tubers (79). They established that the selection of healthy, vigorous plants in a population for replanting would eliminate or minimize this degeneration effect. This experimental evidence helped dispel speculation that the cause of degeneration was due to the fact that potatoes were propagated through vegetative tubers, rather than through sexually generated seeds.

The system for inspection and production of potato seed stocks established in Germany was studied and copied in the United States and Canada. Initial steps in the United States included the passage of the National Plant Quarantine Act of 1912, which prohibited the introduction of potatoes with black wart, a disease caused by the fungus *Synchytrium endobioticum*, which recognized the fact that pests and disease organisms were being distributed throughout the world in association with seed stocks. In 1913, the maritime provinces of Canada and five states in the United States established seed potato certification programs and, in 1914, the Potato Association of America held its first annual meeting. These rapid series of events provided considerable structure to the fledgling seed potato industry and helped to validate its contributions and significance to the entire commercial potato industry. The formal seed industry in North America is now over 80 years old and is concentrated in 16 of the northern most states in the United States and in 10 Canadian provinces. Approximately 270,000 acres of seed potatoes are certified annually (approximately 8% of the total acreage). Although over 100 cultivars are certified annually, 80% of the acreage is comprised of only 10 cultivars.

The participation of growers in formal seed potato certification programs is voluntary, however, participation includes an agreement to abide by standards covering eligibility of seed lots and the maintenance of those cultivars within specified tolerances in order to receive certification (1,2,4,45,79). Seed certification agencies are legally recognized and seed lots can only be labeled "certified" by these agencies. Tolerances are established for the incidence of tuber-borne diseases, for purity, and for the physical condition and quality of the tubers to be shipped. These standards are upheld through the visual inspection process of growing plants and harvested tubers by trained inspectors.

VIRUSES OF POTATO AND METHODS FOR DETECTION

Viruses of Potato

Virus diseases were suspected to be the cause of "potato degeneration" long before the viral etiology was known. Potato leafroll disease under the name "leaf curl" is purported to have caused epidemics as early as 1784 or 1812, more than a century earlier than the identification of the virus itself (6). Potatoes are susceptible to over twenty-five different virus or virus-like diseases (Table 1) throughout the world (10,57). In North America, five viruses (leafroll, virus A, virus S, virus X and virus Y) and potato spindle tuber viroid are of primary concern and have received attention in potato seed certification programs. Potato virus M (PVM) has been rarely encountered and tobacco rattle virus has been restricted to only certain areas of some states in USA and a few backyard gardens in Alberta, Canada   Potato, being a vegetatively propagated crop presents ample potential for the infection and increase in incidence of viruses with every year or generation of field growth, and multiple viral infections can be a common occurrence.   Therefore, virus identification based on visual inspection needs to be substantiated by diagnostics, especially for new cultivars for which symptom expression following viral infection needs substantiation.

Virus Indexing And Potato Seed Certification

In order to prevent virus spread from one area to another area which is free from the virus, indexing of the planting material for virus freedom is an essential step. For viruses, which are not vector-transmitted, this step alone could delay indefinitely the introduction of viruses into new areas. The history of the development of potato virus indexing is a tour-de-force of the progress and the sophistication of the potato seed certification system itself (Table 2). As early as 1908, in order to reduce the "degeneration" of the potato seed crop, the method of "tuber-unit planting" was introduced (85) and adopted to reduce virus levels in potato seed crops (33). It is

still practiced for early generation seed stocks in Canada and the USA (1,2). In tuber-unit planting, all seed pieces of a tuber are planted consecutively in a row and are separated from the adjoining unit by a skip or space. Observation of virus symptoms in any plant of a unit necessitates removal of the entire unit from the seed plot. This practice minimizes the inoculum sources for vectors in the growing crop and the incidence of virus-infected tubers for the subsequent crop. However, some virus-plant combinations exhibit no visible symptoms or only very vague symptoms, which can make detection problematic.

"Tuber-indexing" was added in 1921 (11) as an improvement over the tuber-unit planting. In "tuber-indexing", an eye is removed from a tuber and planted in a greenhouse during winter months. Observation of virus symptom in plants requires removal of the mother tuber from the planting stock. The advantage of the tuber-indexing method lies in the elimination of primary virus sources, which minimizes virus spread by vectors. However, it has the drawback of missing infected tubers when all eyes are not infected and the limitation of space for growing plants (53). A practical modification of tuber-indexing method widely practiced in North America is the "Florida-test" or "Southern-test". In this test, small seed tubers (30-60 g), usually 400 per seed lot, are collected soon after harvest. Tuber dormancy is artificially broken and they are planted either in southern Florida or California, where the winter climate allows for field production of potato plants. Visual inspection is the primary method for detection of infection. Readings for virus disease incidence are made in late January for Florida and late February for California. In both cases plants are inspected approximately 60 days after emergence (1,4). This test has been used effectively for over 50 years to minimize disease problems caused by planting stocks with a high incidence of virus infection.   However, some of the currently popular potato cultivars do not show symptoms or only show transient symptoms when infected with PVY, a factor which can compromise the effectiveness of this test.

Virus Indexing Tests

Callose Test

Among the early dormant tuber tests is the test for PLRV detection. Testing dormant tubers is important because it enables late-season virus

Table 1. Viruses infecting potato: their abbreviation, group and means of spread

Virus name	Abbreviation	Group	Vector/spread
Alfalfa mosaic virus	AlMV	AlMV group	Aphids
Andean potato latent virus	APLV	Tymovirus	Beetles
Andean potato mottle virus	APMV	Comovirus	Beetles
Beet curly top virus	BCTV	Geminivirus	Leafhopper
Cucumber mosaic virus	CMV	Cucumovirus	Aphids
Potato acuba mosaic virus	PAMV	Potexvirus	Aphids
Potato deforming mosaic virus	PDMV	Geminivirus(?)	Whitefly(?)
Potato leafroll virus	PLRV	Luteovirus	Aphids
Potato moptop virus	PMTV	Tobamovirus	Fungi
Potato virus A	PVA	Potyvirus	Aphids
Potato virus M	PVM	Carlavirus	Aphids/ mechanical
Potato virus S	PVS	Carlavirus	Aphids/ mechanical
Potato virus T	PVT	Capilloviruses	True seed
Potato virus V	PVV	Potyvirus	Aphids
Potato virus X	PVX	Potexvirus	Mechanical
Potato virus Y	PVY	Potyvirus	Aphids
Potato spindle tuber viroid	PSTVd	PSTVd group	Mechanical/ true seed
Potato yellow dwarf virus	PYDV	Rhabdovirus	Leafhopper
Potato yellow vein virus	PYVV	?	Whiteflies
Solanum leaf curling virus	SALCV	Geminivirus	?
Tobacco mosaic virus	TMV	Tobamovirus	Mechanical
Tobacco necrosis virus	TNV	Necrovirus	Nematodes
Tobacco rattle virus	TRV	Tobravirus	Nematodes
Tobacco ringspot virus	TRSV	Nepovirus	Nematodes
Tobacco streak virus	TSV	Glarvirus	Pollen/thrips
Tomato black ring virus	TBRV	Nepovirus	Nematodes
Tomato spotted wilt virus	TSWV	Tospovirus	Thrips

Table 2. A timeline for the introduction of test methods for the detection of viruses in potato tubers

Year	Method	Viruses
Growing of Tubers		
1908	Tuber unit planting	LR, mosaic
1921	Tuber indexing	LR, mosaic
1925	Indicator plants	PVX
Cytological		
1955	Callose test	PLRV
Serological		
1934	Serological reaction	PVX
1964	Bentorite flucculation test	PVS, PVM, PVX
1967	Latex agglutination test	PVS, PVM, PVX
1979	Immunosorbent electron microscopy	PLRV
1980	Enzyme-linked immunosorbent assay	PVS, PVM, PVY
1992	Squash-blot immunoassay	PVX, PVY
Nucleic acid based		
1990	Nucleic acid spot hybridization	PVY
1993	Polymerase chain reaction	PVY, PLRV

infections to be detected prior to shipping seed lots. However, virus titer in dormant tubers is less than in growing plants and, therefore, requires sensitive, reliable techniques for detection. Potato tubers infected with PLRV contain large quantities of callose in their phloem (sieve tubes) (13). After staining callose with resorcin blue, it has been possible to distinguish between diseased and healthy tubers (5,38). For reliable results, tubers need to be stored for at least 4 weeks at 10° to 20°C before testing (19). Besides PLRV, the stolbur disease (phytoplasma) also stimulates abnormal callose production (86). However, cultivars vary in callose deposition and some cultivars can not be reliably tested, especially when PLRV infection takes place late in the season.

Indicator Plants

Johnson (39) was the first to report that tuber sap can be used to detect potato virus X. In early studies, tuber macerates were found impractical for PVX detection (15), except for tuber macerates from vascular bundles (17). For large-scale testing tubers were abraded on a coarse piece of sandpaper and the exposed surface was rubbed onto the indicator host (7).

Use of a local lesion host, *Gomphrena globosa*, provided the opportunity to make quantitative observations, and the highest virus concentrations were found in sprout extracts (87). Using this host plant, it was shown that PVX can be detected from various cultivars throughout the year (36).

Besides PVX, tuber sap or cut tuber pieces have been used to detect PVY and PVA (20,68). Both local lesion hosts (20,68) and systemic hosts (43) have been used. Tuber slices or nucleic acid extracts can also be used to detect potato spindle tuber viroid (PSTVd) using *Solanum berthaultii* (69) Specific techniques that have been used to enhance detection include wounding of tubers by cutting twice at four week intervals to increase the detection of PVY^N (20) and using detached leaves of *Physalis angulata* caused PVA lesions to develop several days sooner than on intact leaves (68). However, indicator plants require greenhouse facilities with special environmental conditions for each virus-host combination, a situation which is not amenable to large-scale testing and, therefore, has been limited to testing early generation planting material.

Serological Methods

The first serological tests for virus detection from dormant tubers were those of Gratia and Manil (28) who observed a flocculation reaction with PVX-antiserum and sap from a PVX infected tuber. Later, Scott *et al.* (62) employed bentonite to make the flocculation effect more visible and identified PVX from bud-end extracts of tubers which were not detected by the microprecipitin test (83). This method was extended to PVS, PVM and PVY in later studies (42), however, reliable detection of PVY was not possible.

The serious application of serological tests to systematically detect potato viruses in North American seed lots was pioneered by Shepard (64) and Wright (88) and was applied to detect the latent viruses PVM, PVS and PVX. Initially, the Ouchterlony double-diffusion test was used (63). Several Western states carried out summer testing of leaves for PVX and marked the seed tags as "PVX-tested" (18).

Latex Agglutination Test (LAT)

Latex agglutination test increases virus detectability by adsorbing antibodies onto latex beads. The procedure is fast (<15 min) and consists of only two steps. Test samples and antibody-sensitized latex preparations are mixed in capillary tubes or on microprecipitin plates and shaken vigorously. However, LAT is influenced by virus concentration and more than one antigen concentration is required for tests. It has been shown that LAT is over 1000X more sensitive than double-diffusion tests for Andean potato latent and Andean potato mottle viruses in potato tubers (46) and 500X to 1000X more sensitive than microprecipitin tests (27). LAT has also been investigated for the detection of PVS and PVX in several cultivars (32,44). The sensitivity increase

over microprecipitin tests for elongated viruses like PVS, PVX and PVY is about 100X, however, from tubers LAT underestimates APLV, APMV, PVS, PVX and PVY (27).

Immunosorbent Electron Microscopy

Immunosorbent electron microscopy (ISEM) involves incubating grids coated with carbon films with antibody at 37°C, followed by incubation of coated grids with virus-containing samples at 4°C. ISEM was used to detect PLRV and potato mop-top (PMTV) virus from tuber extracts of different cultivars (54). It was shown to be 1000X more sensitive than conventional methods of electron microscopy.

Enzyme-linked Immunosorbent Assay (ELISA)

The development of ELISA tests for plant virus detection in the late 1970's (16,84) created the first really feasible large-scale application of serological methods for virus detection, including dormant tubers. Immediate attempts were made to apply ELISA to the detection of potato viruses in dormant tubers (21,22,23). It was successful in detecting PVS and PVM, but virus concentration for both declined slightly after 8 weeks tuber storage. No differences were found between bud- and stem-end pieces or storage at 4°C or 20°C (23). For PVX, tubers stored at 4°C for 38-39 weeks could be tested, however, if storage at 4°C was followed by one week of storage at 20°C, detectable virus increased considerably (22). The ELISA procedure has been shown to detect both PVS and PVX in tubers immediately after harvest but it has not been as effective as testing foliage from growing plants. Attempts to detect PVY in dormant tubers have not been satisfactory (21), and artificial breaking of dormancy followed by incubation at room temperature for a few weeks is required before tubers can be tested for PVA and PVY (30,73)

ELISA indexing to supplement visual field inspections, especially for poorly symptomatic cultivars, is practiced in many countries. In the Netherlands alone, about 5×10^6 ELISA tests are performed annually on seed potatoes (37). Broad scale evaluation of antibodies, especially monoclonal antibodies for specificity range in detection of virus strains will help future survey efforts (41).

Squash Blot Immunoassay (SBIA)

A simplified modification of ELISA in which tuber tissues are squash-blotted onto a nylon membrane and processed similar to an ELISA was developed for direct use in the field or in less sophisticated laboratories (12). Squashing samples requires minimal training and can easily be performed by individual farmers prior to shipping to distant laboratories for further processing. Both PVX and PVY have been detected from tubers by SBIA. No detailed large-scale evaluations using tubers, however, have been reported using SBIA.

Nucleic Acid Methods

With the introduction of recombinant DNA technology in the 1970's, the prospect of utilizing the viral genome, not just the coat protein, for the detection of viruses became feasible. Inherent in this approach were the possibilities of using multiple genome segments for making specific or general detection probes, particularly for those pathogens which do not possess a protein coat. The first successful application of a nucleic acid based-method with solid supports applicable to large-scale testing were made for PSTVd by Owens and Diener (52) using sprouted tubers. Earlier studies utilized radioactive labels, e.g. ^{32}P (47, 52), however, non-radioactive molecular probes for PSTVd and other potato pathogens were described (35,49,55). Sensitivity of detection with non-radioactive probes (eg, biotin, photobiotin and digoxigenin labels) have been reported to be as sensitive as radioactive probes (49,55).

Nucleic acid spot hybridization (NASH) has been applied to dormant tubers for the detection of PSTVd using digoxigenin-labelled probes (77). Both cDNA and cRNA (complementary DNA or RNA) probes have been used, however, cRNA probes are more sensitive than cDNA probes (58,77). In the case of PSTVd, multimeric PSTVd forms for test probes were more sensitive than monomeric probes with tetrameric to hexameric multimeric cRNA probes detecting PSTVd in tuber extracts diluted 1:16384 (77).

For the detection of PVY, NASH was evaluated recently (48,51) with nucleic acid extracts suitable for the detection of PVY in leaves but not in dormant tubers (65). However, PVY can be detected in freshly harvested, greenhouse grown tubers of several potato

cultivars if care is taken to maximize nucleic acid extractions. The sensitivity of PVY detection using purified viral RNA has been down to 10-100 pg, detection in tuber extracts has been to a dilution of 1:64, and detection from potato leaves has been to a dilution of 1:1024. Nucleic acids stored at 4°C, -20°C or at -70°C remain suitable for hybridization for over 2 weeks (65). However, when this method has applied to field-grown tubers, PVY has not been detected from certain cultivars and enhancement of the nucleic acid extraction by using DNAse I and proteinase K has been required (66). Further improvements in the sensitivity of PVY detection is possible by using longer cDNA probes as shown for PVY^N and PSTVd (24,77).

Reverse Transcription Polymerase Chain Reaction (RT-PCR)

Reverse transcription polymerase chain reaction (RT-PCR) has been shown to detect PVY and PLRV from dormant tubers and minitubers (8,31,82), however, PVY detection from tubers has been found to be less reliable than by ELISA (8). The RT-PCR method is susceptible to substances present in potato tubers, probably polyphenolics, and; therefore, dilution of nucleic acids or addition of some antioxidant chemicals prior to cDNA synthesis is needed to improve the detectability (66). Application of RT-PCR methods to field-grown tubers of eight potato cultivars and their comparison with plant symptoms or ELISA carried out on plant leaves have shown 95-97% correlation (67).

With the recent progress in developing the nucleic acid-based detection procedures, particularly for dormant tubers, it is now becoming realistic to reliably index potato tubers for the most important viruses prior to planting. This is an approach to post-harvest testing which will be examined closely for economic and routine feasibility over the next few years and which will be compared to the standard winter grow-out tests.

INTEGRATION OF TECHNOLOGY AND PRODUCTION

The utilization of thermotherapy and meristem-tip culture to produce virus-free potato cultivars began in earnest in the late 1960's in North America (50,78). A collection of these virus-free cultivars was established and maintained in Vancouver, British Colombia, Canada and was

Table 3. Limited generation schemes in the United States and Canada

United States seed class	Canadian seed grade	Description
Pre-Nuclear	Pre-Nuclear	*In vitro* stocks maintained with test history for specific pathogens; also for minitubers produced in greenhouses or other protected environments
Nuclear	Pre-Elite	Tubers harvested from first field year; this designation may also be used to minitubers from greenhouses or other protected environments.
Generation 1	Elite I	Tubers from second field year
Generation 2	Elite II	Tubers from third field year
Generation 3	Elite III	Tubers from fourth field year
Generation 4	Elite IV	Tubers from fifth field year
Generation 5	Foundation	Tubers from sixth field year
Generation 6	Certified	Tubers from seventh field year

Recognition of the value of these virus-free stocks led to the adoption of limited generation programs (Table 3) (79).

Two principles are common to all the limited generation systems instituted in North America. Firstly, all seed potato stocks must originate from an in vitro pathogen-tested source. All initial explants must be tested for the major viruses (PVA, PVM, PVS, PVX, PVY, and PLRV), the bacteria *Clavibacter michiganensis* subsp. *sepedonicus* and *Erwinia carotovora*, as well as general microbial contamination (78,79). ELISA has been the standard test for viruses, but nucleic acid-based tests are being evaluated and may be utilized in the future. Potato spindle tuber viroid has been virtually eliminated through the systematic application of nucleic acid hybridization and return gel electrophoresis assays (52,70). In addition, the incidence of lots affected by bacterial ring rot disease or soft rot has also decreased dramatically (4,18,80). The second element common to all programs is that they limit the number of generations that these initial lots can be increased. This stipulation recognizes the fact that these cultivars are still susceptible to infection if inoculated and that multiple generations of increase increase the probability that reinfection will occur. In addition some programs have instituted stricter tolerances for virus disease incidence in the early generation seed lots. Lots exceeding these tolerances are down graded to the next acceptible seed class or are rejected outright from the certification process.

As emphasis on early generation seed stocks has increased, we have seen an evolution in methods to accommodate needs. For example, it has been shown that the efficiency and effectiveness of virus elimination from cultivars can be enhanced by utilizing *in vitro* plantlets, rather than pot grown plants (29,60,81). Advantages include enhanced sanitation, juvenile plant grown which is easier to handle for meristem excision, a requirement for less space and the ability to handle more plants and the ability to control environmental parameters such that treatment stringency can be enhanced. It has been shown that the combined use of heat therapy and chemotherapy on *in vitro* plantlets enables virus-free plants to be obtained from modified nodal cuttings, rather than the smaller meristem tips, which enables the regeneration of plants to be accomplished in 4-6 weeks, rather than 3-6 months (60). In addition, *in vitro* plantlets are increasingly being field transplanted in order to maximize the production capacity of these propagules. This process is being driven both by the need for generation stocks, as well as the recent introduction of genetically altered cultivars with desired resistance traits (14,25,40).

EXAMPLES OF VIRUS DISEASE CONTROL

Potato Virus X

PVX was first known as the healthy potato virus, because it was demonstrated that sap from apparently healthy potato plants caused mosaic symptoms when inoculated to healthy tobacco plants (39). Subsequently, it was found that most potato stocks in North America were uniformly infected with PVX. Severe strains of PVX were eliminated from seed lots through stringent roguing practices, however, mild strains which did not cause discernible symptoms in cultivars persisted. The implementation of thermotherapy and meristem-tip culture made it possible to eliminate PVX from these stocks. By 1980 PVX-free stocks of all major cultivars were available and advanced clones were increasingly being tested and put through therapy procedures to eliminate PVX before introduction into the seed systems. Although there was some resistance to the elimination of PVX from some standard cultivars on the basis that PVX infection conferred cross resistance to other pathogens, these stocks were rapidly phased out of seed systems when it was found that newer cultivars were sensitive to infection with PVX and could rapidly be contaminated by close proximity to these standard cultivars. Although PVX can still be found in seed stocks in North America, frequency of infected lots and the incidence in infected lots has decreased dramatically (18,32). This relatively simple approach to PVX control is possible because PVX does not have an insect vector. If initial seed stocks are free from PVX, continued freedom from PVX relies only on strict sanitation measures on the seed potato farm. Sanitation includes the cleaning of facilities and machinery, as well as strict attention to the order in which seed lots are handled (i.e., early generation seed lots should always be handled before later generation seed lots). The benefits of a good sanitation program certainly extend beyond the impact of reduced incidence of PVX in seed lots as sanitation is a key element in seed certification programs for the management of most tuber-borne pathogens. A number of states in the western USA currently test growing plants for PVX and include "PVX-tested" on seed tags or lots meeting PVX testing tolerances, which is an indicator of seed lot quality.

Potato Spindle Tuber Viroid (PSTVd)

The strategy for PSTVd eradication in Canada was to reduce the inoculum and maintain proper hygiene around the farm. In absence of aerial vectors, PSTVd spread should be minimal. In Canada spindle tuber disease was present in table stock fields at a rate of 4% in 1969-70 in the province of New Brunswick and, in some fields, an infection rate as high as 15% was not uncommon (74). However, a series of improvements in seed potato production systems, in general, and some specifically to individual provinces, reduced the incidence of PSTVd to zero by 1979-80 (72, 75). These improvements included: the development and application of rapid and sensitive test methods; establishment of an "Elite" seed potato production farm to supply initial seed potato stocks to commercial seed growers free from PSTVd and viruses; changes in seed certification regulations to limited generations of potato stocks; adoption of a "Zero tolerance" for PSTVd in seed certification (i.e., preventing the planting of seed potatoes with even trace amount of PSTVd); testing of potato breeding stocks for PSTVd prior to their release to seed growers; enactment of provincial disease eradication "Acts" aimed at PSTVd eradication; and increased emphasis by potato processing companies for their contract growers to plant earlier generation seed classes rather than the minimum "certified" class. In order to reduce PSTVd introduction onto commercial seed farms through newly released cultivars, potato breeders instituted testing for PSTVd of all parental material prior to their use in making crosses. Most of these changes have been common in Canada and USA. As a result PSTVd has been eliminated from seed production areas across the entire region and has not been considered a production risk for more than a decade. In order to ensure continued freedom from PSTVd, leaf and tuber samples from a large number of fields continue to be tested every year and all imported potato seeds must also be thoroughly tested. Canada uses return-polyacrylamide gel electrophoresis for this purpose (70), which is specific to circular RNA molecules.

PSTVd is carried through the pollen and seeds of potato, including wild Solanum species, and can survive in true potato seeds for over 21 years (76). Therefore, possibility of its presence in potato germplasm collections exists. PSTVd has been reported to occur in the tuber-bearing Solanum species of major potato germplasm collections (34). Therefore, the increased dispersal of Solanum species and the long persistence of viroid in *Solanum* true seeds constitutes a continued worldwide risk. Although prompt actions and use of sensitive detection methods have eliminated PSTVd from most germplasm collections (34,56,61) continued vigilance is needed to keep PSTVd out of potato crop in view of recent reports of mixed viroid-virus infections. For example, a high incidence of aphid transmision of PSTVd has been observed in association with PLRV (59), and a new sequence variant of PSTVd from wild *Solanum* species has been found in association with a geminivirus in Australia (9).

Tobacco Veinal Necrosis Strain of PVY (PVY^N)

PVY^N was reported as a recently introduced pest in Canada, and an eradication effort for PVY^N was established (71). The infection level ranged from 0.43 to 1.57% in seedlots in 1990. Since Canada maintains strict records of seed lot multiplication, the system of potato seed tag numbering made it possible to trace back and remove from planting purposes any of the likely sources of infection. In addition, all seedlots found positive in tests (winter or summer), their sister lots and those seedlots in close proximity (buffer lots) were banned for planting. This approach resulted in drastically reduced inoculum potential for 1991 crop. In order to reduce the inoculum further, steps were taken to destroy volunteer potatoes from PVY^N positive fields, and the growing of uncertified potaoes in private gardens was prohibited on Prince Edward Island (3).

The second basic premise for disease eradication was to generate data which would allow proper timing and application of control measures, particularly for a disease like PVY^N. Although infected tubers provide the initial inoculum, these infection foci are few and far apart. Therefore, 4.7 million leaves encompassing 4,000 fields were tested in the 1991 growing season. Only 0.58% of the samples were found positive for PVY^N, which showed a trend to reduced inoculum within a year. The follow up winter and summer testing with 1,000 to 5,000 leaves per field found only occasional PVY^N positive samples. Routine testing of class seeds has failed to detect any case of PVY^N infection since 1993, which suggests that the eradication program has been effective. A similar situation was encountered in 1985 in New Zealand (26). PVY^N was detected in 9% of potato samples in 1985. Three years later no PVY^N was detected in any sample.

SUMMARY AND FUTURE DIRECTIONS

After 80 years seed potato certification programs are still the central management tool for maintaining and producing productive seed stocks. As technology has advanced the utilization of that technology within seed potato programs has also advanced. Knowledge of the ecology and epidemiology of potato viruses enabled seed producers to develop strategies to minimize virus spread within and among seed lots (for example, isolation of seed lots and/or production areas, monitoring aphid populations in seed production areas, and early season vine dessication to avoid rapid spread of viruses in stocks). Further, the systematic adoption of *in vitro* technology coupled with pathogen test procedures, particularly ELISA, has provided the North American industry with the best quality initial seed stocks ever available to the potato industry. The adoption of limited generation systems has minimized reinfection of these seed lots, and has led to the routine certification of seed lots (> 95% of seed lots in most programs now pass certification).

The future as well as the future direction of seed potato certification programs will be interesting. We are now beginning to see the rapid introduction of cultivars which have been genetically altered in order to exhibit a specific physiological trait. Among these traits is virus resistance and cultivars with coat protein and replicase-mediated resistance are now being increased in seed potato programs. Although the impact of transgenic cultivars on potato production generally remains to be seen, they certainly will have an impact on the certification process. In some ways they will increase the need for the certification process as the documentation surrounding these genetically altered cu tivars will become increasingly important because many are proprietary. Since identification of these modified cultivars is not possible morphologically, the initial verification of such cultivars entering programs and the monitoring of these lots through programs will become extremely important. In addition, monitoring the incidence or change in incidence of viral infection within seed lots will become important because seed potato inspectors will be in a unique position to monitor such changes. Interestingly, despite the enhancement in laboratory testing procedures, this latter issue clearly recognizes that a holistic assessment of the plant health status of seed stocks will always require a visual assessment component. Since current expression of these resistances is analogous to single gene resistance, a breakdown of resistance may occur under intense disease pressure through the selection of a virus strain which is a minor component of the general virus strain population, mutation within a virus strain with the resultant variant capable of breaking the resistance, or a recombination event within resistant plants leading to the generation of a new virus. Although we know that these events are possible, we do not know if they are probable under our current production system. We do know that the integration of various production steps to minimize initial virus inoculum and spread of virus inoculum will help to preserve and perpetuate these new resistances. It can be anticipated that seed potato certification programs will remain central to that effort.

REFERENCES

1. Anonymous. 1985. Problems in uniform seed potato certification procedures. Am Potato J. 62:376-386.

2. Anonymous. 1991. Department of Agriculture, Seed Regulations, amendment. Canada Gazette Part II, Vol 120, No. 20.

3. Anonymous. 1991. An act to amend the Plant Health Act (Bill No. 26). Chapter 31, 3rd session, 58th General Assembly, Province of Prince Edward Island, 40 Elizabeth II, 1991.

4. Anonymous. 1992. Seed potato certification: Its purpose, capabilities, and limitations. Am. Potato J. 59:231-236.

5. Baerecke, M.L. 1955. Der nachweis der blattrollinfektion bei kartoffeln durch ein neusses farberfahren. Züchter 25:309-313.

6. Bagnall, R.H. 1988. Epidemics of potato leafroll in North America and Europe linked to drought and sun spot cycles. Can. J. Plant Pathol. 10:193-202.

7. Bald, J.G., and White, N.H. 1942. Potato virus X:

8. The average severity of strain mixtures in three varieties of potato. J. Coun Sci. Industr. Res. Aust. 15:300-306.

8. Barker, H., Webster, K.D., and Reavy, B. 1993. Detection of potato virus Y in potato tubers: a comparison of polymerase chain reaction and enzyme-linked immunosorbent assay. Potato Res. 36:13-20.

9. Behjatnia, S.A.A., Dry, I.B., Krake, L.R., Condé, B.D., Connelly, M.I., Randles, J.W., and Rezaian, M.A. 1996. New potato spindle tuber viroid and tomato leaf curl geminivirus strains from a wild Solanum sp. Phytopathology 86:880-886.

10. Berger, P.H. 1994. Biotechnology and resistance to potato viruses. Pages 535-555 in: Advances in Potato Pest Biology and Management. G.W. Zehnder, M.L. Powelson, R.K. Jansson, and K.R. Raman, eds. APS Press, St. Paul, Minnesota.

11. Blodgett, F.M, Fernow, K. 1921. Testing seed potatoes for mosaic and leafroll (Abstr.) Phytopathology 11:58-59.

12. Bravo-Almonacid, F., Hain, L., and Mentaberry, A. 1992. Rapid immunological detection of potato viruses in plant tissue squashes. Plant Dis. 76:574-578.

13. Brehmer, W.V., and Rochlin, E. 1931. Histologische und mikroche-,ische untersuchungen über pathologische gewebeveränderungen viruskranker kartoffelstauden. Phytopath. Z. 3:471-498.

14. Bryan, J.E. 1988. Implementation of rapid multiplication and tissue culture methods in Third World countries. Am. Potato J. 65:155-207.

15. Burnett, G., and Jones, L.K. 1931. The effect of certain potato and tobacco viruses on tomato plants. Wash. Agr. Exp. Sta. Bull. 259.

16. Clark, M.F., and Adams, A.N. 1977. Characteristics of the microplate method of enzyme-linked immunosorbent assay for the detection of plant viruses. J. Gen. Virol. 34:475-483.

17. David, E., and Stormer, I. 1941. Capsicum annuum als testpflanze fur einige dartoffelviren. Phytopath. Z. 13:532-538.

18. DeBoer, S.H., Slack, S.A. vanden Bovenkamp, G., and Mastenbroek, I. 1996. A role for pathogen indexing procedures in potato certification. Adv. Botanical Research 23:217-242.

19. de Bokx, J.A. 1959. Jaarverslag instituut voor plantenziekten kundig onderziek, Wageningen. 111-113.

20. de Bokx, J.A. 1961. The effect of wounding potato tubers on the activity of virus Y N (tobacco veinal necrosis virus). T. Pl. Ziekten. 68:136-142.

21. de Bokx, J.A., and Maat, D.Z. 1979. Detection of potato virus Y N in tubers with the enzyme-linked immunosorbent assay (ELISA). Meded. Fac. Landbouwwet. Rijksuniv. Gent 44:635-644.

22. de Bokx, J.A., Prinon, P.G.M., and Maat, D.Z. 1980. Detection of potato virus X in tubers with the enzyme-linked immunosorbent assay (ELISA). Potato Res. 23:129-131.

23. de Bokx, J.A., Prion, P.G.M., and Cother, E. 1980. Enzyme-linked immunosorbent assay (ELISA) for the detection of potato viruses S and M in potato tubers. Neth. J. Pl. Path. 86:285-290.

24. Dhar, A.K., and Singh, R P. 1994. Improvement in the sensitivity of PVYN detection by increasing the cDNA probe size. J. Virol. Meth. 50:197-210.

25. Dodds, J.H. 1988. Tissue culture

technology:practical application of sophisticated methods. Am. Potato J. 65:167-180.

26. Fletcher, J.D. 1989. Potato virus Y[N]-host range and incidence in seed potato crops in New Zealand, N.Z.J. Crop Hort. Sci. 17:259-263.

27. Fribourg, C.E., and Nakashima, J. 1984. An improved latex agglutination test for routine detection of potato viruses. Potato Res. 27:237-249.

28. Gratia, A., and Manil, P. 1934. Differentiation serologique des virus X et Y de la pomme de terre chez les plantes infectus on porte uses de ces virus. Compt. Rend. Soc. Biol. 117:490-492.

29. Griffiths, H.M., Slack, S.A., and Dodds, J.H. 1990. Effect of chemical and heat therapy on virus concentrations *in vitro* plantlets. Can. J. of Botany 68:1515-1521.

30. Gugerli, P., and Gehringer, W. 1980. Enzyme-linked immunosorbent assay (ELISA) for the detection of potato leafroll virus and potato virus Y in potato tubers after artificial break of dormancy. Potato Res. 23:353-359.

31. Hadidi, A., Montasser, M.S., Levy, L., Goth, R.W., Converse, R.H., Madkour, M. A. and Skrzeckowski, L.J. 1993. Detection of potato leafroll and strawberry mild-edge luteoviruses by reverse transcription-polymerase chain reaction amplication. Plant Dis. 77:595-601.

32. Hahm, Y., Slack, S.A., and Slattery, R.J. 1981. Reinfection of potato seed stocks with potato virus S and potato virus X in Wisconsin. Am. Potato J. 58:117-125.

33. Harrington, F.M. 1927. Tuber indexing verses tuber-uniting and roguing in seed potato production. Am. Potato J. 9:128-132.

34. Harris, P.S., Miller-Jones, D.N., and Howell, P.J. 1979. Control of potato spindle tuber viroid: The special problems of a disease in plant breeders' material. Pages 231-237 in: Plant Health: The Scientific Basis for Administrative Control of Plant Parasites. D.L. Ebbles, and J.E. King, eds. Blackwell, Oxford, England.

35. Hopp, H.E., Hain, L., Bravo-Almonacid, F., Tazzini, A.C., Orman, B., Arese, A.I., Ceriani, M.F., Saladrigas, M.V., Celnik, R., del Vas, M., and Mentaberry, A.N. 1991. Development and application of non-radioactive nucleic acid hybridization system for simultaneous detection of four potato pathogens. J. Virol. Meth. 31:11-30.

36. Hoyman, W.G. 1951. A method of indexing potatoes for virus X by use of petiole or tuber juice. Am. Potato J. 28:713-721.

37. Huttinga, H. 1996. Sensitivity of indexing procedures for viruses and viroids. Adv. Bot. Res. 23:59-71.

38. Igel, M., and Lange, H. 1995. Verfahren zur frühdiagnose von viruskrankheiten bei pflanzen patentanmeldung bein dt. Patentamt. 18.4. 1953.

39. Johnson, J. 1925. Transmission of viruses from apparently healthy potatoes. Wis. Agr. Exp. Sta. Res. Bull. 63.

40. Jones, E.D. 1988. A current assessment of *in vitro* culture and other rapid multiplication methods in North America and Europe. Am. Potato J. 65:209-220.

41. Jordan, R.L., and Hammond, J. 1991. Comparison and differentiation of potyvirus isolates and identification of strain-, virus-, subgroup-specific and potyvirus group-common epitopes using monoclonal antibodies. Journal of General Virology 72:25-36.

42. Kahn, R.P., Scott, H.A., Bozicevich, J., and Vincent, M.M. 1967. Detection of potato viruses X, M, and S in dormant potato tubers by the bentonite flocculation test. Phytopathology 57:61-65.

43. Keller, E.R., and Bérces, S. 1966. Check-testing for virus Y and leaf-roll in seed potatoes with particular reference to methods of increasing precision with the A6-leaf test for Y. Eur. Potato J. 9:14.

44. Kahn, M.A., and Slack, S.A. 1980. Detection of potato virus S and X in dormant potato tubers by the latex agglutination test. Am. Potato J. 57:213-218.

45. Knutson, K.W. 1988. Implications of new techniques for seed potato certification programs and seed growers. Am. Potato J. 65:229-235.

46. Koenig, R., and Bode, O. 1978. Sensitive detection of Andean potato latent and Andean potato mottle viruses in potato tubers with the serological latex test. Phytopath. Z. 92;275-280.

47. Lakshman, D.K., Hiruki, C., Wu, X.N., and Lang, W.C. 1986. Use of [^{32}P] RNA probes for the dot-hybridization detection of potato spindle tuber viroid. J. Virol. Meth. 14:309-319.

48. LeClerc, D., Eweida, M., Singh, R.P., and AbouHaidar, M.G. 1992. Biotinylated DNA probes for detecting virus Y and aucuba mosaic virus in leaves and dormant tubers of potato. Potato Res. 33:173-182.

49. McInnes, J.L., Habili, N., and Symons, R.H. 1989. Nonradioactive photobiotin-labeled DNA probes for routine diagnosis of viroids in plant extracts. J. Virol. Meth. 23:299-312.

50. Mellor, F.C., and Stace-Smith, R. 1977. Virus-free potatoes by tissue culture. Pages 616-637 in: Applied and Fundamental Aspects of Plant Cell Tissue and Organ Culture. J. Reinert and Y.P.S. Bajaj, eds. Springer-Verlag, Berlin.

51. Nicolaeva, O.V., Morozov, S.Y., Zakhariev, V.M., and Skryabin, K. . 1990. Improved dot-blot hybridization assay for large-scale detection of potato viruses in crude tuber extracts. J. Phytopathol. 129:283-290.

52. Owens, R A., and Diener, T.O. 1981. Sensitive and rapid diagnosis of potato spindle tuber viroid disease by nucleic acid spot hybridization. Science 213:670-672.

53. Richardson, L.T., and Racicot, H.N. 1951. The comparative efficiency of tuber indexing and tuber unit planting in the elimination of virus diseases from seed potatoes. Am. Potato J. 28:765-775.

54. Roberts, I.M., and Harrison, B.D. 1979. Detection of potato leafroll and potato mop-top viruses by immunosorbent electron microscopy. Ann. Appl. Biol. 93:289-297.

55. Roy, B.P., AbouHaidar, M.G., and Alexander, A. 1989. Biotinylated RNA probes for the detection of potato spindle tuber viroid (PSTV) in plants. J. Virol. Meth. 23:149-156.

56. Salazar, L.F. 1989. Potato spindle tuber. Pages 155-167 in: Plant Protection and Quarantine, Vol. II, Selected Pests and Pathogens of Quarantine Significance. R.P. Kahn, ed. CRC Press, Boca Raton, Florida.

57. Salazar, L.F. 1994. Virus detection and management in developing countries. Pages 643-651 in: Advanced Potato Pest Biology and Management. G.W. Zehnder, M.L. Powelson, R.K.

Jannson and K.V. Raman, eds. APS Press, St. Paul, Minnesota.

58. Salazar, L.F., Balbo, I., and Owens R.A. 1988. Comparison of four radioactive probes for the diagnosis of potato spindle tuber viroid by nucleic acid spot hybrization. Potato Res. 31:432-442.

59. Salazar, L.F., Querci, M., Bartolini, I., and Lazarte, V. 1995. Aphid transmission of potato spindle tuber viroid assisted by potato leafroll virus. Fitopatologia 30:56-58.

60. Sanchez, G.E. Slack, S.A., and Dodds, J.H. 1991. Response of selected *Solanum* species to virus eradication therapy. Am. Potato J. 68:299-325.

61. Schwinghamer, M.W., and Scott, G.R. 1986. Survey of New South Wales potato crops for potato spindle tuber viroid with use of a ^{32}P-DNA probe. Plant Dis. 70:774-776.

62. Scott, H.A., Kahn, R.P., Bozicevich, J., and Vincent, M.M. 1964. Detection of potato virus X in tubers by bentonite flocculation test. Phytopathology 54:1292-1293.

63. Shepard, J.F. 1972. Gel-diffusion methods for the serological detection of potato viruses X, S and M. Montana Agr. Expt. Sta. Bull. 662.

64. Shepard, J.B., and Claflin, L.E. 1975. Critical analyses of the principles of seed potato certification. Annu. Rev. Phytopathol 13:271-293.

65. Singh, M., and Singh, R.P. 1995. Digoxigenin-labeled cDNA probes for the detection of potato virus Y in dormant potato tubers. J. Virol. Meth 52:133-143.

66. Singh, M., and Singh, R.P. 1996. Factors affecting detection of PVY in dormant tubers by reverse transcription polymerase chain reaction and nucleic acid spot hybridization. J. Virol. Meth. 60:47-57.

67. Singh, M., Diamond, J.F., Moore, L., DeHaan, T.-L., and Singh, R.P. 1996. Detection of PVY° in dormant tubers by NASH and RT-PCR and its comparison with plant symptoms and leaf-ELISA. Am. Potato J. 73:384-385.

68. Singh, R.P. 1982. *Physalis angulata* as a local lesion host for postharvest indexing of potato virus A. Plant Dis. 66: 1051-1052.

69. Singh, R.P. 1984. *Solanum x berthaultii*, a sensitive host for indexing potato spindle tuber viroid from dormant tubers. Potato Res. 27:163-172.

70. Singh, R.P. 1991. Return-polyacrylamide gel electrophoresis for the detection of viroids. Pages 89-107 in: Viroids and Satellites Molecular Parasites at the Frontier of Life. K. Maramorosch, ed. CRC Press, Boca Raton.

71. Singh, R.P. 1992. Incidence of the tobacco veinal necrotic strain of potato virus Y (PVY^{n}) in Canada in 1990 and 1991 and scientific basis for eradication of the disease. Can. Plant Dis. Surv. 72:113-119.

72. Singh, R.P., and Crowley, C.F. 1985. Evaluation of polyacrylamide gel electrophoresis, bioassay and dot-blot methods for the survey of potato spindle tuber viroid. Can. Plant Dis. Surv. 65:61-63.

73. Singh, R.P., and Somerville, T.H. 1986. Factors affecting the detection of potato virus Y in tubers by enzyme-linked immunosorbent assay (ELISA). Ind. J. Plant Path. 4:75-81.

74. Singh, R.P., Finnie, R.E., and Bagnall, R.H. 1971. Losses due to the potato spindle tuber virus. Am. Potato J. 48:262-267.

75. Singh, R.P., DeHaan, T.-L., and Jaswal, A.S. 1988. A survey of the incidence of potato spindle tuber viroid in Prince Edward Island using two methods. Can. J. Plant Sci. 68:1229-1236.

76. Singh, R.P., Boucher, A., and Wang, R.G. 1991. Detection, distribution and long-term persistence of potato spindle tuber viroid in true potato seed from Heilongjiang, China. Am. Potato J. 68:65-74.

77. Singh, R.P., Boucher, A., Lakshman, D.K., and Tavantzis, S.M. 1994. Multimeric non-radioactive cRNA probes improve detection of potato spindle tuber viroid (PSTVd). J. Virol. Meth. 49:221-234.

78. Slack, S.A. 1980. Pathogen-free plants by meristem-tip culture. Plant Dis. 64:15-17.

79. Slack S.A. 1993. Seed certification and seed improvement programs. In: Potato Health Management. R. C. Rowe, ed. APS Press, St. Paul, Minnesota.

80. Slack, S.A., and French, E.R. 1982. New disease elimination techniques in seed production programs. Proceedings of Research for the Potato in the Year 2000. International Potato Center, Lima, Peru.

81. Slack, S.A., and Tufford, L.A. 1995. Meristem culture for virus elimination from potato. Pages 117-128 in: Methods of Plant Cell, Tissue and Organ Culture and Laboratory Operations. O. Gamberg and G. Phillips, eds. Springer-Verlag, New York.

82. Spiegel, S., and Martin, R.R. 1993. Improved detection of potato leafroll virus in dormant potato tubers and microtubers by the polymerase chain reaction and ELISA. Ann. Appl. Biol. 121:493-500.

83. van Slogteren, D.H.M. 1955. VIII. Serological micro-reactions with plant viruses under paraffin oil. Pages 51-54 in: Proc. Second Conf. Potato Virus Diseases.

84. Voller, A., Bartlett, A., Bidwell, D.E., Clark, M.F., and Adams, A.N. 1976. The detection of viruses by enzyme-linked immunosorbent assay (ELISA). J. Gen. Virol. 33:165-167.

85. Webber, H.J. 1908. Plant-breeding for farmers. Cornell Agri. Expt. Sta. Bull. 251:131-172.

86. Wenzl, H. 1960. Die frühdiagnose der Fadenkeimigkeit bei Kartoffeln. Pflschber. 24:169-180.

87. Wilkinson, R.E., and Blodgett, F.M. 1948. *Gomphrena globosa*, a useful plant for qualitative and quantitative work with potato virus X. Phytopathology 38:28 (Abstr.).

88. Wright, N.S. 1988. Assembly, quality control and use of a potato cultivar collection rendered virus-free by heat therapy and tissue culture. Am. Potato J. 65:181-198.

CHAPTER   20

Virus Certification of Grapevines

G.P. Martelli and B. Walter

CERTIFICATION: THE PRINCIPLES

The consensus is that if the sanitary improvement of any crop is to be obtained, a system of preventive, protective and, often, curative measures has to be established and implemented, encompassing a complex series of interventions currently referred to as "certification".

Speaking of certification, five major questions arise:

What Is It?

Certification can be defined as a procedure whereby candidate mother plants to be used as source of material for propagation, undergo controls and, whenever necessary, treatments to secure trueness-to-type and absence from any number of pathogens, as specified by regulations officially issued, or endorsed, by competent governmental agencies.

There are two major kinds of certification:

Voluntary

A widespread form of certification largely propitiated by the growers' increasing demand for material of known sanitary status in order to establish their plantations, especially if such plants are long-lasting woody crops. Voluntary certification is regimented by regulations issued by a "certifying authority" (i.e. usually a branch of the country's Ministry of Agriculture, or the equivalent) but, by definition, cannot be forcefully imposed.

Compulsory

A type of certification enforced whenever it becomes essential to prevent the dissemination of destructive diseases, whose dispersal through propagative material may establish infection foci that, in turn, may favor explosive vector-mediated spread at a site (e.g., Pierce's disease, tristeza, plum pox).

What Does it Apply To?

In principle, certification can be applied to any cultivated plant species, regardless of whether it is propagated vegetatively (cuttings, buds, tubers, bulbs, setts, offshoots, etc.) or through seeds. Thus, both vegetable and woody crops are liable to enter certification schemes, which, in fact, they do. However, with few exceptions (e.g. potatoes), the most popular and widespread certification programs are those concerning fruit trees, vines, and small fruits. There are no limitations to the kind and number of pathogens that may be considered for exclusion in a certification program. Several species of bacteria, fungi, and nematodes are certification organisms in a number of schemes but, most certainly, graft-transmissible infectious agents like viruses, viroids and intracellular prokaryotes (phytoplasmas and fastidious bacteria) are those of major concern.

Is There a Need for it ?

A progressive, sanitary deterioration of vegetatively propagated crops, especially fruit trees (pome and stone fruits, citrus, and grapevines), has taken place in a recent past. The primary causes of such as an alarming situation reside in:

i)   increased domestic and international demand and trading of nursery products to comply with the requirements of

ii)   expanding world agriculture;

ii)   insufficient or incomplete knowledge of the sanitary problems affecting specific crops;

iii)   presence of hosts that are symptomless carriers of infectious diseases;

iv)   deplorable lack of appropriate sanitary control of propagating material being marketed. All this, together with the inefficiency or absence of quarantine services in many countries has contributed to the generalized dissemination of a number of diseases and pathogens, among which the graft-transmissible ones prevail by far.

Infectious diseases are widespread throughout the world, both in developed and developing countries. For example, an extensive 4-year survey (1984-87) carried out in Southern Mediterranean and Near East countries under the FAO patronage by J. Dunez, G.P. Martelli and A.A. Salibe, unraveled a significant deterioration of the health status of stone fruit trees, grapevines and citrus, due to continued propagation of infected planting material and to the lack of provisions for its sanitary amelioration (1). On the other hand, the European fruit industry is not in a much happier situation for somewhat similar reasons.

The awareness of the disastrous conditions of so many crops over such a wide area, the alarm signals launched by the scientific community, and the increasing demand for planting material with a reasonable health status, has compelled several countries to establish certification programs, or to reorganize the previously existing ones.

What Are the Conditions Needed for its Implementation?

To secure the success of a certification program there are a number of conditions that must be met before venturing into its launch, such as:

i)   Existence of the problem: i.e. occurrence of phytosanitary conditions objectively calling for specific interventions;

ii)   Compelling request by the grower associations so as to create a "political consent";

iii)   Involvement and convinced participation of nurserymen;

iv)   Commitment of governmental authorities to support the program legally, logistically, and, whenever possible, financially;

v)   Adequate legislation: i.e. emanation of regulations that finalize the scheme to be enforced and that regiment the production and marketing of certified material;

vi)   Appointment of the "certifying authority", i.e. a public service entrusted with control duties and delivery of certification labels.

vii)   Unfailing support by scientific institutions;

viii)   Availability of technology for the reliable detection of diseases and their agents and for the effective elimination of diseases and pathogens (sanitation).

How Long Should it Last?

Certification is an integral part of sanitary improvement programs and is often the only way to curb certain diseases of woody crops. Hence, it is by no means a temporary operation. Rather, it is long-lasting endeavor that must continue indefinitely, first to attain the desired health level of the crop taken into the scheme, then to maintain this level through time.

VIRUSES AND OTHER DISEASE AGENTS OF GRAPEVINES

To date, 44 different viruses belonging to five families and sixteen genera (Table 1) have been identified in grapevines. This is indeed the highest number of viral agents encountered in a single woody species. It represents a veritable record, both for the variety of the findings and the relatively short time in which they have occurred. The first isolation of a grapevine virus dates back to the early 1960s (12). Some of these viruses, i.e. cucumber mosaic cucumovirus (CMV), tobacco mosaic tobamovirus (TMV), broad bean wilt fabavirus (BBWV), potato X potexvirus (PVX), artichoke Italian latent nepovirus (AILV), and alfalfa mosaic alfamovirus (AMV) are serious pathogens to many important crops. However, they represent mere scientific curiosities for grapevines, either because they are very rarely found, or because the damage they induce is negligible. By converse, other viruses, especially members of the *Nepovirus, Closterovirus* and *Trichovirus* genera, are veritable pathogens and the agents of diseases

Table 1. Grapevine viruses and their taxonomic position[a]

Family	Genus	Species
Viruses belonging to genera assigned to families		
BROMOVIRIDAE	*Alfamovirus*	Alfalfa mosaic (AMV)
	Cucumovirus	Cucumber mosaic (CMV)
	Ilarvirus	Grapevine line pattern (GLPV)
BUNYAVIRIDAE	*Tospovirus*	Tomato spotted wilt (TSWV) [b]
COMOVIRIDAE	*Fabavirus*	Broadbean wilt (BBWV)
	Nepovirus	Artichoke Italian latent (AILV)
		Arabis mosiac (ArMV)
		Blueberry leaf mottle (BBLMV)
		Grapevine Bulgarian latent (GBLV)
		Grapevine chrome mosaic (GCMV)
		Grapevine fanleaf (GFLV)
		Grapevine Tunisian ringspot (GTRV)
		Peach rosette mosaic (PRMV)
		Raspberry ringspot (RRV)
		Tobacco ringspot (TRSV)
		Tomato ringspot (ToRSV)
		Tomato black ring (TBRV)
		Strawberry latent ringspot (SLRSV)
TOMBUSVIRIDAE	*Tombusvirus*	Petunia asteroid mosaic (PAMV)
		Grapevine Algerian latent (GALV)
	Carmovirus	Carnation mottle (CarMV)
CLOSTEROVIRIDAE	*Closterovirus*	Grapevine leafroll-associated 1 (GLRaV1)
		Grapevine leafroll-associated 2 (GLRaV2)
		Grapevine leafroll-associated 3 (GLRaV3)
		Grapevine leafroll-associated 4 (GLRaV4)
		Grapevine leafroll-associated 5 (GLRaV5)
		Grapevine leafroll-associated 6 (GLRaV6)
		Grapevine leafroll-associated 7 (GLRaV7)
Viruses belonging to genera not assigned to families		
	Sobemovirus	Sowbane mosaic (SoMV)
	Necrovirus	Tobacco necrosis (TNV)
	Potexvirus	Potato X (PVX)
	Tobamovirus	Tobacco mosaic (TMV)
		Tomato mosaic (ToMV)
	Trichovirus	Grapevine A (GVA)
		Grapevine B (GVB)
		Grapevine C (GVC) [c]
		Grapevine D (GVD)
		Grapevine berry internal necrosis (GIBNV) [c]

Table 1. (cont.)

Family	Genus	Species

Viruses belonging to genera not assigned to families (cont.)

	Capillovirus [c]	Unnamed virus
	Furovirus [c]	Grapevine labile rod-shaped (GLRSV)
	unnamed putative new genus [c]	Grapevine fleck (GFkV)
		Grapevine asteroid mosaic (GAMV)
Unclassified viruses		
		Grapevine ajinashika (GAV)
		Grapevine stunt (GSV)

[a] From Martelli (30) modified.
[b] The correct identification of this virus is doubtful.
[c] Tentative assignments.

that have an undoubtedly negative impact on the quality and quantity of the yield.

But why are *Vitis* species so prone to viral infections? The explanation resides in a number of reasons, among which the intrinsic susceptibility of grapevines to viruses, the way in which grapevines are multiplied, and the variety of geographical and climatic environments under which they are grown. In *Vitis*, like in all vegetatively propagated plant species, there is a progressive accumulation of infectious agents which are acquired by the plants with exposure to inoculum in different places and times, and are "clonally" perpetuated with them. This allows the survival and dissemination also of those viruses to which individual vines are little susceptible, but that have found the way to infect them, sometimes quite by chance through the fortuitous inoculation by a passing vector.

Beside viruses, grapevines are susceptible to a number of other intracellular disease agents. Current records report six different viroids, three or more phytoplasmas, and *Xylella fastidiosa*, a xylem-limited fastidious bacterium (10,29).

INFECTIOUS DISEASES OF GRAPEVINES

As compared with the very high number of potential pathogens infecting grapevines, the list of recognized diseases is short (Table 2). The reason is twofold: (i) several of the viruses

recovered may cause latent infections; (ii) the same disease can be induced by multiple agents (Tables 3 and 4).

Grapevine diseases and their impact on the crop have been the object of books and review articles (10,11,16,29,35,44,48) which the readers are referred to for details. What follows is a sketchy description of the diseases that prevail on a worldwide basis both for economic relevance and geographical distribution.

Virus Diseases

Degenerative Diseases and Decline

Nepoviruses are the causal agents of these diseases (Table 3). The degenerative condition caused by European nepoviruses is universally known as fanleaf, whereas the comparable disorders elicited by American nepoviruses are referred to as decline (18,29). Several of the European nepoviruses (GFLV, ArMV, TBRV, GCMV) possess distorting and chromogenic strains that induce malformation of leaves and canes or chrome yellow discolorations of the foliage, respectively. Reduction in vigor and in the quantity and quality of the yield is associated with infection by both types of strains. American nepoviruses evoke responses that vary with the grapevine species and climatic conditions. In cold climates European grapes infected by TRSV or

Table 2. Infectious diseases of grapevines[a]

Major virus diseases

       Degeneration (GFLV and European nepoviruses)
       Decline (American nepoviruses)
       Leafroll (closteroviruses)
       Rugose wood complex (trichoviruses)
       Fleck

Minor virus diseases

       Yellow mottle (alfalfa mosaic alfamovirus)
       Line pattern (grapevine line pattern ilarvirus)
       Stunt (grapevine stunt virus)
       Ajinashika (grapevine ajinashika virus)
       Berry inner necrosis (grapevine berry inner necrosis virus)
       Roditis leaf discoloration (carnation mottle carmovirus and GFLV)

Virus-like diseases (agents unknown)

       Enations
       Vein necrosis
       Vein mosaic
       Summer mottle
       Asteroid mosaic
       Bushy stunt
       Graft-incompatibility disorders

Viroid diseases

       Yellow speckle

Prokaryotic diseases

       Flavescence dorée (elm yellows phytoplasma group)
       Yellows induced by the stolbur phytoplasma group (e.g. Bois noir and Vergilbungskrankheit)
       Yellows induced by the X-disease phytoplasma group
       Pierce's disease (*Xylella fastidiosa*)

---

[a] From Martelli (29), modified

ToRSV exhibit stunted growth, distortion of leaves and canes, low yield, and decline rapidly. In warmer climates, yield but not vigor is affected, and the leaves may show chrome yellow flecking along the veins. Grapevine nepoviruses are disseminated over medium and long distances by propagating material. Their field transmission is mediated by longidorid nematodes, i.e. *Xiphinema* and *Longidorus* (Table 3).

Leafroll

Closteroviruses are the major, if not the only, agents of leafroll. Seven serologically distinct such viruses have been identified so far in leafroll-affected grapevines (Table 4). Although all of these viruses continue to be cautiously called grapevine leafroll-associated (GLRaVs) (8), there is experimental evidence that some (e.g., GLRaV

Table 3. Nepoviruses agents of grapevine diseases and their vectors

Virus	Vector
Agents of degeneration	
Grapevine fanleaf virus (GFLV)	*Xiphinema index*
Arabis mosaic virus (ArMV)	*X. diversicaudatum*
Grapevine chrome mosaic virus (GCMV)	Unknown
Strawberry latent ringspot virus (SLRV)	*X. diversicaudatum*
Raspberry ringspot virus (RRV)	*Longidorus macrosoma* *Paralongidorus maximus* *L. elongatus*
Tomato black ring virus (TBRV)	*L. elongatus, L. attenuatus*
Artichoke Italian latent virus (AILV)	*L. attenuatus, L. fasciatus*
Grapevine Bulgarian latent virus (GBLV)	Unknown
Agents of decline	
Peach rosette mosaic virus (PRMV)	*X. americanum, L. elongatus, L. diadecturus*
Tomato ringspot virus (ToRSV)	*X. californicum, X. rivesi*
Tobacco ringspot virus (TRSV)	*X. americanum*
Blueberry leaf mottle virus (BELMV)	Unknown
Undetermined pathogenicity	
Grapevine Tunisian ringspot virus (GTRV)	Unknown

-1, -2, and -3) induce leafroll-type responses in vines infected by grafting or vectors. This prompted the suggestion that the word "associated" be dropped from the name of the first three GLRaVs of the series (31). Disease symptoms are those typically induced by viruses multiplying in the phloem and affecting its functionality. Leaves are thicker than normal, brittle, with margins rolled downwards and discolored, i.e., yellowish in white-berried cultivars and reddish to deep purple in red-berried cultivars. A great variability in symptom expression is commonly observed in the field, symptomatological variants being probably determined by different viral combinations and differential varietal responses (29, 31).

All grapevine closteroviruses are spread primarily by propagative material but two (GLRaV-3 and probably GLRaV-2) are transmitted by pseudococcid mealybugs and scale insects (Table 4).

Rugose Wood

Rugose wood is a complex disease in which four different disorders denoted rupestris stem pitting, corky bark, Kober stem grooving, and LN33 stem grooving can be recognized and sorted out by graft transmission to *Vitis* indicators (29). In the field, rugose wood is characterized primarily by alterations of the woody cylinder. Affected vines may be dwarfed and less vigorous than normal, have delayed bud opening in spring, and some decline and die within a few years after planting. Grafted vines often show swelling above the bud union and a marked difference between the diameters of scion and rootstock. Sometimes, the

Table 4. Leafroll and rugose wood viruses and their vectors[a]

Disease	Virus	Vector
Leafroll		
	GLRaV-1	Unknown
	GLRaV-3	*Planococcus ficus, Pseudococcus longispinus, Ps. affinis, Pulvinaria vitis*
	GLRaV-4	Unknown
	GLRaV-5	Unknown
	GLRaV-6	Unknown
	GLRaV-7	Unknown
Leafroll and graft incompatibility		
	GLRaV-2	*Ps. longispinus, Ps. affinis*
Rugose wood complex		
Rupestris stem pitting	Unknown	Unknown
Kober stem grooving	GVA	*Pl. citri, Pl. ficus, Ps. longispinus, Ps. affinis*
Corky bark	GVB	*Ps. longispinus, Pl. ficus, Ps. affinis*
LN33 stem grooving	Unknown	Unknown

[a] From Walter and Martelli (48)

bark of the scion above the graft union is exceedingly thick and corky (corky rugose wood) and has a spongy texture and rough appearance. The woody cylinder is typically marked by pits and/or grooves which may occur on the scion, the rootstock, or both. In most cases no specific symptoms are shown by the foliage, but the crop may be reduced.   Four serologically distinct trichoviruses and a possible capillovirus (Tables 1 and 4) were found to be associated with rugose wood-affected vines. However, only for GVA and GVB there is evidence of a likely involvement in the etiology of two of the disorders of the complex, i.e. Kober stem grooving (GVA) and corky bark (GVB) (5). Rugose wood is distributed over long distances by propagative material, whereas its spread at a site is mediated by pseudococcid mealybugs (Table 4).

Fleck

Fleck is latent in European grapes and in most American rootstocks. Symptoms are expressed in *Vitis rupestris*, a self-indexing species, and consist of clearing of the veins of third and fourth order that produce localized translucent spots. Leaf deformity and stunting may also occur (29). Fleck is a widespread and damaging disease (48). Its causal agent is a phloem-limited non mechanically transmissible isometric virus (6), likely belonging to a still undescribed genus (Table 1). No vectors are known, thus spreading is through infected propagating material.

Viroid Diseases

Yellow Speckle

There are six grapevine viroids but only two, denoted grapevine yellow speckle viroid 1 and 2 (GYSVd-1 and GYSVd-2), seem to be pathogenic to grapevines (29, 38). Yellow speckle, the disease they induce, is elusive as its outward expression is conditioned by climatic and, perhaps, varietal factors. Symptoms consist of a few to many minute chrome yellow spots or flecks scattered over part or the whole leaf surface, or gathering along the main veins. The latter condition, known as "vein banding", is enhanced by the synergistic effect resulting from concomitant infection by GFLV and GYSVd-1 (25, 42). Typical yellow speckle may not be too harmful, whereas vein banding has a severe detrimental effect on the yield of certain varieties. As with other viroids, GYSVd does not have a vector but it is perpetuated in propagating material with which it spreads. Mechanical transmission to field-grown vines through surface-contaminated cutting tools is possible, though with a low efficiency (40).

Phytoplasma-induced Diseases

Grapevines are affected by several yellows diseases elicited by different phytoplasmas (Table 2), and called with different names: flavescence dorée, bois noir, Vergilbungskrankheit, Mediterranean, subtropical, or North American grapevine yellows (13). Regardless of the disease and causal agent the symptoms are very similar: leaf rolling, yellowing or reddening of the leaves according to whether the vines are white- or red-berried, necrosis along the veins, incomplete wood ripening, withering of the berries and drying up of the bunches. Severely affected vines may die (29). Phytoplasma infections are spreading in epidemic form in several European countries but vectors have been identified only for flavescence dorée and Vergilbungskrankheit. These are transmitted by the leafhoppers *Scaphoideus titanus* and *Hyalestes obsoletus*, respectively (13,26). Medium and long distance spread is through infected wood.

CERTIFICATION: THE EUROPEAN EXPERIENCE

Historical Background

The amelioration of *Vitis vinifera* varieties through selection started in Germany towards the end of the last century. The first clones were planted in 1900 and were officially recognized in 1921. The selection was visual and based primarily on productivity. The effect of viral infections was not taken into consideration until Muth (34) suggested that "Reisigkrankheit" (= fanleaf disease) was responsible for the variations observed within varietal populations. Thus, in the 1930s clonal and sanitary selection were combined, which led to elimination of vines affected by fanleaf. The totality of grapevine cultivars now grown in Germany has a clonal origin (37).

In France, a specific agency (Section de Sélection et de Contrôle des Bois et Plants de Vigne) was established in 1946 for obtaining fanleaf-free propagating material. The selection was visual and based both on pomological and sanitary characteristics. The first planting of clonally selected vines was established in Alsace in 1969 (48). Clonal and sanitary selection has continued ever since, so that a high proportion of both rootstocks and European grapes now on the market are of clonal origin. When required, selected material undergoes sanitation (heat therapy, meristem tip culture, micrografting) for the elimination of unwanted diseases.

In Italy, interdisciplinary programs for the pomological and sanitary improvement of the country's viticultural industry were initiated in the late 1950s. Since 1976, efforts by individual research institutions were pooled under coordinated research projects supported by governmental and regional authorities (27). Objectives of this undertaking were: (i) clonal and sanitary selection of the main European grape cultivars (wine grapes in particular) and rootstocks grown in the country; (ii) production of disease-free stocks; (iii) technological evaluation of certifiable clones. Also in Italy, a high proportion of commercial propagating material (American rootstocks in particular) now comes from registered clones (41).

In several other European countries (e.g., Switzerland, Hungary, former Czechoslovakia, Yugoslavia, and Soviet Union) comparable activities developed at the initiative of research institutions that started small scale clean stock programs of their own (9). In Switzerland, the implementation of an experimental sanitary improvement program preceded the promulgation of the European Union (EU) Directives discussed in the next section.

Present Status

Infectious diseases of the grapevine occur throughout the Old World, including the viticultural countries of the European Union (EU), i.e. Portugal, Spain, France, Luxembourg, Germany, Austria, Italy, and Greece. In these countries, the incidence of virus and virus-like diseases is high and their spread has been rapid due to the uncontrolled distribution and use of infected scions and rootstocks that took place especially in the post-war period. The alarming sanitary deterioration of the crop, which was only in part counterbalanced by the interventions of individual Members States (France, Germany, Italy), prompted the EU Council to issue in 1968 and 1971 Directives for the improvement of the Union's grapevine industry. Directive 68/93 on the "Marketing of vegetatively propagated material of grapevines" classified propagative material into categories denoted "basic", "certified", and "standard", and contained indications encompassing the sanitary characteristics of mother vineyards destined for the production of these materials. Directive 68/93 was modified by Directive 71/40, which defines the sanitary requirements of current EU certification as follows:

"In the vineyards producing basic material, harmful virus diseases, notably fanleaf and leafroll, must be eliminated. Vineyards producing materials of other categories must be kept free from plants showing symptoms of virus diseases".

Bylaws generated by these Directives were promulgated in EU Member States, except for Austria which has joined the Union only recently. National certification schemes are now under way in all these countries (28). However, whereas in France, Germany and Italy these programs have been in operation for many years, already yielding a substantial number of certified clones, their enforcement in Spain, Greece and Portugal is more recent.

A distinctive trait of EU certification schemes is that they are not mere clean stock programs applied to pomologically uncontrolled mother vines. Rather, only sources true to type and with a well-established clonal nature can be registered and certified. The identification of clones is a lengthy procedure regulated by EU Directive 72/169, and outlined in a recent Resolution (Viti 1/91) of the Office International de la Vigne et du Vin (OIV) (2). Thus, as conceived in EU Member States, clonal and sanitary selection is an interdisciplinary activity requiring the joint effort of viticulturists, virologists and, in the case of wine grapes, technologists.

Candidate clones are selected in vineyards with desirable characteristics in the typical area of cultivation of each variety. Selection is based on varietal conformity, vegetative vigor, bud fertility, quality (e.g. sugar content, titratable acidity) and quantity of the yield, timing and uniformity of ripening, general sanitary conditions. Vines are kept under observation for two to four years and the best performing and least infected ones are chosen as candidate clones. These are grafted onto two different rootstocks and planted in two sister performance plots established in two diverse ecological environments. Candidate clones are kept under scrutiny in these plots for no less than five years. The whole process thus requires 8 to 10 years, or more, if clones must undergo sanitation and be re-indexed for health assessment.

Newly identified clones are described and submitted for registration to governmental authorities, together with a statement defining their sanitary status, issued by recognized laboratories. In France new clones are approved by the "Comité Technique Permanent de la Selection des Plantes Cultivées", in Germany by the "Bundessortenamt", in Italy by the "Comitato Nazionale per la Valutazione delle Varietà di Vite", and in Spain by the "Instituto de Semillas y Plantas de Vivero" which are divisions of the respective Ministries of Agriculture. Approved clones are registered in National Catalogues (28).

Although national certification schemes implemented in EU countries are inspired by, and more or less conform to, EU Directives, they differ to varying extents from one another. Some of these differences pertain to the pomological aspects of clonal selection (e.g., in Germany it lasts longer than in France or Italy), others concern sanitary requirements. In most countries, these requirements are stricter than those contained in EU Directives, which prescribe freedom from fanleaf and leafroll only. However, there is little uniformity among individual schemes as the health status of certified clones may vary from country to country. Thus, for instance, exemption from rugose wood is required in Portugal, France, Italy and in Spain (limitedly to rupestris stem pitting and corky bark) but not in Germany where no tests are made for its detection. France and Germany, contrary to all other countries, require absence of fleck only in roostocks (28).

Although additional differences may exist in the way in which certified material is maintained, propagated and distributed, certain steps are common to the various schemes. Registered clones (primary sources or nuclear stocks) are maintained by the juridical or physical person who owns them ("obtenteur" or conservative breeder) and undergo a first multiplication in specialized outfits (pre-multiplication or foundation blocks). These distribute propagating material to nurseries for the establishment of certified mother vine plots which, in turn, are used for the production of certified budwood, rooted cuttings, or grafted vines for commercial purposes. Official, or officially authorized organizations carry out controls on the health, origin and amount of certified plants and issue certification labels.

CERTIFICATION: THE AMERICAN EXPERIENCE

In North America, grapevine certification schemes are operating in the United States (California, Oregon, Michigan, New York, Washington, Texas, and Virginia) and Canada (British Columbia and Ontario). Unfortunately, the lack of a comprehensive source of information does not facilitate comparison of the various state programs, the best known of which are those implemented in California and New York (21).

The pioneering work of W.B. Hewitt and co-workers (23,24), besides providing the first authoritative scheme for sorting out and classifying virus and virus-like diseases of grapevines, gave California the lead in the knowledge of these disorders and in the strategies for controlling them. In the early 1960s successful attempts were made for freeing vines from GFLV and other viruses by heat therapy and *in vitro* shoot tip culture (17,20). This favored the implementation of the "California Clean Stock Program" established by the University of California in cooperation with the California Department of Food and Agriculture (13,19).

Basic steps of this program are disease recognition, indexing, therapy, registration and certification. Grape selections are tested for freedom from fanleaf and other nepovirus diseases, leafroll, rugose wood (rupestris stem pitting and corky bark), fleck, asteroid mosaic and flavescence dorée. Only selections indexing negative for these diseases enter the certification scheme (21). The program is voluntary and its financial load is shared by participating nurseries. "Foundation blocks" are maintained by the Foundation Plant Material Service (FPMS) of the University of California at Davis, which is also responsible for sanitary checks (indexing and laboratory testing). Propagating material purchased by nurseries from FPMS is used for establishing "registered increase blocks", which are inspected for disease and varietal conformity by the California Department of Food and Agriculture. Wood from registered blocks is destined for the production of certified vines.

The New York State grape certification program became official in 1973. It operates under regulations issued by the New York State Department of Agriculture and Markets, which is responsible for their enforcement and for the inspections to nurseries (15,21). The New York State program is strictly voluntary, as in California, but a fee is charged for plants that qualify for certification. These are vines testing negative for fanleaf, tomato and tobacco ringspot viruses, leafroll, corky bark, rupestris stem pitting, and flavescence dorée.

There are four types of plantings from which material for production of certified vines can be collected: (i) "Nuclear blocks", established with vines derived directly from nuclear stocks (primary sources); (ii) "Foundation blocks I and II", planted with wood propagated from Nuclear blocks and Foundation block I, respectively; (iii) "Registered blocks", established with material originating from any of the above (21).

The Canadian certification scheme is similar to those implemented in California and New York (43). Sanitary standards are also nearly the same, with the exception that the presence of rupestris stem pitting, one of the rugose wood components, is tolerated, on the assumption that its significance is negligible. This difference does not impair the free importation of Canadian certified vines into USA (21).

It is evident from the above, that the EU certification schemes compare rather well with the North American programs. The EU schemes, however, are subsidized to a large extent by public funds and guarantee, in addition to trueness to type and a given sanitary standard, the clonal origin of propagating material. The latter is perhaps the single most significant difference between these programs in Europe and North America. In fact, as pointed out by Meredith and Walker (33) "the term clone certification does not have much meaning in the United States", where no programs exist at the federal or state level for clonal selection. However, clones imported from Europe, directly or via Canada, are recognized and maintain their identification number (21).

PROPOSED SCHEME FOR GRAPEVINE CERTIFICATION IN THE EUROPEAN UNION

The sanitary provisions of the EU Directives discussed above are largely inadequate for two order of reasons. First, they do not guarantee an acceptable sanitary status of propagative material of any category. Second, they are outdated, failing to take into account recent developments of grapevine virology. Furthermore, the Directives do not provide guidelines for the implementation of a standardized certification protocol in the Union. As a consequence, propagative material of *Vitis* currently produced in the EU is not sanitarily uniform, its health status being determined by national certification schemes, which differ from country to country. All this contrasts with the criteria inspiring the agricultural policy of the EU and is prejudicial to the free circulation of this material within the Union.

Faced with such a regrettable situation and sharing the concern of growers, nurserymen and their association (Comité International des Pépinieristes), a group of European grapevine virologists members of the International Council for the Study of Virus and Virus Diseases of the Grapevine (ICVG), reviewed the state-of-the-art of certification in the Union (28) and outlined a scheme mediating between the procedures implemented in EU Member Countries (32). A similar move was made shortly afterwards by the European and Mediterranean Plant Protection Organization (EPPO), which proposed a certification scheme conforming to the same technical criteria used by ICVG virologists (4).

Both protocols took notice that propagative material of grapevines in the EU is classified into categories with a color coding system:

Pre-basic: nuclear stock plants originated from clonal selections and free from all diseases and viruses considered in the scheme (virus-tested).

Basic (white label): material from plants derived by direct propagation from nuclear stocks and grown in purity in specialized propagation plots. Basic material is intended for delivery only to nurseries possessing given qualifications.

Certified (blue label): material or plants for delivery to growers, derived from mother vines grown from basic material by authorized commercial nurseries. The production of new mother vines from certified mother vines is forbidden. To this aim, nurseries must obtain new basic material.

Standard (orange label): material or plants from sources of unknown sanitary status.

Outline of the Scheme

To be certified, grapevine varieties and rootstocks must undergo the following step wise procedure:

1. Identification of candidate clones through selection for pomological and health quality of individual vines.
2. Establishment of candidate clone repositories in soil without nematode vectors. Grape selections can either be grown on their own-roots or can be grafted on virus-tested rootstocks.
3. Assessment of the health status of visually selected candidate clones by indexing and laboratory assays. Selections that do not meet the sanitary requirements of the scheme undergo sanitation and are re-tested for health assessment. If the totality of selections of a given variety or rootstock are expected to be infected, it is advisable to proceed directly with sanitation
4. Establishment of performance plots in which virus-tested candidate clones are evaluated for ultimate selection and identification of clones.
5. Application to governmental authorities for official recognition and registration of clones.
6. Maintenance of registered clones (nuclear stocks, pre-basic material) under

conditions (e.g. insect-proof screenhouse) ensuring freedom from re-infection by soil or aerial vectors. Nuclear stocks are tended by their conservative breeders and checked each year for virus symptoms. ELISA is also used for the detection of certain viruses (e.g. GFLV and ArMV). Re-indexing is advisable if novel or more efficient detection techniques, antisera or indicators become available, or whenever visual inspections suggest tests to be carried out.

7. Multiplication of nuclear stocks in outdoor plantings (propagation blocks) under conditions ensuring freedom from re-infection. Propagation blocks should have a safety distance of 15 to 20 mt from vineyards planted with material of lower category ("certified" or "standard") and be established in soils with no grapevine history or where grapevines were not grown for at least six years. Propagation blocks are the source of "basic material". Mother vines are checked visually each year for virus symptoms and re-indexed regularly so that, according to the size of the plot, all are tested in a 5 to 8 year period.

8. Distribution of basic material to qualified nurseries under official control.

9. Establishment of commercial stands for production of certified material for delivery to growers (certified blocks). These are planted with budwood coming directly from propagation blocks, at a minimum distance of 8 to 10 mt from other vineyards and in soils in which virus-transmitting nematodes are not detected.

10. Certification and labeling. Labels are supplied by the certifying authority which may be either a government agency or an officially recognized organization.

Steps 1-3 are considered to be carried out by a government agency or an official organization; steps 4,6,7,8 and 9 by or under the strict control of an official organization; step 10 under strict control only.

Evidently, certification schemes do not embrace standard material, a category which, in principle, is due to disappear. Trueness to type is guaranteed by all categories. Basic and certified categories guarantee also the sanitary status, as declared for individual scion cultivar, clone or rootstock type. In grafted vines both scion and rootstock must belong to the same category, failing which the resulting grafted combination will have the status of the lower category: i.e., basic/basic = basic; certified/certified = certified; basic/certified or vice versa = certified; basic or certified/ standard or vice versa = standard.

Minimum Sanitary Requirements

To date, only the diseases and viruses listed below are regarded as undesirable in the UE. Their occurrence is incompatible with registration and certification of selected clones:

1. Degeneration, including GFLV and other European nepoviruses.
2. Leafroll
3. Rugose wood complex (rupestris stem pitting, corky bark, LN 33 stem grooving, Kober stem grooving)
4. Fleck
5. All filamentous viruses
6. Yellows diseases

The absence of other disorders, among which some widespread virus-like (vein mosaic, vein necrosis, enations) and viroid diseases (yellow speckle), is optional. Their presence does not impair registration and certification but efforts should be made for their elimination.

Testing for Disease Freedom and Sanitation Procedures

What follows is a brief outline of the diagnostic tests and sanitation procedures currently used for grapevines. These are extensively described in the appendices to the EPPO certification scheme (3) and in other publications (16,29) which readers are referred to for details.

Diagnostic Tests

The type of detection tests, their application and reliability have been the object of extensive debate in the EU. For verifying the efficiency of diagnostic procedures and harmonizing their application, an EU-supported Network of grapevine virologists was established in 1993 (46). This group of experts is still at work but the preliminary results of comparative testing carried out seem to confirm the validity of most of the protocols currently used, with reference to the following main procedures:

Indexing on Vitis Indicators

The use of woody differential hosts is a compulsory step of grapevine certification programs, for there are diseases that can be identified only by the reaction of indicators (Table 5). Inoculation is by: (i) Whip or cleft grafting in the field; (ii) Chip-bud grafting, a technique recommended for detection of rupestris stem pitting. The pits induced by this disease develop on the indicator stem below the grafted chip, and extend basipetally in a band or stripe; (iii) Machine or bench grafting; (iv) Green grafting, a technique to be encouraged because of the distinct advantages over other grafting systems (49).

Inoculation to Herbaceous hosts

Herbaceous indicators detect mechanically transmissible viruses, including those of minor relevance. Their use is complementary to other diagnostic procedures. Sap transmission may be useful for preliminary screening and random testing.

Serological Tests

Serology is to be regarded as a complement to, but not as a substitute for, other diagnostic procedures. However, the use of ELISA is recommended for the identification of GFLV and other European nepoviruses, for GFkV, closteroviruses and trichoviruses antisera to which are available. Source of antigens for ELISA can be grapevine buds, roots, leaves, and cortical scrapings from mature canes. Cortical scrapings are advantageous because they can be used throughout the year without apparent loss of efficiency due to seasonal variation of antigen titer in vegetating organs (47). Furthermore, cortical scraping extracts give low background readings and are much more reliable for detection of closteroviruses in American rootstocks, especially *V. rupestris* and its hybrids (7).

Double-stranded RNAs

Electrophoretic detection of dsRNAs from tissue extracts may complement other diagnostic procedures. It may be useful in evaluating the outcome of sanitation treatments (22).

Molecular Hybridization

Molecular probes and primers for target nucleic acid amplification (PCR) have been developed for several grapevine nepoviruses (29, 45), closteroviruses (31), trichoviruses (5), GFkV (36), viroids (45), and grapevine phytoplasmas (13, 14, 45). Whereas molecular tools have not yet found generalized application for virus detection and identification, they have become the method of choice for phytoplasmas (13).

Sanitation Procedures

The high incidence of infectious diseases in commercial vineyards of EU Member States makes sanitation highly desirable, if not compulsory. The main techniques currently available are:

1. Hot water treatment of dormant canes at 50°C for 45 min, for eliminating prokaryotes.
2. Hot air treatment of vegetating vines at constant temperature of about 38°C, excision and rooting of shoot tips, for eliminating virus and virus-like diseases;
3. Meristem tip culture *in vitro*, for eliminating virus and virus-like diseases. *In vitro* culture can be used in combination with heat therapy or chemotherapy. The consensus is that *in vitro* culture is more efficient than hot air treatment.
4. Micrografting of meristem tips onto *in vitro*-grown seedlings

Regardless of the procedure used, testing of treated material for assessment of its health status must follow, when a suitable period from the end of the sanitation treatment has elapsed.

Transitory Provisions

Advocating a certification scheme that increases the number of diseases considered by EU Directives now in force from two (fanleaf and leafroll) to five (fanleaf, leafroll, rugose wood, fleck, yellows), will raise the problem of the destiny of registered cloned now in circulation This is undoubtedly a touchy issue, especially for some European grape nurserymen who are reluctant to recognize well-established facts, such

Table 5. Main indicators for identification of virus and virus-like diseases of grapevine

Indicator	Disease identified
Vitis rupestris St. George	Degeneration[a], fleck, rupestris, stem pitting, asteroid mosaic
Vitis vinifera Cabernet franc, Pinot noir, or other red-fruited cultivars[b] Kober 5BB LN 33	Leafroll Kober stem grooving Corky bark, LN 33 stem grooving, enations
Vitis riparia Gloire de Montpellier 110 R	Vein mosaic Vein necrosis

[a] In countries where degeneration is caused also by nepoviruses other than GFLV, Siegfriedrebe (FS4 201- 39) may be used as an indicator.
[b] The choice of the most suitable indicator for eafroll depends on climatic conditions of the place of testing.

as the alleged detrimental effects of some of the major virus diseases (4) and the un-deferable necessity for more incisive sanitary controls. This problem, however, could be circumvented by declaring a suitable moratorium (say, 6 to 8 years) during which the old clones will be tested for conformity to the new minimal sanitary standards, to be either re-admitted to registration (directly or after sanitation), or discarded.

CONCLUDING REMARKS

The certification scheme outlined above is tailored to the health situation of the EU viticultural industry as it stands now, with exclusive reference to viruses and phytoplasmas. The diseases considered were all proven to be detrimental in various ways to grapevine production (48) and are by far the most significant for geographical distribution and economic relevance. Yet, there is a number of other pathogen, both fungal (*Phomopsis viticola*, *Eutypa* spp., *Stereum* spp.) and bacterial (*Agrobacterium tumefanciens*, *Xylophilus ampelinus*) that perennate in grapevine woody material and can be disseminated with it. These disease agents were listed as undesirable in the FAO/IBPGR Technical Guidelines for the Safe Movement of Grapevine Germplasm (16) and are included in the EPPO certification scheme (3).

If certification is looked upon from a wider angle, extending to other grape-growing areas of the world, still other diseases and pathogens need to be addressed, many of which are currently classified as quarantine pests for Europe (39). Thus, for instance, *Xylella fastidiosa*, the agent or Pierce's disease, and the American nepoviruses should be banned from certified material, as they actually are in the North American schemes (21). Similarly, it would appear desirable to enforce provisions for freedom from the Japanese diseases "ajinashika" and "stunt" (10, 29).

Grapevine has a worldwide distribution, is exposed to the infection of a variety of local pathogens exotic to other areas of cultivation, and moves tremendous economic (and political) interests. It ensues that current certification is designed to suit the conditions and to protect the interests of somewhat restricted areas. Yet, its promotion from a regional to a worldwide endeavor would be highly desirable.

While waiting for the time to be ripe for a "world certification", European grapevine virologists would be content with the promulgation of an EU scheme which, by taking into account the proposals now on the table, will provide clear and uncontroversial guidelines. Its' implementation may not eliminate all the sanitary problems that affect grapevines in the Union, but it will certainly give a substantial contribution to their solution, much to the benefit of growers, consumers and nurserymen.

REFERENCES

1. Anonymous. 1988. Fruit crop sanitation in the Mediterranean and Near East region, status and requirements. United Nations Development Programme/Food and Agricultural Organization of the United Nations, Rome, Italy.

2. Anonymous. 1991. Programme type pour la réalisation de la selection clonale de la vigne. Bull. OIV 727-728:752-763.

3. Anonymous. 1994. Certification scheme: Pathogen-tested material of grapevine varieties and rootstocks. EPPO Bull. 24:347-367.

4. Anonymous. 1996. Le C.I.P. à nouveau réuni a Genve le 22 Mars. Le Pépiniériste, 107:19.

5. Boscia, D., Minafra, A., and Martelli, G.P. 1996. Filamentous viruses of the grapevine: putative trichoviruses and capilloviruses. in: Filamentous Viruses of Woody Plants. P.L. Monette, ed. Research Signpost, Trivandum, India (in press).

6. Boscia, D., Martelli, G.P., Savino, V., and Castellano, M.A. 1991. Identification of the agent of grapevine fleck disease. Vitis 30: 97-105.

7. Boscia, D., Savino, V., Elicio, V., Jebahi, S.D., and Martelli, G.P. 1991. Detection of closteroviruses in grapevine tissues. Pages 52-57 in: Proc. 10th Meeting ICVG, Volos 1990. I.C. Rumbos, R. Bovey, D. Gonsalves, W.B. Hewitt and G.P. Martelli, eds.

8. Boscia, D., Greif, C., Gugerli, P., Martelli, G.P., Walter, B and Gonsalves, D. 1995. The nomenclature of the leafroll-associated putative closteroviruses. Vitis 34:171-175.

9. Bovey, R. 1989. Control of virus and virus-like diseases of grapevine: production of virus-free propagating material and its performance. Pages 143-152 in: Proc. 9th Meeting ICVG, Kiryat Anavim, 1987.

10. Bovey, R., and Martelli, G.P. 1992. Directory of Major Virus and Virus-like Diseases of Grapevines. Mediterranean Fruit Crop Improvement Council/ International Council for the Study of Viruses and Virus Diseases of the Grapevine, Bari.

11. Bovey, R., Gärtel, W., Hewitt, W.B., Martelli, G.P., and Vuittenez, A. 1980. Virus and Virus-like Diseases of Grapevines. Editions Payot, Lausanne, Switzerland

12. Cadman, C.E., Dias, H.F., and Harrison, B.D. 1960. Sap-transmissible viruses associated with diseases of grape vines in Europe and North America. Nature, London 187:577-579.

13. Caudwell, A. 1993. Advances in grapevine yellows research since 1990. Pages 79-83 in: Extended Abstracts 11th Meeting ICVG, Montreux, 1993.

14. Davis, R.E., and Prince, J.P. 1993. Grapevine yellows diseases: diverse etiologies indicated by new DNA-based methods for pathogens detection and identification - Implications for epidemiology. Pages 93-94 in: Extended Abstracts 11th Meeting ICVG, Montreux, 1993.

15. Deth, P.S. 1973. The New York State grape certification program. N.Y. Food Life Sci. Quarterly 6:5.

16. Frison, E.A., and Ikin, R., eds. 1991. FAO/IBPGR Technical Guidelines For The Safe Movement of Grapevine Germplasm. Food and Agricultural Organization of the United Nations/International Board for Plant Genetic Resources, Rome, Italy.

17. Gifford, E.M., and Hewitt, W.B. 1961. The use of heat therapy and in vitro shoot tip culture to eliminate fanleaf virus from grapevine. Amer. J. Enol. Vitic. 12:129-130.

18. Goheen, A.C. 1977. Virus and viruslike diseases of grapes. HortScience 12:465-469.

19. Goheen, A.C. 1980. The California clean grape stock program. Calif. Agric. 34 (6): 15-16.

20. Goheen, A.C., and Luhn, C.F. 1973. Heat inactivation of viruses in grapevines. Riv. Patol. Veg. 9 (Suppl.): 287-289.

21. Golino, D.A., Pool, R., and Gonsalves, D. 1993. Clean stock and quarantine programs for grapevines in the United States of America. Pages 151-161 in: Viticulture. 73rd General Assembly of the OIV, S. Francisco, 1993.

22. Habili, N., Krake, L.R., Barlass, M., and Rezaian, M.A. 1992. Evaluation of biological indexing and dsRNA analysis in grapevine virus elimination. Ann. Appl. Biol. 121:277-283.

23. Hewitt, W.B. 1954. Some virus and virus-like diseases of grapevines. Bull. Cal. Dept. Agric. 43:47-64.

24. Hewitt, W.B., Goheen, A.C., Raski, D.J., and Gooding, G.V. 1962. Virus diseases of the grapevine in California. Vitis 3: 57-83.

25. Krake, R.L., and Woodham, R.C. 1983. Grapevine yellow speckle agent implicated in the aetiology of vein banding disease. Vitis 22: 40-50.

26. Maixner, M. 1994. Transmission of German grapevine yellows (Vergilbungskrankheit) by the planthopper Hyalestes obsoletus (Auchenorrhyncha: Cixiidae). Vitis 33:103-104.

27. Martelli, G.P., ed. 1977. Utilizzazione dei risultati della selezione della vite da vino. in: Consiglio Nazionale delle Ricerche, Bari, Italy.

28. Martelli, G.P., ed. 1992. Grapevine viruses and certification in EEC countries: State of the art. Quaderno No. 3, Istituto Agronomico Mediterraneo, Bari, Italy.

29. Martelli, G.P., ed. 1993. Graft-transmissible diseases of grapevines. in: Handbook For Detection and Diagnosis. FAO Publication Division, Rome, Italy.

30. Martelli, G.P. 1994. Inquadramento sistematico dei virus delle vite. Pages 245-254 in: Atti Giornate Fitopatologiche 1994, vol. 2. Editrice CLUEB, Bologna, Italy.

31. Martelli, G.P., Saldarelli, P. and Boscia, D. 1996. Filamentous viruses of the grapevine: closteroviruses. in: Filamentous Viruses of Woody Plants. P.L Monette, ed. Research Signpost, Trivandum, India (in press).

32. Martelli, G.P., De Sequeira, O.A., Kassemeyer, H.H., Padilla, V., Prota, U., Quacquarelli, A., Refatti, E., Rüdel, M., Rumbos, I.C., Savino, V., and Walter, B. 1993. A scheme for grapevine certification in the European Economic Community, monograph no. 54. Pages 279-284 in: Plant Health and The European Single Market. D. Ebbels, ed. British Crop Protection Council.

33. Meredith, C.P., and Walker, M.A. 1993. Clone certification in United States. Pages 100-103 in: Viticulture. 73rd General Assembly of the OIV, S.

Francisco 1993.

34.   Muth, F. 1926. Wie selektioniert der Winzer am besten seine Weinberge? Wein und Rebe 8:158-176

35.   Pearson, R.G., and Goheen, A.C , eds. 1988. Compendium of Grape Diseases. American Phytopathological Society Press, St. Paul, MN.

36.   Sabanadzovic, S., Saldarelli, P., and Savino, V. 1996. Molecular diagnosis of grapevine fleck virus. Vitis (in press).

37.   Schöffling, H., and Stellmach, G. 1993. Klon-Züchtung bein Weinreben in Deutschland. Waldkircher Verlag, Germany.

38.   Semancik, J.S. 1993. Current status of research on grapevine viroids. Pages 34-36 in: Extended Abstracts, 11th Meeting ICVG, Montreux, 1993.

39.   Smith, I.M., McNamara, D.G., Scott, P.R., and Harris, K.M. 1992. Quarantine Pests for Europe. CAB International, Wallingford, UK.

40.   Staub, U., Polivka, H., Herrmann, J.V., and Gross, H.J. 1995. Transmission of grapevine viroids is not likely to occur mechanically by normal pruning. Vitis 34:119-123.

41.   Stramaglia, L., and Martelli, G.P. 1993. Legislative measures and procedures for selection and certification of the grapevine in Italy. Pages 1-21 in: Viticulture. 73rd General Assembly of the OIV, S. Francisco, 1993.

42.   Szychowski, J.A., McKenry, M.V., Walker, M.A., Wolpert, J.A., Credi, R., and Semancik, J.S. 1995. The vein-banding disease syndrome: A synergistic reaction between grapevine viroids and fanleaf virus. Vitis 34:229-232.

43.   Uyemoto, J.K. and Welsh, M.F. 1974. Viruses threaten vineyards - nursery certification reduces the threat. N.Y. State Hort. Proc. 119:180-184.

44.   Uyemoto, J., Martelli, G.P., Woodham, R.C., Goheen, A.C., and Dias, H.F. 1978. Grapevine (Vitis) virus and virus-like diseases. in: Plant Virus Slide Series. O.W. Barnett and S.A. Tolin, eds. Clemson University, Clemson, SC.

45.   Walter, B. 1993. Advances in virus disease diagnosis since 1990. Pages 127-130 in: Extended Abstracts, 11th meeting ICVG, Montreux, 1993.

46.   Walter, B. 1996. Harmonisation dans l'Union Européenne des protocoles de dépistage de maladies virales de la vigne: les virologues sont à l'ouvrage. Prog. Agric. Vitic. 112: 460-461.

47.   Walter, B., and Etienne, L. 1987. Detection of grapevine fanleaf viruses away from the period of vegetation. J. Phytopathol. 120:355-364.

48.   Walter, B., and Martelli, G.P. 1996. Sélection clonale de la vigne: sélection sanitaire et sélection pomologique. Influence des viroses et qualité. Bull. OIV (in press).

49.   Walter, B., Bass, P., Legin, R., Martin, C., Collas, A., and Vesselle, G. 1990. The use of green grafting technique for the detection of virus-like diseases of the grapevine. J. Phytopathol. 128:137-145.

CHAPTER 21

Virus Certification of Ornamental Plants - the European Strategy

G. Krczal

The production and trade of ornamentals has constantly increased during recent decades. New production methods, such as *in vitro* propagation, have led to highly specialized businesses which concentrate on the mass production of one or two ornamental species. Outbreaks of pests and diseases can represent very serious risks for businesses, not only through the economic losses of the affected nurseries, but also by inflicting international trade difficulties due to quality and/or quarantine concerns of importing countries. The effects of spreading pests and pathogens by propagation material, especially those which can occur in latent symptoms stages are especially severe since this material will serve as a foundation stock for final mass production in other nurseries within the country or in other countries.

On the other hand, the introduction and rapid spread of dangerous quarantine organisms such as *Frankliniella occidentalis*, *Bemisia tabaci*, *Liriomyza huidobrensis* and tomato spotted wilt tospovirus within the Euro-Mediterranean region showed that traditional plant quarantine measures such as border inspections, phytosanitary certificates and post-quarantine have failed to a certain extent, especially in countries with high trading volumes and/or no geographical isolation.

The existing EC plant health regime, adopted in 1976, has been limited to the consolidation of standards which govern the movement of plants and plant products between EC Member States as well as the import of plants and plant products to the EC from other countries. Inspection of such goods took place prior to export from one Member State and, on a check sampling basis, upon import to another Member State and, finally, upon import to the destined Member State.

The different treatment accorded to plant material in different member states was likely to create trade barriers and thus hinder free movement within the Community. With a view to the single market the barriers were intendend to be removed by setting up Community provisions to replace these laid down by the member states.

The Single European Act (1986), which amended the Treaty of Rome, provided for the creation by January 1993 of an area without internal frontiers and in which the free movement of goods, persons, services and capital is ensured, i.e.a "Single Market". The inspection practices under the 1976 EC plant health regime, described above, were clearly inconsistent with the requirements of the Single Market and therefore the Commission, set out a program of measures necessary to adapt the EC plant health regime.

The main features of this program involved:

i.   alignement of phytosanitary standards for domestic and intra-EC trade;

ii.   transfer of phytosanitary checks from the internal borders to places of production within the EC and to the external borders of the EC for third country products, while affording appropriate protection to certain defined zones within the EC at particular risk from certain harmful organisms for which complete harmonization of standards is not possible throughout the EC ("protected zones");

iii.   establishment of a system of plant passports attached to sensitive goods and attesting to their compliance with Community standards. The passports ensures free circulation within the Community and replaces plant health certificates in intra-EC trade;

iv.   establishment of an EC plant health inspection activity to ensure uniform application of checks by national inspectors and, under certain circumstances, to check imports from third countries.

THE LEGISLATIVE PROGRAM FOR THE ESTABLISHMENT OF THE "SINGLE MARKET"

This large legislative program is now nearing completion. The EC institutions have elaborated the basic texts providing for the adaptation of the existing regime described under the first three points above as well as the mandate for an EC inspection activity. The latter will be carried out by a permanent Office which has now been established by the Commission and is currently being staffed and set up. Concurrently, Member States are adapting their national rules and re-organizing their plant health controls according to the new EC-regulations.

The Commision Directive 77/93/EEC implements protective measures that have to be taken against the introduction into and spread within the Community of harmful or quarantine organisms. Plant material that is traded according to this Directive is labelled with the Plant Passport.

A certificate for ornamentals is only provided, however, if the plants are produced and traded according to the Commission Directive 91/682/EEC concerning the marketing of ornamental plant material. This Directive includes not only aspects of plant health but also of plant quality standards (also trueness to type and varietal purity). It sets out conditions in general to be met by ornamental plant propagating material and ornamental plants and the implementing measures concerning the supervision and monitoring of the suppliers and establishments producing this material.

The system of each of the Directives is based on a new approach of the implementation of quality control, in which the emphasis is laid on holding the supplier responsible. The French word "responsabilation" exactly sums up the intention. The supplier takes all necessary measures to guarantee compliance with the standards laid down by the Directive concerned, at all stages of the production and marketing of propagating and planting material. The responsible official body of the Member State carries out controls and inspections in order to ensure that the supplier meets his obligations.

A number of demands are made upon the supplier before he can be accredited by the responsible official body. Only accredited suppliers may market propagating and/or planting material. In addtion to the demands on the supplier, requirements are made of the propagating and planting material concerning varietal aspects, plant health and quality. These requirements are laid down in detail for schedules for each genus and/or species, with a reference to the plant health conditions laid down in Directive 77/93/EEC. Compared to fruit plants where a traditional certification system with three categories of material is applied (pre-basic, basic and certified, that may be declared virus-free or virus-tested) there is only one category of material, defined as CAC (Conformitas Agraria Communitatis) for ornamental plants.

Propagating and planting material of ornamentals may be only marketed if it meets at least the requirements laid down for CAC material in the schedule concerned, with reference to the variety to which it belongs and accompanied by a document made out by the supplier in accordance with the conditions laid down in the schedule concerned.

Initially, Community rules are established for those genera and species of ornamental plants which are of major economic interest in the Community. A Community procedure for adding further genera and species will be established. Obviously the determination of plant health and quality standards require lenghty and detailed technical and scientific consideration, therefore a procedure for the determination of these has to be laid down.

It is not appropriate to apply these Community rules to plant material of ornamentals that is intended to be exported to third countries, as the rules applicable there may be different from those contained in the Community Regulations. Moreover, provisions will be made for authorizing the marketing within the Community of propagating material and ornamental plants produced in third countries, provided always that they afford the same assurances as material produced in the Community and complying with Community rules. The third-country import regime is necessarily different from the national controls it replaces. It attempts to identify the most critical organisms and hosts and also to take some account of unknown risks. But it will need to be refined through appropriate dispensation where experience shows this to be possible, and it is intended to be compatible at all times with the draft GATT-SPS agreement.

The following genera and species are subject of the rules laid down in Commission Directive 91/682/EEC:

Begonia x hiemalis Fotsch, Pelargonium, Dendranthema x grandiflorum (Ramat.) Kitam., Dianthus caryophyllus L. and hybrids, Euphorbia pulcherrima(Wild ex Kletsch), *Gerbera, Phoenix, Rosa, Citrus* (ornamental), *Malus* (ornamental), *Prunus* (ornamental), *Pyrus* (ornamental), *Lilium, Gladiolus,* and *Narcissus*

The Council Directive 93/49/EEC sets out the conditions to be met by ornamental plant propagating material and ornamental plants in accordance with article 4 of Directive 91/682/EEC. The conditions laid down in this Directive must be regarded as the minimum standard acceptable at this stage, taking into account the current production conditions in the Community; whereas they will progressively be developped and redefined, in order ultimately to achieve high standards of improved quality.

The list of specific harmful organisms and diseases of quality affecting significance comprises for each genera specific insects, mites and nematodes, bacteria, fungi and viruses and virus-like organisms.

The plant material must comply with the relevant plant health conditions laid down in Directive 77/93/EEC and must, at least by visual observation, be substantially free from any harmful organisms and diseases of quality - affecting significance, sings or symptoms thereof, which reduce the usefulness of the propagating material of ornamental plants, and in particular from those listed above.

Any material showing visual signs or symptome of the said harmful organisms or diseases at the stage of the growing crop shall be treated properly, immediately upon appearance or, where appropriate, shall be removed.

In the case of citrus material the following requirements shall also be met:

● it shall be derived from initial material which has been checked and found to show symptoms of the relevant viruses, virus-like organisms or diseases listed (viroids such as exocortis, cachexia

(xyloporosis); infectious variegation disease, citrus leaf rugose, and diseases that induce porosis-like symptoms as porosis, ringspot, cristacortis, impietratura, and concave gum.)

● it shall have been checked and found to be substantially free of such viruses, virus-like organisms or diseases since the beginning of the last cycle of vegetations; and

● in the case of grafting, it shall have been grafted into rootstocks other than those susceptible to viroids.

In the case of flower bulbs the following requirement shall also be met:

● the propagating material shall be derived directly from material which, at the stage of the growing crop, has been checked and found to be substantially free from any harmful organisms and diseases, signs or symptoms thereof particular from those listed.

The supplier's document referred to in Article 11 of Directive 91/682/EEC shall be printed in at least one of the official languages of the community. It shall contain the following information rubrics:

● Indication "EEC quality"

● Indication of EEC member State code

● Indication of responsible official body or its distinguishing code

● Registration or accrediation number

● Name of supplier

● Individual serial, week or batch number

● Date of issue of the supplier's documents

● Botanical name

● Denomination of the variety, where appropriate. In the case of rootstocks, denomination of the variety or its designation

- Denomination of the group of plants, where appropriate

- Quantity

- In the case of imports from third countries under Article 16(2) of Directive 91/682/EEC, the name of the country of harvesting

Suppliers that produce certified material of ornamental plants have to be supervised and monitored by an official body. The supervision and monitoring has to be carried out at least once a year at an appropriate time, in order to ensure continued compliance with the requirements laid down in Directive 91/682/EEC.

In respect of taking samples for analysis, if necessary, in an accredited laboratory, the responsible official body has to monitor the supplier to ensure that the samples are taken in a technically correct manner, at appropriate time. The analysis of the samples has to be carried out by a laboratory which is accredited for that purpose.

The EEC Directives do not recommand method for diagnosis of suspicions samples, all accredited methods are accepted. The European Plant Protection Organization (EPPO) has, however, established standard test protocols for different ornamental plant species that may be applied within these tests.

THE EUROPEAN PLANT PROTECTION ORGANIZATION (EPPO)

The European and Mediterranean Plant Protection Organization (EPPO) is a regional organization established according to the International Plant Protection Convention. One of its' main areas of work is the coordination and harmonization of plant quarantine measures to limit the spread of quarantine pests into or within the region. For this purpose, a "Working Party" has been established which coordinates and designs the work in plant quarantine. Under the "Working Party" a number of Panels of experts from the whole region accomplish the scientific part of the work and draft recommendations which have to be approved by the Executive Committee and Council of EPPO to receive the official status of an EPPO recommendation. Beside its regional work, EPPO is very much involved in the global

efforts to harmonize phytosanitary measures.

Several EPPO Member Governments developed certification schemes for certain ornamental species in order to raise the health standard of the propagation material and, subsequently, to promote trade. Certification schemes in the different countries are, however, mostly not consistent with each other or comparable due to differences in testing methods, sampling techniques or simply because of the pathogens which are tested for. EPPO was, therefore, called by its Member Governments to evaluate exiting national certification schemes and to develop standard schemes from them.

EPPO'S CERTIFICATION SCHEMES

EPPO's work in general and the certification schemes in particular are based on the assumption that Member Governments cooperate and work closely together. The adoption of the EPPO certification schemes (EPPO recommendations are not compulsory for the EPPO Member Governments) in the different EPPO countries could lead to a harmonized, healthy and free market for many ornamentals, not just for the 12 EC Members, but for all 34 EPPO Member Governements. For establishments and nurseries the production of ornamental propagating material according to the high standards of EPPO schemes, even if not compulsory, could lead to a better acceptance of their plant material, world-wide.

The EPPO Panel on "Certification of Ornamentals", composed of experts from nine countries, was founded in 1985 and has since developed certification schemes for five ornamental crops: carnation (2), *Pelargonium* (5), Lily (6), *Narcissus* (7) and chrysanthemum (8). Certification schemes for several other ornamental crops are still under discussion or subject to approval within EPPO (bulbous iris, rose, begonia, impatiens). Beside certification schemes EPPO also developed classifications schemes for tulip bulbs, crocus, bulbous iris, hyacinth, narcissus, fresia and kalanchoe. The difference in the certification/ classification schemes is essentially in the fact that in a certification scheme, certified stock is derived by known filiation steps from individually tested nuclear stock plants. In a classification scheme, the classified stock comes in one step from inspected and randomly tested mother plants whose filiation

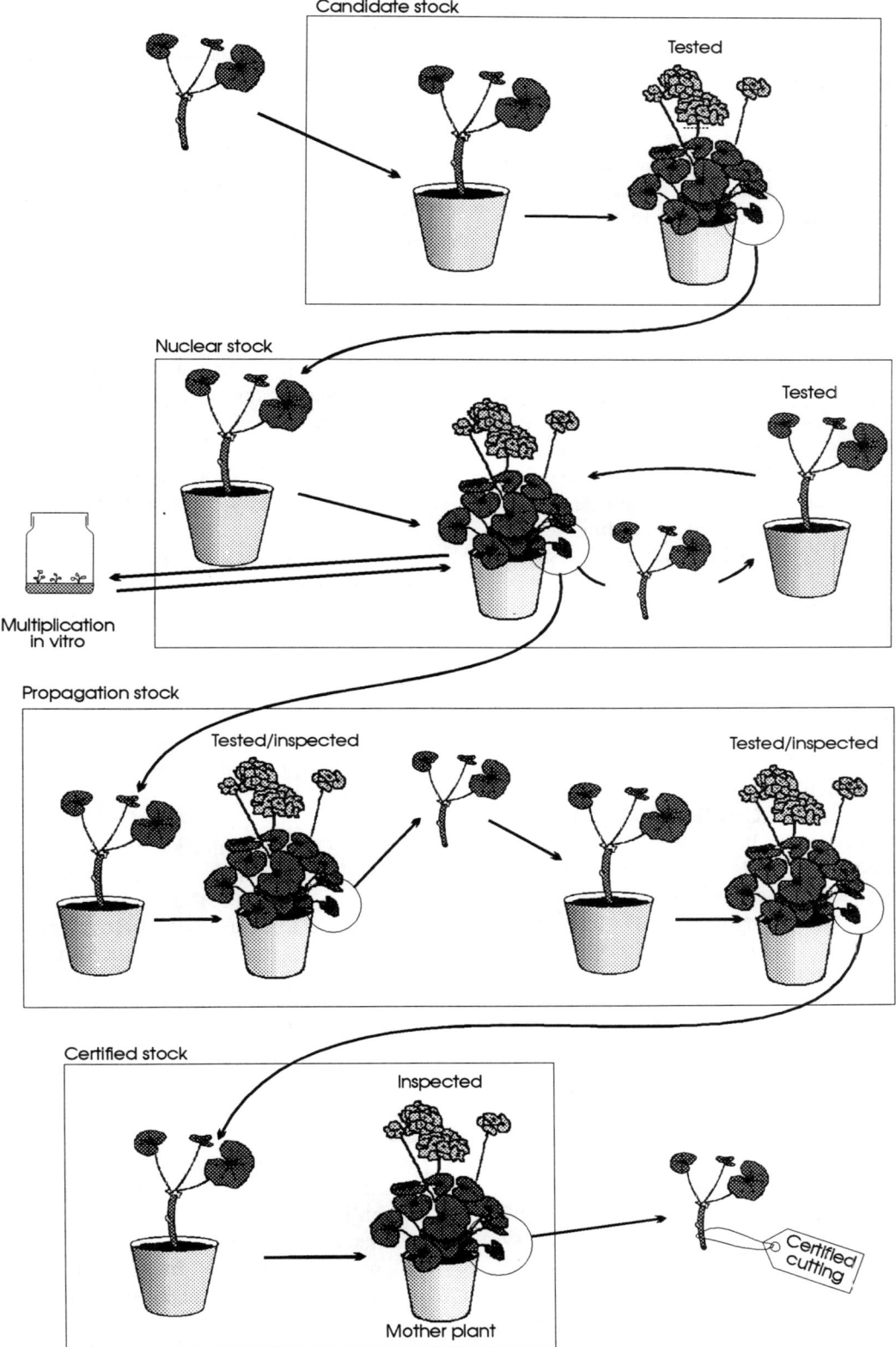

Figure 1. Diagram of the certification scheme for pelargonium as an example for EPPO's certification schemes

is not considered.

A classification scheme may be extended to include several classes, according to the

Propagation Stock

Propagation stock material is, in respect to its health status, a slightly lower category of plants, subject to pathogen tests but not individually. Propagation stock has to be randomly tested for a variety of pathogens and any plant showing viral, bacterial or fungal infections or positive test results should be eliminated. Propagation stock should be kept in suitable greenhouse to avoid re-infection by pathogens and should be separated from any other crop. Cuttings taken from the propagation stock can retain their status if they are tested according to the propagation stock schedule. There is a maximum of filiation steps according to the crop in question for which propagation stock can retain its status. Cuttings which are not tested according to the propagation stock schedule become certified stock.

Certified Stock

Certified stock is the final category within the propagation process of the certification scheme. Certified stock plants are the mother plants from which the final product of the scheme - the certified stock cutting - are taken and sold to any nursery which produces cutflowers or pot plants for consumption. Certified stock plants also have to fulfil certain health standards and must be randomly tested (single plants or pooled samples) according to a defined schedule, less stringent than for propagation stock. Certification will be granted if these health standards are met. Any plants showing visual symptoms of viral diseases or any other diseases should be eliminated. The plants should be kept in separate greenhouses apart from nuclear stock, propagation stock and uncertified material.

One essential part of the certification scheme is the requirement that individual plants within the system can always be traced back through the filiation steps to their original nuclear stock i.e. the number of propagation steps from nuclear stock is known. This makes it possible to detect the origins of diseases as well as to eliminate them more easily. It corresponds to a basic concept of the EPPO schemes, that phytosanitary quality (probability that the material is not infected by any of the pests covered by the scheme) is determined not only by the testing schedule and conditions of isolation, but also by the number of propagation steps.

The propagation steps in the certification schemes require establishments to have different technical know-how. Also the isolation requirements for greenhouses are different for the separate categories. Therefore, it might be useful if the production of nuclear stock, propagation stock and certified stock is separately carried out by establishments which have specialized in the production of one category.

In general the certification scheme should be under the strict control of an official inspection/ certification organization, throughout all of its stages. Establishments involved in the production of nuclear, propagation or certified stock should be officially registered.

SPECIFIC CERTIFACTION SCHEMES

Within the certification schemes already developed by EPPO the different ornamental crops has to be tested for the following viruses with the test methods indicated:

Pelargonium

Candidate material for nuclear stock status has to be tested for and be free of the following viruses:

Cucumber mosaic cucumovirus (CMV)
Pelargonium flower break carmovirus (PFBV)
Pelargonium leaf curl tombusvirus (PLCV)
Pelargonium line pattern virus (PLPV)
Pelargonium ringspot virus (PelRSV)
Tobacco ringspot nepovirus (TRSV)
Tomato black ring nepovirus (TBRV)
Tomato ringspot nepovirus (TomRSV)

The most important of these in practice are: PLPV which has a long incubation period and may therefore be overlooked; PLCV, transmitted mainly by propagation but also by irrigation water, which is tending to disappear from well-managed stocks as it is slow to reinfect; PFBV which is important because of its rapid transmission by mechanical means, irrigation systems and thrips, is particularly damaging in pink cultivars, but latent in other *P. zonale* and *P. peltatum* cultivars and may easiliy be overlooked.

PelRSV is mechanically transmitted and is widespread in old cultivars. CMV, aphidborne and polyphagous, only occurs occasionally in pelargonium. The three nematode-transmitted viruses which have a wide range on other hosts, can fairly easily be kept out of pelargonium stocks. TomRSV does not occur naturally in the EPPO region but is occasionally found in imported ornamentals. It is considered as an EPPO A2 quarantine pest, the main risk being to fruit trees in which it has never been found in Western Europe.

Tomato spotted wilt virus (TSWV) has recently become much more important in the EEPO region, mainly as a result of the introduction of a new thrips vector, *Frankliniella occidentalis*. It can be transmitted by thrips on contaminated implements. Candidate material of pelargonium for nuclear stock status should as far as possible be free from TSWV, by visual inspection for symptoms, control measures taken against its vectors, and use of test methods.

Nuclear stock plants have to be tested individually once a year at least for PFBV, PLPV and TomRSV. Propagation stock plants should be randomly tested for PFBV, same should be done within certified stock.

Test Methods

Cucumber Mosaic Cucumovirus (CMV)

While visual inspection may provide some indication of symptoms (irregular leaf distortion and mottling, severe stunting) and this is sufficient for random checks on propagation and certified stock, nuclear stock (and possibly the first generation of propagation stock) should be tested either by inoculation to *Chenopodium quinoa* or by ELISA.

Pelargonium Line Pattern Virus (PLPV)

The virus causes line and ring patterns or yellow spots on the leaves in spring but is symptomless in many pelargonium cultivars. Since it has a long incubation period, visual examination is insufficient to detect infection, but may be used for random checking of propagation and certified stock. Nuclear stock (and possibly the first generation of propagation stock) should be tested either by inoculation to *C. quinoa* or by ELISA, once a year for nuclear stock. PLPV may be irregularly distributed in the plant and it is therefore preferable to take samples form the lower leaves.

Tomato Ringspot Nepovirus (TomRSV)

Infected plants of many pelargonium cultivars show no conspicuous symptoms, while others show chlorotic yellow rings on the older leaves in spring. Since TomRSV is an EPPO A2 quarantine pest, the testing procedures for it are given in EPPO Quarantine Procedure Nr.27. Essentially, TomRSV can be mechanically inoculated to *C. quinoa* (or other indicators), and its infection to this host can be confirmed by ELISA, or ELISA can be used directly for its detection from pelargonium.

Pelargonium Flower Break Carmovirus (PFBV)

The symptoms are breaking of the flower and mottling on the leaves depending on the variety infected. Plants should be tested for this virus by inoculation to *C. quinoa* or by ELISA.

Pelargonium Ringspot Virus (PelRSV)

Clear ringspot symptoms are caused on *P. peltatum* and line pattern symptoms on *P. zonale*. However, because symptoms are very slow to develop in infected plants (inoculated seedlings develop symptoms only after 15 months), visual inspection is insufficient to detect infection. Plants can be inoculated to *C. quinoa* or can be tested by ELISA.

Tomato Black Ring Nepovirus (TBRV), Tobacco Ringspot Nepovirus (TRSV), Pelargonium Leaf Curl Tombusvirus (PLCV)

Beside visual symptoms, these viruses can all be detected by transmission to *C. quinoa*, as already done for the three more important viruses above. In addition, specific ELISA testing for these viruses is possible.

Chrysanthemum

Candidate material has be tested for and be free of the following viruses and viroids:

Chrysanthemum B carlavirus (CBV)
Tomato aspermy cucumovirus (TAV)
Tomato spotted wilt virus (TSWV)
Chrysanthemum stunt viroid (ChSVd)

All nuclear stock plants have to be tested individually for the above mentionned viruses. Propagation stock plants should be randomly tested for CBV, TAV, TSWV and ChSVd, same has to be done within certified stock, at this stage pooled samples are accepted.

Test Methods

Chrysanthemum B Carlavirus (CBV)

Most cultivars are symptomless when infected, though some show mild leaf mottling and slight loss of flower quality. Therefore, visual examination provides no reliable indications. Nuclear stock can be checked by mechanical inoculation to petunia or by ELISA.

Tomato Aspermy Virus (TAV)

Infected plants are often symptomless, but the virus can induce dwarfing or flower disortion (especially in conjunction with CBV). Visual examination provides no reliable indications. Nuclear stock can be checked by mechanical inoculation to petunia, *Nicotiana tabacum* or *C. quinoa*, or by ELISA. The tests are applied as for CBV above.

Chrysanthemum Stunt Viroid (ChSVd)

ChSVd gives weak or non-specific symptoms, and none at all on certain cultivars, so testing is essential. Three methods are available: 1.) grafting onto a suceptible cultivar, 2) electrophoresis, 3) molecular probe. ChSVd is an EPPO A2 quarantine pest, and the test methods have been presented in EPPO Quarantine Procedure Nr.24 (1).

Tomato Spotted Wilt Virus (TSWV)

Infected plants of chrysanthemum can occassionally show severe symptoms of infection by TSWV. Symptoms are brown necrotic lesions, yellow/brown concentric rings and/or flower distortion. Test methods are available but are not fully reliable, so detection of TSWV should combine visual inspection (also for the thrips vectors) with test methods. TSWV can be transmitted mechanically to petunia or to *N. benthamiana*. It can also be detected by ELISA.

Lily

Candidate material has to be tested for the following viruses:

Arabis mosaic nepovirus (ArMV)
Cucumber mosaic cucumovirus (CMV)
Lily symptomless carlavirus (LSV)
Lily X potexvirus (LXV)
Tobacco rattle tobravirus (TRV)
Tulip breaking potyvirus (TBV)

Of these viruses, the most severe, causing most difficulties in practice, are the aphid-transmitted viruses, LSV and TBV. The nematode-transmitted viruses (nepoviruses, tobraviruses) have a wide host range. They can be excluded from certified stocks without too much difficulty. TRV ist the most important in practice.

Nuclear stock plants must be tested individually for all viruses once a year, except the nemtaode-transmitted one. For propagation stock I (in aphid roof greenhouses) random test on growing plants for TBV and LVX must be negative while infection by LSV should not exceed 1 % (or 2 % for oriental and speciosum hybrids). For progation stock II and certified stock (in the field) records must show that random test by ELISA on bulb samples after harvest show infection by TBV and LVX not exceeding 1 % and by LSV not more than 10 % (for 20 % for oriental and speciosum hybrids).

Test Methods

Arabis Mosaic Nepovirus (ArMV)

The nepovirus ArMV alone induces symptomless infection in lily, so visual inspection provides no reliable indications of infection. It can readily be detected by ELISA and/or ISEM in naturally infected hosts or in inoculated indicator plants.

Cucumber Mosaic Cucumovirus and Tobacco Rattle Tobravirus (CMV, TRV)

These viruses induce no characteristic symptoms in lily. Like the nepoviruses they are best identified by ELISA and/or ISEM in naturally infected hosts or in inoculated indicator plants.

Tulip Breaking Potyvirus (TBV)

Serological procedures (ISEM and ELISA) can be used to detect and identify tulip breaking potyvirus in lily, which can also be inoculated to the indicator plant *Lilium formosanum*.

Lily Symptomless Carlavirus and Lily X Potexvirus (LSV, LXV)

These viruses are best detected and identified in naturally infected lilies and in inoculated indicator plants by ELISA and/or ISEM.

Narcissus

The plants within candidate material should be examined visually at regular intervals for absence of symptoms, and individually tested on 3 occasions, during at least 2 growing seasons for the following viruses:

Narcissus Pseudonarcissus

Arabis mosaic nepovirus (ArMV)
Cucumber mosaic cucumovirus (CMV)
Narcissus latent carlavirus (NaLV)
Narcissus late season yellows potyvirus (NLSYV)
Narcissus mosaic potexvirus (NaMV)
Narcissus tip necrosis tombusvirus (CMV)
Narcissus virus Q (NVQ)
Narcissus white streak potyvirus (NWSV)
Narcissus yellow stripe potyvirus (NYSV)
Raspberry ringspot nepovirus (RRSV)
Strawberry latent ringspot nepovirus (SLRSV)
Tobacco rattle tobravirus (TRV)
Tobacco ringspot nepovirus (TRSV)
Tomato black ring nepovirus (TBRV)

Narcissus Tazetta

Arabis mosaic nepovirus (ArMV)
Carnation latent carlavirus (CLV)
Cucumber mosaic cucumovirus (CMV)
Narcissus degeneration potyvirus (NDV)
Strawberry latent ringspot nepovirus (SLRSV)
Tomato black ring nepovirus (TBRV)

Of these viruses, the most severe, causing most difficulties in practice, are the aphid-transmitted viruses: CMV, NaLV, CLV and potyviruses. The nematode-transmitted viruses (nepoviruses, tobraviruses) have a wide host range and occur only incidentally in bulbs, in which they are relatively unimportant. They can be excluded from certified stocks without too much difficulty. TRV is the most important in practice. NaMV (mechanically transmissible) and NTNV are relatively unimportant.

At least every 3 years, all nuclear stocks plants should be indexed for all viruses listed above. Propagation stock I (greenhouse) must also be free of all viruses. Within propagation stock II (in the field) only 0,05 % of the plants inspected may show severe and 0,5 % mild virus symptoms. If the incidence of pests exceeds these percentages, certification will be refused to the whole lot. The treshold for virus incidence within certified stock plants is 2 % for severe virus symptoms during inspection.

Test Methods

Nepoviruses (ArMV, RRSV, SLRSV, TRSV, TBRV)

The nepoviruses alone induce symptomless infection in narcissus so visual inspection provides no reliable indication of infection. They can readily be detected by ELISA and/or ISEM in naturally infected hosts or in inoculated indicator plants.

Cucumber Mosaic Cucumovirus (CMV), Narcissus Mosaic Potexvirus (NaMV) and Tobacco Rattle Tobravirus (TRV)

These viruses alone induce no characteristics symptoms in narcissus. Like the nepoviruses they are best identified by ELISA and/or ISEM in naturally infected hosts or in inoculated indicator plants.

Narcissus Latent Carlavirus (NaLV) and Narcissus Tip Necrosis Tombusvirus (NTNV)

NaLV induces leaf chlorosis, and NTNV causes characteristic tip necrosis in some narcissus cultivars. Both are best identified by ELISA and/or ISEM in naturally infected narcissus or indicator-plants.

Potyviruses (NWSV, NYSV, NDV, NLSYV)

The potyvirus infecting narcissus usually induce characteristic symptoms, but those of NLSYV and NWSV appear only after flowering. It is not recommended to use indicator plants. No antisera are available for NWSV, which should therefore be identified by visual symptoms and by the detection of filamentous particles unrelated to the other two potyviruses in "decoration tests" by ISEM. Serological procedures (ISEM and ELISA) can be used to detect and identify the other two potyviruses in narcissus.

Carnation

Candidate material for nuclear stock status has to be indexed for the following viruses (4):

Carnation etched ring caulimovirus (CERV)
Carnation latent carlavirus (CLV)
Carnation mottle carmovirus (CarMV)
Carnation necrotic fleck closterovirus (CNFV)
Carnation ringspot dianthovirus (CRSV)
Carnation vein mottle potyvirus (CVMV).

Of these, CRSV causes relatively conspicuous symptoms and has, in consequence, almost disappeared from well-managed production units. The other viruses are often more or less latent and present more difficulties. CarMV may persist at low concentration ("attenuated"), and presents the greatest problems. It is found, in practice, that material from which CarMV has been entirely eliminated by meristem-tip culture is generally also free from the other viruses. CERV, CLV, CNFV and CVMV are aphid-transmitted, while CarMV and CRSV are transmitted by contact.

Nuclear stock plants should individually be retested once a year for CERV and CNFV, and at least twice a year for CarMV. Within propagation stock random tests for CarMV should be performed and may be applied to pooled samples. The presence of CarMV should be also checked in certified stock plants by random testing.

Test Methods

Carnation Mottle Carmovirus (CarMV)

Visual inspection provides no reliable indication of infection. Nuclear stock should be checked by inoculation on *C. quinoa, C. amaranticolor* or *Saponaria vaccaria*. Attenuated forms of the virus may occur, giving few or no local lesions on *C. quinoa*. So a second inoculation, from *C. quinoa* to *C. quinoa*, should be done. However, if *S. vaccaria* is used as the first indicator species, the attenuated forms will produce systemic symptoms.

ELISA testing may be used for nuclear stock and candidate nuclear stock material, in parallel with biological testing. ELISA will not reliably detect attenuated ChaMV. It is recommended to test candidate nuclear stock plants at least 3 times a year by ELISA, with at least one additional test in spring or autumn by biological means. Nuclear stock should be tested twice a year by ELISA, at least 4 month apart, and once by inoculation to indicator plants. For propagating and certified stock, ELISA testing only is applied to appropriate samples. Gel immunodiffusion may be used occasionally as a check.

Carnation Ringspot Dianthovirus (CRSV)

For nuclear stock, the inoculation test to *C. quinoa* or *C. armaranticolor* done for CarMV will also detect CRSV. ELISA testing may also be used.

For propagation and certified stock, material should be inspected for necrotic spots, deformations and stunting of leaves. Symptoms are generally conspicious, and CRSV is now rare in well managed carnation propagation systems.

Carnation Vein Mottle Potyvirus (CVMV)

For nuclear stock, the inoculation test to *C. quinoa* or *C. amaranticolor* done for CarMV will also detect CVMV. ELISA testing may also be used. In addition, visual inspection of the sample of nuclear stock material taken to flowering will allow detection of breaking and malformation of the flowers. Visual inspection of the propagation or certified stock (which is not flowered) will give no usefull information.

Carnation Necrotic Fleck Clostervirus (CNFV)

Nuclear stock material should be tested both by ELISA and by visual inspection of leaves and flowering stems for streaking and purple discoloration. Propagation and certified stock may be visually inspected for these symptoms on leaves.

Carnation Etched Ring Caulimovirus (CERV)

Nuclear stock material should be tested by mechanical inoculation to *S. vaccaria* or *Silene armeria*, and by visual inspections of leaves for necrotic spots or flecks. ELISA testing may also be used.

Propagation and certified stock should be checked by visual inspection.

Carnation Latent Carlavirus (CLV)

For nuclear stock the inoculation test to *C. quinoa* or *C. amaranticolor* for CarMV will also detect CLV. ELISA testing may also be used.

SUMMARY

The main new feature consequent on the Single Market is the need for the most important types of traded plants to be certified, whether for movement within one Member State or between Member States. This will have considerable consequence for producers and official services and will also affect other countries exporting to the EC, for the same standards will apply. For third countries, this aspect may well have more consequence than the harmonization of phytosanitary regulations as such. In so far as this certification includes plant health elements (especially virus-testing) technical standards have to be set, and EPPO's Panels on certification of ornamentals have been preparing such standards for several years.

REFERENCES

1. Anonymous. 1989. Quarantine procedures. Nr. 24. Chrysanthemum stunt viroid-inspection and test methods. Bulletin EPPO Bull. 19:161-164.
2. Anonymous. 1991. Recommendations made by EPPO Council. General scheme for the production of certified pathogen-tested vegetatively propagated ornamental plants. EPPO Bulletin 21:757.
3. Anonymous. 1991. Annual Report for 1990. Bulletin EPPO Bull. 21 (4):731-763.
4. Anonymous. 1991. Certification Scheme: Pathogen-tested material of Carnation. EPPO Bull. 21 (2):279-290.
5. Anonymous. 1992. Certification Scheme: Pathogen-tested material of Pelargonium. EPPO Bull. 22 (2):285-296.
6. Anonymous. 1993. Certification Scheme: Pathogen-tested material of lily. EPPO Bull. 23 (2):215-234.
7. Anonymous. 1993. Certification Scheme: Pathogen-tested material of narcissus. EPPO Bull. 23 (2):225-237.
8. Anonymous. 1993. Certification Scheme: Pathogen-tested material of chrysanthemum. EPPO Bull. 23 (2):239-247.

CHAPTER 22

Virus Certification of Fruit Tree Propagative Material in Western Europe

M. Barba

Breeding programs for the selection of new high quality germplasm are the pre-requisite for qualified and competitive fruit tree production. The production of fruit trees for national and international marketing is continuously changing because of the interest in growing new cultivars which are highly productive and/or resistant to biotic or abiotic stress. In addition, free trade among Western European countries, in the framework of the European common market, has stimulated the production of high quality vegetatively propogated products. Due to the upsurge of fruit tree germplasm exchange at the national and international level during the last few years, a European framework has been established for evaluating technical and operative aspects of production, certification and marketing of vegetativly propagated material. Certification represents an effective way to guarantee the two requirements that propagative material must meet and maintain throughout the different steps of production: trueness to plant cultivar type and sanitary status. Certification is based on the collaboration between scientific and technical organizations and involves - at different levels and with different responsabilities - more than one component. Certification safeguards both the nurseryman, who sells vegetatively propagated plant material, and the grower, who buys the nurserymans products. This chapter deals with the national and international directives issued in western Europe to guarantee the quality of fruit tree propagative material, especially as related to viral disease control.

INTERNATIONAL DIRECTIVES

The European Union (EU) issued directives indicating qualitative and phytosanitary requirements that propagative material must meet in order to be accepted by member countries. In particular, the directive n. 92/34 (9) regulates 22 species and genera of fruit plants (Table 1).

The EU, strongly desiring the immediate initiation of production and commercialization of certified propagative material within the framework of the European market, describes in its' protocols another category of material named CAC (Conformitas Agricolae Communitatis), which satisfies minimumm sanitary conditions such as the absence of quarantine pests and other pathogens that could affect product quality. Absence of pests and pathogens can be ascertained only by visual inspection and random testing under suspicious circumstances. Obviously the CAC category must be used to guarantee not only the sanitary status but trueness to cultivar type of the propagated plant material.

The main differences between certified and CAC material is that the former is produced by filiation in a known number of stages through vegetative propagation from basic material. The absence of pathogens is ascertained by tests and disease control under the responsibility of an official institution, whereas CAC material does not derive from basic material and its' sanitary status must be ascertained by visual inspection or random tests when suspicious circumstances are encountered.

At the present time the EC has not published specific protocols for producing certified propagative material. These protocols are curently under development and will refer to operative schemes previously published by international organizations. Among them, the European Plant Protection Organization (EPPO) is the most important and active in this field. It is an intergovernmental organization with 35 member

Table 1. List of genera and species to which the 92/34 Directive applies

Citrus sinensis	Orange	*Rubus*	Blackberry
Citrus limon	Lemon	*Prunus avium*	Sweet cherry
Citrus reticulata	Mandarin	*Prunus cerasus*	Sour cherry
Citrus paradisi	Grapefruit	*Prunus domestica*	Plum
Citrus aurantifolia	Lime	*Prunus salicina*	Japanese plum
Olea europea	Olive	*Prunus persica*	Peach
Fragaria x *ananassa*	Strawberry	*Prunus armeniaca*	Apricot
Cydonia oblonga	Quince	*Prunus amygdalus*	Almond
Malus pumila	Apple	*Corylus avellana*	Hazelnut
Pyrus communis	Pear	*Juglans regia*	Walnut
Ribes	Redcurrant	*Pistacia vera*	Pistachio

Table 2. EPPO certification schemes being prepared or published

Plant material	Date(s) of publication	Reference
Published schemes		
Fruit trees	1991,1992	1,2,3,4
Grapevine	1994	5
Ribes	1994	6
Rubus	1994	7
Strawberry	1994	8
Under publication		
Citrus		
In preparation		
Olive		
Hazelnut		
Hop		
Vaccinium		

goverments which promotes meetings among specialized researchers from member states with the purpose of integrating several procedures into cetification schemes.

Supported by researchers involved in virological and breeding aspects, EPPO has prepared and published certification schemes to guide growers of specific propagative material (Table 2). When preparing a certification scheme, the following aspects of certification are considered: To define steps to be followed during certification, to describe the category of plant material produced in each step of the scheme, to identify the most dangerous pathogens to be excluded from certified material and to suggest the most reliable methods for their detection.

Among the pathogens, main consideration is given to the systemic ones such as: viruses, phytoplasmas (formerly known as mycoplasma-like organisms or MLOs) and viroids because they are transmitted in a propagative way and cannot be eliminated by pesticide treatments. Furthermore,

Table 3. Operative steps of the Italian certification scheme

Category	Responsibility
Prebasic	Research Institutes, Ministry of Agriculture
Basic	Research Institutes, Ministry of Agriculture
Certified	Nurserymen associations
Certified plant	Private nurserymen

they cannot be detected by visual observation or optical microscopy and, in some cases, they are latent and do not produce any specific symptom in the host plant.

Italy, like other European countries, has followed EPPO recommendations and, on July 2nd, 1991, enacted a law concerning "Constitutive regulation of the voluntary certification service of vegetative propagative material" (D.M. 289) (10). After this "umbrell" law was enacted, specific directives for "stone fruits" (D.M. December, 31, 1992) (11), "strawberry" (D.M. December, 31, 1992) (12), "olive" (D.M. June, 16, 1993) (13), "citrus" (D.M. October, 29, 1993) (14), "pome fruits" (D.M. October, 29, 1993) (15) and "walnut" (D.M. October, 29, 1993) (16) have been published.

SCHEME FOR THE PRODUCTION OF CERTIFIED PROPAGATIVE MATERIAL

As suggested by EPPO, the Italian legislation foresees a certification by filiation; the steps for multiplication are strictly linked allowing, in each stage of the scheme, the identification of the original source of the material under propagation (Table 3). The following levels are considered:

Nuclear Stock (Prebasic Plants)

This is the first step of the certification scheme. The prebasic plants must show, from official checks and tests, varietal characteristics corresponding to those typical of that variety and must have been found free from the pathogens listed in the specific annexes. They are maintained in a screen house, in containers with sterile growth medium, isolated from the soil. Each plant must be checked for trueness to type during its' vegetative stage. Each single plant has to be tested for virus and virus-like organisms, according to the specific annexes. Clonal origin must be identified by a number that must be maintained throughout the propagative production and commercialization process. Plants satisfying the above mentioned conditions are used for production of prebasic plant material.

Propagation Stock (Basic Plants)

Prebasic plant material is used for the production of basic plants. Basic plants are maintained in a screen house and must satisfy the conditions described for maintenance of prebasic plants. Material from these plants can be used as basic material.

Certified Mother Plants

Basic plant material is used for the production of certified mother plants. *In vitro* micropropagation can be used to produce certified mother plants. Every year, at least 10% of these plants must be tested for virus and virus-like diseases according to specific annexes. Mother plants for production of certified plant material must be separated by at least 100 m from any other non-certified plant material and maintained under continuous surveillance to control pests and diseases and the soil must be free of virus transmitting nematodes. These plants are maintained in open air under conditions ensuring freedom from reinfection and trueness to the variety and are used for the production of certified

Table 4. Sanitary control in Italian certification steps

Resposibility	Steps	Control	Level
Scientific Institutions	Prebasic plants	Biological tests ELISA	National
Scientific Institutions	basic plants	Biological tests ELISA	National or Regional
Nurserymen associations	certified mother plants	ELISA	Regional
Nurserymen	Nurseries	Visual inspection	Local

plant material.

Certified Plant Material (Nursery)

Certified plant material is used for production of certified plants. These are kept in plots free of virus transmitting nematodes and separated at least 5 m from other non-certified material plots. Certified plants are visually checked for harmful organisms and diseases that impair quality, or any signs or symptoms thereof. The nurseries are officially inspected by the responsible authority at least once a year. The responsible authority issues the nurserymen the appropriate number of labels for individual labeling of certified plants.

DIAGNOSIS APPLIED TO CERTIFICATION

An effective certification scheme is based on reliable diagnostic methods; these have been improved and optimized, ranging from simple tests (such as biological assays on herbaceous and woody plants) to very sophisticated serological and molecular techniques. Biotechnology contributed enormously to their development by optimizing the methods and increasing the sensitivity of the reagents. Molecular approaches have been very helpful in detecting systemic pathogens (mainly viroids and phytoplasmas) but, until the present, sophisticated, expensive equipment and well trained staff were required. Serology still remains the simplest approach for an effective and rapid diagnosis.

Several serological methods have been performed and used according to the nature of the problem under investigation . In fact, diagnosis is not only applied to certification purposes, but is very helpful in improving knowledge of the taxonomy, ecology and epidemiology of viruses. With few exceptions, enzyme-linked immunosorbent assay (ELISA) is the most frequently used technique. The use of serological tests, particularly ELISA, depends from the level of the scheme (Table 4).

In the first two steps, the sanitary status of the plants must be ascertained not only by ELISA but also by biological tests on woody indicators. This is due to the fact that serological tests, even if universally recognized, useful and reliable, could fail. If a mistake arises at these stages, it could be enormously amplified, generating thousands of infected plants in the following steps. For this reason, to be sure of the sanitary status of a plant belonging to the nuclear stock or to the propagation stock, ELISA is not considered sufficient: in other words, negative serological results must be confirmed by biological tests.

In a propagative center or in a nursery, serological tests are considered reliable and sufficient to guarantee the sanitary status of propagative material, making it possible to control certified material quickly and efficiently. In conclusion, serology improves the practical performance of the certification process, allowing faster and more reliable diagnosis of the most significant pathogens.

Table 5. List of pathogens covered by the scheme for 'virus-tested' certification

Name	Agent	Suggested indicator	Test[a]
ALMOND			
Virus (l)			
Apple chlorotic leaf spot	ACLSV	GF 305	ELISA
Apple mosaic	ApMV	GF 305	ELISA
Plum pox	PPV	GF 305	ELISA
Prune dwarf	PDV	GF 305	ELISA
Prune necrotic ringspot	PNRV	GF 305	ELISA
APRICOT			
Virus (l)			
Apple chlorotic leaf spot	ACLSV	GF 305	ELISA
Apple mosaic	ApMV	GF 305	ELISA
Plum pox	PPV	GF 305	ELISA
Prune dwarf	PDV	GF 305	ELISA
Prune necrotic ringspot	PNRV	GF 305	ELISA
Phytoplasmas			
Apricot chlorotic leafroll	phytoplasma	Luitez	Probe,DAPI
CHERRY			
Virus (l)			
Apple chlorotic leaf spot	ACLSV	GF 305	ELISA
Apple mosaic	ApMV	GF 305	ELISA
Prune dwarf	PDV	GF 305	ELISA
Prune necrotic ringspot	PNRV	GF 305	ELISA
Raspberry ringspot	RRV	GF 305	
PEACH			
Virus (l)			
Apple chlorotic leaf spot	ACLSV	GF 305	ELISA
Apple mosaic	ApMV	GF 305	ELISA
Plum pox	PPV	GF 305	ELISA
Prune dwarf	PDV	GF 305	ELISA
Prune necrotic ringspot	PNRV	GF 305	ELISA
Viroids			
Peach latent mosaic	viroid	GF 305	Probe
PLUM			
Virus (l)			
Apple chlorotic leaf spot	ACLSV	GF 305	ELISA
Apple mosaic	ApMV	GF 305	ELISA
Plum pox	PPV	GF 305	ELISA
Prune dwarf	PDV	GF 305	ELISA
Prune necrotic ringspot	PNRV	GF 305	ELISA

[a] Serological tests do not substitute biological tests.

HOW THE SCHEME WORKS: THE STONE FRUIT CERTIFICATION

Certification of stone fruits is only one aspect of certification programs. It follows the same procedures applied to other species and the effectiveness of the scheme is guaranteed by the filiation. The difference relies on the peculiarity of the agents to be checked.

As for other species, the protocols foresee two levels of sanitary status of propagative material: virus free and virus tested. The difference is mainly based on the number of considered pathogens.

All pathogens listed in Table 5 must be absent from stone fruit propagative material in order to commercialize it as virus tested. Most of the pathogens are viruses, except apricot leaf roll phytoplasma and peach latent mosaic viroid. The detection of the last two pathogens can be performed by biological or molecular tests. As for other species, a good check at the beginning safeguards the final product.

In conclusion, certification can be considered a useful tool when based on the collaboration and coordination among different public and private institutions whose goal is to improve the quality of propagative plant material.

REFERENCES

1.   EPPO. 1991. Certification scheme. Virus-free or virus-tested fruit trees and rootstocks. Part I: basic scheme and its elaboration. EPPO Bulletin 21:267-277.
2.   EPPO. 1992. Certification scheme. Virus-free or virus-tested fruit trees and rootstocks. Part II: tables of viruses and vectors. EPPO Bulletin 22:255-263.
3.   EPPO. 1992. Certification scheme. Virus-free or virus-tested fruit trees and rootstocks. Part III: testing methods for viruses of fruit trees present in the EPPO region. EPPO Bulletin 22:265-275.
4.   EPPO. 1992. Certification scheme. Virus-free or virus-tested fruit trees and rootstocks. Part IV: technical appendices and table of contents. EPPO Bulletin 22:277-283.
5.   EPPO. 1994. Certification scheme. Pathogen-tested material of grapevine varieties and rootstocks. EPPO Bulletin 24:347-367.
6.   EPPO. 1994. Certification scheme. Pathogen-tested material of Ribes. EPPO Bulletin 24:857-864.
7.   EPPO. 1994. Certification scheme. Pathogen-tested material of Rubus. EPPO Bulletin 24:865-873.
8.   EPPO. 1994. Certification scheme. Pathogen-tested strawberry. EPPO Bullètin 24:875-889.
9.   European Community. 1992. Council directive 92/34 of 28 April 1992 on the marketing of fruit plant propagating material and fruit plants intended for fruit production. EC bulletin n. L 157/10 of June 10, 1992.
10.   Ministero dell'Agricoltura e delle Foreste. 1991. Regolamentoistitutivo del servizio di certificazione volontaria del materiale di propagazione vegetale. D.M. n. 290 July 2 1991, G.U. n. 209 of September 6, 1991.
11.   Ministero dell'Agricoltura e delle Foreste. 1992. Norme tecniche per la produzione di materiale di propagazione vegetale certificato delle Prunoidee e dei relativi portinnesti. D.M. December 31, 1992, G.U. n. 15 of January 20, 1993.
12.   Ministero dell'Agricoltura e delle Foreste. 1992. Norme tecniche per la produzione di materiale di propagazione vegetale certificato di fragola (*Fragaria* x *Ananassa*). D.M. December 31, 1992, G.U. n. 15 of January 20, 1993.
13.   Ministero dell'Agricoltura e delle Foreste. 1993. Norme tecniche per la produzione di materiale di propagazione vegetale certificato di olivo. D.M. June 16, 1993, G.U. n. 147 of June 25, 1993.
14.   Ministero dell'Agricoltura e delle Foreste. 1993. Norme tecniche per la produzione di materiale di propagazione vegetale certificato di agrumi. D.M. October 29, 1993, G.U. n. 265 of November 11, 1993.
15.   Ministero dell'Agricoltura e delle Foreste. 1993. Norme tecniche per la produzione di materiale di propagazione vegetale certificato di noce. D.M. October 29, 1993, G.U. n. 265 of November 11, 1993.
16.   Ministero dell'Agricoltura e delle Foreste. 1993. Norme tecniche per la produzione di materiale di propagazione vegetale certificato delle pomoidee. D.M. October 29, 1993, G.U. n. 265 of November 11, 1993.

CHAPTER 23

Virus Certification of Deciduous Fruit Trees in The United States and Canada

G.I. Mink

Worldwide there are more than 200 named diseases of pome and stone fruit crops caused by viruses, viroids, phytoplasmas (formerly called mycoplasma-like organisms), or graft-transmissible agents which are currently uncharacterized but which exhibit many virus-like properties (4,5,8). Tables that list 81 diseases of pome fruits and 136 diseases of stone fruits can be found at the NRSP-5 Web site "HTTP: WWW. TRICITY.WSU.EDU/~GMINK/ NRSPHOME. HTML". Over half of these diseases have been recognized for decades (1,9) and virtually all are described in the literature as being caused by viruses or virus-like agents (8). At least three distinct groups of agents (viruses, viroids, and phytoplasmas) are now known to cause "virus" diseases of deciduous fruit trees. The causal agents of the so-called virus-like diseases remain obscure yet today (8). Regardless of their specific properties, however, all agents referred to as fruit tree "viruses" share one important biological characteristic that distinguishes them from other pathogens that affect fruit trees. When these agents infect trees in commercial orchards, they cannot be eliminated from the trees by any economically feasible means. When infection results in decreased growth, yield or fruit quality, the only option is tree removal.

Application of chemicals, a control procedure that has proven effective against a wide range of other pathogens, has either no effect on fruit tree "viruses" or the methods are prohibitively expensive for use under orchard conditions. Even for those fruit tree "viruses" that are spread by insect or nematode vectors, attempts to control the diseases by the use of chemicals to control the vectors generally have been unsuccessful. The only methods that have proven effective in controlling "virus" diseases of deciduous fruit trees involve some form of pathogen exclusion; programs that prevent "viruses" from occurring in orchards. In the US, Canada, and many other countries, exclusion programs begin with international or regional quarantines to exclude truly exotic "viruses" and include nursery certification programs which are designed to produce and distribute planting stock free of known pathogens. Tree removal as a means to control spread of "virus" diseases within orchards is a method most growers resist until all other alternatives are exhausted.

Trees in most deciduous fruit orchards remain in place for decades where they are exposed to wide range of pathogens for much of that time. Consequently, it is not surprising that many trees eventually become infected with one or more "viruses"; either through natural spread or by graft-inoculation when obsolete cultivars are topworked with newer, but "virus"-infected selections. If wood is taken from infected trees and used for propagation by commercial nurseries, "viruses" can be disseminated rapidly throughout an orchard, a region, or even an entire industry. Programs which provide nurseries and growers with propagation wood and planting stocks that are certified free from known "viruses" can effectively control the majority of "virus" diseases of both pome and stone fruits. Certification is particularly effective against those "viruses" which are transmitted only by grafting. Exclusive use of certified planting stock and propagation materials by the fruit industry can essentially eradicate these group of diseases. Because those "virus" diseases which are disseminated by natural vectors spread slowly under most orchard conditions, growers who establish orchards with certified planting stock and who routinely rogue newly diseased trees can often produce near normal crops even in areas where "virus" diseases are endemic.

FRUIT TREE VIRUS CERTIFICATION SCHEMES IN THE US AND CANADA

An *ad hoc* committee on Certification Standards of the North American Plant Protection Organization (NAPPO) proposed the following definition of a certification scheme: "A domestic program consisting of the maintenance, multiplication, distribution, and production of plant materials intended for release, either domestically or for export, under an officially sponsored certificate attesting to the status of the material" (6). Fruit tree certification programs designed to provide growers with "virus"-free or "virus"-tested propagation materials now exist in several US states (7), Canada (6), and many other counties around the world (8). However, there is considerable variation among "virus" certification programs worldwide in the types of crops certified, the methods used to produce certified trees, and the way in which the programs are funded. These differences reflect geographic differences in the fruit crops grown and the pathogens involved as well as financial and political differences that are unique to a state or country (7). Despite their differences, however, all fruit tree "virus" certification programs worldwide include variations of the following elements(7):

1. A set of published standards that define the purpose, the specific terminology, and the protocols used for certification.
2. Sources of plant material that are determined to be free from specific "viruses" and are true-to-horticultural type.
3. A system for preserving the above material and for adding new items as the old become infected or obsolete.
4. Procedures for increasing the elite materials and distributing them on a commercial scale.
5. A system to monitor established standards.
6. A method to finance most, if not all, operations.

In the US, individual state Departments of Agriculture are empowered with the legal authority to develop and promulgate certification programs. Such actions require that public hearings be held to explain alternatives and to determine the needs and desires of interested growers and commodity groups (7). Once standards for certification are agreed upon and promulgated, the Department of Agriculture or its designated affiliate assumes responsibility for administering the program and, in most cases, undertakes whatever inspection and testing may be required in the regulations. Unlike the US, the certification program in Canada is national in scope and not administered by individual provinces. In both countries participation in fruit tree certification programs is voluntary.

At present a total of 10 US states have fruit tree "virus" certification programs on the books . These are California, Idaho, Michigan, Missouri, Montana, New Jersey, New York, Oregon, Pennsylvania, and Washington. Programs vary considerably not only in the amount of certified plants produced but also in the number of fruit species certified. For example, programs in Idaho, Montana, and New Jersey certify only one or more stone fruit species and may have very few nurseries participating in some years. In contrast, all stone and pome fruit species are certified in California, New York, Michigan, Missouri, Oregon, and Washington where most nurseries participate. The California program also certifies several species of nut trees. In 1991 Washington added certification of ornamental nursery stock of *Malus*, *Pyrus* and *Prunus* to the list of fruiting cultivars of the same species. In Canada all stone and pome fruit species, both fruiting and ornamental are included. Participating Canadian nurseries are located mainly in the fruit growing regions of British Columbia, Ontario, and Quebec.

In most US states fruit tree "virus" certification programs are administered and executed by a unit within the state Department of Agriculture. In California certification of fruit trees and nut trees is a cooperative effort among two entities; the California Department of Food and Agriculture (CDFA) and the Foundation Plant Materials Service (FPMS), a nonprofit entity administered through the University of California, Davis, CA. Certification standards are promulgated by CDFA which is responsible for all pathogen testing done at the nursery level. Pathogen testing of foundation trees; heat therapy of new clones; storage, release and distribution of "virus"-tested materials occurs at the FPMS facility, Davis, CA.

All nursery stock certification programs begin with parental trees that have been indexed free of known "viruses" and examined to be true-to-horticultural type. In the early years of "virus" certification many parent trees were developed by state or federal research scientists located in the individual states. However, in recent years, the number of fruit tree virus workers has declined dramatically nationwide. Changes in priorities for much of the research being conducted in plant virology at US public institutions has shifted emphasis from field oriented problems to those at the molecular level. As a result, development of "virus"-free clones of the fruit tree cultivars used in the US and Canada has become concentrated primarily at three facilities; the National Research Support Project-5 (formerly called Interregional Project-2) located at Prosser, WA;   FPMS at Davis, CA, and the Agriculture and AgriFood Canada, Centre for Plant Health (CPH) located at Sidney, BC. A limited number of commercial cultivars of foreign origin are given heat treatment when introduced through the USDA National Germplasm Resources Laboratory, Plant Germplasm Quarantine Office (PGQO), Beltsville, MD. Heat therapy of some apple and pear germplasm selections is also done at the National Clonal Germplasm Repositories at Geneva, NY and Corvallis, OR, respectively. At all locations "virus"-free clones are produced using a combination of heat therapy and micropropagation techniques and the results verified by a combination of laboratory, greenhouse and field tests

Once "virus"-free sources are available for public distribution, some state certification programs establish foundation block trees in areas isolated from commercial orchards to provide cooperating nurseries with elite propagation materials. Nurseries use budwood from these registered foundation trees to propagate 'mother' blocks from which they obtain enough materials to produce certified trees on a commercial scale. Each state has specific requirements for the establishment and maintenance of foundation and mother blocks. This includes specific protocols for treatment of orchard soils for nematodes, general management of the trees, and the distances required to isolate mother blocks from production blocks.

In all programs, trees to be certified must be observed periodically while in the nursery rows by inspectors from the certifying agency. In some cases representative samples are taken and tested for one or more "viruses" by an approved governmental agency. Nursery inspection and pathogen testing are supported, at least in part, by inspection and testing fees charged to the nursery.

Once trees meet the program requirements of a particular state nurseries are provided with official certification tags provided by the state. Nurseries or individuals who receive certification tags are required to account for all materials produced and sold and to keep such records as may be specified by the certifying agency. In some states the certification tags carry disclaimer clauses in attempts to reduce liability from what has become known as *implied warranty*. Legal actions have established that when a certifying agency places its tag, seal, or other official identification on a product, a performance warranty is implied for virtually all diseases and disorders; even those which may not be detected by the testing procedures used.

THE EVOLUTION OF NRSP-5/IR-2 AND ITS' ROLE IN FRUIT TREE CERTIFICATION

By the mid 1950s over 100 named diseases of pome and stone fruit trees were thought to be caused by "viruses" primarily because their causal agents had not been isolated by conventional techniques and all were transmitted by grafting (1,5). Many of these graft-transmissible agents were reported to be distributed widely through infected nursery stock that had been propagated from infected trees. The fact that many of the infected source trees displayed no conspicuous disease but caused disease when established in certain rootstock/scion combinations caused concern among fruit tree researchers, nurserymen and growers nationwide. Many individuals recognized the need for certification programs that could incorporate "virus" testing prior to propagation and general distribution.

During the 1950s and 1960s, a few fruit-growing states developed certification programs for specific crops using trees found to be free of specific "viruses" by graft-inoculating woody indicator plants. However, it was usually necessary to test large numbers of trees to find a single "virus" free source tree as many of the most popular cultivars were found to be nearly 100 percent infected with certain "viruses". It soon became apparent that no single state regulatory agency or public research institution could afford the resources necessary to develop "virus"-free sources of all of the cultivars needed to satisfy the needs of even one large fruit-growing state like California or Washington to say nothing about fulfilling the needs of the fruit industry nationwide. In 1954, directors of US agriculture experiment stations associated with Land Grant Universities through their Experiment Station Committee on Policy (ESCOP) established a national project designated Interregional Project-2 (IR-2) supported by funds from the Cooperative States Research Service (CSRS) Committee of Nine. The primary objectives of IR-2 were to develop "virus"-free clones of important deciduous fruit tree cultivars and to distribute nuclear amounts of this elite propagation material to CSRS researchers who would supervise its increase and ultimate distribution to nurseries and growers nationwide (3). These "off-the-top" funds provided the first attempt to address a national need in the area of fruit tree virus certification.

During the late 1950s, greenhouse and field facilities for IR-2 were developed at the Washington State University Irrigated Agriculture Research and Extension Center (IAREC), Prosser, WA, with an isolation block for clonal storage located near Moxee, WA. The first "virus"-free propagation wood was released in 1965. In 1992 the program was renamed National Research Support Project 5 (NRSP-5) to recognize the support role this program plays in fruit tree research at the national level. Over its 40+ years of existence NRSP-5/IR-2 has evolved as the primary provider of "virus"-free budwood and seed of deciduous fruit trees to research and certification programs throughout the US. It is also a source of start-up materials for most of the "virus"-free fruit tree programs worldwide.

During the first three decades of operation, virtually all of the "virus"-free clones of the pome and stone fruits released by IR-2 were produced from cultivars which were popular at the time and which had no restrictions on their propagation and distribution. Virus-infected cultivars important to specific fruit-growing areas of the US were submitted by growers to one of four regional technical representatives who served on the IR-2 advisory committee. Because the program had only limited capacity for heat therapy and virus testing, the technical advisory committee established priorities for clonal acceptance. Once clones were accepted they were processed without charge. Industry requests for budwood were channeled through the appropriate regional technical advisor to the IR-2 plant pathologist-in-charge who provided a limited amount of budwood back to the regional technical advisor who, in turn, supervised its increase and distribution among interested nurseries in the region. While this procedure provided the fruit industries of all states with access to "virus"-tested budwood of generally available cultivars, the process was not designed to cope with the rapid changes in cultivar preferences and the move to various forms of plant patent and restricted license agreements that have become the industry norm in recent years.

Since the mid 1980s, most of the new cultivars submitted for virus elimination either have a US patent or are restricted by some form of domestic or international license agreement. Because these items can not be made available for public distribution after heat therapy and virus testing, NRSP-5/IR-2 instituted service fees to cover all expenses involved in processing the various classes of restricted items. The heavy demand for these and for other activities related to development and distribution of "virus" tested fruit cultivars prompted the initiation of several fee-supported programs. These programs which involve direct negotiations between NRSP-5/IR-2 and industry personnel include pathogen elimination from licensed domestic fruit cultivars, pathogen testing of advanced selections from state and federal breeding programs, pathogen elimination, if necessary, from breeding selections, the introduction of cultivars and breeding selections from foreign countries combined with pathogen testing and elimination, "virus" testing for state regulatory agencies and industry organizations, and cultivar identification using isozyme analysis (see NRSP-5 home page for details).

Pathogen-tested clones of patented and restricted cultivars which are developed by NRSP-5/IR-2 are released only to the patent or license holder who, in turn, controls commercial distribution. Despite the limited distribution, most states certify these items with the same standards applied to publicly available cultivars. The trend is toward increased use of patented of restricted cultivars. In most US states and many foreign countries, virtually all of the new apple cultivars that are being produced are being distributed under some type of license agreement.

THE ROLE OF OTHER GOVERNMENT AGENCIES IN FRUIT TREE CERTIFICATION

US federal quarantine regulations list all species of pome and stone fruits as prohibited;

subject to entry into the US only by permit and then only at certain locations (2,10). Permits for importation can be requested through the Permit Unit, Animal and Plant Health Inspection Service (APHIS), Riverdale, MD. Pome and stone fruit germplasm and some commercial cultivars are processed through the National Plant Germplasm Quarantine Center (NPGQC) at Beltsville, MD. Here budwood is indexed for "viruses" by several laboratory, greenhouse, and field tests which are established by APHIS and determined, in part, by the species tested. Items that test free of known "viruses" are passed on to the importer usually within 2-4 years. Items that are found to contain "viruses" are heat treated and then retested. Alternatively, the importer may request budwood from another source be submitted for testing. Clones determined to be free of known "viruses" are released and qualify as a sources for state certification programs.

Pome or stone fruit budwood or propagated trees can be imported into most states under post entry quarantine providing that the material originates from any of several nurseries in Belgium, France, Germany, Great Briton, and The Netherlands that cooperate with APHIS-approved certification programs. Post entry permits allow entrance without a lengthy pathogen testing period. However, all propagants must be grown under minimal isolation in nursery rows and observed for two growing seasons by regulatory inspectors in the receiving state. Materials can be imported in to the US from approved nurseries in Canada without being held in post-entry quarantine.

In Canada, pome and stone fruit budwood from non-approved foreign countries and domestic sources are processed through Centre for Plant Health at Sidney, BC. Protocols for the introduction, testing, heat therapy, and release of plant materials at CPH are quite similar to those used at the NPGQC or NRSP-5/IR-2. As in the US, budwood or trees can be imported into Canada from sources approved by the Plant Protection Division. These sources include nurseries in Belgium, England, France, Germany, The Netherlands, and several US states (CA, MI, MN, MO, MT, NY, OR, PA, TN, SC, WA).

The "virus" certification program for deciduous fruit trees in Canada (6) differs somewhat from those in the US in that 1) the scheme is national in scope being administered by the Plant Protection Division of Agriculture and AgriFood Canada, 2) it is composed of regional, provincial, and national components, 3) it focuses mainly on the phytosanitary quality assuring that plant materials have been tested for specific pathogens, 4) while participation is voluntary, plant materials cannot be exported unless they are produced in accordance with certification guidelines, 5) Regional Plant Protection Division Offices have sole authority for the issuance of phytosanitary certificates for export.

FINANCIAL CONSIDERATIONS

Most US states have found it difficult, if not impossible, to totally finance fruit tree certification programs. As a result, it has been necessary for participants in each state to generate most, if not all, of the necessary funds by use of assessments, fees, or by taxing sales of the final products. In states with few participating nurseries or where the volume of certified plants is low, relatively little money can be generated by any method. In contrast, states with many participating nurseries that sell large volumes of certified trees can usually generate funds to maintain strong programs. The mechanisms by which states funds varies but in the past many of these involved adding some portion of the certification costs to the price of finished trees. Although this method succeeded in raising limited amounts of money, it also established an environment where growers would often select the less expensive, non certified trees, sometimes produced by the same nurseries. In this climate competition would often force nurseries to sell certified trees at noncertified tree price which, in turn, made it difficult to generate sufficient funds to maintain an effective program.

When voluntary nursery certification of fruit trees was first established in Washington state during the 1960s, the legislature, acting upon recommend-ations of the state Horticulture Association, passed a law requiring that all licensed nurseries collect a 1% tax on the gross sale price of *all* fruit trees, fruit tree seedlings, fruit tree rootstocks, and seed sold within or *shipped from the state*. These funds, which are collected by nurseries at the time of sale, are remitted to a special account administered by the state Department of Agriculture. The funds are monitored by a committee of nurserymen and are used for activities related to the tree fruit nursery "virus" certification program. The Washington nursery tax is collected on non certified as well as certified trees propagation materials. Because no premium is added to the cost of certified plants,

Washington nurseries have no incentive to grow non certified trees and there is no savings to growers for buying non certified trees. As a result, the nursery certification program in Washington generates funds that cover all program and administrative costs, the maintenance of foundation plantings, and provides whatever "virus" testing is deemed appropriate. In recent years elements of this funding scheme have been adopted by Oregon and California.

WHAT ABOUT THE FUTURE?

There is a long tradition in US agriculture of grower reliance upon scientists employed by the State Agriculture Experiment Stations and US Department of Agriculture to monitor the occurrence of "virus" diseases, to identify and study the causal agents, to evaluate control procedures, and, in the case of tree fruits, to provide most of the testing services that make "virus" certification possible. In the past these activities were funded mainly by one or more agencies of the state or federal government. However, the level of public funding that made these activities possible has declined significantly over the years. For some agencies, funds to support the development and execution of fruit tree testing have essentially disappeared. Competition for scarce research dollars and the mounting pressures for quick publications at traditional research institutions has discouraged young scientists from addressing problems that require substantial investments of money and time such as those associated with detection and elimination of "viruses" of fruit trees. The long term effects of this redirection of talent is strikingly reflected in the changing levels of participation in regional research activities on fruit tree "viruses" funded by CSRS over the past four decades.

In the mid 1950s when NRSP-5/IR-2 was established, it received technical support from a large number of CSRS-funded fundamental and applied research projects that were coordinated through regional technical committees active in three of the four main geographic regions of the US; Northeast (NE-14), Northcentral (NC-14), and Western (W-22). Each regional committee was composed of 10-20 scientists whose research was funded, in part, by CSRS regional research funds. Annual meetings in each region were attended not only by scientists receiving direct regional research funding but also by fruit tree scientists who were funded by other state and federal agencies. In the 1960s over 150 US and Canadian scientists directed research projects on "virus" diseases of pome and stone fruits. In contrast, a computer search of the CSRS database conducted in late 1994 listed fewer than 20 researchers in the US who receive CSRS support to study "virus" diseases of deciduous fruit trees and these projects are now coordinated through a single regional committee, Western Region Coordinating Committee for Fruit Tree Viruses (WRCC-20).

While state and federal support for research on "viruses" of fruit trees has decreased, there have been a significant increases in funds provided by various industry groups in some states. As might be expected the bulk of industry funds to support research on fruit "viruses" have come from organizations in California and Washington. The California Fruit Tree, Nut Tree, and Grapevine Improvement Advisory Board (IAB) and the Washington State Tree Fruit Research Commission (WSTFC) provide most of the grant funds that support fruit tree virus research in these states. Some funds are also provided by nursery license sur-tax monies generated in Washington and Oregon and by the Central Washington Nursery Improvement Institute (CWNII). While such funds are essential for maintaining research on many applied aspects of fruit tree "virus" diseases, industry funding groups are generally reluctant to fund research on fundamental aspects of those pathogens which cause "virus" diseases. Hence, relatively little research is being done on the molecular aspects of these agents and the diseases they induce.

Despite an overall decline in funds for fruit tree "virus" research over the past several years, it is widely accepted by the fruit industry and by scientists that "virus" certification programs for deciduous fruit tree in the US and Canada have been remarkably successful. The list of named "virus" diseases worldwide has risen from around 100 in the mid 1950s to well over 200(8). Even though there were fewer named fruit tree "virus" diseases in the mid 50s, they were found in most pome and stone fruit orchards throughout North America. Today, while some "virus" diseases are still found in many areas, no single endemic "virus" disease currently threatens the economic viability of any deciduous fruit crop in the US or Canada. Occasionally stone fruit orchards in some areas of California may experience significant losses due to the X-disease phytoplasma which has spread under specific conditions. However, such situations are the exception rather than the rule. In Washington, modification of some management practices combined with the general use of certified planting stock keeps the impact of stone fruit viruses to a minimum. In most states the on y "virus" problems of pome fruits that have been reported in recent years involve orchards that were planted with diseased trees; usually noncertified trees obtained from small nurseries or individuals with no ties to the state certification programs.

The combination of state and national plant quarantine laws, state certification programs, grower education efforts, and continuing research and development by a dwindling group of fruit tree researchers have combined to greatly reduce the impact of "virus" diseases of fruit trees throughout North America. When present day growers weigh the economic risks they face from a variety of sources, most consider the risk of "virus" diseases to be relatively low. They view "virus" disease problems as essentially solved by certification and management. However, it is well to remember that *not one* of the pathogens that created major economic problems for US or Canadian fruit growers in the past has been eradicated. Because of changing consumer preferences and the competition for new international markets, US and Canadian fruit growers and nurseries will continue to find, test, and distribute many new fruit cultivars and rootstocks. If history is any guide to the future most of these selections will be widely distributed before their sensitivity to "viruses' has been determined. Undoubtedly some of these new fruit selections will be highly sensitive and thus adversely affected by individual "viruses" that are benign in cultivars that are currently being grown. If quarantine and certification programs are not continued in combination with supportive research

programs, we can expect that future changes in the fruit industry will create an environment for the resurgence of some of the "virus" diseases that are currently viewed as unimportant.

REFERENCES

1.   Cation, D., G. H. Berkeley, L. C. Cochran, F. P. Cullinan, and R. J. Haskell. 1951. Virus diseases and other disorders with viruslike symptoms of stone fruits in North America, U. S. Depr. Agr., Agr. Handbook 10.

2.   Foster, J. A. 1991. Exclusion of plant pests by inspections, certifications and quarantines. Pages 311-328 in: CRC Handbook of Pest Management in Agriculture, Vol. 1, 2nd ed. D. Pimental, ed. CRC Press, Boca Raton, Fl.

3.   Fridlund, P. R. 1977. The objectives and methods of the IR-2 virus-free deciduous fruit tree repository, HortScience 12:487.

4.   Fridlund, P. R. 1989 Virus and viruslike diseases of pome fruits and simulating noninfectious disorders, Bulletin SP0003, Washington State University, 330 Pages.

5.   Gilmer, R. M., J. D. Moore, G. Nyland, M. F. Welsh, and T. S. Pine. 1976. Virus diseases and noninfectionus disorders of stone fruits in North America, U. S. Dept. Agr., Agr. Handbook 437, 433 Pages.

6.   Lanterman, W. S., R. Johnson, and D. Thompson. 1996. Disease control through crop certification-Woody plants. Canad. J. Plant Pathol. 17:274-277.

7.   Mink, G. I. 1981. Control of plant diseases using disease-free stocks, Vol 1. Pages 317-346 in: Handbook of Pest Management in Agriculture. CRC Press, Boca Raton, FL.

8.   Nemeth, M. 1986. Virus, mycoplasma and rickettsia diseases of fruit trees. Martinus Nijhoff Publishers, 841 Pages.

9.   Posnette, A. F. 1963. Virus diseases of apples and pears. tech. Commun. Bur. Hort. East Malling, 30. 141 Pages.

10. Waterworth, H. E. 1981. Control of plant diseases by exclusion: quarantines and disease-free stocks, Vol 1. Pages 269-296 in: Handbook of Pest Management in Agriculture. CRC Press, Boca Raton, FL.

CHAPTER 24

Indexing for Viruses in Citrus

C.N. Roistacher

The objective of an indexing program for citrus is to detect any citrus graft transmissible pathogens (CGTPs) and thereby assure their elimination from citrus tissue. The ultimate objective is the elimination of these pathogens from propagative budwood and the production and maintenance of disease free primary or foundation block source trees for budwood increase and production of healthy, highly productive citrus groves.

The importance of a good indexing program as a means of identifying causal agents of transmissible diseases cannot be overemphasized. A prime responsibility of those in charge of a certification program is the prevention of the distribution of thousands to millions of trees contaminated with pathogens by propagation of infected budwood throughout a state or country.

INDEXING

Indexing may be defined as any test which will positively and consistently confirm the presence (or absence) of a transmissible pathogen, or identify a disease. The index test should be specific for the pathogen or disease. *Inoculation of plants is the primary means by which most graft transmissible diseases of citrus are diagnosed.* Inoculation is primarily by graft-transmission, but also can be done by vector or by mechanical transmission. Other approaches to indexing are the direct observation of the pathogen in plant cells by light or electron microscopy, microscopic observation of inclusion bodies or gum deposits within the plant cells, antibody-antigen reaction, analysis of pathogen-induced chemical changes, culturing of an organism, and/or observation of pathogen-specific nucleic acids via sequential polyacrylamide electrophoresis (sPAGE), molecular hybridization or polymerase chain reaction. Other indexing techniques involve detec-

tion of changes in water uptake or mineral content of trees as used in testing for blight.

In general, any consistent, measurable or striking change in the morphology or chemistry of a plant brought about by the direct or indirect presence of a pathogen, or the direct observation of the pathogen, its by-products or constituents can be considered as an index.

Certain citrus diseases can be diagnosed by specific field symptoms. However, many graft transmissible disease are symptomless in the major commercial citrus varieties and these can only be detected by indexing.

Although biological indexing is sometimes difficult, involving long term care of plants, it has certain advantages over immuno-response indexing such as ELISA (enzyme-linked immunosorbent assay). Specifically, whereas ELISA depends on titer, biological indexing involves bio-multiplication of a pathogen and can increase the number of particles from extremely few to perhaps billions in the new growth of an indicator plant.

Since plants are the primary means for detecting citrus pathogens, a greenhouse or plant laboratory for the production and maintenance of indicator plants is essential to any successful program for detection and diagnosis of CGTPs. The system recommended for growing citrus indicator plants is the UC system (2,14,15). The plant laboratory is more than a greenhouse. It must be, in every way, as organized and refined as any well run efficient and well equipped laboratory (20).

Some of the techniques used for graft-transmission of citrus pathogens include inoculations by bud graft (buds, blind buds or chip buds), leaf grafts (leaf piece or leaf disc), bark grafts and side grafts. These techniques are described and illustrated in the handbook (23).

Table 1. The major graft transmissible diseases of citrus based on economic impact and transmissibility

Disease	Pathogen

Diseases which induce severe loss and where vector transmission is known or suspected

Tristeza	Citrus tristeza virus
Greening	*Liberobacter sp.*
Stubborn	*Spiroplasma citri*
Blight and related diseases	unknown
Witches' broom of lime trees	Phytoplasma
Citrus variegated chlorosis	*Xylella fastidiosa*
Citrus chlorotic dwarf	Probable virus

Diseases which induce moderate loss are mechanically transmitted and also bud transmitted or vectored by unknown means

Exocortis	Citrus exocortis viroid
Cachexia	Citrus cachexia viroid
Satsuma Dwarf	Satsuma dwarf virus
Tatterleaf	Citrus tatterleaf virus
Infectious variegation	Citrus infectious variegation virus
Transmissible psorosis	Ophiovirus

Diseases which are readily controlled by use of virus-free budwood and are primarily bud transmitted by man

Psorosis-A family	Ophiovirus
Oak-leaf pattern family	Probable virus
-Concave gum	
-Impietratura	
-Cristacortis	
Gummy bark	Probable viroid
Gummy pitting in trifoliate	viroid
Bud union crease	Probable virus

Diseases which are vector transmitted but of minor economic impact

Vein enation	Citrus vein enation virus

THE MAJOR GRAFT TRANSMISSIBLE DISEASES OF CITRUS

The graft transmissible pathogens of citrus have two general effects. Some are destructive and cause severe decline and death of the tree. Others do not kill the tree but can cause severe stunting, loss of vigor, reduced fruit size and reduced yield and may result in extensive long term losses.

The major graft transmissible diseases of citrus are presented based on economic impact and transmissibility as shown in Table 1.

The vector transmitted diseases of tristeza and greening rank as the most serious and destructive of the citrus diseases. Where sour orange is the predominant rootstock and appropriate vectors are present, tristeza can spread rapidly and can destroy the entire industry on sour orange rootstock. Even when tristeza becomes endemic in an area, and tolerant rootstocks replace the sour orange, new stem pitting forms of the CTV can superimpose on the existing infected trees and can decimate and destroy an industry (22). In many areas of the world, where greening has come in, citrus as a viable crop has gone out. It is most important to understand the destructive potential of these two diseases and to develop every strategy to detect, prevent and destroy the pathogens before they can become established within a country or state.

Recently blight and related declines have inflicted severe damage to the citrus industries of many countries, and some diseases such as fruta bolita in northern Argentina, have destroyed the local citrus industries in Misiones Province. New, and quite destructive diseases of citrus have appeared in the past few years such as witches' broom of lime trees in Oman and the United Arab Emirates (4), citrus variegated chlorosis in Brazil and Argentina (12) and citrus chlorotic dwarf in Turkey (7).

Some of the graft and mechanically transmissible diseases of citrus such as tatterleaf, psorosis, ringspot, and satsuma dwarf are considered as moderately destructive. Although they may severely limit crop production and in some cases be quite destructive such as transmissible psorosis in Argentina and Uruguay, these diseases can be readily detected in propagative budwood, can be readily eliminated by shoot tip grafting (29) or thermotherapy and their rapid spread curtailed. The mechanically transmitted viroid diseases such as exocortis and cachexia, are destructive to certain rootstock varieties, and once introduced and established in a country can spread relatively rapidly by mechanical means. These diseases can be readily detected by indexing and should not be permitted to enter the expanding citrus industries of new or older regions.

Other citrus diseases such as concave gum, cristacortis, impietratura, infectious variegation and vein enation are generally less serious compared to the other graft transmissible diseases and are not known to be vector transmitted. Once eliminated from propagative budwood they should not pose further problems.

Certain pathogens, once distributed throughout a region, are difficult to remove, and may remain indefinitely with the citrus industry. Therefore, prevention by rigid quarantine should be a first priority. Detection by indexing, and a program for reducing inoculum either by eradication or a certification program may have

wide benefits.

INDEXING BENEFITS AND OBJECTIVES

The benefits derived from the development of an indexing program will usually far outweigh the costs of development and maintenance of the program. In those countries where no organized program of testing has been established, most old line citrus cultivars are infected with viruses, viroids or other graft transmissible agents. In early studies, reports of indexing trials from Spain, Brazil, Florida, and the Foreign Budwood Importation Program in California, have shown that almost all citrus was infected with one or more CGTPs. Thus, prior to a certification program, it is almost certain that the native citrus in any country will be found heavily infected with one of more CGTPs.

The first benefit of a comprehensive indexing program is the obtaining of an accurate picture of the pathogens which are present in the citrus within the country or state. This knowledge enables the formulation of sound strategies for developing a program to avoid pathogens in propagative budwood.

One indirect benefit of establishing a comprehensive indexing program is its potential for the early discovery of introduced pathogens which could be highly destructive to citrus prior to their dissemination via vector or budwood. For example, the early discovery of the citrus variegated chlorosis in Brazil, and a program for its eradication could have prevented its wide distribution throughout the country. In this world of rapid movement of people or plants by air transportation, budwood from other countries is continually being introduced by uninformed nurserymen and growers. This budwood can readily harbor pathogens potentially destructive to citrus (11). A well equipped and properly staffed diagnostic laboratory would be a first line of defense for the rapid identification of introduced pathogens. If detected early there may be opportunities to eradicate them before they spread.

Perhaps the most rewarding aspect of a comprehensive indexing program is in the production of pathogen-free trees derived by shoot tip grafting *in vitro* or by thermotherapy. Trees so produced are healthy, more uniform, higher yielding with better colored fruit, and ultimately more profitable to the grower. Those countries or states which have well established indexing and certification programs have demonstrated the economic benefits of such programs to the grower. Production of disease free trees requires the knowledge of which CGTPs are present and verification of their elimination after shoot tip grafting or thermotherapy (29). Indexing is imperative in any certification program.

In summary, the ultimate objective or goal of any indexing program should be the development of certified pathogen-free budwood to improve production of the industry and also to have the means for detection of new serious pathogens which are not yet present, may be present or only recently introduced.

CONCEPTS IN DETECTION OF CITRUS GRAFT TRANSMISSIBLE PATHOGENS

Visible Pathogens

Many CGTPs will induce specific symptoms in the field tree, and many of these symptoms are diagnostic for a specific disease. The serious student or individual responsible for indexing should read and study the literature relative to each of the diseases of citrus. Resources are the literature present in all 13 volumes of the proceedings of the International Organization of Citrus Virologists (IOCV). Also, a historical review and bibliography for most of the graft transmissible pathogens of citrus plus shoot tip grafting and thermotherapy is now available (25). Recommended publications illustrating many of the citrus graft transmissible diseases are: Bové and Vogel's Description and Illustration of Citrus Virus and Virus-like Diseases (3); the fourth volume of the Citrus Industry containing the articles by Wallace (31) and others on citrus virus and virus-like diseases and by Calavan *et al.* (6) on registration, certification and indexing of citrus trees and the Compendium of Citrus Diseases (32).

Diseases which may show characteristic and diagnostic symptoms in the field on certain varieties are cachexia, concave gum, cristacortis, exocortis, greening, infectious variegation, satsuma dwarf, psorosis-A, ringspot, stubborn, tatterleaf, tristeza and vein enation-woody gall. However, clear diagnostic symptoms are not always obvious on infected trees in the field. It is very important to realize that most of these same CGTPs are symptomless in various hosts and may show little or no diagnostic evidence of their

presence in the field. Plants showing symptoms of one disease may also be infected with other pathogens. Thus, these CGTPs can unknowingly be spread by bud-propagation, mechanical transmission or by insect transmission throughout a country. These are the 'invisible pathogens' and they can only be detected by indexing.

Invisible Pathogens

Most of the CGTPs exist in an invisible or symptomless state in many citrus cultivars and can only be detected by indexing. Without knowledge of their existence, these symptomless pathogens can be readily spread in propagative budwood and can be widely distributed within a country. The primary sources of infected budwood usually resides within the country, but with modern means of rapid air transportation, new exotic and possibly destructive strains are being brought in by nurseryman and growers (24).

Examples of CGTPs Which Can Be Symptomless in Citrus

1. Certain isolates of citrus tristeza virus (severe stem pitting and seedling yellows strains) which may not show symptoms in field trees of mandarin cultivars on trifoliate orange rootstock. However, when infected mandarins are planted near sweet orange or grapefruit trees, aphids will transmit the severe strains of CTV from the symptomless mandarins into reactive hosts, and this can be extremely destructive to the existing citrus industry (22).

2. Citrus cachexia viroid, which induces no symptoms in grapefruit or sweet orange cultivars, can induce a severe reaction in tangelo or mandarin if infected buds are used as scions or if the viroid is mechanically transmitted to these cultivars.

3. Citrus tatterleaf virus expresses no symptoms in most citrus cultivars including sweet orange, mandarin, sour lemon, Meyer lemon and grapefruit on most rootstocks. However, if virus-infected budwood is used as scion wood and grafted to citrange or trifoliate orange or any hybrids of trifoliate used as a rootstock, a brown bud-union crease will usually be evident, and deep pits and

grooves may develop in the rootstocks and the trees become debilitated and usually decline.

4. Citrus vein enation virus induces no symptoms in most mature citrus trees and can readily be detected only by indexing.

5. The concave gum pathogen induces no symptoms in lemon and grapefruit, but can be quite severe in sweet orange.

6. The psorosis-A pathogen may be present without inducing symptomatic bark scaling in sweet orange for 7 to 15 years. In some rare cases psorosis-A infected sweet orange trees may be symptomless for over 50 years. However, progeny trees propagated with budwood from such symptomless trees will usually show bark scaling symptoms within 7-15 years or even earlier. The psorosis-A pathogen induces no symptoms in infected lemon, sour orange and many varieties of pummelos.

7. Exocortis and related citrus viroids will induce no symptoms in most sweet orange, mandarin and grapefruit scion cultivars on sour orange rootstock. However, when viroid-infected budwood is grafted on rootstocks of trifoliate orange, hybrids of trifoliate, Palestine sweet lime or Rangpur lime, distinct symptoms may occur and these can be severe and debilitating.

CGTPs can also be masked by heat. For example, certain strains of CTV may be symptomless in sweet orange and grapefruit grown under hot conditions. However, when budwood taken from these trees, and propagated in cooler areas, severe stem pitting and stunting may develop in progeny trees.

Pathogen-free Budlines

If a control strategy is adopted based on pathogen-free budwood obtained by shoot tip grafting *in vitro*, with or without thermotherapy. The question is often asked "Why should we index? Won't the shoot tip grafted budwood automatically be pathogen-free"? Shoot tip grafting and thermotherapy, though proven excellent as therapeutic methods for eliminating many pathogens from citrus budwood, will not guarantee 100% success in eliminating all pathogens (5,18,21,29). Actually, 100%

elimination of pathogens by these therapeutic techniques is the exception, and indexing is an absolute necessity if freedom from CGTPs is to be assured. Distribution of budwood after therapy, but without indexing, may result in dissemination of infected material throughout a country.

Once pathogen-free budlines are obtained by therapeutic means, and foundation or mother trees are planted, these foundation-source trees must be periodically reindexed to be sure they have not been contaminated by mechanical transmission or by insect vectors. In countries where vectors of diseases are known to be present, insect-proof screenhouses are needed to protect foundation and mother trees as well as primary increase block trees from reinfection.

Other Developments in Indexing

The use of serology, especially ELISA, for the mass detection of certain pathogens is one of the most exciting and helpful developments in recent years in the field of indexing. The rapidity of the test, and its accuracy for detection of some CGTPs makes it a necessary part of any indexing program. ELISA has proved useful in mapping the distribution of such viruses as tristeza, satsuma dwarf and infectious variegation over large areas and its future is bright for the detection of other pathogens affecting citrus. ELISA, when used in combination with plant biological indexing is highly effective for detection of virus strains which are very mild reacting in plant indicator hosts and which might otherwise escape detection. ELISA and other immuno-assay techniques are available for detection of tristeza, infectious variegation, satsuma dwarf, stubborn and greening. It is important to realize that ELISA should not be used as a substitute for plant indexing in the critical evaluation for presence of pathogens in primary source foundation budwood. Under conditions where temperatures are hot and virus titer is low it has been shown that ELISA will fail to detect (8).

Electrophoretic techniques have been developed for detection of the viroids of exocortis and cachexia and other citrus viroids (8). These electrophoretic methods may become the primary index for some viroids which are difficult to index by other methods. Where the technology is available and laboratory experience proven, sequential polyacrylamide gel electrophoresis (sPAGE) technique for detection of citrus viroids is preferred over the long term citron index (8).

For example, sPAGE can readily detect the citrus cachexia viroid whereas the plant index for cachexia is a long term index taking a year or longer for detection of mild strains. Also, some of the citrus viroids are difficult to detect in citron and sPAGE can readily show their presence or absence.

Identification of viral pathogens by nucleic acid hybridization assays is also promising for some applications, especially for detection of specific strains of a given pathogen and for viroid detection. Development of non-radioactive labels is needed to make this approach more useful for practical application.

The stubborn organism *Spiroplasma citri* can be cultured and also detected by ELISA. Culturing and use of ELISA has replaced plant indexing for this disease.

Other techniques such as SSEM, molecular hybridization, dsRNA, light microscopy with fluorescent antibodies are useful for detections of specific pathogens and may ultimately replace the plant index for some applications. However, as shown in Table 2, plants are still the primary means for the certain detection of many of CGTPs and the need for a well equipped plant laboratory with skilled personnel is evident in any certification program. Indexing, based on plant indicators still remains a valuable adjunct to other procedures, especially where verification of virus freedom is essential. The concept should not be to replace the use of plant indicators but to use these efficiently in a total comprehensive indexing and certification program.

An old proverb perhaps expresses the concept of the need for plant indexing while continuing research for new and more rapid techniques ----

"Be not the first to take the new astride;
Be not the last to cast the old aside".

INDEXING - FACILITIES, METHODS AND CONSIDERATIONS

Strategies and Considerations

How does one get started in an indexing program? This will depend on objectives. If the objective is to survey for the presence or extent of infestation of a particular pathogen or disease within a region i.e. the viruses of tristeza, tatterleaf or satsuma dwarf, and antiserum is available, the use of ELISA would be the preferred strategy for detection. Where antiserum is not available and plants must be used for indexing of field trees over a large area, then other strategies must be used i.e. multiple indexing of a number of trees to larger index plants.

If the objective is to establish a comprehensive indexing and certification program then a plant laboratory is a necessity. Of immediate concern would be the training of individuals to develop and lead a certification program. Also, to develop a system for growing index plants, obtain seed, obtain positive controls, etc. A decision to develop a certification program, including a foundation block with its supportive index facility, must be well supported and adequately financed as a long term program. The program should include policies for education and publicity to inform growers and the public of the dangers of introducing pathogens and the benefits of pathogen-free stock.

Each country or state must establish their individual priorities and goals depending on the seriousness of their immediate problem, the size of their industry, government support and available funding. An important consideration is the selection of a suitable site for the plant laboratory with appropriate temperature conditions. The minimum needs are a plant growth facility and a laboratory equipped for laboratory oriented procedures such as ELISA, polyacrylamide gel electrophoresis and nucleic acid hybridization.

Perhaps the most important ingredient for a successful indexing program is the quality, education, training, dedication and integrity of the personnel responsible for the conduct of the program. A well trained, well paid and dedicated individual may mean the success or failure of a program. Training programs for individuals can be arranged with institutions having established certification programs. In addition, there are courses and workshops in indexing which have been presented in the past by the International Organization of Citrus Virologists and various regional programs for greening and tristeza control (1,11). Programs of training by these organizations are also projected for the future.

Table 2. The minimum number of indicator plants recommended for a comprehensive indexing program and the pathogens which can be detected

Citrus plant	Sdlg. or scion	Minimum no. of plants	Plants[a] per container	Varieties	Viruses detected[b]	
					Primary	Secondary
For detection of cool temperature pathogens						
Lime	Sdlg	4	3	Mexican, Key, etc.	Tr, VE	IV, Ps/RS, TL
Sweet orange	Sdlg	4	3	Pineapple, M. Vinous	Ps/RS, CG	Gr, Imp, Cr, IV, SD, Tr-SP
Sweet orange	Sdlg	2	1	Madam Vinous	Tr-SY, Tr, SP	Imp, Cr, IV, SD, Tr-SP
Mandarin[c]	Sdlg	4	3	PonKan	Gr	Ps/RS, CG, Cr, Imp, IV
Mandarin	Sdlg	4	3	Satsuma	SD	
Tangor	Sdlg	4	3	Dweet	CG, Cr, Imp (OLP)	Ps/RS, Gr
Sour orange	Sdlg	4	3	Standard	Tr-SY, VE	Ps/RS, CG, IV
C. excelsa	Sdlg	4	3	Kalpi lime	TL, Tr	
Citrange	Sdlg	4	3	Rusk, Troyer, Carrizo	TL	SD
Citron	Sdlg	2	3	861, 60-13	IV, Ps/RS	SD, Tr-SP
Grapefruit	Sdlg	4	3	Duncan	Tr-SY, Tr-SP	Ps/RS, Gr
For detection of warm temperature pathogens						
Citron	Scion	5	1	861-S-1/ rough lemon stock	Exocortis & misc. citrus viroids	
Mandarin	Scion	6	1	Parsons/ rough lemon stock	Cachexia	
Sweet orange	Sdlg	5	1	Madam Vinous	Stubborn	

[a] Grown three per container: two are inoculated and one left as a non-inoculated control.

[b] CG = concave gum, Cr = cristacortis, Gr = greening, Imp = impietratura, IV = infectious variegation, Ps/RS = psorosis/ringspot, SD = satsuma dwarf, TL = tatterleaf, Tr = tristeza, Tr-SY = seedling yellows, Tr-SP = stem pitting tristeza, VE = vein enation.

[c] Mandarin and Dweet tangors can be used interchangeably for detection of oak leaf pattern (OLP) inducing pathogens i.e. concave gum, cristacortis and impietratura. Ponkan is recommended for greening, and satsuma is recommended for satsuma dwarf. Other mandarin varieties such as Dancy, Kara or King are excellent indicators.

INOCULATION PROCEDURES FOR DETECTION OF CITRUS GRAFT TRANSMISSIBLE PATHOGENS.

Collection and Storage of Inoculum Tissue

Budwood is the primary inoculum tissue used for most inoculations, however, bark and leaves may also be used. Budwood should not be collected during excessively hot weather since some CGTPs in the perimeter branches of field trees can be temporarily inactivated or severely suppressed by heat (8,27). However, when the season changes and temperature becomes cooler, the pathogen usually will return from its reservoir location in the roots or shaded parts of inner branches. An ice chest should be used for budwood storage when collecting. Clippers should be disinfected by dipping or spraying when moving from tree to tree using a 1% sodium hypochlorite solution (a one to four dilution of the 5.25% commercial household bleach in water). Bark samples can be placed in a small plastic tube with moistened cotton at the bottom and the tube should not be sealed.

Immediately after collecting tissue samples, they should be placed in polyethylene bags to prevent them from drying and immediately put into an ice chest. Be sure to label all samples clearly at the time of collection. Upon arrival at the plant laboratory they should be put directly into a refrigerator at 5° to 6°C. Avoid freezing the inoculum. Budwood can be maintained under refrigeration for up to two weeks or longer, but should preferably be used as soon as possible.

If a tree in the field is selected as a primary candidate whose budwood will be propagated for heat treatment or shoot tip grafting, a budstick should be taken just below or proximal to a well developed and typical fruit. A bud propagation is then made, and the propagation held in the greenhouse. This propagation would then become the primary plant, and inoculum can be taken anywhere from this plant for initial indexing, heat treatment, shoot tip grafting or for use as positive control tissue to test the effectiveness of the heat treated or shoot tip grafted plant.

Inoculation Methods

The most frequently used method for inoculating indicator plants for the detection of most CGTP is by 'bud'-graft inoculation. The term 'bud'-graft includes buds with 'eyes', stem pieces without 'eyes' which are sometimes called blind buds, and also chip-buds. There are other inoculation techniques, most of which are also given and illustrated in the handbook (23). These include side grafts, approach grafts, root grafts, fruit grafts, leaf piece grafts or leaf disc grafts.

Mechanical transmission from citrus to citrus or from citrus to herbaceous plants is done by knife or razor blade slash. The blade is first slashed through the inoculum tissue, and then a single slash is made into the stem of the receptor plant. This procedure is repeated 10 to 25 times per plant. The slashed area of the receptor plant is then wrapped with budding tape. Citron is an excellent donor host as well as a receptor host for mechanical transmission by knife or razor blade slash.

In general, seedlings are preferred as receptor or indicator plants. However, if propagated clonal buds derived from seedling lines are substituted for seedlings, they should be tested and compared against the seedling for their performance as indicators since their performance as budlings may be different than that of seedlings.

Table 2 gives a summary of the minimum number of recommended indicator plants and the pathogens they can detect. The recommended index temperatures and symptomatology, the specific methods of inoculation, the suggested number of indicator plants to use, the preferred inoculum, plant growth, recommended index temperatures, time for development of the first symptoms, and symptoms are given in detail for each of the individual diseases in the handbook for detection and diagnosis (23).

Although 'buds' are used as inoculum for most inoculations, other tissue and techniques i.e. leaf, bark, root, or side grafts should be continually tried and tested to find the most effective means of bringing out maximum symptom expression. This is especially true for any initial indexing of new diseases or diseases of unknown etiology.

Specific clonal selections used as scion propagations rather than seedlings have been found superior as indicators for indexing of certain pathogens i.e. the cachexia, exocortis and exocortis-like citrus viroids. A vigorous rootstock such as rough lemon or Volkamer lemon is recommended as a rootstock under the clonal bud. The forcing of clonal buds is recommended where tristeza is endemic and tristeza-susceptible indicators may show too strong a tristeza reaction, thereby masking symptoms of other pathogens. In many cases tristeza can be filtered out by inoculating trifoliate orange seedlings and using shoots of trifoliate as inoculum. A modification of this technique is to graft an indicator scion bud on a trifoliate or citrange seedling, inoculating the seedling and forcing the indicator bud. In most cases tristeza will be filtered from the new growth of the developing indicator shoot. Some isolates of tristeza can pass through trifoliate or citrange, however most do not.

When testing for the bud union effect of CTV using a sweet orange scion budded on a sour orange rootstock, or for the bud union crease of certain scions on trifoliate or citrange rootstock induced by the tatterleaf virus, propagation of the scion and inoculation of the rootstock can be done simultaneously and the sour orange or trifoliate rootstock seedling is then bent just above the scion bud to promote rapid forcing of that bud (13).

Positive and Negative Controls

It is extremely important that both positive and negative controls be incorporated in each index test. A collection of infected source plants containing mild and severe CGTPs should be developed and maintained in a 'virus bank'. Sweet orange has been found to be an excellent holding or reservoir plant for almost all CGTPs. These reservoir or bank plants should be periodically indexed to be sure the pathogen is present or has not changed. It is important that the mildest strains of pathogens be collected and preserved in the 'virus bank' and these should be used as positive controls for each index test. These positive controls will provide the determining factor as to when an index test should be terminated. The inclusion of negative controls is also very important and should be generously incorporated into every index. The negative control plant gives an indication of possible environmental or insecticidal spray damage, and can show effects other than those induced by pathogens. They also act as a standard for plant size comparison when subtle pathogens or diseases of unknown etiology may stunt index plants, but otherwise show no other leaf symptoms. However, their primary importance is to provide a normal control plant when reading for very mild leaf reactions in the inoculated plants. Thus, the presence of a new pathogen, or a very mild reacting form of a known pathogen can be detected. Although it may appear to be an extra use of seedlings, the presence of a non-inoculated

control plant in each container has been found to be very helpful for a number of reasons but specifically for judging any possible reaction in the two inoculated plants in the same container.

Time of First Symptom Appearance

The time in weeks from inoculation to appearance of the first symptoms under optimum growth and temperature conditions is given in detail in the handbook (23) under the specific section for each of the diseases. During critical flush periods, plants should be observed daily for development of symptoms for certain CGTPs. Symptoms of psorosis, psorosis-like pathogens and concave gum oak-leaf patterns may disappear from the young developing leaves as the leaves matures, and symptoms may not reappear in later flushes. Leaf flecking symptoms are best observed during the first to third flushes of growth. Plants should be watched carefully to catch the growth at maximum unfolding of the leaves for best reading of young leaf symptoms. Different pathogens will show leaf reactions at different times. Records should be carefully maintained for the time of appearance of symptoms with a detailed description of the plant reaction.

Maintaining proper temperatures are extremely important for appearance of some symptoms. If the temperatures are kept too warm, certain 'cool' temperature pathogens may show no symptoms in leaves, or show them very poorly (27,28). If temperatures are kept too cool, symptoms for such warm-temperature diseases as stubborn, cachexia, exocortis or certain citrus viroids may not show or may develop poorly in indicator plants. Also, citron plants used for preparation prior to sPAGE may not build a high titer of viroid under cool conditions. These plants must be held at warm temperatures (9).

The liberal inclusion of mild and severe positive controls give a working indication of the proper time and temperature for symptom appearance. The lack of any symptom development in plants inoculated with these mild positive controls would invalidate the index.

Vigorous growth is important for production of good leaf and stem pitting symptoms. Stem pits are poorly produced in poor unthrifty plants.

Checking Inoculum Survival

Two to three weeks after inoculation, the

wrapping tapes, especially that of plastic mylar, should be removed, the inoculum examined for survival, and the survival recorded. If tapes are cut with a knife or razor blade, these tools should be disinfected in a 1% sodium hypochlorite solution between each cut. When buds are taken from mature wood of a dark colored budstick, or when bark inoculum is used, it is sometimes difficult to tell if the inoculum tissue is dead or alive. A small slice or cut made into the brown bark surface of the inoculum will reveal the bright green color of living tissue beneath, thus indicating that the inoculum is alive. If both inoculum 'buds' are dead, the plant should be re-inoculated or new inoculations made to another plant. Generally, if one of the two inoculum 'buds' are alive, the plant need not be reinoculated providing there are sufficient replications.

Records

A record sheet for each index must be kept. This should include the experiment number, date of inoculation, source of the inoculum, indicator plants used, inoculum survival, reading dates, and a larger space reserved for notes on observations. Records should preferably include temperatures and light conditions under which indexing was done, and any use of artificial lighting (19).

Indexing Using Field Trees

Certain indexes require a longer term for completion for the expression of the mildest symptoms. At such times the inoculated index plants growing in the plant laboratory (greenhouse) needs to be set out in the field, or field trees need to be inoculated and observed. For example, in the long term plant index for cachexia, the mild positive controls may show no symptoms in the greenhouse even after one year. Therefore, it is best to move the indicator plants to the field and plant them at close spacing until the mild controls show positive symptoms. Similarly, certain strains of exocortis or related citrus viroids, may require a field test to show mild bark cracking on their trifoliate or Rangpur lime indicator rootstocks. The testing of sweet orange on sour orange rootstocks for the classical quick decline tristeza reaction may also require an extended period of time for typical tristeza decline symptoms to develop. The testing for cristacortis or bud-union crease also require long term

observation of plants or trees in a screenhouse or in the field. These indexes should be carried out in an environment where temperatures are conducive for best symptom expression. Again, mild and severe positive controls as well as negative controls should be present.

For certain diseases, trees in the field may have to be tested or inoculated to observe specific symptoms i.e. testing for blight, or observing fruit for symptoms of impietratura.

THE PLANT LABORATORY (THE GREENHOUSE)

A greenhouse, screenhouse or a controlled environment structure is necessary for the production of index plants and for indexing. This structure need not be expensive or elaborate. It should provide light, heat and cooling, and be sufficiently well constructed to prevent insect intrusion. An entryway with two doors and a darkened vestibule between is desirable as a preventative measure against insect invasion. The greenhouse covering can be of glass or plastic. Excellent plants can be grown in a simply designed and inexpensive wood structure covered with heavy plastic and containing a good system for heating and cooling. Modern structures are now made with extruded aluminum framing.

The size of the greenhouse will depend upon the amount of indexing and research to be done. There should be at least two or preferably three compartments:

1. A cool temperature room for indexing CGTPs which are best expressed in plants grown under relatively cool temperatures of 24° or 27°C to a maximum not to exceed 30°C during the day, and 18° to 21°C at night.
2. A relatively warm temperature room primarily used for the growing of plants which can be held at flexible temperatures ranging from 30° to 35°C maximum day and 20° to 24°C night.
3. The third room should be a hot room which may be used for preconditioning budwood prior to thermotherapy, and for indexing of diseases such as stubborn, cachexia, exocortis, and various citrus viroids that require hot temperatures for the best symptom expression. Temperatures in this room should be maintained as warm as possible without inducing plant

injury or leaf distortion. A recommended maximum day temperature is 32° to 40°C with minimum night temperatures of about 24° to 27°C. The hot room can also be used for germination of seedlings and growing of plants.

Corrugated plastic sheeting rather than glass is recommended where hail is a problem, and in many respects is preferable to glass since it may be less expensive, does not break and may be easier to construct and maintain.

A double door entryway and vestibule is highly recommended as an insect inspection and trapping area.

Flooring, Benches and Containers

Benches can be made of wood, concrete, wire mesh plastic or any satisfactory container supporting system. A satisfactory bench system in use at the Rubidoux facility at the University of California at Riverside uses 2 x 6 inch (5 x 15 cm) Douglas fir wood boards spaced about 2 cm between boards. The wood is painted or sprayed with a 2% copper naphthenate solution (26) which acts both as a wood preservative and disinfectant. Wood benches can be placed on concrete blocks or on a metal frame or other foundations at a satisfactory working height approximately 80 cm from the ground. There are available plastic bench tops which are quite satisfactory. They are one piece semi-rigid perforated plastic bench top mounted on a wood frame.

Flooring can be of concrete with provisions for drainage. However, gravel flooring with a concrete walkway is highly recommended. Semi course gravel (1-2 cm diameter) should be spread on the ground about 8 to 10 cm thick. This provides good drainage and aids in maintaining sanitation. The greenhouse should be constructed on a well drained soil base. If this is not possible, then supplementary drainage using tile under the flooring should be provided prior to construction.

Plastic containers are recommended for growing citrus plants. A tapered container approximately 15 cm diameter and 15 cm deep has been found very satisfactory over many years of use at Riverside, California and elsewhere. Such a single small lightweight container filled with the UC soil mix will grow three plants to 1 m height readily and without nutritional or other problems. A relatively large number of these small containers can be placed on each bench, but should be adequately spaced to avoid crowding (16). Containers measuring 18 or 20 cm diameter can also be used for larger holding plants. Plastic containers must be pre-tested for their ability to withstand steam (permitting their reuse) since steaming is the preferred method of sterilization. Clay pots are not recommended. Clay will accumulate salts, make heavier pots, are subject to breakage, and must be soaked and washed after each use, which is labor intensive.

Thermographs

It is important to have a thermograph in each room. These should be periodically calibrated against two thermometers for accuracy. At the end of each week, when charts are changed, the maximum and minimum temperature readings for each day should be recorded in a special book. This provides a record for research, and also is a means of noticing any abnormal changes which give warning of heating or cooling unit failure or breakdown.

Supplemental Lighting

Supplemental lighting when applied during winter months will enhance the expression of the OLP symptoms. For example, at Riverside, California at 34°N. latitude, the addition of 5 hours of 40 to 50 foot candles of supplemental light (at the plant level) during the months of October through April induced symptoms in 207 leaves on plants of four indicator varieties compared to only 60 leaves on plants grown without lights (19). In addition to the over 3 fold increase in number of leaves showing symptoms under artificial lights, there was a 32% increase in the total number of leaves produced. Supplemental lighting during the winter months is highly recommended for use in an index plant laboratory. Recent studies have shown that growth of certain citrus seedlings could be significantly enhanced during the winter months with light from 25 watt bulbs, which induced the same growth response as light from 100 watt bulbs (placed 1 m above the bench tops). This has resulted in a considerable savings in electric energy (30).

Heating

Heating can be provided by gas heaters with fans, by steam heat using radiators, or by steam pipes placed along the sides of the structure.

Heat may also be distributed from gas heaters using supplementary fans blowing the heat through perforated plastic tubes. Most gas heaters are placed inside the structure. However, on occasion, ethylene released by faulty heaters can be very damaging to plants. If feasible, gas heaters should be placed outside the structure and the warmed air circulated inside by a fan inside the greenhouse and preferably by forcing the air through large diameter perforated plastic tubes.

Catalogues of the larger companies manufacturing greenhouse structures, heating and cooling equipment should be available in many countries and should be consulted for ideas and costs. The construction of a greenhouse facility is best done through local builders with design and input by those in charge of indexing.

Cooling

There are three general methods for cooling a greenhouse: 1) Bringing in outside air when the outside temperature is cooler than the temperature inside the house. 2) Use of evaporative coolers if the relative humidity is low enough to make evaporative cooling effective. 3) Using refrigeration. There may be other innovative methods that can be used such as a double layered plastic bubble which acts both as an insulator and where cool (or warm) air can be forced between the sandwiched layers. Combinations of any of these methods may be used for economy and efficiency depending upon location.

Air Cooling

The simplest and most economical means of cooling a greenhouse is by bringing in outside air to replace the warm air within. This is best accomplished by fans at one end of the greenhouse which are controlled by thermostats . When temperatures rise, the thermostat is activated, the fans turn on and cooler air is drawn through the greenhouse. A greenhouse so designed as to use the cooling ability of outside air will save much expensive energy and wear on cooling equipment. Air brought in from outside must be screened or filtered to prevent introduction of insects. A protective screen at the air intake with 32 mesh plastic screen should surround the water evaporation cells.

Thermostatic controls are designed to control both heating and cooling. As the temperature warms up inside the greenhouse the

thermostat will activate the fan thus bringing outside air into and through the house forcing the warm air outside. When the temperature rises further the thermostat switches on the water pump which sends water to the cooling cells to begin evaporation cooling.

One way of using outside air to cool a greenhouse is by having vents at the peak of the roof. Vents may be activated mechanically by hand or by a thermostatic controlled motor. When the vents are opened, they permit the warmed inside air to rise and bring in the cooler outside air through filtered vents at the lower sides of the structure. There are many problems associated with this method of air cooling, and though it is present in many older installations, it is generally not recommended for a plant laboratory greenhouse. These vents, if used should be screened with 32 mesh plastic screen.

Evaporator Coolers

The use of evaporator coolers is recommended for most greenhouses where humidity during the summer months is low. An engineering study should be done to calculate the cooling ability of evaporator coolers where humidity is moderate or high during the warm months. Evaporative coolers may prove uneconomical and unsound if the relative humidity is too high. However, in some areas evaporative coolers can be combined with refrigeration for efficient cooling.

A standard commercial evaporator cooler is available in most countries where humidity is low and where homes and buildings are cooled by this means and is relatively inexpensive. If the panels are removed, the inside of such a cooler will show a squirrel cage fan, a water reservoir at the bottom, a water pump forcing water to the top of a pad, usually made of wood fibers or glass wool and the water will drip down from outlets at the top. Such units have been and still are being used at the Riverside, California indexing greenhouses over many years and are very efficient in their cooling ability and in screening out insects. They should be carefully serviced each year by cleaning, painting, and changing the cooling pads. A standby cooler should be available for emergency replacement, as well as spare water pumps, fan belts and drive motor.

A more efficient apparatus for cooling uses cooling cells consisting of rectangular units of specially treated cardboard placed together to form a solid block. Water is pumped from a reservoir tank to a trough above the cells. The water then drips down by gravity over the cardboard cells. The outside air, is forced through the moistened cells by the diminished pressure induced by the fans located at the opposite end of the greenhouse The operation of the fans and water pump is controlled by a thermostat. The cooling cells usually occupy the full length of the outside wall of the greenhouse.

Refrigeration

Refrigeration can be used to supplement evaporator coolers where the relative humidity is too high during the warmer months, where extra cooling capacity is needed as a supplement for the plants in a cool temperature indexing room or for cooling small individual rooms. Small plastic chambers can be built inside of a large greenhouse to give areas of controlled cooling using refrigeration. Refrigeration is recommended in smaller greenhouses for those compartments to be held at cooler temperatures. These units should be designed to be readily removable for repair and replacement, and a spare unit should be held in reserve to replace one that may need repair. Recently in the warm climate of Belize, one area of a screenhouse was blocked off, heavily insulated with 15 cm thick insulation material and a plastic roof placed above. two small refrigeration units kept this room within the temperature range required for indexing of cool temperature pathogens and good symptoms of psorosis and tristeza were observed.

THE UC SYSTEM OF PLANT GROWTH

Since most symptoms of graft transmitted citrus diseases are seen primarily in plants, the plants should be of the highest quality. Therefore, the soil mixture with its balanced supply of micro and macro nutrients is of prime importance, and much emphasis will be given to the soil mixture and plant growth in this appendix. The University of California (or UC) system for producing healthy container grown plants was developed by Dr. Kenneth F. Baker and co-workers and published as Manual 23[1] (2). This system was based on the John Innes system of soil mixes developed in

England. The original objective was to provide a rapidly growing nursery industry in California with the means of producing uniform and healthy plants. This was done by developing a soil mixture of readily available ingredients and nutrients, incorporating a rigid sanitary program at all levels of production -- including clean nursery stock, sanitary greenhouse practices and soil disinfection. The system was modified by Nauer et al. (14,15) for growing citrus by the addition of micronutrients to the artificial mixture. Micronutrients were found to be absolutely essential for the successful growth of most citrus cultivars.

Through strict sanitation practices, the UC system provides a means for the total prevention of soil diseases (especially *Phytophthora* species). It permits fertility control, provides for a renewable, consistent and dependable set of soil ingredients and aids in salinity control. It meets the objective in the production of uniform healthy vigorous citrus and herbaceous plants free of deficiency symptoms, and the production of plant growth flushes containing clear large young developing leaves which are necessary for observing many symptoms.

The modified UC system containing micronutrients for growing citrus and other herbaceous plants has been in use at the Riverside, California indexing facility for over 35 years. In practicing rigid sanitation as recommended by the system, not a single case of soil contamination by *Phytophthora* has appeared in that facility over this time period. Since plants are the 'eyes' with which we see most graft transmissible citrus pathogens, a successful indexing program using plants would be very difficult to maintain if *Phytophthora* were present. Many of the specific index plants recommended as seedling indicators or as rootstocks under specific indicator scions are very susceptible to *Phytophthora*. Since citrus plants take 6 months to one year to reach buddable size, the destruction of these plants by soil-born diseases cannot and should not be tolerated. Every precaution should be taken to prevent *Phytophthora* infection in the plant laboratory. One such precaution is the use of a foot pad containing a soluble copper compound in a foam pad or Bordeaux mixture as a powder for disinfection of shoes and placed at the entrance of the greenhouse.

This manual, now out of print in California, has been re-issued in Australia and is available from Surrey Beatty and Sons PTY. LTD. Chipping Norton, NSW 2170. AUSTRALIA.

UC Mix

Ingredients

The basic soil mixture consists of 50% Canadian peat moss and 50% fine sand, with macro and micro nutrients added to the mix. Canadian peat moss is recommended as the prime ingredient. A trial mixture should first be prepared using equal parts Canadian peat moss and fine sand. Comparative tests can then be made substituting other local or more readily available types of peat[2]. Canadian peat moss has been tested in California and found to be superior to other peat or sphagnum mosses in nutrient retention and kelating ability. Comparative tests can be made to find a substitute for part of the more specific Canadian peat moss. Such ingredients as redwood shavings (if available), other wood shavings complimented with extra nitrogen, other peat mosses, perlite or vermiculite can be tried. The recommended mixture of 50% Canadian peat plus 50% fine sand should always be used as the standard for comparison.

The recommended particle diameters for the fine sand for use in the UC system should range from 0.05 to 0.5 mm. Beach sand 0.5 to 1 mm diameter is not recommended nor is clay. Fine sands can be found in wind blown deposits or as the fine silt separated out as waste material from a sand and gravel company processing pit. A quick and simple test for determining the presence of clay in a proposed sand source is to shake a sample of the test soil in a jar with water. If the sand settles rather rapidly and the water remains relatively clear, it is satisfactory. If clay is present the water will have a muddy appearance and that source preferably should not be used.

The sand should be inert and preferably siliceous. Calcarious or limestone sands should not be used since they may affect the pH. If a good grade silicate sand is not available, consideration should be given to substituting the sand fraction of the mix with vermiculite and perlite in a proportion of 1/2 peat, 1/4 vermiculite and 1/4 perlite or 1/3 of each ingredient. The objective is to obtain an artificial mixture which is consistently reproducible, will absorb and release macro and micro nutrients and will maintain a pH of the drainage water at 5.5 to 6.5.

The ingredients can be mixed together with a shovel on a flat concrete surface. However a small or medium sized electric or gasoline powered concrete mixer is the preferred mixing device. The procedure used at the Riverside, California indexing facility using a medium sized concrete mixer is as follows:

1. A specific number of uniform standard shovel-scoops of soil, peat and redwood shavings (or other substitutes for part of the peat moss) are counted and shoveled into the apron of the concrete mixer.
2. While in the apron, a weighed quantity of macronutrients i.e. phosphate, calcium and magnesium are sprinkled on top of the unmixed ingredients.
3. The soil ingredients plus macronutrients are then dumped into the concrete mixer and thoroughly tumbled.
4. The micronutrients, pre-weighed and mixed together in a package, are first dissolved in a container of water and then poured into the turning mixer.
5. A small quantity of water can be added to the soil while the mixer is turning to bring the soil mixture to a friable moist level. This may be necessary if the soil or peat is too dry.
6. After about 20 minutes of tumbling and mixing, the soil is emptied from the mixer into a trailer fitted with steam pipes on the bottom. The trailer top is covered with a cloth tarpaulin. Containers and flats may be placed on top of the soil in the trailer before covering, or steamed separately[3].
7. The soil mixture is then steamed. The time of steaming will depend on the quantity of steam produced which is proportional to the size and capacity of the boiler. A good general criteria is to continue to steam for about 15 minutes after the steam bellows up the covering tarpaulin. Soil thermometers placed in the

---

J.F. Ballester-Olmos in charge of the citrus indexing program at IVIA Spain has done extensive research on many peats. In a personal communication he suggests that only oligotropic peats should be used. These peats are produced in rain water peat bogs formed in cold regions with high rainfall i.e. Finland, Poland, Germany, USSR, Canada etc. There are two types of oligotropic peats: blond and black. Ballester-Olmos suggests that a good quality peat for a potting mix should be a mixture of both.

The trailer used at the Rubidoux facility in Riverside, California is fitted with 20 mm (3/4 inch) galvanized pipes with 5 mm (3/16 inch) holes spaced 15 to 20 cm (6 to 8 inches) apart. Pipes are spaced 15 cm (6 inches) apart. Holes are located at the bottom of the pipes.

corners of the trailer are helpful in judging the correct period for steaming. One minute at 100°C or 10 minutes at 83°C will kill all known pathogenic organisms. Therefore, oversteaming is a waste of energy, and is unnecessary. Steaming has never been found toxic or harmful to plants grown in a UC mix using Canadian peat moss, redwood shavings, perlite or vermiculite.

Fertilization

The initial mix contains both macro- and micro-nutrients which were added during mixing. The micronutrients are tied up in the peat moss. The peat, which acts as a chelating agent, releases sufficient small amounts of micronutrients to the plant for periods up to 1-2 years. (Nauer *et al.*, 14,15).

Liquid fertilizer is applied with each watering using a proportioning device. There are a number of such devices which inject fertilizer in proportion to the water used. An effective, simple and very inexpensive device is a venturi type siphoner available in most plant nurseries. Immediately after purchase, the siphoner should be calibrated. Many siphoners will vary considerably from the advertized ratio of concentrate to water as printed on the instructions. Calibration is done by putting a measured amount of water (500 or 1000 ml) in a graduated cylinder, then placing the suction end of the siphoner into the graduate and measuring the final amount of water exiting the hose. Allow water to flow and fill a container until the 500 or 1000 ml of measured liquid is siphoned up. Then measure the water in the container and convert the results to a ratio.

Another device which injects a given quantity of liquid fertilizer into the water system at a uniform rate in direct proportion to the water flow is the 'Smith' measuremix proportioner[4]. This is a precision instrument and highly reliable. This devise has been in use at the university of California greenhouses for over 35 years and in Spain and other countries for a considerable period of time with a minimum of upkeep or repair problems. The proportioner is set to deliver at a ratio of 1 to 100. However it should be calibrated in the same manner as for the venturi siphoner.

Information on the Smith measuremix proportioner can be obtained from: Smith Precision Corp. 1299 Lawrence Drive, Newbury Park, California, 91320 U.S.A.

A liquid fertilizer formula based on that given by Nauer *et al.* (15) is as follows:

Mix up the following (dry) parts by weight:

9 parts NH_4NO_3 (Ammonium Nitrate)
3.75 parts $Ca(NO_3)_2$ (Calcium nitrate)
2.75 parts KNO_3 (Potassium nitrate)

(If KNO_3 is difficult to obtain, substitute KCl Muriate of Potash)

This fertilizer should be well mixed and applied at the rate of 67.5 g of mixture to 100 l water (9 oz. per 100 gallons). The fertilizer listed above should be applied at the rate of 67.5 g of mixture per 100 liters (9 oz per 100 gallons) water. Calculate the amount to use and put the proper amount of fertilizer mix in the concentrate tub, add the correct amount of water and stir well. With the UC system of soil mix, fertilize with every watering. The soil should be fertilized directly after mixing and before and <u>immediately</u> after planting since the basic U.C. soil mix contains no nitrogen or potassium.

As a general practice, potted plants in a UC system need to be watered with enough volume periodically to flush out any accumulated salts to prevent salinity build up. It is important that the soil is not filled to the top of the container but a space of 2 to 3 cm is left between the top of the container and the soil level. This will allow for a sufficient volume of water to adequately flush the soil in the container.

It is important to know the pH of the water source being used and also the pH of the leachate should be measured over a period of time. This is done by filling the potted container with tap-water (without fertilizer) and collecting the leachate from the drainage holes at the bottom at the bottom of the container. The proportion of ammonium nitrate to calcium nitrate can be varied and used to adjust the pH of the soil within the container. Nauer *et al.* (15) reported that even though the initial pH of water being used was 7.8, drainage pH dropped to 6.4, 5.9, and 4.4 after 5,7 and 11 months respectively by modifying the nitrogen source. Plants were vigorous and showed little micronutrient deficiency symptoms at the lower pH. A leachate pH of 5.0 to 6.5 is recommended to obtain the best plant growth with minimal micronutrient deficiencies.

PRODUCTION AND CARE OF INDICATOR PLANTS

The indicator plants recommended for indexing are given in Table 2. Seedlings are recommended for indexing for most of the CGTPs. However, in some cases where seedlings are difficult to obtain, or other seedlings are readily available, a clonal propagation can be made with a bud from a selected indicator plant grafted to vigorous growing rootstock seedling. The rootstock is then inoculated, and the indicator bud forced as a scion. This procedure has been successfully used for detection of a number of pathogens. However, a comparison should be made between the seedling and a clonal propagation of the seedling to be sure that the clonal budline will induce clear positive symptoms equal to that of inoculated seedlings. In a number of trials, clonal budlines have not performed as well as seedlings whereas in other trials they have performed equally as well.

Where tristeza is endemic, the use of clonal propagations may be necessary to filter out the tristeza virus to detect or see other pathogens. For example, trifoliate or citrange may be used to filter the citrus tristeza virus so that other pathogens are not masked.

For any consistent long range index program a block of seed source trees containing the desired indicator varieties should be planted as soon as possible to obtain a reliable and consistent source of seed. In the meantime small or large quantities of seed can be obtained from commercial outlets or small quantities can be requested from other Research Stations or the United States Citrus Germplasm repository[5].

Seed Treatment

All fruit collected for seed extraction should be picked as high on the tree as possible, and picking fruit from the ground should be avoided because of the danger of *Phytophthora* infection. After extraction, the seed should be routinely treated against possible contamination by

United States Dept. Agriculture
National Clonal Repository for Citrus
1060 Pennsylvania Ave.
Riverside California, 92507 USA
Telephone (909) 787 4399
FAX (909) 787 4398

Phytophthora by a hot water dip for 10 minutes at 51.6°C (125°F) followed by a short dip in cool water to return the seed to normal temperatures (10). In addition to the hot water treatment the seed should be disinfected with a fungicide to aid in preserving seed during storage and to prevent albinism when the seedlings emerge. The commercial fungicide 'Thiram' as a 75% powder can be used to dust the seed after drying, or the seed can be dipped for 3 minutes in a 1% solution of 8-hydroxy quinoline sulfate, available from most chemical supply houses. The seed is then spread out on paper or on a fine mesh screen and allowed to air dry. It should be turned frequently during drying. Be careful to avoid over-drying. As soon as the last moisture has visibly disappeared, and the surface appears dry, the seed is packaged into small polyethylene bags. The bags are dated, labeled and sealed with a rubber band and then placed in a second small polyethylene bag with a small piece of slightly moist tissue paper placed between the bags. The second bag is then sealed with a rubber band. Seed treated, packaged and sealed in this manner and stored at refrigeration temperatures of 5° to 6°C have maintained excellent viability for up to 3 years (17).

Seed Planting

Seed can be planted in flats of wood or plastic. Redwood, if available is ideal since it is easy to construct, will not decay, will last a long time and can be steam sterilized. Other woods can also be used, but should be dipped or painted with a relatively non-toxic preservative such as copper naphthenate (26). Plastic containers with drainage holes are also quite satisfactory. They should be tested to see if they will withstand steam sterilization so that they can be reused. The sizes of the two types of redwood flats used at Riverside, California for growing seedlings are approximately 40 x 40 cm, and 40 x 20 cm by 14 cm deep.

The sterilized soil is placed in the flat and tamped firm with a flat metal tamper. It is important that the soil is firmly compacted and leveled at about 3 cm from the top of the flat. This permits uniform distribution of water; for if the soil is not level, water will settle at one corner of the flat and the other corner will usually be tested dry and may result in poor seed germination in that corner. A planting board made of thin masonite or plastic with 1.5 cm holes drilled 2.5 cm apart is placed on top of the soil and the seed

Table 3. The time for growth of seedlings from seed to one meter at the Rubidoux greenhouse, University of California, Riverside

Seedling cultivar[a]	No. of seedlings/ container	No. of weeks to reach 1 meter	
		Range[b]	Average
Citron	1	19-28	25
Eureka lemon	1	22-34	27
Rough lemon	1	23-34	28
Citrus excelsa	3	25-30	29
Sour orange	3	26-38	32
Troyer citrange	1	24-42	33
Mexican lime	3	24-44	35
Dweet tangor	3	29-41	36
Duncan grapefruit	3	28-48	40
Pineapple sweet orange	3	48-56	51

[a] Based on 4 to 9 plantings per cultivar over a three year period
[b] Reflects both summer and winter growth

individual placed in each hole. The seed is then lightly pressed into the soil with a dowel After seeding is complete, the planting board is removed and the seed covered with about 1 cm of soil and tamped lightly. Watering should be done using a soft spray perforated sprinkler nozzle on the hose end[6], or hand watered with a watering can with a perforated sprinkler head, until the seedlings emerge.

Speedling trays are satisfactory means of growing individual seedlings. Such trays are available through greenhouse and nursery catalogs. However, tests have shown them to be less efficient than use of flats.

The seeded flats or speedling trays are best kept in the warm growing room but can also be kept in the hot room if rapid forcing is desired. Most seedlings will reach transplantable size in 11 to 19 weeks depending on the variety (Table 3). Sweet orange seedlings take much longer to grow,

An excellent recommended soft-spray sprinkler nozzle made of cast aluminum is the "Dramm" nozzle. The Dramm Company, P.O Box 528, Manitowoc, Wisconsin 54220. U.S.A.

averaging 29 weeks. The seedlings must be periodically and critically examined for off-type, gametic or non-nucellar variants and these must be culled and should not be used since they usually make poor indicators. By permitting seedlings to reach 8 to 15 cm of growth, rather than transplanting when they are too small, the off type variants are more readily detected and can be pulled from the flat.

Insect Control

If possible, the indexing facility should be located away from nearby citrus groves to lower the infestation pressure by insects. Ornamental landscape plants should not be grown directly adjacent to a greenhouse and preferable not too close to the indexing facility since they may provide a host for growth of insects. Personnel should not be exposed to insect infested plants in the field prior to entering the greenhouse. A balanced and thorough insect control program in an indexing facility is of utmost importance. Damage to indicator plants by insects or insecticidal spray can make symptom reading very difficult.

Any insect control program that reduces the amount of insecticide spraying is worthwhile. Insecticide application, although necessary, should be limited and carefully controlled. Spray damage to leaves are usually in the form of circular translucent spots and may confuse symptom readings as well as damage young growth. The presence of adequate non-inoculated control plants should verify any damage to leaves done by insects, spray or other non-viral effects. The insect control program used at the University of California, Riverside index facility is as follows:

1. All plants are critically examined at least once each week for any signs of insects. If insects are found, the area surrounding the infestation is treated by spot spraying with an appropriate insecticide at concentrations usually below those recommended by the manufacturer to avoid chemical damage to the tender young emerging leaves.

2. At least once each week all plants are sprayed with water using a standard pressure hose nozzle with a fine spray attachment. The objective of this water spray program is to control small infestations of mites. A few undetected mites are readily washed off of the leaves by this water spray and usually do not return. This is a preventative measure and it has been found very effective as one means of control and reducing the number of times the entire greenhouse must be sprayed. The combination of periodic inspection, spot spraying and water-spraying have been highly successful in keeping insect infestation under control. The importance of routine periodic preventative inspection for detecting new low level insect infestations cannot be overemphasized.

3. The use of the two-door vestibule entryway for the critical inspection of ones clothes for presence of insects is important. One should avoid wearing green colored clothes during aphid season. Only one door at a time should be opened when bringing in or taking out plants or materials.

4. When spraying with an insecticide becomes necessary, select an insecticide and a dilution which will not spot or

injure plants. New insecticide sprays should first be tested on a few plants. Dilutions should be carefully calculated and double checked to be sure they are correct. A number of effective miticides should be kept on inventory and their use rotated to prevent build up of insect resistance.

In general, aphid infestation in the greenhouse at the Rubidoux indexing facility in Riverside, California has been rare and never a serious problem. Mites are the major pest problem. Soft brown and other scales, whitefly, mealy bugs or thrips can be serious problems. Again, the importance of periodic inspection and good insect control cannot be overstated in the maintenance of an efficient plant laboratory.

The bringing of plant material from other areas or from the field into the greenhouse should be avoided. If the plants must be brought in, they should be carefully examined for pests, then cut back to a minimum number of leaves, given a preventative insecticidal spray and isolated on a separate bench or area until they are shown to be free of pests.

If spraying with insecticides becomes necessary, it may be wise to consider spraying all rooms or houses at one time. If only one room or house is sprayed, reinfestation may readily occur from other areas not sprayed.

REFERENCES

1. Aubert, B., ed. 1987. Regional Workshop on Citrus Greening Huanglungbin Disease. Review and abstracts. FAO-UNDP, Fuzhou, China Dec. 6-12, 1987.
2. Baker, K.F., ed. 1951. The U.C. System for Producing Healthy Container-grown Plants. Calif. Agric. Exp. Sta. Manual 23. Reprinted 1985 by Surrey Beatty and Sons, Chipping Norton, NSW 2170, Australia.
3. Bové, J.M., and Vogel, R., eds. 1980. Description and Illustration of Virus and Virus-like Diseases of Citrus. A collection of color slides. I.R.F.A. SETCO-FRUITS, Paris.
4. Bové, J. M., Zreik, L., Danet, J-L., Bonfils, J., Mjeni, A.M.M., and Garnier, M. 1993. Witches' broom disease of lime trees: Monoclonal antibody and DNA probes for the detection of the Associated MLO and the identification of a possible vector. Pages 342-348 in: Proc. 12th Conf. IOCV. IOCV, Riverside.
5. Calavan, E.C., Roistacher, C.N., and Nauer, E.M.. 1972. Thermotherapy of citrus for inactivation of certain viruses. Plant Dis. Reptr. 56: 976-980.
6. Calavan, E.C., Mather, A.M., and McEachern, E.H. 1976. Registration, certification and indexing of citrus trees. Pages 185-222 in: The Citrus Industry

Vol. 4. W. Reuther, E.C. Calavan and G.E. Carmen, eds. Univ. of Calif. Div. Agric. Sciences.

7.   Çinar, C., Korkmaz, S., and Kersting, U. 1994. Presence of a new whitefly-borne citrus virus disease of possible viral etiology in Turkey. FAO Plant Prot. Bull. 42:73-74.

8.   Dodds, J.A., Jarupat, T.,. Lee, J.G., and Roistacher, C.N. 1987. Effect of strain, host, time of harvest and virus concentration on double-stranded RNA analysis of citrus tristeza virus. Phytopathology 77: 442-447.

9.   Duran-Vila, N., Pina, J.A., and Navarro, L. 1993. Improved indexing of citrus viroids. Pages 202-211 in: Proc. 12th Conf. IOCV. IOCV, Riverside.

10. Klotz, L.J., DeWolfe, T.A., Roistacher, C.N.,Nauer, E.M. and Carpenter, J.B. 1960. Heat treatment to destroy fungi in infected seeds. Plant Dis. Reptr. 44:858-861.

11. Lastra, R, R.F. Lee, M. Rocha-Peña and C.L. Niblett. 1991. Workshop on citrus tristeza virus/ *Toxoptera citricida* in Central America and the Caribbean Basin, CATIE, Turrialba, Costa Rica. May 14-17, 1991.

12. Lee, R.F.,Derrick, K.S., Beretta, M.J.G., Chagas, C.M., and Rossetti, V. 1991. Citrus Variegated Chlorosis: a new destructive disease of citrus in Brazil. Citrus Industry, Oct. 1991, Pages 12-13,15.

13. Nauer, E.M., and Goodale, J.A. 1964. Forcing newly budded citrus. California Citrograph 49: 294- 295, 297.

14. Nauer, E.M., Roistacher, C.N., and Labanauskas, C.K. 1967. Effects of mix composition, fertilization, and pH on citrus grown in UC-type potting mixtures under greenhouse conditions. Hilgardia 38: 557-567.

15. Nauer, E.M., Roistacher, C.N., and Labanauskas, C.K. 1968. Growing citrus in modified UC potting mixtures. Calif. Citrograph 53: 456, 458, 460-461.

16. Nauer, E.M., Holmes, R.C., and Boswel, B.S. 1980. Close spacing in greenhouse inhibits lime seedling growth. HortScience 15: 591-592.

17. Nauer, E.M., and Carson, T. 1985. Packaging citrus seed for long term storage. Citrograph. 70: 229-230.

18. Navarro, L., Roistacher, C.N., and Murashige, T. 1976. Effect of size and source of shoot tips on psorosis-A and exocortis content of navel orange obtained by shoot-tip grafting in vitro., Pages 194-197 in: Proc. 8th Conf. IOCV. IOCV, Riverside.

19. Roistacher, C.N. 1963. Effect of light on symptom expression of concave gum virus in certain mandarins. Plant Dis. Reptr. 47: 914-915.

20. Roistacher, C.N. 1976. Detection of citrus viruses by graft transmission: A review. Pages 175-184. in: Proc. 7th Conf. IOCV. IOCV, Riverside.

21. Roistacher, C.N. 1977. Elimination of citrus pathogens in propagative budwood. I. Budwood selection, indexing and thermotherapy. Proc. Int. Soc. Citriculture 3:965-972.

22. Roistacher, C.N. 1988. Observation on the decline of sweet orange trees in coastal Peru caused by stem pitting tristeza. FAO Plant Protection Bull. 36:19-26.

23. Roistacher, C.N. 1991. Graft-transmissible diseases of citrus. In: Handbook for Detection and Diagnosis. Publication Division Food and Agricultural Organization of the United Nations, Viale delle Terme di Caracalla, 00100 Rome, Italy . 286 pages.

24. Roistacher, C.N. 1993. Arguments for establishing a mandatory certification program for citrus. Pages 18-34 in: Proc. 4th Conf. Int. Soc. Citrus Nurserymen, Johannesburg, South Africa.

25. Roistacher, C.N. 1996. A Historical Review of the Major Graft-transmissible Diseases of Citrus. Food and Agricultural Organization of the United Nations, Regional Office for the Near East. P.O. Box 2223, Cairo, Egypt. 89 pages.

26. Roistacher, C.N., and Baker, K.F. 1954. Disinfecting action of wood preservatives on plant containers. Phytopathology 44: 65-69.

27. Roistacher, C.N., Blue, R.L.,Nauer, E.M., and Calavan, E.C. 1974. Suppression of tristeza virus symptoms in Mexican lime seedlings grown at warm temperatures. Plant Dis. Reptr. 58: 757-760.

28. Roistacher, C.N., and Calavan, E.C.. 1974. Inactivation of five citrus viruses in plants held at warm glasshouse temperatures. Plant Dis. Reptr. 58: 850-853.

29. Roistacher, C.N., L. Navarro, and T. Murashige. 1976. Recovery of citrus selections free of several viruses, exocortis viroid, and *Spiroplasma citri* by shoot tip grafting in vitro. Pages 786-793 in: Proc. 8th Conf. IOCV. IOCV, Riverside.

30. Roistacher, C.N., and Nauer, E.M. 1985. Effect of supplemental light on citrus seedlings in winter. Citrograph 70: 181-182, 196.

31. Wallace, J.M. 1978. Virus and virus like diseases. Pages 67-184 in: The Citrus Industry Vol. IV. W. Reuther, E.C. Calavan and G.E. Carmen, eds. Univ. Calif. Div. Agric. Sciences.

32. Whiteside, J.O., Garnsey, S.M., and Timmer, L.W. 1988. Compendium of Citrus Diseases. APS Press, St. Paul MN 55121, USA.

33. Yang, X., Hadidi, A., and Garnsey, S.M. 1992. Enzymatic cDNA amplification of citrus exocortis and cachexia viroids from infected citrus hosts. Phytopathology 82:279-285.

CHAPTER 25

Virus Certification of Strawberries

S. Spiegel

Cultivated strawberry (*Fragaria* x *Ananassa* Duch), grown commercially in many countries, is considered the most important crop among the 'small (soft) fruits'. Production of strawberries in the world has almost doubled between 1970 and 1994. In the U.S. fresh-fruit market, strawberries are currently the second most valuable crop after apples (1). The trade in fresh and processed fruit, as well as in planting material, has increased in the last decade both within countries and internationally and has highlighted the significance of virus diseases in this crop.

Like many other species, strawberry plants are propagated vegetatively. The majority of planting stocks are being built up from mother plants through runners, and occasionally by *in vitro* propagation (micropropagation). Viruses entering a propagation system at the breeding or germplasm level are being transmitted "vertically" down the propagation chain and are eventually being distributed with planting material in commercial fields. Virus and virus-like diseases do not induce distinct symptoms in most strawberry cultivars, but may reduce considerably both productivity and fruit quality. For convenience, the term 'viruses' will be used here to include virus-like agents such as viroids and phytoplasmas, unless otherwise stated.

Since no practical treatments to cure virus-infected plants in the nursery and/or field are available, controlling virus diseases can be done mainly by using virus-free (tested) planting material (17,18). Procedures concerning production and propagation of high quality, virus-free planting material have been established in 'certification schemes' in strawberry growing countries. This chapter will address mainly virus certification, however, it is emphasized that other pathogens (e.g fungi, bacteria, etc.) should also be included.

VIRUS ELIMINATION AND DETECTION

More than thirty viruses and phytoplasmas have been described in the Genus *Fragaria* (2), three of them recently (8,11,15). Several of the major strawberry viruses which are aphid-borne and those unique to strawberry with unknown vectors are poorly characterized. Few strawberry viruses act singly in their hosts and frequently several viruses are present. With aphid-borne viruses, transmission by vectors contributes to mixed infections which often result in distortion of plant parts including fruit (2).

Virus Elimination Procedures

Reliable virus detection as well as established therapy protocols are prerequisites for production of virus-free strawberry planting material. Shoot-tip culture and thermotherapy have been used either alone or increasingly in combination to eliminate several of the major strawberry viruses. Strawberry mild yellow edge (SMYE)- associated luteovirus (4), strawberry mottle (SMV), strawberry crinkle (SCV) and strawberry vein banding virus (SVBV) and pallidosis agent (PA) have been eliminated with various degrees of success (2), whereas, plants were not cured of June Yellows (JY) using thermotherapy combined with meristem-tip culture (19).

Applying therapy procedures to *in vitro* grown plantlets was reported recently. Strawberry plantlets free of tobacco streak virus (TSV) were recovered from stolon tips excised from infected plants and established on a nutrient medium *in vitro* (16). Addition of 2,4-dioxohexahydro-1,3,5-triazine (DHT) to the nutrient medium on which plantlets infected with SCV and SMV were grown, followed by regeneration of apices, allowed recovery of virus-free plants. SCV and SMV were not eliminated from strawberry, unless both DHT treatment and shoot-tip culture were used (10).

From the information currently available in the literature it seems that variation among viruses and virus isolates, plant cultivars and therapy procedures used by various research groups do not allow conclusions to be made regarding general protocols applicable for most viruses.

Virus Detection Methods

The most reliable detection method applied for the major strawberry viruses, is leaf grafting onto sensitive *F. vesca* and *F. virginiana* clones. This bioassay, known as 'indexing', is time-consuming, requires a temperature controlled greenhouse and can be inaccurate; however, its broad detection spectrum is extremely valuable for poorly characterized viruses (2). Recent developments in diagnostics and characterization of several strawberry viruses have provided serological and molecular assays (3). ELISA has been used successfully to detect SMYE-associated potexvirus (7,12), SVBV(6), *Fragaria chiloensis* ilarvirus (15) and TSV (16). Nematode-borne viruses, known to infect a wide range of hosts, including strawberry, have been detected by ELISA with antisera prepared against these viruses from other hosts (2,3).

Double-stranded RNA analysis was used for detection of SMYE-associated luteovirus (14), SMV (13) and PA (20) and could add information to results obtained from leaf graft bioassays. The polymerase chain reaction (PCR) method, increasingly used for detection of minute quantities of plant viruses in plant tissue, has been applied for detection of the SMYE-associated potexvirus (9) and luteovirus (5).

Determining the virus status of source plants from which planting material is produced is a prerequisite for propagation of virus-free plants. Often, source plants have gone through a therapy step and virus titer may be very low. In these cases, plants need to be grown for a prolonged period of time (preferably through a natural dormancy period) to allow virus titers to reach detection thresholds. The time required to reach reliable virus detection may vary with host plant, virus, and geographical location. Until more information regarding the characterization of viruses is acquired, determining virus status of

strawberry planting material will continue to depend on leaf graft bioassays(3).

CERTIFICATION SCHEMES

Procedures concerning the production and propagation chain of virus-free strawberry plants have been established in various countries in regulations known as 'certification schemes'. The ultimate goal shared by all schemes, regardless of their country of origin, is to ensure that certified planting material obtained by the nursery and/or grower meets the standards required in the relevant program and is a high quality, pathogen-free, and true to type product. Certification schemes are usually organized and administered by government (national, state or province) or government-affiliated bodies. Government control is needed for regulatory measures and for providing a basis for the granting of phytosanitary documentation. Strawberry certification programs are operating, either on a voluntary or a compulsory basis in many countries including the U.S. (California, Florida, Washington); Canada (Nova Scotia, Ontario); Australia (Victoria), France, UK ,Germany, Israel, Italy and The Netherlands.

In general, certification schemes are based on three major principals - freedom of the source plants (nuclear stock) from viruses and other pathogens, prevention of reinfection during two-three subsequent propagation steps and trueness-to-type.

An outline of a typical strawberry certification scheme is based on several consecutive stages of propagation (Figure 1):

Stage 1: *Nuclear stock* is composed of a limited number of healthy clonal plants, derived from selected pre-nuclear plants after a therapy step, if necessary, tested extensively for viruses - preferably for two growing seasons, and evaluated for trueness-to-type. Nuclear stock plants are grown in individual pots on raised benches, in screened enclosures under strict phytosanitary conditions, minimizing the risks of reinfection by vectors. Indexing is repeated several times throughout the year (zero tolerance). From nuclear stock, plant material is taken for growing "foundation stock".

Stage 2: *Foundation stock* is derived directly from nuclear stock plants. Runners, severed from nuclear plants, are potted and maintained under

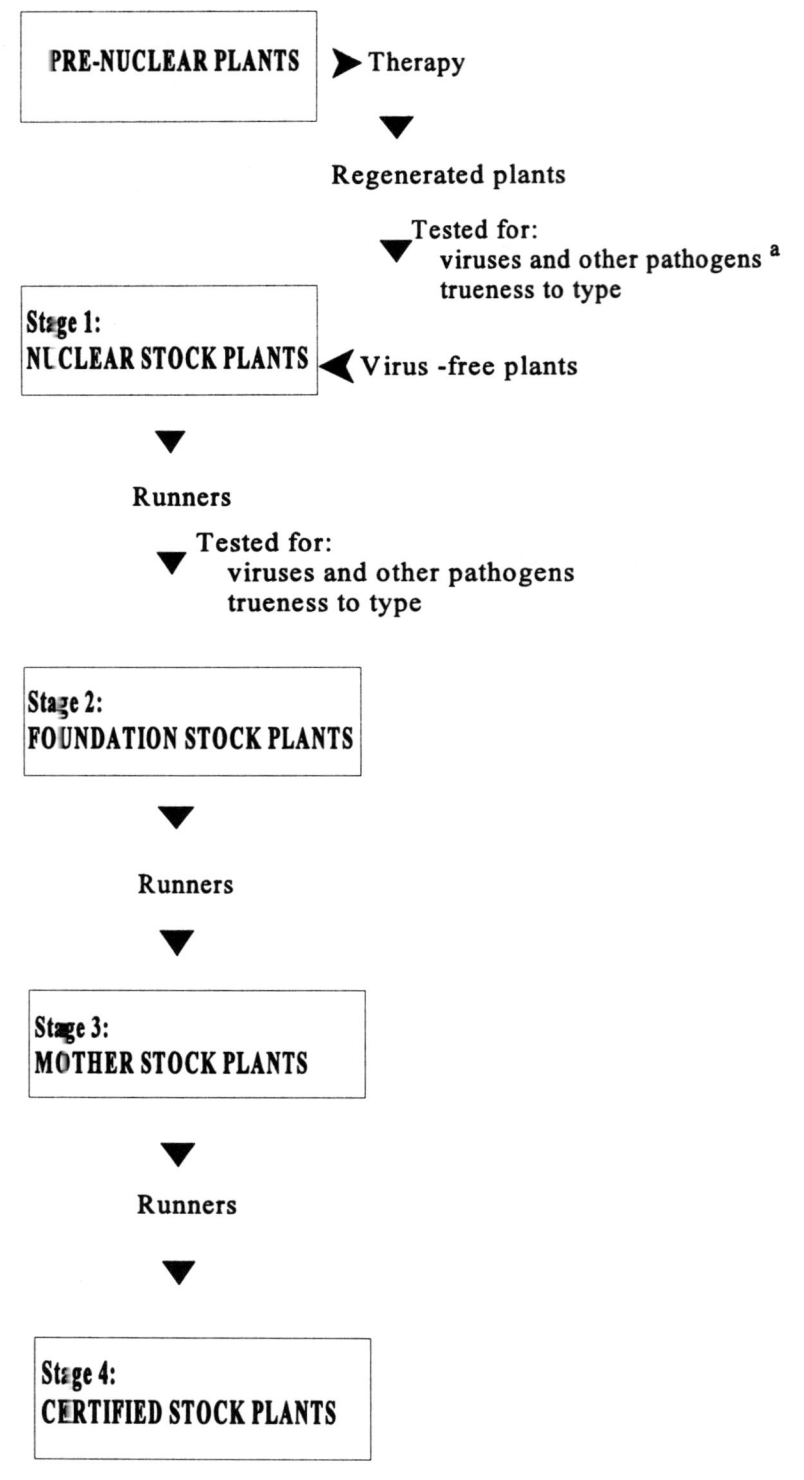

a Testing for viruses is done by leaft graft bioassays, enzyme-linked immunosorbent assays (ELISA), or other appropriate methods. Plants are observed for symptoms of possible other pathogens (e.g. fungi, bacteria). Testing of nuclear stock plants is done on individual plants ,of foundation stock on each clone and of certified stock at random

Figure 1. Stages of production of certified strawberry planting

conditions minimizing reinfection as above. Plants are indexed on a clonal basis rather than each individual plant. This stage of propagation may be done in isolated field plots.

Stage 3: *Mother* (increase) *stock* is derived directly through runners from foundation stock as described in stage 2. This propagation step is usually done in the open field, located at least one mile from any other strawberry field and/or nursery. Plants are randomly checked for quality standards and diseases for which maximum tolerance levels have been defined.

Stage 4: *Certified stock* is derived directly from mother stock. Runners severed from mother stock plants are grown in nurseries for the final mass production of certified planting material. Plants are checked visually and if the required standards are met, plants are harvested and sold as certified stock which will be grown in commercial fruiting fields. It is recommended but not mandatory to grow certified stock under minimum reinfection conditions.

Several points in the basic scheme have been approached differently in various countries. One of the crucial issues concerns the period of time allowed for growing plants at each stage of the scheme before being replaced with new plants. The time limit for nuclear stock plants ranges in various schemes from one to four years. Mother stock plants are being downgraded to certified stock after 2-3 years and new mother plants are derived from foundation stock. Procedures concerning the renewal of nuclear stock plants vary in different programs. Another variation among schemes is the number of propagation stages between nuclear plants to certified stock.

CONCLUSIONS

Certification schemes, operating successfully in strawberry growing countries in various parts of the world, should be periodically evaluated and updated if necessary. Revisions should consider: reports on new pathogens, disease outbreaks and other pests; and new techniques for virus elimination and detection.

It is my opinion that a close cooperation between plant breeders, propagators, growers and plant protection services contribute to a successful scheme regardless of whether the program is voluntary or compulsory.

REFERENCES

1. Berterlson, D. 1995. The U.S. strawberry industry.Commercial Agriculture Division, Economic Research Service, U.S. Department of Agriculture. Statistical Bulletin No. 914.
2. Converse, R.H. 1987. Virus diseases of small fruits. United States Department of Agriculture, Agriculture Handbook No. 63. 277 Pages.
3. Converse, R.H., Adams, A.N., Barbara, D.J., Clark, M.F.,Casper, R., Hepp, R.F., Martin, R.R., Morris, T.J., Spiegel, S., and Yoshikawa, N. 1988. Laboratory detection of viruses and mycoplasma-like organisms in strawberry. Plant Dis.72:744-749.
4. Converse, R.H., and Tanne, E. 1984. Heat therapy and stolon apex culture to eliminate mild yellow-edge virus from Hood strawberry. Phytopathology 74:1315-1316.
5. Hadidi, A., Montasser, M.S., Levy, L., Goth, R.W., Converse, R.H., Madkour, M.A., and Skrzeckowski, L.J. 1993. detection of potato leafroll and strawberry mild yellow edge luteoviruses by reverse transcription-polymerase chain reaction amplification. Plant Dis. 77:595-601.
6. Honetslegrova, J., Mraz, I., and Spak, J. 1995. Detection and isolation of strawberry vein banding virus in the Czech Republic. Acta Hortic. 385:29-32
7. Jawee, A., and Adams, A.N. 1995. Serological detection of strawberry mild yellow edge-associated virus. Acta Hort. 385:98-104.
8. Jelkmann, W., Martin, R.R., Lesemann, D.E., Vetten, H.J., and Skelton, F. 1990. A new potexvirus associated with strawberry mild yellow edge disease. J. Gen. Virol.71:1251-1258.
9. Kaden-Kreuziger, D., Lamprecht, S., Martin, R.R., and Jelkmann, W. 1995. Immunocapture polymerase chain reaction assay and ELISA for the detection of strawberry mild yellow edge associated potexvirus. Acta Hort. 385:33-38.
10. Kondakova, V., and Schuster, G. 1991. Elimination of strawberry mottle virus and strawberry crinkle virus from isolated apices of three strawberry varieties by the addition of 2,4-dioxohexahydro-1,3,5-triazine (5-azadihhydrouracil) to the nutrient medium. J. Phytopathol. 132:84-86.
11. Norrisseau, J.G., Lansac, M., and Garnier, M. 1993. Marginal chlorosis, a new disease of strawberries associated with a bacterium-like organism. Plant Dis. 77:1055-1059.
12. Quail, A.M., Martin, R.R., Spiegel, S.S., and Jelkmann, W. 1995. Development of monoclonal antibodies specific for strawberry mild yellow edge potexvirus. Acta Hortic. 385:39-45
13. Schoen, C.D., and Leone, G. 1995. Towards molecular detection methods for aphid-borne strawberry viruses. Acta Hortic. 385:55-60.
14. Spiegel, S. 1987. Double-stranded RNA in strawberry plants infected with strawberry mild yellow-edge virus. Phytopathology 77:1492-1494.
15. Spiegel, S., Martin, R.R., Leggett, F., Ter Borg, M., and Postman, J. 1993. Characterization and geographical distribution of a new ilarvirus from *Fragaria chiloensis*. Phytopathology 83:991-995.
16. Spiegel, S., Martin, R.R., Ter Borg, M., and Tam, Y. 1995.Uneven distribution of tobacco streak virus in strawberry plantlets grown in vitro. Acta Hortic. 385:122-125.
17. Stace-Smith, R. 1985. Virus free clones through

plant tissue culture. Pages 69-179 in: Comprehensive Biotechnology: The Principles, Applications and Regulations of Biotechnology in Industry, Agriculture and Medicine. M. Moo-Young, ed. Pergamon Press, Oxford, England.

18.   Walkey, D.G.A. 1991. Production of virus-free plants. Pages 270-292 in: Applied Plant Virology. John Wiley & Sons, New York.

19.   Watkins, C.A., McNicol, R.J., Young, K., and

Jones, A.T. 1990. The effect of heat treatment and meristem-tip culture on june yellows in strawberry. Ann. Appl. Biol. 116:489-492.

20.   Yoshikawa, N., and Converse, R.H. 1990. Strawberry pallidosis disease: distinctive dsRNA species associated with latent infections in indicators and in diseases strawberry cultivars. Phytopathology 80:543-548.

Certification for Plant Viruses - an Overview

H.E. Waterworth

In reviewing chapters in this section, it is readily apparent that viruses are recognized as major problems in many agricultural crops, but especially so in vegatatively propagated crops. It is also clear that many programs are in place, particularly in the developed countries, to reduce spread of virus-like pathogens among countries or to ameliorate the detrimental effects of domestic viruses by making virus free planting stock available to growers at reasonable cost.

The need for certified virus-free planting stocks and propagating material, especially the vegetatively propagated crops, became obvious during the mid twentieth century as data became available regarding the worldly distribution of viruses in these crops and the losses they were causing to growers. Because this group of crops, which includes tree fruits, small fruits, tropical tree crops, woody ornamentals and some grasses are vegetatively propagated, virus infections in the source plants become established in the offspring.

The Problem

Losses to virus infections take their toll, not only in the form of obvious symptoms in the plant or its harvested product, but in numerous other ways as well. Viruses take their toll in such diverse ways as reducing bud take or rooting of cuttings when these crops are propagated; for reduced plant vigor, incompatibility between stock and scion variety, reductions in yield, alterations to product quality, reduced life span and/or productivity of orchards, groves, and pastures, and in other ways that go unnoticed. Virus infections also caused plants to be more susceptible to drought and frost and more susceptible to certain fungal pathogens (4,18). Some have been responsible for catastrophic or recurring losses. The literature abounds with examples in which the hundreds of viruses that affect these crops incite their detrimental effects.

The worldly distribution of viruses in vegetatively propagated genera and the frequency of their occurrence in large numbers of samples became apparent as countries developed quarantine programs during international exchanges of germplasm which included testing for virus-like pathogens. It soon became apparent that some viruses are widely distributed or occurred in rather high percentage of samples. For example, 53% of 518 importations of *Malus*, *Pyrus*, *Prunus*, and *Citrus* from ca. 50 countries were infected with one or more viruses (11). Similarly, virus was detected in 65% of 409 accessions of *Vitis* from 27 counties, and in 56% of 424 accessions of *Solanum* from 23 countries. In most cases the imported and established plants displayed no disease symptoms.

In view of this, new strategies to reduce losses by viruses were clearly needed. Control methods until mid century focused on use of chemicals to reduce insect or nematode vector populations, rouging diseased plants, growing disease tolerant varieties, and cross protection. However, by then it was known that some crops remained free of disease for many years when virus free trees were planted in new orchards. This is especially true for viruses spread only by man. Hence, programs were established for the specific purpose of producing and distributing certified virus tested planting stocks. In Chapters 16 and 23, the authors describe some of the oldest and better known certification programs of deciduous tree fruit crops in North America, among which is the Interregional Project no. 2 (IR2/NRP 5)(Mink, Chapter 23) and the quarantine programs of the USA and Canada (Foster and Hadidi, Chapter 16). Similarly well known early virus free certification programs for tree fruit crops were developed in several Western European Countries (6,7,18,19). Certified virus free potato programs were in place in the Netherlands and Scotland by the early 1950s (22). Slack and Singh (Chapter 19) review the history and magnitude of seed potato certification programs in North America.

Scope and Definitions

Some of the authors in this section have defined or described certification programs. The North American Plant Protection Organization defines a certification scheme as: "A domestic program consisting of the maintenance, multiplication, distribution, and production of plant materials intended for release....under an officially sponsored certificate attesting to the status of the material" (13). Freedom from virus infections is not a requirement by this definition. However, the term "certified" planting material can be variously defined. In its broadest sense it includes seed and vegetatively propagated crops in which plants or seeds are distributed between locations within a country, usually by plant nurseries in accordance with regulations of the local governments. Requirements for certification in these programs involve combinations of such activities as fumigation of soils, visual inspection of crops during the growing season for signs of disease and insect infestations, and additional inspections after plants, tubers, roots, or seeds are harvested but usually without tests for virus infections. Each of the US states and several other countries conduct this level of certification activity as a means to provide growers with reasonably pest free planting stock.

"Certified" planting material can also be defined as being free of virus infections based upon tests for all known virus-like pathogens. This kind of certification scheme has been described for various crops such as deciduous tree fruits (Chapters 22 and 23), small fruits (Chapter 25), grapevines (Chapter 20), potatoes (Chapter 19) and ornamental crops (Chapter 21). In his review, Bos (5) describes two types of certification, one of which deals with 'quality' of certified planting material as it is usually applied in commerce. It addresses such items as trueness to type, genetic purity, vitality and levels of allowed indigenous virus infections. The other is labelled 'Certification for Quarantine', meaning that pathogens of quarantine status are not present in germplasm imported to countries. In this review "certified" plant material means that it has been tested for all or specified pathogens and that named virus-like agents were not detected in tested plants. These schemes may be dedicated to certification activities within a country (Chapters 24 and 25 by Roistacher and Spiegel, respectively) or be part of a nation's quarantine program during international exchange of germplasm as described in the chapters by Foster and Hadidi (Chapter 16) and by Frison and Diekmann (Chapter 17).

No attempt has been made in this overview to describe every crop certification program throughout the world, nor the magnitude of each program in terms of crops included and the viruses for which tests are conducted. It would soon be obsolete. Rather, examples are

given as they relate to various points of the discussion. Some public programs are partically funded by private companies or grower associations as described by Mink in Chapter 23. Finally, the details on virus indexing procedures, such as kinds of tests, number of replicates, specific growing conditions for indicators, number of buds to test, names of numbers of indicators to inoculate, and alternative tests, will not be elucidated here. These variables differ among certification programs and crops to be tested, and are widely reported in the literature. In this section, chapters by Maury *et al.*, Slack and Singh, Martelli and Walter, Krczal, Barba and Roistacher (Chapters 18,19,20,21,22,24, respectively) describe testing procedures in detail for the respective crops.

Certification Procedures

Several key components are necessary for a successful plant certification program. Among those offered by Bernhard (3), Hollings (9), Meijneke (80), van der Graaff (24) and by some authors of chapters in this section are:

i. published standards that define the purpose, terminology, and protocols,
ii. selection of the material to be used in the virus free program that is 'true to type',
iii. recognition of the virus symptoms,
iv. availability of reliable indexing methods for the causal pathogens and latent viruses,
v. ability to produce virus free germplasm when only infected plants are available,
vi. enactment of measures to maintain the health of nuclear and foundation stocks, retesting, and adding new items,
vii. effective distribution of their progeny to clean stock programs and growers,
viii. a means to monitor established standards, and
ix. grower acceptance of the virus free materials.

The absence of any of these elements can severely compromise a certification program (5).

Differences Among Certification Programs

As one can see from reviewing the chapters in this section, certification programs among crops and various countries vary considerably according to the objectives of the program, facilities, finances and skills available, and magnitude of the virus problems to be managed. Certification programs are conducted by governments at the national level, in a few countries by other levels of government such as State Departments of Agriculture or by Universities, and by private companies. Essentially no two programs are identical. Among the variables are number of crops included, whether only selected or all viruses are tested for, pre- and post test inspection procedures, manner in which virus free nuclear stock plants and maintained, frequency of retesting, procedures for propagating to volume from tested plants, and marketing/sales of certified plants. The simplest of programs in which certified plants are sold include a single crop wherein tests are conducted for only one of several viruses known to infect that crop, and with only visual inspections for other virus-like disorders. For example, Barba (Chapter 22) mentioned that one of the two kinds of certification programs included in the European Union (EU) protocol, (*Conformitas Agricolae Communitatis*) requires only absence of quarantine pests and pathogens based upon visual inspection and random tests in order to qualify for exchange of germplasm within the EU. Martelli and Walter (Chapter 20), and Spiegel (Chapter 25) describe some of the methods by which therapy is conducted on virus infected grapevines and strawberries.

At the other extreme, programs include elaborate testing protocols some of which are repeated, includes all known viruses, and tested virus free plants may be specially classified such as: nuclear, elite, foundation, and certified stocks. Most of the chapters in this section describe these categories of planting material. Handling of plants in these categories such as degree of isolation, frequency of inspections and retesting, testing of soils for nematodes, is progressively less stringent from nuclear stock to certified stock categories. Most European Economic Community and North American countries include these components in their certification schemes. The best of these require that the plants are propagated from virus-free material of known origin, while others require only that the plants are free from visible disease symptoms (19). Some schemes certify that plants are from virus-tested material and free of important viruses, but known to be infected with other less dangerous viruses.

Most public certification programs involve more than one crop; usually those crops important to that State, country or region of the world. The administration of many to these programs is described by Ebbels (7). The more comprehensive schemes require many more steps than testing for viruses. Among them are fumigation of the soils in which tested plants are grown to eliminate nematodes that spread viruses and taking measures to insure that outdoor sites will not become recontaminated by drainage or irrigation water, and that virus will not be spread among tested plants by such activities as cultivating, digging, and pruning. Some viruses are spread by pollen or insects so that maintaining these trees in areas distant from non-tested plants of the same crop is essential (19). Another difference among certification programs is the frequency of observation of plants being tested both before and after tests are completed. In some programs, plants are examined several times each year for signs of disease that may result from new pathogens, from invasions of pathogens from untested plants, and verification of test results. For these reasons, the more thorough schemes include regularly scheduled retesting as described in chapters by Barba (Chapter 22), Martelli and Walter (Chapter 20), Krczal (Chapter 21) and Mink (Chapter 23). The better schemes also address legal and financial issues such as trueness to type, implied warranties, and marketing and sales. Mink, Martelli and Walter, and Krczal elaborate on these isues in their chapters.

Some Commonly Employed Procedures

All crops in certification programs are known to be hosts of several virus-like pathogens. Among these are pathogens known to be viruses, viroids, phytoplasmas, and rickettsia but also usually includes many uncharacterized infectious agents (18). Because of the differing characteristics among these pathogens and the status of technologies available to detect them, various combinations of tests are usually required in effort to detect all of the virus-like agents known to infect a given plant genus (14,25). Authors of several chapters in this section describe some of the differences between procedures employed to certify various crops.

Among the combinations of tests available to detect virus-like agents are sap transmission to herbaceous genera, serological tests (usually ELISA), electron microscopy of extracted sap, and nucleic acid extractions for PCR and/or hybridization tests. Also usually employed are graft tests to pathogen sensitive genotypes of the same genus as the tested plant because crops such as tree fruits are hosts of 'pathogens' of unknown etiology and therefore no other test is yet available (13,25). For example, more than 100 virus-like diseases affect *Prunus* worldwide but nothing is known about the pathogens associated with about half of these diseases except that they are graft transmissible (18). A similar situation exists for many other crops in certification programs. Several chapters in this sections describe in detail the kinds of tests conducted on various genera in quarantine/ certification programs.

Some varieties or undeveloped genotypes are widely infected with viruses so that the most practical method to obtain pathogen-free germplasm is by therapy. The most widely practiced form of therapy has been hot air treatment of entire infected plants, a technique that has been useful in dealing with pathogens known to be viruses. The reader is reffered to the chapter by Mink *et al.* (Chapter 26) in this book for details on the subject. Some viruses can be eliminated from meristematic tissues of plants grown at elevated temperatures for weeks. Combined with micrografting from treated plants onto disease free stocks, or tissue culturing of excised meristems or growing points, pathogen-free germplasm of most crops can be produced. Tissue culture is also employed to produce large numbers of plants in order to maintain known healthy germplasm for long periods without exposure to re-infection, and to produce virus indicator plants for use in testing operations.

New Technologies for Certification Schemes

In Chapter 17, Frison and Diekman discuss the need for new methods to detect viruses in plant germplasm. Ideal testing methods should provide rapid results, be highly reliable, and require a minimum of labor and space. Developments in molecular detection technology have been disscussed in this book. In addition, Slack and Singh, Spiegel, and Roistacher describe a few of the molecular procedures already in place in some certification schemes. These and other new methods have resulted in more convenient, effective and specific assays which have opened the door to greater use in certification programs.

Certification of Seed Propagated Crops

Well over 100 plant viruses are transmitted via seed from infected mother plants (17). In Chapter 18, Maury *et al.* have discussed the issue in detail by addressing such matters as tolerance limits for grower seed vs seed for research/breeding programs and reliability and sensitivity of test methods. Together with Bos (5), they point out the importance of distinquishing between infection of the embryo versus infection of the seed coat from which transmission to seedlings is not likely to occur. Bos (5) elaborated on the issue of virus transmission in seeds and on procedures to obtain virus free germplasm from a quantity of seed containing virus. Although most countries and each of the 50 U.S. States have seed certification programs and, with few exceptions, virus infections of seeds are not usually addressed.

Seed lot sampling methodologies have been developed to give high levels of test result confidence and laboratory procedures are available to detect most of these pathogens in seed samples. Maury *et al.* discussed the half- dozen or so viruses that are of primary concern in government-managed certification programs. Barley stripe mosaic, pea seed borne mosaic, bean common mosaic, lettuce mosaic and potato viruses are among those for which laboratory tests are conducted in some certification programs. The importance of conducting tests for viruses in seeds during international exchange of germplasm is discussed by Bos (5) and by Maury *et al.* (Chapter 18), however, there is very little information available regarding how widely practiced these tests are in quarantine programs of various countries.

Seeds of a few crops are tested for specific viruses by laboratory procedures to meet import requirements, such as pea or lentil seed destined for Australia or South Africa (8). In other cases, seed must be virus-tested as part of the quarantine requirements of importing countries (5). Such is the case with *Prunus* and true potato seed entering the USA (25). Finally, the International Plant Genetic Resources Institute (IPGRI) has issued guidelines for testing methods to certify seed free of viruses. Among these are legumes, grapevine, small fruits and temperate small grains.

Recommendations for Certification by International Groups

As mentioned in the chapters by Barba (Chapter 22), Martelli and Walter (Chapter 20), and Krczal (Chapter 21), the European and Mediterranean Plant Protection Organization (EPPO) has taken an active role in quarantine and plant certification matters. The EPPO has published a series of reports of recommended procedures to certify various crops as pathogen free. The background and objectives for this activity have also been described by Ebbels (7). These reports are comprehensive and describe in detail the procedures that should be followed to provide maximum assurance that certified plant material is free of all known pathogens.

The IPGRI (formerly IBPGR) has also addressed certification as an important means to reduce spread of virus-like pathogens in plant germplasm shipped among nations of the world. Its' administrative relationship to the International Agricultural Research Centers (IARCs), and to the Consultative Group of International Agricultural Research (CGIAR) is discussed by Frison and Diekmann. According to them the goal of the IPGRI is to "provide relevant information on disease indexing and other procedures that will help to ensure phytosanitary safety when germplasm is moved internationally". The specific recommendations that have been published will improve the health status of germplasm from IARC collections, are listed and discussed by Bos (5).

A third series of reports dealing with production of certified virus free plants and indexing methods are those by the Committee for Cooperation in Fruit Tree and Small Fruit Virus Research of the International Society for Horticultural Science (ISHS). These are published as an appendix to the proceedings of International Symposia on Virus Diseases of Temperate Fruit Crops (2) and are updated every three years during the symposia to consider new technologies that become available and better indicators as they are discovered. However, they include stone fruits, pome fruits and small fruits.

There are also many comprehensive publications that outline in detail appropriate virus indexing procedures. Most also include information about the pathogen and its host range, symptoms and economic importance of the diseases, methods of spread, and geographic distribution of the pathogen. Among these are entire books on pome and stone fruit virus-like pathogens, small fruits, grapevine and citrus.

Role of the Private Sector

The private sector has assumed an appreciable and varied role in the production of certified disease free planting stock. During the era when tests for viruses depended primarily on grafting procedures, certification programs were conducted primarily by public institutions. While governments, public research stations, international germplasm centers, and universities continue with a major role in this regard, the private sector now is also making contributions to the dissemination of certified virus free plants. Their activities in this regard can be categorized into two groups.

Some companies have in-house certification programs which involve the full spectrum of activities such as selection of varieties, conducting the tests, maintaining nuclear stock plants, propagation, and sales of disease free plants usually for private consumption. Most of the genera are the vegetatively propagated fruit trees and herbaceous ornamentals. Among the crops are geranium, orchids, New Guinea impatiens, chrysanthemum, hibiscus, begonia, poinsettia, verbena, and lilies (12,14,20). Some test for a single important virus while others test for all known virus-like pathogens. The number of companies and their product lines continues to expand.

Costs and Benefits Associated with Certified Virus Free Planting Stock or Retail Products

It can be assumed that the cost of producing a certified virus-free items varies considerably among genera and among various programs. Factors such as whether the certification activity is an integral component of other programs, numbers of items produced per year, frequency of field inspections, number and kinds of tests conducted, the genus involved, and whether therapy is needed, whether tests are conducted in-house or under contract by private companies, and many other considerations all contribute to the range in costs of producing a virus free nuclear stock plant. These types of data are not widely available.

Costs of some tests are relatively inexpensive. For example, ELISA tests on 100

330 CERTIFICATION: AN OVERVIEW

samples for a single virus can be obtained for as little as US $6.50, or for 10 viruses about US $16 per sample (23). Other types of tests, such as grafts to virus indicators, nucleic acid extraction-PCR tests for viruses, viroids or phytoplasmas cost more per sample. Added to any test are numerous indirect costs, such as personnel salaries, utilities, equipment, and expendable supplies. Samples that require many kinds of tests, some of which are repeated for many different pathogens, and that require therapy can result in costs of several thousand US dollars per item certified. Virus indexing costs and program funding issues have been described by Mink in Chapter 23.

Regardless of the costs of conducting certification programs, not many would argue that they do not pay for themselves many times over. Nurseries that market potted floral crops are in business to increase profits and if virus free plants were not more valuable than the cost to produce them, they would not undertake certification activities (12,20). For private and public programs, whose products are perennial crops such as trees, the literature abounds with examples describing the increased financial returns (26).

Commercial growers, breeders, germplasm curators, gardeners, nursery workers, and others are rewarded in a diversity of ways when certified virus free germplasm is employed. Scientific and trade journals abound with examples on the virtues of virus free plants.

Growers benefit because crops come into production earlier, yield better, and have more economically productive years. For example, pear trees in England free of vein yellows virus produced 30% more fruit and grew more rapidly (19), and in France, yields were increased from 137 to 203 tons per hectare during an eleven year period when virus-free trees were planted (15). In Korea, Horton et al. (10) attributed potato yield increases from 11 to more than 18 tons per hectare to grower adoption of certified planting stock. Cherries free of tomato ringspot virus yielded 100% more than infected trees (21). Hops, citrus, apples, sweet potatoes, strawberries, sugarcane and grapes are other crops that performed better by planting certified plants.

Breeders and curators involved in germplasm evaluation do not have to contend with the complications, such as Bos (4,5) has outlined, that result when one or more latent virus-like pathogens are present. Virus infections confound experimental results by reducing yield, time of fruit maturation, fruit color, size, texture or nutrient composition; or increasing sensitivity to

other diseases or abiotic stresses. For example, cherry bud survival was reduced as much as 96% when virus was present. Typically curators and breeders have large holdings of genotypes and must be aware of potential contamination of their collections. Many viruses and phytoplasmas spread in nature by pollen, arthropods, nematodes , or by unknown means and could dramatically reduce the usefulness of collections or breeding lines. Isolation as a means to avoid spread of viruses is discussed by Agrios (1).

Nurserymen benefit by working with virus free material because of greater success in propagating varieties (19) and faster growth of plants for sale. Furthermore, certified virus free plants command a higher price in the market place(15).

Home gardeners benefit by planting certified pathogen free plants that obtain higher yields, produce fruit without blemishes caused by viruses as well as perennial crops such as raspberries, citrus, peach and grapevines that will produce fruit for more seasons when free of viruses.

Concluding Remarks

Production of certified crop varieties and acceptance by growers has become a widely utilized method of managing virus-like diseases. And it is reasonable to expect that use of virus free planting stocks will continue to expand in terms of included crops and countries. New molecular and serological virus detection technologies have been recently developed and more are on the horizon. Some will be incorporated into more certification programs to the extent that they are more sensitive, provide more reliable results, are more rapid, less expensive, or save labor. More sensitive techniques will likely result in the discovery of heretofore unknown infectious agents - especially those that are latent under most environmental conditions - and tests for them will be included in some certification programs.

Numbers of private diagnostics companies will continue to expand. Some will sell complete do-it-yourself kits or the component reagents for single or multiple pathogens, while others will conduct tests for plant pathogens for a clientele. We can expect that more non virus-like agents will be included in certification programs--especially bacteria and fungi that are present in very low titers.

Regional and International organizations, such as EPPO, NAPPO, IPGRI, and ISHS have made appreciable contributions to the furtherance of certification and quarantine indexing programs. The IPGRI will continue its role to improve the health status of germplasm during international exchange and in the IARCs according to Frison and Diekmann. We can anticipate that other organizations will continue to make valuable recommendations based upon consensus experience.

Acknowledgements

The author expresses gratitude to Drs. A. Hadidi, E. Civerolo and R.P. Kahn for critical review of this overview.

REFERENCES

1. Agrios, G.N. 1990. Economic considerations. Pages 1-22 in: Plant Viruses, Vol. II - Pathology. C.L. Mandahar, ed. CRC Press, Florida.

2. Anon. 1992. Detection of virus and virus like diseases of fruit trees. Acta Hortic. 309: 407-420.

3. Bernhard, R. 1986. Certification des arbres fruitiers. EPPO Bull. 16: 245-253.

4. Bos, L. 1982. Crop losses caused by viruses. Crop Protection 1: 263-282.

5. Bos, L. 1992. New plant virus problems in developing countries; a corollary of agricultural modernization. Adv. in Virus Res. 41: 349-407.

6. Desvignes, J.C. 1990. Les Virus des Arbres Fruitiers. CTIFL, Paris. 126 Pages.

7. Ebbels, D.L. 1989. Administrative control of pome fruit planting material in countries other than Canada and the United States. Pages 314-323 in: Virus and Virus-like Diseases of Pome Fruits and Simulating Noninfectious Disorders. Wash. State Univ. Coop. Ext. Publ. SP 0003.

8. Hamilton, R.I. 1983. Certification schemes against seed-borne viruses in leguminous hosts, present status and further areas for research and development. Seed Sci & Technol. 11: 1051-1062.

9. Hollings, M. 1965. Disease control through virus-free stock. Annu. Rev. Phytopathol. 3: 367-396.

10. Horton, D.E. 1987. Korea's seed potato program: organization, impact, and issues. Int'l Potato Cntr. for Min. of Agri. and Fisheries, Korea Republic, Lima, Peru. 68 Pages.

11. Kahn, R.P., Waterworth, H.E., Gillaspie, and Foster, J. A. 1979. Detection of viruses or virus like agents in vegetatively propagated plant importations under quarantine in the United States, 1968-1978. Plant Dis. Rptr. 63: 775-779.

12. Klopmeyer, M. 1991. The importance of disease indexing and pathogen-free production. Pages 277-281 in: Ball Redbook, 15th Ed. Ball Publ. Co., Batavia, IL.

13. Lanterman, W.S., Johnson, R., and Thompson, D. 1996. Disease control through crop certification - Woody plants. Can. J. Plant Pathol. 17: 274-276.

14. Lawson, R.H. 1981. Controlling virus diseases in major international flower and bulb crops. Plant Dis. 65: 780-786.

15. Lemoine, J., and Michelesi, J.C. 1990. Latent viruses and agricultural performance. Arboric Fruitiere No. 434: 33-37.

16. Meijneke, C.A.R. 1982. Ideal schemes and associated problems in the production, maintenance, multiplication, distribution and certification of fruit crops. Acta Hortic. 130: 29-31.

17. Mink, G.I. 1993. Pollen and seed transmitted viruses and viroids. Annu. Rev. Phytopathol. 31: 375-402.

18. Nemeth, M. 1986. Virus, mycoplasma and rickettsia diseases of fruit trees. Martinus Nijhoff Publ., Boston. 841 Pages.

19. Posnette, A.F. 1976. Virus/mycoplasm diseases and certifcation schemes for deciduous fruit trees. EPPO Bull. 6: 57-61.

20. Raju, B.C., and Olson, C.J. 1985. Indexing systems for producing clean stock for disease control in commercial floriculture. Plant Dis. 69: 189-192.

21. Ramsdell, D.C., Bird, G.W., Adler, V.A., and Gillett, J.M. 1992. Effects of tomato ringspot virus and Prunus necrotic ringspot virus alone and in combination on the growth and yield of 'Montmorency' sour cherry. Acta Hortic. 309: 111-113.

22. Shepard, J.F., and Claflin, L.E. 1975. Critical analysis of the principles of seed potato certification. Annu. Rev. Phytopathol. 13: 271-293.

23. Sutula, C.L. 1996. Quality control and cost effectiveness of indexing procedures. Adv. in Bot. Res. 23: 279-292

24. van der Graaff, N.A., and Chiarappa, L. 1984. The need for an international approach for the production of virus-free propagation material. EPPO Bull. 14: 373-376.

25. Waterworth, H.E. 1993. Processing foreign plant germ plasm at the national plant germplasm quarantine center. Plant Dis. 77: 854-860.26. Waterworth, H.E., and Hadidi, A. 1997. Economic losses due to plant viruses. (Chap. 1; This book).

CHAPTER 26

Heat Treatment of Perennial Plants to Eliminate Phytoplasmas, Viruses, and Viroids While Maintaining Plant Survival

G. I. Mink, R. Wample, and W. E. Howell

For many years, heat therapy has been the principal method used worldwide to produce perennial plant propagants free from phytoplasmas (previously known as mycoplasma-like organisms, MLO's), viruses, and viroids. While other methods such as tissue culture and chemotherapy are now being used in combination with heat therapy to eliminate specific pathogens, exposure to elevated temperatures remains the basis for most pathogen elimination programs. Although these procedures provide some propagants that are free of phytoplasmas and most viruses, success is decidedly limited when heat sensitive cultivars are infected with heat-tolerant viruses or viroids.

In this report, we briefly review the history of heat therapy and examine what little seems to be known about the effects of elevated temperatures on virus replication and transport. We also examine effects of elevated temperatures on physiological plant processes and plant survival. Finally, we summarize the variety of methods currently in use around the world to produce perennial plants for certification.

HISTORICAL PERSPECTIVE

Heat has been used to eliminate diseases from vegetatively propagated plants for over a century. Nyland and Goheen (48) credit Scots gardeners for pioneering the practice of hot water treatment by immersing bulbs in hot water before planting. They also cite several reports covering a period from 1889 to 1925 in which hot water treatment was used to eliminate sereh disease from sugarcane setts (vegetative seed piece). General interest in the use of heat (either of short exposures to hot water or long exposure to hot air)

to cure virus diseases began in earnest with the work of Kunkel (37-43). He used these techniques to cure several diseases caused by "viruses" in the yellows group. The list of diseases cured included alfalfa witches broom, aster yellows, cranberry false blossom, little peach, red suture, peach rosette, and peach yellows, all of which are now known to be caused by phytoplasmas (MLOs) (47).

It was not until 1949 that the first successful heat treatment of a conventional virus was reported (31,32). By 1957, however, Kassanis (33) listed more than 15 conventional viruses that had been inactivated in plants by heat. Only 12 years later, Nyland and Goheen (48) listed 76 viruses or virus diseases, 2 viroids and 33 phytoplasma-induced diseases that had been inactivated by heat. Although the association of phytoplasmas with diseases in the "yellows group" was just being recognized, these authors pointed out the likelihood that conventional viruses were not inactivated by short exposures to hot water treatments.

Early heat treatment procedures involved maintaining whole plants as long as possible at constant temperatures that ranged from 35-40°C. Although many plants died during treatment, a few survived long enough to provide virus-free propagants. Over the years, numerous procedural modifications have been successfully attempted (48). Some of these modifications include the use of artificial supplemental light, maintenance of reduced root temperatures, preconditioning of plants by short exposures to moderate temperatures, and the use of fluctuating temperatures. In recent years some workers have combined heat therapy with chemotherapy, or apical meristem culture to increase the probability of obtaining virus-free propagants.

Despite the widespread use of heat therapy, relatively little is known about how the process works. This is due in part to the fact that for most perennial plants, once a few virus-free propagants are obtained, enough tissue can be increased to satisfy the needs of an entire industry. Most of the heat-treatment procedures attempted have provided the necessary few virus-free propagants. Consequently, there has been little pressure to understand the fundamental mechanisms involved in pathogen elimination.

EFFECTS OF HEAT ON VIRUSES

Direct Effects

Thermal Inactivation Point

Thermal inactivation point (TIP) is a term that appears frequently in virus characterization literature (48). It is defined as the 10-minute exposure temperature required to inactivate virus in an aliquot of expressed sap or, most commonly, in a neutral buffered tissue extract. While the host-liquid environment and virus isolate used can affect TIP to some extent, most values obtained over time and among laboratories for individual viruses have been remarkably close. Most, but not all, viruses within a taxonomic group have similar TIPs. Values for these groups range from 45°C to 95°C. Because of the ease with which this value can be obtained, it is often one of the first properties determined for an unknown virus that can be transmitted by rub inoculation. For viruses with TIPs below 50°C or above 65°C, these values can assist in identification of the taxonomic group to which an unknown might belong. Although TIP may be used to recognize similar or dissimilar viruses or strains, it has had little relevance in vivo to heat inactivation studies. The TIP determined in vitro is substantially higher than the thermal-death point of the host plant (48).

Inactivation Kinetics

Inactivation of plant viruses generally follows a course of first-order kinetics (20), although inactivation of tobacco necrosis virus appeared to occur at two rates (1). While the thermal inactivation kinetics have been examined for some plant viruses in vitro, no comparable information has been obtained for viruses in vivo.

Attempts have been made to find significance in in vivo half-life determinations made at various temperatures including 38°C (48). These studies have elucidated few specifics about in vivo heat therapy.

Effects of Heat on Viral Replication

Sustained temperatures of 37°C or above completely inhibit multiplication of many viruses (34). In studies with tobacco mosaic virus (TMV), no evidence of viral RNA synthesis was detected when inoculated plants were maintained at 40°C for several hours (11). However, when infected plants were moved from 25°C to 40°C, synthesis of viral double-stranded dsRNA continued at high temperature for a period, slowed, and then stopped (12). During the slowing period, if plants were returned to 25°C, synthesis of virus RNA resumed. However, once dsRNA synthesis stopped at 40°C, the capacity to resume viral RNA synthesis at 25°C was temporarily lost (61). New synthesis occurred only after a 12-hour lag period. In vivo synthesis of three TMV-specific proteins (160, 110 and 17.5 kDa) was inhibited at 40°C in a pattern similar to high temperature inhibition of dsRNA synthesis (13). However, translation of viral mRNA was not temperature-sensitive (13). Although studied less intensively, RNA synthesis for cowpea chlorotic mottle virus (CCMV) was shown to be inhibited at 40°C. In contrast to TMV where dsRNA synthesis continued only briefly at 40°C, synthesis of CCMV dsRNA declined gradually (15). For both TMV and CCMV, synthesis of viral single-stranded ssRNA ceased immediately at 40°C but resumed upon return to 25°C (12,13).

Sudden increases in temperature above 35°C resulted in inhibition of synthesis of most pre-existing plant proteins followed by synthesis of a set of new proteins referred to as heat shock proteins (hsp). In vivo synthesis of TMV-specific proteins was found to proceed independently of heat shock effects (14).

When plants are moved from 25°C to 40°C, RNA synthesis is inhibited for both host and virus (11,15). When shifted from 40°C to 25°C, host RNA synthesis resumes immediately. However, there was a 4-8 hour delay before CCMV RNA synthesis resumed (15) and a 16-20 hour delay before TMV RNA synthesis resumed. This work provides some guidelines for selecting alternating temperatures and time intervals to produce virus-free tissues in temperature-sensitive

Figure 1. Heat sensitive steps in virus replication[a]

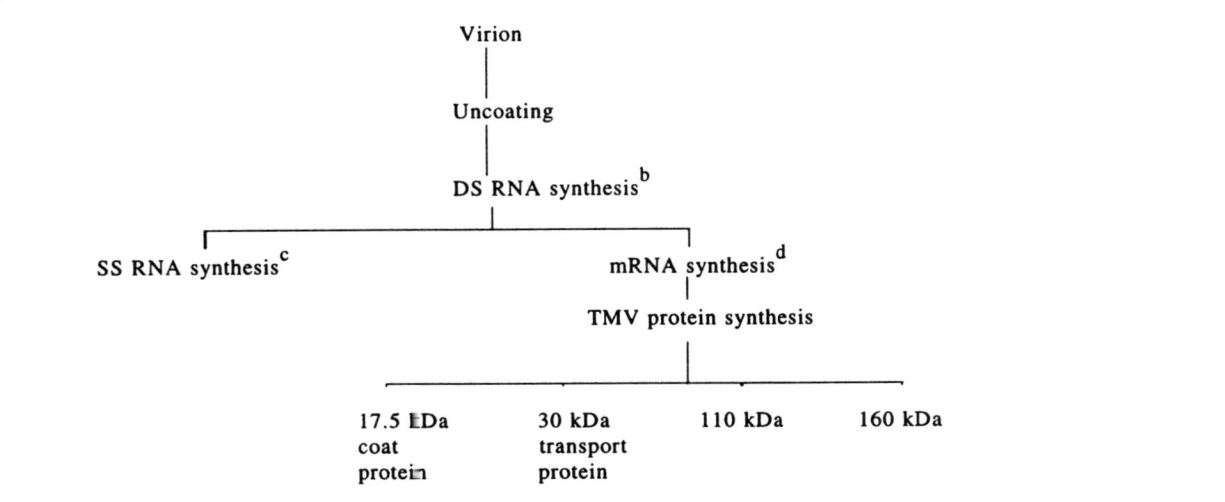

[a] Based on studies by Dawson and colleagues summarized in section on effects of heat on viral replication.
[b] Enzymatic activity decreases to zero during exposure to 40°C. After prolonged exposure, activity does not resume after return to 25°C.
[c] Enzymatic activity ceases abruptly during exposure to 40°C. Activity slowly resumes after return to 25°C.
[d] Enzymatic activity continues after host protein synthesis is repressed by heat shock at 40°C.

plants. Alternating temperatures between 37-40°C and 22-25°C have been used to produce PVS-free potato (46), alfalfa mosaic virus (AMV)- and cucumber mosaic virus (CMV)-free callus cultures of *Nicotiana rustica*, (58,59) and CCMV-free cowpea propagants (13). While alternating temperature allowed sensitive cultivars to be maintained in heat therapy for longer periods of time, these procedures did not significantly increase the number of virus-free propagants obtained from most cultivars (46,58).

Effect of Heat on Virus Movement

Short distance (cell-to-cell) and long distance (tissue-to-tissue) spread of many viruses appears to be mediated by a class of virus-coded proteins collectively referred to as movement or transport proteins (25). The best known of these is the 30 kDa transport protein (TP) used by TMV (16).

Evidence is accumulating that more than mechanism may be involved in short distance movement. According to Hull (25), two likely mechanisms include 1) interactions between a movement protein and plasmodesmata to facilitate passage of a non-viral form of the virus, and 2) passage of intact spherical virus along tubular structures through plasmodesmata. Natural mutants

of viruses and genetically engineered deletion mutants with modified coat proteins exhibit reduced or highly restricted movement within the plant. Although requirements for long-distance spread are less well known, studies reviewed by Hull (25), involving coat protein mutants, suggest that the virus coat protein may facilitate systemic movement.

Information discussed in the preceding section and illustrated in Figure 1 indicates that cessation of ds RNA synthesis by exposure to high temperature would eventually eliminate synthesis of both coat protein and movement protein. This would likely restrict cell-to-cell movement of pre-existing virus.

EFFECTS OF HEAT ON PLANT PROCESSES

General

There is a wide variation in host response to heat stress both within and among plant genera and even within a species. Differences in heat tolerance are recognized by the absolute temperature that plants can tolerate or by the time interval they can survive at some elevated temperature. In addition to genetics, the thermal history of a plant can significantly influence its heat tolerance (60). The mechanism of heat acclimation is not well understood but protein synthesis has been implicated through the use of protein synthesis inhibitors (49) Other changes that can contribute to increased heat tolerance are increases in the level of saturation of membrane fatty acids and changes in the composition of cytoplasmic and organelle fluids that increase osmotic potential and maintain normal metabolism of the cell (60). One consequence of variation in host plant thermotolerance is the wide variety of heat treatment procedures that have evolved to eliminate viruses from plants without killing them (see section on an international survey of heat therapy methods).

Physiological studies of heat stress in plants are usually conducted with plants exposed to relatively short (30 minutes to several hours) time periods (5,44,45,55-57). These studies usually involve either a single, brief exposure (from a few minutes to one hour) or at most a few days exposure to alternating temperatures. There are very few reports where plants, infected or otherwise, have been studied under the long-term heat stress conditions normally used for pathogen elimination. Consequently, little is known about the physiological effects on plants of exposure to temperatures of 35°C and higher for periods of 30 or more days. Table 1 provides a summary of plant responses to elevated temperatures. It is not intended to be an exhaustive list, but rather to introduce interested readers to specific areas. It should be noted that very few of these reports examine plant responses to temperatures above 35°C for periods greater than 48 hours.

Inhibition of Photosynthesis

Photosynthesis appears to be one of the most sensitive, and physiologically important processes affected by heat stress. Short exposure to heat reduces photosynthesis due to changes in chloroplasts rather than as a result of cytoplasmic changes (5). An increase in chloroplast fluorescence caused by a breakdown in the electron transport pathway and studies on other biochemical and biophysical processes (5,60) strongly indicate that the site of initial heat stress response is the multimolecular, chlorophyll containing, protein complex responsible for non cyclic electron transport (photosystem II). Photosystem II, located on the thylakoid membrane system of the chloroplast, is that part of the photosynthetic system which is responsible for

production of oxygen during the splitting of water molecules as a consequence of the absorbance of light by the P680 pigment system (19). This process provides the "reducing power" necessary for the ultimate fixation of carbon dioxide. It is hypothesized that heat stress prevents the necessary charge separation across the thylakoid membrane needed to achieve photolysis of water. Disruption of this system is considered to be irreversible and can result in additional loss of chlorophyll due to photobleaching. Photobleaching results from photoexcitation of pigment molecules in the absence of the electron acceptors and transport system needed to dissipate this energy. Photobleaching is frequently observed in heat treated host plants as a general chlorosis of the younger leaves, which subsequently dehydrate and abscise. The rate at which the processes occur is host plant specific.

In cases of moderate heat stress, there may be an increase in photorespiration, the process where ribulose bisphosphate carboxylase/ oxygenase binds with oxygen instead of carbon dioxide. Photorespiration is associated with low internal carbon dioxide levels, perhaps caused by stomatal closure due to water stress. Photorespiration can be reduced by increasing the level of carbon dioxide. This procedure has been successfully used to reduce detrimental effects during heat therapy of grapevines (36) and high bush blueberry (9). Interestingly, water stress, which is normally accompanied by stomatal closure, has been demonstrated to reduce the loss of photosystem II activity caused by heat stress, and has been proposed as a potential protective adaptation (23). However, the author did not indicate what protective mechanism might be involved. Temperature and photoperiod cycling appear to be effective procedures to reduce the detrimental effects of heat stress on photosynthesis. Periods of reduced temperature and darkness promote increased transport of fixed carbon and provide conditions under which chloroplast can be repaired. In some plants, continuous light may result in starch accumulation in chloroplasts, which, in turn, can reduce photosynthesis. A lack of transport also deprives non-photosynthetic portions of the plant of an energy source. Havaux et al. (24) have also reported that heat induced breakdown of PS II is inhibited by light and increased in the dark. This has important consequences for plant survival during thermotherapy and clearly suggests that temperatures be reduced during dark periods.

Table 1. Plant responses to elevated temperatures

Physiological parameter	Plant response	Reference
Photosynthesis	Loss of electron transport Photooxidation of pigments Increased photorespiration	23,36,54,60
Respiration	Increases with temperature up to a critical point and then declines (this is generally at higher temperatures than where photosynthesis declines).	21
Protein synthesis	Proteolysis and Reduced synthesis of non-stress related proteins Increased synthesis of stress (heat shock) related proteins.	4,8,30,35, 54,56,57
Ribonuclease	Increase.	29
Carbohydrate transport	Increased at moderate temperatures reduced at higher (perhaps at both the chloroplastic and phloem tissue levels).	1
Cell division/growth	Reduced only at very high temperatures (>40C)	2
Plant hormones	Auxin, cytokinins and gibberelins are reduced and abscisic acid and ethylene are increased in short term experiments.	26,27,50
Proline	Short term studies show and increase.	44
Polyphenol-oxidase	Increase.	30

Increased Dark Respiration and Reduced Carbohydrate Translocation

Dark respiration increases with temperature, but is not believed to be a major cause of plant failure during heat stress (21). This is because dark respiration is much more heat resistant than photosynthesis (5). However, there is no quantitative research which has examined a combination of reduced photosynthesis and increased respiration on plant performance over periods of extended heat therapy. When combined with lower carbohydrate transport, this combination of effects would increase the likelihood of failure of non-photosynthetic tissues and organs such as roots and developing shoot apices, thus leading to loss of plant viability (2). Reduced root growth would result in a loss of effective root area for water and nutrient uptake and could eventually lead to water and nutrient stresses during thermotherapy. Therefore the practice of selecting plants with high levels of stored carbohydrate reserves or treating plants in ways to increase these reserves would improve plant survival during heat therapy. Additionally, if photosynthesis continued, reduced phloem transport would result in an accumulation of starch

or other storage products within the chloroplast with potential detrimental effects as mentioned above. As expected, thermal sensitivity of phloem transport varies with plant species and may be more or less of a problem depending upon this (18). However, thermo- and photoperiod cycling are effective procedures for overcoming this problem. It is interesting that in the case of grapes (*Vitis vinifera* cv. Cabernet Sauvignon), a relatively heat tolerant plant, heat stress results in an increase in the lipid fraction in the chloroplast with a reduction in the level of starch (7). This would permit a higher level of carbon storage within the chloroplast and perhaps reduce the physical problem of starch accumulation on chloroplast function. It would be interesting to see if heat tolerant plants show other similar responses.

Protein Synthesis

Heat stress reduces the synthesis of some proteins and increases the synthesis of others. (29,30,35,49,51,57). In general, near lethal heat stress results in accelerated proteolysis of non-stress related proteins and a decrease in other indicators of protein synthesis such as rough

Table 2. Facilities performing virus therapy on perennial plants experimentally or as a service

Lab no.	Facility	Species treated	No. items per yr.
1	D. J. Gumph, Dept. of Plant Pathology - University of California Riverside CA 92521	*Citrus*	25
2	W. E. Howell - IR2/NRSP5, Irrigated Agr. Res. and Ext. Center - WA State University, Prosser WA 99350	*Prunus, Malus, Pyrus, Cydonia*	100
3	Curator, USDA-ARS Plant Genetic Resources Unit - Cornell University Geneva NY 14456	*Malus, Vitis*	0-245
4	Sara Spiegel, Agr. Research Organization - Institute of Plant Protection The Volcani Center, P. O. Box 6, Bet Dagan 50250 Israel	*Prunus*	<1
5	M. Manners, Citrus Institute - Florida Southern College 111 Lake Hollingsworth Drive, Lakeland FL 33801	*Rosa*	50
6	National Germplasm Resources Laboratory-Quarantine Unit USDA-ARS-PSI, Building 580 BARC-East, Beltsville MD 20705	*Malus, Pyrus*	40
7	Carlton Plants, 14301 SE Wallace Road, Dayton OR 97114	*Malus, Pyrus, Prunus*	11
8	D. Thompson, Agriculture Canada - Saanichton Plant Quarantine Station 8801 East Saanich Road, Sidney BC Canada	*Malus, Pyrus, Prunus*	94
9	J. Postman, USDA-ARS, National Clonal Germplasm Repository 33447 Peoria Road, Corvallis OR 97333	*Pyrus, Fragaria, Rubus, Ribes, Corylus, Mentha*	150-300
10	Curator, USDA-ARS, National Clonal Germplasm Repository - Univeristy of California, Davis CA 95616	*Prunus*	30
11	J. Juarez, Tissue Culture Laboratory Instituto Valenciano De Investigaciones Agrarias, Apartado Ofiial 46113 Moncada, Valencia Spain	*Prunus*	
12	Christa Lankes, Institut Fur Obstbau Und Gemusebau der Universitat Bonn Auf dem Hugel 6, 53121 Bonn Germany	*Malus, Rubus*	15
13	B. Di Terlizzi, Instituto Agronomico Mediteranio Via Ceglie 23 Cap 70010, Valenzano (Ba) Italy	*Prunus, Vitis*	55
14	D. Gonalves, Dept. of Plant Pathology - Cornell University, Geneva NY 14456	*Vitis*	<1
15	Head of Virology, Opzoekingsstation van Gorsem Brede Akker 3, B-3800 Saint-Truiden Belgium	*Malus, Pyrus, Prunus*	23
16	G. Jongedijk, NAKB - Dept. of Tree Pathology Johan de Wittlaan 12, 2517 Be Den Haag The Netherlands	*Malus, Pyrus, Prunus*	75
17	G. J. Visser, Directorate Plant and Quality Control - Dept. of Agriculture Private Bag X5015, Stellenbosch 7599 South Africa	*Malus, Pyrus, Prunus, Vitis*	14
18	Elzi Volk, Small Fruits Certification and Virus Elimination Program Dept. of Botany & Plant Pathology - Oregon State University, Corvallis OR 97331	*Fragaria, Rubus*	13

endoplasmic reticulum and the levels of polysomes (30,54). On the other hand, heat stress has been shown to induce the formation of a family of proteins of varying molecular weight and referred to as heat shock proteins (HSP) (57). Heat shock protein synthesis has been found in every organism where it has been looked for and has also been shown to occur in tissue cultured cells (4,62). Not every HSP that has been identified s present in every organism and there are some HSPs that seem to be unique to plants (57). Some classes of HSPs are also present in the absence of heat stress. Both mitochondria and chloroplasts contain such proteins and in each case are associated with the non membrane fraction of the organelle. The presence of these proteins is presumed to confer increased heat tolerance to essential metabolic proteins in these organelles. This is supported by the fact that metabolic pathways in these organelles are functional at temperatures above those where membrane-associated functions, such as electron transport, begin to break down (6). Heat stress induced HSPs are also found in these organelles, as well as other cellular components (57). Although there is a strong correlation between the appearance of HSPs and increased heat tolerance, knowledge of the mode of action is still lacking (57). Furthermore, there is no evidence that HSP synthesis would continue for the duration of a thermotherapy treatment. In fact, Cooper and Ho (10) indicated that HSP formation continued for only 8 hours before declining sharply in continuously heat-stressed maize. Perhaps this is sufficient time for changes in membrane structure and composition to permit acclimation to the higher temperature. Unfortunately this report, like most others, involved heat treatments that lasted for only a few days and thus may not answer questions regarding HSP synthesis during long term thermotherapy. There is a need for additional research in this area.

Cell Division, Cell Growth and Plant Hormones

Although cell division and growth can be reduced directly by heat, available information suggests that these processes are only affected at temperatures above 40°C (21,44). Therefore, growth reductions that occur during thermotherapy near 38°C are most likely due to indirect rather than direct effects of heat. Recognizing how factors such as photosynthesis may influence cell division and growth will result in the use of procedures that will sustain cellular division and growth. Major factors in the regulation of plant growth and development are the plant hormones. Despite the abundance of literature on the presence of these endogenous substances and the effect of exogenously applied growth regulators that approximate them, we know almost nothing about hormone synthesis and activity under the extended conditions of heat therapy. In an earlier report, Itai and Benzioni (27) showed that heat stress caused an initial increase in abscisic acid (ABA) and a decrease in cytokinin-like activity and that this pattern was reversed after 24 to 30 hours. In an attempt to overcome the effect of heat stress by application of either ABA or a cytokinin (kinetin) Itai et al. (28) found that pretreatment with the compounds resulted in greater plant damage than occurred in the controls. Post heat stress treatment with kinetin resulted in a slight improvement. Chen et al. (8) showed that application of gibberellic acid (GA_3) during the heat shock of mung beans increased heat tolerance and stimulated growth of seedlings following heat stress. Other authors have reported that high temperatures reduce cytokinins, auxins and gibberellins and increase ABA and ethylene (26,50,62). However, all of these reports represent short-term heat stress studies. This clearly represents a serious gap in our knowledge of plant physiology and represents a research opportunity.

AN INTERNATIONAL SURVEY OF HEAT THERAPY METHODS

General

Although numerous articles have been published on the use of heat to eliminate viruses from perennial plants, we could find neither a current list of locations where heat therapy was being used, nor a list of current methods used when we began this review. A questionnaire to develop information on current activities in heat therapy of perennial plants was sent to 25 individuals and agencies in five U.S. states and 11 countries. Twenty-three responses were received, 18 from facilities where heat therapy is currently being performed experimentally or as a service. These facilities are listed in Table 2, which is not a comprehensive listing, but rather a sample of the extent to which heat therapy is used for perennial plants.

Table 3. Perennial plant genera undergoing heat therapy worldwide

Genera	No. of facilities	Location no. [a]
Citrus	1	1
Corylus	1	9
Fragaria	2	9,18
Malus	9	2,3,6,7,8,12,15,16,17
Mentha	1	9
Prunus	10	2,4,7,8,10,11,13,15,16,17
Pyrus	8	2,6,7,8,9,15,16,17
Ribes	1	9
Rosa	1	5
Rubus	3	9,12,18
Vitis	4	3,13,14,17

[a] see Table 2 for address of each location

Heat Therapy Methods in Use

Of the 18 currently active heat therapy facilities most are producing virus-free plants of deciduous tree fruit genera (Table 2). Ten facilities are producing virus-free *Prunus* species selections, 9 producing virus-free *Malus*, and 8 producing virus-free *Pyrus*. Virus-free plants of *Vitis*, *Rubus*, *Rosa*, *Ribes*, *Mentha*, *Fragaria*, *Corylus*, and *Citrus* are also being generated at one or more locations (Table 3).

Although all responding facilities use methods based on heat treatment and propagation, specific methods vary considerably among facilities (Table 4). Despite the variations in methodology, all facilities appear to produce virus-free propagants of each species treated (Table 4). Variation among methods are summarized as follows:

Treatment of Plants Before Exposure to Heat

Procedures varied among facilities in how plants are handled prior to exposure to heat. Plants placed in heat vary in age from dormant buds recently inserted onto growing seedlings, to established 2-year-old plants (Table 5). These plants are either placed directly into chambers maintained at temperatures over 36°C or are acclimatized by holding them at temperatures of 28 to 34°C for 2 weeks or more.

Temperature

Three distinct temperature regimes are being used to obtain virus-free tissue. These are 1) constant high temperatures, 2) alternating high and moderate temperatures, and 3) constant warm temperatures (Table 4). Depending upon the laboratory, constant high temperature is maintained somewhere between 36 and 39°C. Specific temperatures may vary somewhat. In the IR2/NRSP5 Heat Therapy Program temperature fluctuates 2 to 3 degrees at the beginning and end of each photoperiod due to equipment design. Where alternating temperatures are used they

Table 4. Temperature regimes and the type of tissues excised in heat therapy programs for perennial plants

Temperature	Excised tissue[a]	No. of labs	Species	Survival (% propagants)	Virus-free propagants	Facility
Constant						
37-39	ST	9	Corylus, Malus	50-80	Yes	2,3,6,9
			Prunus, Pyrus			2,12,13
			Vitus			15,16
38	BUD	2	Rosa, Prunus	25-80	Yes	5,15
38	M	2	Fragaruam, Rubus	[b]	Yes	12,18
36-36.5	ST	2	Malus, Prunus, Pyrus	[b]	Yes	8,15
Alternating[c]						
38-40/30	ST	2	Citrus, Prunus	[b]	Yes	1,17
38/36	ST	1	Malus, Pyrus	60	Yes	7
38/30-36	M	2	Malus, Prunus	[b]	Yes	7,17
38/28	ST	1	Prunus	[b]	Yes	4
36.5/33	ST	1	Prunus	[b]	Yes	8
35/30	M	2	Prunus, Vitis	[b]	Yes	11,17
Alternating[d]						
38/30	ST	1	Pyrus, Mentha Fragaria Ribes, Rubus	[b]	Yes	9
Warm						
30	M	1	Vitis	[b]	Yes	14

[a] ST = shoot tip; M = meristem; BUD = lateral bud & surrounding tissue.
[b] not reported.
[c] 16 h light, high temp; 8 h dark, moderate temp.
[d] alternating 4 h periods of high and moderate temps independent of 16 h/8 h photoperiod.

range from lows of 28 to 36 for periods of 4 to 8 hours to highs of 35 to 40°C for periods of 4 to 18 hours. Warm temperatures of approximately 30°C were used at one facility to treat *Vitis* plants under tissue culture conditions.

Culture of Treated Plants During Treatment

Considerable variation also occurs in the maintenance of the heat treated individuals within heat chambers (Table 4). Several facilities provide 5000 to 30000 lux of artificial light with photoperiods of 16 to 24 hours. Some use available natural light in clear chambers placed in

Table 5. Cultural conditions for plants undergoing heat therapy

Genus	Types of heat-treated plants[a]	Photo period[b]	Lux[b]	Duration of heat treatment (days)
Citrus	bud	16		56-84
Corylus	mo			>20
Fragaria	mo,root,run	16	10000	>20
Malus	bud,mo,1-2yr	14-18	5-30000	20 to >90
Mentha	mo			>20
Prunus	bud,mo,1-2yr,tc	16-18	5-10000	9 to 270
Pyrus	bud,mo,1-2yr	14-18	5-30000	20 to 90
Ribes	mo			>20
Rosa	1yr	24		28
Rubus	mo,root		10000	>20
Vitis	1-2yr,tc	16	10000	30->56

[a] = Method and timing of vegetative propagation: bud = recently bud propagated plant, mo = 2 to 3 months old, 1-2yr = 1 or 2 year old plant, run = runner, tc = plant growing in tissue culture medium.
[b] = artificially supplied light.

greenhouses. Some survey respondents mentioned the use of scheduled fertilizer applications. Others made no mention of such applications. Duration of treatment times ranges from 9 to more than 200 days.

Tissue Used

As with temperatures, the type of tissues excised to obtain virus-free propagation material from the heat treated, infected mother plants varies among facilities. Three general types of tissues are used; 1) shoot tips which may vary in length from 5 to 15 mm depending upon the species and facility, 2) meristems from 0.25 to 0.5 mm in length which may include 1 or 2 leaf primordia taken from shoot tips or auxiliary buds, and 3) lateral stem tissue containing axillary buds (Table 4). The buds and shoot-tips of most species and the meristems of citrus are grafted directly on young virus-free seedlings to initiate growth of a virus-free tree of the clone involved. Meristems of other genera are cultured and rooted on artificial media. Alternatively, some meristems are allowed to elongate on artificial media and then grafted to virus-free seedlings.

Virus

Detection

Response to our questionnaire provided little information about the specific viruses involved at each facility. However, most genera treated appear to contain one or more viral contaminants that are common to that genus. Not surprisingly, for most genera the most frequent viruses or strains of viruses encountered seem to be those that cause few, if any, symptoms on most clones of that genus. For example, for *Malus* and *Pyrus* the preponderance of virus detected in the IR2/NRSP5 Program are the apple "latent" viruses [apple stem grooving virus (ASGV), apple chlorotic leaf spot virus (ACLSV), and apple stem pitting virus (ASPV)]. For *Prunus* certain strains of the Ilar-viruses [(*Prunus* necrotic ring spot virus (PNRSV) and prune dwarf virus (PDV)] are most commonly found. For *Rubus*, raspberry bushy dwarf (RBDV), the raspberry mosaic virus complex and tobacco streak virus (TSV) are the most commonly encountered agents (J Postman, personal communication).

Elimination

Although heat therapy was generally successful at all facilities, the degree of success appears to be dependent on the virus and plant species or even the clone involved. In *Rubus* certain components of the mosaic complex are very heat labile. Other viruses of *Rubus* such as RBDV and TSV are more heat stable (J. Postman, personal communication). Observations based on the heat treatment of over 300 apple clones in the IR2/NRSP5 Program indicate that there is great variation in heat sensitivity among the deciduous fruit tree viruses. ASGV for example is very recalcitrant to heat therapy techniques and requires extensive heat treatment times to increase the probability of obtaining virus-free propagants. Conversely, a high proportion of propagants free of ACLSV are obtained with little, if any, heat treatment. While it has been fairly easy to obtain *Prunus* trees free of PNRSV and PDV through heat therapy, the percentage of virus-free trees obtained by tip propagation alone may be affected only slightly, if at all, by the length of heat treatment.

Our observations and much of the literature on heat therapy suggest that distribution of viruses within plants held for prolonged periods at elevated temperatures is virus dependent. This may reflect differences in temperature sensitivity to critical steps in replicative or transport functions among viruses.

Alternative Methods

Certain alternative methods used by some of the laboratories deserve special mention. In efforts to improve plant growth while limiting viral multiplication, the National Clonal Germplasm Repository, Corvallis, Oregon, alternates temperatures between 38 and 30°C every 4 hours. A 16 hour photoperiod is employed that operates independent of the temperature schedule. The timing of these temperature fluctuations may optimize plant survival while minimizing chances of virus replication (15). In efforts to improve the odds of selecting virus-free tissue for propagation

several heat therapy facilities use meristems and tissue culture rather than shoot tips and grafting in their therapy efforts. With strawberries these meristems are sometimes collected from parent plant runners. At the opposite end of the spectrum two facilities propagate from lateral stem tissue which contains an auxiliary bud. Success in both cases seems to invoke opposing hypotheses. In the first case, small pieces of tissue are selected to 'escape' the virus, which is presumed not to invade the growing point under elevated temperatures. In the second case, tissue is selected with the assumption that the contaminating virus had been eradicated from the excised portion of the plant. Both procedures produce the desired results, a percentage of propagants which are free of virus. When virus-free tissue can be obtained from treated lateral stems, however, it is difficult to discern whether the virus was eradicated or is just erratically distributed. Some facilities attempt to improve plant survival by plant adaptation to heat and by stimulation of heat shock proteins. When treating *Prunus* selections in South Africa, Directorate of Plant and Quality Control (Plant Quarantine Station, Stellenbosch) gradually raises the temperature at specific intervals. After incubating the trees at 28/28°C (day/night) for 2 weeks, the day temperature is increased every third day until 35/28°C is reached within a week. These temperatures are then maintained for two weeks. Day temperature is subsequently increased to 38°C and maintained there for another week at which time the night temperature is increased to 30°C. ELISA tests for ilar-viruses are conducted at each of the several steps and prior to tip excision.

A few facilities are using chemotherapy experimentally, sometimes in combination with thermotherapy (22). The Canadian plant quarantine facility applies weekly sprays of 500 ppm Virazole to some *Prunus* trees through mid summer until tips are taken and grafted on a young rootstock or placed in tissue culture media (22).

Viroids

The discovery of viroids and their apparent preference for warm temperature for replication and accumulation poses questions about the proper therapy for these disease agents. Attempts to obtain viroid-free trees from citrus exocortis viroid-infected individuals by excising and growing meristems from heat treated plants has been very difficult (53), if not impossible (D.

Gumph, personal communication). Similar results have been obtained with potato spindle tuber viroid-infected potato plants (17). Future efforts to obtain viroid-free individuals from these plants will likely be attempted from growth induced at cooler temperatures as was done with grapevine yellow speckle diseased grapevines (3). On the other hand, studies at the IR2/NRSP5 Program indicate that apple scar skin viroid (ASSVd) accumulates optimally at temperatures as low as 18°C with little or no accumulation detected at 38°C (52). A few apple trees apparently free of ASSVd were obtained from infected plants using standard heat treatment at 38°C followed by shoot tip propagation.

SUMMARY

It is clear that many variations of heat therapy or heat applied in combination with other procedures are being used succesfully by scientists worldwide to eliminate phytoplasmas, viruses and viroids from a wide variety of perennial plants. Most of the procedures used appear to have been developed empirically in response to the environmental sensitivities of individual hosts and to some extent the heat tolerance of specific viruses. Despite the fact that all heat therapy programs have achieved some measure of success, virtually nothing has been published from any of these programs that sheds light on the fundamental processes that ultimately provide the pathogen-free tissues. Studies on herbaceous plants have given a brief but tantalizing glance at some possible effects of short exposures to elevated temperatures on virus replication and movement. Herbaceous plants have also provided most of what is known about effects of short duration heat stress on plant-directed physiological processes. It remains to be seen, however, whether or not information generated from studies of herbaceous plants will provide a basis for understanding the processes involved in heat elimination of pathogens from perennial plants.

REFERENCES

1. Babos, P., and Dassanis, B. 1963. Thermal inactivation of tobacco necrosis virus, Virology 20:490-497.
2. Baker, K. F. 1962. Thermotherapy of planting material. Phytopathology. 52:1244-1255.
3. Barlass, M., Skene, G. M., Woodham, R. C., and Krade, L. R. 1982. Regeneration of virus-free grapevines using in vitro apical culture. Ann. Appl. Biol. 101:291-295.
4. Barnett, T., Altschuler, M., McDaniel, C. N., and Mascarenhas, J. P. 1980. Heat shock induced proteins in plant cells. Developmental Genetics. 1:331-340.
5. Berry, J., and Bjorkman, O. 1980. Photosynthetic response and adaptation to temperature in higher plants. Annu. Rev. Plant Physiol. 31:491-543.
6. Bjorkman, O., Badger, M. R., and Armond, P. A. 1980. Response and adapatation of photosynthesis to high temperatures. Pages 233-249 in: Adaptation of Plants to Water and High Temperature Stress. N.C. Turner and P.J. Kramer, eds. John Wiley and Sons, New York.
7. Buttrose, M. S., and Hale, C. R. 1971. Effects of temperature on accumulation of starch or lipid in chloroplasts of grapevine. Planta 101:166-170.
8. Chen, Y., Kamisaka, S., and Masuda, Y. 1986. Enhancing effects of heat shock and gibberellic acid on thermotolerance in etiolated Vigna radiata. I. Physiological aspects on thermotolerance. Physiol. Plant. 66:595-601.
9. Converse, R. H., and George, R. A. 1987. Elimination of mycoplasmalike organisms in cabot highbush blueberry with high-carbon dioxide thermotherapy. Plant Dis. 71:36-38.
10. Cooper, P., and Ho, T. D. 1983. Heat shock proteins in maize. Plant Physiol. 71:215-222.
11. Dawson, W. O. 1976. Synthesis of TMV RNA at restrictive high temperatures. Virology 73-319-326.
12. Dawson, W. O. 1978. Recovery of tobacco mosaic virus RNA replication after incubation at 40°C. Intervirology 9:295-303.
13. Dawson, W. O. 1983. Tobacco mosaic virus protein synthesis is correlated with double-stranded RNA synthesis and not with single-stranded RNA synthesis. Virology 125:314-323.
14. Dawson, W. O., and Boyd, C. 1987. TMV protein synthesis is not translationally regulated by heat shock. Plant Mol. Biol. 8:145.
15. Dawson, W. O., White, J. L., and Grantham, G. L. 1978. Effect of heat treatment upon cowpea chlorotic mottle virus ribonucleic acid replication. Phytopathology 68:1042-1048.
16. Deom, C. M., Olwer, M. J., and Beachy, R. N. 1987. The 30-kilodalton gene product of tobacco mosaic virus potentiates virus movement. Science 237: 389-394.
17. Diener, T. O. 1979. Viroids and Viroid Diseases. John Wiley & Sons, New York. 252 Pages.
18. Farrar, J. F. 1988. Temperature and the partitioning and translocation of carbon. Pages 203-235 in: Plants and Temperature. S.P. Long and F.I. Woodward, eds. The Company of Biologists Ltd., Cambridge.
19. Ghanotakis, D. F., and Yocum, C. F. 1990. Photosystem II and the oxygen-evolving complex. Annu. Rev. Plant Physiol. 41:255-296.
20. Ginza, W. 1968. Inactivation of viruses by ionizing radiation and heat, Vol. 4. Pages 139-209 in: Methods in Virology. K. Maramorosch and H. Koprowski, eds. Acadamic Press.
21. Grace, J. 1988. Temperature as a determinant of plant productivity. Pages 91-107 in: Plants and Temperature. S.P. Long and F.I. Woodward, eds. The Company of Biologists Ltd., Cambridge.
22. Hansen, A. J. 1984. Effect of ribavirin on green ring mottle causal agent and necrotic ringspot virus in Prunus species. Plant Dis. 68:216-218.
23. Havaux, M. 1992. Stress tolerance of

photosystem, II, *In vivo* antagonist c effects of water, heat, and photoinhibition st esses. Plant Physiol. 100:424-432.

24. Havaux, M., Greppin, H., and Strausser, R. J. 1991. Functioning of photosystems I and II, Pea leaves exposed to heat stress in the presence or absence of light. Planta 186:88-89.

25. Hull, R. 1989. The movement of viruses in plants. Annu. Rev. Phytopathol. 27:213-240.

26. Itai, C., Benzioni, A., and Ordin, L. 1973. Correlative changes in endogenous hormone levels and shoot growth induced by heat shock treatment to the root. Physiol. Plant. 29:355-360.

27. Itai, C., and Benzioni, A. 1974. Regulation of plant response to high temperature. Pages 477-482 in: Mechanisms of Regulation of Plant Growth. R.L. Bieleski, A.R. Ferguson, and M.M. Cresswell, eds. Royal Society of New Zealand.

28. Itai, C., Benzioni, A., and Munz, S. 1978. Heat stress: Effects of abscisic acid and kinetin on response and recovery of tobacco leaves. Plant and Cell Physiol. 19:453-459.

29. Johnstone, G. R., and Wade, G. C. 1974. Therapy of virus-infected plants by heat treatment. I. Some properties of tomato aspermy virus and its inactivation at 36 degrees C. Aust. J. Bot. 22:437-450.

30. Johnstone, G. R., and Wade, G. C. 1974. Therapy of virus-infected plants by heat treatment. II. Host protein synthesis and multiplication of tomato aspermy virus at 36 degrees C. Aust. J. Bot. 22:451-460.

31. Kassanis, B. 1949. Potato tubers freed from leafroll virus by heat. Nature 164:881.

32. Kassanis, B. 1954. Heat therapy of virus-infected plants. Ann. Appl. Biol. 41:470-474.

33. Kassanis, B. 1957. Effect of changing temperature on plant virus diseases. Advan. Virus Res. 4:221-241.

34. Kassanis, B. 1957. Some effects of varying temperature on the quality and quantity of tobacco mosaic virus in infected plants. Virology 4:187-190.

35. Key, J. L., Lin, C. Y., and Chen, Y. M. 1981. Heat shock proteins of higher plants. Proc. Natl. Acad. Sci. 78:3526-3530.

36. Kriedemann, P. E., Sward, R. J., and Downton, W. J. S. 1976. Vine response to carbon dioxide enrichment during heat therapy. Aust. J. Plant Physiol. 3:605-618.

37. Kunkel, L. O. 1935. Heat treatment for the cure of yellows and rosetted of peach. Phytopathology 25:24.

38. Kunkel, L. O. 1936. Heat treatments for the cure of yellows and other virus diseases of peach. Phytopathology 26:809-830.

39. Kunkel, L. O. 1936. Peach mosaic not cured by heat treatments. Am. J. Bot. 23:683-685.

40. Kunkel, L. O. 1941. Heat cure of aster yellows in periwinkles. Am. J. Bot. 28:761-769.

41. Kunkel, L. O. 1943. Potato witches' broom transmission by dodder and cure by heat. Proc. Am. Phil. Soc. 86:470-475.

42. Kunkel, L. O. 1949. Studies on cranberry false blossom. Phytopathology. 35:805-821.

43. Kunkel, L. O. 1952. Transmission of alfalfa witch's broom to nonleguminous plants by dodder, and cure in periwinkle by heat. Phytopathology 42:27-31.

44. Levitt, J. 1980. Responses of Plants to Environmental Stresses. Academic Press, New York. 497 Pages.

45. Long, S. P., and Woodward, F. I. 1988. Plants and Temperature. The Company of Biologists Ltd., Cambridge. 415 Pages.

46. Lozoya-Saldana, H., and Dawson, W. O. 1982 The use of constant and alternating temperature regimes and tissue culture to obtain PVS-free potato plants. Am. Potato J. 59:221-230.

47. McCoy, R. E., Caudwell, A., Chang, C.J., Chen, T.A., Chiykowski, L. N., Cousin, M.T., Dale, J.L., de Leeuw, G.T.N., Golino, D.A., Hackett, K.J., Kirkpatrick, B.C., Marwitz, R., Petzold, H., Sinha, R.C., Sugiura, M., Whitcomb, R.R., Yang, I.L., Zhu, B.M., and Seemuller, E. 1989. Plant diseases associated with mycoplasma-like organisms. Pages 546-640 in: The Mycoplasmas, Vol. 5. R. F. Whitcomb and J. G. Tally, eds. Academic Press, N.Y.

48. Nyland, G., and Goheen, A. C. 1969. Heat therapy of virus diseases of perennial plants. Ann. Rev. Phytopathology. 7:331-354.

49. Ougham, H. J., and Howarth, C. J. 1988 Temperature shock proteins in plants. Pages 259-280 in: Plants and Temperature. S.P. Long and F.I. Woodward, eds. The Company of Biologists Ltd., Cambridge.

50. Royal, L. F., Jr., The effects of temperature, water stress and hormones on the germination and early growth of barley (*Hordeum vulgare* L. cv. Himalaya). Dissertation Abstracts International 41(8):2842.

51. Sachs, M. M., and Ho, T. D. 1986. Alteration of gene expression during environmental stress in plants. Annu. Rev. Plant Physiol. 37:363-376.

52. Skrzeczkowski, L. J., Howell, W. E., and Mink, G. I. 1993. Correlation between leaf epinasty symptoms on two apple cultivars and results of a cRNA hybridization probe for rapid detection of apple scar skin viroid. Plant Dis. 77:919-921.

53. Stubbs, L. L. 1969. Apparent elimination of exocortis and yellowing viruses in lemon by heat therapy and shoot tip propagation. Pages 96-99 in: Proc. Conf. Intern. Organ. Citrus Virologists, Italy.

54. Sward, R. J., and Hallam, N. D. 1976. Changes in fine structure of the potato meristem following heat treatment for virus eradication. Aust. J. Bot. 24:597-605.

55. Turner, N. C., and Kramer, P. J. 1980. Adaptation of Plants to Water and High Temperature Stress. John Wiley and Sons, New York. 482 pages.

56. Vierling, E. 1990. Heat shock protein function and expression in plants. Pages 357-375 in: Stress Responses in Plants: Adaptation and Acclimation Mechanisms. R.G. Alscher and J.R. Cumming, eds. John Wiley & Sons, Inc., New York.

57. Vierling, E. 1991. The roles of heat shock proteins in plants, Annu. Rev. Plant Physiol. 42:579-620.

58. Walkey, D. G. A., and Freeman, G. H. 1977. Inactivation of cucumber mosaic virus in cultured tissues of *Nicotiana rustica* by diurnal alternating periods of high and low temperature. Ann. Appl. Bio. 87:375-382.

59. Weis, E., and Berry, J. A. 1988. Plants and high temperature stress. Pages 329-346 in: Plants and Temperature. S.P. Long and F.I. Woodward, eds. TheCompany of Biologists Ltd., Cambridge.

60. White, J. L., and Dawson, W. O. 1978. The effect

of supraoptimal temperatures upon TMV RNA replicase. Intervirology 9:295-303.

61. Wu, M. T., and Wallner, S. J. 1984. Heat stress responses in cultured plant cells heat tolerance induced by heat shock versus elevated growing temperature. Plant Physiol. 75:778-780.

62. Yakushikina, N. I., and Tarasov, S. I. 1982. Growth of maize seedlings with short-term exposure to extreme temperature (Abstr.). Plant Growth Regul. 9:1441.

CHAPTER 27

Virus Elimination by Meristem Tip Culture and Tip Micrografting

G. Faccioli and F. Marani

It is common knowledge that plant viral diseases are widespread around the world and cause serious yield losses; their control is difficult because of the peculiarity of the replication cycle of their agents.

While microorganisms are, in many cases, controlled by therapeutic treatments (e.g. chemicals), for viral diseases prevention is the only really effective measure. With vegetatively propagated species, progressive increase in infection is experienced through the generations, thus it is of basic importance to produce healthy material.

Application of meristem culture began in the early 20th century when trials to grow *in vitro* apical portions of stems and roots were carried out (220). White (286) was the first to hypothesize that plant meristems could be free of viruses. To study virus distribution in plants he grew *in vitro* tomato roots infected with the aucuba strain of tomato mosaic virus (ToMV) sampling successive portions to determine the virus presence by inoculation to *Nicotiana glutinosa*. Few years afterwards Ball (12) demonstrated the morphogenetic potential of apex, regenerating plants from meristems of *Lupinus* and *Tropaeolum*. Morel (172) was the first that, carrying out the aseptic culture of meristems excised from shoot tips, obtained virus-free plantlets from diseased ones, proposing the hypothesis of meristem "immunity" towards viruses. Successive studies by Limasset and Cornuet (144) demonstrated a differential distribution of viruses in systemically infected plants. Using biological and serological methods they established that virus concentration in buds was 150 times lower than in developing leaves. Since meristem tips, excised under the stereoscopic microscope, were found to be mostly free of virus, they also hypothesized that meristem

tips were free of virus. In the same period, Holmes (105) obtained *Dahlia* plants free of tomato spotted wilt virus (TSWV) by grafting meristem tips onto healthy rootstocks. Later, Morel and Martin (176,177) produced healthy plants from some virus-infected cultivars of *Dhalia* and potatoes, by excision and growth of the meristems in a sterile synthetic medium. Their results opened up a new possibility for controlling plant viruses.

When meristem-tip culture was first used to produce virus-free plants, it was assumed that the regenerated plants were healthy since viruses did not invade the meristematic tissues of the bud (127), despite the fact that tobacco mosaic virus (TMV) was reported to be present in meristems of tomato (240). The evidence that carnation mottle virus (CarMV) and carnation ringspot virus (CRSV) can invade carnation meristems (101) was confirmed by other findings, using different techniques. Adopting meristem "squash method" for mechanical inoculation, and electron microscopic observations, it was found that cherry leaf roll virus (CLRV), strawberry latent ringspot virus (SLRSV) and cucumber mosaic virus (CMV) were able to invade meristems of various hosts (279) and tobacco ring spot virus (TRSV) those of bean (63) and tobacco (221). Later on, more precise and sophisticated techniques allowing the detection of TMV, CMV, potato virus X (PVX), potato virus M (PVM), potato virus S (PVS), bean common mosaic virus (BCMV) were used, as it will be described in a following section. Although some viruses e.g. PVX, PVS and PVM are present in apical dome cells of infected plants, virus-free plantlets can be obtained using meristem culture, thus indicating that virus elimination occurs during tissue culture *in vitro* (26,59,78,79).

Different hypotheses have been put forward to answer the questions why in some infected species viruses fail to invade the meristem

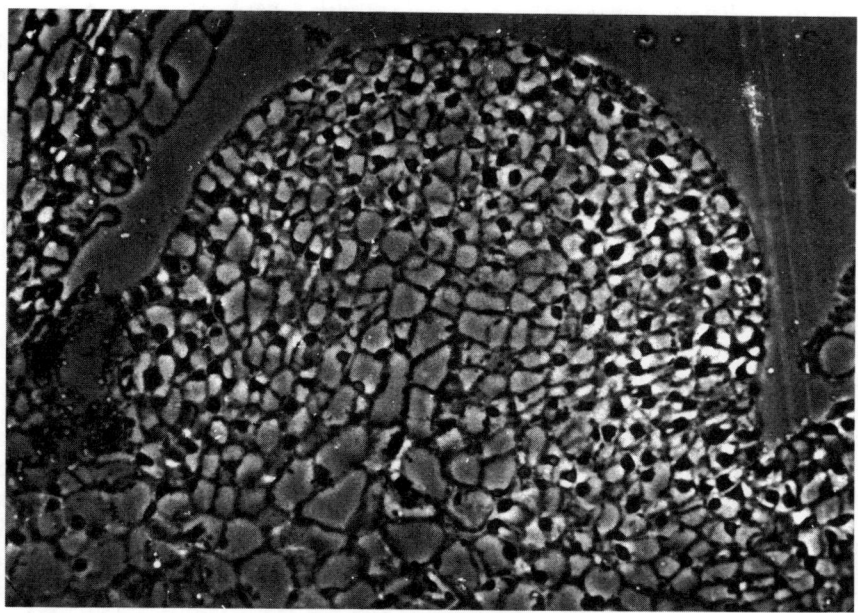

Figure 1. The dome and part of the two first leaf primordia of a potato meristem tip (cv. Majestic).

tip, and how viruses present in the meristem can be eliminated during tissue culture procedure. It was proposed that virus failure to invade meristems may be due either to high auxin concentration in actively dividing meristematic cells (87), or to a competition for nutrients during synthesis active meristematic cells that have enhanced oxidative-reductive processes, the environment is unsuitable for the replication of several viruses.

Various reasons were tentatively given for explaining virus elimination during *in vitro* culture: the action of growth regulators, specially cytokinins (16), the presence of inhibitors as phenolamines (159), the metabolic disruption of enzymes necessary for viral replication, and viral RNA degradation resulting from cell injury during explant excision (164).

Studies on the behavior of plant viruses in tissue cultures have shown that some viruses (e.g. CLRV and arabis mosaic virus: ArMV) are inactivated, whereas others (e.g. TRSV, CMV, TMV and PVX) persist in all or most of the cultures (164). CLRV is, for example, lost in tissue culture after 155 days at 22°C, whereas for CMV elimination, a temperature of 30°C for 66 days is needed. Even in this case we think that several factors influence the persistence of viruses in cultured meristems, but it can perhaps be hypothesized that all events influencing host cell metabolism (e.g. composition of culture media and temperature) affect virus replication within cells.

The reported hypothesis does not satisfactorily explain why some viruses persist during culture, but nevertheless meristem-tip culture has assumed an important and wide practical application in both virus elimination, and multiplication and storage of valuable germplasm under safe conditions. The coupling with thermo- and chemotherapy has also increased its potential for producing virus-free material.

The method, besides its utilization in controlling viral infection, is also valuable in eliminating bacterial and fungal diseases that could be transmitted through simple clonal propagation *in vitro*.

CHARACTERISTICS OF MERISTEM TISSUES

The apical meristem, which is formed during embryo development, is a dome of actively dividing cells located at the apex of shoots and roots (Figure 1). It remains in an active state of division throughout the vegetative phase of the plant forming new tissues and organs (shoots, leaves, roots) and, therefore, has the capability of producing a complete plant *in vivo* and *in vitro*. The totipotency of its undifferentiated cells is at the basis of meristem culture technique, so that meristems cultured on a suitable nutrient medium can be regenerated into plantlets.

Plantlets derived from meristem-tip culture usually retain the genetic characteristics of mother plants, whereas those derived from other tissues such as protoplasts, callus, nucellus or floral primordia may present different kinds of variations (64).

Since the purpose of meristem-tip culture is to eliminate a virus from a selected cultivar, it is of paramount importance to obtain plants that do not differ phenotypically from mother plants.

Virus-free plants can, however, differ "per se" in their external appearance from virus-infected individuals. These new phenotypes may be the result of modified metabolic functions due to virus elimination, or may originate from mutations. Micromutations, reported, for example, in rhubarb and apple after meristem culture (72,73) are difficult to identify, as their expression is often strongly dependent on environmental conditions. Specialized electrophoretic or molecular techniques able to identify variations in isoenzyme patterns, or RFLP, are therefore needed for identification of the mutants.

The culture of meristem usually results in production of plants true-to-type; in fact the frequency of variants seems to be comparable to that observed when using traditional methods of propagation (25). The genetic stability of the meristem depends on the fact that it is the site of storage of genetic information of the plant which, in seed-plants, is transmitted to the germ line in the flower, at the time of transition from vegetative to reproductive phase. Thus the genetic stability of meristem cell lines is the prerequisite for the genetic stability of the germ line in each species (65). Factors of genetic stability in meristems are under the strict control of DNA synthesis, which does not allow extra duplications of DNA responsible for somatic polyploidy, and the continuous cell division which eliminates at least part of the spontaneous chromosomal structural changes.

The security of genetic uniformity in mericlones depends on different factors, but mostly on the use of techniques that avoid the formation of callus (derived from dedifferentiated tissues), which is a common source of genetic variation. Thus, *Asparagus officinalis* plants

obtained from meristem culture are all diploid, whereas plants regenerated from callus and suspension cultures are either diploid or tetraploid, with a predominance of the latter (151,196).

It is well known that chances of mutation increase when adventitious shoot formation occurs in clonal propagation (micropropagation), or in callus culture (somaclonal variations) or when somatic embryogenesis occurs in callus cultures. In fact this embryogenetic pathway, in which plants regenerated from single cells are solid mutants, is of particular significance in mutagenetic studies.

The type of tissue used as starting material influences also the probability of mutation (242). In undifferentiated tissues such as apical meristems, cambium, procambium and pericycle, in which a uniformly diploid state of cells is maintained, the chance of mutations is usually lower than in differentiated tissues, such as pith, where induction of cell division leads to callus cultures. Growth regulators used in growing media especially 2,4-dichlorophenoxyacetic acid (2,4-D), naphtalenacetic acid (NAA) and synthetic cytokinins are also responsible for inducing mutations. 2,4-D, in particular, has been reported as a direct inducer of polyploidization in tissue cultures (258). The induction of higher ploidy is inversely related to concentration of growth regulators.

Nuclear changes, including polyploidy, aneuploidy and chromosomal mutations, are also associated with repeated subculturing *in vitro* and with long-term callus cultures, which can be used in mutant selection (somaclones) (24,69,124).

DETECTION OF VIRUSES IN MERISTEM TIPS

As we have already seen, meristematic tissues of shoots are the plant parts with the lower virus content which, sometimes, may be virus free. Thus even for viruses which invade meristems (dome and/or few leaf primordia) a gradient of increasing virus concentration from the dome to the successive primordia exists, that follows the increasing degree of tissue differentiation. This means that the probability of obtaining virus-free plants is inversely related to the size of the meristem used, but the capability of the meristem to develop into a full plant is directly related to the explant's size. A compromise should, therefore, be found which will assure the largest number of depends on both type of virus and plant-host

Table 1. Viruses detected in apical meristems (0.1-0.5 mm in length)

Virus	Host	Technique	References
CarMV	carnation	"tip-squash" inoculation	101
CRSV	carnation	as above	
CLRV	*Nicotiana rustica* *Chenopodium amaranticolor* *N. clevelandii*	electron microscopy of "tip-squash" electron microscopy of "tip-squash" electron microscopy of "tip-squash"	279
SLRSV	*C.amaranticolor* *Cucumis sativus*	as above	11
CMV	*N. rustica*	as above	11
TRSV	bean tobacco	electron microscopy of embedded tissues electron microscopy of embedded tissues	63 221
PVX	potato	as above	6,137
CMV, TMV, PVX	tomato, tobacco, petunia, potato	fluorescent antibody	179
PVX, PVS, PVM	potato	autoradiography in phase-contrast microscopy	79,227,228
PVX, PVS, PVM	potato	dot-ELISA	59,77, 227
BCMV	bean	autoradiography as above	226

combination (not only species but also cultivar) (59). The viral distribution should then be determined for each individual case of virus-host combination, before carrying out the mass production through the *in vitro* culture of meristem tips (Table 1). One way of checking it, is to determine the percentage of virus-free plantlets obtained after culturing meristems of different size, but since this requires long time it can be substituted by the direct detection of virus in meristem tips and subsequent choice of the explant size which is virus free.

Several methods have been adopted for the purpose. At the beginning rather sophisticated techniques were used, like electron microscopy of tissues (6) which is a time consuming method, or immunofluorescence (179) which is not simpler.

The subsequent use of autoradiography with the optical microscope has already allowed precise and faster virus detection, through the examination of a certain number of meristems in a reasonable amount of time, to carry out a statistical examination of virus presence (26,79,81). The technique is based on the incorporation of ^{3}H-uridine that takes place only in viral RNA, since cellular RNA synthesis is prevented by actinomycin D (AMD) treatments. Plastic-embedded tissues, cut with an ultramicrotome and coated with nuclear emulsion, are developed after two weeks in the dark and examined in a phase contrast microscope: virus localization is determined by the presence of "silver grain" produced by the isotope's radiations hitting the emulsion (Figure 2). However this method was still not simple enough for mass examination, and only with the sensitivity improvement of serological methods, particularly the adoption of dot-ELISA (13) with some minor modifications, the precise and fast virus detection per single meristem was possible. Each meristem ground in 200 µl of ELISA buffer is placed, through plastic templates, on nitrocellulose which is then treated with the usual alkaline-phosphatase-IgG specific conjugate, and stained

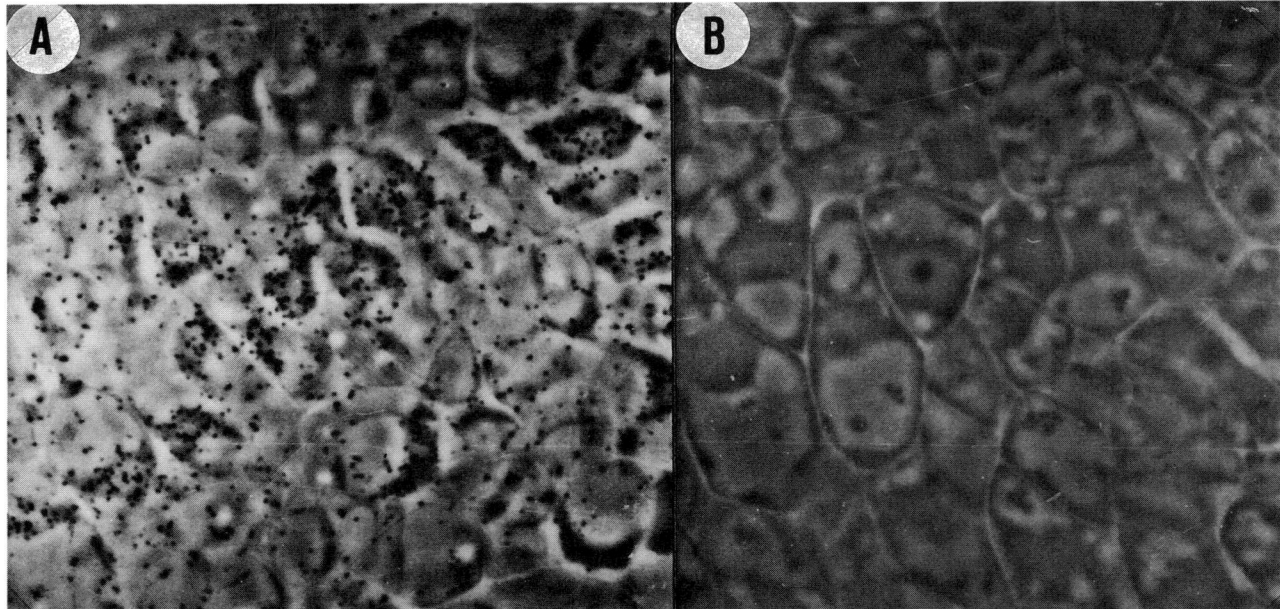

Figure 2. Autoradiograms of ^{3}H-uridine incorporation (silver grains) in meristematic cells of Majestic potato tips (first leaf primordium): samples were given 5 h pulse feeding (50 μC/ml) after a 12 h treatment with actinomycin D (AMD, μg/ml). A = healthy cell; B = PVS-infected cells (Magnification=1450x).

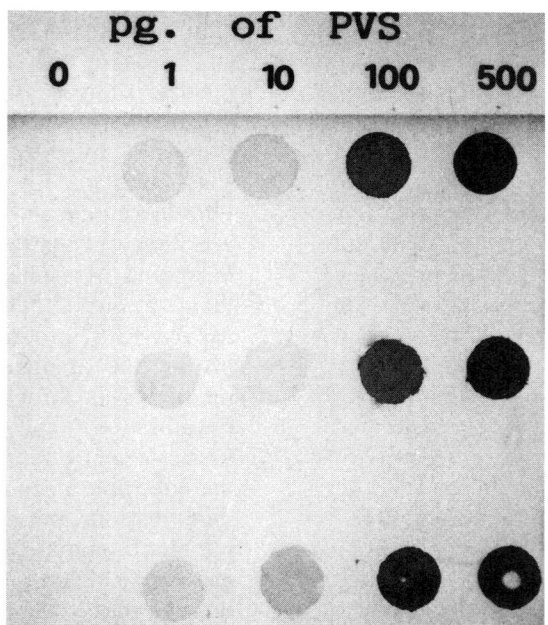

Figure 3. Dot-blot identification of PVS on nitrocellulose strips, virus concentration of samples is expressed in picograms (pg).

with a 1:1 mixture of fast blue BBN salt and naphtol phosphate which develop a blue color. By this technique contents as low as 1 pg of PVX and PVS were detected (Figure 3) (27).

Furthermore, by examining a high number of meristems and culturing *in vitro* an equivalent number of them, it was possible to: (i) establish that also meristems containing a limited amount of virus, could give rise to virus-free plants after culture *in vitro*; (ii) determine for particular virus-plant combinations (PVX, PVY, PVM and potato) the level of virus present in the meristem, below which the progeny would be largely free of virus (27,59,77). That is to say, as already postulated by Quak (24), that minute amounts of virus are lost during the culture *in vitro*: in the above mentioned case, an amount lower than 10 pg.

MERISTEM-TIP CULTURE

Any modern virus-elimination program requires the knowledge of the virus involved and of its distribution into the plant. For virus identification the most used serological techniques such as DAS-ELISA (267) and immunoelectron microscopy (168) could be used, while for virus distribution more sophisticated methods as we have already described are needed. The choice of mother plants to start with, is obviously based on their agronomical characteristics.

Meristem-tip culture consists of three main steps:

-meristem sampling
-meristem culturing and development in the culture medium, i.e. shoot and root formation
-transplanting of plantlets to pots and subsequent virus indexing.

Meristem Sampling

Different names have been used to indicate the explant suitable for virus elimination. The most frequently used are "shoot tip", "tip", "meristem", "meristem tip" (100). The last two names indicate the explant made of meristematic dome and one or two leaf primordia which measures about 100 µm in diameter and 250 µm in length. "Shoot tip" and "tip" indicate part of the shoot apex including the meristem with several leaf primordia and adjacent differentiated stem tissue. This kind of explant is usually removed

from heat treated plants or for clonal propagation if freedom from viruses is not required.

It is quite common the misconception that meristem-tip culture and clonal propagation are synonymous probably because both were indifferently used to obtain virus-free plants. A similarly common mistake is to call meristem-tip culture the culturing of shoots and other tissues, which are not meristematic at all, and should be instead considered as simple clonal propagation. This distinction is important because, as it is well known, non meristematic tissues of virus-infected plant are obviously infected. For instance, the use in propagation by the orchid industry of tips larger than 0.1 mm, which was the meristem size adopted by Morel (173,174) to obtain virus free plants, brought to a rapid increase of orchid viruses which became a major problem for the culture (141).

Meristem-tip explants suitable for virus elimination are about 0.1 mm in diameter and 0.2-0.4 mm in length (apical dome plus one or two couples of leaf primordia), and must be excised under a stereoscopic microscope (8-40 X) with appropriate instruments. These include, forceps, needles and pieces of razor blades with sharp edges, mounted on handles of suitable length, depending on the type of vessels used. Dissection is normally carried out in a sterile room or in a laminar flow cabinet but, if these are not available, also in a normal very clean room. The instruments before every manipulation must be heat sterilized, and the platform of the dissecting microscope must be cleaned with 75% ethanol. The excision phase should be as short as possible, because the light commonly used with stereomicroscope causes desiccation of small explants. Cool light illuminator or glass fibre lamps should be used or, alternatively, dissection could be performed in a Petri-dish, containing a moist sterile filter paper to prevent drying. Before excision of meristem tips from a growing shoot, the external leaves should be removed by hand. The remaining leaves are removed under the stereomicroscope. When the meristem is freed from leaves it is excised together with 1-2 leaf primordia, by 4 cuts perpendicular to the meristematic area, in order to obtain a cube of material from which, with a basal cut, the meristem tip is excised and immediately placed on a nutrient medium to prevent drying.

The first important goal in realizing meristem-tip culture is to avoid the explant's contamination by surface pathogens. Different procedures are reported in the literature on this aspect. Based on our experience, it seems necessary to use different methods according to the source of the meristem tip. If the meristem tip is excised from growing shoots, sterilization may not be strictly necessary. The growing points are, in fact, so well protected by numerous leaf primordia that a careful dissection permits the excision of aseptic explants without any sterilization. Sometimes disinfection treatment may, indeed, increase the incidence of contamination.

In some cases it is useful to grow donor plants in a greenhouse on sterilized soil, and to water the soil rather than leaves.

To reduce contaminant populations, periodic spraying of the plants with systemic fungicides is recommended. For field-grown plants it is advisable to sample small cuttings and grow them in nutrient solutions, because shoots arising from axillary buds are less contaminated than those directly taken from field-grown material.

Meristem tips from underground plant organs, such as tubers, rhizomes, bulbs or corms, from buds or from seeds, should be dipped in 75% ethanol to remove trapped air, sterilized with sodium hypochloride or calcium chloride for 10 min and rinsed three times in sterile water.

Despite these general rules, the operator should adopt the procedure to fit his working conditions trying so as to find the best way to avoid contaminations. An example is given by the meristem-tip excision from garlic: while Bhojwani *et al.* (34) dipped cloves in 95% ethanol and flamed before dissection of sterile bud shoots, in our laboratory sterilization is not used.

Meristem Culturing and Development

Morphogenetic potential of meristems for culturing depends mainly on meristem size, plant species, physiological conditions of the donor plants and meristem position on the plant.

Meristem Size

As per literature, the proper meristem size to be used varies with the authors: some report the total length of meristems used, others the number of leaf primordia. The majority of the workers generally use tips between 0.2-1 mm in length, consisting of meristematic dome and two or more leaf primordia. We think, though, that in order to avoid confusion, the number of primordia should be given rather than dimensions, since these vary remarkably with plant species.

The larger the size of the meristem cultured, the greater is the number of regenerated plants; on the other hand, the number of virus-free plantlets obtainable is inversely proportional to the size of the cultured tips. Explant dimension should, therefore, accomplish both needs: development in a complete plant and virus elimination.

In our experience, the chance for survival of meristems lacking leaf primordia is very limited, as already stated by others (255,272), although some authors reported this possibility for various plants (239,248). Almost always, one or more primordia are included in the explants indicated in the literature as suitable for obtaining a plantlet. *Cassava* meristems generated plantlets only when longer than 0.2 mm, below this size they produced either callus or roots (123). Soybean meristems differentiated into plantlets when longer than 0.4 mm (121). With carnation, within the same cultivar (G.B. Red Shy), the percentage of developing meristems varies from 8 to 21% for dimensions ranging from 0.4 to 1 mm in length (135).

Plant Species

Regenerative potential of meristems is not only related to the species but, whenever investigated, appeared to be connected also with cultivars. This results, for example, from a study on carnation, where yield of developed plantlets ranged from 68 to 83%, for 0.2 mm long meristems of four cultivars, while it was only 8% for 0.4 mm meristems of a 5th cultivar (91,135,149). In general, it is possible to point out that a higher regenerative capability is present in herbaceous plants, rather than in shrubs and trees. However, bulbous plants (iris, gladiolus, lily, fresia) demonstrate a slower meristem growing than dicotyledons, so that they seldom are ready for transfer to soil in less than six months (208), and survival after transplanting to soil is much lower than that of most herbaceous plants.

Physiological Condition of Donor Plant

Generally speaking, the best physiological conditions of donor plants, for meristem sampling, coincide with the highest vegetative activity. The full control of factors affecting this activity would therefore be desirable, which is to say the availability of phytotrons or conditioned greenhouses where light and temperature can be

Figure 4. Rooted shoots of Majestic meristem tips (4 leaf primordia) in the third culture medium (MS + 1 mg/L IAA and 1 mg/L IBA) after a total culture time of 10,12 and 14 weeks respectively (from left to right).

well regulated. Broccoli donor plants, as an example, supply well developing meristems only when grown at 16°C; already at 21°C meristems are unable to develop in tissue culture (5).

Quite often though, donor plants are not kept in these ideal conditions, and especially in these cases seasonal fluctuations are important mainly with reference to light intensity and photoperiod. For example, attention should be paid not to sample from buds already induced to flowering, particularly for ornamental species (*Dahlia*, *Pelargonium*, begonia, *Chrysanthemum*), where flower induction occurs precociously.

In carnation, meristems survive better when excised from plants in early spring and autumn than in winter or summer. Shoots derived from winter culture root more easily than those cultured in summer, but the latter originate more frequently virus-free plants (266). For most potato varieties, meristems isolated in spring and early summer root more readily than those taken later in the year (163,216). In bulbs and corms, meristems are best taken at the end of the dormancy period. For trees, the growth of which is limited to a short period of time in the spring, meristem tips should be excised in this season, or in winter only after some breaking-dormancy treatment. In various *Prunus* species, for examples, this is obtained maintaining stem cuttings at 4°C for almost 6 months before excision of the tips (42).

Meristem Position on the Plant

Although in principle all plant meristems are suitable as starting material, both morphogenetic potential and virus content depend on their position on the stem or shoots, apical or axillary, and in the latter case on the shoot position on the stem.

As far as morphogenetic potential is involved, a wide range of possibilities exist. Meristems from basal buds of asparagus spears, for example, develop into plantlets more easily than meristems of buds positioned near the apex (288). On the contrary, successful culture is obtained only from apical buds in gooseberry (115), while rose meristems that come from the mid part of stems develop faster and more easily (45). In carnation and *Chrysanthemum*, meristem tips excised from apical buds give better results than those derived from axillary buds: out of 3800 Chrysanthemum meristem tips sampled, 32% of those derived from apical meristems developed into complete plants while only 18% of the axillary meristems were successful (102).

Meristem position on the plant is to be taken into account also for its' importance in the phenological stages of of the derived plantlets. More thorough studies in this direction would be desirable, since few are available: see, for example, the enhanced precocity of tomato plantlets (cv. Jumbo) obtained from meristems of the median nodes, with respect to the others (134).

Factors Affecting Meristem Developmment

The development in culture of meristems sampled taking into account the factors mentioned before, will depend mainly on the characteristics of culture media, light, temperature and, to a lesser extent, on secondary factors.

Culture Media

Tissue culture media generally consist of mineral salts, a carbon source of energy, vitamins, growth factors and other minor components.

Meristem development is mainly dependant on hormone composition of the culture medium for they are unable to synthesize growth factors. Since different groups of hormones exert contrasting effects on merisytem differentiation, it is seldom possible to combine all of them in a properly balanced single medium. Several trials were performed in the past trying the single-medium culture for various species, but only a few cases were successful (121), and the yield of developed plantlets was always lower than that obtained using at least two media. The adoption of a first cytokinin-rich medium for shoot development, and of a second auxin-rich medium to stimulate root formation is, therefore, advisable, or even necessary.

In some cases, the use of a medium deprived of hormones to transplant potato shoots

and drain out of them residual cytokinins, proved to be beneficial for increasing rooting in the second medium (G. Faccioli, unpublished). In others, to favor the development of more shoots from a potato meristem, the "multimeristem technique" was adopted (222), based on the use of three media: the first for initial meristem development, a second to induce formation of embryoids from callus, and a third with the usual auxins for rooting (Figure 4) (80,81).

Meristems usually grow well on nutrient media solidified with agar (0.6 - 0.8%), but also liquid media can be used. Meristem development can also be affected by agar concentration and quality (68). Difco Bacto agar as well as alternative products such as Biogel P200 or alginate or Gelrite should be used (208), since some ordinary types of agar could contain organic and inorganic contaminants toxic to meristems (224). When a liquid medium is used, it may be advisable to use growth regulators at a lower concentration since their uptake by the tissues is faster than in agar, and concentration inside the tissues may become too high. Since meristems of some plants (e.g. orchids, fruit-trees or *Pelargonium*) usually release large quantities of phenolic compounds into the medium, this last could become cytotoxic, causing an inhibition of growth. In these cases a liquid medium, where paper bridges (98) or other porous materials are dipped and support the explants, are indicated (208). The use of a liquid medium is recommended for rooting of some woody species (190). Simultaneous combination of solid and liquid media can also be used. In particular cases - e.g. fruit-trees in order to favor the further development of the meristem-derived shoots, due to a better nutrient absorption, it is advisable to pour, on the top of the solid, a liquid medium (double-phase technique) (148). This also avoids possible shoot "vitrification" and increases the yield of axillary shoots in clonal propagation.

The pH of culture medium, which is normally used in the range of 5.0-6.5, is an important variable which affects meristem development mainly according to the plant species. Carnation meristems, for example, cultured at pH 5.5, differentiated 59% of shoots but only 4% when pH was 6.0. On the other hand a pH lower than 5.5 inhibits shoot rooting (255). This factor could probably direct the type of meristem organogenesis, since it has been proved, for example, that at pH of 5.9 tobacco cells also differentiated shoots whereas at 4.9 developed roots (255). This pH effect could be attributed to a different cell uptake of growth factors, vitamins and ammonium ions (48,246). pH may also play a role in amino acids and sugars uptake by the cells (155). The use of buffers, such as Sörensen phosphate, MES [2-(N-morpholino) ethane], sulphonic acid and TRIS [Tris (hydroxymethyl) aminomethane] (38) to control pH falls (generally 0.3-0.5 units) during culture, can give different results according to the medium composition (246). In the preparation of the culture media it has to be taken into account also the fact that an excess of auxins and cytokinins together with a low agar concentration (lower than 6 g/L), high relative moisture and accumulation of gases, may cause the so called "vitrification" or "glassiness". This physiological disorder consists in the formation of hyperhydrated leaves having intercellular spaces filled with water instead of air, and can cause growth inhibition and/or transplant losses up to 60% (129,148,294).

We would finally like to recall the fact that ingredients used for tissue culture media are not always pure enough, and this can create additional problems. Components used in large quantities such as water, sugar and agar, generally give the most frequent inconveniences, but it is also important that growth substances used are obtained from reliable suppliers, and nutrient salts be of research grade.

The time necessary for *in vitro* plantlets production before transplanting, once that the proper growing media are established, greatly varies according to plant species and cultivars. Potato rooted shoots are transplanted after an average time of three months, the early varieties being a bit faster than late ones (Faccioli, *unpublished*). An average time of 4 months is required by carnation and 4 to 6 months are necessary for *Chrysanthemum* and bulbous species (e.g. garlic, lily, *Gladiolum*, *Iris* etc.).

After these general considerations, a more detailed description of the specific importance of the various chemical components of the culture media follows.

Mineral Salts

Different mineral salts at various concentrations have been used and the main mixtures are reported by Bhojwani and Razdan (33), and Gamborg (87).

Differences in the composition lie mainly in the quantity of salts and ions, whereas quality of inorganic nutrients appears to be fairly constant. White's medium, widely used in the earliest plant tissue culture media, includes all the necessary nutrients. Concentrations of various inorganic nutrients, especially of K^+ and NH_4^+, thought to be low in White and Morel media and, therefore, responsible for a poor development and chlorosis of meristem-tip in potato, were later increased (164). On the other hand, Al and Ni were eliminated since they could not be demonstrated as essential. Iron, added in the form of $Fe_2(SO_4)_3$ was firstly replaced by $FeCl_2$ and after by Fe-EDTA (ethylendiaminetetracetic acid), and this greatly improved the availability of this element during pH variations in tissue culturing and possibly of other micronutrients by preventing precipitation (89). Nowadays the media of Murashige and Skoog (MS) (194) and Gamborg *et al.*(B5) (88) are the most widely used. In both recipes, nitrogen is supplied in two forms, nitrate and ammonium compounds. Nitrate is superior to ammonium as source of nitrogen, but when used alone the pH of the medium shifts towards alkalinity. The addition of NH_4 compounds controls this deviation, so that a right NO_3/NH_4 balance is important for a successful development of explants.

The MS medium is widely used since most species grow on it. However its high salt concentration makes it of no use for some salt sensitive species (e.g. *Gerbera*) (209). Rooting of shoots in numerous plant species e.g. *Narcissus* (235), *Malus domestica* (139), *Pelargonium* hybrids (54), *Rubus* spp. (292), requires a lower salt concentration, therefore, half-strength MS salts give better results. B5 medium, originally designed for cells and protoplasts cultures but also effectively used in plant regeneration, differs mainly from MS medium for a much lower amounts of nitrate and especially ammonium (87).

Carbon and Energy Source

The preferred carbon source, for most plant tissue cultures, is sucrose which in the medium is rapidly converted to glucose and fructose, glucose being utilized first. A morphogenetic role was also indicated for sucrose; in *Liliaceae* for example, organogenesis is inhibited by high sucrose levels: when 30 g/L of sucrose were used, 100% of the surviving shoot-tip explants produced bulblets, whereas at 90 g/L sucrose reduced the explants producing bulblets to 10%, while the remaining gave rise to callus. In contrast to shoot induction, rhizogenesis may be stimulated by high sucrose levels (245). Rooting percentage of apple cultivars in media containing sucrose up to 60 g/L was 92 to 100%; rooting percentage dropped to 28-30% without sucrose (291).

A decrease in sucrose concentration (1%) during the rooting phase and increasing of light intensity is beneficial since it stimulates plantlets to shift from an heterotrophic to an autotrophic type of nutrition (193).

Vitamins

Vitamins such thiamine-HCl, biotin, pyridoxin, panthothenic acid and nicotinic acid are often added to the medium at low concentration (0.1- 0.5 mg/L) even if their role is not exactly known. Among them only thiamine-HCl appears to be essential in the culture of most plants (192).

Growth Regulators

Early studies indicated that auxin/cytokinin ratio determined the type and the extent of organogenesis. Following the demonstration by Skoog and Miller (247), cytokinins would cause shoot proliferation and auxins would stimulate root formation. However, no universal rule exists and, since there is great variability in auxin and cytokinin requirements for morphogenesis among taxa and genotypes, appropriate concentrations must be determined for each species, and sometimes even for cultivars.

Since the discovery of auxin/cytokinin responses in plant cell-culture systems, many additional substances influencing growth and morphogenesis, were identified. These include gibberellins, ethylene, and complex organic compounds. All these compounds were shown to influence growth in culture, but the basic biochemical mechanisms by which they alter growth responses are still not understood. We still do not have a proper knowledge of the main events or sites of action of auxin-cytokinin, whether they have a direct action at the gene level or their effect takes place through a series of intermediates.

Growth promoters are generally used at low concentration (0.01-10 mg/L). The commonly used auxins are 3-indoleacetic acid (IAA), 3-indolebutyric acid (IBA), and naphtalenacetic acid (NAA). 2,4-D should be avoided because it generally promotes callusing of the explants instead of morphogenesis.

Cytokinins on the basis of their chemical structure can be subdivided in at least two broad groups of compounds, one having a purine-type ring such as 6-benzyladenine (BA), isopentenyladenine (IPA, 2iP), 6-furfuryl-aminopurine (Kinetin) and zeatin; and the other based on phenylurea structure such as thidiazuron, which shows activity higher or equivalent to that of adenine-based compounds (82).

If meristem-tip explants are 500 µm or more in length they may give rise to entire plants in a medium lacking growth regulators, but generally small amounts of an auxin or a cytokinin or both (0.1-0.5 mg/L) are necessary. Auxins are probably synthesized by the second pair of leaf primordia. Thus the meristematic dome of shoot tips is not autonomous for auxins, and this is demonstrated by the fact that meristem explants of many species (*Coleus blumei*, *Daucus carota*, *Nicotiana tabacum*, *N. glauca*, *Tropaeolum majus*) not including leaf primordia require the presence of exogenous auxins to develop (248).

Two media, as already stated, are usually utilized for meristem development into plantlets, the first to elicit meristem development in a shoot, usually containing cytokinin can also induce axillary-shoots formation both in monocotyledons (bulbous species) and dicotyledons (e.g. *Chrysanthemum*) (F. Marani, unpublished).

The formation of numerous shoots from one meristem does not depend mainly on the species involved, but on the cytokinin level of the medium. These growth substances, in fact, can overcome the apical dominance and induce growth of axillary shoots in the presence of the terminal apex. This process and adventitious shoot formation, which can be repeated several times, are at the basis of clonal propagation (33).

Auxins are generally added to the medium (deprived of cytokinins) in the rooting phase, as most species do not root in the presence of a cytokinin. To obtain full plants, their shoots, about 1 cm in length, have to be transferred to the rooting medium. The most commonly used auxins for rooting are NAA and IBA in the amount of 0.1 to 2 mg/L.

Occasionally, shoots of strawberry (39), gladiolus (109) and *Narcissus* (93) root also in a hormone-free medium. Shoots of certain species developed in culture could be treated as minicuttings and rooted out of the culture. To this purpose the shoot's cut base is treated with a standard rooting powder, or dipped in a IBA solution (100 mg/L) for 1 h and planted in a potting mix. In some woody species (e.g. *Rhododendron* and *Rubus* spp.), probably for the higher oxygen supply in soil with respect to that of agar medium, shoots root better *in vivo* than *in vitro* (138,292); for other species, e.g. *Chrysanthemum*, a large number of shoots is instead lost during *in vivo* rooting due to microbial infections (223).

The role of GA3 seems to be quite variable. In *Dahlia*, at a concentration of 0.1 mg/L it suppresses callus formation favoring growth and differentiation. In *Cassava* its combination with BAP and NAA seems to be essential for obtaining full plants; high concentration could have an inhibitory effect (87). Variable effects were experimented with potato meristem tips of various cultivars: improved growth with some of them , and no effect with others (164). It was also found that rooting of potato meristems on a MS medium lacking GA3 was lower (17%) than on a medium containing it (66%) (164).

Amino Acids and Others

Mixtures of amino acids may be added to the medium, even though meristem tips should be able to synthesize all aminoacids. Adenine sulfate, when added to the medium, can often enhance growth and shoot formation (201). Myoinositol, which is involved in the synthesis of cytoplasmic membrane systems, has also a growth-promoting activity. Addition to the medium of fluoroglucinol was reported to favor rooting in a number of rosaceous fruit-trees (114,293).

Antibiotics

Sudden appearance of contamination by endogenous bacteria may occur in some species (e.g. *Lilium*: F. Marani, unpublished) after months of culture, causing growth inhibition (21). In these cases antibiotics could be used, possibly after having checked their action on both the bacterium and plant tissues for incidental toxic effect. Among antibiotics, gentamicin and tetracyclin seem to be the most effective at a concentration of 100-200 mg/L (61). They could be part of the growing medium or, for their high cost, be used only if contamination occurs.

Minor Components

Cultured tissues release into media a variety of metabolites, some of which are toxic, others inhibit tissue growth (92). These substances usually pass unnoticed, unless they cause discoloration of the medium. Browning of tissues and of the adjacent medium is assumed to be due to polyphenol oxidation resulting in formation of

quinones, which are toxic for tissues (150). A short initial culture period may be beneficial to reduce medium browning, for light promotes phenol oxidation as already proved for *Pelargonium* and potato. Furthermore, to avoid oxidation, antioxidant and adsorbent substances could be used. Ascorbic acid, for example, is routinely added to the media to prevent oxidation in strawberry and fruit-trees cultures (41), other antioxidant, as cystine, thiourea and citric acid may be used (191).

Among adsorbents, activated charcoal (0.5-3%) added to the medium has the beneficial effect of binding phenols, but might also adsorb plant growth regulators and other organic compounds, slowing down explant development (60,84). In a variety of species though, e.g. potato, it hastens meristem growth in shoot, and increases rooting of shoots (83). Polyvinylpyrrolidone (PVP) added to the medium at a concentration of 250 to 1000 mg/L can also adsorb phenol (208).

Finally, transfer of meristem to fresh culture media, one or few days after implanting, has been found to be beneficial since cell compounds and debris remain in the previous medium (e.g. *Pelargonium* and raspberry) (F. Marani, unpublished).

Culture Containers

Glass tubes, 10 x 120 mm in size, closed with caps permitting air exchanges, are the most commonly used containers for meristem culture. Sealing of culture vessels, with the purpose of avoiding contamination, may inhibit effective gas exchanges and cause accumulation of various gases, such as carbon dioxide and ethylene that could adversely affect morphogenesis (262,263).

Dimensions of containers can influence the type of meristem development. It was observed, for example, that under identical cultural condition, *Cassava* meristems could directly differentiate plantlets when cultured in small tubes, but in larger containers they produce callus (125).

Micrografting

When it is not possible to induce meristem development or shoot rooting in culture media, the meristem-tip can be grafted onto a rootstock. In these cases the meristem (0.1 - 0.4 mm) excised from the infected cultivar, is aseptically grafted onto the vascular ring of a decapitated virus-free rootstock, such as a seedling or a shoot propagated in culture. The procedure consists of four steps: preparation of rootstock and scion materials, grafting, culture *in vitro* of grafted plantlets, and transfer to soil.

Micrografting technique is important mainly for woody species, since in this group of plants meristem-tip culture is often difficult. The procedure was first used to eliminate different viruses from *Citrus* spp. as an alternative to the use of nucellar seedlings (195,198,230). Citrus plants regenerated through *in vitro* grafting did not show the long juvenile phase of plants obtained from nucellar tissue (198).

Micrografting technique was also applied to several fruit-tree species with variable results. Success in grafting varied markedly according to the species: 45-70% for peach (*Prunus persica*), 45% for almond (*P. amygdalus*), 40% for Japanese plum (*P. salicina*) "Red Beauty", 26% for "Stanley" (*P. domestica*) and 8% for "Pollizo" (*P. insititia*) European plums, and 10-15% for apricot (*P. armeniaca*) (120).

Survival in peach upon transplanting in soil was reported to be around 60-70% by Navarro *et al.* (199), but only 18% by others (184), although a similar success in grafting was experienced (157,183). In grapevine survival after transplanting was also low (about 10%) with respect to grafting survival (about 60%) (158).

The percentage of grafting success in *Citrus* was around 30-50% (198), and an increase up to 63 to 91% was experienced when a pretreatment of both meristems and seedlings with BAP was employed (254).

An improvement of micrografting results was also obtained when some drops of citric and ascorbic acids (100 mg/L) were added to just grafted grapevine meristem, to reduce oxidation, and few drops of IAA (1 mg/L) were delivered 24 hours later, to stimulate meristem growth (158). The application of DIECA (Na-diethyldithio-carbamate) (1g/L) on the cut surfaces to prevent oxidation, and of a cytokinin (10 mg/L), to stimulate development, were also reported in various species and had favorable effects (112). In the case of peach an application of zeatin (10 mg/L) allowed a 44% success instead of 22% (183).

Micrografting is a very useful technique also because it speeds up meristem growth, thus peach plants ready for indexing could be available two months after grafting (199). Finally, micrografting technique may also be used, for an early detection and analysis of incompatibility mechanisms in *Prunus* heterografts (112,113).

Temperature

Although temperature requirement for growth of cultured meristems are supposedly similar to those of mother plants, most species are satisfactorily grown at a constant temperature of 20-24°C.

Temperature need is often similar within the same botanical family but is mainly dependent on the species, and the best development is obtained taking into account this parameter. A constant temperature of 26°C together with a 16 h photoperiod at 3 to 4 Klux is satisfactory for meristem development of various legumes (chickpea, soybean, cowpea, bean, peanut), sweet potato and coffee (121). Similar temperatures are adopted for shoot-tip cultures of various *Rosaceae* such as apple, strawberry (25-27°C) and (22-28°C) and *Pyrethrum* (25°C), whereas lower temperatures are required for *Brassica oleracea* groups (20-24°C), and carnation (22°C) (257). Bulbous ornamental plants, in particular iris, require even lower temperatures (10-12°C) to form shoots (28). For some species a decrease in temperature after shoot formation is beneficial for rooting, e.g. cauliflower, and *Pyrethrum* (18 and 17°C respectively: 280,282), and rose (5°C for 1 week: 130).

Alternated temperatures, simulating day and night conditions in nature, assure in some case a much better meristem development. *Cassava* and tomato meristem, for example, generated 100% shoots when exposed to 20 and 15°C alternating temperatures using a 16 h photoperiod at 4 Klux, whereas only 40 to 50% was the yield of shoots obtained at 26°C constant temperature (122). In pea, 100% meristems differentiated shoots if grown at 20°C in daytime and 15°C at night (124), but 80% was the yield for those cultured at 26°C (126). Diurnal cycles are also used for *Citrus* spp. (16), chestnut (270), grape (55) (see review by Styer and Chin, 257). Alternated temperatures seems to be beneficial also for roots formation. In *Chondrilla juncea* maximum rooting of shoots was obtained at 21-27°C in daytime and 16-22°C at night. It appeared that the higher day temperature was essential for cambium formation, and the lower temperature necessary for differentiation of cambium into root primordia (192). Conversely, meristems of species as cowpea do not develop well at alternated temperature (20 to 15°C): only 25-30% regenerate shoots (122).

Interaction of temperature effects with light influence and hormone composition of the substrate is also to be taken into account.

Besides growing temperatures, for some groups of plants like the bulbous ones, a lowering of temperature before transplanting is useful to avoid dormancy, thus for certain *Lilium* cultivars for example, exposure at 5°C for a prescribed period will allow plantlets and bulbs produced in tissue cultures to develop right away upon transfer to soil (192). Gladiolus corms and plantlets do not develop in soil unless previously exposed at 2°C for 4 to 6 weeks (99).

Light

Light affects meristem organogenesis in a complex way through its various characteristics, involving exposure time (photoperiod), intensity and wavelength, but few studies are available on the separate effect of these parameters. Most of the times, a 14 to 18 hours of daylight with an intensity of 1 to 4 Klux of a cool white fluorescent lamp, either alone or implemented by enrich-red fluorescent light (Gro-lux) are adopted, even though a differentiated use of these parameters according to plant species and to type of wanted morphogenesis will improve, or even elicit, meristem development.

Light requirement for morphogenesis varies among taxa (108), and although differences normally exist between cultured tissues (heterotrophic) and autotrophic whole plant, one may suppose that the former have similar requirements as the latter. It is safe, therefore, to assume that photoperiod requirement of tissue culture resembles, somehow, that of the whole plant (191). Because photoperiod and light intensity are interdependent requirements, they are often reported together in the literature for different plant species, whereas wavelength requirement are similar for all plant species and, supposedly, cultured tissue.

According to Murashige (192), the suggestion of various experiments on shoot formation carried out on *Plumbago* spp. (201) and *Brassica* spp. (154), is that short exposure periods are sufficient when relatively high light intensity is used, e.g. 9 hours at 4 Klux for *Brassica*, and vice versa. For various species belonging to different families a light photoperiod of 16 hours with a light intensity of 1-4 Klux is suggested for shoot formation: *Brassica oleracea* (var. *capitata*, *gemmifera*, *botrytis*), *Dianthus caryophyllus*, strawberry, *Prunus*, *Rosa*, potato and monocotyledons such as banana and *Asparagus* spp. (257). In some of these plants it is beneficial to deliver a lower light intensity at shoot initiation and to increase it later for shoot development: 1 to 3 Klux in globe artichoke, 1-4 to 10 Klux in *Dianthus caryophyllus* (257). For root formation, a decrease in light or even a dark period is beneficial: *Chrysanthemum cinerariaefolium* likes a decrease from 3 to 1.4 Klux, while *Rosa* and strawberry require full dark, which is not surprising since roots are underground organs and, like potato tubers and bulbs, are favored by very low light intensity besides a reduced photoperiod (257).

As it is well known, photosynthesis in plants occurs preferentially at specific wavelength bands where chlorophyll a and b absorb it, i.e. in the red (680-650 nm respectively) and blue (440-470 nm) regions. In tissue culture it seems that these two light bands preferentially direct the morphogenesis toward formation of adventitious roots (143) and shoots (236,237), respectively. In pea, however, lateral root formation is inhibited by red light, and the inhibition is negated by far-red light; the phenomenon is phytochrome mediated and reversible (85). It seems, therefore, that formation of lateral and adventitious roots have different requirement.

Various experiments were carried out to determine the type of lamps to be used in different species. For potato meristems, the effect of three different light regimes were investigated (206): sources rich in red light, sources approximating daylight, and a 2:1 combination of both. Under mixed light, the best rooting and most vigorous plants were obtained. For pea and cowpea meristems it was found that the critical point of light spectrum for shoot development is in the blue region, while root initiation is stimulated by red light (121,122). Therefore "Gro-lux" lamps or combination of white fluorescent and incandescent lamps should be ideal for regeneration of whole plants (121). In our experience Fluora 77R (Osram) with prevalent band emission in the red and blue regions (3 Klux), using a 16 h photoperiod, resulted in a complete and satisfactory morphogenetic response by 4 leaf primordia meristems of various potato cultivars (79).

Transplanting of Plantlets

When shoots have developed satisfactory root systems, they are transferred from culture vessels to well-drained soil. Careful attention must be taken during transplanting, in order to preserve roots from damage: removal of agar medium sticking on them, to avoid future contaminations by fungi and bacteria, should be done by gentle washing. The most important requirement in successive operations, will be the maintenance of plants under a very high humidity (90-100%) for the first 10-15 days. Plantlets are then generally covered with clear plastic bags or kept under intermittent mist, right after potting. The humidity is gradually decreased to allow plants adjustment to normal conditions, and 4-6 weeks later they are ready to grow under normal greenhouse conditions. Plants forming resting organs such as tubers, corms, rhizomes and bulblets should be induced to produce these organs in culture, in order to overcome this critical phase.

The most important condition for successful transplanting is to maintain the plants under a very high humidity since the cuticle is generally poorly developed in plants grown in test tubes and when transplanted to soil water loss through cuticular evaporation usually occurs (259). During the first few hours of acclimatization, plantlets undergo significant water stress: stomata do not properly operate since in tissue culture closure mechanism is not operative, poor vascular connections are generally present between shoots and roots (285), and roots of plantlets grown *in vitro* are deprived of hairs.

INDEXING

After transferring regenerated plantlets from the culture medium to soil, they must be indexed to ensure that they are virus-free. Methods for virus testing depend on the species involved and on the infecting virus. Methods of grafting indicator plants, transmission to susceptible hosts by mechanical sap inoculation, electron microscopic examination of sap material and various serological tests have been used.

Increased availability of antisera against plant viruses and the introduction of ELISA in its various forms, and of immunoelectron microscopy (ISEM) have provided methods of greater sensitivity. Other methods used, in particular for viroid detection, are nucleic acid extraction from infected plant, and identification by polyacrylamide gel electrophoresis (PAGE) and nucleic acid hybridization. Plantlets are usually indexed a few times to make sure that they are virus-free. This is because, after meristem-tip culture, the concentration of virus may fall below the level of detection and time is needed for the virus build up. For mother plants kept in the

greenhouse, periodic testing should be carried out before any further multiplication. ELISA is the serological technique most used today. In its' more prevalent version, DAS (double antibody sandwich, 57), it requires the preparation or purchase of a different antibody-conjugate for each virus and takes about two days to provide results (268,269). Several samples can be examined on a microtitre plate, and quite a few of them can be processed together and examined with automatic readers. In our experience this method is rather sensitive, being able to detect various potato virus (PVY, PVX, PVS, PVM) at very low concentrations (approximately 1 ng) (59,78,228).

Several "indirect" ELISA methods have also been developed in order to use one universal antibody conjugate for all viruses. The "CIq" method involves trapping of anti-virus IgG in the immunoplate with the CIq portion of the serum complement and anti-rabbit IgG produced in another animal conjugated with a detection enzyme (264,265). In a similar method (PAS-ELISA) wells are first coated with protein-A which will trap IgG. Protein-A conjugated with enzyme is again used as revealing reagent. In a third method, the F(ab')2 part of IgG globulin is used, instead of the entire IgG, together with a conjugate of protein A and detection enzyme which specifically binds to the Fc region of IgG (15).

Other applications, involve the use of fluorescent or radiolabelled antibodies which confer greater sensitivity but are less used in practice. An improvement in sensitivity was obtained using a dot-ELISA technique on nitrocellulose membrane (13), by which we have detected, as already reported in another section, as low as 1 pg of potato viruses. The only problem when using plant sap material instead of meristem, is that the green pigment can interfere with proper staining reactions, but this can be overcome by diluting sap and "bleaching" nitrocellulose supports, after completion of the reaction, with sodium hypochlorite.

A practical version of ELISA that can be carried out with the same sensitivity of DAS-ELISA is the "squash blot" (169), that we have modified by using commercial PVC strips to directly trap the virus and carry out the reaction as in dot-ELISA (77). The method is fast, simple and labor saving in detecting PVY, PVX, PVS, and PVM but not PLRV.

Direct use of DAS-ELISA on *in vitro* material to detect viruses, was reported to be successful by various researchers.   necrotic ringspot virus (PNRSV) in sweet cherry was detected in leaves or cultured shoot (249), and alfalfa mosaic virus (AMV) (274), grapevine viruses (17), PVX and PVS (86), and prune dwarf virus (PDV) (23) were similarly detected. In our experience it is not always safe to test only the material *in vitro* for the reason specified above. In the case of PVS, for example, some potato plantlets that were virus free *in vitro* by DAS- and dot-ELISA when tested one month after transplanting appeared to be virus-infected. Double checking is, in our opinion, highly desirable.

Immunosorbent electron microscopy (ISEM) is also a very sensitive technique that allow, by antibody coating of electron microscopic grids, the trapping of limited virus amounts (168), and the specific identification of different viruses by the so-called "decoration" procedure. This consists in exposing virus particles to an antiserum so that the antibodies stick to the particles in a readily visible way (167). To exemplify the enhanced sensitivity of this method, with respect to ordinary electron microscopy, the results of Allen and Anderson (4), who found a 105 increase in concentration of LSV particles on antibody coated grids, can be reported.

Nucleic acid hybridization has been used to detect viruses in the nanogram range (58) by spotting infected sap on nitrocellulose and detecting the virus by hybridization with a probe (cDNA or cRNA) complementary to it. The use of ^{32}P or digoxigenin labeled probes in conjunction with exposure of the hybridization membrane to X-ray film are required for detection (22). In our experience molecular hybridization for PVX and PLRV detection is comparable to DAS-ELISA but less efficient than dot-ELISA . It can surely find a proper application on viroid detection (14) for obvious reasons, being more sensitive than PAGE electrophoresis of viroidal RNAs (118,119,181).

Recently the adoption of biotin-labelled probes and of the system fluorescein conjugated with avidin (140), has overcome the safety problems due to the use of radioactive probes.

For a more complete review of virus detection methods see Jones and Torrance (116), and related chapters in this book.

VIRUS ELIMINATION

Besides all factors regulating morphogenetic process of meristems cultured *in vitro* previously described, virus elimination

depends on meristem size, type of virus and virus strain, plant species and cultivar, physiological condition of mother plants and meristem position on it. Among these factors, of great importance for determining success in virus elimination is the meristem size to be taken in relation to species and virus involved. Observations reported by Mori and Hosokawa (179) and by Faccioli *et al.* (77,78) clearly showed that in each plant-virus combination exists a specific apical "immunity" regarding a certain meristem size. Thus before starting a successful elimination program, virus distribution in meristem tip have to be precisely determined. Although this survey is the most correct way to adopt, only few researchers followed it. Most work done in this field was based on the empirical approach of using meristems of various sizes, and often very small, with the result of having very low percentages of development. Thus results obtained must be considered taking this behavior into account.

Meristem and Micrografting Culture

A correlation between meristem size and virus elimination has been found in carnation by various authors. The percentage of virus-free plantlets obtained, culturing meristems of size ranging from 0.1 to 1.0 mm, varied from 67% to nil (256). Similar results were obtained by Kowalska (135) who with 3 groups of carnation meristems (0.4, 0.4-0.7, 0.7-1.0 mm) found an increase of infection (6 to 100%) proportional to meristem size. Various elimination results were reported for CarMV, also in relation to the cultivar: 0.2 mm-long meristems belonging to 4 cultivars gave elimination percentage ranging from 15 to 84% (91,149). In carnation it was also found that some viruses were more easily eliminated than others. With meristems smaller than 0.1 mm, elimination percentages for CarMV, carnation latent virus (CLV), CRSV and carnation vein mottle virus (CVMV) were 10%, 71%, 83% and 89%, respectively (256). Complete elimination for CarMV was obtained in different carnation cultivars using longer meristems (0.2 mm) after 8 weeks at 38°C (91), and 95% elimination from CarRSV with 3 mm meristems following heat treatment at 36°C for 4 weeks.

For chrysanthemum infected with filamentous viruses, low elimination percentages: 10% (152) and 13% (103) were reported, whereas 40% elimination was obtained by Dunez and Monsion (74) and 98.5% by Hollings and Stone

(103) after heat treatment.

A high percentage (100%) of gladiolus plantlets free from both bean yellow mosaic virus (BYMV) and CMV was obtained culturing apical meristems of 0.5-0.7 mm (146). BYMV elimination varying from 63 to 100%, in relation to the gladiolus cultivar has also been reported (30).

Meristem tip culture was also suitable to produce lilies free of lily symptomless virus (LSV): 53 to 100% elimination depending on the cultivar (28).

Sometimes particular techniques were coupled with meristem culture to improve results, as in *Cymbidium* orchids infected with Odontoglossum ringspot virus (ORSV): virus-free plantlets were obtained by treating excised apical meristems with virus-specific antiserum (110); in the experiment, meristems 0.1-0.3 mm in length, immersed for 1 hour in ORSV-antiserum (1:2048), yielded 52.3% plants free of virus.

Early reports on potato virus elimination indicated a differential behavior: excising meristems of 2 leaf primordia assured a 100% of regenerated plantlets free of PLRV, PVY and PVM, but 1/3 still contained PVX and PVS (1). By culturing 0.1 mm meristems with or without a leaf primordium, high elimination percentage of PVX and PVS (95%) was obtained, even though only 20 plantlets were regenerated from 196 meristems (128). Among potato viruses, potato virus A (PVA) and PVY are easily eliminated from 85-90% of excised meristems smaller than 0.3 mm long (178), whereas other potato viruses such as PVX and PVS were not eliminated, probably because of their capability to invade meristems. With four leaf primordia meristems, PLRV and PVY were totally eliminated from potato (79,81).

In garlic the percentage of plantlets free from filamentous viruses varied from 70 to 85% depending on selections, which were also found to be differently suitable for *in vitro* meristem culture, since survival in some selections was lower than 30%, and in others higher than 60% (31). Different results in elimination (5 of 7 selections were completely free), were obtained by Bhojwani *et al.* (34) with similar meristem survival rate.

In the attempts to eliminate cassava mosaic disease in *Manhiot esculenta*, it was found that up to 60% of the plantlets regenerated from meristems, 0.4 mm in length, were virus-free, while all the plants obtained from meristems exceeding 0.4 mm were still infected (125).

Pea seed-borne mosaic virus (PSbMV) was eliminated with high frequency (90-100%) from more than 100 breeding lines of *Pisum sativum*, by culturing shoot apical meristems 0.4-0.5 mm in length (122).

In strawberry, 100% elimination was reported for different viruses when meristems excised were 0.1-0.3 mm and culture medium contained BAP (0.1 mg/L) as cytokinin (41). Using this meristem size, strawberry crinkle virus (SCrV) was completely eliminated (40), whereas for 0.5-1 mm meristems, elimination was 56% (156).

An inverse correlation between meristem size and percentage elimination was also found by McGrew (162) working on pallidosis: elimination was 100% for 0.2 mm meristems, whereas it was 95% and 50% for 0.5 and 1 mm meristems respectively.

For different *Citrus* species, shoot-tip grafting was effective to produce plants free of tristeza (CTV), exocortis (CEVd), psorosis, concave gum, impietratura, xyloporosis, vein enation and stubborn, making it possible to eliminate several agents that could not be eliminated by thermotherapy. Using shoot tip meristems of 3 leaf primordia, 80-100% of plants were freed from most diseases (197). Similar results were reported from Italy: 85% elimination from CEVd, 88% from a mild strain of it and 87% from psorosis (253,254). Plants free from CTV were readily obtained by shoot tip grafting (198).

A particular type of apical culture, called "fragmented shoot-apex" was adopted by Barlass *et al.* (17) to regenerate virus-free grapevines. This involved, fragmenting with a scalpel shoot apices containing 2 to 3 leaf primordia (1 mm in length) and culturing in a liquid medium the clumps of derived cells, which developed into leaf-like structures. A 100% success in elimination was obtained growing fragments at 27°C/day and 20°C/night for grapevine leafroll virus (GLRV), yellow speckle, fleck and summer mottle diseases, but not for Grapgvine fanleaf virus (GFLV) (17). GLRV in particular, was eliminated using 0.2 to 0.3 mm meristems (111).

Micrografting on grapevine seedling *in vitro* (19,20,200,202,271) or on shoots propagated and rooted *in vitro* (9), was useful in obtaining disease-free grapevines.

Virus elimination in fruit-trees was obtained both with meristem-tip culture (41,42) and shoot-tip micrografting *in vitro* (71,112,199). Much work has been done for *Prunus* species (42,217), while only few reports deal with apple virus elimination (107).

Elimination depends not only on the type of virus involved but also on the virus strain: in general closterovirus are easily eliminated (90-100%), probably because of their phloematic location, whereas PNRSV and PDV are more difficult to be eliminated (199). The use of meristem tips 0.3-0.8 mm in length, resulted in the complete elimination of apple chlorotic leaf spot virus (ACLV), partial elimination of PDV (90%) and non elimination of PNRSV from two cultivars of *Prunus cerasus* (41). Use of micrografting technique in peach, resulted in a highly variable elimination of different strains of PNRSV (0-92%); for PD, 0% and for ACLSV 100% elimination were obtained using this technique (199). These differences in the successful elimination of PNRSV and PDV strains may be due to a different ability to enter meristem tissues. A correlation between growth conditions of mother plants from which meristems are excised and virus elimination by micrografting was found in peach (199).

Grafting technique *in vitro*, often associated with thermotherapy, was applied to different fruit-tree species such as apricot and plum (157), peach (183), cherry (72), apple (107) and grapevine (111) to obtain virus-free plants. In fact, infected peaches grown in greenhouse at 25-33°C gave better results than plants kept at 10-27°C. In the first case, elimination of PNRSV varied from 60 to 77%, in the second from 5 to 20%; for PDV a 73-95% elimination vz. 0-73%, for the above mentioned cobnditions, were obtained, while for ACLV no differences were found (199). For cherry , virus-free plants were obtained by applying high cytokinin concentrations to the medium after 8 to 9 subcultures in order to induce intensive vegetative multiplication of sweet cherry explants infected with PNRSV, PDV and ACLV (70). These results suggest that hormone balance in the media may have an inhibitory effect on virus replication as suggested by Quak (214). The recovery of virus-free plants, using a simple method of intensive shoot propogation, seems to be possible provided that this procedure does not favor mutations. The main results of virus elimination obtained by meristem-tip culture are summarized in Table 2.

Thermo- and Chemotherapy

To improve pathogen elimination, other techniques such as thermo- and chemotherapy coupled with meristem-tip culture can be used.

Thermotherapy can be carried out on mother plants, before meristem excision, or during the *in vitro* culture of meristem tips. The first type of application is less damaging and has also the advantage of allowing the use of somewhat larger meristems, which more easily develop into plants, since the treatment contrast virus multiplication (44,251,283).

Although mechanisms providing virus elimination due to heat treatment are not completely known, it seems that the most important are the slowing down or cessation of virus replication and/or virus transport in the plant (232). It was also supposed that alterations of cellular biochemistry due to heat bring about the formation of o-quinones, activation of ribonucleases, inhibition of replicases and reduction in ribosome number (160).

Better percentages of pathogen elimination were obtained in several cases, when the meristem tip alone did not give satisfactory results. In strawberry, for example, SMYEV elimination was about 82% when a heat treatment at 36°C for six weeks was used against a 25% without it (0.3-0.8 meristem length) (188). For strawberry latent A and strawberry mottle viruses and a mixed infection of the two, an increase in virus-free material produced was found, from 17 to 71%, 35 to 100% and 0 to 17%, respectively, when 0.4-0.8 mm meristem tips were heat treated for 5 weeks at 38°C (250).

With this procedure several potato cultivars were freed from PVS and PVX (147,251). High yield of PVS-free Majestic plants (93%) was obtained from meristem tips of 4 leaf primordia, sampled from plants heat treated at 35°C for 26 days. The treatment caused a reduction in virus content to less than 10 pg in the majority of meristems (85%), while untreated meristems contained normally from 10 to 100 pg of virus (27). Longer heat treatment did not improve yield of virus-free plants and caused a reduction in the percentage of developed plants.

In general, one can say that effectiveness of heat treatment is related to its length, but up to a certain point. In *Chrysanthemum*, for example, increasing exposure from 10 to 30 days resulted in an increase of virus-free plants from 9 to 90%, but a longer treatment (40 days or more) did not improve the results and caused a marked decrease in the number of developing meristems (95). Thus, the shorter treatment was more effective.

Direct heat treatment in tissue culture is also effective, although meristem tips are more

Table 2. Virus-free plant species obtained by meristem-tip culture and tip micrografting

Plant species	Virus eliminated[a]	References
Allium sativum (garlic)	GMV	33
	OYDV	218
	GYSV	34
Ananas sativus (pineapple)	unspecified	33
	mosaic	33
Armoracia rusticana	TuMV	33
A. lapathifolia (horse radish)	CaMV,TuMV	33
Asparagus officinalis	AV2	32
	unspecified	33
Brassica oleracea (cauliflower)	CbBRSV	33
	TuMV, CaMV	33
Buddleia davidii	CMV	33
Caladium hortulanum (aroid)	DsMV	33
Chrysanthemum spp.	CCMVd	33,203
	CSVd	33,137
	TAV	33,74
	CVB	33,152
	CVMV	33
	complex of viruses	33
Citrus spp.	CEVd, psorosis	253
	CTV, CVV, psorosis,	197,198
	concave gum, vein	197,198
	enation, yellow vein	197,198
	sweet mottle, cachexia,	197,198
	CEVd	197,198
Colocasia esculenta	DsMV	33
Cymbidium spp.	CyMV	33,110
Dahlia spp.	complex of viruses	33
	DMV,TAV,CVMV,CVB	33
Daphne spp.	ArMV,CMV,RRV	33
Daphne odora	DVS	33
Dianthus barbatus	CLV,CarMV,CRSV,CVMV	256
D. caryophyllus (carnation)	CERV	33
	CLV	33,256
	CarMV	33,135,149,256
	CRSV,CVMV	33,256
	complex of viruses	33
	unidentified	33
Forsythia intermedia	unspecified	33

Table 2. (cont.)

Plant species	Virus eliminated[a]	References
Fragaria spp.	Complex of viruses	33
(strawberry)	SCrV	33,166
	latent A	33
	latent C	33
	mottle	33,166,188
	pallidosis	162,188
	SMYEV	33,188
	SVBV	33,166
	yellow virus complex	33
	vein chlorosis	33
Freesia spp.	FrMV	33
	freesia virus 1	33
	phaseolus virus 2	33
Gladiolus spp.	BYMV,CMV	146
	BYMV	30
	unidentified virus	30
Glycine max	SMV	33
(soybean)		
Hippeastrum spp.	mosaic	33
(amaryllus)		
Humulus lupulus	HpLVd	33
(hop)	necrotic ringspot	33
Hyacinthus spp.	HyMV,LSV	33
Hydrangea macrophylla	HRSV	33
Ipomoea batatas	SPFMV	33
(sweet potato)	internal cork	33
	rugose mosaic	33
	unidentified	33
Iris spp.	IMV,	33
	IMMV,ISMV,NLV,BYMV	27
	unidentified	33
Lilium spp.	CMV	33
	HyMV,LSV,LMV	33
	latent	33
	LMV	33
	LSV	29
	unidentified	33
Lolium multiflorum	RGMV	33
(ryegrass)		
Malus spp.	latent viruses	33
(apple)	ASGV	107
Malus pumila	ACLSV	33

Table 2. (cont.)

Plant species	Virus eliminated[a]	References
Manihot spp. (cassava)	ACMV,CBSV mosaic	33 123
Musa spp. (banana)	CMV, unidentified	33
Musa acuminata x M. balbisiana	mosaic	33
Narcissus tazetta	ArMV narcissus degeneration NYSV, NLV	33 33 207
Nasturtium officinale (watercress)	CMV,CaMV,TuMV	33
Nerine spp.	nerine latent unidentified	33 33
Nicotiana rustica	CMV, AMV CLRV, ArMV,TRSV,CMV	33 33,278
N. tabacum	TMV	33
Ornithogalum	OrMV	33
Pelargonium spp.	CMV,TBRV TomRSV unidentified	33 33 33
Petunia spp.	TMV TNV	33 275
Polyanthes tuberosa	mosaic	33
Pisum sativum	PsbMV	121
Prunus spp.	ACLV, PDV, PNRSV PDV,PNRSV PNRSV	41,70,107,199 184 249
Rheum rhaponticum (rhubarb)	TRV,CMV,CLRV,SLRV, TuMV	33 33
Ribes grossularia (gooseberry)	vein banding	33
Rubus idaeus (raspberry)	mosaic	33
Saccharum officinarum (sugarcane)	mosaic	33
Solanum melongena	EMCV	33
Solanum tuberosum (potato)	PLRV paracrinckle PAMV PVA PVG	33,81 33 33 33 33

Table 2. (cont.)

Plant species	Virus eliminated[a]	References
S. tuberosum (cont.)	PVM	33,227
	PVS,PVX	33,79
	PVY,PVX,PLRV	33,80
	PSTVd	145
Vitis vinifera (grapevine)	GFLV	111
	GLRV, yellow speckle,	17
	fleck, summer mottle	17
Zinziber officinale (ginger)	mosaic	33,283

[a] ACMV = African cassava mosaic virus, AMV = alfalfa mosaic virus, ACLV = apple chlorotic leafspot virus, ASGV = apple stem grooving virus, ArMV = arabis mosaic virus, AV2 = asparagus virus 2 BYMV = bean yellow mosaic virus, CbBRSV = cabbage black ringspot virus, CBSV = cassava brown streak virus, CERV = carnation etched ring virus, CLV = carnation latent virus, CarMV = carnation mottle virus, CRSV = carnation ringspot virus, CVMV = carnation vein mottle virus, CaMV = cauliflower mosaic virus, ChLRV = cherry leafroll virus, CSVd = chrysanthemum stunt viroid, CVB = chrysanthemum virus B, CCMVd = chrysanthemum chlorotic mottle viroid, CEVd = citrus exocortis viroid, CTV = citrus tristeza virus, CVV = citrus variegation virus, CMV = cucumber mosaic virus, CyMV = cymbidium mosaic virus, DVS = dafne virus S, DMV = dahlia mosaic virus, DsMV = dasheen mosaic virus, EMCV = eggplant mottle crinkle virus, FrMV = freesia mosaic virus, GMV = garlic mosaic virus, GYSV = garlic yellow streak virus, GFLV = grapevine fanleaf virus, GLRV = grapevine leafroll virus, HpLVd = hop latent viroid, HyMV = hyacinth mosaic virus, HRSV = hydrangea ringspot virus, IMMV = iris mild mosaic virus, IMV = iris mosaic virus, ISMV = iris severe mosaic virus, LMV = lily mosaic virus, LSV = lily symptomless virus, NLV = narcissus latent virus, NYSV = narcissus yellow stripe virus, OYDV = onion yellow dwarf virus, OrMV = ornithogalum mosaic virus, PsbMV = pea seed-borne mosaic virus, PAMV = potato acuba mosaic virus, PLRV = potato leafroll virus, PSTVd = potato spindle tuber viroid, PVA = potato virus A, PVM = potato virus M, PVS = potato virus S, PVX = potato virus X, PVY = potato virus Y, PNRSV = prunus necrotic ringspot virus, PDV = prune dwarf virus, RRV = raspberry ringspot virus, RGMV = ryegrass mosaic virus, SMV = soybean mosaic virus, SCrV = strawberry crinkle virus, SLRSV = strawberry latent ringspot virus, SMYEV = strawberry mild yellow edge virus, SVBV = strawberry vein banding virus, SPFMV = sweet potato feathery mottle virus, TMV = tobacco mosaic virus, TNV = tobacco necrosis virus, TRV = tobacco rattle virus, TAV = tomato aspermy virus, TBRV = tomato black ringspot virus, ToRSV = tomato ringspot virus, TuMV = turnip mosaic virus.

sensitive to heat. With this type of application CMV and AMV were eliminated, in 4-5 days at 30°C or in 9 days at 45°C, from meristem tips of Nicotiana rustica (274,276). GLRV and the agent of grape vein necrosis were also similarly eliminated (189). GFLV was eliminated from culture of grapevine fragmented shoot-apex by heat treatments at 35°C, while cultures held at normal temperatures (25°C day/20°C night) were all infected (17). The same temperature was used for four weeks to free shoot-tip cultures of sweet cherry from PNRSV; elimination was complete but only 40% of tips survived (249).

Since prolonged exposure to heat may damage host tissue (275), the use of alternated temperature proved useful in various cases by giving an increased elimination effect. Inactivation of CMV in Nicotiana rustica tissue was obtained by heating daily at 40°C for 16 h and at 22°C for 8h, or for two days at 40°C followed by two at 35°C (278). More recently this method was

successfully applied to grapevine (170) and Narcissus (46) to eliminate GFLV and ArMV, respectively. In the later case, viruses were eliminated from 0.2 mm shoot tip cultures held 6 h at 22°C and 18 h at 39°C for 40 days.

Low temperature could also be utilized to free tissues from viruses before meristem sampling - PVA and PVY were eliminated from potato tissues grown at 5°C (185) - but even more from viroids which develop well at high temperature: above 30°C and under high light intensity they accumulate in high concentration (182,231). PSTVd was eliminated from potato by cooling at 5°C for 3-6 months with a yield of 53%, probably consequent to a slowing down of its replication (145). Comparatively, two cycles of heat treatment at 33-35°C and subsequent meristem tip culture yielded a much lower percentage of free plantlets (2.4-6%) (25). CSVd was also effectively eliminated by low temperature treatments and meristem tip culture from Chrysanthemum (203).

The use of meristem tips of increasing size (0.2, 0.25, 0.5 mm), produced also in this case, a reduction of viroid elimination: 66.7, 22 and 14%, respectively (202,204).

Chemotherapy can also be applied to both meristems cultured *in vitro*, or to plants before meristem sampling. In the first case the chemicals are directly incorporated into the culture medium, and can affect meristem growth quite a bit. Several substances have being tried for the purpose (for a review, see Schuster (234), but only few of them have shown an exploitable effect. The most used antiviral substance is the synthetic analog of guanosine, Ribavirin=Virazole (1-b-D - ribofuranosyl - 1,2,4 - triazole -3- carboxamide). Incorporated into the media in the amount of 10 to 50 mg/L, it has been quite effective against infections of various potato viruses (PVX,PVY,PVS,PVM) in potato and tobacco (51,52,53,131,132), and also against CMV in meristematic culture of *Nicotiana rustica* (244), when used at 50-100 mg/L, although incubation of 127 days was required to free the culture. Effective elimination of ACLV (100%), in apple meristem tissues cultured for two months on a substrate containing 10-20 µM ribavirin, was experienced with reduced phytotoxic effects (slight chlorosis) (96).

Results for some bulbous viruses were less promising. For example LSV and tulip breaking virus (TBV) were partially eliminated (61.4% against 35.4% of the control) from meristems adventitiously formed on infected bulb-scale explants cultured on 10-40 µM virazol (non toxic dose), and only for one variety out of two of *Lilium longiflorum* (35). The presence of this chemical did not reduce the percentage of HyMV present in plants derived from meristems adventitiously formed on bulb-scale tissue culture of hyacinth, to which up to 40 µM of virazol was added (36).

One problem that arises with ribavirin is the phytotoxic effect that this substance can exert on plant tissue at active doses, which vary with plant species. This normally results in a noticeable increase of culture time, and therefore requires many transplantings to fresh media, but it may also affect meristem development. For example with concentration of 190 mg/L for elimination of PVX in potato culture, development of meristems required 6-8 months (131). Regeneration of potato sprouts from meristematic cultures was delayed from 6 to 8 weeks, for ribavirin concentration ranging from 10 to 50 mg/L (53).

Tentatives to overcome toxicity problems were carried out by supplying for two weeks ribavirin to hydroponic cultures of potato stem cuttings infected by PVX, PVS and PVM, before meristem sampling, with positive results (79). Culturing meristems of four leaf primordia, sampled from cuttings treated with 5 mg ribavirin + 5 mg of DHT (2,4-dioxo-hexahydro-1,3,5-triazine), 90% of meristems generated plantlets and 74% of the progeny was PVS-free. Dot-ELISA determinations showed that about 90% of the meristems had less than 10 pg of virus. Use of either one of the two chemicals alone (10 mg/L DHT or 5 mg/L ribavirin) had a lower effect, yielding only a 67% of meristem having less than 10 pg of PVS. Similar results were experienced for PVM, where around 69% of virus-free plantlets against 39% of the untreated control, were obtained. The simultaneous use of the two chemicals was beneficial, because at a concentration higher than 5 mg/L ribavirin had toxic effect, delaying also meristem development, whereas DHT is a base analog that had no toxic effect at the dose-range normally used (59). Cyanoguanidine, another base analog, gave similar results (61.4% free plantlets vz 34.7%) against the same virus (233).

Strawberry mottle virus and SCrV were absent in about 65 to 74% plantlets of 3 different strawberry cultivars, regenerated from 0.5 mm apices grown on Gamborg's medium containing 0.2 mg/L of DHT (133). 2-thiouracil did also increase from 29.8 to 71.2% the yield of plantlets free from SCrV when administered to strawberry apices (133). For a review on these chemicals see Hecht and Diercks (97).

Other chemicals were used both *in vitro* or hydroponic culture of PVS-infected potatoes with limited success. Cycloleucine (0.1-0.5 mM) was toxic both for cultured meristems and stem cuttings, giving only around 3-11% of developed plantlets, 33% of which were virus-free, with respect to 15% of the controls. 3-Deazauridine (0.05-0.1 mM) gave slightly better but still not practical results (59,78).

Other antimetabolites such as 8-azaguanine (Aza G), 5-fluorouracil (FU), 2-thiouracil (TU) and p-fluorophenylalanine (FPA) were used by Pennazio (205) in the culture media of PVX-infected meristem tips: none of them was effective at doses that did not block meristem development. Another almost ineffective antimetabolite was amantadine that added to the culture media (100 mg/L) to free culture of CSVd gave only 10% success (106). Against *Narcissus* viruses, unsatisfactory results were obtained using

6-azouracil and vidarabine *in vitro* for 3 narcissus cultivars, because both were toxic, and only in one cultivar they reduced narcissus virus Q (NVQ) content (207).

It is finally interesting to notice that when none of the three eradicating methods, either alone or combined by the two, were successful in eradicating a virus, all three together could be applied with good results. This was the case of peanut mottle virus (PeMoV) in *Arachis hypogaea*. Meristem-tip culture of heat treated plants at 35°C for 45 days generated infected plants, and the same was true for tips grown on medium supplemented with ribavirin (5 to 20 mg/L); when all three procedure were used though, 80 to 100% virus-free progenies were obtained. This suggest that quite a few "difficult" virus infection could be probably eliminated in this way (56).

Effects of Virus Elimination

Virus elimination through meristem-tips can exert various effects on the derived cultures and their production. Besides the supposed increase in yield of virus-free plants derived from virus infected material, other aspects such as variations in susceptibility (or resistance) towards viruses and other pathogens, and variations of genetic characteristics should be carefully taken into account.

The real benefit in obtaining virus-free plants is the increase in quantity and quality of the products obtained, since viruses cause great losses. Their effects are different in relation to the virus, the species and cultivar, and may be related to plant growth, number and quality of flowers, weight and number of fruits, tubers and bulbs, etc. We therefore think that, reporting few examples of crop losses due to virus infections that could be eliminated by meristem-tip culture, will be helpful in understanding the potentiality of this technique in improving crop yields.

In tulips, TBV may cause a reduction in bulb weight from 5 to 40-50% depending on the severity of infection and the cultivar; TNV may cause about 10% decrease in bulbs weight. In bulbous iris, yield difference between virus-free and double infected plants by IMMV and NLV, ranged between 24 to 34% (8). In TBV-infected lily bulbs, weight may be reduced from 10 to 35%; LSV-free plants of cultivar Enchantment were 2/3 taller than the infected ones, their production was doubled and flowers were 12% bigger (165).

Flowering of virus-free *Lilium*, was delayed by 6 days and required, when grown in pots, 2-3 times more growth retardants. Virus-free narcissus plants produced significantly more flowers of a better quality (47). In rhubarb the effect of virus elimination was an increase of petiole yield by 60-90% (277).

More accurate estimates of yield losses were made infecting healthy plants with known viruses and comparing their production with that of virus-free plants of the same clone (164). Potato virus-free clones for example, compared with several others PVS-infected, produced 10% more and this was attributed to a higher number of tuber produced (127). PVX- and PVA-free clones exhibited a production 60% higher than infected clones (93). For some potato cultivars, crop losses due to necrotic strain of PVX were around 50%, and those due to PVS around 10-25% (26).

Viruses can often greatly reduce growth and yield of fruit crops and induce failure of bud grafts (62). Apple mosaic virus (ApMV), for example reduces production of apple cvs. Allington Pippin and Cox's Orange by 30% (211), ACLV that of pears up to 40% (213), and viruses of plum by 23-82% (212).

Some viruses adversely affect also the quality of fruits, causing sometimes a reduction of sugar content and taste (62). Others, like PNRSV, can greatly reduce hop yield, often without induction of visible symptoms, causing 20% reduction of brewing material (200).

PNRSV can also induce deleterious effects on roses, reducing the size and the number of blooms, increasing the number of deformed flowers, causing early leaf fall and loss of vigor (261). It has also been found that virus infections can severely affect pasture crops. In USA, double infection by AMV and BYMV reduced forage yields of white clover cultivars by 23-55%, their seed yields by 29-54% and their longevity from 10-20 years to 3-6 years (18,136).

When plants are virus-infected they normally react to a secondary infection differently than healthy one. Thus freeing plants from a primary infection may result in a different susceptibility (or resistance) towards other viruses or pathogens. First of all, considering infections due to strains of the same virus, it is well known that mild strains of a virus usually cross-protect against the virulent ones (161). Acquired resistance against virus-infection is also a common phenomenon which can protect plants from successive infections (76,161,225). In the above mentioned cases, therefore, plant free from viruses may become more susceptible to virus infections. On the other hand several examples are known, demonstrating that often a plant infected by a virus is more susceptible to another, or the infections of the two together damages the host more than the sum of individual effects of the two; classical example is that of PVX-PVY mixed infection in potato (67). Interactions between viruses and fungi also influence susceptibility of the host. Examples of increased resistance to fungus infections of virus-infected plants are numerous, and in these cases an increased susceptibility of the cured plant is experienced. PVX and PVY infected potatoes, for example, are less susceptible to *Phytophthora infestans* attacks (186). PVX infected tubers are less susceptible to *Fusarium roseum* (dry rot) than the comparable healthy ones (117). Many reports however indicate that virus infections increase susceptibility to fungus infections (47), and in these cases elimination give also the benefit of increasing resistance towards fungal infections. Russel (229) showed that infection by beet mild yellowing virus (BMYV) increased susceptibility of sugarbeet towards *Alternaria* spp. infection. Prior infection of barley yellow dwarf (BYDV), predispose wheat or barley to the infection of *Cladosporium* spp. and *Verticillium* spp. (3); asparagus infected by asparagus virus 2 (AV2) had increased susceptibility to *Fusarium* crown and root rot (75).

Plants obtained from meristem-tip culture generally show little or no genetical differences from the original virus-infected sources. Minor physiological changes can, however, occur in regenerated plants following virus removal: In apple, minor variations in the timing of flowering and fruiting were detected (49). In rhubarb, major changes with regard to the low temperature requirement to break dormancy were detected (50), and in garlic, a 10 day longer vegetative period (F. Marani, unpublished) were observed. A careful selection of the horticulturally most suitable clones should be done to prevent the distribution of "vitrovariants" i.e. of plants differing from the original source because of somaclonal variation. This phenomenon may result from changes in chromosome numbers, gene mutations, mutations of characteristics controlled by chloroplastic or mitochondrial DNA, or variations in the copy number of a gene. All these modifications arise mainly in the passage of plant tissues through the callus phase, but they can also arise during mass propagation following meristem culture where chance mutations are not evident. A trueness-to-type control is therefore of greatest importance

before using virus-free material as mother plants.

Some examples of somaclonal variations are reported in the literature for chrysanthemum regenerated from long-term cultures (260), grapevine (94,180), potato (241) and strawberry (43). These variations are probably frequent, but not much considered; in any case it should not be forgotten that somaclonal variations can be exploited for breeding purpose (238).

MAINTENANCE OF VIRUS-FREE STOCKS AND REINFECTION

Virus-free plants obtained from meristem-tip culture can become easily reinfected; to prevent any contamination, careful procedures based on the knowledge of virus epidemiology and type of crop should be used.

Virus-free stocks (mother plants) should be grown in sterilized soil, to avoid contamination by nematode- and fungus-transmitted viruses, in virus- and vector-free greenhouses. Successive multiplications of the material (two or three) have to be carried out in insect-proof screenhouses, and bulk multiplication in isolated areas, where chances of reinfection are minimal for the absence of virus sources and vectors. In any case, mother plants should be controlled singly and frequently, while plants multiplied in nurseries or fields for commercial purpose should be periodically indexed, randomly or in groups.

Strict hygienic measures should be adopted mainly for contact-transmitted viruses and, in any case, it is safe to store healthy stocks material also *in vitro*.

To reduce chances of reinfection in the field, pesticide treatments could be used according to the virus vectors. To reduce spread of aphid-transmitted non persistent viruses, since normal treatments are ineffective, mineral oils should possibly be employed. In bulb crops, mineral oils alone (73,243) or in combination with synthetic pyrethroids have been quite effective (7). For some plants like potatoes or tomatoes, though, decrease in production due to tissue asphyxia, caused by the oil coat, could be experienced (G. Faccioli, unpublished).

As we already mentioned, plants regenerated by meristem-tip culture can be stored *in vitro* rather than in insect-proof greenhouses, or in other protecting structures. The *in vitro* storage avoids reinfection and saves space.

Storage of nuclear stocks using low temperatures is possible for long periods of time.

Virus-free cultures of strawberry, for example, were stored up to 6 years at 4°C in the dark, adding every 3 months one or two drops of fresh medium to replace the water loss due to evaporation, and to supply nutrients (187). Possibility of storage (i.e. length of storage and percentage of survival) was tested at 2°C in 19 strawberry cultivars and varied from 9 to 27 months depending on the cultivar (66). Grape plants have been stored for over 15 years at 9°C by yearly transfer to fresh medium (175), and potato shoot-tip cultures are used to maintain germplasm collection in Peru (284). The use of minimal nutrients in the medium, the addition of inhibitors (i.e.abscissic acid) and the lowering of temperature conditions could reduce growth rates during conservation, and therefore transplanting frequency. For both potato and sweet-potato these conditions allowed growth reduction in culture, and the storage of the material for at least 1 year.

More recently, cryopreservation using liquid nitrogen has been applied to plant cells (10), and adapted to different types of tissue and species. Its' success depends on tissue differential possibility by withstanding freezing temperatures.

Meristems are preferred to callus cultures for long-term preservation of germplasm, because they regenerate plants more easily, are genetically more stable and withstand sudden freezing. Potato plants of three cultivars were regenerated for example from meristems freeze-preserved (-196°C) for 24 months (11), and one lily cultivar from meristems undergoing an adaptation period of five days at 5°C before freezing (38). Although some success has been obtained, before adopting the tool as a general preservation method, the high rate of meristem survival and the absence of genetic changes occurring during long-term storage should be ensured.

When virus-free stocks regenerated from meristem-tip cultures are grown in the field, reinfection with mechanically or vector transmitted viruses is a permanent threat. This is particularly true for bulbous ornamentals (lily, tulip, gladiolus, narcissus, hyacinth, iris), carnation, chrysanthemum and horticultural crops such as potato and strawberry, for which sanitation programs are operative in different countries.

Data on reinfection of virus-free carnations, for example, indicate a 75% reinfection by CarMV after six months, when plants were grown together with ordinary infected commercial stocks in greenhouse: CarMV is known to be easily transmitted by contact (102). In subsequent surveys on carnation reinfection, only a few CarMV-infected plants were detected in the first 10 months, and over the whole period of 16 months the rate of reinfection was very slow, reaching only 35% (104,290). Older carnations appeared much less susceptible than young plants to infection, and these observations show the importance of excluding the virus in the first few months of culture.

Surveys on virus reinfection of carnation, chrysanthemum and gladiolus crops indicated an average reinfection of carnation with CarMV of 0.95% (19.84% if plants were grown in conjunction with ordinary infected stocks), and of chrysanthemum with CVB and TAV of 9% and 7% respectively. In gladiolus plants, grown among infected gladiolus, reinfection averaged 16.5% with BYMV and 0.002% with CMV after one growing season, whereas plants grown 0.5-2 km away from infected crops did not become infected by either virus (171). These results indicate that for non-persistent aphid-borne viruses, such as TAV, CVB, BYMV, CMV, long distance spread seems to be non operative.

The reinfection rate of virus-free hops with hop mosaic virus (HpMV), which is transmitted by aphids in a non-persistent manner, was 54% after the first year and 87% after three years, when healthy plants were grown in a field near one of the infected plants (2).

For virus-free rhubarbs grown isolated, the reinfection percentage with TuMV was 10% after three years (281). In fruit-trees reinfection was reported to occur only slowly: plum pox, little cherry and chat fruit diseases have spread into nursery plants five years after releasing the tested clones (210). The reinfection from plum pox virus (PPV) and peach latent mosaic viroid (PLMVd), which are aphid transmitted, and from PNRSV and PDV, which are pollen transmitted, should be prevented using drastic pruning and flower abscission by chemical treatments, respectively (90).

CONCLUSIONS

From experiments of Morel in 1948, the use of meristem tips to obtain virus-free material has been extended to most herbaceous species, since culture *in vitro* has been developed for the majority of them. With the advent of meristem micrografting, also grapevine and fruit-trees have been successfully cultured *in vitro*, so that the technique can now be applied to almost all crops of interest (Table 2).

While for some years the applications of meristem-tip culture has been based on the assumption of meristem "immunity" towards viruses, successive researches have shown that the presence of viruses in meristematic cells is strictly dependent on the particular virus-plant species and even cultivar. Until recently, the technique was mostly applied rather empirically for virus elimination, by culturing meristems of different sizes and determining, at the end of the culture, whether or not elimination was successful, and to what extent.

With the development of more sensitive methods and devices for practical virus detection in meristem (e.g. autoradiography in phase contrast microscopy, dot-ELISA), it is now possible to establish the most appropriate size of meristem to be used, and the eventual need of using complementary techniques as chemo- and thermotherapy before or during meristem culture.

Numerous viruses in several hosts have been eliminated so far, rendering meristem-tip culture essential for the production of virus-free material.

A further important development comes from the possibility of storing cultures *in vitro* at low temperatures (4-6°C), which allows the mass conservation of virus-free material in limited space and a reduced need of frequent transplanting. Even more perspectives are open by the new technique of meristem cryopreservation, which will give the opportunity of conserving germplasm at a large scale for many years.

For all these reasons, meristem-tip culture, which is not yet fully exploited, will furnish one of the best ways to control the spread of virus infections in plants in the future.

ACKNOWLEDGMENTS

The authors wish to thank Prof. G.P. Martelli, Dipartimento Protezione delle Piante dalle Malattie, University of Bari, Italy, for his critical reading of the manuscript and Miss Annalisa Zanotti for her patience and diligence in typing the manuscript.

REFERENCES

1. Accatino, P. 1966. Papa corahila libre de virus mediante cultivo de meristemas. Agric.Tec. 26: 34-39.
2. Adams, A.N., Barbara, D.J., Manwell, W.E., and Thresh, J.M. 1979. Hop mosaic virus (HMV). East Malling Res. Stn.Rep. 1978: 103.
3. Ajayni, 0., and Dewar, A.M. 1983. The effects of barley yellow dwarf virus, aphids and honeydew on *Cladosporium* infection of winter wheat and barley. Ann.Appl.Biol. 102: 57-65.
4. Allen, T.C., and Anderson, W.C. 1980. Production of virus-free ornamental plants in tissue culture. Acta Hortic. 110: 245-251.
5. Anderson, W.C., and Carstens, J.B. 1977. Tissue culture propagation of broccoli, *Brassica oleracea* (Italica group) for use in F1 hybrid seed production. J.Am.Soc.Hortic.Sci. 102: 69-73.
6. Appiano, A., and Pennazio, S. 1972. Electron microscopy of potato meristem tips infected with potato virus X. J.Gen.Virol. 14: 273-276.
7. Asjes, C.J.1985. Control of field spread of non-persistent viruses in flower-bulb crops by synthetic pyrethroid and pirimicarb insecticides, and mineral oils. Crop Protection 4: 485-493.
8. Asjes, C.J.1990. Production for virus freedom of some principal bulbous crops in the Netherlands. Acta Hortic. 266: 517-529.
9. Ayuso, P. 1985. Le microgreffage appliqué à la régéneration sanitaire de la vigne. (Abstr.). Pages 191-192 in: Proc. Moet-Hennessy Coll. Amélioration de la Vigne et Culture *in vitro*, Paris. France.
10. Bajaj, Y.P.S. 1976. Regeneration of plants from cell suspensions frozen at -20, -70 and - 196°C. Physiol.Plant. 37: 263-268.
11. Bajaj, Y.P.S. 1981. Regeneration of plants from potato meristems freeze-preserved for 24 months. Euphytica 30: 141-145.
12. Ball, E. 1946. Development in sterile culture of stem tips and subjacent regions of *Tropaeolum majus* L. and *Lupinus albus* L. Am.J.Bot. 33: 301-318.
13. Banttari, E.E., and Goodwin, P.H. 1985. Detection of potato viruses S,X and Y by enzyme-linked immunosorbent assay on nitrocellulose membranes. Plant Dis. 69: 202-205.
14. Bar-Joseph, M., Seger, D., Blickle, W., Yesodi, V., Franck, A., and Rosner, A. 1986. Application of synthetic DNA probes for the detection of viroids and viruses. Pages 13-23 in: Development in Applied Biology 1. Developments and Applications in Virus Testing. R.A.C. Jones and L. Torrance, eds. Association Applied Biologists, Lavenham Press Ltd., Sudbury, UK.
15. Barbara, D.J., and Clark, M.F. 1982. A simple indirect ELISA using F(ab') fragment of immunoglobulin. J.Gen.Virol. 58: 315-322.
16. Barlass, M., and Skene, K.G.M. 1982. *in vitro* plantlet formation from *Citrus* species and hybrids. Sci .Hortic. 17: 333-341.
17. Barlass, M., Skene, K.G.M., Woodham, R.C., and Krabe, L.R. 1982. Regeneration of virus-free grapevines using *in vitro* apical culture. Ann.Appl.Biol. 101: 291-295.
18. Barnett, O.W., and Diachum, S. 1984. Virus diseases of clovers. Pages 235-268 in: Clover Science and Technology, Madison, ASA-CSSA-SSSA, USA.
19. Bass, P., and Legin, L. 1981 Thermothérapie et multiplication *in vitro* d'apex de vigne. Application à la séparation ou à l'elimination de diverses maladies de type apical et à l'evaluation de dégâts. C.R.AcadAgric.Fr. 67: 922-933.
20. Bass, P., and Vuittenez, A. 1977. Amélioration de la thermotherapie des vignes virosÈes en moyen de la culture d'apex sur milieux nutritifs ou par greffage sur vignes de semis, obtenue aseptiquement

in vitro. Ann. Phytopathol. 9: 539-540.

21.   Bastiaens, L. 1983. Endogenous bacteria in plants and their implications in tissue culture. Med. Fac. Landbouw. Rijksuniv. Gent. 48: 1-11.

22.   Baulcombe, D., Flavell, R.B., Boulton, R.A., and Jellis, G.J. 1984. The sensitivity and specificity of a rapid nucleic acid hybridization method for detection of potato virus X in crude sap samples. Plant Pathol. 33: 361-370.

23.   Baumann, G., Casper, R., and Kornkamhaeng, P. 1984. Detection of prune dwarf virus by ELISA in meristem-propagated sour cherry plants during *in vitro* culture. Phytopathol.Z. 110: 168-171.

24.   Bayliss, M.W. 1973. Origin of chromosome number variation in cultured plant cells. Nature (London) 246: 529-530.

25.   Beauchesne, G. 1982. Appearance of plants not true to type during *in vitro* plant propagation. Pages 268-302 in: Variability in Plants Regenerated from Tissue Culture. E.D. Earle and Y. Demarly, eds. Praeger Publishers, New York.

26.   Beemster, A.B.R., and Rozendaal, A. 1972. Potato viruses: properties and symptoms. Pages 115-143 in: Viruses of Potatoes and Seed-Potato Production. J.A.de Bokx, ed. Cent.Agric.Publ.Doc., Wageningen, Netherlands.

27.   Bertaccini, A., and Bellardi, M.G. 1991. Shoot-tip culture in the production of virus-indexed Dutch iris. Adv.Hortic.Sci. 5: 23-26.

28.   Bertaccini, A., and Marani, F. 1986. Individuation of growing media for shoot-tip culture of Dutch *Iris*. (Abstr.) Proc.6th Intern.Congress of Plant Tissue and Cell Culture Minneapolis, USA: 238.

29.   Bertaccini, A., and Marani F. 1987. Produzione di *Lilium* esenti da virus latente mediante coltura *in vitro* di apici meristematici. La Difesa delle Piante 10: 163-170.

30.   Bertaccini, A., and Marani, F. 1986. BYMV-free clones of eight gladiolus cultivars obtained by meristem-tip culture. Acta Hortic. 177: 299-308.

31.   Bertaccini, A., Marani, F., and Borgia, M. 1986. Shoot-tip culture of different garlic lines for virus elimination. Riv. Ortoflorofrutt. Ital. 2: 97-105.

32.   Bertaccini, A., Marani, F., and Passarelli, V. 1984. Risanamento dell'asparago dall'infezione da virus 2. Atti Giornate Fitopatologiche: 437-447.

33.   Bhojwani, S.S., and Razdan, M.K. 1996. Plant Tissue Culture: Theory and Practice, a Revised Edition. Elsevier, Amsterdam. 767 Pages.

34.   Bhojwani, S.S., Cohen, D., and Fry, P.R. 1982. Production of virus-free garlic and field performance of micropropagated plants. Sci.Hortic. 18: 39-43.

35.   Blom-Barnhoorn, G.J., and van Aartrijk, J. 1985. The regeneration of plants free of LSV and TBV from infected *Lilium* bulb-scale explants in the presence of virazole. Acta Hortic. 164: 163-168.

36.   Blom-Barnhoorn, G.I., van Aartrijk, J., and van der Linde, P.C.G. 1986. Effect of virazole on the production of hyacinth plants free from hyacinth mosaic virus (HMV) by meristem culture. Acta Hortic. 177: 571-574.

37.   Bonga, J.M., and Durzan, D.J. 1982. Tissue Culture in Forestry. Nijhoff and Junk Publishers, The Hague. 420 Pages.

38.   Bouman, H., and de Klerk, G.J. 1990. Cryopreservation of lily meristems. Acta Hortic. 266: 331-337.

39.   Boxus, P. 1976. Rapid production of virus-free strawberry by *in vitro* culture. Acta Hortic. 66: 35-38.

40.   Boxus, P. 1981. Commercial production of strawberry plants produced by meristem culture and micropropagation. Pages 310-348 in: Les Floralies Internationale de Montreal, Colloques Scientifiques, Vol. 5 Perspec. Hortic. Arbres Fruitiers et Petits Fruits.

41.   Boxus, P. 1984. Assainissement des arbres fruitiers et du fraisier par culture de mÈristÈmes. Parasitica 40: 139-155.

42.   Boxus, P., and Quoirin, M. 1974. La culture de meristemes apicaux de quelques especes de . Bull.Soc.R.Bot.Belg. 107: 91-101.

43.   Boxus, P.,Damiano, C., and Brasseur, E. 1984. Strawberry. Pages 453-486 in: Handbook of Plant Cell Culture. Vol. 3. Crop Species. P.V. Ammirato, D.A. Evans, W.R. Sharp and Y. Yamada, eds. Macmillan Publishers, New York.

44.   Boxus, P., Quorin, M., and Laine, J.M. 1977. Large scale propagation of strawberry plants from tissue culture. Pages 130-143 in: Plant Cell, Tissue and Organ Culture. J. Reinert and Y.P.S. Bajaj, eds. Springer-Verlag, Berlin.

45.   Bressan, P.H., Kim, Y.J., Hyndman, S.E., Hasegawa, P.M., and Bressan, R.A. 1982. Factors affecting *in vitro* propagation of rose. J. Am. Soc. Hortic. Sci. 107: 979-990.

46.   Brunt, A.A. 1980. A review of problems and progress in research on viruses and virus diseases of narcissus in Britain. Acta Hortic. 110: 23-30.

47.   Brunt, A.A. 1986. Methods of producing and some benefits of growing virus-free plants. Pages 55-65 in: Proc.Symp. Healthy Planting Material: Strategies and Technologies, Mono. 33.

48.   Butenko, R. 1968. Plant Tissue Culture and Plant Morphogenesis. Israel Programme for Scientific Translations, Jerusalem, Isreal, 291 Pages

49.   Campbell, A.I. 1974. Cited by Walkey, D.G.A. 1978. *in vitro* methods for virus elimination. Pages 245-254 in: Frontiers in Plant Tissue Culture. T.A. Thorpe, ed. University Calgary Press, Canada.

50.   Case, M.W. 1973. Rhubarb. 15th Ann.Report., Stockbridge House E.H.S., UK.

51.   Cassells, A.C. 1987. *in vitro* induction of virus-free potatoes by chemotherapy. Pages 40-50 in: Biotechnology in Agriculture and Forestry. Vol. 3. Potato.Y.P.S. Bajaj, ed. Springer-Verlag, Berlin.

52.   Cassells, A.C., and Long, R.D. 1980. The regeneration of virus-free plants from cucumber mosaic virus and Potato virus Y infected tobacco explant cultures in the presence of virazole. Z.Naturforsch. 35: 350-351.

53.   Cassells, A.C., and Long, R.D. 1982. The elimination of potato viruses X,Y, S and M in meristem and explant cultures of potato in the presence of virazole. Potato Res. 25: 165-173.

54.   Cassels, A.C., Minas, G., and Long, R. 1980. Culture of *Pelargonium* hybrids from meristems and explants: chimeral and beneficially-infected varieties. Pages 125-130 in: Tissue Culture Methods for Plant Pathologists. D.S. Ingram and J.P. Helgeson, eds. Blackwell Scientific Publishers, Oxford.

55.   Chee , R., and Pool , R.M. 1982. The effects of growth substances and photoperiod on the development of shoot apices of Vitis cultured *in vitro*. Sci. Hortic. 16: 17-27.

56.   Chen, W.Q., and Sherwood, J.L. 1991. Evaluation

of tip culture, thermotherapy and chemotherapy for elimination of peanut mottle virus from *Arachis hypogaea*. J.Phytopathol. 132: 230-236.

57.   Clark, M.F., and Adams, A.N. 1977. Characteristic of the microplate method of enzyme-linked immunosorbent assay for the detection of plant viruses. J.Gen.Virol. 34: 475-483.

58.   Coleman, M.C., and Powell, W. 1992. Virus elimination and testing. C6, Pages 1-12 in: Plant Tissue Culture Manual. Kluwer Academic Publisher, Dordrecht, Netherlands.

59.   Colombarini, A. 1990. Individuazione dei virus PVS e PVM in patata e risanamento mediante coltura di meristemi, chemio- e termoterapia. Dr.degree Thesis, University of Bologna, Italy. 65 Pages.

60.   Constantin, M.J., Henke, R.R., and Mansur, M.A. 1977. Effect of activated charcoal on callus growth and shoot organogenesis in tobacco. *in vitro* 13: 293-296.

61.   Cornu, D., and Michel, M.F. 1987. Bacteria contaminants in shoot cultures of avium L. Choice and phytotoxicity of antibiotics. Acta Hortic. 212: 83-86.

62.   Cropley, R. 1979. The production and practical benefits of virus-free propagating material of fruit crops. Pages 121-127 in: Plant Health. D.L. Ebbels and J.E. King, eds. Blackwell Scientific Publishers, Oxford.

63.   Crowley, N.C., Davidson, E.M., Francki, R.I.B., and Owusu, G.K. 1969. Infection of bean root-meristems by tobacco ringspot virus. Virology 39: 322-330.

64.   D'Amato, F. 1977. Cytogenetics of differentiation in tissue and cell cultures. Pages 343-357 in: Applied and Fundamental Aspects of Plant Cell, Tissue and Organ Culture. J. Reinert and Y.P.S. Bajaj, eds. Springer-Verlag, Berlin.

65.   D'Amato, F. 1978. Chromosome number variation in cultured cells and regenerated plants. Pages 287-295 in: Frontiers of Plant Tissue Culture. T.A. Thorpe, ed. Univ.Calgary Press, Canada.

66.   Damiano, C. 1977. La ripresa vegetativa di piantine di fragola provenienti da colture *in vitro*. Frutticoltura 39: 3-7.

67.   De Bokx, J.A. 1972. Test plants. Pages 102-110 in: Viruses of Potatoes and Seed-Potato Production. J.A. De Bokx and I.P.H. Van der Want eds. Cent.Agric.Publ.Doc., Wageningen, Netherlands.

68.   Debergh, P.C. 1983. Effects of agar brand and concentration on the tissue culture medium. Physiol. Plant. 59: 270-276.

69.   Demoise, C.F., and Partanen, C.R. 1969 Effects of subculturing and physiological condition of medium on the nuclear behaviour of a plant tissue culture. Am. J. Bot. 56: 147-151.

70.   Deogratias, J.M., Dosba, F., and Lutz, A. 1989. Eradication of prune dwarf, necrotic ringspot virus, and apple chlorotic leaf spot virus in tissue cultured sweet cherry. Can. J. Plant Pathol. 11: 332-336.

71.   Deogratias, J.M., Lutz, A., and Dosba, F. 1986. *in vitro* micrografting of shoot tips from juvenile and adult *avium* (L.) and *persica* (L.) Batsch. to produce virus-free plants. Acta Hortic. 193: 139-145.

72.   Deogratias, J.M., Lutz, A., and Dosba, F. 1986. Microgreffage d'apex de cerisiers(*avium* L.) multipliés *in vitro* en vue de l'élimination de trois types de particules virales (CLSV, PDV, NRSV). Fruits 41: 675-680.

73.   Deutsch, M., and Loebenstein, G. 1967. Field experiments with oil sprays to prevent yellow mosaic virus in irises. Plant Dis. Rep. 51: 318-319.

74.   Dunez, J., and Monsion, M. 1968. Possibilités de regeneration des chrysanthémes contaminés par les virus de l'aspermie et de la mosaique. Ann.Epiphyt. 19 (H.S.): 165-175.

75.   Evans, T.A., and Stephens, C.T. 1989. Increased susceptibility to *Fusarium* crown and root rot in virus-infected asparagus. Phytopathology 79: 253-258.

76.   Faccioli, G., and Capponi, R. 1983. An antiviral factor present in plants of *Chenopodium amaranticolor* locally infected by tobacco necrosis virus: 1 Extraction, partial purification, biological and chemical properties. Phytopathol.Z. 106: 289-301.

77.   Faccioli, G., and Colombarini, A. 1990. Risanamento della patata da seme dai principali virus a mezzo coltura di meristemi ed eventuali termoterapia e chemioterapia. Agricoltura Ricerca 115: 19-22.

78.   Faccioli, G., and Colombarini, A. 1996. Correlation of potato virus S and virus M contents of potato meristem tips with the percentage of virus free plantlets produced *in vitro*. Potato Res. 39:129-140.

79.   Faccioli, G., and Rubies-Autonell, C. 1982. PVX and PVY distribution in potato meristem tips and their eradication by the use of thermotherapy and meristem-tip culture. Phytopathol.Z. 103: 66-76.

80.   Faccioli, G., Rubies-Autonell, C., and Resca, R. 1984. Autoradiographic detection of PLRV, PVY and PVX in potato meristem tips and production of virus-free material by meristem-tip culture.(Abstr.) Potato Res. 27: 102.

81.   Faccioli, G., Rubies-Autonell, C., and Resca, R. 1988. Potato leafroll virus distribution in potato meristem tips and production of virus-free plants. Potato Res. 31: 511-520.

82.   Fellman, C.D., Read, P.E., and Hosier, M.A. 1987. Effects of thidiazuron and CPPU on meristem formation and shoot proliferation. HortScience 22: 1197-1200.

83.   Fridborg, G., and Eriksson, T. 1975. Effects of activated charcoal on growth and morphogenesis in cell cultures. Physiol.Plant. 34: 306-308.

84.   Fridborg, G., Pederson, M., Landstron, L., and Eriksson, T. 1978. The effect of activated charcoal on tissue cultures: adsorption of metabolites inhibiting morphogenesis. Physiol.Plant. 43: 104-106.

85.   Furuya, M., and Torrey, J.G. 1965. The reversible inhibition by red and far-red light of auxin-induced lateral root initiation in isolated pea roots. Plant Physiol. 39: 987-991.

86.   Gallenberg, D.J., and Jones, E.D. 1985. Detection of potato viruses X and S in tissue culture plantlets. Am.Potato J. 62: 118-119.

87.   Gamborg, O.L.1984. Plant cell cultures: nutrition and media. Pages 18-26 in: Cell Culture and Somatic Cell Genetics of Plants. Vol. 1. I.K. Vasil, ed. Academic Press, New York.

88.   Gamborg, O.L., Miller, R.A., and Ojima, K. 1968. Nutrient requirements of suspension cultures of soybean root cells. Exp.Cell.Res. 50: 151-158.

89.   Gamborg, O.L., Murashige, T., Thorpe, T.A., and Vasil, I.K. 1976. Plant tissue culture media. *in vitro* 12: 473-478.

90.   Giunchedi, L., and Poggi Pollini, C. 1984. Principali malattie da virus e virus-simili delle drupacee in

Italia. Inf.tore Fitopatol. 34: 75-94.

91. Goethals, M., and van Hoof, P. 1971. Régénération des oeillets par la culture de méristèms combinée à la thermotherapie. Parasitica 27: 36-41.

92. Gould, J.H., and Murashige, T. 1985. Morphogenic substances released by plant tissue cultures. I. Identification of berberine in Nandina culture medium, morphogenesis, and factors influencing accumulation. Plant Cell Tissue Org.Cult. 4: 29-42.

93. Gregorini, G., and Lorenzi, R. 1974. Meristem tip culture of potato plants as a method of improving productivity. Potato Res. 17: 24-33.

94. Grenan, S., and Truel, P. 1983. Observations sur un aspect de la variabilité constatée au cours de la multiplication végétative de variétés de vigne issues de semis de *Vitis vinifera* L. Agronomie 3: 675-680.

95. Hakkaart, F.A., and Quak, F. 1964. Effect of heat treatment of young plants on freeing chrysanthemums from virus B by means of meristem culture. Neth.J.Plant.Pathol. 70: 154-157.

96. Hansen, A.J., and Lane, W.D. 1985. Elimination of apple chlorotic leafspot virus from apple shoot cultures by ribavirin. Plant Dis. 69: 134-135.

97. Hecht, H.,and Diercks, R. 1978. Wirkung antiviraler Agentien auf das Kartoffel-Y-Virus in intekten Kartoffelpflanzen. II. Antimetaboliten und andere antivirale Substanzen. Landwirtsch. Jahrb., Munchen, 55: 433-457.

98. Heller, R. 1953. Recherches sur la nutrition minerale des tissus vegétaux cultivés *in vitro*. Ann.Sci.Bot.Veg.Paris 14: 1-223.

99. Hildebrandt, A.C. 1971. Growth and differentiation of single plant cell and tissues. Pages 71-93 in: Les Cultures de Tissues de Plants, Colloq. Int , CNRS.

100. Hollings, M. 1965. Disease control through virus-free stock. Annu.Rev. Phytopathol. 3: 367-396.

101. Hollings, M., and Stone, O.M. 1964. Investigation of carnation viruses. I. Carnation mottle. Ann.Appl.Biol. 53: 103-118.

102. Hollings, M., and Stone, O.M. 1968. Techniques and problems in the production of virus-tested planting material. Sci.Hortic. 20: 57-72 .

103. Hollings, M., and Stone, O.M. 1968. Heat treatment, meristem-tip culture and the production of virus-tested carnation and chrysanthemum. Rep.Glass.Crops Res.Inst. 1967: 106-107.

104 Hollings, M., Stone, O.M., and Smith, D.R. 1972. Productivity of virus-tested carnation cloned and the rate of reinfection with virus. J.Hortic.Sci. 47: 141-149.

105. Holmes, F.O. 1948. Elimination of spotted wilt from dahlia. Phytopathology 38: 314.

106. Horst, R.K., and Cohen, D. 1980. Amantadine-supplemented tissue culture medium: a method for obtaining chrysanthemums free of chrysanthemum stunt viroid. Acta Hortic. 100: 315-319.

107. Huang, S.C., and Millikan, D.F. 1980. *in vitro* micrografting of apple shoot tips. HortScience 15: 741-743.

108. Hughes, K.W. Ornamental species. Pages 5-139 in: Cloning Agricultural Plants via *In Vitro* Techniques. B.V. Conger, ed. CRC Press, Inc. Boca Raton, USA.

109. Hussey, G. 1977. *in vitro* propagation of some members of Liliaceae, Iridaceae and Amaryllidaceae. Acta Hortic. 78: 303-309.

110. Inouye, N. 1983. Effect of antiserum treatment on the production of virus-free Cymbidium by means of meristem culture. Nogaku Kenkyu 60: 123-126.

111. Iri, M., Shimura, T., Togawa, H., and Ueno, K.

1982. Elimination of grapevine viruses by meristem tip culture. Pages 807-808 in: Plant Tissue Culture, A. Fujiwara, ed. Maruzen, Tokyo.

112. Jonard, R., Hugard, J., Macheix, J.J., Martinez, J., Mosella, C.L., Poessel, J.L., and Villemur, P.1983. *in vitro* micrografting and its applications to fruit science. Sci. Hortic. 20: 147-159.

113. Jonard, R., Lukman, D., Schall, F., and Villemur, P. 1990. Early testing of graft incompatibilities in apricot and lemon trees using *in vitro* techniques. Sci.Hortic. 43: 117-128.

114. Jones, O.P. 1976. Effect of phloridzin and phloroglucinol on apple shoots. Nature (London) 262: 392-393.

115. Jones, O.P., and Vine, S.J. 1969. The culture of gooseberry shoot tips for eliminating virus. J.Hortic.Sci. 43: 289-292.

116. Jones, R.A.C., and Torrance L. 1986. Developments in Applied Biology 1: Developments and applications in virus testing. Association Applied Biologists, Lavenham Press Ltd., Sudbury, UK. 300 Pages

117. Jones,E.D., and Mullin, J.M. 1974. The effect of potato virus X on susceptibility of potato tubers to *Fusarium roseum* "Avenacearum". Am.Potato J. 51: 209-215.

118. Jordan, R., and Dodds, J.A. 1985. Double-stranded RNA in detection of diseases of known and unproven viral etiology. Acta Hortic. 164: 101-108.

119. Jordan, R.L., Heich, G.A., Dodds, J.A., and Ohr, H. 1983. Rapid detection of sunblotch viroid RNA and virus-like double stranded RNA in multiple avocado samples. (Abstr.) Phytopathology 73: 791.

120. Juarez, J., Arregui, J.M., Deogratias, J.M. and Navarro, L. 1990. Shoot-tip grafting *in vitro* in temperate stone-fruit trees. (Abstr.) Proc. 7th Inter.Congress Plant Tissue and Cell Culture, Amsterdam: 146.

121. Kartha, K.K. 1981. Meristem culture and cryopreservation. Methods and applications. Pages 181-211 in: Plant Tissue Culture, T.A. Thorpe, ed. Academic Press, New York.

122. Kartha, K.K. 1981. Tissue culture technique for virus elimination and germplasm preservation. Genet.Eng.Crop Improv., Rockfeller Found.Conf. 1980: 123-141.

123. Kartha, K.K., and Gamborg, O.L. 1975. Elimination of cassava mosaic disease by meristem culture. Phytopathology 65: 826-828.

124. Kartha, K.K., Gamborg, O.L., and Constabel, R. 1974. Regeneration of pea (*Pisum sativum* L.) plants from shoot apical meristems. Z.Pflanzenphysiol. 72: 172-176.

125. Kartha, K.K., Gamborg, O.L., Constabel, F., and Shyluk, J.P. 1974. Regeneration of cassava plants from apical meristem. Plant Sci.Lett. 2: 107-113.

126. Kartha, K.K., Leung, N.L., and Gamborg, O.L. 1979. Freeze-preservation of pea meristems in liquid nitrogen and subsequent plant regeneration. Plant Sci.Lett. 15: 7-15.

127. Kassanis, B. 1965. Therapy of virus infected plants. J.R. Agric.Soc.Engl. 126: 105-114.

128. Kassanis, B., and Varma, A. 1967. The production of virus-free clones of some British potato varieties. Ann.APagesBiol. 59: 447-450.

129. Kevers, C., Coumans, M., CoumansóGilles, M.F., and Gaspar, Th. 1984. Physiological and biochemical events leading to vitrification of plants *in vitro*. Physiol.Plant. 61: 69-74.

130. Khosh-Khui, M., and Sink, K.C. 1982. Rooting-enhancement of *Rosa hybrida* for tissue culture propagation. Sci.Hortic. 17: 371-376.

131. Klein, R.E., and Livingston, C.H. 1982. Eradication of potato virus X from potato by ribavirin treatment of cultured potato shoot tips. Am.Potato J. 59: 359-365.

132. Klein, R.E., and Livingston, C.H. 1983. Eradication of potato viruses X and S from potato shoot tip cultures with ribavirin. Phytopathology 73: 1049-1050.

133. Kondakova, V., and Schuster, G. 1991. Elimination of strawberry mottle virus and strawberry crinkle virus from isolated apices of three strawberry varieties by the addition of 2,4-Dioxohexahydro-1,3,5triazine (5-Azadihydrouracil) to the nutrient medium. J. Phytopathol. 132: 84-86.

134. Kouadio, J., and Phan, C.T. 1987. Heterogeneity of progeny in meristem culture. Acta Hortic. 212: 411-417.

135. Kowalska, A. 1974. Freeing carnation plants from viruses by meristem tip culture. Phytopathol.Z. 79:301-309.

136. Kreitlow, K.W., and Hunt, O.J. 1958. Effect of alfalfa mosaic and bean yellow mosaic viruses on flowering and seed production of ladino white clover. Phytopathology 48: 320-321.

137. Krylowa, N.V., Stepanenko, V.I., and Reifuran, V.G. 1973. Potato virus X in potato apical meristems. Acta Virol. 17: 172.

138. Kyte, L., and Briggs, B. 1979. A simplified entry into tissue culture production. Proc.Int.Plant Prop.Soc. 29: 90-95.

139. Lane, W.D. 1978. Regeneration of apple plants from meristem tips. Plant Sci. Lett. 13: 281-285.

140. Lange, L. 1986. The practical application of new development in test procedure for the detection of viruses in seed. Pages 269-287 in: Development in Applied Biology 1: Developments and Applications in Virus Testing. R.A.C. Jones and L. Torrance, eds. Association Applied Biologists, Lavenham press Ltd., Sudbury, UK.

141. Langhans, R.W., Horst, R.K., and Earle, E.D. 1977. Disease-free plants tissue culture propagation. HortScience 12: 149-150.

142. Larkin, P.J., and Scowcroft, W.R. 198_. Somaclonal variation a novel source of variability from cell cultures for plant improvement. Theo.Appl.Genet. 60: 197-214.

143. Letouze, R., and Beauchesne, G. 1969. Action d'eclairements monochromatiques sur la rhizogenese de tissue topinambour. C.R.Acad.Sci. Paris, 269: 1528-1531.

144. Limasset, P., and Cornuet, P. 1949. Recherche de virus de la mosaïque du tabac dans les méristèms des plants infectées. C.R. Acad.Sci. (Paris) 228: 1971-1972.

145. Lizarraga, R.E., Salazar, L.F., Roca, W.M.T., and Schildeó Reutschler, L. 1980. Elimination of potato spindle tuber viroid by low temperature and meristem culture. Phytopathology 70: 754-755.

146. Logan, A.E., and Zettler, F.W. 1985. Rapid *in vitro* propagation of virus-indexed gladioli. Acta Hortic. 164: 169-175.

147. Mac Donald, D.M. 1973. Heat treatment and meristem culture as a means of freeing potato varieties from viruses X and S. Potato Res. 16: 263-269.

148. Maene, L.J., and Debergh, P.C. 1985. Liquid medium additions to established tissue cultures to improve elongation and rooting *in vivo*. Plant Cell Tissue Org. Cult. 5: 23-33.

149. Maia, E., Beck, D., and Gaggelli, D. 1969. Obtention de clones d'oeillets méditerranens indemnes de virus.é Ann.Phytopathol. 1 (H.S.): 311-319.

150. Maier, V., and Metzlier, D.M. 1965. Quantitative changes in date polyphenols and their relation to browning. J.Food Sci. 30: 80-84.

151. Malnassy, P., and Ellison, J.H. 1970. Asparagus tetraploids from callus tissue. HortScience 5: 444-445.

152. Marani F. 1972. Ricerche sulla coltura di meristemi per la produzione di crisantemi virus-esenti. Inf.tore Fitopatol. 7-8: 3-8.

153. Marani, F., and Bertaccini, A. 1980. Risanamento dell'infezione virale dell'aglio mediante coltura di apici meristematici. Atti Giornate Fitopatologiche: 23-30.

154. Margara, J. 1969. Étude des facteurs de la neoformation de bourgeons en culture *in vitro* chez le chou-fleur (*Brassica oleracea* L., var. botrytis). Ann. Physiol. Veg. 11: 95-112.

155. Margara, J. 1982. Bases de la multiplication végétative. Les méristèmes et l'organogenèse. INRA, Paris. 262 Pages.

156. Martelli, G.P., and Bar-Joseph, M. 1991. Closterovirus group. Pages 345-347 in: Classification and Nomenclature of Viruses. R.I.B. Francki, C.M. Fauquet, D.L. Knudson and F. Brown, eds. Archives of Virology Suppl. 2, Springer-Verlag, Wien.

157. Martinez, J., Hugard, J., and Jonard, J. 1979. Sur le differentes combinaisons de greffages des apex realizés *in vitro* entre pecher (*Prunus persica* Batsch), apricotier (*Prunus armeniaca* L.) et myrobalan (*Prunus cerasifera* Ehrh.). C.R.Acad.Sci. 288: 759-762.

158. Martino, C. 1985. Il microinnesto *in vitro* della vite. Petria 2, Suppl.1: 17-25.

159. Martinó Tanguy, J. 1985. The occurrence and possible function of hydroxycinnamoyl acid amides in plants. Plant Growth Regul. 3: 381-399.

160. Matta, A., and Pennazio, S. 1985. Impiego di mezzi fisici contro le malattie parassitarie delle piante. Pages 187-207 in: Metodi Alternativi alla Lotta Chimica nella Difesa delle Piante Agrarie. G. Goidanich and P. Baronio, eds. Comune di Cesena, Italy.

161. Matthews, R.E.F. 1991. Plant Virology. 3rd ed. Academic Press, New York. 835 Pages

162. Mc Grew, J.R. 1980. Meristem culture for production of virus-free strawberry. Pages 80-85 in: Proc.Conf. Nursery Production Fruit Plants through Tissue Culture, Applications and Feasibility. R.H. Zimmerman, ed. Agric. Res. Sci. Educ. Admin., USDA, Beltville.

163. Mellor, F.C. and Stace-Smith, R. 1969. Development of excised potato buds in nutrient medium. Can.J.Bot. 47: 1617-1621.

164. Mellor, F.C., and Stace-Smith, R. 1977. Virus-free potatoes by tissue culture. Pages 616-635 in: Applied and Fundamental Aspects of Plant Cell, Tissue and Organ Culture. J. Reinert and Y.P.S. Bajaj eds. Springer-Verlag, Berlin.

165. Menhenett, R. and Hanks, G.R. 1982. The responses of virus-free and virus infected lily "Enchantment" to the retardant ancymidol, chloromequat chloride,

mepiquat carbamate. Sci.Hortic. 17: 61-70.

166. Miller, R.W., and Belkengren, R.O. 1963. Elimination of yellow edge, crinkle, and vein banding viruses and certain other virus complexes from strawberries by excision and culturing of apical meristems. Plant Dis.Rep. 47: 298-300.

167. Milne, R.G. 1988. Quantitative use of the electron microscope decoration technique for plant virus diagnostics. Acta Hortic. 234: 321-329.

168. Milne, R.G., and Luisoni, E. 1977. Rapid immunoelectron microscopy of virus preparations. Methods Virol. 6: 265-281.

169. Mitchell, D.J., Powell, P.J., and Rose, D.G. 1990. Squash blot, a novel device for sampling plant viruses.(Abstr.) Potato Res. 33: 144-145.

170. Monette, P.L. 1986. Elimination in vitro of two grapevine nepoviruses by an alternating temperature regime. J.Phytopathol. 116: 88-91.

171. Moran, J.R., and Wilson, J.M. 1985. Rates of reinfection with virus in commercial carnation, chrysanthemum and gladiolus crops. Acta Hortic. 164: 325-332.

172. Morel, G. 1948. Recherches sur la culture associée de parasities obligatoires et de tissus végétaux. Ann. Epiphyt. 1: 123-234.

173. Morel, G.M. 1960. Producing virus-free cymbidiums. Am.Orchid Soc.Bull. 29: 495-497.

174. Morel, G.M. 1964. Tissue culture - a new means of clonal propagation of orchids. Am.Orchid Soc.Bull. 33: 473-478.

175. Morel, G. 1975. Meristem culture techniques for the long-term storage of cultivated plants. Pages 327-332 in: Crop Genetic Resources for Today and Tomorrow. O.H. Frankel and J.G. Hawkes, eds. Cambridge University Press.

176. Morel, G., and Martin, C. 1952. Guerison de dahlias atteints d'une maladie a virus. C.R.Acad.Sci. (Paris) 235: 1324-1325.

177. Morel, G., and Martin, C. 1955. Guerison de pommes de terre atteints de maladie a virus. C.R.Acad.Agr.Fr. 41: 472-475.

178. Morel, G., Martin, C., and Muller, J.F. 1968. La guerison des pommes de terre atteints de maladies a virus. Ann.Physiol.Veg. 10: 113-139.

179. Mori, K., and Hosokawa, D. 1977. Localization of viruses in apical meristems and production of virus-free plants by means of meristem and tissue culture. Acta Hortic. 78: 389-396.

180. Morini, S., Marzialetti, P., and Barbieri, C. 1985. in vitro propagation of grapevine. Riv.Ortoflorofrutt.Ital. 69: 385-396.

181. Morris, T.J., and Dodds, J.A. 1979. The isolation and analysis of double stranded RNA from virus-infected plants and fungal tissue. Phytopathology 69: 854-858.

182. Morris, T.J., and Smith, E.M. 1977. Potato spindle tuber disease. Procedures for the detection of viroid RNA and certification of disease-free potato tubers. Phytopathology 67: 145-150.

183. Mosella, C.L., Macheix, J.J., and Jonard, R. 1979. Sur les améliorations apportées aux techniques de microgreffage des apex in vitro chez les arbres fruitieres du picher (Prunus persica Batsch). C.R. Acad. Sci. 289:505-508.

184. Mosella, C.L., Signoret, P.A., and Jonard, R. 1980. Sur la mise au point de techniques de microgreffage d'apex en vue de l'élimination de deux types de particules virales chez le picher (Prunus persica Batsch, C.R.Acad.Sci. 290: 287-290.

185. Moskovets, S.N., Gorbarenko, N.I., and Zhuk, I.P. 1973. The use of the method of the culture of apical meristems in the combination with low temperature for the sanitation of potato against mosaic virus. Sel-skokhozyaiastvennaya Biol. 8: 271-275.

186. Muller, K.O.L. and Munro, J. 1951. The reaction of virus infected potato plants to Phytophthora infestans. Ann. Appl. Biol. 38: 765-773.

187. Mullin, R.H., and Schlegel, D.E. 1976. Cold storage mantainance of strawberry meristem plantlets. Hortic. Sci. 2: 100-101.

188. Mullin, R.H., Smith, S.H., Frazier, N.W., Schlegel, D.E., and Mc Call, S.R. 1974. Meristem culture freed strawberries of mild yellow edge, pallidosis and mottle diseases. Phytopathology 64: 1425-1429.

189. Mur, G. 1979. Thermothérapie de variétés de Vitis vinifera par la méthode de culture in vitro. Prog. Agric. Vitic. 96: 148-151.

190. Murashige, T. 1973. Sample preparation of media. C. Plant Cultures. Pages 698-703 in: Tissue Culture Methods and Applications.P.F. Kruss,Jr. and M.K. Patterson, eds. Academic Press, New York.

191. Murashige, T. 1974. Plant propagation through tissue cultures. Annu.Rev.Plant Physiol. 25: 135-166.

192. Murashige, T. 1977. Manipulation of organ initiation in plant tissue cultures. Bot. Bull. Acad. Sin. 18: 1-24.

193. Murashige, T. 1978. The impact of plant tissue culture on agriculture. Pages 15-26 in: Frontiers of Plant Tissue Culture. T.A. Thorpe, ed. Univ. Calgary Press, Canada.

194. Murashige, T., and Skoog, F. 1962. A revised medium for rapid growth and bioassays with tobacco tissue cultures. Physiol. Plant. 15:473-497.

195. Murashige, T., Bitters, W.B., Rangan, T.S., Naver, E.M., Roistacher, C.N., and Holliday, P.B. 1972. A technique of shoot apex grafting and its utilization towards recovering virus-free citrus clones. HortScience 7: 118-119.

196. Murashige, T., Shabde, M.N., Hasegawa, P.M., Takatori, F.H., and Jones, J.B. 1972. Propagation of asparagus through shoot apex culture. I.Nutrient medium for formation of plantlets. J.Am.Soc.Hortic.Sci. 97: 158-161.

197. Navarro, L., and Juarez, J. 1977. Elimination of citrus pathogens in propagative budwood. II. in vitro propagation. Proc.Int.Soc. Citriculture 3: 973-987.

198. Navarro, L., Roistacher, C., and Murashige, T. 1975. Improvement of shoot tip grafting in vitro for virus-free Citrus. J.Am.Soc.Hortic.Sci. 100: 471-479.

199. Navarro, L., Llacer, G., Cambra, M., Arrequi, J.M., and Juarez, J. 1982. Shoot-tip grafting in vitro for elimination of viruses in peach plants (Prunus persica Batsch). Acta Hortic. 130: 185-192.

200. Neve, R.A. 1979. Hop diseases: the risks and consequences of spread in a vegetatively-propagated crop. Pages 155-161 in: Plant Virus Epidemiology. R.T. Plumb and J.A. Thresh, eds. Blackwell Scientific Publishers, Oxford.

201. Nitsch, C., and Nitsch, J.P. 1967. The induction of flowering in vitro in stem segments of Plumbago indica L. The production of vegetative buds. Planta 72: 355-370.

202. Paduch-Cichal, E., and Kryczynski, S. 1987. A low temperature therapy and meristem-tip culture for eliminating four viroids from infected plants. J.Phytopathol. 118: 341-346.

203.   Paludan, N. 1980. *Chrysanthemum* stunt and chlorotic mottle. Establishment of healthy *Chrysanthemum* plants and storage at low temperature of *Chrysanthemum*, carnation, campanula and *Pelargonium* in tubes. Acta Hortic. 110: 303-313.

204.   Paludan, N. 1985. Elimination of viroids in chrysanthemum by low temperature treatment and meristem-tip culture. Acta Hortic. 164: 181-186.

205.   Pennazio, S. 1973. Effects of four antimetabolites on PVX inhibition in infected potato tips cultured on artificial substrate. Riv. Patol.Veg. 9: 3-10.

206.   Pennazio, S., and Redolfi, P. 1973. Factors affecting the culture *in vitro* of potato meristem tips. Potato Res. 16: 20-29.

207.   Phillips, S. 1990. The efficacy of four antiviral compounds in the elimination of narcissus viruses during meristem-tip culture. Acta Hortic. 266: 531-538.

208.   Pierik, R.L.M. 1987. *in vitro* Culture of Higher Plants. Martinus Nijhoff Publishers, Dordrecht, Netherlands. 321 Pages.

209.   Pierik, R.L.M., Steegmans, H.H.M. and Marelis, J.J. 1973. *Gerbera* plantlets from *in vitro* cultivated capitulum explants. Sci.Hortic. 1: 117-119.

210.   Posnette, A.F. 1977. Virus diseases of woody plants. Sci.Hortic. 29: 7-13.

211.   Posnette, A.F., and Cropley, R. 1959. The reduction in cropping caused by apple mosaic. Rep.East Malling Res.Station 1958. 89-90.

212.   Posnette, A.F., and Cropley, R. 1970. Decline and other effects of five virus infections on three varieties of plum. Ann.Appl.Biol. 65: 111-114.

213.   Posnette, A.F., and Cropley, R. 1973 The effect of viruses on growth and cropping of pear trees. Ann.Appl.Biol. 73: 39-43.

214.   Quak, F. 1961. Heat treatment and substances inhibiting virus multiplication in meristem culture to obtain virus-free plants. Adv.Hortic.Sci.Appl.l: 144-148.

215.   Quak, F. 1965. Infection of tobacco-callus tissue with tobacco mosaic virus and multiplication of virus in such tissue. Proc.Int.Conf.Plant Tissue Culture, Berkeley, USA: 513-519.

216.   Quak, F. 1977. Meristem culture and virus-free plants. Pages 598-615 in: Applied and Fundamental Aspects of Plant Cell, Tissue and Organ Culture. J. Reinert and Y.P.S. Bajaj, eds. Springer-Verlag, Berlin.

217.   Quoirin, M., and Lepoivre, P. 1977. Etude de milieux adaptes aux cultures *in vitro* de *Prunus* . Acta Hortic. 78: 437-442.

218.   Rabiti, A.L., Marani, F., Poggi Pollini, C., Giunchedi, L., and Pizzi, L. 1990. Applicazione della tecnica immunoenzimatica (ELISA) per la individuazione di un potyvirus dell'aglio. Atti Giornate Fitopatologiche: 139-146.

219.   Randles, J.W. 1991. Luteovirus group. Pages 309-311 in: Classification and Nomenclature of Viruses. Archives of Virology Suppl. 2. R I.B. Francki, C.M.Fauquet, D.L. Knudson and F. Brown, eds. Springer-Verlag, Wien.

220.   Robbins, W. 1922. Cultivation of excised root tips and stem tips under sterile conditions Bot.Gaz. 73: 376-390.

221.   Roberts, D.A., Christie, R.G., and Archer, M.C. 1970. Infection of apical initials in tobacco shoot meristems by tobacco ringspot virus. Virology 42: 217-220.

222.   Roca, W.M., Espinoza, N.O., Roca, M.R., and Bryan, J.E. 1978. A tissue culture method for the rapid propagation of potatoes. Am . Potato J . 55: 691-701.

223.   Roest, S., and Bokelmann, G.S. 1975. Vegetative propagation of *Chrysanthemum morifolium* Ram. *in vitro*. Sci.Hortic. 3: 317-330.

224.   Romberger, J.A., and Tabor, C.A. 1971. The *Picea abies* shoot apical meristem in culture. I. Agar and autoclaving effects. Am. J. Bot. 58: 131-140.

225.   Ross, A.F. 1961. Systemic acquired resistance induced by localized virus-infections in plants. Virology 14: 340-358.

226.   Rubies-Autonell, C., and Faccioli, G. 1985. Distribution of bean common mosaic and bean yellow mosaic viruses in bud tips of *Phaseolus vulgaris* L. Phytopathol.Mediterr. 24: 241-244.

227.   Rubies-Autonell, C., Faccioli, G., and Colombarini, A. 1989. Detection of PVM in potato meristem tips and preliminary results on its eradication.(Abstr.) Potato Res. 33: 142.

228.   Rubies-Autonell, C., Resca, R., Faccioli, G., and Colombarini, A. 1987. PVS and PVX rapid detection in potato meristem tips and their eradication by the use of thermo- and chemotherapy.(Abstr.) Proc.7th Congress Mediterr.Phytopathol. Union, Granada, Spain: 78-80.

229.   Russel, G.E. 1966. The control of *Alternaria* species on leaves of sugar beet infected with yellowing viruses. II. Experiments with two yellowing viruses and virus-tolerant sugar beet. Ann.Appl.Biol. 57: 425-434.

230.   Russo, F., and Starrantino, A. 1973. Ricerche sulla tecnologia dei microinnesti nel quadro del miglioramento genetico sanitario degli agrumi. Ann.Ist.Sperim.Agrum. 6: 209-222.

231.   Sanger, H.L., and Ramm, K. 1975. Radioactive labelling of viroid RNA. Pages 229-252 in: Modification of the Information Content of Plant Cells.R. Markham, D.R. Davies, D.A. Hopwood and R.W. Horne, eds. North. Holland Elsevier, Amsterdam.

232.   Savino, V. 1991. Lotta ai virus: risanamento. Pages 133-139 in: Aspetti Biologici e Molecolari dei Virus delle Piante. Assoc.Genetica Ital. and CNR-RAISA, Cortona, Italy.

233.   Schuster, G. 1982. Antiphytovirale Verbindungen mit Guanidinstruktur. Phytopathol.Z. 103: 77-86.

234.   Schuster, G. 1988. Synthetic antiphytoviral substances. Appl.Virology Res. 1: 265-283.

235.   Seabrook, J.E.A., Cumming, B.G., and Dionne, L.A. 1976. The *in vitro* induction of adventitious shoot and root apices on *Narcissus* (daffodil and narcissus) cultivar tissue. Can.J.Bot. 54: 814-819.

236.   Seibert, M. 1973. The effects of wavelength and intensity on growth and shoot intiation in tobacco callus. *in vitro* 8: 435-438.

237.   Seibert, M., Wetherbee, P.J., and Job, D.D. 1975. The effects of light intensity and spectral quality on growth and shoot initiation in tobacco callus. Plant Physiol. 56: 130-139 .

238.   Semal, J., and Lepoivre, P. 1990. Application of tissue culture variability to crop improvement. Pages 301-315 in: Plant Tissue Culture. Applications and Limitations. S.S. Bhojwani, ed., Elsevier, Amsterdam.

239.   Shabde, M., and Murashige, T. 1977. Hormonal requirements of excised *Dianthus caryophyllus* L.

shoot apical meristem in vitro. Am. J. Bot. 64: 443-448.

240. Sheffield, F.M.L. 1942. Presence of virus in the primordial meristem. Ann.Appl.Biol. 29: 16-17.

241. Shepard, J.F. 1980. Mutant selection and plant regeneration from potato mesophyll protoplasts. Pages 185-219 in: Genetic Improvement of Crops, Emergent Techniques. I. Rubenstein, B. Gegenbach, R.L. Phillips and C.E. Green, eds. University of Minnesota Press, Minneapolis, USA.

242. Sheridan, W.F. 1975. Plant regeneration and chromosome stability in tissue culture. Pages 263-295 in: Genetic Manipulation with Plant Material. L. Ledoux, ed. Plenum Press, London.

243. Simons, J.N., and Zitter, T.A. 1980. Use of oils to control aphid-borne viruses. Plant Dis. 64: 542-546.

244. Simpkins, I., Walkey, D.G.A., and Neely, H.A. 1981. Chemical suppression of virus in cultured plant tissues. Ann.Appl.Biol. 99: 161-169.

245. Skirvin, R.M. 1980. Fruit crops. Pages 51-139 in: Cloning Agricultural Plants via in vitro Techniques. B.V. Conger, ed. CRC Press, Inc. Boca Raton, USA.

246. Skirvin, R.M., Chu, M.C., Mann, M.L., Young, H., Sullivan, J., and Fermanian, T. 1986. Stability of tissue culture medium pH as a function of autoclaving, time, and cultured plant material. Plant Cell Rep. 5: 292-294.

247. Skoog, F., and Miller, C.0. 1957. Chemical regulation of growth and organ formation in plant tissues cultured in vitro. Symp.Soc.Exp.Biol. 11: 118-131.

248. Smith, R.H., and Murashige, T. 1970. in vitro development of isolated shoot apical meristems of angiosperms. Am.J.Bot. 57: 562-568.

249. Snir, I., and Stein, A. 1985. in vitro detection and elimination of necrotic ring spot virus in sweet cherry (avium L.). Riv.Ortoflorofruttic.Ital. 69: 191-194.

250. Sobcykiewicz, D., 1979. Heat treatment and meristem culture for the production of virus-free strawberry plants. Acta Hortic. 95: 79-82.

251. Stace-Smith, R., and Mellor, F.C. 1968. Eradication of potato viruses X and S by thermotherapy and axillary bud culture. Phytopathology 58: 199-203.

252. Stace-Smith, R., and Mellor ,F. 1970. Eradication of potato spindle tuber virus by thermotherapy and axillary bud culture. Phytopathology 60: 1857-1858.

253. Starrantino, A. 1991. Il microinnesto in vitro degli agrumi. Petria 2, Suppl. 1: 27-35.

254. Starrantino, A., and Caruso, A. 1988. The shoot-tip grafting technique applied in citriculture. Acta Hortic. 227: 101-103.

255. Stone, O.M. 1963. Factors affecting the growth of carnation plants from shoot apices. Ann.Appl.Biol. 52: 199-209.

256. Stone, O.M. 1968. The elimination of four viruses from carnation and sweet william by meristem-tip culture. Ann.Appl.Biol. 62: 119-122.

257. Styer, D.J., and Chin, C.K. 1983. Meristem and shoot-tip culture for propagation, pathogen elimination and germplasm preservation. Hortic.Rev. 5: 221-277.

258. Sunderland, N. 1977. Nuclear cytology. Pages 177-205 in: Plant Tissue and Cell Culture. H.E. Street, ed. Blackwell Scientific Publishers, Oxford.

259. Sutter, E., and Langhans, R.W. 1979. Epicuticular wax formation on carnation plantlets regenerated from shoot tip culture. J.Am.Soc.Hortic.Sci. 104: 493-496.

260. Sutter, E., and Langhans, R.W. 1981. Abnormalities in Chrysanthemum regenerated from long term cultures. Ann.Bot. 48: 559-568.

261. Thomas, B.J. 1984. The effect of necrotic ringspot virus on field grown roses. Ann.Appl. Biol. 100: 129-134.

262. Thomas, D.D.S., and Murashige, T. 1979. Volatile emissions of plant tissue cultures. I. Identification of the major components. in vitro 15: 654-658.

263. Tisserat, B., Murashige, T. 1977. Effects of ethephon, ethylene, and 2,4-dichlorophenoxyacetic acid on asexual embryogenesis in vitro. Plant Physiol. 60: 437-439.

264. Torrance, L. 1980. Use of bovine C1q to detect plant viruses in an enzyme-linked immunosorbent-type assay. J.Gen.Virol. 51: 219-232.

265. Torrance, L. 1981. Use of bovine C1q enzyme-linked immunorbent assay to detect plant viruses and their different serological strains. Ann.Appl.Biol. 88: 291-299.

266. van Os, H. 1964. Production of virus-free carnations by means of meristem culture. Neth.J.Plant Pathol. 70: 18-26.

267. van Regenmortel, M.H.V., Dekker, E.L., Dore, I., Porta, C., Weiss, E., and Burckard, J. 1988. Recent advances in serodiagnosis of plant virus diseases. Acta Hortic. 234: 175-183.

268. van Shadewijk, A.R. 1986. Detection of tulip breaking virus and lily symptomless virus in lily bulbs by means of ELISA. Acta Hortic. 177: 121-128.

269. van Shadewijk, A.R., Derks, A.F.L.M., Lemmers, M.E,C , Hollinger. Th.C 1988. Detection of iris mild mosaic virus in bulbs and leaves of bulbous iris by ELISA. Acta Hortic. 234: 199-206.

270. Vieitez, A.M., and Vieitez, M.L. 1980. Culture of chestnut shoots from buds in vitro. J. Hortic. Sci. 55: 83-84.

271. Vuittenez, A., Bass, P., and Ateb, N. 1974. Application de la culture de fragments d'organes in vitro pour l'étude des virus chez la vigne et autres plantes hôtes expérimentales ligneuses ou herbacées. (Abstr.) Ann. Phytopathol. 6: 498.

272. Walkey, D.G.A. 1968. The production of virus-free rhubarb by apical tip culture. J. Hortic.Sci. 43: 283-287.

273. Walkey, D.G.A. 1972. Production of apple plantlets from axillary-bud meristems. Can.J.Plant Sci. 52: 1085-1087.

274. Walkey, D.G.A. 1976. High temperature inactivation of cucumber and alfalfa mosaic viruses in Nicotiana rustica cultures. Ann.Appl.Biol. 84: 183-192.

275. Walkey, D.G.A. 1978. in vitro methods for virus elimination. Pages 245-254 in: Frontiers of Plant Tissue Culture. T.A. Thorpe, ed. International Association for Plant Tissue Culture, University Calgary Press, Canada.

276. Walkey, D.G.A., and Cooper, V.C. 1972. Some factors affecting the behaviour of plant viruses in tissue culture. Physiol. Plant Pathol. 2: 259-264.

277. Walkey, D.G.A., and Cooper, V.C. 1972. Comparative studies on the growth of healthy and virus infected rhubarbs. J. Hortic. Sci. 47: 37-41.

278. Walkey, D.G.A., and Freeman, G.H. 1977. Inactivation of cucumber mosaic virus in cultured tissues of Nicotiana rustica L. by diurnal alternating periods of high and low temperature. Ann. Appl. Biol. 87: 375-382.

279. Walkey, D.G.A., and Webb, M.J.W. 1968. Virus in

plant apical meristems. J.Gen.Virol. 3: 311-313.

280. Walkey, D.G.A., and Woolfitt, J.M.G. 1970. Rapid clonal multiplication of cauliflower by shake culture. J. Hortic. Sci. 45: 205-206.

281. Walkey, D.G.A., Creed, c., Delaney, H., and Whitwell, J.D. 1981. Studies on the reinfection and yield of virus-tested and commercial stocks of rhubarb cv. Timperley Early. Plant Pathol. 31: 253-261.

282. Wambugu, F.M.,and Rangan, T.S. 1981. *in vitro* clonal multiplication of pyrethrum (*Chrysanthemum cinerariaefolium* Vis.) by micropropagation. Plant Sci. Lett. 22: 219-226.

283. Wang, P., and Hu, C. 1980. Regeneration of virus-free plants through *in vitro* culture. Pages 61-99 in: Advances in Biochemical Engineering. Vol. 18 Plant Cell Cultures II. A. Fiechter, ed. Springer-Verlag, Berlin.

284. Westcott, R.J., Henshaw, G.G., and Roca, W.M. 1977. Tissue culture storage of potato germplasm: culture initiation and plant regeneration. Plant Sci. Lett. 9: 309-315.

285. Wetzstein, H.Y., and Sommer, H.E. 1982. Leaf anatomy of tissue-cultured *Liquidambar styraciflua* (Hamamelidaceae) during acclimatization. Am.J.Bot. 69: 1579-1586.

286. White, P.R. 1934. Multiplication of the viruses of tobacco and aucuba mosaic in growing excised tomato roots. Phytopathology 24: 1003-1011.

287. Wu, J.H., Hildebrandt, A.C., and Riker, A.J. 1960. Virus-host relationships in plant tissue culture. Phytopathology 50: 587-594.

288. Yang, H.J., and Clore, W.J. 1973. Rapid vegetative propagation of asparagus through lateral bud culture. HortScience 8: 141-143.

289. Zaid, A., and Tisserat, B.H. 1983. *in vitro* shoot tip differentiation in Phoenix dactylifera L. Date Palm J. 2: 163-182.

290. Zandvoort, R. 1973. The spread of carnation mottle virus in carnation in glasshouse. Neth.J.Plant.Pathol. 79: 81-84.

291. Zimmerman, R.H. 1982. Factors affecting *in vitro* propagation of apple cultivars. Acta Hortic. 131: 171-178.

292. Zimmerman, R.H., and Broome, O.C. 1980. Micropropagation of thornless blackberries. Pages 23-26 in: Proc.Conf. on Nursery Production of Fruit Plant through Tissue Culture, Applications and Feasibility. Agric.Res.Sci. Educ.Admin., USDA., Beltsville.

293. Zimmermann, R.H., and Broome, O.C. 1981. Phloroglucinol and *in vitro* rooting of apple cultivar cuttings. J.Am.Soc.Hortic.Sci. 106: 648-652.

294. Ziv, M., Meir, G., and Halevy, A.H. 1983. Factors influencing the production of hardened glaucous carnation plantlets *in vitro*. Pl. Cell Tiss. Cult. Org. 2:55-65.

CHAPTER 28

Advanced Diagnostic Tools as an Aid to Controlling Plant Virus Diseases

R. R. Martin

Accurate disease diagnosis combined with sensitive, rapid and early detection of plant viruses is critical for effective management of most crop systems. Appropriate control procedures can only be applied effectively if the disease is correctly identified and distribution in an area or crop is known. There have been several major improvements in virus detection over the past two decades. Serological detection was greatly improved with the application of the enzyme-linked immunosorbent assay (ELISA) to plant virus detection in the mid-1970's (20). This was enhanced further with the development of monoclonal antibody technology and its application to a large number of plant viruses in the 1980's. Similarly, nucleic acid hybridization has been used successfully for the detection of many plant viruses and is the method of choice for detection of viroids. Cloning of plant viral nucleic acids and the development of nonradioactive detection methods have increased the utility of nucleic acid hybridization for virus detection. Most recently, the development of polymerase chain reaction (PCR) has greatly improved the sensitivity and utility of hybridization and other nucleic acid based assays. Immunocapture PCR combines the advantages of serology and PCR into a very sensitive method of detection (17,44,77,79,99).

Monoclonal antibodies, nucleic acid hybridization and PCR provide the potential for the development of diagnostic reagents with desired specificities. Strain, virus or group specific reagents have been developed for viruses in several plant virus groups (15,25,42,45,47,53,-64,80,84,86,87,89,98). There is still much work needed to develop diagnostic reagents for many of the plant viruses. Equally important are the further application of these technologies to recalcitrant viruses, their use in certification schemes, the study of viral epidemiology, mechanisms of virus transmission and detection of viral pathogens in transgenic plants that contain part of one or more viral genomes.

In this chapter I will review recent advances in virus detection and diagnosis and the implications of transgenic plants on virus detection, certification and plant quarantine. Some of this material has been reviewed elsewhere in more detail (18,23,59,76). I will not review polymerase chain reaction or its application to plant virology since this is the subject of another chapter in this book.

REQUIREMENTS FOR DETECTION AND DIAGNOSIS

The requirements for the specificity and sensitivity of detection will vary in different situations. In a clean plant program, the manager wants to ensure that the starting material is free of known viruses. In this situation, where much depends on the virus status of relatively few individual plants, several tests will likely be employed and the virus status of every plant will be determined. Serological or nucleic acid-based tests often can be employed for detection of well characterized viruses known to occur in the crop but tests that detect a wide range of viruses are also desirable. Mechanical transmission to selected herbaceous indicator plants, electron microscopic examination of leaf dips, double-stranded RNA (dsRNA) analysis and/or grafting may be desirable to ensure that the nuclear material (also referred to as mother block, foundation, elite; all refer to the few plants that are the basis of all plant material in a clean plant propagation scheme) is virus-free. New material being added to such a program may be put through a virus eradication program to enhance the likelihood that the material will be free of viruses not reported previously in the crop and free of new strains of a virus which may not be detected with available antibodies or nucleic acid probes.

I will use certification to refer to the later stages of a clean plant program and consider it distinct from nuclear stock. The reason for this separation is that at the later stages of such programs many more plants are tested but they are usually tested for fewer viruses or at least fewer tests are employed to ascertain their virus status. At this stage of a clean plant scheme a predetermined percentage of the plants are usually virus-indexed. In a plant certification scheme it is important to standardize the tests in such a way that results from one laboratory can be compared with those from other facilities. This requirement to meet standards may slow the introduction of a new test to such a system. As an example, suppose that a certification program has a tolerance limit that will allow up to 1% virus infection based on test X. Then test Y is developed which is 1000 time more sensitive than X. A rigorous comparison between the results obtained with test X and Y must be made before test Y is introduced as a standard test. The level of virus infection in a crop may be reduced, or the tolerance limit may be changed when a more sensitive test is introduced into a certification scheme.

In plant quarantine it is necessary to ensure that plant material is free of quarantinable viruses prior to its release to industry or breeding programs. Quarantine programs often work with only a few plants of a given cultivar but the level of indexing on these few plants is as complete as is possible. Maximum sensitivity is the goal of virus detection in quarantine programs. These programs often use an agreed upon test for each specific virus. The test may be grafting or mechanical transmission onto a specific indicator plants, an ELISA, hybridization assay of PCR for specific viruses (27). Detection of viruses of woody plants will often require graft transmission tests to indicator plants. When grafting is required for a single virus in a crop it might be more efficient to employ graft transmission tests for all viruses of quarantine significance since the work is already being done for the one virus. Thus, quarantine facilities may not opt for the "newer laboratory" techniques until they are available for all quarantinable viruses capable of infecting that crop.

Quarantine officials may also require extreme specificity in cases where only one strain of a virus is on a quarantine list. Such a case occurred recently with potato virus Y (PVY) in Canada (68). In the USA, only the necrotic strain of PVY, PVY^N, is on the quarantine list and it causes a severe disease in tobacco, tomato and pepper crops. It was necessary to ensure that seed potatoes were free of PVY^N, whereas a low level of infection with PVY^O was acceptable for movement of seed potatoes from Canada to the USA.

When there are several sources of plant material it might be easier to destroy all material that is infected with a particular virus rather than trying to determine which strain of a virus is present. There is a resistance breaking strain of raspberry bushy dwarf virus (RBDV) that occurs in Europe and is not known to not occur in North America. The only way to differentiate the two strains is to graft onto resistant varieties of raspberry (5). Both strains are readily detected, but can not be differentiated serologically and in practice any material being imported to North America from Europe that tests positive for RBDV is destroyed rather than risk introducing the resistance breaking strain of the virus. Recently, a RT-PCR test has been developed to differentiate these two strains of RBDV (7).

Studies in epidemiology require a sensitive assay in which a large number of samples can be tested quickly usually for a single or a few viruses. These studies may involve monitoring of vectors for the presence of a plant virus, monitoring a large number of plants over time to assess the movement of a virus in a region or within a field or monitoring of weeds and indigenous vegetation for various viruses to determine potential sources of inoculum (12,14,29,30,72,78,93). The results of the monitoring often are the basis of management decisions such as the application of a pesticide to reduce vector populations, top killing of seed potatoes to ensure virus incidence does not exceed a certain level or no action required if the vector is not viruliferous. To date most such tests make use of ELISA or dot immuno-binding assay (DIBA) since they are suited to handling large numbers of samples in a short period of time. Nucleic acid based tests also work very well for dot or squash blots in these situations, but many laboratories doing epidemiological studies are not equipped to work with the radioisotopes with which most nucleic acid probes are labelled. However, as nonradioactive probes are more widely accepted, dot blot assays for nucleic acids will become more commonly used (12,78,81,100).

Sensitivity becomes a problem in each of the above examples since recent infections can be difficult to detect regardless of the method of detection used. Vectors that have acquired virus several weeks previous to testing, or have only fed on an infected source plant for a short period of time may have a virus concentration below detectable levels. These vectors could still play an important role in the epidemiology of persistently transmitted viruses. Aphids that do not vector or are an inefficient vector of a specific luteovirus can still harbor it (37). Transmission efficiency of the vector must also be considered in studies on virus epidemiology where vectors are being assayed for virus.

SEROLOGICAL DETECTION

Gel diffusion tests were the most commonly used serological tests prior to the introduction of ELISA. Gel diffusion will not be reviewed here, for a discussion on various types of gel diffusion assays readers are referred to (2,48). Most serological assays today make use of a solid phase support: ELISA on microtiter plates, DIBA on membranes, and immunospecific electron microscopy (ISEM) on electron microscope grids. One big advantage of using solid supports is the lack of a problem with optimal concentrations as must be considered when using gel diffusion assays (2). These tests are much more sensitive than gel diffusion assays and use less antibody (20,26,70). Several hundred thousand samples may be indexed in some certification programs and the amount of antibody required per test can be an important consideration.

ELISA and DIBA

ELISA and DIBA are currently the most widely used methods of serological detection of plant viruses. There have been relatively few changes in ELISA or DIBA for virus detection since the application of monoclonal antibodies (McAbs) (46,59,76). ELISA has been reviewed elsewhere (18,21,23) and the general outline will not be covered here. Rather, some of the principles of the assays will be discussed. These assays are carried out on a solid phase [usually plastic, nitrocellulose (3) or filter paper (40)] where each component of the test is applied successively and the reaction between virus and antibody is detected by enzymatic hydrolysis of a substrate that results in a color change or light emission.

Assays carried out on nitrocellulose or filter paper are referred to as dot immunobinding assays. The DIBA is about as sensitive as ELISA (70) and offers the advantage that sample preparation can be very simple. A few microliters of sap can be spotted onto the membrane or a freshly cut edge of a leaf, petiole or stem can be pressed gently against the membrane which absorbs small amounts. The latter approach is referred to as tissue blotting and also provides some information on the location of the virus in the tissue, e.g. phloem (50). The other advantage of the DIBA is that it is readily adapted to field situations and applications in areas with minimal laboratory facilities (40,70). The membranes can be taken to the field and plant tissue or insects blotted directly onto the membrane. Both of these assays can be performed without any specialized equipment.

Polyclonal antisera often contain anti-host antibodies which may result in elevated background readings in ELISA or DIBA. When virus concentrations are very low the resulting background readings may prevent a differentiation between infected and healthy samples. The background signal may be high enough that an antiserum is not useable in an ELISA without some type of cross absorbtion procedure to remove antibodies to plant antigens. The cross absorbtion can be accomplished easily by diluting the detecting antibody in buffer containing filtered plant sap of the test species being used. The conjugate should be incubated with the leaf sap 15 min to 2 h before adding to the ELISA plate (67). This is not always necessary, since with good polyclonal antisera (those with minimal nonspecific antibodies) overnight (or longer) incubation with substrate can give excellent results. Part of the problem with nonspecific reactions in ELISA can be overcome by reducing the amount of virus used for the immunization of the animal used for antiserum production. Often 1 to 10 mg of virus is used per injection when immunizing rabbits, whereas 50 to 100 μg of virus per injection is adequate (97). Increasing the amount of virus used to immunize a rabbit 10- to 50-fold higher than necessary results in little (if any) increase in the titer of the antiserum obtained but may increase by 10- to 50-fold the amount of plant contaminants. The increased contaminants may then be at a level where they will induce an immune response. In the long run, it is worth the extra effort to ensure that virus preparations are highly pure before injecting into an animal for production of polyclonal antibodies. The result will be an antiserum that will give much better results in ELISA.

The ELISA procedure is quite adaptable and many variations on a test will likely give suitable results. A standardized test protocol should be used in certification schemes to ensure consistent results between laboratories or from year to year in the same facility. Standardization of an ELISA protocol may include factors such as manufacturer of microtiter plate used in the assay, identification of a specific monoclonal or polyclonal antiserum, part of plant to be sampled and at what stage of plant development sampling should take place, list of buffers to be used at each step of the procedure, the length and temperature of each incubation, how long the substrate should develop before absorbance values are taken, and what should be used as positive and negative controls.

ELISA tests used in plant virology usually have two or three steps that use an antibody. The antibody applied directly to the microtiter plate is usually referred to as the trapping or coating antibody, since its purpose is to selectively bind the antigen of interest to the plate. The coating antibody is applied as purified IgG diluted in an appropriate buffer (usually carbonate or PBS). The second antibody (often the same source of IgG as the coating antibody) is conjugated with an enzyme in the standard double antibody sandwich (DAS) ELISA and is referred to as the conjugate or detecting antibody. In triple antibody sandwich (TAS) ELISA, the coating is done in the same manner as in DAS-ELISA but the second antibody (primary antibody) is specific for the antigen of interest and produced in a different animal than the trapping or coating antibody. The primary antibody is then followed by a conjugated antibody (secondary antibody) that is specific for antibodies produced by the animal that was the source of the primary antibody. For example, if the primary antibody was a McAb produced in a mouse, then the secondary antibody might be a rabbit-anti-mouse conjugate. I generally prefer to have a conjugate made in the same host species as the coating antibody to prevent cross reaction between conjugate and coating antibody.

ELISA procedures have been divided into direct and indirect methods (62). Direct ELISA methods involve conjugation of the antiviral antibody to an enzyme which subsequently is used to indicate the formation of an antibody-antigen complex. Indirect methods use a conjugate that is specific for the primary antibody. These conjugates are often antibodies specific for the detecting antibody ie. rabbit anti-mouse, goat anti-mouse, goat anti-rabbit, or protein A conjugates which are specific for the Fc fragment (C-terminal half of the heavy chains of an IgG molecule, this half has no activity as an antibody but carries most of the carbohydrate moiety of an IgG molecule) of the antibody. One advantage of indirect systems is that a single conjugate, available commercially, can be used to detect a number of different viruses. When protein A enzyme conjugates are used to detect primary antibody the Fc fragment of the trapping antibody must be cleaved (4).

When using McAbs for virus detection or diagnosis, the substrate can usually be incubated overnight to increase the sensitivity of the assay. In this way, McAbs can be used to detect virus in individual aphids using a standard ELISA protocol (73,94). In our laboratory we routinely take absorbance readings 1-2 h after adding substrate and again after overnight incubation at room temperature. With some virus-host systems readings can be taken after 15 minutes. The time of incubation is not critical.

The most important aspect of developing an assay is to maximize the absorbance of infected/healthy samples. A good polyclonal or monoclonal antiserum can be used to develop an assay that does not require statistical analysis to differentiate between known positive and negative samples. If such a test is available, samples that give borderline results should be retested. It must be remembered however, that even with the best of detection methods, recently infected samples may well give a borderline result. The application of statistics to determine when an absorbance value in ELISA represents a positive result has been presented elsewhere (92).

Tissue Blotting

Tissue blotting for virus detection has some advantages over ELISA. It is similar to ELISA in that the sample is blotted onto a solid support (nitrocellulose or filter paper), then probed with antibodies specific for the virus (50,52,70). A second antibody that is conjugated and specific for the first antibody can then be used to detect the presence of the virus. The enzyme hydrolysis of the substrate usually results in a precipitate that can be visualized. One disadvantage of using precipitating substrates is that they can be obscured by pigment from the plant sap, reducing the sensitivity of the assay. Recently developed chemiluminescent subsrates which emit light upon hydrolysis are not affected by plant pigments and therefore tend to have greater sensitivity. Light emitting substrates the advantage of providing a permanent copy of the results since they can be easily recorded using X-ray film. Nitrocellulose filters with precipitating substrate can be fragile and the color of the precipitate tends to fade over time. It may be necessary to follow tissue blotting with fluorescence or electron microscopy to localize the virus to subcellular sites. Nucleic acid probes can also be used to identify pathogens in tissue blots (17,51,78,81,88).

Immnuospecific Electron Microscopy (ISEM)

In ISEM the virus is visualized with an electron microscope. The virus can be decorated with antiserum before staining. Decoration is useful for examining relationships between viruses and looking at mixed infections, especially when the viruses have similar morphology. ISEM is not well suited to the testing of large numbers of samples but for small numbers of samples it can be much quicker than ELISA. ISEM is expensive and not always available.

Decoration may not be suitable to identify Tospo- and Rhabdovirus particles since they may have irregular shapes if part of the membrane is not present. The membrane proteins on these particles make it difficult to be sure that the halo is composed of antibody. Gold labelling of the antibody greatly enhances one's confidence in the identification of the virus because the gold particles are uniform and very easily recognized in the electron microscope.

NUCLEIC ACID HYBRIDIZATION ASSAYS

Detection of pathogens by nucleic acid hybridization is based on the specific pairing between the target nucleic acid sequence (denatured DNA or RNA) and a complementary nucleic acid probe to form double stranded nucleic acids. Thus, either RNA or DNA sequences may be used as probes. The kinetics of hybridization and the stability of the duplex is reviewed elsewhere (13). If the probe is labelled with a detectable marker, e.g., ^{32}P, a protein or an enzyme, then the formation of the duplex can be assayed after removal of unhybridized probe. For detection of plant viruses, hybridizations are usually carried out on solid filter supports (1) where the target nucleic acids are immobilized and the labelled nucleic acid probe is allowed to hybridize to them. Nitrocellulose and charged nylon are the most commonly used filters for hybridization. Nylon based filters are easier to handle and can be rehybridized several times (36). Methods have been developed to covalently bind nucleic acids to filters. This prevents the filter-bound nucleic acids from being washed off during hybridization. These include uv-induced-binding (19) and alkali treatment. Hybridization reactions are affected by probe concentration, base composition, type of probes (RNA or DNA), temperature, ionic strength, pH, viscosity, levels of formamide in the hybridization solution, and degree of mismatching (1,75).

The use of nucleic acid hybridization assays for virus detection has been reviewed recently (18,54). The greatest improvements in hybridization assays in recent years have been the advances in nonradioactive detection systems. Several methods of labelling nucleic acids have been developed. Incorporation of modified nucleotides such as biotin-11-UTP (63) digoxigenin tagged UTP (11) which are then detected by streptavidin or an anti-digoxigenin antibody, respectively. Modification of probe by direct binding of biotin to probe (69,85) again detected with streptavidin. A third approach has been to develop antibodies to chemically modified DNA (39) where the antibodies are used to detect probes modified in the same manner.

Digoxigenin labelled dUTP appears to be the most widely used nonradioactive tag for labelling probes for detection of plant viruses. Digoxigenin is linked via a spacer arm to dUTP and then incorporated into probe with the same enzymes used to make ^{32}P-labelled probes (32). Antibodies specific for the digoxigenin and conjugated to an enzyme (usually alkaline phosphatase or horse radish peroxidase) are then used to complex with the bound probe. Either a precipitating or a light emitting substrate is then used to visualize the presence of the probe. Many biotechnology supply companies now market kits for tagging probes with nonradioactive labels.

There are two types substrates that are widely used for detection of nonradioactive probes. These are the same ones used in DIBA the precipitating and chemiluminescent substrates. With the precipitating substrates, detection of nucleic acids in clean samples was approximately as sensitive as with ³²P labelled probes. However when relatively crude plant extracts were used, the pigmentation from the sample interfered with the assay. Chemiluminescent technology is not affected by the plant pigments since the signal is light emmission rather than color cevelopment. Several systems claim sensitivity equal to that obtained with radioactive probes. The improved sensitivity will lead to the adoption of this technology in laboratories and testing facilities that previously did not use nucleic acid based tests because of the radioactive labels. With the development of nonradioactive probes tissue- and squash blot assays should become more widely used in studies on virus epidemiology and ecology. The sample preparation is easy and now facilities not equipped to handle radioisotopes will be able to perform these assays (12,15,51,78,81).

During the last 5-6 years many researchers have shown that nonradioactive probes can be used successfully for virus detection and can be substituted for ³²P without any loss of sensitivity (24,28,34, 41,61,74,81).

Prior to the development of nonradioactive probes nucleic acid hybridization technology did not have much of an impact in testing of field samples for virus infection. PCR is becoming used in many diagnostic laboratories suggesting that the problem with the application of nucleic acid based detection was the required use of isotopes and the lack of a quantifiable signal (8).

The sensitivity of detection of plant viruses by nucleic acid hybridization is roughly similar to that of ELISA (6,18,34,74). It was found that hybridization assays were more sensitive than ELISA with subterranean clover stunt virus (SCSV) when purified virus was used (18) however, when infected tissues were used, ELISA was found to be more sensitive: the pathogen was detectable at a sap dilution of 1/625 while hybridization assays had a dilution limit of 1/125. The sensitivity of nucleic acid hybridizaiton and ELISA were similar with barley yellow dwarf and cherry leaf roll viruses (34,74). However, nucleic acid hybridization was more sensitive than ELISA in detecting tomato ringspot virus from bark or root tissue of infected nectarine trees (84).

Comparison of sensitivity between different assays should be made using plant extracts rather than purified virus. Another consideration is the ease of carrying out an assay. If two methods of detection are sensitive enough to meet the needs of an experiment, then the method that is simpler to carry out likely will be used. The expertise of the worker will determine which test is simpler. Someone who has extensive experience working with nucleic acids will be more inclined to use PCR or hybridization assays rather than ELISA; the converse is true for someone who is more familiar with serology than nucleic acid based tests.

There are many viruses of woody plants that have only been described in terms of symptomatology (22,38). It is likely that nucleic acid based detection will be available before serological methods are developed for these viruses. Developments in cloning viral specific dsRNAs of plant viruses (57,58,100) make it possible to readily make cDNA from dsRNA templates. Thus, for viruses where dsRNA can be extracted from diseased tissues, probes for nucleic acid hybridization or sequence information required to develop a PCR test are realistic short term goals. Once the sequences are known it will be possible to prepare antibodies to the coat proteins of these viruses for use in serological assays if this is desired.

DETECTION BASED ON MORE TRADITIONAL METHODS

Diagnostic tools are not available to detect many of the viruses of woody plants with serological or nucleic acid based tests because it has not been possible to purify the viruses. It is still necessary to carry out graft transmissions or mechanical inoculations onto herbaceous hosts for these viruses. Attempts to improve certification schemes for tree fruit or small fruit crops is limited because these tests are labor intensive, and especially with grafting, require substantial amounts of space to grow test plants. It may be several years after grafting before the results of the test are obtained.

The introduction of new germplasm from collections made in the wild present the possibility that undescribed viruses will be encountered (91). New viruses encountered will not be on any quarantine list but may still pose a biological risk. This material should be assayed with a broad spectrum test such as mechanical transmission, grafting or dsRNA analysis. Quite often we tend to be concerned about the viruses that we are familiar with and forget those that have not yet been described.

VIRUS DETECTION IN PLANTS TRANSGENIC FOR A VIRAL GENE

Hundreds of different plants have been transformed with viral sequences (9,33,71) since the development of the first transgenic plants carrying the viral coat protein (83). Incorporation of viral genes into a plant could interfere with interpretation of results obtained with standard virus testing procedures. Virus coat protein gene expressed in a plant may well give positive ELISA results for the transgenic plant because of the expressed protein. Transforming plants with any viral sequences could interfere with the detection of virus by nucleic acid hybridization. ELISA and nucleic acid hybridization will still be useful for monitoring virus infection in transgenic plants carrying a viral gene or sequences. It may require that specific probes be used for detection. As transgenic plants reach commercialization it will be necessary to know how the inserted gene may interfere with virus detection.

Low expression levels of the coat protein in transgenic plants may result in an increased background signal, reducing the reliability of an ELISA test. It may be possible to use antibodies to other viral proteins when coat proteins are expressed in plants. Monoclonal antibodies to the helper-component protease (HC-Pro) of PVY resulted in ELISA tests with the same specificity as antibodies to PVY coat protein. The HC-Pro PVY McAbs could be used to divide PVY into subgroups that matched the grouping based on coat protein McAbs (16). When nonstructural genes are used to transform plants, ELISA based on coat protein will still be a useful tool to monitor virus infection. Potyviruses replicate via a polyprotein which means all proteins should be produced in equimolar amounts. Therefore, it should be feasible to use antibodies to nonstructural proteins for detection without loss of sensitivity. Where coat protein is expressed from a subgenomic messenger RNA it is likely that some of the nonstructural proteins will be present in much lower amounts than the coat protein. With these viruses, ie. luteo, sobemo, carmo, tombus etc. it is likely that ELISA tests based on nonstructural proteins will not be as sensitive as those based on coat protein.

It might appear that the use of monoclonal antibodies that only react with intact viral particles (60,95) would be useful for virus detection in transgenic plants. Not so for two reasons. First, some viruses are known to form empty particles during infection. Similar empty particles have been observed from the expressed coat protein of arabis mosaic virus in transgenic plants (10). Alternately, if the expressed protein can encapsidate RNA from other viruses, then these antibodies could react with the transencapsidated particle. There are several examples of transgenic plants where the expressed coat protein has encapsidated RNA from other viruses (31,65).

Transformation of plants with untranslatable viral sequences, either through the removal of the AUG start codon (82), through the introduction of a stop codon immediately downstream of the start codon (66), or through the use of antisense constructs (60,96) should not interfere with the use of standard virus detection protocols based on ELISA. There are no viral proteins expressed in these transgenic plants. What will be required for hybridization based detection of virus in these plants is the use of a probe that is specific for a region of the virus other than the transformed gene (90). Successful PCR based detection will require the use of a set of virus specific primers with at least one primer outside of the incorporated gene.

Commercialization of transgenic plants with viral genes or sequences will impact certification schemes. However, since the exact viral sequences that have been incorporated are known it will be possible and relatively straight forward to develop a test that will be suitable for any transgenic plant. It may become a necessity for a certification program to incorporate more than one type of test into its scheme. If all tests were previously based on serology, it might be necessary to incorporate a PCR or hybiridization test with transgenic plants.

THE SIGNIFICANCE OF A TEST RESULT

In situations where a test result has significant biological or economic impact one should use more than one type of test to confirm the presence or absence of a virus. The recent incidence of PVY[N] in Canada and its implications for shipping seed potatoes to the USA is an excellent example of where the test result can have considerable economic impact and a confirmatory test could potentially save a lot of money. The effect of mixed infections confounded the results of the bioassay leading to a large percentage of false positives. As the movement of agricultural products between countries increases, quarantine restrictions based on plant pathogens will continue to be important. For example, several *Prunus* spp. were found to be infected with virus isolates which cross-reacted with plum pox potyvirus (PPV) antisera in ELISA (43,55). In this instance the impact of the test result was very significant since plum pox is a quarantine status virus not known to occur in North America. Plum pox is a member of the potyviridae which produce diagnostic inclusions in infected hosts. The prunus virus isolates did not cause these inclusions, their coat proteins were atypical of members of the potyviridae (49,56) and RT-PCR tests with oligo-nucleotides specific to the 3' or 5' noncoding regions of plum pox virus did not amplify any fragments (43,56). It is now thought that these prunus virus isolates are not plum pox virus despite their reaction with antisera specific to plum pox virus. Reliance solely on the initial test results would have had very serious implications

Another important consideration when using diagnostic tests that are not based on biological activity is the significance of the result. For example, many people are aware that PCR has been used to amplify DNA from insects embedded in amber for millions of years. This positive "test" for insect DNA does not show that the insect is living. A similar situation could arrive when using laboratory tests to index plants for viruses. It has been shown that biologically active prunus necrotic ringspot virus is slowly lost from *Prunus pennsylvanica* seed (35). Would ELISA, PCR or hybridization give positive results when biologically the virus was no longer seed transmitted? In the case of biologcial tests it is also important to know that the symptoms observed are due to the virus in question rather than a mixed infection as was the case with some of the PVY^N testing mentioned above

The legal implications of a decision to refuse a shipment of plant product based on a positive test which does not consider biological activity is an issue that needs to be discussed. If a shipment of grain that has been fumigated with methyl bromide is assayed for a specific fungal pathogen by PCR, it would be possible to get a positive result based on non-living fungal tissue. Similarily, the results of a positive ELISA test for plum pox potyvirus could be used to turn back a shipment of *Prunus* planting stock, when in fact the material was free of plum pox potyvirus. Where does the liability of the testing agency end in such cases? These questions need to be addressed.

REFERENCES

1. Anderson, M.L.M., and Young, B.D. 1985. Quantitative filter hybridization. Pages 73-111 in: Nucleic Acid Hybridization, A Practical Approach. B.D. Hames and S.J. Higgins eds. IRL Press, Oxford.
2. Ball, E.M. 1990. Agar diffusion techniques, introduction. Pages 97-100 in: Serological Methods for Detection and Identification of Viral and Bacterial Plant Pathogens. R. Hampton, E. Ball and S. DeBoer eds. APS Press, St. Paul.
3. Banttari, E.E., and Goodwin, P.H. 1985. Detection of potato viruses S, X, Y, by enzyme-linked immunosorbent assay on nitrocellulose membranes (dot-ELISA). Plant Dis. 69:202-205.
4. Barbara, D.J., and Clark, M.F. 1982. A simple indirect ELISA using F(ab')^2 fragments of immunoglobulin. J. Gen. Virol. 58:315-322.
5. Barbara, D.J., Jones, A.T., Henderson, S.J., Wilson, S.C., and Knight, V.H. 1984. Isolates of raspberry bushy dwarf virus differing in *Rubus* host range. Ann. Appl. Biol. 105:49-54.
6. Barbara, D.J., Dawata, E.E., Ueng, P.P., Lister, R.M., and Larkins, B.A. 1987. Production of complementary DNA clones from the MAV isolate of barley yellow dwarf virus. J. Gen. Virol. 68:2419-2428.
7. Barbara, D.J., Morton, A., Spence, N.J., and Miller, A. 1995. Rapid differentiation of closely related isolates of two plant viruses by polymerase chain reaction and restriction fragment legth polymorphism analysis. J. Virol. Methods 55:121-131.
8. Barnes, L.W. 1994. The role of plant clinics in disease diagnosis and education: A North American perspective. Annu. Rev. Phytopathol. 32:601-609.
9. Beachy, R.N., Loesch-Fries, S., and Tumer, N.E. 1990. Coat protein-mediated resistance against virus infection. Annu. Rev. Phytopathol. 28:451-474.
10. Bertioli, D.J., Harris, R.D., Edwards, M.L. Cooper, J.I., and Hawes, W.S. 1991. Transgenic plants and insect cells expressing the coat protein of arabis mosaic virus produce empty virus-like particles. J. Gen. Virol. 72:1801-1809.
11. Boehringer Mannheim. 1993. The DIG system user's guide for filter hybridization. Biochemica 90 Pages.
12. Boulton, M.I., and Markham, P.G. 1986. The use of squash-blotting to detect plant pathogens in insect vectors. in: Developments and Applications in Virus Testing: Proceedings of a Conference at the University of Cambridge, 10-12 April, 1985. R.A.C. Jones and L. Torrance, eds. Univ. of Cambridge Press, Cambridge, UK. 300 Pages.
13. Britten, R.J., and Davidson, E.H. 1985. Hybridization strategy. Pages 3-15 in: Nucleic Acid Hybridization, A Practical Approach. B.D. Hames and S.J. Higgins, eds. IRL Press, Oxford.
14. Bulger, M.A., Stace-Smith, R., and Martin, R.R. 1990. Transmission and field spread of raspberry bushy dwarf virus. Plant Dis. 74:514-517.
15. Candresse, T., Macquaire, G., Lanne, M., Bousalem, M., Quiot-Douine, L., Quiot, J.B., and Dunez, J. 1995. Analysis of plum pox virus variability and development of strain-specific PCR assays. Acta Hortic. 386:357-365.

16. Canto, T., Ellis, P., Bowler, G., and Lopez-Abella, D. 1995. Production of monoclonal antibodies to potato virus Y helper component-protease and their use for strain differentiation. Plant Dis. 79:234-237.

17. Chevalier, S., Greif, C., Clauzel, J.M. Walter, B., and Fritsch, C. 1995. Use of an immunocapture-polymerase chain reaction procedure for the detection of grapevine virus A in Kober stem grooving-infected grapevines. J. Phytopathol. 143:369-373.

18. Chu, P.W.G., Waterhouse, P.M., Martin, R.R., and Gerlach, W.L. 1989. New approaches to the detection of microbial plant pathogens. Biotechnology and Genetic Engineering Reviews 7:45-111.

19. Church, G.M., and Gilbert, W. 1984. Genomic sequencing. Proc. Natl. Acad. Sci. USA 81:1991-1995.

20. Clark, M.F., and Adams, A.N. 1977. Character-istics of the microplate method of enzyme-linked immunosorbent assay for the detection of plant viruses. J. Gen. Virol. 34:475-483

21. Clark, M.F., and Bar-Joseph, M. 1984. Enzyme immunosorbent assays in plant virology. Methods in Virology 7:51-85.

22. Converse, R.H., ed. 1987. Virus Diseases of Small Fruits. USDA-ARS Agriculture Handbk. #631.

23. Converse, R.H., and Martin, R.R. 1990. ELISA for plant viruses. Pages 179-196. in: Serological Methods for Detection and Identification of Viral and Bacterial Plant Pathogens. R. Hamptom, E. Ball, and S. DeBoer, eds. APS Press, St. Paul, MN.

24. Crespi, S., Accotto, G.P. Caciagli, P., and Gronenborn, B. 1991. Use of digoxigenin-labelled probes for detection and host-range studies of tomato yellow leaf curl geminivirus. Res. Virol. 142:283-288.

25. D'Arcy, C.J., Torrance, L., and Martin, R.R. 1989. Discrimination among luteoviruses and their strains by monoclonal antibodies and identification of common epitopes. Phytopathology 79:869-873.

26. Derrick, K.S. 1990. Serologically specific electron microscopy (SSEM). Pages 313-319 in: Serological Methods for Detection and Identification of Viral and Bacterial Plant Pathogens. R. Hampton, E. Ball and S. DeBoer, eds. APS Press, St. Paul.

27. Diekmann, M., Frison, E.A., and Putter, R. eds. 1994. FAO/IPGRI Technical Guidelines for the Safe Movement of Small Fruit Germplasm. Food and Agriculture Organization of the United Nations, Rome/International Plant Genetic Resources Institute, Rome.

28. During, K. 1991. Ultrasensitive chemiluminescent and colorigenic detection of DNA, RNA, and proteins in plant molecular biology. Anal. Biochem. 196:433-438.

29. Ellis, P.J. 1992. Weed hosts of beet western yellows virus and potato leafroll virus in British Columbia. Plant Dis. 76:1137-1139.

30. Ellis, P., and Stace-Smith, R. 1993. Beet western yellows virus is not an important component of potato leafroll disease in Canada and the United States. Plant Dis. 77:718-721.

31. Farinelli, L., Malnoae, P., and Collet, G.F. 1992. Heterologous encapsidation of potato virus Y strain O (PVYO) with the transgenic coat protein of PVY stran N (PVYN) in Solanum tuberosum cv. Bintje. BioTechnology 10:1020-1025.

32. Feinberg, A.P., and Vogelstein, B. 1983. A technique for radiolabelling DNA resriction endonuclease fragments to high specific activity. Anal. Biochem. 132:6.13.

33. Fitchen, J.H., and Beachy, R.N. 1993. Genetically engineered protection against viruses in transgenic plants. Annu. Rev. Microbiol. 47:739-763.

34. Fouly, B.M., Domier, L.L., and D'Arcy, C.J. 1992. A rapid chemiluminescent detection method for barley yellow dwarf virus. J. Virol. Methods 39:291-298.

35. Fulton, R.W. 1964. Transmission of plant viruses by grafting, dodder, seed and mechanical innoculation. pages 39-97. in: Plant Virology. M.K. Corbett and H.D. Sisler, eds. Univ. of Florida Press, Gainsville, Florida.

36. Gatti, R.A., Concanon, P., and Salser, W. 1984. Multiple use of Southern blots. BioTechniques 2:148-155.

37. Gildow, F.E. 1993. Evidence for receptor-mediated endocytosis regulating luteovirus acquisition by aphids. Phytopathology 83:270-277.

38. Gilmer, R.M., Moore, J.D., Nyland, G., Welsh, M.F., and Pine, T.S., eds. 1976. Virus Diseases and Noninfectious Disorders of Stone Fruits in North America. USDA-ARS Agriculture Handbk. #437.

39. Gratzner, H.C. 1982. Monoclonal antibodies to 5-bromo- and 5-iododeoxyuridine: A new reagent for detection of DNA replication. Science 218:474-475.

40. Haber, S., and Knapen, H. 1989. Filter paper sero-assay (FiPSA): a rapid, sensitive technique for sero-diagnosis of plant viruses. Can. J. Plant Pathol. 11:109-113.

41. Haber, S., Wakarchuk, D.A., Cvitkovitch, S., and Murray, G. 1992. Diagnosis of flame chlorosis, a viruslike disease of cereals, by detection of disease-specific RNA with digoxigenin-labeled RNA probes. Plant Dis. 76:590-594.

42. Hadidi, A., and Hammond, R.W. 1989. Construction of molecular clones for identification and detection of tomato ringspot and arabis mosaic viruses. Acta Hortic. 235:223-230.

43. Hadidi, A., and Levy, L. 1994. Accurate identification of plum pox virus and its'differentiation from Asian prunus latent potyvirus in Prunus germplasm. EPPO Bull. 24:633-643.

44. Hadidi, A., Levy, L. and Podleckis, E.V. 1995. Polymerase chain reaction technology in plant pathology. Pages 167-187 in: Molecular Methods in Plant Pathology. R.P. Singh and U.S. Singh, eds. CRC Press, Boca Raton, FL.

45. Hadidi, A., and Powell, C.A. 1991. Complementary DNA cloning and analysis of RNA 2 of a Prunus stem-pitting isolate of tomato ringspot virus. Molecular and Cellular Probes 5:337-344.

46. Halk, E.L. and DeBoer, S.H. 1985. Monoclonal antibodies in plant disease research. Annu. Rev. Phytopathol. 23:321-350.

47. Hammond, J., and Jordan, R.L. 1991. Monoclonal antibodies against potyvirus-associated antigens, hybrid cell lines producing these antibodies, and use therefore. Patent application, U.S. Patents and Trademarks Office, Washington, D.C. 20231.

48. Hampton, R., Ball, E., and DeBoer, S., eds. 1990. Serological Methods for Detection and Identification of Viral and Bacterial Plant Pathogens. APS Press, St. Paul, MN. 389 Pages.

49.   Hari, V., Abdel Ghaffar, M.H., Levy, L., and Hadidi, A. 1995. Asian prunus latent virus: An unusual potyvirus detected in germplasm from East Asia. Acta Hortic. 386:78-82.

50.   Holt, C.A. 1992. Detection and Localization of Plant Pathogens. Pages 127-137 in: Tissue Printing. P.D. Reid, R.F. Pont-Lezica, E.D. Campillo, and R. Taylor, eds. Academic Press, New York.

51.   Hooftman, R., Arts, M-J., Shamloul, A.M., Van Zaayen, A., and Hadidi, A. 1996. Detection of chrysanthemum stunt viroid by reverse transcription-polymerase chain reaction and by tissue blot hybridization. Acta Hortic. 432:120-128.

52.   Hsu, H.T., and Lawson, R.H. 1991. Direct tissue blotting for detection of tomato spotted wilt virus in Impatiens. Plant Dis. 75:292-295.

53.   Huguenot, C., Furneaux, M.I., and Hamilton, R.I. 1994. Capsid protein properties of cowpea aphid-borne mosaic virus and blackeye cowpea mosaic virus confirm the existence of two major subgroups of aphid-transmitted, legume-infecting potyviruses. J.Gen. Virol. 75:3555-3560.

54.   Hull, R. 1988. Rapid-diagnosis of plant virus infections by spot hybridization. Trends Biotechnol. 2:88-91.

55.   James, D., Thompson, D.A., and Godkin, S.E. 1994. Cross reactions of an antiserum to plum pox virus. EPPO Buletin 24:605-614.

56.   James, D., Godkin, S.E., Eastwell, K.C., and MacKenzie, D.J. 1996. Identification and differentiation of Prunus virus isolates that cross react with plum pox virus and apple stem pitting virus antisera. Plant Dis. 80:536-543.

57.   Jelkmann, W., Martin, R.R., and Maiss, E. 1989. cDNA cloning of four plant viruses from dsRNA templates. Phytopathology 79:1250-1253.

58.   Jelkmann, W., Maiss, E., and Martin, R.R. 1992. Sequence of a potexvirus associated with strawberry mild yellow edge virus. J. Gen. Virol. 73:475-479.

59.   Jordan, R.L. 1990. Strategy and techniques for the production of monoclonal antibodies; monoclonal antibody applications for viruses. Pages 55-85. in: Serological Methods for Detection and Identification of Viral and Bacterial Plant Pathogens. R. Hampton, E. Ball, and S. DeBoer, eds. APS Press, St. Paul, Mn.

60.   Kawchuk, L.M., Martin, R.R., and McPherson, J. 1991. Sense and antisense RNA-mediated resistance to potato leafroll virus in Russet Burbank potato plants. Mol. Plant-Microbe Interact. 4:247-253.

61.   Kimpton, C.P., Corbitt, G., and Morris, D.J. 1989. Detection of CMV DNA using probes labeled with digoxigenin. J. Virol. Methods 24:335-346.

62.   Koenig, R., and Paul, H.L. 1982. Variants of ELISA in plant virus diagnosis. J. Virol. Methods 5:113-125.

63.   Langer, P.R., Waldrop, A.A., and Ward, D.C. 1981. Enzymatic synthesis of biotin-labelled polynucleotides: Novel nucleic acid affinity probes. Proc. Natl. Acad. Sci. USA 75:6633-6637.

64.   Langeveld, S.A., Dore, J.M., Memelink, J., Derks, A.F.L.M., van der Vlugt, C.I.M., Asjes, C.J., and Bol, J.F. 1991. Identification of potyviruses using the polymerase chain reaction with degenerate primers. J.Gen. Virol. 72:1531-1541.

65.   Lecoq, H., Ravelonandro, M., Wipf-Scheibel, C., Monsion, M., Raccah, B., and Dunez, J. 1993. Aphid transmission of a non-aphid-transmissible strain of zucchini yellow mosaic potyvirus from transgenic plants expressing the capsid protein of plum pox potyvirus. Mol. Plant-Microbe Interact. 6:403-406.

66.   Lindbo, J.A., Silva-Rosales, L., and Doughtery, W.G. 1993. Pathogen derived resistance to potyviruses: working, but why? Semin. Virol. 4:369-379.

67.   Lister, R.M. 1978. Application of the enzyme-linked immunosorbent assay for detecting viruses in soybean seed and plants. Phytopathology 68:1393-1400.

68.   MacDonald, J.G., Kristjansson, G.T., Singh, R.P., Ellis, P.J., and McNab, W.B. 1994. Consecutive ELISA screening with monoclonal antibodies to detect potato virus Y^N. Am. Potato J. 71:175-183.

69.   McInnes, J.L., Habili, N., and Symons, R.H. 1989. Non-radioactive, photobiotin-labelled DNA probes for the routine diagnosis of viroids in plant extracts. J. Virol. Methods 23:299-312.

70.   Makkouk, K.M., Hsu, H.T., and Kumari, S.G. 1993. Detection of three plant viruses by dot-blot and tissue-blot immunoassays using chemiluminescent and chromogenic substrates. J. Phytopathol. 139:97-102.

71.   Martin, R.R. 1995. Alternatives to the use of virus coat protein for engineering virus resistance in plants. Acta Hortic. 385:18-28.

72.   Martin, R.R., and Bristow, P.R. 1988. A carlavirus associated with blueberry scorch disease. Phytopathology 78:1636-1640.

73.   Martin, R.R., and Ellis, P. 1986. Detection of potato leafroll and beet western yellows viruses in aphids. in: Proc. of Workshop on Epidemiology of Plant Virus Diseases, Orlando, Florida, August 6-8, 1986.

74.   Mas, P. Sanchez-Navarro, J.A., Sanchez-Pina, M.A., and Pallas, V. 1993. Chemiluminescent and colorigenic detection of cherry leaf roll virus with digoxigenin-labeled RNA probes. J. Virol. Methods 45:93-102.

75.   Meinkoth, J., and Wahl, G. 1984. Hybridization of nucleic acids immobilized on solid supports. Anal. Biochem. 138:267-284.

76.   Miller, S.A. and Martin, R.R. 1988. Molecular diagnosis of plant disease. Annu. Rev. Phytopathol. 26:409-432.

77.   Minafra, A., and Hadidi, A. 1994. Sensitive detection of grapevine virus A,B, or leafroll-associated III from viruliferous mealybugs and infected tissue by cDNA amplification. J. Virol. Methods 47:175-188.

78.   Navot, N., Ber, R., and Czosnek, H. 1989. Rapid detction of tomato yellow leaf curl virus in squashes of plants and insect vectors. Phytopathology 79:562-568.

79.   Nemchinov, L., Hadidi, A., Candresse, T., Foster, J.A., and Verderevskaya, T.D. 1995. Sensitive detection of apple chlorotic leaf spot virus from infected apple or peach tissue using RT-PCR, IC-RT-PCR, or multiplex IC-PCR. Acta Hortic. 386:51-62.

80.   Nemchinov, L., Hadidi, A., Maiss, E., Cambra, M., Candresse, T., and Damsteegt, V. 1996. Sour cherry strain of plum pox potyvirus (PPV): molecular and serological evidence for a new subgroup of PPV strains. Phytopathology 86:1215-1221.

81.   Podleckis, E.V., Hammond, R.W., Hurtt, S., and Hadidi, A. 1993. Chemiluminescent detection of

potato and pome fruit viroids by digoxigenin-labeled dot blot and tissue blot hybridization. J. Virol. Methods 43:147-158.

82. Powell, P.A., Sanders, P.R., Tumer, N., Fraley, R.T., and Beachy, R.N. 1990. Protection against tobacco mosaic virus infection in transgenic plants requires accumulation of coat protein rather than coat protein RNA sequences. Virology 175:124-130.

83. Powell-Abel, P., Nelson, R.S., De, B., Hoffman, N., Rogers, S.G., Fraley, R.T., and Beachy, R.N. 1986. Delay of disease development in transgenic plants that express the tobacco mosaic virus coat protein gene. Science 232:738-743.

84. Powell, C.A., Hadidi, A., and Halbrendt, J.M. 1991. Detection of tomato ringspot virus in nectarine trees using ELISA and transcribed RNA probes. HortScience 26:1290-1292.

85. Reisfeld, A., Rothenberg, J.M., Bayer, E.A., and Wilchek, M. 1987. Non-radioactive hybridization probes prepared by the reaction of biotin hydrazide with DNA. Biochem. Biophys. Res. Comm. 142:519-526.

86. Robertson, N.L., French, R., and Gray, S.M. 1991. Use of group-specific primers and the polymerase chain reaction for the detection and identification of luteoviruses. J. Gen. Virol. 72:1473-1477.

87. Rojas, M.R., Gilbertson, R.L., Russell, D.R., and Maxwell, D.P. 1993. Use of degenerate primers in the polymrase chain reaction to detect whitefly-transmitted geminiviruses. Plant Dis. 77:340-347.

88. Romero-Durban, J., Cambra, M., and Duran-Vila, N. 1995. A simple imprint-hybridization method for detection of viroids. J. Virol. Methods 55:37-47.

89. Rose, D.G., and Hubbard, A.L. 1986. Production of monoclonal antibodies for the detection of potato virus Y. Ann. Appl. Biol. 109:317-321.

90. Smith, O.P., Damsteegt, V.D., Keller, C.J., Beck, R.J., and Hewings, A.D. 1993. Detection of potato leafroll virus in leaf and aphid extracts by dot-blot hybridization. Plant Dis. 77:1098-1102.

91. Spiegel, S., Martin, R.R., Legget, F., ter Borg, M., and Postman, J. 1993. Characterization and geographical distribution of a new ilarvirus from *Fragaria chiloensis*. Phytopathology 83:991-995.

92. Sutula, C.L., Gillet, J.M., Morrissey, S.M., and Ramsdell, D.C. 1986. Interpreting ELISA data and establishing the positive-negative threshold. Plant Dis. 70:722-726.

93. Thomas, P.E., Hang, A.N., Reed, G., Gilliland, G.C., and Reisenauer, G. 1993. Potential role of winter rapeseed culture on the epidemiology of potato leaf roll disease. Plant Dis. 77:420-423.

94. Torrance, L. 1987. Use of enzyme amplification in an ELISA to increase sensitivity of detection of barley yellow dwarf virus in oats and in individual vector aphids. J. Virol. Methods 15:131-138.

95. Tremaine, J.H., Ronald, W.P., and MacKenzie, D.J. 1985. Southern bean mosaic virus monoclonal antibodies: reactivity with virus strains and with the virus antigen in different conformations. Phytopathology 75:1208-1212.

96. van der Wilk, R., Willink, D.P.L., Huisman, M.J., Huttinga, H., and Goldbach, R. 1991. Expression of the potato leafroll luteovirus coat protein gene in transgenic potato plants inhibits viral infection. Plant Molec. Biol. 17:431-439.

97. Van Regenmortel, M.H.V. 1982. Serology and Immunochemistry of Plant Viruses. Academic Press, New York. 302 Pages.

98. Varveri, C., Ravelonandro, M., and Dunez, J. 1987. Construction and use of a cloned cDNA probe for the detection of plum pox virus in plants. Phytopathology 77:1221-1224.

99. Wetzel, T., Candresse, T., Macquaire, G. Ravelonandro, M., and Dunez, J. 1992. A highly sensitive immunocapture polymerase chain reaction method for plum pox potyvirus detection. J. Virol. Methods 39:27-37.

100. Winter, S., Purac, A., Leggett, F., Frison, E.A., Rossel, H.W., and Hamilton, R.I. 1992. Partial characterization and molecular cloning of a closterovirus from sweet potato infected with the sweet potato virus disease complex from Nigeria. Phytopathology 82:869-875.

CHAPTER 29

Identification and Detection of Recalcitrant Temperate Fruit Crop Viruses Using dsRNAs and Diffusion Antisera

W. Jelkmann

With the introduction of the microplate method of enzyme-linked immunosorbent assay (ELISA) to plant virology in the late 1970's (2,68), an important objective was achieved by specifically testing, with high sensitivity, a large number of samples for the presence or absence of a virus in a relatively short time. Subsequently, ELISA protocols have been modified in a number of ways to allow the detection of viruses from different plant hosts and tissues as well as to improve the sensitivity and reliability of detection. Tests have been performed as direct ELISA with enzymes covalently linked to the antiviral immunoglobin, or as indirect ELISA with enzymes coupled to a molecule that detects the antiviral immunoglobin (4,23). Among the temperate fruit crop viruses to be detected with ELISA using polyclonal antisera in the 1970's were plum pox virus (PPV), arabis mosaic virus (ArMV), apple mosaic virus (ApMV) (3), and apple chlorotic leafspot virus (ACLSV) (13). The number of temperate fruit crop viruses detected by ELISA using polyclonal antisera rapidly increased soon after this. In addition, in the early 1980's viral monoclonal antibodies (McAbs) were prepared and used in ELISA for the detection of more than 30 different plant viruses, including important viruses that infect temperate fruit crops such as prune dwarf virus (PDV), prunus necrotic ringspot (PNRS) and raspberry bushy dwarf (RBDV) (67). In subsequent years, continuous progress was achieved in detecting several other temperate fruit crop viruses with ELISA using viral polyclonal or monoclonal antibodies.

The detection of the causal agents of several important viral diseases and diseases of uncertain etiology which infect temperate fruit crops, however, is still largely based on biological indexing. This is because the viruses associated with these diseases are difficult to purify by conventional purification methods. Thus, the availability of viral antigen for the production of antibodies and of viral RNA for cDNA cloning and probe production is limited or nil. These diseases include little cherry, apple stem pitting, and related diseases in pome fruit as well as strawberry crinkle, strawberry mottle, and strawberry mild yellow edge. Viral RNA templates for cDNA synthesis are generally isolated from purified virion preparations. Those temperate fruit crop viruses that could be purified in sufficient quantities were cloned and sequenced (8,15,45,50,54,73) and can be detected by molecular hybridization methods.

An alternative method to provide viral RNA template for cDNA synthesis involves extraction and isolation of viral dsRNA from infected cells (7,56). Although this method has not become widely used for routine virus detection, it has been supportive for obtaining the starting material for virus characterization. It has been of particular value to better characterize recalcitrant temperate fruit crop viruses and their strains. Table 1 shows temperate fruit crop viruses for which viral dsRNA has been extracted from infected cells.

cDNA SYNTHESIS FROM dsRNA AND MOLECULAR CLONING

Molecular cloning of dsRNA has long been a well established method (1,9,16,25,63). It is essentially done following the protocols that are used for cDNA synthesis from single stranded

Table 1. Recalcitrant virus or virus-like diseases of temperate fruit crops for which associated viral dsRNAs were reported

Virus	Accession no.	Reference
Apple stem pitting virus [a]	D21829	
Pear vein yellows [a]	D21828	34, 60
Cherry green ring mottle virus [a]		74
Cherry mottle leaf virus [a]		[b]
Cherry necrotic rusty mottle		55
Cherry twisted leaf		28
Cherry virus A[a]	X82547	32
Little cherry virus	X93351; Y10237	10,22,32,47
Raspberry leafspot		44
Strawberry june yellows		69
Strawberry mild yellow edge [a]	D01227	51, 64
Strawberry mottle virus		49
Strawberry pallidosis		72

[a] Serological detection by immunosorbent electron microscopy (ISEM) or ELISA has been reported.
[b] D. James, personal communication.

RNA (ssRNA) (18). However, before the (+)- and/or (-) strand of the dsRNA is subjected to reverse transcription a denaturing step is required. This can be achieved by incubation in 90% dimethyl sulfoxide (DMSO) at 65°C for 30 min, incubation at room temperature in 20mM methylmercuric hydroxide (MeHg) for 10 min or boiling for 5-10 min. Whereas initially the cloning from large amounts of dsRNA was described the technique demonstrated to be useful for the cloning from less than microgram quantities of dsRNA "replicative form" templates purified from virus-infected fruit plants (37). Initiation of first strand synthesis of cDNA is generally done by the use of random hexamers or for polyadenylated dsRNAs by the use of oligo(dT) primers or chimeric oligo(dT) primers containing a polylinker sequence. Alternatively, homopolymeric tailing of dsRNA with rATP by poly(A)-polymerase prior to cDNA synthesis has been performed. Second strand synthesis has largely been done according to (18). Another choice is homopolymeric tailing

of first strand cDNA initiated with a specific or chimeric (dT)-primer followed by PCR amplification with the same chimeric primer. Cloning of dsRNA has also been achieved after PCR amplification with two specific primers when the region of interest was flanked by known sequences. Degenerate oligonucleotides derived from known sequences and corresponding to sequence motifs have been used to amplify sequences of related viruses (19,32,37,39, 40,52,59,65).

Largely depending on the amount of double-stranded cDNA a variety of methods have been described for molecular cloning. As for most of the recalcitrant temperate fruit viruses only small amounts of cDNA became available, homopolymeric tailing after second strand reaction andannealing with complementary tailed plasmid vectors has shown to be useful. Alternatively the ligation of linkers or blunt-end cloning have been described. Molecular cloning has been done in *Escherichia coli* or lambda systems.

MOLECULAR CHARACTERIZATION

Application of cDNA synthesis to dsRNA and molecular cloning followed by nucleotide sequence analysis revealed partial or complete viral genomes of several filamentous recalcitrant viruses of temperate fruit crops.

The genome of the newly identified strawberry mild yellow edge associated potexvirus (SMYEaV) has been obtained (35,36). This virus is the causal agent of the disease as demonstrated by inoculation of *Fragaria vesca* 'Alpine' indicator seedlings with an *in vivo* infectious cDNA clone (46). Whereas investigated strains of SMYEaV are transmitted by *Chaetosiphon fragaefolii*, aphid transmission does not occur after single infection with the infectious cDNA. The result supports a previous hypothesis that SMYEaV is transmitted by heterologous encapsidation with the suspected strawberry mild yellow edge (SMYE) luteovirus (5).

The genome of apple stem pitting virus (ASPV) has been determined and appeared to be most closely related to that of potexviruses, but with a larger coat protein of Mr 44 kDa (31). ASPV is recognized as a latent virus in commercial apple varieties and causes xylem pits in the stem of Virgina Crab and epinasty and decline in Spy 227. There are several reports on related diseases in pear and quince, but only pear vein yellows disease has been proven to be caused by an isolate of ASPV (31, 48). The nucleotide sequence of the filamentous virus causing cherry green ring mottle (GRMV) and pitting of the stem of sweet and shirofugen cherry encodes for a potexvirus like genome bearing a very close relationship to that of ASPV. However the coat protein has a size of 32 kDa which is in the range as typically observed with carlaviruses (75).

Molecular cloning of cDNA obtained from the high molecular weight dsRNA purified from a European isolate (UW1) of little cherry disease (LCD) unraveled the genome of a monopartite closterovirus (LChV) (33,41). Due to a mixed infection in the sample a hitherto undescribed capillovirus, cherry virus A (CVA), was identified and its genome characterized (32). While little cherry disease has a major impact on sweet cherry production and has a worldwide distribution, CVA currently cannot be related to a described disease in cherry. Preliminary data (W. Jelkmann, unpublished) suggest that CVA is widely distributed but latent in sweet cherry. The

development of an LCD specific cDNA clone has also been reported from the Canadian isolate LC5, obtained after transmission of a clostero-like virus by the mealybug *Phenacoccus aceris* from symptomatic *Prunus avium* (10).

VIRUS IDENTIFICATION AND DETECTION

The establishment of cDNA clone libraries of the above described filamentous recalcitrant viruses and their molecular characterization has provided the basics for a sensitive nucleic acid based detection. Moreover, amino acid alignments of the putatively encoded gene products of these viruses with data base entries allowed the identification of structural and non structural viral proteins. Such genes have been subcloned into procaryotic and eucaryotic expression systems and diffusion proteins have been produced. Diffusion protein antisera to the bacterially expressed coat proteins have been obtained for SMYEaV (36,38,61), ASPV(34), CVA (32) and LChV (41). These antisera have been of various qualities and differed in their usefulness for serological assays. Diffusion antisera that have been prepared for SMYEaV as fusion proteins to either protein A (pRIT2T; Pharmacia), glutathione S-transferase (pGEX; Pharmacia)or a tag of six histidines (pQE type IV; Quiagen) were able to trap and decorate virus particles in immunosorbent assays (ISEM). They also reacted specifically with the coat protein when used as probes in immunoblots. However, only an antiserum to the fusion protein from the pQE-system was specific to SMYEaV in an indirect plate trapping ELISA. It also was suitable in a TAS-ELISA for trapping in combination with a McAb as second antibody and in immunocapture RT-PCR (IC-RT-PCR) (38,61). Monoclonal and polyclonal antisera to SMYEaV useful in ELISA were also prepared after partial virus purification from *Chenopodium quinoa* co-infected with a second virus (30). The diffusion protein antisera to ASPV and CVA reacted in ISEM and immunoblots but failed to detect virus particles in DAS-ELISA (31,32). However, the ASPV antiserum produced with pQE identified the virus readily in a plate trapped ELISA with infected *Nicotiana occidentalis* and with some limitations from apple and pear leaves. Although diffusion protein antisera have been widely used for the identification of structural virus proteins there are few reports on their use in DAS-ELISA. While a diffusion antiserum to the coat protein of citrus tristeza virus (CTV) reacted in an indirect ELISA and was suitable in ELISA for trapping (58), an antiserum to the nucleoprotein of tomato spotted wilt tospovirus (TSWV) also reacted in DAS-ELISA (66). The expression of recombinant proteins in insect cells by baculovirus systems (12,53) and yeast systems (27) provide alternatives to *Escherichia coli* and have been used to express structural virus proteins. Moreover, all three systems allowed for potyviruses the assembly into virus-like particles (12,26,27). Such systems have the potential to yield virus particles in high titers and which may subsequently be purified for antigen production. Moreover, antibodies to remaining impurities may not react with plant proteins resulting in low background reactions.

Hybridization to purified dsRNA in slot blots or Northern blots has been reported for SMYEaV (51), ASPV/PVYV (34), CVA (32) and LChV (11).Although these methods allow much faster virus testing versus the use of indicator plants, there are several disadvantages. The extraction of dsRNA is labor intensive and lacks high sensitivity. Moreover, the use of radiolabeled probes is not suitable for routine testing. The application of hybridization for the above mentioned viruses has lead to a better characterization rather than providing a rapid and sensitive diagnostic test. The highest potential to develop sensitive detection for routine purposes can be expected by the application of already existing PCR methods as for SMYEaV or other fruit viruses or viroids for which woody plant tissue based PCR has been shown (17,20-22,42, 62,70,71) (see also chapter by Candresse *et al.* in this volume). PCR tests for ASPV/PVYV and LChV, based on total nucleic acid extracted from woody tissue, have been shown to be sensitive. Furthermore, primers have been selected that allow for the detection of a wide range of isolates (W. Jelkmann, unpublished). Alternatively, ELISA tests using diffusion protein antisera as trapping antibodies in combination with MaAbs to be developed have the potential for sensitive routine testing such as shown for SMYEaV (61).

CONCLUSIONS

Methods for the molecular characterization of recalcitrant temperate fruit viruses have been described. The application of cDNA synthesis from small amounts of dsRNA followed by molecular cloning and genome analysis has been shown to be of particular interest when virion ssRNA is unavailable. Once partial or complete genomic information has been described the currently available technology allows the development of routine detection methods that can be applied to woody plants. A promising tool is the production of diffusion protein antisera and their use in ELISA. However, most of the antisera described had limitations in DAS-ELISA tests. When available, a combination with a second antibody, i.e. McAbs, were able to overcome recorded problems such as specificity or sensitivity. The development and application of PCR based detection methods follow the same principles as methods widely described for viruses that have been molecularly characterized using ssRNA templates. As the PCR technique has become more common and reached a high level of sensitivity for the detection of woody host viruses, their use in certification schemes can be supported. Moreover, the PCR detection of multiple virus infection is also feasible.

REFERENCES

1. Cashdollar, L.W., Esparza, J., Hudson, G.R., Chmelo, R., Lee, P.W.K., and Joklik, W.K. 1982. Cloning the double-stranded RNA genes of reovirus: sequence of the cloned S2 gene. Proc. Natl. Acad. Sci. 79:7644-7648.

2. Clark, M.F., and Adams, A.N. 1977. Characteristics of the microplate method of enzyme-linked immunosorbent assay for the detection of plant viruses. J.Gen.Virol. 34:475-483.

3. Clark, M.F., Adams, A.N., Thresh, J.M. and Casper, R. 1976. The detection of plum pox and other viruses in woody plants by enzyme- linked immunosorbent assay (ELISA). Acta Hortic. 67:51-57.

4. Clark, M.F., and Bar-Joseph, M. 1984. Enzyme immunosorbent assays in plant virology. Pages 51-85 in: Methods in Virology, Vol. 7. K. Maramorosch and H. Koprowski, eds. Academic Press, New York.

5. Converse, R.H., Martin, R.R. and Spiegel, S. 1987. Strawberry mild yellow-edge. Pages 25-29 in: Viruses of Small Fruits , Agriculture Handbook No. 631. R.H. Converse, ed. United States Department of Agriculture, Washington, D.C.

6. Dodds, J.A. 1986. The potential for using double-stranded RNAs as diagnostic probes for plant viruses. Pages 71-84 in: Developments and Applications in Virus Testing. R.A.C. Jones and L. Torrance, eds. Association of Applied Biologists, Wellesbourne, Warwick   CV35 9EF.

7. Dodds, J.A., Morris, T.J., and Jordan, R.L. 1984. Plant viral double-stranded RNA. Annu. Rev. Phytopathol. 22:151-168.

8. Dood, S.M., and Robinson, D.J. 1984. Nucleotide sequence homologies among RNA species of strains of tomato black ring virus and other nepoviruses. J.Gen.Virol. 65:1731-1740.

9. Dyall-Smith, M.L., Ellemann, T.C., Hoyne, P.A.,

Holmes, I.H., and Azad, A.A. 1983. Cloning and sequence of UK bovine rotavirus gene segment 7: marked sequence homology with simian rotavirus gene segment 8. Nucleic Acids Res. 11:3351-3363.

10. Eastwell, K.C., and Bernardy, M.G. 1996. Association of high molecular weight double-stranded RNA with little cherry disease. Can. J. Plant Pathol. 18:203-208.

11. Eastwell, K.C., Bernardy, M.G., and Li, T.S.C. 1996. Comparison between woody indexing and a rapid hybridisation assay for the diagnosis of little cherry disease in cherry trees. Ann. Appl. Biol. 128:269-277.

12. Edwards, S.J., Hayden, M.B., Hamilton, R.C., Haynes, J.A., Nisbet, I.T., and Jagadish, M.N. 1994. High level production of potyvirus-like particles in insect cells infected with recombinant baculovirus. Arch. Virol. 136:375-380.

13. Flegg, C.L., and Clark, M.F. 1979. The detection of apple chlorotic leafspot virus by a modified procedure of enzyme-linked immunosorbent assay (ELISA). Ann. Appl. Biol. 91:61-65.

14. Friedlund, P.R., ed. 1989. Virus and virus-like diseases of pome fruits and simulating noninfectious disorders. Cooperative Extension College of Agriculture and Home Economics, Washington State University, Pullman, Washington, 330 pages.

15. German, S., Candresse, T., Lanneau, M., Huet, J.C., Pernollet, J.C., and Dunez, J. 1990. Nucleotide sequence and genomic organization of apple chlorotic leaf spot closterovirus. Virology 179:104-112.

16. Gorziglia, M., Cashdollar, L.W., Hudson, G.R., and Esparza, J. 1983. Molecular cloning of a human rotavirus genome. J.Gen.Virol. 64:2585-2595.

17. Griesbach, J.A. 1995. Detection of tomato ringspot virus by polymerase chain reaction. Plant Dis. 79:1054-1056.

18. Gubler, U., and Hoffman, B.J. 1983. A simple and very efficient method for generating cDNA libraries. Gene (Amst.) 25:263-269.

19. Habili, N., and Rezaian, M.A. 1995. Cloning and molecular analysis of double stranded RNA associated with grapevine leafroll disease. Ann. Appl. Biol. 127:95-103.

20. Hadidi, A., Levy, L., and Podleckis, E.V. 1995. Polymerase chain reaction technology in plant pathology. Pages 167-187 in: Molecular Methods in Plant Pathology. R.P. Singh and U.S. Singh, eds. CRC Press, Boca Raton, FL.

21. Hadidi, A., and Yang, X. 1990. Detection of pome fruit viroids by enzymatic cDNA amplification. J. Virol. Methods 30:261-270.

22. Halpern, B.T., and Hillman, B.I. 1996. Detection of blueberry scorch virus strain nj2 by reverse transcriptase polymerase chain reaction amplification. Plant Dis. 80:219-222.

23. Hamilton, R.I., Dodds, J.A,. and Raine, J. 1980. Some properties of a nucleic acid associated with little cherry disease. Acta Phytopathol. Acad. Sci. Hun. 15:75-77.

24. Hampton, R., Ball, E., and de Boer, S., eds. 1990. Serological Methods for Detection and Identification of Viral and Bacterial Plant Pathogens - A Laboratory Manual. Am. Phytopathol. Soc. St. Paul, Minnesota, USA.

25. Imai, M., Richardson, M.A., Ikegami, N., Shatkin, A.J., and Furuichi, Y. 1983. Molecular cloning of double-stranded RNA virus genomes. Proc. Natl. Acad. Sci. 80:373-377.

26. Jagadish, M.N., Huang, D., and Ward, C.W. 1993. Site-directed mutagenesis of a potyvirus coat protein and its assembly in Escherichia coli. J. Gen. Virol. 74:893-896.

27. Jagadish, M.N., Ward, C.W., Gough, K.H., Tulloch, P.A., Whittaker, L.A., and Shukla, D.D. 1991. Expression of potyvirus coat protein in Escherichia coli and yeast and its assembly into virus-like particles. J. Gen. Virol. 72:1543-1550.

28. James, D. 1992. Isolation of a high molecular weight dsRNA associated with twisted leaf disease in cherry. Can. J. Plant Pathol. 14:281-284.

29. James, D., and Mukerji, S. 1993. Mechanical transmission,identification, and characterization of a virus associated with mottle leaf in cherry. Plant Dis. 77:271-275.

30. Jawee, A., and Adams, A.N. 1995. Serological detection of strawberry mild yellow edge-associated virus. Acta Hortic. 385:98-104.

31. Jelkmann, W. 1994. Nucleotide sequences of apple stem pitting virus (ASPV) and of the coat protein gene of a similar virus from pear associated with pear vein yellows disease and their relationship with potex- and carlaviruses. J. Gen. Virol. 75:1535-1542.

32. Jelkmann, W. 1995. Cherry virus A: cDNA cloning of dsRNA, nucleotide sequence analysis and serology reveal a new plant capillovirus in sweet cherry. J. Gen. Virol. 76:2015-2024.

33. Jelkmann, W., Fechtner, B., and Agranovsky, A.A. 1997. Little cherry virus: complete genome structure and phylogenetic analysis of a mealybug transmissible closterovirus. J. Gen. Virol. (in press).

34. Jelkmann, W., Kunze, L., Vetten, H.J., and Lesemann, D.E. 1992. cDNA cloning of dsRNA associated with apple stem pitting disease and evidence for the relationship of the virus-like agents associated with apple stem pitting and pear vein yellows. Acta Hortic. 309:55-62.

35. Jelkmann, W., Maiss, E., and Martin, R.R. 1992. The nucleotide sequence and genome organization of strawberry mild yellow edge-associated potexvirus. J. Gen .Virol. 73:475-479.

36. Jelkmann, W., Martin, R.R., Lesemann, D.-E., Vetten, H.J., and Skelton, F. 1990. A new potexvirus associated with strawberry mild yellow edge disease. J. Gen. Virol. 71:1251-1258.

37. Jelkmann, W., Martin, R.R. and Maiss, E. 1989. Cloning of four plant viruses from small quantities of double-stranded RNA. Phytopathology 79:1250-1253.

38. Kaden-Kreuziger, D., Lamprecht, S., Martin, R.R., and Jelkmann, W. 1995. Immunocapture polymerase chain reaction assay and ELISA for the detection of strawberry mild yellow edge associated potexvirus. Acta Hortic. 385:33-40.

39. Karasev, A.V., Nikolaeva, O.V., Koonin, E.V., Gumpf, D.J., and Garnsey, S.M. 1994. Screening of the closterovirus genome by degenerate primer-mediated polymerase chain reaction. J. Gen. Virol. 75:1415-1422.

40. Karasev, A.V., Nikolaeva, O.V., Mushegian, A.R., Lee, R.F., and Dawson, W.O. 1996. Organization of the 3'-terminal half of beet yellow stunt virus genome and implications for the evolution of closteroviruses. Virology 221:199-207.

41. Keim-Konrad, R., and Jelkmann, W. 1996. Genome analysis of the 3' terminal part of the little cherry

disease associated dsRNA reveals a monopartite clostero-like virus. Arch. Virol. 141:1437-1451.

42. Kinard, G.R., Scott, S.W., and Barnett, O.W. 1996. Detection of apple chlorotic leaf spot and apple stem grooving viruses using RT-PCR. Plant Dis. 80:616-621.

43. Koganezawa, H. and Yanase, H. 1990. A new type of elongated virus isolated from apple trees containing the stem pitting agent. Plant Dis. 74:610-614.

44. Kurppa, A., and Martin, R.R. 1986. Use of double-stranded RNA for detection and identification of virus diseases of *Rubus* species. Acta Hortic. 186:51-62.

45. Lain, S., Riechmann, J.L., Mendez, E. and Garcia, J.A. 1988. Nucleotide sequence of the 3' terminal region of plum pox potyvirus RNA. Virus Res. 10:325-342.

46. Lamprecht, S., and Jelkmann, W. 1996. Synthesis of a full-length *in vivo* infectious cDNA clone of strawberry mild yellow edge associated potexvirus (SMYEaV). Phytopathology. Abstract 394A, 1996 APS meeting.

47. Legrand, G., and Verhoyen, M. 1986. Use of different methods to detect little cherry disease in ornamental cherry trees. Acta Hortic. 193:283-289.

48. Leone, G., Lindner, J.L., Jongedijk, G,. and van der Meer, F.A. 1995. Back transmission of a virus associated with apple stem pitting and pear vein yellows, from *Nicotiana occidentalis* to apple and pear indicators. Acta Hortic. 386:72-77.

49. Leone, G., Lindner, J.L. and Schoen, C.D. 1995. Unstable infectivity and abundant viral RNAs associated with strawberry mottle virus. Acta Hortic. 385:76-85.

50. Maiss, E., Timpe, U., Brisske, A., Jelkmann, W., Casper, R., Himmler, ,G., Mattanovich, D. and Katinger, H.W.D. 1989. The complete nucleotide sequence of plum pox virus RNA. J. Gen. Virol. 70:513-524.

51. Martin, R.R., Jelkmann, W., Spiegel, S., and Converse, R.H. 1989. Molecular cloning of the dsRNA associated with strawberry mild yellow-edge virus. Acta Hortic. 236:111-116.

52. Mawassi, M., Mietkiewska, E., Gofman, R., Yang, G., and Bar-Joseph, M. 1996. Unusual sequence relationships between two isolates of citrus tristeza virus. J. Gen. Virol. 77:2359-2364.

53. McGowan, E.M., Hayden, M.B., Edwards, S.J., Pye, D., Love, D.N. and Whalley, J.M. 1994. Expression and characterisation of equine herpesvirus 1 glycoprotein H using a recombinant baculovirus. Arch. Virol. 137:389-395.

54. Meyer, M., Hemmer, O., Mayo, M.A., and Fritsch, C. 1986. The nucleotide sequence of tomato black ring virus RNA-2. J. Gen. Virol. 67:1257-1271.

55. Moore, D.L., and Cameron, H.R. 1986. Double-stranded RNA isolated from necrotic rusty mottle-diseased cherries. Acta Hortic. 193:307-310.

56. Morris, T.J., and Dodds, J.A. 1979. Isolation and analysis of double-strande RNA from virus-infected plant and fungal tissue. Phytopathology 69:854-858.

57. Németh, M. 1986. Virus, Mycoplasma and Rickettsia Diseases of Fruit Trees. Martinus Nijhoff Publishers, Dordrecht, Boston, Lancaster.

58. Nikolaeva, O.V., Karasev, A.V., Gumpf, D.J., Lee, R.F., and Garnsey, S.M. 1995. Production of polyclonal antisera to the coat protein of citrus tristeza virus expressed in *Escherichia coli*:

Application for immunodiagnosis. Phytopathology. 85:691-694.

59. Pappu, H.R., Karasev, A.V., Anderson, E.J., Pappu, S.S., Hilf, M.E., Febres, V.J., Eckloff, R.M.G., McCaffery, M., Boyko, V., Gowda, S., Dolja, V.V., Koonin, E.V., Gumpf, D.J., Cline, K.C., Garnsey, S.M., Dawson, W.O., Lee, R.F., and Niblett, C.L. 1994. Nucleotide sequence and organization of eight 3' open reading frames of citrus tristeza closterovirus genome. Virology 199:35-46.

60. Paunovic, S. 1995. Double-stranded RNA associated with fruit deformation of quince. Acta Hortic. 386:45-50.

61. Quail, A.M., Martin, R.R., Jelkmann, W., and Spiegel, S. 1995. Development of monoclonal antibodies specific for strawberry mild yellow edge potexvirus. Acta Hortic. 385:39-45.

62. Rowhani, A., Maningas, M.A., Lile, L.S., Daubert, S.D., and Golino, D.A. 1995. Development of a detection system for viruses of woody plants based on PCR analysis of immobilized virions. Phytopathology. 85:347-352.

63. Skipper, N. 1983. Synthesis of a double-stranded cDNA transcript of the killer toxin-coding region of the yeast M1 double-stranded RNA. Biochem. Biophys. Res. Comm. 114:518-525.

64. Spiegel, S. 1987. Double-stranded RNA in strawberry plants infected with strawberry mild yellow-edge virus. Phytopathology. 77:1492-1494.

65. Tian, T., Klaassen, V.A., Soong, J., Wisler, G., Duffus, J.E., and Falk, B.W. 1996. Generation of cDNAs specific to lettuce infectious yellows closterovirus and other whitefly transmitted viruses by RT-PCR and degenerate oligonucleotide primers corresponding to the closterovirus gene encoding the heat shock protein 70 homolog. Phytopathology. 86:1167-1173.

66. Vaira, A.M., Vecchiati, M., Masenga, V,. and Accotto, G.P. 1996. A polyclonal antiserum against a recombinant viral protein combines specificity with versatility. J. Virol. Methods 56:209-219.

67. Van Regenmortel, M.H.V. 1986. The potential for using monoclonal antibodies in the detection of plant viruses. Pages 89-101 in: Developments and Applications in Virus Testing. R.A.C. Jones and L. Torrance, eds. Association of Applied Biologists, Wellesbourne, Warwick CV35 9EF.

68. Voller, A., Bartlett, A., Bidwell, D.E., Clark, M.F., and Adams, A.N. 1976. The detection of viruses by enzyme-linked immunoassay for factor VIII related antigen. J. Gen. Virol. 33:165-167.

69. Watkins, C.A., Jones, A.T., Mayo, M.A., and Mitchell, M.J. 1990. Double-stranded RNA analysis of strawberry plants affected by June yellows. Ann. Appl. Biol. 116:73-83.

70. Wetzel, T., Candresse, T., Macquaire, G., Ravelonandro, M., and Dunez, J. 1992. A highly sensitive immunocapture polymerase chain reaction method for plum pox potyvirus detection. J. Virol. Methods 39:27-37.

71. Wetzel, T., Candresse, T., Ravelonandro, M., and Dunez, J. 1991. A polymerase chain reaction assay adapted to plum pox potyvirus detection. J. Virol. Methods 33:355-365.

72. Yoshikawa, N., and Converse, R.H. 1990. Strawberry pallidosis disease: distinctive dsRNA species associated with latent infections in indicators and in diseased strawberry cultivars. Phytopathology. 80:543-548.

73. Yoshikawa, N., Sasaki, E., Kato, M., and Takahashi, T. 1992. The nucleotide sequence of apple stem grooving capillovirus genome. Virology 191:98-105.

74. Zagula, K.R., Aref, N.M., and Ramsdell, D.C. 1989. Purification, serolgy and some properties of a mechanically transmissible virus associated with green ring mottle disease in peach and cherry. Phytopathology. 79:451-456.

75. Zhang, Y.-P., Uyemoto, J.K., and Kirkpatrick, B.C. 1996. Nucleotide sequence and molecular characterization of sour cherry green ring mottle virus. Phytopathology APS/MSA 1996 annual meeting. Abstract 391.

CHAPTER 30

Detection and Identification of Plant Viruses and Viroids Using Polymerase Chain Reaction (PCR)

T. Candresse, R.W. Hammond and A. Hadidi

Following the development of the polymerase chain reaction (PCR) in 1983 by K. Mullis and collaborators (81,111), its' immense potential for use in the diagnostic field was rapidly exploited. First harnessed for the detection of genetic diseases such as sickle cell anaemia (112), PCR was rapidly adapted to the detection of pathogens such as human hepatitis B and rhino viruses (28,63). Since these early efforts and with the introduction of automated DNA thermal cyclers in 1989, the applications of PCR have grown in an exponential manner as researchers developed new adaptations of the original technique (48). These include the coupling of PCR with a preliminary reverse transcription step, thus allowing the amplification of RNA sequences in a cDNA form.

In the diagnostic field, the importance of the PCR stems from its' enormous amplification power (10^6 to 10^9 fold amplifications can routinely be obtained during a 3-4 hour reaction), thus allowing, at least in theory, the synthesis of a detectable product from a single target molecule. As a consequence, PCR may yield detection assays with a sensitivity far superior to all previously developed assays. Another very important quality of PCR is its' high level of specificity which, when using carefully designed primers, allows for the discrimination of sequences differing by single nucleotide mutations. This high specificity makes PCR a prime candidate for the development of tests which allow not only the detection, but also the identification, of specific genetic sequences and pathogens. Following the initial publications demonstrating the feasibility of the adaptation of PCR to the detection of human viruses, a number of research teams rapidly tried to adapt this technique to the detection of plant viruses and viroids, with the first reports appearing in 1990

(40,109,138). In the landmark paper by Hadidi and Yang (40), they stated: "In summary, our results demonstrate a successful approach for using RT-PCR to directly detect plant pathogenic viroids in total nucleic acids from infected pome fruit trees. This feature not only increases the speed of detection but also makes the construction of a probe and hybridization unecessary. Our results also suggest the potential utility of the PCR technique in detecting other viroids, plant viral satellite RNAs, plant viruses, and possibly other plant pathogens. In view of its' high sensitivity, the PCR method could also be useful in many other plant pathological studies, including the study of pathogen-host interaction." Since then, the number of papers reporting the adaptation of PCR to the detection of plant viruses and viroids has grown steadily. To date, although several reviews have addressed the use of PCR for the detection of plant pathogens (37,44,98,133), there have been no specific efforts to exclusively review the applications of PCR to the field of plant virology. Although it is clear that other areas of molecular plant virology have greatly benefited from the introduction of PCR, this chapter will only concentrate on those applications specifically applicable to the field of detection and identification of plant viruses and viroids.

VIRUS OR VIROID/HOST COMBINATIONS FOR WHICH PCR-BASED DETECTION ASSAYS HAVE BEEN DEVELOPED

Since the initial reports in 1990 of the feasibility of adapting PCR for the detection of plant viruses and viroids (40,109,138), over 100 publications have described the development of PCR detection assays for plant viruses and viroids

Table 1. Viruses for which PCR-based detection systems have been reported

Virus	Host/Vector	Reference
dsDNA viruses		
Caulimoviruses		
Cauliflower mosaic	*Brassica* (leaves), aphids	73
Badnavirus		
Rice tungro bacilliform	Rice (leaves), leafhoppers	130
ssDNA viruses		
Geminiviruses		
Tomato yellow leaf curl	Tomato	22,84
	Whiteflies	84
Maize streak and geminiviruses of grasses	Corn or grass (leaves)	109
Variety of subgroup I viruses	Corn, sugarcane, grasses (leaves)	47
Variety of whitefly transmitted geminiviruses	Variety of dicot hosts	20
	Whiteflies	20,78
	Variety of dicot hosts	104
Unassigned virus group		
Subterranean clover stunt	Legumes (leaves or seeds)	4
Banana bunchy top	Banana (leaves)	46,52,117
	Banana (tissue culture)	117
	Aphids	117
Faba bean necrotic yellows	Faba bean (leaves)	Shamloul *et al.* (unpublished)
dsRNA viruses		
Reoviruses		
Fiji disease	Sugarcane (leaves)	124,125
ssRNA viruses		
Potexviruses		
Cymbidium mosaic	Orchids (leaves)	70
Tobraviruses		
Tobacco rattle	*Nicotiana* and *Narcissus* (leaves)	76,102
	Nematodes	136
Nepoviruses		
Grapevine fanleaf	Grapevine(leaves)	9,92,107
	Grapevine (leaves,bark, roots and shoots)	106
Cherry leafroll	Walnut (twigs)	92
	Walnut (buds)	7
	Walnut (leaves)	107
Tomato ringsppot (serotypes)	Herbaceous and woody tissues	33
Closteroviruses		
Citrus tristeza	Orange (twigs)	92
	Orange (leaves)	107
Grapevine leafroll associated virus III	Grapevine (bark and petioles)	79
	Mealybugs	79

Table 1. (cont.)

Virus	Host/Vector	Reference

ssRNA viruses (cont.)

  Trichoviruses

Apple chlorotic leaf spot	Rosaceous (fruit tree tissues)	54
Grapevine A and B	Grapevine (bark and petioles)	79
Grapevine A	Mealybugs	79
	Grapevine	79,80

  Capilloviruses

| Apple stem grooving | Rosaceous (fruit tree tissues) | 54 |

  Alfamoviruses

| Alfalfa mosaic | Legumes (leaves or seeds) | 4 |

  Ilarviruses

Apple mosaic	Cucumber and rose tissues	107
Prunus necrotic ringspot	Peach (leaves)	128,129
	Cucumber and Cherry (leaves)	107
Prune dwarf	Cherry and Plum (bark, young green buds, base of leaf petioles, pollen)	96
	Peach	107

  Potyviruses

Pea seedborne mosaic	Pea (leaves,pollen,seeds)	57
Bean yellow mosaic	Legumes (leaves or seeds)	4
	Broad bean (leaves)	92
	Gladiolus (leaves and corms)	105
	Gladiolus (corms)	139
	Gladiolus (leaves)	138
Clover yellow vein	Legumes (leaves or seeds)	4
Tulip and Lily infecting potyviruses	Tulip (leaves)	19
Variety of potyvirids	Monocot and dicot hosts	29
Plum pox	Plum (bark and buds)	59
	Peach, plum and apricot (leaves and anthers)	65,69
	Sour cherry (leaves, flowers, seeds, fruit, roots, bark)	90
	Plum, peach and apricot (leaves and bark)	142,143
Potato Y	Potato (leaves)	31,66
	Dormant tubers	121
	Aphids	66,122
Soybean mosaic	Soybean (leaves)	94
Barley yellow mosaic	Barley (leaves)	120
Sweet potato feathery mottle	Sweet potato (leaves)	15,16
Sugarcane mosaic	Sugarcane (leaves)	124
Zucchini yellow mosaic	Squash	2,132

  Cucumoviruses

Cucumber mosaic	Legumes (leaves or seeds)	4
	Pepper (leaves)	92
	Lupin (seeds)	144
	Nicotiana and cucumber (leaves)	18

Table 1. (cont.)

Virus	Host/Vector	Reference
ssRNA viruses (cont.)		
Sobemoviruses		
Subterranean clover mottle	Legumes (leaves or seeds)	4
Tombusviruses		
Artichoke mottled crinkle	Artichoke (leaves)	3
Tobamoviruses		
Pepper mild mottle	*Nicotiana* (leaves)	92
Pepper mild mottle	Pepper (leaves)	131
Paprika mild mottle	Pepper (leaves)	131
Tobacco mosaic	*Nicotiana*	24
Luteoviruses		
Subterranean clover redleaf	Legumes (leaves or seeds)	4
Potato leafroll	Potato (leaves and/or tubers)	38,92,115,127
	Aphids	38,122,123
	Physalis (leaves)	101
Strawberry mild yellow edge	Strawberry (leaves)	38
Beet western yellows	Oilseed rape (leaves)	51
	Physalis (leaves)	101
Beet mild yellowing	Sugar beet (leaves)	51
Barley yellow dwarf serotypes	Wheat and oat (leaves)	101
	Aphids	13
Tospoviruses		
Tomato spotted wilt	Pepper (leaves)	92
	Nicotiana, Datura, tomato	82,83
	Thrips	134
Groundnut ringspot	*Nicotiana rustica*	21
Tomato chlorotic spot	*Nicotiana rustica*	21
Furoviruses		
Beet necrotic yellow vein	*Chenopodium* (leaves)	108
	Sugar beet (roots)	60
Beet soilborne mosaic	*Chenopodium* (leaves)	108
Wheat soilborne mosaic	Wheat (leaves)	97
"Idaeoviruses"		
Raspberry bushy dwarf	*Chenopodium* (leaves)	2
	Arctic bramble	58

Table 2. Viroids for which PCR-based detection systems have been reported

Viroid	Host	Reference
Group I		
Avocado sunblotch	Avocado (leaves)	R. Schnell, personal communication
Peach latent mosaic	Peach, nectarine, plum, cherry, apricot (leaves, bark, fruit)	30,35
Group II		
Subgroup: potato spindle tuber viroid		
Chrysanthemum stunt	Chrysanthemun (leaves)	45
Citrus exocortis	Sweet and sour orange (leaves, bark, fruit)	67,137,146
	Grapefruit	6
	Grape (leaves)	140
	Citron, herbaceous hosts	27
Potato spindle tuber	Potato (tubers, leaves, pollen, true seeds)	118
Subgroup: coconut cadang cadang viroid		
Coconut cadang cadang	Coconut	43
Subgroup: hop stunt viroid		
Cucumber pale fruit	Cucumber (leaves)	146
Plum dapple	Plum (bark)	39
Grapevine variant	Grapevine (leaves)	39,140
Citrus viroid IIa	Citrus (leaves)	27,39,67
Citrus viroid IIb (cachexia or xyloporosis)	Citrus (leaves)	27,39,67,137
Subgroup: apple scar skin viroid		
Apple scar skin	Apple, pear (leaves, bark, fruit)	40,147
	Apple, pear (seed, flowers, leaves,bark and roots)	36
Dapple apple	Apple (leaves, bark, fruit)	40,147
	Apple (seeds, flowers, leaves,bark, roots)	36
Grapevine yellow speckle	Grapevine (leaves)	99,140
Pear rusty skin	Pear (leaves, bark, fruit)	40,147
Australian grapevine	Grape (leaves)	140

belonging to all major virus and viroid groups. Tables 1 and 2 give tentative lists of the viruses and their genera (Table 1) as well as viroids (Table 2) for which PCR detection assays have been reported. From these data, it is clear that the technique has now been adapted to the detection of viroids (circular ssRNA molecules consisting of 246 to 371 nucleotides) and also to the detection of both DNA and RNA viruses (with the inclusion of a reverse transcription step before the PCR itself for viroids and RNA viruses) possessing both single- or double-stranded genomes. It is also clear that PCR amplification of whole viroid genomes and of parts of viral genomes has been successfully achieved from a variety of plants covering both monocot and dicot hosts, and using varied samples including woody or lignified material (7,10-12,17,35-42,45,52-56,58, 59,65,69,77,79,80,85-93,96,98,99,106,107,117-119,128,129,137,140,142,143,145-147). In addition to these examples, but outside of the diagnostic field, amplification of a number of other plant viruses has been successfully achieved from many other sources for cloning and/or characterization purposes during research projects focusing on the molecular biology of plant viruses. Thus, although the development of an efficient PCR-based detection assay may still require an important time investment, especially during the assay validation period, there is no doubt that the use of PCR constitutes a potential alternative to existing plant virus and viroid detection techniques.

DETECTION OF PLANT VIRUSES IN THEIR VECTORS

In addition to allowing p ant virus detection in infected host plants, PCR has also been adapted to the detection of viruses in their vectors (see Table 1). The vectors analyzed in these studies cover a range of insects such as aphids (13,38,46,66,73,117,122,123), mealybugs (79), whiteflies (20,78,84), thrips (134) and leafhoppers (130). In addition, successful detection of viruses in their nematode vectors has also been reported (136). Although sample preparation protocols may have to be optimized, amplification from vectors does not appear to present more significant problems than amplification from an infected host-plant. In each of the models studied, the virus appears to be retained for extended periods by the vector (circulative or semi-persistent viruses), a fact which might indicate that relatively high titers of the virus are to be found in the vector. In contrast to the circulative and semi-persistent viruses, few reports have appeared on the detection of non-persistent or stylet-borne viruses in their vectors (66,122,123). The great sensitivity of the PCR-based assay has usually translated into the ability to detect the virus in extracts prepared from a very small number of vectors (usually 1-3 vectors are sufficient to get a positive signal). These findings have obvious, important applications for the study of the epidemiology of plant viruses, especially since in some cases it has been shown that the PCR assays can be performed on vectors that have been stored in various ways for extended periods of time (20,78,134). However, care should be exercised when analyzing the results of such experiments, since it has been shown that positive amplification of the virus does not parallel the viruliferous state of the vector as identical signals could be obtained from aphids containing an aphid transmissible or a non-aphid transmissible isolate of cauliflower mosaic virus (73). A similar lack of correlation between the PCR status of adult thrips and the ability to transmit tomato spotted wilt virus has also been reported (134). In contrast, however, a correlation was established between PCR assays and the ability of aphids to transmit

banana bunchy top virus (46).

THE PCR AMPLIFICATION REACTION AND ITS VARIANTS

Besides the PCR, a number of techniques allowing exponential nucleic acid sequences have been described, such as the ligase-chain reaction (LCR, 5), the self-sustained sequence replication (3SR, 34), the strand displacement amplification (SDA, 141) or the probe amplification system based on the Qβ replicase (71). Thanks to its simplicity of use and to its sensitivity, PCR has proven to be by far the most successful of these techniques. Similarly, although a number of thermoresistant DNA polymerases are now available, the *Taq* polymerase isolated from *Thermus aquaticus* is by far the most frequently used enzyme for PCR assays. However, in a few cases, other thermoresistant enzymes have been used and sometimes compared to the *Taq* polymerase. The use of such enzymes has at least two potential applications: i) the use of enzymes possessing both a reverse transcriptase (RT) and a DNA polymerase activity, allowing the use of a single enzyme in RT-PCR assays, and ii) the use of DNA polymerases having a 3' to 5' exonuclease (proof-reading) activity, which results in an increased fidelity of the amplification process (75,115). Although there are reports demonstrating the applicability of PCR employing *Taq* DNA polymerase for the assessment of viral genome variability (1), the use of proof-reading thermoresistant polymerases has clear implications for PCR-based assays aimed at the study of the molecular variability of plant viruses.

In a number of cases the viral genome consists of RNA and all known plant viroids consist of RNA molecules, which require the introduction of a preliminary reverse transcription (RT) step before the PCR amplification process (RT-PCR). The two most frequently used enzymes to perform this RT step are the reverse transcriptases (RTases) isolated either from avian myeloblastosis virus (AMV-RT) or from Moloney murine leukemia virus (MMLV-RT). As indicated in the previous paragraph, some thermoresistant DNA polymerases also possess a reverse transcriptase activity, thus allowing the use of a single enzyme for the entire RT-PCR procedure. Such a system, employing the *Tth* polymerase from *Thermus thermophilus*, has been developed for tobacco mosaic virus amplification (24). However, in parallel assays, a recombinant version of the *Tth* polymerase (*rTth*) proved less efficient than the MMLV-RT, resulting in decreased sensitivity of the assay (125). Even if the MMLV-RT appears to be the most frequently used enzyme, there are few reports providing direct comparison of the efficiency of the various RTases for plant RNA virus amplification. In the case of sugarcane mosaic potyvirus, Smith and Van de Velde (124) indicate that MMLV-RT provides a better sensitivity than a recombinant version of the same enzyme devoid of RNAseH activity (MMLV Superscript, BRL). Similarly, without providing precise data, Nolasco *et al.* (92) report that MMLV-RT generally performed better than AMV-RT in a two step RT-PCR assay.

The most frequently used approach to perform a RT-PCR assay is to run a RT reaction under conditions optimal for RTase activity and then to add a fraction of the RT reaction medium to a standard PCR reaction. Another alternative is to simply add the PCR reagents to the RT reaction. In this case, sometimes referred to as 'single-tube' RT-PCR, the reaction conditions are sequentially adapted to the enzyme used. Such an approach has successfully been used by several authors (92,142,143). Yet another possibility is to use intermediate reaction conditions allowing the simultaneous activity of both the RTase and the *Taq* polymerase. All components of the RT and of the PCR reaction can thus be added together, without the need for further manipulations once the RT reaction is started. This form of RT-PCR is sometimes referred to as 'one-tube, one manipulation' or 'one step' RT-PCR (116). This last format has the advantage of requiring less hands-on time to set up an assay, and also to reduce the possibility of false positive results through contaminations, since it involves only a single series of pipeting. One step RT-PCR has been successfully utilized for the detection of potato spindle tuber viroid (PSTVd) from total nucleic acids obtained from infected tomato leaf tissue (118). In the case of apple chlorotic leaf spot trichovirus, use of such a 'one step' protocol has resulted in increased sensitivity of detection (Candresse *et al.*, unpublished results).

Besides these fairly straightforward versions, a very large number of techniques deriving from the original PCR protocol have been presented (see for example Innis *et al.*, 48; Erlich *et al.*,25; Hadidi *et al.*, 37). While few of these derivatives have already been applied to the detection of plant viruses, many of them offer a great potential in this domain. For example, the case of the 'nested' PCR which involves two rounds of PCR amplification, the second of which is performed using primers that hybridize within the fragment amplified during the first round of PCR (100,135). This approach offers the advantage of increasing both the specificity and the sensitivity of the PCR assay. It has recently been used to detected tobacco rattle virus in single viruliferous nematodes, a case where the virus could not be reproducibly detected using a single round of RT-PCR (136). The increased sensitivity of nested PCR assays was also demonstrated in the case of the detection of grapevine fanleaf nepovirus from grapevine tissues (9). Another potentially very interesting version of PCR is the single primer amplification approach described by Lambden *et al.*, (61), which allows, in theory, the amplification of viruses for which no sequence information is yet available.

SAMPLE PREPARATION FOR PCR-BASED DETECTION ASSAYS

It is clear that a number of parameters, such as magnesium ion concentration, primer annealing temperature, choice of primers (see below), "hot starting" the PCR reaction (14), number of PCR cycles, presence of various additives to increase specificity (109,113), and inclusion of an RNA denaturation step before reverse transcription (125,143), can dramatically influence the final sensitivity of a PCR assay. In order to achieve maximal sensitivity of detection, these parameters should therefore be carefully optimized. Such optimization is, however, usually very specific for a given target/primer pair combination and is therefore considered to be outside the scope of this chapter.

Another extremely important parameter is the technique adopted for sample preparation before PCR or RT-PCR amplification. This part is often the most labor intensive stage of the detection assay, and therefore may very significantly contribute to the overall economy of the process. In addition, plant samples often pose specific problems not encountered when treating samples from animal or human origin. Outside of the use of direct tissue printing techniques for sample preparation (93), PCR obviously requires a grinding step. Plant samples are very frequently rich in ill-identified compounds (polysaccharides, phenolics...) that have the potential to interfere with the RT or with the PCR reaction. Indeed, there are now a number of reports indicating that crude plant extracts, or crude nucleic acid extracts derived from them, contain inhibitors of the PCR reaction (3,7,38,60,70,78,95,106,139). In some cases, a similar observation has also been made with extracts prepared from insect vectors of plant viruses (73). A very large number of sample preparation protocols have been developed to try to overcome these problems. Unfortunately, some of these protocols are very tedious or lengthy and may use costly chemicals or consumables, which may be acceptable at the research laboratory level, but often prove incompatible with the large scale application of the technique to agricultural products of low added value.

The simplest approach to overcome problems linked with the presence of potential inhibitors of the PCR reaction is to initially immobilize nucleic acid targets on membranes (93,107) or prepare crude plant homogenates in simple buffers (or even in sterile water) and then dilute the extract. Given the great sensitivity of the PCR, it is frequently possible to dilute out the inhibitors while retaining sufficient quantities of pathogen nucleic acids to allow successful amplification. Depending on the particular virus/host combination tested, widely differing dilution ratios have been reported to provide sensitive amplification. These range from 10-fold [cucumber mosaic cucumovirus in cucumber: (18); plum pox potyvirus in peach: (143)] to intermediate values in the 10^2-10^3 range [cauliflower mosaic caulimovirus in *Brassica*: (73); grapevine A capillovirus in grapevine : (79,80); artichoke mottled crinkle tombusvirus in artichoke: (3)] to extreme cases where 10^6-10^9 dilution factors have to be used in order to get successful amplification [*Cymbidium* mosaic virus in orchids: (70)]. Although appealing in their simplicity, these sample preparation protocols have the disadvantage of lowering the overall detection sensitivity when it is expressed as the lowest quantity of virus detectable in a given amount of plant tissue. For viruses reaching only relatively low titers in their hosts, the dilution step may ultimately lead to sensitivity levels close to those of existing diagnostic techniques, even if their detection limit expressed as the minimal quantity of virus detectable per assay is dramatically lower than that of existing assays (see below).

The other, and probably most widely applied approach, is the use of more or less purified total plant nucleic acids (TNA) or total plant RNAs from virus or viroid infected tissue as the starting material in the PCR or RT-PCR process. A large variety of protocols have now

been published for the preparation of these purified nucleic acid preparations. Although several protocols not involving solvent extractions have been reported (95,104,106,129,144), the most frequently used approach is to grind the plant material in an extraction buffer (sometimes by using liquid nitrogen to reduce the plant material to a powder first) and then to perform one or more rounds of phenol and/or chloroform extraction to remove plant proteins and contaminating substances. The nucleic acids are then recovered by ethanol or isopropanol precipitation, resuspended in sterile water and used for amplification. Additional purification steps are also frequently added, since even solvent extraction may not always succeed in removing interfering substances from nucleic acid preparations. These include steps such diverse procedures as lithium chloride precipitation of high molecular weight RNAs (4,24,101), polyadenylated RNA purification by affinity chromatography on oligo-dT magnetic beads (60), commercial nucleic acid minicolumn chromatography (38,59,146) and gel filtration (139).

Thus, although many different protocols may provide effective ways to prepare nucleic acid preparations free from interfering substances, some host plants or tissues may require the use of more extensive purification schemes, with ensuing cost increases. Sample preparation is therefore usually a step in which both optimization and simplification of protocols must be attended to, if the goal is to develop a practical detection assay.

There are several interesting alternatives to the use of crude extracts or of total nucleic acid preparations. One is the use of the GeneReleaser™ matrix (65,66,69). This commercially available resin is used directly on crude plant homogenates and is apparently able to remove interfering substances from a variety of host plants including tobacco, apple, peach, apricot, grapevine and periwinkle (69). A second alternative is to use an immunoaffinity step to purify viral particles from a crude plant extract, in a fashion reminiscent of first steps of the well-known ELISA assay, before pursuing the PCR or RT-PCR assay. This approach, initially developed for animal viruses [antigen-capture PCR: (49)], has been adapted to plant viruses under the name 'immuno-capture PCR' (IC-PCR), and shown to be extremely efficient (8,9,73,79,88,92,143) (see below). A third alternative, similar to IC-PCR, is 'print capture' PCR (PC-PCR) which allows viral detection from infected plants without grinding the samples (93). A number of proteins may be used for the capture phase, thus obviating the need for virus-specific immunoglobins.

TECHNIQUES USED FOR PCR-BASED ASSAY RESULT ANALYSIS

Once the PCR amplification has taken place, several strategies are available to reveal the presence of the amplified DNA fragment(s). Agarose gel electrophoresis followed by ethidium bromide staining is by far the simplest and most frequently used technique. A variation of this technique, which bypasses the need for agarose gel electrophoresis, has been developed by Nolasco et al., (92) who used a fluorimeter to directly measure the amount of amplified DNA in the presence of the fluorochrome Hoechst 33258. However, more complex approaches have also been used, mostly because they increase the efficiency of visualization of the amplified DNA, with corresponding increases in the sensitivity of detection of the pathogen. One relatively simple appproach is the use of polyacrylamide gel electrophoresis followed by silver nitrate staining (40,69). Another possible technique is molecular hybridization of the amplified DNA with a labeled probe, following gel electrophoresis and transfer to a membrane. Both radioactively labeled (40,51,138) and non-radioactive probes have been used (35,68,86,96,118,125), and direct comparison with agarose or polyacrylamide gel electrophoresis alone indicates that molecular hybridization may increase the sensitivity of detection 10 to 25-fold (40,51,138). Besides providing increased sensitivity, molecular hybridization also offers the extra advantage of providing an independent test of the specificity of the amplification reaction (18,101). It does, however, significantly increase the amount of work required per test and is not easily amenable to automation.

Alternative approaches include the use of labeled deoxynucleotide precursors in the PCR reaction so that a labeled PCR product is synthesized, thus potentially simplifying the detection assay and improving its sensitivity. Biotin labeling of the PCR product in this way, followed by a membrane spot assay, has been developed for the sensitive detection of plum pox potyvirus in bark samples from infected fruit trees (59). Fluorescently labeled nucleotides or oligonucleotide primers have also been used, in conjunction with laser-excited fluorescence detection on an automated DNA sequencer, for the efficient detection of grapevine leafroll virus in grapevine tissues (55,56). Besides probably increasing the sensitivity of detection, these techniques, or derivatives thereof, offer the extra advantage of the possibility of automation should high throughput assays be required.

An interesting system aimed at increasing the sensitivity of detection of bean yellow mosaic potyvirus in gladioli has also been applied by Rosner et al., (105). In this case, the use of PCR primers containing the sequence of the phage T7 RNA polymerase promoter allowed subsequent amplification of the signal by in vitro transcription of the DNA synthesised during the PCR reaction. A ten to fifty-fold signal amplification was achieved in this way, with final signal detection achieved by either molecular hybridization or agarose gel electrophoresis.

A new proceedure which increased the sensitivity of PCR detection of potato leafroll virus has been described (115). It consists of pre-PCR immunocapture of the virus from tuber extracts by paramagnetic beads carrying an antiserum against the virus, reverse transcription of a specific portion of the viral coat protein open reading frame and a fluorogenic 5' nuclease detection assay using the TaqMan™ System (Applied Biosystems, Foster City, CA) to detect the amplified transcript.

A number of options are therefore available for the final detection of the amplified DNA resulting from the PCR amplification. As seen frequently in the diagnostic field, selection of a given technique will probably reflect a compromise between sensitivity of detection and simplicity. In this respect, the potential of some of the techniques described above to be automated is a very important factor that may, in the long run, prove crucial to the development of routine PCR-based detection assays.

FINE-TUNING OF THE SPECIFICITY LEVEL OF PCR-BASED DETECTION ASSAYS

Besides its extreme sensitivity, one of the major advantages of the PCR-based assays is the ability, through careful selection of the primers, to fine tune the specificity level of the amplification procedure. At one end of the spectrum, primers allowing the amplification of genomic sequences from many members of a virus group have been described for potyvirus (15,16,19,29,62), luteoviruses (38,101), geminiviruses

(20,78,84,104,109) and, to a lesser extent, to tombusviruses (3). Primers of such wide specificity are usually targeted to conserved regions of the viral genome that are likely to be strongly conserved between members of a genus. In addition, the primers used are frequently degenerate at variable positions, with reported degeneracy levels varying between 2-4 fold (101) and 250-1000 fold (109). Instead of using degenerate primers, some authors have also used primers containing inosine, which can base pair with any other base (62), although this may severely affect the PCR overall sensitivity. The use of such degenerate or inosine containing primers may also require specific optimization on a case-by-case basis, including the balancing of the concentrations of the two primers. However, broad specificity amplification offers a number of advantages for some studies, including the ability to amplify viruses for which no sequence information is yet available.

At the other range of the specificity spectrum are reports of primers showing extremely precise specificity and thus allowing discrimination of viral isolates or strains within a virus species (103,114). So far, this possibility has not been highly exploited in the case of plant viruses. Candresse *et al.*, (12) have, however, been able to design two primer pairs allowing the specific amplification of the two major serotypes, D and M, of plum pox potyvirus. In addition, Nemchinov and Hadidi (87) and Nemchinov *et al.* (85) have designed primers specific for the amplification of the cherry serogroup of plum pox virus (PPV-C). Strain-specific polymerase reactions for discrimination of prunus necrotic ringspot virus isolates have been developed and employ primers containing single nucleotide mismatches (42). More frequently, discrimination between amplified viral isolates has been achieved by further processing of the amplified DNA fragments. Such processing steps include restriction fragment length polymorphism (RFLP) analysis as well as sequencing in cloned or uncloned form (32,64). The RFLP approach, while quite simple, has allowed the differentiation both of different viruses amplified using polyvalent primers (51,108,131), and of isolates or strains within a given virus (8,17,60,86,143). Such narrow specificity primers or discriminating techniques have obvious important applications for the study of the molecular variability of plant viruses, and as tools in epidemiological studies.

However, for detection purposes, the ideal primers should show an intermediate level of specificity allowing both the detection of all isolates of the virus and absence of amplification from samples containing related viruses. Unfortunately, gaining an understanding of the reactivity of a primer pair will often require extensive testing against a variety of viral isolates, in a fashion reminiscent of the evaluation of a monoclonal antibody. Results reported in the literature are frequently concerned only with the evaluation of PCR against a single viral isolate, the sequence of which was used for primer design. As a consequence, even if PCR-based detection assays have now been reported for a large number of viruses, there are in fact far fewer viruses for which the specificity of the primer pair used has been extensively evaluated to insure that all isolates of the virus will be properly amplified [see for example De Blas *et al.* (18); Robinson (102; Wetzel *et al.* (143); Bousalem *et al.* (8); Levy and Hadidi (66)]. Although quite tedious, such extensive testing of a primer pair is, and will continue to be in the future, a prerequisite to the translation into diagnostic practice of PCR-based assays. Failure to appreciate this will almost certainly considerably limit the usefulness of any given assay.

EVOLVED FORMATS FOR PCR-BASED DETECTION ASSAYS

As described above, one variant of PCR, the immuno-capture PCR (IC-PCR), seems particularly well-suited for the large scale development of PCR-based assays for the detection of plant viruses (9,73,79,88,92,142). Although it has the drawback of requiring an antiserum against the virus to be detected, IC-PCR offers both the advantage of providing a very simple way of preparing the plant extracts before amplification, and of affording increased sensitivity over other techniques (79,142). Indeed, most of the plant inhibitors are likely removed during the washes that follow the immunocapture step. The immunocapture itself can be carried out in separate tubes or ELISA plates but can also be directly carried out in the Eppendorf tubes in which the PCR reaction is to be performed, thus eliminating the need for further transfer of the capture particles (142). Transfer can also be eliminated by performing the PCR amplification directly in the ELISA plate that has been used for the immunocapture step (92). An even more interesting twist of the technique is to perform PCR amplification in plates that have already been through a complete ELISA assay. This offers the advantage of testing by PCR only those samples found negative in a preliminary ELISA screening, without apparently compromising the sensitivity of the PCR amplification procedure (92).

Besides providing an efficient way to remove interfering plant substances, IC-PCR allows the processing of large volumes of plant extract (100-200 μl, equivalent to 10-50 mg of starting plant material), thus increasing the sensitivity of the technique as expressed in quantity of virus per amount of infected plant. By comparison, techniques involving the dilution of crude plant extracts will usually involve the testing of 1 mg or less of plant material per assay, while those involving total nucleic acids extracts routinely use the equivalent of 1-4 mg of tissue, with very few reported protocols showing the testing of amounts equivalent to those used in IC-PCR (4). As a result, in direct comparisons between IC-PCR and other techniques, a significant increase in sensitivity is usually associated with the use of IC-PCR (79,142). In testing extracts from mealybug insect vectors, IC-PCR has, however, been reported to be somewhat less sensitive than a total nucleic acid procedure, although it yielded less non-specific amplification products (79).

Another very interesting approach is the development of 'Multiplex' PCR, a technique which allows the simultaneous amplification of several viroids or viruses using a mixture of specific primer pairs. Although posing specific amplification problems linked with the necessity to find PCR conditions allowing the efficient amplification of all the viroids or viruses tested, this strategy offers the advantage of dramatically reducing the diagnostic cost. The feasibility of this technique has been demonstrated by studies involving both animal viruses [see for example Soler et al. (126)], plant viroids (67), and plant viruses. In the latter case, both simultaneous amplification of closely related viruses using a single polyvalent primer pair (16,108) and simultaneous detection of unrelated viruses using a mixture of virus-specific primer pairs (79,124) have been reported. In one particularly remarkable example, Bariana et al. (4) reported the development of an assay allowing the simultaneous detection of five seedborne viruses from legume seeds, using a mixture of nine primers. All five viruses could be detected efficiently without apparent cross-interference. A negative effect on the efficiency of amplification of the simultaneous presence of the two potyviruses targeted was

however noted, possibly due to the fact that these two viruses were amplified using primer pairs composed of a specific primer and of a degenerate primer common to both viruses. Attempts to include another primer pair for the amplification of plant 18S RNA as an internal control of the PCR reaction in this already complex assay proved unsuccessful due to spurious amplifications (4).

Thus, Multiplex PCR offers the possibility of cutting down the costs of indexing by reducing the number of assays to perform if detection of several different viruses or viroids is needed. It also opens the possibility of introducing internal amplification controls by targeting plant sequences for amplification. Recent published and unpublished results from A. Hadidi's and T. Candresse's laboratories, respectively, indicate that Multiplex PCR can be coupled with a preliminary immunocapture step (88; Candresse, unpublished).

SENSITIVITY OF THE PCR-BASED ASSAYS AS COMPARED WITH OTHER TECHNIQUES

In theory, the PCR has the possibility to permit amplification and detection of a single target sequence (25,48,101). Even if the introduction of a notoriously inefficient reverse transcription is required to detect viroids or RNA viruses, detection assays based on PCR technology should have a sensitivity far greater than that of other currently existing detection techniques such as the ELISA assay for viruses or dot blot molecular hybridization for viruses and viroids. Achieving maximal efficiency requires, however, optimization of many parameters and may also require, in many cases, the efficient removal of inhibitory plant substances. In particular, properties of the target nucleic acid molecule and of the primers may very significantly affect the sensitivity of detection. In one extreme example, detection sensitivity of sugarcane mosaic potyvirus and of Fiji disease fijivirus in identical sugarcane total nucleic acid extracts was shown to differ by three orders of magnitude (124). Use of specially designed computer programs (74,110) may in some cases help to select the most efficient primer pair.

There are many ways in which the sensitivity of a PCR based assay can be evaluated, often making comparisons between different publications or with other diagnostic techniques complicated. One possibility, which may, however, not reflect actual diagnostic situations, is to estimate the minimal amount of purified target virus or viral nucleic acid which will still give a positive amplification signal. Extremely large variations in sensitivity have been reported using this criterion, ranging from 100 pg (38) to 0.1 fg of purified viral RNA (125). More representative values are usually in the more restricted 1-100 fg range (18,54,57,70,142). These values are in turn roughly equivalent to the detection of approximately 200-20,000 viral particles. Although such figures give a precise estimate of the sensitivity of the PCR procedure, they do not take into account the inhibitory effects of plant substances on the overall sensitivity, and therefore do not provide an adequate estimate of the assay sensitivity under 'real life' indexing conditions.

More representative values may be obtained by directly comparing the sensitivity of a PCR assay with those of other techniques by performing the assays on successive dilutions of an infected plant extract. Again, it should be kept in mind that in order to reflect a real life situation, the dilutions should be prepared using healthy plant extracts rather than buffer. As for the absolute specificity estimated on purified viral components, an extremely large variation in the sensitivity of PCR-based assays is observed using this second criterion. Reported values range from two-fold better than an ELISA assay (82) to a 10^9-fold difference observed in one system using dilutions of the infected plant extract in water (7). Following dilution in plant sap, the higher reported increase in sensitivity over an ELISA assay is 10^5-fold (4). Hundred to thousand-fold increases in sensitivity are, however, more frequently observed under these conditions (18,142). Dilution of the infected plant extract, while being less representative of actual indexing, usually gives sensitivity increases over the ELISA assay in the 10^2-10^4-fold range (70,125,127,138).

CONCLUSION : POSSIBILITIES FOR FURTHER IMPROVEMENT

Whatever the precise sensitivity figures, it is now clear that the introduction of PCR technology in the field of plant virus and viroid diagnosis has fulfilled its promises and that PCR has demonstrated its' potential to be adapted to detection assays providing very significant improvements in sensitivity when compared to already available assays. There is, however, still room for much improvement. There is also a need

for further work in order to translate many of the currently available laboratory assays into *bona fide* diagnostic techniques that can be performed by the indexing laboratory or extension service. One of the characteristics of plant disease diagnostics is the low value added of the samples to be indexed. One of the primary selection criteria for a diagnostic technique is therefore its cost. Although chemicals, enzymes and equipment may significantly affect the cost of an assay, the cost of the labor involved in performing the assay usually represents an important fraction of the total cost. In this respect, there is still much room for improvement of the PCR-based assays, to move from labor intensive, laboratory scale assays to routine large-scale indexing. Several lines to reduce the hands-on time have already been explored, such as simplification of sample preparation, use of one-tube one-manipulation RT-PCR formats, development of multiplex PCR, etc. These various aspects need to be developed further. Recent results on the sensitive simultaneous detection of plum pox potyvirus and apple chlorotic leaf spot trichovirus, using a Multiplex IC-PCR system in a 'one-manipulation' format indicate the feasibility of such approaches (Candresse *et al.*, unpublished results). Some versions of the PCR assays also seem to be particularly well-suited for automation (55,56,92,115), a development which could also significantly affect the economy of PCR-based indexing assays.

Another aspect that has not yet been fully integrated into the currently reported assays is the introduction of measures specifically aimed at preventing false positives due to contamination problems. Given their extreme sensitivity, PCR-based assays are particularly prone to such problems. Several techniques have been developed to try to overcome this problem (23,26,50,72) but, for the most part, do not yet appear to be integrated fully in the field of plant virus an viroid detection. Although this might not pose a problem when the assays are performed on a reduced, laboratory scale, failure to integrate contamination prevention techniques upon scaling up of the assays is bound to cause much disappointment in the long run.

Lastly, it should be kept in mind that while it is relatively easy to develop a PCR-based assay performing well in the laboratory on the strain whose sequence was used to design the primers, such a result is still far off from a useful diagnostic assay. Indeed, much work is then needed before one can be confident of the validity (by extensive comparison with existing assays) and the general applicability (by confrontation with the variability of the pathogen to be detected) of the assay. This second phase, which could aptly be named the 'development' phase by analogy with the 'research and development' vocabulary of industrial research, is often not pursued as actively as it should be. As a consequence, many published assays may ultimately not be translated into practice. Although the research laboratories are frequently neither prepared nor equipped to perform such 'development' tasks, this is probably the greatest challenge with which they are confronted. Only efforts aimed at the validation of a new PCR-based assay will, in the long run, insure that it becomes a generally accepted and used detection method.

ACKNOWLEDGEMENT

We wish to thank Dr. Sandra Kofalvi for reviewing the manuscript.

REFERENCES

1. Almond, N., Jones, S., Heath, A.B., and Kitchin, P.A. 1992. The assessment of nucleotide sequence diversity by the polymerase chain reaction is highly reproducible. J. Virol. Methods 40:37-44.

2. Barbara, D.J., Morton, A., Spence, N.J., and Miller, A. 1995. Rapid differentiation of closely related isolates of two plant viruses by polymerase chain reaction and restriction fragment length polymorphism. J. Virol. Methods 55:121-131.

3. Barbarossa, L., Grieco, F., Iosco, P., and Gallitelli, D. 1994. Use of the polymerase chain reaction and sandwich hybridization for detecting artichoke mottled crinkle tombusvirus in artichoke. J. Phytopathol. 140:201-208.

4. Bariana, H.S., Shannon, A.L., Chu, P.W.G., and Waterhouse, P.M., 1994. Detection of five seedborne legume viruses in one sensitive multiplex polymerase chain reaction test. Phytopathology 84:1201-1205.

5. Barringer, K.J., Orgel, L., Wahl, G., and Gingeras, T.R. 1990. Blunt-end and single-strand ligations by *Escherichia coli* ligase : influence on an *in vitro* amplification scheme. Gene 89:117-122.

6. Ben-Shaul, A., Guang Y., Mogilner, N., Hada, R., Mawassi, M., Gafny, R., and Bar-Joseph, M. 1995. Genomic diversity among populations of two citrus viroids from different graft-transmissible dwarfing complexes in Israel. Phytopathology 85:359-364.

7. Borja, M.J., and Ponz, F. 1992. An appraisal of different methods for the detection of the walnut strain of cherry leafroll virus. J. Virol. Methods 36:73-83.

8. Bousalem, M., Candresse, T., Quiot-Douine, L., and Quiot, J.B. 1994. Comparison of three methods for assessing plum pox virus variability : further evidence for the existence of two major groups of isolates. J. Phytopathol. 142:163-172.

9. Brandt, S., Himmler, G., and Kattinger, H. 1993. Anwendung der immunocapture polymerase chain reaction (IC/PCR) für den nachweis von rebviren aus holzigem material. Mitteilungen Klosterneuburg 43:143-147.

10. Candresse, T., Kofalvi, S.A., Lanneau, M., and Dunez, J. 1997. A PCR-ELISA proceedure for the simultaneous detection and identification of prunus necrotic ringspot and apple mosaic ilaviruses. 17th International Symposium on Virus and Virus-like Diseases of Temperate Fruit Crops, Bethesda, MD (Abstr.).

11. Candresse, T., Lanneau, M., Grasseau, N., Macquaire, G., German, S., Malinovsky, T., and Dunez, J. 1995. An immunocapture PCR assay adapted to the detection and the analysis of the molecular variability of apple chlorotic leaf spot virus. Acta Hortic. 386:136-147.

12. Candresse, T., Macquaire, G., Lanneau, M., Bousalem, M., Wetzel, T., Quiot-Douine, L., Quiot, J.B., and Dunez, J. 1994. Detection of plum pox potyvirus and analysis of its molecular variability using immunocapture-PCR. EPPO Bull. 24:585-594.

13. Canning, E.S.G., Penrose, M.J., Barker, I., and Coates, D. 1996. Improved detection of barley yellow dwarf virus in single aphids using RT-PCR. J. Virol. Methods 56:191-197.

14. Chou, Q., Russel, M., Birch, D.E., Raymond, J., and Bloch, W., 1992. Prevention of pre-PCR and primers-dimerizations improves low-copy number amplification. Nucleic Acids Res. 20:1717-1723.

15. Colinet , D., and Kummert, J. 1993. Identification of a sweet potato feathery mottle virus isolate from China (SPFMV-CH) by the polymerase chain reaction with degenerate primers. J. Virol. Methods 45:149-159.

16. Colinet, D., Kummert, J., Lepoivre, P., and Semal, J. 1993. Identification of distinct potyviruses in mixedly infected sweetpotato by the polymerase chain reaction with degenerate primers. Phytopathology 84:65-69.

17. Crescenzi, A., d'Aquino, L., Comes, S., Nuzzaci, M., Boscia, D., Piazolla, P., and Hadidi, A. 1997. Characterization of the sweet cherry isolate of plum pox potyvirus. Plant Dis. (in press).

18. De Blas, C., Borja, M.J., Saiz, M., and Romero, J., 1994. Broad spectrum detection of cucumber mosaic virus (CMV) using the polymerase chain reaction. J. Phytopathol. 141:323-329.

19. Dekker, E.L., Derks, A.F.L.M., Asjes, C.J., Lemmers, M.E.C., Bol, J.F., and Langeveld, S.A. 1993. Characterisation of potyviruses from tulip which cause flower breaking. J. Gen. Virol. 74:881-887.

20. Deng, D., McGrath, P.F., Robinson, D.J., and Harrison, B.D. 1994. Detection and differentiation of whitefly-transmitted geminiviruses in plants and vector insects by the polymerase chain reaction with degenerate primers. Ann. Appl. Biol. 125:327-336.

21. Dewey, R.A., Semorile, L.C., and Grau, O. 1996. Detection of *Tospovirus* species by RT-PCR of the N-gene and restriction enzyme digests of the products. J. Virol. Methods 56:19-26

22. Di Martino, M.T., Albanese, G., Di Silvestro, I., and Catara, A. 1993. Rapid detection of tomato yellow leaf curl virus in plants by polymerase chain reaction. Riv. Pat. Veg. 3:35-40.

23. Dougherty, R.M., Phillips, P.E., Gibson, S., and Young, L. 1993. Restriction endonuclease digestion

eliminates product contamination in reverse transcribed polymerase chain reaction. J. Virol. Methods 41:235-238.

24. Drygin Y.F., Korotaeva, S.G., and Dorokhow, Y.L. 1992. Direct RNA polymerase chain reaction for TMV detection in crude cell extracts. FEBS Lett. 309:350-352.

25. Erlich, H.A., Gelfand, D., and Sninsky, J.J. 1991. Recent advances in the polymerase chain reaction. Science 252:1643-1651.

26. Fox, J.C., Azit-Khaled, M., Webster, A., and Emery, V.C. 1991. Eliminating PCR contamination : is UV irradiation the answer. J. Virol. Methods 33:375-382.

27. Francis, M.I., Szychowski, J.A., and Semancik, J.S. 1995. Structural sites specific to citrus virus groups. J. Gen. Virol. 76:1081-1089.

28. Gama, R.E., Hugues, P.J., Bruce, C.B., and Stanway, G. 1988. Polymerase chain reaction amplification of rhinovirus nucleic acids from clinical material. Nucleic Acids Res. 16:9346.

29. Gibbs, A., and Mackenzie, A. 1997. A primer pair for amplifying part of the genome of all potyvirids by RT-PCR. J. Virol. Methods 9-16.

30. Giunchedi, L., Dougdoug, K., Gentit, P., Nemchinov, L., Poggi-Pollini, C., and Hadidi, A. 1997. Plum spotted fruit: a disease associated with peach latent mosaic viroid. 17th International Symposium of Virus and Virus-like Diseases of Temperate Fruit Crops, Bethesda, MD (Abstr.).

31. Glais, L., Kerlan, C., Tribodet, M., Tordo, V.M-J., Robaglia, C., and Astier-Manifacier, S. 1996. Molecular characterization of potato virus Y^N isolates by PCR-RFLP. Europ. J. Plant Pathol. 102:655-662.

32. Grant Lewis, J., Chang, G.J., Lanciotti, R.S., and Trent, D.W. 1992. Direct sequencing of large flavivirus PCR products for analysis of genome variation and molecular epidemiological investigations. J. Virol. Methods 38:11-24.

33. Griesbach, J.A. 1995. Detection of tomato ringspot virus by polymerase chain reaction. Plant Dis. 79:1054-1056.

34. Guatelli, J.C., Whitfield, K.M., Kwoh, D.Y., Barringer, K.J., Richman, D.D., and Gingeras, T.R. 1990. Isothermal, in vitro amplification of nucleic acids by a multienzyme reaction modeled after retroviral replication. Proc. Natl. Acad. Sci. USA 87:1874-1878.

35. Hadidi, A., Giunchedi, L., Shamloul, C., Poggi-Pollini, and Amer, A.M. 1997. Occurrence of peach latent mosaic viroid in stone fruits and its transmission with contaminated blades. Plant Dis. 81:154-158.

36. Hadidi, A., Hansen, A.J., Parish, C.L., and Yang, X. 1991. Scar skin and dapple apple viroids are seed borne and persistent in infected apple trees. Res. Virol. 142:289-296.

37. Hadidi, A., Levy, L., and Podleckis, E.V. 1995. Polymerase chain reaction technology in plant pathology. Pages 167-187 in: Molecular Methods in Plant Pathology. R.P. Singh and U.S. Singh, eds. CRC/Lewis Press, Boca Raton, FL.

38. Hadidi, A., Montasser, M.S., Levy, L., Goth, R.W., Converse, R.H., Madkour, M.A., and Skrzeckowski, L.J. 1993. Detection of potato leafroll and strawberry mild yellow edge luteoviruses by reverse transcription-polymerase chain reaction amplification. Plant Dis. 77:595-601.

39. Hadidi, A., Terai, Y., Powell, C.A., Scott, S.W., Desvignes, J.C., Ibrahim, L.M., and Levy, L. 1992. Enzymatic cDNA amplification of hop stunt viroid variants from naturally infected fruit crops. Acta Hortic. 309:339-344.

40. Hadidi, A., and Yang, X. 1990. Detection of pome fruit viroids by enzymatic cDNA amplification. J. Virol. Methods 30:261-270.

41. Halpern, B.T., and Hillman, B.I. 1996. Detection of blueberry scorch virus strain nj2 by reverse transcriptase polymerase chain reaction amplification. Plant Dis. 80:219-222.

42. Hammond, R.E., Crosslin, J.M., Howell, W.E., and Mink, G.I. 1997. The design of strain-specific polymerase chain reactions for the discrimination of prunus necrotic ringspot virus isolates. 17th International Symposium of Virus and Virus-like Diseases of Temperate Fruit Crops, Bethesda, MD (Abstr.).

43. Hanold, D., and Randles., J.W. 1991. Coconut cadang-cadang disease and its' viroid agent. Plant Dis. 75:330-335.

44. Henson, J.M., and French, R. 1993. The polymerase chain reaction and plant disease diagnosis. Annu. Rev. Phytopathol. 31:81-109.

45. Hooftman, R., Arts, M-J., Shamloul, A.M., Van Zaayen, A., and Hadidi, A. 1996. Detection of chrysanthemum stunt viroid by reverse transcription-polymerase chain reaction and by tissue blot hybridization. Acta Hortic. 432:120-128.

46. Hu, J.S., Wang, M., Sether, D., Xie, W., and Leonhardt, K.W. 1996. Use of polymerase chain reaction (PCR) to study transmission of banana bunchy top virus by the banana aphid (Pentalonia nigronervosa). Ann. Appl. Biol. 128:55-64.

47. Hughes, F.L., Rybicki, E.P., and Von Wechmar, B.M. 1992. Genome typing of southern African subgroup I geminiviruses. J. Gen. Virol. 73:1031-1040.

48. Innis, M.A., Gelfand, D.H., Sninsky, J.J., and White, T.J. 1989. PCR protocols : a guide to methods and applications. Academic Press, San Diego.

49. Jansen, R.W., Siegl, G., and Lemon, S.M. 1990. Molecular epidemiology of human hepatitis A virus defined by an antigen-capture polymerase chain reaction method. Proc. Natl. Acad. Sci. USA 87:2867-28

50. Jinno, Y., Yoshiura, K., and Niikawa, N. 1990. Use of psoralen as extinguisher of contaminated DNA in PCR. Nucleic Acids Res. 18:6739.

51. Jones, T.D., Buck, K.W., and Plumb, R.T. 1991. The detection of beet western yellows virus and beet mild yellowing virus in crop plants using the polymerase chain reaction. J. Virol. Methods 35:287-296.

52. Karan, M., Harding, R.M., and Dale, J.L. 1994. Evidence for two groups of banana bunchy top virus isolates. J. Gen. Virol. 75:3541-3546.

53. Karasev, A.V., Nickolaeva, O.V., Koonin, E.V., Gumpf, D.J., and Garnsey, S.M. 1994. Screening of the closterovirus genome by degenerate primer-mediated polymerase chain reaction. J. Gen. Virol. 75:1415-1422.

54. Kinard, G.R., Scott, S.W., and Barnett, O.W. 1996. Detection of apple chlorotic leafspot and apple stem grooving using RT-PCR. Plant Dis. 80:616-621.

55. Knorr, D.A., Blasband, A.J., Rowhani, A., and Golino, D.A., 1993. Fluorescence-based PCR assay

for the detection of grapevine fanleaf virus. Phytopathology 83-1397.

56. Knorr, D.A., Rowhani, A., and Golino, D.A., 1993. Fluorescence-based PCR assay for detection of grapevine fanleaf virus. Am. J. Enol. Vitic. 44, 352.

57. Kohnen, P.D., Dougherty, W.G., and Hampton, R.O., 1991. Pea seedborne mosaic virus (PSbMV) detection using the polymerase chain reaction. Phytopathology 81:1155.

58. Kokko, H.I., Kivinea, M., and Karenlampi, S.O. 1996. Single step immunocapture RT-PCR for the detection of raspberry bushy stunt virus. BioTechniques 20:842-846.

59. Korschineck, I., Himmler, G., Sagl, R., Steinkellner, H., and Kattinger, H.W.D. 1991. A PCR membrane spot assay for the detection of plum pox virus RNA in bark of infected trees. J. Virol. Methods 31:139-146.

60. Kruse, M., Koenig, R., Hoffmann, A., Kaufmann, A., Commandeur, U., Solovyev, A.G., Savenkov, I., and Burgermeister, W. 1994. Restriction fragment length polymorphism analysis of reverse transcription-PCR products reveals the existence of two major strain groups of beet necrotic yellow vein virus. J. Gen. Virol. 75:1835-1842.

61. Lambden, P.R., Cooke, S.J., Caul, E.O., and Clarke, I.N. 1992. Cloning of noncultivable human rotavirus by single primer amplification. J. Virol. 66:1817-1822.

62. Langeveld, S.A., Dore, J.M., Memelink, J., Derks, A.F.L.M., Van der Vlugt, C.I.M., Asjes, C.J., and Bol, J.F. 1991. Identification of potyviruses using the polymerase chain reaction with degenerate primers. J. Gen. Virol. 72:1531-1541.

63. Larzul, D., Guigue, F., Sninsky, J.J., Mack, D.H., Brechot, C., and Guesdon, J.L. 1988. Detection of hepatitis B virus sequences in serum by using *in vitro* amplification. J. Virol. Methods 20:227-237.

64. Lee, E., Nesterowicz, A., Marshall, I.D., Weir, R.C., and Dalgarno, L. 1992. Direct sequence analysis of amplified dengue virus genomic RNA from cultured cells, mosquitoes and mouse brain. J. Virol. Methods 37:275-288.

65. Levy, L., and Hadidi, A. 1993. Rapid and simple preparation of samples for PCR from tissue containing viruses, viroids and MLO's. Phytopathology 83:1378.

66. Levy, L., and Hadidi, A. 1994. A simple and rapid method for processing tissue infected with plum pox potyvirus for use with specific 3' non-coding region RT-PCR assays. EPPO Bull. 24:595-604.

67. Levy, L., Hadidi, A., and Garnsey, S.M. 1992. Reverse transcription-polymerase chain reaction assays for the rapid detection of citrus viroids using multiplex primer sets. Proc. Int. Soc. Citriculture 2:800-803.

68. Levy, L., Kolber, M., Tokes, G., Nemeth, M., and Hadidi, A. 1995. 3' non coding region RT-PCR detection and molecular hybridization of plum pox potyvirus in anthers of infected stone fruit. Acta Hortic. 386:331-339.

69. Levy, L., Lee, I.M., and Hadidi, A. 1994. Simple and rapid preparation of infected plant tissue extracts for PCR amplification of virus, viroid and MLO nucleic acids. J. Virol. Methods 49:295-304.

70. Lim, S.T., Wong, S.M., Yeong, C.Y., Lee, S.C., and Goh, C.J. 1993. Rapid detection of cymbidium mosaic virus by the polymerase chain reaction (PCR). J. Virol. Methods 41:37-46.

71. Lizardi, P.M., Guerra, C.E., Lomeli, H., Tussie-Luna, I., and Kramer, F.R. 1988. Exponential amplification of recombinant-RNA hybridization probes. Biotechnology 6:1197-1202.

72. Longo, M.C., Berninger, M.S., and Hartley, J.L. 1990. Use of uracyl DNA glycosylase to control carry-over contamination in polymerase chain reactions. Gene 93:125-128.

73. Lopez-Moya, J.J., Cubero, J., Lopez-Abella, D., and Diaz-Ruiz, J.R. 1992. Detection of cauliflower mosaic virus (CaMV) in single aphids by the polymerase chain reaction (PCR). J. Virol. Methods 37:129-138.

74. Lowe, T., Sharefkin, J., Yang, S.Q., and Dieffenbach, C.W. 1990. A computer program for selection of oligonucleotide primers for polymerase chain reaction. Nucleic Acids Res. 18:1757-1761.

75. Lundberg, K.S., Shoemaker, D.D., Adams, M.W.W., Short, J.M., Sorge, J.A., and Mathur, E.J. 1991. High-fidelity amplification using a thermostable DNA polymerase isolated from *Pyrococcus furiosus*. Gene 108:1-6.

76. MacFarlane, S.A. 1996. Rapid cloning of uncharacterised tobacco rattle virus isolates using long template (LT) PCR. J. Virol. Methods 56:91-98.

77. MacKenzie, D.J., McLean, M.A., Mukerji, S., and Green, M. 1997. Improved RNA extraction from woody plants for the detection of viral pathogens by reverse transcription-polymerase chain reaction. Plant Dis. 81:222-226.

78. Mehta, P., Wyman, J.A., Nakhla, M.K., and Maxwell, D.P. 1994. Polymerase chain reaction detection of viruliferous *Bemisia tabaci* (Homoptera:Aleyrodidae) with two tomato-infecting geminiviruses. J. Econ. Entomol. 87:1285-1290.

79. Minafra, A., and Hadidi, A., 1994. Sensitive detection of grapevine virus A, B, or leafroll associated III from viruliferous mealybugs and infected tissue by cDNA amplification. J. Virol. Methods 47:175-188.

80. Minafra, A., Hadidi, A., and Martelli, G.P. 1992. Detection of grapevine closterovirus A in infected grapevine tissue by reverse transcription-polymerase chain reaction. Vitis 31:221-227.

81. Mullis, K.F., Faloona, F., Scharf, S., Saiki, R., Horn, G., and Erlich, H. 1986. Specific enzymatic amplification of DNA *in vitro* : the polymerase chain reaction. Cold Spring Harbor Symp. Quant. Biol. 51:263-273.

82. Mumford, R.A., Barker, I., and Wood, K.R. 1994. The detection of tomato spotted wilt virus using the polymerase chain reaction. J. Virol. Methods 46:303-311.

83. Mumford, R.A., Barker, I., and Wood, K.R. 1996. An improved method for the detection of *Tospoviruses* using the polymerase chain reaction. J. Virol. Methods 57:109-115.

84. Navot, N., Zeidan, M., Pichersky, R., and Czosnek, H. 1992. Use of the polymerase chain reaction to amplify tomato yellow leaf curl virus DNA infected plants and viruliferous whiteflies. Phytopathology 82:1119-1202.

85. Nemchinov, L., Crescenzi, A., Hadidi, A., Piazzolla, P., and Verderevskaya, T.D. 1997. Present status of the new cherry subgroup of plum pox virus (PPV-C). A chapter in this book.

86. Nemchinov, L., and Hadidi, A. 1996. Characterization of the sour cherry strain of plum

pox virus. Phytopathology 86:575-58J.

87. Nemchinov , L., and Hadidi, A. 1997. Polymerase chain reaction detection of plum pox virus - cherry (PPV-C) subgroup using PPV-C specific primers. 17th International Symposium of Virus and Virus-like Diseases of Temperate Fruit Crops, Bethesda, MD (Abstr.).

88. Nemchinov, L., Hadidi, A., Candresse, T., Foster, J.A., and Verderevskaya, T.D. 1995. Sensitive detection of apple chlorotic leaf spot virus from infected apple or peach tissue using RT-PCR, IC-RT-PCR, or multiplex IC-RT-PCR. Acta Hortic. 386:51-62.

89. Nemchinov, L., Hadidi, A., Maiss, E., Cambra, M., Candresse, T., and Damsteegt, V. 1995. Sour cherry strain of plum pox potyvirus (PPV): molecular and serological evidence for a new subgroup of PPV strains. Phytopathology 86:1215-1221.

90. Nemchinov, L., Hadidi, A., and Verderevskaya, T.D. 1995. Detection and partial characterization of plum pox isolate from infected sour cherry. Acta Hortic. 386:226-237.

91. Nemchinov, L. Hadidi, A. Verderevskaya, T.D., Howell, W.E., and Mink, G.I. 1997. Polymerase chain reaction detection of apple stem pitting virus from herbaceous and pome fruit hosts. 17th International Symposium of Virus and Virus-like Diseases of Temperate Fruit Crops, Bethesda, MD (Abstr.).

92. Nolasco, G., de Blas, C., Torres, V., and Ponz, F., 1993. A method combining immunocapture and PCR amplification in a microtiter plate for the detection of plant viruses and subviral pathogens. J. Virol. Methods 45:201-218.

93. Olmos, A., Dasi, M.A., Candresse, T., and Cambra, M. 1996. Print-capture PCR: a simple and highly sensitive method for the detection of plum pox virus (PPV) in plant tissues. Nucleic Acids Res. 24:2192-2193.

94. Omunyin, A., Hill, J.H., and Miller, W.A. 1996. Use of unique RNA sequence specific oligonucleotide primers for RT-PCR to detect and differentiate soybean mosaic virus strains. Plant Dis. 80:1170-1174.

95. Palmans, E., Höfte, M., de Rijke, M., and Van Haute, E. 1993. Detection of potato leafroll virus via polymerase chain reaction. Med. Fac. Landbouww. Univ. Gent 58:1077-1084.

96. Parakh, D.R., Shamloul, A.M., Haddi, A., Scott, S.W., Waterworth, H.E., Howell, W.E., and Mink, G.I. 1995. Detection of prune dwarf ilarvirus from infected stone fruits using reverse transcription-polymerase chain reaction. Acta Hortic. 386:421-430.

97. Pennington, R.E., Sherwood, J.L., and Hunger, R.M, 1993. A PCR-based assay for wheat soilborne mosaic virus in hard red winter wheat. Plant Dis. 77:1202-1205.

98. Podleckis, E.V., and Hadidi, A. 1995. New trends in the detection of fruit tree viroids. Acta Hortic. 386:606-610.

99. Rezaian, M.A., Krake, L.R., and Golino, D.A. 1992. Common identity of grapevine viroids from U.S.A. and Australia revealed by PCR analysis. Intervirology 34:38-43.

100. Rimstad, E., and Ueland, K. 1992. Detection of feline immunodeficiency virus by a nested polymerase chain reaction. J. Virol. Methods 36:239-248.

101. Robertson, N.L., French, R., and Gray, S.M. 1991. Use of group specific primers and the polymerase chain reaction for the detection and identification of luteoviruses. J. Gen. Virol. 72:1473-1477.

102. Robinson, D.J., 1992. Detection of tobacco rattle virus by reverse transcription and polymerase chain reaction. J. Virol. Methods 40:57-66.

103. Rodriguez, A., Martinez-Salas, E., Dopazo, J., Davila, M., Saiz, J.C., and Sobrino, F. 1992. Primer design for specific diagnosis by PCR of highly variable RNA viruses : typing of foot-and-mouth disease virus. Virology 189:363-367.

104. Rojas, M.R., Gilbertson, R.L., Russell, D.R., and Maxwell, D.P. 1993. Use of degenerate primers in the polymerase chain reaction to detect whitefly-transmitted geminiviruses. Plant Dis. 77:340-347.

105. Rosner, A., Stein, A., Levy, S., and Lilien-Kipnis, H. 1994. Evaluation of linked PCR-transcription amplification procedure for bean yellow mosaic virus detection in gladioli. J. Virol. Methods 47:227-236.

106. Rowhani, A., Chay, C., Golino, D.A., and Falk, B.W. 1993. Development of a polymerase chain reaction technique for the detection of grapevine fanleaf virus in grapevine tissue. Phytopathology 83:749-753.

107. Rowhani, A., Maningas, M.A., Lile, L.S., Daubert, S.D., and Golino, D.A. 1995. Development of a detection system for viruses of woody plants based on PCR analysis of immobilized virions. Phytopathology 85:347-352.

108. Rush, C.M., French, R., and Heidel, G.B. 1994. Differentiation of two closely related furoviruses using the polymerase chain reaction. Phytopathology 84:1366-1369.

109. Rybicki, E.P., and Hughes, F. 1990. Detection and typing of maize streak virus and other distantly related geminiviruses of grasses by polymerase chain reaction amplification of a conserved viral sequence. J. Gen. Virol. 71:2519-2526.

110. Rychlik, W., and Rhoads, R.E. 1989. A computer program for choosing optimal oligonucleotides for filter hybridization, sequencing and in vitro amplification of DNA. Nucleic Acids Res. 17:8543-8551.

111. Saiki, R.K., Gelfand, G.H., Stoffel, S., Scharf, S.J., Higuchi, R., Horn, G.T., Mullis, K.B., and Ehrlich, H.A. 1988. Primer-directed enzymatic amplification of DNA with a thermostable DNA polymerase. Science 239:487-491.

112. Saiki, R.K., S., Scharf, S.J., Faloona, F., Mullis, K.B., Horn, G.T., and Ehrlich, H.A. 1985. Enzymatic amplification of β-globin genomic sequences and restriction site analysis for diagnosis of sickle cell anaemia. Science 230:1350-1354.

113. Sarkar, G., Kalpener, S., and Sommer, S.S., 1990. Formamide can dramatically improve the specificity of PCR. Nucleic Acids Res. 18:7465.

114. Scherba, G., Jin, L., Schnitzlein, W.M., and Vodkin, M.H. 1992. Differential polymerase chain reaction for detection of wild type and a vaccine strain of Aujerszky's disease (pseudorabies) virus. J. Virol. Methods 38:131-144.

115. Schoen, C.D., Knorr, D., and Leone, G. 1996. Detection of potato leafroll virus in dormant potato tubers by immunocapture and a fluorogenic 5' nuclease RT-PCR assay. Phytopathology 86:993-999.

116. Sellner, L.N., Coelen, R.J., and Mackenzie, J.S.

1992. A one-tube, one manipulation RT-PCR reaction for detection of Ross river virus. J. Virol. Methods 40:255-264.

117. Shamloul, A.M., El-Afifi, S.I., and Hadidi, A. 1995. Sensitive detection of banana bunchy top virus from infected leaves, tissue cultures and viruliferous aphids using polymerase chain reaction. Phytopathology (Abstr.) 85:632.

118. Shamloul, A.M., Hadidi, A., Zhu, S-F., Singh, R.P., and Sagredo, B. 1997. Sensitive detection of potato spindle tuber viroid using RT-PCR and identification of a viroid variant naturally infecting pepino plants. Can. J. Plant Pathol. 19:89-98.

119. Shamloul, A.M., Minafra, A., Hadidi, A., Giunchedi, L., Waterworth, H.E., and Allam, E.K. 1995. Peach latent mosaic viroid: nucleotide sequence of an Italian isolate, sensitive detection using RT-PCR and geographic distribution. Acta Hortic. 386:522-530.

120. Shi, N-N., Chen, J., Wilson., T.M.A., MacFarlane, S.A., Antoniw, J.F., and Adams, M.J. 1996. Single-strand conformation polymorphism analysis of RT-PCR products of UK isolates of barley yellow mosaic virus. Virus Research 44:1-9.

121. Singh, M., and Singh, R.P. 1996. Factors affecting detection of PVY in dormant tubers by reverse transcription polymerase chain reaction. J. Virol. Methods 60:47-57.

122. Singh, R.P., Kurz, J., and Boiteau, G. 1996. Detection of stylet-borne and circulative potato viruses in aphids by duplex reverse transcription-polymerase chain reaction. J. Virol. Methods 59:189-196.

123. Singh, R.P., Kurz, J., Boiteau, G., and Bernard, G. 1995. Detection of potato leafroll virus in single aphids by the reverse transcription-polymerase chain reaction and its' potential epidemiological application. J. Virol. Methods 55:133-143.

124. Smith, G.R., and Van de Velde, R. 1994. Detection of sugarcane mosaic virus and Fiji disease virus in diseased sugarcane using the polymerase chain reaction. Plant Dis. 78:557-561.

125. Smith, G.R., Van de Velde, R., and Dale, J.L. 1992. PCR amplification of a specific double stranded RNA region of Fiji disease from diseased sugarcane. J. Virol. Methods 39:237-246.

126. Soler, C., Allibert, P., Chardonnet, Y., Cros, P., Mandrand, B., and Thivolet, J. 1991. Detection of human papillomavirus types 6, 11, 16 and 18 in mucosal and cutaneous lesions by the multiplex polymerase chain reaction. J. Virol. Methods 35:143-157.

127. Spiegel, S., and Martin, R.R. 1993. Improved detection of potato leafroll virus in dormant potato tubers and microtubers by the polymerase chain reaction and ELISA. Ann. Appl. Biol. 122:493-500.

128. Spiegel, S., Rosner, A., and Stein, A. 1994. Detection of Prunus necrotic ringspot virus in peach by the polymerase chain reaction. Phytoparasitica 22:178-179.

129. Spiegel, S., Scott, S.W., Bowman-Vance, V., Tam., Y., Galiakparov, N.N., and Rosner A. 1996. Improved detection of prunus necrotic ringspot virus by the polymerase chain reaction. Europ. J. Plant Pathol. 102:681-685.

130. Takahashi, Y., Tiongco, E.R., Cabauatan, P.Q., Koganezawa, H., Hibino, H., and Omura, T. 1993. Detection of rice tungro bacilliform virus by the polymerase chain reaction for assessing mild infection of plants and viruliferous vector leafhoppers. Phytopathology 83:655-659.

131. Tenllado, F., Garcia-Luque, I., Serra, M.T., and Diaz-Ruiz, J.R. 1994. Rapid detection and differentiation of tobamoviruses infecting L-resistant genotypes of pepper by RT-PCR and restriction analysis. J. Virol. Methods 47:165-174.

132. Thomson, K.G., Dietgen, R.G., Gibbs, A.J., Tang Y.C. Liesack, W., Teakle, D.S., and Stackebrandt, E. 1995. Identification of zucchini yellow mosaic potyvirus by RT-PCR and analysis of sequence variability. J. Virol. Methods 55:83-96.

133. Torrance, L. 1992. Developments in methodology of plant virus detection. Neth. J. Pl. Pathol. 98:21-28.

134. Tsuda, S., Fujisawa, I., Hanada, K., Hidaka, S., Hogo, K., Kameya-Iwaki, M., and Tomaru, K. 1994. Detection of tomato spotted wilt virus S RNA in individual thrips by reverse transcription and polymerase chain reaction. Ann. Phytopathol. Soc. Japan 60:99-103.

135. Tuke, P.W., Luton, P., and Garson, J.A. 1992. Differential diagnosis of HTLV-I and HTLV-II infections by restriction enzyme analysis of 'nested' PCR products. J. Virol. Methods 40:163-174.

136. Van der Wilk, F., Korsman, M., and Zoon, F. 1994. Detection of tobacco rattle virus in nematodes by reverse transcription and polymerase chain reaction. Europ. J. Plant Pathol. 1:109-122.

137. Villalobos, W., Rivera, C., and Hammond, R., 1997. Occurrence of citrus viroids in Costa Rica. J. Tropical Biologia (in press).

138. Vunsh, R., Rosner, A., and Stein, A. 1990. The use of the polymerase chain reaction (PCR) for the detection of bean yellow mosaic virus in gladiolus. Ann. Appl. Biol. 117:561-569.

139. Vunsh, R., Rosner, A., and Stein, A. 1991. Detection of bean yellow mosaic virus in gladioli corms by the polymerase chain reaction. Ann. Appl. Biol. 119:289-294.

140. Wah, Y.F.W.C., and Symons, R.H., 1997. A high sensitivity RT-PCR assay for the diagnosis of grapevine viroids in field and tissue culture samples. J. Virol. Methods 63:57-69.

141. Walker, G.T., Little, M.C., Nadeau, J.G., and Shank, D.D. 1992. Isothermal in vitro amplification of DNA by a restriction enzyme/DNA polymerase system. Proc. Natl. Acad. Sci. USA 89:392-396.

142. Wetzel, T., Candresse, T., Macquaire, G., Ravelonandro, M., and Dunez, J. 1992. A highly sensitive immunocapture polymerase chain reaction method for plum pox virus detection. J. Virol. Methods 39:27-37.

143. Wetzel, T., Candresse, T., Ravelonandro, M., and Dunez, J. 1991. A polymerase chain reaction assay adapted to plum pox virus detection. J. Virol. Methods 33:355-365.

144. Wylie, S., Wilson, C.R., Jones, R.A.C., and Jones M.G.K. 1993. A polymerase chain reaction assay for cucumber mosaic virus in lupin seeds. Aust. J. Agric. Res. 44:41-51.

145. Xie, W.S., and Hu, J.S. 1995. Molecular cloning, sequence analysis and detection of banana bunchy top virus in Hawaii. Phytopathology 85:339-347.

146. Yang, X., Hadidi, A., and Garnsey, S.M. 1992. Enzymatic cDNA amplification of citrus exocortis and cachexia viroids from infected citrus hosts. Phytopathology 82:279-285.

147. Zhu, S.F., Hadidi, A., Hammond, R.W., Yang, X., and Hansen, A.J. 1995. Nucleotide sequence and

secondary structure of pome fruit viroids from dapple apple diseased apples, pear rusty skin diseases pears and apple scar skin symptomless pears. Acta Hortic. 386:554-559.

CHAPTER 31

Ultra Microscopic Detection of Plant Viruses and Their Gene Products

V. Hari and P. Das

The electron microscope has been used as a tool for virus diagnosis and for analyzing the structure of viruses (8,17,22-24,28). Plant viruses often exist in sufficient concentrations in infected tissues that they can be extracted and examined for particle structure and morphology by standard electron microscopy. Modern methods developed in the last decade now permit us to diagnose virus infections using small amounts of virus-infected tissues. In this review, we will describe various electron microscopic procedures that are useful for the detection of plant viruses in infected tissues and will also include a brief description of Atomic Force Microscopy. We will not, however, go into details with regard to purification of viruses, preparation of sample support grids, shadowing methods, thin sectioning procedures and the actual process of operation of the microscopes. For this, the reader is referred to various text books and reviews (13,17,22-24). This chapter will focus on: 1. conventional electron microscopy. 2. immune electron microscopy. 3. immunogold electron microscopy. 4. sphere-linked immuno diagnostic assay. 5. *in situ* electron microscopic methods and 6. atomic force microscopy.

CONVENTIONAL METHODS

All those methods that are currently available for the direct visualization of virus particles under the EM are included in this category. These include: leaf dip methods, tissue squash methods, tissue exudate methods and purified virus. All these methods involve the release of virus particles from virus-infected cells and examining them under the EM after shadowing with metals (8,28, or more commonly, staining the particles with a suitable stain. A number of different stains are currently available and are listed in Table 1. Diagnosis of the virus involved presupposes a knowledge of the structural features of different viruses (13,23). Descriptions regarding the morphological features that would be helpful in diagnosis are to be found in various books, reviews (13,23) and in different volumes of the CMI/AAB descriptions of plant viruses.

Leaf Dip Methods

This method was originally described by Hitchborn and Hills (18). The method is mainly used for the detection of viruses in leaf tissues and is particularly used as a rapid means of diagnosing those viruses (e.g. tobacco mosaic, TMV) that reach high concentrations in leaves.

Procedure

1. Load a drop of negative stain such as uranyl acetate or phosphotungstate on a carbon-coated grid.

2. Take a forceps with a fine tip similar to the one used to grip the grid and tear off a piece of epidermal tissue from the underside of the virus-infected leaf. This process is easy if you first make a shallow cut into the mid rib or a prominent leaf vein. Grip the cut end with the forceps and tear the epidermis from the leaf. The torn exposed surface of the strip is then dipped into the droplet of stain on the grid.

3. After a minute, draw off the excess fluid from the grid with a filter paper and dry the grid.

4. Examine the grid under the EM.

An alternative procedure involves simply

Table 1.   A list of commonly used negative stains for electron microscopy and their properties

---

Ammonium molybdate

Used as a 2-5% solution in water. The pH of a 5% solution is 5.0-5.5. The pH can be adjusted with ammonium hydroxide to get a pH range of 5-8. The stain is most commonly used in the pH range of 7.0-7.4 (16,17). It gives good contrast when staining membranes in thin sections and can be used as an alternative to phosphotungstate staining of virus particles.

Phosphotungstic acid (PTA)

This was the first stain to be used as a negative stain. Phosphotungstic (dodecatungstophosphoacid) is dissolved in water to give a concentration of 1-2%. The dissolved acid is then titrated with sodium or potassium hydroxide to give a pH in the range of 5-8. The stain is stable at room temperatures in light unless CO_2 from air dissolves in it in which case, the solution might become turbid. The stain tolerates non-volatile buffers such as phosphate and it migrates less than uranyl acetate under the electron beam (16). PTA does not however give as good a negative stain contrast as uranyl acetate and also has the disadvantage of disrupting the structure of many plant viruses (16).

Uranyl acetate

This yellow colored salt is dissolved at a concentration of 0.2-2.0% in distilled water in which it is soluble up to a concentration of about 4%. The stain solution has a pH of 3.0-5.5 and is unstable (precipitate formation) above a pH of 6.0. The stain should be stored in a dark bottle away from light and is stable for about 3-4 weeks. The stain gives excellent contrast but is not very suitable for directly adding to tissue extracts since precipitation of proteins occur due to the low pH of uranyl acetate (UA). As a consequence, if UA is used to stain crude sap, the grids have to be washed a few times to reduce particulate precipitates.

Uranyl formate

This stain should be prepared preferably fresh each time before use since it degrades rapidly. However, it can be stored up to a week if kept in a dark bottle away from light. The degraded stain has a brown color as opposed to a clear yellow color for the fresh stain. The solid salt is dissolved in water to give concentrations of 1-2%, the pH is about 3-0 but can be titrated to a pH of about 4.0-5.0 with ammonium hydroxide. The stain is better than UA for carrying out high resolution studies because of its greater penetration properties (20). As with UA, the stain should be added to the virus preparation on the grid and not directly mixed with the virus.

Uranyl acetate-EDTA

A combination of UA and EDTA can be used to stain virus particles at neutral pH. The mixture is prepared by adding 0.1% ammonium acetate and 0.5% EDTA to a 1% solution of UA.

Other Stains

Methylamine tungstate (16); Gold thio glucose, Lanthanum acetate, Lithium tungstate, Sodium silico tungstate, tungsto borate, uranyl nitrate , uranyl sulfate and uranyl oxalate are other salts used as stains (16).

dipping the cut edge of a leaf disc or wedge into the stain on the grid and proceeding to process the sample as above for microscopy. We have also found that viruses such as TMV can easily be detected in droplets of exudate released from wounds. Here, the petiole or stem is cut and the droplet of exudate that comes out is examined after staining.

Leaf Squash and Extraction Methods

Small amounts of tissue can be examined for presence of virus by this procedure. One of the disadvantages of the method is that a lot of cellular debris would be deposited on the grid and discrimination of isometric viruses becomes difficult since some of them can easily be confused with ribosomes. Grind a small piece of tissue with one or two drops of stain or buffer on a glass slide using a glass rod. Alternatively, place a piece of tissue on a glass slide with one or two drops of stain or buffer. Place another glass slide on top of the first slide and gently macerate the tissue by rotating the slide such that the tissue wedged between the slides is macerated. The top slide is turned loose and the tissue extract is then examined after loading the extract on to grids and staining if not previously extracted in stain.

Partially Purified Samples

Many viruses particularly those viruses which have a low titer in tissues need to be processed further before they can be used for virus diagnosis. Herein, 1-10 g of tissue are powdered in liquid nitrogen and the powder is re suspended in a suitable buffer. Alternatively, the samples may be ground up directly in buffer. The powder is clarified by heating at about 60 °C and centrifuged at low speeds in a microcentrifuge or a clinical centrifuge. The heating step may be omitted if you suspect the virus to be thermolabile. The clarified sample may then be centrifuged at high speeds in an ultra centrifuge and the pelleted material may be examined in the EM after suspending the pellets in a small volume of buffer.

IMMUNOELECTON MICROSCOPY

Leaf-dip Serology

This is a modification of the leaf-dip procedure wherein, cut leaf preparations or epidermal strips are dipped into virus-specific antiserum (11). In this procedure, a drop of virus-specific antiserum diluted in 0.001 M ammonium acetate is placed on a carbon-coated collodion film on a grid. A small piece of cut tissue is dipped into the drop as in the leaf-dip procedure. Alternatively, an epidermal strip can also be similarly dipped into the drop. The drop is incubated for about 5-10 min at 40 °C and dried. The grid is then stained with 1:3 mix of vanadyl molybdate:phospotungstate and examined in the EM. Depending on the virus examined, virus particles were found to be clumped and coated. The advantage of the method over the conventional leaf-dip method lies in the fact that in addition to observing virus particle morphology, one can also directly visualize the serological relationships of the virus under observation. The concentration of ammonium acetate or the type of buffer used while diluting the serum seems to influence the results (1). Thus, it was observed that no virus particles were found when leaf strips were dipped in drops containing 0.02-0.04 ammonium acetate or 0.0015-0.006 M phosphate buffer with 0.007-0.028M sodium chloride. Also, incubating the leaf dips overnight at 2°C also resulted in absence of rods (1). The reason for some of these findings have not been investigated further. Anyone planning to use this technique should try various buffers and incubation conditions before using this method for diagnostic purposes.

Immunosorbent Electron Microscopy (ISEM)

This method is a further development of the procedure described above and was originally described by Derrick (11). The method has also been variously described as serologically specific electron microscopy (SSEM), solid phase immune electron microscopy (SPIEM). The method makes use of the trapping of antiserum-specific virus particles onto EM grids that have been coated by the specific antiserum. Samples that are serologically related to the antibodies on the coated grids get concentrated on the grids enabling the detection of virus in samples containing low amounts of virus. Further, the method allows the investigator to characterize the serological relationships of viruses when only small amounts of sample are available. The actual procedure involves coating carbon-coated EM grids with diluted antiserum to a specific virus. A drop of the sample virus preparation being examined is placed on the coated grid and incubated for 30 seconds-4 h. In many instances, incubation times of 30 seconds is sufficient to trap virus particles. The grids are rinsed in appropriate buffers and stained before examination in the EM. Various parameters such as buffer used to dilute the antiserum, the time and temperature of incubation influence the results (25). In many instances, crude sap from presumably virus-infected plants could be used to diagnose the presence of viruses. However, in cases such as plum pox virus, grapevine fanleaf virus or in general, viruses extracted from plants of the Rosaceae, grapevine or poplar, best results are obtained only when the buffer used for extraction of virus contained 2% polyvinyl pyrrolidone or 2.5% nicotine as protectants.

An alternative ISEM procedure involving the use of grids pre-coated with protein A was introduced by Shukla and Gough, (29) and Leesemann and Paul (21). In this procedure, grids are first coated with a solution of protein A prior to coating with the specific antiserum for ISEM. The main advantage seems to be that serum proteins other than the antibodies in the antiserum do not bind to the grids and antibodies from low titer antisera become concentrated on the grids due to the trapping of the antibodies by the protein A.

Immunogold Decoration Techniques

The ISEM technique has been further modified by incorporating various decoration techniques in particular gold conjugated antibodies (3). The method enhances the ability to detect and characterize viruses. Virus particles in solution from purified preparations or plant sap are loaded on to carbon coated EM grids. The grids are dried by removing the excess fluid by means of a filter paper. A drop of blocking buffer (TTBS = 0.1 M Tris pH 7.0 containing 0.1% tween 20, 0.65% NaCl) containing 1% bovine serum albumin (BSA) is placed on the grid. This ensures that sites on the grid not occupied by the virus or other plant proteins are blocked so that non-specific binding of antibodies added in the next step does not take place. The blocking buffer is removed with the aid of a filter paper and a drop of antiserum diluted in TBS (Y) is placed on the grid. After a 1 h incubation time at room temperature (ca. 22 °C) the antiserum is removed with the aid of a filter paper and the grid is rinsed several times in TBS and then in blocking buffer. The sample on the grids is now incubated in a drop of secondary antibody conjugated with 10-15 nm gold particles.

After further rinses in TBS, the samples are dried and viewed in a EM. Figure 1 shows the results of such an Immunogold decoration experiment. We have found that a mixture of at least two different types of viruses can be coated on to duplicate grids and examined for serological specificity by using two separate antisera.

SPHERE-LINKED IMMUNODIAGNOSTIC ASSAY (SLIDA)

This method is based on the basic principles involved in ELISA (15). The method has been successfully used for the detection of TMV, TEV, PVX, PVY, BMMV and MDMV (10,15). It has also been used for detection of bacterial antigens and animal viruses such as human immuno-deficiency virus (HIV) antigen (unpublished). The method involves the covalent binding of disaggregated virus antigen on to plastic spheres (micro spheres) thereby creating what amounts to "artificial spherical viruses" and then reacting the virus antigen-coated spheres with specific antisera followed by gold-conjugated secondary antisera. The presence of virus antigen is indicated by the presence of gold-labeled secondary antibody on the spheres. The main advantages of the method are: 1. The method does not depend on the morphology of the virus since any type of virus antigen can be tested. 2. Virus structural and non-structural proteins can be tested. Thus for example a tissue extract containing non-structural proteins of the virus such as the CI protein of potyviruses or the 126 KD protein of TMV can be covalently attached to the microspheres and examined for presence of CI or 126 KD protein using specific antisera to these proteins. 3. Spheres can be coated with an extract from a plant that is infected by more than one virus and aliquots of these spheres can then be exposed to different antisera to test for viruses that react specifically to that particular antiserum.

Procedure

1.   Total proteins from purified, partially purified or infected plant tissues are prepared by the acetic acid method (12). Samples in buffer are mixed with two volumes of cold acetic acid and centrifuged to remove insoluble material. The acetic acid extract is dialyzed against water to remove the acid and then against

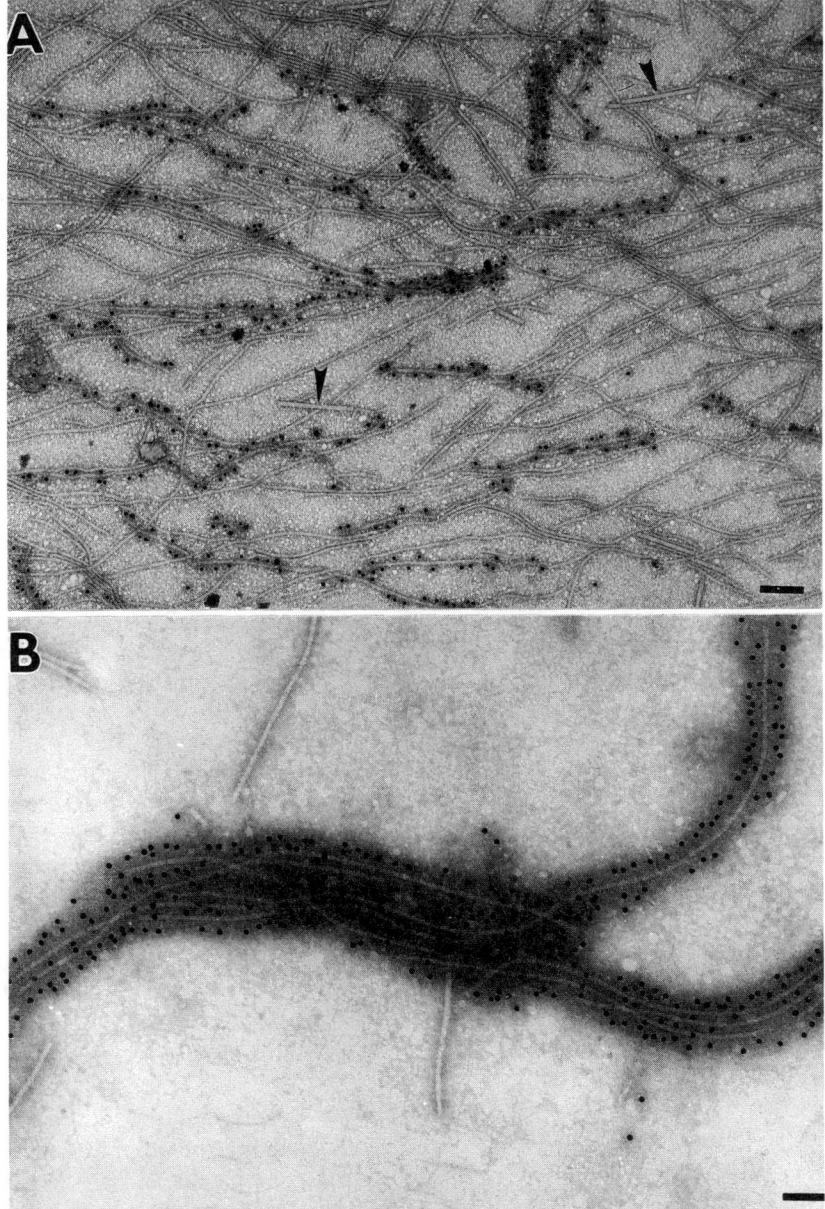

Figure 1. Immunogold decoration of a mixture of tobacco etch virus (TEV), tobacco mosaic virus (TMV) and potato virus PVX (panel A) and of maize dwarf mosaic virus (MDMV) (Panel B). Panel A shows a mixture of TMV, TEV and PVX exposed to antiserum to TEV. Note the specific labeling of TEV. Arrow shows TMV particles. Panel B shows a sample of MDMV exposed to an antiserum to MDMV. Magnification: A= bar equals 180 nm and B= bar equals 120 nm.

a suitable buffer (0.1 M Tris pH 7.0). Alternatively, the acetic acid extract may be neutralized to pH 7.0 and used to coat virus antigen on to the spheres. Other methods that dis-aggregate virus or virus-coded proteins can also be tried but we have not tried methods other than the acetic acid procedure.

2. About 0.1 ml of a suspension containing approximately 1.1 X 10^{11} plastic spheres (0.9 μm) or 3.3 x 10^{10} (0.5 μm) covaspheres (Duke Scientific Corp, Palo Alto, Calif.) are sonicated to remove clumping of the spheres. The covaspheres are plastic microspheres that have carboxyl groups on their surfaces so that the amino groups of proteins can form covalent bonds with the spheres when they are incubated together. The sonicated spheres are mixed with about 20 μg of virus protein or an extract of the virus proteins in water or 0.1 M phosphate buffer pH 7.0 prepared as in step 1, sonicated as before and incubated for 75 min at room temperature. This step allows the covalent bonding of the virus antigens to the spheres. Although, we used about 0.1 ml of the sphere suspension in our experiments, we see no reason why much more diluted samples of the spheres cannot be used for the experiments.

3. The spheres from step 2 are centrifuged for 5 min in a microcentrifuge at full speed in order to pellet the spheres.

4. The protein-coated spheres are suspended in TBS buffer (20 mM Tris, 500 mM NaCl, pH 7.5) containing 5% BSA for 1 h at room temperature.

5. The spheres from step 4 are centrifuged in a micro centrifuge as before or at 10,000 rpm in a Sorvall RC2 B using an SS34 rotor. The pellets are resuspended in the above buffer.

6. About 10 μl of antigen-coated suspension is sonicated and loaded on to parlodion-coated/carbon-coated 200 mesh nickel grids. The excess liquid is drawn off with a filter paper and the grids are allowed to dry at room temperature.

7. A drop of blocking buffer (TBS + 5% BSA) is placed on the grid in order to block regions of the grid which do not contain the spheres.

8. The grids are then incubated for 1 h with appropriate dilutions of pre-immune or antigen-specific rabbit antiserum (the antisera are diluted in antibody buffer: 20 mM Tris pH 7.5; 500 mM NaCl, 0.1% tween 20 and 1% BSA).

9. The grids are rinsed for 30 sec with a jet of TTBS buffer (20 mM Tris pH 7.5; 500 mM NaCl, 0.1% tween 20) using a wash. The rinsed grids are then dipped in the above buffer (TTBS) for 5 min. The process of rinsing and dipping are repeated 3 times.

10. The grids are filter-dried and a drop of 15 nm gold-conjugated goat anti-rabbit antiserum (1:25 dilution in an antibody buffer) is placed on the grid. The dilution of the GAR-gold serum will vary depending on the titer of the commercially available serum. The grids are incubated in the secondary serum for 20 min.

11. The grids are now washed 3 times with a squirt of TTBS using a wash bottle and then dipped several times in a beaker of double distilled water. The grids are dried and examined in an EM and photographed.

12. The number of gold particles per sphere are counted manually using the photographs. The number of gold particles gives an estimate of the titer of the antiserum if the spheres were exposed to several dilutions of the primary serum. It also gives an estimate of the antigen concentration in a tissue or virus extract.

Comments: The spheres that have been coated once with an extract of virus-infected tissue can be stored for at least 6 months and aliquots of the extract can be used to test for different viruses if the test tissue is being tested for different viruses. Figure 2 shows the results of some experiments involving SLIDA.

IN SITU METHODS

Thin Sectioning

Detailed procedures for thin sectioning and examination of these sections for virus and virus gene products are described elsewhere (22) and we will not attempt to repeat these here.

Immuno-gold Methods

These methods have been used by a

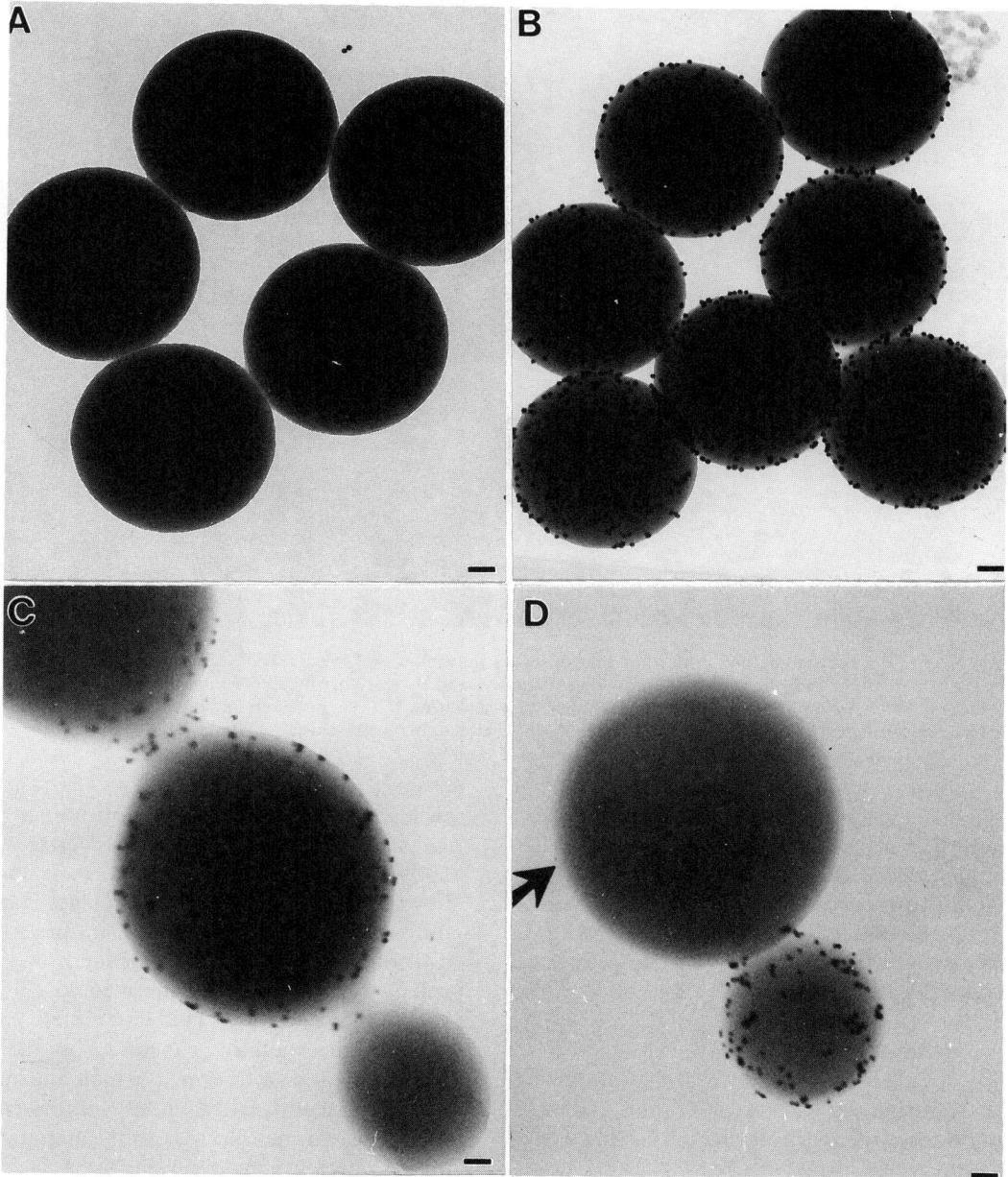

Figure 2.   SLIDA of TMV and TEV capsid proteins. Disaggregated capsid protein samples of TMV or TEV were coated on to covaspheres and examined in the EM after incubating the coated spheres in pre- immune rabbit sera (panel A) rabbit antibodies to TMV capsid (panel B). Panel C shows a mixture of large spheres coated with TEV capsid and small spheres coated with TMV capsids and incubated with antiserum to TEV. Panel D shows a similar mixture of spheres as in panel C but incubated with antiserum to TMV capsid.   Magnifications bar = 90nm (A); 90 nm (B); 105 nm (C); 105 nm (D).

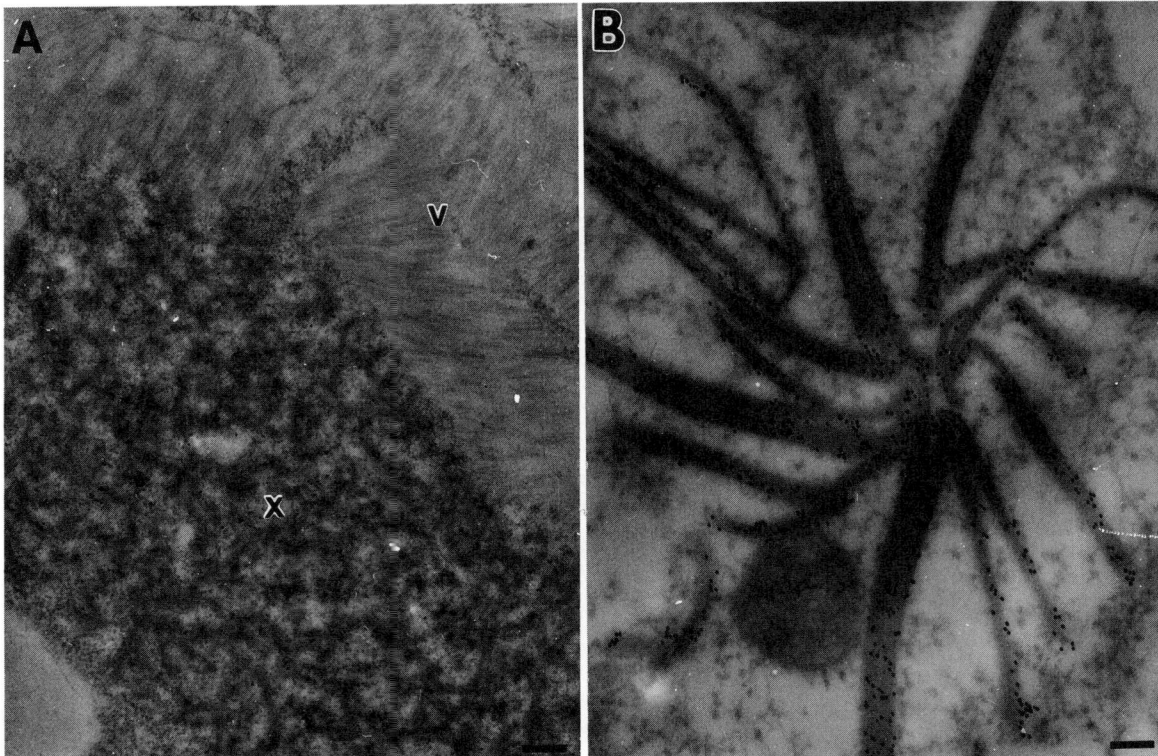

Figure 3. Immunogold electron microscopy of tain sections of leaves infected with TMV, or TEV. Thin sections of Lowicryl embedded samples were subjected to immunogold electron microscopy. Panels A and B, respectively, show immunogold labeling of X-bodies in TMV-infected cells, CI bodies in TEV-infected cells. The primary sera used were respectively, an antiserum to the 126 K protein of TMV (A) and an antiserum to the CI protein of TEV (B). Magnifications, bar= 225 nm (A);150 nm (B).

number of investigators for the detection and localization of various viruses and their gene products in infected cells. In this chapter, we will outline some proceedures that have been successfully used for detecting and localizaing TMV, TEV, PVY, MDMV and the ncn-structural proteins coded by these viruses (2-4,9).

Tissue Processing for Lowicryl K11

Leaf strips of 2 X 0.5 mm from systemically infected leaves are vacuum infiltrated with cold phosphate-citrate buffer (2-4,9). The tissue pieces are fixed overnignt in 1% glutaraldehyde in phosphate- citrate buffer, washed three times in the same buffer and dehydrated in a graded series of alcohol. The dehydrated tissue is then passed through a graded

series of methyl acetate as follows: 30% methyl acetate for 30 min, 70% methyl acetate for 30 min; 100% methyl acetate for 30 min; 1:1 mixture of Lowicryl K11 and methyl acetate for 60 min followed by 100% Lowicryl resin for 24-48 h. The infiltrated tissue is then embedded in Lowicryl in gelatin capsules and polymerized at -20 °C for 24 h using a UV light The latter step is done by placing the gelatin capsules in the -20 °C compartment of a freezer in which a hand held UV light is placed. Ultra violet polymerization can also be done at room temperature but low-temperature polymerization is recommended in order to reduce antigen modifications. The blocks are hardened by placing the capsules at room temperature for 24 h after they had polymerized. Thin sections are cut with the aid of a diamond knife and the sections placed on

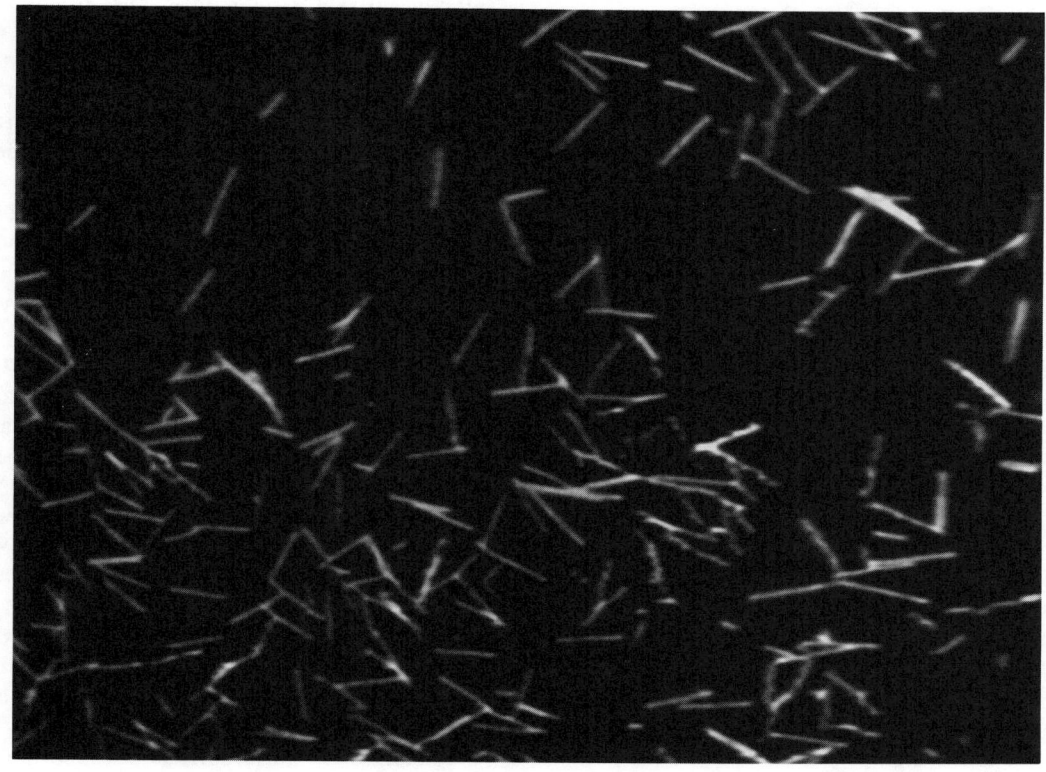

Figure 4. Purified preparations of TMV were spread on freshly cleaved mica and imaged by AFM under ambient conditions. The figure was captured on a computer screen and photographed. Magnification bar = 500 nm.

parlodion-coated grids for viewing in the EM. Figure 3 shows immunogold labeled cells of plant leaves infected with TEV or TMV and embedded in Lowicryl.

Tissue Processing for LR White

1. Cut tissue into small pieces as above.
2. Fix in 2% paraformaldehyde, 2.5% glutaraldehyde in 0.1 M phosphate or cacodylate buffer pH 7.4 containing 0.5% sucrose for 2 h at 22 and then for 14 h at 14.
3. Wash tissue pieces 5 times in the above buffer for 60 min each wash.
4. Dehydrate the tissue in a graded series of alcohol as above.
5. Infiltrate the tissue at room temperature with 1:1 mixture of LR white and 100% ethanol for 4 h.
6. Infiltrate in 100% LR White for 16-24 h Some types of plant tissue such as corn might need longer infiltration times.
7. Transfer the tissues to gelatin capsules containing LR White. Make sure that air bubbles are not trapped.   Seal the capsules and polymerize at 50 °C for 24-48 h.

Immunogold Labeling

The thin sections embedded in Lowicryl or LR White as the case may be are transferred to parlodion-carbon coated grids. The sections are then incubated with virus protein-specific antibody (primary serum) for 1-3 h at room temperature. The sections are washed in TTBS buffer containing 0.01 % BSA and incubated with gold-conjugated secondary antibody for 20 min, washed in TTBS and then stained with uranyl acetate and post-stained with lead citrate. The sections are viewed in the EM and photographed.

ATOMIC FORCE MICROSCOPY/ SCANNING FORCE MICROSCOPY

This is a method that has not yet found use in plant pathology but protocols are now being developed which suggest that this type of microscopy could become an important tool for studying virus structure and for diagnostic procedures. The Atomic Force Microscope (AFM) also known as a Scanning Force Microscope is a high resolution profilometer capable of resolving

atomic dimensions (5-7,27,30). The basic operational principle involves the response of an extremely sharp probe to highly localized forces between the probe and sample (30). The probe is attached to a flexible cantilever that is deflected as it responds to the forces between the probe and the specimen and these deflections are monitored optically from the feedback loop and photodiode arc and the output is recorded and digitized on a computer. The images derived can be observed on the computer screen, saved on a hard or floppy disc and printed with the aid of color or black and white printers. Compared to EM, the AFM has the advantage of obtaining higher resolution real time three dimensional images and to image virus and other molecules in liquid. Thus, imaging surfaces by AFM has proven to be a powerful tool in a number of fields because many materials under certain conditions are not accessible by any other technique in the size range from < 1-50 nm (27). Reliable length measurements of DNA and excellent pictures of TMV particles have been obtained by different investigators (19,30,31). AFM is also being applied to other areas of biological research (14,26). Figure 4 is a topographic image of TMV particles obtained by AFM. Protocols for visualizing virus particles, are now being developed and it is hoped that AFM will become an important tool for not only the diagnosis of virus infections but also the study of virus structure at high resolution.

ACKNOWLEDGMENTS

The following former and current students have contributed to the development of SLIDA and immunogold techniques described here: David Baunoch, Pritam Das, Mary Browning, and Mamdouh Abdel-Ghaffar. Figure 4, showing TMV particles through AFM was contributed by Dr. Gang-Yu Liu of the Chemistry Dept. at Wayne State University.

REFERENCES

1.   Ball, E.M., and Brakke, M.K. 1968. Leaf-dip serology for electron microscopic identification of plant viruses. Virology 36: 152-155.
2.   Baunoch, D.A., Das, P., Browning, M.E., and Hari, V. 1991. A temporal study of the expression of the capsid, cytoplasmic inclusion and nuclear inclusion proteins of tobacco etch potyvirus in infected plants. J. Gen. Virol. 72: 487-492.
3.   Baunoch, D.A., Das, P., and Hari, V. 1988. Intracellular localization of TEV capsid and inclusion proteins by immunogold labeling. Journal of Ultrastructure and Molecular Structure Research.

99: 203-212.

4. Baunoch, D.A., Das, P., and Hari, V. 1990. Potato virus Y helper protein is associated with amorphous inclusions. J. Gen. Virol. 71: 2479-2482.

5. Binnig, G. 1992. Force microscopy. Ultramicroscopy 42-44 : 7-15.

6. Binnig, G., Gerber, C., Stoll, E., Albrecht, T.R., and Quate, C.F. 1987. Atomic resolution with atomic force microscopy. Europhysd. Lett. 3: 1281-1286.

7. Binnig, G., Quate, C.F., and Gerber, C. 1986. Atomic force microscope. Phys. Rev. Lett. 56: 930-933.

8. Backus, R.C. and Williams, R.C. 1950. Use of spraying methods and of volatile suspending media in the preparation of specimens for electron microscopy. J. Appl. Phys. 21: 11-15.

9. Das, P. and Hari, V. 1992. Intracellular distribution of the 126K/183Kand capsid proteins in cells infected by some tobamoviruses. J. Gen. Virol. 73: 3039-3043.

10. Das, P. and Hari, V. 1993. Detection of Antibodies to the HIV-1 p25/24 and gp120 proteins by sphere-Linked immunodiagnostic assay (SLIDA). BioTechniques 14: 774-780.

11. Derrick, K.S. 1973. Quantitative assay for plant viruses using serologic specific electron microscopy. Virology 56: 652-653.

12. Fraenkael-Conrat. H. 1957. The degradation of tobacco mosaic virus with acetic acid. Virology 4: 1-4.

13. Francki, R.I.B., Fauquet, C.M., Knudson, D.L., and Brown, F. 1991. Classification and nomenclature of viruses. in: Fifth Report of the International Committee on Taxonomy of Viruses. Arch. Virol. Supplement 2, Springler Verlag, Wein, Austria.

14. Haberle, W., Horber, J.K.H., Ohnesorge, F., Smith, D.P.E., and Binnig, G. 1992. In situ investigations of single living cells infected by viruses. Ultramicroscopy 42-44: 1161-1167.

15. Hari, V., Baunoch, D.A., and Das, P. 1990. Sphere-linked immunodiagnostic assay (SLIDA): An electron microscopic method for detecting specific antibodies. BioTechniques 9: 342-350.

16. Hayat, M.A., and Miller, S.E. 1990. Virological methods. Pages 51-155 in: Negative Staining. McGraw Hill Publishing, New York.

17. Hill, S.A. 1984. Electron microscopy. Pages 130-153 in: Methods in Plant Virology. T.F. Preece, ed. Blackwell Scientific Publications, Oxford, England.

18. Hitchborn, J.H., and Hills, G.J. 1965. The use of negative staining in the electron microscopic examination plant viruses in crude extract. Virology 27: 528-540.

19. Imai, K., Yoshimura, K., Tomitori, M., Nishikawa.,

O., Kokawa, R., Yamamoto, M., Kobayashi, M., and Ikai, A. 1993. Scanning tunnelling microscopy of T4 bacteriphage and tobacco mosaic virus. Japan J. Appl. Phys. 32: 2962-2964.

20. Leberman, L. 1965. Use of uranyl formate for as a negative stain. J. Mol. Biol. 13:606.

21. Lessemann, D.E., and Paul, H.L. 1980. Conditions for the use of protein A in combination with the Derrick method of immunoelectron microscopy. Acta. Horticultura 110: 119-128.

22. Martelli, G.P., and Russo, M. 1984. Use of thin sectioning for visualization and identification of plant viruses. Pages 143-216 in: Methods in Virology, Vol. III. K. Maramorosch and H. Kaprowski, eds. Acad. Press, NY.

23. Matthews, R.E.F. 1993. Plant Virology, A Text Book. Acad. Press, New York. 897 Pages

24. Milne, R.G. 1972. Electron microscopy of viruses. Pages 76-126 in: Principles and Techniques in Plant Virology. C.I. Kado, and O. Aggarwal, eds. Van Nostrand Reinhold Co., NY.

25. Milne, R.G., and Lessemann, D.E. 1984. Immunosorbent electron microscopy in plant virus studies. Pages 85-100 in: Methods in Virology, Vol. VIII. K. Maramorosch and H. Kaprowski, eds. Acad. Press, New York.

26. Mulhern, P.J., Blackford, B.L., Jericho, M.H., Southam, G., and Beveridge, T.J. 1992. AFM and STM studies of the interaction of antibodies with the S-layer sheath of the archebacterium Methanospirillum hungatei Ultramicroscopy 42-44: 1214-1221.

27. Ohnosorge, F., and Binnig, G. 1993. True atomic resolution by atomic force microscopy through repulsive and attractive forces. Science 260: 1451-1456.

28. Pinteric, L., and Taylor, J. 1962. The lowered drop method for the preparation of specimens of partially purified virus lysates for quantitative electron microscopic analysis. Virology 18: 359-371.

29. Shukla, D.D., and Gough, K.H. 1979. The use of protein A from Staphylococcus aureus in immune electron microscopy for detecting plant virus particles. J. Gen. Virol. 45: 533-536.

30. Vesenka, J., Guthold, M., Tang, C.L., Keller, D., Delaine, E., and Bustamente, C. 1992. Substrate preparation for reliable imaging of DNA molecules with the scanning force microscope. Ultramicroscopy. 42-44: 1243-1249.

31. Zenhausen, F., Adrian, M., Emch, R., Taborelli, M., Jobin, M., and Descouth, P. 1992. Scanning force microscopy and cryoelectron microscopy for tobacco mosaic virus as test specimen. Ultramicroscopy 42-44: 1168-1172.

CHAPTER 32

Detection and Management of Plant Viroids

R.P. Singh and A.K. Dhar

Viroids are the smallest autonomously replicating disease-causing agents at the lowest level in the hierarchy of life. Biochemically, they are low molecular weight single-stranded circular RNA of 246 to 375 nucleotides and cause diseases only in higher plants (21,98). The unique structural features of viroids combined with their apparent inability to code for any proteins, have made them ideal objects for molecular research in understanding basic life processes (61). Since viroid pathogenicity is essentially governed by their nucleotide sequence and structural conformation of RNA, understanding of the host-pathogen interactions at the cellular level may immensely contribute to our knowledge of plant viroid maladies and their management.

Since the first demonstration that spindle tuber disease of potatoes is caused by free RNA molecules (20,122), the so-called viroid, many plant maladies of once uncertain etiology have been described as diseases caused by viroids (Table 1). Some of them cause enormous damage to the host crop, e.g. cadang-cadang disease, which affects coconut and oil palm plantations in the Philippines (48). Others with no apparent effect on their primary host, do considerable damage to other crops situated near the infected host species. Such latent hosts may be significant in plant quarantine and germplasm introductions, both in determining the epidemiology of viroid diseases and in formulating management strategies. This chapter will attempt to provide an update of the viroid literature useful in the management of crops and the prevention of viroid infection and spread.

IMPACT OF VIROID DISEASES ON AGRICULTURE

Symptomatic Plants

Their biochemical novelty and extraordinary biological potency have made viroids a valuable tool for molecular biological research. But their potential as plant pathogens and their economic impact on agricultural production should not be overlooked (40). Diseases like potato spindle tuber, coconut cadang-cadang, chrysanthemum stunt and hop stunt have all had a major impact on production (48,66,129,159). Although presently the potato spindle tuber viroid (PSTVd) is of rare occurrence in North America, the infection rate of 25-90% was observed in the 1920s (118). PSTVd can cause as much as 65% reduction in yield of tubers (129). Even in the 1990s, in Heilongjiang province of the Peoples Republic of China, PSTVd has been very widespread and efforts to eradicate it are currently being actively pursued (137). In 1982, there was an outbreak of potato spindle tuber disease in potato breeding lines in Australia (110) which was subsequently brought under control. Hop stunt disease was detected in 17% of the total acreage of hops in Japan's Fukushima Prefecture in 1968 with some gardens having as high as 60% infection (159).

During 1946-47, chrysanthemum stunt disease was widespread throughout the United States and Canada and in some greenhouses 50-100% of plants were infected (11). The chrysanthemum industry suffered a major loss within a few years after its first appearance (67). An outbreak of chrysanthemum stunt viroid (CSVd) was reported in 1990 from Brazil, which occurred during a rapid multiplication of plants by tissue culture plantlets (27). However, the most dramatic losses are caused by coconut

Table 1. Viroids, their molecular size and indicator plants

Diseases (viroids)	Main-Host	Family	Indicator/Host	No. of nucleotides
Group I				
Avocado sunblotch (ASBVd)	Avocado	Lauraceae	*Persea americana* cv. 'Haas'; 'collison'	246-251[a]
Carnation stunt associated (CarSAVd)	Carnation	Caryophyllaceae	*Dianthus caryophyllus*	275[b]
Peach latent mosaic (PLMVd)	Peach	Rosaceae	*Prunus persica*	336-337[b]
Group II				
Subgroup: PSTVd				
Chrysanthemum stunt (CSVd)	Chrysanthemum	Compositae	*Chrysanthemum morifolium* 'Mistletoe'	354-356[a]
Citrus exocortis (CEVd)	Citrus	Rutaceae	*Citrus medica* 'Etrog	370-375[a]
Columnea latent (CLVd)	Columnea	Gesneriaceae	--[i]	370-372[a]
Potato spindle tuber (PSTVd)	Potato	Solanaceae	*Lycopersicon esculentum* 'Sheyenne'; 'Rutgers' *Scopolia sinensis* *Solanum berthaultii*	356-360[c]
Tomato apical stunt (TASVd)	Tomato	Solanaceae	Same as for PSTVd	360[a]
Tomato planta macho (TPMVd)	Tomato	Solanaceae	*L. esculentum* 'Rutgers'	360[a]
Subgroup: CCCVd				
Coconut cadang cadang (CCCVd)	Coconut	Palmae	*Cocos nucifera*	246-247[a]
Coconut tinangtaja (CtiVd)	Coconut	Palmae	*Cocos nucifera*	254[a]
Hop latent (HLVd)	Hop	Cannabaceae	--[i]	256[a]

Table 1. Continued

Diseases (viroids)	Main-Host	Family	Indicator/Host	No. of nucleotides
Subgroup: HSVd				
Cucumber pale fruit (CPFVd)	Cucumber	Cucurbitaceae	*Cucumis sativus* 'Suyo'	303[a]
Hop stunt (HSVd)	Hop	Cannabaceae	*Cucumis sativus* 'Suyo'	297[a]
Citrus viroid IIa (CVIIa)	Citrus	Rutaceae	*Cucumis sativus* 'Suyo'	302[d]
Citrus viroid IIb (CVIIb)(Cachexia)	Citrus	Rutaceae	*Cucumis sativus* 'Suyo'	299[d]
Subgroup: ASSVd				
Apple scar skin (ASSVd)	Apple	Rosaceae	*Malus pumila*	329-330[e,f]
Dapple apple (DAVd)	Apple	Rosaceae	*Malus pumila*	331[f]
Australian	Grapevine	Vitaceae	'Indo'	369[a]
Citrus bent leaf (CBLVd)	Citrus	Rutaceae	--[i]	318[g]
Grapevine yellow speckled				
(GYSVd-1)	Grapevine	Vitaceae	--[i]	366-368[a]
(GYSVd-2)	Grapevine	Vitaceae	--[i]	363[a]
Pear blister canker (PBCVd)	Pear	Rosaceae	*Pyrus communis* 'Fieudière'	315[b]
Pear rusty skin (PRSVd)	Pear	Rosaceae	*Pyrus communis*	334[f]

Table 1. Continued

Diseases (viroids)	Main-Host	Family	Indicator/Host	No. of nucleotides
Subgroup: CBVd				
Coleus blumei				
(CBVd)-1	Coleus	Lamiaceae	-[i]	248[a]
(CBVd)-2	Coleus	Lamiaceae	-[i]	299[h]
(CBVd)-3	Coleus	Lamiaceae	-[i]	302[h]

[a] From the data of McInnes and Symons (75).
[b] From the data of Hernandez et al. (52,53,54), Minafra et al. (77), and Shamloul et al. (115).
[c] From the data of Herold et al. (55).
[d] From the data of Levy and Hadidi (69).
[e] From the data of Hashimoto and Koganezawa (51), Puchta et al. (89), and Yang et al. (160).
[f] From the data of Zhu et al. (162).
[g] From the data of Ashulin et al. (3).
[h] Personal communication from H.L. Sänger.
[i] Symptomless plants.

cadang-cadang viroid (CCCVd) which has destroyed over 30 million coconut trees since 1926 in the Philippines. Over 500,000 trees are dying each year because of CCCVd (48).

Non-symptomatic Plants

Besides causing recognizable symptoms in host plants, there are several examples of viroids causing widespread infection without symptoms. Columnea latent viroid (CLVd) which does not manifest any symptoms in its natural host, *Columnea erythrophae* Deene ex Houll (22), when inoculated into potato cv. Saco, the viroid produces symptoms similar to those of the type strain of PSTVd. Similarly, a viroid latently carried in hops, hop latent viroid (HLVd), occurs worldwide in practically all hop cultivars (90). A strain of hop stunt viroid (HSVd) has been isolated from symptomless plants of grapevines in Japan (105), having been detected in plants introduced into Japan from France, Germany, Austria, Hungary and the United States. In one commercial planting of hop in the United Kingdom, HLVd was detected in as high as 89% of the samples. Although described as a latent viroid it is considered to be a potentially serious pathogen (5). A widely distributed and apparently symptomlessly carried viroid of Coleus (CbVd) has been reported from Brazil (31), Canada (136), Germany (146), and India (92). A high incidence (16 to 68%) of CbVd has been encountered in commercial seed lots available from various countries (92,136). Peach latent mosaic disease, sometimes symptomatic, recognized only in the 1970s, has been shown to be present in France in 20% of the peach cultivars introduced from the USA and in 60% of those from Japan. The disease appears frequently and universally in France (30). Recently, a viroid has been isolated from the symptomless pouchflower plant (*Nematanthus wettsteinii*), a commonly-grown ornamental. When the viroid was mechanically transmitted to potato, tomato or *Scopolia sinensis* they developed symptoms similar to those caused by PSTVd (140).

Beneficial Effects

Viroid infection is generally regarded as harmful to plants whether or not it produces recognizable symptoms. Stunting is one of the common manifestations of viroid pathogenesis. Efforts have been made to make use of this aspect

of viroid-altered growth habit for beneficial purposes. In citrus orchards, the harvesting, spraying and other grove care operations constitute the major cost of production which can be minimized considerably with smaller trees having more compact canopies. Citrus exocortis viroid (CEVd) infected citrus trees propagated on susceptible root stock not only have stunted growth but also tend to bear fruit at an early age (71). Interestingly, the longevity of the tree as well as fruit size and quality is not adversely affected when orange trees are grafted on trifoliate root stock (34). In Australia, trials of high density plantings of citrus trees deliberately dwarfed by CEVd inoculation have been conducted (34). However, the major concern over such deliberate use of CEVd is the possibility of a severe viroid mutant arising from an increased reservoir of infected plants. Moreover, the possible health hazards, if any, of CEVd-infected fruits have not been evaluated. Theoretically, a more feasible approach will be to understand the stunting mechanism of viroid infection and to apply that understanding instead of using intact infectious viroid (40).

Viroids are essentially confined to the host nuclei and remain in equilibrium with the host metabolic pathways without exerting dramatic changes in all host species. Semancik proposed the use of viroid as an "effective exogenously introduced regulatory molecule...." because of their relative simplicity in the expression of biological potential (113). In fact, it has been shown that a viroid infection, by hardening the host cell wall, actually has the beneficial effect of better preventing fungal and bacterial infection (113).

Biological Properties of Viroids

Macroscopic Symptoms

A number of plant diseases which are now known to be caused by viroids were previously attributed to viruses. Many of the host responses to the presence of viroids resemble either virus infections or physiological stresses. Nevertheless, the appearance of stunting seems to be one of the primary consequences of the biological expression of viroid RNA. Necrosis of vascular tissues, epinasty and rugosity of leaves, internode shortening, and deformation of fruits and tubers are some of the other symptoms produced as a result of viroid infection. PSTVd, CEVd and tomato apical stunt viroid (TASVd) in *S. sinensis* cause small dark brown necrotic lesions under experimental conditions (116,118).

Variation in the symptomatology of a viroid disease has been shown among other causes due to the duplication of a certain part of viroid sequence. In the case of CEVd a dramatic moderation of symptom expression in *Gynura aurantiaca* D.C. has been attributed to a CEVd variant containing a 92-nucleotide repeat sequence (114).

Cytopathological Symptoms

In addition to macroscopic symptoms, distinct cytopathic changes have been observed in viroid-infected tissues. Aberrations in cell wall structures and chloroplast organization have been reported for several host-viroid combinations. Increased frequency of plasma membrane proliferations, the so-called plasmalemmasomes or paramural bodies, have been reported in *G. aurantiaca* infected with CEVd (111), PSTVd-infected tomato leaf tissues (49), and avocado sunblotch viroid (ASBVd)-infected avocado (16). However, in recent years, contradictory observations have been presented regarding the presence of plasmalemmasomes in healthy and viroid-infected tissues. For example, Wahn *et al.* (154) did not observe any difference in the frequency of plasmalemmasomes in CEVd-infected and healthy *G. aurantiaca*. Similar findings have been recorded in healthy and CEVd- and HSVd-infected cucumber leaf cells (64).

Disorganization of chloroplast structure, distortion of thylacoid membrane, paucity of grana, disappearance of tonoplast, breakdown of outer chloroplast membrane and eventual release of chloroplast content in cytoplasm have been observed in viroid-infected tissues. Chloroplast disintegration has also been noticed in plants that do not display visible symptoms as in the case of cucumber pale fruit viroid (CPFVd)- and HSVd-infected tomato (64). In ASBVd-infected avocado, chlorosis constitutes a major symptom expression. Ultrathin sections through the chlorotic tissues demonstrated swelling of chloroplasts and formation of crystalline bodies (16). Accumulation of electron dense material in viroid-infected plants has been observed in CSVd-infected chrysanthemum (65) and PSTVd-infected *S. sinensis* (84). Viroid-host cell interaction has also been studied using cell-suspension culture. Photosynthetically active callus and cell-suspension cultures derived from

healthy and PSTVd-infected tomato cells have revealed that PSTVd-infected cells grow slowly, are morphologically different in size and shape and form tight cell aggregates. PSTVd infection also caused starch accumulation in chloroplasts, deformation of chloroplast envelope and irregular plasmalemmasomes at the cell membrane (149).

Transmission and Spread of Viroids

All viroids are transmissible readily by mechanical means and are highly contagious. Viroids are perpetuated through unlimited generations, provided the initial propagating material is infected with viroids (40), in crops commercially propagated by vegetative means like avocado, chrysanthemum, citrus, hops and potatoes. With PSTVd, mechanical transmission by contact with farm implements is mainly responsible for the spread of the disease in potatoes (73). In the case of CEVd and PLMVd, contaminated budding knives and other tools have been implicated (37,115).

PSTVd and CbVd are transmitted through both pollen and true seeds (123,136). But the percentage of transmission may be highly variable with infection reaching as high as 80% (123,138). However, since true potato seeds (TPS) are not used in North America and Europe for commercial potato production, pollen and seed transmission of PSTVd is of greater concern in breeding and experimental studies. But in China and other developing nations where TPS is being promoted for commercial production, PSTVd-infected seed may pose a real threat. Seed transmission has also been demonstrated in ASBVd (155), CCCVd (48), and CEVd (39).

Evidence also exists for experimental arthropod transmission of viroid. For example, de Bokx and Piron (19) demonstrated 5% aphid transmission of PSTVd in a nonpersistent manner by *Macrosiphum euphorbiae*. Similarly, TASVd can be transmitted by *Aphis craccivora* (156). Tomato planta macho (TPMVd) is transmitted by *Myzus persicae*. The TPMVd persists in the vector for at least 8 days, and transmission efficiency may go as high as 97% (35).

Biochemical Changes in Viroid-infected Plants

Proteins

Viroid infection does not cause any major

qualitative or quantitative changes in the nucleic acid profiles of the host, which would indicate a greater coordination of viroid and host nucleic acid synthesis. Nevertheless, some viroid-induced changes in host proteins have been demonstrated for different viroid-host combinations. For example, in PSTVd-infected tomato leaves two proteins were detected (155 kDa and 195 kDa) (161). However, similar proteins were found to have accumulated in response to tobacco mosaic virus infection (153). Similarly, drastic increases in the concentration of two low molecular weight host-coded proteins CEV-P$_1$ (12 to 13.7 kDa) and CEV-P$_2$ (16.3-18 kDa) have been detected in CEVd-infected G. aurantiaca, Etrog citron, potato and tomato (28,112). Interestingly, it was observed that the molecular weights of these two proteins depend on the host species (15). In uninfected G. aurantiaca identical proteins termed senescence proteins (SEN-P$_1$ and SEN-P$_2$), were detected as apparent products of natural senescence (113). Hadidi (42) demonstrated that in PSTVd-infected leaves the concentration of a 33 kDa protein significantly increased. He also showed that a protein of similar size from infected tissue bound to PSTVd in vitro. Camacho and Sänger (12) showed that in tomato plants infected with different viroids (PSTVd, CEVd, CSVd, CPFVd) dramatic increases in the concentration of the protein P14 in order of Mr 14 kDa (P14) are obvious. However, accumulation of P14 was observed in response to many fungal and viral infections, indicating the induction of P14 is a general response to microbial stress. The complete amino acid sequence of P14 has been determined and comparison with the published amino acid sequences of several related pathogenesis proteins revealed no sequence homology, thereby characterizing P14 as a structurally novel type of protein (72).

Although different proteins have been implicated as a consequence of viroid pathogenesis, the biological significance of these disease-associated proteins remains to be determined. For the P$_1$-protein of CEVd-infected G. aurantiaca, ribonuclease activity was assigned as a potential function. However, consistent information is lacking to attribute a definite function to this protein (28).

Hormones

The symptom expression manifested by viroid pathogenesis suggests disturbances in phytohormonal metabolism. Significant decrease in gibberellin concentration as a result of CEVd infection in G. aurantiaca has been reported (47). Delay in root initiation has been attributed to reduction in diffusible auxin like substances from the apical bud (29). Delay or almost complete inhibition of root initiation has also been described for CSVd-infected chrysanthemum (57), HSVd-infected hop plants (158) and CEVd-infected tomato (23).

Structure and Replication

Viroids are single-stranded, covalently-closed circular RNA molecule containing 246 to 375 nucleotides (61). Due to extensive regions of intra-molecular complementarity, the viroid molecule under native conditions appear as double-stranded in the electron microscope. Pioneering studies with PSTVd have shown that this rod-like molecule undergoes a structural transition with increasing temperature (97). Initially, transition occurs at the premelting regions of viroid molecule termed premelting (PM), PM1, PM2, and PM3. In a highly cooperative transition at 77°C all base pairs are disrupted with concomitant formation of three stable hairpins - I, II and III. Hairpins I and III dissociate at 90°C whereas hairpin II melts at 100°C giving rise to a circular RNA without any base pairing. This thermal transition is a reversible process. The discovery that the integrity of hairpin II seems to be indispensable for viroid replication suggests a promoter function for it (97). This structural transition characteristic has been utilized for viroid assay procedure (107,108,126).

On the basis of sequence similarity and processing of oligomeric replication intermediates into monomeric circular progeny molecules, viroids have been classified into two major groups. Avocado sunblotch viroid group includes avocado sunblotch, carnation stunt associated, and peach latent mosaic viroids. The viroids in this group self-cleave through a hammerhead structure. Potato spindle tuber viroid group in which members share a highly conserved central domain (see below) and require a trans factor present in host cell nuclei for processing of the oligomeric templates during replication. Viroid molecules have been assigned certain functional regions or domains (60,61) (Figure 1).

1) The **central domain** is a highly conserved region among most viroids. It consists of a bulged helix, flanked on both sides by an inverted repeat which can form a stem loop

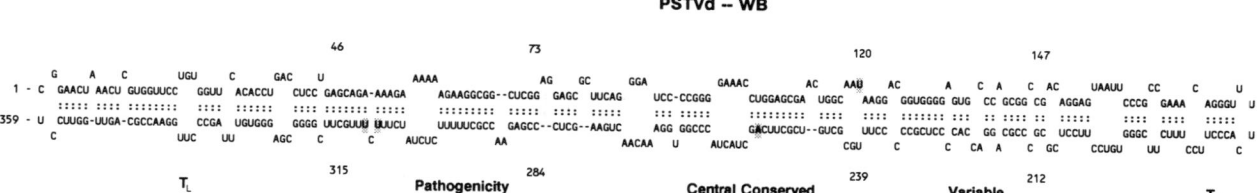

Figure 1. Nucleotide sequence and five structural domains of a mild strain of potato spindle tuber viroid (from Singh *et al.* (142)).

structure. The postulated function of this domain involves processing of the viroid replicative intermediates. 2) The **pathogenicity domain** includes a long adenine-dominated oligopurine sequence of 15 to 17 nucleotides in one strand and an oligo (U_{4-7}) sequence in the opposite strand. Sequence changes in this domain are involved in the alteration of symptom expression and host-range of viroids. 3) The **variable domain** is on the right-hand side of the central conserved region, and as its name implies, sequence variants show most differences in this region. A significant sequence relationship in this region between viroids is the presence of the oligopurine: oligopyrimidine helix. 4) The **terminal domain (left)** consists of an end loop on the left side of the molecule. It is characterized by a strictly conserved sequence of CCUC. 5) **terminal domain (right)** on the right-hand end of the viroid molecule, is characterized by the strictly conserved sequence of CCUUC. The sequences in the terminal domains are postulated to facilitate binding of DNA-dependent RNA polymerase II.

Viroids are autonomously-replicating molecules and do not code for their replicase or a subunit of the replicating enzyme (103). It is apparently conceivable that a viroid RNA molecule could be translated *in vivo* into a small polypeptide that might be capable of converting a host enzyme into viroid-specific replicase. A case in point is Qβ, an RNA bacteriophage, whose phage-coded protein combines with three other host factors to create Qβ- specific replicase (59). In fact, the coding capacity of a viroid could result in a protein of about 120 amino acids (98). However, attempts to demonstrate mRNA activity of viroid RNA using a cell-free *in vitro* protein synthesizing system or using *Xenopus*

oocytes failed to yield viroid specific polypeptides (18,46). The lack of ribosome binding sites, the absence of "cap" structure, the nearest neighbor sequences, the stable secondary structure and the circularity of the genome offer additional, circumstantial evidence against mRNA activity. The logical conclusion one can infer is that viroid replication depends on host polymerases.

Different workers have proposed replication models for viroid RNA. Essentially, the replication scheme involves a rolling circle mechanism in which the circular viroid (+) RNA serves as a template for synthesizing a multimeric complementary sense, linear molecule, (-) RNA. This complementary strand then serves as template for the production of a multimeric viroid (+) RNA strand, which is cleaved to yield a unit-length molecule. These unit- length molecules are finally circularized, giving the circular monomeric viroid (+) RNA as progeny (10,58,81). Conceptually, the process of cleavage of multimeric viroid RNA into monomeric progeny RNA is analogous to splicing precursor RNAs and joining exons to form functional RNAs. In ASBVd the mechanism of self-cleavage has been extensively studied. A characteristic secondary structure called a "hammerhead" seems to be a prerequisite (32,33). Whereas most satellite RNAs follow a self-cleaving process, viroids other than ASBVd group require a *trans* factor present in cell nuclei (151).

DETECTION OF VIROIDS

Viroids have been isolated from over twelve plant families. Rapid and specific detection of viroids is a prerequisite for their management. Since viroids lack a protein coat, diagnostic approaches based on antigenic

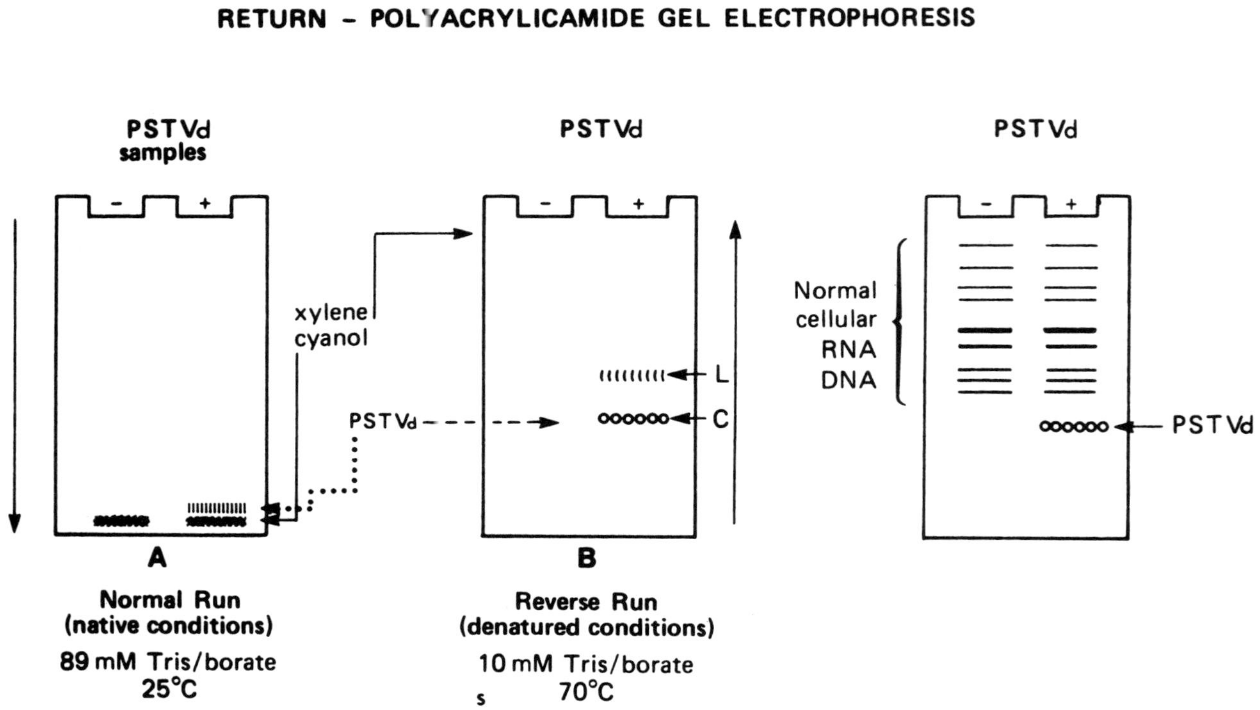

RETURN – POLYACRYLICAMIDE GEL ELECTROPHORESIS

Figure 2. A diagrammatic representation of return-polyacrylamide gel electrophoresis procedure, using PSTVd-infected (+), and healthy (-) nucleic acids samples. (A) First run; (B) second run; and the last one, depicting the stained gel bands (from Singh, R.P., Viroids and Satellites: Molecular Parasites at the Frontier of Life, Chapter 5, 1991 with permission).

properties are not applicable. Ideally, the detection procedures should be simple and sensitive enough to detect very low levels of the pathogen. In view of the potential danger posed by some viroids like PSTVd and CCCVd, detection methods which can easily be applied for a large number of samples are desirable. Detailed accounts of the diagnosis of viroid diseases have been given (56,75,121,150). The present review will provide a comprehensive account of procedures for detecting viroids with a greater emphasis on recent developments, based on molecular biological approaches.

Indicator Plants

The commonly-used and time-consuming method for viroid diagnosis is based on their symptom expression on indicator plants (Table 1). For example, symptom development of ASBVd can take from 2 months to 2 years after inoculation of susceptible avocado seedlings (86). However, a long incubation period and the lack of suitable indicator plants often limit this biological method of detection. Moreover, symptom expression on the host plant is influenced by temperature. For example, an increase in temperature from 20°C to 30° C reduces the incubation period of CPFVd in cucumber from 76 to 12 days and from 38 days at 21°C to 17 days at 33°C for HSVd in cucumbers (106,152). ASBVd symptoms develop faster in a hot (30/28°C day/night) than in a cool (20/18°C day/night) environment (17). Symptom expression of PSTVd is enhanced by high light intensity (650 μ Em^{-2}S^{-1}, sodium vapor lamp), high temperature (24-39°C) (36) and host nutrition (68,130).

Not all viroid infections can be detected visually, as some viroids remain symptomless on some hosts. In addition, several indicator plants like tomato, *Scopolia* sp., *Gynura* sp. etc., react similarly to many viroids, making identification of a specific viroid difficult (118).

Mixed infections of virus and viroid are of common occurrence. Double infection of PVY and PSTVd may produce new disease symptoms and without PSTVd being encapsidated by PVY particles (127,141).

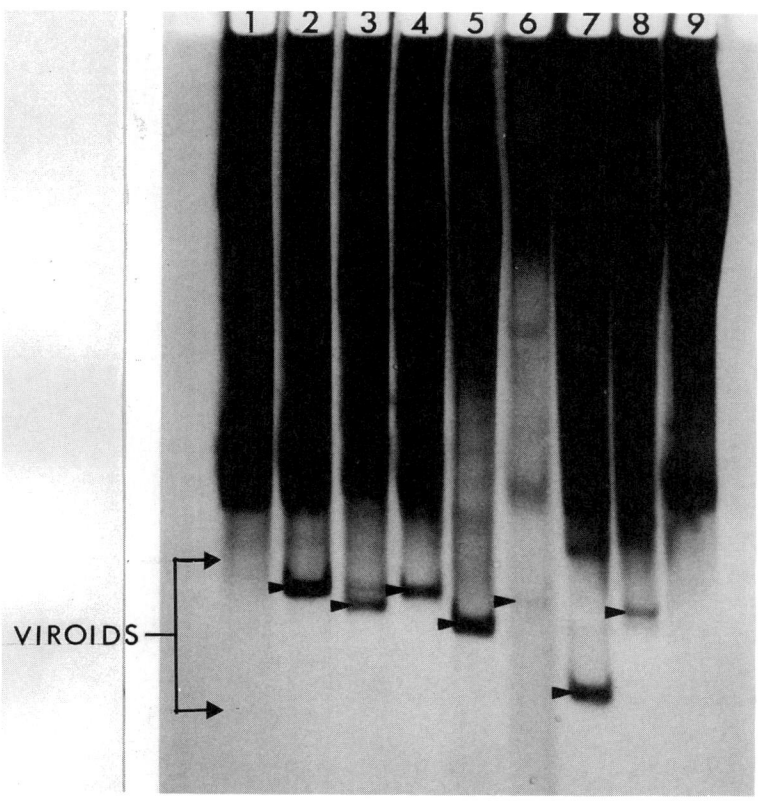

Figure 3. Detection of viroids of varying sizes, by R-PAGE from different plant species. Viroids affecting potato (lane 2), tomato (3), citrus (4), chrysanthemum (5), apples (6), Coleus (7), and pouchflower (8), lanes 1 and 9 contained nucleic acids from healthy plants (from Singh, R.P., Viroids and Satellites: Molecular Parasites at the Frontier of Life, Chapter 5, 1991 with permission).

Polyacrylamide Gel Electrophoresis

Intramolecular base-pairing which leads to unique secondary structural properties has been used in the detection and purification of viroids (97). A method of two-dimensional electrophoresis was developed by Schumacher *et al.* (107) for detection of viroids because of their genome circularity and conformational transitions. A modified method, called Return-polyacrylamide gel electrophoresis (R-PAGE) (108,126), essentially uses the thermodynamic properties of viroids. In the first step of electrophoresis, nucleic acid extracts from healthy and viroid-infected plants are electrophoresed on a nondenaturing polyacrylamide gel. Both circular and linear forms of the viroid RNA separate from the host nucleic acids but viroids migrate together during the first run. In the second step electrophoresis is carried out under denaturing conditions (using a boiled, low-salt buffer) and with the polarity of migration reversed. Under these conditions the double-stranded structure of the viroid is transformed into an unfolded single-stranded circle, with much slower mobility than other nucleic acids of the same size. The viroid RNA therefore forms a distinct band and remains separated from normal cellular RNA (Figure 2). Using R-PAGE, PSTVd has been detected in single true potato seeds, in minute seeds of Coleus, and dormant potato tubers (128,133,136), and other viroids from different plant parts (Figure 3). R-PAGE has also been used in the identification of viroid strains (126) and in the cross-protection studies (62,135). Because of its rapidity, simplicity, and specificity to circular RNAs, R-PAGE can routinely be used to detect viroids and can accommodate a fairly large number of samples. It can be used especially in the initial stages of multiplication of clean

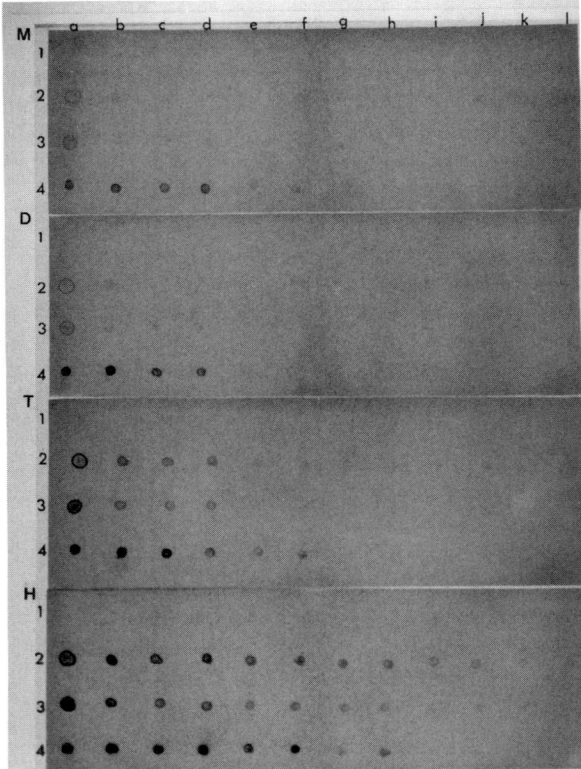

Figure 4. Detection of PSTVd using various sizes of RNA probes; (M) monomer; (D) dimer; (T) tetramer; and (H) hexamer. Row #1, nucleic acid extracts from healthy potato leaf; #2 mild PSTVd from potato leaf; #3 mild PSTVd from tuber; and #4 purified PSTVd. Rows 1 to 3 contained 1 ul of sample diluted 1:2 (a), 1:512 (b), 1:1024 (c-d), 1:2048 (e-f), 1:4096 (g-h), 1:8192 (i-j), and 1:16384 (k-l). Row 4 was spotted with purified viroid, ·12 ng (a-b), 1.2 ng (c-d), 120 pg (e-f), 12 pg (g-h), 1.2 pg (i-j), and 0.12 pg (k-l). [From Singh *et al.* 1994 (143)]

propagating materials and general screening of the quarantine material.

Occasionally, small double-stranded RNAs are also detected in the range typical of viroids by the R-PAGE (144). This type of contamination can be eliminated by conducting R-PAGE procedure, where samples on the first gel are applied after boiling and rapid cooling. Thus, the ds-RNAs in boiled samples remain linear and separated, while viroids form rod-like molecules on rapid cooling and thus result in slower and differential migration (9).

Nucleic Acid Hybridization

Nucleic acid hybridization tests for detecting specific DNA or RNA sequences have gained wide acceptability in recent years. Although the term hybridization was originally proposed by Spiegelman (147) to describe DNA-RNA hybrids, today the term hybridization includes the formation of DNA-DNA, DNA-RNA, or RNA-RNA complexes.

Biological properties of DNA or RNA molecules are essentially governed by the sequences of four component nucleotides. Since the nucleotide sequences of these molecules are largely conserved, any procedure which can readily differentiate between sequences will be useful in the specific identification of a particular nucleic acid. Since viroids have no protein coat, a detection procedure based on an immunological approach is ruled out. Diagnostic approaches based on nucleic acid hybridization are not only highly specific but also applicable for routine testing of large number of samples (56,75, 150).

Important aspects for nucleic acid hybridization are the choice of the compound used to label the nucleic acid and the subsequent detection of the label. Generally ^{32}P is used to label nucleic acids. But the problems associated with the use of radioactive labels (viz. a short half-life thus restricting the shelf life of a labeled probe; possible health hazards; and disposal of radioactive wastes) necessitate a suitable alternative. Different non-radioactive labels are now available for specific detection of a target nucleic acid (63). Of them, the use of biotin and digoxigenin-labeled probes is gaining greater attention. In the case of a biotinylated probe, the nucleic acid is labeled with biotin, which is then detected by streptavidin or avidin conjugated to alkaline phosphatase. With digoxigenin (DIG)-labeled probes the particular nucleotide is linked via a spacer arm to the steroid hapten digoxigenin, which is then detected by anti-DIG-alkaline phosphatase. An enzyme-catalyzed color reaction with 5-bromo-4-chloro indolyl phosphate (x-phosphate) and nitroblue tetrazolium salt (NBT) produces an insoluble color precipitate which indicates the hybrid molecule.

Owens and Diener (80) employed a radio-labeled cDNA probe to detect PSTVd, using a solid support (a nitrocellulose membrane). More recently, dot-blot hybridization has been used for ASBVd (7), PSTVd (102), and CSVd (13) with a full-length monomeric cDNA clone. A synthetic oligonucleotide probe has also been used to detect ASBVd (6,94), HSVd (104), CEVd (104), and PSTVd (157). Sano et al. (104) demonstrated that synthetic probes directed towards the central conserved region of the viroid have broader specificity, detecting members of the same group whereas probes directed towards variable regions could be strain-specific. However, the sensitivity and specificity of detection can be further improved by an RNA probe prepared by using an RNA polymerase system (56,76). An RNA probe has several advantages over a DNA probe viz., strand-specific probe can be made; the RNA:RNA hybrid is more stable than DNA:DNA or DNA:RNA hybrids, a non-specifically bound probe can be removed by digestion with RNAse A, which is highly specific for single-stranded RNA; and a riboprobe has a higher sensitivity than a nick-translated DNA probe. ^{32}P-labeled c-RNA probes have been successfully used for the detection of pome fruit viroids (45). A biotinylated RNA probe has been used for detection of PSTVd (100). However, the biotinylated cDNA probe is 15 times less sensitive than ^{32}P-labeled probe and cRNA probes are 6 times less sensitive (14). Digoxigenin-labeled cRNA probe of PSTVd or ASSVd, on the other hand, was found to be as sensitive as a ^{32}P-labeled

probe (87). In our laboratory (143) multimeric cRNA probes, labeled with digoxigenin have shown a 4-fold higher sensitivity than monomeric probes (Figure 4).

Polymerase Chain Reaction (PCR)

PCR is a highly specific and versatile method of DNA amplification using thermostable DNA polymerase from *Thermus aquaticus* (Taq polymerase) or *Pyrococeus furiosus* (Pfu polymerase). To amplify a specific DNA segment, two primers flanking the target DNA are designed. The amplification procedure involves denaturation of the target nucleic acid, followed by annealing the primers to their complementary sequences and amplification of the target region using DNA polymerase. These three steps are repeated in a thermocycler until sufficient amount is produced (2).

Cross-hybridization with a non-homologous viroid is a major limitation of viroid detection using the nucleic acid hybridization test. The situation is further complicated when more than one viroid is present in a given host, as observed in grapevine (95), and citrus (26). Reverse-transcription, followed by PCR (RT-PCR) has been used to detect apple scar skin viroid (ASSVd) group viroids from pome fruit trees (43) and CEVd, CVIIa, CVIIb, plum dapple, peach latent mosaic, grapevine yellow speckled, and PSTVd (44). Using viroid specific primers in RT-PCR, it has been possible to differentiate grapevine yellow speckled viroid1 (GYSVd1) and GYSVd2, which could not be distinguished by nucleic acid hybridization using a cDNA probe (96). Although nucleotide sequencing is the most definitive approach for viroid identification, its application has limitations. In that direction, PCR can provide a practical alternative.

MANAGEMENT OF VIROID DISEASES

Since the discovery of viroid in the early 1970s (20,122), considerable progress has been made in understanding the biochemical novelty of these pathogens. A great deal of research has also been done in improving viroid detection techniques. However, management of viroid diseases, the ultimate goal of farmers/growers, has received comparatively lesser attention. Although efforts to eradicate viroids from commercially-important crops have achieved considerable success in some cases (e.g. PSTVd from potato) (120), in others destruction of diseased plants is the only available practical solution at present (e.g. cadang-cadang disease in coconut palms) (48). With increasing knowledge of the biochemistry of viroids and the molecular basis of their interaction with hosts, opportunities to control viroid diseases are becoming possible.

Prevention of New Viroids From Germplasm Introduction

One of the primary strategies to control viroid diseases is to prevent introduction of the pathogen from germplasm collections. Interest in this approach has renewed, in view of the large-scale movement of plant materials and changes in agricultural practices in recent years. Because viroids have a high seed transmission potency and are extremely contagious, it becomes difficult to eradicate most viroids once they are introduced into germplasm collections. This is particularly important in an initial breeding program. For example, the outbreak of PSTVd in potato breeding lines in Australia in 1983 (110); the isolation of a mild PSTVd strain from some wild *Solanum* germplasm collections in India (82), and the discovery of a PSTVd isolate widely prevalent in the potato breeding lines in China posed threats to the breeding programs (142). Sequence analyses of the last two PSTVd isolates (Indian and Chinese) indicated that they are identical to a mild strain of PSTVd found in North America (142). Considering that the PSTVd has not been detected in wild *Solanum* sp. growing in South America, where potato originated, or from countries not importing seed potatoes from North America (4), it appears that the primary center of distribution of PSTVd may have been the sources in North America, and the viroid spread to Asia and other South American countries has taken place through movement of seed potatoes or germplasm. However, characterization of PSTVd from widely different geographical locations is needed before achieving a final conclusion.

Use of Clean Propagating Material

Planting of viroid-free materials is a prerequisite for the management of viroid disease. This is especially important for vegetatively propagated material, where viroids can perpetuate over the generations through contaminated initial material. Indexing of propagating material for viroid detection is therefore essential. Regular indexing of seed potatoes as well as field samples helped to eradicate PSTVd from Canada (120). Seed certification schemes have also been successfully employed for large scale production of viroid-free planting material in chrysanthemum (67). Since most viroids are generally not vectored by insects, use of clean propagating material ensures an obvious additional advantage compared to the aerial vectored viral diseases.

Disinfection of Equipment

Their highly contagious nature, coupled with the apparent ability of viroid RNA to withstand extreme conditions, ensure that viroids can spread quickly, by means of contaminated tools used in various tissue culture and field operations. It has been reported that viroids remain infective on contaminated surfaces for a long time (7). Transmission via contaminated tools may pose a serious threat if there is disease source in the field. Disinfecting tools, knives, and other equipment with 0.25% sodium or calcium hypochlorite is very effective (38,99). Heat treatment or use of detergents may not be very effective, especially with brief exposure (134). PSTVd has been found to remain infective up to 3 h at 120 C.

Use of Resistant Plant Material

Incorporation of resistance in cultivated crop plants is one of the most preferred methods of control for plant viruses. However, with plant viroids not much success has been achieved, mainly because of the paucity of resistant donors.

Singh and Slack (124) tested 81 tuber-bearing *Solanum spp* for resistance to mild and severe strains of PSTVd. The reactions varied widely from symptomless (tolerant) to very severe symptoms, but no immune plants were found. However, the use of tolerant cultivars may not be practically feasible because, once infected, they will continue to serve as a reservoir for inoculum. In a later study Singh (119) reported two clones of *S. berthaultii* resistant to PSTVd on sap inoculation but not by grafting. Although the graft-inoculated plants remained symptomless, viroids could be detected by bioassay on *S. sinensis* and by dot-blot tests.

Cross-protection

Cross-protection has been defined as the phenomenon by which a plant infected with a strain of pathogen is protected against subsequent challenge inoculation with a more virulent strain. The concept of cross-protection has been favorably exploited in the management of plant virus diseases like citrus tristeza, cocoa swollen shoot, papaya ring spot, etc. However, in plant viroid diseases cross-protection is less effective. When tomato plants with a prior infection by a mild strain of PSTVd are challenge-inoculated with a severe strain, symptom development by the challenge strain is delayed by 48 days (62). However, presence of both strains could be detected. The severe strain was found to replicate faster than the mild strain and eventually replaced the latter (62). But increasing the gap between protecting and challenge inoculations decreased the number of doubly-infected plants (140). Cross-protection of PSTVd strains is a host-dependent phenomenon. In contrast to tomato, cross-protection was found to be complete in certain potato cultivars but not in others (135).

More recently (79), it has been demonstrated that tomato plants pre-inoculated with non-necrogenic satellite RNA of cucumber mosaic virus (CARNA-5) produce milder symptoms when challenge-inoculated with a severe strain of PSTVd. Dot blot, polyacrylamide gel electrophoresis and northern hybridization revealed a reduced level of viroid RNA accumulation although the satellite RNA concentration remained unchanged (79).

ELIMINATION OF VIROIDS FROM INFECTED PLANTS

Thermotherapy, Meristem Culture and Nucleic Acid Inhibition

Viroids are warm-climate pathogens, and replicate at higher rates under high temperature conditions (118) except ASSVd (145). Meristem culture techniques, in combination with heat therapy, have been used to eliminate plant viruses from plant tissues and organs (91). However, application of these techniques for viroid elimination has yielded a very low proportion of PSTVd-free potato (148). PSTVd-infected potato plants, however, have been freed of viroids by a combination of cold treatment and meristem culture (70). Recently, additional viroids or viroid-like diseases (chrysanthemum stunt, chrysanthemum chlorotic mottle, and cucumber pale fruit) were eradicated using prolonged low-temperature therapy (83). Although viroid-free meristems can be obtained from infected plants grown for 3-6 months at 6-10 C, because this procedure is very slow, it can only be used to recover plantlets from valuable and rare germplasms.

Shoot-tip culture has been successfully used to obtain CEVd-free citrus and tomato (24). CHCMVd and CSVd from chrysanthemum (85), and GVYSVd1, and HSVd from grapevine (*Vitis vinifera* L. cv cabernet Sauvignon) (25) have been eliminated using meristem shoot-tip culture.

Procedures which are effective for one viroid, may not be applicable to other viroids or to the same viroid in other host plants. For example, meristem culture in combination with low temperature is effective in eliminating several viroids (83), but low temperature does not enhance the efficiency for producing HSVd-free hop plants (78). Shoot-tip culture has failed to eliminate GVYSVd1 and GVYSVd2 from several grapevine cultivars (41). Similarly, the heat treatment, which is not effective for the elimination of PSTVd may be very effective for the elimination of ASSVd in apples (145). ASSVd from Chinese pears, however, was eliminated by in vitro thermo-therapy and apical meristem culture (88).

The antisense DNA (complementary to the pathogenicity domain) of PSTVd has been shown to cause significant reduction of PSTVd infection at whole plant or single cell levels. The inhibition was strongly sequence specific and more efficient than corresponding antisense RNA approach (74).

Chemotherapy and Meristem Culture

So far, no chemical treatment is available for the elimination of a viroid disease. Two antibiotics, tetracycline and penicillin failed to cure CCCVd from coconuts (93). An aqueous emulsion containing 1% piperonyl butoxide and 0.01% Triton X-100 only protects against PSTVd infection in the potato and *S. sinensis* if viroid inoculation of the sprayed leaves takes place within four days (117,131). Treatment of *G. aurantiaca* plants, infected by CEVd with 300-1400 mg/C concentrations of ribavirin (1-B-D-ribofuranosyl-1,2,4-triazole-3-carboxamide), a broad spectrum antiviral agent, produced almost complete remission of symptoms on newly developed leaves (8). Ribavirin has been successful in reducing PSTVd concentration in *S. sinensis* plants, but in shoot tip culture or cuttings grown from ribavirin-treated plants, viroid concentration rapidly reappears (Singh - unpublished data).

EXAMPLES OF VIROID ERADICATION

Potato Spindle Tuber Disease: Eradication From Canada

Potato spindle tuber disease, which was recognized on the North American continent in the early 1920s and was widespread in most Canadian provinces, has been completely eliminated from commercial potato production in Canada (120). The disease was present in tablestock fields at a rate of up to 4% in 1969-70 in the province of New Brunswick, and in some fields, an infection rate of as high as 15% was not uncommon (129). However, by 1979-80, it had decreased to an undetectable level and was not detected by sensitive methods of detection (125,132). The elimination of PSTVd from Canada was achieved by a series of improvements in the potato seed production system. These measures included: development and application of rapid and sensitive test methods; establishment of an "Elite" seed potato farm to supply the seed potato industry with the highest classes of disease-free (virus and viroid-free) seed; programs to produce virus/viroid-free nuclear material by thermotherapy and axillary bud culture; improvement in seed certification regulations; adoptation of "zero tolerance" for PSTVd in seed certification, i.e. preventing the planting of seed potato with even trace amounts of PSTVd; increased emphasis on the testing of potato breeding stocks for PSTVd prior to their release to seed growers; enactment of provincial disease eradication "Acts" aimed at PSTVd eradication; and the increased emphasis by processing companies for their contract growers to plant higher classes of seeds, rather than the minimum recommended for processing. In order to ensure continued freedom from PSTVd, leaf and tuber samples from a large number of fields are tested regularly during every growing season and all imported potato samples are thoroughly tested in a plant quarantine laboratory for the presence of any viroid-like molecules by general type of detection method like R-PAGE (121).

Potato Spindle Tuber Viroid Outbreaks Eradicated From Germplasm

PSTVd is carried through the pollen and seeds of potato. It has been reported to occur in the tuber bearing *Solanum* species of the major potato germplasm collections, eg. the Commonwealth Potato Collection in Scotland (50), the International Potato Center in Peru (50), the Inter-Regional Potato Project at Sturgeon Bay, Wisconsin, U.S.A. (50); and in potato breeding lines in Australia (110). The increasing dispersal of *Solanum* species and the long persistence of viroid infection in *Solanum* botanical seeds constitutes a continued worldwide risk to potatoes. However, prompt action and use of sensitive detection procedures have eliminated PSTVd from most germplasm collections (50,101,109). The field outbreak of PSTVd in South Australia was similarly controlled and eradicated (109).

CONTAINMENT OF VIROID OUTBREAKS

Surveys to determine the occurrence of viroid infection, using general procedures like R-PAGE (121), which detects viroid in various ornamental, horticultural and agricultural plants, should be carried out.

Plant quarantine regulations should then be imposed on diseases, which do not occur in a particular area. Strict enforcement of these regulations can prevent the introduction of viroids.

Detection of a viroid disease outbreak in a new area should not be considered a threat to that area or to a crop, unless infection is widespread. Since most viroids are not readily transmitted by arthropod vectors, elimination of infected material from propagation and the application of hygienic procedures should reduce the risk of further spread. Continued monitoring of plants for a few years can eradicate the viroid from an area, as discussed above.

ACKNOWLEDGEMENTS

The authors would like to thank Richard Anderson for editorial assistance for the manuscript, and T.H. Somerville for the technical assistance in preparing the photographs.

REFERENCES

1. Allen, R.M. 1968. Survival time of exocortis virus of citrus on contaminated knife blades. Plant Dis. Reptr. 52:935-939.

2. Arnheim, N., and Erlich, H. 1992. Polymerase chain reaction strategy. Annu. Rev. Biochem. 61:131-156.

3. Ashulin, L, Lachman, O., Hadas, R., and Bar-Joseph, M. 1991. Nucleotide sequence of a new viroid species, citrus bent leaf viroid (CBLVd) isolated from grapefruit in Israel. Nucleic Acid Res. 19:4767.

4. Avila, A.C., Singh, R.P., Dusi, A.N., Fonseca, M.E.N., and de Castro, L.A.S. 1990. Lack of evidence of the presence of potato spindle tuber viroid in the main potato crop in Brazil. Fitopatol. Bras. 15:186-189.

5. Barbara, D.J., Mortow, A., and Adams, A.N. 1990. Assessment of U.K. hops for the occurrence of hop latent and hop stunt viroids. Ann. Appl. Biol. 116:265-272.

6. Bar-Joseph, M., Segev, D., Twizer, S., and Rosner, A. 1985. Detection of avocado sunblotch viroid by hybridization with synthetic oligonucleotide probes. J. Virol. Methods 10:69-73.

7. Barker J.M., McInnes, J.L., Murphy, P.J., and Symons, R.H. 1985. Dot-blot procedure with ^{32}P DNA probes for the sensitive detection of avocado sunblotch and other viroids in plants. J. Virol. Method 10:87-98.

8. Belles, J.M., Mansen, A.J., Granell, A., and Conejero, V. 1986. Antiviroid effects of ribavirin on citrus exocortis viroid infection in *Gynura aurantiaca* DC. Physiol. Mol. Plant Pathol. 28:61-65.

9. Beuther, E., Wiese, U., Lukaes, N., van Slobbe, W.G., and Riesner, D. 1992. Fatal yellowing of oil palms: Search for viroids and double-stranded RNA. J. Phytopathol. 136:297-311.

10. Branch, A.D., and Robertson, H.D. 1984. A replication cycle for viroids and other small infectious RNA's. Science 223:450-455.

11. Brierley, P. 1953. Virus diseases of chrysanthemum. Pages 596-601 in: Plant Diseases: The Yearbook of Agriculture. U.S. Government Printing Office, Washington, D.C.

12. Camacho Henriquez, A., and Sänger, H.L. 1982. Analysis of acid extractable tomato leaf proteins after infection with a viroid, two viruses and a fungus and partial purification of the pathogenesis-related protein P^{14}. Arch. Virol. 74:181-193.

13. Candresse, T., Macquaire, G., Monsion, M., and Dunez, J. 1988. Detection of chrysanthemum stunt viroid CSVd using nick translated probes in a dot-blot hybridization assay. J. Virol. Methods 20:185-193.

14. Candresse, T., Macquaire, G., Brault, V., Monsion, M., and Dunez, J. 1990. ^{32}P-and biotin labelled *in vitro* transcribed cRNA probes for the detection of potato spindle tuber viroid and chrysanthemum stunt viroid. Research Virol. 141:97-107.

15. Conejero, V., Picazo, I., and Segado, P. 1979. Citrus exocortis viroid (CEV): protein alterations in different hosts following viroid infection. Virology 97:454-456.

16. Da Graca, J. V., and Martin, M.M. 1981. Ultrastructural changes in avocado leaf tissue infected with avocado sunblotch. J. Phytopathol. 102: 185-194.

17. Da Graca, J.V., and Van Vuuren, S.P. 1987. Use of high temperature to increase the rate of avocado sunblotch symptom development in indicator seedlings. Plant Dis. 65:46-47.

18. Davies, J.W., Kaesberg, P., and Diener, T.O. 1974. Potato spindle tuber viroid. XII. An investigation of viroid RNA as a messenger for protein synthesis. Virology 61:281-286.

19. de Bokx, J.A., and Piron, P.G.M. 1981. Transmission of potato spindle tuber viroid by aphids. Neth. J. Plant Pathol. 87:31-34.

20. Diener, T.O. 1971. Potato Spindle Tuber Virus. IV. A replicating, low molecular weight RNA. Virology 45:411-428.

21. Diener, T.O. 1983. Viroids. Adv. Virus. Res. 28:241-283.

22. Diener, T.O. 1987. Columnea latent. Page 297 in: The Viroid. T.O. Diener, ed. Plenum Press, N.Y.

23. Duran-Vila, N., and Semancik, J.S. 1982. Effects of exogenous auxins on tomato tissue infected with the citrus exocortis viroid. Phytopathology 72:777-781.

24. Duran-Vila, N., and Semancik, J.S. 1986. Shoot-tip culture and the eradication of viroid-RNA. Scientia Hortic. 29:199-203.

25. Duran-Vila, N., Juarez, J., and Arregui, J.M. 1988. Production of viroid-free grapevines by shoot tip culture. Am. J. Ecology Viticulture 39:217-220.

26. Duran-Vila, N., Roistacher, C.N., Rivera, Bustamante, R., and Semancik, J.S. 1988. A definition of citrus viroid group and their relationships to the exocortis disease. J. Gen. Virol. 69:3069-3080.

27. Dusi, A.N., Fonseca, M.E.N., and de Avila, A.C. 1990. Occurrence of a viroid in chrysanthemum in Brazil. Plant Pathol. 39:636-637.

28. Flores, R., Chroboczek, J., and Semancik, J.S. 1978. Some properties of CEV-P$_1$ protein from citrus exocortis viroid-infected *Gynura aurantiaca* DC. Physiol. Plant Pathol. 13:193-201.

29. Flores, R., and Rodriguez, J.L. 1981. Altered patternof root formation on cuttings of *Gynura aurantiaca* infected by citrus exocortis viroid. Phytopathology 71:964-966.

30. Flores, R., Hernandez, C., Desvignes, J.C., and Llaur, G. 1990. Some properties of the viroid inducing peach latent mosaic virus disease. Res. Virol. 141:109-118.

31. Fonseca, M.E.N., Boiteux, L.S., Singh, R.P., and Kitajima, E.W.A. 1989. A small viroid in *Coleus* species from Brazil. Fitopatol, bras. 14:94-96.

32. Forster, A.C.,and Symons, R.H. 1987. Self-cleavage of plus and minus RNAs of a virusoid and a structural model for the active site. Cell 49:211-220.

33. Forster, A.C., Jeffries, A.C., Sheldon, C.C., and Symons, R.H. 1987. Structural and ionic requirements for self cleavage of virusoid RNA and trans self-cleavage of viroid RNA. Cold Spring Harbor Symp. Quant. Biol. 52:249-259.

34. Fraser, L.R., and Broadbent, P. 1979. Virus and Related Diseases of Citrus in New South Wales. 77 Pages. Surrey Beatty & Sons, Chipping Norton, N.S.W.

35. Galindo, A.J. 1988. Ecology and vector of tomato planta macho viroid (TPMV). Pages 28-33 in: Abstracts of Papers, Yamanishi Viroid Disease

Workshop, "Possible Viroid Etiology and Detection". 2nd meeting of the International Viroid Working Group, August 16-19, 1988 Yamanashi, Japan.

36. Grasmick, M.E., and Slack, S.A. 1985. Symptom expression enhanced and low concentration of potato spindle tuber viroid amplified in tomato with high light intensity and temperature. Plant Dis. 69:49-91.

37. Garnsey, S.M., and Jones, J.W. 1967. Mechanical transmission of exocortis virus with contaminated budding tools. Plant Dis. Reptr. 51: 410-413.

38. Garnsey, S.M., and Whidden, R. 1971. Decontamination treatments to reduce the spread of citrus exocortis virus (CEV) by contaminated tools. Proc. Fla. State Hort. Soc. 84:63-67.

39. Garnsey, S.M., and Weathers, L.G. 1972. Factors affecting mechanical spread of exocortis virus. Pages 105-111 in: Proc. 5th Conf. Int. Org. Citrus Virol. W.C. Price, ed. Univ. Florida Press, Gainesville.

40. Garnsey, S.M., and Randles, J.W. 1987. Biological interactions and agricultural implications of viroids. Pages 127-160 in: Viroids and Viroid like Pathogens. J.S. Semancik, ed. CRC Press, Boca Raton, Florida.

41. Habili, N., Krake, L.R., Barlass, M., and Rezaian, M.A. 1992. Evaluation of biological indexing and dsRNA analysis in grapevine virus elimination. Ann. Appl. Biol. 121:277-283.

42. Hadidi, A. 1988. Synthesis of disease-associated proteins in viroid-infected tomato leaves and binding of viroid to host proteins. Phytopathology 78:575-578.

43. Hadidi, A., and Yang, X. 1990. Detection of pome fruit viroids by enzymatic cDNA amplification. J. Virol. Methods 30:261-270.

44. Hadidi, A., Levy, L., and Podleckis, E.V. 1995. Polymerase chain reaction technology in Plant Pathology. Chapter 13 in: Molecular Methods in Plant Pathology. R.P. Singh and U.S. Singh, eds. CRC Press, Boca Raton, USA.

45. Hadidi, A., Huang, C., Hammond, R.W., and Hashimoto, J. 1989. Homology of the agent associated with dapple apple disease to apple scar skin viroid and molecular detection of these viroids. Phytopathology 80:263-268.

46. Hall, T.C., Wepprich, R.K., Davies, J.W., Weathers, L.G., and Semancik, J.S. 1974. Functional distinctions between the ribonucleic acids from citrus exocortis viroid and plant viruses. Cell-free translation and aminoacylation reactions. Virology 61:486-492.

47. Hanks, R.W., and Feldman, A.W. 1972. Changes in amounts of auxin like growth promoter, gibberellin, and inhibition in citrus infected with exocortis virus. Pages 244-250 in: Proceedings of the 5th Conference of the International Organization of Citrus Virologists. W.C. Price, ed. University of Florida Press, Gainesville.

48. Hanold, D., and Randles, J.W. 1991. Coconut cadang-cadang disease and its viroid agent. Plant Dis. 75:330-335.

49. Hari, V. 1980. Ultrastructure of potato spindle tuber viroid-infected tomato leaf tissue. Phytopathology 70:385-387.

50. Harris, P.S., Miller-Jones, D.N., and Howell, P.J. 1979. Control of potato spindle tuber viroid: The special problems of a disease in plant breeders'

material. Pages 231-237 in: Plant Health: The Scientific Basis for Administrative Control of Plant Parasites. D.L. Ebbels and J.E. King, eds. Blackwell, Oxford, England.

51. Hashimoto, J., and Koganezawa, H. 1987. Nucleotide sequence and secondary structure of apple scar skin viroid. Nucleic Acids Res. 15:7045-7052.

52. Hernandez, C., and Flores, R. 1992. Plus and minus RNAs of peach latent mosaic viroid self-cleave in vitro via hammerhead structure. Proc. Natl. Acad. Sci. 89:3711-3715.

53. Hernandez, C., Daros, J.A., Elena, S.F., Moya, A., and Flores, R. 1992. The strands of both polarities of a small circular RNA from carnation self-cleave in vitro through alternative double and single-hammerhead structures. Nucleic Acids Res. 20:6323-6329.

54. Hernandez, C., Elena, S.F., Moya, A., and Flores. R. 1992. Pear blister canker viroid is a member of the apple scar skin subgroup (apscaviroids) and also has sequence homology with viroids from other subgroups. J. Gen. Virol. 73:2503-2507.

55. Herold, T., Hass, B., Singh, R.P., Boucher, A., and Sänger, H.L. 1992. Sequence analysis of five new isolates demonstrates that the chain length of potato spindle tuber viroid (PSTVd) is not strictly conserved but as variable as in other viroid. Plant Mol. Biol. 19:329-333.

56. Hiruki, C. 1991. Use of complementary RNA as diagnostic probes for viroids. Pages 79-88 in: Viroids and Satellites: Molecular Parasites at the Frontier of Life. K. Maramorsch, ed. CRC Press, Boca Raton, Florida.

57. Horst, R.K., Langhans, R.W., and Smith, S.H. 1977. Effects of chrysanthemum stunt, chlorotic mottle, aspermy and mosaic on flowering and rooting of chrysanthemums. Phytopathology 67:9-14.

58. Ishikawa, M., Meshi, T., Ohano, T, Okada, Y., Sano, T., Uyeda, I.,and Shikata, E. 1984. A revised replication cycles for viroids: The role of longer than unit length RNA in viroid replication. Mol. Gen. Genet. 196:421-428.

59. Kamen, R.I. 1975. Structure and function of the Q replicase. Pages 203-234 in: RNA Phage. N.D. Zinder, ed. Cold Spring Harbor Laboratory, Cold Spring Harbor, N.Y.

60. Keese, P., and Symons, R.H. 1985. Domains in Viroids: Evidence of intermolecular RNA rearrangements and their contribution to viroid evolution. Proc. Natl. Acad. Sci, U.S.A. 82:4582-4586.

61. Keese, P., and Symons, R.H. 1987. The structure of viroids and virusoids. Pages 1-47 in: Viroids and Viroid-like Pathogens. J.S. Semancik, ed. CRC Press, Boca Raton, Florida.

62. Khoury, J., Singh, R.P., Boucher, A., and Coombs, D.H. 1988. Concentration and distribution of mild and severe strains of potato spindle tuber viroid in cross-protected tomato plants. Phytopathology 78:1331-1336.

63. Kricka, L.J., ed. 1992. Non-isotopic DNA probe technique, Academic Press, California.

64. Kojima, M., Murai, M., and Shikata, E. 1983. Cytopathic changes in viroid-infected leaf tissue, cucumbers, tomatoes. J. Fac. Agr. Hokkaido Univ. 61:219-224.

65. Lawson, R.H., and Hearon, S.S. 1971.

Ultrastructure of chrysanthemum stunt virus infected and stunt free Mistletoe chrysanthemum. Phytopathology 61:653-656.

66. Lawson, R.H. 1987. Chrysanthemum stunt. Pages 247-259 in: The Viroid. T.O. Diener, ed. Plenum Press, N.Y.

67. Lawson, R.H. 1987. Controlling virus diseases in major international flowers and bulb crops. Plant Dis. 65:780-786.

68. Lee, C.R.,and Singh, R.P. 1972. Enhancement of diagnostic symptoms of potato spindle tuber virus by manganese. Phytopathology 62:516-520.

69. Levy, L., and Hadidi, A. 1993. Direct nucleotide sequencung of PCR-amplified DNAs of the closely related citrus viroids IIa and IIb (cachexia). Pages 180-186 in: Proceedings of the XIIth International Conference on Citrus Virology.

70. Lizarraga, R.E., Salazar, L.F., Roca, W.M., and Schilde-Rentschler, L. 1980. Elimination of potato spindle tuber viroid by low temperature and meristem culture. Phytopathology 70:754-755.

71. Long, K., Fraser, L., Broadbent, P., and Duncan, J. 1977. Dwarfing of citrus trees by inoculation with a pathogen. Proc. Int. Soc. Citricult. 3:866-868.

72. Lucas, J., Camacho Henriquez, A., Lottspeich, F., Henscher, A., and Sänger, H.L. 1985. Amino acid sequence of the "pathogenesis-related" leaf protein P[14], from viroid-infected tomato reveals a new type of structurally unfamiliar proteins. EMBO J. 4:2745-2749.

73. Manzer, F.E., and Merriam, D. 1961. Field transmission of the potato spindle tuber virus and virus X by cultivating and hilling equipment. Am. Potato J. 38:347-352.

74. Matousek, J., and Rakouský, S. 1993. Antisense DNA inhibits infection of potato spindle tuber viroid. Folia Biologica 39:87-100.

75. McInnes, J., and Symons, R.H. 1991. Comparative structure of viroids and their rapid detection using radioactive and nonradioactive nucleic acid probes. Pages 21-58 in: Viroids and Satellites: Molecular Parasites at the Frontier of Life. K. Maramorsch, ed. CRC Press, Boca Raton, Florida.

76. Melton, D.A., Krieg, P.A., Rebagliati, M.R., Maniatis, T., Zinn, K., and Green, M.R. 1984. Efficient in vitro synthesis of biologically active RNA and RNA hybridization probes for plasmids containing a bacteriophage SP6 promoter. Nucleic Acids Res. 12:7035-7056.

77. Minafra, A., Shamloul, A.M., Hadidi, A., and Giunchedi, L. 1993. Detection and characterization of peach latent mosaic viroid by polymerase chain reaction (PCR) technology. 6th International Congress of Plant Pathology, Montreal, Quebec, Canada. Page 319.

78. Momma, T., and Takahashi, T. 1983. Cytopathology of shoot apical meristem of hop plants infected with hop stunt viroid. J. Phytopathol. 106:272-280.

79. Montasser, M.S., Kaper, J.M., and Owens, R.A. 1991. First report of potential biological control of potato spindle tuber viroid disease by virus-satellite combination. Plant Dis. 75:319.

80. Owens, R.A., and Diener, T.O. 1981. Sensitive and rapid diagnosis of potato spindle tuber viroid disease by nucleic-acid hybridization. Science 213:670-672.

81. Owens, R.A., and Diener, T.O. 1982. RNA intermediates in potato spindle tuber viroid replication. Proc. Natl. Acad. Sci. USA 79:113-117.

82. Owens, R.A., Khurana, S.M.P., Smith, D.R., Singh, M.N., and Garg, I.D. 1992. A new mild strain of potato spindle tuber viroid isolated from wild Solanum spp. in India. Plant Dis. 76:527-529.

83. Paduch-Cichal, E., and Kryczynski, S. 1987. A low temperature therapy and meristem-tip culture for eliminating four viroids from infected plants. J. Phytopathol. 118:341-346.

84. Paliwal, Y.C., and Singh, R.P. 1981. Cytopathological changes induced by potato spindle tuber viroid in Scopolia sinensis. Can. J. Bot. 59: 677-682.

85. Paludan, N. 1984. Elimination of chrysanthemum chlorotic mottle and chrysanthemum stunt viroids, 6th Int. Symp. Virus Diseases of Ornamental Plants. Cornell University, Ithaca, N.Y.

86. Palukaitis, P., Rakowski, A.G., Alexander, D. McE., and Symons, R.H. 1981. Rapid indexing of the sunblotch disease of avocados using a complementary DNA probe to avocado sunblotch viroid. Ann. Appl. Biol. 98:439-449.

87. Podleckis, E.V., Hammond, R.W., Hurtt, S.S., and Hadidi, A. 1993. Chemiluminescent detection of potato and pome fruit viroids by digoxigenin-labeled dot blot and tissue blot hybridization. J. Virol. Methods 43:147-158.

88. Postman, J.D., and Hadidi, A. 1995. Elimination of apple scar skin viroid from pears by in vitro thermo-therapy and apical meristem culture. Acta Hortic. 386:536-543.

89. Puchta, H., Luckinger, R., Yang, X., Hadidi, A., and Sänger, H.L. 1990. Nucleotide sequence and secondary structure of apple scar skin viroid (ASSVd) from China. Plant Molecular Biology 14:1065-1067.

90. Puchta, H., Ramm, K., and Sänger, H.L. 1988. The molecular structure of hop latent viroid (HLV), a new viroid occurring worldwide in hops. Nucl. Acids Res. 16:4197-4216.

91. Quak, F. 1977. Meristem culture and virus-free plants. Pages 598-615 in: Plant Cell Tissue and Organ Culture. J. Reinert and Y.P.S. Bajaj, eds. Springer Verlag, New York.

92. Ramachandran, P., Kumar, D., Varma, A., Pandey, P.K., and Singh, R.P. 1992. Coleus viroid in India. Current Sci. 62:271-272.

93. Randles, J.W., Boccardo, G., Retuerma, M.L., and Rillo, E.P. 1977. Transmission of the RNA species associated with cadang-cadang of coconut palm, and the insensitivity of the disease to antibiotics. Phytopathology 67:1211-1216.

94. Rey, M.E.C., and Moses, A.V. 1993. Detection of avocado sunblotch viroid using cDNA and synthetic oligonucleotide probes. Phytophylactica 25:257-262.

95. Rezaian, M.A., Koltunow, A.M., and Krake, L.R. 1988. Isolation of three viroids and a circular RNA from grapevine. J. Gen. Virol. 69:413-422.

96. Rezaian, M.A., Krake, L.R., and Golino, D.A. 1992. Common identity of grapevine viroids from U.S.A. and Australia revealed by PCR analysis. Intervirology 34:38-43.

97. Riesner, D. 1990. Structure of viroids and their replication intermediates. Are thermodynamic domains also functional domains? Seminars in Virol. 1:83-99.

98. Riesner, D., and Gross, H.J. 1985. Viroids. Annu. Rev. Biochem. 54:531-564.

99. Roistacher, C.N., Calavan, E.C., and Blue, R.L. 1969. Citrus exocortis virus - chemical inactivation on tools, tolerance to heat and separation of isolates. Plant Dis. Reptr. 53:333-336.

100. Roy, B.P., Abouhaider, M.G., and Alexander, A. 1989. Biotinylated RNA probes for the detection of potato spindle tuber viroid (PSTV) in plants. J. Virol. Methods 23:149-156.

101. Salazar, L.F. 1989. Potato spindle tuber. Pages 155-167 in: Plant Protection and Quarantine, Vol. 2, Selected Pests and Pathogens of Quarantine Significance. R.P. Kahn, ed. CRC Press, Boca Raton, Florida.

102. Salazar, L.F., Balbo, I., and Owens, R.A. 1988. Comparison of four radioactive probes for the diagnosis of potato spindle tuber viroid by nucleic acid spot hybridization. Potato Res. 31:431-442.

103. Sänger, H.L. 1987. Viroid Function: Viroid Replication. Pages 117-166 in: The Viroid. T.O. Diener, ed. Plenum Press, N.Y.

104. Sano, T., Kudo, H., Sugimoto, T., and Shikata, E. 1988. Synthetic oligonucleotide hybridization probes to diagnose hop stunt viroid strains and citrus exocortis viroid. J. Virol. Methods 19:109-120.

105. Sano, T., Ohshima, K., Hataya, T., Uyeda, I., Shikata, E., Chou, T.-G., Meshi, T., and Okada, Y. 1986. A viroid resembling hop stunt viroid in grapevines from Europe, te United States and Japan. J. Gen. Virol. 67: 1673-1678.

106. Sasaki, M., and Shikata, E. 1977. On some properties of hop stunt disease agent, a viroid. Proc. Japan Acad. 63B:109-112.

107. Schumacher, J., Randles, J.E., and Riesner, D. 1983. A two-dimensional electrophoretic technique for detection of circular viroids and virusoids. Anal. Biochem. 135:228-295.

108. Schumacher, J., Meyer, N., Riesner, D., and Weidemann, H.L. 1986. Diagnostic procedure for detection of viroids and viruses with circular RNAs by "return" gel electrophoresis. J. Phytopathol. 115:332-343.

109. Schwinghamer, M.W., and Scott, G.R.. Survey of New South Wales potato crops for potato spindle tuber viroid with use of the ^{32}P-DNA probe. Plant Dis. 70:774-776.

110. Schwinghamer, M.S., Scott, G.K., Mallinson, F.K., Tesoricro, L.A., and Morrison, W.L. 1983. Potato spindle tuber viroid: an extensive infection in the New South Wales potato breeding program, Abstracts. 4th Int. Cong. Plant Pathol. Melbourne. Abstracts, 122.

111. Semancik, J.S., and Vanderwoude, W.J. 1976. Exocortis viroid: Cytopathic effects at the plasma membrane in association with pathogenic RNA. Virology 69:719-726.

112. Semancik, J.S., Conejero, V., and Gerhart, J. 1977. Citrus exocortis viroid: Survey of protein synthesis in Xenopus laevis oocytes following addition of viroid RNA. Virology, 80:218-221.

113. Semancik, J.S., and Conejero-Tomas, V. 1987. Viroid pathogenesis and expression of biological activity. Pages 71-126 in: Viroids and Viroid like Pathogens. J.S. Semancik, ed. CRC Press, Boca Raton, Florida.

114. Semancik, J.S., Szychowski, J.A., Rakowski, A.G., and Symons, R.H. 1994. A stable 463 nucleotide variant of citrus exocortis viroid produced by terminal repeat. J. Gen. Virol. 75:727-732.

115. Shamloul, A.M., Minafra, A., Hadidi, A., Giunchedi, L., Waterworth, H.E., and Allam, E.K. 1995. Peach latent mosaic viroid: nucleotide sequence of an Italian isolate, sensitive detection using RT-PCR and geographic distribution. Acta Hortic. 386:522-530.

116. Singh, R.P., A local lesion host for potato spindle tuber virus. 1971. Phytopathology 61:1034-1035.

117. Singh, R.P. 1977. Piperonyl butoxide as a protectant against potato spindle tuber viroid infection. Phytopathology 67:933-935.

118. Singh, R.P. 1983. Viroids and their potential danger to potatoes in hot climates. Can. Plant Dis. Surv. 63:13-18.

119. Singh, R.P. 1985. Clones of Solanum berthaultii resistant to potato spindle tuber viroid. Phytopathology 75:1432-1434.

120. Singh, R.P. 1988. Occurrence, diagnosis and eradication of the potato spindle tuber viroid from Canada. Pages 37-50 in: Viroids of plants and their detection, International Seminar, August 12-20. 1986 Warsaw Agricultural University Press, Warsaw, Poland.

121. Singh, R.P. 1991. Return-polyacrylamide gel electrophoresis for the detection of viroids. Pages 89-118 in: Viroids and Satellites: Molecular Parasites at the frontier of Life. K. Maramorsch, ed. CRC Press, Boca Raton, Florida.

122. Singh, R.P., and Clark, M.C. 1971. Infectious low-molecular weight ribonucleic acid from tomato. Biochem. Biophys. Res. Commun. 44:1077-1083.

123. Singh, R.P., and Finnie, R.E. 1973. Seed transmission of potato spindle tuber metavirus through the ovules of Scopolia sinensis. Can. Plant Dis. Surv. 53:153-154.

124. Singh, R.P., and Slack, S.A. 1984. Reactions of tuber-bearing Solanum species to infection with potato spindle tuber viroid. Plant Dis. 68:784-787.

125. Singh, R.P., and Crowley, C.F. 1985. Evaluation of polyacrylamide gel electrophoresis, bioassay, and dot-blot methods for the survey of potato spindle tuber viroid. Can. Plant Dis. Surv. 65:61-63.

126. Singh, R.P., and Boucher, A. 1987. Electrophoretic separation of a severe from mild strains of potato spindle tuber viroid. Phytopathology 77:1588-1591.

127. Singh, R.P., and Somerville, T.H. 1987. New disease symptoms observed on field-grown potato plants with potato spindle tuber viroid and potato virus Y infections. Potato Res. 30:127-132.

128. Singh, R.P., and Boucher, A. 1988. Comparative detection of mild strains of potato spindle tuber viroid from dormant potato tubers by return-polyacrylamide gel electrophoresis and nucleic acid hybridization. Potato Res. 31:159-166.

129. Singh, R.P., Finnie, R.E., and Bagnall, R.H. 1971. Losses due to the potato spindle tuber virus. Am. Potato J. 48:262-267.

130. Singh, R.P., Lee, C.R., and Clark, M.C. 1974. Manganese effect on the local lesion symptom of potato spindle tuber 'virus' in Scopolia sinensis. Phytopathology 64:1015-1018.

131. Singh, R.P., Michniewiuz, J.J., and Narang, S.A. 1975. Piperonyl butoxide, a potent inhibitor of potato spindle tuber viroid in Scopolia sinensis. Can. J. Biochem. 53:1130-1132.

132. Singh, R.P., DeHaan, T.-L., and Jaswal, A.S. 1988. A survey of the incidence of potato spindle tuber viroid in Prince Edward Island using two testing methods. Can. J. Plant Sci. 68:1229-1236.

133. Singh, R.P., Boucher, A., and Seabrook, J.E.A. 1988. Detection of the mild strains of potato spindle tuber viroid from single true potato seed by return electrophoresis. Phytopathology 78:663-667.

134. Singh, R.P., Boucher, A., and Somerville, T.H. 1989. Evaluation of chemicals for disinfection of laboratory equipment exposed to potato spindle tuber viroid. Am. Potato J. 66:239-246.

135. Singh, R.P., Boucher, A., and Somerville, T.H. 1990. Cross-protection with strains of potato spindle tuber viroid in the potato plant and other Solanaceous hosts. Phytopathology 80:246-250.

136. Singh, R.P., Boucher, A., and Singh, A. 1991. High incidence of transmission and occurrence of a viroid in commercial seeds of *Coleus* in Canada. Plant Dis. 75:184-187.

137. Singh, R.P., Boucher, A., and Wang, R.G. 1991. Detection, distribution and long-term persistence of potato spindle tuber viroid in true potato seed from Heilongjiang, China. Am. Potato J. 68:65-74.

138. Singh, R.P., Boucher, A., and Somerville, T.H. 1992. Detection of potato spindle tuber viroid in the pollen and various parts of potato plant pollinated with viroid-infected pollen. Plant Dis. 76:951-953.

139. Singh, R.P., Khoury, J., Boucher, A., and Somerville, T. 1989. Characteristics of cross-protection with potato spindle tuber viroid strains in tomato plant. Can. J. Plant Pathol. 11:263-267.

140. Singh, R.P., Lakshman, D.K., Boucher, A., and Tavantzis, S.M. 1992. A viroid from *Nematanthus wettsteinii* plants closely related to the *Columnea* latent viroid. J. Gen. Virol. 73:2769-2774.

141. Singh, R.P., Boucher, A., Wang, R.G., and Somerville, T.H. 1992. Potato spindle tuber viroid is not encapsidated *in vitro* by potato virus Y particles. Can. J. Plant Pathol. 14:18-21.

142. Singh, R.P., Singh, M., Boucher, A., and Owens, R.A. 1993. A mild strain of potato spindle tuber viroid from China is similar to North American isolates. Can. J. Plant. Path. 15:134-138.

143. Singh, R.P., Boucher, A., Lakshman, D.L., and Tavantzis, S.M. 1994. Multimeric non-radioactive cRNA probes improve detection of potato spindle tuber viroid (PSTVd). J. Virol. Methods 49:221-234.

144. Singh, R.P., de Avila, A.C., Dusi, A.N., Boucher, A., Trindade, D.R., van Slobbe, W.G., Ribeiro, S.G., and Fonseca, M.E.N. 1988. Association of viroid-like nucleic acids with the fatal yellowing disease of oil palm. Fitopatol. bras. 13:392-394.

145. Skrzeczkowski, L.J., Howell, W.E., and Mink, G.I. 1993. Correlation between leaf epinasty symptoms on two apple cultivars and results of cRNA hybridization for detection of apple scar skin viroid. Plant Dis. 77:919-921.

146. Spieker, R.L., Haas, B., Charng, Y.-C., Freimuller, K., and Sänger, H.L. 1990. Primary and secondary structure of a new viroid 'species' (CbVd1) present in the *Coleus blumei* cultivar 'Bienvenue'. Nucl. Acids Res. 18: 3998.

147. Spiegelman, S. 1964. Hybrid nucleic acids. Sci. Am. 221:48-65.

148. Stace-Smith, R., and Mellor, F.C. 1970. Eradication of potato spindle tuber virus by thermotherapy and axillary bud culture. Phytopathology 60:1857-1858.

149. Stöcker, S., Guitton, M.C., Barth, A., and Mühlbach, H.P. 1993. Photosynthetically active suspension cultures of potato spindle tuber viroid infected tomato cells as tools for studying viroid-host cell interaction. Plant Cell Reports 12:597-602.

150. Symons, R.H. 1984. Diagnostic approaches for the rapid and specific detection of plant viruses and viroids. Pages 93-124 in: Plant Microbe Interactions, Molecular and Genetic Perspectives, Vol. 1. T. Kosuge and E.W. Nester, eds. Macmillian Publishing Company, New York.

151. Tsagris, M., Tabler, M., Muhlbach, H.P., and Sänger, H.L. 1987. Linear, oligomeric potato spindle tuber viroid RNAs are accurately processed *in vitro* to the monomeric circular viroid when incubated with a nuclear extract from healthy potato cells. EMBO J. 6:2173-2183.

152. Van Dorst, H.J.M. and Peters, D. 1974. Some biological observations on pale fruit, a viroid-incited disease of cucumber. Neth. J. Plant. Pathol. 80:85-96.

153. Van Loon, L.C. 1985. Pathogenesis related proteins. Plant Mol. Biol. 4:111-116.

154. Wahn, K., Rosenberg-de Gomez, F., and Sänger, H.L. 1980. Cytopathic changes in leaf tissue of *Gynura aurantiaca* infected with the viroid of citrus exocortis disease. J. Gen. Virol. 49:355-365.

155. Wallace, J.M., and Drake, R.J. 1962. A high rate of seed transmission of avocado sunblotch virus from symptomless trees and origin of such trees. Phytopathology 52:237-241.

156. Walter, B. 1981. Un viroide de la Tomate en Afrique de l'Oquest: identite avec la viroide du "Potato Spindle Tuber"? CR. Acad. Sci. Paris 292:537-542.

157. Welnicki, M., Skrzeczkowski, J., Soltynska, A., Jonczyk, P., Markiewicz, W., Kierzek, R., Imiolczyk, B., and Zagorski, W. 1989. Characterization of synthetic DNA probe detecting potato spindle tuber viroid. J. Virol. Methods 24:141-152.

158. Yaguchi, S., and Takahashi, T. 1984. Survival of hop stunt viroid in hop garden. J. Phytopathol. 109:32-44.

159. Yamamoto, H., Kagami, Y., Jurokawa, M., Nishimura, S., Ukawa, S., and Kubo, S. 1973. Studies on hop stunt disease in Japan. Rept. Res. Lab. Kirin Brewery Co. Ltd. 16:49-62.

160. Yang, X., Hadidi, A., and Hammond, R.W. 1992. Nucleotide sequence of apple scar skin viroid reverse transcribed in host extracts and amplified by the polymerase chain reaction. Acta Hortic. 309:305-309.

161. Zaitlin, M., and Hariharasubramanian, V. 1972. A gel electrophoretic analysis of protein from plants infected with tobacco mosaic and potato spindle tuber viruses. Virology 47:296-305.

162. Zhu, S.F., Hadidi, A., Hammond, R.W., Yang, X., and Hansen, A.J. 1995. Nucleotide sequence and secondary structure of pome fruit viroids from dapple apple diseased apples, pear rusty skin diseased pears and apple scar skin symptomless pears. Acta Hortic. 386:554-559.

CHAPTER 33

Present Status of Controlling Barley Yellow Dwarf Virus

P. A. Burnett and R.T. Plumb

The disease known as barley yellow dwarf (BYD) has been recognized for more than 40 years (50) as being caused by a virus, transmitted in the persistent manner by aphids. Since that first report, our knowledge of the disease and especially of its causal agents has increased considerably. It soon became apparent that it was not a disease with a single causal virus, and strains of the barley yellow dwarf viruses (BYDVs) have now been divided into five groups according to their vector specificity (Table 1). These five groups have now been divided into two subgroups based on serological relationships, cytopathology and nucleic acid sequences (26,44,66,77). These two subgroups should probably be identified and described as two distinct viruses, possibly BYDVs for the MAV, PAV, and RMV strains and cereal yellow dwarf virus for the RPV and SGV strains (21).

Knowledge of the vector, epidemiology and host reaction are critical for the control of these viruses by breeding, cultural or chemical methods. Exploitation of the opportunities available through transformation and recombinant DNA technologies are, however, dependent on a detailed knowledge of the structure and function of the virus.

While the causal agent was tentatively identified as a virus some forty years ago, the disease has been recognized for much longer (4,5,20) and could well have co-evolved with the development of cultivated Poaceae. The expansion of areas sown to small grain cereals, which are especially vulnerable to infection by BYD-inducing viruses, may also have encouraged the spread of these viruses and their aphid vectors to new areas and to wild hosts not previously exposed.

Table 1. Biological properties of five strains of barley yellow dwarf luteoviruses that have been used in the classification of these viruses

Strain	Vector[a]	Reaction in oats
RMV	*Rhopalosiphum maidis*	Weakly virulent
RPV	*Rhopalosiphum padi*	Weakly virulent
MAV	*Macrosiphum avenae*	Moderately virulent
PAV	*Rhopalosiphum padi* *Macrosiphum avenae*	Strongly virulent
SGV	*Macrosiphum avenae* *Schizaphis graminum*	Weakly virulent

[a] Vector refers to the most efficient vector(s) for each of the strains (44)

The viruses which cause BYD infect a wide range of members of the Poaceae, but infect no other plant families (20). All major cereal crops are susceptible, including bread wheat (*Triticum aestivum* L.), durum wheat (*Triticum turgidum* L.), corn (*Zea mays* L.), barley (*Hordeum vulgare* L.), oat (*Avena sativa* L.), rice (*Oryza sativa* L.), rye (*Secale cereale* L.), and sorghum (*Sorghum bicolor*), and these crops show clear symptoms of the BYDV infection. Most of these plants show discoloring and stunting (both of shoots and roots), and sometimes die, especially in the competitive environment of crops. Barley leaves usually turn a bright yellow; the discoloration begins at the leaf tip and margin, and moves down the entire leaf. Oat leaves turn red, purple, orange, or tan depending on the environment and cultivar. Infected leaves of wheat, triticale, and rye usually turn yellow and sometimes purple. BYD-infected corn leaves turn red, purple, or yellow, while infected rice leaves exhibit a yellow or orange discoloration. Symptoms may on occasion be confused with other factors such as nitrogen or phosphorus deficiency, moisture or temperature stress, and aster yellows (20).

However, many grasses show few or no symptoms (20), although those species most closely investigated are usually those used within agriculture and grazing, haying, or silaging can remove plants before symptoms appear. Thus, in cereal breeding, infected plants were unlikely to have been selected and, even if they had been, the amount of their seed would probably have been less than that from uninfected plants. Consequently, even in the absence of any knowledge of the condition or its causes, unconscious selection would likely have eliminated very susceptible lines. In addition, as reported below, experience would have demonstrated the best dates on which to sow crops within the constraints imposed by local weather conditions. Thus, if crops sown at a particular time were consistently dwarfed and discolored while crops sown earlier or later remained green there would be a move to the sowing date that resulted in the greenest crop.

Given the above points, why are crops that are easily infected and killed by early infection still grown? Part of the reason is undoubtedly that more intensive agriculture has highlighted the problem. In a crop short of nitrogen, as many primitive crops likely were, it is not easy to diagnose BYD infection on symptoms alone. Furthermore, large areas of genetically uniform crops provided good environmental conditions for the growth of aphids and the viruses they transmit. Priorities for breeding have also changed. In the past 200 years, the developed world has changed from an essentially agrarian society, in which most people produced their own food, to one in which a very small proportion of the population (less than 5% in much of the "North") provides food for the majority who live in towns or cities. For most of that period, population has been increasing rapidly, causing recurrent famines in many less 'developed' parts of the world. Political uncertainties and wars have also added an emphasis on maximizing yield. The resulting displacement of traditional practices, learned by experience but limited by hand- or horse-power, have also contributed to disease problems. In the 1990s, therefore, BYD is probably more of a problem than ever before. Worldwide BYD is the most widespread and damaging virus disease of small grain cereals, and its control is therefore a high priority in many parts of the world, although the methods used differ.

METHODS OF CONTROL

Because BYD is caused by aphid-transmitted viruses, there are several opportunities and restrictions on methods available for their control. The methods can be broadly categorized as avoidance, tolerance or resistance, and control of the vectors. Infection, once it occurs in a plant, cannot be reversed or restricted by external intervention, in contrast to the use of fungicides to control fungal pathogens.

Avoidance

Avoidance of infection, usually by sowing crops at appropriate times or by growing crops where BYD is rare, is probably the 'traditional' way of coping with BYD. In the UK, late-sown (April/May) spring crops were so frequently infected that they were expected to turn yellow and acquired the name of 'cuckoo barley,' because their drilling and/or emergence coincided with the return of migrant cuckoos. Changing sowing dates does not prevent infection but can delay it until later growth stages when plants are more resistant to infection and less susceptible to damage.

In some regions sowing dates of autumn-sown crops are sufficiently flexible to avoid infection by conditions that would normally kill them (Table 2). However, the onset of conditions

Table 2. Incidence (%) of BYD and the effects of a single aphicide spray on yield of winter barley sown at different times in 1989[a]

Sowing Date	BYD (%) Untreated	Yield (t/ha)[b]	
		(-) Cypermethrin	(+) Cypermethrin
5 September	32.2	4.42	5.95
18 September	11.9	6.51	7.77
29 September	8.2	7.29	7.75
9 October	5.5	7.40	7.69
18 October	2.6	7.02	7.42

[a] Modified from Plumb and Johnstone (55)
[b] Standard error deviation = 0.295

that would prevent sowing is unpredictable and the larger the area to be sown, the earlier seeding has to start to ensure its completion. There is also a potential yield benefit from earlier autumn sowing, although crops can be sown too early, but this potential often has to be protected in other ways, often by chemical treatments (Table 2).

Generally, spring-sown crops are sown as soon as soil conditions permit. Early spring sowing in the UK usually avoids early BYD infection as the first aphid vectors usually fly in May, by which time a February- or March-sown crop is well developed. The sources of the aphids that spread BYD differ among regions. In cool, moist, often maritime areas where summer drought is uncommon, there are two sources of infection: external or internal. External sources can be perennial grass swards, or for autumn-sown crops, late-harvested corn, from which aphids fly carrying virus to infect newly sown cereal crops. Internal sources can be grass swards, stubble regrowth, or volunteers on land destined to be sown to cereals. These sources can provide a 'green bridge' between two susceptible crops if the 'green crop' is not destroyed by desiccation or cultivation before the next susceptible crop has emerged. A dense grass sward that has been established for several years is often a persistent source of aphids and BYDVs, and may not be

killed immediately by cultivation. If a seed bed is quickly prepared on the plowed surface, aphids can move from the dying grass onto the newly-emerged cereal. The importance of stubble regrowth as a source depends on amount of infection in the crop from which it derived, but volunteers have to be infected before they can act as a source. In areas where the green bridge is a problem, volunteers and stubble of autumn-sown barley present the greatest risk because this crop is usually the first cereal harvested. Summer rainfall has a pronounced influence on both internal and external sources. Sufficient rain is required to keep grass swards green, stimulate the germination of shed grain, and encourage regeneration of stubbles if these are to be potential sources of BYD. Where there is prolonged summer drought, as in 'Mediterranean' regions then these sources are less important and aphids depend upon longer-lived crops, such as corn, or irrigated crops to survive. It is often more difficult for aphids to survive hot, dry summers than cold, wet winters.

External sources of infection can be suppressed by changing sowing date or controlling aphids. Controlling internal sources is generally straightforward, by either early and effective cultivation, or the use of a broad spectrum herbicide or desiccant. There is evidence from the

UK that, when internal sources are not a problem, crops sown on plowed land have more BYD than those sown on land minimally tilled (36). Apparently, minimal tillage provides some natural control by allowing more of the aphids' natural enemies to survive.

Tolerance and Resistance

Epiphytotics of BYD are irregular in both time and space; BYD can occur in a given year in one area and not in an area immediately adjacent, and may subsequently not occur for several years. Yield losses, however, can be severe and some regions of the world are probably affected each year. According to recent reviews (39,52), natural infection can cause losses of approximately 10-25% and complete crop loss is not uncommon. Oat and barley are generally more severely affected than wheat but disease severity in a crop depends on a number of factors including cultivar, viruses involved, time of infection, vector numbers and environment. Because of the wide host range of the BYDVs among the grasses (20), tolerance or resistance is a logical choice for control of BYD and in some regions is the only option as treatment is not economical and modifying sowing date is not a practical option.

When researchers begin to screen germplasm for field resistance to a disease they often do not know whether the absence of disease symptoms is due to resistance, tolerance, or lack of infection (7). In the literature, BYD-infected plants that exhibit mild symptoms have often been referred to as "resistant" but true resistance, i.e., decreased viral replication in infected plants (19), has not often been assessed. Although the term "resistance" has been widely used in the literature, the word "tolerance" is more appropriate to indicate any field resistance. Host reactions have been categorized as ranging from tolerant to susceptible, depending on symptoms, usually on a scale of 0-9: 0 indicates no symptoms, 9 is highly susceptible (7,11,14,56). Symptoms alone may not be sufficient for determining field tolerance to BYD, but they are usually a good first step (7,11,14,57). Field tolerance to BYD has been termed "slow yellowing" by Fox et al. (23) or selection for a "stay green" character by Burnett (6). Various methods have been used to manipulate the screening conditions to enhance BYD symptom expression (7).

Barley

The early screening work for barley lines that exhibited field tolerance to BYD is summarized by Bruehl (4) and Rochow (65). C.W. Schaller at the University of California, Davis, USA (69) found field tolerant lines in barleys from Ethiopia during the Californian epiphytotic of BYD in 1951 and began an active breeding program (56). The field tolerance was traced to a major semidominant gene Yd_2 (64) located on chromosome 3 (68) and this gene is now known to map close to the centromere on the long arm (18).

The Californian program has released many cultivars incorporating this material (7). Genetic analysis of other Ethiopian barleys, summarized by Burnett et al. (7), indicated that the same Yd_2 gene conditions tolerance although the possibility of closely linked genes or a series of alleles cannot be excluded.

Some 'reversals' of tolerance to BYD have been noted (67). However, as the range of virus isolates that cause BYD is large (39,80) and isolates classified as the same serotype sometimes have different biological properties (40), lines should be selected for local use. Some of these 'reversals' may need further study but, in general, field tolerance bestowed by the Yd_2 is easy to use and has remained stable.

Other sources of resistance to BYD other than Yd_2 have been utilized (7). It seems that sufficient tolerant material does exist in barley to provide suitable parental material for breeding programs in most areas of the world (7,67). Burnett et al. (7) lists 27 cultivars of barley, many with Yd_2, that exhibit tolerance to BYD. However, local conditions and expectations of yield affect whether they are used as there is still a loss of yield compared with healthy plants.

The ICARDA-CIMMYT barley breeding program in Mexico has used the Yd_2 gene in crossing programs. Additionally, the winter barley 'Post' has been used as a parent. Many of these lines exhibit resistance in Mexico, Ecuador and Colombia (N.E. Vivar, ICARDA-CIMMYT barley breeder, Mexico, pers. comm.). Recently, Vivar (78), using selection sites in Kenya and Mexico, identified barley lines resistant to both BYDV strains from both subgroups. Additionally, he reported the presence of a single gene in the barley line Shyri, different from the Yd_2 gene present in Atlas 68, that provides resistance to MAV.

The presence of the Yd_2 reduces virus

concentration when plants are infected with MAV and PAV strains, but this is not so when plants are infected with the RPV strain (31,73). This means that Yd_2 bestows resistance only to some strains of the BYDVs, usually those in subgroup 1. Holloway and Heath (32) found that plants with Yd_2 contain a unique, constitutively expressed polypeptide, and have developed a system for the assessment of the presence of Yd_2 that could aid selection of early generation breeding material. DNA markers could also be used (30). Such markers would be very useful especially to researchers who do not have access to artificial inoculation methods. Enzyme-linked immunosorbent assays (ELISA) are a useful tool for assessing resistance conferred by Yd_2 to PAV (S. Haber and W. Legge, pers. comm. in 7) and have been used to assess the virus titres of four Mexican strains of BYDVs, Mex-PAV, Mex-MAV, Mex-RPV and Mex-RMV in a number of barley lines (63). The lines that had previously exhibited field tolerance to BYD also had reduced virus titres of some of the virus strains. By crossing some of these lines, it may be possible to produce lines that are tolerant to both subgroup 1 and subgroup 2 strains of BYDVs (63). Because Yd_2 decreases the virus titre against some BYD strains, it may be correct to describe it as bestowing true resistance to these viruses.

Recent work in the UK on Spring barley has shown that some cultivars, notably Chad, Cork and Optic, have less infection than other cultivars in field trials. With controlled inoculation in the greenhouse, infected plants of these cultivars suffered as much yield loss as others when infected, but were more difficult to infect. This suggested that the 'field resistance' was mediated through the vector rather than the virus. Subsequent work has been inconclusive but if the vector effect is confirmed, it would be the first known example of BYD control through resistance to the vector, and, if this character could be combined with the Yd_2 gene, both effects could be enhanced.

Oat

Work on evaluating oat germplasm for field tolerance to BYD has been well summarized (4,7,33,65). Many oat breeders observed oat lines that appeared to exhibit tolerance during the 1959 outbreak of BYD in the U.S. (Plant Disease Reporter Supplement 262, 1959). In the U.S., the program at the University of Illinois at Urbana-Champaign has been prominent and other programs in North America are those in Indiana, Iowa, North Dakota, Wisconsin, Manitoba and Québec.

A source of "tolerance" in Albion (C1 729) found by the Illinois group has been widely used. Burnett et al. (7) list 32 cultivars of oat that exhibit tolerance to BYD, including the lines Otee, Ogle, and Hazel from the Illinois program. This review also listed many other sources of BYD tolerance in oat. Good progress has been made in identifying sources of resistance to BYD in oat and incorporating these into agronomically suitable cultivars. It appears that one to four genes may be involved. Good sources of resistance have also been found in other *Avena* species of various ploidy levels (12).

Wheat

Much of the work reported on sources of tolerance in wheat to BYD has been summarized (7,57,60). Compared with oat and barley, little early screening on wheat has been done (4,65). A variety that has received much attention because of its tolerance to BYD in California is Anza (59). Crosses with Anza generated transgressive segregation for tolerance (56,60,75) and Qualset et al. (60) and Topcu (75) showed quantitative inheritance of tolerance to BYD. Recently, Singh et al. (72) reported the tolerance in Anza and nine other wheats to be due to a common partially recessive and partially dominant gene, *Bvd1*. It is thought that there is a genetic association of this gene with the genes *Lr34* and *Yr 18* for adult resistance to leaf rust and stripe rust, respectively (71).

The winter wheat variety Caldwell and the winter wheat germplasm Elmo have both been released by Purdue University, U.S. (49,51). Burnett et al. (7) summarized much of the other work on winter wheats and found that there are good tolerant sources available, so there seems to be strong between-site variation. For example, a line NS 87914 that appears extremely tolerant in India appears susceptible in Mexico, and Anza appears susceptible in China.

Chile has had an active program on screening for BYD tolerance (61) and good sources of tolerance in South American wheats have been found. Tolbay ("Tolerant to Barley Yellow Dwarf") shows little yield loss to BYD infection in Chile and in California (56,62). Thirty-four wheat lines that have exhibited good tolerance to BYD in Chile for three years and other promising material are under test (62).

Comeau and St. Pierre (16) include a list of wheats that have exhibited good tolerance in Québec. Some very tolerant wheats have been produced by selection in both Canada and Mexico (13).

Chromosome 3 of barley, containing Yd_2, has been added to wheat (46,47), and the resistance to BYD has been expressed under field conditions, but no lines are available commercially. Limited sources of BYD tolerance exist in durum wheat (9) but some promising lines are listed by several investigators (8,16,56,58,62). Cheour et al. (10) found that ELISA was useful in screening for resistance to BYD in durum wheat. Virus levels in wheat lines, measured by ELISA, have indicated tolerance rather than resistance. However, Banks et al. (2) have found a line from Ecuador that showed resistance, and Lorens et al. (41) found heritable variation in PAV-RNA concentration in wheat, suggesting that both tolerance and resistance may exist. Recently, Comeau et al (17) reported on an interspecific-derived durum that passes very good BYD tolerance.

Recently Haber et al. (28) have reported on indoor studies with BYD that predict the performance of lines in large-scale outdoor tests under BYD disease pressure. These procedures will be valuable to breeding programs that are aiming to produce BYD-tolerant material. In a preliminary application of this approach, S. Haber, Agriculture and Agri-Food Canada, Winnipeg, Canada (pers. comm.), found that in crosses made between parents that possess different parameters of good performance, progeny could be selected that were much more tolerant to BYD than either parent.

Other Poaceae and Grasses

Sources of BYD tolerance are available on winter and spring triticale, rice, corn, a range of other grass species, and interspecific derivatives, but there is little evidence of these being used consciously in breeding programs to control BYD (7). In rice, the tolerance to BYD (giallume in rice) appears to be conditioned by one incomplete dominant gene (1). More promising approaches are seen in the work of Comeau et al. (15), who derived wheat-like lines with immunity or resistance to BYD from an Agrotricum amphiploid, and of Xin et al. (81), who identified an amphiploid of wheat, Thinopyrum intermedium (Host) "Zhong 5," which ELISA assays showed to have good resistance to BYD. This resistance was

incorporated into bread wheat in Australia (2), and, more recently, at CIMMYT (B. Skovmand and L. Bertschinger, CIMMYT Mexico, pers. comm.). Comeau et al. (15) derived wheat-like lines with resistance to BYD from an Agrotricum amphiploid found to be resistant by others. The resistance of this line [Oklahoma (OK) 721 1542] was derived from hybrids of wheat with Thinopyrum ponticum (Qin Chen, AAFC Lethbridge Research Centre, pers. comm.) Many of these interspecific materials are promising resources for breeding for true resistance to BYD and are likely to lead to the rapid incorporation of new sources of resistance into wheat.

Chemical Control of Vectors

Cereal-feeding aphids are easy to kill, with no confirmed reports of resistance to the chemicals that have now been widely used for their control. However, the economics and timing of treatments are critical to the success of this method of control, and experience with the development of resistance in other virus vectors, especially Myzus persicae, shows that its use should be based on need and the likelihood of an economic return. There is little point in treating crops too early, before aphids arrive, because effectiveness of any treatment may have disappeared before any aphids appear. Equally, there is little point in treating crops late in growth after a virus has spread and when any subsequent infection will cause little damage. However, treatment at this time to prevent direct feeding damage may be appropriate. Therefore, timing is critical and this needs to be coupled with an expectation, or forecast, of how much virus would spread and how much damage would be caused if treatment were not applied. Consequently, a rational approach to the chemical control of BYDV vectors should be based on good knowledge of the epidemiology of BYD.

The use of insecticides has been greatest in Western Europe where, in many regions, potential yields of autumn-sown wheat and barley are high (6-10 t/ha) and the cost of chemical is small relative to the value of the crop. Thus, for winter wheat, the yield increase required to repay the cost of chemical and its application (application costs are often ignored as the insecticide is frequently applied with herbicide) can be as little as 0.1 t/ha. With the modest expectation of a 6 t/ha crop, this cost can be recouped with a less than 2% increase in yield.

For a 10 t/ha crop, the yield increase is only 1%. Such increases are very difficult to measure in field experiments. Consequently, many farmers prefer not to risk infection, especially as a severe attack of BYD could cause yield losses of 1-2 t/ha in wheat and twice as much in barley. While there is a strong economic argument for the prophylactic approach, its use may diminish because of decreasing margins on cereal crops; increasing pressure to decrease pesticide inputs on environmental grounds; risk of adverse effects on predators and parasites of aphids; and risk that aphids may become insensitive to pesticides.

The basis of a more refined approach is knowledge of vector phenology and virus epidemiology. In autumn, the principal migrant aphid is *Rhopalosiphum padi*, but in northern regions, *R. padi* has a sexual as well as an asexual cycle. This has a profound effect on BYD epidemiology, as males and sexual females (gynoparae) are produced on hosts susceptible to BYD and thus may have acquired virus, but their biology minimizes the likelihood of its transmission because they are unlikely, naturally, to feed on cereals, as they are migrating to their primary host. Consequently, assessing the risk of infection depends not only on a measure of vector populations but also a quantitative assessment of their biological characteristics and infectivity. Both of these assessments are now being incorporated into forecasting systems that guide insecticide applications (29,35,54).

Crops sown at the beginning or even slightly before the autumn migration starts are most at risk because (i) they are exposed to aphid vectors for the whole of the autumn migration, and (ii) a larger proportion of the early autumn migrants are parthenogenetic forms capable of feeding on cereals. The later the crops are sown, the shorter the period they are exposed to aphids, and the larger the proportion of aphids that are non-cereal-feeding. Consequently, the benefits of aphicides are greatest on the earliest-sown crops (Table 2). But not in all years is treatment beneficial on even the earliest-sown crops, and the sowing date after which treatment is unnecessary, the earliest safe-sowing date (42), can differ from year to year. Nevertheless, based on such evidence, the proportion of crops routinely treated can be decreased. Clearly, in the example given (Table 2), the yield of the earliest-sown crop was also decreased by factors other than BYD.

For spring-sown crops, the effect of sowing date on BYD incidence is the reverse of that for crops sown in autumn; latest-sown crops suffer the most damage (34). However, because the negative effect of late sowing on spring-sown crops is well known, most crops are sown as soon as conditions permit, and usually escape infection. Even when conditions dictate late sowing, there is no good evidence that insecticide treatment is an economic method of control (43).

The effectiveness of any aphicide treatment is determined by the sensitivity of the target and the timeliness of the treatment. So far, there is no evidence that aphid vectors of BYDVs have developed resistance to any insecticides. This may be because autumn-applied insecticides kill aphids before they die naturally over winter, so any selection for tolerance is lost. However, treatment is often also used to control direct feeding on ears and this adds a further selection pressure. The sexual generation or *R. padi* provides possibility of recombination to produce resistant biotypes. Consequently, the use of synthetic pyrethroid insecticides on cereals in spring and summer was prohibited in the UK until recently.

The correct timing of treatments, now fairly well defined, is at the end of the autumn migration (24,25). A spray applied at this time obviously does not prevent the primary infection resulting from the autumn migration, but it does stop the secondary spread of virus within the crop. Primary infection causes little yield loss; it is the secondary spread that causes damage.

Some pesticides are more effective than others (3). This is not related to aphid sensitivity but to the persistence and action of the chemicals in relatively cool conditions at the time of application. Thus, synthetic pyrethroid insecticides have generally given better results than organophosphorus or carbamate compounds when applied in autumn.

In most years, if treatment is not applied soon after the end of the autumn migration, then it will be too late to apply, because (i) the weather and conditions deteriorate quickly and treatment is either impractical or ineffective, and (ii) most of the virus spread has already occurred and there is little benefit to be gained. Only in some years, especially those when the autumn migration poses little threat because of limited number and infectivity of primary migrants, and the following winter is so mild that aphids are not killed, can a late winter or early spring (February/March) spray be beneficial. In most years, if crops are not treated in autumn, there is no point spraying them to control aphids transmitting BYDVs the following spring and autumn.

The comparatively good results obtained from spray treatments is due to the epidemiology of the disease and the relative importance of primary and secondary spread of BYD. Various attempts have been made to control BYD vectors using pre-drilling soil-applied pesticides (53), but methods have been too unwieldy and expensive, especially when sprays applied after sowing are very effective. Recently, a promising new seed-applied pesticide, imidacloprid NTN 33893 (70), has been developed and is widely used on cereals in France. Gourmet *et al.* (27) showed that this product may increase yeild of oat and wheat in Illinois, even for cultivars considered tolerant to BYD. Of course, such a treatment has to be applied in advance of infection and may, as a consequence, be unnecessary.

Biological Methods

Although much effort has been devoted to exploiting predators and parasites as control agents of cereal aphids, most of this effort has been directed at ameliorating the direct feeding damage they cause. Consequently, the emphasis has been on keeping populations below a damage threshold, usually measured in terms of numbers of aphids per shoot, ear, or plant. At the stages of plant growth when such populations occur, damage by BYD is likely to be small and reduction in aphid population, that may limit direct damage, are unlikely to stop virus spread. Introduced or natural biological control agents rarely decrease BYD incidence and spread, except where a vector or vectors have been introduced to an area without, or with only part, of their accompanying predators and parasites. In these cases the introduction of parasites and predators markedly reduced aphid populations and associated BYD infection (82); however, as reported above, method of tillage can influence virus incidence by its relative effect on predator populations (36).

FUTURE CONTROL

In the future, transformation and regeneration of cereals will become a standard procedure. Much progress has already been made with corn and rice, and transformed wheat (38,76), oats (74) and barley (37) have been regenerated. Much of the knowledge of viral genetic sequences, including those for coat protein that could be incorporated in transgenic plants, already exists (45,48). Consequently, by the time transformation

and regeneration of the small grain cereals becomes commonplace, we should be in a good position to exploit this knowledge. It should also be possible to benefit from the knowledge gained from transformed dicotyledonous plants. There are numerous options for such methods of control, including coat protein insertion, which has been achieved for barley with two BYDV-PAV isolates (79); antisense sequences; and ribozymes. For some BYDV isolates (RPV), satellite virus mediated control may be possible. However, we do not know how effective and stable such systems will be, and whether they can be incorporated and expressed in genetic backgrounds which meet all the other criteria that modern cereal cultivars have to satisfy. The prospects look promising, but the time scale is uncertain, and public reaction and legislative controls may slow the introduction of such novel sources of resistance. Experience with broad-leaved species is likely to result in a set of criteria under which transformed cereals will have to operate.

It seems unlikely that transgenic cultivars of small grain cereals will be widely grown within five years. It is probable that such plants will at first carry resistance to herbicides and modified quality traits rather than resistance to pests and disease. Resistance to BYD is likely to be an early target for breeders. Therefore, it is likely that virus resistance through recombinant DNA technology will be available within 10 years. In that time, conventional plant breeding may also have made progress, although no gene as effective as the Yd_2 gene in barley has been identified for wheat and oats. Before such methods become available, control will depend on the cultural and chemical methods, none of which may have value for much longer. Changes in pricing structure and government policy and legislation will change the relative benefits of these methods and the resources devoted to their production and implementation.

One way forward is to make current methods more accurately targeted, to decrease the quantity of pesticides applied to crops by developing more effective (and cheaper) materials, and to develop novel control strategies. Aphids can be deterred from feeding on cereals by using extracts of naturally-occurring compounds such as polygodial (22). In autumn-sown cereals (22), repeated applications of this material controlled BYD almost as effectively as a well-timed spray of synthetic pyrethroids, though at greater cost. One approach is to use behavior-modifying chemicals to deter aphids from feeding on the majority of a crop, and/or an aggregation pheromone to attract them to a localized area where they can then be sprayed with aphicide or exposed to biological methods of control under favorable conditions. This minimizes the area treated with no diminution in control.

While much of this is uncertain, what is certain is that BYD will not go away. A way needs to be found to protect crops against the damage it causes, now and for the future.

ACKNOWLEDGEMENTS

The authors wish to thank R.A. Butts, H.H. Janzen, A. Comeau and S. Haber for reviewing this work and making helpful suggestions.

REFERENCES

1. Baldi, G., Moletti, M., and Osler, R. 1990. Inheritance of resistance to giallume in rice. Pages 429-433 in: World Perspectives on Barley Yellow Dwarf. P.A. Burnett, ed. CIMMYT, Mexico, D.F., Mexico.

2. Banks, P., Xu, S.J., Wang, R.C., and Larkin, P.J. 1993. Varying chromosome composition of 56-chromosome wheat × Thinopyrum intermediate partial amphiploids. Genome 36:207-215.

3. Barrett, D.W.A., Northwood, P.J., and Horellou, A. 1981. The influence of rate and timing of autumn applied pyrethroid and carbamate insect cide sprays on the control of barley yellow dwarf virus in English and French winter cereals. Pages 405-412 in: Proceedings of the British Crop Protection Conference, Pests and Diseases, Vol. 2

4. Bruehl, G.W. 1961. Barley Yellow Dwarf. Monograph No. 1. The American Phytopathological Society. 52 Pages.

5. Burnett, P.A., ed. 1984. Barley Yellow Dwarf, a Proceedings of the Workshop. CIMMYT, Mexico, D.F., Mexico, 209 Pages.

6. Burnett, P.A. 1990. Breeding for resistance to barley yellow dwarf. Pages 179-193 in: Proceedings Aphid-Plant Interactions: Populations to Molecules. D.C. Peters, J.A. Webster, and C.S. Chlouber, eds. Stillwater, OK, USA.

7. Burnett, P.A., Comeau, A., and Qualset, C.D. 1995. Host plant tolerance or resistance for control of barley yellow dwarf. Pages 321-343 in: Barley Yellow Dwarf Virus: 40 Years of Progress. C.J. D'arcy and P.A. Burnett, eds. APS Press, St. Paul, MN.

8. Burnett, P.A., and Mezzalama, M. 1992. The Centro Internacional de Mejoramiento de Maiz y Trigo (CIMMYT) barley yellow dwarf program Pages 33-45 in: Barley Yellow Dwarf in West Asia and North Africa. A. Comeau and K.M. Makkouk, eds. ICARDA, Aleppo, Syria.

9. Chéour, F., Comeau, A., and Asselin, A. 1989. Genetic variation for tolerance of resistance to barley yellow dwarf virus in durum wheat. Euphytica 40:213-220.

10. Chéour, F., Comeau, A., and Asselin, A. 1993. Barley yellow dwarf virus multiplication and host plant tolerance in durum wheat. J. Phytopathol. 139:357-366.

11. Chicaiza, O. 1989. Effectiveness of the visual symptom score in selecting germplasm for resistance to the barley yellow dwarf virus disease in wheat (Triticum aestivum L.). M.S. thesis, University of California, Davis, USA. 47 Pages.

12. Comeau, A. 1984. Barley yellow dwarf virus resistance in the genus Avena. Euphytica 33:49-55.

13. Comeau, A., Burnett, P.A., McKenzie, R., Haber, S., and St-Pierre, C.A. 1993. Promising sources of barley yellow dwarf virus tolerance for bread wheat. Page 179 in: Sixth International Congress of Plant Pathology, Abstract 9.5.12.

14. Comeau, A., Collin, J., and Chéour, F. 1992. Barley yellow dwarf virus symptoms and ELISA data in relation to biomass and yield loss. Pages 155-168 in: Barley Yellow Dwarf in West Asia and North Africa. A. Comeau and K.M. Makkouk, eds. ICARDA, Aleppo, Syria.

15. Comeau, A., Makkouk, K.M., Ahmad, F., and St-Pierre, C.A. 1994. Bread wheat × Agrotricum crosses as a source of immunity and resistance to the PAV strain of barley yellow dwarf luteovirus. Agronomie 2:153-160.

16. Comeau, A., and St-Pierre, C.A. 1992. The Agriculture Canada/Laval University barley yellow dwarf program. Pages 55-60 in: Barley Yellow Dwarf in West Asia and North Africa. A. Comeau and K.M. Makkouk eds. ICARDA, Aleppo, Syria.

17. Comeau, A., St-Pierre, C.A., Maës and Collin, J. 1997. Interspecific hybridyzation of wheat: recent progress and prospects for the future. Page 45 in: Biotechnology, Plant Breeding and Food Security. AUPELF-UREF, Paris (in French).

18. Collins, N.C., Partridge, M.G., Ford, C.M., and Symons, R.H. 1996. The Yd_2 gene for barley yellow dwarf resistance maps close to the centromere on the long arm of barley chromosome 3. Theoretical and Applied Genetics 92:858-864.

19. Cooper, J.I., and Jones, A.T. 1983. Responses of plants to viruses: proposals for use of terms. Phytopathology 73:127-128.

20. D'arcy, C.J. 1995. Symptomatology and host range of barley yellow dwarf virus. Pages 9-28 in: Barley Yellow Dwarf Virus: 40 Years of Progress. C.J. D'arcy and P.A. Burnett, eds. APS Press, St. Paul, MN.

21. D'arcy, C.J. 1997. Letter and documents resulting from the discussions at the Luteovirus meeting at Cirencester earlier this year. personal communication.

22. Dawson, G.W., Griffiths, D.C., Pickett, J.A., Plumb, R.T., Woodcock, C.M., and Chang, Z.-N. 1988. Structure activity studies on aphid alarm pheromone derivatives and their field use against transmission of barley yellow dwarf virus. Pesticide Science 22:17-30.

23. Fox, P.M., Tola, J., and Chicaiza, O. 1990. Barley yellow dwarf in the Andean countries of South America. Pages 25-28 in: World Perspectives on Barley Yellow Dwarf. P.A. Burnett, ed. CIMMYT, Mexico, D.F., Mexico.

24. George, K.S. 1974. Damage assessment aspects of cereal aphid attack in autumn- and spring-sown cereals. Ann. Appl. Biology 77:67-74.

25. George, K.S. 1975. The establishment of economic

thresholds with particular reference to cereal aphids. Pages 79-85: Proceedings of the British Insecticide and Fungicide Conference 8th Vol. 1.

26. Gill, C.C., and Chong, J. 1979. Cytopathological evidence for the division of barley yellow dwarf virus isolates into two subgroups. Virology 95:59-69.

27. Gourmet, C., Kolb, F.L., Smyth, C.A., and Pedersen, W.L. 1996. Use of imidacloprid as a seed-treatment insecticide to control barley yellow dwarf virus (BYDV) in oat and wheat. Plant Dis. 80:136-141.

28. Haber, S., MacKenzie, R.I.H., and Townley-Smith, T.F. 1997. A robust model that predicts how barley yellow dwarf (BYD) affects yield in spring wheat. Can. J. Plant Pathol. 18:(in press)

29. Harrington, R., Mann, J.A., Plumb, R.T., Smith, A.J., Taylor, M.S., Foster, G.N., Holmes, S.J., Masterman, A.J., Tones, S.J., Knight, J.D., Oakley, J.N., Barker, I., and Waters, K.F.A. 1994. Monitoring and forecasting for BYDV control - the way forward? Pages 197-206 in: Sampling to Make Decisions, Aspects of Applied Biology, Vol. 37.

30. Hayes, P., Prehn, D., Vivar, H., Blake, T., Comeau, A., Henry, I., Johnston, M., Jones, B., Steffenson, B., and St.-Pierre, C. 1996. Multiple disease resistance loci and their relationship to agronomic and quality loci in a spring barley population. J. Quantitative Trait Loci (a reviewed internet journal) http://probe.nalusda.gov:8000/otherdocs/jqtl/jqtl19 96-02/jqtl22.html:abstract:A scald resistance gene is closely linked to BYDV resistance Yd_2.

31. Herrera, G.F. 1989. Interactions between host plants and British isolates of barley yellow dwarf virus. Ph.D. thesis, University of London. 232 Pages.

32. Holloway, P.J., and Heath, R. 1992. Identification of polypeptide markers of barley yellow dwarf virus resistance and susceptibility genes in non-infected barley (Hordeum vulgare) plants. Theor. Appl. Genet. 85:346-352.

33. Jedlinski, H. 1984. The genetics of reistance to barley yellow dwarf virus in oats. Pages 101-105 in: Barley Yellow Dwarf, a Proceedings of the Workshop. P.A. Burnett, ed. CIMMYT, Mexico, D.F., Mexico.

34. Jenkyn, J.F., and Plumb, R.T. 1983. Effects of fungicides and insecticides applied to spring barley sown on different dates in 1976-9. Ann. Appl. Biology 102:421-433.

35. Kendall, D.A., and Chinn, N.E. 1990. A comparison of vector populations indices for forecasting barley yellow dwarf virus in autumn-sown cereal crops. Ann. Appl. Biology 116:87-102.

36. Kendall, D.A. Chinn, N.E., Smith, EB.D., Tidboald, C.M., Winstone, L., and Western, N.M. 1991. Effects of straw disposal and tillage on spread of barley yellow dwarf virus in winter barley. Ann. Appl. Biology 119:359-364.

37. Lazzeri, P.A. 1995. Stable transformation of barley via direct DNA uptake: electroporation - and PEG-medicited protoplast transformation. Pages 95-106 In: Plant Gene Transfer and Expression Protocols (Methods in Molecular Biology, Vol. 49). H. Jones, ed. Humana Press, Totowa.

38. Lazzeri, P.A., Barro, F., Connell, M.E., Rasco-Gaunt, S.M., Tatham, A.S., Shewry, P.R., and Barcelo, P. 1996. Development of a robust wheat transformation system. In: Proceedings IBC Conference on Biotechnology in Agriculture, London.

39. Lister, R.M., and Ranieri, R. 1995. Distribution and economic importance of barley yellow dwarf. Pages 29-53 in: Barley yellow dwarf virus: 40 years of progress. C.J. D'arcy and P.A. Burnett, eds. APS Press, St. Paul, MN.

40. Lister, R.M., and Sward, R.J. 1988. Anomalies in serological and vector relationships of MAV-like isolates of barley yellow dwarf virus from Australia and the USA. Phytopathology 78:766-770.

41. Lorens, G.F., Falk, B.W., and Qualset, C.O. 1989. Inheritance of resistance to barley yellow dwarf virus detected by northern blot analysis. Crop Sci. 29:1076-1081.

42. Lowe, A.D. 1967. Avoid yellow dwarf by late sowing. N.Z. Wheat Review 10:59-63.

43. Mann, J.A., Carter, N., and Plumb, R.T. 1992. The epidemiology of barley yellow dwarf virus in spring-sown cereals and the potential for its control. Page 205 in: 5th International Plant Virus Epidemiology Symposium, Bari, Italy.

44. Martin, R.R. and D'Arcy, C.J. 1995. Taxonomy of barley yellow dwarf viruses. Pages 203-214 in: Barley Yellow Dwarf Virus: 40 Years of Progress. C.J. D'arcy and P.A. Burnett, eds. APS Press, St. Paul, MN.

45. Mayo, M.A., and Ziegler-Graaf, V. 1996. Molecular Biology of Luteoviruses. Advances in Virus Research 46:413-460.

46. McGuire, P.E. 1984. Status of an attempt to transfer the barley yellow swarf virus resistance gene Yd_2 of barley to hexaploid wheat. Pages 113-119 in: Barley Yellow Dwarf, a Proceedings of the Workshop. P.A. Burnett ed. CIMMYT, Mexico, D.F., Mexico.

47   McGuire, P.E., and Qualset, C.O. 1990. Transfer of the Yd_2 barley yellow dwarf virus gene from barley to wheat. Pages 476-481 in: World Perspectives on Barley Yellow Dwarf Virus. P.A. Burnett ed. CIMMYT, Mexico, D.F., Mexico.

48. Miller, W.A., and Young, M.J. 1995. Prospects for genetically engineered resistance to barley yellow dwarf viruses. Pages 345-369 in: Barley Yellow Dwarf - 40 Years of Progress. C.J. D'Arcy and P.A. Burnett, eds. APS Press, St. Paul, MN.

49. Ohm, H.W., Patterson, F.L., Carrigan, L.L., Shaner, G.E., Foster, J.E., Finney, R.E., and Roberts, J.J. 1981. Registration of Elmo common wheat germplasm. Crop Sci. 21:803.

50. Oswald, J.W., and Houston, B.R. 1951. A new virus disease of cereals, transmissible by aphids. Plant Dis. Rep. 11:471-475.

51. Patterson, F.L., Ohm, H.W., Snaer, G.E., Finney, R.E., Gallun, R.L., Roberts, J.J., and Foster, J.E. 1982. Registration of Caldwell wheat. Crop Sci. 22:691-692.

52. Pike, K.S. 1990. A review of barley yellow dwarf virus grain yield losses. Pages 356-361 in: World Perspectives on Barley Yellow Dwarf. P.A. Burnett, ed. CIMMYT, Mexico, D.F., Mexico.

53. Plumb, R.T. 1981. Chemicals in the control of cereal virus diseases. Pages 136-145 in: Strategies for the Control of Cereal Disease. J.F. Jenkyn and R.T. Plumb, eds. Blackwell Scientific Publications, Oxford.

54. Plumb, R.T. 1986. A rational approach to the control of barley yellow dwarf virus. J. Royal Agr. Soc. England 147:162-171.

55.   Plumb, R.T., and Johnstone, G.R. 1995. Cultural, chemical and biological methods for the control of barley yellow dwarf. Pages 307-319 in: Barley Yellow Dwarf - 40 Years of Progress. C.J. D'Arcy and P.A. Burnet,t eds. APS Press, St. Paul, MN.

56.   Qualset, C.O. 1992. Developing host plant resistance to barley yellow dwarf virus: an effective control strategy. Pages 115-129 in: Barley Yellow Dwarf in West Asia and North Africa. A. Comeau and K.M. Makkouk, eds. ICARDA, Aleppo, Syria.

57.   Qualset, C.O., Lorens, G.F., Ullman, D.E., and McGuire, P.E. 1990. Genetics of host plant resistance to barley yellow dwarf virus. Pages 368-382 in: World Perspectives on Barley Yellow Dwarf. P.A. Burnett, ed. CIMMYT, Mexico, D.F., Mexico.

58.   Qualset, C.O., McGuire, P.E., Vogt, H.E., and Topcu, M.A. 1977. Ethiopia as a source of resistance to the barley yellow dwarf virus in tetraploid wheat. Crop Sci. 17: 527-529.

59.   Qualset, C.O., Vogt, H.E., and Borlaug, N.E. 1984. Registration of Anza wheat. Crop Sci. 24:827-828.

60.   Qualset, C.O., Williams, J.C., Topcu, M.A., and Vogt, H.E. 1973. The barley yellow dwarf virus in wheat: importance, sources of resistance and heritability. Pages 465-470 in: Proceedings of the Fourth International Wheat Genetics Symposium, Missouri Agricultural Experimental Station, Columbia, MO, USA.

61.   Ramirez, I. 1990. A review of barley yellow dwarf virus in the southern cone countries of South America. Pages 29-33 in: World Perspectives on Barley Yellow Dwarf. P.A. Burnett, ed. CIMMYT, Mexico, D.F., Mexico.

62.   Ramirez, I., Zerene, M., and Cortazar, R. 1992. The barley yellow dwarf virus program in Chile. Pages 47-54 in: Barley Yellow Dwarf in West Asia and North Africa. A. Comeau and K.M. Makkouk, eds. ICARDA, Aleppo, Syria.

63.   Ranieri, R., Lister, R.M., and Burnett, P.A. 1993. Relationships between barley yellow dwarf virus titer and symptom expression in barley. Crop Sci. 33:968-973.

64.   Rasmusson, D.C., and Schaller, C.W. 1959. The inheritance of resistance in barley to the yellow-dwarf virus. Agron. J. 51: 661-664.

65.   Rochow, W.F. 1961. The barley yellow dwarf virus disease of small grains. Adv. Agron. 13:217-248.

66.   Rochow, W.F., and Duffus, J.E. 1981. Luteoviruses and yellows diseases. Pages 147-170 in: Handbook of Plant Virus Infections and Comparative Diagnosis. E. Kurstak, ed. Elsevier/North Holland, Amsterdam.

67.   Schaller, C.W. 1984. The genetics of resistance to barley yellow dwarf virus in barley. Pages 93-99 in: Barley Yellow Dwarf, a Proceedings of the Workshop. P.A. Burnett, ed. CIMMYT, Mexico, D.F., Mexico.

68.   Schaller, C.W., Qualset, C.O., and Rutger, J.N. 1964. Inheritance and linkage of the Yd_2 gene conditioning resistance to barley yellow dwarf virus disease in barley. Crop Sci. 4: 544-548.

69.   Schaller, C.W., Rasmusson, D.C., and Qualset, C.O. 1963. Sources of resistance to the yellow dwarf virus in barley. Crop Sci. 3: 342-344.

70.   Schmeer, H.E., Bluett, D.J., Meredith, R., and Heatherington, P.J. 1990. Field evaluation of imidacloprid as an insecticidal seed treatment in sugar beet and cereals with particular reference to virus vector control. Pages 29-36 in: Proceedings of the Brighton Crop Protection Conference, Pests and Diseases Vol. 1.

71.   Singh, R.P. 1993. Genetic association of gene $Bdv1$ for tolerance to barley yellow dwarf virus with genes $Lr34$ and $Yr18$ for adult plant resistance to rusts in bread wheat. Plant Diseases 77:1103-1106.

72.   Singh, R.P., Burnett, P.A., Albarran, M., and Rajaram, S. 1993. $Bvd1$: A gene for tolerance to barley yellow dwarf virus in bread wheats. Crop Sci. 33: 231-234.

73.   Skaria, M., Lister, R.M., Foster, J.E., and Shaner, G. 1985. Virus content as an index of symptomatic resistance to barley yellow dwarf virus in cereals. Phytopathology 75: 212-216.

74.   Somers, D.A, Rines, H.W., Gu, W., Kaeppler, H.E., and Bushnell, W.R. 1992. Fertile transgenic oat plants. Bio/Technology 10:1589-1594.

75.   Topcu, M.A. 1975. The inheritance of resistance to barley yellow dwarf virus in wheat. Ph. D. thesis, University of California, Davis, CA, USA.

76.   Vasil, V., Castillo, A.M., Fromm, M.E., and Vasil, I.K. 1992. Herbicide resistant fertile transgenic wheat plants obtained by microprojectile bombardment of regenerable embryonic callus. Bio/Technology 10:667-674.

77.   Vincent, J.R., Lister, R.M., and Larkins, B.A. 1991. Nucleotide sequence analysis and genomic organization of the NY-RPV isolate of barley yellow dwarf virus. J. Gen. Virol. 72: 2347-2355.

78.   Vivar, H.E. 1996. Barley: wide adaptation. Pages 260-266 in: Proceedings of 5th International Oat Conference and 7th International Barley Genetics Symposium--Invited Papers. G. Scoles and B. Rossnagel, eds.

79.   Wan, Y., and Lemaux, P.G. 1994. Generation of large numbers of independantly transformed fertile barley plants. Plant Physiology 104:37-48.

80.   Webby, G.N., Lister, R.M., and Burnett, P.A. 1993. The ocurrence of barley yellow dwarf viruses in CIMMYT bread wheat nurseries and associated cereal crops during 1988-1990. Ann. Appl. Biol. 123: 63-74.

81.   Xin, Z.Y., Brettell, R.I.S., Cheng, Z.M., Waterhouse, P.M., Appels, P., Banks, P.M., Zhou, G.H., Chen, X., and Larkin, P.J. 1988. Characterization of a potential source of barley yellow dwarf virus resistance for wheat. Genome 30:250-257.

82.   Zuniga, E. 1990. Biological control of cereal aphids in the southern cone of South America. Pages 362-367 in: World Perspectives on Barley Yellow Dwarf. P.A. Burnett, ed. CIMMYT, Mexico.

CHAPTER 34

Present Status of Controlling Rice Tungro Virus

H. Koganezawa

Modern rice varieties with high yield potential were adopted in the late 1960s and the early 1970s to avert the impending food shortages in Asian developing countries (15). Modern rice varieties have relatively short stems, are early maturing and have high fertilizer responsiveness. Because these early modern varieties were highly susceptible to tungro and as its vectors, the epidemic outbreaks of tungro in the late 1960s and the early 1970s caused devastating production losses in South and Southeast Asia (82). Before the introduction of modern varieties, diseases showing tungro-like symptoms have been known under different names such as "mentek" (also called "penyakit habang") in Indonesia, "penyakit merah" in Malaysia, "dwarf or stunt" in the Philippines and "yellow orange leaf" in Thailand (62). Tungro now occurs in most South and Southeast Asian countries (62) and also in China (95)·

To stabilize rice yield, the breeding programs at the International Rice Research Institute (IRRI) as well as in national research institutions have been designed to develop varieties resistant to tungro (39). Various integrated pest management schemes were also implemented. Although several control methods against tungro were proposed and applied, its epidemics have occurred periodically in the past two decades. The disease appears in areas where tungro has not been recorded or has remained endemic at low levels and immediately becomes prevalent throughout whole regions. Among rice pathogens, tungro causes the highest yield loss. Thus, scientists give high priority to tungro in rice research activities (23).

CAUSAL VIRUSES

Tungro is a composite disease caused by two viruses, rice tungro bacilliform virus (RTBV) and rice tungro spherical virus (RTSV). RTBV particles are bacilliform, 100-300 nm in length, and 30-35 nm in width. RTBV contains a circular double-stranded DNA (27,37). The genome analysis revealed that RTBV has some similarities to caulimoviruses and is more closely related to *Commelina* yellow mottle virus (64). RTBV is a proposed member of a newly recognized group of badnaviruses with nonenveloped bacilliform plant viruses that contain circular double-stranded DNA. RTSV particles are isometric and 30 nm in diameter. RTSV contains single-stranded RNA (27,37) and belongs to the maize chlorotic dwarf virus group.

Both viruses multiply independently in rice plants. In rice plants, RTSV particles are restricted to the phloem tissues (17), while RTBV particles are present in both phloem and xylem cells (83). The most conspicuous symptoms in plants infected with both RTBV and RTSV are stunting and yellow or yellow-orange discoloration. Plants infected with RTBV alone develop similar but milder symptoms than those caused by double infection, whereas RTSV alone causes no clear symptoms, except very mild stunting (24, 28). Grain yield reduction is as high as 100% in doubly infected plants, and 40% in highly susceptible varieties infected with RTSV alone (20). RTSV alone is widespread in the Philippines (6) and probably in other South and Southeast Asian countries and southern China. RTSV was once epidemic in southern Japan causing serious damage. Its causal agent was called rice waika virus (18).

EPIDEMIOLOGY OF TUNGRO

Transmission

The tungro viruses are transmitted by *Nephotettix virescens, N. cincticeps, N.*

nigropictus, N. malayanus, N. parvus, and *Recilia dorsalis* (48). The transmission efficiency differs depending on the vector species and the colonies (34,48). *N. virescens* is the most efficient vector. The role of other species may be small, at least in tungro epidemic areas, either because they are indigenous to temperate regions, rice is not their host, or their transmission efficiency is low. Actually, in several regions of Indonesia and Muda area of Malaysia, the period when tungro became a big problem is consistent with the period when *N. virescens* replaced *N. nigropictus* and became predominant in rice fields (32,81).

Green leafhoppers transmit tungro viruses in semi-persistent manner. It is reported that the minimum acquisition and inoculation feeding periods for *N. virescens* are 5 min and 7 min, respectively (48). *N. virescens* retains tungro infectivity for 2 to 6 days but loses infectivity after molting (48). Leafhoppers readily acquire RTSV from plants infected with RTSV alone, but do not acquire RTBV from plants infected with RTBV alone. Leafhoppers acquire RTBV only when exposed to RTSV-infected plants before feeding on RTBV-infected plants (24,29). RTSV-infected plants begin to serve as source for helper factor of transmission 36 hr after inoculation and as virus source for tungro 72 hr after inoculation (11). Tungro becomes widespread within a short period because of its semi-persistent transmission manner and its short latent period in plants.

Leafhoppers are considered sedentary and less migratory than the delphacids (42). Migratory flight of *N. virescens* over moderate distance of 5-30 km had been observed (66). Viruliferous leafhoppers were caught in the middle of a lake 3 km apart from the nearest rice field (3). It is unclear how the moderate-distance migration of vectors plays a role in tungro spread over wide areas.

Alternative Hosts

The host range of tungro viruses among weeds has been studied in several countries with conflicting results. Tungro viruses caused symptoms on several weeds in some studies, but did not on the same hosts in other studies (38). Actually tungro viruses can infect several Graminaceous weeds other than rice, but its transmission rate from weeds to rice is extremely low. Likewise, *N. virescens*, the main vector, is a typical monophagous species restricted to rice. Therefore it is considered to be very rare that

infected weeds play an important role as disease sources. Wild rices are considered to play a role as disease source in Thailand (31). Natural infection of tungro in wild rices was observed in the Philippines (89). A major disease source is rice plants; that is standing and volunteer crops (89), ratoons (1), and stubbles (59, 89). Tungro occurrence is associated with areas where irrigation facilities are well-developed and where water remains even in the dry season (31). Volunteer crops and ratoons can survive there in off-seasons.

Disease Development

The role of nursery in the spread of tungro has been controversial. The necessity to control the disease in nurseries was emphasized in some reports (50,56,59). On the other hand, negligible infection in nurseries was demonstrated in other reports (35,87,90). Tungro infection is usually low in seedbed where the seedbed period is 3-4 weeks. Rice seedlings, however, may be infected in the nursery, where the seedbed period is 5-6 weeks for long-duration varieties like in India and also when the seedbed is raised near rice crops at the middle growth stage.

Infection mainly occurs after transplanting through immigrant green leafhoppers. At early growth stages, rice plants are vulnerable to tungro infection (49). The spread of tungro in the field is faster in younger plants than in older ones (80). The infection of plants with RTSV alone precedes infection of RTBV in fields. Although discernible symptoms do not appear, RTSV-alone infection reaches the maximum level about one month after transplanting. RTBV infection increases subsequently and tungro symptoms appear (87). The early infection of plants with RTSV indicates the movement of vectors largely infective with RTSV alone soon after transplanting. This can be explained by the independent transmission of RTSV (24) and the widespread occurrence of RTSV in fields where little or no RTBV occurs (6).

Most tungro endemic areas in tropical Asia have a tropical monsoon climate. The time of onset and end of the rains greatly influence rice cropping patterns, hence the vector populations and tungro incidence (47,55,59,84). Vectors appear to multiply more rapidly on wet season crop than on the dry season crop, and the outbreaks of tungro usually occur mainly in the rainy season. The patterns of population fluctuation of *N. virescens* vary from region to region, but are closely related to rice cropping patterns (12, 93). Generally the number of insects increases rapidly on younger rice plants in the vegetative phase but decreases after the reproductive phase. The peak population usually appears during the middle stage of rice in the wet season (82).

Cause of Epidemics

Tungro epidemic has always been sporadic. The upsurge of tungro is associated with several factors.

In the Philippines, severe outbreaks occurred around the 1950s (74). Later tungro outbreaks occurred in 1971 (45, 50) and 1987 (4). The high-yielding variety IR8 was released from IRRI in 1966 and soon adopted by Philippine farmers. In 1971, IR8 became susceptible to *N. virescens* and vector population became very high (50). In 1987, prolonged drought in Mindanao resulted in late transplanting (4).

In Indonesia, "mentek" disease has been a major problem in rice production in Java since the 19th century, and serious crop losses due to the disease were recorded in 1921 and 1936. The disease declined after the release of resistant varieties in 1940 to 1942 (82). IR8 and IR5 were introduced to Indonesia in 1967, and subsequently Pelita (produced in the national breeding program) in 1971. These varieties were highly susceptible to tungro. During 1972 - 1975, tungro caused serious damage on rice in South Sulawesi (54). Although continuous data on leafhopper population are not available, a light trap caught more than 100,000 leafhoppers within 4 weeks in 1975 (90). An extremely long and severe dry season was recorded in 1972 (90).

In Malaysia, a serious outbreak of tungro (known as penyakit merah) occurred in 1969 in North Krian district where rice fields were planted throughout the year. The outbreak was attributed to early buildup and high numbers of leafhoppers and late planting due to the introduction of double cropping (46). During 1981 - 1983, tungro spread to Penang, Kedah, and Perlis following extension of double cropping in the farmer's fields (32,33). Factors contributing to increase in tungro incidence in the Muda area after 1981 were high populations of green leafhopper and the change of species composition from *N. nigropictus* to *N. virescens*. Another factor was the staggered planting in 1981 - 1983 caused by unusual dry

weather, which resulted in water shortage and delayed planting (7,32).

In Thailand, a severe outbreak of tungro (known as yellow orange leaf) occurred in 1966 and about 660,000 ha of rice fields were affected moderately to severely (44). The population of green leafhopper in 1966 was 6 times higher than that of the previous year (41). Severe tungro incidence continued in 1968 - 1971, declined in 1972 and 1973, and again occurred severely in 1974. The peak incidence during this period occurred in 1969. In 1967, it was dry until September, and infection was low in early planted rice, but severe in late plantings (92). In a light trap study from 1969 to 1974 at the Bangkhen Experimental Station, the number of green leafhoppers was highest in 1969 and 1970, lowest in 1971 and 1972, and ranked moderately in 1973 and 1974. In 1972, the drought weather prevailed continuously during May to August (35). Tungro-resistant new varieties RD1 and RD3 were released in 1969, which might have contributed to the decline of tungro incidence in 1970 to 1973, but in 1974, RD1 became susceptible (92). There was an outbreak again in 1979. Morinaka *et al.* (58) attributed the occurrence to high population of green leafhopper in the field.

In India, the high-yielding variety Taichung Native 1 was introduced and widely planted during 1968 - 1969, and then IR8 and Jaya were released (15). Subsequently a severe outbreak of tungro occurred in Uttar Pradesh and Bihar (36). In Tamil Nadu, severe outbreaks occurred in 1978 - 1981 and 1984. In all cases leafhopper populations were very high compared with those during non-epidemic years (91). In 1990, tungro outbreaks occurred in Andhra Pradesh and Orissa (5). In this case, heavy rain and floods led to late planting in many places and the leafhopper population was quite high (A. Anjaneyulu, personal communication).

In summary, widespread occurrence of tungro is associated mainly with three factors; 1) introduction of new susceptible varieties; 2) late planting due to adverse weather, hence the vulnerable growth stage of rice coinciding with peaks of vector population; 3) high vector population, although it is not clearly understood how the population increases several folds only during certain seasons.

CULTURAL CONTROL

Elimination of Disease Source

As rice plants are the main source of tungro viruses, roguing or removal of affected plants are recommended in farmers' brochures to reduce the chances of disease spread by vectors. Tungro incidence in rogued fields is lower than in nonrogued fields, but roguing does not increase yield (16). Roguing may not be a practical procedure. Ratoons and stubbles in the affected rice fields serve as potential sources of virus inoculum and of vectors. Destruction of all ratoons and stubbles in endemic areas by ploughing, burning or herbicide application is feasible and recommendable.

Fallow Period and Synchronized Cropping

The tungro epidemic area is limited to where rice cropping is not synchronized and where rice grows continuously throughout the year. Tungro incidence and vector populations are high in asynchronous cropping areas (7,51). The disease source is limited to infected rice plants and *N. virescens* is monophagous. Viruliferous green leafhoppers lose infectivity within a week. If a fallow period can be implemented over a wide area, it is possible to reduce potential disease sources and vector populations. However, it is adversely possible that natural enemy populations are more drastically reduced than vectors during a fallow period. Therefore, vector populations may increase rapidly in the crop season after a fallow period. Synchrony of cropping is a prerequisite for a fallow period and synchrony itself is as effective as a fallow period (69).

There are several cases in which tungro was successfully controlled by introducing fallow period and synchronized cropping. During 1979 wet season there was a severe occurrence of tungro in Tuaran district, Sabah, East Malaysia. The district authorities discontinued the supply of irrigated water, forcing the farmers to refrain from planting rice in the next dry season of 1980. The incidence of tungro declined in the subsequent main season of rice cropping (96). In the Northern Plain of West Java, Indonesia, where irrigation system was completed years before and where synchronized cropping is accomplished by the control of water supply by the local government, tungro has not been a problem for the past 20 years. In other areas such as West and the Central Ronboku district of West Tengara and part of Central Java, where synchronized cropping was implemented, tungro incidence dropped sharply without exception (84).

Synchronized cropping is very difficult to implement due to difficulty in water control and labor shortage. It also requires the leadership of the authorities and coordination among farmers. Areas that are breeding grounds for tungro epidemics are those with staggered transplanting of rice. Such areas are distributed in Mindanao in the Philippines, Bali in Indonesia, Bangladesh, and West Bengal in India. Unfortunately, most of these areas have special considerations which do not allow the implementation of synchronized cropping. For example, farmers in Bali disregard tungro occurrence because rice is not a major part of their income.

Adjustment of Planting Time

Appropriate planting time when vector population is low was recommended. In India, disease incidence was lowest during January and February (rabi), and June and July (kharif), which correspond with the beginning of normal transplanting. When seedlings are transplanted late or early, tungro incidence becomes high (59,78). However, a shift in the pattern of vector population growth was observed in Indonesia, following a change in planting dates (69).

Broadcasting

Tungro incidence is lower in direct seeded rice than in transplanted rice (88, 90). The low tungro incidence in broadcast rice also shows the promising role of this cultural practice not only in alleviating the scarce manpower in farm operations but also in managing tungro. Higher tungro infection was observed in direct seeded rice culture in flooded conditions than in nonflooded conditions (43). Plant spacing is known to affect the spread of the disease: the rate of disease spread was slower under closer spacings than under wider spacings (79). The lower infection of tungro in direct seeded rice may be due to the narrow spacing of rice plants and to less water which attracts vectors.

USE OF RESISTANT VARIETIES

The use of resistant varieties is one of the

most important and economical components of a tungro management strategy. According to Sogawa (82), breeding of rice varieties resistant to "mentek" disease was initiated in Indonesia as early as 1934. During 1940 to 1942, resistant cultivars such as Bengawan, Peta, Intan, and Tjeremas were released. The need for varieties resistant to tungro and its main vector *N. virescens* was recognized at an early stage of the breeding program at IRRI, and routine screening tests have been used since the late 1960s (40). The International Rice Testing Program (now called International Network for Genetic Evaluation of Rice) was initiated in 1975 to test rice germplasm under a wide range of agroclimatic and cultural conditions, including evaluation for tungro resistance. The national research systems introduced tungro-resistant varieties bred in IRRI, and conducted the breeding of resistant varieties through crosses with traditional varieties. The All-India Coordinated Rice Improvement Project (AICRIP) initiated an intensive screening program to identify suitable donors for tungro resistance (57,76). Pankhari 203, Latisail, Kataribhog, and Ambemohar 159 were used as donor parents. However, the type of resistance in varieties bred in most national institutions is not clear due to lack of facilities to detect tungro viruses.

Vector Resistance

The breeding strategy for tungro-resistant variety at IRRI has been based almost exclusively on incorporating vector resistance. Most IRRI crosses made after 1969 had at least one parent with resistance to *N. virescens*. All IR varieties, except IR22 and all IR lines named by the Philippine Government, are rated as having green leafhopper resistance at the time of release. Of the seven known genes for vector resistance, four have been incorporated into improved varieties (39). The main donors are Ptb18, Gam Pai 30-12-15, and Ptb33. Such rice cultivars resistant to vector adequately escape tungro infection in the field under light to moderate tungro and vector pressure (22,30).

For decades the introduction of vector-resistant varieties has been the major strategy of controlling tungro. The vector-resistant varieties helped to minimize yield losses that are likely to have occurred because of tungro disease. However, the breakdown of vector resistance occurred after a few consecutive seasons of intensive cultivation of formerly resistant cultivars

(14,35,54). The cause of the breakdown is the overwhelming appearance of new virulent vectors (14). To reduce the probability of the vector adapting to a specific variety, a varietal rotation technique was devised and implemented in Indonesia (see below).

Virus Resistance

The variety TKM6 from India was found resistant to stem borers, bacterial leaf blight, and tungro. It was used as a good parent in the breeding program of IRRI. TKM6 was crossed with a selection from Peta x Taichung Native 1, and IR20 was selected from the progeny and released in 1969. IR20 was later found resistant to RTSV infection. IR26, IR30, and IR40, which also have TKM6 in their parentage, also show resistance to RTSV (25). IR20 was grown in 1973 to 1974 in farmers' fields, but was soon replaced because of its susceptibility to brown plant hopper and poor grain quality. Malaysian variety MR81 was developed from the cross of MR24/IR36, which are both susceptible to tungro. MR81 shows resistance to RTSV infection and may inherit the resistance from Pankhari 203 (33). MR81 is recommended for tungro-prone areas of Malaysia.

These varieties show high infection rates in artificial inoculation tests because of their susceptibility to RTBV infection, but disease development is slow in fields (69). There have been many varieties reported as resistant to RTSV infection (10,26). Shahjahan *et al.* (75) reported that RTSV resistance in Utri Merah was controlled by a single recessive gene, whereas resistance in Kataribhog and Pankhari 203 is by three complementary recessive genes.

In spite of previous reports, sources for resistance to RTBV infection have not been found in rice germplasm (Koganezawa, unpublished data), but several varieties show symptomatic resistance or tolerance (9,20,26). Tolerant varieties like Utri Merah do not show conspicuous symptoms and yield losses. The RTBV concentration in some tolerant varieties is too low to be detected by ELISA (9,86). These varieties seem to be good source for resistance, because virus spread in fields is expected to be slow.

Durability of Resistance

The isolation of tungro virus strains was reported in India (2) and the Philippines (67).

However, these reports were made before the discovery that tungro is caused by two distinct viruses. Actually tungro M-strain of the Philippines was revealed to be infected with RTSV only by electron microscopical study (68). Now, several efforts are being made to find out the strains of each virus. Dahal *et al.* (13) compared varietal reaction to three isolates from the Philippines, Malaysia, and India. There were variations in symptoms severity and transmission profile among the three isolates. Variety TKM6 reacted differently to RTSV isolates from India and the Philippines. In the Philippines, there is a RTSV strain virulent to several varieties which were previously identified as resistant. There are also several strains of RTBV which induce only mild symptoms (8). These facts indicate that resistance to the virus may not be durable in the same way as resistance to the vector. A survey of strains that can attack resistant varieties is needed prior to the deployment of resistant varieties. Among varieties so far tested in several regions, variety Utri Merah which originated in Indonesia, always showed tolerance. The virus multiplication is suppressed in infected Utri Merah plants. Efforts to incorporate the tolerant genes in Utri Merah to improved varieties are being made at IRRI. The combination of vector- and virus resistances may confer more durability to rice varieties.

CHEMICAL CONTROL OF VECTORS

Extensive studies on screening of effective insecticides and on their application procedure have been made in the last two decades (56). The most effective insecticides were carbofuran (71,77,94), a systemic granular insecticide and cypermethrin (53,70), a synthetic pyrethroid having repellent action. Other insecticides reported to be effective in controlling tungro were deltamethrin, ethofenprox (53), acephate, bendiocarb, carbaryl, and isoprocarb (72). Some of the insecticides mentioned above cause the resurgence of brown planthopper, the most important rice pest. Recently buprofezin, an insect growth regulator and slow-acting insectistatic compound, is recommended to control brown leafhopper. Buprofezin is effective in controlling tungro when used in combination with knockdown-type insecticides (52).

Some systemic insecticides are absorbed by the roots, when applied into the root-zone of rice plants. Root-zone application technique was first developed at IRRI in 1972 using medical gelatin capsules. Later, insecticides were incorporated in mud balls. This technique proved more effective than conventional broadcasting application (63,65,73). Incorporating carbofuran into the soil is also as effective as the root-zone application (73).

Because the presence of a relatively small number of vectors may cause substantial spread of tungro disease in the field, applying insecticide on susceptible cultivars is sometimes not effective in controlling tungro especially when vector populations are very high (54). Moreover, in most cases it is not cost-effective (22). It is important that farmers are taught to monitor tungro incidence, especially at the early stage of crop growth, so that appropriate insecticide can be applied at the right time. In Indonesia, a predetermined economic control threshold helped farmers to judge the situation and take immediate control action against tungro (85). The weekly thresholds were determined from the relationships between the percentage of infected hills with leaf discoloration and yield losses. The control thresholds for the economic one of 10% yield loss are 0.05%, 0.2%, 0.8%, and 1.5% of diseased hills at 2, 3, 4, and 5 weeks after transplanting, respectively. If the percentage of infected hills in young rice stages is more than 4 times as large as the economic threshold, severely tungro will occur (85).

INTEGRATED TUNGRO CONTROL

Muda Area in Malaysia

The Muda area, located in the north of Peninsular Malaysia, is a principal granary of rice. In the area, double-cropping was first implemented in 1970, and by 1974, full double cropping cycle was completed on more than 90% of the total area. The increase in cropping intensity through the change-over from single to double cropping, together with the large-scale planting of high-yielding, fertilizer-responsive rice and a large quantity of fertilizer, resulted in the change of occurrences of rice pests. The presence of tungro was not confirmed until 1981, although tungro-like symptoms had been observed in the Muda area during the single cropping as well as the early double cropping days. In 1981, 5,884 ha, mainly in the southern part of the Muda area, were affected with tungro. The disease subsequently inflicted more severe damage in 1982 and 1983 over the entire Muda area. To overcome the disease, a proposal was formulated to improve the planting schedule in the Muda area. Large-scale campaigns were organized by MADA (19,32,61). The proposal involved 1) large-scale planting of vector-resistant or moderately resistant varieties, e.g. IR42, MR71, MR73, MR77, and MR84; 2) institution of one-month fallow period in January/February to break continuous cropping by controlling water supply; 3) destruction of inoculum sources by applying herbicide to kill ratoons and volunteer crops, by burning of straws and stubbles, and by dry plowing soon after the second season harvesting; and 4) judicious and timely applications of insecticides for vector control. Carbofuran granules are applied in the nursery beds. For standing crops, carbamate insecticides such as BPC, MIPC, and carbaryl are applied as early as possible when disease symptoms are detected.

As a result of the concerted efforts of several organizations, the population density of leafhoppers had declined drastically since 1983 and tungro incidence significantly declined after 1984. From 1986 to the present (1994), tungro symptoms were undetected over the entire Muda area. A field survey carried out in early 1985 revealed that the relative population density of *N. virescens* after rice harvest was in the descending order of volunteer seedlings in the unplowed field > ratoons in the unplowed fields > volunteer seedlings in the plowed fields > grasses in the surrounding of fields. This observation further substantiated the fact that the fallow period together with dry plowing is effective in reducing green leafhopper populations and lowering tungro infection (32). The wide cultivation of IR42, which is resistant to green leafhopper, also contributed to the decline of vectors and tungro.

South Sulawesi in Indonesia

In Indonesia, tungro (as mentek hebane) was reported as early as the 19th century. As in other countries, tungro became a problem in Indonesia in the 1960s after dwarf and high-yielding new varieties were cultivated intensively. There was no serious outbreak of the disease in Sulawesi until 1972, but during 1972 - 1975, tungro caused damage on 100,000 ha of rice in South Sulawesi. After these years there were repeated outbreaks (84,90).

To solve the tungro problem, an integrated management scheme was developed and first practiced in 1983 in the provinces of South and Central Sulawesi (54). The strategy had three components: 1) appropriate planting time for wet and dry seasons, 2) varietal rotation according to vector resistant gene, and 3) use of insecticides. In the areas under the scheme, the wet and dry season crops are initially planted when the *N. virescens* population is low. From the studies on the fluctuation of vector, the best time to plant rice at Maros, South Sulawesi, is December to January for wet season and June to July for dry season. Planting time is further adjusted, depending on the time for cultivars to reach maturity, so that harvesting occurs almost simultaneously. In 1985, 90% or more fields were planted following the recommended planting schedule (69). The varietal rotation technique was devised in order to avoid the emergence of virulent biotypes of *N. virescens* on vector-resistant varieties. Commercial rice cultivars were grouped into four based on resistance to *N. virescens* and reaction to tungro using *N. virescens* populations that have been maintained on *N. virescens*-resistant IR26, IR42, IR54, or susceptible Pelita. Varieties under the different categories are deployed in an appropriate rotation cycle depending on the tungro situation in each area. Effective insecticides were selected and applied to affected fields to reduce vector density.

After the scheme was implemented , tungro incidence drastically dropped and the use of insecticide was reduced. Since 1986, tungro occurred sporadically in small areas. In the survey conducted during 1987 - 88, tungro incidence on vector-resistant varieties IR54 and IR42 was higher than that on susceptible variety Cisadane. The varietal rotation seems to have a limited role in reducing tungro incidence. The reason for the success of the scheme was attributed to synchronized cropping as a result of revised planting schedule among the various components (69).

CONCLUSION

Tungro is the most complicated virus disease, which involves two viruses and their strains as well as vectors and their biotypes. Although many papers about varietal reaction to tungro have been published, consistent results have not always been obtained. Breeding of resistant varieties will continue to be the main component in tungro management research, but there is a possibility that resistant varieties will succumb to tungro as a result of new virus strains and vector biotypes. Conferring durable resistance to improved varieties is still beyond our sight. However, the success of tungro control in Malaysia and Indonesia indicates that synchronized cropping, together with fallow period, is the most important component in tungro management known so far.

REFERENCES

1. Ali, M. A., and Miah, S. A. 1990. Presence of rice tungro virus in the first ratoon crop of tungro infected main crop of rice. Bangladesh J. Bot. 19:155-159.

2. Anjaneyulu, A., and John, V. T. 1972. Strains of rice tungro virus. Phytopathology 62:1116-1119.

3. Anonymous 1991. Serological techniques for virus disease monitoring. Page 201 in: Program Report 1990. International Rice Research Institute, P.O. Box 933, Manila, Philippines.

4. Anonymous 1987. Tungro. Page 36 in: Annual Report 1986, BPI (Bureau of Plant Industry). San Andres, Manila, Philippines.

5. Anonymous 1991. Outbreaks of pests and diseases: India - Tungro virus on rice. Q. Newsl. Asia Pac. Plant Prot. Comm. 33(4):20.

6. Bajet, N. B., Aguiero, V. M., Daquioag, R. D., Jonson, G. B., Cabunagan, R. C., Mesina, E. M., and Hibino, H. 1986. Occurrence and spread of rice tungro spherical virus in the Philippines. Plant Dis. 70:971-973.

7. Bottenberg, H., Litsinger, J. A., Loevinsohn, M. E., and Kenmore, P. 1991. Impact of cropping intensity and asynchrony on the epidemiology of rice tungro virus in Malaysia. J. Plant Prot. Trop. 7:103-116.

8. Cabauatan, P. Q., Cabunagan, R. C., and Koganezawa, H. 1995. Biological variants of rice tungro viruses in the Philippines. Phytopathology 85:77-81.

9. Cabunagan, R. C., Flores, Z. M., Coloquio, E. C., and Koganezawa, H. 1993. Virus detection in varieties resistant to tungro (RTD). Int'l. Rice Res. Notes 18(1):22-23.

10. Cabunagan, R. C., Flores, Z .M., and Koganezawa, H. 1993. Resistance to rice tungro spherical virus (RTSV) in rice germplasm. Int'l Rice Res. Notes 18(1):24-25.

11. Chowdhury, A. K., Teng, P. S., and Hibino, H. 1990. Production of helper component in rice tungro virus (RTSV)-infected plants. Int. Rice Res. Newsl. 15:14.

12. Cook, A.G., and Perfect, T. J. 1989. Population dynamics of three leafhopper vectors of rice tungro viruses, Nephotettix virescens (Distant), N. nigropictus (Stal) and Recilia dorsalis (Motschulsky) (Hemiptera: Cicadellidae), in farmers' fields in the Philippines. Bull. Entomol. Res. 79:437-451.

13. Dahal, G., Dasgupta, I., Lee, G., and Hull, R. 1992. Comparative transmission of, and varietal reaction to, three isolates of rice tungro virus disease. Ann. Appl. Biol. 120:287-300.

14. Dahal, G., Hibino, H., Cabunagan, R. C., Tiongco, E. R., Flores, Z. M., and Aguiero, V. M. 1990. Changes in cultivar reaction to tungro due to changes in "virulence" of the leafhopper vector. Phytopathology 80:659-665.

15. Dalrymple, D. G. 1986. Development and spread of high-yielding rice varieties in developing countries. Bureau for Science and Technology, Agency for International Development, Washington, D.C.

16. Estano, D. B., and Shepard, B. M. 1989. Effect of roguing on rice tungro virus (RTV) incidence and rice yield. Int'l Rice Res. Newsl. 14(6):22.

17. Favali, M. A., Pellegrini, S., and Bassi, M. 1975. Ultrastructural alterations induced by rice tungro virus in rice leaves. Virology 66:502-507.

18. Furuta, T. 1977. Rice waika, a new virus disease, found in Kyushu. Rev. Plant Prot. Res. 10:70-80.

19. Habibuddin, H., Takita, T., and Ho, N. K. 1987. Research and management of tungro disease in Peninsular Malaysia. Pages 86-91 in: Proc. Workshop on Rice Tungro Virus, Maros Research Institute for Food Crops, Maros, Indonesia.

20. Hasanuddin, A., and Hibino, H. 1989. Grain yield reduction, growth retardation, and virus concentration in rice plants infected with tungro-associated viruses. Trop. Agric. Res. Ser. 22:56-73.

21. Heinrichs, E.A., and Rapusas, H. 1983. Correlation of resistance to the green leafhopper, Nephotettix virescens (Homoptera: Cicadellidae) with tungro virus infection in rice varieties having different genes for resistance. Environ. Entomol. 12:201-205.

22. Heinrichs, E. A., Rapusas, H. R., Aquino, G. B., and Palis, F. 1986. Integration of host plant resistance and insecticides in the control of Nephotettix virescens (Homoptera: Cicadellidae), a vector of rice tungro virus. J. Econ. Entomol. 79:437-443.

23. Herdt, R.W. 1991. Research Priorities for rice biotechnology. Pages 19-54 in: Rice Biotechnology. G. S. Khush and G. H. Toenniessen, eds. International Rice Research Institute, P.O. Box. 933, Manila, Philippines.

24. Hibino, H. 1983. Transmission of two rice tungro-associated viruses and rice waika virus from doubly or singly infected source plants by leafhopper vectors. Plant Dis. 67:774-777.

25. Hibino, H., Daquioag, R. D., Cabauatan, P. Q., and Dahal, G. 1988. Resistance to rice tungro spherical virus in rice. Plant Dis. 72:843-847.

26. Hibino, H., Daquioag, R. D., Mesina, E. M., and Aguiero, V. M. 1990. Resistances in rice to tungro-associated viruses. Plant Dis. 74:923-926.

27. Hibino, H., Ishikawa, K., Omura, T., Cabauatan, P. Q., and Koganezawa, H. 1991. Characterization of rice tungro bacilliform and rice tungro spherical viruses. Phytopathology 81:1130-1132.

28. Hibino, H. Roechan, M., and Sudarisman, S. 1978. Association of two types of virus particles with Penyakit habang (tungro disease) of rice in Indonesia. Phytopathology 68:1412-1416.

29. Hibino, H., Saleh, N., and Roechan, M. 1979. Transmission of two kinds of rice tungro associated viruses by insect vectors. Phytopathology 69:1266-1268.

30. Hibino, H., Tiongco, E. R., Cabunagan, R. C., and Flores, Z. M. 1987. Resistance to rice tungro-associated viruses in rice under experimental and natural conditions. Phytopathology 77:871-875.

31. Hino, T., Wathanakul, L., Nabheerong, N., Surin,

P., Chaimongkol, U., Disthaporn, S., Putta, M., Kerdchocchai, D., and Surin, A. 1974. Studies on rice yellow orange leaf virus disease in Thailand. Tech. Bull. Tropical Agriculture Research Center, Japan, No.7.

32. Hirao, J., and Ho, K. 1987. Status of rice pests and their control measures in the double cropping area of the Muda irrigation scheme, Malaysia. Trop. Agric. Res. Ser. 20:107-115.

33. Imbe, T., and Habibuddin, B. H. 1989. Studies on breeding for resistance to tungro disease of rice in Malaysia. Tropical Agriculture Research Center (TARC), Japan and Malaysian Agricultural Research and Development Institute (MARDI), Malaysia.

34. Inoue, H. 1986. Vector specificity of leaf- and planthoppers in rice virus transmission. Trop. Agr. Res. Ser. 19:220-228.

35. Inoue, H., and Ruay-Aree, S. 1977. Bionomics of green rice leafhopper and epidemics of yellow orange leaf virus diseases in Thailand. Trop. Agri. Res. Ser. 10:117-121.

36. John, V. T. 1968. Identification and characterization of tungro, a virus disease of rice in India. Plant Dis. Reptr. 52:871-875.

37. Jones, M., Gough, K., Dasgupta, I., Subba Rao, B. L., Cliffe, J., Qu, R., Shen, P., Kaniewska, M. B., Davies, J. B., Beachy, R. N., and Hull, R. 1991. Rice tungro disease is caused by an RNA and A DNA virus. J. Gen. Virol. 72:757-761.

38. Khan, M. A., Hibino, H., Aquiero, V. M., and Daquioag, R. D. 1991. Rice and weed hosts of rice tungro-associated viruses and leafhopper vectors. Plant Dis. 75:926-930.

39. Khush, G. S. 1989. Multiple disease and insect resistance for increased yield stability in rice. Pages 79-92 in: Progress in Irrigated Rice Research. International Rice Research Institute, P.O. Box 933, Manila, Philippines.

40. Khush, G. S. 1977. Disease and insect resistance in rice. Adv. Agron. 29:265-341.

41. King, T. H. 1968. Occurrence and distribution of diseases and pests of rice and their control in Thailand. FAO Pl. Prot. Bull. 16:41-44.

42. Kiritani, K. 1979. Pest management in rice. Annu. Rev. Entomol. 24:279-341.

43. Koganezawa, H., Pablico, P. P., Cabunagan, R. C., Tiongco, E. R., Cabangon, R., Tuong, T. P., and Yamauchi, M. 1993. The relationship between tungro infection and water level in direct seeded rice field. Int'l Rice Res. Notes 18(2):28.

44. Lamey, H. A., Surin, P., Disthaporn, S., and Wathanakul, S. 1967. The epiphytotic of yellow orange leaf disease of rice in 1966 in Thailand. FAO Plant Proc. Bull. 15:67-69.

45. Lapis, D. B. 1991. The incidence of rice tungro and some contributory factors for its outbreaks in the Philippines. Pages 87-93 in: Proc. National Conference and Workshop on Integrated Pest Management in Rice, Corn and Selected Major Crops. National Crops Protection Center, UPLB, College, Laguna, Philippines.

46. Lim, G. S. 1972. Studies on penyakit merah disease of rice III. Factors contributing to and epidemic in North Krian, Malaysia. Malaysian Agr. J. 48:278-294.

47. Lim, G. S., and Heong, K. L. 1977. Habitat modification for regulating pest population of rice in Malaysia. MARDI Report 50:1-28.

48. Ling, K. C. 1972. Rice virus diseases. International Rice Research Institute, P.O. Box 933, Manila, Philippines,142 Pages.

49. Ling, K. C., and Palomar, M. K. 1966. Studies on rice plants infected with the tungro virus at different ages. Philippine Agr. 50:165-177.

50. Ling, K. C., Tiongco, E. R., and Flores, Z. M. 1983. Epidemiological studies of rice tungro. Pages 249-257 in: Plant Virus Epidemiology: The spread and control of insect borne viruses. R. T. Plumb and J. M. Thresh, eds. Blackwell Scientific Publications, Oxford.

51. Loevinsohn, M. E., and Alviola, A. A. 1991. Effect of asynchronized rice planting on vector abundance and tungro (RTD) infection. Int. Rice Res. Newsl. 16(5):20-21.

52. Macatula, R. F., Mochida, O., and Litsinger J. A. 1988. Using mixtures of buprofezin and cypermethrin or deltamethrin for green leafhopper (GLH) and rice tungro virus (RTV) control. Int'l Rice Res. Newsl., 13(4):38-39.

53. Macatula, R. F., Valencia, S. L., and Mochida, O. 1987. Evaluation of 12 insecticides against green leafhopper for preventing rice tungro virus disease. IRRI Rice Res. Pap. Ser. 128:1-9.

54. Manwan, I., Sama, S., and Rizvi, S. A. 1985. Use of varietal rotation in the management of tungro disease in Indonesia. Indonesian Agric. Res Devlop. J. 7(3/4):43-48.

55. Miah, S. A., Rahman, M. M., and Nahar, M. A. 1984. Seasonal ariation of the incidence of rice tungro disease and its vector *Nephotettix virescens*. Bangladesh J. Bot. 13:12-15.

56. Mochida, O., Valencia, S. L., and Basilio, R. P. 1986. Chemical control of green leafhoppers to prevent virus diseases, especially tungro disease, on susceptible/intermediate rice cultivars in the tropics. Trop. Agri. Res. Ser. 19:195-208.

57. Mohanty, S. K., Bhaktavatsalam, G., Anjaneyulu, A., and Krishnamurty, A. 1990. Management of rice tungro disease through host resistance. Oryza 27:191-195.

58. Morinaka, T., Tsurumachi, M., Putta, M., Chettanachit, D., Patirupanusara, T., Parejarearn, A., and Disthaporn, S. 1986. Seasonal changes of incidence of rice viruses and their insect vectors in Thailand. Pages 84-93 in: Virus Diseases of Rice and Legumes in the Tropics. T. Kajiwara and S. Konno, eds. Tropical Agriculture Research Center, Japan (Tech. Bull. No. 21).

59. Mukhopadhyay, S., Chowdhury, A. K., and Chakrabarty, S. K. 1987. Epidemilology and control of rice tungro virus disease in West Bengal. Pages 142-153 in: Proc. Nat'l Seminar on Rice hoppers, Hopperborne Viruses and Thier Integrated Management. S. Mukhopadhyay and M. R.Ghosh, eds. Plant Virus Research Center, Bidhan Chandra Krishi Viswaavidyalaya, Mohnpur, West Bengal, India.

60. Mukhopadhyay, S., Nath, P. S., Sarkar, T. K., Sarkar, S., and Mukhopadhyay, S. 1988. Interrelationship between rainfall and rice green leafhopper population in West Bengal. Indian J. Agric. Sci. 58:34-38.

61. Nozaki, M., Wong, H. S., and Ho, N. K. 1984. A new double cropping system proposed to overcome instability of rice production in the MUDA irrigation area of Malaysia. JARQ (Japan Agricultural Research Quarterly) 18:60-68.

62. Ou, S. H. 1985. Rice diseases. Commonwealth

Mycological Institute, Kew, UK.

63. Pathak, M. D., Encarnacion, D., and Dupo, H. 1974. Application of insecticides in the root zone of rice plants. India J. Pl. Prot., 1(2):1-16.

64. Qu, R., Bhattacharyya, M., Laco, G. S., Kochko, A. de, Subba Rao, B. L., Kaniewska, M.B., Elmer, J.S., Rochester, D.E., Smith, C.E., and Beachy, R.N. 1991. Characterization of the genome of rice tungro bacilliform virus: Comparison with Commelina yellow mottle virus and caulimoviruses. Virology 185:354-364.

65. Rapusas, H .R., Heinrichs, E. A., Aquino, G. B., and Basilio, R. P. 1986. Root-zone application of insecticides for control of the rice tungro virus vector. J. Plant Prot. Trop. 3:111-119.

66. Riley, J. R., Reynolds, D. R., and Farrow, R.A. 1987. The migration of *Nilaparvata lugens* Stal Delphacidae and other Hemiptera associated with rice during the dry season in the Philippines: a study using radar, visual observations, aerial netting and ground trapping. Bull. Entomol. Res. 77:145-169.

67. Rivera, C. T. and Ou, S. H. 1967. Transmission studies of the two strains of rice tungro virus. Plant Dis. Reptr. 51:877-881.

68. Saito, Y. 1976. Interrelationship among waika disease, tungro and other similar disease of rice in Asia. Trop. Agr. Res. Ser. 10:129-135.

69. Sama, S., Hasanuddin, A., Manwan, I., Cabunagan, R. C., and Hibino, H. 1991. Integrated management of rice tungro disease in South Sulawesi, Indonesia. Crop Protection 10:34-40.

70. Satapathy, M. K., and Anjaneyulu, A. 1984. Use of cypermethrin, a synthetic pyrethroid, in the control of rice tungro virus disease and its vector. Trop. Pest Manage. 30:170-178.

71. Satapathy, M.K., and Anjaneyulu, A. 1989. Experimental epidemics of tungro and its vectors in nursery beds under different pesticide treatments. Int. J. Trop. Pl. Dis. 7:137-150.

72. Satapathy, M. K., and Anjaneyulu, A. 1989. Management of tungro virus disease by application of wettable powder and flowable insecticides. Trop. Pest Manage. 35:41-47

73. Satapathy, M. K., and Anjaneyulu, A. 1991. Rice tungro management by different methods of carbofuran application. Oryza 28:377-387.

74. Serrano, F. B. 1957. Rice "accep na pula" or stunt disease - a serious menace to the Philippine rice industry. Philipp. J. Sci. 86:203-230.

75. Shahjahan, M., Imbe, T., Jalani, B. S., Zakri, A. H., and Othman, O. 1991. Inheritance of resistance to rice tungro spherical virus in rice (*Oryza sativa* L.). Pages 247-254 in: Rice Genetics II, Proc. 2nd Int'l Rice Genetics Symp. International Rice Research Institute, P.O. Box 933, Manila, Philippines.

76. Shastry, S. V. S., Freeman, W. H., Seshu, D. V., and John, V. T. 1971. Some investigations on resistance to rice tungro virus. Indian J. Gen. Pl. Breed. 31:536-542.

77. Shukla, V. D., and Anjaneyulu, A., 1980. Evaluation of systemic insecticides for control of rice tungro. Plant Dis. 64:790-792.

78. Shukla, V. D., and Anjaneyulu, A. 1981. Adjustment of planting date to reduce rice tungro disease. Plant Dis. 65:409-411.

79. Shukla, V. D., and Anjaneyulu, A. 1982. Plant spacing to reduce rice tungro incidence. Plant Dis. 65:584-586.

80. Shukla, V. D., and Anjaneyulu, A. 1989. Spread of tungro virus disease in different ages of rice crop. J. Pl. Dis. Protect. 88:614-620.

81. Siwi, S. S., and Roechan, M. 1983. Species composition and distribution of green rice leafhopper, *Nephotettix* spp. and spread of tungro virus disease in Indonesia. Pages 263-276 in: Proc 1st Int'l Workshop on Leaf and Planthopper of Economic Importance. W.J. Knight, N.C. Pant, T.S. Robertson, and M.R. Wilson, eds. Commonwealth Institute of Entomology, Kew, U.K.

82. Sogawa, K. 1976. Rice tungro virus and its vectors in Tropical Asia. Rev. Plant Protec. Res. 9:21-46.

83. Sta. Cruz, F. C., Koganezawa, H., and Hibino, H. 1993. Comparative cytology of rice tungro viruses in selected rice cultivars. J. Phytopathology 138:274-282.

84. Suzuki, Y. 1991. Occurrence of the rice tungro disease and its vector, green leafhoppers in Indonesia. Plant Protection (Tokyo) 45:377-380 (in Japanese).

85. Suzuki, Y., Astika, I. G. N., Widrawan, I. K. R., Gede, I. G. N., Raga, I. N., and Soeroto 1992. Rice tungro disease transmitted by the green leafhopper: its epidemiology and forecasting technology. JARQ (Japan Agricultural Research Quarterly) 26:98-104.

86. Takahashi, Y., Tiongco, E. R., Cabauatan, P. Q., Koganezawa, H., Hibino, H., and Omura, T. 1993. Detection of rice tungro bacilliform by polymerase chain reaction for assessing mild infection of plants and viruliferous vector leafhoppers. Phytopathology 83:655-659.

87. Tiongco, E. R., Cabunagan, R. C., Flores, Z. M. Hibino, H., and Koganezawa, H. 1993. Serological monitoring of rice tungro disease development in the field: Its implication in disease management. Plant Dis. 77:877-882.

88. Tiongco, E. R., Cabunagan, R. C., Flores, Z. M., and Mew, T. W. 1990. Tungro (RTV) incidence in direct seeded and transplanted rice. Int'l Rice Res. Newsl. 15(1):30.

89. Tiongco, E. R., Flores, Z. M., Koganezawa, H., and Teng, P. S. 1993. Inoculum sources of rice tungro viruses. Philippine Phytopathol. 29:30-41.

90. Van Halteren, P. 1979. The insect pest complex and related problems of lowland rice cultivation in South Sulawesi, Indonesia. Meded. Lanbouwhogeschool Wageningen 79:1-111.

91. Vidhyasekaran, P. 1990. Forecasting of rice tungro epidemic. Pages 119-125 in: Environmental ecology and aerobiology. S. T. Tilak, ed. Today & Tomorrow's Printers & Publishers, New Delhi, India (Recent Res. Ecol. Environ. Pollut. 3).

92. Watahnakul, L., and Weerapat, P. 1977. Yellow orange leaf virus disease in Thailand. Trop. Agr. Res. Ser. 10:165-169.

93. Widiarta, I. N., Suzuki, Y., Sawada, Y., and Nakasuji, F. 1990. Population dynamics of the green leafhopper, *Nephotettix virescens* Distant (Hemiptera: Cicadellidae) in synchronized and staggered transplanting areas of paddy fields in Indonesia. Res. Popul. Ecol. (Japan) 32:319-328.

94. Wilkins, R. M., Batterby, S., Heinrichs, E. A., Aquino, G. B., and Valencia, S. L. 1984. Management of the rice tungro virus vector *Nephotettix virescens* (Homoptera: Cicadellidae) with controlled-release formulation of carbofuran. J. Econ. Entomol. 77:495-499.

95. Xie, L. H., Lin, Q. Y., Zhu, Q. L., Lai, G. B., Chen,

N. Z., Huang, M. J., and Chen, S. M. 1983. On the occurrence and control of rice tungro in Fujian. J. Fujian Agric. Coll. 12:275-284 (in Chinese).

96.   Yamaguchi, T. 1984. Overlapping of cropping season may cause severe occurrence of some diseases in the tropics. Jap. J. Trop. Agric. 28(4):246-248, 252.

CHAPTER   35

Present Status of Controlling Rice Stripe Virus

R. Kisimoto and Y. Yamada

The occurrence of rice stripe virus (RSV) disease in Japan was documented as early as the 1900s in the Kanto area, central Japan (1,22). It showed an epidemic pattern, beginning with small-scale occurrences at core sites followed by a great increase in terms of area and severity. After severe outbreaks for a few years, the disease declined to a chronic status. Several RSV epidemics have occurred intermittently in this area, but the factors that initiated the epidemic have not been clarified. On the other hand, when new rice-growing practices were introduced, for example, drill seeding of rice in standing wheat fields as a labor-saving measure in Okayama in the 1940s (47), or early transplantation of ordinary rice cultivars in mid to late May to increase the grain yield, as practiced widely in western Japan in the 1960s (14), movement of the small brown planthopper (SBPH), *Laodelphax striatellus* Fallén (Delphacidae, Homoptera), the vector of RSV, from wheat to nearby young rice plants or into recently transplanted paddies was favored, and severe RSV outbreaks followed. The RSV outbreaks in Okayama ceased within a few years after the abandonment of drill seeding, but in western Japan the outbreaks continued for more than 10 years after the practice declined.

Trials for controlling RSV have been carried out from various viewpoints, such as changes in rice-growing practices, ecological and chemical control of vectors and breeding of resistant cultivars. Recently, virological studies of RSV have made great progress, and RSV and related viruses presenting peculiar filamentous particles were recognized as a new group named tenuivirus in 1987. Trials to breed transgenic cultivars resistant to RSV are now underway. In this chapter, the present status of RSV control by conventional but practical procedures currently available and modern biotechnological trials is presented and discussed.

PLANT-VIRUS-VECTOR COMPLEX

RSV control is based on an understanding of the complex between three biological entities that differ considerably from each other; the rice plant, RSV and the vector.

Geographical Distribution of SBPH and RSV

SBPH is distributed widely in Asia and Europe but occurs predominantly in temperate East Asia. In tropical countries such as the Philippines, North Sumatra and Indochina, it may be found on upland rice (45). On the other hand, the distribution of RSV is, so far, confined to temperate East Asia; Japan, Korea, Vladivostok and China, where the life cycle of SBPH is intimately associated with paddy rice.

In Hokkaido, the northern boundary for rice-growing area, RSV was first detected in 1968 and a rice-RSV-SBPH complex was for the first time established; before this, rice and SBPH had long shown a close but simple relationship. RSV epidemics occurred in 1972, 1977-1979 and 1985-1986 (27). In Taiwan, RSV was first recorded in 1969 and an epidemic occurred in 1985-88 (3). In these two cases, damage due to sap-sucking by SBPH was also a problem for rice production.

SBPH also transmits rice black-streaked dwarf virus (RBSDV) to rice, barley and maize in areas similar to RSV. However, so far, no special control measure has been introduced except for that on maize.

RSV

The appearance of RSV in electron microscopy is pleomorphic, due partly to artifacts (38). Purified RSV presents filamentous particles 3-8 nm wide with 4 different contour lengths of

2,110 nm, 840 nm, 610 nm and 510 nm, corresponding to each of the RNA molecules. The particle 8 nm wide is considered to be a supercoiled helix of a circular filament 3 nm wide. The nucleoprotein (CP) of RSV is a single protein of molecular weight 32 kDa, which was proved to be encoded on RNA 3 (8). It is also noteworthy that a large quantity of non-capsid protein, specific to RSV-infected rice plant tissue (s-protein: SP), is found in susceptible cultivars (21). However, the amount is much reduced in tolerant rice cultivars and wheat, and extremely low in corn. The SP is known to be encoded on RNA 4 (8), and is detectable also in RSV-infective SBPH by the ELISA method (39). Crystalline inclusion bodies observed in RSV-infected rice cells probably correspond to SP. In ultrathin sections of RSV-infected plant and vector tissue, no filamentous particles are evident, but the viroplasmic structures found in the cytoplasm of infected cells react positively with RSV antiserum (39).

RSV types presenting different symptoms, milder or severer than the ordinary ones, have occurred in laboratory stocks (18). Type SV (severe) produces pronounced yellowing of leaves in wheat but spotted mild yellowing in rice, whereas BW (bleached white) produces whitish-yellow stripes with a dark green background, the stripes appearing along the veins, in wheat and rice. Rice plants infected with SV survive for a long time, but those infected with BW are severely stunted and wither rapidly. The latent period in wheat and vector is shortest for SV, followed by N (ordinary) and BW. The proportion of transovarially infective members in a family produced by an infective female differs considerably according to the type, as described below.

Host Range of RSV and Symptoms

Japanese upland rice and most cultivars of *indica* rice have been shown to be resistant to RSV, but when infested heavily by viruliferous vectors they present symptoms of mild yellow spots or short stripes scattered over the leaf. Twisting and curling of young leaves which are typical symptoms of RSV on susceptible rice plants, do not develop. Cereals infected experimentally with RSV other than rice are *Hordeum vulgare*, *Triticum aestivum*, *Avena sativa*, *Secale cereale*, *Setaria italica*, *Panicum miliaceum*, *Echinochloa utilis*, *Zea mays*, forage crops, such as *Lolium multiflorum*, *L. perenne* and *Phleum pratense*, and gramineous weeds commonly grown in and around paddy and upland farms, such as *Beckmannia syzigachne*, *Digitaria adscendens*, *D. violascens*, *Echinochloa crus-galli*, *Eragrostis multicaulis*, *Poa annua* and *Setaria viridis* (34). On these plants, greenish-yellow to whitish-yellow stripes on the leaf appear after long latent periods and infected plants survive much longer than infected rice plants, except for *Zea mays*, which shows very severe symptoms with or without simultaneous infection with RBSDV. Common weeds, such as *Agropyron tsukushiense*, *Eleusine indica*, *Eragrostis ferruginea*, *Pennisetum alopecuroides*, *Phalaris arundinacea* and *Zoysia japonica*, are not infected (34).

Host Plants of SBPH

SBPH also propagates on wheat, barley and Italian rye grass in spring and on upland rice, *Echinochloa crus-galli*, *Digitaria adscendens* and *D. violascens* in summer, although RSV occurrences on these crops and weeds are negligible because of the low temperature in winter or tolerance to RSV in summer. SBPH often lands on young corn plants and transmits RSV, but it cannot propagate on corn.

Most tenuiviruses are transmitted by several sibling or related species of planthoppers, but only a single species, common and predominant on the target crop, is economically significant. Potential vectors of RSV, such as *Unkanodes sapporona* Matsumura, *U. albifascia* M. and *Terthron albovittatum* M., have their own specific host plants and practically none of them infest cereals in the field.

SBPH occasionally causes damage to rice plant by sucking the plant sap when its density increases, although RSV occurs at much lower densities. Consequently, the economic threshold for control is much lower than that as a sap sucker. It is therefore important to monitor the occurrence of the vector in terms of the proportion of infective individuals in a given local population as well as the density.

Ability of SBPH to Acquire RSV from Infected Plants

The probability that SBPH acquires RSV by feeding on infected plants is low compared to the probability that the same vector acquires RBSDV. Rearing of SBPH on RSV-infected rice plants for two years, except in the winter when it is reared on weeds, produced colonies in which 16.7-25% of members were infective (48). However, selective breeding in the laboratory at 25°C for 9 generations produced a highly acquisitive line, and 50-60% of vectors became infective within 4 days. On the other hand, in low-ability lines produced at the 11th to 13th selective generations, less than 10% of vectors acquired RSV within 8 to 11 days (15). With RBSDV, 4% of field-collected vectors acquired the virus within 30 min; 67-70% within 1 day and 90% within 3 days (34). Geographical variations in the acquisitive ability of SBPH have not been reported. A SBPH colony collected at Davao, the Philippines, in 1972 transmitted RSV in laboratory trials (R. Kisimoto, unpublished).

Ability of SBPH to Transmit RSV to Plants and Offspring Through the Ovary

SBPH can transmit RSV to rice plants by inoculative access within a minimum period of 3 min, but practically significant transmission usually occurs through access for longer than 10-30 min (34, 46). Therefore, vectors should be removed from the host plant within a short time to avoid infection. The inoculative ability is high in the middle to old nymphal stages, but decreases at the adult stage and males scarcely inoculate (18).

RSV is transmitted transovarially to a high proportion of the offspring, i.e., 75.5-95% in Okayama stock of SBPH (46), 96-100% in Tokyo and Saitama stocks (34), and 90% in Fukuyama stock (31), but it is also true that the proportion fluctuates considerably according to RSV type and among individual vector females. In Zentsuji stock transmitting type SV, the average proportion of infective members in 23 families extending over 8 generations was 94.3% for males and 98.5% for females, while in a stock transmitting type N the average of 22 families was 84.3% for males and 90.4% for females. With type BW the average of 67 families was as low as 35.1% for males and 34.4% for females (18).

SBPH females showing a higher ability to inoculate plants tend to produce families with higher proportions of infective members. For example, in a SBPH stock transmitting type N, all the females which inoculated 4 to 5 test plants in 5 successive inoculative feedings of 4 days each on a wheat seedling produced infective families and average proportion of infective members was

86.3%. On the other hand, 4 females (20%) out of the 20 that inoculated only 1 plant produced non-infective families (none of the family members were infective), and in the remaining 16 families the average proportion of infective members was 40.6%. Ten (43.5%) out of the 23 females that belonged to an infective family but did not inoculate any plant themselves produced non-infective families, and in the remaining families the average proportion of infective members was only 10.9%. As no infective individuals have appeared in their descendants, the females and their families showing no infectivity were considered to have lost RSV completely. However, members of these non-infective families became RSV-infective by acquisitive access to infected plants at a similar proportion to non-infective families collected from the field. These findings strongly suggest that the concentration of RSV differs according to individual vectors and that it reflects the ability to inoculate plants and the rate of transovarial passage. The RSV concentration may decrease on average along with the generation passage of vector unless individuals having high ability for transovarial transmission are selected out. The rate of transovarial passage tends to be high when the female acquires RSV from an infected plant (18).

ECOLOGY OF THE VECTOR AND RSV OCCURRENCE IN RELATION TO RSV CONTROL

Life Cycle of the Vector

In temperate East Asia, SBPH overwinters mostly as a 4th stage nymph in diapause on weeds or withered host plants in and around the paddy fields where it has grown. The survival rate of SBPH during winter may affect the population size invading paddies in the next season. Field sanitation measures in fall through spring, such as turning over the paddy soil, weeding of levees and insecticide application, may lower vector density but the effects have not been well estimated. The short-winged forms (brachypter: flightless) predominates among males and females in the overwintering population.

In central to western Japan at ca 32-36°N, overwintered SBPHs move to wheat and barley to propagate for one generation. From late May to mid-June when the crops ripen, long-winged (macropter) males and females of the first generation emerge and leave the host plant. They fly actively on fine days in the late afternoon and land on preferentially on recently transplanted paddy fields. An increase of immigrants due to natural factors or better synchronization between SBPH dispersal and rice transplantation triggers an epidemic (see below). The population of SBPH usually remains low during summer through fall, but when the proportion of infective individuals is high, transmission by the second generation is important in paddies transplanted still later (23).

In Hokkaido at ca 44°N, SBPH has two generations per year. Overwintered SBPHs move directly into nursery beds and recently transplanted paddies from mid-May to early June (27).

At altitudes between the two, the generation process of SBPH is not synchronized well with rice plants at the early growth stage, and consequently RSV scarcely occurs (4).

In the southern part of the temperate region, for example in Chejiang, China, SBPH has 6 generations per year and macropters in the first generation immigrate into the first rice crop from late May to early June, and those in the 3rd generation appearing from mid-July to early August immigrate into seedling beds and recently transplanted paddies of the second rice crop (28). In the subtropical region, for example in Taiwan, SBPH grows throughout the year in paddies. The population of SBPH increases mainly in two periods, one in early May to late June (peak in early June) on the first rice crop, and the other in late September to early October on the second crop. RSV infection of the first crop mainly occurs 21 to 60 days after transplanting (3). The nymphal diapause of SBPH in subtropical and tropical colonies is less obvious.

Dispersal of SBPH and RSV Infection

Experimentally, SBPH may continue wing-beating over several hours. The distance over which SBPH disperses depends on the wind, and under ordinary climatic conditions during the dispersal period in central Japan it may disperse for several kilometers. Considerable numbers of SBPH have been trapped by wind-borne tow nets on the East China Sea at 126°E and 31°N, ca 400 km from the nearest mainland in late June to early July on the warm and humid SW monsoon. In these cases two important sap-suckers on rice plant, the brown planthopper, *Nilaparvata lugens* Stl, (also a vector of rice grassy stunt and rice ragged stunt in the tropics) and the white-backed

planthopper, *Sogatella furcifera* Horváth, and many other small insects were trapped together. A few SBPH individuals gave a positive haemagglutination reaction using RSV antibody-sensitized blood cells (0/146 in 1977, 5/154 in 1978, 2/187 in 1979, 2/200 in 1980) (17). On Ishigaki Is., ca 200 km east of Taiwan, immigrant SBPH appears with other planthoppers in May to June carried by SW winds. In 1987, the population infectivity of SBPH collected on May 19 was estimated to be 9.9% (39/394) and on June 10-12 5.5% (17/310) (R. Kisimoto, unpublished), a severe outbreaks of RSV occurring one month later (40).

RSV transmission to rice plants takes place mainly during the first 30 to 35 days after transplanting. When rice plant are transplanted 2-3 weeks before the peak of SBPH dispersal, immigrants land on young rice plants at the tillering stage and transmit RSV, but when the plants are transplanted later, immigrant SBPH decreases and RSV occurrence is reduced (16). In this case, transmission by the nymphs produced by immigrants may occur. Late transplanting is usually beneficial for avoiding RSV infection, but it is not necessarily acceptable by farmers.

Monitoring of Vector Density

Vector density sometimes fluctuates enormously for reasons that are still unclear, although climatic factors such as warm and fine weather in spring when the first generation propagates may favor egg-laying and the survival of nymphs.

During Overwintering

The density of overwintering SBPH can be estimated by ordinary net-sweeping but the efficiency of this method is very much influenced by temperature and wind. Tapping the sunny hibernacula by hand agitates the nymphs and induces them to leap on a white towel placed on the south side of the hibernacula, the nymphs then being collected with an aspirator. Sucking hibernacula with an insect collecting vacuum may allow nymph collection at a standardized efficiency.

During the Dispersal Period

Several types of equipments and method are available for monitoring of vector density (19).

Wind-borne Tow Net (WBTN)

SBPH flying in the air is trapped by a wind-borne tow net, 1 m in diameter and 1.7 m in depth, set at a height of 8-20 m. In years of high density, more than 2,000 SBPH may be trapped in a dispersal season. As many other flying insects and spiderlings are trapped together, it is recommended to empty the net as frequently as possible because trapped arthropods are very fragile.

Yellow Pan Water Trap (YPWT)

A galvanized iron or plastic water pan ca 60 cm in diameter (depending on choice for easy handling) colored lemon-yellow, not orange-yellow, is set inside a paddy field at a distance of 5 m or more from the edge to avoid a possible edge effect, and the height is adjusted to the rice canopy level. The pan is half-filled with water containing 0.1% liquid detergent. Planthoppers are trapped predominantly in late afternoon. All the trapped insects are removed daily using a meshed spoon and sorted carefully under a binocular. Addition of a few drops of formaldehyde solution keeps the catches from spoiling for a few days. Water in the pan is renewed every few days to prevent it from becoming soiled.

Visual Counting

The number of planthoppers on each rice hill can be counted from above with the naked eye. Several rows are systematically sampled. Males tend to be restless and counts are less reliable than those for females.

Correlation between catches obtained by WBTN, YPWT and visual counting are significant and these methods have been proved useful for estimation of vector density (19).

SURVEYS OF THE PROPORTION OF INFECTIVE INDIVIDUALS IN A LOCAL SBPH POPULATION : POPULATION INFECTIVITY

The probability that a rice plant is infected by RSV depends on the number of infective vectors on a host plant at the susceptible stage. The number of infective vectors is the product of vector density and population infectivity.

Relationship Between Population Infectivity and RSV Occurrence

The infectiveness of SBPH can be estimated precisely, individually and efficiently by the haemagglutination reaction using RSV antibody-sensitized blood cells or latex. A single observer can test more than several hundred vectors per day. From a stochastic viewpoint, more than 300 vectors should be tested per population when the population infectivity is around 5-10%. Live SBPH, preferably older than the 4th nymphal stage, either in or not in diapause, should be used.

In fields transplanted at various times around the period of SBPH immigration, the relationship between the number of infective SBPH (number of SBPH trapped by YPWT during the immigration period x the average population infectivity in the same period) and the percentage of infected plants (3 plants per hill) at Konosu, Saitama, was obtained when RSV occurrence was at epidemic status. The percentage of infected plants rises at a continually decreasing rate (Figure 1).

Category of RSV Prevalence Based on Population Infectivity

The relationship between local population infectivity and the percentage of infected hills in paddies transplanted at ordinary times was surveyed in Fukuoka (western Japan) (20), Hyogo (western-central Japan)(10), and Saitama (central Japan) (20). In localities where the population infectivity was lower than ca 4%, no economically significant RSV infection occurred, whereas RSV infection exceeding 20% of infected hills (about 5% yield loss) tended to occur in localities where the local population infectivity exceeded 7-8% (Figure 2). In central to northern Honshu along the coast of the Sea of Japan, where no economically significant RSV occurrences have been recorded, the population infectivities were lower than 2-3% (R. Kisimoto, unpublished). From these facts the prevalence of RSV can be classified by population infectivity as chronic (population infectivity lower than 4%), epidemic (higher than 7-8%) and transient (between the two). It is recommended that the criteria should be modified according to the economic injury level adopted in the locality, as well as local climatic conditions and rice-growing practices.

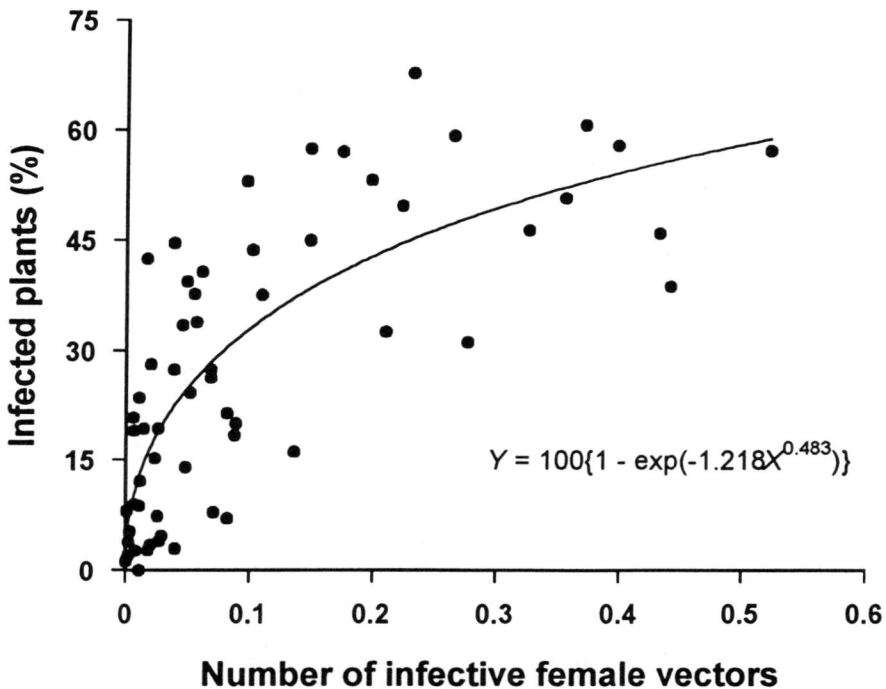

Figure 1. Relationship between the number of infective SBPH (females)collected by a yellow pan water trap set in each plot transplanted at various date from mid-May to late July and the percentage of RSV-infected plants at Konosu (1973-1985).

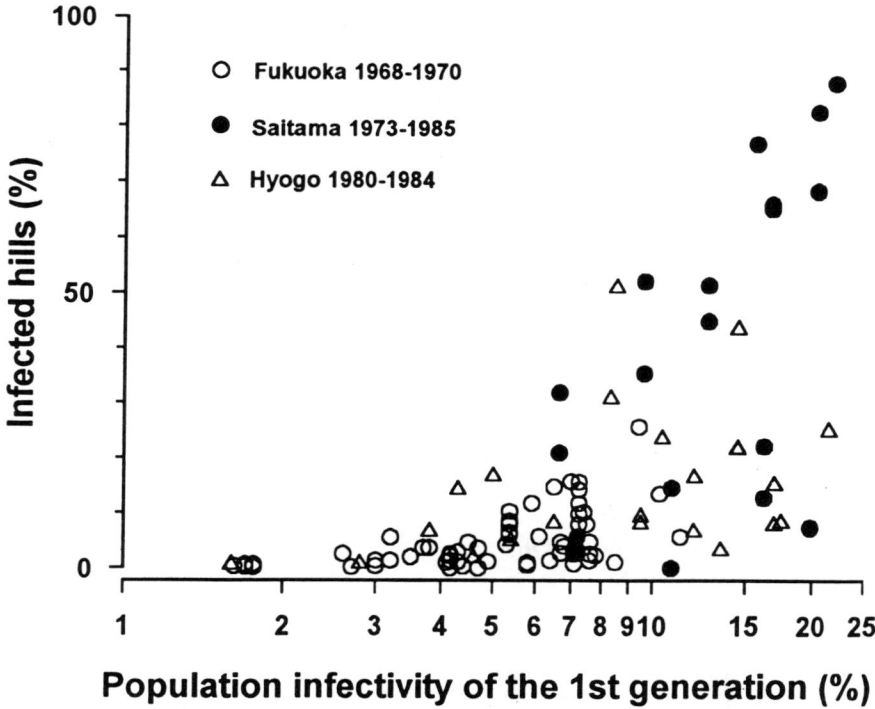

Figure 2. Relationship between local population infectivity and the proportion of infected hills in paddy fields transplanted at the ordinary time.

A MODEL DESCRIBING RSV EPIDEMIC

The population infectivity is controlled by three factors, i.e., a decrease from one generation to the next resulting from incomplete transovarial passage of RSV, an increase resulting from the change of non-infective individuals into infective ones by acquisition of virus from an infected plant, and intermixing of vector populations with different population infectivities. The following formula describing the population infectivity of each vector generation starting from the overwintering generation has been proposed (20).

$$P_n = v P_{n-1} + (1- v P_{n-1}) [1- \exp (- mwH)] \quad (a)$$

For this, the third factor was neglected. where P_n (n = 0, 1,..., 5) is the population infectivity of the nth generation, 0 the overwintering generation and 5 the overwintering generation for the next year; v is the rate of transovarial passage from one generation to the next; H is the proportion of infected plants that is finally achieved in the year; m is the proportion of H which is attained at each generation, being tentatively evaluated as 0.5 at the second generation, 1.0 at the 3rd and 4th generations, and 0 at the 1st and 5th generations; w is the relative efficiency of the acquisitive access. In this equation $v P_{n-1}$ is the portion of P_n due to transovarial passage; $(1- v P_{n-1})$ is the proportion of non-infective vectors at the beginning of the generation, and $[1 - \exp (-mwH)]$ is the probability that a non-infective member will become infective during the generation.

A submodel describing H was proposed as follows,

$$H = 1 - \exp (- aNP_1) \quad (b)$$

in which a is the coefficient of inoculative access of an infective vector and N is the density of vectors. The value of a was estimated at 3.548 using H_s for H (H_s : statistics on the degree of disease occurrence in Saitama for 1973-1980) and N_{tn} (migrant density index) for N [(the total number of vectors caught by a WBTN during the first generation) x (total rice area in Saitama in ha)$^{-1}$ x 10^2], and P_1 (population infectivity of the first generation within the Konosu area). Parameter v in equation (a) was estimated to be 0.9431 by comparing population infectivities of the overwintering and the first generation within the Konosu area in 1973-1982, as no apparent

virus sources in host plants were available during the generations. Parameter w was estimated by the method of least squares for non-linear regression. The estimated value for w, which makes $\sum[P_5 \text{(observed)} - P_5 \text{(estimated)}]^2$ minimum for the P_5 - values of 1973-1981, was 3.109.

As shown in Figure 3, the model well described an RSV epidemic which occurred at Konosu from 1973 to 1982, when the introduction of RSV-resistant cultivars began. The epidemic was initiated by unusual increases of vectors in the first generation in the two successive years of 1977 and 1978. RSV infection, which had been at a rather low level, increased greatly after 1978 followed by a steep increase in the population infectivity from 5-7% to 15-20% in 2 to 3 years due to the severe RSV outbreak. Statistics of the Agricultural Pest Forecasting Program in 1983 showed that 10.6% of the total rice area in Saitama was severely infected by RSV (more than 20% of hills infected) and 34% were damaged to various extents. A rice-growing area with the radius of ca 50 km, covering Saitama, Tochigi, Gunma and Ibaraki, has followed the same process. The population infectivity did not show any sign of decrease even after the vector density became reduced to the former level.

Year to year fluctuation of population infectivity in the first generation (P_1) was traced in relation to the vector density in the first generation (N_{tn}) (Figure 4). Curve CN (critical level of N) shows N_{tn} which induces neither an increase nor decrease in P_1 in the following year for various P_1 levels. Other contour lines show ranges of P_{tn} inducing an increase or decrease in P_1 by a specified population infectivity. From 1973 to 1976, the data points, showing 8-10% of P_1 and about 1.1 of N_{tn}, are scattered around curve CN, which means P_1 may not shift if N_{tn} remains level. After two years P_1 has increased to another equilibrium level of around 20%. This figure suggests that it is necessary to suppress vector density to a level of 0.5 or still lower for years if we expect to lower P_1 from 20% to less than 10%. In Figure 4, other CN curves (dotted), expected when RSV-resistant cultivars (almost immune: see below) are introduced at certain proportions, are also presented.

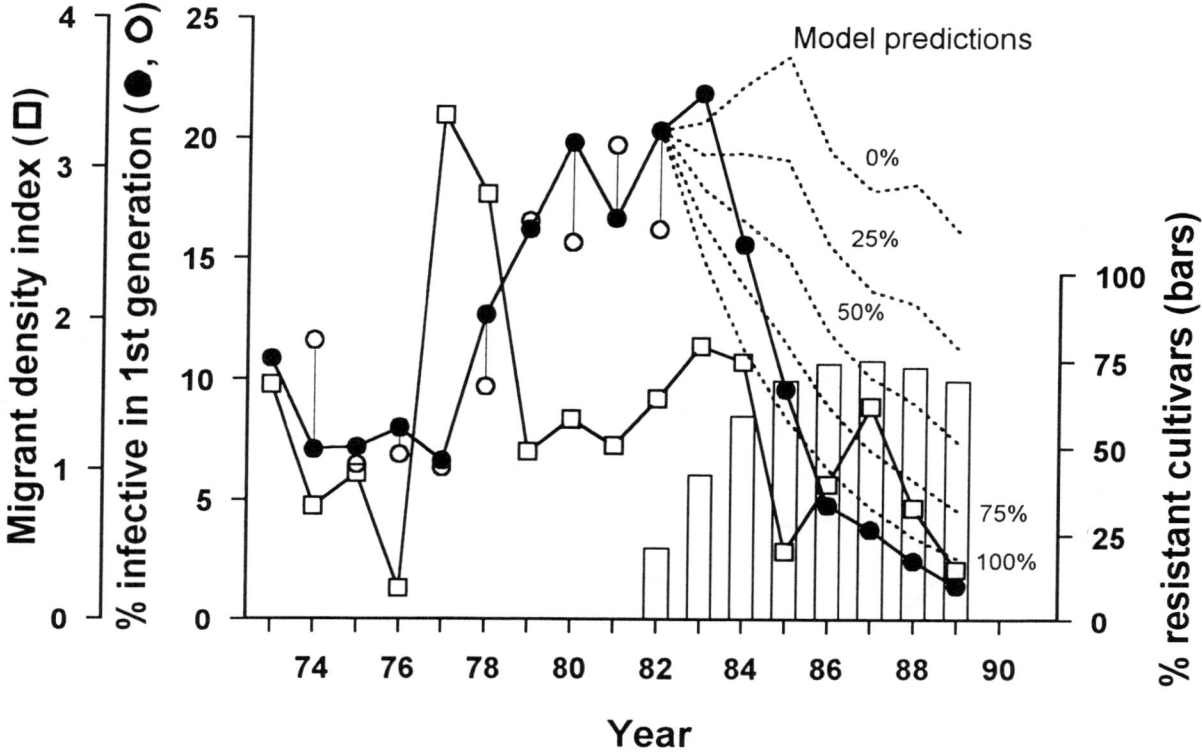

Figure 3. Annual trend of migrant density index (clear squares) and populationinfectivity of the first generation of vectors surveyed at Konosu, and the proportion of acreage of RSV-resistant cultivars introduced since 1982 in Saitama. Population infectivities represented by solid circles are the field surveyed, and represented by clear circles are estimated from the model. Dotted lines show destiny of the population infectivity predicted by the model, assuming that resistant cultivars are grown at each specified proportion.

STRATEGIES FOR CONTROLLING RSV

Insecticidal Control of Vectors

The most instantaneous effect would be achieved if vectors could be suppressed by insecticides to an insignificant level during the first few weeks of rice growing, when rice plants are susceptible to RSV. Many trials aimed at controlling RSV by dusting, spraying or soil application of granules of conventional organophosphorus and carbamate insecticides effective against hemipterans have been attempted, but the effects were not always sufficiently successful (12). Granule application to seedling boxes just before transplantation was tested in the expectation of long-lasting effects in suppressing SBPH, but RSV occurrences were not always suppressed to a sufficient degree (Figure 5, clear circles). In Gunma, a regional cooperative program for control of RSV was initiated. In a rice area of 2,000 ha, 3-4 aerial applications of conventional insecticides from April through July in addition to application of granules to individual paddies to suppress SBPH, and 4-5 regional cooperative applications against all rice insects during the rice season, were conducted from 1976 to 1982, but RSV occurrence was not significantly suppressed (11).

A recently developed granular insecticide containing imidacloprid as an active ingredient showed a long-lasting effect, suppressing SBPH and other rice insects, when it was applied to seedling boxes.

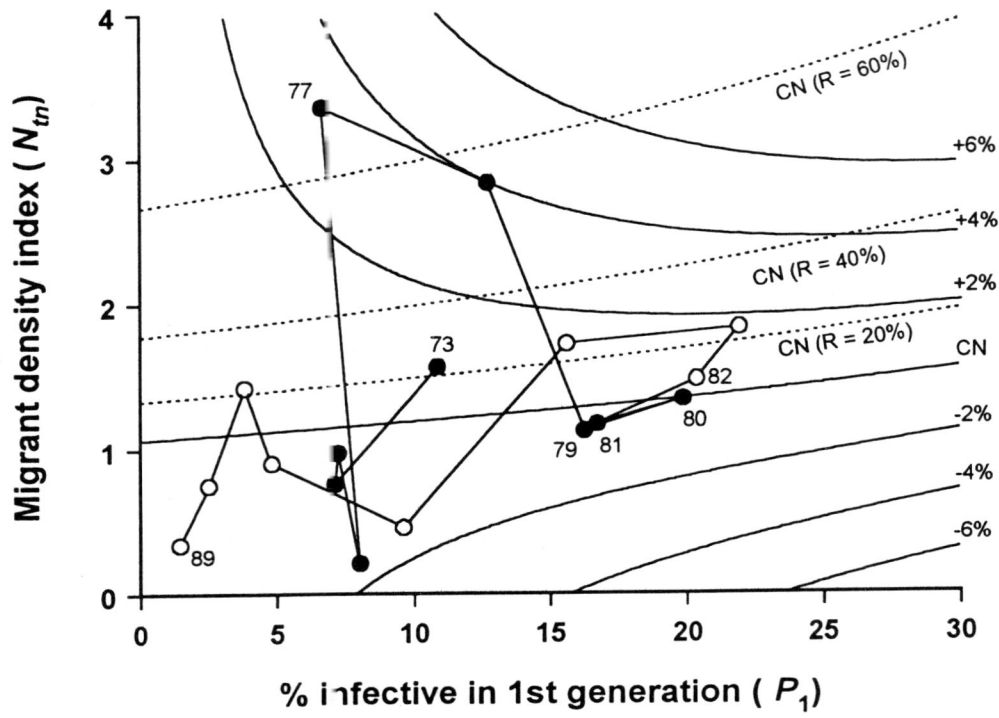

Figure 4. Yearly fluctuation in the population infectivity in the first generation along the migrant density index surveyed at Konosu since 1973. Solid circles represent the values before introduction of resistant cultivars and clear circles those after introduction (cf. Figure 3). Curve CN means the level of the migrant density index inducing neither an increase nor decrease in the population infectivity in the following year, and contour lines show those which induce an increase or decrease by each specified population infectivity. Dotted lines represent CN curves estimated from the model, assuming that resistant cultivars are grown at each specified proportion.

Its RSV-controlling effect was tested cooperatively at various localities (Figure 5, solid triangles) (13). In certain cases RSV occurrences were apparently suppressed to insignificant levels. Further field tests are needed to clarify the effects when RSV is at epidemic status.

RSV-resistant Cultivars, Genetics and Breeding

All the Japanese paddy rice cultivars tested by the seedling test method were proved to be susceptible, whereas most Japanese upland rice cultivars are highly resistant or almost immune, with few exceptions (43). In cultivars collected outside Japan, most *indica* rice cultivars were resistant, but *japonica* rice cultivars were susceptible (29,30,41). RSV resistance of Japanese upland rice is controlled by two pairs of complementary dominant genes, St_1 and St_2 (43), while that of *indica* rice is controlled by a single incomplete dominant gene, St_3 (44), the symbol of which was changed to St_2^i, since St_3 was proved to be an allelomorph of St_2. The different level of resistance in *indica* cultivars may be due to various expressions of multiple alleles of St_2^i. Cultivars of the *indica* type were considered to be more beneficial than Japanese upland cultivars during the breeding of RSV-resistant cultivars once the resistance gene was implicated in Japanese paddy cultivars.

Breeding of RSV-resistant cultivars was started in 1962, when an epidemic of RSV

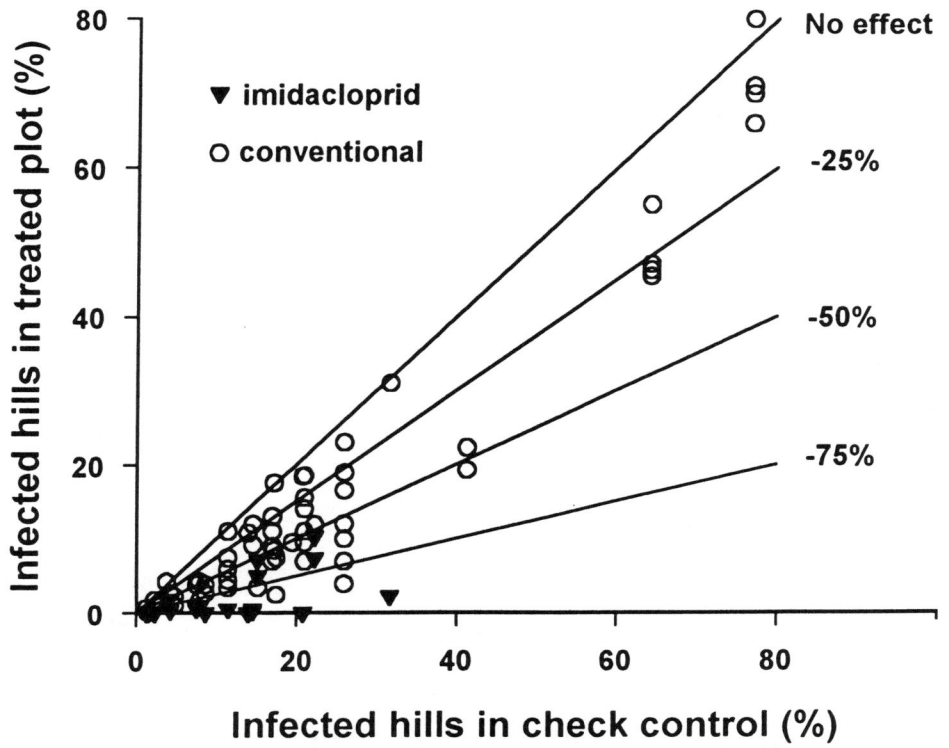

Figure 5. Effect of granular insecticide application for suppression of RSVoccurrence. Insecticides were applied to seedling boxes just before transplanting. Clear circles correspond to applications ofconventional insecticides and solid triangles correspond to granules containing imidacloprid (after 12, 13).

occurred in western Japan. A highly resistant parental line St No.1, was found among lines bred through 5 backcrosses (B_5F_7) between Norin 8, a Japanese paddy cultivar, as a recurrent parent, and Modan, a Pakistani *indica* cultivar, as a donor (29). From the same crossing, a related resistant line, Chugoku 31, was found in the B_5F_8 generation (36). The original crossing was made in 1951, aiming at resistance to rice blast, and in the last 1 to 2 generations selection was made on the basis of RSV resistance.

Mineyutaka (37), the first RSV-resistant cultivar, was selected from the cross between St No.1 and Sachikaze and released in 1972, 20 years after the original crossing. Musashikogane (25), the second RSV-resistant cultivar was bred in the lineage of St No.1 and released in 1980. It is now widely grown in the Kanto area. From a breeding

line of KC 89 (*indica/japonica* hybrid, *indica* type), which came from crossings including the *indica* cultivars Mudgo and IR 8, and Japanese paddy cultivars, a new cultivar, Akenohoshi, was released in 1984 (35). The crossings were aimed primarily at super-high yielding ability. Other RSV- resistant cultivars bred thereafter were mostly derived from St No.1, i.e., Himeminori (1981), Hoshinohikari (1982), Aoisora (1983), Tamaminori (1984), Tsukinohikari (1985), Asanohikari (1987), Aoinokaze (1989), Heiseimochi (1989) and Akanezora (1991). Those from Chugoku 31 were Nadanishiki (1981) and Tamahonami (1985), and one from KC 89 was Hoshiyutaka (1987).

When heavily inoculated at the 2-3-leaf stage, St No.1, Chugoku 31 and other resistant cultivars became infected at a considerable

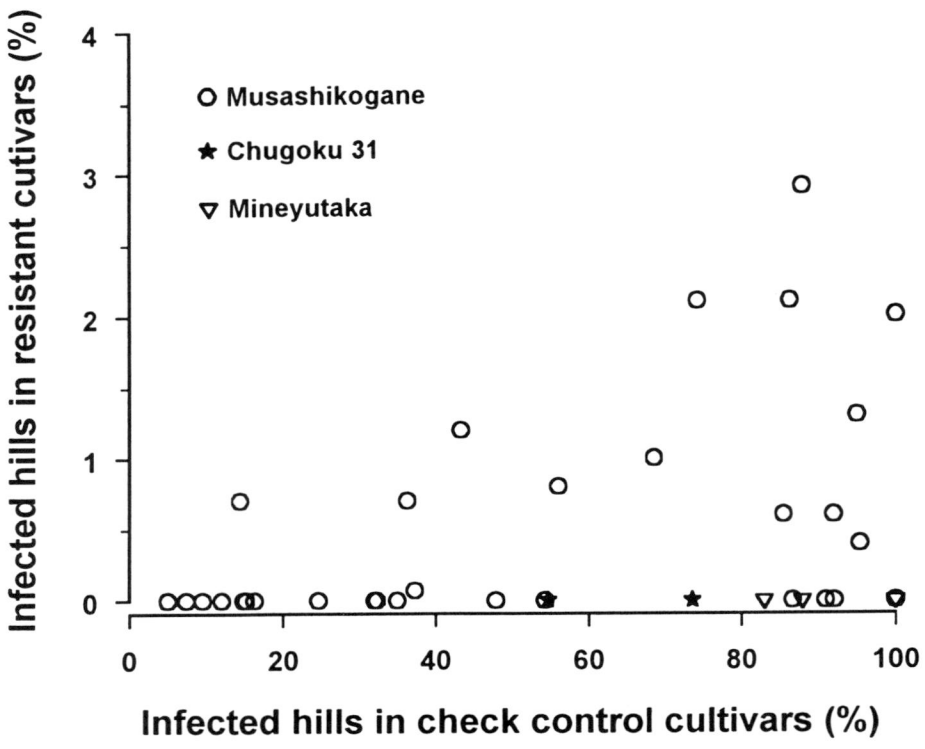

Figure 6. Percentage infected hills of resistant cultivars when susceptible check control cultivars were infected severely in the field. Check control susceptible cultivars are Nipponbare for Musashikogane, Chuseishinsenbon for Mineyutaka, and Norin 29 for Chugoku 31 (after 26, 36, 37).

frequency, but disease symptoms were much less severe, sometimes masked as the host plant grew, than those on susceptible cultivars (36,37). In the field, however, RSV infections on the resistant cultivars were suppressed to insignificant levels even when RSV was at epidemic status (Figure 6). A few hills of Musashikogane showed RSV symptoms when inoculated at the 5-leaf stage in the field, but 73% of the infected hills survived and produced normal heads. The dilution end-point of RSV in the infected plant estimated using RSV antiserum was at a level comparable to that of the susceptible Nipponbare, but it decreased as the plants grew (26).

Effect of Introducing Resistant Cultivars

Since 1982, the RSV control strategy has changed in the Kanto area, and the acreage of resistant cultivars such as Musashikogane, Tsukinohikari and Aoisora, has greatly increased (Figure 3). The proportion of infective vectors in the area began to decrease in 1984 when the resistant cultivars increased to 60 % or more of the total rice area, and was reduced to a level of 2-3% in 3-4 years, lower than the level before the epidemic began. Thus, the RSV epidemic was over.

In Figure 3, the P_1 after 1983 is predicted assuming that resistant cultivars are grown at a constant proportion for successive years. In this case, the values of N_{tn} and the following equation was used,

$$H = (1 - R) [(1 - \exp(-aNP_1))] \qquad (c)$$

in which **R** is the proportion of resistant cultivars. The observed values of P_1 decreased at a higher rate than predicted. A similar massive decrease of population infectivity was experienced also at Zentsuji and Chikugo (western Japan), when RSV epidemic was at the final stage (20). Analyses of equations (a) and (c) enabled the expected *CN* curves to be drawn in Figure 4 when resistant cultivars were grown at a certain proportion of the total rice area. The *CN* curves indicate that when resistant cultivars are grown in 40% or more of the rice area the population infectivity will decrease in most years, but that at 20% the decrease is insufficient.

It has been pointed out with some concern that SBPH has increased on Musashikogane to a level 1.5-5 times as high as that on Nipponbare, a popular susceptible cultivar (24). However, as discussed already in Figure 3, the population infectivity was soon suppressed to an insignificant level after introduction of RSV-resistant cultivars, due not to suppression of vector density but to deprivation of virus sources in plants from the same area. On Musashikogane the green leafhopper, *Nephotettix cincticeps* Uhler, a vector of rice dwarf virus, increased 1.5-3 times (24), and the white-backed planthopper, *Sogatella furcifera* Horváth, has, in certain cases, increased to an unprecedented level and caused hopperburn or heavy sooty molds (6). Rice diseases, such as rice blast, rice bacterial leaf blight and rice sheath blight, also increased on Musashikogane (5). Countermeasures against these pest insects and diseases are needed.

It may be a problem that most of the released RSV-resistant cultivars originate from a single or closely related gene source, and that once RSV variants virulent to the resistant cultivars occur, all the resistant cultivars will be exposed to a crisis of simultaneous breakdown. Breeding of resistant cultivars having different resistance genes is therefore needed. Fortunately, no such virulent RSV variants have occurred so far even in those areas where more than 80% of the total rice fields are covered by resistant cultivars.

Transgenic Cultivars Resistant to RSV

Coat protein (CP)-mediated resistance to virus infection has been developed in various monocot crops (2). Production of transgenic rice plant has become possible by electroporation of plasmids into rice protoplasts (32,33). A plasmid pLAN150 was constructed as follows (7);

| CaMV 35S pro | int | RSV-CP | nos ter |

in which CaMV 35S pro means the 35S promoter of cauliflower mosaic virus, int: the first intron of the caster bean catalase gene, RSV-CP: a cDNA of RSV RNA3, nos ter: a nopaline synthase polyadenylation signal. The plasmid was digested at the *Hind* III site, which is located immediately upstream from the CaMV 35S promoter, resulting in a linear vector. This plasmid was cotransformed with the hygromycin B resistance plasmid as a selection marker. Protoplasts prepared from the embryo of the mature seed of two common rice cultivars highly susceptible to RSV, Nipponbare and Kinuhikari, were used. Transformed protoplasts were selected in medium containing hygromycin B (30 g/ml). Integration of the CP gene was detected in 139 out of 238 hygromycin-resistant calli.

RSV resistance of the primary transformants was tested at the 4th leaf stage (20 cm high). RSV was inoculated using virulent SBPH for 2-3 days. In 3 clones expressing CP, 4 individual plants for each, no RSV symptoms appeared, whereas all the plants of 2 clones which did not express CP, 3 plants each, presented the symptoms. Among progeny of 2 CP-expressing lines, 21% and 38% of tested individual plants presented symptoms, whereas 94% in a CP non-expressing line and 82% in a non-transgenic line did so. Accumulation of SP protein, which reflects RSV multiplication, was tested in the 11 progeny plants. Eight weeks after inoculative access, a nontransformed rice plant and a progeny plant of a transgenic plant which lacked the CP gene showed SP accumulation, whereas 9 out of 10 plants expressing the CP and presenting no symptoms, showed no SP accumulation, except for 1 plant showing a trace amount of SP (7). These facts show that the CP gene was introduced into two common rice cultivars, inherited by the next generation, and expressed CP-mediated RSV resistance under laboratory conditions. Growth tests for safety in closed and open fields had been carried out by 1994. Field tests for RSV resistance under severe RSV outbreaks are needed before the lines can be extended.

CONCLUSIONS

The process of RSV epidemic was traced through surveys of vector density during the period of immigration and of the proportion of infective vectors in a local population. It was shown that once RSV occurrence had attained a high level induced by an unusual increase of the vector, then population infectivity also increased. The increased population infectivity was not reduced to the former level in a short period even when the vector density was lowered to the former level, and RSV occurrence lasted for years. In this situation vector control using conventional insecticides was mostly unsuccessful in suppressing the occurrence of RSV. Up to now, only the introduction of RSV-resistant cultivars accounting for over 40% of the total rice area has been successful. However, several worrying problems remains to be solved, i.e., an increases of the vector density in addition to increases of other diseases and insect pests. Grain quality competition between cultivars resistant and susceptible but of higher quality is also a problem. Breeding of RSV-resistant cultivars having genes different from those available at present will also be necessary. Transgenic rice cultivars may provide new approaches to solving these problems efficiently and simultaneously.

ACKNOWLEDGMENTS

We acknowledge support from the Ministry of Education, Science and Culture, Japan (Grant-in-Aid for Scientific Research No. 62760045).

REFERENCES

1. Amano, E. 1933. On the rice stripe disease. Byochugai Zasshi (J. Plant Prot. Japan) 20: 634-638.
2. Beachy, R. N., Loesch-Fries, S., and Tumer, N. E. 1990. Coat protein-mediated resistance against virus infection. Annu. Rev. Phytopathol. 28:451-474.
3. Chen, C. C., and Ko, W. F. 1986. Studies on the time of rice stripe virus infection and field experiments on disease control. Res. Bull. Taichung District Agric. Improv. Sta. (Taiwan) 12:51-59.
4. Fuji, S., and Kojima, M. 1989. Ecological aspect of rice stripe disease in Niigata Prefecture. Proc. Assoc. Plant Prot. Hokuriku 37:12-14.
5. Fujita, Y., and Noda, S. 1986. Rice diseases in a rice stripe disease resistant cultivar. Proc. Kanto-Tosan Plant Prot. Soc. 33:38-39.
6. Hara, E., and Saito, M. 1984. A relationship between rice varieties and population density of the white-backed planthopper, *Sogatella furcifera* Horváth. Proc. Kanto-Tosan Plant Prot. Soc. 31:109.
7. Hayakawa, T., Zhu, Y., Itoh, K., Kimura, Y., Izawa, T., Shimamoto, K., and Toriyama, S. 1992. Genetically engineered rice resistant to ricestripe virus, an insect-transmitted virus. Proc. Natl. Acad. Sci. USA. 89:9865-9869.
8. Hayano, Y., Kakutani, T., Hayashi, T., and Minobe,Y.1990. Coding strategy of rice stripe virus: major nonstructural protein is encoded in viral RNA segment 4 and coat protein in RNA complementary to segment 3. Virology 177:372-374.
9. Hsieh, C. Y. 1973. Transmission of rice stripe virus by *Laodelphax striatellus* Falln in Taiwan. Plant Prot. Bull. (Taiwan) 15:153-162.
10. Hyogo Agricultural Experiment Station. 1982, 1983, 1984, 1985. Annual Report on Insect Pest Control for 1981, 1982, 1983, 1984. 46 Pages, 45 Pages, 30 Pages, 37 Pages.
11. Iizuka, K., Ohzeki, K., Tamura, M., Nishida, N., Kakinuma, K., and Taguchi, T. 1987. Control of rice stripe virus disease by insecticides and resistant cultivars. Nogyo oyobi Engei (Agriculture and Horticulture) (Japan) 62:740-746.
12. Japan Plant Protection Association. 1977-1985.Reports on Insecticidal Control of Rice Insects. 572 Pages (1977), 665 Pages (1978), 685 Pages (1979), 509 Pages (1980), 482 Pages (1981), 453 Pages (1982), 489 Pages (1983), 530 Pages (1984), 554 Pages (1985).
13. Japan Plant Protection Association. 1992. Special Report of the Cooperative Experiments on the Seedling Box Application of Admire Granule. 156 Pages.
14. Kiritani, K. 1983. Changes in cropping practices and the incidence of hopper-borne diseases of rice in Japan. Pages 239-247 in: Plant Virus Epidemiology. R. T. Plumb and J. M. Thresh, eds. Blackwell Scientific Publications, Oxford.
15. Kisimoto, R. 1967. Genetic variation in the ability of a planthopper vector; *Laodelphax striatellus* (Falln) to acquire the rice stripe virus. Virology 32:144-152.
16. Kisimoto, R. 1969. Ecology of insect vectors, forecasting, and chemical control. Pages 243-255 in: The Virus Diseases of the Rice Plant. The International Rice Research Institute, ed. The Johns Hopkins Press, Baltimore, MD.
17. Kisimoto, R. 1981. Development, behaviour, population dynamics and control of the brown planthopper, *Nilaparvata lugens* Stl. Rev. Plant Prot. Res. (Japan) 14:26-58.
18. Kisimoto, R. 1986. Transovarial infectivity of three variants of rice stripe virus transmitted by the small brown planthopper, *Laodelphax striatellus* Fallén. Pages 115-132 in: Proc. Intern. Symp. Transmission ofPlant and Animal Viruses by Vectors. J. Hidaka and N. Sako, eds., Fukuoka, Japan.
19. Kisimoto, R. 1991. Long-distance migration of rice insects. Pages 167-195 in: Rice Insects: Management Strategies. E. A. Heinrichs and T. A.Miller, eds. Springer-Verlag, New York.
20. Kisimoto, R., and Yamada, Y. 1986. A planthopper-rice virus epidemiology model: rice stripe and small brown planthopper, *Laodelphax striatellus* Fallén. Pages 327-344 in: Plant Virus Epidemics, Monitoring, Modelling and Predicting Outbreaks. G. D. McLean, R. G. Garrett and W. G. Ruesink, eds. Academic Press Australia, North Ryde, N.S.W.
21. Kiso, H., and Yamamoto, T. 1973. Infection and symptom development in rice stripe diseases, with

special reference to disease specific protein other than virus. Rev. Plant Protect. Res. (Japan) 6:75-100.

22. Kuribayashi, K. 1931. Studies on the rice stripe disease. Bull. Nagano Agric. Expt. Stn. 2:45-69.

23. Matsui, T., Ueda, Y., and Ino, M. 1983. Immigration of the second generation adults of the small brown planthopper into late transplanted rice fields. Proc. Kanto-Tosan Plant Prot. Soc. 30:107-108.

24. Murakami, M., and Kanda, T. 1986. Occurrence of insect pests in rice stripe disease resistant cultivar. Proc. Kanto-Tosan Plant Prot. Soc. 33:186-187.

25. Niwayama, T., and Shiobara, H. 1981. A new rice cultivar 'Musashikogane' bred in Saitama Prefecture. Nogyo Gijutsu (Agriculture Technology) (Japan) 36:465-468.

26. Noda, S., Kanda, T., and Murakami, M. 1986. Occurrence of rice stripe disease and concentration of rice stripe virus in rice stripe diseaseresistant cultivar. Proc. Kanto-Tosan Plant Prot. Soc. 33:36-37.

27. Okuyama, S., and Kajino, Y. 1980. Studies on the control of the small brown planthopper transmitting rice stripe disease, I. Disease occurrence and the rate of infective vectors. Hokuno (Northern Agriculture) (Japan) 47(7):10-22.

28. Ruan, Y. L., Chiang, W. L., and Lin, R. F. 1981. Studies on the rice virus vector, small brown planthopper Laodelphax striatellus Fallén. Acta Entomol. Sinica 24:283-290.

29. Sakurai, Y., and Ezuka, A. 1964. The seedling test method of varietal resistance of rice plant to stripe virus disease. 2. The resistance of various varieties and strains of rice plant by the method of seedling test. Bull. Chugoku Natl. Agric. Expt. Stn. A 10:51-70.

30. Sakurai, Y., Ezuka, A., and Okamoto, H. 1963. The seedling test methods of varietal resistance of rice plants to stripe virus disease (part 1). Bull. Chugoku Natl. Agric. Expt. Stn. A 9:113-125.

31. Sakurai, Y., Ezuka, A., Yunoki, T., and Morinaka, T. 1965. The seedling test method of varietal resistance of rice plant to stripe virus disease. 3. Study on virus-free planthoppers found in progenies of viruliferous ones. Bull. Chugoku Natl. Agric. Expt. Stn. A 11:145-154.

32. Shimamoto, K. 1991. Transgenic rice plants. Pages 1-15 in: Plant Gene Research, Molecular Approaches to Crop Improvement. E. S. Dennis and D. J. Llewellyn, eds. Springer-Verlag, Wien.

33. Shimamoto, K., Terada, R., Izawa, T., and Fujimoto, H. 1989. Fertile transgenic rice plants regenerated from transformed protoplasts. Nature 338:274-276.

34. Shinkai, A. 1962. Studies on insect transmission of rice virus diseases in Japan. Bull. Natl. Instit. Agric. Sci. Series C 14:1-112.

35. Shinoda, H., Toriyama, K., Fujii, K., Shibata, M., Yamamoto, T., Sekizawa, K., Ogawa, T., Okamoto, M., and Yamada, T. 1989. A high-yielding rice variety 'Akenohoshi'. Bull. Chugoku Natl. Agric. Expt. Stn. 4:13-27.

36. Toriyama, K., Sakurai, Y., Washio, O., and Ezuka, 1966. A newly bred rice line, Chugoku No. 31, with stripe disease resistance transferred from an Indica variety. Bull. Chugoku Natl. Agric. Expt. Stn. A 13:41-51.

37. Toriyama, K., Washio, O., Sakurai, Y., Ezuka, A., Shinoda, H., Sakamoto, S., Yamamoto, T., Morinaka, T., and Sekizawa, K. 1972. 'Mineyutaka', the first rice variety possessing stripe disease resistance. Bull. Chugoku Natl. Agric. Expt. Stn. A 21:1-19.

38. Toriyama, S. 1986. Rice stripe virus: prototype of a new group of viruses that replicate in plants and insects. Microbiolog. Sci. 3:347-351.

39. Toriyama, S. 1986. Viruses and ambisense RNA genomes of tenuivirus. Protein, Nucleic Acid and Enzyme 37:2467-2473.

40. Tsurumachi, M., and Yasuda, K. 1989. Long-distance immigration and infestation of the rice planthoppers and the rice leaffolder in Okinawa. Res. Rep. Tropical Agric. (Tsukuba) 66:50-54.

41. Washio, O., Ezuka, A., Sakurai, Y., and Toriyama, K. 1967. Studies on the breeding of rice varieties resistant to stripe disease. I. Varietal difference in resistance to stripe disease. Japan. J. Breeding 17:91-98.

42. Washio, O., Ezuka, A., Toriyama, K., and Sakurai, Y. 1968. Testing method for, genetics of and breeding for resistance to rice stripe disease. Bull. Chugoku Natl. Agric. Expt. Stn. A 16:39-197.

43. Washio, O., Toriyama, K., Ezuka, A., and Sakurai, Y. 1968. Studies on the breeding of rice varieties resistant to stripe disease. II. Genetic study on resistance to stripe disease in Japanese upland rice. Japan. J. Breeding 18:96-101.

44. Washio, O., Toriyama, K., Ezuka, A., and Sakurai, Y. 1968. Studies on the breeding of rice varieties resistant to stripe disease. III. Genetic studies on resistance to stripe in foreign varieties. Japan. J. Breeding 18:167-172.

45. Wilson, M. R., and M. F. Claridge. 1991. Handbook for the identification of leafhoppers and planthoppers of rice. C.A.B. International, UK. 142 pp.

46. Yamada, W., and Yamamoto, H. 1955. Studies on the rice stripe disease of rice plant. I. On the virus transmission by an insect, Delphacodes striatella Fallén. Spec. Bull. Okayama Pref. Agric. Expt. Stn. 52:93-112.

47. Yamada, W., Shiomi, T., and Yamamoto, H. 1955. Studies on the stripe disease of rice plant. II. On the disease occurrence in Okayama Prefecture and its control. Spec. Bull. Okayama Pref. Agric. Expt. Stn. 52:113-124.

48. Yasuo, S., Ishii, M., and Yamaguchi, T. 1965. Studies on rice stripedisease. I. Epidemiological and ecological studies on rice stripe disease in Kanto-Tosan district in central part of Japan. J. Centr. Agric. Expt. Stn. 8:17-108.

CHAPTER 36

Epidemiology and Control of Maize Streak Disease

V.D. Damsteegt and E.C.K. Igwegbe

Maize streak disease is one of the most important corn diseases in sub-Saharan Africa and neighboring countries. The disease was first recorded in South Africa as "mealie variegation" in 1901 (31). It is now found in most corn producing areas of Africa (22,26) and is considered one of the most destructive virus diseases of gramineous crop plants in the world (40). Maize streak virus (MSV) is endemic in Africa and southeast Asia (88) on corn, a congenial host (26).

Yield losses caused by MSV can be controlled in part by agronomic practices although more practical and stable control would be by use of tolerant or resistant varieties. Several breeding programs have made important strides in developing high yielding, well-adapted, MSV resistant lines for different ecological zones. These national and international breeding programs will be discussed later.

During the First Eastern, Central, and Southern Africa Regional Maize Workshop in 1985, 11 of 19 African countries considered MSV a significant constraint on corn production (22, 32). Maize streak virus also is economically important in other crops such as sugarcane in South Africa (105), Egypt (1), India (74); finger and pearl millet in India (12,13); barley, oats, rye, and wheat in South Africa (6,105), and rice in Nigeria (68).

HISTORICAL IMPORTANCE OF CORN AND MAIZE STREAK VIRUS IN AFRICA

Corn (*Zea mays* L.) is a widely grown gramineous crop in the tropics. Corn was introduced into West Africa in the 16th century by the Portuguese who brought it from the West Indies and Central and South America (26). In many countries corn has become a major cereal crop and an important component of animal and human diets. Despite the significance of corn as a staple food crop, low yields across Africa, averaging only 1 ton/ha, are caused by diseases, insect pests, and low-input subsistence farming. As is true in every culture, farmers fail to commit resources to crops with high probabilities of failure (22).

Maize streak virus disease has played a major role in corn production in Africa although the importance of MSV on corn varies from place to place. In the past 25 years, the increasing popularity of corn in Africa has transformed it from a garden or backyard crop to a major field crop in much of west Africa (22). This temporal and spatial buildup of the crop has lead to greater ecological opportunities for the virus and its leafhopper vectors. Increased prevalence of corn production is directly correlated with the increases in MSV disease (81). Additionally, the increase in irrigation practices throughout much of eastern and southern Africa has increased cropping of related grain crops which facillitates vector buildup and widens the ecological window for epidemics of MSV disease (22).

The disease occurs in the forest and savannah zones, from sea level to 1800 meters (4). Disease incidence and destructiveness vary from year to year and from season to season (84). The magnitude of yield loss due to MSV infection is dependent on weather, vector population densities, percent carryover inoculum, and growth stage of crops at the time of infection (4). Severe outbreaks are often associated with late plantings or second season cropping (4,22). Yield reductions of 100% have occurred during severe epidemics (22).

SYMPTOMATOLOGY

Symptoms of MSV infection in susceptible plants begin as interveinal, circular to oval, nearly white spots 0.5 to 2.0 mm in diameter, near the base of the newest expanding leaf. The spots may be scattered over the entire leaf or confined to a few adjacent interveinal areas. As the leaves expand, spots become more numerous and elongate until long streaks become apparent (5,18,22,38). Highly susceptible corn genotypes may exhibit widespread, almost complete chlorotic streaking of all new leaves following infection, often associated with morphological teratology, including leaf margin splitting, leaf tip twisting, necrosis of emerging leaf, and reduced leaf size, tassel sterility, and shoot stunting (5,18,25). Streaks are broad, pale, cream-colored to light green against a darker green background with a glassine appearance in the first affected leaves (95). Tolerant corn lines exhibit more scattered, discontinuous streaks with less plant stunting and discoloration. Symptoms of MSV in wheat, oats, barley, rye, rice, and wild grass hosts are similar to those found in corn (19).

CAUSAL VIRUS(ES)

The causal agent of maize streak was first visualized in nuclei of infected leaves by Sylvester et al. (106) and Plasvic and Maramorosch (73). Whitcomb and Davis (115) postulated from indirect evidence that it was probably a small RNA virus but later evidence (40) demonstrated that MSV was a DNA virus with a circular, single-stranded genome. The virions are present as singlet (20 nm), but, more commonly doublet (20 X 30 nm) particles from which Harrison (40) proposed the name geminivirus.

MSV is a monopartite geminivirus containing a single 2.7 - 2.9 kb circular, single-stranded DNA (ssDNA) genomic component (53,63) making it the smallest known plant virus. The single genomic component resembles the A component of bipartite geminiviruses (53). Three separate MSV genomic sequences have been determined (42,54,63).

The genome organization of MSV is similar to that of other geminiviruses (53,93) with major open reading frames (ORFs) diverging from a large intergenic region (LIR) and a bidirectional pattern of transcription (28,61, Figure 1). Of seven potential ORF's only four, two (V1 and V2) in the virion (+) sense, and two (C1 and C2) in the complementary (-) sense, are thought to be expressed. The remaining three ORFs are not conserved among the monopartite geminiviruses (53,54,62). The ducts encoded by V1 and V2 have been detected: V1 product is a 10.9 kDa protein possibly involved in disease development (streak width) (8,88) and the V2 product is the 26.8 kDa capsid protein involved in systemic spread (55,56,75). V1 and V2 gene products do not affect replication.

Autonomous replication is controlled by sequences located in the two intergenic regions (large intergenic region [LIR] and small intergenic region [SIR]) and in the two overlapping complementary ORFs (56). Mutants containing insertions or deletions in the coat protein gene (V2) produced dsDNA although virion ssDNA was not detected (9). Symptom development has been shown to be linked to a potential hairpin structure in the LIR with a conserved sequence TAATATTAC in the loop (88). Differences in host range and symptom development between a severe Nigerian isolate (MSV-Ns) and a mild isolate (MSV-Nm) were linked to three nucleotide changes between the isolates (8) which did not affect amino acid sequences. One change at nucleotide 40 (nt 40) in the V1 gene affected streak width while severity of chlorosis, streak length, latency, and host range were related to a base change at nucleotide 2473 (nt 2473) in the LIR.

MSV isolated from many different hosts have been considered as forms, isolates, strains, or viruses (21,60,58,72,73,105). Although some cross-protection has been demonstrated (105), the various isolates have been referred to as separate, closely related viruses which have been host adapted. Markham et al.(57) have shown that the vector species plays a major role in host range determination. Although different isolates or strains could be efficiently transmitted to different plant hosts by species of Cicadulina when the insects were injected with virus (57), different vector species commonly favored specific host species in nature.

At least 4 strains of MSV recovered from Panicum maximum, Digitaria setigera, Saccharum officinarum, and Zea mays have been identified by polyclonal and monoclonal (MAbs) antibodies (21,70,71,72). Two of the host-adapted strains have now been recognized as separate viruses: sugarcane streak virus (SSV) (43,44), and Panicum streak virus (PSV) (11). At least 15 MAbs have been raised against a Nigerian corn isolate; one internal epitope was recognized by all

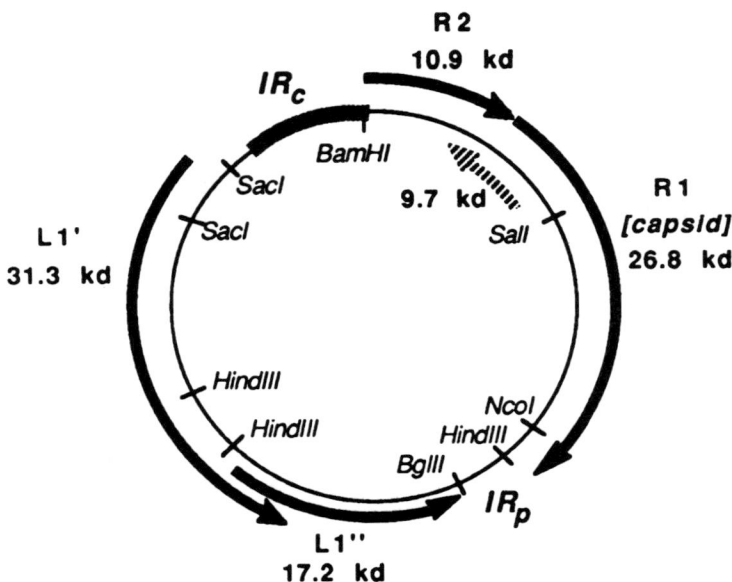

Figure 1. Genomic organization of MSV. Diagrammatic representation of the single genomic component of MSV-S based on sequence analysis of infectious, cloned viral DNA (54). Arrows indicate limits and directions of potential ORFs encoding predicted proteins as shown (kd). '+' strand (R1, R2) and '-' strand (L1', L1") ORFs have been named according to the convention adopted for the bipartite geminiviruses and to reflect homologies with corresponding ORFs in the A component of the bipartite viruses, where these exist. Identification of the capsid gene (R1) is by homology to a Nigerian isolate of MSV (61). (Reprinted from Lazarowitz *et al.* (56).

MAbs. Sap extracts were more reliable than virus preparations as antigen sources (21,71).

Although only 45% nucleotide sequence similarity is shared among grass infecting monopartite geminiviruses, the sequenced isolates of MSV (MSV-K, MSV-N, and MSV-S) are 98% similar (10). The nucleotide sequences of the small intergenic region (SIR) and the coat protein gene of 12 MSV isolates from different geographic locations showed divergence of only 10.9% at the nucleotide level and 2.0% at the amino acid level (10).

The polymerase chain reaction (PCR) technique (87) has been developed for typing strains, mapping genomes, phylogenetic analyses and taxonomy (10, 86). During the past 15 years MSV has been used extensively to provide rapid and sensitive markers for the transfer of DNA to cereals (7).

TRANSMISSION

All MSV strains are transmitted by species in the genus *Cicadulina*, Jassidae leafhopper species of the grasslands in the tropics (67). Storey's early studies with maize streak transmission and the relationship of this virus with its leafhopper vectors are classics in plant virology (94 - 103). The principal vector of all strains of MSV is *Cicadulina mbila* (Naude), although 8 additional *Cicadulina* spp. have been implicated as vectors. The eight species are: *C. storeyi* China, *C. latens* Fennah, *C. parazeae* Ghauri, *C. bipunctata* (Melichar),(= *C. bipunctella zeae* China; Heller & Linnavuori, 1968; Vilbaste, 1976), *C. arachidis* China, *C. similis* China, *C. ghauri* and *C. anestae* Van Rensburg (5,15,16,27,45,67,85,92,108).

Most of the detailed information on vector-virus relationships comes from Storey's classic studies of MSV transmission by *C. mbila*

(99,102,103). Acquisition access periods are generally less than an hour and can be as short as five minutes. Latent periods can be as little as 6 hours although 12-24 hours are more commonly reported. Following the latent period, transmission can be accomplished in 5 minutes although greater transmission rates occur when the inoculation access period (IAP) is increased up to 24-48 hours. All life stages of *C. mbila* can acquire and subsequentially transmit MSV intermittently for the life of the individual insects; females (86%) are better vectors than males (26%) (20,95,96,110). Individual insects can transmit following moults; no evidence exists for transovarial passage. Temperature affects virus acquisition and transmission (20) with 30°C optimum. Ability to transmit is controlled by a single, dominant, sex-linked gene (97, 98). Genetically "active" and "inactive" forms are found in all species of *Cicadulina* except *C. parazeae*. "Inactive" forms can be rendered "active" by puncturing the gut lining (102).

HOST RANGE

Hosts of MSV are species within the Gramineae. Extensive lists of naturally-infected genera, species and cultivars have been published. Bock *et al.* (6) reported that in Africa MSV had been recorded in species in the tribes Andropogoneae (*Cymbopogon, Imperata, Saccharum, Rottboellia*), Eragrosteae (*Dactyloctenium, Diplachne, Eleusine, Eragrostis, Leptochloa, Setaria*), Sporoboleae (*Sporobolus*), Zoysieae (*Tragus*), Maydeae (*Euchlaena, Zea*), Hordeae (*Triticum*),and Aveneae (*Avena*). Gorter (35) reported that species of *Urochloa, Chloris, Lolium, Hordeum,* and *Secale* were susceptible as seedlings. Rose (78) found naturally infected species of *Panicum, Pennisetum,* and *Sorghum.* Damsteegt (19) demonstrated the potential for MSV infection in 54 of 138 annual and perennial grasses within 45 additional graminaceous species in the tribes Hordeae (*Aegilops, Agropyron, Elymus*), Aveneae (*Holcus*), Agrostideae (*Agrostis, Calamagrostis*), Chlorideae (*Schedonnardus*), Andropogoneae (*Schizachyrium, Sorghastrum*), Paniceae (*Echinochloa*), Oryzeae (*Oryza*), and Glycerieae (*Glyceria*). No species were reported in the subfamily Bambusoideae (5, 19) Corollary crops such as wheat, barley, rye, oats, and rice were susceptible to MSV; two perennial *Zea* species, *Z. perennis* (Hitchc.) Reeves & Mangelsd. (tetraploid) and *Z. diploperennis* Iltis, Doebley, &

Guzman (diploid) were immune to infection (19,72). The virus host range is directly correlated to the host range of the insect vectors (57). Damsteegt (19), however, reported that several annual and perennial grass species were good food and oviposition hosts for *C. mbila*, but not hosts of MSV.

Several MSV related diseases have been described in corollary crops. Wheat stunt (33) in South Africa and India (91), bajra streak (90) in India, Uba cane streak (58), and sugar cane streak in Egypt (1).

EPIDEMIOLOGY

The prevalence and distribution of MSV is absolutely associated with the vector population which is influenced by rainfall, temperature, and availability of host plants (79). In forest areas of West Africa with bi-modal rainfall distribution and two cropping seasons, the second crop is more severely affected and in savannah areas with unimodal rainfall distribution the incidence is higher in later plantings. Irrigated crops usually suffer more (25,78). In Zimbabwe, the populations are lowest during the dry winter season (May to August), rise sharply in August - November during a hot dry season, and continue high during the wet season from December to April (78).

The size of *Cicadulina* populations is mainly determined by the numbers of host grasses suitable for oviposition and development of nymphs. Population cycles follow a distinct pattern with young annual grasses and cereals being colonized by predominantly female populations (78, 81). A predominance of females is found in populations of *Cicadulina* which settle on cereals. This is also true of other leafhoppers which disperse over long distances, e.g. *Empoasca fabae* (48), *Circulifer tenellus* (52), and *Macrosteles fascifrons* (59) because females fly further than males from their breeding sites.

Patterns of streak infection in corn fields conform closely with seasonal distributions of different *Cicadulina* vectors (77,80). Females tend to accumulate more in edge rows while males are found uniformly throughout the fields (probably because males are more flighty) (76). Females of *C. mbila* have been shown to be better vectors than males, in part because they show a strong preference for settling on the chlorotic regions of MSV infected corn where the virions are found (41).

Considerable information is available on how constant and fluctuating temperatures affect fecundity, development time, survival (longevity), and ability to transmit strains of MSV (1,2,14,20,79,81,85). All *Cicadulina* species survived and reproduced readily at temperatures ranging from 20 - 30°C although *C. arachidis* survived the longest (2). Constant low temperatures increased leafhopper development time and longevity and reduced fecundity (20, 66). In at least one study (2) the distribution of *C. ghaurii* in Nigeria appeared to be affected more by relative humidity than temperature. The most efficient vector of MSV was *C. mbila*, closely followed by *C. storeyi* and *C. ghaurii*. Okoth and Dabrowski (69) and Asanzi *et al.* (2) found *C. arachidis* a poor vector of MSV.

Yield losses are dependent on stage of plant growth at the time of infection (107). Van Rensburg and Kuhn (112) reported total loss, 50% loss, and negligible loss from plants infected at 1, 3, and 8 weeks after planting. Fajemisin *et al.* (25) reported that early losses occurred from lack of plant survival and later yield losses resulted from fewer harvestable ears, poorly filled ears, shorter plants, narrower stem diameter (weak stems), smaller leaf area and smaller tassels. Mzira (64) found a closer relationship between yield loss and time of infection than between yield loss and disease incidence. Guthrie (39) found correlation between cob weight and grain weight and suggested use of cob weight as a measurement of yield loss.

DISEASE MANAGEMENT

Control strategies for MSV are interdependent on cultural practices, sources of resistance in corn, sources of resistance in *Cicadulina* spp., and populations of different *Cicadulina* spp. (22). Cultural practices utilized in different countries have included field sanitation, early planting, chemical treatment for vector control, rogueing of scattered infected plants, and breeding for resistant cultivars (22,36,81).

Each ecological area experiences a different combination of epidemiology factors. Outbreaks of MSV often can be associated with drought conditions or irregular early rains. Early season crops may escape severe losses due to low vector populations while later crops suffer severe losses because of high vector populations which favor early MSV spread (22,81).

Natural sources of the virus exist in indigenous grasses from which the vectors emigrate to susceptible corn fields. However, there is evidence that natural grasslands harbor only limited sources of virus strains virulent to corn (81). Storey (100) believed that infected plants within previously planted gramineous crops were the major sources of maize streak. This is true for corn crops planted after the flight season from grasses is over and for irrigated corn crops. In areas where irrigation has extended the growing season of corn or associated gramineous crops, *Cicadulina* populations build up and quickly spread MSV into new corn fields (81).

The control strategy used depends on whether vectors are flying from grass hosts or previously planted cereals (81). Dispersal of leafhoppers occurs when colonized hosts mature or dry (81). If vectors are coming from grass hosts there are fewer viruliferous individuals. If they are coming from other infected cereals they usually have flown short distances and are more viruliferous. These natural flight seasons are determined by ambient temperatures, host plant species and available moisture, all of which affect developmental time of leafhoppers (20,113). Although maize streak epidemics may occur under several different environmental scenarios, most occur in or near irrigated corn crops.

Cultural and Chemical Control

Several cultural practices have been recommended for control of maize streak disease such as i) planting downwind of earlier gramineous crops (34), ii) leaving barriers of 10 meters of bare ground around the crop (34), iii) rotating crops in time and place (65), iv) selectively using insecticides (81), v) planting streak resistant corn cultivars (81), vi) eradicating scattered infected plants (80,81), and vii) planting all susceptible crops concurrently (65). In specific cases insecticides have been used to reduce vector populations with accompanying reduction in percentage of streak infection (80).

Chemicals for control of MSV vectors have been insecticidal sprays, seed treatments, and soil systemic insecticides (109). Carbofuran applied as a 10% granular formulation in the planting furrow together with the seed proved very effective in controlling *C. mbila* populations (109). In greenhouse tests, carbofuran was the only insecticide that provided a rapid knockdown of *Cicadulina* vectors and prevented MSV transmission up to 46 days after treatment (111). Carbofuran also gave protection against other major insect pests, thereby limiting additional applications of pesticides (114). However, soil systemic insecticides may be six times more expensive than conventional foliar insecticides. In general, insecticides are expensive prophylactics and the most reliable control is the use of resistant cultivars (25,29,36,83).

Breeding for Resistance to Virus

Sources of Resistance

Tolerant or resistant sources of corn germplasm have been known for many years. Fielding (29) identified the first source of resistance in corn in a variety called Peruvian Yellow. Rose (82) identified another source in the variety Arkells Hickory resulting in new varieties called P X H and SA 31 (25). Gorter (36) reported on several resistant varieties from Rhodesia (Zimbabwe) and Storey and Howland (104) described the incompletely-dominant, single-gene resistance mechanism of the P X H crosses.

East Africa's contributions to the pool of resistant germplasm began in the 1970's with the ultimate discovery and development of a source of resistance called "La Revolution". This source was found on Reunion Island (24) and has remained very effective. Damsteegt (19) also found Revolution to be a good source of resistance from among more than 500 corn varieties and lines.

The national plant breeding program in South Africa has produced several tolerant or resistant lines (30). Two Vaalharts lines were susceptible as young seedlings but were somewhat resistant in the six-to-eight week stage (19). A Vaalharts composite VHCY, developed by Engelbrecht (23), was used as the non-recurrent parent in crosses with US inbreds Mo17 and B73 in 1989 to develop streak-resistant inbred lines with improved combining ability for yield (51). Lines were tested for yield with and without streak, under high and low moisture, and in the presence of other stressors. From 68 S5 lines characterized in 1995, a total of 15 superior inbred lines were ready for release in 1996 (51). Nearly all MSV-resistant hybrids sold in South Africa prior to 1992 were yellow-grained (3). Since the bulk of corn produced in Africa is consumed by humans and white corn is preferred, there is a need to produce resistant white-grained hybrids (3).

Breeding Programs

In 1975 scientists at the International Institute of Tropical Agriculture (IITA) initiated a method of mass rearing *Cicadulina* spp. for artificial infestation of corn germplasm (17,22). The corn research unit at IITA in cooperation with Centro Internacional de Majoramiento de Maize y Trigo (CIMMYT) Mexico, developed an international program integrating the national corn breeding programs from all areas of Africa. Sources of resistance were identified in the varieties TZ-Y (Tropical Zea-Yellow) derived from a cross between Tuxpeno Planta Baja from CIMMYT and a source from East-Africa. After six generations a highly resistant line designated IB32 was selected for studying mode of inheritance. Crosses with four susceptible inbred lines (B14, B68, B73, and Mo17) produced segregating lines without distinct reaction classes indicating quantitative inheritance with three major genes (49) and additional modifier genes affecting growth characteristics (4). The resistance of "La Revolution" is dominant monogenic (49). Using the resistance from IB32 and Revolution, IITA and CIMMYT scientists in collaboration with national programs have developed over 115 different hybrids and inbreds with stable resistance to MSV. These have been adapted to different ecological habitats and consumer preferences in Africa (49). National programs in South Africa, Zimbabwe, Kenya, Zaire, Nigeria, and other areas in cooperation with the international center at Ibadan (IITA) have continued to screen corn germplasm for sources of usable resistance. Yield of resistant hybrids in the absence of MSV has continued to be a problem in widespread acceptance of new hybrids. It is obvious that any new resistant hybrid needs to compete with present-day commercial hybrids in non-MSV years. To control natural infections of MSV in susceptible cultivars a systemic carbamate must be applied at a cost equal to 7% of the yield. Therefore, any new hybrid must yield at least 93% of the susceptible hybrid it would replace in the absence of MSV. Many areas of Africa have low yield potential and resistant cultivars would eliminate the need for expensive pesticides (3).

New resistant inbreds and hybrids are being released from national programs. In 1992, MSIRI 3B, developed by the Mauritius Sugar Industry Research Institute, was released in Mauritius as a new source of resistance different from that used at IITA. It may be especially useful in the Islands of the Indian Ocean (37,50). In the absence of MSV it outyielded the composite variety from which it was selected by 15 - 30%.

Breeding for Insect Resistance

Resistant corn cultivars have been released on a large scale in the past 10 years (3,22) with the emphasis on resistance to the virus. In screening programs symptomless plants are discarded because they may include escapes. This practice needs to be re-examined for it could exclude genotypes which were less preferred by the insect vectors (46,47).

Jones (46) classified insect resistance mechanisms into four main types: interference with location of host plants, interference with initial settling and feeding behavior of vectors, interference with sustained feeding behavior of vectors, specific interference with vector transmission of virus. For insect resistance to be effective with MSV, prolonged probing must be precluded by some mechanism (47).

Kairo *et al.* (47) determined that certain resistant corn genotypes were less preferred by *C. mbila* than susceptible genotypes which added to their resistant phenotype. Three vector characteristics (settling, probing, and oviposition) were compared on three resistant cultivars and one susceptible cultivar. On one, two, and three week-old seedlings, *C. mbila* showed a distinct preference for settling on the streak susceptible cultivar H512. There were almost three times as many probing marks on one resistant cultivar than on the susceptible cultivar. Oviposition was not significantly different on the four cultivars although the nymphs were not allowed to reach maturity on the oviposition hosts. Overall the settling characteristic was significantly different in the resistant cultivars and other characteristics were less consistent.

CONCLUSIONS

Maize streak is the most damaging and economically important virus disease of corn in the world. It affects corn production in most sub-Saharan African corn growing areas, the neigboring countries and India. The ssDNA monopartite geminivirus has been the focus of intensive research in genome organization and function (at least three MSV isolates cloned, sequenced, and characterized). The MSV is able

to provide a rapid and sensitive marker for the transfer of DNA to cereals (7).

Several *Cicadulina* spp. are obligate vectors of MSV isolates; *C. mbila* is the principal corn vector. All known isolates are infectious to corn although several host-adapted strains exist.

Many sources of resistance have been identified in national breeding programs. During the last 20 years, mass rearing of leafhoppers and artificial inoculations of genetic lines have expanded the scope of resistance breeding. Collaborative research between international centers (IITA and CIMMYT) and national breeding programs has provided lines with stable resistance to both the virus and vector in widely divergent ecological zones. There is a continued urgent need to incorporate some of the known resistance genes into white-grained corn.

Extensive testing of several MSV-resistant hybrids is currently being undertaken in several countries in Africa. The ultimate goal is to identify and encourage widescale cultivation of MSV-resistant corn hybrids by farmers to remove MSV as a major limiting factor in corn production in Africa.

ACKNOWLEDGEMENTS

The authors wish to thank EMBO Journal and S. Lazerowitz for use of published Figure 1 and to M. Bonde, G. Gillaspie and G. Milbrath for constructive reviews.

REFERENCES

1. Ammar, E.D. 1975. Biology of the leafhopper *Cicadulina chinai* Ghauri (Homoptera, Cicadellidae) in Giza, Egypt. Z. Angew. Entomol. 79:337-345.

2. Asanzi, M.C., Bosque-Perez, N.A., Nault, L.R., Gordon, D.T., and Thottappilly, G. 1995. Biology of *Cicadulina* species (Homoptera:Cicadellidae) and transmission of maize streak virus. Afr. Entomol. 3:173-179.

3. Barrow, M.R. 1992. Development of maize hybrids resistant to maize streak virus. Crop Protect. 11:267-271.

4. Bjarnason, M. 1986. Progress in breeding for resistance to the maize streak virus disease. Pages 197-207 in: To Feed Ourselves. Proc. 1st Eastern, Central, Southern Africa Regional Maize Workshop. B. Gelaw, ed. CIMMYT, Mexico, D.F. Lusaka, Zambia, Mar 10-17, 1985.

5. Bock, K.R. 1974. Maize streak virus. CMI/AAB Descriptions of Plant Viruses No. 133. 4 Pages.

6. Bock, K.R., Guthrie, E.J., and Woods, R.D. 1974. Purification of maize streak virus and its relationship to viruses associated with streak diseases of sugar cane and *Panicum maximum*. Ann. Appl. Biol. 77:289-296.

7. Boulton, M.I., and Davies, J.W. 1990. Monopartite geminiviruses: Markers for gene transfer to cereals. Aspects Appl. Biol. 24:1-8.

8. Boulton, M.I., King, D.I., Donson, J., and Davies, J.W. 1991. Point substitutions in a promoter-like region and the V1 gene affect the host range and symptoms of maize streak virus. Virology 183:114-121.

9. Boulton, M.I., Steinkellner, H., Donson, J., Markham, P.G., King, D.I., and Davies, J.W. 1989. Mutational analysis of the virion-sense genes of maize streak virus. J. Gen. Virol. 70:2309-2323.

10. Briddon, R.W., Lunness, P., Chamberlin, L.C.L., and Markham, P.G. 1994. Analysis of the genetic variability of maize streak virus. Virus Genes 9:93-100.

11. Briddon, R.W., Lunness, P., Chambelin, L.C.L., Pinner, M.S., Brundish, H., and Markham, P G. 1992. The nucleotide sequence of an infectious insect-transmissible clone of the geminivirus *Panicum* streak virus. J. Gen. Virol. 73:1041-1047.

12. Choudhary, G.G., Singh, G., and Bhatnagar, G.C. 1980. Losses in yield components of wheat variety Lal Bahadur caused by streak virus disease. Indian Phytopathol. 33:604.

13. Choudhary, G.G., Singh, G., and Bhatnagar, G.C. 1980. Reactions of pearl millet lines to Pennisetum strain of maize streak virus. Indian J. Mycol. Plant Pathol. 10:71.

14. Dabrowski, Z.T. 1985. The biology and behaviour of *Cicadulina triangula* in relation to maize streak virus resistance screening. Insect Sci. Applic. 6:417.

15. Dabrowski, Z.T. 1987. *Cicadulina ghaurii* (Hem., Euscelidae): Distribution, biology and maize streak virus (MSV) transmission. J. Appl. Entomol. 103:489-496.

16. Dabrowski, Z.T. 1987. Comparative studies of *Cicadulina* leafhoppers in West Africa. Pages 35-39 in: Proc. 2nd Int. Workshop on Leafhoppers and Planthoppers of Economic Importance, Brigham Young University, Provo, Utah. M. R. Wilson and L. R. Nault, eds. CAB Int. Inst. Entomol, London.

17. Dabrowski, Z.T. 1987. Some parameters affecting suitability of *Cicadulina* species for resistance screening to maize streak virus (MSV). Insect Sci. Applic. 8:757-764.

18. Damsteegt, V.D. 1981. Exotic virus and virus-like diseases of maize. Pages 110-123 in: Virus and Virus-like Diseases of Maize in the United States. D.T. Gordon, J.K. Knoke, and G.E. Scott, eds. Southern Cooperative Series Bulletin 247, June 1981. 210 Pages.

19. Damsteegt, V.D. 1983. Maize streak virus: I. Host range and vulnerability of maize germplasm. Plant Dis. 67:734-737.

20. Damsteegt, V.D. 1984. Maize streak virus: Effect of temperature on vector and virus. Phytopathology 74:1317-1319.

21. Dekker, E.L., Pinner, M.S., Markham, P.G., and Van Regenmortel, M.H.V. 1988. Characterization of maize streak virus isolates from different plant species by polyclonal and monoclonal antibodies. J. Gen. Virol. 69:983-990.

22. Efron, Y., Kim, S.K., Fajemisin, J.M., Mareck, J.H., Tang, C.Y., Dabrowski, Z.T., Rossel, H.., Thottappilly, G., and Buddenhagen, I.W. 1989. Breeding for resistance to maize streak virus: A multidisciplinary team approach. Plant Breeding 103:1-36.

23. Engelbrecht, G.C. 1973. Die genetika van weerstandbiedendheid teen streepsiektevirus by *Zea mays*. Unpublished D. Sc. (Agric.) Thesis, Univ. Orange Free State, Bloemfontein.

24. Etienne, J., and Rat, B. 1973. Le stripe: Une maladie importante du mais a la Reunion. L'Agron. Trop. 28:11-17.

25. Fajemisin, J.M., Kim, S.K., Efron, Y., and Alam, M.S. 1984. Breeding for durable disease resistance in tropical maize with special reference to maize streak virus. FAO Plant Prot. Bull. 55:49-71.

26. Fajemisin, J. M., and Shoyinka, S. A. 1976. Maize streak and other maize virus diseases in West Africa. Pages 52-60 in: Proc. Int'l Maize Virus Dis. Colloq. and Workshop. L.E. Williams, D.T. Gordon, and L.R. Nault, eds. Aug 16-19, 1976, OARDC, Wooster, OH. .

27. Fennah, R.G. 1960. A new species of *Cicadulina* (Homoptera:Cicadellidae) from East Africa. Ann. Mag. Nat. Hist. (Series 13) 11:757-758.

28. Fenoll, C., Black, D.M., and Howell, S.H. 1988. The intergenic region of maize streak virus contains promoter elements involved in rightward transcription of the viral genome. EMBO J. 7:1589-1596.

29. Fielding, J. 1933. Field experimental work on rotation crops. Pages 10-14 in: Empire Cotton Growing Assoc. Progr. Rep., 1931-1932.

30. Fourie, A.P., and Pienaar, J.H. 1983. Breeding for resistance to maize streak virus: A report on the Vaal Harts breeding programme. Tech. Commun.-S. Afr. Dept. Agric. Pretoria. The Department 182:41-50.

31. Fuller, C. 1901. Mealie variegation. First Report Gov. Entomol. Natal 1899-1900:17-18.

32. Gelaw, B. 1986. CIMMYT's maize improvement role in East, Central, and Southern Africa. Page 208 in: To Feed Ourselves, A Proc. 1st Eastern, Central and Southern Africa Reg. Maize Workshop. B. Gelaw, ed. CIMMYT, Mexico, D.F. Lusaka, Zambia, Mar. 10-17, 1986.

33. Gorter, G.J.M.A. 1947. Wheat stunt - a new cereal disease. Farm. S. Afr. 22:29-32, 44.

34. Gorter, G.J.M.A. 1951. Streak disease of maize - helpful measures for its prevention. Farm. S. Afr. 26:361-62, 64.

35. Gorter, G.J.M.A. 1953. Studies on the spread and control of the streak disease of maize. Sci. Bull. Dept. Agric. For. Union S. Afr. No. 341. 20 Pages.

36. Gorter, G.J.M.A. 1959. Breeding maize for resistance to streak. Euphytica 8:234-240.

37. Govinden, N., and Rummun, K. 1996. Registration of MSIRI 3B streak-resistant maize germplasm. Crop Sci. 36:824.

38. Guthrie, E.J. 1976. Virus diseases of maize in East Africa. Pages 62-66 in: Proc. Intl. Maize Virus Dis. Colloq. and Workshop. L.E. Williams, D.T. Gordon, and L.R. Nault, eds. Aug 16-19, 1976, Wooster, OH.

39. Guthrie, E.J. 1978. Measurement of yield losses caused by maize streak disease. Plant Dis. Rep. 62:839-841.

40. Harrison, B.D. 1985. Advances in geminivirus research. Annu. Rev. Phytopathol. 23:55-82.

41. Heathcote, R.J. 1975. The fecundity of *Cicadulina mbila* (Naude) in relation to maize streak virus. MSc Thesis. Univ. Rhodesia, Salisbury. 69 Pages.

42. Howell, S.H. 1984. Physical structure and genetic

organization of the genome of maize streak virus (Kenyan isolate). Nucleic Acids Res. 12:7359-7375.

43. Hughes, F.L., Rybicki, E.P., and Kirby, R. 1993. Complete nucleotide sequence of sugarcane streak monogeminivirus. Arch. Virol. 132:171-182.

44. Hughes, F.L., Rybicki, E.P., Kirby, R., and von Wechmar, M.B. 1991. Characterization of the sugarcane streak agent as a distinct geminivirus. Intervirology 32:19-27.

45. Ishihara, T. 1969. Families and genera of leafhopper vectors. Pages 235-254 in: Viruses, Vectors, and Vegetation. K. Maramorosch, ed. Interscience Publishers, N.Y.

46. Jones, A.T. 1987. Control of virus infection in crop plants through vector resistance: A review of achievements, prospects, and problems. Ann Appl. Biol. 111:745-772.

47. Kairo, M.T.K., Kiduyu, P.K., Mutinda, C.J.M., and Empig, L.T. 1995. Maize streak virus: evidence for resistance against *Cicadulina mbila* Naude, the main vector species. Euphytica 89:109-114.

48. Kieckhefer, R. W. 1962. Some factors affecting populations of the potato leafhopper, *Empoasca fabae* (Harris). PhD Thesis, Univ. Wisc., Madison. 92 Pages.

49. Kim, S-K., Efron, Y., Fajemisin, J.M., and Buddenhagen, I.W. 1989. Mode of gene action for resistance in maize to maize streak virus. Crop Sci. 29:890-894.

50. Kim, S-K., Efron, Y., Khadr, F., Fajemisin, J., and Lee, M.H. 1987. Registration of 16 maize streak virus-resistant tropical maize parental inbred lines. Crop Sci. 27:824-825.

51. Kuhn, H.C., and Van Rensburg, J.B.J. 1995. Release of streak-resistant maize inbred lines. S. Afr. Tydskr. Plant Grond 12:180-181.

52. Lawson, F.R., Chamberlain, J.C., and York, G.T. 1951. Dissemination of the beet leafhopper in California. U.S. Dept. Agric. Tech. Bull. No. 1030. 59 Pages.

53. Lazarowitz, S.G. 1987. The molecular characterization of geminiviruses. Plant Mol. Biol. Rep. 4:177-192.

54. Lazarowitz, S.G. 1988. Infectivity and complete nucleotide sequence of the genome of a South African isolate of maize streak virus. Nucleic Acids Res. 16:229-249.

55. Lazarowitz, S.G., and Pinder, A.J. 1989. Molecular genetics of maize streak virus. Pages 167-183 in: Molecular Biology Plant-Pathogen Interactions, UCLA Symposia on Molecular and Cellular Biology, Vol. 101. Mar. 26-Apr. 11, 1988, Steamboat Springs, CO.

56. Lazarowitz, S.G., Pinder, A.J., Damsteegt, V.D., and Rogers, S.G. 1989. Maize streak virus genes essential for systemic spread and symptom development. EMBO J. 8:1023-1032.

57. Markham, P.G., Pinner, M.S., and Boulton, M.I. 1984. The transmission of maize streak virus by leafhoppers, a new look at host adaptation. Bull. Soc. Entomol. Suisse 57:431-432.

58. McClean, A.P.D. 1947. Some forms of streak virus occurring in maize, sugar-cane and wild grasses. Sci. Bull. Dept. Agric. Union S. Afr. No. 265. 39 Pages.

59. Meade, A.B. 1962. The origin and development of populations of the six-spotted leafhopper, *Macrosteles fascifrons* (Stal), on an area basis. PhD Thesis, Univ. Minn., Minneapolis. 130 Pages.

60. Mesfin, T., Bosque-Perez, N.A., Buddenhagen, I.W., Thottappilly, G., and Olojede, S.O. 1992. Studies of maize streak virus isolates from grass and cereal hosts in Nigeria. Plant Dis. 76:789-795.

61. Morris-Krsinich, B.A.M., Mullineaux, P.M., Donson, J., Boulton, M.I., Markham, P.G., Short, M.N., and Davies, J.W. 1985. Bidirectional transcription of maize streak virus DNA and identification of the coat protein gene. Nucleic Acids Res. 20:7237-7256.

62. Mullineaux, P.M., Boulton, M.I., Bowyer, P., Van Der Vlugt, R., Marks, M., Donson, J., and Davies, J. W. 1988. Detection of a non-structural protein of Mr 11,000 encoded by the virion DNA of maize streak virus. Plant Mol. Biol. 11:57-66.

63. Mullineaux, P.M., Donson, J., Morris-Krsinich, B.A.M., Boulton, M.I., and Davies, J.W. 1984. The nucleotide sequence of maize streak virus DNA. EMBO J.3:3063-3068.

64. Mzira, C.N. 1984. Assessment of effects of maize streak virus on yield of maize. Zimbabwe J. Agric. Res. 22:141-149.

65. Mzira, C.N. 1984. Cultural control of maize streak virus in wheat by spacing and time of planting. Zimbabwe Agric. J. 81:189-192.

66. Nault, L.R., and Madden, L.V. 1985. Ecological strategies of *Dalbulus* leafhoppers. Ecol. Entomol. 10:57-63.

67. Nielson, M.W. 1968. The leafhopper vectors of phytopathogenic viruses (Homoptera:Cicadellidae). Taxonomy, biology and virus transmission. USDA Tech. Bull. No. 1382:1-386.

68. Obi, I.U. 1987. Disease and pest problems of late season maize (*Zea mays* L.) in relation to time of planting on the Nsukka plains of south-eastern Nigeria. E. Afr. Agric. For. J. 53:1-11.

69. Okoth, V.A.O., Dabrowski, Z.T., Thottappilly, G., and Van Emden, H.F. 1987. Comparative analysis of some parameters affecting maize streak virus (MSV) transmission of various *Cicadulina* spp. populations. Insect Sci. Applic. 8:295-300.

70. Peterschmitt, M., Quiot, J.B., Reynaud, B., and Baudin, P. 1992. Detection of maize streak virus antigens over time in different parts of maize plants of a sensitive and a so-called tolerant cultivar by ELISA. Ann. Appl. Biol. 121:641-653.

71. Pinner, M.S., and Markham, P.G. 1990. Serotyping and strain identification of maize streak virus isolates. J. Gen. Virol. 71:1-6.

72. Pinner, M.S., Markham, P.G., Markham, R.H., and Dekker, E.L. 1988. Characterization of maize streak virus: description of strains; symptoms. Plant Pathol. 37:74-87.

73. Plavsic-Banjac, B., and Maramorosch, K. 1972. Electronsko-mikroskopska dijagnoza oboljenja indijske patuljaste zutice rize, proliferacije sandalovog drveta i crticavosti afrikanskog kukuruza. Acta. Biol. Iugosl. Ser. B. Mikrobiol. 9:201-211.

74. Raychaudhuri, S.P., Seth, M.L., Renfro, B.L., and Varma, A. 1981. Principal maize virus diseases in India. Pages 69-77 in: Proc. Int. Maize Dis. Colloq. and Workshop, 16-19 Aug., 1976. L.E. Williams, D.T. Gordon, and L.R. Nault, eds. Ohio Agric. Res. Dev. Cent., Wooster.

75. Roberts, I.M., Robinson, D.J., and Harrison, B.D. 1984. Serological relationships and genome homologies among geminiviruses. J. Gen. Virol. 65:1723-1730.

76. Rose, D.J.W. 1971. The biology and dispersal of *Cicadulina* spp. PhD Thesis, Univ. London, England. 359 Pages.

77. Rose, D.J.W. 1973. Distances flown by *Cicadulina* spp. (Hem., Cicadellidae) in relation to distribution of maize streak disease in Rhodesia. Bull. Entomol. Res. 62:497-505.

78. Rose, D.J.W. 1973. Field studies in Rhodesia on *Cicadulina* spp. (Hem., Cicadellidae), vectors of maize streak disease. Bull. Entomol. Res. 62:477-495.

79. Rose, D.J.W. 1973. Laboratory observations on the biology of *Cicadulina* spp., with particular reference to the effects of temperature. Bull. Entomol. Res. 62:471-476.

80. Rose, D.J.W. 1974. The epidemiology of maize streak disease in relation to population densities of *Cicadulina* spp. Ann. Appl Biol. 76:199-207.

81. Rose, D.J.W. 1978. Epidemiology of maize streak disease. Annu. Rev. Entomol. 23:259-282.

82. Rose, F.M. 1938. Rotation crops. Pages 21-25 in: Emp. Cotton Grow. Corp. Prog. Rep., 1936-1937.

83. Rose, F.M. 1941. Rotation crops. Pages 27-31 in: Empire Cotton Growing Assoc. Progr. Rep., 1939-1940.

84. Rossel, H.W., and Thottappilly, G. 1985. Virus diseases of important food crops in tropical Africa. IITA Publication Series. 61 Pages.

85. Ruppel, R.F. 1965. A review of the genus *Cicadulina* (Hemiptera, Cicadellidae). Mich. State Univ. Mus. Biol. Ser. 2:385-428.

86. Rybicki, E.P., and Hughes, F.L. 1990. Detection and typing of maize streak virus and other distantly related geminiviruses of grasses by polymerase chain reaction amplification of a conserved viral sequence. J. Gen Virol. 71:2519-2526.

87. Saiki, R.K., Gelfand, D.H., Stoffel, S., Scharf, S.J., Higuchi, R., Horn, G.T., Mullis, K.B., and Erlich, H.A. 1988. Primer-directed enzymatic amplication of DNA with a thermostable DNA polymerase. Science 239:487-491.

88. Schneider, M., Jarchow, E., and Hohn, B. 1992. Mutational analysis of the 'conserved region' of maize streak virus suggests its involvement in replication. Plant Mol. Bio. 19:601-610.

89. Scott, D.B., and Van Niekerk, H.A. 1991. Resistance in wheat to maize streak virus. S. Afr. Tydskr. Plant Grond 8:110-111.

90. Seth, M.L., Raychaudhuri, S.P., and Singh, D.V. 1972. Bajra (pearl millet) streak: A leafhopper-borne cereal virus in India. Plant Dis. Rep. 56:424-428.

91. Seth, M. L., Raychaudhuri, S.P., and Singh, D.V. 1972. Occurrence of maize streak virus on wheat in India. Curr. Sci. (Bangalore) 41:684.

92. Soto, P.E. 1978. A new vector of maize streak virus. East African Agric. and For. J. 44:70-71.

93. Stanley, J., and Davies, J.W. 1985. Structure and function of the DNA genome of geminiviruses. Pages 191-218 in: Molecular Plant Virology, Vol II. J.W. Davies, ed. CRC Press, Boca Raton, FL.

94. Storey, H.H. 1924. Streak disease, an infectious chlorosis of sugarcane, not identical with mosaic disease. Pages 132-144 in: Rep. Imp. Bot. Congr., London, 1924.

95. Storey, H.H. 1925. The transmission of streak disease of maize by the leafhopper *Balclutha mbila* Naude. Ann. Appl. Biol. 12:422-439.

96. Storey, H.H. 1928. Transmission studies of maize streak disease. Ann. Appl. Biol. 15:1-25.

97. Storey, H.H. 1931. The inheritance by a leafhopper of the ability to transmit a plant virus. Nature 127:928.

98. Storey, H.H. 1933. The inheritance by an insect vector of the ability to transmit a plant virus. Proc. R. Soc. Lond. Ser. B. 112:46-60.

99. Storey, H.H. 1933. Investigations of the mechanism of the transmission of plant viruses by insect vectors. -I. Proc. R. Soc. Lond. Ser. B. 113:463-485.

100. Storey, H.H. 1936. Virus diseases of East African plants. IV. A survey of the viruses attacking the Gramineae. East Afr. Agric. J. 1:333-337.

101. Storey, H.H. 1937. A new virus of maize transmitted by *Cicadulina* spp. Ann. Appl. Biol. 24:87-94.

102. Storey, H.H. 1938. Investigations of the mechanism of the transmission of plant viruses by insect vectors. II. The part played by puncture in transmission. Proc. R. Soc. Lond. Ser. B. 125:455-477.

103. Storey, H.H. 1939. Investigations of the mechanism of the transmission of plant viruses by insect vectors. III. The insect's saliva. Proc. R. Soc. Lond. Ser. B. 127:526-543.

104. Storey, H.H., and Howland, A.K. 1967. Inheritance of resistance in maize to the virus of streak disease in East Africa. Ann. Appl. Biol. 59:429-436.

105. Storey, H.H., and McClean, A.P.D. 1930. The transmission of streak disease between maize, sugar cane and wild grasses. Ann. Appl. Biol. 17:691-719.

106. Sylvester, E.S., Richardson, J., and Nickel, J L. 1973. An additional note on virus-like particles associated with maize streak disease. Plant Dis. Rep. 57:414-416.

107. Van Rensburg, G.D.J. 1981. Effect of plant age at the time of infection with maize streak virus on yield of maize. Phytophylactica 13:197-198.

108. Van Rensburg, G.D.J. 1983. Southern African species of the genus *Cicadulina* China (Homoptera: Cicadellidae) with descriptions of new species. Dept. Agric., Rep. S. Afr, Entomol. Memoir No. 57.

109. Van Rensburg, G.D.J., and Giliomee, J.H. 1989. Comparative efficacy of pre- and post-emergence application of insecticides for simultaneous control of the maize leafhopper, *Cicadulina mbila*, and the stalk borers, *Busseola fusca* and *Chilo partellus*, on maize. Phytophylactica 21:399-402.

110. Van Rensburg, G.D.J., and Giliomee, J.H. . 1990. A comparison of females and males of *Cicadulina anestae* and *C. mbila* (Homoptera:Cicadellidae) as vectors of maize streak virus. Phytophylactica 22:241-243.

111. Van Rensburg, G.D., and Giliomee, J.H. 1991. Chemical prevention of maize streak virus transmission by the maize leafhopper, *Cicadulina mbila* (Homoptera:Cicadellidae). Phytophylactica 23:141-144.

112. Van Rensburg, G.D.J., and Kuhn, H.C. 1977. Maize streak disease. Farm. S. Afr. Maize Series, No. E. 4 Pages.

113. Van Rensburg, G.D.J., Pringle, K. ., and Giliomee, J.H. 1991. Effect of temperature and vector numbers on maize streak virus incubation and symptom severity in maize. Phytophylactica 23:145-147.

114.   Van Rensburg, G.D.J., Van Rensburg, J.B.J., and Giliomee, J.H. 1991. Towards cost effective insecticidal control of the maize leafhopper, *Cicadulina mbila*, and the stalk borers, *Busseola fusca* and *Chilo partellus*. Phytophylactica 23:137-140.

115.   Whitcomb, R.R., and Davis, R.E. 1970. Mycoplasma and Phytarboviruses as plant pathogens persistently transmitted by insects. Annu. Rev. Entomol. 15:405.

CHAPTER 37

Present Status of Management of Sugarcane Mosaic Virus

R.K. Jain, G.P. Rao and A. Varma

Sugarcane mosaic virus (SCMV) is one of the oldest definitive members of the potyvirus genus and potyviridae family (65). SCMV was first reported in 1919 from the United States by Brandes (15). However, the disease syndrome (named as yellow stripe or Gelestrepenziekte) associated with it were first recognized in 1892 in Java (now Indonesia) (53) and subsequently in Taiwan by Miyake in 1915 (17). The importance of SCMV was realized later when it was inadvertently introduced into Argentina from Java through POJ canes (4). The import of new varieties, vegetative propagation of the host and the absence of conspicuous symptoms subsequently resulted in its introduction into Brazil, Cuba, Puerto Rico, USA (Louisiana) and other cane growing countries during 1916-1925 (43,77). The widespread presence of SCMV was also recognized in India (19,20,69). SCMV is now known to occur in majority of the sugarcane growing areas of the world (56).

THE VIRUS

SCMV particles are flexuous, rods of 730-755 nm long and 13 nm wide and composed of a single polypeptide species of 28,500 - 35,000 Daltons consisting of 264 - 328 amino acid residues surrounding a single stranded, positive sense RNA species (43,80). Recent studies on the comparison of the amino acid sequences of coat proteins of SCMV isolates have indicated that the sequences in the N-terminal region correlate with host range of the isolates (90). Its genome is 9766 nucleotide long [SCMV-JG (Aust)] and has an organization similar to those of other potyviruses (33). It is moderately immunogenic and antisera usually have titres of 1:256 (80). It is readily transmitted by grafting, mechanical inoculation and a number of aphids in a non-persistent manner (43).

ECONOMIC LOSSES

It causes mosaic diseases in sugarcane, corn, sorghum and other Poaceous species worldwide (80). It has resulted in considerable economic losses in sugarcane and failure of commercial clones in several countries. Yield losses of 30-40% and sometimes 60-80% have been recorded in the western hemisphere (27,38,43). Although early studies in India indicated that even 100% infection reduced yields by only 10%, yet considering the vast area under sugarcane, even less than 10% loss in yield is enormous (84). SCMV is also responsible for yield losses of 10-30% and 10-50% in China and South Africa, respectively (18,26).

The extent of loss depends on clonal susceptibility, time of infection, pathogenic variability and interaction with other diseases. Yield reduction of 3-33% and 9-17% in the clones CO281 and CO290, respectively, have been recorded (77). Losses tend to be greater in ratoon crops (31-50%) than in the plant cane crop (17%) of the clone CP44-101 (1). Yield losses are amplified when SCMV predisposes the host plant to other diseases, such as ratoon stunting and pythium root rot (39,40).

PATHOGENIC VARIABILITY

SCMV strains, namely A,B,C and D were first identified using sugarcane clones CP29-291, CP31-294 and CO281 by Summers (76). Subsequently, other strains, E,F.G,H,I,J,K,L, and M were identified in the United States on the clones CP31-294 and CP31-558 and sweet

Table 1. World distribution of SCMV strains

Country	Strain												
	A	B	C	D	E	F	G	H	I	J	K	L	M
United States	A	B	C	D	E	F	G	H	I	J	K	L	M
Argentina		+											
Australia	+									+			
Cameroon				+									
Colombia		+	+		+								
Dominican Republic	+	+		+									
Egypt				+									
Iran		+											
Jamaica			+										
Japan	+	+						+	+				
Kenya	+												
Mexico		+		+									
Pakistan	+					+							
Puerto Rico	+	+		+									
South Africa	+			+						+			
Spain		+		+									
Taiwan	+	+		+									

sorghum cv. 'Rio' (3,42,78,81,92) . Strains of SCMV occurring in other countries including India have also been recognized. But, direct attempts to identify strains relative to the United States strains were made only from 17 countries (Table 1) (2,30-32). Hence, to ascertain world distribution of SCMV strains, comparative studies have to be undertaken at one place on standard differentials. Nevertheless, limited studies indicate that strains A,B and D are widely prevalent (Table 1).

A major breakthrough in the identification of SCMV strains was achieved recently. Studies on N-terminal serology, cytoplasmic inclusion morphology, amino acid sequence and peptide profiling of coat proteins, 3' non-coding nucleotide sequences homology and differential hosts reaction led some virologists to speculate that SCMV strains, previously considered as single virus, represented four distinct viruses, namely Johnsongrass mosaic (JGMV), corn dwarf mosaic (MDMV), sorghum mosaic (SrMV) and SCMV (Table 2) (28,44,48,49,63,64,83). In view of these

Table 2. Grouping of SCMV strains into four distinct viruses

JGMV	MDMV	SCMV	SrMV
SCMV-JG(Aust)	SCMV-JG(US)	MDMV-B(US)	SCMV-H(US)
MDMV-O(US)	MDMV-A(I)	SCMV-A(US)	SCMV-I(US)
MDMV-KS1(US)	MDMV-A(US)	SCMV-B(US)	SCMV-M(US)
	MDMV-A(YU)	SCMV-D(US)	
	MDMV-481(GER)	SCMV-E(US)	
	MDMV-D(US)	SCMV-SC(Aust)	
	MDMV-E(US)	SCMV-BC(Aust)	
	MDMV-F(US)	SCMV-Sabi(Aust)	
		SCMV-ISIS(Aust)	
		SCMV-Brisbane(Aust)	
		SCMV-Bundabey(Aust)	
		SCMV-522(GER)	

JGMV-Johnsongrass mosaic virus; MDMV-Corn dwarf mosaic virus; SrMV-Sorghum mosaic virus; Aust-Australia; GER-Germany; I-Italy; US-United States; YU-Yugoslavia. Source: (44,64)

Table 3. Approaches to differentiate four viruses of the SCMV subgroup

Approach	JGMV	MDMV	SCMV	SrMV
Host Reaction				
Sorghum (Atlas)	M	M	NS+ST	RL
Johnsongrass	M	M	NI	NI
Oat	MM	NI	NI	NI
Sugarcane	NI	NI	M	M
Cytoplasmic inclusion morphology	Scrolls only	Scrolls+ Needles like	Scrolls+ Laminated aggregates	Scrolls+ Amorphous inclusions
Coat protein amino sequence homology	52-57%	57-71%	56-71%	52-70% acid
3'-non coding region nucleotide sequence homology	44-45%	44-57%	44-61%	45-61%

M: Mosaic; MM: Mild Mosaic; NI: No Infection ; NS: Necrotic Streak; RL: Red Leaf; ST: Stripes. Source: (44,64,83)

Table 4. Cure of seed-cane transmitted pathogens by heat therapy

Disease	Pathogen	Systems of heat therapy				
		HWT	SHWT	HAT	MHAT	AST
Red Rot	*Colletotrichum falcatum*	-	-	54°C/ 8 h	54°C/ 2 h	50-52°C
Downy Mildew	*Sclerospora sacchari*	50°C/ 3 h	45,52°C 1 h each at 24 h interval	-	-	-
Smut	*Ustilago scitaminea*	55°C/ 10 min or 52°C/ 8 min or 52°C/ 45 min	-	-	-	-
Ratoon stunting	*Clavibacter* xyli pv. *xyli*	50°C/ 2 h	-	54°C/ 8 h	54°C/ 2 h	51°C/ 4 h
Leaf scald	*Xanthomonas albilineans*	50°C/ 3 h	-	-	54°C 2 h	-
Grassy shoot	Phytoplasma	50°C/ 2 h	-	54°C/ 8 h	54°C/ 2 h	50°C/ 1 h
White leaf	Phytoplasma	55°C/ 10 min	-	-	-	-

HWT = Hot water treatment; SHWT = Serial hot water treatment; HAT = Hot air treatment; MHAT = Moist hot air treatment; AST = Aerated steam treatment. Source: (71)

findings, taxonomic status of SCMV strains occurring in various countries need to be reascertained and the availability of various approaches to distinguish the four viruses in the SCMV subgroup (Table 3) now makes it feasible.

MANAGEMENT

Various practices, such as avoidance of sources of infection, avoidance or control of vectors, cultural control, genetical resistance and use of biotechnological approach have been suggested for the management of potyviruses (85).

Management of SCMV is also possible through these practices, but the task gets complicated by vegetative nature, long duration and ratooning practice of sugarcane crop.

Avoidance of Sources of Infection

Infected planting material, alternate host and vectors are the primary sources of infection (43) and their avoidance provide an ideal strategy for SCMV management.

Use of Virus-free Planting Material

Since sugarcane is a vegetatively propagated crop and SCMV is a seed-cane transmitted potyvirus (80), use of virus free planting material is imperative for raising a healthy crop as well as ratoons.

Critical studies in India and elsewhere have clearly documented that heat therapy, i.e. treating the seed-cane at 50-55°C is useful in curing seed-cane transmitted pathogens of sugarcane (71) (Table 4) and eventually a "three tier seed program" was formulated (70). But, none of these physical methods, hot water treatment (HWT), aerated steam treatment (AST) and moist hot air treatment (MHAT) were suitable for curing seed-cane SCMV infection. Thus, heat therapy with entirely different time-temperature combinations suitable for reducing the frequency of SCMV infection were worked out (Table 5). Steib and Cifuentses (75) obtained good control of SCMV by AST at high temperature of 56-57°C for 2-3 h. According to Waterworth and Kahn (88), three sequential exposures of cutting at 24h intervals to hot water for 20 min at 52, 57 and 57°C, respectively, were successful in eliminating SCMV-H and -I from germplasm. In another study, Benda (10) reported that cures from SCMV-H and -I were obtained at maximum temperature in treatment series of 54.8°C for 10.5 min each and 57.3°C when infected with SCMV-A, -B and -D. Recently, Benda et al. (11) suggested still higher temperature treatment series to cure infected seed pieces from SCMV-A, -D,-F and -H (Table 5). However, lack of a single time-temperature combination, lack of absolute control, reduction in seed-cane germinability, uneconomical cost-benefit ratio and difficulties in treating planting material in bulk are major deterrents for large scale application of heat therapy.

Meristem culture alone or in combination with heat therapy has also been found useful in generating SCMV free plants (36,45,59,62,88). Over 90% success was obtained in CO740 when shoot apices of 0.05 - 1.5 mm long from SCMV infected plants were used for tissue culture (36). In another study, complete cure was claimed when tissue culture was combined with heat therapy (88). Numerous chemicals, like phenolic compounds, hemeopathic drugs, thiouracil and 2,4 - dinitrophenol have been found to inhibit SCMV infectivity (54,61,66-68). Besides, application of plant extracts, such as seed extracts of *Capsicum annum*, *Cassia auriculata*, *Chenopodium* *amaranticolor*, *Datura metal*, *D. stramonium*, *D. inoxia*, and *Phytolacca americana* have shown considerable promise in inhibiting SCMV infection (51,52). However, the scope of these approaches in a long duration crop like sugarcane is limited as maintaining and keeping plants SCMV free in the field would be difficult.

A useful approach to produce virus-free commercial seed-cane is to select cane from a healthy crop raised in virus-free areas, as suggested for the production of pea seeds free from pea seed borne mosaic virus (PSbMV) in India (86). But, such seed-canes require careful testing as they may not be free from infection as was found for cassava infected with Indian cassava mosaic virus (46) and permissible "tolerance limit" for commercial seed-cane should be determined. Biotechnological methods like enzyme-linked immunosorbent assay (ELISA) (21), immunosorbent electron microscopy (ISEM) (50), immunocapture polymerase chain reaction (89), etc. should find a place in programs for developing SCMV free planting materials. Diagnosis based on such methods would also ensure risk free movement of germplasm. To minimize the risk in germplasm exchange, ELISA and ISEM are used for SCMV detection at quarantine stations in Columbia, France, Mauritius, Morocco, Senegal, South Africa and USA (5).

Avoidance of Alternate Hosts

The natural host range of SCMV is restricted to the members of Poaceae. A number of cultivated and wild grasses which serve as natural reservoir of SCMV have been identified (9,12,14,57,60,79,82) (Table 6). Of these, corn and sorghum are of most importance for SCMV spread to sugarcane as they are planted in areas adjacent to sugarcane (80). Eradication of these hosts at frequent intervals during the cropping season yields good results (13,43).

Vector Control

A number of aphid species are important in the field spread of SCMV (Table 7). Of these, *Dactynotus ambrosiae*, *Melanaphis sacchari* and *Rhopalosiphum maidis* are considered efficient vectors (13,43). The virus is more readily transmitted to or from hosts such as corn and sorghum than to or from sugarcane (80).

The success of vector control measures

Table 5. SHWT with various time-temperature combinations developed for curing SCMV seed-cane infection

Treatment	Strain	Clone	% cure	% inhibition in germination	Reference
52,57.3,57.3° C at 24 h intervals, 20 min each	A,B,D	CP65-357	90	-	10
54.8,57.3,57.3°C at 24 h intervals, 7 min each	A,B,D	CO419	65	20	25
51.5,60,60°C at 24 h intervals for 20, 12.5 and 25 min respectively.	A,D	BL4	51	-	11
51.5,60, 60° C at 24 h intervals for 20, 12.5 and 22.5 min respectively	F	NCO310	100	-	11
52,54.8,54.8°C at 24 h intervals for 20 min each	H,I	CP65-357	90	-	10
52,57,57C° at 24 h intervals for 20 min each	H,I	NCO310 CP52-68 CP65-357	15(100) 41(100) 33(100)	28(83) 30(38) 86(50)	88
51.5,60,60°C at 24 h intervals for 20, 12.5 and 22.5 min respectively.	H	NCO310	100	-	11

Values in parentheses represent effect of heat therapy in combination with tissue culture

depends on the availability of information on vector population dynamics. High population of aphids throughout the long period in fall and spring has been attributed to maximum SCMV spread and incidence in Louisiana (USA) (8). Bailey and Fox (6) observed rapid spread of SCMV in late summer to early fall in South Africa. Similarly, peak activity of *R. maidis* and maximum SCMV incidence were observed in July and August in India (55).

Chemical Control

Though insecticidal control of vectors is commonly practiced for the management of viral diseases, it has not found its' place in SCMV management, of the sugarcane crop. Besides, Charpentier (16) reported that the application of insecticides failed to prevent vectors from spreading SCMV.

Table 6. Wild and cultivated grasses as natural reservoirs of SCMV

Host Species	Strain
Arundinaria gigantea	A
Digitaria didactyla	SCMV-BC
Echinocloa spp.	A,F
Eleusine indica	A,F
Eragrostis ferruginea	Kazekusa mosaic virus group
Imperata cylindrica var.*koenigii*	Chigaya mosaic virus group
Miscanthus sinensis	Susuki mosaic virus group
Pennisetum purpureum	A,F
P. typhoides	A,F
Sorghum halpense	SCMV-JG
Sorghum vulgare	A,F
Stenotaphrum secundatum	E
Urochloa mosambicensis	SCMV-Sabi
Zea mays	A,F

BC: Blue Couch Grass Strain; JG: Johnsongrass Strain; Sabi: Sabi Grass Strain

Table 7. Aphid vectors responsible for SCMV spread in the field

From	To		
	Sugarcane	Corn	Sorghum
Sugarcane	1-14	3,8,10-13	3,8,10-13
Corn	3,8,10-13	3,8,10,12,13	
Sorghum	3,8,10-13	3,8,10-13	

1: *Acyrthosiphon pisum* ; 2: *Amphorophora sonchi*; 3: *Aphis gossypii*; 4: *Aphis nerii*; 5: *Carolinaia cyperi*; 6: *Dactynotus ambrosiae*; 7: *Hysteroneura setariae*; 8: *Lipaphis pseudobrassicae*; 9: *Melanaphis idiosacchari*; 10: *Melanaphis sacchari*; 11: *Myzus persicae*; 12: *Rhopalosiphum maidis*; 13: *Rhopalosiphum rufiadominalis*; 14: *Schizaphis graminum*. Source: (13,43)

Biological Control

Management of SCMV through biological control of vectors has not received the attention it deserves. Rhizvi and Bhargava (58) reported parasitization of aphid spp. colonies by a number of predators (Table 8). However, vector control would depend on those predators which can survive on the surface of cane, corn and sorghum particularly on the young top region.

Cultural Control

Since SCMV has host range mainly restricted to the members of Poaceae only (80), modification of cultural practices, such as roguing, isolation of crops in time and space and crop rotation offer considerable promise in impeding

Table 8. Predators on aphid spp. colonies

Predators	Aphid Spp.
Coccinellid grubs	A. gossypii L. pseudobrassicae M. sacchari M. persicae R. maidis
Epilachna beetle	A. gossypii L. pseudobrassicae M. persicae R. maidis
Trioxys sp.	M. sacchari

SCMV incidence.

Roguing of Diseased Plants

The traditional practice of roguing from fields has been extremely useful in minimizing SCMV spread and maintaining virus free seed nurseries (7,13,23,47). However, for roguing to be economical and effective, it should be commenced early and repeated 3-4 times in the growing season (43). Chemicals like Erbon or Glytoc have been used for roguing to replace digging out of plants (24.41).

Planting Time

Selection of suitable planting time so as to avoid the coincidence of susceptible growth stage with the peak activity of infective aphids helps in escaping infection (43). Avoidance of planting of susceptible cane crops in late spring to mid summer in South Africa gave good control of SCMV (7).

Crop Rotation

The practice of crop rotation after plant and ratoon crops with rice, wheat, and green manure crops (especially Sesbania spp.) is in vogue (74). The effect of this cropping system on SCMV incidence needs to be worked out.

Use of Resistant Clones

In the absence of proper chemical control, use of resistant clones is the pragmatic and environmentally sound approach to manage SCMV. Breeding for SCMV resistance has been the main objective of most sugarcane improvement programs in various countries (43). Though resistance to SCMV essentially comes from wild canes, their genetic basis of resistance is not known. S. spontaneum and S. barberi have been mainly employed in breeding program to impart resistance in cultivated clones (34,73). A number of clones have been identified as tolerant/resistant to SCMV in many countries (Table 9).

Transgenic Plants

Coat-protein mediated resistance is the most effective non-conventional strategy to impart resistance to specific viruses (87). There are numerous reports of transfer of viral coat protein genes conferring resistance including SCMV (37). Recently, transient expression of the SCMV coat protein gene in sugarcane protoplast has been reported (29,72).

CONCLUSIONS

No single approach can provide effective and long lasting management of SCMV. Judicious integration of use of virus free seed-cane, appropriate cultural practices and resistant clones

Table 9. Clones identified as tolerant/resistant to SCMV

Country	Clones
1. Cuba	B4362, B42231
2. India	B.O.91,B.O.70,B.O.101,CO212,CO453,CO1148,COS802, 837,7918,8407,8408,8432,85233,CoSe93232,92231,92234, 92422,92423,92426, 92427,92428,92430, 92437,93231,92439
3. United States	LCP 86-454,LCP 87-17, LCP 87-491,L88-46,L88-63, CP88-702, CP88-739,CP88-764,CP88-769,CP70-321,CP74-383,CP79-318,L CP82-89,LHo 83-153, LCP 85-384,HoCP 85-845
4. Phillippines	Phil 6723, Phil 7495
5. South Africa	N 12
6. Taiwan	ROC1,ROC3,ROC5,ROC6,ROC7,ROC8,F160,F177

(developed through conventional breeding or genetic transformation) can only provide ideal control of SCMV. For the production of virus free seed-cane on large scale, SCMV free areas need to be identified. Similar approach has been successful in generating PsbMV-free germplasm and peanut stripe virus-free peanut germplasm (35,91). Though use of resistant clones is the most economical preposition, fundamental studies on identification of SCMV resistant genes in these clones is necessary. Further , for obtaining the optimum advantage of available resistance in clones, appropriate cultural practices suitable for particular ecosystem need to be followed. In the future, transgenic sugarcane expressing SCMV coat protein gene may form an integral part of environmentally safe sugarcane crop improvement program.

REFERENCES

1. Abbott, E.V. 1960. Studies on mosaic problem in Louisiana. Sugar Bull. 39(2):23-27.
2. Abbott, E.V., and Stokes, I.E. 1966. A World survey of sugarcane virus strains. Sugar Y Azucar. 61:27-29.
3. Abbott, E.V., and Tippett. R.L. 1966. Strains of sugarcane mosaic virus. U.S. Dept. Agric. Res. Serv. Tech. Bull. 1340. 25 Pages.
4. Artschwager, E., and Brandes, E.W. 1958., Sugarcane (Saccharum officinarum). U.S. Dept. Agric. Handbook 122. 307 Pages.
5. Autrey. L.J.C., Dhayan, S., and Sullivan, S. 1989. An evaluation of quarantine procedures adopted for sugarcane in sixteen countries. Proc. Int. Soc. Sugar Cane Technol 20(2):776-795.
6. Bailey, R.A., and Fox, P.H. 1980. The susceptibility of sugarcane varieties to mosaic and the effect of planting date on mosaic incidence in South Africa. Proc. S. Afr. Sugar Technol. Assoc. 64:1-7.
7. Bailey, R.A., Cronje, C.P.R., and Bechet, G.R. 1992. Integrated control of sugarcane mosaic. Int. Sugar J. 94:68.
8. Benda, G.T.A. 1969. Sugarcane mosaic in Louisiana: Some aspects of a chronic problem. Proc. Am.Soc. Sugar Cane Technol. 16:61-81.
9. Benda, G.T.A. 1970. Sugarcane mosaic virus from Arudinaria gigantea, a Bamboo. Plant Dis. Rep. 54:815-816.
10. Benda, G.T.A. 1972. Control of sugarcane mosaic by serial heat treatment. Proc. Int. Soc. Sugar Cane Technol. 14:55-960.
11. Benda, G.T.A., Mock, R.G., and Gillaspie, A.G. 1989. Control of sugarcane mosaic by serial heat treatment. II. The pattern of cure at high temperatures. Sugar Cane 2:6-11.
12. Bhargava, K.S. 1971. Investigations on virus diseases of sugarcane in relation to sugar industry. Rept. PL480 grant No FG.IN.274. Project No.A7-CR 112. 203 Pages.
13. Bhargava, K.S. 1975. Sugarcane Mosaic - retrospect and prospects. Indian Phytopathol. 28(1):1-11.
14. Bhargava, K.S., Joshi, R.D,. and Rishi,N. 1969. Elephant grass (Pennisetum purpureum) a natural host of sugarcane mosaic virus. Sugarcane Pathol. Newsl. 3:15.
15. Brandes, E.W. 1919. The mosaic disease of sugarcane and other grasses. U.S. Dep. Agric. Dep. Bull. 829. 26 Pages.

16.   Charpentier, L.J. 1956. Systemic insecticide studies for control of vectors and sugarcane mosaic in Louisiana. J. Econ. Entomol. 49:413-414.

17.   Chen, C.T., Yang, S.L., and Deng, T.C. 1990. Studies on sugarcane mosaic and maize dwarf viruses in Taiwan. Taiwan Sugar 37:9-15.

18.   Chiu, W. F. 1988. The economic impact of filamentous plant viruses- China. Pages 387-391 in : The Filamentous Plant Viruses. R.G. Milne, ed. Plenum Press, New York.

19.   Chona, B.L. 1944. Sugarcane mosaic and its control. Indian Farming 4:178-181.

20.   Chona, B.L. 1958. Some diseases of sugarcane reported from India in recent years. Indian Phytopathol. 11:1-9.

21.   Clark, M.F., and Adams, A.N. 1977. Characteristics of the microplate method of enzyme-linked immunosorbent assay for the detection of plant viruses. J .Gen. Virol. 34:475-483.

22.   David, H., Easwaramoorty, S., Jayanthi, R., and Mukunthan, N. 1992. Integrated pest management in Sugarcane. Pages 77-96 in : Farming Systems and Integrated Pest Management. J.P. Verma and A. Varma, eds. Malhotra Publishing House, New Delhi, India.

23.   Edgerton, C.W. 1958. Mosaic. Pages 227-248 in : Sugarcane and Its' Diseases. C.W Edgerton, ed. Louisiana State Univ. Press, Baton Rouge.

24.   Fanguy, H.P., and Garrison, D.D. 1977. Roguing sugarcane mosaic using glyphorate (Abstr.). Proc. Am. Soc. Sugar Cane Technol. 6(NS):102.

25.   Farrag. S.H., and Kandarsamy, T.K. 1980. Control of sugarcane mosaic virus by heat therapy. Proc. Int. Soc. Sugar Cane Technol. 17(2):1536-1543.

26.   Fauquet, C., and Wechmar, M. B. Von. 1988. The economic impact of filamentous plant viruses - Africa. Pages 379-384 in : The Filamentous Plant Viruses. R.G. Milne, ed. Plenum Press, New York.

27.   Forbes, I.L., and Steib, R.J. 1964. Controlling sugarcane mosaic. Sugar J. 27:51-52.

28.   Frenkel, M.J., Jilka, J., McKern, N.M., Strike, P.M., Clark, J.M., Jr., Shukla, D.D., and Ward, C.W. 1991. Unexpected sequence diversity in amino-terminal ends of the coat proteins of strains of sugarcane mosaic virus. J. Gen. Virol. 72:237-242.

29.   Gambley, R.L., Bryant, J.D., and Smith, G.R. 1994. Microprojectile transformation of sugarcane meristems and callus for SCMV resistance. Page 21 in: Proc. 1SSCT Pathology Workshop, 4ᵗʰ.

30.   Gillaspie, A.G., Mock, R.G., and Smith, F.F. 1978. Identification of sugarcane mosaic virus and characterization of strains of the virus from Pakistan, Iran and Cameroon. Proc. Int. Soc. Sugar Cane Technol. 16: 347-355.

31.   Gillaspie, A.G., and Mock, R.G. 1979. Recent survey of sugarcane mosaic virus strains from Colombia, Egypt, and Japan. Sugarcane Pathol. Newsl. 22:1-23.

32.   Gillaspie, A.G., Chen, C.T., Mock, R.G., and Harris, R.W. 1980. Sugarcane mosaic virus strains in Taiwan. Proc. Int. Soc. Sugar Cane Technol. 17:1505-1509.

33.   Gough, K.H., and Shukla, D.D. 1993. The nucleotide sequence of Johnsongrass mosaic potyvirus genome RNA. Intervirology 136:181-192.

34.   Grisham, M.P., Burner, D.M., and Legendre, B.L. 1992. Identification of resistance to sugarcane

35.   mosaic virus among wild relatives of sugarcane. Int. Sugar J. 94:67.

35.   Hampton, R.O., Kraft, J.M., and Muehlbauer, F.J. 1993. Minimizing the threat of seedborne pathogens in crop germplasm - elimination of pea seedborne mosaic virus from the USDA - ARS germplasm collection of Pisum sativum. Plant Dis. 77:220-224.

36.   Hendre, R.R., Mascarenhas, A.F., Nadgir, A.L., Pathak, M., and Jagannathan, V. 1975. Growth of mosaic virus-free sugarcane plants from apical meristems. Indian Phytopathol. 28:175-178.

37.   Hull, R., and Davies, J.W. 1992. Approaches to nonconventional control of plant virus diseases. Critical Reviews in Plant Sciences 11:17-33.

38.   King, N.C. 1955-56. Cane diseases and weed control. Ann. Rept. Expt. Stat. S. Afr. Sugar Assn. Pages 25-30.

39.   Koike, H. 1977. Diseases as a factor influencing sugarcane yields in Louisiana during the last decade. Proc. Am. Soc. Sugar Cane Technol., 6(NS):178-181.

40.   Koike, H., and Yang, S. 1971. Influence of sugarcane mosaic virus strain H and Pythium graminicola on growth of sugarcane. Phytopathol. 61:1090-1092.

41.   Koike, H., and Tippett, R.L. 1972. Rogueing mosaic diseased stubble in seed plots with herbicide. Sugar Bull. 50(13):10-11.

42.   Koike, H., and Gillaspie, A.G. 1976. Strain M, a new strain of sugarcane mosaic virus. Plant Dis. Rep. 60:50-54.

43.   Koike, H., and Gillaspie, A.G. 1989. Mosaic. Pages 301-322 in : Diseases of Sugarcane - Major Diseases. Ricaud et al., eds. Elsevier Science Publishers, Amsterdam.

44.   Lesemann, D.E., Shukla, D.D., Tosic, M., and Huth, W. 1992. Differentiation of the four viruses of the sugarcane mosaic virus subgroup based on cytopathology. Pages 353-361 in: Potyvirus Taxonomy, Arch. Virol. Suppl. 5. O.W. Barnett, ed. Springer-Verlag, New York.

45.   Leu, L.S. 1972. Freeing sugarcane from mosaic virus by apical meristem culture and tissue culture. Ann. Rept. Taiwan Sugar Expt. Stn. 57:57-63.

46.   Malathi, V.G., Varma, A., and Nambisan, B. 1989. Detection of Indian cassava mosaic virus by ELISA. Curr. Sci. 58:149-150.

47.   McClean, A.P.D. 1932. Mosaic disease in sugarcane and its control in South Africa. Proc. Int. Soc. Sugar Cane Technol. 4:27.

48.   McKern, N.M., Whittaker, L.A., Strike, P.M., Ford, R.E., Jensen, S.G., and Shukla, D.D. 1990. Coat protein properties indicate that maize dwarf mosaic virus - KSI is a strain of Johnsongrass mosaic virus. Phytopathol. 80:907-912.

49.   McKern, N.M., Shukla, D.D., Toler, R.W., Jensen S.G., Tosic, M., Ford, R.E., Leon, O., and Word, C.W. 1991. Confirmation that the sugarcane mosaic virus subgroup consists of four distinct potyviruses by using peptide profiles of coatproteins. Phytopathol. 81:1025-1029.

50.   Milne, R.G., and Lesemann, D.E. 1984. Immunosorbent electron-microscopy in plant virus studies. Pages 85-101 in : Methods in Virology, Vol. VIII. K. Maramorosch and H. Koprowski, eds. Academic Press, Orlando, FL.

51.   Molina, M., Peralta, E.L., and Sierra, P. 1986. Inhibitory affect of Phytolacca americana extracts

on sugarcane mosaic virus. Revista de Proteccion Vegetal 1(1):15-20.

52. Molina, M., and Leon, O. 1991. Antiviral activity of leaf extracts of *Bougainvillea spectabilis*, *Capsicum annuum*, *Datura metel* and *Datura stramonium* on sugarcane mosaic virus. Revista de Proteccion Vegetal 6(2-3):150-155.

53. Musschenbroek, S.C. Van. 1892. Beschrijving vantot dusueree in West Java onbokende Rietziekten, Soerabajasche Verreinging van suiker Fabrikanten, Circulative No.42, Bijlage. Pages 113-118.

54. Prakash, J., and Joshi, R.D. 1977. 2,4 - dinitrophenol - an inhibitor of sugarcane mosaic virus. Int. Sugar J. 79:303.

55. Rao, G.P., Singh, S.P., and Upadhyaya, P.P. 1994. Population dynamics of aphids in relation to ecological parameters and spread of sugarcane mosaic virus in two districts of Uttar Pradesh, India. Pages 151-158 in : Current Trends in Sugarcane Pathology. Rao et al., eds. Int. Books and Periodicals Supply Service, Delhi, India.

56. Ricaud, C., Egan, B.T., Gillaspie, A.G., and Hughes, C.G. 1989. Diseases of Sugarcane - Manor Diseases. Elsevier Science Publishers, Amsterdam. 399 Pages.

57. Rishi, N. 1972. An alternate host for sugarcane mosaic. Sugarcane Path. Newsl. 9:1972.

58. Rizvi, S.M.A., and Bhargava, K.S. 1973. Predators and parasites of vectors of sugarcane mosaic. Sugarcane Pathol. Newsl. 10:18-19.

59. Roth, G. 1973. The elimination of the virus diseases "streak" and "mosaic" by thermotherapy and tissue culturing, Abst. No. 901. 2nd Int. Cong. Plant Pathol., Minnesota.

60. Saladini, J.L., and Zettler, F.W. 1972. Characterisation of strain E of sugarcane mosaic virus infecting St. Augustine grass. Plant Dis. Rep. 56:885-889.

61. Shah, S.S. 1972. Sugarcane agriculture, contribution of Sugarcane Breeding Institute. Indian Farming 22:38-43.

62. Shaheen, M.S., and Mirza, M.S. 1989. In vitro regeneration of virus free sugarcane plants. Int. Conf. on in vitro selection and propagation of economic plants, 13-26 March, Peshawar, Pakistan.

63. Shukla, D.D., Tosic, M., Jilka, J., Ford, R.E., Toler, R.W. and Langham, M.A.C. 1989. Taxonomy of potyviruses infecting maize, sorghum and sugarcane in Australia and the United States as determined by reactivities of polyclonal antibodies directed towards virus-specific N-termini of coat proteins. Phytopathology 79:223-229.

64. Shukla, D.D., Frenkel, M.J., McKern, N.M., Ward, C.W., Jilka, J., Tosic, M., and Ford, R.E. 1992. Present status of the sugarcane mosaic subgroup of potyviruses. Pages 363-373 in : Potyvirus Taxonomy, Arch. Virol. Suppl. 5. O.W. Barnett, ed. Springer-Verlag, New York.

65. Shukla, D.D., Ward, C.W., and Brunt, A.A. 1994. The Potyviridae. CAB International, Wallingford, U.K. 516 Pages.

66. Shukla, K. 1978. Studies on some aspects of a mosaic disease of sorghum from Gorakhpur. Ph.D. Thesis. Univ. Ghorakhpur, India, Pages 155-156.

67. Shukla, K., and Joshi, R.D. 1979. Influence of some phenolic acids on the infectivity of sugarcane mosaic virus. Sugarcane Pathol. Newsl. 23:19-21.

68. Shukla, K., and Joshi, R.D. 1982. Inhibition by some homeopathic drugs of sugarcane mosaic virus in sorghum. Sugarcane Pathol. Newsl. 29:48-50.

69. Singh, K. 1971. Virus diseases of sugarcane and the seed programme. Advances in Agriculture (Kanpur) 1:69-87.

70. Singh, K. 1977. Sugarcane diseases and the three tier seed programme. Sugar News 9:81-89.

71. Singh, R.P., and Agnihotri, V.P. 1987. Thermotherapy of sugarcane for disease control. Rev. Trop. Pl. Path. 4:305-330.

72. Smith, G.R., Ford, R., Frenkel, M.J., Shukla, D.D., and Dale, J.L. 1992. Transient expression of the coat protein of sugarcane mosaic virus in sugarcane protoplasts and expression in *Escherichia coli*. Arch. Virol. 125:15-23.

73. Srinivasan, K.V., and Chenulu, V.V. 1956. A preliminary study of the reaction of Saccharum spontaneum variant to red rot, smut, rust and mosaic. Proc. Int. Soc. Sugar Cane Technol. 9:1097-1107.

74. Srivastava, S.C., Johari, D.P. and Gill, P.S. 1988. Manual of sugarcane production in India. ICAR, New Delhi. 194 Pages.

75. Steib, R.J., and Cifuentses, O.W. 1977. Use of aerated steam as a possible method for the control of sugarcane mosaic and ratoon stunting disease. Sugarcane Pathol. Newsl. 18:24-27.

76. Summers, E.M. 1936. An investigation of types or strains of the mosaic virus of sugarcane in Louisiana. Iowa State Coll. J. Sci. 11:118-120.

77. Summers, E.M. 1943. Effect of sugarcane mosaic caused by different strains of virus in yield of susceptible sugarcane varieties. Sugar Bull. 21:181-183.

78. Summers, E.M., Brandes, E.W., and Rands, R.D. 1948. Mosaic of sugarcane in the United States, with special reference to strains of the virus. U.S. Dept. Agric. Tech. Bull. No. 955. 124 Pages.

79. Teakle, D.S., and Grylls, N.E. 1973. Four strains of sugarcane mosaic virus infecting cereals and other grasses in Australia. Aust. J. Agric. Res. 24:465-477.

80. Teakle, D.S., Shukla, D.D., and Ford, R.E. 1989. Sugarcane Mosaic Virus. AAB Description of Plant Viruses, No. 342.

81. Tippett, R.L., and Abbott, E.V. 1968. A new strain of sugarcane mosaic virus in Louisiana. Plant Des. Rep. 52:449-451.

82. Toriyama, S., and Yora, K. 1972. Virus diseases of wild grasses and cereal crops in Japan. Univ. Tokyo Press. 68 Pages.

83. Tosic, M., Ford, R.E., Shukla, D.D., and Jilka, J. 1990. Differentiation of sugarcane, maize dwarf, johnsongrass and sorghum mosaic viruses based on reactions of oat and some sorghum cultivars. Plant Dis. 74:549-552.

84. Varma, A. 1988. The economic impact of filamentous plant viruses - the Indian Subcontinent. Pages 371-376 in : The Filamentous Plant Viruses. R.G. Milne, ed. Plenum Press, New York.

85. Varma, A. 1993. Integrated management of plant viral diseases. Pages 140-157 in : Crop Protection and Sustainable Agiculture. Wiley, Chichester (Ciba Foundation Symposium 177).

86. Varma, A., Khetarpal, R.K., Vishwanath, S.M., Kumar, D., Maury, Y., Sharma, B., and Tyagi, M.C. 1991. Detection of pea seedborne mosaic virus in commercial seeds of pea and germplasm of pea and lentil. Indian Phytopathol. 44:107-112.

87. Varma, A., and Sinha, S.K. 1992. Sustainable development through long term biotechnological alternatives in agriculture. Pages 44-57 in : Proceedings of the International Seminar on Impacts of Biotechnology in Agriculture and Food in Developing Countries. Committee on Science and Technology in Developing Countries (COSTED), Madras.

88. Waterworth, P. and Kahn, R.P. 1978. Thermotherapy and aseptic bud culture of sugarcane to facilitate the exchange of germplasm and passage through quarantine. Plant Dis. Rep. 62:772-776.

89. Wetzel, T., Condresse, T., Macquaire, G., Ravelanandro, M., and Denez, J. 1992. A highly sensitive immunocapture polymerase chain reaction method for plum pox potyvirus detection. J. Virol. Methods 39:27-37.

90. Xiao, X.W., Frenkel, M.J., Teakle, D.S.,Ward, C.W., and Shukla, D.D. 1993. Sequence diversity in the surface-exposed amino-terminal region of the coat proteinsof seven strains of sugarcane mosaic virus correlates with their host range. Arch. Virol. 136 : 381-387.

91. Zettler, F.W., Elliott, M.S., Purcifull, D.E., Mink. G.I., Gorbet, D.W., and Knauft, D.A. 1993. Production of peanut seed free of peanut stripe and peanut mottle viruses in Florida. Plant Dis. 77:747-749.

92. Zummo, N., and Stokes, I.E.. 1973. Sugarcane mosaic strain K : A new strain of sugarcane mosaic virus in Meridian, Mississippi. Sugarcane Pathol. Newsl. 10:16-17.

CHAPTER 38

Present Status of Controlling Cucumber Mosaic Virus

D. Gallitelli

Cucumber mosaic virus (CMV) is the type species of the genus *Cucumovirus* in the family *Bromoviridae* (98) and is recognized as the causal agent of disease of many important crops throughout the temperate and some of the tropical regions of the world. CMV has probably the largest host range of any virus i.e. over 800 plant species, representing no less than 85 botanical families. Recently, CMV was associated with tomato diseases of epidemic scale in several countries of the Mediterranean basin. During epidemics, at least two new disease phenotypes appeared; one of which was co-determined by a variant of CMV satellite RNA (satRNA) and the other was caused by CMV alone (see below). This situation renewed the scientific interest for: (i) obtaining new information on the eco-epidemiology of CMV and its satRNA; (ii) searching for a match between data collected in the "natural laboratory" with those previously obtained experimentally; (iii) setting up new strategies for controlling virus epidemics. Indeed, most of these control strategies would have been only of scientific interest if CMV outbreaks were not the cause of enormous economic losses. Because of this, their application into practice is more and more exploited.

'NEW DISEASES' CAUSED BY CMV

Tomlinson (120) listed CMV as the first virus of economic importance in celery, cowpea, cucurbit, lettuce, pepper and tomato crops. Limitedly to vegetable crops, CMV appears to be the most important virus in Eastern China, Croatia, France, Egypt, Greece, Israel, Italy, Japan, Poland, Portugal, Sweden and in the northeastern United States. In other countries, CMV ranks 2nd or 3rd (120). Palukaitis *et al.* (90) reviewed CMV-

induced diseases in banana and pasture legumes and Tricoli *et al.* (123) listed CMV among the most damaging viruses to squash production. In the Mediterranean basin the economic importance of CMV, correlates with recurrent outbreaks in canning tomato crops (42). After the tomato necrosis disease in Alsace (France) in 1972, tomato diseases induced by CMV became suddenly important in Italy and Spain in 1987 (41, 63) where the virus caused up to 100% crop losses in the main production areas. To date, a number of reports indicate that CMV populations, with or without satRNA, are well established in tomato cultivated areas in the Mediterranean where CMV is frequently found in mixed infections with alfalfa mosaic alfamovirus (AMV), potato Y potyvirus (PVY), potato M carlavirus (PVM) and tomato spotted wilt tospovirus (TSWV).

In addition to the well known fern leaf/shoestring and tomato necrosis (synonym: tomato lethal necrosis) diseases (68), tomato fruit necrosis and tomato top stunting, two previously unrecorded disorders were observed during CMV epidemics.

Tomato Fruit Necrosis

A disease of tomato fruits caused by CMV and distinguishable from the physiological disorder known as 'internal browning' (10). In the field, the symptoms consist of more or less extended internal necrosis of the fruits. In most cases, the plants are vigorous and do not show symptoms on the foliage, except for occasional mild discolorations of the leaves. Sometimes, however, the plants show also the fern leaf/shoestring condition typically induced by ordinary CMV strains. Fruit tissues affected by necrosis are the mesocarp and particularly those surrounding the base of the pedicel. Externally, brown soft blotches correspond to altered tissues. From tomato plants with symptomless foliage and fruit necrosis, a subgroup I CMV strain (CMV-Tfn) was consistently isolated. CMV-Tfn supported a 390-ribonucleotide satellite named Tfn-satRNA (25, 27). This satellite RNA was not present in plants showing both fruit necrosis and fern leaf/shoestring. Its occurrence was apparently influent in determining the fruit necrosis condition, but prevented the appearance of phylimorphism on the leaves. Crescenzi *et al.* (25) demonstrated that the disease was induced by a specific CMV strain that affected the vascular system of fruit pedicels. Comparable vascular alterations were not caused by the ordinary CMV strains that induce fern leaf/shoestring symptoms.

Tomato Top Stunting

This disease is characterized by a marked shortening of the internodes of apical shoots that confers a compact and bushy appearance to the vegetation. Leaflets are small, with the blade rolled upward as their middle vein curves downward and twists. Older leaves have a normal size and exhibit a mild mosaic. The crop is much reduced and unmarketable as fruits are few, small-sized and mature unevenly. The disease phenotype was co-determined by a subgroup I CMV strain (CMV-TTS) supporting a 339 nucleotide satRNA of a non-necrogenic phenotype (47). CMV-TTS artificially deprived of its satRNA induced ordinary fern-leaf and shoestring symptoms.

Similar diseases were also described in Spain under the name of internal browning and curl-stunt (63). Infected plants did not contain Tfn-like satRNA which correlates positively with the observation that under Spanish conditions, fruit necrosis was always associated with fern-leaf/shoestring symptoms.

VIRUS PROPERTIES

The molecular biology and biochemistry of CMV was recently reviewed by Palukaitis *et al.* (90). Thus, only a brief update of the most important CMV features will be given.

CMV is a virus with polyhedral virions, all the same size and morphologically undistinguishable from each other, constructed by 180 subunits (Mr ranging from 24,159 to 24,300) and exhibiting icosahedral symmetry (T=3). Virus particles encapsidate three linear plus-sense single-stranded genomic RNAs (RNA1, RNA2, RNA3) and the subgenomic RNA (RNA4) transcribed from RNA3 which codes for the viral capsid protein (CP). RNA1 and RNA2 are encapsidated in distinct particles whereas RNA3 and RNA4 are encapsidated together (78). Hence, CMV is a multi component virus for which inoculation by the three types of particles is required to infect plants.

The complete nucleotide sequence of six CMV strains (30,51,52,60,72,73,81,86,88,-96,97,102,103 and GenBank accession No. D10209) has been determined. RNA1 is monocistronic coding for a single product often designated 1a protein, that is thought to have both methyltransferase and RNA helicase activity. Until recently (34), also RNA2 was considered to be monocistronic coding for one protein, designated 2a protein, containing conserved amino acid sequence motifs (GDD) of viral replicase (RNA-dependent RNA polymerase = RdRp) proteins. Both 1a and 2a proteins function as the virus-encoded component of the specific CMV-replicase complex isolated from CMV-infected tobacco plants (53). Within the three dimensional structure of this complex, the functional regions for methyltransferase, helicase and polymerase activity are located in positions accessible to viral RNA template with which they interact during replication (54).

In addition to the 2a protein, RNA2 of CMV-Q encodes another product with unknown function, designated 2b protein, that overlaps the C-terminal portion of the major 2a gene (34). The 2b protein which is translated by a subgenomic RNA designated RNA4A, does not seem to be essential for systemic spreading of the virus in infected plants nor for replication of CMV satRNA (33, 34). The open reading frame (ORF) 2b but not the subgenomic RNA4A has been found in CMV-Ixora (81) and CMV NT9 (60).

RNA3 is also bicistronic encoding the 3a protein in the 5' proximal half of the molecule and containing the ORF for the CP in the 3' proximal half. The CP cistron is translated via the subgenomic RNA4. The 3a protein, also designated MP, is involved in viral movement (24, 33, 71, 127). Unlike MPs of most of other plant viruses, CMV MP is not located in the cell wall but, rather in the cytosol of infected plants. Although it requires the cooperation of CP for viral cell-to-cell movement, it does not induce formation of tubular structures protruding from cell membranes, as reported for other plant viruses with a CP-mediated movement (12, 71, 77).

Some CMV strains encapsidate a fifth RNA known to consist of a mixed population of molecules of about 300 nucleotides derived from the imperfectly conserved 3'-terminal region of genomic RNAs 2 and 3 (11). Since RNA5 contains the tRNA-like structure that is likely to be involved in replication of CMV (90) and other viral species in the *Bromoviridae* family (98) it is possible that CMV RNA5 plays a role in viral replication (11).

Strains of CMV can be divided into two subgroups, I (or WT) and II (or S), on the basis of their sequence similarity and serological relationships. In addition, strains of subgroup I do not appear to encapsidate the subgenomic RNA4A (56, 81).

Some isolates of CMV encapsidate a satellite RNA (satRNA), often referred to as CARNA5 (for Cucumber Mosaic Virus Associated RNA5) (68), a small RNA molecule completely dependent on viral genome for its replication and spread but that does not supply the helper virus with any essential function (21, 55, 107, 134). The effects of CMV satRNA in viral pathogenesis have been repeatedly reviewed (67 and references therein). The majority of CMV satRNAs contain 332-335 nt. Another group of these molecules, however, has a size ranging from 368 nt up to the 405 nt of the newly described KN satRNA (58). SatRNAs have attracted much scientific and practical interest as they exist in nature as several variants capable of attenuating or aggravating disease symptoms induced by the helper virus. This effect is particularly relevant in tomato where the CMV infection phenotype may be asymptomatic or lethal depending on the satRNA. Appearance of specific symptoms like tomato lethal necrosis or tobacco chlorosis was associated with certain nucleotides occurring at specific sites and to other sequence elements present in the proper sequence context (122). The host species, its cultivar, the temperature and the helper virus strain are other critical parameters leading to induction of a particular phenotype (39, 70, 130, 135). Recent studies have shown that helper strain can also drive preferential replication or amplification of one satRNA variant over another (81, 105).

EPIDEMIOLOGY

The existence of a large number of CMV isolates that cover a very broad host range suggests that the multi component nature of the virus does not act as a biological constraint to its ecological success (38). Rather, the consensus is that reassortment of intact genomic segments from multi component viruses (pseudo-recombination) is a parasexual process leading to generation of new variants able to extend the host range to new plant species and/or to implement other specific functions. Other sources of variation are, true recombination (RNA-RNA recombination) which was recently demonstrated for CMV (35) and the production of mixed-subunit capsids from two different cucumoviruses (17) that may lead to radical recognition changes in insect-mediated transmission. Viable pseudo-recombinants have been prepared *in vitro* and used to assign specific functions to individual segments of the genome (90 and references therein). Symptoms induced in particular hosts and insect-mediated transmission map in RNA3 (91, 93, 142) whereas specific helper functions for replication, disease phenotype induction, systemic movement and accumulation of satRNA in host plant are encoded by RNAs 1 and 2 (19,81,83,106). Albeit postulated several years ago, the natural occurrence of pseudo-recombinants in the genus *Cucumovirus* was demonstrated only recently in bean distortion mosaic cucumovirus (BDiMV), found in Chile (131).

CMV can be acquired and transmitted by more than 80 species of aphids with very short probes on infected plant. By analogy with potyviruses, CMV particles probably bind to the anterior portion of the food canal of the vector because of interactions between the surface domains of the virion, the host and the vector (93). The rate of transmission can be affected by the species of aphid, of the donor and the recipient plant, and by the virus strain (92). Some CMV strains are not or are poorly transmissible by aphids. These differences in transmissibility are probably due to few differences in amino acid composition of the virus shell, although it is not yet known how they affect transmissibility (92, 93)

CMV is also transmitted by seeds of several plant species. Among cultivated crops, seed transmission of CMV seems particularly relevant for species in the *Leguminosae* and, to a lesser extent, in some cucurbits. Hampton and Francki (49) studied seed transmission of CMV in *Phaseolus vulgaris* using pseudo-recombinants prepared *in vitro* by exchanging genomic RNAs of a non seed-transmitted strain (CMV-Le) and the Pg strain which is transmitted in 20 to 50% of the seeds. None of the six possible recombinants was transmitted with the same efficiency of the Pg strain and, interestingly, seed-transmission, appeared to map on RNA1. In turn, this suggests the existence of some relationships between RNA1-coded product of the viral replicase complex and seed-transmission. Yang *et al.* (137) reported that CMV can be replicated in spinach reproductive organs (ovary, anthers and endosperm tissues) and seeds. Hampton and Francki (49) postulated that seed-transmission in *P. vulgaris* could be dependent on replication after the virus entered the embryo. Lack of host-encoded products of the viral replicase complex in embryonic cells could account for the absence of seed-transmission in a number of plants that host CMV infections.

The rate of seed-transmission in weed species remains largely unknown although seeds as well as rhizomes of certain plant species seem fundamental for perpetuation of CMV strains in cultivated areas (100). Crescenzi *et al.* (26) have shown that during CMV outbreaks in tomato in two Italian regions (Basilicata and Campania), weeds played a key role in determining epidemics of subgroup II isolates which prevailed in wild plants throughout winter. Since CMV isolates of the subgroup II were associated with lethal necrosis disease in tomato, their absence in summer correlated positively with the natural disappearance of some weeds in this season and with the self-destructive nature of lethal necrosis. Therefore, in such condition only the weeds were able to maintain (probably in their seeds) the inoculum until next autumn. The situation appeared different with subgroup I strains for which distinct crops rather than weeds represented inoculum sources. Cultivated areas rather than weeds seemed also fundamental in the epidemiology of CMV-satRNA during viral epidemics in Italy, because: (I) satRNA sequences were found only in few weeds (26) and (ii) heterogeneity among satRNA variants isolated in Italy from 1988 to 1993 did not vary significantly from one epidemic episode to the other (F. Grieco, C. Lanave and D. Gallitelli, unpublished results). In addition, a phylogenetic study showed that a satRNA variant from tobacco appeared to be the common ancestor of most of the necrogenic variants isolated in Italy. This suggests that a crop, rather than a weed, could have been the *focus* leading to amplification and subsequential epidemic spread of CMV-satRNA (48, F. Grieco, C. Lanave and D. Gallitelli, unpublished results).

CONTROL

Diagnosis

Control of plant virus diseases capitalizes on the availability of sensitive, reliable and rapid methods of diagnosis. Both serological and molecular techniques have been adapted and successfully applied to diagnosis of CMV in order to: (i) allocate newly recognized field isolates in one of the two currently recognized viral subgroups, (ii) identify CMV and its satRNA as etiological agents of specific plant disease and (iii) obtain timely virus detection for quarantine and certification purposes. Although most CMV strains are poorly immunogen, efficient antisera can be raised using aldehyde-stabilized CMV particles. However, if the test is addressed to strain differentiation, also the reacting antigen must be fixed because strain-differentiating antigenic determinants appear to be present only on the surface of intact virus particles (90 and references therein). Ziegler *et al.* (143) have selected antibody fragments (single-chain FV=scFV) from a library of synthetic fragments displayed on the surface of a filamentous bacteriophage that specifically recognize CMV particles in sap of infected plants. Selection of scFVs from phage display library, overcomes recourse to animal immunization and several other difficulties such as the poor immunogenicity that represents a constraint for the production of polyclonal sera and monoclonal antibodies. ScFV specific to CMV performed well either in plate trapped antigen ELISA (PTA-ELISA) (121) or in immunoblotting using lupin seed extracts. The fragments reacted with CMV strains of both subgroups but there is no information available on their use for strain differentiation.

ELISA is routinely used in CMV basic research and diagnosis (29, 31, 74). Recently, direct antigen coated ELISA (DAC-ELISA) has been used for assessment of virus-free accessions of cowpea seed lots (44). The viral assay was carried out with leaf disks excised from the top leaflets of three trifoliolate leaves of cowpea seedlings. However, a great effort is currently being made in minimizing manipulation of the samples collected. Immunoassay techniques have been adapted to field use to overcome complicated procedures for extracting the virus from plant tissue. Ideally, the test itself should be done in the field but methods that overcome at least sample transfer to the laboratory, would already represent a great practical improvement. Squash blot, tissue print or whole leaf press are examples of such applications. Riedle (99) has blotted CMV-infected leaves of cucumber plants by pressing them between two pieces of nitrocellulose membrane. The method allowed immunodetection of CMV and, at the same time, virus localization in specific leaf tissues. A high concentration of virus particles was found in the leaf hair. This may imply that dispersal of CMV particles could also occur by aerosol produced by hair breakage, as with contact-transmitted viruses.

Immunological techniques, albeit practical and convenient, often are less sensitive than molecular methods. In a comparative study for detection of CMV in banana plants in Hawaii, Hu et al. (61) have shown that the dot blot assay was 100 times more sensitive than ELISA. The reverse-transcription polymerase chain reaction (RT-PCR) was even more sensitive than dot blot and therefore was recommended for examining samples that gave inconclusive responses in ELISA and dot blot. This was particularly true with CMV-Ohau, a strain that was undetectable in banana by ELISA but resulted ELISA-positive after mechanical transmission to squash.

Nucleic acid hybridization technology has been successfully applied to CMV diagnosis as it permits to overcome some limitations of immunological techniques. Among the mixed-phase hybridization formats, (62), dot blot represents the most suitable for routine testing (40). Since the first report of CMV outbreaks in tomato crops in 1988 in many Mediterranean countries, the dot blot test has been applied extensively to analyze heterogeneity of CMV populations consequent to epidemic events (26, 63, 74, 104). Dot blot was particularly useful for correlating a specific disease phenotype with the presence or absence of CMV satRNA in naturally infected plant tissues (26,41,63,69, D. Skoric , M. Krajacic , L. Barbarossa, F. Cillo, F. Grieco, A. Saric and D. Gallitelli, unpublished results). Other recent applications of nucleic acid hybridization to CMV diagnosis and strain differentiation with subgroup-specific riboprobes (89) are reported for Kava dieback disease (31) and in diseases occurred in ornamental plants (36). Localization of CMV or its satRNA in infected plant tissues was successful with *in situ* hybridization (25) and by printing sections of tomato fruits and stems onto Nylon membranes (F. Cillo and D. Gallitelli, unpublished results). Effective differentiation of CMV isolates has also been obtained with RT-PCR. Rizos *et al.* (101) used double stranded RNA

replicative intermediates extracted from infected leaf tissue as template for RT-PCR amplification with two CMV-specific primers flanking the CMV-CP gene. The amplified products were then subjected to endonuclease digestion using *Msp*I, and the isolates were divided into two groups on the basis of the restriction pattern obtained. The two groups correlated positively with the current subgroups I and II. The RT-PCR approach was used to screen banana plants affected by infectious chlorosis disease (111). Finally, Anderson *et al.* (3) proposed differentiation of CMV isolates on the basis of identification of an *Eco*RI site present only in RNA 4 of subgroup II (S) isolates. Such differentiation, including identification of distinct satRNA variants can also been obtained by RNase protection assay (RPA) (5,6). Although undoubtedly effective and highly sensitive, RT-PCR and other methods based on specific sequences identification seem laborious for routine testings when hundred of samples are to be analyzed in a short time. ELISA and dot blot hybridization are more suitable for this purpose; especially the latter, considering that blotted membranes prepared in far apart places, can easily be mailed to the same central laboratory for hybridization. A dot blot hybridization system using digoxigenin-labeled riboprobes and chemiluminescent detection was developed for the diagnosis of CMV and additional five viruses in the sanitary certification of about 45 million tomato seedlings in Italy (40, 108).

Control of Vectors

Insecticide sprays are effective means for controlling aphid populations but they do not act fast enough to prevent inoculation of those viruses that, like CMV, are transmitted during the probing phase on the host plant. However, properly applied treatments can reduce disease incidence by killing the aphid before it has time to move to another plant. In practice, the efficacy of insecticide sprays is greatly reduced because: (i) viruliferous aphids are dispersed and survive long enough to transmit the virus, and (ii) the low aphid specificity of CMV transmission allows aphids either to acquire from or to transmit the virus to plant species on which they probe but do not feed. Effective aphid control and consequential low disease incidence can be achieved if winged aphid forms are kept at low levels. In several instances it has been observed that CMV epidemics correlate positively with extraordinary high aphid movement. In temperate regions like those of the Mediterranean basin, this usually occurs after mild winters or after high spring rainfall followed by rapid temperature increase. The conditions greatly favor reproduction within aphid colonies that rapidly reach a crowded stage leading to emergence of numerous winged forms. Such colonies often develop on weeds that can be also CMV reservoirs and therefore their destruction is a valid means for controlling and preventing virus outbreaks. This is particularly true with annual crops. In this case, weeds should be eliminated from the field at least one week before transplanting. This measure proved effective in reducing incidence of CMV in celery, cucumber and lettuce (100 and reference therein) and tomato (D. Gallitelli, unpublished observation). To be really effective, this measure should be applied over large areas where CMV outbreaks correlate with infected weed sources rather than with the crop itself (26).

Another possible control approach is to avoid the physical contact between aphids and plants. This is relatively easy to achieve in nurseries where the plants can be grown under screenhouses equipped with double doors, insect-proof nets and yellow sticky traps. Although simple in principle, in practice this control measure could cause serious problems in reducing ventilation, thus leading to a damaging increase of temperature in the screenhouse. Elimination of this problem by forced ventilation and cooling may be very expensive. Reflective mulches have been used under several circumstances to reduce incidence of non-persistently transmitted aphid-borne viruses in the field. Summers et al. (112) made a comparison of the efficiency of spray mulches, film mulches and nets - the latter being either suspended on wire hoops to form tunnels or placed on soil surface - in protecting zucchini squash from non persistently-transmitted aphid-borne viruses. Mulching performed better than netting because it covered more than 60% of soil surface. Silver mulches were significantly more effective in repelling Aphis gossypii than other variously colored (white, brown, green, black) mulches. The repelling power of each color should be tested for each species of aphid before suggesting its use in large scale agriculture (112 and references therein). In addition, silver netting used as tunnels was initially as effective as other silver mulches in repelling aphids but after removal for the first fruit harvest, the plants became infested by aphids and showed viral symptoms very soon. Also plants protected by silver mulches became infected but only in a later stage when canopy cover significantly decreased the efficiency of mulches in repelling aphids. However, older plants can tolerate virus infection better than young ones, thus yielding marketable fruits. Finally, spray mulches appear more convenient than polyethylene film mulches for they are biodegradable, thus eliminating the problem of disposal at the end of the crop. Cucurbits possess a strong enough structure to support unwoven tissue laid directly onto the plant leaves even at the seedling stage. Tomassoli et al., (117) showed that this early protection from incoming viruliferous aphids delayed CMV infection in a highly infested area. As for net tunnels, the infection was not totally escaped because the coverage had to be periodically removed for routine agricultural practices and fruit harvesting.

Agronomic Practices

The potential of controlling CMV diseases through agronomic practices must also be considered in particular for vegetable crops. The removal of weed hosts of CMV at least one week before transplanting has already been mentioned although in some cases it must be considered unpractical depending upon the position of the field and neighboring crops. Early sowing of lettuce in New York proved effective in preventing CMV diseases (100) and avoided overlapping crops. Late transplanting of tomato seedlings may also be useful for escaping or reducing CMV incidence in Southern Italy. Late transplanting avoids the overlap of tomato growing season with celery which may host heavy infections of CMV and its satRNA (26) and allows to escape high aphid flight peaks. Regular irrigation coupled with mulching reduced incidence of tomato fruit necrosis in Southern Italy. This probably correlates with the damage caused by the virus to the vascular system of infected plants that cannot resist the physiological stress induced by an inconstant soil humidity (25, D. Gallitelli, unpublished observation).

Genetic (Conventional) Resistance

An environmentally friendly approach to control plant viruses is to confer resistance by introgression of a gene or genes usually found in wild species into a commercial variety. For CMV, resistant or tolerant varieties are available for some crops including spinach, pepper and some cucurbits. Nono-Womdim *et al.* (87) studied multiplication and systemic invasion of CMV in susceptible and tolerant lines of pepper. Lines Perennial and Vania were tolerant to CMV infection as Vania inhibited virus movement to uninoculated leaves and Perennial viral multiplication. The latter feature could be associated with lack of host-encoded products of CMV-replicase complex. Vania and Milord, another bell pepper variety that inhibits CMV movement, were also evaluated under field conditions showing appreciable levels of tolerance to CMV infections. Through recurrent selection, tolerance has been introduced into commercial F1 hybrids that are also resistant to common strains of PVY and tobacco mosaic tobamovirus (TMV). New CMV-tolerant pepper lines were recently produced by Hobbs *et al.* (59).

New accessions of *Cucurbita pepo* and *C. maxima* were also evaluated for resistance to CMV on the basis of their ability to react with local or systemic symptoms to viral infection (76). The accessions showed variable levels of tolerance although accessions from *C. maxima* were generally more tolerant than those obtained for *C. pepo*. This was also demonstrated for some accessions of *C. pepo*, previously reported as resistant to CMV infection, that expressed severe symptoms on the cotyledons but not on upper leaves. Tolerance to CMV (inhibition of systemic spread) was also reported for some lines of melon (2).

Non-conventional Resistance

Since CMV-resistant varieties will probably take long time before being commercially available, a different approach based on pathogen-derived resistance has evolved very rapidly. This approach, often referred to as "non conventional resistance", was conceptually proposed by Sanford and Johnston (109) as based on the transfer of a mild strain of the pathogen or some of its genes into an appropriate plant species. The pathogen can simply be inoculated onto the host or the latter can be genetically engineered with pathogen's gene(s). This strategy shows indisputable advantages over the 'conventional' resistance for the unlimited availability of genes as they are carried by the pathogen itself. Wilson (132) has reviewed current knowledge on different aspects of pathogen-derived resistance and Tepfer (115) has made an exhaustive analysis of the risks associated with this type of approach.

Cross-protection

Cross-protection is a mechanism whereby infection of a mild strain of a virus (protecting strain) prevents the subsequent infection or the expression of infection by a virulent strain (challenging strain). Cross-protection (often referred to as pre-immunization or vaccination) has been successfully used in several instances in the biocontrol of CMV-induced diseases. Cross-protection proved particularly useful in tomato because resistant or tolerant varieties are not available and it represents an immediate type of response against recurrent CMV outbreaks as those occurring in many Mediterranean countries. For CMV, cross-protection can be obtained using a satRNA-free mild strain or with a mild strain supporting a satRNA with ameliorative phenotype. Because of concern about safety in releasing large amounts of satRNA in the environment, the satRNA-free approach should be restricted to areas where there are no risks of outbreaks driven by CMV strains supporting necrogenic satRNAs. In these areas, the protective strain itself is likely to replicate the necrogenic satRNA with dramatic consequences (67). Therefore the ameliorative satRNA approach which can be only suggested under conditions characterized by a low natural presence of satRNA becomes imperative in places where natural necrogenic variants exist. Ecological risks associated with cross-protection using ameliorative satRNAs have been exhaustively reviewed by Tepfer (115) and by Jacquemond and Tepfer in this book. Proposed mechanisms of satellite-mediated cross protection have been repeatedly described (65,66,67,116). Cross protection has been used in small experimental fields in Italy and U.S.A (43, 82) but after preliminary experimentation (110, 133, 140) the technology was transferred to commercial fields in China and Japan where fruits from cross-protected plants are already on the market. In the several instances in which cross-protection with ameliorative satRNA has been used, no adverse effects were observed.

One of the most undesirable effects is the potential reassortment or true recombination of genome segments between protecting and challenging strains. Different opinions exist about complete inhibition of replication of the challenging strain and its satRNA in cross-protected plants but although investigated in several instances there is no definitive answer to this fundamental question.

Cillo and Gallitelli (18) obtained

preliminary results that seem to exclude any recombination and/or reassortment between the two strains. A CMV pseudo-recombinant was assembled, characterized and used as a biocontrol agent in Apulia (Southern Italy). The helper virus was obtained *in planta* by inoculation of tomato seedlings with CMV-S (subgroup II) genomic RNAs 1+2, CMV-Tfn (an Italian isolate belonging to subgroup I) genomic RNA-3, and a biologically active transcript of a cDNA clone of Tfn-satRNA (a 390 ribonucleotide satellite supported by CMV-Tfn). This pseudo-recombinant, named CMV-ST, was produced because naturally attenuated strains of CMV have not yet been found Italy and for: (i) monitoring pseudo-recombinant virus and its satellite RNA with either CMV subgroup and satellite-specific probes; (ii) obtaining a virus with CP coded by the RNA-3 of an Italian isolate of CMV that could be recognized by specific monoclonal antibodies; (iii) using a satellite RNA present in nature in Italy.

In greenhouse experiments, CMV-ST protected tomato plants against CMV-PG, an Italian isolate supporting a necrogenic satRNA. Experimentally, CMV-ST did not induce detectable symptoms in zucchini squash, cucumber, pepper, lettuce, eggplant, spinach and some of the most widely used tomato cultivars and hybrids in Italy and, for unknown reasons, it was not transmitted from tomato to tomato by *Aphis fabae*, *Macrosiphon euphorbiae* and *Myzus persicae*. CMV-ST spread systemically throughout the plant and was still clearly detectable by ELISA and molecular hybridization five months after inoculation. In tomato plants it reached a maximum concentration within seven days from inoculation then decreased rapidly to very low levels. Moreover, the maximum and minimum CMV-ST titer was about four times lower than that of the parental isolate CMV-Tfn. Apparently, CMV-ST was able to prevent infection by CMV-PG for neither the challenging strain or "new pseudo-recombinants" were recovered by mechanical or aphid transmission from challenge-inoculated plants. However, since CMV-ST is a pseudo-recombinant strain the possibility was tested that parental strains could reconstitute or that other reassortants could arise upon infection of the pre-immunized plants. The results obtained have shown that reconstitution of parental strains was indeed possible only when the pseudo-recombinant and the challenging strain were inoculated in mixture or simultaneously onto two different cotyledons. Such reconstitution was not observed when plants were challenged three days

after pre-immunization either on pre-immunized cotyledons or true leaves. In the latter case the challenging virus seemed totally unable to initiate the infection as suggested by the failure of its detection with routine diagnostic methods and lack of multiplication following six sub-inoculations onto tomato seedlings. The possibility that true recombination could take place between satRNAs of the protecting and challenging strains was excluded by RPA analysis both in pre-immunized plants and in their progeny (F. Cillo, L. Barbarossa and D. Gallitelli, unpublished results). Such possibility is now being investigated for genomic RNAs.

Experimental fields with tomato seedlings pre-immunized with CMV-ST were established in 1994 and 1995 in Apulia. Seedlings were transplanted in the field three weeks after pre-immunization i.e. when the virus titre was already stabilized at low levels. In both years, natural CMV infections were too low to evaluate the protective effect of CMV-ST however the trial provided the following information: (i) reduction in fruit yield caused by CMV-ST infection was of about 5%; (ii) fruit ripening and their color were more homogeneous than in control plants; (iii) no statistical differences in pH/soluble solids were observed between control and pre-immunized plants; (iv) the nucleotide sequence of ten Tfn-CARNA5 variants amplified by PCR from tissues collected from pre-immunized plants just before harvesting (middle of August) showed only one base change at position 173 in one of the variants, thus indicating that the molecule retained its primary nucleotide sequence also after five months exposure to natural conditions; (v) CMV-ST was not transmitted to weeds either growing within or in neighboring experimental fields (20).

Transgenesis

Most of the CMV-encoded genes, including different satRNA variants have been stably introduced into plant genomes with the objective of understanding fundamental functions of the viral genome or conferring suitable levels of pathogen-derived resistance to virus infection. The latter approach seems particularly useful from a practical point of view and some of the transgenic plants have also been tested in small-scale experiments in the greenhouse or grown in the field. It seems that, to date, only plants expressing CMV CP or satRNA sequences have been tested in the field. In China, transgenic tobacco plants expressing the Ra variant of CMV satRNA (32) have already been released to farmers but it is expected that a number of other species expressing other CMV genes will reach this stagein the near future. Apparently, no adverse effects could be seen in these large-scale releases. Except for the Chinese experience, other releases of transgenic plants were of experimental size.

Gonsalves *et al.* (46) tested cucumber plants of cultivar Poinsett 76 expressing the CP gene of a non-aphid transmissible strain of CMV under field conditions. Transgenic plants were compared with the genetically-resistant plants of cultivar Marketmore 76. Mechanically-infected plants interspersed among test plants provided the inoculum that was dispersed by natural aphid population. The results of the trial showed that transgenic cucumbers had a level of resistance comparable to that of genetically-resistant plants. However when plants entered senescence, CMV infection was 35% in transgenic line , 62% in Markemore 76 and 85% in non-transformed Poinsett 76 plants. Yields of transgenic lines were higher than those of other plants.

Tomassoli *et al.* (118, 119) and Barba *et al.* (7) compared transgenic UC82B tomato plants transformed with the CP gene of CMV-D with other UC82B plants expressing the CP of Italian CMV isolates of subgroup I (CMV-22) and II (CMV-PG). Two experimental sites were chosen in a highly CMV-infected area near Naples and in a low-infected area near Rome. In the latter case, mechanically inoculated plants were used as inoculum source but in both sites the natural aphid population assured spread of challenging inocula. Plants expressing CP genes of the two Italian isolates were nearly 100% resistant to CMV infection, thus performing better than the plants expressing the CP of CMV-D.

Multiple CP constructs were also used to transform squash. Tricoli *et al.* (123) observed suitable levels of resistance in squash lines transformed with CP of CMV and of two potyviruses, watermelon mosaic virus 2 (WMV-2) and zucchini yellow mosaic (ZYMV). The squash line ZW-20 (37) expressing CPs of WMV-2 and ZYMV has already obtained approval for commercialization whereas the CZW-3 line which in addition to potyvirus CPs, expresses also the CP of CMV is still under trial. In the same paper, the authors announced attempts to introduce the CP gene of papaya ringspot potyvirus, to develop a transgenic squash line with resistance to the four major cucurbit viruses. This multiple resistance is expected to be incorporated into commercial squash cultivars by back-crossing (123).

Simpler ways to obtain transgenic resistance against multiple viruses were recently suggested. Cooper *et al.* (23) reported that a defective movement protein of tobacco mosaic tobamovirus (TMV) expressed in tobacco plants produced a significant delay in symptoms appearance and reduced systemic spread of tobacco rattle tobravirus, tobacco ringspot nepovirus, AMV, peanut chlorotic streak caulimovirus and CMV. This was probably due to presence of similar functions in movement proteins of a number of distinct plant viruses (23). Watanabe *et al.* (129) obtained resistance mediated by a yeast-derived RNase specific for double stranded RNAs. Formation of double-stranded replicative intermediates is a common step of the life cycle of plant viruses with positive strand RNA genome. Transformed plants showed resistance against TMV, PVY and CMV.

Additional reports on field releases of plants expressing CMV satRNA as a transgene are by Wai *et al.* (128) who tested in U.S.A. satellite transgenic lines of tomato UC82B and Lichun for resistance to CMV by artificial challenging, whereas Colombo *et al.* (22) tested two tomato lines of the cultivar 'San Marzano' transformed with the Ra variant of satRNA (124) in a highly CMV-infected area of Southern Italy. In both cases, a delay in symptom appearance was observed and about 50% increase in yield, compared with non transformed plants.

Other approaches to confer transgenic resistance to CMV are evolving very rapidly although they are still at laboratory level.

The ability to out-compete viral genome for replication, thus leading to suppression of disease symptoms, ideally poses the satRNA-mediated resistance as the most immediate and fascinating approach (80,138). However, for unknown reasons mRNA of satRNA of the DNA transgene is less abundantly expressed than other CMV-derived transgenes. Transgenic plants attain significant titers of satRNA following CMV infection. Only then the mature satRNA out-competes the viral replication. In fact transgenic plants expressing satRNA often undergo recovery after an initial phase in which they show symptoms. The reverse is true with the CP gene. In most cases, plants expressing viral CP show a delay in symptom development for a period of time after which the virus overcomes the protection conferred by the CP gene. Yie *et al.* (139) obtained high levels of resistance to CMV in a commercial tobacco cultivar expressing both CP and satRNA the two transgenes being active, respectively, in an early and late stage of infection. In spite of the encouraging results obtained by Yie *et al.* (139) this type of combined resistance may still find some limitations in the strain-specificity conferred by the CP gene. From comparison of CMV CP amino acid sequences i.e. about 95% identity within the subgroup versus 80% identity between subgroups (28, 94), two different levels of protection can be expected from the expressed gene, against strains of the homologous and heterologous subgroup. Coat protein genes from subgroup I CMV strains afforded good levels of protection of transgenic tobacco plants against other strains of the same subgroup, whereas the CP gene of subgroup II strain CMV-WL had a broader spectrum of protection, also against isolates of the heterologous subgroup (85). High levels of protection against CMV strains of both homologous and heterologous subgroup were also obtained in the transgenic tomato line G-80 expressing CMV-WL CP gene (136). These results indicate that CMV-WL CP gene and probably CP genes of other subgroup II strains, would be the most useful under a variety of field conditions (45, 46) and for a combined resistance with satRNA.

Until recently (113), replicase-mediated resistance against CMV infection was attained by transformation of tobacco plants with a cDNA clone containing a truncated form of CMV-Fny (subgroup I) RNA2 (4). In transgenic plants, the deleted RNA2 was translated into a product of 70 kDa i.e. smaller than the 2a protein (94 kDa) expressed by the complete RNA2 of CMV-Fny. The resistance was effective against the same virus and other subgroup I strains but was overcome by subgroup II strains and some other subgroup I strains found in Asia (141). The sequence responsible for breaking (CMV-K) replicase-mediated resistance by one of these strains (CMV-K) was located on RNA2 (57).

Potential mechanisms of replicase-mediated resistance have been proposed by Carr and Zaitlin (14), Baulcombe (9), Hellwald and Palukaitis (56) and Mueller *et al.* (84). The so-called 'sense suppression mechanism' (79) as well as the protein-mediated inhibition for building up a functional replicase complex (13) do not seem to apply to the CMV model. Hellwald and Palukaitis (56) showed that viral RNA may be targeted by two independent mechanisms. One mechanism leads to RNA-RNA recognition between the transgene RNA and that of the challenging virus in the region of the "stem-loop" structures of 5' and 3' untranslated regions of RNA2. The other

mechanism inhibits viral movement in which RNA2 seems also to be involved. Single nucleotide differences in the proposed secondary structure of such regions of K-CMV and its ability to overcome inhibition of translocation in transgenic plants were proposed to account for the breakage of replicase-mediated resistance by this CMV strain (56). Suzuki *et al.* (113) obtained transgenic tobacco plants resistant to CMV infection by expressing a functional form of viral replicon (i.e. RNA1 and RNA2). The result was totally unexpected since transgenic tobacco plants expressing AMV RNA1 and 2 not only were not resistant to virus challenge but complemented RNA1 and/or RNA2- deficient inocula (114, 125, 126). The plants expressing CMV replicon showed also resistance to strain C of tomato aspermy cucumovirus (TAV). A resistance mechanism was proposed whereby the excess of RNA1 and RNA2 expressed in transgenic plants could affect the rate of encapsidation of RNA 3 and 4 by capturing most of the CP newly synthesized by RNA4, thus leading to exposure of both RNA3 and RNA4 to RNase attack (113).

Finally, plant transformation with antisense RNA constructs received much attention some years ago (28, 95). F. Cellini and coworkers (personal communication) introduced an antisense RNA obtained from CMV-1 RNA3 into W38 tobacco plants. The antisense RNA contained a 3'-half domain of RNA3 spanning from part of the subgenomic RNA promoter of the intercistronic region to the tRNA-like domain. One tobacco line (T1) showed a level of resistance that exceeding that previously reported (28, 95). The interference of the intercistronic region in antisense configuration with the production of minus-strand RNA proposed as a possible resistance mechanism (F. Cellini, personal communication).

Quarantine Inspections and Sanitary Programs

The main task of quarantine is to prevent the entry of pests into areas where they do not occur or are not widely distributed. Because of its worldwide distribution on a broad host range, quarantine inspections may appear hardly applicable to CMV. However, they can be useful for preventing: (i) introduction of new CMV strains that may also harbor harmful satRNAs and that may jeopardize the benefits gained by transgenesis and/or conventional resistance; (ii) virus spread in areas with low incidence of CMV diseases; (iii) reintroduction of virus inocula in areas where CMV incidence was significantly lowered by integrated disease management.

Quarantine inspections should be directed primarily to seed germplasm and new accessions of plant species in which the virus may be seedborne. Hampton et al. (50) suggested that such new accessions should be assayed for seedborne viruses and that first-generation seed increase should be carried out in an insect-proof screenhouse. Often the sanitary status of mother plants is evaluated by visual inspection to detect specific virus infections. In several instances this approach proved unreliable. Gillaspie et al. (44) found that 40 out of 60 cowpea seed lots from asymptomatic mother plants representing distinct genotypes, contained one or more seedborne viruses, including CMV. The test was done by ELISA on pre-introduction seed lots from Botswana, India and Kenya. A similar approach was also used in Taiwan (16) for the production of virus-free stocks of asparagus bean (*Vigna unguiculata* ssp. *sesquipedalis*) by roguing virus-infected seedlings after ELISA detection of blackeye cowpea mosaic virus and CMV. In the fields planted with virus-free seedlings incidence of virus diseases was less than 50% i.e. significantly lower than that usually recorded in the same areas. Ornamental plants, especially those propagated by bulbs and rhizomes may constitute important virus reservoirs. Often those plants are kept in the same screenhouse where seedlings of vegetable crops are grown. Kaminska (64) reported that 65% of lily bulbs tested in Poland harbored CMV infection.

Like other preventive control measures, quarantine inspections may also have drawbacks. For example, selection of seed mother plants for production may lead to loss of diversity. Significant loss of diversity was reported as consequence of programs for selection of virus-free accessions of beans and peas (1, 75). On the other hand the seedborne virus itself may be responsible for loss of diversity by killing or severely affecting mother plants. However, in considering the relative influence of the two opposite problems, one should first evaluate if a CMV seedborne infection may constitute the *focus* for new infections spread by aphids over large areas. Therefore, loss of diversity could be an unavoidable cost to be paid to ensure the introduction and subsequent use of virus-free germplasm.

Sanitary certification is a natural complement to quarantine measures for releasing virus-free propagation material to the farmers. Voluntary certification of tomato seedlings was launched in Apulia (Southern Italy) in agreement with Directive n. 33/92 of the European Union, issued on 24 August 1992. Nurserymen entering the certification program had to accept technical rules issued by Apulian Regional Extension Service for Plant Protection and strictly enforced throughout. In particular, greenhouses were equipped with insect-proof nets and double doors, weeds were accurately controlled in the surroundings of the greenhouses, insecticide sprays were frequently applied and inspections by authorized personnel were carried out periodically. Certified seedling lots were tagged with a blue label listing the viruses they had been tested for.

Sanitary certification required the timely and quick processing of a large number of samples, which, in turn, required the use of a rapid and sensitive detection method such as the multiple chemiluminescent dot-blot procedure devised in our laboratory (40,108). How to sample millions of tomato seedling was the first problem. To reduce errors due to biases introduced by the operator, we opted for a systematic sampling design. Tomato samples were collected from nursery seedling lots at the four/six true leaf stage. For a systematic and reproducible sampling, 24 sub-samples of leaf tissue were collected with a simple home-made instrument along a W-shaped sampling pattern (8) for each two square meters of polystyrene trays in which seedlings (2300 plants) were grown. Each sample (1.2 g) was collected directly into the plastic bag in which it was then processed. In principle, unrestricted random sampling where each individual plant has the same chance of being chosen should be preferred but, as pointed out by Barnett (8), the systematic sampling also has a form of randomness because the first sample is selected in a random manner and only the following samples are taken at a fixed distance. Barnett (8) proposed sampling patterns that would be appropriate for different survey situations. The W-shaped pattern would be recommended for large fields because most of the field would be covered and because unlike other patterns it does not amplify edge effects that in a greenhouse could be particularly relevant.

The use of a mixture of riboprobes for multiple detection of CMV, TSWV, AMV, PVY, tomato mosaic tobamovirus (ToMV) and tomato yellow leaf curl geminivirus (TYLCV) was an important achievement. The six probes did not interfere with one another, and did not cross-hybridize with heterologous target nucleic acid nor gave aspecific reactions with healthy sap. The possibility of detecting up to six viruses in a one-step assay saved time and allowed to carry out the complete protocol in three days, from sample collection to release of the results. Frequency of virus infections was negligible. Insect-proof nets together with insecticide sprays used routinely by nurserymen, strongly reduced the chance of infection by insect-transmitted viruses. ToMV was the only virus recorded over two years of observations. The infection, however, was very low and more than likely, seed-borne.

CONCLUDING REMARKS

CMV, known for many years as a threatening plant pathogen, continues to cause serious crop losses and has become more difficult and expensive to control. While the search for a more effective integrated disease management and for new sources for genetic resistance have made remarkable strides, is the pathogen-derived resistance approach that is gaining momentum, although being based on mechanisms that are still unclear.

The cross-protection strategy may find application as an immediate response to particularly dramatic situations like sudden epidemic events. The concern for potential risks consequent to deliberate release of a mild CMV strain carrying an ameliorative satRNA should be alleviated by the availability of readily monitorable strains which are not aphid-transmissible. However, not less risky is the tendency to assess resistance levels in field trials by transplanting plants infected with harmful CMV strains to serve as inoculum sources to be spread by aphids. Cross-protection has other disadvantages, as it needs continuous testing of different batches of the protecting strain to check its stability and the mechanical inoculation of millions of seedlings if applied over large areas.

It has been discussed that to be really effective the mild CMV strain should support a satRNA with ameliorative phenotype. The use of CMV satRNA as a biological control agent is the object of hot dispute because of the possibility of its conversion from the ameliorative into a necrogenic form. Observations on a number of satRNA variants indicate that, under natural conditions, ameliorative satRNA variants tend to maintain their molecular stability, lending support to the notion that this conversion is remote. A cDNA copy of satRNA stably inserted in the plant genome is considered less subjected to sequence changes that might alter its biological behavior. However, this is only partially true because following viral infection, the satRNA escapes the plant transcription system becoming completely dependent on and amplified by the viral replicase. This possible drawback could be alleviated only if the titer of satRNA in transgenic plants is high enough to out-compete viral replication from the very beginning of the infectious process. Efforts for increasing the satRNA expression in transgenic plants and for a better assessment of the environmental risks associated with its use are expected in the near future.

CP is the other transgene that conferred high resistance to CMV under field conditions. Constraints generated by strain-specificity were overcome by either using two CP genes derived from strains of both virus subgroups or the CP gene of a strain with a broader spectrum of activity. Compared with satRNA, CP-mediated resistance is believed to represent a less harmful approach, although a pollen-mediated transfer of the transgene to compatible plant species and hetero-encapsidation of nucleic acid of a different virus are considered as two possible problems. Transcapsidation, that could also be the consequence of a pollen-mediated transfer, extinguishes its deleterious effects as soon as the hetero-encapsidated viral nucleic acid codes its own coat protein in the new host. A third and probably more serious problem could be recombination between the transgene and the genome of any virus that infects the transformed plant. Since this is a common problem with the other pathogen-derived genes, studies to assess risk of recombination between genome of distinct viruses are in progress in many laboratories.

In the near future, an increasing number of reports claiming multiple resistance against distinct viruses is expected. Although looking promising, most of these approaches have not yet attained the field trial phase where the transgenic plants will be faced with several uncontrollable factors and with the few virus particles transmitted by a probing aphid.

ACKNOWLEDGMENTS

I wish to thank prof. G.P. Martelli for helpful comments and for revising the text.

REFERENCES

1. Alconero, R., Weeden, N.F., Gonsalves, D. and Fox, D.T. 1985. Loss of genetic diversity in pea germplasm by the elimination of individuals infected by pea seedborne virus. Ann. Appl. Biol. 106: 357-364.

2. Alvarez, J. 1993. Resistencia al CMV (virus del mosaico del pepino) en melon. Phytoma-Espana 50: 66-70.

3. Anderson, B., Boyce, P.M., and Blanchard C.L. 1995. RNA 4 sequences from cucumber mosaic virus subgroups I and II. Gene 161: 293-294.

4. Anderson, J.M., Palukaitis, P., and Zaitlin, M.1992. A defective replicase gene induces resistance to cucumber mosaic virus in transgenic tobacco plants. Proc. Nat. Acad. Sci. USA 89: 8759-8763.

5. Aranda, M.A., Fraile, A., and García-Arenal, F. 1993. Genetic variability and evolution of the satellite RNA of cucumber mosaic virus during natural epidemics. J. Virol. 67: 5896-901.

6. Aranda, M.A., Fraile, A., García-Arenal, and Malpica, J.M. 1995. Experimental evaluation of the ribonuclease protection assay method for the assessment of genetic heterogeneity in populations of RNA viruses. Arch. Virol. 140: 1373-1383.

7. Barba, M., Tomassoli, L., Ilardi, V., Kaniewski, W., Mitsky, T., and Layton, J. 1995. Piante di pomodoro geneticamente trasformate per la resistenza al virus del mosaico del cetriolo: tre anni di risultati. Petria 5 (suppl. 1): 70-72.

8. Barnett, O. 1986. Surveying for plant viruses: design and consideration. Pages 147-166 in: Introduction to Plant Disease Epidemiology. C.L. Campbell, and L.V. Madden, eds. Wiley Interscience, NY.

9. Baulcombe, D. 1994. Replicase-mediated resistance: a novel type of virus resistance in transgenic plants? Trends in Microbiology 62: 18-21.

10. Blancard, D. 1988. Maladies de la Tomate, Observer, identifier, lutter. Pages 1-211 in: Revue Horticole. INRA, Paris.

11. Blanchard, C., Boyce, P., and Anderson, B.J. 1996. Cucumber mosaic virus RNA5 is a mixed population derived from the conserved 3'-terminal regions of genomic RNAs 2 and 3. Virology 217: 598-601.

12. Boccard, F., and Baulcombe, D. 1993. Mutational analysis of cis-acting sequences and gene function in RNA3 of cucumber mosaic virus. Virology 193: 563-578.

13. Brederode, F.T., Taschner, P.E.M., Posthumus, E., and Bol, J. 1995. replicase-mediated resistance to alfalfa mosaic virus. Virology 207: 467-474.

14. Carr, J.P., and Zaitlin, M. 1993. Replicase-mediated resistance. Seminars in Virology 4: 339-347.

15. Carr, J.P., Gal-On, A., Palukaitis, P., and Zaitlin, M. 1994. Replicase-mediated resistance to cucumber mosaic virus in transgenic plants involves suppression of both virus replication in the inoculated leaves and long-distance movement. Virology 199: 439-447.

16. Chang, C.A., Yang, T.T., and Tsan, T.M. 1995. Production and application of virus-free seed for control of virus diseases of asparagus bean in Taiwan. Pages 43-44 in: Proc. 6th International Plant Virus Epidemiology Symposium: Epidemiological Aspects of Plant Virus Control. Ma'ale Hachamisha (Jerusalem), Israel, April 23-24, 1995.

17. Chen, B., Randles, J.W., and Francki, R.I.B. 1995. Mixed-subunit capsids can be assembled in vitro with coat protein subunits from two cucumoviruses. J. Gen. Virol. 76: 971-973.

18. Cillo, F., and Gallitelli, D. 1995. La preimmunizzazione nella difesa del pomodoro contro il CMV: valutazione dei fattori di rischio. Petria 5 (suppl. 1): 69-70.

19. Cillo, F., Barbarossa, L., Grieco, F., and Gallitelli, D. 1994. Lethal necrosis, fruit necrosis and top stunting: molecular-biological aspects of three cucumber mosaic virus-induced diseases of processing tomatoes in Italy. Acta Hortic. 376: 369-376.

20. Cillo, F., Tamba, M.L., Borgatti, S., Cerato, C., Di Candilo, M., Barbarossa, L., Di Franco, A., Grieco, F., Mileti, V., and Gallitelli, D. 1995. Biological properties of a satellite-containing pseudorecombinant of cucumber mosaic virus and its' use for tomato protection in Italy. Page 47 in: Proc. 6th International Plant Virus Epidemiology Symposium, Ma'le Hachamisha (Jerusalem), April 23-28, 1995.

21. Collmer, C.W., and Howell, S.H. 1992. Role of satellite RNA in the expression of symptoms caused by plant viruses. Annu. Rev. Phytopathol. 30: 419-442.

22. Colombo, M., Monti M.M., Valanzuolo, S., and Cassani, C. 1995. Induzione di resistenza al CMV in pomodoro 'San Marzano' mediata da RNA satellite. Petria 5 (suppl. 1): 68.

23. Cooper, B., Lapidot, M., Heick, J.A., Dodds, J.A. and Beachy, R. 1995. A defective movement protein of TMV in transgenic plants confers resistance to multiple viruses whereas the functional analog increases susceptibility. Virology 206: 307-313.

24. Cooper, B., Scmitz, I., Rao, A.L.N., Beachy, R.N., and Dodds, J.A. 1996. Cell-to cell transport of movement-defective cucumber mosaic and tobacco mosaic viruses in transgenic plants expressing heterologous movement protein genes. Virology 216: 208-213.

25. Crescenzi, A., Barbarossa, L., Cillo, F., Di Franco, A., Vovlas, N., and Gallitelli, D. 1993. Role of cucumber mosaic virus and its satellite RNA in the etiology of tomato fruit necrosis in Italy. Arch Virol. 131: 321-333.

26. Crescenzi, A., Barbarossa, L., Gallitelli, D. and Martelli, G.P. 1993. Cucumber mosaic cucumovirus populations in Italy under natural epidemic conditions and after a satellite-mediated protection test. Plant Dis. 77: 28-33

27. Crescenzi, A., Grieco, F., and Gallitelli, D. 1992. Nucleotide sequence of a satellite RNA of a strain of cucumber mosaic virus associated with tomato fruit necrosis. Nucleic Acids Res. 20: 2886.

28. Cuozzo, M., O'Connel, K.M., Kaniewski, W., Fang, R.X., Chua, N.H., and Tumer, N.E. 1988. Viral protection in transgenic tobacco plants expressing the cucumber mosaic virus coat protein or its antisense RNA. Bio/technology 6: 549-557.

29. Daniels, J., and Campbell, R.N. 1992. Characterization of cucumber mosaic virus isolates from California. Plant Dis. 76: 1245-1250

30. Davies, C., and Symons, R.H. 1988. Further implications for the evolutionary relationships between tripartite plant viruses based on cucumber mosaic virus RNA 3. Virology 165: 216-224.

31. Davis, R.I., Brown, J.F., and Pone, S.P. 1996. Causal relationship between cucumber mosaic cucumovirus and kava dieback in the south pacific. Plant Dis. 80: 194-198.

32. Devic, M., Jaegle, M. and Baulcombe, D. 1989. Symptom production on tobacco and tomato is determined by two distinct domains of the satellite RNA of cucumber mosaic virus (strain Y). J. Gen. Virol. 70: 2765-2774.

33. Ding, B., Li, Q., Nguyen, L., Palukaitis, P., and Lucas, W.J. 1995. Cucumber mosaic virus 3a protein potentiates cell-to-cell trafficking of CMV RNA in tobacco plants. Virology 207: 345-353.

34. Ding, S.W., Anderson, B.J., Haase, H.R., and Symons, R.H. 1994. New overlapping gene encoded by the cucumber mosaic virus genome. Virology : 593-601.

35. Fernandez-Cuartero, B., Burgyan, J., Aranda M.A., Salanki, K., Moriones, E., and García-Arenal, F. 1994. Increase in the relative fitness of a plant virus RNA associated with its recombinant nature. Virology 203: 373-377.

36. Flasinski, S., Scott, S.W., Barnett, O.W., and Sun, C. 1995. Diseases of *Peperomia, Impatiens* and *Hibbertia* caused by cucumber mosaic virus. Plant Dis. 79: 843-848.

37. Fuchs, M., and Gonsalves, D. 1995. resistance of transgenic hybrid squash ZW-20 expressing the coat protein genes of zucchini yellow mosaic virus and watermelon mosaic virus 2 to mixed infection by both potyvivruses. Bio/Technology 13: 1466-1473.

38. Fulton, R.W. 1908. Biological significance of multicomponent viruses. Annu. Rev. Phytopathol. 18: 131-146.

39. Gal-On, A., Kaplan, I., and Palukaitis, P. 1995. Differential effects of satellite RNA on the accumulation of cucumber mosaic virus RNAs and their encoded proteins in tobacco vs. zucchini squash with two strains of CMV helper virus. Virology 208: 58-66

40. Gallitelli, D., and Saldarelli, P., 1996. Molecular identification of phytopathogenic viruses. Pages 57-79 in: Methods in Molecular Biology 50, Species Diagnostics Protocols: PCR and Other Nucleic Acid Methods. J.P Clapp, ed. Humana Press, NJ.

41. Gallitelli, D., Di Franco, A., Vovlas, C., and Kaper, J.M. 1988. Infezioni miste del virus del mosaico del cetriolo (CMV) e di potyvirus in colture ortive di Puglia e Basilicata. Inf. Fitopatol. 12: 57-64.

42. Gallitelli, D., Martelli, G.P., Gebre-Selassie, K., and Marchoux, G. 1995. Progress in the biological and molecular studies of some important viruses of *solanaceae* in the Mediterranean. Acta Hortic. 412: 503-514.

43. Gallitelli, D., Vovlas, C., Martelli, G.P., Montasser, M.S., Tousignant, M.E., and Kaper, J.M. 1991. Satellite-mediated protection against cucumber mosaic virus: II. Field test under natural epidemic conditions in southern Italy. Plant Dis: 75: 93-95.

44. Gillaspie, A.G. Jr., Hopkins, M.S., Pinnow D.L., and Hampton, R.O. 1995. Seedborne viruses in pre-introduction cowpea seed lots and establishment of virus-free accessions. Plant Dis. 79: 388-391.

45. Gonsalves, D., and Slightom, J.L. 1993. Coat protein-mediated protection: analysis of transgenic plants for resistance in a variety of crops. Seminars in Virology 4: 397-405.

46. Gonsalves, D., Chee, P., Provvidenti, R., Seem, R., and Slightom, J.L. 1992. Comparison of coat protein-mediated and genetically-derived resistance in cucumbers to infection by cucumber mosaic virus under field conditions with natural challenge inoculations by vectors. Bio/Technology 10: 1562-1570.

47. Grieco, F., Cillo, F., Barbarossa, L., and Gallitelli D. 1992. Nucleotide sequence of a cucumber mosaic virus satellite RNA associated with a tomato top stunting. Nucleic Acids Res. 24: 6733

48. Grieco, F., Lanave, C., and Gallitelli, D., 1995. Evolutionary dynamics of cucumber mosaic virus satellite RNA variants isolated during natural epidemics in Italy. Page 19 in: Proc. 6th International Plant Virus Epidemiology Symposium. Ma'le Hachamisha (Jerusalem), April 23-28, 1995.

49. Hampton, R.O., and Francki, R.I.B. 1992. RNA-1 dependent seed transmissibility of cucumber mosaic virus in *Phaseolus vulgaris*. Phytopathology 82: 127-130.

50. Hampton, R.O., Kraft, J.M., and Muehlbeuer, F.J. 1982. Importance of seedborne viruses in crop germplasm. Plant Dis. 66: 977-978.

51. Hayakawa, T., Mizukami, M., Nakamura, I., and Suzuki, M. 1989. Cloning and sequencing of RNA-1 cDNA from cucumber mosaic virus strain O. Gene 85: 533-540.

52. Hayakawa, T., Mizukami, M., Nakajima, I., and Suzuki, M. 1989. Complete nucleotide sequence of RNA 3 from cucucmber mosaic virus (CMV) strain O: comparative study of nucleotide sequences and amino acid sequences among CMV strains O, Q, D, and Y. J. Gen. Virol. 70: 499-504.

53. Hayes, R.J., and Buck, K.W. 1990. Complete replication of a eukaryotic virus RNA *in vitro* by a purified RNA-dependent RNA polymerase. Cell 63: 363-368.

54. Hayes, R.J., Pereira, V.C.A., McQullin, A., and Buck, K.W. 1994. Localization of functional regions of the cucumber mosaic virus RNA replicase using monoclonal and polyclonal antibodies. J. Gen. Virol. 75: 3177-3184.

55. Hayes, R.J., Toush, D., Jacquemond, M., Pereira, V.C., Buck, K.W., and Tepfer, M. 1992. Complete replication of a satellite RNA *in vitro* by a purified RNA-dependent RNA polymerase. J Gen Virol 73: 1597-1600.

56. Hellwald, K.H., and Palukaitis, P. 1995 Viral RNA as a potential target for two independent mechanisms of replicase-mediated resistance against cucumber mosaic virus. Cell 83: 937-946.

57. Hellwald, K.H., and Palukaitis, P. 1994. Nucleotide sequence and infectivity of cucumber mosaic cucumovirus (strain K2) RNA2 involved in breakage of replicase-mediated resistance in tobacco. J. Gen. Virol. 75: 2121-2125

58. Hidaka, S., and Hanada K. 1994. Structural features unique to a new 405-nucleotide satellite RNA of cucumber mosaic virus inducing tomato necrosis. Virology 200: 806-808.

59. Hobbs, H.A., Black, L.L., Dufresne, D.J., and Valverde, R.A. 1995. resistance in *Capsicum annuum* (pepper) line to geographically diverse cucumber mosaic virus isolates. Phytopathology 85: 1192.

60. Hsu, Y.H., Wu, C.W., Lin, B.Y., Chen, H.Y., Lee, M.F., and Tsai, C.H. 1995. Complete genomic RNA sequences of cucumber mosaic virus strain NT9 from Taiwan. Arch. Virol. 140: 1841-1847.

61. Hu, J.S., Li, H.P., Barry, K., Wang, M., and Jordan,

R. 1995. Comparison of dot blot, ELISA, and RT-PCR assays for detection of two cucumber mosaic virus isolates infecting banana in Hawaii. Plant Dis. 79: 902-906.

62. Hull, R. 1993. Nucleic acid hybridization procedures. Pages 267-271 in: Diagnosis of Plant Virus Diseases. R.E.F. Matthews, ed. CRC Press, Boca Raton.

63. Jordá, C., Alfaro, A., Aranda, M.A., Moriones, E., and García-Arenal, F. 1992. Epidemic of cucumber mosaic virus plus satellite RNA in tomatoes in Eastern Spain. Plant Dis. 76: 363-366.

64. Kaminska, M. 1995. Some aspects of virus infection of lilies in Poland. Page 44 in: Proc. 6th International Plant Virus Epidemiology Symposim: Epidemiological Aspects of Plant Virus Control. Ma'ale Hachamisha, (Jerusalem), Israel, April 23-24, 1995.

65. Kaper, J.M. 1993. Satellite-mediated symptom modulation: an emerging technology for the biological control of viral crop disease. Microb. Releases. 2: 1-9

66. Kaper, J.M. 1993. Viral satellites, molecular parasites for plant protection. Pages 134-143 in: Pest Management: Biologically Based Technologies. R.D. Lumsden and J.L. Vaughn, eds. American Chemical Society, Washington, DC.

67. Kaper, J.M. 1995. Role of satellites in viral pathogenesis: nested parasistic nucleic acids competing for expression. Pages 377-392 in: Pathogenesis and Host Specificity in Plant Diseases, Histopathological, Biochemical, Genetic and Molecular Bases, Vol. III Viruses and Viroids. R.P. Singh, U. Singh and K. Kohmoto, eds. Pergamon Press.

68. Kaper, J.M., and Waterworth, H.E. 1981. Cucumoviruses. Pages 257-332 in: Handbook of Plant Virus Infections and Comparative Diagnosis. E. Kurstak, ed. Elsevier/North-Holland Biomedical Press.

69. Kaper, J.M., Gallitelli, D., and Tousignant, M. 1990. Identification of a 334-ribonucleotide viral satellite as principal aetiological agent in a tomato necrosis epidemic. Res. Virol.141: 81-95.

70. Kaper, J.M., Geletka, L.M., Wu, G.S., and Tousignant, M.E. 1995. Effect of temperature on cucumber mosaic virus satellite-induced lethal tomato necrosis is helper virus strain dependent. Arch. Virol. 140: 65-74.

71. Kaplan, I.B., Shintaku, M.H., Li, Q., Zhang, L., Marsh, L.E. and Palukaitis, P. 1995. Complementation of virus movement in transgenic tobacco expressing the cucumber mosaic virus 3a gene. Virology 209, 188-199.

72. Kataoka, J., Masuta, C., and Takanami, Y. 1990. Complete nucleotide sequence of RNA 1 of cucumber mosaic virus Y strain. Ann. Phytopathol. Soc. Japan 56: 501-507.

73. Kataoka, J., Masuta, C., and Takanami, Y. 1990. Complete nucleotide sequence of RNA 2 of cucumber mosaic virus Y strain. Ann. Phytopathol. Soc. Japan 56: 495-500.

74. Kearney, C.M., Zitter, T.A., and Gonsalves, D. 1990. A field survey for serogroups and the satellite of cucumber mosaic virus. Phytopathology 80: 238-1243.

75. Klein, R.E., Wyatt, S.D., and Kaiser, W.J. 1990. Effect of diseased plant elimination on genetic diversity and bean common mosaic virus incidence in Phaseolus vulgaris germplasm collection. Plant Dis. 74: 911-913.

76. Lebeda, A., and Kristkova, E. 1995. Screening of cucurbita spp. germplasm for resistance to cucumber mosaic virus. Pages 154-157 in: Proc. 8th Conference on Virus Diseases of Vegetables, Prague, Czech Republic, July 9-15, 1995.

77. Li, Q., and Palukaitis, P. 1996. Comparison of the nucleic acid and NTP-binding properties of the movement protein of cucumber mosaic cucumovirus and tobacco mosaic tobamovirus. Virology 216: 71-79.

78. Lot, H., and Kaper, J.M. 1976. Physical and chemical differentiation of three strains of cucumber mosaic virus and peanut stunt virus. Virology 74: 209-222.

79. Matzke, M.A. and Matzke, A.J., 1995. How and why do plants inactivate homologous (trans)genes? Plant Physiol. 107: 679-685

80. McGarvey, P., and Kaper, J.M. 1993. Transgenic plants for conferring virus tolerance - satellite approach. Pages 277-296 in: Transgenic Plants. S.D. Kung and R.E. Wu, eds. Academic Press, New York, U.S.A.

81. McGarvey, P., Tousignant, M., Geletka, L., Cellini, F., and Kaper, J.M. 1995. The complete sequence of a cucumber mosaic virus from Ixora that is deficient in the replication of satellite RNAs. J. Gen. Virol. 76: 2257-2270.

82. Montasser, M., Tousignant, M., and Kaper, J.M. 1991. Satellite-mediated protection of tomato against cucumber mosaic virus: I. greenhouse experiments and simulated epidemic conditions in the field. Plant Dis. 75: 86-92.

83. Moriones, E., Diaz, I., Fernandez-Cuartero, B. Fraile, A., Burgyan, J. and García-Arenal, F. 1994. Mapping helper virus functions for cucumber mosaic virus satellite RNA with pseudo-recombinants derived from cucumber mosaic and tomato aspermy viruses. Virology 205: 574-577.

84. Mueller, E., Gilbert, J., Davenport, G., Brigneti, G., and Baulcombe, D.C. 1995. Homology-dependent resistance: transgenic virus resistance in plants related to homology-dependent gene silencing. Plant J. 7: 1001-1013.

85. Namba, S. Ling, K., Gonsalves, C., Gonsalves, D., and Slightom, J.L. 1991 Expression of the gene encoding the coat protein of cucumber mosaic virus (CMV) strain WL appears to provide protection to tobacco plants against infection by several different CMV strains. Gene 107: 181-188.

86. Nitta, N., Masuta, C. Kuwata, S., and Takanami, Y. 1988. Comparative studies on the nucleotide sequence of cucumber mosaic virus RNA 3 between Y strain and Q strain. Ann. Phytopathol. Soc. Japan 54: 516-522.

87. Nono-Womdim, K. Gebre-Selassie, K. Palloix, A. Pochard, E., and Marchoux, G. 1993. Study of multiplication of cucumber mosaic virus in susceptible and resistant Capsicum annuum lines. Ann. Appl. Biol. 122: 49-56.

88. Owen, J., Shintaku, M., Aeschleman, P., Tahaar, S.B., and Palukaitis, P. 1990. Nucleotide sequence and evolutionary relationships of cucumber mosaic virus (CMV) strains: CMV-RNA 3. J. Gen. Virol. 71: 2243-2249.

89. Owen, J., and Palukaitis, P. 1988. Characterization of cucumber mosaic virus. I. Molecular heterogeneity mapping of RNA3 in eight CMV

strains. Virology 166: 495-502.

90. Palukaitis, P., Roossink, M., Dietzgen, R.G., and Franchi, R.I.B. 1992. Cucumber mosaic virus. Adv. Virus Res. 41: 281-348.

91. Perry, K.L., and Francki, R.I.B. 1992. Insect-mediated transmission of mixed and reassorted cucumovirus genomic RNAs. J. Gen. Virol. 73: 2105-2114.

92. Perry, K.L., Zhang, L., and Palukaitis, P. 1995. Differential transmission of cucumber mosaic virus by two aphids: mutation in the coat protein restore transmission by *Aphis gossypii* but not by *Myzus persicae*. Phytopathology 85: 1143

93. Perry, K.L., Zhang, L., Shintaku, M.H., and Palukaitis, P. 1994 Mapping determinants in cucumber mosaic virus for transmission by *Aphis gossypii*. Virology 205: 591-595.

94. Quemada, H., Kearney, C., Gonsalves, D., and Slightom, J.L. 1989. Nucleotide sequences of the coat protein genes and flancking regions of cucumber mosaic virus strains C and WL RNA 3. J. Gen. Virol. 70: 1065-1073.

95. Rezaian, M.A., Skene, K.G.M., and Ellis, J.G. 1988. Anti-sense RNAs of cucumber mosaic virus in transgenic plants assessed for control of the virus. Plant Mol. Biol. 11: 463-471.

96. Rezaian, M.A., Williams, R.H.V., and Symons, R.H. 1985. Nucleotide sequence of cucumber mosaic virus RNA 1. Presence of a sequence complementary to part of the viral satellite RNA and homologies with other viral RNAs. Eur. J. Biochem. 150: 331-339.

97. Rezaian, M.A., Williams, R.H.V., Gordon, K.H.J., Gpuld, A.R., and Symons, R.H. 1984. Nucleotide sequence of cucumber mosaic virus RNA 2 reveals a translation product significantly homologous to corresponding proteins of other viruses. Eur. J. Biochem 143: 277-284.

98. Ribicki, E.P. 1995. Family *Bromoviridae*. Pages 450-457 in: Virus Taxonomy, 6th report of the International Committee on Taxonomy of Viruses. F.A. Murphy, C.M. Fauquet, D.H.L Bishop, S.A. Ghabrial, A.W. Jarvis, G.P. Martelli, M.A. Mayo and M.D. Summers, eds. Springer-Verlag, Wien, New York.

99. Riedle, M. 1995. Localization of cucumber mosaic cucumovirus in infected cucumber plants by whole leaf press blotting. Pages 162-167 in: Proc. 8th Conference on Virus Diseases of Vegetables, Prague, Czech Republic, July 9-15, 1995.

100. Rist, D.L., and Lorbeer, J.W. 1988. Occurrence and overwintering of cucumber mosaic virus and broad bean wilt virus in weeds growing near commercial lettuce fields in new York. Phytopathology 79: 65-69.

101. Rizos, H., Gunn, L.V., Pares, R.D., and Gillings, M.R. 1992. Differentiation of cucumber mosaic virus isolates using the polymerase chain reaction. J. Gen. Virol 73: 2099-103.

102. Rizzo, T.M., and Palukaitis, P. 1988. Nucleotide sequence and evolutionary relationships of cucumber mosaic virus (CMV) strains: CMV RNA 1. J. Gen. Virol. 69: 1777-1787.

103. Rizzo, T.M., and Palukaitis, P. 1989. Nucleotide sequence and evolutionary relationships of cucumber mosaic virus (CMV) strains: CMV RNA 2. J. Gen. Virol. 70: 1-11.

104. Rodríguez-Alvarado, G., Kurath, G. and Dodds, A. 1995. Heterogeneity in pepper isolates of cucumber mosaic virus. Plant Dis. 79: 450-455.

105. Roossinck, M.J., and Palukaitis, P. 1995. Genetic analysis of helper virus-specific selective amplification of cucumber mosaic virus satellite RNAs. J. Mol. Evol 40: 25-29.

106. Roossink, M., and Palukaitis, P. 1991. Differential replication in zucchini squash of cucumber mosaic virus satellite RNA maps to RNA-1 of the helper virus. Virology 181: 371-373.

107. Roossink, M.J., Sleat, D., and Palukaitis, P. 1992. Satellite RNAs of plant viruses: structure and biological effects. Microbiol. Rev. 56 (2): 265-279.

108. Saldarelli, P., Barbarossa, L., Grieco, F., Gallitelli, D., and Martelli, G.P., 1995. Molecular hybridization analysis with chemiluminescent riboprobes applied to sanitary certification of tomato in Italy. Page 29 in: Proc. 6th International Plant Virus Epidemiology Symposium, Ma'le Hachamisha (Jerusalem), Israel, April 23-28, 1995.

109. Sanford, J.C., and Johnston, S.A. 1985. The concept of parasite-derived resistance: Deriving resistance genes from the parasite's own genome. J. Theor. Biol. 113: 395-405.

110. Sayama, H., Sato, T., Kominato, M., Natsuaki, T., and Kaper, J.M. 1993. Field testing of a satellite-containing attenuated strain of cucumber mosaic virus for tomato protection in Japan. Phytopathology 83: 405-410.

111. Singh, Z., Jones, R.A.C,. and Jones, M.G.K. 1995. Identification of cucumber mosaic virus subgroup I. Isolates from banana plants affected by infectious chlorosis disease using RT-PCR. Plant Dis. 79: 713-716.

112. Summers, C.G., Stapleton, J.J., Newton, A.S., Duncan, R.A., and Hart, D. 1995. Comparison of sprayable and film mulches in delaying the onset of aphid-transmitted virus diseases in zucchini squash. Plant Dis. 79: 1126-1131.

113. Suzuki, M., Masuta, C., Takanami, Y., and Kuwata, S. 1996. Resistance against cucumber mosaic virus in plants expressing the viral replicon. FEBS Lett. 379: 26-30.

114. Taschner, P.E.M., van der Kuyl, A.C., Neeleman, L., and Bol J.F. 1991. Replication of an incomplete alfalfa mosaic virus genome in plants transformed with replicase gene. Virology 181: 445-450.

115. Tepfer, M. 1993. Viral genes and transgenic plants. Bio/Technology 11: 1125-1132.

116. Tien, P., and Wu, G.S. 1991. Satellite RNA for the biocontrol of plant disease. Adv. Virus Res. 39: 321-339.

117. Tomassoli, L., Cupidi, A., and Barba, M. 1993. Control of zucchini yellow mosaic virus in zucchini crop. Petria 3 (suppl. 1): 81-82.

118. Tomassoli, L., Ilardi, V., Kaniewski, W., and Barba, M. 1995. Piante di pomodoro geneticamente modificate per la resistenza al virus del mosaico del cetriolo. Inf. Agr. 51(4): 49-52.

119. Tomassoli, L., Kaniewski, W., Ilardi, V. and Barba, M. 1995. Coat protein mediated resistance of transgenic tomato to cucumber mosaic virus under field conditions. Page 56, in: Proc. 6th International Plant Virus Epidemiology Symposium, Ma'ale Hachamisha (Jerusalem), 23-28 April, 1995.

120. Tomlinson, J.A. 1987. Epidemiology and control of virus diseases of vegetables. Ann. Appl. Biol. 110: 661-681.

121. Torrance, L. 1992. Serological methods to detect plant viruses: production and use of monoclonal

antibodies. Pages 7-33 in: Techniques, for Rapid Detection of Plant Pathogens. J.M. Duncan and L. Torrance, eds. Blackwell Scientific Publications, Oxford.

122.   Tousignant, M., and Kaper, J.M. 1993. Cucumber mosaic virus-associated RNA 5. XIII. Opposite necrogenicities in tomato of variants with large 5' half insertion/deletion regions. Res. Virol. 144: 349-360.

123.   Tricoli, D.M., Carney, K.J., Russell, P.F., MacMaster, J.R., Groff, D.W., Hadden, K.C., Himmel, P.T., Hubbard, J.P., Boeshore, M.L., and Quemada, H.D. 1995. Field evaluation of transgenic squash containing single or multiple virus coat protein gene constructs for resistance to cucumber mosaic virus, watermelon mosaic virus, and zucchini yellow mosaic virus. Bio/Technology 13: 1458-1465.

124.   Valanzuolo, S., Catello, S. Colombo, M., Dani, M.M., Monti, M.M., Uncini, L., Petrone, P., and Spigno, P., 1994. Cucumber mosaic virus resistance in transgenic San Marzano tomatoes. Acta Hortic. 376: 377-386.

125.   van der Kuyl, A.C., Neeleman, L., and Bol, J.F. 1991. Deletion analysis of cis and trans-acting elements involved in replication of alfalfa mosaic virus RNA3 in vivo. Virology 183: 731-738.

126.   van Dun, C.M.P., van Vloten-Doting, L., and Bol, J. F. 1988. Expression of alfalfa mosaic virus cDNA1 and 2 in transgenic tobacco plants. Virology 163: 572-578.

127.   Vaquero, C., Turner, A.P., Demangeat, G., Sanz, A., Serra, M.T., Roberts, K., and Garcia-Luque, I. 1994. The 3a protein from cucumber mosaic virus increases the gating capacity of plasmodesmata in transgenic tobacco plants. J. Gen. Virol. 75: 3193-3197.

128.   Wai, T., Stommel, J.R., Tousignant, M.E., and Kaper, J.M. 1995. Field test of satellite-transgenic tomato resistance against CMV. Phytopathology 85: 1192.

129.   Watanabe, Y., Ogawa, T., Takahashi, H., Ishida, I., Tacheuchi, Y., Yamamoto, M., and Okada, Y. 1995. Resistance against multiple plant viruses in plants mediated by a double stranded-RNA specific ribonuclease. FEBS Lett. 372: 165-168.

130.   White, J.L., Tousignant. M.E., Geletka, L.M., and Kaper, J.M. 1995. The replication of a necrogenic cucumber mosaic virus satellite is temperature-sensitive in tomato. Arch Virol 140: 53-63.

131.   White, P.S., Morales, F., and Roossink, M. 1995. Interspecific reassortment of genomic segments in the evolution of cucumoviruses. Virology 207: 334-337.

132.   Wilson, T.M.A. 1993. Strategies to protect crop plants against viruses: Pathogen-derived resistance blossoms. Proc. Nat. Acad. Sci. USA 90: 3143-3141.

133.   Wu, G., Kang, L., and Tien, P. 1989. The effect of satellite RNA on cross-protection among cucumber mosaic virus strains. Ann. Appl. Biol. 114: 489-496.

134.   Wu, G., Kaper, J.M., and Jaspars, E.M.J. 1991. Replication of cucumber mosaic virus satellite RNA in vitro by an RNA-dependent RNA polymerase from virus-infected tobacco. FEBS Lett. 292: 213-216.

135.   Wu, G., Kaper, J.M., Tousignant, M.E., Masuta C., Kuwata, S., Takanami, Y., Pena, L., and Diaz-Ruiz, J.R. 1993. Tomato necrosis and the 369 nucleotide Y satellite of cucumber mosaic virus: factors affecting satellite biological expression. J. Gen. Virol. 74 (Pt. 2), 161-168.

136.   Xue, B., Gonsalves, C., and Provvidenti, R. 1994. Development of transgenic tomato expressing a high level of resistance to cucumber mosaic virus strains of subgroups I and II. Plant Dis. 78: 1038-1041.

137.   Yang, Y., Kim, K.S., and Anderson, E.J. 1995. Ultrastructural studies on seed transmission of cucumber mosaic virus in spinach. Phytopathology 85: 1184.

138.   Yie Y., and Tien, P., 1993. Plant virus satellite RNAs and their role in engineering resistance to virus diseases. Seminars in Virology 4: 363-368.

139.   Yie Y., Zhao, F., Zhao, S.Z., Liu, Y.Z. , Liu, Y.L., and Tien, P. 1992. High resistance to cucumber mosaic virus conferred by satellite RNA and coat protein in transgenic commercial tobacco cultivar G-140. Molecular Plant-Microbe Interactions 5: 460-465.

140.   Yoshida, K., Goto, T., and Iizuka, N. 1985. Attenuated isolates of cucumber mosaic virus produced by satellite RNA and cross-protection between attenuated isolates and virulent ones. Ann. Phytopath. Soc. Japan 51: 238-242.

141.   Zaitlin, M., Anderson, J.M., Perry, K.L., Zhang, L., and Palukaitis, P. 1994. Specificity of replicase-mediated resistance to cucumber mosaic virus. Virology 201: 200-205.

142.   Zhang, L., Hanada, K., and Palukaitis, P. 1994. Mapping local and systemic symptom determinants of cucumber mosaic cucumovirus. J. Gen. Virol. 75: 3185-3191.

143.   Ziegler, A., Torrance, L., Macintosh, S.M., Cowan, G.H., and Mayo, M.A. 1995. Cucumber mosaic cucumovirus antibodies from a synthetic phage display library. Virology 214: 235-238.

CHAPTER 39

Present Status of Controlling Bean Common Mosaic Virus

F. J. Morales

Bean common mosaic virus (BCMV) is one of the earliest plant viruses described as a pathogen of an economically important plant species, *Phaseolus vulgaris* (28). Its original name, 'bean mosaic virus', was changed in 1934 to 'bean common mosaic virus' (BCMV) to differentiate it from bean yellow mosaic virus (20). The characteristic foliar symptom induced by BCMV consists of a mosaic of well-defined dark and light green areas, often observed in a vein-banding pattern (Figure 1 A).

However, BCMV can induce other symptoms in mosaic-affected plants, such as leaf curling, blistering, dwarfing, and chlorosis (3,16). A different syndrome of BCMV consisting of top necrosis and subsequent plant death (Figure 1 B), was first described by Grogan and Walker in 1948 (7).

This systemic necrosis reaction is first observed in young trifoliolate leaves, affecting the vascular system. As the vein necrosis progresses, the leaf dies and the necrosis spreads systemically down the petiole and stem, till it reaches the roots. Hence the name "black root", also given to this syndrome (11). In mature plants bearing pods, these may be affected by local lesions or systemic necrosis.

As a member of the potyvirus group, BCMV consists of filamentous particles about 750 nm long and 14 nm wide, containing single-stranded RNA. BCMV is transmitted by mechanical inoculation, by several aphid species in a nonpersistent manner, and through seed and pollen. The virus induces the formation of cylindrical 'pinwheel' inclusions in the cytoplasm of infected cells (16).

Currently, there is a proposal to consider BCMV as one of two species of the genus Potyvirus, family Potyviridae, that cause bean common mosaic or 'black root' symptoms. Three necrosis-inducing strains previously considered as BCMV, are now considered a new virus species called 'bean common mosaic necrosis virus', as discussed below in more detail.

PATHOGENIC VARIABILITY OF BCMV

The first BCMV isolate studied came to be known as the 'Type strain' or US 1 (7). When bean cultivars 'Michelite' and 'Robust', possessing resistance to the Type strain, became infected in 1939, the first BCMV variant was recorded as the New York 15 or US 2 strain (23).

Two other CMV variants described in 1961, in the United States, included the 'Idaho' (US 3), and 'Western' (US 4) strains (5,25), which belong to the same pathogenicity group as the Dutch 'Colana' (NL 6) strain (6). In 1962, the BCMV-Florida (US 5) was detected by Zaumeyer and Goth (32). This strain was observed to display a higher level of virulence in certain bean genotypes, than either the Type or New York 15 strains.

In 1964, a new BCMV strain (US 6) was shown by Silbernagel to be seed-borne in cultivar Red Mexican 35, previously considered as resistant to all mosaic-inducing BCMV strains. The 'Mexican' BCMV strain induced symptoms as severe or worse than the Florida strain in some cultivars (24). This strain had been previously described from the Netherlands as the 'Great Northern' (NL 4) strain (10).

Concerning the BCMV strains capable of inducing the hypersensitive 'black root' reaction, several observations made by various researchers between 1943 and 1977, demonstrated that there were BCMV strains that could elicit the reaction

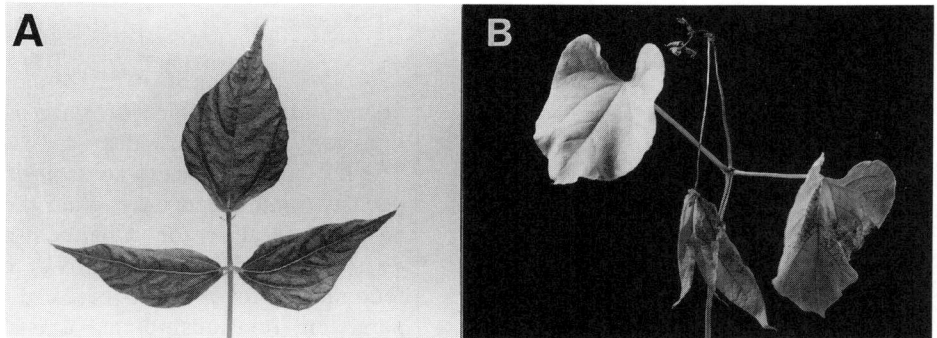

Figure 1. Mosaic (A) and systemic necrosis (B) symptoms induced by BCMV and BCMNV in *Phaseolus vulgaris* L.

and others which could not (6,9). Also, it was shown that for systemic necrosis to occur, it was necessary to maintain inoculated test plants at high temperatures (30-32°C) for some strains (temperature-dependent), such as NL2 and NL6, whereas for other strains the hypersensitive reaction could take place at moderate temperatures (17-26°C) (9). The most pathogenic temperature-independent strains are NL3 and NL5, first detected by Hubbeling in the Netherlands in 1963 and 1972, respectively (9,10).

Serological tests using monoclonal antibodies have been used to divide BCMV strains into serotype A, which includes temperature-independent BCMV strains NL3, NL5, and NL8; and serotype B, which includes the non-necrotic BCMV strains Type, Florida, NY 15, NL4, and NL7, and the temperature-dependent necrotic strains, NL2, and NL6 (31). Also, peptide profiles of coat protein digests have confined the existence of two distinct potyviruses among the known BCMV strains (15). Recently, it has been proposed that BCMV NL3, NL8, and NL5 should be strains of a new species called 'bean necrosis mosaic virus' (BCMNV), while the rest of the BCMV strains maintain their current denomination. The proposal further recommends that some related azuki bean, cowpea, peanut, and soybean potyviruses should be considered as strains of BCMV (F.J. Morales and BCMV working Group, unpublished data). This proposal has apparently been accepted by the International Committee on Taxonomy of Viruses (unpublished data).

Most of the current methodology implemented at the International Center of Tropical Agriculture (Centro Internacional de Agricultura Tropical, CIAT) to genetically control bean common mosaic, is based on Drijfhout's unique contribution to the understanding of the genetic interaction between *P. vulgaris* and BCMV (6). Since the division of BCMV into two separate viruses, BCMV and BCMNV, does not modify the genetic improvement or screening methodologies followed at CIAT, we prefer to deal with the disease, bean common mosaic, in generic terms, than with the viral pathogens as separate entities. After all, it must be taken into account that both BCMV and BCMNV include strains capable of inducing mosaic or systemic necrosis symptoms depending on the bean genotype inoculated.

INTEGRATED BEAN COMMON MOSAIC MANAGEMENT

Exclusion

One of the first measures that can be taken to control a plant virus is exclusion, that is, to avoid the introduction of an exotic virus or virus strain into a region free of the pathogen. Unfortunately, BCMV and BCMNV are some of the most ubiquitous viruses known, found wherever beans are grown in the world (16). The reason for its universal distribution is its high incidence of transmission in the seed produced by susceptible bean cultivars. On an average, 30-50% BCMV/BCMNV transmission rates in *P. vulgaris* seed can be expected in most susceptible landraces (17).

Nevertheless, the 10 BCMV/BCMNV strains described by Drijfhout are not equally distributed in the world (6). In Latin America, the center of origin of *P. vulgaris*, the Type strain predominates, followed by the Florida strain, NL6, and NL4. The main necrotic strain, BCMNV NL3, is also present in Latin America but at a low incidence (F. Morales, unpublished data).

In North America, the Type, NY-IS, NL6

and Florida BCMV strains, and more recently, the BCMNV necrosis-inducing strains NL3 and NL8, have been detected (13,22). In Europe and particularly in the Netherlands, the Type (NL1), NL2, NL3, NL4, NL5, NL6, and NL8 BCMV and BCMNV strains have been reported (6).

In east Africa, the necrotic BCMNV NL3 strain predominates in most countries, followed by BCMV NL6, NL4 and Type strains, and by BCMNV NL8 (26). The high incidence of BCMNV NL3 in East Africa is noteworthy. Some virologists speculate that BCMNV NL3 is probably a potyvirus of cultivated or wild legumes in Africa (27), but it is also possible that NL3 may have been introduced into Africa in infected navy bean seed from Europe (17). A similar situation is found in Chile, the only country in South America where navy beans have been imported from Europe (17).

Thus, we can conclude that, while the exclusion of certain BCMV or BCMNV strains, such as the necrosis-inducing strain BCMNV NL3, is still possible for some countries where these strains have not been reported yet, the deficient quarantine regulations and facilities found in most bean-producing countries, will eventually lead to the worldwide distribution of most mosaic- and necrosis-inducing strains of BCMV and BCMNV.

Eradication

Once viruses are introduced in a given production region, it is usually difficult to eradicate them due to the existence of alternative plant hosts and foci of wild or cultivated species of the primary host. However, in the case of BCMV/BCMNV, it could be possible to eradicate the virus or reduce the primary inoculum by eliminating common mosaic-susceptible bean genotypes in a given germplasm collection (14) or production region, since there are few plant species recognized as important reservoirs of these viruses in nature.

The physical eradication (roguing) of BCMV/BCMNV-infected plants in the field, is not recommended since even low populations of aphid vectors can spread the virus in the field faster than symptom development, so that systemically infected but symptomless plants can act as efficient virus reservoirs for secondary spread of the viruses in susceptible bean fields.

Protection

Protection is a control measure to prevent or minimize the attack of plant pathogens, and it is usually accomplished by the use of chemicals. The use of pesticides to control viral diseases, has been recommended only when the causal viruses have a natural vector. Chemical control of aphid vectors of BCMV or BCMNV, however, is not recommended in most cases due to the non-persistent relationship that exists between these viruses and their aphid vectors (3,16). That is, aphid vectors are capable of acquiring and transmitting these viruses in a matter of seconds before any insecticide can act. Also, aphids rarely colonize bean plants in the main production regions of the world, with the exception of east Africa and some Asian countries (F.J. Morales, unpublished). In these cases, where the aphid vectors are not migrating into bean fields but, rather, originate within the crop, chemical control could be feasible to prevent significant feeding damage and early virus attack.

The application of oil formulations has been shown to be effective against stylet-borne or non-persistent viruses (30), but their cost and requirements to obtain an effective coverage, have discouraged their use.

Genetic Control

Breeding for common mosaic resistance in *P. vulgaris*, has been the main objective of most bean improvement programs around the world. The genetics of resistance to BCMV in *P. vulgaris* was initially studied by Pierce (21) in 1935, who described two types of genetic resistance: dominant, as found in 'Corbett Refugee', and recessive, first observed in cultivars 'Robust' and 'Great Northern 1'. In 1950, Ali (1) proposed the existence of two dominant genes, one required for virus infection and another conditioning a hypersensitive necrosis reaction. Additionally, he described the recessive alleles of those genes conditioning susceptibility or resistance to mosaic (depending on the virus strain). Andersen and Down in 1954 (2), and Petersen in 1958 (19), complemented this information for different BCMV strains, confining the existence of both dominant and strain-specific recessive resistance to BCMV.

However, it was not till 1978, when Drijfhout published his work on the genetic interaction between *P. vulgaris* and BCMV (6),

Table 1. Genetic interaction between selected *Phaseolus vulgaris* cultivars and potyvirus species inducing bean common mosaic

Differential cultivar	Resistance genes	BCMV/BCMNV strain and pathogenicity genes[a]						
		Type PO	Fla NL6 $P1.1^2$	NY15 NL2 P1.2	NL3 NL5 $P1.1^2 2$	NL4 $P1.1^2.2^2$	NL7 P1	NL8 P2
Dubbele Witte	I^+	+	++	++	++	+	+	+
Imuna	I^+bc-u bc-1		++	++	++	+	+	
RG-B	I^+bc-u bc-1^2		++		++	+		
Michelite	I^+bc-u bc-2			++	++			+
Pinto 114	I^+bc-u bc-1 bc-2			++	++			
Great Northern 31	I^+bc-u bc-1^2 bc-2^2					+		
Widusa	I		$+^t$		++			+
Jubila	I bc-1		$+^t$	$+^t$	++			
Topcrop	I bc-1		$+^t$	$+^t$	++			
Amanda	I bc-1^2				+			

[a] Reaction to BCMV/BCMNV: + = mosaic in I^+I^+ cultivars or systemic necrosis in II cultivars; t =temperature-dependent; blank spaces = no reaction. Source: E. Drijfhout 1978 (6).

that the existence of five strain-specific, recessive resistance genes, bc-1, bc-1^2, bc-2, bc-2^2, and bc3, and one strain-unspecific gene, (bc-u), required for expression of the latter genes, was revealed. Genes bc-1, bc-2, and their alleles, had a gene-for-gene relationship with four genes for pathogenicity present in BCMV and BCMNV strains. No BCMV or BCMNV strain has been found to attack the bc-3 gene (Table 1)

BREEDING FOR MONOGENIC DOMINANT RESISTANCE TO BCMV/BCMNV

The majority of Latin American bean landraces do not possess any BCMV/BCMNV strain-specific resistance genes. For the most part, these virus-susceptible bean genotypes show common mosaic symptoms and various degrees of plant malformation, mainly leaf curling, under field conditions. These symptoms are caused by the chronic systemic invasion of BCMV or BCMNV. As it can be observed in Table 1, there is always a strain or groups of strains capable of systemically infecting those cultivar groups which only possess recessive resistance to BCMV/BCMNV. It is for this reason that the incorporation of strain-specific recessive genes in susceptible cultivars has not been a widely practiced breeding strategy.

On the other hand, the dominant necrosis (1) gene has been shown to protect *P. vulgaris* genotypes against chronic systemic infection, so

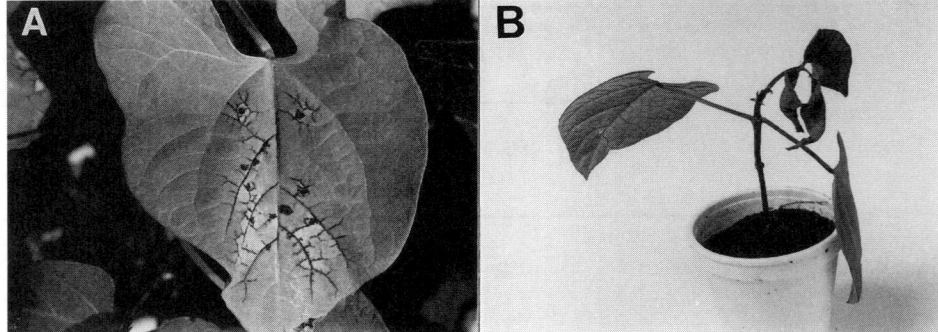

Figure 2. Vein (A) and/or top necrosis (B) observed in bean genotypes possessing monogenic dominant resistance, when artificially inoculated with necrotic BCMV/BCMNV strains.

that no mosaic, plant malformation or seed transmission can occur. Furthermore, screening for a single dominant gene is a rather simple process and, therefore, most of the improved bean cultivars in the world possess dominant resistance to BCMV or BCMNV.

Screening Methodology

The first step in breeding for monogenic dominant resistance involves the selection of suitable parents known to possess the dominant necrosis gene. This information is available at the Genetic Resources Unit of CIAT for more than 22,000 bean accessions from all over the world. Once the hybridization between an 1 gene donor parent and the mosaic-susceptible genotype to be improved, has been achieved, the Fl seed is used to generate F2 nurseries or populations, in which individual plant selections can be made for desirable agronomic traits. At this stage, approximately 25% of the F2 plant population should be susceptible to BCMV, assuming that no other resistance genes are present. These plants may show mosaic symptoms in the field if aphid vector populations are high. Planting BCMV-infected bean seed around these nurseries, helps increase the incidence of BCMV in common mosaic-susceptible F2 plants.

Once the symptomless individual selections are made in F2 plant populations, part of the seed (8-10 seeds) harvested from individual plants, is set aside for a progeny test with selected BCMV/BCMNV strains. In this test, the F3 plants obtained can be inoculated in trays or individual pots, preferably under controlled green- or screenhouse conditions. The selection of virus strains depends on the availability of BCMV or BCMNV strains capable of inducing necrosis in I gene plants. If available, BCMNV NL3, has been the strain of choice for the progeny test, since it can elicit local lesions and vein necrosis in inoculated leaves of I gene plants (Figure 2 A). The observation of local lesion and/or vein necrosis in inoculated primary leaves, or top necrosis (Figure 2 B) in bean seedlings inoculated at the primary leaf stage, is a clear indication that a plant possesses monogenic dominant resistance. Thus, depending upon the number of plants showing necrosis, out of the total number inoculated, it is possible to select lines homozygous for the I gene (100% vein or top necrosis). The presence of over 50 % mosaic-affected plants in a given line would indicate that the F2 plant selected was homozygous susceptible, whereas the presence of approximately 25 % mosaic-affected plants in an inoculated line suggests that the individual F2 plant selection was heterozygous for the I gene. Once identified, these heterozygous lines can be further selected on an individual basis to repeat the progeny test.

In case a necrosis-inducing BCMV or BCMNV strain is not available for screening purposes, a mosaic-inducing BCMV strain can be used for the progeny test. In this case, only mosaic-resistant (symptomless) lines should be selected, and it must be kept in mind that it takes a combination of mosaic-inducing BCMV strains, such as NY15 and NL4, to detect most non-I-gene genotypes. Furthermore, an infectivity (inoculation of BCMV susceptible bean genotypes) or serological assay would be necessary to exclude the possibility of symptomless infections.

PROBLEMS ASSOCIATED WITH THE INCORPORATION OF MONOGENIC DOMINANT RESISTANCE

Genetic Linkage

Temple and Morales demonstrated in 1985, that the red-mottled seed color in a bean genotype was linked to BCMV susceptibility, while darker purple-mottle grain types were closely linked to the dominant necrosis I gene (29). The existence of genetic linkage between certain seed colors and susceptibility to BCMV, has been observed for the 'azufrados' or 'canarios' (yellow), most light red types, and the Mexican 'Flor de Mayo' genotypes. The linkage problem has been partially circumvented by selecting intermediate seed colors, which are still commercially acceptable (F. J. Morales, unpublished data).

Increased Susceptibility to Other Viruses

Recently, Morales and Castaño reported on the higher degree of susceptibility observed in dominant 1 gene bean cultivars challenged by some comoviruses belonging to the "bean severe mosaic virus" complex (18). Bean plants possessing dominant resistance, usually show partial necrosis in leaves and stems, and are severely malformed including the pods. Plant death has been observed following severe attacks of these beetle-transmitted viruses in El Salvador and Guatemala. The combination of the dominant 1 gene and recessive genes, (such as bc-2^2), known to protect the I gene against necrosis-inducing strains of BCMV, does not diminish the severity of the symptoms induced by these comoviruses. On the contrary, bean genotypes possessing a combination of dominant and recessive genes for resistance to BCMV, react with severe necrosis and plant malformation when infected with bean severe mosaic comoviruses.

Susceptibility to Necrosis-inducing BCMV/BCMNV Strains

The commercial cultivation of dominant 1 gene bean genotypes has been a successful approach to the control of BCMV/BCMNV in Latin America for several decades. However, in another important bean production zone of the world, east Africa, dominant 1 gene bean genotypes have often been severely attacked by necrosis-inducing strains, particularly by BCMNV NL3. There has been considerable speculation concerning the nature of this phenomenon but, so far, the main difference between Latin America and east Africa remains the significantly higher incidence of necrosis-inducing BCMNV strains in Africa, while mosaic-inducing BCMV strains, such as the Type strain, predominate in Latin America.

Considering the fact that beans are often the main food staple of thousands of small farmers in east Africa, the deployment of monogenic dominant resistance into this region, has been discontinued until the necrosis I gene can be protected by recessive genes effective against necrotic BCMV/BCMNV strains, as explained in the next section.

BREEDING FOR MULTIGENIC RESISTANCE TO BCMV/BCMNV

Although it was already known that certain bean genotypes were resistant to both common mosaic and systemic necrosis, Drijfhout was the first scientist to demonstrate that the combination of the dominant 1 gene with strain-specific recessive genes, such as bc-2^2, resulted in genotypes that were resistant to the common mosaic and black root syndromes, when challenged by any of the known BCMV/BCMNV strains (6). For instance, bean plants possessing the combination 1 and bc-2^2 only react with pinpoint local lesions (Figure 3 A) on artificially inoculated primary leaves (6).

The role of bc-1^2 is not clear either, but this gene is suspected to confer 1 gene bean genotypes partial resistance to some necrosis-inducing BCMNV strains. Artificially inoculated bean plants suspected to possess the I bc-1^2 gene combination, react with limited vein necrosis on the inoculated primary leaves, but most necrosis-inducing BCMV/BCMNV strains do not become systemic, with the exception of BCMNV-NL5 (F.J. Morales, unpublished data).

Although the genetics of multigene BCMV/BCMNV resistance is not fully understood, since other recessive genes may be involved in the I gene protection mechanism, the Great Northern 31 group of bean cultivars (Table 1) is the main source of the recessive genes bc-u, bc-1^2, and bc-2^2, which protect the dominant 1 gene against necrosis-inducing strains of BCMV and

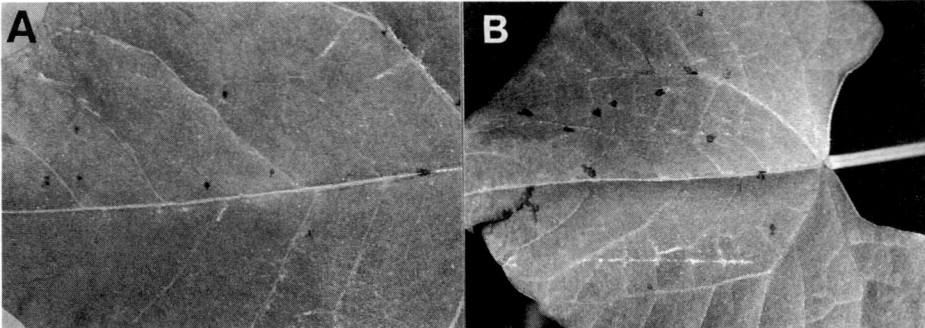

Figure 3. Pin-point local lesions (A) and enlarged point local lesions (13) induced by BCMNV and necrotic BCMV strains in bean genotypes possessing dominant and recessive resistance genes effective against all BCMV/BCMNV strains.

BCMNV.Besides the pin-point local lesions, other types of local lesions, such as enlarged point local lesions (Figure 3 B) have been observed in F3 plants obtained from crosses between I genotypes and bean cultivars possessing the bc-u, bc-1^2, and/or bc-2^2, genes. These black root-resistant plants can be detected in the F3 generation by individual selection of inoculated F2 plants showing pin-point local lesions or no symptoms at all. Obviously, some of these symptomless plants may be 'escapes', that is, artificially-inoculated susceptible plants which were not infected.

The simultaneous incorporation of common mosaic and black root resistance in *Phaseolus vulgaris* is thus possible, albeit at a lower efficiency. While approximately 25% of the F3 lines obtained from simple crosses between mosaic-susceptible and dominant 1 gene progenitors, should be homozygous resistant, less than 1 % of the F2 plants obtained from the hybridization of Royal Red (I) and Red Mexican 35 (bc-1^2,bc-2^2) were resistant to both mosaic and systemic necrosis (4).

For the screening process, it is possible to start the artificial inoculation work at the F2 generation, using BCMNV-NL3, selecting both symptomless and local lesion plants for a subsequent progeny test. As mentioned above, it is possible to obtain local-lesion plants in the F3 derived from symptomless F2 plants. The genetic basis for this observation has not been elucidated.

BREEDING FOR IMMUNITY TO BCMV/BCMNV

As explained before, the bc-3 gene is the only recessive gene for which a corresponding pathogenic BCMV or BCMNV strain is not known. Unfortunately, the bc-3 gene was originally available only in an experimental line, IVT 7213, bred by Drijfhout in the Netherlands (6). Consequently, its adaptation to tropical and semitropical environments was poor, preventing its use in breeding programs outside temperate regions. Later on, Drijfhout, in collaboration with the Bean Program of CIAT, initiated a breeding project designed to incorporate the bc-3 gene in tropically-adapted bean genotypes. Currently, several experimental bean lines possessing the bc-3 gene and good adaptation to the tropics, are available at CIAT for breeding purposes. The combination of the 1 gene and bc-3 is also possible to obtain more durable resistance to all BCMV/BCMNV strains.

The symptomless nature of bean genotypes possessing the bc-3 gene, following their inoculation with BCMV or BCMNV, may be a minor drawback to the use of this gene, as it requires an additional test for all symptomless plants, to differentiate between resistant and symptomless (but BCMV-susceptible) bean genotypes. This test can be conducted using specific antisera to the virus strain used for inoculation or by a suitable infectivity assay. Recently, molecular markers (RAPD) linked to the bc-3 gene in Andean bean genotypes have been identified in the United States (12).

CURRENT SITUATION OF BCMV IN THE WORLD

Latin America

Considerable progress has been achieved in releasing BCMV/BCMNV-resistant bean cultivars in Latin America. For the past 15 years, all of the breeding lines produced by CIAT in collaboration with national agricultural research programs, have resistance to bean common mosaic. Also, some national programs, such as INIA in Chile, have a long tradition of breeding for BCMV/BCMNV resistance.   In other Latin American countries, such as Brazil, and Venezuela, most of the black-seeded bean cultivars grown, possess monogenic dominant resistance to BCMV and BCMNV.

Nevertheless, there is still a large number of common mosaic-susceptible bean landraces grown all over Latin America by traditional small farmers, who do not have access to seed of improved bean cultivars, or who are forced to respond to strong market/consumer preferences for beans of a given seed shape, color, and culinary characteristics, often found in BCMV/BCMNV susceptible landraces.

East Africa

Most of the beans grown by small farmers in east African countries, such as Kenya, Rwanda, Burundi, Zaire, and Uganda, consist of varietal mixtures, the majority of which are susceptible to BCMV/BCMNV. These mixed bean populations, however, exhibit considerable stability to several biotic and abiotic factors which could cause severe losses in genetically homogenous bean cultivars. Also, it is apparent that the incorporation of monogenic dominant resistance in east African bean cultivars is a potentially dangerous approach, due to the high incidence of necrosis-inducing strains, mainly BCMNV NL3, in east Africa. Some of these strains are also suspected of having alternative wild legume hosts, such as *Cassia*, *Crotalaria*, *Rhynchosia*, *Macroptillum*, and *Vigna* species in east Africa (27).

Consequently, considerable efforts are being made by the regional CIAT projects in collaboration with national programs to introduce mosaic- and black root-resistance in varietal mixtures and bean cultivars grown in east Africa.

This strategy can only be justified in the absence of any detrimental side effects that might jeopardize the genetic stability of the varietal mixtures currently grown by small-scale farmers in east Africa.

North America

Although some of the main bean cultivars grown in America possess monogenic dominant resistance to BCMV/BCMNV, there are still a number of bean genotypes, such as pinto and some navy and red-seeded cultivars, which only possess partial recessive resistance. The BCMV/BCMNV situation in North America was under control until 1982, when new strains, namely BCMV NL4, and BCMNV NL3 and NL8 made their appearance in the U.S. (13,22,8). Considerable research has been conducted in recent years to produce bean cultivars that possess multigenic resistance to all known strains of the viruses.

Asia

Some Asian countries, such as Turkey, Iran, India, and China grow *P. vulgaris*.   The majority of the landraces found in these countries are susceptible to BCMV, expressing mosaic and severe plant malformation. The BCMV/BCMNV incidence in countries, such as Iran and Turkey, where beans are grown under dry, irrigated conditions, is high (often 100%) due to the proliferation of aphid vectors (F.J. Morales, unpublished data).

Progress in the incorporation of BCMV/BCMNV resistance into Asian cultivars, has been slow due to lack of human resources and technical expertise in the area. Furthermore, some of the seed types grown in the above mentioned countries, such as cranberries, are difficult to improve for their BCMV/BCMNV resistance due to seed color-virus susceptibility linkage problems. Specific breeding projects are now underway to exploit recessive gene resistance, namely bc-3, for the improvement of red-seeded, cranberry and other Asian bean genotypes.

CONCLUSIONS AND OUTLOOK

There is no doubt that both the mosaic- and necrosis-inducing strains of BCMV and BCMNV can be genetically controlled . This is possible thanks to the work of several scientists, particularly Dr. E. Drijfhout, who elucidated the genetic interaction between BCMV/BCMNV and *P. vulgaris* (6). Soon after this valuable information was available, CIAT developed screening methodologies to select bean genotypes resistant to both mosaic and systemic necrosis induced by any strain. These methodologies have been successfully implemented and, consequently, we could say that BCMV/BCMNV should no longer be a production constraint anywhere in the world. Unfortunately, the lack of adequate virology facilities and bean breeding expertise in most developing, bean-producing countries, have delayed the adoption of the technology available to genetically control the various manifestations of bean common mosaic.

Many scientists in developed and developing nations have continued working on the molecular characterization of BCMV/BCMNV strains, although the outcome of this research might not significantly advance the current screening or genetic improvement strategies. The development of molecular markers for BCMV/BCMNV resistance genes in *P. vulgaris*, on the other hand, is expected to expedite the selection of BCMV/BCMNV-resistant bean genotypes.

REFERENCES

1. Ali, M.A. 1950. Genetics of resistance to the common bean mosaic virus (bean virus 1) in the bean (*Phaseolus vulgaris* L). Phytopathology 40:69-79.

2. Andersen, A.L., and Down, E.E. 1954. Inheritance of resistance to the variant strain of the common bean mosaic virus. (Abstr.) Phytopathology 44:481.

3. Bos, L. 1971. Bean common mosaic virus. CMI/AAB. Descriptions of Plant Viruses No.73.

4. Caceres, S.R., and Morales, F.J. 1985. Incorporación de genes de resistencia dominante y recesiva a cepas del virus del frijol (*Phaseolus vulgaris* L.). Acta Agron. 35:7-20.

5. Dean, L.L., and Wilson, V.E. 1959. A new strain of common bean mosaic in Idaho. Plant Dis. Reptr. 43:1108-1110.

6. Drijfhout, E. 1978. Genetic Interaction between *Phaseolus vulgaris* and Bean Common Mosaic Virus with Implications for Strain Identification and Breeding for Resistance. Ctr. Agric. Publ. Doc. Wageningen.

7. Grogan, R.G., and Walker, J.C. 1948. The relation of common mosaic to black root of bean. J. Agric. Res. Washington D.C. 77:315-331.

8. Hamptom, R.O., Silbernagel, M.J., and Burke, D.W. 1983. Bean common mosaic virus strains associated with bean epidemics in the northwestern United States. Plant Dis. 67:658-661.

9. Hubbeling, N. 1963. Complicaties bij de toetsing van bonerassen op resistentie tegen *Phaseolus* virus 1 ten gevolge van het voorkomen van afwifkende virusstammen. Mededelingen van de Landbouwhogeschool en de Opzoekingstations van de Staat te Gent 28:1025-1033.

10. Hubbeling, N. 1972. Resistance in beans to strains of bean common mosaic virus. Mededelingen van de Faculteit Landbouwwetenschappen, Rijksuniversiteit Gent 37:458-466.

11. Jenkins, W.A. 1940. A new virus disease of snap beans. J. Agric. Res. Washington. D.C. 60:279-288.

12. Johnson, W.C., and Gepts, P. 1994. Two new molecular markers linked to bc 3. Annu. Rep. Bean Improv. Coop. 37:206-207.

13. Kelly, J.D., and Saettler, A.W. 1976. Severe virus outbreaks affect Michigan dry bean production. Annu. Rept. Bean Improv. Coop. 29:9-10.

14. Klein, R.E., Wyatt, S.D., and Kaiser, W.J. 1986. Virus eradication from a bean germplasm collection. Annu. Rept. Bean Improv. Coop. 29:37.

15. McKern, N.M., Mink, G.I., Barnett, O.W., Mishra, A., Whittaker, L.A., Silbernagel, M.J., Ward, C.W., and Shukla, D.D. 1992. Isolates of bean common mosaic virus comprising two distinct potyviruses. Phytopathology 82:923-929.

16. Morales, F.J., and Bos, L. 1988. Bean common mosaic virus. CMI/AAB Descriptions of Plant Viruses. No. 337.

17. Morales, F.J., and Castaño, M. 1987. Seed transmission characteristics of selected bean common mosaic virus strains in differential bean cultivars. Plant Dis. 71:51-53.

18. Morales, F.J., and Castaño, M. 1992. Increased disease severity induced by some comoviruses in bean genotypes possessing monogenic dominant resistance to bean common mosaic potyvirus. Plant Dis. 76:570-573.

19. Petersen, H.J. 1958. Beiträge zur Genetik von *Phaseolus vulgaris* L. anf Infektion mit *Phaseolus* Virus 1 Stamm Voldagsen, Zeitschrift für Pflanzenzüchtung 39:187-224.

20. Pierce, W.H. 1934. Virues of the bean. Phytopathology 24:87-115.

21. Pierce, W.H. 1935. The inheritance of resistance to common bean mosaic in field and garden beans. Phytopathology 25:875-883.

22. Provvidenti, R., Silbernagel, M.J., and Wang, W.J. 1984. Local epidemic of NL-8 strain of bean common mosaic virus in bean fields of western New York. Plant Dis. 68:1092-1094.

23. Richards, B.L., and Burkholder, W.H. 1943. A new mosaic disease of beans. Phytopathology 33:1215-1216.

24. Silbernagel, M.J. 1966. Mexican common bean mosaic virus, a new seedborne mosaic of beans. Annu. Rept. Bean Improv. Coop. 9:36.

25. Skotland, C.B., and Burke, D.W. 1961. A seedborne bean virus of wide host range. Phytopathology 51:565-568.

26. Spence, N.J., and Walkey, D.O. 1991. Identification of strains of bean common mosaic virus occurring in different regions of Africa. Annu. Rept. Bean Improv. Coop. 34:5-6.

27. Spence, N.J., and Walkey, D.O. 1991. Bean common mosaic virus related isolates in legume weeds and other non-*Phaseolus* hosts in Africa. Annu. Rept. Bean Improv. Coop. 34:7-8.

28. Stewart, V.B., and Reddick, D. 1917. Bean mosaic. Phytopathology 7:61.

29. Temple, S.R., and Morales F.J. 1986. Linkage of dominant hypersensitive resistance to bean common mosaic virus to seed color in *Phaseolus vulgaris*.

Euphytica 35: 331-333.

30. Vanderveken J.J. 1977. Oils and other inhibitors of nonpersistent virus transmission. Pages 435-450 in: Aphids as Virus Vectors. K.F. Harris and K. Maramorosch, eds. Academic Press, New York.

31. Wang, W.Y., Mink, G.I., and Silbernagel, M.J. 1985. A broad spectrum monoclonal antibody prepared against bean common mosaic virus. (Abstr.) Phytopathology 75:1352.

32. Zaumeyer, W.J. and Goth, R.W. 1962. A new strain of common bean mosaic virus. (Abstr.) Phytopathology 52:758.

CHAPTER 40

Epidemiology and Control of Faba Bean Necrotic Yellows Virus

K.M. Makkouk, H.J. Vetten, L. Katul, A. Franz and M.A. Madkour

Faba bean necrotic yellows virus (FBNYV), which affects legume production in many countries of West Asia and North Africa (WANA), has isometric particles, 18 nm in diameter, contains a multipartite circular ssDNA genome and is transmitted by aphids (e.g. *Acyrthosiphon pisum* Harris and *Aphis craccivora* Koch.) in the persistent manner. Together with banana bunchy top virus (BBTV) (9, 22), coconut foliar decay virus (CFDV) (41, 42) and subterranean clover stunt virus (SCSV) (6, 11) it seems to form a separate group of ssDNA plant viruses. It has been reported to naturally infect faba bean (*Vicia faba* L.) in Algeria, Egypt, Ethiopia, Jordan, Lebanon, Morocco, Syria and Turkey (17, 29). During the growing season of 1991/1992, a severe virus epidemic affected faba bean crop in Middle Egypt leading to yield losses of over 90%. Surveys conducted in the following two years indicated that the epidemic of 1992 was most likely caused by FBNYV (35).

Symptoms induced by FBNYV on the cool-season legume crops chickpea (*Cicer arietinum* L.), faba bean (*Vicia faba* L.), lentil (*Lens culinaris* Medik.) and pea (*Pisum sativum* L.) are mostly yellowing, stunting and poor pod-setting. Such symptoms are very similar to those induced by bean leaf roll luteovirus (BLRV) (15). For many years it was assumed that BLRV was the most economically important virus affecting legumes and the cause for several diseases such as chickpea stunt in chickpea, lentil yellows in lentil or bean leaf roll in faba bean (24, 33). After the characterization of FBNYV and development of diagnostic reagents for its detection (29), it became evident that in most instances in the past where the symptomatology and aphid transmission of a legume virus disease was considered as evidence for BLRV infections, probably FBNYV was the causal agent. The relative importance of FBNYV and luteoviruses (e.g. BLRV) varies from one country to another, but results of recent surveys (34, 35) suggest that FBNYV has a wide distribution and occurs at higher incidence than BLRV in some WANA countries.

FBNYV is taxonomically closely related to SCSV reported in Australia and New Zealand, and causes similar symptoms in clovers, faba bean, French bean and other legumes (11, 19). A lesser studied virus, known for a long time in Japan, is milk vetch dwarf virus (MDV) (25, 44). Further investigations are proving that FBNYV is very similar to MDV and may be a synonym of this virus (17). More recently, another distinct virus, chickpea chlorotic dwarf geminivirus was also found to infect chickpea and faba bean in India (23) and the Sudan (36) causing symptoms similar to those produced by FBNYV on these crops. It is known that viruses which belong to the same taxonomic group and share host ranges have similar ecologies, but those which belong to different taxonomic groups could also have similar ecologies. This is the case with FBNYV which is taxonomically different from luteoviruses. Since FBNYV and legume luteoviruses are transmitted by almost the same aphid species, are phloem-limited in their host plants, cause nearly identical symptoms and are not seed-transmitted, they appear to have similar ecologies. Because of this similarity and because luteoviruses are better studied than FBNYV and related viruses, information generated in relation to the control of luteoviruses are useful in the development of possible control measures for FBNYV.

EPIDEMIOLOGY

Sources of Inoculum

The host range of FBNYV seems to be largely restricted to the family Fabaceae (17). Many food and forage legume crops such as *Cicer arietinum*, *Lathyrus* spp., *Lens culinaris*, *Medicago* spp., *Pisum sativum*, *Trifolium* spp., *Vicia faba* (many cvs.) and many *Vicia* forage species are susceptible to FBNYV infection (17). More recently, the virus was found to naturally infect cowpea (*Vigna unguiculata*) and French bean (*Phaseolus vulgaris*) (16), both summer legumes, widely grown in the WANA region. In addition to the above legume crops, leguminous weeds such as *Melilotus officinalis* L., *Tetragonolobus purpureus* Moench, *Lathyrus* spp., *Medicago* spp. and *Trifolium* ssp. were also found naturally infected with FBNYV in Egypt and Syria (17).

Most of these susceptible legume species are winter crops, except cowpea and French bean which seem to be important over-summering hosts of the virus. When aphid vectors are abundant, the virus can be spread from overlapping winter to summer legumes and again to winter legumes. Severe epidemics can occur when environmental conditions permit the build-up of the aphid-vector populations which have access to virus sources.

Virus spread in any region not only depends on vector activity and population build-up and on virus titres in sources of infection but also on the number and proximity of these source plants. Thus a nearby infected crop provides a much higher inoculum pressure than a few infected weed plants (8). In Egypt, a FBNYV-susceptible crop such as Egyptian clover (*Trifolium alexandrinum* L.), which is planted in large areas and often next to faba bean fields, probably plays a role in FBNYV spread to the faba bean crop. Quantitative studies have not been conducted yet to determine the relative importance of such crops versus weeds as sources of FBNYV infection. However, final infection pressure to which a given crop (e.g. faba bean) is subject to, is of course, determined by inoculum potential and vector pressure.

Transmission

Only the aphid species *A. pisum*, *Aphis craccivora* and *A. fabae* Scop. have so far been reported as FBNYV vectors. The efficiency of FBNYV transmission by aphids was found to be high for the first two species and very poor for *A. fabae* (16, 29). The information available on FBNYV aphid transmission indicates that FBNYV has the typical features of a persistently transmitted virus which is circulated but does not multiply in the vector insects. Aphids require long acquisition and inoculation feeding periods for becoming efficient vectors, and FBNYV persists in the aphids for almost their entire life (Franz, unpublished). Similar to all phloem-limited viruses, FBNYV is not known to be transmitted by seed or mechanical means. Since all FBNYV hosts reported so far are propagated by true seeds, the only method of natural spread for this virus is by aphid vectors.

To understand the dynamics of epidemic build-up, information on the efficiency with which FBNYV is acquired by aphids from wild hosts as compared with legume crops is essential. If crop plants are more important than wild plants as a source of virus inoculum this should lead to agricultural practices different from those needed if wild plants are the most important sources.

Preliminary observations in faba bean fields in Egypt indicated that aphid vectors, transmit FBNYV to a few plants in the field early in the season (September-October), often close to the field border. These foci enlarge as adjacent plants become infected. When weather conditions favor aphid population build-up and activity, these single-plant foci increase to patches of 50-100 infected plants within 2-3 weeks. This type of spread suggests that faba bean is a favorable host for both virus and vectors and that the spread of FBNYV mainly depends on colonizing aphid species within the crop in a fashion typical of persistently transmitted viruses.

Environmental conditions play an important role in FBNYV spread. High incidence levels have been observed in regions characterized by mild winters e.g. Middle Egypt (Beni Suef and Minia governorates), the Jordan Valley of Jordan and the coastal areas of Syria and Turkey (Makkouk, unpublished). In all these regions winter temperatures rarely fall below 5°C permitting the aphid vector to overwinter parthenogenetically. When temperatures rise the aphids become active, multiply and spread the virus.

CONTROL

Disease control is based on the results

from ecological and epidemiological studies. Control measures can be one of three major categories: (i) those aimed directly at the virus, (ii) those aimed directly at the vector, and (iii) integrated approaches, which combine all possible components in a way to complement each other and be applied at the farmers' level as one control package.

Methods to Control the Virus

Reducing Sources of Infection

A significant decrease in disease incidence can be expected if sources of infection are eliminated from within or near crops. In practice, this is not easy, and is rather impossible to eliminate all weed hosts of the virus. Prior to advocating such an approach for FBNYV control, more information on the relative importance of wild legumes as sources of FBNYV infection is needed.

In Egypt and Syria higher FBNYV incidence is commonly observed at the edges of faba bean fields. This phenomenon suggests that optical stimuli, such as colour contrasts between the crop canopy and the bare soil or between two different crops (2) attract the alate aphids carrying FBNYV. Although pronounced edge effects caused by SCSV in French bean in Australia were ascribed to large numbers of aphids migrating onto French bean seedlings from adjacent SCSV-infected faba bean plants (18), it is generally accepted for persistently transmitted viruses that there is no relationship between the levels and patterns of disease incidence in a field and the distance of the infection sources. Therefore, it will be difficult to determine the relative importance of wild legumes as sources of FBNYV infection just by looking at plants in the vicinity of faba bean fields.

Results of a field trial at ICARDA showed that highest yield losses and even complete crop failure occur when faba bean plants are inoculated with FBNYV-carrying aphids at a very young stage and that infections by FBNYV at later growing stages are economically less important (17). Similarly in Middle Egypt and the coastal areas of Syria many faba bean fields when sown early in the fall (September or early October) suffered high incidence of FBNYV, sometimes reaching 100%. Although early infections by FBNYV are undoubtlessly serious in most WANA countries, there are also observations in Egypt that

infections at later stages (flowering or podding) can also cause serious losses, re-confirming the need for further studies. Such FBNYV epidemics which result most likely from high aerial and on-ground (plant) populations of aphids as well as from abundant sources of infection can be prevented by proper timing of sowing dates, avoiding periods of peak aphid populations and by spraying the fields with appropriate insecticide once or twice during the period when viruliferous aphids introduce the virus to the crop.

Another measure to decrease disease incidence is to eliminate sources of infection from within the crop. Roguing of FBNYV-infected faba bean eliminates or reduces primary infection foci inside the field. Significant reduction in virus incidence was noticed (El-Hadi and Hilmi, personal communication) in a number of fields in Egypt when roguing was practised 2-3 times during the growing season. More experimentation on this aspect is still required to evaluate the economic impact of roguing on crop yield. On the other hand, a scattered distribution of initially infected plants throughout a crop was observed in some faba bean fields in Egypt suggesting that FBNYV had been introduced from a distant source. This type of spread makes the reduction of infection through management of cropping patterns alone, unlikely.

Selection and Breeding for Virus Resistance

Classical Breeding

In preliminary screening for FBNYV resistance using 200 faba bean pure lines, no resistance was found (Makkouk, unpublished). Further work, however, is planned to screen the remaining available faba bean germplasm.

When lentil lines were tested for FBNYV resistance, several lines such as ICARDA ILL 213, 291, 6198, 6193 and 6245 were found highly resistant after artificial field inoculation by viruliferous aphids (38). Resistance was expressed by low virus incidence and extremely low (less than 5%) yield loss. These results were obtained from a single environment and require confirmation. Such preliminary data suggest that lentil genotypes vary widely in their reaction to FBNYV infection as compared to faba bean lines.

Novel Approaches for Virus Resistance

Breeding for resistance has generally aimed at incorporating resistance genes found in wild species into established cultivars. In the case of FBNYV, no sources of resistance in *Vicia faba* have been identified so far. If sources of resistance are not readily available, the concept of pathogen-derived resistance may be used to circumvent disease (43). During the last decade, virus resistance has been introduced successfully into plants using viral coat protein (21) or replicase sequences of several plant RNA viruses (10). Only few reports are available on pathogen-derived resistance against DNA viruses, e.g. geminiviruses (13,31,45). Since the replication strategy of FBNYV seems to be similar to that of geminiviruses (6), there is an obvious potential for genetic engineering to produce FBNYV-resistant faba bean. An indispensable prerequisite for generating transgenic plants is the establishment of a regeneration system for *Vicia faba* which has been recently developed (3).

Faba bean and FBNYV appear to be a promising model for assessing the potential of pathogen-derived resistance to a DNA virus in faba bean by using various FBNYV genes. Since the replicase gene of FBNYV has been already characterized (30) and the capsid protein (CP) gene has been identified (Katul, unpublished), these two viral genes appear to be suitable candidates to be used for transforming faba bean. It is not known if protection against FBNYV can be obtained only by expressing CP in plants or if expression of a truncated CP or non-translatable form of the CP gene will also confer similar levels of resistance. Transformation for faba bean with FBNYV genes depends not only on the availability of FBNYV genes and a regeneration system for *V. faba* but also on the use of suitable and highly efficient promoters, either supplied by the FBNYV genome itself or by using well-established promoters, such as the CaMV 35S and the PR1 promoters, which can be linked to suitable FBNYV genes.

Although the molecular details of the replication mechanism of ssDNA viruses, such as FBNYV and SCSV, have not yet been elucidated, it appears valid to assume that their overall replication strategy is similar to that of the geminiviruses (32). Several features of the genome structure and the replication-associated proteins (Rep proteins) of the group of small isometric ssDNA viruses (6, 22, 30) are very similar to those of the geminiviruses: (i) an inverted repeat sequence with the potential to form a hairpin-like structure; (ii) an AT-rich sequence within the loop of the hairpin, very similar to the conserved nonamer sequence that is the origin of geminivirus

plus-strand replication; and (iii) a putative replicase sharing essential amino acid motifs with the replicase of geminiviruses.

Therefore, the development of a pathogen-derived resistance strategy for FBNYV should concentrate on approaches that have already been shown to be successful for geminiviruses (1,5,14,37). Particularly *in vitro* engineered replicase genes hence appear to be the most promising candidates for generating transgenic faba bean plants with high levels of FBNYV resistance.

Methods Directed Towards Avoidance of Vectors or Reducing Their Incidence

Chemical Control

Since FBNYV is a persistently transmitted virus whose spread depends predominantly on colonizing aphid species which require long acquisition and inoculation feeding periods for becoming efficient vectors, FBNYV spread should theoretically be controlled efficiently by using systemic aphicides against the virus vector. The efficacy of vector control by chemical means is governed by the persistence of the aphicide in the plant and the time intervals at which they are applied. Systemic insecticides have been shown to reduce the incidence of persistently transmitted viruses. In Tasmania, infection of faba bean by subterranean clover red leaf virus at harvest varied from 13 to 23% in plots sprayed 3 to 5 times with demeton-S-methyl compared with 31 to 84% in unsprayed plots (27). Pyrethroids, such as deltamethrin, which has a rapid knock-down effect on insects, were found to also reduce primary infections in faba bean crops (26).

In Egypt, farmers often spray insecticides to control aphids as they are a pest in faba bean crops. In developing countries, however, the application of 3-5 sprays is mostly uneconomical and can have adverse environmental effects. Therefore, proper timing of a single spray or a minimum number of sprays to prevent the build-up of high aphid population and the movement of viruliferous aphids within a crop should be economical and given priority. In Britain, farmers apply a single aphid spray early in November to cereal crops at risk to prevent secondary spread of barley yellow dwarf luteovirus (40).

The use of systemic insecticides such as imidacloprid (Gaucho^R) as a seed-dressing was effective at experimental level in reducing the incidence of BYDV in cereals (Bertschinger, personal communication). The seed treatment offered reasonable protection from BYDV for 2-3 months after sowing. Such a treatment may prove to be economical and less hazardous to the environment. The feasibility and applicability of such chemicals to protect legumes from infection with persistently transmitted viruses such as FBNYV has been only recently initiated.

Sowing Date

In both Syria and Egypt, faba bean crops planted early in September are often severely attacked leading to 100% FBNYV infection. In such circumstances, farmers plough the crop under and replant another crop. Delaying sowing until October or November often resulted in lower levels of virus infection, most probably due to the reduced vector populations and activity. More studies are still required to establish the most suitable sowing date.

Breeding for Vector Resistance

Some decades ago genetic resistance to the vector was advocated (39), but not much progress has been made in this area. The availability of cheap and effective chemicals in the 1950s and 1960s reduced interest in investigations of vector resistance in plants. However, the development of insect resistance to pesticides and the public awareness of environmental hazards resulting from their heavy use has generated new interest in breeding for insect resistance in plants (28).

Resistance to aphid vectors has been reported in a number of legume crops. Gunasinghe *et al.* (20) reported the presence of soybean genotypes resistant to the aphid species *Aphis citricola* van der Goot, *Myzus persicae* (Sulzer) and *Rhopalosiphum maidis* Fitch and that such resistance reduced the spread of soybean mosaic virus in the field. Amin (4) also reported resistance to groundnut bud necrosis virus through crop resistance to the thrips vector *Frankliniella schultzei* (Trybom). In soybean the mechanism of resistance is directed against the vector due to leaf pubescence, whereas the mechanism of resistance in groundnut is not known.

Resistance of faba bean to the aphids *A. craccivora*, *A. fabae* and *Acyrthosiphon pisum* has been reported (12). In Egypt, over 1000 lines were screened for aphid resistance, and 36 were

classified resistant. The faba bean line BPL 23 was resistant to both *A. craccivora* and *A. fabae* (7). Whether or not the use of such cultivars could reduce FBNYV spread in the field awaits further evaluation.

Integrated Approach

Each of the control measures mentioned provides only partial control, but combining genetic resistance, cultural practices, and chemical sprays is expected to lead to improvements. The use of host resistance, whether obtained by classical breeding or genetic engineering, and one or two well-timed sprays coupled with optimal planting date and early roguing of virus-infected plants could offer reasonable and economic control and stabilize cool-season legume production. However, to develop a sustainable package of integrated control, more work is needed on the ecology and epidemiology of FBNYV in the different regions where the virus causes damage to cool-season legumes. If we obtain a better understanding of the distribution, natural hosts, and vector transmission of FBNYV as well as the interrelationship between epidemic build-up and climatic conditions, better control strategies can be developed.

The seriousness of FBNYV infection of cool-season legumes, especially faba bean, in some countries of WANA requires a multi-disciplinary approach to exploit all possible control measures to achieve sustainable yields in areas where epidemics due to FBNYV infection are likely to occur.

REFERENCES

1. Aaziz, R. 1994. Resistance dérivée du pathogene: Une proteine negative dominante protege les plantes de *Nicotiana bentamiana* contre le tomato yellow leaf curl virus (TYLCV). DEA de Phytopathologie, Université Paris VI/XI, INA Paris-Grignon.

2. A'Brook, J. 1964. The effect of planting date and spacing on the incidence of groundnut rosette disease and of the vector *Aphis craccivora* Koch at Mokwa, Northern Nigeria. Ann. of Appl. Biol. 54:199-208.

3. Abu-Salha, A.E., Soliman, M., Jacobsen H.J., and Madkour, M.A. 1994. Regeneration and marker proteins for embryogenic differentiation pattern in *Vicia faba* L. Egypt. J. Genet. Cytol. (in press).

4. Amin, P.W. 1985. Apparent resistance of groundnut cultivar Robut 33-1 to bud necrosis disease. Plant Disease 69:718-719.

5. Bendahmane, M., and Gronenborn, B. 1996. Resistance engineering against tomato yellow leaf curl virus (TYLCV) using antisense RNA and ribozyme. Plant Mol. Biol. (in press).

6. Boevink, P., Chu, P.W.G., and Keese, P. 1995. Sequence of subterranean clover stunt virus DNA: affinities with the geminiviruses. Virology 207:354-361.

7. Bond, D.A., Jellis, G.J., Rowland, G.G., Le Guen, J., Robertson, L.D., Khalil, S.A., and Li-Juan, L. 1994. Present status and future strategy in breeding faba bean (*Vicia faba* L.) for resistance to biotic and abiotic stresses. Pages 592-616 In: Expanding the Production and Use of Cool Season Legumes. F.J. Muelbauer and W.J. Kaiser, eds. Kluwer Academic Publishers, The Netherlands.

8. Bos, L., and Makkouk, K.M. 1994. Insects in relation to virus epidemiology in cool season food legumes. Pages 305-332 In: Expanding the Production and Use of Cool Season Legumes. F.J. Muelbauer and W.J. Kaiser, eds. Kluwer Academic Publishers, The Netherlands.

9. Burns, T.M., Harding, R.M., and Dale, J.L. 1995. The genome organization of banana bunchy top virus: analysis of six ssDNA components. J. Gen. Virol. 76:1471-1482.

10. Carr, J. P., and Zaitlin, M. 1993. Replicase-mediated resistance. Semin. Virol. 4:339-347.

11. Chu, P.W.G., and Helms, K. 1988. Novel virus-like particles containing circular single-stranded DNAs associated with subterranean clover stunt disease. Virology 167:38-49

12. Clement, S.L., El-Din Sharaf, N., Weigand, S., and Lateef, S.S. 1994. Research achievements in plant resistance to insect pest of cool season food legumes. Pages 290-304 In: Expanding the Production and Use of Cool Season Legumes. F.J. Muelbauer and W.J. Kaiser eds. Kluwer Academic Publishers, The Netherlands.

13. Day, A.G., Bejerano, E.R., Buck, K.W., Burrell, M., and Lichtenstein, C.P. 1991. Expression of antisense viral gene in transgenic tobacco confers resistance to the DNA virus tomato golden mosaic virus. Proc. Natl. Acad. Sci. USA 88:6721-6725.

14. Desbiez, C., David, C., Mettouchi, A., Laufs, J., and Gronenborn, B. 1995. Rep protein of tomato yellow leaf curl geminivirus (TYLCV) has and ATPase activity required for viral DNA replication. Proc. Natl. Acad. Sci. USA 92:5640-5644.

15. Duffus, J.E., Falk, B.W., and Johnstone, G.R. 1990. Luteoviruses - one system, many variations. Page 511 In: World perspectives on barley yellow dwarf. P.A. Burnett, ed. CIMMYT, Mexico, D.F., Mexico.

16. Franz, A., Makkouk, K.M., and Vetten, H.J. 1995. Faba bean necrotic yellows virus naturally infects Phaseolus bean and cowpea in the coastal area of Syria. J. of Phytopathol. 143:319-320.

17. Franz, A., Makkouk, K.M., Katul, L., and Vetten, H.J. 1996. Monoclonal antibodies for the detection and differentiation of faba bean necrotic yellows virus isolates. Ann. Appl. Biol. 128(2):255-268.

18. Garrett, R.G., and McLean, G.D. 1983. The epidemiology of some aphid-borne viruses in Australia. Pages 199-209 in: Plant Virus Epidemiology. R.T. Plumb and J.M. Thresh, eds. Blackwell Scientific Publications, Oxford, U.K.

19. Grylls, N.E. and Butler, F.C. 1959. Subterranean clover stunt, a virus disease of pasture legumes. Australian J. Agric. Res. 10:145-159.

20. Gunasinghe, U.B., Erwin, M.E., and Kampmeier, G.E. 1988. Soybean leaf pubescence affects aphid vector transmission and field spread of soybean mosaic virus. Ann. Appl. Biol. 112:259-272.

21. Hackland, A.F., Rybicki, E.P., and Thomson, J.A. 1994. Coat protein-mediated resistance in transgenic plants. Arch. Virol. 139:1-22.

22. Harding, R.M., Burns, T.M., Hafner, G., Dietzgen, R.G., and Dale, J.L. 1993. Nucleotide sequence of one component of the banana bunchy top virus genome contains a putative replicase gene. J. Gen. Virol. 74:323-328.

23. Horn, N.M., Reddy, S.V., Roberts, I.M., and Reddy, D.V.R. 1993. Chickpea chlorotic dwarf virus, a new leafhopper-transmitted geminivirus of chickpea in India. Ann. Appl. Biol. 122:457-469.

24. Horn, N.M. 1994. Viruses involved in chickpea stunt. Ph.D. Thesis. 137 Pages. Wageningen, The Netherlands.

25. Inouye, T., Inouye N., and Mitsuhata, K. 1968. Yellow dwarf of pea and broad bean caused by milk-vetch dwarf virus. Ann. Phytopathol. Soc. Japan 34:28-35.

26. Johnstone, G.R. 1984. Control of primary infections of subterranean clover red leaf virus, a luteovirus, in a broad bean crop with the synthetic pyrethroid deltamethrin. Australasian Plant Pathol. 13:55-56.

27. Johnstone, G.R. and Rapley, P.E.L. 1981. Control of subterranean clover red leaf virus in broad bean crops with aphicides. Ann. Appl. Biol. 99:135-141.

28. Jones, A.T. 1987. Control of virus infection in crop plants through vector resistance: a review of achievements, prospects and problems. Ann. Appl. Biol. 111:745-772.

29. Katul, L., Vetten, H.J., Maiss, E., Makkouk, K.M., Lesemann, D.E., and Casper, R. 1993. Characterisation and serology of virus-like particles associated with faba bean necrotic yellows. Ann. Appl. Biol. 123:629-647

30. Katul, L., Maiss, E., and Vetten, H.J. 1995. Sequence analysis of a faba bean necrotic yellows virus DNA component containing a putative replicase gene. J. Gen. Virology 76:475-479.

31. Kunik, T., Salomon, R., Zamir, D., Navot, N., Zeidan, M., Michelson, I., Gafni, Y., and Czosnek, H. 1994. Transgenic tomato plants expressing the tomato yellow leaf curl capsid protein are resistant to the virus. Bio/Technology 12:500-504.

32. Lazarowitz, S.G. 1992. Geminiviruses: genome structure and gene function. Crit. Rev. Pl. Sci. 11:327-349.

33. Makkouk, K.M., Bos, L., Azzam, O.I., Kumari, S.G., and Rizkallah, A. 1988. Survey of viruses affecting faba bean in six Arab Countries. Arab J. Pl. Protection 6:53-61.

34. Makkouk, K.M., Kumari, S.G., and Al-Daoud, R. 1992. Survey of viruses affecting lentil (*Lens culinaris* Med.) in Syria. Phytopathol. Medit. 31:188-190.

35. Makkouk, K.M., Rizkallah, L., Madkour, M., El-Sherbeeny, M., Kumari, S.G., Amriti, A.W., and Solh, M.B. 1994. Survey of faba bean (*Vicia faba* L.) for viruses in Egypt. Phytopathol. Medit. 33:207-211.

36. Makkouk, K.M., Dafalla, G.A., Hussein, M.M., and Kumari, S. 1995. Chickpea chlorotic dwarf geminivirus affecting faba bean and chickpea in the Sudan. J. Phytopathology 143:465-466.

37. Mettouchi, A. 1992. Mutants dominants negatifs dans le site de liasison de NTP de la protein CI du virus de la courbure et du jaunissement des feuilles

de la tomate. DEA de Microbiologie/Virologie Generale, Institut Pasteur.

38. Mouhanna, A. 1994. Survey of virus diseases of wild and cultivated legumes in the coastal region of Syria. M.Sc. thesis, University of Tishreen, Lattakia, Syria.

39. Painter, R.H. 1951. Insect Resistance in Crop Plants. New York, MacMillan. 520 Pages.

40. Plumb, R.T. 1983. Barley yellow dwarf virus - a global problem. Pages 185-198 in: Plant Virus Epidemiology: The Spread and Control of Insect-Borne Viruses. R.T. Plumb and J.M. Thresh, eds. Blackwell Scientific Publication, Oxford, UK.

41. Randles, J.W., and Hanold, D. 1989. Coconut foliar decay virus particles are 20-nm icosahedra. Intervirology 30:177-180.

42. Rohde, W., Randles, J.W., Langridge, P., and Hanold, D. 1990. Nucleotide sequence of a circular single-stranded DNA associated with coconut foliar decay virus. Virology 176:648-651.

43. Sanford, J. C., and Johnston, S.A. 1985. The concept of parasite-derived resistance: deriving resistance genes from the parasite's own genome. J. Theor. Biol. 113:395-405.

44. Sano, Y., Isogai, M., Satoh, S., and Kojima, M. 1993. Small virus-like particles containing single-stranded DNAs associated with milk vetch dwarf disease in Japan. 6th International Congress on Plant Pathology, Montreal, Canada, Abstract no. 17.1.27.

45. Stanley, J., Frischmuth, T., and Ellwood, S. 1990. Defective viral DNA ameliorates symptoms of geminivirus infections in transgenic plants. Proc. Natl. Acad. Sci. USA 87:6291-6295.

CHAPTER 41

Control Measures for the Economically Important Peanut Viruses

D.V.R. Reddy

The measures designed to control virus diseases should lead to reduction of sources of inoculum within and outside the crops, facilitate reduction in the transmission by the principal vector(s) and minimize crop losses. Research on the influence of adopting various control practices for virus diseases on other economically important diseases and pests has received very little attention. As in the case of other plant diseases, breeding of resistant cultivars is preferred. Research into epidemiology, which include identification of principal vector(s), determination of virus-vector relationships, ecology of the vector, alternative hosts, survival of the virus during the off season and determination of the periods when infection is most likely to occur, is essential for formulation of cultural practices. This will lead to information on the development of epidemics and the control measures which can be recommended to the farmers. Field trials to test the effectiveness of various cultural practices should be performed under conditions adopted by and practical to the farmer. Determination of cost effectiveness of these practices should be given high priority.

I would like to emphasize that precise identification of the causal virus(es) is preliminary to application of control measures. In the case of peanut virus diseases, symptoms caused by some of the potyviruses, for example peanut mottle virus, peanut stripe virus and peanut green mosaic virus, resemble each other. Symptoms produced by very different viruses such as peanut clump virus and groundnut rosette virus to a large extent resemble one another. Symptoms caused by the two tospoviruses, peanut bud necrosis virus, and tomato spotted wilt virus, are difficult to distinguish. Therefore identification on the basis of symptoms alone can be misleading.

Although precise losses due to viruses infecting peanut have not been estimated, it is apparent from published reports that groundnut rosette virus, peanut bud necrosis tospovirus, tomato spotted wilt tospovirus, peanut stripe potyvirus and cucumber mosaic cucumovirus can cause economic losses. Therefore, so far, research has only been focused on these viruses. For the other virus diseases reported on peanuts (8,9) research was restricted to characterizing the causal virus and production of diagnostic aids.

The wide annual variation in the incidence of virus diseases, especially groundnut rosette, peanut bud necrosis and tomato spotted wilt viruses, has been reported. Therefore losses caused by these viruses will vary substantially from year to year. Losses due to viruses such as peanut clump virus, being soil-borne, occur consistently in the same area with more or less the same incidence, and hence can be predicted with some precision.

In this chapter information currently available on the various options for the control of peanut virus diseases is given. The viruses are grouped on the basis of their geographical distribution.

ASIA-PACIFIC REGION

The major viruses in this region are peanut bud necrosis tospovirus (PBNV), Indian peanut clump furovirus (IPCV), peanut stripe potyvirus (PStV) and cucumber mosaic cucumovirus (CMV).

Tospoviruses

Two tospoviruses have been reported from this region. They are PBNV and tomato spotted wilt virus (TSWV). PBNV is widely distributed in South Asia and also occurs in Southeast Asia. It is by far the economically most important of all viruses occurring in peanuts in South Asia. TSWV was reported from Australia. However, its distribution in Asia is currently not known. The occurrence of PBNV in Australia has so far not been reported. PBNV and TSWV can only be distinguished by using group specific polyclonal antisera (9).

PBNV is transmitted by *Thrips palmi* in India and TSWV by a range of *Thrips* vectors, including *Frankliniella occidentalis*, *F. schultzei*, *F. fusca*, *Thrips tabaci* and *T. tenuicornis*. Both viruses have extremely wide host ranges. Cultural practices can be effective in reducing the incidence of TSWV and PBNV. These include sowing so that the presence of seedlings coincide with the lowest population of vectors, use of seed of high quality treated with a seed protectant to avoid seedling fungal diseases, sowing at the recommended rate and spacing. It is extremely important to ensure adequate soil moisture to facilitate good germination and seedling establishment. Maintenance of optimum plant population (ca 2-3,000,000 per hectare) and vigorous crop growth to develop a close canopy can greatly reduce incidence of PBNV or TSWV. Though many weeds have been shown to act as reservoir hosts, their removal to reduce the source of inoculum is not considered to be practical. Additionally, removal of infected plants, especially those infected during early stages of crop growth, can create gaps in the field which can lead to increase in the incidence.

Good sources of field resistance have been identified for TSWV and PBNV (e.g. Southern Runner for TSWV; ICGV 86029, ICGV 86031 and ICGV 86388 for PBNV). Several cultivars with vector resistance to PBNV, such as ICGS 44 and ICGS 11, have been released in India.

In field trials conducted in India and Thailand, spraying with insecticides, to reduce *Thrips* population, resulted in increasing the disease incidence. Therefore if insecticides are to be applied to control peanut pests, especially sucking insects, their choice and number of sprays should be carefully planned to reduce their influence on the incidence of PBNV or TSWV.

Cultivation of PBNV or TSWV field-resistant cultivars, under conditions to maintain optimum plant population and vigorous plant growth, can lead to very good control of these two virus diseases. High incidence of PBNV can be expected in peanut crops grown adjacent to PBNV susceptible crops protected with insecticides in areas where irrigation is available throughout the year. All the peanut genotypes field-resistant to PBNV or TSWV are either medium-duration or long-duration types. Short-duration virus resistant cultivars are yet to be released. Since a number of PBNV and TSWV genes are available for incorporation into peanut through non-conventional means, this approach should be explored for producing short-duration PBNV or TSWV resistant peanut cultivars.

Potyviruses

Peanut stripe (PStV) and peanut mottle (PeMoV) are the two most widely distributed potyviruses in the Asia-Pacific region. PeMoV is known to be present in all the major peanut growing countries in the region. PStV is by far the most widely distributed of all the viruses occurring in peanut crops in Southeast Asia and China. PeMoV and PStV are transmitted through peanut seeds and have assumed global quarantine importance. Frequency of PeMoV seed transmission in peanut is less than that of PStV. Therefore the quarantine risk of an accidental introduction of PStV is considerably greater than that of PeMov. From the monitoring done on crops in institutional plantings where PStV was introduced through exotic germplasm (1,5), it is evident that it can become established within a span of four years.

Both the viruses can be distinguished in serological tests. Polyclonal antisera produced either from purified viruses or cloned viral coat proteins can be used to identify and diagnose PStV and PeMoV, in addition to host reactions on plants such as *Pisum sativum* and *Phaseolus vulgaris* (cv. Topcrop) (9).

Both PeMoV and PStV resistance in cultivated peanut could not be found. Tolerance to PeMoV (no significant yield reduction due to infection) in the genotype ICG 5043 (NcAc 2240) has been identified. However, at least one high yielding peanut genotype, ICG 89336, with tolerance to PeMoV is currently available.

Evidence from studies of the epidemiology of the two viruses show that the primary source of inoculum comes from infected seeds. Therefore, growing genotypes in which the viruses are not seed-transmitted is likely to contribute to the gradual reduction in incidence.

In peanut genotypes ICG 2716 (EC 76446, 292) ICG 7013 (NcAc 17133RF) and ICG 1697

(NcAc 17090), PeMoV was not seed-transmitted. Fortunately these genotypes possess a high degree of resistance to peanut rust, which is one of the important fungal diseases of peanut. The seed of advanced breeding lines from these crosses have been tested for frequency of PeMoV transmission. High yielding and agronomnically acceptable material which do not transmit PeMoV have emerged, e.g. ICGS 65 and ICGS 76.

Unfortunately attempts to locate genotypes which do not transmnit PStV have not been successful. The common practice of using seeds from the previous seasons' crop assures the continuous presence of PStV. Therefore seeds should be obtained from peanut crops free of PStV. "Mulching" with transparent plastic sheet has been shown to reduce the incidence of PStV (12). However data on the economics of adopting this measure for PStV control are not available. The genomes of PStV and PeMoV have been sequenced and putative genes have been identified. Efforts are currently being made to utilize viral genes to incorporate resistance to PStV into peanut. Resistance to both the potyviruses has been located in wild *Arachis* species.

Indian Peanut Clump Furovirus

Peanut clump virus disease has been reported to occur in the Indian subcontinent and in West Africa. The virus occurring in India is referred to as Indian peanut clump virus (PCV). The IPCV has been shown to occur as several serologically distinct isolates. The virus is seed-borne and soil-borne. It is transmitted by a soil fungus (*Polymyxa* sp). IPCV isolates have been grouped into three serotypes. They all produce very similar symptoms on peanut.

A collaborative project between ICRISAT and Catholique University in Louvain in Belgium, funded by the Belgian Government has been established to study the epidemiology of IPCV. The *Polymyxa* sp. prefers cereal crops for its multiplication and infects non-preferred hosts such as peanut only when preferred hosts are not available. Therefore, rotation or intercropping with susceptible cereal crops such as wheat, barley, sorghum, finger millet, and pearl millet leads to high disease incidence. Currently the effects of temperature and soil moisture on the fungus lifecycle are being studied (P. Delfosse, personal communication). It was earlier demonstrated that non-preferred crops such as peanut, grown when the ambient temperatures were lower than 25 °C,

escaped the disease (10).

IPCV incidence can also be reduced by the application of soil biocides such as dibromochloropropane and carbofuran. They are not recommended because of their hazardous nature and high cost.

Over 9000 peanut genotypes have been screened for resistance to IPCV under natural conditions and none was found to be resistant. The genome of one of the IPCV isolates has been fully sequenced and the prospects are excellent for utilizing viral genes to induce resistance into peanut by nonconventional methods.

Cucumoviruses

The two cucumoviruses which are economically important in China are cucumber mosaic virus (CMV) and peanut stunt virus (PSV). Severe epidemics due to these viruses have been recorded in many provinces of northeast China. Both viruses are transmitted by many aphid species and through seeds (12). Primary source of inoculum for CMV was shown to be seed-transmitted plants. Therefore, peanut crops raised from virus-free seed showed very low CMV. However, plots planted with virus-free seed should be separated by at least 100 meters from commercial fields with high CMV incidence. Application of insecticides is not recommended.

AFRICA

Only three virus diseases are currently known to be economically important on peanuts in Africa. Of the three, groundnut rosette disease is one of the most important diseases of peanuts in Africa south of the Sahara.

Groundnut Rosette Disease

Groundnut rosette disease (GRD) was reported nearly 90 years ago and since then many epidemics have been recorded in Africa. Losses due to the epidemic which occurred in 1975 were estimated to exceed 250 million US dollars in Nigeria alone. The two major types of GRD's referred to as chlorotic and green are recognized. Wide variation in symptoms especially due to chlorotic rosette has been observed. GRD is caused by a complex of two viruses and a satellite RNA. One of the viruses is a typical luteovirus, the other a single stranded positive sense RNA virus, called groundnut rosette virus (GRV). The satellite depends on GRV for its replication and it also modulates symptom production. GRV and its satellite RNA depend on the luteovirus for aphid transmission. Diagnostic aids are available for all the three components involved in GRD.

Meaningful epidemiological studies can be conducted now because of the recent progress on the identification of the causal agents of GRD. It was demonstrated conclusively that the disease can be effectively controlled by cultural practices. These include sowing early in the season and at high density, as recommended in the case of PBNV. Excellent sources of resistance to GRD were found in peanut germplasm which originated in Africa. Many long-duration and medium-duration peanut cultivars, have been bred using the long duration resistant sources e.g. RG1, RMP 12, RMP 91, RRI/6 and ICGV-SM 90704. All the currently known rosette-resistant cultivars are susceptible to the luteovirus, but resistant to GRV and its satellite. Early maturing rosette resistant genotypes have been identified recently and will be used to produce early maturing GRD resistant cultivars. One of the major recent breakthroughs in the control of GRD is the development of several high-yielding, Short-duration rosette-resistant breeding lines which include ICGV-SMs (94581, 94584, 94586 and 94587 (P. Subrahmanyam, personal communication).

By growing rosette resistant cultivars and adopting recommended cultural practices, it should be possible to control GRD. However, many factors which include non-availability of good quality seed of GRD-resistant cultivars and lack of understanding of epidemiology of GRD are some of the important reasons why epidemics due to GRD continue to occur throughout Africa. Since the coat protein gene of the luteovirus has been fully sequenced, prospects for producing short duration GRD-resistant peanut cultivars using nonconventional approaches are good.

Peanut Clump Furovirus

Peanut clump virus (PCV) has been reported from Senegammbia, Burkina Faso, Niger and Cote d'Ivoire in West Africa and from South Africa. Occurrence of PCV in other parts of Africa is currently not known. One of the isolates of PCV has been fully sequenced and differences in the genome organization between IPCV and PCV are known.

Data from epidemiological research currently being conducted in India on IPCV are

also expected to provide clues for formulating cultural practices to control PCV. Since the virus is seed-transmitted in cereal crops, seed obtained from crops grown on infested soils should not be used for planting. Intercropping and crop rotation with susceptible cereal crops such as sorghum and pearl millet should be avoided. Crops which are not preferred by the vector and which are resistant to PCV such as cowpea and sunflower are recommended for cultivation on infested soils.

Peanut Mottle and Peanut Stripe Potyviruses

PeMoV is widely distributed in Africa. However, losses due to PeMoV have not been estimated. Precise surveys for the occurrence of PStV have been conducted only in some of the countries in West Africa. Its occurrence on farmers' fields in Africa has so far not been reported. Therefore plant quarantine should take every care to avoid entry of PStV into Africa. Control measures recommended above for the two viruses are not applicable.

NORTH AND SOUTH AMERICA

TSWV and PeMoV are the two most widely distributed viruses on these two continents.

Tospoviruses

Bud necrosis disease caused by TSWV is currently recognized as a serious threat to production of peanut in North America. TSWV was recorded more than 20 years ago in the USA. Severe damage to peanut crops in Texas was reported in 1986, 1990 and 1991, with yield reduction up to 95% (2). In Georgia TSWV was noted at a low incidence until 1986. However, its incidence has increased dramatically since 1986 (3). PBNV has so far not been reported from this continent. In 1994 crop losses in Georgia were estimated at 26 million dollars (J.W. Demski, personal communication).

F. occidentalis and *F. fusca* have been shown to be the principal vectors in southeastern United States (11).

The control measures suggested above for tospoviruses are applicable. The disease occurred mainly due to primary infection from adult thrips which migrated from outside reservoirs. Insecticide application, as expected, did not influence the disease incidence (3). Resistance to

TSWV was noted in the peanut cultivar Southern Runner (4).

PeMoV was first reported from the USA and is widely distributed in North and South America. An economic loss of 5 to 6 % to peanut in Georgia has been reported.

PStV distribution appears to be restricted to institutional productions in the USA. The virus has also been reported recently from Brazil. Its occurrence is suspected in many countries in the region because of germplasm imports from countries which are known to harbor PStV. Many aphid species have been shown to transmit PeMoV and PStV. Options for the control of PeMoV and PStV are given above.

Potyviruses

PeMoV was first reported from the USA and is widely distributed in North and South America. An economic loss of 5 to 6% to peanuts in Georgia has been reported.

PStV distribution appears to be restricted to institutional productions in the USA. The virus has also been reported recently from Brazil. Its occurrence is suspected in many countries in the region becuase of germplasm imports from countries which are known to harbor PStV. Many aphid species have been shown to transmit PeMoV and PStV. Options for the control of PeMoV and PStV are given above.

CONCLUSIONS

Effective control strategies for virus diseases are best based on knowledge of the causal viruses and their principal vectors. All the currently recognized economically important viruses occurring in the major peanut growing countries have now been characterized and diagnostic aids developed. Therefore meaningful surveys for their distribution can be undertaken and studies on epidemiology for some of them have been initiated. Screening for locating sources of disease resistance can be carried out with precision. Several peanut viruses are seed-transmitted and often observed to occur at a higher incidence in institutional productions than in farmers' fields. Many countries in Africa and some in the Asia-Pacific region do not have well-equipped virus laboratories with the result that these viruses escape plant quarantine controls. Therefore provision of diagnostic aids and technology transfer to scientists of the National Agricultural Research Systems in these countries should be given high priority.

Sources of resistance in cultivated peanut is only available for PBNV, GRD and TSWV. The genomes of viruses such as PeMoV, PStV and PCV and IPCV have been sequenced, therefore permitting utilization of the genes of these viruses in the production of transgenic peanut plants. Recently, methods to transform and regenerate peanut plants have been reported (6,7). However, they need to be evaluated with cultivars other than those utilized in these experiments. Additionally the procedures need further refinements to improve the efficiency of frequency of transformation.

Changes in the cultivars grown in various crop production systems and factors which lead to massive multiplication and movement of vectors can cause rapid changes in incidence and severity of virus diseases. Therefore, monitoring the incidence of widely distributed viruses, which are currently considered to be of minor importance, is essential. A very good example is cowpea mild mottle (CMMV), currently known to be present in many peanut growing countries in the Asia-Pacific region and in Africa, albeit at a low incidence. The virus is transmitted by the whiteflies *Bemisia tabaci* and peanuts grown adjacent to CMMV susceptible whitefly preferred crops such as soybeans and cowpea are likely to be severely affected.

ACKNOWLEDGEMENTS

I am extremely grateful to Drs. R.J. Shepherd (University of Kentucky, Lexington), J.A. Wightman, J.M. Lennè and P. Subrahmanyam (ICRISAT) for many valuable comments. Submitted as Journal Article No. JA 1856 by ICRISAT.

REFERENCES

1. Basu, M.S. 1996. Current research on peanut stripe virus and measures taken to contain its spread in India. in: Proc. of Working Group on Groundnut Viruses in the Asia-Pacific Region. D.V.R. Reddy and C.L.L. Gowda, eds. ICRISAT, Patancheru, India (In press).

2. Black, M.C., Andrews, T.D., and Smith, O.D. 1993. Interplot interference in field experiments with spotted wilt disease of peanut. Page 65 in: Proc. Am. Peanut Res. Educ. Soc. Vol. 25, APRES, USA.

3. Camann, MA., Culbreath, A.K., Pickering, J., Todd, J.W., and Demski, J.W. 1995. Spatial and temporal patterns of spotted wilt epidemics in peanut. Phytopathology 85:879-885.

4. Culbreath, A.K. Todd, J.W., Gorbet, d.W., and Demski, J.W. 1993. Spotted wilt apparent disease progress in the component lines of Southern Runner cultivar. Peanut Science 20:81-84.

5. Demski, J.W., and Lovell, G.R. 1985. Peanut stripe virus and the distribution of peanut seed. Plant Dis. 69:735-738.

6. Li, Z., Cheng, M., Xing, A., Jarret, R.L., Pittman, R., and Demski, J.W. 1996. Peanut transformation research in Griffin, GA. in: Proc. Working group on Groundnut Viruses in the Asia-Pacific Region. D.V. R. Reddy and C.L.L. Gowda, eds. ICRISAT, Patancheru, India (In press).

7. McKentley, A.H., Moore, G.A. Doosdar, H., and Niedz, R.P. 1995. Agrobacterium-mediated transformation of peanut (Arachis hypogaea L.) embryo axes, and the development of transgenic plants. Plant Cell Reports. 14:699-703.

8. Reddy, D.V.R. 1991. Groundnut viruses and virus diseases: distribution, identification and control. Review of Plant Pathology 70:665-678.

9. Reddy, D.V.R., and Demski, J.W. 1996. Virus diseases. in: Compendium of Peanut Diseases. D.H. Porter, D. H. Smith, R. Rodriguez-Kabana and P. Subrahmanyam, eds. American Phytopathological Society, St. Paul, MN. USA (In press).

10. Reddy, D.V.R., Nolt, B.L., Hobbs, H.A., Reddy, A.S., Rajeshwari, R., Rao, A.S., Reddy, D.D.R., and McDonald, D. 1988. Clump virus in India: Isolates, host range transmission and management. Pages 239-246 in: Developments in Applied Biology Vol. 2. Viruses with Fungal Vectors. J.I. Cooper and M.J.C. Asher, eds. Association of Applied Biologists, UK.

11. Todd, J.W., Culbreath, A.K., Demski, J.W., and Beshear, R.J. 1990. Thrips as vectors of TSWV. Page 81 in: Proc. Am. Peanut Res. Educ. soc. Vol. 22. APRES, USA.

12. Xu, Z., Zongyi, Z., Kunrong, C., Jinxang, C., and Reddy, D.V.R. 1996. Current research on groundnut virus diseases in China. in: Proc. of Working Group on Groundnut Viruses in the Asia-Pacific Region. D.V.R. Reddy and C.L.L. Gowda, eds. ICRISAT, Patancheru, India (In press).

CHAPTER 42

Approaches for Controlling Tomato Spotted Wilt Virus

J. Cho, R.F.L. Mau, S-Z Pang, M. Wang, C. Gonsalves, J. Watterson, D.M. Custer and D. Gonsalves

Tomato spotted wilt tospovirus (TSWV) seriously affects production of food and ornamental crops worldwide. First described in southern Australia on tomato by Brittlebank (8), TSWV is now widespread in temperate and subtropical regions throughout the world (10, 18). The virus has an extensive host range which includes over 200 dicotyledonous and 8 monocotyledonous plant species (7,11,6). Important economic crops affected include ornamentals (e.g. chrysanthemum, dahlia, gloxinia, cineraria, impatiens, lily), vegetables (e.g. tomato, pepper, lettuce), agronomic crops (e.g. peanut, tobacco) and fruits (e.g. papaya, pineapple).

Although crop damage caused by TSWV occurs on a worldwide basis, the economic effect of this virus has been documented thoroughly in Hawaii. In Hawaii, we have been combating this virus disease since it was first described in 1929 (30). TSWV has been devastating to the agricultural industry. At first, losses in lettuce and tomato were tolerated because epidemics occurred only during the summer months, however, the problem increased forcing several growers out of production and others to produce alternative non-susceptible crops. On the Hawaiian island of Maui, TSWV started causing problems in lettuce in 1980 with 5% to 10% losses, 4 to 5 years later, epidemics were observed with losses of 50% to 75%. Today, TSWV commonly causes losses of 25% to 50%. On the island of Hawaii we saw a similar scenario with TSWV losses of 5% to 10% in 1984; two years later epidemics with losses over 50% were observed; and presently 25% to 50% are common. A look at some of Hawaii's production statistics (3) shows that lettuce and tomato production has decreased by over 65% and 35%, respectively, since 1984 when TSWV began

causing major epidemics statewide (10,11).

Control efforts of TSWV were initiated soon after it was discovered. Over the years, two general approaches have been utilized to control TSWV. These include the use of TSWV resistant cultivars and the use of management tactics that serve to reduce the incidence of the virus. Recently, the concept of "pathogen-derived resistance" (54) has been applied to control virus diseases. The most prevalent approach, referred here as engineered protection, has been by the development of transgenic plants which express the coat protein gene of the virus that one wants to control. In this chapter, we will discuss the efforts and effects of conventional genetic resistance, disease management, and engineered protection for controlling TSWV.

PREVIOUS ATTEMPTS IN BREEDING FOR TSWV RESISTANCE

Resistance in tomato was first reported by Samuel, *et al.* (53) in a wild *Lycopersicon* species, *Lycopersicon pimpinellifolium* Mill. Since that report several other sources of resistance have been identified in other *Lycopersicon* species, including *L. esculentum* Mill., *L. hirsutum* Humb. & Bonpl., and *L. peruvianum* (L.) Mill. *Lycopersicon pimpinellifolium* in particular has been utilized extensively as a genetic resource for several disease resistances including TSWV, *Fusarium* wilt and *Stemphyllium solani* (24). However, in the case of TSWV resistance tomato lines derived from such crosses have not been effective in the field. In 1945, Kikuta and Frazier (34), using TSWV resistance found in German sugar derived originally from an interspecific cross between *L. esculentum* and *L. pimpinellifolium*, developed a TSWV-resistant cultivar which they named Pearl Harbor. Unfortunately, Holmes (28) found that Pearl Harbor was susceptible when grown in New Jersey. Holmes further reported resistance found in two *L. esculentum* cultivars, Rey de los Tempranos and Manzana, which originated from Argentina. These lines have been found to be susceptible to TSWV isolates from other areas.

Unsuccessful previous attempts to develop useful TSWV-resistant tomato lines were partly due to several factors including the use of mechanical inoculations in the evaluation of germplasm; the lack of sensitive virus detection methods, the inappropriate evaluation of germplasm for disease resistance, the use of defective TSWV mutants in inoculation experiments, and the presence of numerous biological strains that can overcome the resistance. Resende *et al.* (51) demonstrated that TSWV readily develops defective interfering RNA mutants upon mechanical passage in plants. These mutants types (defective particles) included: 1) morphological mutants unable to produce membrane glycoproteins and not able to form envelope particles; and 2) isolates with deleted internal regions of the large (L) RNA. Further, they found a consistent association between the generation of defective L RNA mutants and symptom attenuation. For example, tobacco plants infected by 'wild or normal' type TSWV isolate with full length L RNA usually developed severe necrosis, whereas a mild mosaic or mottling was induced in plants infected with isolates containing defective L RNA species in addition to the complete genome. Development of defective virus RNAs affecting pathogenicity is not unique to TSWV and has been described previously for a few other plant viruses including tomato bushy stunt virus (26,36), turnip crinkle virus (38), and cymbidium ringspot virus (9).

Discovery of the development of defective mutants from mechanical maintenance of the virus could explain earlier work by Best (7) and Norris (42) in which four major TSWV strain groups were described based upon differences in plant host reactions. These strain groups designated as tip blight, necrotic, ringspot, and mild differed from each other in the degree of disease severity. Perhaps the tip blight strain represented the 'wild-type' while the less severe strains were defective mutants. In our TSWV resistance screening programs we find that mechanically maintained TSWV isolates readily lose their ability to induce severe tip blight systemic infection in tomato plants. The use of these kinds of isolates can lead to identification of genetic resistance only to attenuated virus strains which may not have any practical application in 'real farm' conditions.

In our work, we have observed a difference in the level of resistance in *L. pimpinellifolium* lines when inoculated with mechanically maintained isolates as compared with wild or thrips maintained isolates. Plants inoculated with mechanically maintained isolates initially become systemically infected by the virus, however, the rapidly growing plant readily outgrows any visual signs of virus infection and generally no detectable virus particles can be found in the developing plant tissues based upon ELISA. On the other hand, we find that *L. pimpinellifolium* plants inoculated with TSWV isolates naturally maintained by vector thrips are systemically invaded by the virus, stunted and unable to out grow the virus. We have also observed this phenomenon in greenhouse studies with susceptible *L. esculentum* lines including Rey de los Tempranos, previously found resistant to certain mechanically maintained isolates (28). Interpretation of the ability to out grow mechanically maintained TSWV isolates as genetic resistance by early investigators (14,42) may explain why previously developed TSWV-'resistant' materials have not held up in the field.

TSWV is relatively labile *in vitro* in comparison with other plant viruses and inconsistent mechanical transmission results have been experienced by several investigators (40,42). Local infections may or may not be observed on inoculated leaf tissues of tomato. In our screening program, local infections on inoculated tissues are generally not observed with susceptible tomato plants, however, 10 to 14 days later typical TSWV symptoms are observed on the upper systemically infected leaves. Because of the difficulties associated with mechanical inoculations, susceptible escapes may be misidentified as resistant or immune.

SCREENING FOR TSWV RESISTANCE IN TOMATO: A CASE STUDY IN HAWAII

Even with the difficulties associated with using mechanically maintained TSWV isolates in a resistance screening program, this procedure is preferred because it is fast, less labor intensive, and permits screening of large numbers of plants.

We have found that it is possible to use mechanically maintained TSWV isolates effectively to identify useful resistance in tomato and, based on preliminary studies, also in pepper. Unfortunately, we have found that this screening program is not applicable to several other crops which require naturally or thrips maintained inoculum for effective evaluation for genetic resistance. These crops include lettuce, chrysanthemum, and peanut.

Several measures were incorporated into our screening and selection program in order to ensure the identification of TSWV resistance that could be used in the development of commercial tomato. We have chosen to use a conservative definition of useful resistance to evaluate tomato lines in our program. Detection of virus either visually and/or by double sandwich (DAS) ELISA (11, 23) beyond the sites of initial inoculation has been classified as a susceptible reaction. Secondly, we have virtually eliminated susceptible escapes by: 1) using an inoculum source containing actively growing, high titer of virus; 2) inoculating leaves of tomato plants at the 3 to 5 leaf stage of development and a second time 3 to 4 days later; 3) evaluating plants weekly for virus infection and holding plants in the greenhouse for at least 6 weeks after infection; 4) routinely assaying inoculated plant materials for the presence of TSWV by DAS ELISA to verify infection; and 5) confirming disease resistance by screening the selfed progeny of previously selected resistant plants for TSWV resistance. Finally, plants found to be resistant after the rigid tests outlined above are then screened for resistance by vector thrips inoculation in greenhouse and field tests.

Development of TSWV Resistant Commercial Tomato

In 1985, we screened some of Petoseed Company's inbred tomato lines for TSWV resistance. These lines originated from intra- and intergeneric crosses between *L. esculentum* and suspected TSWV-resistant tomato lines including *L. peruvianum*, *L. pimpinellifolium*, and *L. esculentum* var. Platense. Initial evaluations were made by mechanical inoculations in the greenhouse and by transplanting seedlings to Hawaiian farms experiencing high TSWV disease incidence. From these preliminary studies, three lines from *L. peruvianum* crosses appeared to be highly resistant; whereas, all other lines were susceptible.

Resistance associated with the *L. peruvianum* crosses was expected, since previous work by early investigators demonstrated a high level of resistance to TSWV in certain lines of *L. peruvianum* (14,15,42). *L. peruvianum* has been identified as a potential source of resistance to several diseases including tomato mosaic virus (2), root knot nematode (20), as well as a source of variation for form and adaptability (24). However, until recently genetic incompatibility barriers with cultivated tomato (*L. esculentum*) (24,27,52,55) had prevented the large scale use of *L. peruvianum* as a resistance source. Fortunately, techniques to overcome these barriers have been identified (24, 55) and used successfully by Petoseed Company for the introgression of TSWV-resistance into *L. esculentum* lines used in our development of commercial TSWV-resistant tomatoes (J. Watterson, unpublished).

In our screening programs, we find that TSWV resistance derived from *L. peruvianum*, differs from *L. pimpinellifolium* in that inoculated plants generally exhibit only a localized hypersensitive infection on inoculated leaves without becoming systemically infected by the virus. Further, the expression of TSWV resistance by the necrotic localized plant response can be readily determined 7 to 10 days after inoculation and has facilitated screening segregating populations for resistance.

Because resistance appeared to be useful for commercial application, in early 1986 we initiated a cooperative project with Petoseed Company Inc., to verify TSWV resistance under greenhouse conditions, to determine the inheritance of TSWV resistance and to develop TSWV-resistant commercial fresh market tomatoes.

Inheritance of TSWV Resistance in Tomato

Genetic resistance to TSWV was demonstrated to be conferred by a single dominant gene (Table 1). In these studies, one (SW 307) of the three TSWV resistant lines identified in our preliminary work was selected for further genetic studies. SW 307 was an F_7 generation selection, which originated from an interspecific cross between a Petoseed Company's breeding line and a TSWV resistant *L. peruvianum* line. This line was crossed with TSWV-susceptible tomato parent lines (P48,P49) and the F_1, F_2, and BC (backcross)

Table 1. Segregation of resistance and susceptibility to tomato spotted wilt tospovirus (TSWV) in progenies derived from crosses of susceptible (P1, P2) and resistant (SW 307) *Lycopersicon esculentum* line; these observations were confirmed by DAS-ELISA

Generation	No. plants Resistant	No. plants Susceptible	Expected ratio	c_2	P
P48	0	20			
P49	0	20			
SW 307	20	0			
P48 x SW 307					
F_1	20	0			
F_2	102	39	3:1	0.53	0.48
BC [(P1 x SW 307) x P1]	46	44	1:1	0.04	0.86
P49 x SW 307					
F_1	14	2			
F_2	33	15	3:1	1.0	0.34
BC (P1 x SW 307) x P1	28	24	1:1	0.08	0.40

generations were developed for genetic analyses.

Both susceptible parent plants (P48, P49) were susceptible to mechanical inoculations with TSWV. Susceptible parents generally did not exhibit obvious disease reactions on inoculated leaves but developed systemic infections characteristic of TSWV symptoms in the upper developing leaves and shoots. Occasionally a faint brown discolored ringspot was observed in the inoculated susceptible leaf tissues. Systemic infections were characterized by vein clearing, chlorotic mottling and stunting of growth. Sometimes a few of the systemically infected plants appeared to outgrow the infection and did not exhibit signs of the disease, however, these tissues were always found to test positive for the presence of the virus by DAS-ELISA.

On the other hand, all resistant SW 307 plants developed necrotic localized leaf spots generally surrounded by a chlorotic halo on the inoculated leaves 7 to 10 days after inoculation. Visual signs of infections were confined to the inoculated leaf tissues, whereas systemic infections did not occur.

All F_1 hybrid plants from crosses between resistant (SW 307) and susceptible parents (P48, P49) developed localized necrotic infections on the inoculated leaves, but the virus generally failed to develop systemic infections, indicating that resistance was dominant. Occasionally, systemic infections were observed in a few of the resistant

F_1 hybrid plants, however, symptoms differed from reactions associated with susceptible plants. Infections were dark brown necrotic spots on leaves associated with the vascular tissues (veins), and longitudinal dark brown necrotic streaks along the stems, petioles, and vascular tissues of leaves. Further, preliminary observations in the screening of other populations suggested that this type of systemic invasion occurs more frequently in F_1 heterozygous populations and less frequently in homozygous populations, indicating that resistance may be additive.

When the F_2 populations from both crosses were screened for resistance approximately 25% (a ratio of 3 resistant to 1 susceptible) of the plants became systemically infected, indicating that resistance was controlled by one gene (monogenic). Plants of backcrosses generated by crossing the F_1 generation with either susceptible parent (P48, P49) segregated in a ratio of one resistant to one susceptible (1:1) providing additional confirmation to inheritance of a single dominant gene for TSWV resistance.

Response of TSWV-Resistant Tomatoes to Vector Inoculations

Western flower thrips (WFT), *Frankliniella occidentalis* (Pergande), was used in vector transmission studies as described previously

(60). Thrips were obtained from virus-free laboratory colonies which were reared and maintained on green bean pods. A TSWV isolate (MT) isolated from a naturally infected tomato plant on the island of Maui was used in all vector transmission experiments. Each replication consisted of five plants each of P1 (a susceptible tomato line), SW 307 (F$_7$ generation material derived from an interspecific cross between *L. esculentum* and *L. peruvianum*), and PSR 55289 (experimental TSWV-resistant hybrid derived originally through crosses between P1 and SW 307). This experiment was replicated 9 times. Plants were maintained in the greenhouse and observed weekly for disease symptoms.

In *F. occidentalis* transmission tests, 51% of the susceptible parent (P1) plants (n=42), none of the homozygous resistant parent (SW 307) plants (n=45), and 5.6% (2 out of 48 plants) of F$_1$ generation plants (PSR 55289) inoculated by WFT became systemically infected by TSWV. Infections were significantly different based on Tukey's studentized range test ($P > 0.001$) between the susceptible parent (P1) and the resistant (homozygous and heterozygous) lines. Infected P1 plants exhibited typical susceptible reactions observed in mechanical inoculations with the virus. Systemic infections observed in the resistant heterozygous (PSR 55289) plants were similar to the local infections that developed on leaf tissues of the TSWV-resistant lines observed after mechanical inoculations.

Development of TSWV-Resistant Fresh Market Tomato

Successful development of TSWV-resistant hybrids for commercial fresh market tomato production can be attributed to the collaborative efforts between the University of Hawaii and Petoseed Company, Inc. Traditional backcrossing programs were employed using SW 307 as one source of TSWV resistance in the development of resistant commercial tomatoes. Because these programs involve lengthy breeding and selection to reduce linkage drag, in order to accelerate development time, duties were divided between the University of Hawaii and Petoseed Company, Inc. to take advantage of the expertise at the individual locations.

The accelerated program initiated in 1986 involved an annual crossing-selfing-selection cycle for the introgression of resistance into advanced horticultural tomato lines. In Hawaii: 1) F$_2$ generation materials from genetic crosses made by Petoseed in Woodland, California between SW 307 and Petoseed inbred parental lines by mechanical inoculations; 2) Resistant survivors are transplanted to the field for further selection; 3) Pollen of recurrent parents was sent from Woodland to Hawaii for crossing onto resistant plants; 4) Hybrid and open pollinated seed were harvested from resistant single plant selections; and 5) Open pollinated seed from resistant single plant selections were evaluated in Hawaii to confirm resistance. In Woodland: 1) Horticultural selections, backcrossing and selfing were made from resistant hybrid lines developed in Hawaii; and 2) Seed from selected individuals segregating for TSWV resistance sent back to Hawaii for resistance screening.

Field Evaluation TSWV-Resistant Fresh Market Tomato

By 1989, TSWV-resistant selections appeared sufficiently advanced horticulturally that experimental hybrid combinations were generated for field evaluations. Seven experimental hybrids were selected and evaluated for horticultural suitability and disease resistance on Maui in August 1989.

We were able to compare horticultural suitability of the 7 resistant hybrids with our growers' standard susceptible hybrid (Celebrity), because of low TSWV disease pressure during the experiment. Less than 1% of the Celebrity plants as compared with none of the experimental hybrid lines were infected during this trial. The 7 experimental hybrids had horticultural characters similar to Celebrity. Characters included vigorous determinate plant growth, adequate fruit yield and quality for the commercial market, firm fruit, acceptable fruit taste, and uniform development of fruit color when ripened with ethylene. Comparison of fruit yield is summarized in Table 2. Fruit yields from all 7 hybrids were equal or higher than Celebrity.

Two hybrids, PSR 55289 and PSR 55689, were selected from this trial for further evaluation by growers. Both hybrids appeared to have more desirable horticultural characters suitable for possible commercial production in Hawaii. In September 1990, approximately two kilograms of each hybrid were produced by Petoseed Company and sample seed quantities (50 to 60 grams) distributed to over thirty Hawaii growers statewide. The objectives of the distribution were

Table 2. Comparison of seven TSWV-resistant hybrid tomato lines and a standard susceptible tomato hybrid, Celebrity, for average fruit production (weight in kg) per plant grown at Pulehu, Maui (total of six harvests)

Tomato hybrid	Grade 1[a]	Grade 2[a]	Total marketable[b]	Total yield[c]
PSR 55289	1.8 bc	2.7 b	4.3 bc	5.4 bc
PSR 55389	2.1 a	3.0 ab	5.2 ab	6.2 ab
PSR 55689	1.9 ab	2.8 ab	4.8 bc	5.7 abc
PSR 55789	1.3 c	2.4 b	3.9 c	47 c
PSR 55989	1.4 c	2.9 ab	4.4 bc	5.7 abc
PSR 56089	1.7 abc	3.0 ab	4.8 bc	6.0 ab
PSR 56289	1.9 ab	3.6 a	6.0 a	6.9 a
Celebrity	1.3 c	6.6 ab	4.3 bc	5.5 bc

[a] Grade 1 and 2 = Department of Agriculture tomato market grade standards. Letters next to numbers correspond to statistical significance at 5% level.
[b] Total marketable = combined tomato marketable fruit grades including 1, 2, and offgrade
[c] Total yield includes total marketable grades and unmarketable culls.

to determine suitability of these lines for commercial production and whether resistance would be adequate on the 'real farm' situation.

Farm observations of the TSWV-resistant tomato hybrids showed that both lines were significantly resistant to TSWV (Table 3) when compared with standard susceptible lines (i.e. Celebrity, Floramerica, Main Pack) grown by Hawaii farmers. In general, TSWV incidence associated with the resistant lines was low and provided adequate protection under all but severe disease pressure. Under severe TSWV pressure as observed on Farm 5, nineteen percent of the resistant PSR 55689 became infected. However, this was considerably lower than the susceptible hybrid at 80%.

Systemic TSWV infections associated with the resistant F_1 hybrids exhibited symptoms similar to what we observed in greenhouse inoculation studies. Disease symptoms were also observed on the fruit of resistant lines. Fruit that developed from systemically infected resistant lines were generally malformed, small in size, exhibited dark brown necrotic concentric and broken ringspots on the fruit surface, and dark brown necrotic spots associated with the internal vascular tissues of the fruit. Dark necrotic spots and rings were also observed associated with completely healthy resistant F_1 hybrids. Symptoms were similar to the external fruit symptoms associated with systemically infected plants, however differed in that fruits were normal in size and shape, and by

the lack of virus symptoms on all other plant parts including the internal parts of the fruit. TSWV could be detected by ELISA throughout the systemically infected plant including the internal and external parts of the fruit, and the fruit peduncle. In contrast, TSWV could only be detected by ELISA from the external surface fruit tissues but not the internal fruit tissues nor peduncle or other plant tissues associated with plants exhibiting only fruit symptoms. We believe that these latter fruit symptoms develop from inoculation of the fruit surface by viruliferous thrips feeding on flower and young fruit tissues.

In our farm observations, we have noticed that the incidence of TSWV infections in the resistant lines are higher in the row of plants closest to the infected susceptible lines, which serve as a tremendous source for the virus and viruliferous thrips. On the other hand, on farms where only resistant tomato lines are planted systemic infections are generally less than 1%. We believe that there will be considerable reduction in TSWV incidence in tomato when an adequate supply of TSWV-resistant tomato seed becomes available and growers are not forced to mix planting susceptible and resistant materials. In this regard, Hawaii farmers initiated a seed production contract with Petoseed Company Inc. and this year will be the first to produce TSWV-resistant hybrid tomatoes commercially.

Table 3. Comparison of TSWV resistance between TSWV-resistant (PSR 55289, PSR 55689) and - susceptible (Celebrity, Floramerica, Main Pack) tomato lines grown on Hawaiian farms experiencing TSWV epidemics

Farm number	Location	Observation date	Tomato line	No. plants	TSWV infection (%)
1	Maui	May 1990	Celebrity	50	>50
2	Maui	Dec 1990	Celebrity	1000	>50
			PSR 55289	1000	<1
3	Maui	Jan 1991	Celebrity	300	>50
			PSR 55289	300	<1
4	Maui	Jan 1991	Celebrity	500	15
			PSR 55289	400	<1
5	Maui	Jan 1992	Celebrity	500	80
			PSR 55689	500	19
6	Maui	Jan 1992	Celebrity	500	30
			PSR 55289	500	<1
7	Hawaii	Jan 1991	Floramerica	500	100
			PSR 55289	500	0
8	Oahu	Feb 1992	Main Pack	450	75
			PSR 55289	450	0

Epilogue

Traditional breeding methods were used in the successful development of TSWV resistant tomatoes for commercial production in Hawaii. Based on preliminary evaluations, this resistant material appears to be broadly resistant to TSWV in many different geographical regions. Based on preliminary work this resistance appears to be holding up in several southern states in the USA., South Africa, Spain, Italy, and Argentina. Factors contributing to the success of this program can be attributed to the stringent tests implemented in our breeding and selection program. These tests included: our greenhouse inoculation procedures, verification of infection by DAS-ELISA, elimination of susceptible escapes, and verification of TSWV resistance to vector inoculation in the greenhouse and field.

Kikuta and Frazer (34) were the first to release a commercial tomato with TSWV resistance. However, resistance in their line proved to be ineffective very quickly. Reasons for unsuccessful development may be explained by their use of *L. pimpinellifolium* as a genetic source of resistance and evaluating their materials for resistance only under natural field conditions. In our evaluations, we have found *L. pimpinellifolium* to be susceptible in greenhouse inoculations, however, find this species to be partly field resistant (37). Field resistance has been observed in the evaluation for resistance of many crops to a number of insect transmitted plant viruses (31) including many wild *Lycopersicon* species (12,56). In view of this phenomenon, it is critical to verify TSWV infection of screened materials when evaluating for resistance.

Breeding to develop commercial TSWV resistant tomatoes was accelerated by combining the expertise at the University of Hawaii and a commercial seed company (Petoseed Company, Inc.), however, development time was relatively slow and six years of intensive breeding and selection were required before TSWV-resistant seed could be produced commercially. In addition, further breeding and selection are required to improve horticultural characters and for the development of other improved hybrids and tomato types.

Recombinant DNA technologies have allowed for accelerating the process of identifying, selecting, and breeding improved plant cultivars. Molecular DNA markers, initially isoenzymes, then restriction fragment length polymorphisms (RFLPs), and most recently random amplified polymorphic DNAs (RAPDs), have revolutionized the ease of mapping the genome of several plant species. Identification of molecular markers linked to disease resistance genes would be extremely useful in accelerating traditional breeding programs by providing a means to monitor presence of the gene in exotic germplasms and in segregating plant populations developed through traditional breeding programs (58,63). In this regard, molecular markers tightly linked with TSWV-resistance have been identified in tomato lines developed from genetic crosses with SW 307 (Brommonschenkel, Tanksley and Cho, unpublished) and individual TSWV-resistant plants with minimum linkage drag have selected. Future use of these plants and identified linked markers will allow us to accelerate introgression of TSWV resistance into other commercial tomatoes with minimum linkage drag.

MANAGEMENT OF TSWV

Tomato spotted wilt virus has caused significant disease losses in several crops including lettuce, tomato and peppers in Hawaii since the early 1980s. Efforts to develop and implement TSWV management strategies to reduce economic losses from the disease has been complicated by the occurrence of several insect vector species and their relationship with possibly different TSWV isolates and the extensive plant hosts that can serve as reservoirs for both virus and vector. The lack of adequate information about vector-virus and vector-host relationships made it impossible to make precise recommendations relating to control of vectors, control of alternate plant hosts, and modifications in cultural management practices to effectively reduce economic TSWV losses in lettuce. Therefore, research was initiated in an effort to effect logical control measures.

Although seven thrips species (60) were

confirmed as vectors of TSWV by Ullman *et al.*, this was prior to the establishment of the Tospovirus genus. TSWV is one of several viruses in the group and it is vectored by four species. They are western flower thrips, *Frankliniella occidentalis* (Pergande), tobacco thrips, *F. fusca* (Hinds), common blossom thrips, *F. schultzei* (Trybom), and onion thrips, *Thrips tabaci* Lindeman. All of the species are found in Hawaii.

Thrips tabaci was reported to be the principal vector associated with TSWV epidemics when the disease was first reported in tomato in 1915. However, *T. tabaci* does not appear important in present epidemics. Instead, *F. occidentalis* has become the predominant vector species. This may indicate a specific relationship between thrips vector species and TSWV isolate. For example, Paliwal in 1976 (43) demonstrated that tobacco thrips (*F. fusca*) transmitted two Canadian TSWV isolates more efficiently than *F. occidentalis* and that onion thrips (*T. tabaci*) were not able to efficiently transmit either isolate. Thrips transmission studies in our laboratory indicated that *F. occidentalis*, *F. schultzei*, and *T. tabaci* larvae could readily acquire TSWV isolates associated with Hawaiian epidemics based on tests by ELISA, however, transtadial passage and effective transmission by adult insects only occurred with *F. occidentalis*. This phenomenon reinforces the suggestion of a specific virus-vector relationship.

In general, TSWV-vector thrips are able to colonize and complete their life cycle on a large number of plant hosts (4,5,57,64,65). The choice of plant host by adult vector thrips plays an important role in the development and perpetuation of the disease cycle, because of a vector-virus relationship unique among the plant viruses. In this relationship, TSWV acquisition which occurs only by feeding larval stages of the vector can result in effective transmission by the larvae and developing adults. TSWV has an extensive host range (7,11,16) which include several hundred plant species.

In Hawaii, *F. occidentalis* colonize and develop on a large number of crop and non-crop plant hosts present in the farm ecosystem. In our vegetable ecosystem over 44 different susceptible weed hosts have been identified (11,64,65). Laboratory tests were conducted to evaluate the relative importance of the different weed species in perpetuating TSWV epidemics in the field (4, 5). Important reservoirs should meet two criteria: 1) support larval development since this is the stage where virus acquisition occurs, and 2) support adult feeding and oviposition. Larval development on the leaves of 17 different plant hosts were compared. Of the hosts tested, buckwheat (*Fagopyrum esculentum* Moench) was the most suitable with 73% of the larvae developing into adults; lettuce was second with 55%; then a group including cheeseweed (*Malva parviflora* L.), sowthistle (*Sonchus oleraceus* L.), green Amaranth [*Amaranthus hybridus* (L.)], cabbage (*Brassica campestris* L.), and Jimson weed (*Datura stramonium* L.) third with a range of 16% to 35%; a fourth group including Nasturtium (*Tropaeolum majus* L.), burdock (*Articum lappa* L.), Peruvian daisy (*G. parviflora* Cav., apple of *Peru* [*Nicandra physalodes* (L.) Gaertn], and verbena (*Verbena litoralis* HBK.) were relatively unsuitable with a range of 2.5% to 9%; and finally a completely unsuitable fifth group where no development occurred including Flora's paintbrush [*Emilia sonchifolia* (L.) DC.], golden crown's bread [*Verbesina enceloides* (Cav.) Benth & Hook. ex Gray], and statice [*Limonium latifolium* (Sm.) Ktze.].

In further laboratory tests, adult *F. occidentalis* were given a choice of five different plant species in vegetative or flowering stages to evaluate feeding and oviposition preference (5). These plants included lettuce, cheeseweed, burdock, jimson weed, and golden crown's beard, which were selected because they are abundant, and have been shown to serve as TSWV and vector reservoirs in Hawaiian farm lands. Thrips preferences differed for vegetative and flowering stages of plant development. In feeding trials during vegetative growth, feeding damage on the leaves of lettuce was significantly higher than the other four plants evaluated. Successful oviposition by adult thrips measured by the number of larvae that developed after a four day feeding exposure to adult *F. occidentalis* show a higher average number developing on lettuce (50.1), followed by burdock (29.7) and cheeseweed (19.9), with the lowest numbers developing on jimson weed (6.3) and golden crown's beard (5.0). A high correlation was found between leaf feeding damage and successful oviposition (r = 0.83; df =23; P < 0.0001). We used this comparison to assist in ordering the preference of these host plants to thrips with lettuce being the most preferred, burdock and cheeseweed next, and jimson weed and golden crown's beard last. In feeding trials during flowering, the mean feeding damage on leaves indicated that lettuce again was most preferred,

cheeseweed second, jimson weed third and golden crown's beard the least preferred. On the other hand, analysis of oviposition preference showed significantly more larvae emerging from the foliage and flowers of colonized cheeseweed, followed by golden crown's beard, lettuce and jimson weed.

This kind of laboratory analyses should be useful in evaluating the relative contribution to the TSWV epidemic by the multitude of plant hosts that can be found in the farm ecosystem. Our studies indicate the following conclusions: 1) lettuce is highly attractive to *F. occidentalis* colonization, which may initiate the TSWV epidemic by enhancing thrips movement from nearby weeds to cropped fields; 2) the high suitability of lettuce both for oviposition and larval development by thrips makes secondary spread and perpetuation within the crop feasible; 3) the suitability of non-crop weeds such as cheeseweed and jimson weed for thrips development make them important alternative hosts and possible outside sources of primarily infection; 3) the importance of burdock and golden crown's beard in the spread of TSWV is less important than previously thought since one was less preferred for feeding and both supported only a small proportion of the larval population to develop into adults; 4) the vegetative stages of apple of Peru, Flora's paint brush, nasturtium, Peruvian daisy, statice, tomato, tobacco, verbena were relatively unsuitable for larval development and therefore would not contribute much in TSWV spread and 5) the high correlation between the amount of feeding injury and development of larvae indicate that feeding preference may provide a relative measure of a plant host's importance as a *F. occidentalis* host and possibly source of viruliferous thrip.

Additional tests were done to evaluate the potential of lettuce, burdock, cheeseweed, jimson weed, and golden crown's beard as acquisition and inoculation hosts for TSWV by *F. occidentalis* (5). Although some of these plant hosts were shown to be suitable for thrips development, the success of acquiring and transmission of TSWV by vector thrips is perhaps one of the major requirements in generating epidemics in the field.

The efficiency of larval thrips to acquire TSWV differed significantly among the plant hosts tested. Based on ELISA, a higher proportion of the larvae acquired TSWV at higher titers after feeding on TSWV infected jimson weed (80.9%), intermediate on lettuce (54.5%) and lowest on burdock (35.9%). Transtadial passage from larval stages to adults was detected within cohorts which had acquired TSWV from all three plant hosts. Although the percentage of passage was low, cohorts that acquired TSWV from jimson weed (11.4%) and lettuce (8.2%) was nearly 3 to 4 times higher than burdock (2.9%). Host plant susceptibility, its ability to support high levels of virus, and the internal distribution of the virus in the plant have been suggested as factors affecting acquisition efficiency of thrips (18, 60). Although jimson weed, lettuce and burdock are systemically infected by TSWV, direct immunoblotting of systemically infected leaves showed that virus distribution was evenly distributed in jimson weed leaves but unevenly distributed in burdock (60) and may explain the differences observed in our studies.

TSWV transmission efficiency varied with different combinations of acquisition and inoculation hosts. Highest transmission occurred to jimson weed when thrips acquired TSWV from lettuce and jimson weed. Further, thrips which acquired TSWV from jimson weed could transmit the virus to lettuce and burdock, indicating that this plant host may be an important virus reservoir in the field. On the other hand, burdock was found to be an unsuitable host for TSWV acquisition and of the thrips acquiring the virus transmitted the virus poorly or not at all. These tests indicated that burdock does not play a significant role in perpetuating the TSWV disease cycle in spite of the high numbers of adults associated with this crop in Hawaiian farms.

A partial understanding of the virus-vector-plant host relationships assisted in the development of cultural field practices that could be used to minimize the occurrence of TSWV losses in susceptible crops. None of the management practices suggested are sufficiently effective by themselves; however, it is possible to integrate these practices in order to significantly reduce disease losses. Effective management strategies are preventative and include the following proven techniques to reduce crop loss:

1. Protect seedlings: Seedlings are particularly vulnerable to infection of TSWV by thrips when left unprotected. Thrips are attracted to light green color of young seedlings. Seedlings should not be grown at the farm site where cultivation of TSWV-susuceptible crops are being grown. If this is not possible seedlings can be grown under cover such as in a protected greenhouse or through the use of fine screen coverings. A regular insecticide maintenance program should be implicated to reduce the chances of thrips buildup in the nursery area.

2. Avoid Sequential Planting: Growers are advised to minimize planting sequentially in the same field area. In Hawaii, this tactic is not always practical because of limited land availability and generally small farm size. Therefore it is suggested that growers continue to grow susceptible crops as long as disease incidence is not present and implement management strategies upon onset of the disease in the crop.

3. Crop Placement: Planting of susceptible crops in areas where the incidence of TSWV is high should be avoided. The planting of these areas with non-susceptible crops or fallowing for a period of approximately three weeks to allow thrips to migrate out of the area will reduce the threat of TSWV and again allow cultivation of susceptible.

4. Elimination of TSWV Crop Reservoirs: Plow harvested or abandoned susceptible crops immediately to eliminate virus reservoirs and thrips food sources. This will cause thrips to migrate out of the area.

5. Weed Management: There are many weeds which are hosts of TSWV and act as reservoirs for the virus. Therefore, weed management of these reservoirs within and around the farm is important. It is important that the timing for the elimination of important weed reservoirs coincides with periods when thrips populations are low to minimize the build up and movement of thrips into the crop.

6. Minimize Cultivation: Cultivation of weeds within the crop stimulates adult thrips to move. This practice should be minimized particularly in fields where disease is present to reduce TSWV spread.

7. Interplant Susceptible With Non-Susceptible Crops: It has been difficult to control TSWV epidemics once it has become established in the crop. Only changing to a non-susceptible crop under these conditions will eliminate the epidemic. However, this generally has not been an economically viable option for growers in Hawaii. The interplanting of a non-susceptible inter-crop can be used as a strategy to minimize TSWV losses by alternating blocks of susceptible crops with blocks of non-susceptible crops. There are several non-susceptible crop options such broccoli, cauliflower, cabbage, squash, cucumber, onion, carrots, radish and sweet potato.

Three management strategies has been used effectively to reduce TSWV losses in the Hawaiian farm ecosystem. Crop placement has been successfully used by several lettuce growers to significantly reduce TSWV. When disease reached levels in excess of 50% in an area where lettuce had been sequentially planted, crop placement of new lettuce transplants at a new uncultivated field resulted on several instances in a reduction of disease incidence to less than 10%. For example, a grower on the island of Hawaii was able to significantly reduce his TSWV losses by moving no more than 200 m to an uncultivated, isolated field, which was away from the line of site from a field where disease incidence was greater than 60%. On Maui, another grower has effectively used this same tactic by moving his lettuce operation from one farm site to another site located approximately 2 miles away when disease incidence reaches levels from 50% to 60%. By utilizing this tactic, both growers were able to produce lettuce sequentially for several planting. A few growers have reduced disease losses by reducing cultivation in fields where TSWV occurred. The third effective practice was to separate susceptible crops with blocks of non-susceptible crops such as broccoli, cabbage, and cauliflower.

ENGINEERED PROTECTION

Pathogen-derived resistance was first proposed by Sanford and Johnston (54) and demonstrated by Powell-Abel et al. (1) to be effective against viral infection in transgenic tobacco expressing the coat protein gene of tobacco mosaic virus. Coat protein mediated protection has since been reported for many other plant-virus combinations (for review, see 25, and the chapter related to this subject in this book). Recently, other forms of genetically engineered resistance have been described, which include expression of antisense RNAs, defective interfering molecules, nonstructural gene sequences and untranslatable coat protein coding sequences (for review, see 29,61, and related chapters in this book). Despite the increasing reports of pathogen-derived resistance under greenhouse conditions, field tests to ascertain that pathogen-derived resistance also operates under field conditions and that the transgenic plants resist virus infection by natural vectors are limited. A few studies have shown that transgenic potatoes and tomatoes are resistant to PVY (33), PVX (32,33) and TMV (41) under field conditions following mechanical inoculation of the challenge viruses. Gonsalves et al., (22) showed that

transgenic cucumbers are resistant to cucumber mosaic virus (CMV) under field conditions where disease pressure is exerted by inoculations from the virus vector throughout the course of the crop growth cycle. Recently, however, a transgenic squash with resistance to watermelon mosaic virus 2 and zucchini yellow mosaic virus has been commercialized (17,59).

Engineered Protection Against Tospoviruses

Engineered resistance to tospoviruses was first reported by Gielen et al. (19) and Mackenzie and Ellis (39), showing that transgenic tobacco plants expressing the nucleocapsid (N) gene of TSWV were resistant to infection by the homologous isolate. De Haan et al. (13) have subsequently suggested that the protection they previously observed in transgenic plants expressing the translatable N gene was primarily due to the presence of N transgene transcripts and not due to the N transgene protein. Their protection was effective against mechanical and vector inoculation by TSWV isolates and not effective against other tospoviruses that share considerable nucleotide sequence homology in their N genes to TSWV (13). Our independent studies, on the other hand, have shown that N gene sequence-derived resistance to tospoviruses is complex and appears to be attributed to both N protein accumulation and N transgene silencing (Table 4) (45,47).

N protein-mediated resistance to tospoviruses was clearly shown by Pang et al. (45, 47, 48). Transgenic tobacco and N. benthamiana plants that express translatable or untranslatable N gene of the TSWV lettuce (TSWV-BL) isolate were produced. Analyses of these transgenic plants showed that plants that accumulated high amounts of this protein but not this RNA showed a broad spectrum but lower levels of protection not only against homologous isolate but also against distantly related impatiens necrotic spot tospoviruses (INSV-Beg and INSV-LI). This N protein-mediated protection can be overcome by increasing inoculum strength, which is typical for coat protein-mediated protection. Interestingly, the protection was not effective to closely related TSWV-10W nor to the GRSV-BR, a tospovirus isolate whose N gene shares 78% nucleotide sequence identity with that of TSWV-BL.

Mechanisms for N protein-mediated protection are still not clear. It is possible that the

Table 4. The comparison of resistance to different tospoviruses in transgenic plants expressing various forms of the TSWV-BL N gene

Tospovirus	Homology to TSWV-BL N gene			PoN	N		mN		asN	
	nt	aa identity	aa similarity		low	high	low	high	low	high
TSWV-BL	100	100	100	S	I	R	I	S	R	S
TSWV-10W	99.0	98.1	98.4	S	I	S	I	S	R	S
INSV-Beg	60.4	53.9	68.8	S	S	R	S	S	S	S
INSV-LI	62.1	54.3	68.8	S	S	R	S	S	S	S
TSWV-B	77.6	79.5	90.3	S	S	S	S	S	S	S

The nucleotide sequences of the tospovirus N genes are reported by Pang et al. (44-46) or our unpublished results. Percent identity was obtained using the BESTFIT program of the GCG package. Reactions of transgenic tobacco or N. benthamiana plants containing promoterless (PoN) N gene coding sequence of TSWV-BL or expressing low and high levels of the intact (N), untranslatable (mN) or antisense (asN) N gene sequence of TSWV-BL to inoculation with the tospoviruses are reported by Pang et al. (45,47,48) or ourunpublished results. Reaction of transgenic plants to inoculation: I, immune (independent of inoculum strength); R, resistant (dependent on inoculum strength); S, susceptible (no appreciable levels of resistance). Since INSV isolates only induce local lesions on tobacco, transgenic N. benthamiana plants were also used to ensure the accuracy of the inoculation results.

N protein produced in transgenic plants, as proposed for resistance to tobacco mosaic virus, prevent the uncoating of the invading homologous virus, thus disrupting the virus infection cycle (6,29). For protection against the distantly related INSV-Beg or -LI isolates, the N protein produced in transgenic plants may serve as a dysfunctional N protein when incorporated into a distantly related, attacking virus particle, such as INSV-Beg (47). The resulting heteroencapsidated virus may lose some important biological properties required for replication or spread. If this is true, then N protein-mediated protection against distantly related tospoviruses could only be expected against virus isolates that cannot functionally exchange their own N proteins with those produced in transgenic plants, and in transgenic plants which produce sufficient levels of "dysfunctional" N protein to compete with the N protein of the attacking virus (47). The lack of N protein-mediated protection against infection by the more related GRSV-Br suggests that the TSWV-BL N protein can form functional heteroencapsidated TSWV-B viral particles (47); they share 90% amino acid identity (48).

Transgenic tobacco plants that express either translatable or untranslatable N gene were protected against homologous and closely related isolates but not against distantly related tospoviruses (13,47), suggesting the involvement of the transgene RNA in protection. Nuclear run off transcription analyses revealed this RNA-mediated protection against tospoviruses (13,47) was actually due to the N transgene silencing (Pang et al., unpublished data). Moreover, analyses of transgenic lettuce plants expressing the N gene of TSWV-BL showed that they were protected against lettuce-infecting TSWV isolates via an N transgene silencing mechanism activated by its overexpression (Pang et al., unpublished data). In a transgenic lettuce line, post-transcriptional gene silencing was activated at a relatively earlier developmental stage in homozygous than in hemizygous progenies. As a result, the homozygous progenies generally showed a uniform suppression of N protein accumulation and consequently high levels of virus resistance in all leaves of the silenced plants. In contrast, responses of the hemizygotes to virus inoculation depended on the time of inoculation in relation to plant development. The small hemizygous plants were not silenced and expressed high levels of N gene. Inoculation at this developmental stage generally induced resistance phenotypes which is similar to the ones of N protein-mediated protection. As these plants further developed the young upper leaves became silenced, resulting in chimeric plants consisting of the unsilenced lower leaves and the silenced upper leaves. When the lower leaves of these plants were deliberately inoculated, the inoculated leaves produce reduced number of local lesions but systemic infection did not occur. When the upper silenced leaves were deliberately inoculated, neither local lesions nor systemic infection were produced. Thus, the level of virus resistance depends on the timing of N gene silencing, which could be shifted to a much earlier developmental stage in homozygous transgenic plants and could be affected by environmental factors such as the time of the year when plants were grown and whether plants were grown in a greenhouse or in the field.

Similarly, transgenic tobacco expressing antisense N transgene displayed various levels of protection against homologous and closely related tospoviruses only when antisense N transgene RNA accumulated at low levels (Table 4) (47), indicating a mechanism of post transcriptional transgene silencing. However, the level of the protection was relatively low compared to the one mediated by sense N transgene silencing as it depends on both the concentration of inoculum and physiological stage of test plants. The protection is effective to mechanical inoculation at low inoculum concentrations and the level of protection increased dramatically as test plants matured. Thus, the antisense transgene silencing-mediated protection might provide sufficient protection under field conditions, especially against viruses that invade the target crops late in the growing season.

When resistance to distantly related viruses is considered, the N protein-mediated protection may be the best choice since the sense and antisense transgene silencing-mediated protections were effective only to homologous and closely related (more than 90% nucleotide sequence identity) isolates. Because the level of sense transgene silencing-mediated protection is very high and can not be overcome by increasing inoculum strength, it may be desirable to engineer plants with more than one untranslatable tospovirus N genes, N gene fragments or other genomic sequences (to reduce "gene load") to

obtain a wide spectrum and a high level of protection against all tospoviruses that are present in the field environment. In fact, Prins et al. (49) recently showed that transgenic tobacco expressing the N genes of three different tospoviruses were indeed resistant to these viruses.

The concept of pathogen-derived resistance opens a whole range of new ways for tospovirus control. Implementation of these strategies should increase our ability to combat plant viral diseases.

Engineered Protection of Commercial Crops

The tomato (*L. esculentum*) is widely consumed throughout the world (62). However, while demand for this popular vegetable has steadily increased, so has the incidence of TSWV infections. There is a need to develop more TSWV-resistant lines, for although resistant cultivars have been developed by classical breeding, maintaining that resistance can be problematic (see previous section).

We have produced transgenic tomato resistant to TSWV by transferring the N gene of TSWV-BL (45) into 'Geneva 80' cotyledon explants via *Agrobacterium tumefaciens*-mediated gene transfer (21). 'Geneva 80' tomato was selected for transformation because it regenerates easily, and is resistant to tobacco mosaic virus, *Verticillium wilt,* and *Phytophthora infestans* (50). ELISA analysis of R_1 transgenic plants showed that 98% of the R_1 and R_2 tomato seedlings contained the NPT II gene, suggesting that multiple insertions of this gene had occurred. However, only 18% produced detectable levels of the N protein in ELISA tests.

In Greenhouse tests, R_1 plants from nine transgenic lines were resistant to TSWV-BL (21). Less than 5% of these plants showed symptoms, compared to 96% of the controls (Table 5). R_2 populations from three R_1 lines were inoculated (Table 6) with the homologous TSWV-BL, the TSWV-91 isolate, and the distantly related GRSV-BR tospovirus. None of the plants inoculated with TSWV-BL became infected; 16% of those inoculated with TSWV-91 became infected; and 48% of those inoculated with GRSV-BR became infected. Kim et al (35) have also produced TSWV-resistant tomato, although a high percentage of their transgenic plants were susceptible to TSWV.

To develop multiple virus resistant

Table 5. Reactions of R_1 seedlings from fruit of R_0 transgenic lines against TSWV-BL

Test Plants[b]	Reactions[a]	
	ELISA+/Tested	Infected/Tested
Experiment 1		
T13-1	NT	0/13
Control	NT	3/3
Experiment 2		
T13-9	6/20	0/20
T13-10	5/25	0/25
T13-11	1/22	0/22
T13-12	4/29	1/29
T13-13	3/22	0/22
Control	NT	20/21
Experiment 3		
T13-2	8/20	6/20
T13-3	5/25	2/25
T13-4	6/25	0/25
Control	NT	20/21

[a] Enzyme linked-immunosorbent assays (ELISA) were done on plants prior to inoculation using TSWV-BL antiserum. Test plants were inoculated with 1/20 dilution of TSWV-BL infected Havana 523 tobacco. Symptoms were recorded up to 36-45 days after inoculation. Plants were regarded as infected if they showed symptoms. NT=not tested.
[b] Test plants from each subline (e.g. T13-2) originated from seeds of individual fruits of the R_0 T13 plant. Control plants were Geneva 80 that were not transformed. All transgenic plants were positive for NPT II.

tomato, we made a cross between a transgenic TSWV-resistant R_1 line T13-1-7 and transgenic CMV-resistant line (19). This CMV-resistant line, TT5-007-11,was developed using 'Geneva 80' and is homozygous for the coat protein gene of CMV (50, 62). Progenies of the cross were inoculated with TSWV-BL, CMV-China or a mixture of both inocula (Table 7). Four percent (1 out of 24 plants) of TSWV-BL inoculated seedlings became infected; 18% (2 out of 11 plants) of CMV-China inoculated seedlings became infected; and 27% (3 out of 11 plants) of the seedlings inoculated with both TSWV-BL and CMV-China became infected. Although the transgenic TSWV-CMV resistant plants have yet to be tested in the field, the results in the greenhouse infectivity tests are very promising. In summary, genetic engineering is a rapid method of incorporating desirable genes into tomato, without altering other genes that already exist in a plant. Studies are continuing in order to

facilitate the future use of tomatoes with engineered viral resistance for commercial farms, home gardens, as well as subsistence farming.

REFERENCES

1. Abel, P.P., Nelson, R.S.,De,B., Hoffmann, N., Rogers, S.G., Fraley, R.T., and Beachy, R.N. 1986. Delay of disease development in transgenic plants that express the tobacco mosaic virus coat protein gene. Science 232: 738-743.
2. Alexander, L.J. 1963. Transfer of a dominant type of resistance to the four known Ohio pathogenic strains of tobacco mosaic virus (TMV) from *Lycopersicon peruvianum* to *L. esculentum*. Phytopathology 53: 869.
3. Anonymous. 1991. Statistics of Hawaiian Agriculture, 1990. 100 Pages.
4. Bautista, R.C., and Mau, R.F.L. 1994. Preferences and development of western flower thrips (Thysanoptera: Thripidae) on plant hosts of tomato spoted wilt tosovirus in Hawaii. Environ. Entomol. 23:151-1507.

Table 6. Reactions of R₂ seedlings from R₁ line T13-1 to inoculations with TSWV-BL, TSWV-91, and GRSV-BR *Tospoviruses*

Test Plants[b]	Reactions[a]	
	ELISA+/Tested	Infected/Tested
TSWV-BL		
T13-1-9	2/8	0/8
T13-1-7	0/8	0/8
T13-1-11	0/9	0/9
Control	NT	8/8
GRSV-BR		
T13-1-9	0/8	2/8
T13-1-7	2/8	5/8
T13-1-11	0/9	5/9
Control	NT	8/8
TSWV-91		
T13-1-9	1/8	1/8
T13-1-7	2/8	2/8
T13-1-11	3/9	1/9
Control	NT	8/8

[a] Enzyme linked-immunosorbent assays (ELISA) were done on plants prior to inoculation using TSWV-BL antiserum. Test plants were inoculated with 1/20 dilution of virus infected Havana 523 tobacco or *Nicotiana benthamiana*. Symptoms were recorded up to 36-45 days after inoculation. Plants were regarded as infected if they showed symptoms. NT=not tested.

[b] Test plants from each subline (e.g. T13-1-11) originated from seeds of individual fruits of the R₁ T13-1 plant. Control plants were Geneva 80 that were not transformed. All transgenic plants were positive for NPT II.

5. Bautista, R.C., Mau, R.F.L., Cho, J.J., and Custer, D.M. 1995. Potential of timato spotted wilt tospovirus plant hosts in Hawaii as virus reservoirs for transmission by *Frankliniella occidentalis* (Thysanoptera: Thripidae). Phytopathol. 85: 953-958.

6. Beachy, R.N., Loesch-Fries, S., and Tumer, N.E. 1990. Coat protein-mediated resistance against virus protection. Annu. Rev. Phytopathol. 85:953-958

7. Best, R.J. 1968. Tomato spotted wilt virus. Pages 65-145 in: Advances in Virus Research, Vol.13. K. M. Smith and M. A. Lauffer, eds. Academic Press, New York.

8. Brittlebank, C. C. 1919. Tomato diseases. J. Agric. Victoria Aust. 17: 213-235.

9. Burgyan, J., Grieco, F., and Russo, M. 1989. A defective interfering RNA molecule in Cymbidium ringspot virus infections. J. Gen. Virol. 70: 235-239.

10. Cho, J. J., Mau, R. F. L., German, T. L., Hartmann, R. W., Yudin, L. S., Gonsalves, D., and Provvidenti, R. 1989. A multidisciplinary approach to management of tomato spotted wilt virus in Hawaii USA. Plant Dis. 73: 375-383.

11. Cho, J. J., Mau, R. F. L., Gonsalves, D., and Mitchell, W. C. 1986. Reservoir weed hosts of tomato spotted wilt virus. Plant Dis. 70: 1014-1017.

12. Costa, A. S. 1944. Observacoes sobre vira-cabeca en tomateiros. Bragantia 4: 589-608.

13. De Haan, P., Gielen, J.J.L., Prins, M., Wijkamp, I. G., Van Schepen, A., Peters, D., Van Grinsven, M.Q.J.M., and Goldbach, R. 1992. Characterization of RNA-mediated resistance to tomato spotted wilt virus in transgenic tobacco plants. Bio/Technology 10: 1133-1137.

14. Finlay, K. W. 1952. Inheritance of spotted wilt resistance in the tomato. I. Identification of strains of the virus by the resistance or susceptibility of tomato species. Aust. J. Sci. Res. 5: 303-314.

15. Finlay, K. W. 1953. Inheritance of spotted wilt resistance in the tomato. II. Five genes controlling spotted wilt resistance in four tomato types. Aust. J. Biol. Sci. 6: 153-163.

16. Francki, R. I. B., and Hatta, T. 1981. Tomato spotted wilt virus. Pages 491-512 in: Handbook of Plant Virus Infections and Comparative Diagnosis. E. Kurstak, ed. Elsevier/North Holland Biomedical Press, Amsterdam

17. Fuchs, M., and Gonsalves, D. 1995. Resistance of transgenic hybrid squash ZW-20 expressing the coat protein genes of zucchini yellow mosaic virus and

Table 7. Reactions of R_1 seedlings from cross of transgenic TSWV resistant line T13-1-7 and CMV resistant line T55-007-11 to inoculations with TSWV-BL and CMV-China

T13-1-7 x T55-007-11[a]	No Infected/No Tested[b]
TSWV-BL inoculated-Exp. 1	
Transgenic	1/13
Control	8/8
TSWV-BL inoculated-Exp 2	
Transgenic	0/11
Control	7/10
CMV-China inoculated-Exp 2	
Transgenic	2/11
Control	10/10
CMV-China & TSWV-BL inoculated-Exp 2	
Transgenic	3/11
Control	10/10

[a] Parent TSWV resistant line was an R_2 plant from line TSWV T13-1-7 expressing the TSWV-BL nucleocapsid protein gene while the parent CMV resistant line was a homozygous trangenic line TT5-007-11 expressing the CMV-WL gene (50, 62).

[b] Test plants were inoculated with 1/10 dilution of TSWV-BL infected *Nicotiana benthamiana* and/or CMV-China infected tomato. Symptoms were recorded up to 45 days after inoculation. Control plants were Geneva 80 that were not transformed. Plants were regarded as infected if they showed symptoms. NT=not tested.

watermelon mosaic virus 2 to mixed infections by both potyviruses. Bio/Technology 13: 1466-1473.

18.   German, T. L., Ullman, D. E., and Moyer, J. W. 1992. Tospoviruses: Diagnosis, molecular biology, phylogeny, and vector relationships. Annu. Rev. Phytopathol. 30: 315-348.

19.   Gielen, J.J.L., De Haan, P., Kool, A.J., Peters, D., Van Grinsven, M.Q.J.M. and Goldbach, R.W. 1991. Engineered resistance to tomato spotted wilt virus, a negative-strand RNA virus. Bio/Technology 9: 1363-1367.

20.   Gilbert, J.C., and McGuire, D.C. 1956. Inheritance of resistance to severe root-knot from *Meloidogyne incognita* in commercial type tomatoes. Proc. Am. Soc. Sci. 68: 437-442.

21.   Gonsalves, C., Xue, B., Pang, S.-Z., Provvidenti, R., Slightom, J.L. and Gonsalves, D. 1993. Breeding multiple virus-resistant transgenic tomatoes. Abstracts of 6th International Congress of Plant Pathology, Montreal, Canada, July 28-August 6, 1993. 190 Pages.

22.   Gonsalves, D., Chee, P., Provvidenti, R., Seem, R., and Slightom, J.L. 1992. Comparison of coat protein-mediated and gentically-derived resistance in cucumbers to infection by cucumber mosaic virus under field conditions with natural challenge inoculations by vectors. Bio/Technology 10: 1562-1570.

23.   Gonsalves, D., and Trujillo, E.E. 1986. Tomato spotted wilt virus in papaya and detection of the

virus by ELISA.   Plant Dis. 70: 501-506.

24.   Gradziel, T.M., and Robinson, R.W. 1991. Overcoming unilateral breeding barriers between *Lycopersicon peruvianum* and cultivated tomato, *Lycopersicon esculentum*. Euphytica 54: 1-9.

25.   Grumet, R. 1994. Development of virus resistant plants via genetic engineering. Plant Breeding Reviews 12: 47-79.

26.   Hillman, B.I., Carrington, J.C. and Morris, T.J. 1987. A defective interfering RNA that contains a mosaic of a plant virus genome. Cell 51: 427-433.

27.   Hogenboom, N.G. 1972. Breaking breeding barriers in *Lycopersicon*. 1. The genus *Lycopersicon*, its breeding barriers and the importance of breaking these barriers. Euphytica 21: 221-227.

28.   Holmes, F.O. 1948. Resistance to spotted wilt in tomato. Phytopathology 38: 467-473.

29.   Hull, R., and Davies, J.W. 1992. Approaches to nonconventional control of plant virus diseases. Crit. Rev. Plant Sci. 11: 17-33.

30.   Illingworth, J.F. 1931. Yellow spot of pineapple in Hawaii. Phytopathology 21: 865-880.

31.   Jones, A.T. 1987. Control of virus infection in crop plants through vector resistance: a review of achievements, prospects and problems. Ann. Appl. Biol. 111: 745-772.

32.   Jongedijk, E., Deschutter, A.A.J.M., Stolte, T., Vandenelzen, P.J.M., and Cornelissen, B.J.C. 1992. Increased resistance to potato virus-X and

preservation of cultivar properties in transgenic potato under field conditions. Bio/Technology 10: 422-429.

33. Kaniewski, W., Lawson, C., Sammons, B., Haley, L., Hart, J., Delannay, X., and Tumer, N.E. 1990. Field resistance of transgenic Russet Burbank potato to effects of infection by potato virus X and potato virus Y. Bio/Technology 8: 750-754.

34. Kikuta, K. and Frazier, W.A. 1946. Breeding tomatoes for resistance to spotted wilt in Hawaii. Proc. Am. Soc. Hort. Sci. 47: 271-276.

35. Kim, J.W., Sun, S.S.M. and German, T.L. 1994. Disease resistance in tobacco and tomato plants transformed with the tomato spotted wilt virus nucleocapsid gene. Plant Dis. 78: 615-621.

36. Knorr, D.A., Mullin, R.H., Hearne, P.Q., and Morris, T.J. 1991. De novo generation of defective interfering RNAs of tomato bushy stunt virus by high multiplicity passage. Virology 181: 193-202.

37. Kumar, N.K.K., Ullman, D.E., and Cho, J.J. 1993. Evaluation of germ plasm for tomato spotted wilt tospovirus resistance by mechanical and thrips transmission. Plant Dis. 77: 938-941.

38. Li, X.H., Heaton, L.A., Morris, T.J., and Simon, A.E. 1989. Turnip crinkle virus defective interfering RNAs intensify viral symptoms and are generated de novo. Proc. Natl. Acad. Sci. (USA) 86: 9173-9177.

39. MacKenzie, D.J., and Ellis, P.J. 1992. Resistance to tomato spotted wilt virus infection in transgenic tobacco expressing the viral nucleocapsid gene. Molecular Plant-Microbe Interactions 5: 34-40.

40. Milbrath, J.A. 1939. Tomato tip-blight virus. Phytopathology 29: 156-168.

41. Nelson, R.S., McCormick, S.M., Delannay, X., Dube, P., Layton, J., Anderson, E.J., Kaniewska, M., Proksch, R.K., Horsch, R.B., Rogers, S.G., Fraley, R.T., and Beachy, R.N. 1988. Virus tolerance, plant growth, and field performance of transgenic tomato plants expressing coat protein from tobacco mosaic virus. Bio/Technology 6: 403-409.

42. Norris, D.O. 1946. The strain complex and symptom variability of tomato spotted wilt virus. Aust. Counc. Sci. Ind. Res. Bull. 202: 51.

43. Paliwal, Y.C. 1976. Some characteristics of the thrips vector relationship of tomato spotted wilt virus in Canada. Can. J. Bot. 54: 402-405.

44. Pang, S.-Z., Bock, J.H., Gonsalves, C., Slightom, J.L., and Gonsalves, D. 1994. Resistance of transgenic *Nicotiana benthamiana* plants to tomato spotted wilt and Impatiens necrotic spot tospoviruses: Evidence for the involvement of the N protein and the N gene RNA in resistance. Phytopathology 84: 243-249.

45. Pang, S.-Z., Nagpala, P., Wang, M., Slightom, J.L., and Gonsalves, D. 1992. Resistance to heterologous isolates of tomato spotted wilt virus in transgenic plants expressing its nucleocapsid protein gene. Phytopathology 82: 1223-1229.

46. Pang, S.Z., Slightom, J.L., and Gonsalves, D. 1993. The biological properties of a distinct tospovirus and sequence analysis of its RNA. Phytopathology 83: 728-733.

47. Pang, S.Z., Slightom, J.L., and Gonsalves, D. 1993. Different mechanisms protect transgenic tobacco against tomato spotted wilt and impatiens necrotic spot tospoviruses. Bio/Technology 11: 819-824.

48. Pang, S.Z. Bock, J.H., Gonsalves, C., Slightom, J.

49. L., Gonsalves, D. 1994. Resistance of transgenic *Nicotiana benthamiana* plants to tomato spotted wilt and impatiens necrotic spot tospoviruses: Evidence of involvement of the N protein and N gene RNA in resistance. Phytopathology. 84: 243-249.

49. Prins, M., De Haan, P., Luyten, R., Van Veller, M., Van Grinsven, M.Q.J.M., and Goldbach, R. 1995. Broad resistance to tospoviruses in transgenic tobacco plants expressing three tospoviral nucleoprotein gene sequences. Molecular Plant Microbe Interactions 8: 85-91.

50. Provvidenti, R., and Gonsalves, D. 1995. Inheritance of resistance to cucumber mosaic virus in a transgenic tomato line with the coat protein gene of the white leaf strain. J. Heredity 86: 85-88.

51. Resende, R.D.O., De, H.P., Avila, A.C.D., Kitajima, E.W., Kormelink, R., Goldbach, R., and Peters, D. 1991. Generation of envelope and defective interfering rna mutants of tomato spotted wilt virus by mechanical passage. J. Gen. Virol. 72: 2375-2384.

52. Rick, C.M. 1963. Barriers to interbreeding in *Lycopersicon peruvianum*. Evolution 17: 216-2321.

53. Samuel, G., Bald, J.G. and Pittman, H.A., 1930. Investigation on 'spotted wilt' of tomatoes. Aust. Counc. Sci. Ind. Res. Bull. 44: 64 pp.

54. Sanford, J.C., and Johnston, S.A. 1985. The concept of parasite-derived resistance - Deriving resistance genes from the parasite's own genome. J. Theor. Biol. 113: 395-405.

55. Smith, P.G. 1944. Embryo culture of a tomato species hybrid. Phytopathology 34: 413-416.

56. Smith, P.G., and Gardner, M. W. 1951. Resistance in tomato to the spotted wilt virus. Phytopathology 41: 257-260.

57. Stobbs, L.W., Broadbent, A.B., Allen, W.R., and Stirling, A.L. 1992. Transmission of tomato spotted wilt virus by the westernflower thrips to weeds and native plants found in Southern Ontario. Plant Dis. 76: 23-29.

58. Tanksley, S.D., Young, N.D., Paterson, A.H., and Bonierbale, M.W. 1989. RFLP mapping in plant breeding: new tools for an old science. Bio/Technology 7: 257-264.

59. Tricoli, D.M., Carney, K.J., Russell, P.F., McMaster, J.R., Groff, D.W., Hadden, K.C., Himmel, P.T., Hubbard, J.P., Boeshore, M.L., and Quemada, H.D. 1995. Field evaluation of transgenic squash containing single or multiple virus coat protein gene constructs for resistance to cucumber mosaic virus, watermelon mosaic virus 2, and zucchini yellow mosaic virus. Bio/Technology 13: 1458-1465.

60. Ullman, D.E., Cho, J.J., Mau, R.F.L., Hunter, W.B., Westcott, D.M., and Custer, D.M. 1992. Thrips-tomato spotted wilt interactions: Morphological, behavior and cellular components influencing thrips transmission. Pages 195-239 in: Advances in Disease Vector Research. K. F. Harris, ed. Springer, New York

61. Wilson, T.M.A. 1993. Strategies to protect crop plants against viruses: Pathogen-derived resistance blossoms. Proc. Natl. Acad. Sci. (USA) 90: 3134-3141.

62. Xue, B., Gonsalves, C., Provvidenti, R., Slightom, J.L., Fuchs, M., and Gonsalves, D. 1994. Development of transgenic tomato expressing a high level of resistance to cucumber mosaic virus strains

of subgroup I and II. Plant Dis. 78: 1038-1041.

63.   Young, N.D. 1990. Potential applications of map-based cloning to plant pathology. Physiol. Mol. Pl. Pathol. 37: 81-94.

64.   Yudin, L.S., Cho, J.J. and Mitchell, W.C. 1986. Host range of western flower thrips *Frankliniella occidentalis* Thysanoptera Thripidae with special reference to *Leucaena glauca*. Environ. Entomol. 15: 1292-1295.

65.   Yudin, L.S., Tabashnik, B.E., Cho, J.J., and Mitchell, W.C. 1988. Colonizations of weeds and lettuce by thrips Thysanoptera Thripidae. Environ. Entomol. 17: 522-526.

CHAPTER 43

Epidemiology and Management of Tomato Yellow Leaf Curl Disease

M. K. Nakhla and D. P. Maxwell

The most serious disease of tomatoes throughout the Mediterranean region, the Middle East, and tropical regions of Africa is tomato yellow leaf curl disease (29,78,104). It is caused by a geminivirus transmitted by *Bemisia tabaci* (Gennadius).

Tomato yellow leaf curl disease was first reported in Israel in 1939 in association with outbreaks of *B. tabaci* [K. J. Avidov, 1944 cited in Cohen and Antignus (23)], and has since spread to every region in the Mediterranean. It has been reported in Egypt (104), Jordan (5,77), Lebanon (79), Saudi Arabia (83), and Cyprus (57). Czosnek *et al.* (34) used a tomato yellow leaf curl virus (TYLCV)-specific probe to detect the virus in Turkey, Cape Verde, Senegal, The Sudan, and Tanzania. Recently, TYLCV has been found throughout the Dominican Republic (94) and in Jamaica (W. McLaughlin *et al.*, personal communication).

TYLC disease has become the major factor limiting tomato production during summer, fall, and winter in the Mediterranean; and losses range from 28-92%, depending on the age of the plants at the time of infection and the percentage of plants infected (4,78,79,104). In Egypt, 100% of the fall-grown tomato plants are usually infected and production losses may reach 80% (89), so tomato production during the autumn is unprofitable. In the Jordan Valley in the 1990/1991 seasons, all field-grown tomato plants were severely infected with TYLCV and heavy losses occurred (62).

SYMPTOMS

Affected tomato plants are stunted; the shoots are erect and have short internodes; the leaves are small, curled, leathery, and chlorotic;

and most of the flowers drop prematurely. The leaflets which appear soon after infection are cupped down, while those that develop subsequently are strikingly chlorotic and curl upward (29,78,93). Affected plants produce either no fruit or a few small fruits, depending on the stage of development at the time of viral infection. Young plants lose vigor after infection, and rarely produce any marketable fruits (25,89).

MOLECULAR CHARACTERIZATION OF TYLCV

TYLCV from the Eastern Mediterranean

The geminate shape of the Eastern Mediterranean TYLCV was first observed in 1980 in thin sections of infected tomato tissue (122); and virus particles, 20 X 30 nm, were purified by Czosnek *et al.* (32). The viral coat protein encapsidates a monopartite, circular, single-stranded DNA genome of approximately 2.8 kb (100).

The double-stranded replicative form of TYLCV, isolated from field-infected tomatoes in Israel, was cloned (100) and shown by nucleotide sequence analysis to consist of 2,787 nucleotides which encode six open reading frames, two on the virion strand and four on the complementary strand (Figure 1). No second, or DNA-B, component was found; so, unlike all previously described, whitefly-transmitted geminiviruses (36), which contained two components, this Eastern Mediterranean TYLCV was monopartite. The cloned component was analogous to the DNA-A components of bipartite geminiviruses, yet it coded for all of the functions of the complete virus, including whitefly transmission from

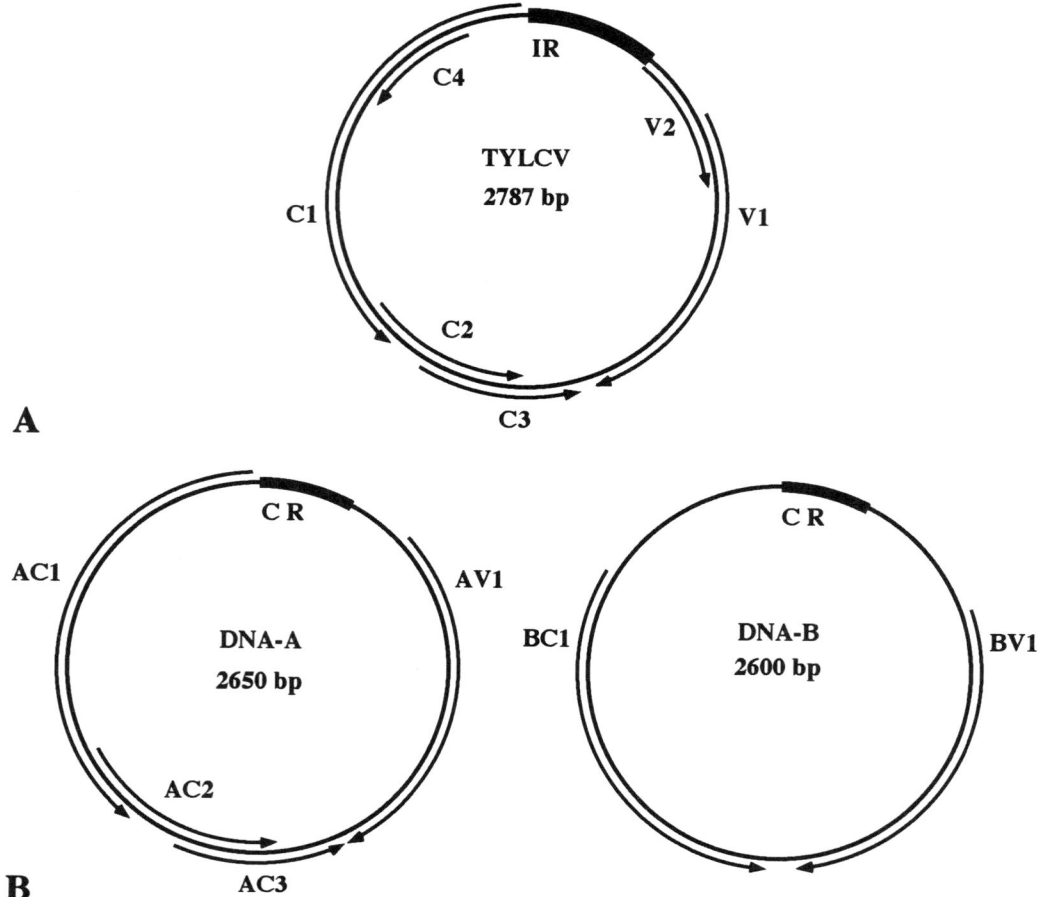

Figure 1. Diagrammatic representation of the replicative forms of the monopartite geminivirus, tomato yellow leaf curl virus (A) and of a typical bipartite geminivirus (B) transmitted by whiteflies (*Bemisia tabaci* or *B. argentifolii*). Open reading frames (C1/AC1, V1/AV1, etc) are represented by arrows and the common region (CR) or intergenic region (IR) containing the origin of replication by a solid bar.

infected plants. Nakhla *et al.* (96) used three different polymerase chain reaction (PCR)-based procedures to clone a TYLCV isolate from Egypt. No DNA-B component was detected for this isolate either, and dimeric constructs of the cloned TYLCV-DNA, introduced into tomato plants by *Agrobacterium*-mediated inoculation, caused severe, typical disease symptoms. The progeny produced in these plants are transmissible by the vector, *B. tabaci* (96). Recently, full-length genomes of two more Eastern Mediterranean-type TYLCV isolates, one from the Dominican Republic (94) and one from Jamaica (W. McLaughlin *et al.*, personal communication), have been cloned by a PCR-based procedure, and both clones infected tomatoes when radicles were inoculated by particle gun acceleration (M.K. Nakhla *et al.*, unpublished).

Diversity of TYLCVs and Tomato Leaf Curl Geminiviruses (TLCVs)

Several distinct geminiviruses have now been identified as the causative agents of the many diseases originally thought to be the result of infection by one or two viruses. These diseases were simply called either tomato yellow leaf curl (TYLC) or tomato leaf curl (TLC) disease, and the viruses were correspondingly called tomato yellow leaf curl virus (TYLCV) or tomato leaf curl virus (TLCV). Considerable confusion has resulted (117). This chapter outlines the current understanding and nomenclature of the TYLC diseases.

In 1990, Rochester *et al.* (119) reported cloning a bipartite TYLCV from Thailand (TYLCV-Tha) and its sequence has been published

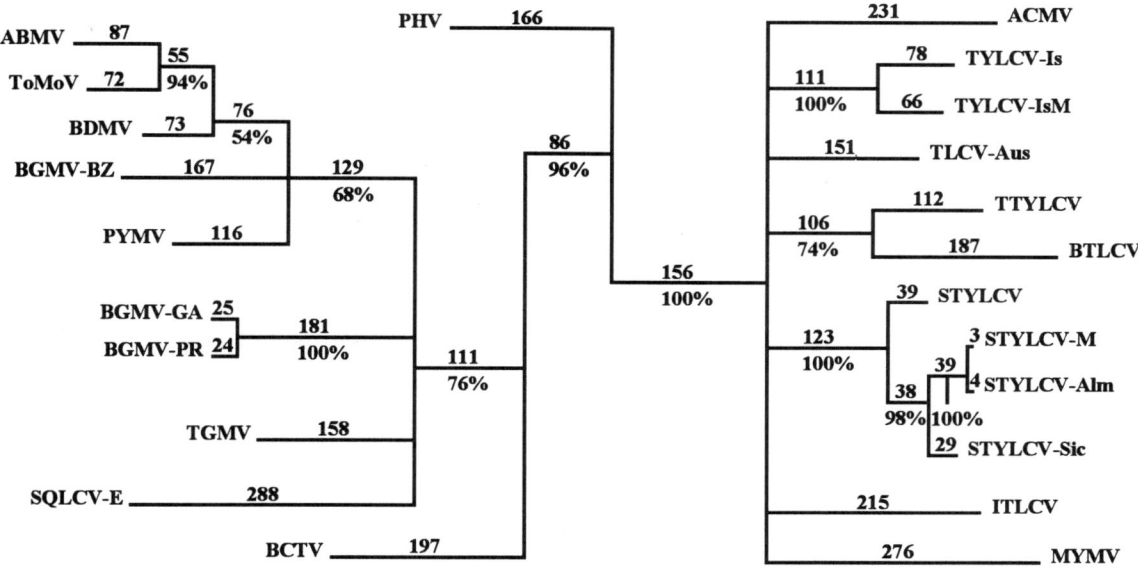

Figure 2. Cladogram showing relationships among one leafhopper-transmitted geminivirus (beet curly top geminivirus (BCTV, No. M24597) and 22 whitefly-transmitted geminiviruses, based on the total nucleotide sequences of the AC1/C1 ORF. Sequences were analyzed by the Phylogenetic Analysis Using Parsimony software developed by D. L. Swofford, using Heuristic tree construction, and branch strength was tested by constructing 100 trees by the bootstrap with branch swapping. The number of nucleotide changes are noted on top of the horizontal lines, and the percentage of trees with a given branch are below the horizontal lines. Vertical distances are arbitrary, and horizontal distances are in proportion to the number of nucleotide differences between branch nodes. Western Hemisphere geminiviruses: abutilon mosaic virus (AbMV, No. X15983), tomato mottle virus (ToMoV, No. L14460), bean dwarf mosaic virus (BDMV, No. M88179), bean golden mosaic geminivirus type I from Brazil (BGMV-BZ, No. M88686), potato yellow mosaic virus (PYMV, No. D00940), BGMV type II from Guatemala (BGMV-GA, No. M91604), BGMV type II from PortoRico (BGMV-PR, No. M10070), tomato golden mosaic virus (TGMV, No. K02029), squash leaf curl virus-extended host range (SQLCV-E, No. M38183), and pepper huasteco virus (PHV, No. D00940). Eastern Hemisphere geminiviruses: African cassava mosaic virus (ACMV, No. X17095), tomato yellow leaf curl virus from Israel (TYLCV-Is, No. X15656), TYLCV mild strain from Israel (TYLCV-IsM, No. L27708 and X76319), tomato leaf curl virus from Australia (TLCV-Aus, No. S53251), TYLCV from Thailand (TTYLCV, S. Attathom, personal communication), TLCV from Bangalore, India (BTLCV, No. L11746), TYLCV from Sardinia (STYLCV, No. X61153), TYLCV from Spain, Murcia (STYLCV-M, No. Z25751), TYLCV from Spain, Almeria (STYLCV-Alm, No. L27708), TYLCV from Sicily (STYLCV-Si, No. Z28390), TLCV from India (ITLCV, No. U15015), and mungbean yellow mosaic virus (MYMV, No. D14703).

recently (118). In 1991, two research groups independently showed that the TYLCV from Israel (100) and another from Sardinia, Italy (TYLCV-Sar; 65) are monopartite. Since then, five more isolates have been cloned and sequenced and shown to be monopartite: the TYLCV from Egypt (95,96, 97), which is nearly identical to TYLCV from Israel, two isolates from Spain (103; GenBank No. L277081), an isolate from Sicily (GenBank No. Z28390), and a mild isolate from Israel, maintained on *Datura stramonium* since 1969 (9).

Nucleotide sequence comparisons show that the TYLCV isolates which have been characterized are at least three distinct geminiviruses. This separation is based on the suggestion by Padidam *et al.* (106) that geminiviruses with nucleotide differences greater than 10% in homologous open reading frames (ORFs) should be considered separate

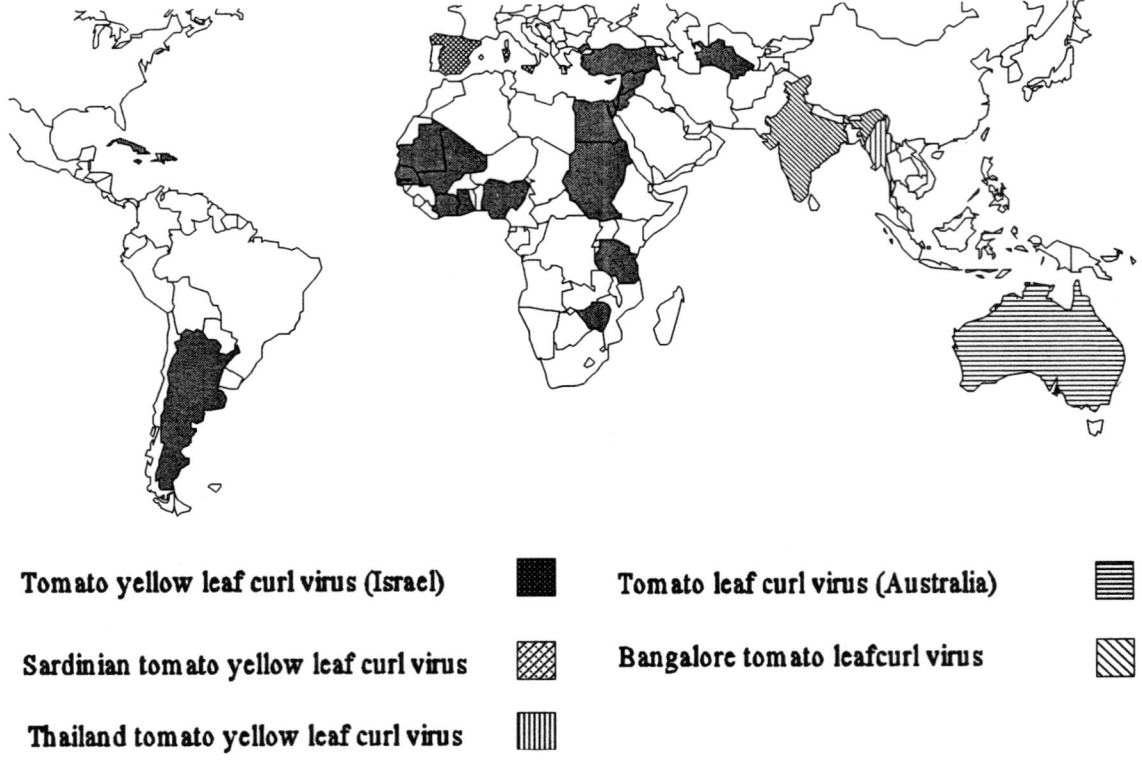

Figure 3. Distribution of tomato yellow leaf curl-like geminiviruses. Viruses were identified either by specific viral DNA probes or by sequencing of cloned viral DNA. Tomato yellow leaf curl virus (Israel) (33,85,94,95; H. Czosnek, *personal communication*); Sardinian tomato yellow leaf curl virus (65, Z25751, L27708, Z28390); Thailand tomato yellow leaf curl virus (118); tomato leaf curl virus (Australia) (40); Indian tomato leaf curl virus (107), and Bangalore tomato leaf curl virus (20); Taiwan tomato leaf curl virus in Taiwan (Chiang and associates, *personal communication*).

geminiviruses. Computer-assisted phylogenetic analyses also place these TYLCV isolates in three different branches (Figure 2). Geminivirus nomenclature is not clear because some isolates, now differentiated by molecular characterization, still carry identical names. Rochester *et al.* (117) proposed including the country of origin in the name (e.g., STYLCV, Sardinian tomato yellow leaf curl virus). A group of identical (i.e., nucleotide sequence identities greater than 90%) geminiviruses would all carry the name of the isolate in that group described first. Each isolate described after the initial one would have an additional "isolate identification code" (e.g., STYLCV-Alm, an isolate nearly identical to the Sardinian one but from Almeria, Spain). Since the isolate from Israel was the very first TYLCV described, it would be called simply TYLCV, without a country of origin designation. The isolate from Egypt, whose nucleotide sequence is nearly identical (>90%) would be TYLCV-Eg1.

The mild isolate from Israel would be TYLCV-mild isolate, since its nucleotide identity to the TYLCV-severe isolate is 98%. The two isolates from the Caribbean, nearly identical to the , isolate from Israel, would be TYLCV-DR1 (Dominican Republic, 94) and TYLCV-Jam1 (Jamaica, W. McLaughlin *et al.*, personal communication). The isolates from the western Mediterranean are nearly identical to each other (in the C1 ORF), but are less than 80% identical to TYLCV so they would have the designation, STYLCV, after the isolate from Sardinia which was described first (65). The isolates from Murcia and Almeria, Spain would be STYLCV-M (103) and STYLCV-Alm (GenBank. No. L277081) and the isolate from Sicily would be STYLCV-Sic (GenBank No. Z28390). The isolate from Thailand becomes TTYLCV since it has < 80% nucleotide identity with either TYLCV or STYLCV (118).

Some other tomato-infecting geminiviruses

of the Eastern Hemisphere, which produce leaf curling symptoms similar to those of TYLCV, have historically been called tomato leaf curl virus (TLCV). Based upon nucleotide identity of greater than 80% for the C1 ORF, these TLCVs separate into four distinct viruses. The monopartite TLCV isolate from Australia (40) was the first one molecularly characterized and is designated TLCV. Two TLCVs from India have recently been cloned and sequenced: one, collected from northern India, is bipartite and is called Indian TLCV (ITLCV; 107) and the other, from southern India (Bangalore), is monopartite and is called BTLCV (20 and O. Chatchawankanphanich *et al.*, personal communication). The TLCV from Taiwan (TTLCV, 48) appears to be monopartite and distinct from the other TLCVs (B.-T. Chiang *et al.*, personal communication). Serological differences, as well as nucleotide identities, support the distinction between TYLCV and BTLCV (92).

The many different geminiviruses that cause TYLC and TLC diseases around the world (Figure 3) represent considerable genetic diversity, therefore recombinant DNA strategies for engineering TYLCV- or TLCV-resistant tomatoes must take this into account. This genetic diversity also has implications for the evaluation of germplasm for resistance. *Lycopersicon chilense* (133) has resistance to TYLCV, but it is not known if the same level of resistance will be expressed against other tomato-infecting geminiviruses from the Eastern Hemisphere. Since the isolates of TYLCV from Egypt and Israel are closely related, *L. chilense*-resistant germplasm should be considered when breeding tomatoes for Egypt and other regions of the Eastern Mediterranean.

The tomato-infecting geminiviruses in the Eastern Hemisphere are clearly distinct from those which have originated in the Western Hemisphere (Figure 2). All the Western Hemisphere WFT geminiviruses identified so far are bipartite and have nucleotide identities < 80% with homologous regions of the Eastern Hemisphere geminiviruses. The Eastern Hemisphere and Western Hemisphere tomato-infecting geminiviruses have been reviewed by Green and Kalloo (47).

Methods for Detection of TYLCV

At least 10 distinct whitefly-transmitted geminiviruses are know to infect tomatoes

worldwide (47). Because of the widespread occurrence of epidemics associated with these geminiviruses and their potential threat to tomato production, rapid and specific procedures for geminivirus detection are needed in both *B. tabaci* and plants to aid in epidemiological and disease management studies.

Traditionally, serological methods have been the primary means of virus detection and diagnosis. This approach has met limited success with the whitefly-transmitted geminiviruses (50,92), because it has been difficult to obtain sufficient quantities of highly purified coat protein for production of antisera. TYLCV has only been partially purified and no high titer antisera are available (21,31,33). As would be expected, tests with polyclonal antisera have shown that whitefly-transmitted geminiviruses are antigenically related (74,116). Monoclonal antibodies (Mabs) have been produced against the purified particles of the whitefly-transmitted geminivirus, African cassava mosaic virus (130) and Indian cassava mosaic virus (2). With the availability of these Mabs, B.D. Harrison *et al.* have initiated more detailed studies of the serological relationships among whitefly-transmitted geminiviruses (50,129). They found that some epitopes are shared by particles of different viruses, whereas others are unique to individual viruses; and from these results, Mabs profiles are emerging for the different geminiviruses and for the distinct geminiviruses form different geographical regions, i.e., geminiviruses from the same geographical area tend to share more epitopes regardless of their hosts than do the viruses from the same host but from different geographical areas. Macintosh *et al.* (75) used Mabs raised against ACMV or ICMV to examine the relationship among the different tomato yellow leaf curl viruses in Europe and to design detection methods.

More recently, nucleic acid hybridization techniques have been used for detection of whitefly-transmitted geminiviruses. These methods are versatile and have many applications in epidemiology and disease management studies. Generally, young leaf samples can be squashed onto nylon membranes and these membranes hybridized with general or specific viral DNA probes at low or high stringency hybridization conditions (46). Taking advantage of these features, Czosnek *et al.* (34) carried out the first study on the geographical distribution of the TYLCV-like viruses. Tomato squash blot membranes were prepared from samples collected in countries of the Mediterranean basin, the

Americas, western Africa, and southeast Asia. These membranes were hybridized with a specific DNA probe for TYLCV. Viral nucleic acids can be specifically and easily detected in many organs of the infected plant, e.g. leaves, stems, flowers, roots, and young fruits (95). Geminivirus can also be detected in an individual whitefly (99). Nucleic acid hybridization can also be used to estimate the geminiviral titer in a large number of plants, and this has been useful in evaluation of tomato germplasm for resistance (133).

The polymerase chain reaction (PCR) provides a sensitive and specific technique for detection and identification of geminiviruses (101,120). Polymerase chain reaction methods amplify a specific DNA fragment that lies between two regions of known nucleotide sequence which anneal to synthetic oligonucleotide primers. The PCR primers may be specific for distinct geminiviruses as was the case for TYLCV (94,101) or degenerate primers (120), which can amplify fragments from many distinct geminiviruses. Because of the sensitivity of PCR methods, detection of geminiviruses in individual whiteflies is easy (86,101). Navot et al. (101) found that they could detect TYLCV in whiteflies by mixing DNA from a single viruliferous whitefly with DNA from whiteflies reared on uninfected plants. After a 1:1,000 dilution, the PCR signal obtained with TYLCV specific primers was still very strong. Also the viral DNA was amplified from insect DNA diluted in water up to 10^{-9}. When two pairs of specific primers are used for distinct geminiviruses, mixed infections can be detected in plants and TYLCV and tomato mottle geminivirus (ToMoV) were found in individual whiteflies (86). In the future, PCR methods will have many applications in diagnosis (see 94,112) and epidemiological studies of geminiviruses and their vectors. With PCR and specific primers, it is possible to identify a specific geminivirus associated with a plant or vector in less than 6 hours.

TYLCV TRANSMISSION

The vector for TYLCV is the tobacco or sweetpotato whitefly *Bemisia tabaci* (29,77,79,93). The damage to tomato crops by TYLCV has always been directly correlated with fluctuations in the whitefly populations. It has recently become possible to identify different biotypes of *B. tabaci* by isozyme analysis (108). The most prevalent biotype, B-biotype or silver leaf whitefly, has been

designated as a new species, *Bemisia argentifolii* (12).

No means of transmission of TYLCV other than *B. tabaci* (or *B. argentifolii*) has been found. The spread of TYLCV between tomato plants by contact seems unlikely since the virus cannot be mechanically transmitted from tomato to tomato (29,77,93). Transmission between tomato plants by natural root grafting has not been reported; TYLCV is not transmissible through seed (77,93,102); and tomato plants do not become infected when grown in soil collected from around TYLCV-infected plants (77). Therefore, the spread of TYLCV in tomato fields appears to be due entirely to the whitefly vector.

The interaction between TYLCV and its vectors, *B. tabaci* and *B. argentifolii*, has been studied extensively. A single whitefly is able to transmit the virus, and the rate of transmission increases as the population density of the vector increases (80,87). Cohen and Nitzany (29) reported that female whiteflies transmit TYLCV more efficiently than males; and both adults and larvae can acquire the virus from infected plants (29,87). The acquisition and inoculation thresholds are between 15 and 30 min (29,87,136); however, the rate of transmission increases with longer acquisition-access periods (23,80,87,136). The latent period is between 21 and 24 h (29,80,87). Viruliferous adult *B. tabaci* retain TYLCV for 10 to 20 days after the end of the acquisition-access period (23,80,136). The virus is retained through the molt from larva to adult (29,87), but not through the egg to the progeny insect (29).

Whiteflies carrying TYLCV progressively lose their ability to transmit the virus, in spite of repeated or prolonged acquisition on an infected plant. The ability to transmit TYLCV can be restored only after virus transmission has ceased and the whiteflies have been allowed a subsequent acquisition-access period. This unique phenomenon in virus-vector interaction was termed "periodic acquisition" (25).

More recently, molecular detection methods have been used to detect TYLCV in whiteflies during transmission experiments (8,32,87,99,136). TYLCV-specific probes allowed detection of viral DNA in *B. tabaci* after a 1 h acquisition-access period (87), and the concentration of viral DNA in the vector increased with the length of the acquisition-access period (P. Mehta *et al.*, personal communication). When viruliferous *B. tabaci* were maintained on cucumbers, a nonhost for TYLCV, the detectable amount of viral DNA per whitefly increased by about 60% during a 4-day period (87). *Trialeurorles vaporariorum*, the greenhouse whitefly and a nonvector, acquired TYLCV during a 24 h acquisition-access period; and when these viruliferous adults were maintained on cucumbers, the detectable viral DNA per insect remained the same over a period of 4 days (P. Mehta *et al.*, personal communication). One explanation for this difference could be that virus replication occurs in a vector (*B. tabaci*) but not in a nonvector (*T. vaporariorum*). Zeidan and Czosnek (137) suggested that TYLCV DNA replicates in the insect vector shortly after virus acquisition. In contrast, Cohen and Antignus (23) detected no viral replicative forms (dsDNA) in extracts of viruliferous whiteflies, so they concluded that the virus does not multiply in its whitefly vector. Thus, more information using different experimental approaches is needed to resolve this question.

The acquisition of TYLCV by nonvector insects (*Myzus persicae* and *T. vaporariorum*) indicates that virus-vector specificity among geminiviruses lies in the hemocoel-salivary gland barrier rather than in the gut-hemocoel barrier (23). It is known that the capsids of all the whitefly-transmitted geminiviruses have one or more antigen epitopes in common, and it has been suggested that these may be the determinants of vector specificity (116). The important role of the coat protein in insect transmission of geminiviruses was shown by the substitution of the coat protein ORF of the leafhopper-transmitted geminivirus, beet curly top virus, for the coat protein ORF of the whitefly-transmitted geminivirus, ACMV. This engineered virus could then be transmitted by the leafhopper vector (16). *B. tabaci* failed to acquire and transmit virions of TYLCV with a coat protein altered by UV irradiation, or naked ssDNA, or cloned dsDNA of TYLCV (8). Similarly, *B. tabaci* did not transmit bean golden mosaic geminivirus from beans infected with mutants which did not produce the coat protein (10). Molecular biological methods are available to specifically define the amino acid motif associated with transmission of geminiviruses by their vectors.

A technique based on the polymerase chain reaction (PCR) has been developed to detect TYLCV in individual vectors rapidly and accurately (86,101). Using PCR and specific primers Navot *et al.* (101) detected TYLCV DNA in a mixture of one viruliferous whitefly plus 1,000 whiteflies reared on uninfected plants, and in a 1x10⁻⁹ dilution of DNA from one whitefly. Also, PCR methods were used to detect both TYLCV and ToMoV-FL in one adult *B. tabaci* (86). In the future, the use of PCR methods will provide many opportunities for elegant epidemiological field experiments.

HOST RANGE

Host range studies were carried out under greenhouse conditions using artificial inoculation by viruliferous *B. tabaci* (23,29,58,80,93,102). The following plant species were found to be susceptible to TYLCV : *Datura stramonium, Datura bernhardii, Solanum nigrum Hyoscyamus desertorum, Lycopersicon esculentum* (all cultivars), *Lycopersicon hirsutum, Lycopersicon peruvianum, Lycopersicon pimpinellifolium, Nicotiana sylvestris, Nicotiana benthamiana, Nicotiana glutinosa, Chaerogphyllum* sp., *Cynanchum acutum, Lens esculenta* 'Moench', *Malva nicaensis, Malva parviflora, Nicotiana tabacum*, 'Samsun', 'Havana 423', and 'Yellow Special Flue Cured Virginia', *Phaseolus vulgaris* 'Bulgarit', and *Sonchus oleraceus*.

Unlike the original Israeli isolate, a TYLCV isolate from Jordan did not infect *P. vulgaris, M. parviflora,* and *N. benthamiana* (80). Recently McLaughlin *et al.* (personal communication) cloned a TYLCV DNA fragment from naturally infected hot peppers from Jamaica. Also, samples of *D. stramonium* from the Dominican Republic hybridized with a TYLCV-specific probe at high stringency (94).

Over 15,000 plant samples were collected from Egypt and assayed in our laboratory with a TYLCV (from Egypt) full-length DNA probe at low stringency to establish a database for future epidemiological and control studies. Fourteen plant species reacted positively with this probe using squash blot hybridization tests: *L. esculentum, Capsicum annuum, Capsicum frutescens, N. tabacum* cv. 'White Burley', *N. benthamiana, D. stramonium, S. nigrum, Arachis hypogea, Gossypium hirsutum, Corchorus trictorius, Hibiscus syriacus, Malva arvensis, Sesamum indicum,* and *Vitis vinifera*. Since the hybridization was done at low stringency, these plants were infected either with TYLCV or other related geminiviruses (A. Sabek *et al.*, personal communication).

EPIDEMIOLOGY

Several factors, either singly or in combination, have contributed to the widespread incidence of TYLC diseases. Among these factors are high vector populations, susceptibility of tomato cultivars, overlapping of successive tomato crops, migration of the vector from mature tomato, cotton, and other crops near harvest to newly plante᷉ tomatoes, and infection of tomato seedlings in unprotected nurseries.

The incidence of TYLC disease is positively correlated with the population size of its whitefly vector *B. tabaci* (26,59,82). During spring and early summer the populations of *B. tabaci* are very small; so, when sowings are made in February-April in Egypt, Cyprus, Lebanon, Jordan, and Israel, the incidence of TYLC disease is negligible. The populations then increase sharply during late summer and fall to peak during September and mid-October when sowings are made for fall tomato crops. TYLCV infection is then very high in Egypt (82), Cyprus (59), Lebanon (79), Israel (102), Jordan (77), and Saudi Arabia (83). Subsequently, during November and December, the whitefly populations gradually decrease (26,60,77,82).

Other crops can serve as reservoirs for the whiteflies which eventually infest tomato. Most of the whitefly population builds up in adjacent fields of vegetables or field crops planted earlier than tomato fields. In Egypt the infection pressure of TYLC disease was measured as the percentage of tomato seedlings which became infected when distributed as bait plants in fields of various crops throughout the Fayoum governorate. Infection pressure is quite low at the beginning of the growing season; it increases to a high at the end of September; and, after the middle of October, drops. The infection pressure was approximately three times higher in tomato fields than in other crop fields (84 and H. Mazyad *et al.*, personal communication). Prior studies showed that early in the growing season a few infections in tomato fields are apparently caused by inoculation from external sources. These infections spread rapidly within the tomato field (23,58,82). In the fall season, however, when the population of the whiteflies is very high, there are two ways of virus spread: from sources outside the field and from sources within the field (82).

In many countries almost all tomato plants are infected by TYLCV soon after transplanting, so there may be numerous sources of inoculum. In Egypt older tomato fields and *D. stramonium* are considered the main sources of the virus (82,93). In the Jordan Valley it was found that winds carry viruliferous whiteflies from TYLCV-infected *C. acutum* into the tomato-growing area. The resulting primary infection is followed by a secondary spread. *M. parviflora, Malva nicaensis,* and *D. stramonium* also serve as natural reservoirs for TYLCV (23,102). In Cyprus, tobacco and *D. stramonium* are the only other likely reservoirs of TYLCV. The practice of year round, overlapping tomato crops provides a constant source of inoculum, and this seasonal carry-over appears to be greatest from overwintered tomato crops, the main source of inoculum for the primary infection of new tomato plantings in the spring (58).

DISEASE MANAGEMENT STRATEGIES

Several methods, such as insecticide treatments and various cultural practices, have been tested for control of TYLC disease, but none of these methods alone was effective. Instead, a satisfactory level of disease management can apparently be obtained only through an integrated approach which combines cultural practices and insecticide treatments (23,58,60,82). Those practices found to be effective are the following:

Eliminating or Reducing Sources of Initial Inoculum

The spread of TYLCV in Egypt was successfully controlled by rouging infected plants and removing all overwintered tomato crops early in the spring prior to the emergence of adult whiteflies. In such treated fields virus was spread only from sources outside the field (82). To be effective, inoculum reduction measures must be applied on an area-wide basis and, preferably, where there is little or no tomato production in greenhouses, often the site of adult whitefly activity and active virus spread the year round. When these measures were followed in Cyprus for 3 consecutive years, primary virus spread to spring plantings was almost completely prevented, and secondary spread to summer plantings was reduced from 40-50% to less than 5% (58).

Eradication of volunteer tomatoes and weeds during the summer in the Jordan Valley significantly reduced the population of viruliferous whiteflies, and thus, reduced the incidence of TYLC disease during the following fall (5).

Unfortunately, it is not easy to eliminate common weed hosts such as *D. stramonium, C. acutum,* and *M. nicaensis.* In the Arava region in Israel no virus-infected natural hosts which could serve as a bridge between seasons are found, yet TYLC incidence in this region can reach 100% by the end of the summer. Since whiteflies were trapped as far as 16 km from plant sources of TYLCV, the virus may be introduced yearly into the region by long distance, wind-carried, viruliferous whiteflies from the east or the northern part of the Jordan Valley (23). Unfortunately, the importance of long distance movement of viruliferous *B. tabaci* adults is not known for most agricultural situations.

Production and Use of Healthy Transplants

Early seedling infection reduces tomato yield and enhances the secondary spread of the virus in the field. In Egypt, Cyprus, and the Jordan Valley, virus-free tomato seedlings were produced successfully both in insect-proof propagation centers and under 500 mesh cheesecloth tunnels (Moustafa *et al.*, personal communication; 7,58,84). A range of protective cultural strategies was evaluated to develop methods for the production of healthy tomato seedlings in Egypt (Moustafa *et al.*, personal communication, and M.K. Nakhla *et al.*, unpublished): 1) seedbeds under low tunnels of cheesecloth, 2) open seedbeds surrounded with sticky yellow plastic polyethylene sheets, 3) open seedbeds treated with insecticides, and 4) open seedbeds treated with insecticides and intercropped with squash. In fall plantations, the production of seedlings under low tunnels markedly reduced the number of infected plants observed in the first few weeks after transplanting. However, in summer plantations, the incidence of TYLCV-infected plants was very low and no differences in infection rate or tomato yield were seen between the different methods of seedling protection (Moustafa *et al.*, personal communication, and M.K. Nakhla, unpublished).

Date of Planting

Manipulation of the planting date for open-field tomatoes in Egypt and Cyprus effectively reduced whitefly infestation and TYLCV infection. Numerous experiments and observations showed that TYLCV infection in greenhouses and in low tunnels is reduced or completely avoided by delaying transplanting to the winter months. In the open-field crops this can be achieved by transplanting as early in the spring as possible. The objective in both cases is to avoid the exposure of young plants to high whitefly populations (Moustafa *et al.*, personal communication; 59,62,78,82).

Chemical Management of the Vector Populations

In the last two decades, attempts to reduce the incidence of TYLCV-infected plants by chemical control of the vector have been extensive, but control of *B. tabaci* to a level that results in a significant reduction in virus infection has been difficult (123). Some of the factors causing the difficulties in managing whiteflies are: i) a wide host range which exceeds 500 species (45); ii) the presence of immature stages and adults on the underside of leaves; iii) the extreme mobility of adults; and iv) the ability of *B. tabaci* to rapidly develop resistance to most classes of existing insecticides. Many conventional insecticides such as organophosphates, carbamates, and pyrethroides effectively reduced whitefly populations (60,78,82,102,132); however, they provided only partial TYLC disease control, even when sprayed as frequently as 2-3 times a week. In spite of insecticide spraying, the incidence of TYLC-infected plants and populations of the whitefly vector have increased in the last few years. At the end of summer and during autumn, Israeli farmers spray daily. Disease incidence in treated fields reaches 25-50%; but, when high whitefly populations occur, 100% infection is common (23). Continued excessive use of these compounds not only failed to control whiteflies but also had a negative impact on natural enemies and on the environment. In a very short time insecticide resistance in *B. tabaci* develops (38). The whiteflies have become resistant to organophosphorous insecticides (72,113) and all other insecticides used for their control (38); therefore, insecticide resistance management systems are needed to extend the effective period for current and new materials.

Because of massive outbreaks of TYLC disease in Egypt during the growing seasons of 1980 and 1989, a national management program was developed. In the Fayoum area, large-scale insecticidal control of the vector successfully reduced losses in the fall growing seasons of 1981,

1990, and 1991. The tomato crop was sprayed almost weekly for 8 to 10 weeks either by surface or aerial equipment, and damage caused by the virus dropped markedly from losses of 80-100% to 30-35% (82; M.K. Nakhla *et al.*, personal communication)

At the beginning of the 1990's, relatively selective insecticides, such as buprofezin, pyriproxyfen, diafenthiuron, and imidacloprid, with novel modes of action and activities were found to be very effective for controlling the developmental stages of *B. tabaci* in cotton and other crops. J. J. Pacheco-Covarrubias (105) found a clear difference among the several nymphal whitefly instars in their susceptibility to imidacloprid. The N3 and N4 nymphs were 24 and 82% more tolerant than N1 and N2 nymphs.

In an attempt to delay the onset of *B. tabaci* resistance to novel insecticides as well as to efficiently apply conventional ones, an integrated pest management-insecticide resistance management strategy with an extensive resistance monitoring program was implemented in 1987 in cotton fields in Israel. The main principles of the strategy are: restricting the use of the insecticides to one pest-generation period, alternating compounds with different modes of action (preferably selective ones), and minimizing the number of treatments according to a provisional action threshold (56).

Non-insecticidal products and oils can effectively reduce both whitefly numbers (infestation) and the percentage of infected plants. Puri *et al.* (114) found that four laundry detergent powders (Nirma, Rin, Surf, and Wheel) and two natural oils (cottonseed and neem) markedly reduced the number of adult whiteflies as well as nymphs on treated cotton plants. Sharaf and Allawy (124) reported two- to three-fold increases in tomato yield and a reduction in TYLC disease incidence when the insecticides permethrin, methidathion, and pirimphos-methyl were applied with mineral oils. The advantages of mixing oils with insecticides to control other whitefly-transmitted viruses have also been reported (125,126).

Experience indicates that chemical measures can at best provide only partial control of TYLC disease; however, in potential epidemic cases, wide-scale chemical control of *B. tabaci* might be recommended for the early stages of plant growth (82). Because whiteflies rapidly develop resistance to insecticides, the benefits of these chemicals may decrease greatly in successive 28,82; M.K. Nakhla, unpublished). The

development of resistance (38) and the loss of natural predators and parasites following repeated application of insecticides contribute to control problems. Consequently, the frequent application of insecticide sprays cannot be justified either economically or ecologically. Insecticides should be applied only infrequently and as a supplement to cultural management. Suggestions for improving chemical control include the use of insecticide combinations (38), insecticides with new modes of actions (55), oils, detergents, and other natural products (17).

Cultural Practices

Yellow Traps and Crop Mulches

Covering the soil of tomato fields with sawdust, fresh wheat straw, or yellow polyethylene sheets markedly reduced *B. tabaci* populations and, consequently, the incidence of TYLC disease in Israel (24,27,28,64). Whiteflies are attracted to the yellow color of the mulches and are killed there by the heat (28). Mulch was most effective in seedbeds and at early stages of growth, and delayed the spread of TYLCV for a relatively short period of 2-4 weeks (22,27). The fading of the mulch color and changes in the ratio of canopy to mulch area probably caused the reduction in the control effect (23).

Intercropping of Plant Species

Interplanting with rows of a species, which is a good host for the vector but a non-host for the virus, might attract the vector and hence reduce the incidence of the disease in tomato. TYLCV infection was delayed during the first 2 months in Jordan when tomato was interplanted with cucumber (planted 4 weeks before tomato transplantation) (3,4)

Physical Barriers

In general, any materials fine enough to block insect passage (infiltration) yet provide adequate light and ventilation can be used to exclude whiteflies. In Egypt it was found that tomatoes grown under low mesh tunnels or in screen houses had lower infection rates during the first 3 weeks after transplanting (84). For the protection of open-field plants, various types of light covers either placed directly on the plants or supported as low tunnels, have been successfully tested. They include various types of spunbonded, nonwoven fabrics (e.g., Lutradur of 17-30 g/m^2). Coarse mesh shading nets (e.g., AgronetP of 15-20 g/m^2) are also occasionally used but always in combination with frequent insecticide sprays on both the crop and the nets themselves (58,115). In the Jordan Valley, the use of muslin tunnels for the first 30 days after transplanting successfully delayed the onset of TYLC disease, yet allowed sufficient ventilation (6). For cultivation in insect-proof screenhouses various types of fine-mesh screens, such as perforated polyethylene, polypropylene sheets (Agryl), and fine mesh nets (especially 50 mesh) are used (15,24,58,59).

Breeding for Resistance to TYLCV

The most effective and environmentally sound management would be the planting of resistant or tolerant lines, so breeding TYLCV-resistant plants is probably the most important long-term goal. Unfortunately, all commercially acceptable tomato cultivars tested have been susceptible to TYLCV (63,81,109).

Wild *Lycopersicon* species were screened for their response to the virus, and certain accessions of *L. pimpinellifolium, L. cheesmanii, L. hirsutum, L. peruvianum,* and *L. chilense* were found to have naturally occurring resistance, which varied from partial to complete (29,52,53,54,63,-81,93,109,110,133). The disease may be expressed in different accessions with varying degrees of symptom severity. The difficulties in visual scoring for resistance, in addition to the year-to-year variation in the severity of the TYLC epidemic, can produce conflicting results regarding the response of the accessions to TYLCV. The putative existence of different strains of TYLCV may also cause different reactions (9,133). Determination of the resistance level in wild tomato types was difficult but new molecular detection approaches may provide better means of screening wild species for resistance. The squash-blot procedure (99) was used to screen *Lycopersicon* accessions for resistance to TYLCV. Three categories of response to TYLCV infection were defined: 1) susceptibility - plants contain viral DNA and develop symptoms of the disease, 2) tolerance - plants contain detectable amounts of viral DNA but are symptomless, and 3) resistance - the virus is not detectable by squash blot hybridization tests and the plants are symptomless (133). In another study, viral DNA accumulation correlated positively with symptom severity (121).

Programs to develop TYLCV-resistant cultivars were begun in 1974 in Israel and more recently in Egypt, Lebanon, Jordan, and France using several wild species, including *L. pimpinellifolium, L. peruvianum, L. hirsutum,* and *L. cheesmanii* (51,53,54,61,62,68,69,78,-91,109,110,135). The breeding program in France is developing improved populations derived from resistant wild tomatoes: *L. pimpinellifolium* (LA121; LA3737; LA1478; LA1582, 'Hirsute'), *L. hirsutum* (PI129157 H2, LA1777), *L. peruvianum* (CMV sel. INRA), and *L. cheesmanii* (LA1401) (69,70,71). These four species are used in recurrent selection to combine the resistance elements of the various progenitors to obtain improved plant populations for breeding lines. Screening for resistance is done by natural infection under conditions that result in high infection rates in susceptible tomato varieties. Using seeds of the selected plants, INRA scientists carry out backcrosses or intercrosses with less affected commercial varieties. The commercial varieties found to be less affected by the virus during severe epidemics are Colombian, Roza, Progress No. 1 (United Arab Emirates, Senegal), Lignon C8-6, Lignon C20-5 (Mali, Senegal, Cuba), Rowpack (Cape Verde), VF 145 B 7879 (Egypt), Anahu (Sudan), and EC 104395 (India, Sudan, United Arab Emirates) (69,70).

A similar approach has been used in Egypt; however, tomato seedlings are artificially inoculated by viruliferous whiteflies before transplantation to the field. Interspecific crosses are made with *L. peruvianum, L. cheesmanii,* and *L. pimpinellifolium* using a backcross method. Six lines with superior tolerance and yield have been selected (91).

Tolerance Derived from L. pimpinellifolium

Infection with TYLCV does not seriously affect the growth of *L. pimpinellifolium,* and symptoms appear much later (8-10 weeks after inoculation) than in the sensitive *L. esculentum* cultivars (3-4 weeks). This long latent period could contribute to better plant growth and yield, and this tolerance has been introduced into commercial cultivars. Unfortunately, the selected lines do not have high enough yields (78).

Accessions of *L. pimpinellifolium* varied in their reactions: LA373 exhibited the least severe TYLC disease symptoms; LA121, LA1579, LA1589, and LA1690 were heterogeneous for disease symptoms; and all plants of LA411, LA1256, LA1370, LA1583 and LA1634 exhibited severe symptoms. The virus was detected in all symptomless plants (52). Tolerance in *L. pimpinellifolium* accessions LA121 and LA373 appeared as a partially recessive, poorly heritable quantitative character (53). In contrast, an incomplete dominant single gene was associated with accession LA121 (109). Kasrawi (61) has described a monogenic, dominant gene for resistance in the *L. pimpinellifolium* Hirsute-INRA and LA1478.

Tolerance Derived from L. cheesmanii and L. hirsutum

All tested accessions of *L. cheesmanii* (LA1401), *L. hirsutum* (LA386, LA1295, LA1352, LA1393, LA1691) and *L. hirsutum* f. *glabratum* (LA1252, LA1624) were highly resistant to TYLCV. None of the tested plants exhibited TYLC symptoms, and grafting experiments indicated that none carried the virus (52). A very high level of resistance was recently found in one accession of *L. hirsutum*, LA1777 (41,89). Inheritance of resistance to TYLCV was studied in the interspecific crosses, *L. esculentum* cv. 'UC 82' x *L. cheesmanii* LA1401 and *L. esculentum* cv. 'VF145-B7879' x *L. hirsutum* LA386. Parents, F_2 and F_3 plants, and backcross populations were artificially inoculated with TYLCV prior to transplanting and were later evaluated under field conditions. The reactions of parents, F_2 and F_3 plants, and backcrosses of resistant F_2 plants to UC 82, indicated that resistance derived from *L. cheesmanii* seems to be recessive (54). A polygenic recessive pattern in *L. cheesmanii* was described by Kasrawi (62). The resistance derived from *L. hirsutum* is dominant and controlled by more than one gene (54). In Egypt a promising tolerant line (line 44) was derived from crosses between the commercial cultivar, Pakmor, and the wild species, *L. cheesmanii*. The selection and evaluation of the plants in segregating generations following each backcross led to the development of the tolerant line (91).

Tolerance Derived from L. peruvianum

All tested accessions of *L. peruvianum* (LA372, LA452, LA462, LA1274, LA1333, LA1373, CMV sel INRA) and *L. peruvianum* f. *humifusum* (LA385) were highly resistant and did not produce any TYLC disease symptoms (52). Kasrawi *et al.* (63) reported a high level of resistance in *L. peruvianum* LA385 and *L. peruvianum* f. *humifusum*, however back-indexing

showed that they were infected with TYLCV. Pilowsky and Cohen (110) indicated that tolerance in *L. peruvianum* was controlled by five recessive factors. Resistance in *L. peruvianum* CMV sel INRA to the transmission of the virus via whiteflies was broken when inoculations were made at high temperatures (mostly over 42°C during daytime). Plants of this accession were susceptible to graft inoculations but they were symptomless (81). The disadvantage of working with *L. peruvianum* is the difficulty of obtaining interspecific hybrids with *L. esculentum* because of incompatibility. This problem was overcome by culturing immature seeds *in vitro* and by the "pollen mixture" technique (78). A highly TYLCV-resistant F_4 line was developed from the interspecific cross of *L. esculentum* 'Mortelglan' X *L. peruvianum* CMV sel. INRA and this line was used in the production of F_2 BC1 plants (51). A tolerant line (# 53) was derived from the interspecific cross of UC 82 x *L. peruvianum* (91). The tolerant F_1 hybrid, TY-20, was developed in Israel using *L. peruvianum* PI126935 as the source of tolerance (110).

Tolerance Derived from L. chilense

Recently, a new source of very high resistance has been identified in *L. chilense* LA1969 (133). Regardless of whether a plant is inoculated by viruliferous *B. tabaci* or by grafting, *L. chilense* does not show symptoms. Furthermore, only 3% of the plants contain viral DNA detectable by the molecular hybridization technique (133,135). It is, however, detectable by back-grafting LA1969 onto a sensitive cultivar (41). The main TYLCV resistance gene (*Ty-1*) has been introgressed into *L. esculentum* and the *Ty-1* gene mapped to chromosome 6 using RFLP analysis (134). The *Ty-1* gene was shown to reduce long-distance virus translocation in plants (88).

Commercial Tomato Hybrids with Tolerance to TYLCV

Recently, several TYLCV-tolerant hybrids have been released by the private sector. The first one available was the F_1 hybrid, TY-20, which was released in 1988 in Israel (Hazera Seed Co.) for open-field cultivation (111). *L. peruvianum* PI126935 was the source of this tolerance (110). When infected with TYLCV, the leaves of young TY-20 plants exhibit only a mild interveinal chlorosis. In mature plants, the leaflets usually become slightly cupped. When early infection is prevented, the plants give an acceptable yield in spite of TYLCV infection (1,111). DNA assays indicate that the tolerance mechanism operates through the reduction of the multiplication and/or spread of the virus in the plants (121). Fragette (41) reported that resistance is expressed only in old plants. Therefore it is recommended that TY-20 seedlings be produced in insect-proof screenhouses, with an intensive insecticide spraying program. Insecticide should also be applied for the first month after transplanting.

Other commercial TYLCV-tolerant hybrids are Fiona F1 (=E437), Jackal F1 (=E438) (Sluis and Groot, Holland), TOP 21 F1 (Clause), Saria (Petoseed), Tycoon F1, Tymoor F1, Tyking F1, Tydal F1, Tyger F1, and Tygold F1 (Royal Sluis47). In Egypt the hybrids TY-20, BB234, BB235, and Typhoon were found to be tolerant and are recommended for the fall season (90).

Host Plant Resistance to the Vector

Several accessions of wild *Lycopersicon* spp., such *as L. hirsutum* LA771418, LA1777, *L. hirsutum* f. *glabratum* LA407, PI34418, and *L. pennellii* LA716, have been used to incorporate resistance to *B. tabaci* into tomatoes (13,14,66). This resistance is usually based either on the density of the leaf trichomes (127) or on the sticky excretion from the leaf hairs (35). *Lycopersicon hirsutum*, unlike other species, has type Vic glandular trichomes and the exudate from these glands entraps whiteflies before they can transmit the virus (19).

Recombinant DNA Approaches for Engineering Tomatoes with Resistance to TYLCV

Molecular approaches have the potential for providing solutions to crop improvement problems that have not been solved by more traditional methods (42,131). Development of antiviral strategies for engineering plants with resistance has progressed more rapidly with RNA viruses than with any other group of plant pathogens (42). Some strategies include transgenic plants with the following constructions: i) coat protein for RNA viruses (42); ii) truncated coat protein gene for RNA viruses (73); iii) 54-kDa component of the replicase complex of tobacco mosaic virus (18); iv) truncated viral replicase gene (39); v) proteinase of a potyvirus

(76); and vi) defective movement protein of tobacco mosaic virus (30). Recently, the first genetically engineered, virus-resistant plant, Freedom II squash (Asgrow Seed Co.), was approved by the United States Department of Agriculture for commercial production. This squash has been engineered with the coat protein gene from zucchini yellow mosaic potyvirus (H. Quemada, personal communication; 98). Even though molecular approaches have been used with relative ease to create plants with resistance to RNA viruses (42), unfortunately, this is not the case for DNA plant viruses.

For geminiviruses, Stanley et al. (1990; reviewed in Frischmuth and Stanley, 43) have shown that N. benthamiana, transformed with a single copy of a tandem repeat of subgenomic DNA-B of ACMV, had reduced symptoms and lower titers of the wild-type DNA-A and DNA-B components. The symptom reduction was specific to ACMV and did not occur in this transgenic plant when inoculated with tomato golden mosaic geminivirus (TGMV). Frischmuth and Stanley (44) obtained similar results when N. benthamiana was transformed with a partial repeat of subgenomic DNA of the monopartite geminivirus, beet curly top virus (BCTV). Also, Stenger (128) transformed N. benthamiana with a tandem repeat of a defective-interfering (DI), subgenomic DNA from the Logan strain of BCTV and these plants showed reduced symptoms when challenged with the Logan strain, but not with the CFH or Worland strains. Only the Logan strain was able to mobilize and replicate an episome from the integrated tandem repeat DI DNA.

Antisense approaches against the replication-associated protein have also shown promise. Bejarano and Lichtenstein (11) and Day et al. (37) reported that transgenic N. tabacum plants with TGMV antisense-AC1 ORF constructions had reduced symptoms after inoculation with TGMV, and replication was reduced. This resistance was not expressed against ACMV but replication of BCTV was reduced (11). They (11) suggested that the difference between the replication of ACMV and BCTV was due to the high identity (82%) of a 280-nucleotide region within the AC1 ORFs of TGMV and BCTV, whereas there was not a similar region of high nucleotide identity between the AC1 ORFs of TGMV and ACMV. The use of mutant replication-associated proteins of geminiviruses to interfere with wild-type replication has been proposed by Hanson et al. (49) and is currently being evaluated for tomato-infecting geminiviruses (Nakhla *et al.*, unpublished).

More recently, Kunik *et al.* (67) reported that some plants of the F₁ interspecific tomato hybrid, *L. esculentum* X *L. pennellii*, transformed with the coat protein ORF of TYLCV expressed delayed symptoms and recovery when plants were inoculated by whiteflies fed on TYLCV-infected plants. All plants that showed this reduced symptom development had expressed viral coat protein and the transgenic plants that developed wild-type symptoms accumulated coat protein RNA, but no coat protein. Similarly, beans transformed with the coat protein ORF of bean golden mosaic geminivirus accumulated coat protein RNA, were generally as susceptible as nontransformed beans, and did not accumulate coat protein (O. Azzam *et al.*, personal communication).

CONCLUSIONS

The seriousness of geminiviral diseases of tomatoes in many countries requires collaboration to exploit all possible management measures. Each strategy investigated so far has limited efficiency and provides only partial control, but combinations can be expected to lead to improvements in disease management.

The application of multiple management measures for reducing incidence of TYLCV is dramatically illustrated by the current situation in the Dominican Republic. Symptoms of TYLC disease were first noticed on tomatoes in the spring of 1992 (J.K. Brown and J. Bird in Green and Kallo, 47) and severe epidemics occurred in the 1992-1993 and 1993-1994 growing seasons with losses of about 80%. An Eastern Hemisphere-type geminivirus (112) and, more specifically, TYLCV (94) was detected in these plants. Multiple management strategies were implemented before the fall of 1994. This included a government imposed host-free period, the use of the new systemic insecticide, imidacloprid, for treatment of seed beds and transplanted seedlings, the production of tomato seedlings in isolated areas away from potential sources of inoculum, a time of transplanting tomatoes delayed to correspond to lower vector populations, the use of the most tolerant cultivars, and the applications of insecticides. At least in part because of these multiple approaches, TYLC disease incidence was less than an estimated 20% in most fields in January 1995. These approaches

have merit for other tomato growing areas and highlight the importance of tomatoes as possibly the major source of inoculum for TYLCV (R.L. Gilbertson and D.P. Maxwell, personal communication).

Effective management of future TYLC disease epidemics will depend on the rapid and accurate identification of the causal agent. The use of polymerase chain reaction methods to characterize TYLCV and the use of specific DNA probes for the detection of TYLCV-like geminiviruses in both tomato seedlings and potential virus reservoirs should help in studying the epidemiology of the virus, in designing disease management programs, in breeding for disease resistance, and in developing antiviral strategies involving recombinant DNA approaches. In addition, the development of tolerant cultivars both by traditional approaches and genetic engineering methods, the careful use of insecticides, and the management of plantings to avoid inoculum have promise for the immediate future.

ACKNOWLEDGMENTS

The authors wish to express their gratitude to Martha D. Maxwell for her editorial services. This review was supported in part by the College of Agricultural and Life Sciences, University of Wisconsin-Madison, and by Asgrow Seed Co. and Petoseed Co.

REFERENCES

1. Adler, U., Omar, S., Kern, J., and Ben-Nur, Z. 1989. Autumn tomatoes in Bet Shean Valley. Hassadeh 70:230-232.
2. Aiton, M. M., and Harrison, B. D. 1989. Monoclonal antibodies to Indian cassava mosaic geminivirus (ICMV). Report of the Scottish Crop Research Institute for 1988, p. 175.
3. Al-Musa, A. 1981. Incidence, economic importance and prevention of tomato yellow leaf curl virus in Jordan. Page 47 in: Abstracts of the 1981 International Workshop on Pathogens Transmitted by Whiteflies. Oxford, England.
4. Al-Musa, A. 1982. Incidence, economic importance and control of tomato yellow leaf curl in Jordan. Plant Dis. 66:561-563.
5. Al-Musa, A. 1986. Tomato yellow leaf curl virus in Jordan: epidemiology and control. Dirasat 13:199-208.
6. Al-Musa, A., Nazer, I. K., and Sharaf, N. 1987. Effect of combined agricultural treatments on whitefly population and incidence of yellow leaf curl virus. Dirasat 14:127-134.
7. Al-Musa, A., Sharaf, N., and Qasem, S. 1982. Low cost and effective methods for production of tomato

transplants free from tomato yellow leaf curl virus. Dirasat 9:27-32.

8. Antignus, Y., Adler, O., and Cohen, S. 1991. The use of a molecular probe to study the interaction of TYLCV with its vector *B. tabaci*. Proceedings of the Third International Congress of Plant Molecular Biology, Tucson, Arizona. Abstr. #1160.

9. Antignus, Y., and Cohen, S. 1994. Complete nucleotide sequence of an infectious clone of a mild isolate of tomato yellow leaf curl virus (TYLCV). Phytopatology 84:707-712.

10. Azzam, O., Frazer, J., De la Rosa, D., Beaver, J. S., Ahlquist, P., and Maxwell, D. P. 1994. Whitefly transmission and efficient ssDNA accumulation of bean golden mosaic geminivirus require expression functional coat protein. Virology 204:289-296.

11. Bejarano, E. R., and Lichtenstein, C. P. 1994. Expression of TGMV antisense RNA in transgenic tobacco inhibits replication of BCTV but not ACMV geminiviruses. Plant Mol. Biol. 24:241-248.

12. Bellows, T. S., Jr., Perring, T. M., Gill, R. J., and Headrick, D. H. 1994. Description of a species of *Bemisia* (Homoptera: Aleyrodidae). Ann. Entomol. Soc. Am. 87:195-206.

13. Berlinger, M. J., and Dahan, R. 1987. Breeding for resistance to virus transmission by whiteflies in tomatoes. Insect Sci. Appl. 8:783-784.

14. Berlinger, M. J., Dahan, R., Berlinger, O. C., and Mordechi, S. 1989. Honeydew excretion as a possible tool to screen tomato resistance to virus transmission by *Bemisia tabaci*. Pages 121-131 in: Proceedings of the Working Group "Breeding for Resistance to Insects and Mites", Marcelin, Switzerland.

15. Berlinger, M. J., Mordechi, S., Liper, A., Piper, A., Katz, J., and Levav, N. 1991. The use of nets to prevent the penetration of *Bemisia tabaci* into greenhouses. Hassadeh 71:1579-1583.

16. Briddon, R. W., Pinner, M. S., Stanley, J., and Markham, P. G. 1990. Geminivirus coat protein gene replacement alters insect specificity. Virology 177:84-94.

17. Butler, G. D., Jr., Puri, S. N., and Henneberry, T. J. 1991. Plant-derived oil and detergent solutions as control agents for *Bemisia tabaci* and *Aphis gossypii* on cotton. Southwest Entomol. 16:331-337.

18. Carr, P. J., and Zaitlin, M. 1993. Replicase-mediated resistance. Sem. Virol. 4:339-347.

19. Channarayappa, G. S., Muniyappa, V., and Frist, R. H. 1992. Resistance of *Lycopersicon* species to *Bemisia tabaci*, a tomato leaf curl virus vector. Can. J. Bot. 70:2184-2192.

20. Chatchawankanphanich, O., Chiang, B.-T., Green, S. K., Singh, S. J., and Maxwell, D. P. 1993. Nucleotide sequence of a geminivirus associated with tomato leaf curl from India. Plant Dis. 77:1168.

21. Chiemsombat, P., Murayama, A., and Ikegami, M. 1991. Tomato yellow leaf curl virus in Thailand and tobacco leaf curl virus in Japan are serologically identical. Ann. Phytopath. Soc. Japan. 57:595-597.

22. Cohen, S. 1981. Control of whitefly vectors of viruses by non-conventional means. Page 51 in: Abstracts of the 1981 International Workshop on Pathogens Transmitted by Whiteflies, Oxford, England.

23. Cohen, S., and Antignus, Y. 1994. Tomato yellow leaf curl virus (TYLCV), a whitefly-borne geminivirus of tomatoes. Pages 259-288 in:

24. Cohen, S., and Berlinger, M. J. 1986. Transmission and cultural control of whitefly-borne viruses. Agric. Ecosyst. Environ. 17:89-97.

25. Cohen. S., and Harpaz, I. 1964. Periodic, rather than continual acquisition of a new tomato virus by its vector, the tobacco whitefly (*Bemisia tabaci* Gennadius). Entomol. Exp. Appl. 7:155-166.

26. Cohen, S., Kern, J., Harpaz, I., and Ben-Joseph, R. 1988. Epidemiological studies of the tomato yellow leaf curl virus (TYLCV) in the Jordon Valley, Israel. Phytoparasitica 16:259-270.

27. Cohen, S., and Melamed-Madjar, V. 1978. Prevention by soil mulching of the spread of tomato yellow leaf curl virus transmitted by *Bemisia tabaci* (Gennadius) (Homoptera:Aleyrodidae). Israel. Bull. Entomol. Res. 68:465-470.

28. Cohen, S., Melamed-Madjar, V., and Hameiri, J. 1974. Prevention of the spread of tomato yellow leaf curl virus transmitted by *Bemisia tabaci* (Gennadius) (Homoptera:Aleyrodidae). Israel. Bull. Entomol. Res. 64:193-197.

29. Cohen, S., and Nitzany, F. E. 1966. Transmission and host range of the tomato yellow leaf curl virus. Phytopathology 56:1127-1131.

30. Cooper, B., Lapidot, M., Heichk, J. A., Dodds, J. A., and Beachy, R. N. 1995. A defective movement protein of TMV in transgenic plants confers resistance to multiple viruses whereas the functional analog increases susceptibility. Virology 206:307-313.

31. Credi, R., Betti, L., and Canova, A. 1989. Association of a geminivirus with a severe disease of tomato in Sicily. Phytopathol. Medit. 28:223-226.

32. Czosnek, H., Ber, R., Antignus, Y., Cohen, S., Navot, N., and Zamir, D. 1988. Isolation of the tomato yellow leaf curl virus, a geminivirus. Phytopathology 78:508-512.

33. Czosnek, H., Ber, R., Navot, N., Zamir, D., Antignus, Y., and Cohen, S. 1988. Detection of tomato leaf curl virus in lysates of plants and insects by hybridization with a viral DNA probe. Plant Dis. 72:949-951.

34. Czosnek, H., Navot, N., and Laterrot, H. 1990. Geographical distribution of tomato yellow leaf curl virus. A first survey using a specific DNA probe. Phytopathol. Medit. 29:1-6.

35. Dahan, R. 1985. *Lycopersicon pennellii* as a source for resistance to the tobacco whitefly *Bemisia tabaci* in tomato. M. S. thesis. Ben Gurion University of the Negev, Be'er Sheva, Israel.

36. Davies, W., and Stanley, J. 1989. Geminivirus genes and vectors. Trends Genet. 5:77-81.

37. Day, A. G., Bejarano, E. R., Buck, K. W., Burrell, M., and Lichtenstein, C. P. 1991. Expression of an antisense viral gene in transgenic tobacco confers resistance to the DNA virus tomato golden mosaic virus. Proc. Natl. Acad. Sci. USA 88:6721-6725.

38. Dittrich, V., Uk, S., and Ernst, G. H. 1990. Chemical control and insecticide resistance of whiteflies. Pages 263-286 in: Whiteflies: their bionomics, pest status, and management. D. Gerling, ed. Intercept Ltd., Andover, Hants, UK.

39. Donson, J., Kearney, C. M., Turpen, T. H., Khan, I. A., Kurath, G., Turpen, A. M., Jones, G. E., Dawson, W. O., and Lewandowdki, D. J. 1993. Broad resistance to tobamoviruses is mediated by a modified tobacco mosaic virus replicase transgene.

Advances in Disease Vector Research. Vol. 10. Springer-Verlag, New York.

Mol. Plant-Microbe Interact. 6:635-652.

40. Dry, I. B., Rigden, J. E., Krake, L. R., Mullineaux, P. M., and Rezaian, M. A. 1993. Nucleotide sequence and genome organization of tomato leaf curl geminivirus. J. Gen. Virol. 74:147-151.

41. Fargette, D. 1991. Quelques proprietes de la resistance varietale a l'enroulement de la tomate. Pages 47-49 in: Resistance of the Tomato to TYLCV. Proceedings of the seminar of EEC contract DGXII-TS2-A-055 F (CD) partners. H. Laterrot, and C. Trousse, eds. INRA-Stationd'Amelioration des Plantes Maraicheres, Montfavet-Avignon, France.

42. Fitchen, J. H., and Beachy, R. N. 1993. Genetically engineered protection against viruses in transgenic plants. Annu. Rev. Microbiol. 47:739-763.

43. Frischmuth, T., and Stanley, J. 1993. Strategies for the control of geminivirus diseases. Sem. Virol. 4:329-337.

44. Frischmuth, T., and Stanley, J. 1994. Beet curly top virus symptom amelioration in *Nicotiana benthamiana* transformed with a naturally occurring viral subgenomic DNA. Virology 200:826-830.

45. Gameel, O. I. 1972. A new description, distribution and hosts of the cotton whitefly, *Bemisia tabaci* (Gennadius) (Homoptera:Aleyrdidae). Rev. Zool. Bot. Africa 86:50-64.

46. Gilbertson, R. L., Hidayat, S. H., Martinez, R. T., Leong, S. A., Faria, J. C., Morales, F., and Maxwell, D. P. 1991. Differentiation of bean-infecting geminiviruses by nucleic acid hybridization probes and aspects of bean golden mosaic in Brazil. Plant Dis. 75:336-342.

47. Green, S. K. and Kalloo, G. 1994. Leaf curl and yellowing viruses of pepper and tomato: an overview. Asian Vegetable Research and Development Center. Technical Bulletin No. 21, 51 p.

48. Green, S. K., Sulyo, Y., and Lesemann, D. E. 1987. Leaf curl virus on tomato in Taiwan Province. FAO Plant Prot. Bull. 35:62.

49. Hanson, S. F., Gilbertson, R. L., Ahlquist, P. G., Russell, D. R., and Maxwell, D. P. 1991. Site-specific mutations in codons of the putative NTP-binding motif of the AL1 gene of bean golden mosaic geminivirus abolish infectivity. (Abstr.) Phytopathology 81:1247.

50. Harrison, B. D., Muniyappa, V., Swanson, M. M., Roberts, I. M., and Robinson, D. J. 1991. Recognition and differentiation of seven whitefly-transmitted geminiviruses from India, and their relationships to African cassava mosaic and Thailand mungbean yellow mosaic viruses. Ann. Appl. Biol. 118:299-308.

51. Hassan, A. A., Laterrot, H., Mazyad, H. M., Moustafa, S. E., and Nakhla, M. K. 1987. Use of *Lycopersicon peruvianum* as a source of resistance to tomato yellow leaf curl virus. Egypt. J. Hort. 14:173-176.

52. Hassan, A. A., Mazyad, H. M., Moustafa, S. E., and Nakhla, M. K. 1982. Assessment of tomato yellow leaf curl virus resistance in the genus *Lycopersicon*. Egypt. J. Hortic. 9:13-116.

53. Hassan, A. A., Mazyad, H. M., Moustafa, S. E., Nassar, S. H., Nakhla, M. K, and Sims, W. L. 1984. Genetics and heritability of tomato yellow leaf curl virus tolerance derived from *Lycopersicon pimpinellifolium*. Pages 81-87 in: EUCARPIA, Tomato working group 9th meeting, Wageningen,

The Netherlands.

54. Hassan, A. A., Mazyad, H. M., Moustafa, S. E., Nassar, S. H., Nakhla, M. K., and Sims, W. L. 1984. Inheritance of resistance to tomato yellow leaf curl virus derived from *Lycopersicon cheesmanii* and *Lycopersicon hirsutum*. HortScience 19:574-575.

55. Hemmeberry, T. J., and Butler, G. D., Jr. 1992. Whiteflies as a factor in cotton production with specific reference to *Bemisia tabaci* (Gennadius). Pages 674-683 in: Proc. Beltwide Cotton Production and Research Conference, National Cotton Council of America, Memphis, TN.

56. Horowitz, A. R., and Ishaaya, I. 1994. Chemical control of *Bemisia tabaci*-management and application. *Bemisia* Newsletter 8:37-38.

57. Ioannou, N. 1985. Yellow leaf curl and other virus disease of tomato in Cyprus. Plant Pathol. 34:428-434.

58. Ioannou, N. 1987. Cultural management of tomato yellow leaf curl disease in Cyprus. Plant Pathol. 36:367-373.

59. Ioannou, N., and Hadjinicolis, A. 1991. Epidemiology and control of tomato yellow leaf curl virus in Cyprus. Pages 3-5 in: Resistance of the Tomato to TYLCV. Proceedings of the Seminar of EEC contract DGXII-TS2-A-055F (CD) Partners. H. Laterrot, and C. Trousse eds. INRA-Station d'Amelioration des Plantes Maraicheres, Montfavet-Avignon, France.

60. Ioannou, N., and Iordannou, N. 1985. Chemical control of the whitefly *Bemisia tabaci* (Genn.) and its effect on tomato yellow leaf curl virus., Agricultural Research Institute, Nicosia, Cyprus, Technical Bulletin No. 68, 8 p.

61. Kasrawi, M. A. 1989. Inheritance of resistance to tomato yellow leaf curl virus (TYLCV) in *Lycopersicon pimpinellifolium*. Plant Dis. 73:435-437.

62. Kasrawi, M. A. 1991. Tomato production and tomato yellow leaf curl viruses in Jordan. Pages 14-16 in: Resistance of the Tomato to TYLCV. Proceedings of the seminar of EEC contract DGXII-TS2-A-055 F (CD) partners. H. Laterrot, and C. Trousse eds. INRA-Station d'Amelioration des Plantes Maraicheres, Montfavet-Avignon, France.

63. Kasrawi, M. A., Suwwan, M.A., and Mansour, A. 1988. Sources of resistance to tomato yellow leaf curl virus in *Lycopersicon* species. Euphytica 37:61-64.

64. Kern, J., Decco, Z., Cohen, S., and Ben-Joseph, R. 1991. Reduction of damage from TYLCV and tobacco whitefly in tomatoes by yellow plastic mulch. Hassadeh 71:864-868.

65. Kheyr-Pour, A. M., Bendahmane, V. M., Accotto, G. P., Crespi, S., and Gronenborn, B. 1991. Tomato yellow leaf curl virus from Sardinia is a whitefly-transmitted monopartite geminivirus. Nucleic Acids Res. 19:6763-6769.

66. Kisha, S. A. 1984. Whitefly, *Bemisia tabaci*: infestations on tomato varieties and a wild *Lycopersicon* species. Ann. Appl. Biol. 104:124-129.

67. Kunik, T., Salomon, R., Zamir, D., Navot, N., Zeidan, M., Michelson, I., Gafni, Y., and Czosnek, H. 1994. Transgenic tomato plants expressing the tomato yellow leaf curl virus capsid protein are resistant to the virus. Bio/Technology 12:500-504.

68. Laterrot, H. 1986. La tomate. Interet et utilisation

des especes sauvages pour la creation varietale. P. H. M. Revue Horticole 295:3-7.

69. Laterrot, H. 1991. Materiel vegetal en cours de selection pour la resistance au TYLCV. Pages 44-46 in: Resistance of the Tomato to TYLCV. Proceedings of the seminar of EEC contract DGXII-TS2-A-055 F (CD) partners. H. Laterrot and C. Trousse ed. INRA-Station de Amelioration des Plantes Maraicheres, Montfavet-Avignon, France.

70. Laterrot, H. 1992. Resistance genitors to tomato yellow leaf curl virus (TYLCV). Tomato Leaf Curl Newsletter 1:2-3.

71. Laterrot, H., and Makkouk, K. M. 1983. Selection for partial resistance to tomato yellow leaf curl virus (TYLCV). Tomato Genetics Cooperative Rep. 33:4-5.

72. Lemon, R. W. 1992. Resistance around the globe. IRAC Resistant Pest Management Newsletter. 4:5-8.

73. Lindbo, J. A., and Dougherty, W. G. 1992. Pathogen-derived resistance to a potyvirus: immune and resistance phenotypes in transgenic tobacco expressing altered forms of a potyvirus coat protein nucleotide sequence. Mol. Plant-Microbe Interact. 5:144-153.

74. Luisoni, E., Milne, R. G., and Vecchiati, M. 1992. Purification and serology of tomato yellow leaf curl geminivirus. Pages 56-58 in: Recent Advances in Vegetable Virus Research. 7th Conference ISHS, Vegetable Virus Working Group, Athens, Greece.

75. Macintosh, S., Robinson, D. J., and Harrison, B. D. 1992. Detection of three whitefly-transmitted geminiviruses occurring in Europe by tests with heterologous monoclonal antibodies. Ann. Appl. Biol. 121:297-03.

76. Maiti, I. B., Murphy, J. F., Shaw, J. G., and Hunt, A. G. 1993. Plants that express a potyvirus proteinase gene are resistant to virus infection. Proc. Natl. Acad. Sci. USA 90:6110-6114.

77. Makkouk, K. M. 1978. A study on tomato viruses in Jordan Valley with special emphasis on tomato yellow leaf curl. Plant Dis. Reptr. 62:259-262.

78. Makkouk, K. M., and Laterrot, H. 1983. Epidemiology and control of tomato yellow leaf curl virus. Pages 315-321 in: Plant Virus Epidemiology. R. T. Plumb and J. M. Thresh, eds., Blackwell Scientific Publications, Oxford, UK.

79. Makkouk, K. M., Shehab, S., and Majdalan, S. E. 1979. Tomato yellow leaf curl: incidence yield losses and transmission in Lebanon. Phytopathol. Z. 96:263-267.

80. Mansour, A., and Al-Musa, A. 1992. Tomato yellow leaf curl virus: host range and virus-vector relationships. Plant Pathol. 41:122-125.

81. Mazyad, H. M., Hassan, A. A., Nakhla, M. K., and Moustafa, S. E. 1982. Evaluation of some wild Lycopersicon species as sources of resistance to tomato yellow leaf curl virus. Egypt. J. Hort. 9:241-246.

82. Mazyad, H. M., Nakhla, M. K., El-Amrety, A. A., and Dos, S. A. 1986. Further studies on the epidemiology of tomato yellow leaf curl virus in Egypt. Acta Hortic. 190:121-130.

83. Mazyad, H. M., Omar, F., Al Taher, K., and Salah, M. 1979. Observations on the epidemiology of tomato yellow leaf curl disease on tomato plants. Plant Dis. Reptr. 63:695-698.

84. Mazyad, H. M., Peters, D., and Maxwell, D. 1994. Tomato yellow leaf curl virus in Egypt: epidemiological and management aspects. 1st International Symposium on Geminiviruses, Almeria, Spain.

85. McGlashan, D., Polston, J. E., and Bois, D. 1994. Tomato yellow leaf curl geminivirus in Jamaica. Plant Dis. 78:1219.

86. Mehta P., Wyman, J. A., Nakhla, M. K., and Maxwell, D. P. 1994. Polymerase chain reaction detection of viruliferous Bemisia tabaci (Homoptera:Aleyrodidae) with two tomato-infecting geminiviruses. J. Econ. Entomol. 87:1285-1290.

87. Mehta P., Wyman, J. A., Nakhla, M. K., and Maxwell, D. P. 1994. Transmission of tomato yellow leaf curl geminivirus by Bemisia tabaci (Homoptera:Aleyrodidae). J. Econ. Entomol. 87:1291-1297.

88. Michelson, I., Zamir, D., and Czosnek, H. 1994. Accumulation and translocation of tomato yellow leaf curl virus (TYLCV) in a Lycopersicon esculentum breeding line containing the L. chilense TYLCV tolerance gene Ty-1. Phytopathology 84:928-933.

89. Moustafa, S. E. 1991. Tomato cultivation and breeding program for tomato yellow leaf curl virus. Pages 6-8 in: Resistance of the Tomato to TYLCV, Proceedings of the seminar of EEC contract TS2-A-055 F (CD) partners. H. Laterrot, and C. Trousse, ed. INRA-Station de'Amelioration des Plantes Maraicheres, Montfavet-Avignon, France.

90. Moustafa, S. E., and Hassan, A. A. 1994. Tomato cultivar evaluation with emphasis on tomato yellow leaf curl virus tolerance in Egypt. Tomato Leaf Curl Newsletter 5:3.

91. Moustafa, S. E., and Nakhla, M. K. 1990. An attempt to develop a new tomato variety resistant to tomato yellow leaf curl virus (TYLCV). Assiut J. Agric. Sci. 21:167-184.

92. Muniyappa, V., Swanson, M. M., Duncan, G. H., and Harrison, B. D. 1991. Particle purification, properties and epitope variability of Indian tomato leaf curl geminivirus. Ann. Appl. Biol. 118:595-604.

93. Nakhla, M. K., El-Hammady, M., and Mazyad, H. M. 1978. Isolation and identification of some viruses naturally infecting tomato plants in Egypt. Pages 1042-1051 in: 4th Conf. Pest Control, NRC, Cairo, A.R.E.

94. Nakhla, M. K., Maxwell, D. P., Martinez, R. T., Carvalho, M. G., and Gilbertson, R. L. 1994. Widespread occurrence of the Eastern Mediterranean strain of tomato yellow leaf curl geminivirus in tomatoes in the Dominican Republic. Plant Dis. 78:926.

95. Nakhla, M. K., Mazyad, H. M., and Maxwell, D. P. 1993. Molecular characterization of four tomato yellow leaf curl virus isolates from Egypt and development of diagnostic methods. Phytopathol. Medit. 32:163-173.

96. Nakhla, M. K., Mehta, P., Wyman, J. A., Stout, J., and Maxwell, D. P. 1994. Construction of infectious clones of tomato yellow leaf curl geminivirus from polymerase chain reaction-amplified fragments. The American Society for Virology 13th Annual Meeting July 1994, Madison, WI.

97. Nakhla, M. K., Rojas, M. R., McLaughlin, W., Wyman, J., Maxwell, D. P., and Mazyad, H. M.. 1992. Molecular characterization of tomato yellow leaf curl virus from Egypt. Plant Dis. 76:538.

98. Namba, S., Ling, K., Gonsalves, C., Slightom, J. L., and Gonsalves, D. 1992. Protection of transgenic plants expressing the coat protein gene of

watermelon mosaic virus II or zucchini yellow mosaic virus against six potyviruses. Phytopathology 82:940-946.

99. Navot, N., Ber, R., and Czosnek, H. 1989. Rapid detection of tomato yellow leaf curl virus in squashes of plants and insect vectors. Phytopathology 79:562-564.

100. Navot, N., Pichersky, R., Zeidan, M., Zamir, D., and Czosnek, H. 1991. Tomato yellow leaf curl virus: a whitefly-transmitted geminivirus with a single genomic component. Virology 185:151-161.

101. Navot, N., Zeidan, M., Pichersky, E., Zamir, D., and Czosnek, H. 1992. Use of the polymerase chain reaction to amplify tomato yellow leaf curl virus DNA infected plants and viruliferous whiteflies. Phytopathology 82:1199-1202.

102. Nitzany, F. E. 1975. Tomato yellow leaf curl virus. Phytopathol. Medit. 14:127-129.

103. Noris, E., Hidalgo, E., Accotto, G. P., and Moriones, E. 1994. High similarity among the tomato yellow leaf curl virus isolates from the West Mediterranean Basin: the nucleotide sequence of an infectious clone from Spain. Arch. Virol. 135:165-170.

104. Nour-Eldin, F., Mazyad, H. M., and Hassan, M. S. 1969. Tomato yellow leaf curl virus disease. Agric. Res. Rev. 47:49-54.

105. Pacheco-Covarrubias, J. J. 1995. Response of ninfal instars of *Bemisia* spp. to imidacloprid. Page 52 in: Proc. of Third Annual Progress Review of The 5-Year National Research and Action Plan for Development of Management and Control Methodology for Silverleaf Whitefly, San Diego, CA.

106. Padidam, M., Beachy, R. N., and Fauquet, C. M. 1995. Classification and identification of geminiviruses using sequence comparisons. J. Gen. Virol. 76:249-263.

107. Padidam, M., Beachy, R. N., and Fauquet, C. M. 1995. Tomato leaf curl geminivirus from India has a bipartite genome and the coat protein is not essential for symptom development. J. Gen. Virol. 76:25-35.

108. Perring, T. M., Cooper, A. D., Rodriguez, R. J., Farrar, C. A., and Bellows, T. S., Jr. 1993. Identification of a whitefly species by genomic and behavioral studies. Science 259:74-77.

109. Pilowsky, M., and Cohen, S. 1974. Inheritance of resistance to tomato yellow leaf curl virus in tomatoes. Phytopathology 64:632-635.

110. Pilowsky. M., and Cohen, S. 1990. Tolerance to tomato yellow leaf curl virus derived from *Lycopersicon peruvianum*. Plant Dis. 74:248-250.

111. Pilowsky, M., Cohen, S., Ben-Joseph, R., Shlomo, A., Chen, L., Nahon, S., and Krikun, J. 1989. TY-20 a tomato cultivar tolerant to tomato yellow leaf curl virus. Hassadeh 69:1212-1215.

112. Polston, J. E., Bois, D., Serra, C. A., and Concepcion, S. 1994. First report of tomato yellow leaf curl-like geminivirus in the Western Hemisphere. Plant Dis. 78:831.

113. Prabhaker, N., Coudriet, D. L., and Toscano, N. C. 1988. Effect of synergists on organophosphate and permethrin resistance in sweetpotato whitefly (Homoptera:Aleyrodidae). J. Econ. Entomol. 81:34-39.

114. Puri, S. N., Bhosle, B. B., Ilyas, M., Butler, G. D., Jr., and Henneberry, A. T. J. 1994. Detergents and plant-derived oils for control of the sweetpotato whitefly on cotton. Crop. Prot. 13:45-48.

115. Reyd, D. 1993. Nonwovens in agriculture: a way to prevent insect transmitted viruses. Tomato Leaf Curl Newsletter. 3:2.

116. Roberts, I. M., Robinson, D. J., and Harrison, B. D. 1984. Serological relationships and genome homologies among geminiviruses. J. Gen. Virol. 65:1723-1730.

117. Rochester, D. E., Beachy, R. N., and Fauquet, C. M. 1993. Geminivirus nomenclature: the need to set taxonomic standards. Arch. Virol. 132:221-224.

118. Rochester, D. E., DePaulo, J. J., Fauquet, C. M., and Beachy, R. N. 1994. Complete nucleotide sequence of the geminivirus tomato yellow leaf curl virus, Thailand isolate. J. Gen. Virol. 75:477-485.

119. Rochester, D. E., Kositratana, W., and Beachy, R. N. 1990. Systemic movement and symptom production following agroinoculation with a single DNA of tomato yellow leaf curl geminivirus (Thailand). Virology 178: 520-526.

120. Rojas, M. R., Gilbertson, R. L., Russell, D. R., and Maxwell, D. P. 1993. Use of degenerate primers in the polymerase chain reaction to detect whitefly-transmitted geminiviruses. Plant Dis. 77:340-347.

121. Rom, M., Antignus, Y., Gidoni, D., Pilowsky, M., and Cohen, S. 1993. Accumulation of tomato yellow leaf curl virus DNA in tolerant and susceptible tomato lines. Plant Dis. 77:253-257.

122. Russo, M., Cohen, S., and Martelli, G. P., 1980. Virus-like particles in tomato plants affected by the yellow leaf curl disease. J. Gen. Virol. 49:209-213.

123. Sharaf, N. 1986. Chemical control of *Bemisia tabaci*. Agric. Ecosyst. Environ. 17:111-127.

124. Sharaf, N. S., and Allawy, T. F. 1981. Control of *Bemisia tabaci* Genn., a vector of tomato yellow leaf curl virus disease in Jordan. J. Plant Dis. Prot. 87:123-131.

125. Singh, S. J., Sastry, K. S. , and Sastry, K. S. 1975. Effect of alternate spraying of insecticides and soil on the incidence of tomato leaf curl virus. Pesticides, India 9:45-46.

126. Singh, S. J., Sastry, K. S., and Sastry, K. S. 1979. Efficacy of different insecticides and oil in the control of leaf curl virus disease of chilies. J. Plant Dis. Prot. 86:253-256.

127. Snyder, J. C., and Carter, C. D. 1985. Trichomes on leaves of *Lycopersicon hirsutum*, *L. esculentum* and their hybrids. Euphytica 34:53-64.

128. Stenger, D. C. 1994. Strain-specific mobilization and amplification of a transgenic defective-interfering DNA of the geminivirus beet curly top virus. Virology 203:397-402.

129. Swanson, M. M., Brown, J. K., Poulos, B. T., and Harrison, B. 1992. Genome affinities and epitope profiles of whitefly-transmitted geminiviruses in the Americas. Ann. Appl. Biol. 121:285-296.

130. Thomas, J. E., Massalski, P. R., and Harrison, B. D. 1986. Production of monoclonal antibodies to African cassava mosaic virus and differences in their reactivities with other whitefly-transmitted geminiviruses. J. Gen. Virol. 67:2739-2748.

131. Wilson, T. M. A. 1993. Strategies to protect crop plants against viruses: pathogen-derived resistance blossoms. Proc. Natl. Acad. Sci. USA 90:3134-3141.

132. Yassin, A. M. 1975. Epidemics and chemical control of leaf curl virus disease of tomato in the Sudan. Exp. Agric. 11:161-165.

133. Zakay, Y., Navot, N., Zeidan, M., Kedar, N.,

Rabinowitch, H., Czosnek, H., and Zamir, D. 1991. Screening *Lycopersicon* accessions for resistance to tomato yellow leaf curl virus: presence of viral DNA and symptom development. Plant Dis. 75:279-281.

134. Zamir, D., Michelson, I., Zakay, Y., Navot, N., Zeidan, M., Sarfatti, M., Eshed, Y., Harel, E., Pleban, T., Van-Oss, H., Kedar, N., Rabinowitch, H., and Czosnek, H. 1994. Mapping and introgression of a tomato yellow leaf curl virus tolerance gene Ty-1. Theor. Appl. Genet. 88:141-146.

135. Zamir, D., Zakay, Y., Zeidan, M., and Czosnek, H. 1991. Combating the tomato yellow leaf curl viruses in Israel: the agrotechnical and genetics approaches. Pages 9-14. in: Resistance of the Tomato to TYLCV. Proceedings of the seminar of EEC contract DGXII-TS2-A-055 F (CD) partners. H. Laterrot, and C. Trousse, ed. INRA-Station d'Ameliioration des Plantes Maraicheres, Montfavet-Avignon, France.

136. Zeidan, M., and Czosnek, K. 1991. Acquisition of tomato yellow leaf curl virus by the whitefly *Bemisia tabaci*. J. of Gen. Virol. 72:2607-2614.

137. Zeidan, M., and Czosnek, K. 1994. Replication of tomato yellow leaf curl virus in its insect vector, the whitefly *Bemisia tabaci*. *Bemisia* Newsletter 8:52.

CHAPTER 44

Present Status of Controlling Potato Leafroll Virus

U. Jayasinghe and L.F. Salazar

Among the more than twenty-seven viruses that have been reported to infect potatoes, potato leafroll (PLRV), potato virus Y (PVY), and potato virus X (PVX) are the most damaging as well as being distributed worldwide. Yield reduction in potatoes by PLRV may reach 80-90% in susceptible cultivars, but an even greater loss may be expected when PLRV occurs in simultaneous infections with PVX or PVY (30).

PLRV is a Luteovirus (30) having isometric particles with c. 26 nm in diameter. The virus is restricted to the phloem sieve tubes and companion cells of infected plants (6,40) and can be transmitted by more than ten different aphid species, of these, *Myzus persicae* Sulz. its most efficient natural vector (10,22).

As with other potato viruses, production and use of healthy planting materials, control of vectors, and use of resistant cultivars are the most important measures for PLRV control. In developing countries, planting resistant cultivars is the preferred alternative because healthy planting materials are in short supply and insecticide applications are very costly. Developed country interest in the use of virus resistance has also increased during recent years, because environmental concern over the use of pesticides has increased.

PLRV RESISTANCE

Resistance to PLRV has been reported to be controlled polygenically by minor genes (11,37). This type of resistance has been found in both cultivated and wild *Solanum* species, though most determinations have been through field observations and in greenhouse experiments which use aphids to inoculate the virus (16,23,26,38,41). Until the development of sensitive methods of virus detection such as the enzyme-linked immunosorbent assay (ELISA) (15) only resistance to virus infection was identified. This is a relative type of resistance which can only be determined in field exposure trials by comparing the percentage of infection in the cultivar tested with that of a known susceptible cultivar in the trial (16).

The improvement in the sensitivity of the methodology for detection of PLRV allowed us to identify different components contributing towards the general resistance to the virus such as: resistance to virus infection, resistance to virus multiplication, hypersensitivity or intolerance, resistance to virus translocation, vector antixenosis and antibiosis

In cultivars exhibiting resistance to PLRV, one or a combination of two or more of these components may be present (Table 1).

Resistance to Virus Infection

Potato genotypes that exhibit this component of the resistance do not become easily infected when inoculated with viruliferous aphids. Resistance to infection can be considered in two aspects: 1) Direct resistance where the resistance of the genotype is acting on the virus; and 2) indirect resistance where the resistance to PLRV is due to the resistance of the genotype to the vector.

Many attempts have been made to determine the levels of resistance in potato genotypes under greenhouse conditions (14). However, the results obtained in these experiments did not reflect the resistance of the genotypes under field conditions. Field exposure trials greatly depend on the natural aphid population, and trials frequently have to be repeated to obtain reliable results. The need to perform field trials to

Table 1. Reaction of some selected potato genotypes to components of resistance to PLRV

Genonotype	Resistance to Infection	Resisistance to Multiplication	Resistance to Translocation	Antixenosis	Symptoms Upon Infection
DTO-28	No	No	No	No	Yes
Serrana	Yes	No	No	No	Yes
ariva	Yes	No	No	No	Yes
P. crown	No	Yes	No	No	Yes
B-71-240.2	Moderate	Yes	No	Yes	No
T. condemayta	Yes	No	No	Yes	?
S. acaule	Yes	Yes	Yes	Yes	?

determine PLRV resistance increases the time frame of a breeding program by several years. Considering the time needed and the cost of field exposure trials, attempts were made at CIP (Centro Internacional de la PaPa = the International Potato Center) to develop a valid greenhouse screening method. This method has allowed us to obtain results comparable to those from field exposure trials (24). The method consists of inoculating five plants of the clone to be tested with either 25 or 50 viruliferous M. persicae aphids and allowing them to feed for three days. Two to three weeks later, plants are tested by ELISA for PLRV infection.

Under the conditions of the method described above, clones which have resistance to PLRV infection do not become infected when inoculated with 25 or 50 viruliferous aphids. Those with moderate resistance to PLRV are infected with 50 but not with 25 aphids, whereas the susceptible ones are infected with 25 aphids. Recent field evaluations confirmed the greenhouse results, but it was found that moderate resistance was not exhibited under field conditions (20,23). This was probably due to the inability to control the aphid population in field exposure trials to the extent that it can be done in the greenhouse.

Our experience has shown that the results from trials testing for resistance to virus infection obtained in one region are not applicable to other regions. The most important reasons for this lack of resistance stability seem to be the variability of aphid populations, the transmission efficiency of different aphid species and biotypes, and the influence of environmental conditions. Variability in aphid populations is a common factor. For instance, in certain potato growing areas of India

(28) and in Ica, a town located 340 Km south of Lima, Peru, the populations of M. persicae are negligible in comparison to other aphid vector species. However, the extremely high population of less efficient vectors such as Aphis gossypii compensate for the low population of M. persicae and cause considerable spread of the disease. The efficiency of PLRV transmission by different biotypes of aphids of the same species was also found to vary greatly between localities in one region (9).

Temperature seems to be the most important environmental factor that affects efficiency of transmission by aphids. Cervantes (13) found that high temperature regimes (above 30°C) during the incubation period prevented the transmission of PLRV by M. persicae, but not as dramatically during the transmission and acquisition periods of the virus.These results, however, do not indicate whether the effect of high temperature is on the virus or on the vector. Some observations about the elimination of the virus from tubers stored under high temperature regimes might suggest that effect on the virus is the main reason.

Resistance to Virus Multiplication

By means of ELISA, it is possible to accurately determine the concentration of PLRV in infected tissue. In comparative experiments with a number of potato clones under greenhouse conditions, it was found that in a known susceptible clones such as DTO-28 (Figure 1) the PLRV concentration reached a maximum 2-3 weeks following inoculation. Then, it drops and maintains a lower level for the rest of the

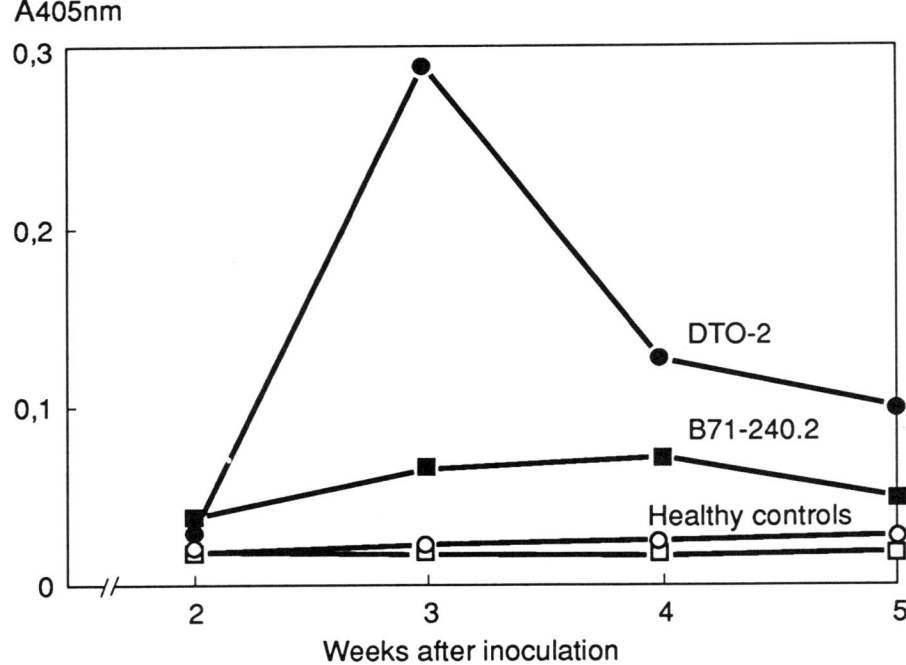

Figure 1. Concentration of PLRV in a susceptible (DTO-2) and a resistant (B71-240.2) potato clone over a period of five weeks after inoculation.

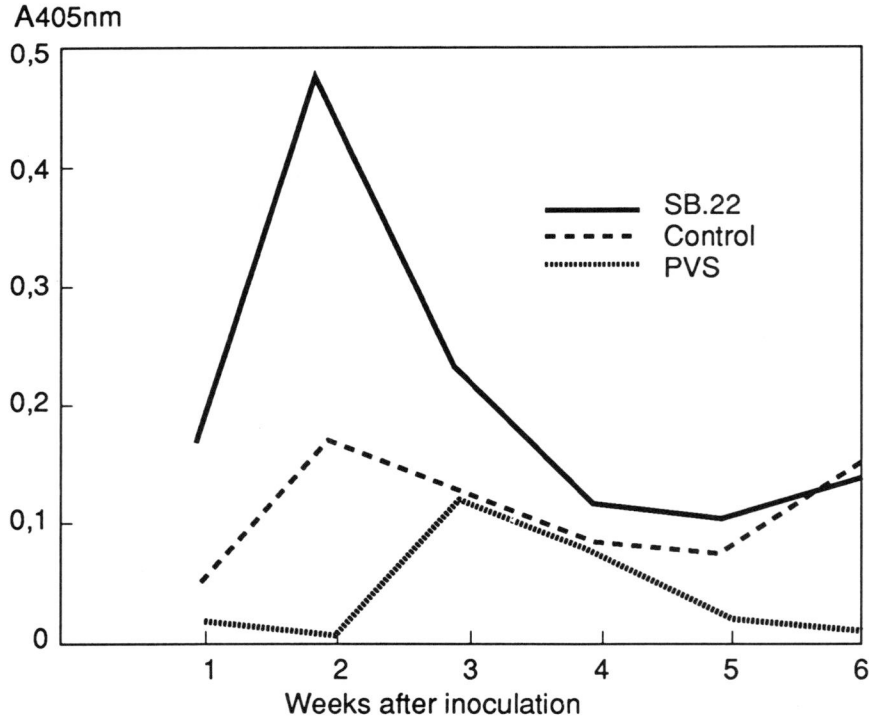

Figure 2. Increase of PLRV concentration in plants of the potato clone B71-240.2 also infected with SB 22 (potato yellowing virus, 19) but not in those infected with PVS. Control plants were only inoculated with PLRV.

vegetative cycle. In some other clones, notably B-71-240.2, the maximum concentration of PLRV is attained much later and this concentration is 5 to 10 times less than in the susceptible clones. This characteristic is constant and does not seem to be affected by environmental conditions. The low concentration of the virus in clones B-71-240.2 is confirmed by the very low amount of virus recovered from infected tissue following a purification protocol. Clones having this characteristic are said to have resistance to virus multiplication. They do not show symptoms under field conditions and behave as though they are tolerant to PLRV . In addition, these clones are poor sources of the virus (6). In the case of B-71-240.2, no resistance to virus infection has been demonstrated indicating that the two components of the resistance (infection and multiplication) are independent mechanisms. In fact, Barker (2,5) reported that resistance to virus multiplication is governed by a single dominant gene in *S. tuberosum*.

As in the case of resistance to infection, the component of resistance to virus multiplication is also modified by interaction with other viruses. In plants of clone B-71-240.2 infected with potato yellowing virus (19) or, the concentration of PLRV is higher than in plants infected with PLRV alone or with PVS (Figure 2) PVX and PVY also increase the concentration of PLRV. Barker (3) reported that in plants infected with both PVX and PLRV a larger portion of companion cells also become infected with PLRV. In single infections, PLRV very rarely infects these cells. Other work has shown that in plants infected with PVX or PVY, PLRV can infect mesophyll and parenchyma cells and reaches a higher concentration than in plants infected with PLRV alone (4). The increase of PLRV in doubly infected plants may have important epidemiological implications and might have been the reason that the breaking of the resistance of cultivars developed and tested in other countries. For instance, previous results have shown that the levels of resistance to PLRV infection in cv. Mariva are higher than in cv. Serrana. Under field conditions, however, cv. Serrana maintains its resistance to PLRV whereas cv. Mariva is easily degenerated by PLRV. Upon examination of plants, it was found that all Mariva plants infected with PLRV were also infected with either PVX or PVY or both. The reason for this difference is easily understood because cv. Serrana shows high levels of hypersensitivity to both PVX and PVY. When cv. Serrana is grown in warmer regions where hypersensitivity genes do not operate, it becomes infected with PLRV due to pre-infection of plants with PVX or PVY (36,39).

Hypersensitivity

Primarily or secondarily PLRV-infected plants of certain Polish and German cultivars, such as Apta and Carla, react with wilting, necrosis of the sprouts, premature death of the plants, and lack of tuber setting (12). These symptoms are different from those seen in the field with *S. tuberosum* ssp. *tuberosum* or ssp. *andigena*. Phloem necrosis can occur in certain clones infected with PLRV (17). However, in clones with hypersensitivity genes to PLRV, phloem necrosis is severe and inmediate. These clones are also considered "self eliminating" because the severe necrosis of infected plants and sprouts cause tubers not to germinate. Hypersensitivity to PLRV is controlled by a single dominant gene which is modified by minor genes (44).

Resistance to Virus Translocation

Upon infecting a plant cell, a virus first multiplies to a determined concentration and then invades neighboring cells through the plasmodesmata and distant tissues by translocation through the vascular tissues of the plants. Initial PLRV infection occurs in few companion cells and later invades distant tissue using the sieve tubes. Experiments conducted at CIP indicate that the initial direction of PLRV translocation in a potato plant is always towards the apical bud of the plant irrespective of age or of the infection site. Later it moves down to the roots. The rate of translocation depends on the speed at which the virus multiplies in the phloem companion cells and its entry into the sieve tubes. Therefore, in potato cultivars having resistance to PLRV multiplication, a slower translocation rate also seems to occur.

Experiments conducted under screenhouse conditions at CIP did not show any difference in the rate of PLRV translocation in cultivated potato. However, in wild *Solanum* species such as in *S. acaule* in which a strong resistance to virus multiplication has been found, PLRV translocation is also strongly retarded.

Antibiosis and Antixenosis

These two components of resistance are available in the plant and affect the vector and not the virus directly. Antibiosis includes all adverse effects exerted by the plant which alter the biology of the aphid vector, including any factor which reduces aphid growth rates, the number of nymphs produced, or even aphids death. The advantage of this component of resistance is primarily its negative effect on the aphid population build-up within a given crop by preventing or reducing the spread of the disease. The insect trapping glandular trichomes in some potato species (21) belong to this type of resistance.

Antixenosis, or no preference, is the repelling action of the plant toward the insect. This may be due to toxins or volatile compounds or non-glandular trichomes present in the plant. Field or screenhouse-grown plants which exhibit this component of resistance usually do not allow aphids to feed on them; or, if the effect is not too strong, a few aphids may feed on the plants but aphid establishment is prevented. In the Peruvian cultivar Tomasa Condemayta this resistance component it is very strongly exhibited. Aphids placed on these plants move away immediately. No aphids can be detected in field grown plants and over many years of observation we have very rarely found any PLRV-infected plants. However, it is common to observe plants infected with PVY, which indicates that aphids do probe on Tomasa Condemayta plants, but do not remain in them long enough to transmit PLRV. Studies using aphid feeding monitors and cytological studies indicate that during feeding aphids find it difficult to reach the phloem (25) the site of PLRV establishment and multiplication. Results obtained at the Scottish Crop Research Institute, Scotland (H. Barker, personal communication, 1992) showed that antixenosis in cv. Tomasa Condemayta operates shortly after initial aphid contact and can be overcome by forced confinement of aphids on this cultivar.

In limited experimentation antixenosis in cv. Tomasa Condemayta has been shown not to be a heritable character. However, in some accessions of S. acaule having this type of resistance, the characteristic is transmitted to the progeny of crosses between S. acaule and S. tuberosum. It is probable that the antixenosis in this wild species is different in nature from that of cv. Tomasa Condemayta.

RESISTANCE COMPONENTS AND THE OVERALL RESISTANCE TO PLRV

Analyses of wild *Solanum* species and cultivated potatoes that exhibit resistance to PLRV indicate that they contain one or more of the resistance components described above. These results have also shown that the level of resistance conferred by each component individually is rather low in comparison to the levels of resistance in potato to PVX and PVY. Therefore, a genotype that possesses more than one of the components of resistance exhibits a higher level of overall resistance to PLRV than a genotype having only one component. *S. acaule* is one of the few genotypes which seems to have all the components of resistance known for PLRV (Table 1).

Different components of resistance are usually inherited independently such as resistance to infection and antixenosis in *S. acaule,* or the resistances to multiplication and infection of several genotypes. Because resistance to virus multiplication results in limited virus accumulation, genotypes having this type of resistance are poor sources of inoculum which in turn reduces virus spread. The combination of resistance to infection and to multiplication is, therefore, seen as a possibility for developing virtually PLRV- immune genotypes.

Due to modification in the levels of resistance to PLRV by pre-infecting plants with other viruses-notably PVX and PVY, breeding for resistance to PLRV should be based on populations with immunity to PVX and PVY. A number of promising progenitors with immunity to PVX and PVY, including some having a triplex loci for immunity to these viruses, are now available at CIP. Those genotypes having immunity genes in a triplex condition will reduce the screening efforts for both viruses. The cultivars that serve as the best sources of PLRV resistance now have been crossed with those progenitors having immunity to PVX and PVY (34).

The importance of the vector in the control of the disease has already been discussed. Usually the vector population in cooler, high altitude climatic regions of the world are lower than in the warmer, low altitude areas. The efficiency of the same species of aphids to transmit PLRV in these two regions can also be very different (10). World potato growing regions

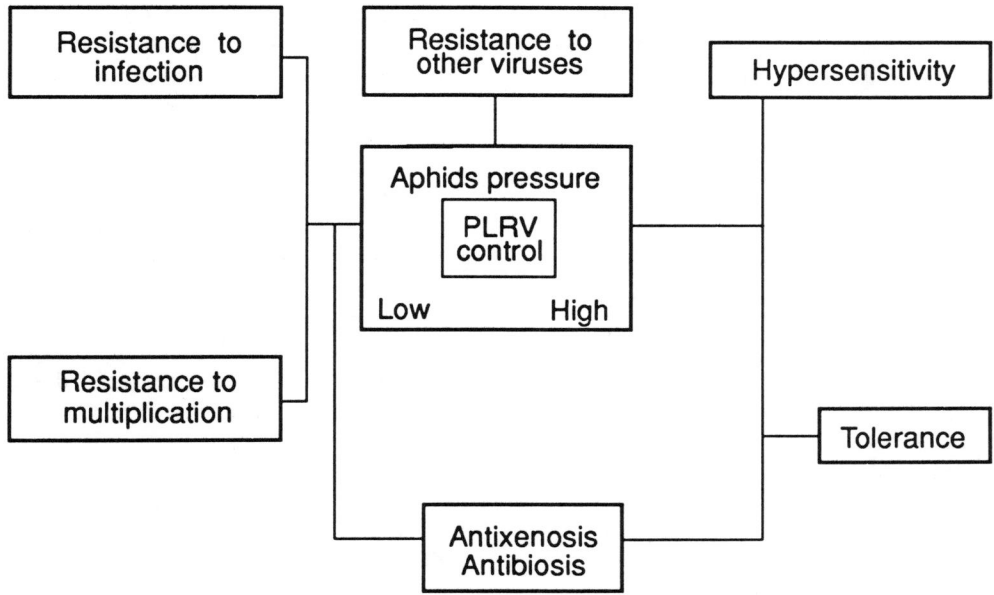

Figure 3. The use of different components of the resistance to control PLRV are dependent on the aphid pressure in the region.

can be broadly divided into low aphid and high aphid population areas. In regions where the aphid population is low, adequate control of PLRV can be achieved using genotypes having a combination (Figure 3) of hypersensitivity and aphid resistance (antixenosis and antibiosis). In regions where the aphid population is high, a different strategy is required because the use of hypersensitivity genes can lead to destruction of the whole crop due to the necrosis developed in these cultivars. A combination of resistance to infection, multiplication, and antixenosis in one genotype will provide adequate control of PLRV disregarding the aphid pressure in the region.

The success of a breeding program for resistance to viruses greatly depends on the methodology adopted to screen for resistance (29). This is especially important for PLRV which in comparison to other viruses has a tendency to remain in low concentration in the affected tissue. The sensitivity of the detection techniques used as well as their efficiency and cost, are determining factors in the success of developing resistance to PLRV.

ENGINEERING RESISTANCE TO PLRV

Transformation of plants with sequences encoding virus coat proteins has been used in several plant species to confer resistance to a large number of plant viruses (8). The knowledge recently accumulated on the genome organization of PLRV (31,33,42) has been useful for developing PLRV-resistant transgenic plants. Indeed, potato plants transformed with sequences encoding the coat protein of PLRV showed virus resistance (27,43), although little or no coat protein was synthesized.

The use of other viral genes such as viral replicase or protease genes may also confer resistance to PLRV. Likewise, the discovery of self-cleavage reactions in RNA catalyzed by an RNA molecule (18) opened interest to synthesizing ribozymes designed to cleave virus nucleic acid sequences. Bahner *et al.* (1) synthesized two ribozymes that cleaved the positive strand of PLRV RNA within the regions known to encode the viral coat protein and the predicted RNA polymerase gene. These ribozymes have been shown to work in vitro and their use for the development of transgenic potato plants resistant to PLRV might now be possible.

The mechanism underlying transgenic resistance is not well known. However, Barker *et al.* (7) have shown that transgenic and host-mediated types of resistance (resistance to virus multiplication) restrict virus multiplication and distribution in a similar fashion.

SEED PRODUCTION

Production of virus-free seed potatoes is, at present, the most adequate method of virus control. Among the potato viruses, PLRV and PVY are the most important (35), requiring the most stringent levels of elimination at different seed multiplication stages. Mother plants and first stages of basic or elite seed should be totally virus-free, and only 1-3% infection is allowed in later seed stock multiplications in most seed programs.

Virus detection technology is a critical factor in seed multiplication. Symptom observation is not sensitive enough to detect latent infections or to detect PLRV in situations where virus-cultivar-environment interactions make symptoms indistinguishable or cause different diseases. This is especially important during the production of basic or elite seed. Fortunately, today there are highly sensitive detection methods, hence ELISA is playing a very important role in PLRV detection although nucleic acid spot hybridization (NASH) and polymerase chain reaction might offer better possibilities for detection in the future.

In large extensions of seed lots, as is the case for certified grade seed, virus detection is carried out by visual observation for symptoms. Therefore, well trained inspectors are fundamental to eliminating PLRV-infected plants, as well as observing adequate cultural practices and vector control to prevent infection.

CONCLUDING REMARKS

Controlling viruses which are well adapted to a crop is not an easy task. Because PLRV spreads by aphids and most currently grown cultivars are susceptible, PLRV acquires more importance than other aphid-transmitted viruses in potato. Application of insecticides helps to reduce spread of PLRV. But this practice to be effective in seed lots must be applied at short intervals, a process which increases the cost of production. In

addition, the world environment protection attitude calls for reduction of pesticide use. Under this circumstances genetic resistance seems to be the most appropriate approach even though complete immunity has not been achieved. Transgenic resistance for PLRV has not yet demonstrated to be better than host - mediated resistance. In fact the most strongly expressed forms of host resistance to viruses are superior to transgenic resistance (32). Other disadvantages of preferring transgenic resistance over host-mediated resistance are the high cost of transformation and the uncertainties on the durability and effectiveness against diverse strains of the virus. In addition there is a risk in that, if protein is produced, transgenically expressed coat protein might encapsidate heterologous, possibly pathogenic, RNA (32). This aspect should deserve detailed research since Salazar *et al.* (39) have found that PLRV virions act as helpers of PSTVd transmission by aphids.

We feel that the available types and levels of host - mediated resistance to PLRV are adequate and sharing them among different breeding programs in the world should be encouraged.

ACKNOWLEDGMENTS

The authors wish to thank Miss Violeta Caballeo for typing the manuscript and Eng. Carlos Chuquillanqui for technical assistance.

REFERENCES

1. Bahner, I., Lamb, J., Mayo, M.A., and Hay, R.T. 1990. Expression of the genome of potato leafroll virus: readthrough of the coat protein termination codon *in vivo*. J. Gen. Virol. 71: 2251-2256.
2. Barker, H. 1987. Multiple components of the resistance of potatoes to potato leafroll virus. Ann. Appl. Biol. 111:641-648.
3. Barker, H. 1987. Invasion of non-phloen tissue in *Nicotiana clevelandii* by potato leafroll luteovirus is enhanced in plants also infected with potato Y potyvirus. J. Gen. Virol. 68: 1223-1227.
4. Barker, H. 1987. Potato leafroll virus multiplies in non-phloem of *Nicotiana clavelandii*. Page 164 in: Meeting of the Virology Section of the European Association for Potato Research, Netherlands 30 (1).
5. Barker, H., and Harrison, B.D. 1985. Restricted multiplication of potato leafroll virus in resistant potato genotypes. Ann. Appl. Biol. 107: 205-212.
6. Barker, H., and Harrison, B.D. 1986. Restricted distribution of potato leafroll virus antigen in resistant potato genotypes and its effects on transmission of the virus by aphids. Ann. Appl. Biol. 109: 595-604.
7. Barker, H., Reavy, B., Kumar, A., Webster, K., and Mayo, M.A. 1992. Restricted virus multiplication in

potatoes transformed with the coat protein gene of potato leafroll luteovirus: similarities with a type of host-mediated resistance. Ann. of Appl. Biol. 120: 55-64.

8. Beachy, R.N., Loesh-Fries, S., and Turner, N.E. 1990. Coat protein-mediated resistance against virus infection. Annu. Rev. Phytophol. 28:451-474.

9. Brignetti, G. 1991. Eficiencia de diferentes especies de áfidos en la transmision del virus del enrollamiento de la hoja de la papa (PLRV), Tesis Biol., Universidad Nacional Agraria, Lima, Perú. 141Pages.

10. Brigneti, G., and Jayasinghe, U. 1992. Efficiency of different aphid species in transmitting potato leafroll virus. Fitopatologia 27: 1-7.

11. Brown, C.R. 1984. Breeding virus-resistant potato cultivars for developing countries. CIP Circular, Peru. 12:1.

12. Butkiewicz, H. 1978. Pages 5-37 in: Intolerance to Potato Leafroll Virus (PLRV) Ocurring in Potato Plants. Ziemniak, Poland.

13. Cervantes, M. 1988. Efecto de la alta temperatura en la transmisión del virus del enrollamiento de la hoja de la papa y su localización intracelular en el áfido *Myzus persicae* ulzer, Tesis, Biol., presentada en la Universidad Ricardo Palma, Lima, Perú. 78 Pages.

14. Chuquillanqui, C., and Jones, R.A.C. 1980. A rapid technique for assessing the resistance of families of potato seedlings to potato leafroll virus. Potato Research, 23: 121-128.

15. Clark, M.F., and Adams, A.N. 1977. Characteristics of the microplate method of enzyme-linked inmunosorbent assay for the detection of plant viruses. J. of Gen. Virol. 34: 475-483.

16. Davidson, T.M.W. 1973. Assessing resistance to leafroll in potato seedlings. Potato Research 16: 99-108.

17. Douglas, D.R., and Pavek, J.J. 1972. Net necrosis of potato tubers associated with primary, secondary and tertiary infection of leafroll, Am. Potato J. 49(9): 330-333.

18. Foster, A.C., and Symons, R.H. 1987. Self-cleavage of plus and minus RNA's of a virusoid and a structural model of active sites. Cell 49: 211.

19. Fuentes, S. 1992. Identificación, características y distribución de un virus baciliforme aislado de papa, Tesis M. Sc, presentada en la Universidad Nacional Agraria, Lima. 184 Pages.

20. Garcilazo, J.M. 1990. Estabilidad de la resistencia al virus del enrollamiento de la hoja de la papa, Tesis Ing. Agr. presentada en la Universidad Nacional Agraria San Luis Gonzaga, Ica, Perú. 85 Pages.

21. Gibson, R.W. 1971. Glandular hair providing resistance to aphids in certain wild species. Ann. Appl. Biol. 68:113-119.

22. Harrison, B.D. 1958. Ability of single aphids to transmit both avirulent and virulent strains of potato leafroll virus. Virology 6: 278-286.

23. Hooker, W.J. 1982. Personal communication.

24. Ingante, P.H. 1991. Estabilidad de la resistencia al PLRV enclones y variedades de papa expuestas por segunda vez en campos, Tesis Ing. Agr., Universidad Nacional San Luis Gonzaga, Ica, Peru. 92 Pages.

25. International Potato Center (CIP). 1985. Thrust IV, Control of virus and virus-like diseases, Ann. Rep., Lima, Perú.

26. Jayasinghe, U. 1990. Variability of, and resistance to potato leafroll virus (PLRV). Pages 141-153 in: Control of Viruses and Virus-Like Diseases of Potato and Sweetpotato. Lima, Peru.

27. Kawchuk, L.M., Martin, R.R., and McPherson, J. 1990. Resistances in transgenic potato expressing the potato leafroll virus coat protein gene. Molecular Plant-Microbe Interactions 3(7): 301-307.

28. Khurana, S.M.P. 1989. Personal communication.

29. Mackinnon, J.P. 1967. Greenhouse evaluation of potato seedlings for leafroll virus resistance. Am. Pot. J. 44(8): 309-315.

30. Matthews, R.E.F. 1982. Classification and nomenclature of viruses: Fourth report of the International Committee on Taxonomy of Viruses, Karger, Switzerland. 17: 1-199.

31. Mayo, M.A., and Barker, H. 1993. Personal communication.

32. Mayo, M.A. 1993. Personal communication.

33. Mayo, M.A., Robinson, D.J., Jolly, C.A., and Hyman, L. 1989. Nucleotide sequence of potato leafroll luteovirus RNA, J. Gen. Virol. 70: 1037-1051.

34. Mendoza, H.A., Fernandez-Northcote, E., Jayasinghe, U., Salazar, L.F., Galvez,R., and Chuquillanqui, C. 1989. Breeding for resistance to potato viruses Y,X, and leafroll: Research strategy, selection procedures, and experimental results. Control of viruses and virus-like diseases of potato and sweet potato , Lima, Peru. 155-171.

35. Peters, D. and Jones, R.A.C. 1981. Potato leafroll virus, Pages 68-70 in: Compendium of Potato Diseases. W.J. Hooker, ed. The Amer. Phytopathology. Soc., St. Paul, MN

36. Querci, M., Owens, R.A., and Salazar, L.F. 1996. Encapsidation of potato spindle tuber viroid (PSTVd) by potato leafroll virus particles is responsible for aphid transmission of PSTVd. Pages 312-313 in: Proceedings of the 13th Triennal Conference of the European Association for Potato Research (abstract).

37. Ross, H. 1958. The breeding of virus-resistant potatoes. Eur. Potato J. 1(4):1-19.

38. Ross, H. 1978. Methods for Breeding Virus Resistant Potatoes. Pages 93-114 in: Planning Conference on Developments in the Control of Potato Viruses. International Potato Center, Lima, Perú.

39. Salazar, L.F., Querci, M., Bartolini, I., and Lazarte, V. 1995. Aphid transmission of potato spindle tuber viroid assisted by potato leafroll virus. Fitopatologia 30(1):56-58.

40. Shepardson, S. 1980. Ultrastructure of potato leaf phloem infected with potato leafroll virus. Virology 105(2): 379-392.

41. Simpson, G.W., Bonde, R., Merriam, D., Akeley, D.F., Manzer, F.E., and Hovey, C.L. 1952. Procedure in field testing potato seedlings for leafroll resistance. Bulletin No. 502-University of Maine, Maine Agricultural Experiment Station. 14 Pages.

42. Van der Wilk,F., Huisman, M.S., Cornelissen, B.J.C., Huttinga, H., and Goldbach, R. 1989. Nucleotide sequence and organization of potato leafroll virus genome RNA, FEBS Letters, 245: 51.

43. Van der Wilk, F., Posthumus-Lutke Wielink, D., Huisman, M.J., Huttinga, H., and Goldbach, R. 1991. Expression of the potato leafroll virus coat protein gene in transgenic potato plants inhibits

virus infection. Plant Molecular Biology. 17(3): 431-439.

44.   Zadina, J., and Novak, F. 1983. The inheritance of extreme intolerance to potato leafroll virus. Sbornit Üvtiz Genetica a Slecnteni 19: 189-194.

CHAPTER   45

Present Status of Controlling Mechanically and Non-persistently Aphid-transmitted Potato Viruses

S.M.P. Khurana and I.D. Garg

Potato had played a significant role in the industrial revolution of many European countries (82) and it has the potential of feeding hundreds of millions in other parts of the world due to its wide adaptability and high yield. A native of South America, it is now cultivated in many parts of the world under diverse climatic conditions. Spread of potato cultivation to more and more parts of the world has been enabled through advances in research in breeding, seed production, pest and disease management, and the production technology as well as postharvest practices. Advances in healthy seed production programs have overcome some of the major constraints in potato produciton. Thus both area under potato produciton and yield per unit area have doubled in many parts of Asia (including India, Pakistan and China) and Africa during the last two decades (28).

Potato crop is a host to over two dozen plant viruses under natural conditions. Due to vegetative mode of propagation, virus infection (s) having once taken place go on increasing in incidence and intensity with successive planting of the same (infected) stocks which ultimately leads to a state referred to as 'running out' or 'degeneration' when yields are almost half and not very economical. The major constraint of healthy seed in potato production is mitigated by adopting an organized system which involves selection of virus-free clones/stocks and their multiplication in a manner to prevent their reinfection. This system has been further supported by research in other fields namely breeding, virus-detection, pest and disease management, etc. (54-56).

NATURE AND EPIDEMIOLOGY OF VIRUSES

Important mechanically and/or non-persistently aphid transmitted potato viruses indude, potato virus X (PVX), potato virus S (PVS), potato virus A (PVA), potato virus Y (PVY) and potato virus M (PVM). Besides these important viruses of wide distribution, a few others belonging to this category, namely tobacco mosaic virus (TMV), alfalfa mosaic virus (AlMV), potato aucuba mosaic virus (PAMV), Andean potato mottle virus (APMV), potato virus T (PVT), potato virus V (PVV) and Andean potato latent virus (APLV) are of limited distribution and significance. All these viruses cause diseases collectively called mosaics and are mainly responsible for degeneration of potato stocks. In addition, potato spindle tuber viorid (PSTVd) is of great significance as a disease agent under warm climates. It is spread by mechanical contact and is freely true-seed transmitted (99).

The diseases called mosaics, caused by any one or none of these viruses, are characterized by a wide range of symptoms from semi latent/transient and typical mosaic to general chlorosis accompanied by various types of leaf discoloration and deformations, namely mild mosaic (mottle), yellow mosaic, veinal mosaic, mosaic with necrotic spots, rugose mosaic, leaf drop streak, crinkle, chlorotic spots, veinal necrosis, etc. Mixed infection by more than one virus in the field is the rule rather than the exception. PVS causes symptomless infection in potato and happens to be ubiquitous in distribution. PVX causes imperceptible to distinctly mild mosaic depending on the virus strain, potato cultivar and the ambient conditions. Various types of symptoms due to viruses are given in Table 1 and some of their charactersitics are summarized in Table 2.

Mechanically Transmitted Viruses

PVX

It is the type member of the potex group of plant viruses. Virions are elongated particles with average dimensions of 515 x 13 nm.

The virus is readily mechnaically transmitted and is widely distributed in potato growing areas of the world. Biologically, four strains of PVX are recognized (69). Besides, a resistance breaking strain was recognized and designated-as PVX-HB with ability to infect PVX immune hybrids/cvs like Saco, Saphir and USDA seedling 41956 (68). PVX also infects *Gomphrena globosa*, a local lesion indicator plant to PVX-HB, systematically and symptomlessly.

PVS

It belongs to the carlavirus group, with flexuous virions measuring ca. 650 x 12 nm. At least two strains are recognized on the basis of systemic reaction on *Chenopodium* species, the type strain and the Andean strain. Isolates of the typical strain from North America and Europe cause local lesions on *Chenopodium* species whereas the Andean isolate causes systemic symptoms on *C. quinoa*, *C. amaranticolor* and *C. rubrum*. The Andean isolate may sometime cause chlorosis and bronzing of leaves in European potato cultivars (83). The Andean strain is readily aphid-transmitted (101).

PVT

It is a member of the capillovirus group, its' virions are also flexuous, it measures ca. 687 x 12 nm and it is known to occur in some South American regions. It is readily sap transmitted, and is seed transmitted in *Nicandra physaloides*, *Datura stramonium* and *Solanum demissum* (85, 86). Many solanaceous hosts get infected but do not express symptomsupon inoculation with PVT. Potato cultivar King Edward develops slight vein necrosis and chlorosis after 12 days of grafting. Many isolates of the virus are known to occur. It is serologically related to apple stem grooving virus.

APMV

It is a member of the comovirus group with isometric virions ca. 28 nm in diameter. It is a multicomponent virus and the components sediment as three components with sedimentation coefficients of 53(T), 93(M) and 112S(B) and contain, respectively, 0%, 27%, and 34% single-stranded, plus sense RNA.

It is readily transmitted by contact between plants, by carriage on farm implements and through seed. The virus has been found in Andean regions at elevations of 2000-4000 m above sea level and is suspected to be present in germplasm material originating from the Andes.

APMV infection induces mild mottle upon primary infections whereas symptoms are rather severe in secondary (tuber borne) infections.

APLV

APLV occurs in some parts of South America in the Andes at 2000 to 4000 m above sea level and is serologically related to *Ononis* yellow mosaic and *Dulcmara* viruses. Virions are isometric ca. 30 nm in diameter. Purified virus preparations consist of two types of particles: one containing the nucleoprotein and the other have only protein (empty protein shell). Obviously, only ones with nucleoprotein are infectious (33). The virus is readily transmitted by contact between plants. It is transmitted through true seed only at a low frequency, and only poorly by potato flea beetle (*Epitrix* sp.). Primary infections show only chlorotic netting of minor leaf veins. Secondary infection often cause mild mosaic. Depending on the potato cultivar, infections may also cause rugosity and severe mosaic. Two serologically distinct strains are recognized.

PSTVd

The causal agent of the disease 'potato spindle tuber' was considered a virus until 1967 when it was suggested that it is a naked nucleic acid (27). It is a small single stranded RNA with 359 nucleotides and with a high degree of intramolecular complementarity (39).

Three main strains, designated as severe, intermediate and mild, are recognized on the basis of severity of symptoms produced in potato and tomato (87). these strains differ by 6 to 10 nucleotide exchanges.

In spite of its wide experimental host range, PSTVd generally affects potato in nature.

Table 1. Symptomatic key for important potato viruses

Latent or Faint Mosaic

PVX	Latent in many vars. or interveinal to barely perceptible mosaic. Var. Craigs Defiance shows top necrosis.
PVS	Usually latent or very mild or barely perceptible mottle & faint vein banding.
PVT	Often symptomless but may produce mild mottle or slight vein necrosis/chlorotic spots.

Mild Mosaic

PVM	Latent in some varieties or mild to severe mosaic, slight leaf chlorosis, deformations & wavy margins of leaflets.
PVA	Mosaic, faint mottling sometimes leaf distortion, top necrosis in some vars. rarely rugosity, shiny leaves.
PSTVd	Very obscure and unusual. Plants rather erect and spindly growth, stunted with (often curling) leaves darker green than healthy. Leaflets twisted, petioles subtended at 45° angle from the main stem. Tubers cylindrically long with tapered ends. Eyes numerous, conspicious and with distinct eye browing of tuber flesh over each eye.
APLV	Mild mosaic or mottle and chlorotic/necrotic netting of minor veins. Leaf deformation and mosaic in sensitive vars. under cool conditions.

Severe mosaic

Green types

PVY	Symptoms vary with strains/variety. Mild or severe mosaic and veinal necrosis and leafdrop streak. Plants stunted. Rugosity, twisting of leaves with slight inverted cupping of leaflets. Older leaves often collapse and drop or show mild mosaic.
APMV	Mild mottle (primary symptom); severe mottle or mosaic (secondary symptom); top necrosis and stunting, leaf systemic necrosis.

Yellow types

PAMV	Bright yellow spots interspersed in the green lamina & stunting.
Calico- AlMV	Bright yellow markings or complete yellowing of leaflets.

Table 2. Some characteristics of mechanically and non-persistently aphid-transmitted potato viruses

Virus name	Virus group	Virion size	Strains (variability)	Seroaffinity	Transmission	Host range
PVX	Potex	515x13nm	Many	-	Mechanical	Wide
Potato aucuba mosaic virus (PAMV)	-do-	580x12nm	Few	White clover mosaic virus	Mechanical; aphid	Restricted
PVS	Carla	650x12nm	-do-	PVM,carnation latent virus	-do-	-do-
PVM	-do-	-do-	-do-	PVS,carnation latent virus	-do-	-do-
PVA	Poty	730x15nm	-do-	PVY	Aphid	Solanaceae
PVY	-do-	730x11nm	Many	-	-do-	Wide
PVV	-do-	760x10nm	Three	PVY^c-AB	-do-	Solanaceae
PVT	Capillo	680x12nm		Apple stem grooving virus	Mechanical, seed	Wide
APLV	Tymo	Ca.30nm; isometric	Two	-	Mechanical, beetles(?), seed	-do-
APMV	Como	28nm, isometric	Few	Eggplant mosaic virus	Mechanical,flea beetle, seed	-do-
AlMV	Alfamo	58,52,42x18nm (baciliform)	Many	-do-	Aphid	Wide
Cucumber mosaic virus (CMV)	Cucumo	30nm; isometric	-do-	-	-do-	-do-
TMV	Tobamo	300nm	-do-	-	Mechanical	-do-
PSTVd	Viroid	Small RNA molecule, 357-361 nucleotides long	Three	-	Mechanical,seed	-do-

PSTVd vary greatly depending on many factors like potato cultivar, strain of PSTVd and environmental conditions. Severely affected plants may be stunted with upright growth due to a more acute angle between stems and petioles. Leaflets show a degree of undulation and size reduction. Tubers may become spindly and reduced in size and may show growth cracks when plants have infection of severe strain of PSTVd. There is also a decrease in eye depth or raising of eyebrows.

Mild to severe yield losses due to PSTVd infection have been reported. Singh *et al.* (100) estimate the decrease in tuber yields of the variety Saco which ranged between 17-24% for mild strains and 64% for a severe strain. Pfannenstiol and Slack (77) observed a non-significant tuber yield reduction, 46% during the first year, up to 87% in the second year and up to 97% in the third year, caused by a severe strain in fourteen potato cultivars.

PSTVd is readily transmitted through mechanical contact (66), sap inoculation (10), infected tools and farm equipment (64, 66) and vertically through pollen and/or seed (45). Recent studies at The International Potato Center in Lima, Peru, have indicated the possibility of dissemination of PSTVd through the female nematode *Meloidogyne incognita*. The viroid may be carried on the surface of the female nematode, on the cuticle of juveniles, inside the egg mass and on the surface of eggs extracted from PSTVd infected plants (4).

PSTVd has been reported to occur in potato fields in northern and northeastern US, Canada (26), the former USSR (58), China (59) and in potato collections in several other countries. But, PSTVd has not been found to be present in the Andean potato cultivars (1).

PSTVd is of great quarantine significance particularly for the warm climate zones where it invokes severe and early symptoms (40) due to high rate of viroid multiplication in warm climates (88). Its' introduction through symptomless cultivars and through the seed is really a serious threat for the countries not well equipped for PSTVd detection. Though it is not transmitted through insect vectors, its' highly stable nature and high rate of seed transmission mitigate this limitation and enable it to spread fast (52,99).

PSTVd is controlled by following the same measures of control applicable for the mechanically transmitted potato viruses X and S. However, in contrast to PVX and PVS, PSTVd cannot be controlled by growing potato through true potato seed (TPS) due to its seed-borne

nature. Therefore, it is imperative to first free the stocks/clones from PSTVd before taking up potato production through TPS. Materials found to carry PSTVd should be destroyed, if dispensable. Elimination of infected materials and the application of phytosanitary procedures help in reducing the risk of recontamination. Tools and operators' hands should be disinfected with 10% solution of calcium hypochlorite (2).

Non-persistently Aphid Transmitted and Mechanically-transmitted Potato Viruses

PVM

It belongs to the carlavirus group, the virions measure ca. 650 x 12 nm and it causes various mottle, mosaic, crinkling and rolling symptoms in leaves and stunting of shoots. It has a narrow host range and several biologically distinguishable strains are detected. Diseases caused by PVM in potato are often called leaf-rolling mosaic, interveinal mosaic or paracrinkle depending on the virus isolate and the potato cultivar. Mild strains produce a mosaic between the veins of the leaflet tips and some deformation of the leaves. This deformation is in the form of twisting of leaflet tips, sometimes severe and sometimes rolling, especially of the top leaves. Symptoms are produced only if young plants get infected but are masked at higher temperatures(ca. 24°C).

PVM is not easily transmitted like PVX or PVX-HB but, of course, it is readily sap and aphid transmitted by the efficient vector *Myzus persicae*.

PVA

It belongs to the potyvirus group, is a flexuous rods measuring ca. 730 x 15 nm and has a narrow host range confined mainly to Solanaceae. It is transmitted efficiently by the aphid *M. persicae* with an acquisition period of 20 seconds and a retention period of 20 minutes (63).

Symptoms induced by PVA depend on the potato cultivar and ambient conditions, comprize mild mosaic, roughness of the surface, waviness of the leaf margins, or no symptoms. Some cultivars show hypersensitive response in the form of top necrosis. The yield losses due to PVA infection may go up to 40% (22).

Isolates of PVA differing in the severity of symptoms produced on potato have been encountered and classed into four groups, viz. very mild, moderately severe and severe (62). PVA is widespread in some cultivars. In combination with PVX, it usually causes a severe disease known as potato crinkle. Similarly it results in a severe 'rugose' disease in combination with PVY. Many cultivars are hypersensitive (field resistant) to PVA and upon inoculation by stem and tuber-core grafting respond with top necrosis.

PVY

It is the type member of the potyvirus group. Like PVA, virions of PVY are flexuous rods measuring ca. 730 x 11 nm. PVY infection may depress the tuber yield up to 80% depending on the virus strain and the potato cultivar. Combined infection with PVA, PVX and/or PVS may cause severe disease that sometimes may destroy the entire crop of a susceptible variety.

Isolates of PVY show great variability in host response and serological relationships. Isolates of PVY are subdivided into the ordinary (PVY^0), the stipple streak (PVY^c) and tobacco veinal necrosis (PVY^N) strain subgroups (21,47). PVY^0 and PVY^N differs from PVY^c in being able to overcome dominant hypersensitivity gene N present in many potato cultivars (16,19). The virus incites symptoms which may be mild to severe mottle, streak on leaf-drop streak with necrosis along the veins of the underside of the leaflets, and stipple-streak. The primary symptoms may be necrosis, mottling or yellowing of leaflets, leaf dropping, and sometimes premature death. Necrosis, which starts as spots or rings on the leaflets, may cause their collapse. Secondarily infected plants are dwarfed and brittle, with leaves crinkled and/or badly puckered.

PVV

PVV is a potyvirus distantly related serologically to PVY. It causes severe necrotic reaction when inoculated to potato cultivars carrying PVY hypersensitive gene N^c. Originally it was classified as belonging to PVY strain group c (16) which is called Y-GI isolate. Fribourg and Nakashima (28) showed that PVY isolates designated as UF, Y^c-GI and Y^c -AB are representatives of a third poty virus (distinct from PVY and PVA) which they called potato virus V (PVV).

PVV infection in potato cultivars 'Estima' (Irish) and 'Farmosa' (Dutch) is often symptomless but faint mosaic and mild rugosity symptoms may sometimes be expressed. Infected plants of potato cv. Home Guard, another Irish variety, may show mild mosaic and rugosity with some necrosis of lower leaves (46).

PVV is more closely related serologically to Peruvian tomato virus (PTV) and wild potato mosaic virus (WPMV) (Fernandez-Northcote, personal communication).

PAMV

It is a member of the potexvirus group. Virions are flexuous rods measuring about 580 x 11nm. The virus is of less economic significance and is known to occurr in Europe and North America. It causes bright yellow flecking especially of lower and middle leaves of certain cultivars. The symptoms are especially clear in young plants.

PAMV is readily sap-transmitted. It may also spread through contact between infected and healthy plants under natural conditions. Some strains are transmitted by *M. persicae* in the presence of a helper virus (PVA or PVY) (50). Strains of PAMV can be differentiated on the basis of symptoms on different potato cultivars.

AlMV

It is a multipartite virus which is the type member of the alfamo group. Of the four types of virions, three are bacilliform measuring ca. 60, 48 and 36 nm in length and ca. 18 nm in diameter. The fourth component is ellipsoidal ca. 28 x 18 nm. The smallest particles contain the subgenomic mRNA which encodes the viral capsid protein.

Though worldwide in distribution, this virus has little significance to potatoes. Besides, its transmission through tubers, it is also transmitted through sap and non-persistently through the aphid species *Aphis nasturtii* Kaltenbach, *Aulacorthurn solani* (Kaltenbach), *Macrosiphum euphorbiae* and *M. persicae* (22). The virus spreads from lucerne to potatoes by migrant aphids from lucerne (9).

Two main groups of strains have been recognized namely potato calico and potato tuber necrosis strains. The main symptoms comprise yellow leafspots with sometimes whole of the leaflets becoming yellow usually accompanied by nec'rnsis and leaf deformation. If necrosis is not there, symptoms are irregular bright yellow blotches on leaves. Secondary infection causes calico markings and slight dwarfing.

SPREAD OF VIRUSES

Planting of infected potato tubers is the major source of spread of potato viruses. Easily mechanically transmitted viruses, namely PVX, PVS and PAMV, spread from plant to plant through contact of foliage or roots of the diseased plant with that of the healthy plants under field conditions. These viruses are very stable and spread among potato fields through contaminated farm machinery, clothes and hands of farm workers. A high rate of infection (54-93%) was observed when PVX-contaminated cultivating and hilling equipment were used (64). Different types of surfaces like painted or unpainted wood, iron, rubber, jute, soil, human skin, etc. can retain infective PVX for different lengths of time (110). Besides infected potatoes, many wild plant species may act as reservoirs of PVX (31, 98). Though their precise role in the spread of PVX to potatoes has not been worked out, they might play an important role in the epidemiology of the virus due to its highly contagious nature and high stability. PVX is not transmitted through aphids, consequently, virus spread occurs through contact of contaminated objects/animates with the healthy potato plants.

PVS is readily and mainly transmitted through contact except the Andean strain which is also transmitted by aphids. PVS is very stable and use of contaminated knives may cause 25 to 97% infection of PVS in different cultivars (109). PVS has a narrow host range and only a few other plant species are known to harbor it under natural conditions (98).

PVM, PVA, PVY, PVV and AlMV are spread by various aphid species both within and between potato fields. PAMV, in the presence of PVA and/or PVY and the Andean strain of PVS (which is also transmitted by aphids), is also spread non-persistently by aphids to some extent. Consequently aphids spread most of these viruses effectively from infected to healthy potato plants within the potato field, and between potato fields which are not separated by very large distances. Though many aphid species can transmit these viruses (76), green peach aphid, *M. persicae,* is the most efficient vector (56, 96). The precise role of different aphid species in the spread of these viruses under natural conditions is still not clearly understood. Virus spread by aphids depends on many factors like host preference, efficiency of the vector, retention of infectivity, location of virus-source, crop age, stage of vector development, climatic conditions (particularly the temperature), vector behavior during migration, etc. Attempts have been made to co-relate the aphid population dynamics to the virus spread (106) by calculating vector pressure from the number (N) of aphids trapped in the trap and their vector efficiency. A model, when devised and developed to simulate the spread of PVY in the field, will be of great value to forecast the likely impact of virus spread to make decision whether to adopt aphid control or not. Such a model would take into account the contribution of factors like vector efficiency, mature-plant resistance, planting date, proportion of diseased plants in the crop and their removal, temperature, etc. in the virus spread.

CONTROL

Conventional Methods

Different potato viruses adopt different strategies of spread and perpetuation, thus, while easily mechanically transmitted viruses like PVX, PVS, PAMV and APMV spread mainly by contact of healthy plant leaves with infected leaves, or with contaminated farm machinery, livestock and farm workers, the non-persistently aphid-transmitted potato viruses use aphids as their vehicle for spread. Though the sources/reservoirs of these viruses are mainly infected potato plants within the field itself, many other plant species are also known to harbor these viruses. Due to different modes of virus spread used by these viruses, different control strategies have been employed. Many of them are employed as a package referred to as Integrated Pest Management. Various components of this package are discussed hereunder.

Exclusion

There are many potato viruses which have been reported to occur only in some specified regions of the world and not others. Such viruses, viz. Andean strain of PVS, APMV and APLV occur mainly in the Andean region of South America and their spread to other parts of the world can be checked through quarantine regulations. Similarly, PVV, so far, is restricted in distribution. Many countries have sophisticated and well staffed quarantine laboratories to intercept such viruses which are otherwise non-existent in those countries. Now commonly employed method of germplasm exchange to avoid

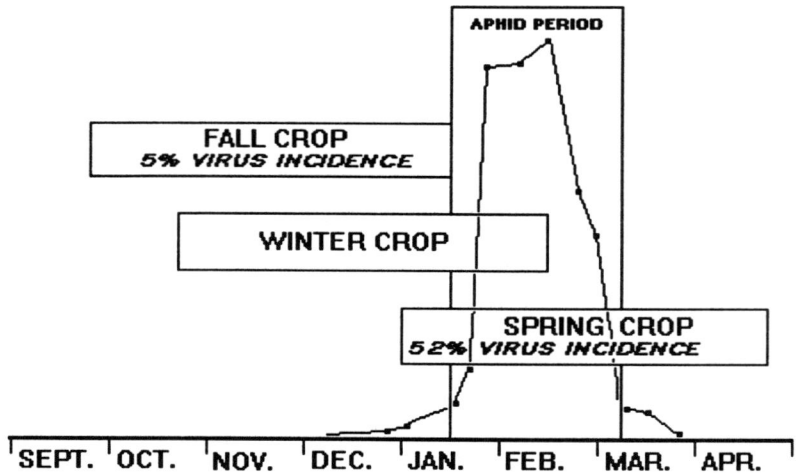

Figure 1. Aphid build-up in relation to early, main and spring crops in northwest plains of India

virus/viroid spread is through *in vitro* cultured plantlets rather than tubers.

Healthy Seed Production

Main reservoirs/sources of potato viruses are the infected tubers used as seed. Obviously, the virus disease management means selection and multiplication of virus-free seed tubers. This is the main strategy of virus disease management adopted by many countries throughout the world. There has been considerable improvement in the health standards of potato stocks during the last one decade mainly due to recent improvements in the virus-detection technology and also adoption of new technology by many countries in Asia and Africa. The techniques like enzyme-linked immunosorbent assay (ELISA), latext agglutination test (LAT), immune electron microscopy (IEM) and nucleic acid spot hybridization (NASH) have revolutionized the quality of virus detection in enormously large potato stocks which when found free from viruses with one or more of these techniques are further multiplied as per the guidelines prescribed by the seed-production agencies in different countries depending on the climatic conditions. Generally, virus-free stocks are multiplied in high hills exposed to heavy wind currents which minimize the vector pressure. Alternatively, they are multiplied in regions having climate suitable for potato crop production

and either lacking the aphid vector or having aphid population below a certain critical level at least during the crop period. Thus, in India, virus-free potato stocks referred to as breeders' seed are multiplied in high hills above 2000 m in the northwest during summer and in the Indo-Gangetic plains during fall (September/October to December/January) when vector pressure is very low (Figure 1). Haulm destruction of potato seed crop is broadcast as soon the level of *M. persicae* reaches 20 aphids per 100 compound leaves from 33 plants selected at random. This critical level in the northwest plains is reached by the 4th week of December (95,97).

Another method of raising virus-free potato stocks is through *in vitro* multiplication. This method is rapidly gaining importance and many countries with tropical climate are following this (28). Virus free seed potatoes thus produced are free from more than one virus(es). But, cost of such virus-free seed production is quite high and consequently calls for multiplication for virus free seed right up to certified seed production stage. Where potato crop production is often more economical by even using seed carrying some tolerable level of virus infection. Due to high cost of seed production, most of the seed programs are either run by the government agencies or have been highly subsidised since healthy seed is the backbone of potato industry.

Clonal selection is the basis of healthy

potato seed production and a complete seed production program has the following three tiers: (i) Clonal selection, (ii) Basic seed production and (iii) Certified seed production. Clonal selection involves the selection of healthy looking and true-to-type plant(s) desirable for a particular cultivar and tested for its health status and if found virus-free, it is further multiplied as basic seed by adopting all possible measures to prevent virus infection during multiplication. All the basic seed plants are tested individually in the field for their virus-freedom and those found free are multiplied further. Freedom of plants from viruses is checked at every stage of multiplication. Such a virus-free stock forms the basic or nucleus seed. Certified seed is raised from the basic seed by multiplying it either on government-owned farms or on registered growers' farms under direct supervision of the experts/field technicians and inspected by an independent national agency for the purpose.

Seed potatoes can be grown only where the growing conditions are ideal and virus infection pressure and incidence of tuber diseases are low. A minimum of 70 days growing period is required for a reasonable tuber yield. Besides the ideal climatic conditions, proper isolation of the seed crop is required. Diseased plants are rogued out as soon as detected. However, tolerance limits for the pests and diseases in seed tubers vary according to the region and keeping in view the interests of the growers of the region and also the ability of seed growers to meet these requirements (Tables 3 and 4).

Northwest Europe and North America have extensive and complete seed production programs and even growers of ware poatoes use fresh lots of certified seed every year. Application of improved techniques has increased the availability of quality seed in many countries. Several private companies are now generating pre-basic seed tubers in Argentina and Brazil thereby cutting down their imports of quality seed drastically. Improved seed production has grown by more than 80% during the last decade in Central America and the Caribbean countries. Similarly, countries in East Africa: Burundi, Ethiopia, Kenya, Rwanda, Madagascar, Malawi, Mauritius and Tanzania have seed programs supplying potato seed to farmers with technical support from the International Potato Center in Lima, Peru (28). A basic seed production scheme has been launched by the Korean Republic Government which involves the use of tissue culture for producing, storing and rapidly multiplying virus-free planting material. Coupled with the use of improved techniques for field multiplication, inspection, and pathogen testing of the foundation stock, registered and certified seeds have resulted in a sharp reduction in incidence of virus diseases in farmers fields and a spectacular increase in the yields (44).

The Seed Production Program in India (Figure 2) was initially confined to the northwest hills having high altitude (above 2500 m). Subsequent research work at Central Potato Research Institute (CPRI) in Shimla revealed the potential of Indo-Gangetic plains as seed-production areas during fall/winter. CPRI produces healthy seed potato called breeders' seed, which is further multiplied on the state government-owned farms, university farms and also in small quantities on private seed farms. Further multiplication and distribution is handled by the State Seed Corporations (SSC) and National Seeds Corporation (NSC) (93).

Use of true (botanical) seed collected from homogeneous populations of potato plants holds promise as the starting point of healthy seed. Potato production from homogeneous seedling populations of true seed was started in 1972. True seed was produced in Inner Mongolia and used for raising small tubers (20-50g) in the southwest mountainous region. The small tubers were then used for the production of commercial crop from which large tubers were consumed and small ones reused as seed for the following crop. This was done about five times so that only once in five years true seed had to be brought in to refresh the seed tubers for the ware crops (102). The *in vitro* multiplication of healthy seed potato stocks was started in 1978 in China.

The main constraint on expansion of potato cultivation to additional areas in India as well as many other parts of the world has been non-availability of quality seed for ware crop production. In India, attempts have been initiated to supplement the supply of quality seed by (i) involving some of the state agricultural universities in breeders' seed production, (ii) rapid multiplication of stocks through *in vitro* culture techniques, and (iii) development of suitable true non-existent potato seed (TPS) lines. Research on TPS has been undertaken jointly by CPRI, Shimla and International Potato Center (CIP), Lima (Peru).

The first attempted use of TPS for commercial potato production was made in India during 1950s but was abandoned due to heterogeneity in crop maturity and non-uniformity of the produce. CIP organized the 19th World

Table 3. Field tolerance (no. of affected plants in %) for basic and certified seed, as applied in four countries

Disease	Country							
	Canada[a]		France[a,c]		The Netherlands[a,c]		U.K.[b]	
				Seed type[d]				
	B	C	B	C	B	C	B	C
Severe mosaic, including leaf-roll, stipple streak and crinkle	0.25	1.0	0.33	1.0	0.09	0.25	0.1	2.0
Mild mosaic	0.25	1.0	0.33	1.0	0.09	2.0	0.5	5.0
Erwinia sp.	0.1	2.0	0.5	1.0	nil	5 plants/ha	0.5	2.0
Varietal impurity (control)	0.0	0.1	0.1	0.2	0.0	0.0	0.05	0.5

[a] Three official inspections for basic seed and two for certified seed.
[b] Two official inspections for basic seed and one for certified seed.
[c] The data presented for basic and certified seed concern those for the grades E and A, repsectively (grades E and A are generally marketed)
[d] B = basic seed; C = certified seed

Table 4. Tolerance limits of virus incidence in different stages of seed production in India (53)

Virus diseases	Stage of crop growth	Minimum permissible limit (%) in		
		F-I	F-II	Certified
Mild mosaic (1st and 2nd inspection)	Before haulm cutting	1.0	2.0	3.0
Severe mosaic, leafroll and yellows	-do-	0.5	0.75	1.0
Total viruses[a]	-do-	1.0	2.0	3.0

[a] Of the two inspections, the higher virus percentage will be considered for purposes of the specified limits of tolerance.

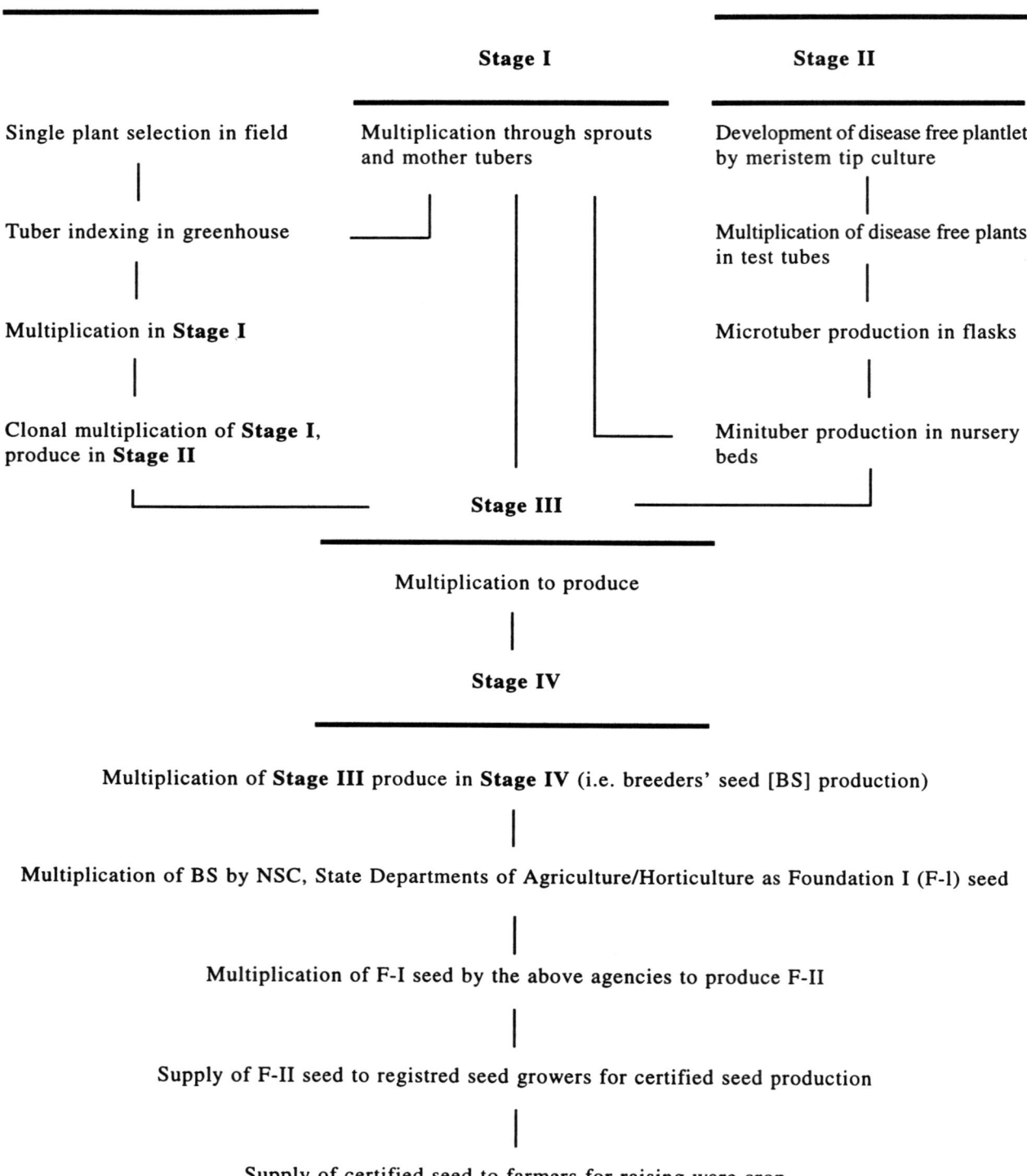

Conventional methods

Stage I

Stage II

Unconventional methods

Single plant selection in field

Multiplication through sprouts and mother tubers

Development of disease free plantlet by meristem tip culture

Tuber indexing in greenhouse

Multiplication of disease free plants in test tubes

Multiplication in **Stage I**

Microtuber production in flasks

Clonal multiplication of **Stage I**, produce in **Stage II**

Minituber production in nursery beds

Stage III

Multiplication to produce

Stage IV

Multiplication of **Stage III** produce in **Stage IV** (i.e. breeders' seed [BS] production)

Multiplication of BS by NSC, State Departments of Agriculture/Horticulture as Foundation I (F-1) seed

Multiplication of F-I seed by the above agencies to produce F-II

Supply of F-II seed to registred seed growers for certified seed production

Supply of certified seed to farmers for raising ware crop

Figure 2. Proceedure for seed production used in India (53)

Planning Conference on Potato Production from TPS in 1979 in Manila (Philippines) for generating interest in this technology in the countries of South Asia. This was followed by a Regional Workshop on TPS for Potato Production in 1985 in Kathmandu (Nepal) where scientists from Bangladesh, Bhutan, India, Nepal, Pakistan and Sri Lanka presented the results of their research (37). The use of TPS offers the following advantages: (i) reduced incidence of disease since only a few disease agents are known to be true seed transmitted; (ii) very low cost of the seed and convenience in its storage and transport, and (iii) genetic diversity in the TPS crop ensures a good degree of protection from being wiped out by a disease. Besides, potato production through TPS also spares more farm area for ware crop production since traditional tuber-seed production requires much larger area than the seedling tubers (tuberlet) production from TPS. However, TPS-based potato production has disadvantage which include (i) greater requirement of labor and time, and (ii) potential non-uniformity in the produce. But, TPS based production technology has been improved and promising TPS families, namely TPS-2 (OP), TPS-3(OP), TPS-7(OP), HPS4/3, HPS1/13, HPS7/III, HPS1/IIi, HPS2/III, etc. have been selected as their successful performance in the field trials demonstrated (Table 5). It has been found that transplanting seedlings in the field and use of seedling tubers as seed are successfull alternative approaches for commercial potato production. Further improvements in the technology would make it a technology of the near future.

Breeding for Virus Resistance

Breeding for resistance to viruses is one of the most important methods of virus-disease management. This is the cheapest and most effective means of controlling virus diseases in potato growing, and the method which can/helps avoid the excessive use of costly and environment-polluting pesticides. This is also the method of choice for areas where potatoes, especially seed potatoes, are grown without the help of a well organized, costly seed certification scheme.

Though breeding for resistance to virus infection can be done at an empirical level with a little knowledge of the genetic basis, a more responsible exploitation of genetic resources requires thorough knowledge of the genetic mechanism(s) of the virus-resistance. Responses of potato plants to various viruses may be grouped

under various categories, viz. (i) tolerance to virus where the virus multiplies in the plant without producing perceptible symptoms, (ii) resistance to the virus multiplication; (iii) resistance to the virus-spread within the plant, and (iv) immunity to the virus. Plants may also escape infection by a virus if they either repel or do not act as a host to the virus vector(s). The type of resistance which prevents virus spread within the plant is often controlled by one dominant gene which confers the ability to localize the virus through a response called hypersensitivity (HS) or hypersensitive reaction (HR). Breeding for virus-tolerant varieties is not desirable for cultivation in the areas earmarked for seed-production since that will allow carrying of viruses symptomlessly. Tolerance may be of value where virus-free stocks can not be maintained/produced and import of healthy seed is expensive. The commonly used resistance mechanism in potato breeding programs is of the extreme, HS type. Another type of resistance referred to as 'resistance to infection' is of a quantitative type since it is appraised quantitatively by the number of plants of a cultivar getting infected upon exposure to the virus. This type of resistance is effective against all strains of the virus involved, and it is inherited polygenically. It is also strongly influenced by environmental conditions and, under heavy infection pressure, all individuals of the cultivar may become infected. Age of the potato plant also influences resistance since older parts are not as easily infected as young plants (mature plant resistance) (8).

HS resistance is characterized by the development of necrosis in parenchymatous tissues in the area of virus invasion and is often expressed as necrotic local lesions on the leaves. HS is inherited in a monogenic dominant fashion and the gene involved is coded N. There are strong indications that extreme resistance is an enhanced form of HS. Extreme resistance is distinguished from the HS by a graft test. A scion of the clone to be tested is grafted on to the virus-infected stock. Severe top necrosis in the scion indicates HS while no scion reaction or formation of only small, inconspicuous necrotic spots indicates extreme resistance denoted by the single dominant gene R (7).

Breeding for Resistance to PVA and PVY

Many potato cultivars possess a high level of resistance to infection by PVY. Many Dutch potato cultivars possessing the dominant

Table 5.   Tuber yields (t/ha) in three successive clonal generations of 8 TPS progeny families and the cultivar Kufri Bahar (37)

TPS progeny	Clonal generation		
	C_1 (1985-86)	C_2 (1986-87)	C_3 (1987-88)
TPS-2(OP)	27.9	26.0	28.6
TPS-3(OP)	22.0	25.1	28.8
TPS-7(OP)	33.8	28.7	34.6
HPS 4/3	27.2	29.1	24.1
HPS 7/Ill	34.3	33.2	33.6
HPS 1/13	33.4	32.6	37.3
HPS I/Ill	35.8	35.7	32.9
HPS 2/Ill	36.8	36.7	30.5
Kufri Bahar	36.1	32.2	34.7

Table 6. Interrelationship of genes for hypersensitivity and strain groups of PVX (15)

Cultivar	Genotype	Strain group[a]			
		1	2	3	4
Arran Banner	nx nb	s	s	s	s
Epicure	Nb nb	R	s	R	s
Arran Victory	nx Nb	R	R	s	s
Craig's Defiance	Nx Nb	R	R	R	s

[a] R=hypersensitive, s=susceptible

resistance gene *Nc* are hypersensitive to PVY but susceptible to some strains of Y and Y^0 subgroups. Similarly, some potato cultivars carrying the resistance gene *Na* are hypersensitive to PVA while some cultivars possess resistance to both PVA and PVY. Extreme resistance to PVY has been located in a number of wild *Solanum* spp. and in *S. tuberosum* spp. *andigena*, which is very closely related to potato. *S. stoloniferum* carries extreme resistance to PVY and PVA and there are six resistance genes for these viruses at three different loci and are differentiated by the phenotypic expression of plants after inoculation with PVY or PVA (16). Potatoes possessing extreme resistance to all strains of PVY and PVA are bred by employing the *Ry (sto)* gene from *S. stoloniferum* and the *Ry (adg)* gene from *S. tuberosum* spp. *andigena* (24,29). Ten cultivars mainly with extreme resistance to PVY have been registered in Europe. Two of these cultivars also possess extreme resistance to PVX.

Breeding for Resistance to PVX

Genes conferring resistance to PVX have been encountered in many species of *Solanum* including *S. tuberosum*. Resistance genes denoted by *Nb* and *Nx* are strain-specific and have been used to differentiate PVX strains (Table 6)(15,16). Cultivars having various combinations of these genes are used to distinguish different strain groups of PVX. Extreme resistance to PVX was first observed by Schultz and Raleigh (91) in the clone USDA 41956. This gene denoted by *Rx(adg)*

is found in many accessions of *S. tuberosum* ssp. *andigena* and also in the progenitor CPC 1673 (Commonwealth Potato Collection) very often used in Dutch breeding programs for nematode resistance. Another extreme resistance gene *Rx(ac1)*, is present in *S. acaule* (84). Twenty-three European potato cultivars carry extreme resistance to PVX but, the extreme resistance is temperature sensitive and demonstrable multiplication of PVX occurs at 20°C or lower (23). PVX evokes necrotic lesions 0.5 mm in diameter at 20°C and 3 mm or less at 8°C, on the leaves of graft-inoculated plants. Tubers derived from graft-inoculated plants developed necrosis under certain conditions of storage. However, the plants grown from such tubers were virus free.

An unusual strain of PVX, designated as PVX-HB, was found in germplasm collection from Bolivia. It is capable of invading hypersensitive and extremely resistant potato cultivars including the USDA clone 41956(68). It systemically infects British cultivars carrying the *Nb* or *Nx* gene. Resistance to PVX-HB has since been located in *S. sucrense* (13). Immunity to PVX has been incorporated in potato cultivars Kufri Lalima (CIP 800971), (released in India), San Juan (CIP 800977), Anett (CIP 800981), Nata (CIP 701131), Santanlalla (CIP 701171), Curipaamba (CIP 703350) and Clones BW-4 (CIP 379690.8), 381371.81, 384327.42 and 384329.21(3).

Recently, resistance to PVX^c and PVX-HB in *S. sucrense* has been shown to give a differential hypersensitive reaction to these viruses (4). Plants grafted with PVX-HB or PVX-GUA suffered severe apical necrosis while plants grafted with PVX^C reacted with the production of only small necrotic lesions. The virus could not be detected in grafted plants. However, co-infection of PVX-HB or PVX^C with PVY broke down the resistance to PVX. Combined resistance to PVX^C, PVX-HB, PVX-GUA, PVY and PSTVd has been found in two accessions of *Solanum candolleanum*. Immunity to PVX and PVY is also transmitted through protoplast hybridiztion and the somatic hybrid shows profuse flowering and improved male fertility under CIP, Lima (Peru) conditions (4).

Breeding for Resistance to PVS

Potato cv. Saco is field immune to PVS. This resistance, inherited as a recessive, is also present in the Canadian cultivar Jemseg (5). Hypersensitive type resistance is found in *S. tuberosum* ssp. *andigena* and in *S. megistacrolobum*; the gene responsible for this resistance is denoted by N_5 (5). Resistance to PVS was also found in several accessions of *S. brevidens* in Helsinki (Finland) (3).

Therapy

Valuable germplasm lines and potato cultivars can be made virus-free through the following procedure(s):

Meristem Tip Culture

Tissue culture techniques have proved useful in the production of virus-free plants. This involves, plantlet regeneration, using sterile culture techniques, from excised short time (meristem tip culture). The process is based on the observation that the virus content of plant meristems is usually very low. Consequently, meristems provide the best source of tissue for the production of virus-free plants (32, 80, 103). The success for getting total elimination of virus actually depends on the size of the meristem used. The chances of isolating virus-free tissue are best when the smallest possible tip is used (less than 0.5 mm), however, survival of the smaller tips in culture is quite limited. This problem may be overcome by first growing the infected plant for several weeks at 36°C before cutting the meristem tip. Heat treatment also reduces the virus concentration in the infected plant and thus enhances the chances of recovery of virus-free plantlets (103). Chemotherapy using anti-

metabolites and antiviral chemicals, viz. Virazole (108), has also been reported to improve the recovery of virus-free plantlets from larger meristem sections.

Potato viruses differ in their sensitivity to heat. Multiplication of PVX is inhibited above 30°C while PSTVd multiplication is inhibited below 10°C. Heat treatment also eliminates PVY from small stocks of seed potato (80).

Murashige and Skoog's culture medium (70) (with or without modifications) is used universally. Meristems, placed on the medium are cultured under fluorescent light (2000 to 4000 lux, 16th/day) at 22-25°C (80). When plantlets are 2-4 cm long and have some roots, they are transferred to small pots containing a sieved compost mixture. Leaf samples are drawn when plantlets attain a height of 20-30 cm and tested for viruses. Success of virus elimination varies considerably depending on the virus. Chances of PVY and PVA elimination were 70% when meristem tips were used while the chances of eliminating PVX were hardly 10% (79). However, when excised meristems of not more than 0.1 mm long were used, only 20 out of 196 developed into plantlets, but 19 of them were virus-free (51).

Elimination of PVS is most difficult. It was eliminated from certain cultivars only when the mother plant had been treated with 2-thiouracil prior to meristem-tip excision (79). Other antimetabolites, namely 8-azaguanine, 5-fluorouracil, 2-thiouracil or p-fluorophenylalanine, used by Pennazio (74) for the elimination of PVX through meristem-tip culture, greatly inhibited plant growth. The few plantlets obtained on a medium containing 2-thiouracil were not completely free from PVX. However, incorporation of Virazole (ribavirin) into the culture medium has been reported to result in a higher percentage of virus-free progeny plants from virus infected explants than were produced in comparable cultures without Virazole (14). Other viruses which could be eliminated include PVX, PVM and potato leaf roll virus (PLRV).

PVX and PVS were eliminated by growing infected plants of potato cultivar White Rose under continuous light, with air temperatures fluctuating between 35-37°C and soil temperatures between 30-32°C. Axillary buds were excised from these plants at different intervals and cultured. The proportion of PVX-free plants reached 50% after 8 weeks and almost 100% after 18 weeks. Elimination of PVS is difficult with this treatment (104). Even elimination of PVX was rather difficult with heat treatment in certain potato cultivars having mixed virus infection (74).

Elimination of PSTVd from infected potato seed stocks is also difficult. However, the rate of PSTVd elimination can be enhanced by taking small meristems from plants kept for several months at 5-8°C under suitable light intensity and humidity (60).

The precise mechanism of elimination of virus through meristem-tip culture is still not known. A discussion of the probable mechanism(s) involved is given by Stace-Smith (103). It appears as if three main processes act in a concerted manner in the elimination of the virus. They are physical restriction of movement of the virus to the meristems due to lack of vascular/plasmodesmatal connections; successful out-competition of the virus by the host, possibly resulting from the rapid cell division of the meristematic cells; and production of antiviral substances. In addition, eradication of the virus may result directly due to the excision and the medium constituents and or culture process (65).

Elimination of potato viruses through chemo- and thermo-therapy without the meristem tip culturehas been attempted (38). Potato plantlets, established in vitro, were exposed to an alternating four hour cycle of 35-31°C and ribavirin (20 mg/L) therapy for 4 weeks followed by ELISA testing: Plantlets were then further propagated via nodal cuttings at room temperature, without exposure to ribavirin. The virus-free plantlets were then grown out as mature plants for a final virus assay. Ribavirin alone or in combination with heat not only reduced PVM, PVS and PVX titres (10-and 60 fold) but also freed many plants from these viruses. Similarly, Conrad (17) has reported elimination of PVS in 10% or more plantlets as well as PVX, PVY and PLRV from over 100 potato varieties by the use of ribavirin. Infected nodal segments were cultured on agar solidified medium containing ribavirin. The desirable potato stocks, after freeing them from viruses, need proper 'preservation' to serve as the starting material for various purposes like seed production, breeding, etc. They have to be preserved in such a manner which avoids their reinfection by viruses and maintains their viability. The method of choice is in vitro preservation of these stocks. The rate of growth of the culture is retarded by lowering the temperature or by the application of chemicals like N-dimethyl succinamic acid (67). Such cultures need transfer to fresh culture medium once every 1 to 2 years.

Cultural Practices

Cultural practices have a profound bearing on the spread of virus diseases. Some management practices bring together large communities of crop plants and weeds associated with their cultivation. Some of these crop plants and the weeds happen to be the reservoirs of viruses and their vectors. Factors like date of planting, isolation, roguing of volunteers, virus-infected plants at the appropriate time, disinfection of tools and farm equipment, killing of vines at the right time, etc. play an important role in the spread and dissemination of potato viruses. Thus, reinfection of healthy potato stocks during seed multiplication is effectively checked by (i) selecting a field free from soil borne diseases and pests, (ii) selecting a planting date which ensures a minimum of 75 days free from aphid vectors and the period is suitable for crop growth and tuberization; (iii) Soil application of systemic granular insecticides at the time of planting to keep the vectors under control; (iv) adopting to full (blind) earthing up at planting time and using herbicides to control weeds thereby minimizing the chances of spread of mechanically transmitted viruses and PSTVd; (v) inspection of the crop to remove off-types and virus-infected plants; (vi) sufficient isolation of the field from ware potato crop; (vii) killing of the vines as soon as vector population tends to reach the critical level, and preventing regrowth to avoid infection through new growth; (viii) proper storage of the produce, etc.

Insecticides and Oilsprays

Insecticides are often effective against the spread of persistently aphid-transmitted viruses like PLRV but not against the spread of non-persistently aphid transmitted viruses (42). Sometimes, use of conventional insecticides even increases the incidence of the virus within the crop, presumably due to increased probings by the agitated vector. Non-persistently aphid transmitted viruses are transmitted quickly even by the short duration probings of less than a minute by the vector. Consequently, only the newer class of insecticides, called synthetic pyrethroids, which can quickly knock down the vector, hold some promise. The pyrethroid (Karate:ICI, PP 321) application gave maximum control of the aphid vectors *Macrosiphum euphorbiae* and *Myzus persicae* and halved the incidence of PVY (75).

Since the first demonstration by Bradely (11) on the effectiveness of mineral oil sprays in reducing the infection of PVY, many others have reported similar results. Prevention of PVY spread by the oil sprays has been reported to vary from 50% to over 90%. Giving six to nine sprays, with a total volume of 100 to 150 liters, starting from the moment the plants emerged, often resulted in reduction of PVY spread by 70% or more (89, 90). However, the use of mineral oils has not become popular for virus control in potatoes due to high cost of the oil and also labor required for several sprays.

Insect Repellents

Exploitation of alarm pheromones, which are secreted by aphids in response to attack by predatory insects and which cause them to disperse rapidly, is being investigated. Derivatives of the common alarm pheromone (E)-B-farnesene (a sequiterpene hydrocarbon) have been found to inhibit acquisition of beet yellow virus and PVY by the aphids (20). The same pheromone has been found in the glandular leaf hairs of the wild potato *Solanum berthaultii* which inhibits colonization by aphids (34). Similarly, the active antifeedant molecule from *Polygonum hydropiper* is the sequiterpene dialdehyde polygodial (71) which decreases the aphid colonization and acquisition of PVY (35).

Non-conventional Methods

Production of Virus-resistant Plants Through Genetic Engineering

Cross-prntection or virus induced resistance in a plant, infected with one virus, to a second infection by the same or a related virus, is a long standing observation (94). The precise mechanism of this phenomenon is still not understood. It is possible that this type of resistance may be operative at one or more different levels during the process of virus infection which involves viral entry, uncoating, replication, maturation, cell-to-cell movement and symptom development. Interference with any of these stages can, in principle, result in the plant being resistant to the virus. Genetic engineering has recently been employed to understand this mechanism and evolve new scientific method(s) for virus disease control.

Genetic engineering employing the tools of plant genetic transformation has made possible the introduction of selected genes into plants to confer novel phenotypes. These techniques have broken the species/genus/phylum barriers for the transfer of genes as different plant virus genes have been introduced into plants to confer novel phenotypes (81). The first example of genetically engineered virus resistance was reported by Powell-Abel *et al.* (78). Plant virus nucleic acid sequences, particularly the ones encoding the coat protein (CP) gene, have been found useful in the development of virus-resistant plants (6).

Virus Coat-protein Gene Mediated Protection

Important commercial potato cultivars Bientje, Escort (43), and Russet Burbank (57) have been transformed with and express the PVX CP-gene. Transgenic potato plants accumulated PVX-CP at 0.05 to 0.2% of total plant protein. Inoculation of PVX-CP(+) Bintje or Escort with PVX caused a 20-to 50-fold less PVX accumulation-than PVX-CP(-) plants, and PVX-CP(-) plants showed slower symptom development (43). Plants accumulating more PVX-CP were more resistant to PVX and PVX-CP (+) Russet Burbank plants did not accumulate any PVX at three different inoculum concentrations tested (57). Like potato plants expressing PVX-CP gene, those expressing the PVY-CP gene also showed resistance to PVY infection and accumulated PVY-CP equivalent to 0.01-0.05% (W/W) of leaf protein (57). Plant lines of Russet Burbank potatoes expressing the CP genes of both PVX and PVY were resistant to inoculation with PVX and PVY simulateneously (57).

Russet Burbank potato lines CP(+) for PVX and PVY and CP(-) have been compared for their performance in the field. Out of the four CP(+) lines tested, two were significantly protected from PVX infection while three lines were not protected from infection by PVY upon simultatneous inoculation by both viruses (49). Tuber yields at maturity in uninoculated plots were the same for all the lines [i.e. CP(+) and CP(-)]. In contrast, tuber yields of all inoculated lines, except 303 CP(+)whose yield was not affected by virus inoculation, were markedly reduced.

Three years of field trials with the transformed potato cultivars Bientje and Escort, which express the PVX-CP gene, have shown a large decrease in PVX incidence among clonal progeny obtained from these cultivars previously infected with PVX. Based on the evaluation of 50 defined morphological characteristics, 81.8% of the Escort cultivar and 17.9% of the Bientje cultivar derived transgenic clones proved to be true to type. The results are taken to demonstrate the commercial feasibility of improving potato cultivars by selectively adding new traits while preserving intrinsic properties (48).

The mechanism underlying the CP-mediated resistance is still not clear. It is likely that CP-mediated resistance is affected at several stages depending on the virus-host interaction. One of the stages may be the early event(s) in infection, e.g. uncoating of the introduced virus into the cell and/ or replication of the released viral RNA. Such a phenomenon may be operative in the case of virus infections where CP-mediated resistance is overcome if viral RNA is used as inoculum rather than virions, e.g. TMV (72), AlMV (61,105) and tobacco rattle virus (105). However, incoulation with PVX-RNA did not overcome CP-mediated resistance against PVX (41). The other stage where CP-mediated resistance may play role is the cell-to-cell and long distance virus transport within the plant. Prevention or delaying of systemic spread of viral infection in CP(+) plants as compared to CP(-) plants has been observed in many studies (6).

Unlike the link between the amount of CP accumulation and the degree of virus resistance in the transgenic plants reported by many researchers (6), Vlugt and his associates (107) have reported that the resistance in transgenic tobacco plants showed the expression of the CP gene only at the transcription level and not at the translation level. Their results strongly suggested that the resistance observed in the transgenic plants was principally based on the presence of PVY-CP RNA sequences rather than on the accumulation of viral CP. CP-mediated resistance against PVS in *Nicotiana debneyii* plants expressing the PVS-CP gene has been demonstrated (62). The transgenic plants expressing the PVS CP gene were highiy resistant to subsequent infection by the ME strain of PVS as evidenced by the absence of symptom development and a lack of virus accumulation in hoth the inoculated and upper uninoculated leaves. Like plants expressing the PVX-CP gene, plants expressing the PVS CP gene were also protected when inoculated with PVS RNA.

Replicase Gene Mediated Resistance

Genetically engineered resistance has also been demonstrated in plants expressing the read-through portion of the putative replicase gene from TMV (36). The transgenic tobacco lines expressing the putative replicase gene of PVX (ORF[1]) were resistant to PVX infection when inoculated with either PVX virions or PVX RNA (12). Analyses of lines containing various portions of the ORF gene demonstrated that resistance is conferred to plants by expressing approximately the first half of the ORF gene. The line expressing the untranslated leader and first 674 codons of ORF was highiy resistant to PVX infection. On the other hand, lines expressing either approximately the third or fourth quarter of the ORF gene, which contain the conserved nucleotide triphosphate (NTP) binding motif and the glycine/aspartic acid (GDD) motif, respectively are not protected from PVX infection. When performance of the most resistant ORF[1](+) line was compared with the most resistant CP(+) line in a resistance test, the best ORF[1](+)line was more resistant to PVX infection than the best transgenic line expressing the PVX-CP gene.

Other Methods of Resistance Incorporation

Other techniques of incorporating virus-resistance in plants through genetic engineering, viz. satellite RNAs, defective interfering nucleic acids and antisense RNAs, have not been exploited in potato obviously due to either the lack of association of satellite RNAs/defective interfering nucleic acids with the potato viruses discussed here or weak resistance imparted by the antisense RNAs.

CONCLUSIONS

The list of mechanically and non-persistently aphid transmitted viruses infecting potatoes under natural conditions has remained unchanged despite the generation of enormous amounts of information about their molecular biology during recent years. Availability of better virus detection technology and *in vitro* rapid mu!tiplication of stocks has enabled many countries to expand potato production.

The sources of resistance to many of the viruses have been identified and resistance to PVX, PVA and PVY have been incorporated in many potato cultivars. Raising of potato crop from botanical seed, at least in some parts of the world, does not appear to be an unrealizable dream. More region-based information needs to be generated regarding the interactions between the virus isolates, vector behavior and host resistance in order to evolve virus forecasting systems.

Though preliminary results of incorporation of virus resistance in plants through their transformation with different viral genes have been encouraging, full impact of the use of such plants on the viruses in particular and the environment in general, is still to be understood.

REFERENCES

1.   Anonymous. 1981. Annual Report International Potato Center (CIP) for the year 1980. CIP, Lima, Peru. 198 Pages.

2.   Anonymous. 1983. Annual Report International Potato Center (CIP) for the year 1982. CIP, Lima, Peru. 148 Pages.

3.   Anonymous. 1991. Annual Report International Potato Center for the year 1990. CIP, Lima, Peru. Pages 53-58.

4.   Anonymous. 1992. Annual Report, International Potato Center for the year 1990. CIP, Lima, Peru. Pages 61-69.

5.   Bagnall, R.H. 1988. Role of plant resistance in control of the potato viruses. Pages 70-76 in: Potato Pest Management in Canada. G. Boileau, R.P. Singh and R.H. Parry, eds. Canada Agric. NB.

6.   Beachy, R.N., Loesch-Fries, S., and Tumer, N.E. 1990. Coat protein-mediated resistance against virus infection. Annu. Rev. Phytopathol. 28:451-474.

7.   Beekman, A.G.B. 1987. Breeding for resistance. Pages 162-174 in: Viruses of Potatoes and Seed-Potato Production. J.A. deBokx and J.P.H. van der Want, eds. Wageningen, The Netherlands.

8.   Beemster, A.B.R. 1987. Virus translocation and mature-plant resistance in potato plants. Pages 116-125 in: Viruses of Potatoes and Seed-Potato Production, 2nd Ed. J.A. deBokx, and J.P.H. van der Want, eds. Pudoc, Wageningen, The Netherlands.

9.   Beemster, A.B.R., and deBokx, J.A. 1987. Survey of properties and symptoms. Pages 94 in: Viruses of Potatoes and Seed Potato Production, 2nd Ed. J.A. deBokx, and J.P.H. van der Want, eds. Pudoc, Wageningen, The Netherlands.

10.   Bonde, R., and Merriam, D. 1951. Studies on the dissemination of potato spindle tuber virus by mechanical inoculation. Am. Potato J. 28:558-560.

11.   Bradley, R.H.E. 1963. Some ways in which a paraffin oil impedes aphids transmission of potato virus Y. Canadian J. Microbiol. 9:369-380.

12.   Braun, C.J., and Hemenway, C.L. 1992. Expression of amino-terminal portions or full-length viral replicase genes in transgenic plants confers resistance to potato virus X infection. Plant Cell 4:735-744.

13.   Brown, C.R., Salazar, L.F., Ochoa, C., and Chuquillanqui, C. 1986. Strain-specific immunity to PVX[HB] is controlled by a single dominant gene. Proc. 9th Triennial Conf. Eur. Assoc. Pot. Res. Interlaken. Pages 249-250 (Abstract).

14.   Cassels, A.C., and Long, R.D. 1982. The elimination of potato viruses X, Y, S and M in meristem and explant cultures of potato in the presence of Virazole. Potato Res. 25:165-173.

15.   Cockerham, G. 1955. Strains of potato virus X. Proc. 2nd Conf. Potato Virus Diseases. Lisse-Wageningen. Pages 89-92.

16.   Cockerham, G. 1970. Genetical studies on resistance to potato viruses X and Y. Heredity 25:309-348,

17.   Conrad, P.L. 1991. Potato virus S-free plants obtained using antiviral compounds and nodal segment culture of potato. Am. Pot. J. 68:507-513.

18.   Costerveld, P. 1987. Inspection and grading. Pages 204-214 in: Viruses of Potatoes and Seed-Potato Production. J.A. deBokx and J.P.H. van der Want, eds. Pudoc, Wageningen.

19.   Davidson, T.M.W. 1980. Breeding for resistance to virus diseases of the potato (*Solanum tuberosum*) at the Scottish Plant Breeding Station. Report Scottish Plant Breeding stn. Pages 100-108.

20.   Dawson, G.W., Gibson, R.W., Griffiths, D.C., Pickett, J.A., Rice, A.D., and Woodlock, C.M. 1982. Aphid alarm pheromone derivatives affecting settling and transmission of plant viruses. J. Chem. Ecology 8:1377-1388.

21.   deBokx, J.A., and Huttinga, H. 1981. Potato virus Y. CMI/AAB Descriptions of plant viruses, No. 242. 6 Pages.

22.   deBokx, J.A. 1987. Biological properties. Pages 58-82 in: Viruses of Potatoes and Seed-Potato Production. J.A. deBokx and J.P.H. Van der Want, eds. Pudoc, Wageningen,.

23.   Delhey, R. 1974. Zur Natur der extremen Virusresistenz bei der Kartoffel. I. Das X-Virus. Phytopathol. Z. 80:97-119.

24.   Delhey, R. 1975. Zur Natur der extremen Virusresistenz beider kartoffel. II. Das Y-Virus. Phytopathol. Z. 82:163-168.

25.   Dhingra, M.K., Khurana, S.M. P., and Lakhanpal, T.N. 1987. Effect of virazole and 2, 4-D on the growth of potato virus X content of potato leaf callus. J. Indian Potato Assoc. 14:100-103.

26.   Diener, T.O. Viroids and Viroid Diseases. John Wiley & Sons, New York.

27.   Diener, T.O., and Raymer, W.B. 1967. Potato spindle tuber virus: plant virus with properties of a free nucleic acid. Science 158:378-381.

28.   Dodds, J.H., and Horton D. 1990. Impact of biotechnology on potato production in developing countries. Agbiotech. News and Information 2:397-400.

29.   Fermandez-Northcote, E.N. 1982. Prospects for stability of resistance to potato virus Y. Research for the potato in the year 2000. Proc. of the International Congress in Celebration of the 10th Anniversary of the International Potato Center. Lima, Peru. Page 82.

30.   Fribourg, C.E., and Nakashima, J. 1984. Characterization of a new potyvirus from potato. Phytopathology 74:1363-1369.

31.   Garg, I.D., Khurana, S.M. P., and Singh, M.N. 1990. Common weed reservoirs of potato viruses X,S and Y in Shimla hills. Plant Dis. Res. 5:84-86.

32.   Garg, G.K., Singh, U.S., Khetrapal, R.K., and Kumar, J. 1988. Application of tissue culture in plantpathology. Pages 83-119 in: Experimental and Conceptual Plant Pathology. W.M. Hess, R.S. Singh, U.S. Singh and D.J. Waber, eds. Cordon and Breach, New York.

33.   Gibbs, A.J., Hacht-Poinar, E., Woods, R.D., and McKee, R.K. 1966. Some properties of three related viruses: Andean potato latent, Dulemara mottle and Ononis yellow mosaic. J. Gen. Microbiol. 44:177-193.

34.   Gibson, R.W., and Pickett, J.A. 1983. Wild potato repels aphids by release of aphid alarm pheromone. Nature 302:608-609.

35.   Gibson, R.W., Rice, A.D., and Sawicki, R.M. 1982. Effects of the pyrethroid deltamethrin on the

36. acquisition and inoculation of viruses of by *Myzus persicae*. Ann. Appl. Biol. 100:49-54.

36. Golomboski, D.B., Lomonossoff, G.P., and Zaitlin, M. 1990. Plants transformed with a tobacco mosaic nonstructural gene sequence are resistant to the virus. Proc. Natl. Acad. Sci., USA. 87:6311-6315.

37. Grewal, J.S. (ed.) 1992. True potato seed for potato production in India, Revised Edn. Tech. Bull. 24. CPRI, Shimla. 16 Pages.

38. Griffiths, H.M., Slack, S.A., and Dodds, J.H. 1990. Effect of chemical and heat therapy on virus concentrations *in vitro* potato plantlets. Can. J. Bot. 68:1515-1521.

39. Gross, H.J., Domdey, H., Lossow, C., Jank, P., Raba, M., Alberty, H., and Sänger , H.L. 1978. Nucleotide sequence and secondary structure of potato spindle tuber viroid. Nature 273:203-208.

40. Gross, R.W. 1930. The symptoms of potato spindle tuber and unmottled curly dwarf on the potato. Nebr. Agric. Exp. Stn. Res. Bull. 47:39.

41. Hemenway, C., Fang, R.X., Kaniewski, J.J., Chua, N.H., and Tumer, N.E. 1988. Analysis of the mechanism of protection in transgenic plants expressing the potato virus X coat protein or its antisense RNA. EMBO J. 7:1273-8.

42. Hille Ris-Lambers, D., Reestman A.J., and Schepers, A. 1953. Insecticides against aphid vectors of potato viruses. Netherlands J. Agr. Sci. 1:188-201.

43. Hoekema, A., Huisman, M.J., Molendijk, L., van der Elzen, P.J.M., and Cornelissen, B.J.C. 1989. The genetic engineering of two commercial potato cultivars for resistance to potato virus X. Bio/Technology 7:273-278.

44. Horton, D. 1988 Potatoes, Production, Marketing and Programmes for Developing Countries. Westview Press, Boulder, Colorado, USA. 243 Pages.

45. Hunter, D.E., Darling, D.H., and Beale, W.L. 1969. Seed transmission of potato spindle tuber virus. Am. Potato J. 46:247-249.

46. Jones, R.A.C., and Fuller, N.J. 1984. Incidence of potato virus V in potato stocks in England and Wales. Plant Pathol. 33:595-597.

47. Jones, R.A.C., Fribourg, CE., and Slack, S.A. 1981. Potato Virus and Virus-like Diseases, Set No. 2. O.W. Barnett and S.A. Tollin, eds. Clemson University, South Carolina. 59 Pages.

48. Jongedijk, E., Schutter, A.A.J.M., de Stolte, T., van den Eljen, P.J.M., and Cornelissen, B.J.C. 1992. Increased resistance to PVX and preservation of cultivar properties in transgenic potato under field conditions. Bio/Technology 10:422-429.

49. Kaniewski, W.K., Lawson, C., Sammons, B., Haley, L., Hart, J. Delanny, X., and Tumer, N.E. 1990. Field resistance of transgenic Russet Burbank potato to effects of infection by potato virus X and potato virus Y. Bio/Technology 8:750-754.

50. Kassanis, B., and Govier, D.A. 1971. New evidence on the mechanism of aphid transmission of potato virus C and potato aucuba mosaic viruses. J. Gen. Virol. 10:99-101.

51. Kassanis, B., and Varma, A. 1967. The production of virus-free clones of some British potato varieties. Ann. Appl. Biol. 59:447-450.

52. Khurana, S.M.P. 1990. True potato seed: certification and quality control. Pages 120-127 in: Commercial Adoption of True Potato Seed Technology - Prospects & Problems. P.C. Gaur, ed.,

53. CPRI, Shimla.

53. Khurana, S.M.P. 1992. Potato Viruses and Viral Diseases. Tech. Bull. No. 35. CPRI, Shimla. 23 Pages

54. Khurana, S.M.P., and Garg, I.D. 1992. Potato mosaics. Pages 148-164 in: Plant Diseases of International Importance, Vol. II. H.S. Chaube, *et al.*, eds. Prentice Hall, NH (USA).

55. Khurana, S.M.P., and Garg, I.D. 1993. New techniques for detection of viruses and viroid. Pages 529-566 in: Advances in Horticulture. K.L. Chadha and J.S. Grewal, eds. MPH, New Delhi.

56. Khurana, S.M.P., Singh, M.N., and Garg, I.D. 1992. Potato viruses-detection and control. Pages 177-196 in: Farming Systems and Integrated Pest Management. J.P. Verma and A. Varma, eds. MPH, New Delhi.

57. Lawson, C., Kaniewski, W., Haley, L., Rozman, R., Newell, C., Sanders, P., and Tumer, N.E. 1990. Engineering resistance to mixed virus infection in a commercial potato cultivar. Resistance to potato virus X and potato virus Y in transgenic Russet Burbank. Bio/Technology, 8:127-134.

58. Leonteva, Y.A. 1964. Indentification of potato gothic. Izv. Kinkyshev Khoz. Inst. 14:279-287.

59. Liu, A. 1980. Experimental results on separation of potato spindle tuber viroid (PSTV-RNA). Acta Phytopathol. Sin. 10:103-105.

60. Lizarraga, R.E., Salazar, L.F., Roca, W.M., and Schilde-Rentschler, L. 1980. Elimination of potato spindle tuber viroid by low temperature and meristem culture. Phytopathology 70:754-755.

61. Loesch-Fries, L.S., Merlo, D., Zinnen, T., Burhop, L., Hall, K. *et al.* 1987. Expression of alfalfa mosaic virus RNA4 in transgenic plants confers virus resistance. EMBO J. 6:1845-1851.

62. Machenzie, D.J. and Tremaine, J.H. 1990. Transgenic *Nicotiana debnejii* expressing viral coat protein are resistant to potato virus S infection. J. Gen. Virol. 71:2167-2170.

63. MacLachlan, D.S., Larson, R.H., and Walker, J.C. 1954. Potato virus A. Am. Potato J. 31:67-69.

64. Manzer, F.E. and Merriam, D. 1961. Field transmission of the potato spindle tuber virus and virus X by cultivating and hilling equipment. Am.Potato J. 38:346-348.

65. Mellor, F.C., and Stace-Smith, R. 1977. Virus-free potatoes by tissue culture. Pages 617-635 in: Applied and Fundamental Aspects of Plant Cell, Tissue and Organ culture. J. Reinert and Y.P.S. Baja, eds. Springer-Verlag, Berlin.

66. Merriam, D., and Bonde, R. 1954. Dissemination of potato spindle tuber by contaminated tractor wheels and by foilage contact with diseased plants (abstr.). Phytopathology 44:111.

67. Mix, G. 1982. *In vitro* preservation of potato material. Plant Genetics Resources Newsletter 51:6-8.

68. Moreira, A., Jones, R.A.C., and Fribourg, C.E. 1980. Properties of a resistance-breaking strain of potato virus X. Am. Appl. Biol., 95:93-103.

69. Munro, J. 1981. Potato virus X. Page 125 in: Compendium of Potato-Diseases. W.J. Hooker, ed. Am Phytopathol. Soc., Minnesota, USA.

70. Murashige, T., and Skoog, F. 1962. A revised medium for rapid growth and bioassays with tobacco tissue cultures. Physiol. Pla. 15:473-497.

71. Nakanishi, K., and Kubo, I. 1977. Studies on warburganal, muzigadial and related compounds.

Israel J. Chem. 16:28-31.

72. Nelson, R.S., Powell-Abel, P., and Beachy, R.N. 1987. Lesions and virus accumulation in inoculated transgenic tobacco plants expressing the coat protein gene of tobacco mosaic virus. Virology 158: 126-132.

73. Owens, R.A., Khurana, S.M.P., Smith, D.R., Singh, M.N., and Garg, I.D. 1992. A new mild strain of potato spindle tuber viroid isolated from wild *Solanum* spp. in India. Plant Dis. 76:527-529.

74. Pennazio, S. 1971. Potato therapy : meristem tip culture combined with thermotherapy (Italian). Reviesta dell' Ortoflorofrutticoltura Italian, 5:446-452.

75. Perrin, R.M., and Gibson, R.W. 1985. Control of some insect-borne plant viruses with the pyrethroid rate. International Pest Control 27:142-145.

76. Peters, D. 1987. Spread of viruses in potato crops. Pages 126-145 in: Viruses of Potatoes and Seed-Potato Production, 2nd Ed. J.A. de Bokx and J.P.H. van der Want, eds. Pudoc, Wageningen.

77. Pfannenstiel, M.A., and Slack, S.A. 1980. Response of potato cultivars to infection by potato spindle tuber viroid. Phytopathology 70:1015-1017.

78. Powell-Abel, P.A., Nelson, R.S., De, B., Hoffman, N., Rogers, S.G., Fraley, R.T., and Beachy, R.N. 1986. Delay of disease development in transgenic plants that express the tobacco mosaic virus coat protein gene. Science 232:738-743.

79. Quak, F. 1961. Heat treatment and substances inhibiting virus multiplication in meristem culture to obtain virus-free plants. Adv. Hort. Sci. 1:144-148.

80. Quak, F. 1987. Therapy of individual plants Pages 151-161 in: Viruses of Potatoes and Seed-Potato Production, 2nd Ed. J.A. deBokx, and J.P.H. van der Want, eds. Pudoc, Wageningen.

81. Reavy, B., and Mayo, M.A. 1992. Genetic engineering of virus resistance. Pages in: Plant Genetic Manipulation for Crop-Protection. A.M.R. Gatehouse, V.A. Hilder and D.J. Boulter, eds. CAB International, Wallingford (U.K.).

82. Rhodes, R.E. The incredible potato. Natl. Geographic 161:68-694.

83. Rose, D.G. 1983. Some properties of an unusal isolate of potato virus S. Potato Res. 26:49-

84. Ross, H. 1954. Die Veverbung der 'Immunitat' gegen das X-Virus in tetraploidem *Solanum acaule*. Proc 9th-Intern. Congr. Genetics; Bellagio, Italy (1953). caryologia 6 (supplement) : 1128-1132.

85. Salazar, L.F., and Harrison, B.D. 1978. Host range, purification and properties of potato virus T. Ann. Appl. Biol., 89:223-235.

86. Salazar, L.F., and Harrison, B.D. 1978. Potato virus T. CMI/AAB Description of Plant Viruses No. 187. 4 Pages.

87. Sänger, H.L. 1982. Biology, structure functions and possible origin of viroids. Pages 368-454 in: Encyclopedia of Plant Physiology, New Series, Vol, 14 B. Nucleic Acids and Proteins in Plants. B. Parthier and D. Boutler, eds. Springer-Verlag, Berlin.

88. Sänger, H.L. and Ramm, K. 1974. Radioactive labelling of viroid RNA, in modification of the information content of plant cells. Proc. 2nd John Innes. Symp. R. Markham, *et al.*, eds. Norwich, England. 250 Pages.

89. Schepers, A., Bus, C.B., and Styszko, L. 1978. Effects of application of mineral oil on seed potatoes (Abstr.).. Pages 269-270 in: Proc. 7th Trien, Conf. Eur. Assoc. Pot. Res., Warsaw.

90. Schepers, A., Harrewijn, P., and Bus, C.B. 1984. Non-insecticidal treatments to reduce aphid development and infection with stylet-borne virus diseases in seed potatoes (Abstr.). Pages 204-206 in: Proc 9th Trien Conf. Eur. Assoc. Pot. Res. Interlaken.

91. Schultz, E.S., and Raleigh, W.P. 1933. Resistance of potato to latent mosaic. Phytopathology, 23:32.

92. Semancik, J.S. 1979. Small pathogenic RNA in plants-the viroids. Annu. Rev. Phytopathol. 17:461.

93. Shekhawat, G.S., Grewal, J.S., and Verma, S.C., eds. 1992. Potato in India, Revised Edn. Tech Bull. CPRI, Shimla. 62 Pages.

94. Sherwood, J.L. 1987. Mechanisms of cross-protection between plant virus strains. Pages 136-150 in: Plant Resistance to Viruses, CIBA Foundation Symposium No. 133. Wiley, Chichester.

95. Singh, M.N., Khurana, S.M.P., Nagaich, B.B., and Agrawal, H.O. 1981. Epidemiological studies on potato viruses Y and leafroll in sub-tropical India. Pages 89-90 in: Proc. Intern. Epidemiological Conf. Oxford.

96. Singh, M.N., Khurana, S.M.P., Nagaich, B.B., and Agrawal, H.O. 1982. Efficiency of *Aphis gossypii* and *Acyrthosiphon pisum* in transmitting potato viruses leafroll and Y. Pages 289-293 in: Potato in Developing Countries. B.B. Nagaich, *et al.*, eds. Indian Potato Assoc., Shimla.

97. Singh, M.N., Nagaich, B.B., and Agrawal, H.O. 1984. Spread of viruses Y and leafroll by aphids in potato fields. Indian Phytopathol. 37:241-251.

98. Singh, R.P. 1988. Role of weeds in potato virus spread. Pages 355-362 in: Potato Pest Management in Canada. G. Bioiteau, R.P. Singh and R.H Parry, eds. Canada Agric. N.B.

99 Singh, R.P. 1989. Plant viroids-a biochemical novelty. Pages 259-288 in: Plant Viruses, Vol. I, structure and Replication. C.L. Mandahar, ed. CRC Press, FL.

100. Singh, R.P., Finnie, R.F., and Bagnall, R.E. 1971. Losses due to the potato spindle tuber virus. Am. Potato J. 48:262-264.

101. Slack, S.A. 1983. Identification of an isolate of the Andean strain of potato virus S in North America. Plant Dis. 67:786-789.

102. Song, B.F. 1984. Use of true potato seed in China. Circular International Potato Centre, 12(2):6-7.

103. Stace-Smith, R. 1985. Virus-free clones through plant tissue culture. Pages 169-179 in: Comprehensive Biotechnology. C.W. Robinson and R.A. Howell, eds. Oxford Pergamon.

104. Stace-Smith, R., and Mellor, F.C. 1968. Eradication of potato viruses X and S by thermotherapy and axillary bud culture. Phytopathology 58:199-203.

105. van Dun, C.M.P., Bol, J.F., and van Vloten-Doting, L.1987. Expression of alfalfa mosaic virus and tobacco rattle coat protein genes in transgenic plants confers virus resistance. EMBO J. 6:1845-1851.

106. van Harten, A. 1983. The relation between aphid flights and the spread of potato virus Y^N (PVY^N) in the Netherlands. Potato Res. 26:1-15.

107. Vlugt, R.A., van der, Ruiter, R.K,. and Goldbach, R. 1992. Evidence for sense RNA-mediated protection to potato virus Y^N in tobacco plants transformed with the viral coat protein cistron. Plant. Mol. Biol. 20:631-637.

108. Walkey, D.G.A. 1987. Production of virus-free plants by tissue culture. Pages 151-161 in: Tissue

Culture M ethods for Plant Pathologists. D.S. Ingram and J.P. Helgeson, eds. Pudoc, Wageningen.

109. Wetter, C. 1971. Potato virus S. CMI/AAB Descriptions of plant viruses No. 60. 3 Pages..

110. Wright, N.S. 1974. Retention of infectious potato virus X on common surfaces. Am Potato J. 51:251-255.

CHAPTER 46

Present Status of Controlling Conventional Strains of Plum Pox Virus

H. Kegler and W. Hartmann

Plum Pox disease was detected in 1910 in Macedonia and was first described by Atanasoff in 1932 (2). The disease thereafter was observed on apricot in Bulgaria and on peach trees in Hungary and Germany (9, 56).

Other than the northern regions, plum pox disease is more or less widely distributed throughout Europe and has become one of the most destructive virus diseases of fruit trees. In severely affected areas about 90-100 % of the plum trees can be infected. Yield losses may amount to 90-100 % in susceptible cultivars.

Recently, plum pox disease on cherry has been described in Europe. This disease and its' causal agent are presented in the next chapter.

PLUM POX VIRUS

Virus and Virus Strains

Plum pox potyvirus (PPV) has filamentous particles 750 nm long and 15 nm in diameter. It is a single-stranded RNA virus with a molecular weight of 3.5×10^6 Da. Protein inclusions of the pinwheel type are present in the cytoplasm of infected leaves and fruits. The nucleotide sequence of the viral RNA has been determined (38). Sequence differences among PPV strains have been detected and seem to be spread in a uniform manner on the genome (43).

Two groups of PPV strains were identified on the basis of immunological properties, the size of capsid protein, the restriction profile of the cDNA of the corresponding nucleic acid and the sequence of the nucleic acid. These two groups are the D- and M-strains which show different epidemiological behavior: M-strains seem to spread more readily in nature than D-strains and seem to cause more severe infections in peach

orchards (34). Virus isolates from occidental Europe normally corresponded to the D-strains whereas those from oriental Europe showed a profile similar to the M-strains.

Symptoms

PPV damages the fruits of sensitive cultivars of plums, apricots, peaches and nectarines. Affected plum fruits are deformed and show deeply engraved rings, irregular lines and poxes on the surface. The flesh turns brownish red and is saturated by gum. Damaged fruits drop prematurely and become unsuitable both for direct consumption and for industrial processing (Figure 1). Several plum cultivars, however, develop only colored spots, rings or lines appearing on the epidermis.

Colored rings and bands appear on the skin of apricot fruits. The discolored areas sometimes turn brown and become sunken. Fruit symptoms of peaches and nectarines are mostly restricted to the skin where pale rings and diffuse bands appear before maturation (Figure 2).

Most of the susceptible plum, apricot, peach and nectarine cultivars show leaf symptoms which appear as pale or yellowish green rings, spots or leaf mottling (Figure 3).

Distinct, highly sensitive plum cultivars like 'Ortenauer', aside from fruit and leaf symptoms, show bark splitting and cancers on the shoots that are brittle and degrade (Figure 4) (22). Infected trees decline within only a few years.

Hosts

PPV can infect cultivated, ornamental and wild species of *Prunus*, such as *P. armeniaca, P. cerasifera, P. domestica, P. glandulosa, P.*

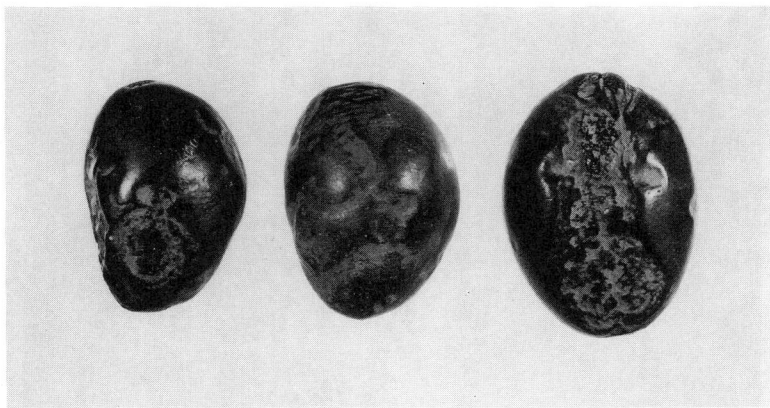

Figure 1. Plum fruits with plum pox symptoms

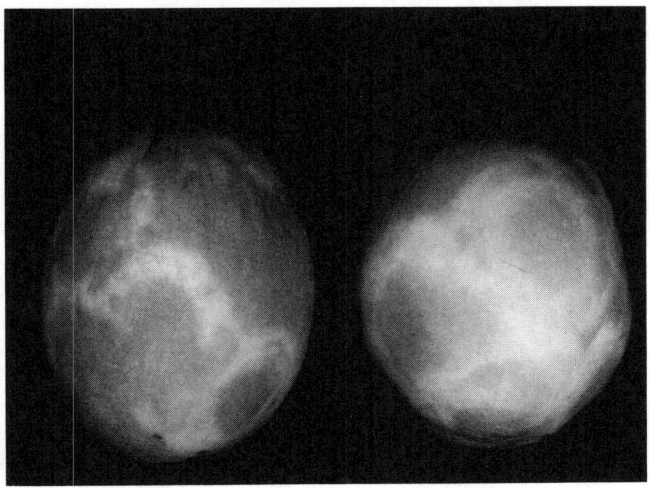

Figure 2. Peach fruits with plum pox symptoms

insititia, *P. persica*, *P. persica* var. *nectarina*, *P. spinosa*, *P. salicina* and *P. tomentosa*. The virus was also detected in naturally infected sour cherry trees (*P. cerasus*) in Moldova and was found to be not identical with PPV from plums (41). Weedy annual plants can carry the virus, but natural transmission between herbaceous plants and *Prunus* species has never been proved.

Natural Virus Transmission

PPV is naturally spread by aphid vectors. *Brachycaudus helicrysi*, *B. cardui*, *Myzus persicae* and *Phorodon humuli* transmit the virus most efficiently (3,26,35). PPV is a non-persistently transmissible virus, the acquisition time amounts to only 5-10 min and the vectors remain infectious for only a few hours, however, this short period may be enough for spreading the virus over long distances by passive flight of the aphids. The efficiency of the vectors in transmitting PPV varies according to the susceptibility of the cultivars. Only short acquisition periods and only few migrant aphids have been proved to be sufficient to transmit the virus after feeding on highly susceptible cultivars. More resistant

Figure 3. Plum leaf (left) and apricot leaf (right) with plum pox symptoms

Figure 4. Bark splitting on a sensitive plum cultivar

Figure 5. PPV induced chlorotic local lesions on *C. foetidum*

cultivars need longer acquisition periods and higher aphid densities for successful PPV transmission (23,40). Furthermore, the virus strains may differ in their vector transmissibility to distinct host plants, especially to peach (39).

The rapidity of natural PPV spread in orchards depends on the distance of the healthy trees to the infection sources. It was established that within a 100 m diameter from an infected individual tree, 48-100 % of the previously healthy trees became infected within 10 years (24). However, plum trees which were grown at a distance of 500 m from an infection source remained healthy.

Virus Detection

PPV is irregularly distributed in the infected tree, therefore, visual inspection of the plants during the growing season is an essential measure for evaluating the state of health.

Biological, serological and/or molecular biological techniques can be used in order to test symptomless trees for latent infections.

Biological Testing

A suitable indicator plant for detecting PPV by biological testing in the greenhouse is *Chenopodium foetidum*. The mechanical inoculation is carried out with crude leaf extracts which were prepared from young leaves mixed with a stabilizing mixture (1 : 4 / w : v) of 0.015 M Na-diethyl-dithiocarbamate, 0.015 M N'N'-diphenylthiourea and 0.03 M phosphate-buffer, pH 7.5 (33).

After 5-10 days, *C. foetidum* reacts with chlorotic or necrotic local lesions on the inoculated leaves (Figure 5). Biological indexing, however, is not as reliable as serological testing.

Serological Testing

PPV is a serologically active antigen but antiserum production is not feasible for every virus strain, due to the different serological aptitude of the strains (58). Among other factors, the difference in serological activity may be due to the different accumulation of the PPV strains in the host plants. "Necrotic strains" attain much higher titers than "yellow strains" (47). The PPV titer also depends on the host plant. The most suitable species for virus multiplication regarding antiserum production are *Nicotiana clevelandii* and *N. occidentalis* (32).

Considering all these factors mentioned above, it should be possibleto receive diagnostic antisera which are suitable for detecting PPV in order to produce virus-free propagation materials and for evaluating virus resistance.

The leaf extracts for ELISA are prepared with PBS tween-buffer + 2 % PVP (1 : 10 / w : v) (8), using the microvariant of ELISA (53) with the simultaneous addition of conjugates and antigens (16). Because young leaves have lower virus titers, leaves of the base or middle regions of the shoots should be used for diagnosis. The most suitable period for serological detection of PPV, under middle European conditions, is from the end of May to early August (19).

PPV was also diagnosed by ISEM in plum and peach trees with leaf and fruit symptoms (46) as well as in the bark of infected trees without symptoms.

Molecular Techniques

A highly sensitive assay, based on polymerase chain reaction amplification (PCR) of cDNA synthesized from the viral RNA of antibody-captured viral particles, was developed for PPV detection (66). The reaction, called immunocaptured/PCR (IC/PCR), yields a specific 243-bp product. The immunocapture step, by allowing the use of large sample volumes and by the viral particle pre-purification it achieves, dramatically increases the sensitivity of the assay. As few as 8000 target viral particles per ml of plant extract could be detected by IC/PCR. When compared with direct PCR, molecular hybridization using ^{32}P-labelled cRNA probes and ELISA this result corresponds to a 250-fold, 625-fold and 5000-fold increased sensitivity, respectively. The high sensitivity of IC/PCR was confirmed during an indexing trial with field samples collected from naturally infected apricot trees. Another sensitive RT-PCR assay for the detection of PPV is that described for the 3' non-coding region of the viral genome. This assay has been successfully utilized in establishing that the potyvirus infecting sour cherry trees in Europe is indeed a strain of PPV.

METHODS OF CONTROLLING PLUM POX DISEASE

Some countries started the controlling of plum pox disease by destroying all obviously infected trees within a growing region. But the results of these procedures proved to be insufficient because newly planted healthy trees became infected within a short time by several natural infection sources among the wild growing host plants of the genus *Prunus*. Long term experiments demonstrated that different measures applied together are most effective.

Choice of Growing Region

A most important prerequisite for the health and yield effectiveness of plum, apricot and peach orchards is the proper choice of the orchard location. In regions where PPV is an endemic pathogen, cultivars sensitive to this virus should not be grown near infection sources. Such sources may be severely affected orchards, single infected trees in gardens or near the roads as well as latent virus carriers among several *Prunus* species. The minimal distance from infected plants should be not less than 500 m (22).

A protecting effect against spontaneous infections by PPV from outside the orchards may be obtained by growing non-host-plants around susceptible stone fruit crops. For this matter pome fruit species and nut trees are suitable. Because most of the economically important plum cultivars are sensitive to PPV, orchards of these cultivars should be "cut through" by other cultivars which are not as productive, but resistant to PPV. These resistant cultivars may act as 'barriers" against a fast spread of PPV within the orchard.

Plant Quarantine

Many examples exist which show that PPV, in the past, has been introduced into previously sharka-free regions by virus-infected budwood, rootstocks or young nursery trees. To prevent this from occurring, the exchange of plant materials is controlled by stringent regulations in most countries. Since PPV is one of the most dangerous plant viruses, PPV-free plant material must be guaranteed by an official phytosanitary certificate for international exchange of susceptable plant material

Because PPV is the object of quarantine, all imported host material (except seeds) should come from a field subjected to inspection during the growing-season. Inspection should also include examination of the immediate vicinity of the field. Budwood, rootstocks and nursery plants should be derived from tested mother plants and should be certified as virus-free.

Not only the international exchange of plant material, but also the exchange of plant material within a country should heed the danger of bringing PPV into previously sharka-free areas. Therefore, different forms of inland-quarantine measures have been introduced in several countries. Introduced plant material, regardless of PPV-free certification, should be transferred and kept under quarantine until investigators have established its' health status. Special attention should be given to ensure proper isolation of the quarantine blocks.

Production of Virus-free Plant Materials

One of the most important prerequisites for minimizing the natural spread of plum pox disease within orchards is the planting and growth of virus-free trees. Therefore, care must be taken to use virus-free propagation materials in the nurseries. In order to achieve this goal, most countries established programs for indexing mother plants, selecting virus-free individuals and propagating virus-free budsticks and rootstocks.

The selection of virus-free source material starts by selecting individual plants which are characterized by some important properties: trueness of the cultivar, high productivity and freedom from disease symptoms. Plants which fulfill these prerequisites are kept and propagated by budding while simultaneously performing biological indexing in an insect-free screenhouse covered by dense (0.6 mesh) nets. Additionally, aphids in the greenhouse should be controlled by frequent spraying with insecticide.

Virus testing is performed using a procedure which is divided into two steps:

Preliminary Testing

Biological testing in the greenhouse by using herbaceous test plants and/or serological testing by ELISA as well as ISEM and IC/PCR techniques are used as methods for preliminary testing.

The reliability of the pre-testing methods is highly dependent on the virus titer and the degree of the systemic virus spread in the host plant. Therefore, highly susceptible cultivars with higher virus titers and a more complete systemic virus spread are desirable.

Preliminary testing enables the detection of virus-infected plants in a short time and makes the main testing more effective.

Main Testing

Main testing is carried out in the field with woody indicators. They are most reliable but very costly, time-consuming and labor-intensive. Suitable indicators for detecting PPV are 'Italian Prune' (*P. domestica*) and seedlings of *P. tomentosa*. The indicator plants can be inoculated by chip budding, grafting or budding. Chlorotic spots, rings or bands and leaf deformations appear on the leaves after inoculation and sometimes bark splitting develops if the source plants were infected.

By the end of the testing period, healthy progeny of indicator mother trees are selected. The trees can be then transferred from the screening nursery to foundation blocks or nuclear stocks (42). They should be protected from spontaneous infections (a minimum distance of 500 m) and by intensive insecticide spraying. Plantings should be established in an area where plum pox disease is not endemic. The mother plants of the nuclear stocks should be serologically re-indexed annually.

Heat Therapy

In case no healthy individuals could be selected by testing, heat therapy either alone or combined with apical meristem culture is applied. For heat therapy treatments young trees growing in pots are put into a heat chamber at temperatures of 37°C for 20-40 days. Conditions within the chamber should enable a rapid growth to attain a dual effect: inhibition of virus multiplication by high temperatures and a decrease in virus concentration due to intensive plant growth. After some weeks the apical portion of the shoots (short tips about 10-20 mm) are cut off and grafted to the healthy potted seedling rootstocks. Smaller portions of the apical tips (a maximum of 2 mm long) are placed on nutrient agar containing macro- and micro-nutrients, vitamins, plant growth hormones and sugars after surface disinfection. The cultures are kept in sterile, closed containers in air-conditioned rooms at a suitable temperature (about 24°C) and light intensity (about 10.000 lux). Later the plantlets are transferred to greenhouse conditions. Determination of the virus-free status of these plantlets must be established by testing as described above.

Health Control Within the Orchards

Young orchards must be monitored in the

years after planting. Diseased trees should be destroyed and replaced by healthy trees. After 5-8 years, plum trees seem to become less endangered because virus spread in the orchards takes place slower and the virus often remains limited to single parts of the crowns (22).

Spring and fall are especially dangerous times for the transmission of PPV (35). Chemical control of virus vectors may reduce the spread of the virus within orchards. Controlling the vectors is particularly important in fall when the aphids change from herbaceous to woody host plants for laying winter eggs. This takes place in middle European countries from the end of September to the end of October. Therefore, late ripening plum cultivars can be treated after harvest. The most suitable time for applying insecticides should be announced by specialists who continuously investigate the migration of the aphids. Among the suitable insecticides, Pirimor has proven long term effectiveness and it spares useful insects. It should be arranged to apply insecticides over a large area (22).

GROWING PPV RESISTANT CULTIVARS

Virus diseases which are spread rapidly in nature can not be sufficiently controlled solely by the measures mentioned above. If economic damages are severe, growth of virus resistant cultivars is the most effective way to control virus diseases.

Types of Resistance

A knowledge of resistance mechanisms is essential for effective inspection and breeding for resistance. In every host-virus-combination different types of resistance may occur. Both qualitative and quantitative virus resistance have been established in plum, peach and apricot against PPV (28).

Qualitative virus resistance is mostly mono- or oligogenically controlled. This type of resistance is complete because the virus is localized by hypersensitive reaction of the host plant around the infection sites. Qualitative resistance often is not durable because new pathotypes may arise and break the resistance.

Quantitative resistance is mainly polygenically controlled and is not complete because the virus does not become localized ("non-localized resistance"). The extent of this type of resistance is variable and is characterized by different traits, such as extension of the incubation period, a smaller infection rate, reduced virus titer in the host tissues, an incomplete systemic virus spread in the host plant, milder or no symptoms, and smaller or no yield losses are the most important traits. All of these traits must be investigated in order to evaluate the type and grade of resistance. The manifestation of resistance is affected by a range of factors; e.g., the virulence of virus strains, infection dose, the age of the plants and environmental conditions, particularly the temperature.

Screening for PPV Resistance

One of the most important prerequisites for screening for virus resistance is the choice and characterization of suitable virus strains. Virus strains should be endemic in the region where the cultivars to be bred are grown. Strain virulence should adequately match the level of resistance which can be expected from the genotypes to be investigated: Highly resistant genotypes should be chosen for highly virulent virus strains and vice versa. The virulence can be characterized and evaluated by the severity of symptoms they cause in different cultivars and by the virus concentration in the host plant.

Screening for PPV-resistance can take place in two steps (27):

Preliminary Screening

Testing for PPV-resistance can be initially carried out in the greenhouse and/or in the field. In the greenhouse, genotypes can be investigated by double grafting on seedling rootstocks. A budstick of the virus source is combined with a budstick of the potted rootstock. Highly susceptible genotypes show leaf mottling and high PPV titers in the leaves about 8 weeks after grafting. Genotypes classified as highly quantitatively resistant to or possibly immune to PPV show no leaf symptoms and no or very low virus accumulation. Hypersensitive genotypes, however, react with tip necrosis and/or necrotic leaf spotting. ELISA extinction values of these genotypes differ rather strongly depending on the strength of the necroses (29).

Preliminary screening can also be carried out by crown-grafting. Plum, apricot, peach or nectarine trees of five year old, highly susceptible cultivars should be infected with a suitable PPV strain by chip budding. Two years later, these "infection stocks" are grafted with every 3-5 scions of the genotypes to be investigated. More than one genotype can be grafted on one infection stock if there are sufficient graftable shoots. Leaf symptoms of the grafted genotypes can be evaluated during the season after grafting and the fruit reactions one or two seasons later. Hypersensitive genotypes react with tip, leaf and bark necrosis as described by the greenhouse test. Partially (quantitative) resistant genotypes show very mild or no leaf symptoms and an intrinsic lower virus titer in the leaves. Cultivars susceptible to PPV are used in the experiments as standards for comparison.

The fastest for the detection of reactions of apricot cultivars to PPV infection were obtained using aphid transmissions of PPV-M strain. Different levels of resistance or tolerance were found (12).

Main Screening

In a second step, genotypes with traits of PPV-resistance are to be investigated by infection experiments with young trees under field conditions. Experimental infections are induced by chip budding or grafting on infected rootstocks and natural infections are induced by aphid vectors. The infection rates, severity of symptoms on leaves and fruits, virus concentration in the leaves, and fruit yield and quality are to be investigated and measured over the course of 5 years.

The severity of the symptoms can be visually evaluated by different scores:

1 = no symptoms
2 = mild symptoms
3 = "normal" symptoms
4 = severe symptoms
5 = very severe symptoms, including necroses.

A disease index (DI) can then be calculated with these scores (65):

$$DI = \frac{(n_1 m_1 + n_2 m_2 + n_3 m_3 + \ldots + n_n m_n)}{N \times m} \times 100$$

n_1 = number of damaged leaves/fruits
$m_1 - m_5$ = grade of damaging (scores)
N = number of leaves/fruits investigated

m = maximal damaging

Using reflectance spectroscopy it is possible to differentiate between the leaf colors of healthy and infected plants and to indicate the gradual differences in leaf discoloration occurring in sensitive and resistant genotypes of plums (31). By utilizing this technique it is possible to establish an exact relationship between symptom intensity and relative virus titer in the leaves.

The virus titer is measured by ELISA as described above. Low virus titer is not only an important indicator for virus resistance but is also important for PPV ecology: Cultivars with low virus titer in the leaves are poor sources for virus acquisition by aphids and, thus, counteract virus spread in orchards.

The yield is measured in terms of the development of usable fruit on infected trees as compared to that of uninfected trees of the same genotype. The smaller the difference between the yields from infected and non-infected trees, the higher the level of (quantitative) PPV-resistance.

Evaluation of PPV Resistance

The degree of leaf and fruit injuries caused by the virus is estimated by the symptoms induced after artificial and natural infections. A correlation between the degree of leaf injury and virus titer (5) as well as a correlation between leaf and fruit symptoms were found. There are exceptions, therefore, the fruit symptoms are the most important criteria for selection.

The resistance has to be evaluated with regard to the type of resistance that exists. Qualitative PPV-resistance is evaluated alternatively: A genotype remains healthy or becomes infected. Hypersensitive genotypes may react with tip necroses (Figure 6). PPV-M strains remain localized to the necrotic zone of the shoot tip and no symptoms appear on the other leaves and shoots. PPV-D strains, however, in addition to the tip necrosis cause chlorotic and/ or necrotic spots on the leaves and the virus is not localized (17). This phenomenon confirms that qualitative virus resistance is virus strain specific.

Quantitative PPV-resistance is very important for resistance breeding because it is durable. However, we should notice that this type of resistance is relative, therefore, gradual differences in the level of cultivar resistance must be proven. In most of the resistant plum cultivars this type of resistance was established by

Figure 6. Tip necrosis on a hypersensitive plum genotype ('K 4') after graft inoculation with PPV-M

evaluation of fruit and leaf symptoms. Genotypes with milder or no any symptoms were characterized as resistant or tolerant.

Field resistant genotypes do not become infected in PPV affected orchards although they are susceptible when graft inoculated. From the breeders point of view, this type of resistance may be very useful especially in countries with a high phytosanitary standard. The trees may remain healthy in the orchard and there will be no negative influence on their productivity, similar to virus-free nursery trees with other types of resistance. For example, field resistance was found in 'Bühler Frühzwetsche', as well as other cultivars. Some field resistant cultivars were found less susceptible to aphid vectors (57). Observations showed that leaf color may play an important role in attractiveness of the plant to virus vectors. Light green leaves are preferred by aphids and may support virus infection.

Breeding PPV-resistant Cultivars

The efficiency of fruit breeding is relatively low because of the long juvenile period of seedlings (4-8 years). In the past, twenty to twenty-five years were needed for producing new cultivars. In order to shorten the juvenile period, pre-selection methods and regional testing after the first selection very efficiently abridge the cycle of breeding by 10-15 years. A quick solution to the sharka problem through breeding is impossible, but there are some opportunities for maintaining plum cultivation in PPV affected areas.

Immediate Measures

The first step for the maintenance of plum cultivation in PPV affected areas is to look for cultivars proven resistant or tolerant in other regions or countries.

In a second (or parallel) step, the PPV-resistance should be checked by testing the cultivars under the specific regional conditions with the spectrum of virus strains occurring in the region. By utilizing this screening process, recommending resistant cultivars for cultivation is possible after 5-6 years.

A special breeding program which combines PPV-resistance with other desired properties can be started in a second step.

Genetic Resources

Looking for genetic resources involves first the inspection of naturally infected orchards (10, 25). The resistant cultivars 'Stanley', 'Monfort', 'Burbank' and 'Ontario' were found this way. The aim of this screening was to find resistant cultivars only for plum production and not for breeding., Different cultivars were examined for PPV resistance by artificial inoculation for the same purpose (1, 11, 20, 29, 64). In Moldova, the highly resistant cultivars 'Pamjatj Vavilova' and 'Vengerka postnjaja', which exceeded the standard, 'Stanley', in the level of resistance to the development of fruit symptoms, were detected (4). Other cultivars also seem to be suitable for crossing experiments and commercial growing. However, the evaluation of PPV resistance of certain cultivars from different sources is not always identical. Different virus strains and climatic conditions as well as false cultivar identities may be reasons for this discrepancy.

Wild or primitive cultivars may be useful as genetic resources as well. they are only indirectly useful for practical breeding because back crossings with adapted parents are necessary. A direct crossing for breeding cultivars is not advisable because, as a rule, the progenies are not productive (55). The use of wild forms in PPV-resistance breeding is only interesting if immune cultivars are found. Resistant hybrids are received more quickly when highly resistant cultivars with good fruit characters like 'Cacanska Naibolja', 'Stanley' and others are used.

Several clones of *P. cerasifera*, *P. insititia* and *P. mariana* and interspecific hybrids (*P. mariana* x *P. cerasifera*) were tested against PPV strains D and M. Two clones of *P. cerasifera* and two hybrids showed less susceptibility than susceptible cultivars (14).

Breeding Programs for PPV Resistance

Specific programs for PPV resistance breeding were elaborated in different countries. Immediately after 'Scoldus' and 'Butilcovidna' had been assessed as sharka resistant (59), a breeding program was started in Yugoslavia with these and several other cultivars: 'Scoldus No. 1' and 'Zh'lta Butilcovidna' were used as immune parents whereas 'Opal', 'Stanley' and 'Ca-canska Naibolja' were used as partially (quantitative) resistant parents. 'Cacanska Lepotica', 'Large Sugar Prune' and 'Ruth Gerstetter' served to tolerant parents and sources of resistance/tolerance. A total of 12 crosses resulted in 367 seedlings from which 56 showed complete resistance. Most of them (20) derived from a cross 'Large Sugar Prune' x 'Zh'lta Butilcovidna' (48).

The results obtained so far are encouraging only in regard to the transfer of resistance to hybrid progenies, but not always encouraging when one examines fruit size and quality (6). Breeding of plums with round fruits seems more easier. Resistant hybrids were received from crossings with 'Tuleu Gras', 'Althans' and 'Anna Späth' (11).

An evaluation of more than 1,300 plum hybrids was carried out in Moldova (4). According to fruit symptoms the greatest number of tolerant seedlings was found in progenies from 'Kirke' plum.

A program was also started for resistance breeding in Germany (21). Special aspects were resistant cultivars with blue color, oval or oblong shape and good fruit quality. Among others, the plum cultivars 'Stanley', 'President' and 'Cacanska Naibolja' were chosen as resistant parents for crossing experiments and 'Bühler Frühzwetsche' as a cultivar with field resistance. 'Italian Prune' and 'Valor' were used as parents with good fruit qualities. Out of more than 50 progenies, selected for good fruit characters, 25 % showed only mild or no fruit symptoms.

Special markers for PPV-resistance were detected for shortening the breeding cycle. A positive correlation between phenolic compounds in the leaves of plums and PPV-resistance of the cultivars was found (54). Using the phenolic content in leaves of young hybrids as a marker for pre-selection for resistance to PPV seems to be possible.

Further investigation demonstrated a significant correlation between the contents of steroid glycosids in plum buds and the degree of PPV infection of the plum tree. This correlation was considered in the evaluation of resistance of plum cultivars to PPV (36).

Furthermore, molecular genetic methods like RFLP-marker and DNA-fingerprinting probably may be suitable for early diagnosis of

PPV-resistance in *P. domestica.*

Genetic Engineering

Because of the requirement of a long span of time, improvement of fruit trees by traditional breeding methods is a tedious effort. The progress in molecular biology allowed to support this process by creating genetic engineered resistance. It has been established that viral coat protein interferes with un-coating of the challenge virus and subsequently prevents infections in plants. This approach has been used to incorporate the coat protein gene of PPV into plants. The PPV coat protein was engineered at the N-terminal amino acid by inserting a translational start site and encoding a protein of 36 kDa. This PPV coat protein was inserted into a binary vector mated to *Agrobacterium tumefaciens* by the triparental plasmid transfer method. The PPV coat protein gene was proved stable in *Nicotiana benthamiana* and *N. xanthii*. Progenies of regenerated plants exhibited a variety of different levels of resistance (51).

A system was also developed which facilitated the transfer of foreign genes into apricot cultivars (7). Cotyledonary explants excised from immature embryos were placed on 2 different regeneration media following transformation with *A. tumefaciens* containing various binary plasmids, pBinGUSint, carrying β-glucuronidase (GUS) as the marker gene and pBinPPVm, carrying the coat protein gene of PPV. The regeneration rate was highest for immature embryos that had been collected between 68 and 89 days after full flowering. The marker gene GUS was used for optical evaluation of the efficiency of the transformation system. The coat protein gene of PPV was used to introduce coat protein mediated resistance against PPV.

Furthermore, gene transfer experiments were initiated in *P. domestica* (52). This species was transformed by inoculating hypocotyl slices with *A. tumefaciens* containing the binary plasmid pGGPPVCP which includes the neophosphomycin transferase II (NPTII), GUS and the PPV CP genes with its T-DNA region. Analysis of the selected transformants showed the integration of the engineered PPV CP gene.

It remains to be seen to what extent the PPV resistance of transgenic *Prunus* plants is heritable, stable and effective in controlling different PPV strains under natural conditions.

PPV-resistant Cultivars

Several plum, apricot, peach and nectarine cultivars were bred or selected with different types or mechanisms of PPV-resistance.

The plum hybrid 'K 4' and other progenies from 'Kirke' plum are characterized by complete resistance because they react with hypersensitive tissue necrosis around the infection sites which prevent or inhibit a systemic spread of PPV (strain M). However, some progenies of 'K 4' show hypersensitive necrotic reactions but no localization of the virus (30).

The plum cultivars 'Scoldus No. 1', 'Jelta covidna' and 'Butilcovidna' were described as completely PPV-resistant or immune (49).

Among many other plum cultivars described as quantitatively resistant or tolerant to different degrees, the following cultivars should be mentioned because of their economic importance: 'Anna Späth', 'Bessarabische Zwetsche', 'Bühler Frühzwetsche', 'Cacanska Naibolja', 'Cacanska Rana', 'Cacanska Lepotica', 'Czar', 'Elena', 'Ersinger', 'Graf Althans', 'Herman', 'Katinka', 'Nancymirabelle', 'Ontario', 'Opal', 'Oullins Reneklode', 'President', 'Ruth Gerstetter', 'Stanley', 'St. Hubertus', 'Tuleu Gras', 'Valjevka' and 'Valo' (20, 22, 45). Some of the resistant or tolerant cultivars ('St. Hubertus', 'Ortenauer') can develop fairly marked symptoms under unfavorable conditions (44).

The apricot cultivars 'Harlayne' and 'Stella' were described as PPV-immune because no virus was detectible after grafting onto diseased rootstocks (15, 60). Other cultivars like 'Badani', 'Early Orange', 'Harcot' and 'Nanno' show symptoms only near the infection site and were classified as highly resistant to PPV. 'Veecot' and 'Paviot' are PPV-tolerant cultivars (13,18). Resistance was more frequently found in hybrids of 'Stark Early Orange' than with 'Henderson' as a parent (15).

Different types of resistance were also described in peach. Localized resistance was detected in the peach clones 'TS 1' to 'TS 10' whereas aphid resistance (to *Myzus persicae* and *M. varians*) occurs in the selections 2605 and 2678. The peach cultivars 'Andross', 'Coronado', 'Jame Berta', 'Ranger' and 'Rio Oso Gem' seem to be PPV-tolerant (37,61).

Nectarine cultivars tolerant or less sensitive to PPV are 'Crimson Gold', 'Fantasia', 'May Grand', 'Delicious', 'Stark Sunglo', 'Nectared No. 4, No. 6 and No. 8', 'Charlampidis

No. 2' and 'Red Gold' (62, 63).

REFERENCES

1. Albrechtova, L., Karesova, R., and Pluhar, Z. 1989. Zur Bewertung der Resistenz von Pflaumensorten und-hybriden gegenüber dem Scharkavirus (plum pox virus). Z. Pflanzen-krankh. Pflanzenschutz 96: 455-463.

2. Atanasoff, D. 1932. Sarka po slivite, edna nova virus a bolest. Jb. Univ. Sofia, Agron. Fak. 11: 49-70.

3. Atanasoff, D. 1935. Old and new virus diseases of trees and shrubs. Phytopath. Z. 8: 197-223.

4. Bivol, T.F., Ignat, V.T., Kukurusak, E., and Kegler, H. 1987. Experiments on resistance of plum varieties and hybrids in Moldavia. Arch. Phytopathol. Pflanzenschutz 23:443-449.

5. Bivol, T.., Meyer, U., and Verderevskaja, T.D. 1988. Nachkommenschaftsprüfung von Pflaumenhybriden auf Scharka- resistenz. Arch. Züchtungsforschg. 18: 365-392.

6. Bivol, T.F., Verderevskaja, T.D., Kukurusak, E.A., and Lahmatova, I.T. 1989. The initial heritability of plum hybrids susceptibility to sharka (plum pox) virus. Rec. Results in Plant Virology, Inst. Phytopathologie Aschers-leben: 7.

7. Câmara Machado, M.L. Da, Câmara Machado A. Da, Hanzer, V., Weiß, H., Regner, F., Steinkellner, H., Mattanovich, D., Plail, R., Knapp, E., and Kalthoff, B. 1992. Regeneration of transgenic plants of Prunus armeniaca containing the coat protein gene of plum pox virus. Plant Cell Rep. 11: 25-29.

8. Clark, M.F., and Adams, A.N. 1977. Characteristics of the microplate method of enzyme-linked immunosorbent assay for the detection of plant viruses. J. Gen. Virology 34: 475-483.

9. Christoff, A. 1937. Maladies à virus des arbres fruitiers. Publ. Bulg. Plant Prot. Service 29: 23.

10. Christoff, A. 1951. The control of plum diseases and enimies. Proc. dec. Nation. Fruitgr. Conf. 1949: 53-61.

11. Cociu, V., Minoiu, N., Roman, R., Gheorgiu, E., Isac, M., and Popescu, I. 1984. Cercetari, privind obtinerea de soiuri noi de prun resistente la virusu varsatul prunului (plum pox virus). Prool. Gent. Teoret. si Aplic. 16:59-67.

12. Dosba, F., Denise, F., Maison, P., Massonie, G., and Audergon, J. M. 1991. Plum pox virus resistance of apricot. Acta Hortic. 293: 569-579.

13. Dosba, F., Lansac, M., Audergon, J. M., Maison, P., and Massonie, G. 1988. Tolerance to plum pox virus in apricot. Acta Hortic. 235: 275-281.

14. Dosba, F., Lansac, M., Eyquard, J. P., Bonnet, A., and Salesses, G. 1994. Behavior towards PPV interspecific hybrids and plum varieties. Acta Hortic. 359:136-144.

15. Dosba, F., Orliac, S., Dutrannoy, F., Maison, P., Massonie, G., and Audergon, J. M. 1992. Evaluation of resistance to plum pox virus in apricot trees. Acta Hortic. 309: 211-220.

16. Flegg, C.L., and Clark, M.F. 1979. The detection of apple chlorotic leaf spot virus by a modified procedure of enzyme-linked immunosorbent assay. Ann. Appl. Biol. 91: 61-65.

17. Fuchs, E., Grüntzig, M., Kegler, H., Krczal, G., and Avenarius, U. 1994. A biological test for characterization plum pox virus strains. Proc. Int. Conf. Plant Virology Apriltsi, Troyan, Bulgaria: 11.

18. Giunchedi, L., 1986. Piante ospiti e sintomatologia della vaiolatura delle drupaceo. Inform. Fitopatologia 7: 14-19.

19. Grüntzig, M., Fuchs, E., and Kegler, H., 1986. Untersuchungen zum Nachweis des Scharka-Virus (plum pox virus) in Pflaumenbäumen. Arch. Phytopathol. Pflanzenschutz 22:441-449.

20. Hamdorf, G. 1984. Zur Toleranz von Pflaumen-, Zwetschen- und Mirabellensorten gegenüber dem Scharka-Virus. Mitt. Biol. Bundesanst. Land- u. Forstwirtsch., Berlin-Dahlem, H. 223: 328.

21. Hartmann, W. 1989. Züchtungsarbeiten bei Pflaumen und Zwetschen in Hohenheim. Erwerbsobstbau 31: 77-79.

22. Hartmann, W. 1990. Das Scharkaproblem im Zwetschenbau und die Frage der Sortentoleranz bzw. -resistenz. Obstbau 15: 350-354.

23. Jordovic, M. 1967. Simptomi virozne encije na liscubreskve. Zast. Bilja 18: 143-145.

24. Jordovic, M. 1968. Effect of sources of infection on epidemiology of sharka (plum pox) virus disease. Tagungsber. DL DDR, Berlin, 97: 301-308.

25. Jovicevic, B. 1967. Reakcija nekih sorti sliva spontano inficiranich virosom sarke posredst vektora. Zast. Bilja 18: 415-420.

26. Kassanis, B., and Sutic, D. 1965. Some results of recent investigations on sharka (plum pox) virus disease. Zast. Bilja 16: 335-340.

27. Kegler, H. 1990. Resistenz gegen das Scharka-Virus (plum pox virus). Arch. Gartenbau 38: 499-517.

28. Kegler, H., Fuchs, E., and Grüntzig, M. 1986. Different types of resistance to plum pox in plums. Acta Hortic. 193: 201-205.

29. Kegler, H., Grüntzig, M., and Fuchs, E. 1994. A glasshouse test for detecting resistance of plum genotypes to plum pox virus. Acta Hortic. 359: 152-158.

30. Kegler, H., Grüntzig, M., and Schimanski, H.-H. 1991. Zur Resistenz der Pflaumenhybride K 4 und ihrer F_1-Nachkommen gegen das Scharka-Virus der Pflaume (plum pox virus). Nachrichtenbl. Dt. Pflanzenschutzd. 43: 102-106.

31. Kegler, H., Meyer, U., and Berka, K. 1990. Messung virus-bedingter Blattverfärbungen mit Hilfe der UV-VIS Remissionsspektroskopie. Jenaer Rundschau 35: 36-38.

32. Kegler, H., and Proll, E. 1989: Nicotiana occidentalis 37 B als Viruswirt. Arch. Phytopathol. Pflanzenschutz 25: 435-443.

33. Kegler, H., Verderevskaja, T.D., and Bivol, T.F. 1977. Untersuchungen zur Schnelldiagnose von Obstviren. 1. Biologische Testung durch mechanische Virusübertragung. Arch. Gartenbau 25: 171-178.

34. Krczal, G. 1993. Characterization of PPV strains occurring in Germany by PCR and RFLP. Proc. EPPO Conf. Sharka, Bordeaux, France: 35.

35. Krzcal, H., and Kunze, L. 1972. Untersuchungen zur Übertragung des Scharkavirus durch Blattläuse. Mitt. Biol. Bundesanst. Land- u. Forstwirtsch., Berlin-Dahlem, H. 114:71-83.

36. Lahmatova, I.T., Kintia, P.K., Verderevskaja, T.D., Bilkey, N.D., and Juravel, A.M. 1994. Plum resistance to plum pox virus in Moldova. Some biochemical aspects. Int. Conf. Fundamental and Applied Problems in Phytovirology, Yalta, Ukraina: 23.

37. Maison, P., and Massonie, G. 1982. Premières observations sur la specifité de la resistance du pêcher à la transmission aphidienne du virus de la sharka. Agronomie 2:681-684.

38. Maiss, E., Timpe, V., Brieske, A., Jelkmann, W., Casper, R., Himmler, G., Mattanovich, D., and Kattinger, H.W. D. 1989. The complete nucleotide sequence of plum pox virus. J. gen. Virology 70: 513-524.

39. Marenaud, C., Mazy, K., and Chauffurin, M. 1976. Observations sur la diffusion et la détection du virus de la sharka. Phytoma 280: 20-24.

40. Minoiu, N. 1973. Vectori virusului varsatului la prun Prunus virus 7 Christ. Anala Inst. Cercetari Pentru, Prol. Plantelor 9: 49-56.

41. Nemchinov, L., Hadidi, A., and Verderevskaja, T.D. 1994. Plum pox virus in sour cherry: Confirmation of its identity by RT-PCR, cloning and nucleotide sequencing of the genomic 3'non-coding region. Int. Conf. Fundamental and Applied Problems in Phytovirology, Yalta, Ukraina: 11.

42. Németh, M. 1986. Virus, mycoplasmas and rickettsia diseases of fruit trees. Akadémia Kiadó Budapest. 841 Pages.

43. Palkovics, L., Burgyán, J., and Balázs, E. 1993. Comparative sequence analysis of four primary structure of plum pox strains. Virus Genes 7: 339-347.

44. Petruschke, M. 1990. Scharkatolerante Sorten für einen wirtschaftlichen Pflaumen- und Zwetschenanbau. Erwerbs- obstbau 32: 80-83.

45. Petruschke, M., and Schröder, M. 1994. Pflaumen- und Zwetschenanbau mit Scharka-fruchttoleranten Sorten. Lan- desanstalt für Pflanzenschutz: 1-19.

46. Polák, J. 1989. Diagnosis of plum pox virus in infected symptomless trees of apricot, peach and *Prunus cerasifera* ssp. myrobalana by ELISA and ISEM. Acta Hortic. 235: 299-303.

47. Rankovic, M. 1978. Podognost nekin metoda u preciscavanju sojeva virus sarke sljive. Zast. Bilja 29: 179-195.

48. Rankovic, M., Ogasanovic, D., and Paunovic, S. 1994. Breeding of plum cultivars resistant to sharka (plum pox) disease. Acta Hortic. 359: 69-74.

49. Rankovic, M., and Paunovic, S. 1988. Further studies on the resistance of plums to sharka (plum pox) virus. Acta Hortic. 235: 283-290.

50. Rankovic, M., Paunovic, S., and Dulic-Markovic, I. 1994. Currant situation and future trends in solving sharka problem in SR Yugoslavia. Proc. XVI Inter. Symp. Fruit tree Virus Dis., Rome, Italy: 55.

51. Ravelonandro, M., Monsion, M., Teycheney, P., Delbos, R. P., and Dunez, J. 1992. Transgenic tobacco plants that contain the plum pox virus (PPV) coat protein gene. Acta Hortic. 309: 191-196.

52. Ravelonandro, M., Scorza, R., Callahan, A., Cordts, J., Monsion, M., Fuchs, M., Delbos, R., Bachelier, J., and Dunez, J. 1993. Transgenic plants in the control of plum pox virus. Proc. Conf. Sharka, Bordeaux, France: 59.

53. Richter, J., Kleinhempel, H., Döring, U., and Augustin, W. 1978. Zur Empfindlichkeit des Nachweises von Pflanzenviren mit einer Mikrovariante des "enzyme-linked immuno-sorbent assay" (ELISA) bei Verwendung von PVC-Tiefzieh-blistern als Trägern. Arch. Phytopathol. Pflanzenschutz 15: 361-366.

54. Rühl, K. 1990. Phenolische Inhaltsstoffe in Blättern verschiedener Pflaumen- und Zwetschensorten (*Prunus domestica* L., *P. insititia* L.) in Beziehung zur Resistenz gegenüber Scharkaviren. Diss. Universität Hohenheim.

55. Schnell, F.W. 1985. Zuchtmethodische Prinzipien und Probleme bei vegetativer Vermehrung. Vortr. Pflanzenzüchtung, H. 8: 5.

56. Schuch, K. 1962. Die Scharka, eine für Zwetschen und Aprikosen gefährliche Viruskrankheit. Ges. Pflanzen 14:126-128.

57. Simeone, A. M. 1982. Indagime sulla sensibilita varietale del susino ad alcumi principali parasiti. Fruticoltura 12: 112.

58. Sutic, D. 1973. Neke seroloske osobine sojeva virusa sarke. Mikrobiologija 10: 199-206.

59. Sutic, D., and Rankovic, M. 1981. Resistance of some plum cultivars and individual trees to plum pox (sharka) virus. Agronomie 1: 617-622.60. Syrgianidis, G. D. 1980. Selection of two apricot varieties to sharka virus. Acta Phytopathol. Acad. Sci. Hung. 15: 85-87.

61. Syrgianidis, G. D., and Mainou, A. C. 1985. Étude de la sensibilité à la maladie de la sharka (plum pox) de 33 variété de pêcher et de nectarines. Agric. Res. 9: 207-209.

62. Syrgianidis, G.D., and Mainou, A. C. 1989. Susceptibility of some nectarine cultivars to plum pox virus (sharka) virus. Acta Hortic. 235: 121-124.

63. Syrgianidis, G.D., and Mainou, A.C. 1991. Deux nouvel-les variétés d'abricotier résistantes à la maladie à virus de la Sharka (Plum pox). Issues de croisements. 2ème rencontre de groupe abricotier. Avignon: 6.

64. Trifonov, D. 1974: Cuvstvitelnost na dvanadeset rasnovid-no sti na sorta kjustendilska sinja sliva sprjamo virus na bolestta sharka-plum pox virus. Virusni Bol. Rast.: 131-137.

65. Verderevskaja, T.D., Bivol, T.F., Kegler, H., and Kukurusak, E. A. 1984. Ustuicivost sortov i gibridov slivik virusu sarki. Genetika, Immuniteta i Selekzija sel.-chos. Rast. Ust. Moldavii, Kishinev: 99-107.

66. Wetzel, T., Candresse, T., Macquaire, G., Ravelonandro, M., and Dunez, J. 1992. A highly sensitive immunocapture polymerase chain reaction method for plum pox potyvirus detection. J. Virol. Methods 39: 27-37.

CHAPTER 47

Present Status of the New Cherry Subgroup of Plum Pox Virus (PPV-C)

L. Nemchinov, A. Crescenzi, A. Hadidi, P. Piazzolla, and T. Verderevskaya

Plum pox (Sharka) disease has been recognized for years as the most important viral disorder of stone fruit trees in Europe and the Mediterranean region. The losses caused by plum pox virus (PPV) may be dramatic - in some cases they may reach 100% of fruit yield in susceptible cultivars (33,38). PPV also occurs in Egypt, Syria and Chile (15). Recently, it has been tentatively diagnosed in India (4). PPV has never been reported from North America and extreme quarantine measures to avoid its' introduction to the continent have been implemented.

Traditionally, among stone fruits, cherry trees have been considered resistant to PPV infection (16,33). Experimental transmission of several isolates of PPV from plum, peach, or apricot to sweet cherry, *Prunus avium* L., and to *P. mahaleb* L. showed that graft inoculated plants were ELISA negative for PPV, but plants inoculated through aphid transmission were ELISA-positive. The virus remained localized to the infection sites and became undetectable (16).

Reports from Moldova in 1989 (19) and Italy in 1994 (12) have claimed that sour cherry , *P. cerasus*, and sweet cherry trees may be naturally infected with PPV. Subsequently, the Moldovian sour cherry isolate of PPV (PPV-SoC) had been reported to differ from other PPV isolates which affect plum, peach and apricot in Moldova by its' transmission to a wide range of herbaceous hosts and by its' symptomatology (20). PPV-SoC and the plum isolate of PPV, however, could not be differentiated by ELISA or ISEM (20). Molecular characterization of PPV-SoC demonstrated that the virus is a prototype of a new cluster of PPV isolates and that its' unusual biological behavior is based on the specific properties of the viral genome (28-31). The identity of an isolate of PPV initially discovered by immunological findings in Italy was also established by molecular investigations using polymerase chain reaction (PCR) and nucleotide sequencing (8-14). This isolate has been termed PPV-sweet cherry (PPV-SwC).

Thus, utilizing PCR-technology with PPV-specific primers in combination with other sensitive tools such as restriction fragment length polymorphism (RFLP), molecular hybridization and nucleotide sequencing, it was demonstrated that the potyviruses affecting sour cherry trees in Moldova and sweet cherry trees in Italy are indeed isolates of PPV. It has been suggested and then demonstrated that both isolates differ in their genomic organization as compared to other known isolates of PPV which allows PPV-SoC or PPV-SwC to infect unusual natural hosts such as sour cherry or sweet cherry. Actually, the recently obtained 3'-terminal nucleotide sequences of PPV-SoC and PPV-SwC showed a high level of nucleotide divergence in the N-terminus of the viral CP. These changes might be responsible for the extension of the typical host range of PPV to cherries. The unique serological reactivity of both PPV-SoC and PPV-SwC has also been demonstrated (10,31).

Apparently, the information available on characterization of PPV-SoC and PPV-SwC suggests the establishment of a new subgroup of PPV, namely PPV-Cherry (PPV-C), which significantly differs from the conventional D, M and El-Amar strains of PPV (10,11,31). The possibility exists that PPV-C may be a potential source of PPV dissemination in many countries as present certification and quarantine regulations do not require testing for PPV in cherry tissue.

Figure 1. PPV-SoC symptoms on sour cherry leaves (a) and fruits (b). PPV symptoms on leaves of graft-inoculated sweet cherry cv. Mazzard (c).

Figure 2. PPV-SwC symptoms on naturally infected sweet cherry leaves (a) and fruits (b).

SYMPTOMS ON CHERRY

PPV-SoC

Infected sour cherry trees show characteristic chlorotic ringspot symptoms on leaves; depressions, necrosis and rings on fruits (Figure 1a and b). Fruit rings gradually disappear during ripening. Similar leaf symptomatology is observed on sweet cherry leaves after graft transmission of PPV-SoC to sweet cherry (Figure 1c).

PPV-SwC

Naturally PPV-infected sweet cherry trees in southern Italy display the symptomatology defined as apical necrosis of sweet cherry (12). Diseased plants show diffuse necrosis starting from the apical portion and spreading to the whole branch. Fruits, subjected to heavy drops, show chlorotic and necrotic ring spots or notches. No symptoms appear on the seeds. The symptoms appear in their maximum intensity during the spring and disappear in summer. Similar symptomatology is observed on seedlings of *P. avium*, cultivars "Ferrovia" and "Sunburst", and *P. mahaleb*. The same symptoms are observed on different clones of micropropagated plants of *P. cerasus* after grafting or mechanical transmission.

THE CAUSAL VIRUS

Filamentous virus particles 700-800 nm in length were observed in leaf dip preparations of infected sour cherries, and cylindrical inclusions, characteristic for Genus Potyvirus, were observed in ultrathin sections of cherry leaves (19). Using ISEM, the flexuous virus particles in sour cherry leaf tissue were decorated with PPV antiserum (20).

PPV-SoC has been detected by PPV-specific PCR assay in infected flowers, including stamens, filaments, anthers, pollen and pistils (28,30). PPV-SoC-infected seeds from different sour cherry cultivars were also positive in PCR tests. Electron microscopy of sour cherry anthers revealed PPV-encoded "pinwheel" inclusion bodies which were located in the anther wall (tapetum cells) (28).

The nucleotide sequence of the 3' 1360 nucleotides of PPV-SoC has been determined (31). A significantly low degree of similarity has been found between the sequenced region of PPV-SoC

containing a portion of NIb (putative replicase) gene, the entire coat protein gene and the 3' non-coding region (3' NCR), and the corresponding segments from other PPV isolates. This is mostly due to the high nucleotide divergence in the N-terminal part of PPV-SoC CP sequence. The nucleotide variations in this fragment have considerable effect on the predicted amino acid sequence. Nucleotide sequencing of the 3'-terminal end of PPV-SwC genome showed a high degree of homology with the corresponding fragment of PPV-SoC (10,14). Cambra *et al.*, (6) established 21 PPV serogroups on the basis of the reaction pattern of 81 PPV isolates against nine monoclonal antibodies (MCA), and Asensio *et al.*, described 23 serogroups when using 85 PPV isolates against the same MCAs (1). PPV-SoC gave positive reaction only against MCAs 5B and 4DB7 (31). MCA 5B represents a common and frequent epitope in the coat protein of all PPV isolates . The fact that PPV-SoC failed to react against MCAs 4DG5 and 4DG11, which are described as D-type specific (1,6) and it was not recognized by MCA A, which is M-type specific (5), indicates that PPV-SoC is serologically representative of an uncommon type of PPV-neither M type or D type. Very similar serological reactivity was demonstrated for PPV-SwC (10). It reacted positively in ELISA tests against MCA 5B (general MCA against all PPV isolates) but did not react either against MCA 4DG5 (anti PPV-D) or MCA AL (anti PPV-M), thus confirming the hypothesis that PPV-SoC and PPV-SwC belong to a new serological subgroup of PPV strains.

Based on the present available data on molecular organization and serological reactivity, PPV-C is a unique strain of PPV which differs from other conventional strains of PPV. The major differences that cause the serological specificity and the unusual biological properties may be due to the distinct structure of the N-terminal domain of PPV-C coat protein. This region of the potyvirus genome is implicated in the study of viral taxonomy, host range, vector transmission and movement functions, and tentatively plays an important role in potyviral pathogenesis (34).

APHID TRANSMISSION

Transmission of PPV by aphids depends on the interaction of two viral-encoded proteins, the CP and the Helper Component (HC), (2,3,26). Several non-aphid transmissible PPV isolates with altered properties of the coat protein and HC have been reported (26,27). Substitution in the conserved triplet of amino acids (DAG) near the N terminus of the CP was determined to greatly affect aphid transmission (2,3).

All aphid transmissible isolates of PPV show conserved DAG block close to the N-terminus of CP. Since this amino acid sequence was found in PPV-SoC CP, we suggested that the virus is naturally transmitted by aphid vectors. To verify our molecular data we used one of the most active PPV natural vectors, the green peach aphid *Myzus persicae*, in attempts to transmit virions from PPV-SoC-infected *N. benthamiana* and sweet cherry to *N. benthamiana*, *N. clevelandii* and plum, *P. domestica* cv "Stanley" (31). The acquisition access period ranged from 1 min to 16 h. The distinctive symptoms (mild mosaic) on *Nicotiana* and Stanley plum leaves appeared 2-3 weeks and 6-7 weeks, respectively, after viral transmission with aphids.   Eleven out of 27 *Nicotiana* plants became infected (1/3 *N. clevelandii* and 10/24 *N. benthamiana*) and two out of three Stanley plums became infected. Acquisition access periods of 5 min to 16 h were successful. The transmission type was stylet-born and only the aphids which made short probes before moving were actually transmitting. Molecular hybridization of tissue extracts from leaves with symptoms to a DIG-labeled cRNA probe specific only for members of PPV-C subgroup or to a cRNA probe specific for all PPV isolates, revealed strong hybridization signals in both cases. RT-PCR assays using primers for the 3' non-coding region of PPV genome were positive. Thus, PPV-SoC is aphid transmissible.

PPV-SwC has been successfully transmitted in a non-persistent manner by *Aphis fabae* and *Myzus persicae* from *N. benthamiana* to the same plant species and to *P. avium* and *P. mahaleb* (10).

HOST RANGE

PPV-SoC

PPV-SoC was graft transmitted to sweet cherry, *P. mahaleb* and peach GF-305 developing chlorotic spots on leaves of sweet cherry cv. Mazzard (Figure 1,c) and vein yellows with necrosis on leaves of *P. mahaleb*; peach GF-305 expressed typical shoestring symptomatology of PPV (31), and *P. tomentosa* developed vein-associated chlorotic pattern which was milder than that of PPV-D or PPV-M symptoms on the same

host (V. Damsteegt, personal communication). The symptoms, however, were similar to those induced by PPV-El-Amar. PPV-SoC was also graft transmitted to different cultivars of *P. domestica*; susceptible cultivars later showed mild mosaic (20). PPV-SoC has been aphid transmitted to plum cv. Stanley, *N. benthamiana* and *N. clevelandii*. Plants were systemically infected.

The virus was mechanically transmitted to different herbaceous plants, including three species of *Chenopodium* (*C. foetidum*, *C. murale*, and *C. foliosum*) on which it gives local lesions; 5 species of *Nicotiana* (*N. clevelandii*, *N. rustica*, *N. occidentalis B37*, *N. acuminata*, and *N. benthamiana*) and *Petunia hybrida*, on which systemic symptoms appear (20).

PPV-SwC

PPV-SwC is mechanically and/or graft transmitted from woody plants to *P. avium*, *P. cerasus*, *P. mahaleb*, *P. cerasifera*, *P. persicae*, *P. domestica*, *P. tomentosa* and *P. armeniaca*. Chlorotic spots, distortion and vein yellowing with necrosis developed on infected leaves. The symptoms appear about 15 days after mechanical inoculation. The isolate was mechanically transmitted to several herbaceous plants and induced the following symptoms: systemic chlorotic and necrotic mottling on *N. benthamiana*, *N. glutinosa* and *N. occidentalis*; systemic faint mosaic on *N. tabacum* cv "Burley 64"; local chlorotic and necrotic lesions followed by systemic mottling on *N. clevelandii*; infection was latent on *N. sylvestris*. *Chenopodium foetidium*, *C. quinoa*, *C. murale* and *C. amaranticolor* were not infected.

GEOGRAPHIC DISTRIBUTION AND EPIDEMIOLOGY

PPV was originally suspected in the cherry orchard collection at the Horticultural Research Institute, Kishinev, Moldova, in 1989 (19). Subsequent evaluation of the sour cherry collection using ELISA, ISEM, molecular hybridization, and PCR, showed that 6 cultivars, including "Kistevaya", "Orlovskaya konservnaya", "Rannneaya 2", "Podbel'skii", "Pandy", and "8-34-3", were infected with PPV (20,28,30). About 3% of the 1500 trees in the orchard were PPV positive; the diseased orchard has been destroyed to prevent the spread of PPV-SoC in the area. PPV-SoC has also been reported on sour cherry trees in the Tambov and Lipetsk regions of Russia (20). PPV-SwC was found by biological and serological assays in a 15 year-old commercial orchard of sweet cherry cv "Ferrovia" located in the Apulia region of southern Italy.

Recently, screening of sweet and sour cherry trees in Bulgaria for PPV using DAS-ELISA demonstrated that a high percentage of cherry trees were infected with PPV (36,37). PPV has also been recently detected by ELISA and PCR in cherry trees in Hungary (21). The exact identity of the Hungarian cherry isolate has been established based on the sequencing results which suggest that it possess the same distinct structure features characteristic of the PPV-C subgroup (32).

These findings show that the PPV-C subgroup may be more widespread in Europe than originally thought. PPV-C was experimentally transmitted to many other *Prunus* spp. but conventional strains of PPV failed to systemically infect cherries, thus, PPV-C is potentially more epidemic than the conventional strains because of its wider host range (8,20,28). Natural infection of plum, peach, and apricot with PPV-C is highly possible but it may be undistinguishable from conventional strains of PPV. PPV-C specific cRNA probe, PCR, and monoclonal antibodies may serve to differentiate between PPV-C and conventional subgroups of PPV

In spite of the fact that there are several observations on the occurrence PPV-C in different countries, the status of the disease worldwide and the damage it may cause, however, has yet to be established

VIRAL DETECTION AND IDENTIFICATION

In addition to diagnosis of PPV-C from early symptoms, PPV-C may be detected by routine assays which are commonly used for general PPV isolates. The assays may include biological tests on established woody and herbaceous indicators, ELISA and ISEM using polyclonal PPV antiserum. A polyclonal antiserum has been obtained against a peptide corresponding to amino acids 1-14 of the N-terminal region of the coat protein of PPV-SwC. Polyclonal antibodies specifically detected PPV-SwC in DAS-ELISA, Western immunoblot and immuno-electron microscopy plus decoration, but the antiserum did not react with D and M isolates of PPV or with PPV-EL-Amar strain (A. Crescenzi *et al.*,

unpublished). This antiserum may be used in diagnosis, epidemiological investigations and the study of the distribution of PPV-C in susceptible *Prunus* species. PCR may also be used for diagnosis of PPV-C. Screening for PPV using DNA primers for the 3' NCR (18,24) or the carboxyterminal region of PPV CP (39) is to date PPV-specific and reliable for detecting PPV in stone fruits (Figure 3a and b). It is even more sensitive in combination with immunocapture of viral particles prior to the RT-PCR (40). The sensitivity of the latter useful technique may be additionally increased by using protein A before the application of antiserum (L. Nemchinov, unpublished) (Figure 3,a).

Along with the detection of PPV, the basic necessity to identify a given isolate may arise because of its' different epidemiological behavior. A highly specific detection method for the members of the PPV-C subgroup has been developed based on the distinct properties of the N-terminal region of PPV-SoC CP sequence. A 447 bp DNA fragment containing 303 bp and 144 bp from the CP gene and NIb gene, respectively, was amplified from the cloned 1360 nt 3'-terminal fragment of the PPV-SoC genome and then subcloned into a suitable plasmid vector, containing a promoter for DNA-dependent RNA polymerase. This fragment does not have any continuous sequence homology (more then 10-15 nt) with corresponding regions of the rest of PPV isolates (31). A non-radioactive DIG-labeled cRNA probe was generated from the subclone and used in dot-blot hybridization assay with several PPV isolates including PPV-D strain from France, PPV-M from Greece, PPV-El Amar from Egypt, PPV isolates from Romania. Additionally, DIG-labeled cRNA probe derived from the conserved C-terminal region of PPV CP gene was used for hybridization to ascertain that tested samples were infected with PPV. PPV-SoC cRNA probe hybridized strongly to samples from PPV-SoC infected tissues only. The probe did not hybridize to samples infected with other isolates of PPV. In contrast, PPV cRNA probe for the conserved region of PPV CP gene hybridized strongly to all isolates of PPV, including those of PPV-SoC (31). To verify the identity of PPV-SwC, this probe was used in dot-hybridization assays with PPV-infected sweet cherry samples from Italy (Figure 4). Positive results proved the close relationship between these two isolates and suggested a high sequence homology between the 5'-end of the CP gene of each isolate, which is not typical for PPV. Based on the distinct

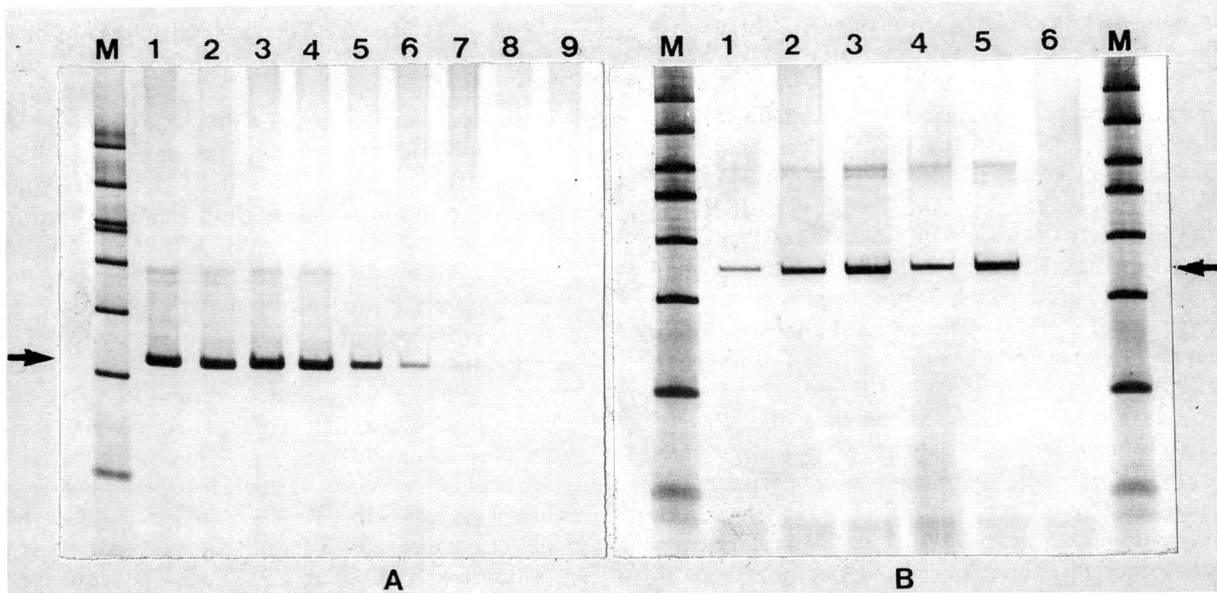

Figure 3. Application of PCR-technology for the detection of PPV. A. IC-RT-PCR with PPV-specific 3' non-coding region primers (18,24) and PPV-SoC infected samples. Protein A was used prior the applying of antiserum for capturing of viral particles. Dilutions: 10^{-1} - 10^{-7}, lanes 1-7, respectively; lane 8 represents amplification from healthy extract, lane 9 Amplification with H_2O as a control. B. IC-RT-PCR with PPV-specific DNA primers for the carboxy-terminal region of PPV coat protein gene (39). PPV cDNA was amplified from: lane 1, PPV-SoC; lanes 2 and 3, PPV-D-infected apricot and plum, respectively; lanes 4 and 5, PPV-D-infected peach; lane 7, uninfected peach. M-DNA markers (1000bp, 700bp, 525bp, 500bp, 400bp, 300bp, 200bp, 100bp).

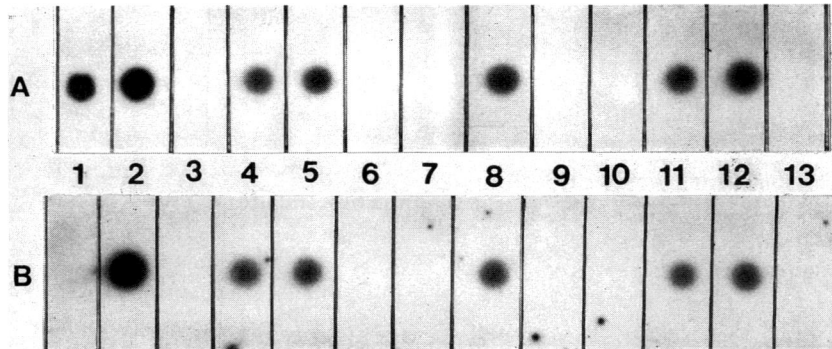

Figure 4. Dot-hybridization of PPV-SoC and PPV-SwC infected tissue extracts with: A. DIG-labeled cRNA probe derived from the conserved C-terminal region of PPV-D coat protein gene. B. PPV-SoC-specific DIG-labeled cRNA probe: 1, PPV-D; 2, PPV-SoC; 3-12, sweet cherry samples from Italy, 13, non-infected tissue.

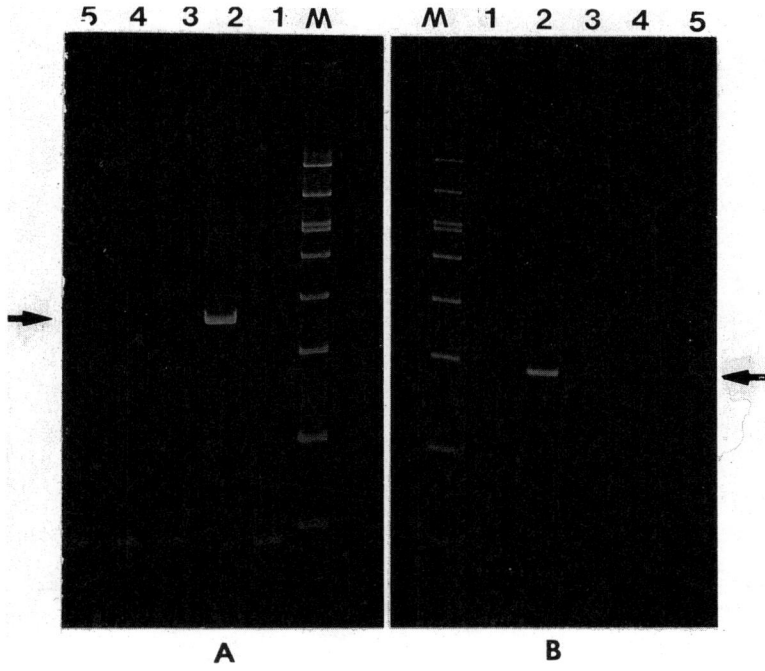

Figure 5. IC-RT-PCR amplification of PPV cDNA using two sets of PPV-SoC -specific DNA primers. M, DNA markers (1000bp, 700bp, 525 bp 500 bp, 400 bp, 300 bp, 200 bp, 100 bp). 1,PCR products from PPV-SoC-infected tissue prepared with GeneReleaser™ (25); 2-4 PCR products from PPV-SoC, PPV-D, PPV-M-infected tissues, respectively prepared using immunocapture of virus particles; 5, amplification from non-infected tissue. A. HSoC-1/CSoC-1, 259 bp. B. HSoC-2/CSoC-2, 193 bp (see Table 1).

features of PPV-SoC CP sequence, we designed two sets of primers specific for amplification of PPV-SoC-derived fragments only, which are located within the region corresponding to the viral-specific cRNA probe (Table 1, Figure 5). Preliminary results using these primers in highly sensitive immunocapture PCR with different isolates of PPV belonging to all known subgroups/serotypes of the virus demonstrated their specificity (L. Nemchinov and A. Hadidi, unpublished). Thus, the cRNA probe and PPV-SoC-specific PCR primers may be used for studying the identity, epidemiology and geographic distribution of members of PPV-C subgroup world-wide.

DISEASE CONTROL

As mentioned above, the status of PPV-C world-wide, along with its' economic significance, has not yet been established. Logically, all basic control measures which are effective against PPV, namely: I.) use of virus-free, certified cherry plant material for propagation; ii.) assaying for PPV in stone fruit germplasm, including cherry germplasm, in quarantine; iii.) early detection of the disease and destroying infected trees; iv.) growing non-susceptible biological barriers around orchards and, v.) use of resistant cultivars, etc., may be successfully applied to control PPV-C.

The genetic engineering approach using PPV-C-derived genes in designing recombinant constructs for plant transformation could be a promising step in the development of pathogen-derived resistance strategy. The production of transgenic cherry plants containing PPV-SoC-derived CP sequence is currently under investigation (M. Laimer daCâmara-Machado, personal communication). PPV-SwC-derived CP sequences are also being used in order to produce transgenic plants resistant to PPV infection (A. Crescenzi, unpublished).

Table 1. PCR primers specific for the cherry strains of PPV

Primer	Sequence	Product size
Primer Set A		
HsoC-1/upstream pos. 99-120[a]	TAACAATGCTGGGGATGAGCTC	
CsoC-1/downstream pos. 340-358	ATGGACTTGTCACTACTGG	259 bp
Primer Set B		
HsoC-2/upstream pos. 249-267	TCCACCATTCCCAAATCTG	
CsoC-2/downstream pos. 423-441	TACATCTCGATCCTTCCTC	193 bp

[a] Position number indicates nucleotide position on PPV-SoC genome (31).

POSSIBLE PATHWAYS OF PPV-C EVOLUTION

Preliminary results show that the PPV-C genome, except its' unique N-terminal domain of CP, may represent characteristic PPV RNA. cRNA probes specific for the different regions of PPV-D type, reacted strongly to PPV-C-infected tissues (L. Nemchinov, unpublished). Thus, the exceptional properties of PPV-C are most likely determined by the observed changes in the N-terminus of viral coat protein.

What is the nature of such genetic hyper-diversity in this region of the PPV-C genome? It seems that simple mutations, deletions, frame shifts and partial duplications, which often represent the main source of molecular variability in RNA viruses (17), were not involved in its' formation. Most probably, the N-terminus of PPV-C CP had arisen *de novo* as a result of natural recombination event between PPV-like virus and another unknown viral RNA or host cellular nucleic acid. However, database searches have shown that no sequence more similar to the N-terminal region of PPV-SoC CP than those of other PPV isolates has been found - if the relatively long nucleotide and/or amino acid sequence fragments (more than 25-30 nt or 10 amino acids) were submitted to the database.

Interestingly, when submitting selected short sequences from PPV-SoC which greatly differ in nucleotide composition from the corresponding regions of general PPV isolates, and were most probably involved in host/viral or interviral gene transfer and recombination, some similarities between the viral sequences and plant cellular sequences were observed (L. Nemchinov, unpublished observations).

Thus, hypothetically, PPV-C may have diverged from an ancestor of PPV-consensus type by genetic recombination with uncharacterized RNA virus (viruses) or host nucleic acid which contained nucleotide sequence similar or identical to those found in N-terminus of CP coding region of PPV-C. Hence, strains of PPV-C group may represent evolved isolates of the existing PPV which, one way or another, went through recombination processes that clearly provided selective advantages for the virus under other natural conditions and led to the expansion of the viral host range.

Although it has been assumed by many investigators that, theoretically, illegitimate recombination occurs in most RNA viruses and is a driving force of their evolution, this conclusion was indirectly predicted by comparative analysis of available sequences or experimentally induced (7,22,23,35), but rarely observed under natural conditions. Based on our investigation, we suggest that modern filamentous plant viruses affecting stone fruits undergo consecutive, naturally occurring evolution by nonhomologous recombination between distantly related viral and/or host RNA sequences. Resulting recombinants may be little-known or newly discovered pathogens with unusual biological characteristics.

REFERENCES

1. Asensio, M., Gorris, M.T., Sanz, A., Camarasa, E., Perez, E., Carbonell, E.A., and Cambra, M. 1995. Characterization and detection of plum pox virus using monoclonal antibodies. Acta Horticulturae 386:354-356.

2. Atreya, P., Atreya, D., and Pirone, T. 1991. Amino acid substitutions in the coat protein results in loss of insect transmissiblity of a plant virus. Proc. Natl. Acad. Sci. USA 88:7887-7891.

3. Atreya, P.L., Lopez-Moya, J.J., Chu, M., Atreya,C.D., and Pirone, T.P. 1995. Mutational analysis ot the coat protein N-terminal amino acids involved in potyvirus transmission by aphids. J. Gen. Virol. 76:265-270.

4. Bhardwaj, S.V., Thakur, P.D., Kohosla, K., and Sharma, D.R. 1995. Detection of plum pox virus in India. Acta. Hortic. 386:237-241.

5. Boscia, D., Zeramdini, H., Cambra, M., Gorris, M.T., Potere, O., Myrta, A., Di Terlizi, B., and Savino, V. 1997. Production and characterization of a monoclonal antibody specific to the serotype M of a plum pox potyvirus. Eur. J. Plant Pathol. In press.

6. Cambra, M., Asensio, M., Gorris, M.T., Perez, E., Camarasa, E., Garcia, J.A., Moya, J.J., Lopez-Abella, D., Vela, C., and Sanz, A. 1994. Detection of plum pox potyvirus using monoclonal antibodies to structural and non-structural proteins. EPPO Bull. 24:569-577.

7. Cervera, M.T., Riechmann, J.L., Martin, M.T., and Garcia, J.A. 1993. 3'terminal sequence of the plum pox virus PS and 06 isolates: evidence for RNA recombination within the potyvirus group. J. Gen. Virol. 74:329-334.

8. Crescenzi, A., D'Aquino, L., Nuzzaci, M., Comes, S., and Piazzolla, P. 1996. Un nuovo ceppo del virus della vaiolatura del susino (PPV) isolato da ciliegio dolce e trasmissibile a diverse specie del genere Prunus. Implicazioni epidemiologiche. Convegno annuale S.I.Pa.V. September 26-27, 1996.

9. Crescenzi, A., d'Aquino, L., Comes, S., Nuzzaci, M., Piazzolla, P., and Hadidi, A. 1996. Characterization of the sweet cherry isolate of plum pox potyvirus. Page 42 in: Middle European Meeting '96 on Plum Pox, Budapest, 2-4 October, 1996.

10. Crescenzi, A., d'Aquino, L., Comes, S., Nuzzaci, M., Boscia, D., Piazolla, P., and Hadidi, A. 1997. Characterization of the sweet cherry isolate of plum pox potyvirus. Plant Dis. 81:711-714.

11. Crescenzi, A., Nemchinov, L., Piazolla, P., and Hadidi, A. 1996. Sweet and sour cherry isolates of plum pox potyvirus: Prototypes of a new group of PPV. Page 145 in: X[th] International Congress of Virology. Jerusalem, August 11-16, 1996.

12. Crescenzi, A., Nuzzaci, M., Levy, L., Piazzola, P., and Hadidi, A. 1994. Infezioni di sharka su ciliegio dolce in Italia meridionale. Inf. Agrar. 34:73-75.

13. Crescenzi, A., Nuzzaci, M., Levy, L., Piazzolla, P., and Hadidi, A. 1995. Plum pox virus in sweet cherry. Acta Hortic. 386:219-225.

14. Crescenzi, A., Piazzola, P., and Hadidi, A. 1996. Nucleotide sequence of the 3'-terminal region of the sweet cherry isolate of plum pox potyvirus. Page 120 in: X[th] International Congress of Virology, Jerusalem, August 11-16, 1996.

15. Diekmann, M., and C.A.J. Putter. 1996. FAO/IPGRI Technical Guidelines for the safe movement of germplasm - No. 16: Stone fruits. Food and Agriculture Organization of the United Nations, International Plant Genetic Resourses Institute, Rome.

16. Dosba, F., Maison, P., Lansac, M., and Massonie, G. 1987. Experimental transmission of plum pox virus (PPV) to Prunus mahaleb and Prunus avium. J. Phytopathol. 120:199-204.

17. Gibbs, A.J., Calisher, C.H., Garcia-Arenal, F., eds. 1995. Molecular Basis of Virus Evolution. Cambridge University Press. 603 Pages.

18. Hadidi, A., and Levy, L. 1994. Accurate identification of plum pox virus and its' differentiation from Asian prunus latent potyvirus in Prunus germplasm. EPPO Bull. 24:633-643.

19. Kalashyan, Y.A., and Bilkey, N.D. 1989. Identifikatsiya virusa sharki na vishne. Plant Virology. (Abstr.) Page 106 in: Proceedings of the 10th Conference of the Czeshoslovak Plant Virologists.

20. Kalashyan, Y.A., Bilkey, N.D., Verderevskaya, T.D., and Rubina, E.V. 1994. Plum pox virus on sour cherry in Moldova. EPPO Bull. 24:645-649.

21. Kölber, M., Nemeth, M., Papp, E., Kiss, E., Pocsai, E., Hangyál, R., Tõkés, G., Krizbai, L., Bereczki, Zs., Szõnyegi, S., Pete, A., Vollent, Á., Takács, M., Bencze, E., Merõ, F., Hajnóczy, GY., and Imre, P. 1997. Five-year study for determination of eventual occurrence of plum pox virus in cherry cultivars in Hungary. Page 132 in: Abstracts of the 17th International Symposium on Virus and Virus-like Diseases of Temperate Fruit Crops, June 23-27, Bethesda, Maryland.

22. Lai, C.M. 1992. RNA recombination in animal and plant viruses. Microbiol. Rev. 56:61-79.

23. Lai, C.M. 1995. Recombination and its' affect on viruses with RNA genomes. Pages 119-132 in: Molecular Basis of Virus Evolution. Gibbs, Calisher and Gaecia-Arenal. Cambridge University Press.

24. Levy, L., and Hadidi, A. 1994. A simple and rapid method for processing PPV-infected tissue for use with PPV-specific 3' non-coding region RT-PCR assays. EPPO Bull. 24:595-604.

25. Levy, L., I-M Lee, and Hadidi, A. 1994. Simple and rapid preparation of infected plant tissue extracts for PCR amplification of virus, viroid, and MLO nucleic acids. J. Virol. Meth. 49:295-304.

26. Lopez-Moya, J.J., Canto, T. Diaz-Ruiz, J.R. and Lopez-Abella, D. 1995. Transmission by aphids of a naturally non-transmissible plum pox virus isolate with the aid of potato virus Y helper component. J. Gen. Virol. 76:2293-2297.

27. Maiss, E., Timpe, U., Brisske, A., Jelkmann, W., Casper, R., Himmler, G., Mattanovich, D., and Kattinger, H.W.D. 1989. Complete nucleotide

sequence of plum pox virus RNA. J. Gen. Virol. 70:513-524.

28. Nemchinov, L., and Hadidi, A. 1996. Characterization of the sour cherry strain of plum pox virus. Phytopathology 86:575-580.

29. Nemchinov, L., Hadidi, A., and Verderevskaya, T. 1994. Plum pox virus in sour cherry : confirmation of its identity by RT-PCR, cloning and sequencing of the genomic 3' non-coding region. (Abstr.) Page 11 in: International Conference on Fundamental and Applied Problems in Phytovirology, Yalta, Ukraine.

30. Nemchinov, L., Hadidi, A., and Verderevskaya, T. 1995. Detection and partial characterization of plum pox isolate from infected sour cherry. Acta Hortic. 386:226-237.

31. Nemchinov, L., Hadidi, A., Maiss, E., Cambra, M., Candresse, T., and Damsteegt, V. 1996. Sour cherry strain of plum pox potyvirus (PPV):molecular and serological evidence for a new subgroup of PPV strains. Phytopathology 86:1215-1221.

32. Nemchinov, L., Kölber, M., Nemeth, M., and Hadidi, A. 1997. Molecular evidence for the occurrence of plum poxvirus - C subgroup in Hungary. Page 54 in: Abstracts of the 17th International Symposium on Virus and Virus-like Diseases of Temperate Fruit Crops, June 23-27, Bethesda, Maryland.

33. Nemeth, M. 1986. Virus, Mycoplasma, and Rickettsia Diseases of Fruit Trees. Akademia Kiado, Budapest. 841 Pages.

34. Shukla, D., Ward, C.W., and Brunt, A. 1994. The Potyviridae. CAB International, Wallington, Oxon, UK.

35. Simon, A.E. and Bujarsky, J.J. 1994. RNA-RNA recombination and evolution in virus-infected plants. Annu. Rev. Phytopathol. 32:337-362.

36. Topchiiska, M. 1991. Detection of plum pox virus by ELISA in Prunus spp. at the different stages of their development. in: XVth International Symposium on Virus and Virus Diseases of Temperate Fruit Crops, Vienna, July 8-13, 1991.

37. Topchiiska, M. 1996. Plum pox virus in some *Prunus* spp. in Bulgaria. Page 27 in: Middle European Meeting '96 on Plum Pox, Budapest, 2-4 October 1996.

38. Verderevskaya, T.D., and Marinesku, V.G. 1985. Virus and mycoplasma diseases of fruit trees and grapevine. (In Russian) Published by 'Shtiintsa', Kishinev, Moldova.

39. Wetzel, T., Candresse, T., Ravelonandro, M., and Dunez, J. 1991. A polymerase chain reaction adapted to plum pox virus detection. J. Virol. Methods 33:355-365.

40. Wetzel, T., Candresse, T., Ravelonandro, M., and Dunez, J. 1992. A highly sensitive immunocapture polymerase reaction method for plum pox potyvirus detection. J. Virol. Methods 39:27-37.

CHAPTER 48

Control Strategies for Citrus Tristeza Virus

S.M. Garnsey, T.R. Gottwald and R.K. Yokomi

Citrus is a major fruit crop in many tropical and subtropical regions throughout the world and numerous varieties of sweet oranges, mandarins, grapefruits, lemons and limes are grown for domestic consumption and export. These have been selected for high fruit quality and yield and are propagated by grafting to rootstock cultivars selected for resistance to soil-borne diseases and adaptability to different soil conditions. In addition to a number of fungal and bacterial pathogens, citrus is affected by numerous graft-transmissible pathogens (46). These include several types of fastidious prokaryotes, a spiroplasma, viruses, viroids and virus-like agents of undetermined etiology (38). Citrus tristeza virus (CTV) is one of the most important pathogens affecting citrus and is the focus of extensive and diverse types of control efforts. The diversity of CTV, the crop, and the ecosystems where citrus is grown create a challenging and complex disease control problem.

Diseases Caused by CTV

Tristeza (which means sadness in Spanish and Portuguese) was the name originally used to describe the rapid and widespread decline and death of millions of trees on sour orange rootstock in Argentina and Brazil following introduction and spread of CTV in the 1930s (37,45). Association with an aphid-transmitted viruslike agent was established in 1946 (4). CTV actually caused two distinct diseases that are economically important. One is the decline of trees grafted on sour orange (*Citrus aurantium* L.) rootstocks as originally described. The second is a stem pitting disease of limes [*C. aurantifolia* (Christm.) Swing and *C. latifolia* Tan.], grapefruit (*C. paradisi* Macf.), and oranges [*C. sinensis* (L.) Osbeck] which debilitates trees and reduces yields. Some CTV isolates cause neither decline nor stem pitting, some cause

both, and others cause either one or the other disease.

Tristeza decline is caused by a CTV-induced phloem necrosis in the bark of the sour orange rootstock just below the budunion (Figure 1C). Movement of carbohydrates to the root system from the canopy is restricted or prevented. Once carbohydrate reserves in the root system are exhausted, new fibrous roots cannot be generated, and trees decline when the existing fibrous root system degenerates (45). This decline is often fast and lethal once symptoms begin to appear 1 to 2 years after infection (Figure 1A,B), and the disease is often called "quick decline". Sometimes trees decline gradually, but in either case, the affected trees rapidly become non-productive and must be replaced. Decline outbreaks have occurred in California, Florida, Israel, Peru, Spain, Venezuela and other locations (37,41). In spite of the hazards of CTV, sour orange has been a highly popular rootstock for citrus because of its vigor and ability to produce high quality fruit; and its tolerance to phytophthora foot rot and many other citrus viruses, viroids and virus-like pathogens (37). It is adaptable to a wide range of soils and climates, and in the absence of CTV, trees on sour orange are vigorous, and can be productive for more than 100 years. We estimate that over 2 million acres of citrus on sour orange rootstock (approximately 200 million trees) still exist worldwide, and these remain at risk to CTV-induced decline.

Stem pitting can be lethal to certain highly sensitive cultivars, but more typically, affected trees are stunted, and produce fruit of reduced size and quality. The productive life of trees affected by stem pitting is often only 5 to 15 years (26). Stem pitting results from a virus-induced disruption of the normal differentiation of cambial cells in different areas of the stem. When localized, discrete pits form in the trunk, branches

Figure 1. Disease symptoms caused by citrus tristeza virus (CTV). A) Aerial shot of a citrus grove on sour orange rootstock in Southwest Florida showing extensive tree loss from on epidemic of tristeza quick decline. Note empty spaces where trees were removed the previous year, and trees currently in various stages of decline. Over 50 percent of the trees were lost within several years. B) Close view of citrus tree in final stages of quick decline. Note fruit from current season is still present. C) Close up of budunion of citrus tree suffering from quick decline. Note yellow-brown discoloration on the surface of the wood and on the inner bark indicative of the virus-induced phloem necrosis below the budunion which triggers decline. (Trees which decline more gradually may have many small holes [honeycombing] in the inner bark face just below the union.) D) Close up budunion area of a trunk of a red grapefruit tree on rough lemon rootstock infected by a stem pitting isolate of CTV. Note severe pitting in the grapefruit portion of the trunk and the absence of pitting in the rough lemon portion which is tolerant. E) Closeup of a stem of sweet orange infected with a severe stem pitting isolate of CTV. The bark has been removed to show the general disruption of cambial differentiation over most of the stem. There are many small pits and an accumulation of gum. F) Fruit from a red grapefruit tree infected with a severe stem pitting isolate of CTV. Small, misshapen fruit are typical from trees with severe pitting in small twigs.

and roots where xylem cells fail to develop as they do in surrounding unaffected tissue (Figure 1D). In some cases the affects on cambial differentiation are more generalized, and nearly the whole stem or branch is affected. In this case, the bark is abnormally thick and the wood is brittle and often impregnated with gum (Figure 1E). While the individual discrete pits are more easily observed, they have less effect on tree vigor and fruit production than the diffuse pitting which severely affects fruit yields (Figure 1F). Limes and grapefruit are highly susceptible to stem pitting and are severely affected in many countries. Sweet oranges which are more tolerant but also are severely affected by some isolates (26,45).

In the long term, the stem pitting disease caused by some isolates of CTV is even more serious than decline. In contrast to CTV-induced decline, stem pitting is not a budunion disease, but affects all parts of the sensitive cultivar. Sensitive cultivars are affected, even when grafted on tolerant or immune rootstocks (Figure 1D).

Virus Properties

CTV is a member of the closterovirus group (1,26), and the complete sequence and genetic organization of one isolate has recently been determined (25,34). The virus particle is approximately 10-12 X 2,000 nm and is highly flexuous. It is easily sheared by mechanical forces, but is relatively stable otherwise. The genome is composed of a single positive sense RNA of 19,256 nucleotides which encodes 12 open reading frames that potentially code at least 17 proteins (34). One of these is the virus coat protein of approximately 25 kD (1). Isolates of CTV vary widely in symptom expression, but biological functions have yet to be associated with a specific part of the genome. Serological relationships have been established among nearly all known isolates using polyclonal antisera to different isolates and highly conserved epitopes have been identified (14,36). However, current research on molecular characterization of different isolates suggests that although there is a relatively high level of conservation in the coat protein gene, there is less homology among different isolates in other regions (Karasev and Hilf, personal communication). The concept that CTV may actually be a group of related viruses is currently gaining favor. There is also apparently some variation among the RNA species which constitute

a single isolate, and isolates whose properties appear stable when propagated in the same host may show segregation when exposed to selection pressure (30). Defective RNAs have also been recognized recently (29). These are apparently quite common and may affect symptom expression of some isolates.

Despite the recently recognized diversity at the molecular level, most CTV isolates share a number of common properties. All are phloem-limited, and the host range of all is essentially confined to citrus and citrus relatives. The only non rutaceous hosts reported so far are some species of *Passiflora* (2). Most CTV isolates are readily transmitted by grafting from hosts in which they replicate well. None are easily transmitted mechanically, but they can be transmitted experimentally by stem slash inoculation. Seed transmission is unknown. Most isolates are aphid transmitted, but efficiency varies (1,49).

Vectors

Six species of aphids have been described as vectors of CTV. *Toxoptera citricida* (the brown citrus aphid), *Aphis gossypii* (the melon aphid) and *Aphis spiraecola* (the spirea aphid) are the principal vectors and vary in transmission efficiency, host range and distribution (37). The brown citrus aphid is the most efficient vector of CTV (49). It is common in most citrus regions worldwide, but is still absent from the continental U.S., Mexico, and the Mediterranean Basin. Introduction of the brown citrus aphid into those areas has become more likely after its recent movement into the Arabian peninsula, Central America, and into many caribbean islands, including Cuba (37). Citrus is the primary host for the brown citrus aphid and it develops large populations on new growth. *T. aurantii*, a closely related species, is not an efficient vector. *T. aurantii* is distributed worldwide, and has a wide host range, but is not found in high populations on citrus. It is found in many countries where the brown citrus aphid is absent, and it has sometimes been mistakenly identified as the brown citrus aphid. This has created some confusion about distribution of the latter (49).

The melon aphid can transmit some CTV isolates with high efficiency, but in general, is much less efficient than the brown citrus aphid (2). The melon aphid is distributed worldwide and has a wide host range, but it prefers nonrutaceous

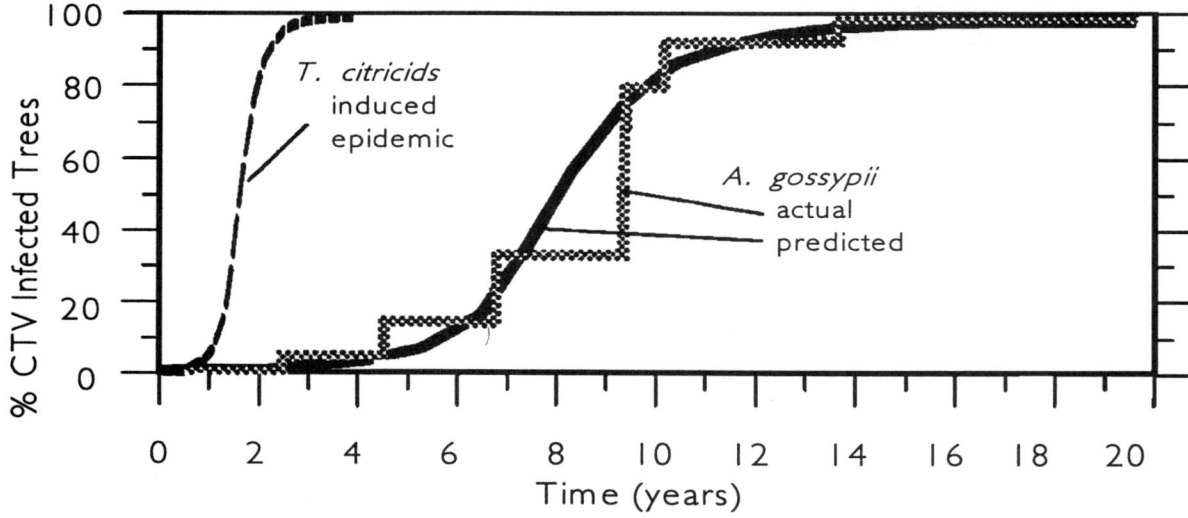

Figure 2. Comparative increase of CTV infection in field situations when vectored by the brown citrus aphid (*Toxoptera citricida*), an efficient vector, and by the melon aphid (*Aphis gossyppii*), a less efficient vector. Data for the brown citrus aphid was taken from test plots in Costa Rica and the Dominican Republic. Data for the melon aphid was taken from surveys and experimental plots in Spain, Florida and California. Initial infection levels were less than 1%. Note "stairstep" progression in infection with the melon aphid which is believed to correspond to periodic heavy aphid migrations.

hosts and only occasionally colonizes citrus. Recently, however, researchers in Spain, California, and Corsica have noted melon aphid colonies more frequently in citrus, and this may reflect higher aphid populations in surrounding crops. The melon aphid tends to migrate through citrus, only stopping to feed. These migrations generally occur in the spring and fall, and may vary in intensity from year to year. A "stair step" progression in CTV infections may result, with plateaus associated with periods of low melon aphid activity and brief rises associated with periods of aphid migration (Figure 2).

The spirea aphid is the least efficient of the three principal vectors in experimental tests, but is abundant on citrus. It is distributed world wide and also has a wide host range.

Host Reactions

Citrus and citrus relatives show various reactions to CTV. Most citrus species and cultivars are susceptible to infection, but many are tolerant and do not show obvious symptoms. Mandarins (*C. reticulata* Blanco) are especially tolerant to stem pitting isolates of CTV. Small acid limes (*C. aurantifolia*) are among the most sensitive hosts, and show vein clearing, stem

pitting and stunting when infected by most isolates of CTV. The genetic control for the decline syndrome and for stem pitting are apparently distinct (13).

Some citrus and citrus relatives are highly resistant or immune to infection, (15). In some cases, the resistance is specific to certain CTV isolates, but trifoliate orange [*Poncirus trifoliata* (L.) Raf.] is immune to a wide range of CTV isolates, and has been used as a breeding parent for rootstocks, and for development of CTV immune scions (3,22). A number of citrus species and hybrids with resistance or tolerance to CTV can be used to produce grafted trees tolerant to CTV-induced decline, but most of these rootstocks are susceptible to other diseases or are less well adapted to a wide range of soil conditions, and growers often choose to use sour orange in spite of the threat of CTV-induced decline (8).

CTV Detection

The traditional method to detect CTV infection in symtomless hosts has been to graft-inoculate lime plants and look for vein clearing symptoms in new flushes of growth (38). However, severity of symptoms in limes may not be correlated to symptom severity in other hosts.

accurate identification and differentiation of decline and stem pitting is still done by graft-inoculation to indicator hosts for the specific syndrome. For example, to determine if an isolate can cause decline, a grafted combination of sweet orange on sour orange is inoculated, and to test for stem pitting, grape fruit and sweet orange seedlings are inoculated. Many isolates produce a seedling yellows (SY) reaction in sour orange, lemon and grapefruit seedlings. An SY response is commonly considered as an indication for a severe isolate, but some stem pitting isolates and some decline isolates do not cause SY. A standard host range has been devised which measures decline, SY and stem pitting effects of CTV (13).

In recent years serological methods, especially ELISA, have been used increasingly for detection of CTV (14,36). Polyclonal antibodies detect nearly all isolates and ELISA provides a rapid and inexpensive means for detection. High accuracy is achieved if tissue samples are collected and processed appropriately. Monoclonal antibodies have also been developed. Some are reactive to epitopes which are widely conserved among diverse CTV isolates and can provide nearly universal detection (36). Others are to epitopes that are more isolate specific, and one (MCA13) has been widely used to differentiate mild isolates from those that induce decline and stem pitting in Florida (35). however, no monoclonals have been described which are specifically associated to either decline or stem pitting isolates as a group. Application of different serological procedures for CTV detection has recently been reviewed (36). Recently, PCR has been used for experimental applications and may become more widely used where higher levels of sensitivity or specificity are needed (33).

CTV Epidemiology

The epidemiology of CTV is very complex, and because of this development of disease, management strategies are difficult to develop. Unlike field crops, the crop cycle for citrus is long, and trees can be exposed repeatedly to infection, often by different isolates. Often several cultivars of differing susceptibility are grown in the same area. CTV isolates are quite diverse in severity, and aphid vectors differ in distribution, efficiency and abundance. Citrus, CTV and the brown citrus aphid all apparently originated in Asia. The earliest movement of

citrus from Asia was by seed, and citrus plantings in new areas were tristeza-free. Subsequently, as methods for shipping improved, citrus plants or budwood were moved between countries. It was then that CTV and the brown citrus aphid were inadvertently moved to new regions and epidemics of tristeza occurred (2). In most cases the spread of CTV has occurred through a combination of propagating activities by man and by aphid transmission (2).

Early information of CTV movement was based on observation of diseased trees and data collection was usually begun only after an epidemic was well in progress and early events were unclear. With the advent of rapid detection techniques, such as ELISA, it is now possible to monitor infection rates in symptomless trees and to do large scale surveys repetitively over time to gather both temporal and spatial data. Information is now being gathered on CTV movement in several areas where CTV is just now beginning to spread or where virus-free trees have been planted in infected areas and the location of initial inoculum can be determined.

CTV Control Efforts

Control strategies for CTV differ according to the incidence and severity of CTV isolates in each area, and with the cultivars and rootstocks used. There is no single control strategy for CTV that will be applicable in all situations. Growers with producing trees on sour orange rootstocks worry most about the threat of tree decline and how to conserve existing groves on sour orange. In many other citrus growing areas sour orange is no longer used because of CTV decline, and the focus in on control of stem pitting.

Several approaches have been used to control damage from CTV (2,37,45). The choice of approach is dependent on CTV incidence. In areas where CTV is rare, and natural spread has been limited, the emphasis has been on avoiding the virus. Quarantines and certification programs have been used to prevent introduction and propagation of infected plants into new areas. Eradication programs have also been used where infection levels are low or localized to suppress further natural spread of CTV. In areas where CTV is endemic and the chances of natural infection are high, the approach is to control losses in production. Tristeza decline has been avoided or overcome by substitution of tristeza tolerant rootstocks for sour orange. Mild strain cross protection has been used to control stem pitting in sensitive scions.

Intensive research on the biological, serological and molecular characterization of CTV over the past few years, plus development of new molecular and genetic engineering technology is expanding our understanding of CTV, and will lead to new control options. New information on the epidemiology of CTV and on vector biology is also being developed that will improve suppression programs and will allow more skillful disease management.

Control options for several different commonly encountered situations are presented. Prevention techniques are emphasized for areas where CTV is still absent or rare. Development of eradication or suppression schemes is covered for areas where only a limited infection is present. Disease management strategies are discussed for areas where the virus is endemic and efficient vectors are present. In addition to traditional approaches for controlling diseases causes by CTV, application of emerging technologies is also discussed.

Only selected references have been cited, but additional relevant literature on CTV and its control, can be accessed through these. An appendix is included with more specific information on detection, therapy and certification methods for CTV and other graft transmissible pathogens of citrus.

PREVENTING DAMAGE WHERE CTV INCIDENCE IS LOW

The value of disease prevention cannot be over-emphasized. Often, attention is first given to CTV disease control only after CTV problems are well developed. At this point, control options are limited and often expensive.

CTV is still absent or rare in many citrus growing areas, and simple and inexpensive preventative efforts can still be used to avoid introduction of CTV and subsequent damage. Much has been learned from observing CTV-induced problems in various countries, and this information can be applied to prevent similar problems from occurring elsewhere. The approaches for preventing CTV damage presented in this section mostly involve some form of regulatory action.

Exclusion and Quarantines

It is important to prevent importation of CTV-infected citrus into areas where CTV is absent or rare (26). Many citrus growing regions, states and countries, have strict quarantines on importation of live citrus tissue. However, quarantines must be enforced to be useful. In many countries, extensive international travel and commerce makes it difficult to thoroughly inspect all baggage and cargo for live plant materials. Even regional quarantines between or within countries are often difficult to enforce, especially when there are long borders and multiple entry points. Fortunately, in the case of virus diseases such as CTV, movement of fruit from infected trees poses little hazard. Seed are generally also less strictly regulated because most graft-transmissible citrus pathogens are not seed transmitted.

The greatest hazards for introduction of CTV is actually within the citrus industry. Citrus growers and horticulturists are constantly in search of new varieties, and are often anxious to import and propagate cultivars from other areas or countries which may offer a competitive advantage, or which have other interesting attributes. materials imported illegally by growers or nurserymen are especially hazardous because these will generally be grown in a commercial citrus area, and may be rapidly disseminated by propagation. Commercial growers or horticulturists are unlikely to deliberately import budwood from plants visually affected by CTV or other pathogens. However, if the cultivar is tolerant to CTV, the infection will go unrecognized. Introduction of severe stem pitting isolates of CTV into several areas has occurred through importation of CTV-infected mandarin cultivars (41). These are highly tolerant to CTV and appear healthy, but become a source of inoculum for secondary spread of CTV (37,41). For this reason, quarantines are usually effective only when coupled with extensive educational efforts among citrus growers and nurserymen to promote support and understanding for the need for regulation.

Illegal importation of propagating materials can also be reduced by establishing practical, rapid and safe methods to legally introduce cultivars from other countries. Most large citrus growing areas have mechanisms for legal introduction of new cultivars to meet grower needs and to reduce the pressure for illegal importation. Budwood is imported under permit, and checked for presence of CTV and other pathogens under quarantine. if necessary, therapy treatments may be used to eliminate graft-transmissible pathogens, including CTV (appendix 1). Indexing for most diseases can be completed within the time period required to verify fruiting characteristics and to produce to produce needed for primary release (38). Initial field plantings of introduced cultivars should be made in isolated areas and the propagation observed for any unusual characteristics.

Comprehensive indexing for graft-transmissible pathogens of citrus and therapy treatments to eliminate pathogens requires substantial investment in laboratory and greenhouse facilities, well trained and experienced personnel, and long term and stable financial support. Countries with small citrus industries, or those with limited resources, may find that cooperation with those countries which already have good certification programs may be the most practical and inexpensive way to quickly establish a basic bank of virus-free materials. Guidelines for safe international movement of citrus germplasm have been formulated and published by the International Board for Plant Genetic Resources (24).

Certification Programs

A vital aspect for preventing spread of CTV is to prevent its movement and the increase of infected trees by propagation (37). Most past epidemics of CTV were apparently begun or enhanced initially by unknowing propagation and movement of infected plants over long distances (2).

Nearly all citrus is clonally propagated by budding select scion cultivars to nucellar seedlings of a rootstock cultivar. Budwood is often taken from a mature tree and used directly in the nursery. Several thousand buds can be taken from a single tree. This budwood may be amplified in a "budwood increase nursery" prior to propagating commercial trees. Several thousand buds taken from a single budwood source tree can be increased a hundred-fold within a few months by a commercial nurseryman. These buds can then produce several hundred thousand trees ready for sale within another 6 to 12 months. Most graft-transmissible pathogens, including CTV, are easily propagated by grafting, so if the parent source of budwood is infected, nearly all propagations will be infected. Commercial nurseries often sell trees

over a large region or even to international markets, and a CTV infection in a single budwood tree can be spread over hundreds of miles within a few months.

Propagation of CTV-infected trees can be easily prevented by regular indexing of the primary budwood source trees and budwood increase blocks. This indexing can be done inexpensively by serological tests such as ELISA in a matter of days (see appendix). Parent sources of virus-free budwood must be protected from infection by being housed under screen or by being located in areas remote from sources of possible infection. Budwood source trees should be checked regularly for infection. Field nurseries should not be located near large plantings of infected trees, although infection rates in nurseries are normally low. Use of a CTV-immune rootstock, such as Carrizo or Troyer citrange (*C. sinensis* X *P. trifoliata*) or trifoliate orange, removes the threat that the rootstock seedlings could potentially become infected in the field prior to budding.

Even in areas where CTV is already present, it is critical to avoid further increasing the inoculum level, and the foci of infection by propagation of CTV-infected trees. Use of virus-free propagating sources will not prevent spread of CTV by vectors from existing infected trees, but it can greatly reduce the rate that existing plantings are lost since dissemination of CTV from a single focus of infection by aphids will be much slower than if infected trees are moved over long distances, and multiple foci of infection are established.

Budwood certification is also essential to avoid damage from other graft-transmissible pathogens when new rootstocks are used to replace sour orange. Most rootstock alternatives to sour orange are susceptible to other viruses and viroids. Sudden shifts from sour orange to other rootstocks using untested budwood has been disastrous in a number of countries, as new plantings on CTV-tolerant rootstocks failed because of other graft-transmissible pathogens. Identification of CTV in new areas should intensify budwood certification efforts and evaluation of alternative rootstocks (37).

Eradication and Suppression Programs

Once CTV is introduced, the second line of defense may be to reduce or remove the inoculum that has been established.

Many citrus growing areas are still largely free of CTV, and most of these areas with little history of CTV infection still have many or most trees on sour orange. Nearly all countries or regions with a long history of citriculture probably have at least a few CTV-infected trees. Often, the CTV-infected trees will be present only in variety collections or door yard plantings while commercial plantings remain CTV-free. These CTV-infected trees may be introductions made prior to the discovery of CTV, and although no natural spread may have occurred, they remain a threat if CTV is transmitted to a more favorable donor host or if the brown citrus aphid is introduced. Timely surveys and eradication of a few infected trees may suffice to avoid a future epidemic.

In other cases, localized areas within large commercial regions may have been contaminated by introduction of CTV-infected trees and some natural spread from these sources may have occurred. This is a more complicated situation and one that may demand an immediate response. Once such CTV infections are recognized, eradication of infected trees is frequently proposed to reduce the possibility of further natural spread. Eradication or suppression efforts against CTV have been conducted with variable success in several areas, including, Israel, Spain, California and Mexico (2,26,37).

Several factors affect the probable success of eradication or suppression programs. When the number of infected trees and the number of infection foci is small, and the rate of natural spread is low, the chances for success are good if resources for rapid and extensive surveys and tree removal are available. When infection is already widespread, and natural spread is common, the chances for success are poor.

In the past, decisions to establish eradication programs were made without comprehensive data on the extent and distribution of CTV infection, on the rate of natural spread and on projected rates of increase. It is now technically feasible to estimate the current distribution and percentage of infected trees, and to forecast future increases in infection (11,19,20). Data on CTV increase rates and spatial spread from Florida, California, Spain, Costa Rica, the Dominican Republic, and Trinidad has been collected and examined (6,20,31, S.M. Garnsey and T.R. Gottwald, unpublished). This data base can be divided into two broad categories: 1) Areas where CTV spread is by the melon aphid, and/or the spirea aphid, and 2) areas where the

brown citrus aphid is the predominant vector. The data from individual CTV epidemics were analyzed, and mathematical models were developed to describe disease increase over time (19). These models can now be used to forecast CTV increase under similar conditions in the future. Models based on data taken from areas such as Spain and Florida, where the melon aphid is the predominant CTV vector, have shown that CTV incidence normally progresses from low levels (~5%) to high levels (~95%) in 8-15 years. In contrast, when the brown citrus aphid was present, (Costa Rica and the Dominican Republic), a similar increase occurred in only 2-4 years (Figure 2). This indicates that CTV suppression would be much more difficult where the brown citrus aphid is present.

The increase of virus infection compounds over time. In the early stages of a CTV epidemic (when CTV infection is below 1%) the increase in CTV infected trees from one year to the next may not be very noticeable. Once the number of CTV-infected trees in a grove reaches the 5-10% infection level, the increase in infected trees is more dramatic because the infection rate increases and is operating on a larger number of trees. A disease curve based on Florida data, applied to a 10,000 tree grove predicts only 50 infected trees through year four. However, by year six 370 trees are infected and by year seven, over 1000 are infected.

The acceleration of infection and the total number of infected trees is highly important to the operation of a suppression program. As natural spread progresses, greater numbers of trees must be removed to offset new infections, and once infection increase reaches the logarithmic phase, tree removal sufficient to offset new infections becomes unfeasible.

Due to the large areas involved, and the lag period between when trees are infected by CTV and the subsequent detection of the infection, tree removal programs realistically are suppression programs since it is difficult to detect and remove every CTV-infected tree. If successfully designed, a suppression program can maintain CTV at low levels with a realistic input of effort and expense. Figure 3 demonstrates the theoretical effect of various tree removal rates on a CTV epidemic. CTV suppression will be achieved only if tree removals offset or exceed the number of newly infected trees. There is a certain threshold beyond which a tree removal strategy becomes cost prohibitive in labor and manpower, even if theoretically possible. This threshold must be

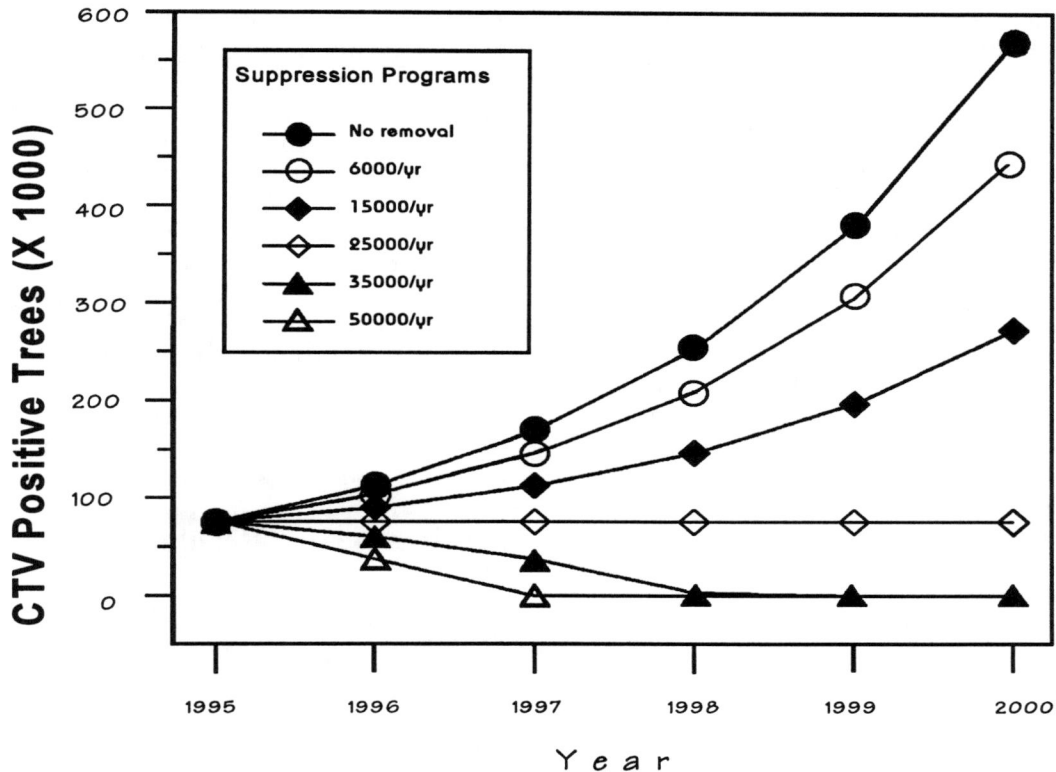

Figure 3. Effect of various tree removal rates on suppression of CTV infection. Figure is from a model developed for a tree removal project in California. Surveys indicated approximately 75,000 CTV infected trees in an area of approximately 80,000 ha. An infection increase of 50% per year of existing infected trees is assumed. If 25,000 of the exisiting trees are removed, then a 50% increase in the remaining 50,000 creates essentially a static situation. Increasing the initial tree removal to 50,000 rapidly reduces the population of infected trees and lowers overall tree removals compared less aggressive initial approaches.

determined for individual areas in terms of the number of trees that it is economically feasible to remove per year. If the number of tree removals is not sufficient to offset new infections and to decrease the number of CTV-infected trees in a timely manner, then tree removal as a control strategy will fail.

Survey Techniques

Accurate information on the distribution and incidence of CTV-infected trees are highly important for decisions to implement a CTV eradication or suppression program, and for program operation. The problem is that accurate infomation on infection incidence is hard to develop through limited surveys when the incidence is very low. Suppression programs for large areas usually will not be implemented if the

level of infection exceeds 5%. In many cases, the initial levels will be less than 1% or will rapidly drop below that point once the program begins. This means that suppression programs need survey data when infection percentages are so low that accuracy is difficult to achieve without extensive sampling. A variety of limited survey techniques have been employed in the past (7,19), but these were designed for quantitative data, such as the number of infections per plant, and are inaccurate when applied to binary data such as the +/- confirmation from ELISA normally determined for virus infected plants (23, Hughes and Gottwald, unpublished).

The need to gather an accurate sample, and the need for the sampling process to be simple and practical pose a conflict that is still being resolved. some initial information on patterns of spread from exisiting sources of infection (aggregated or random) may be essential for accurate estimates of infection. Fortunately, extensive data sets for CTV are available for analysis and are being used to develop new sampling protocols.

Survey operations normally occur in at least two phases. Initial surveys usually are geared to testing a relatively low percentage of trees over the entire area. Some initial sampling may also be biased toward testing the most suspected cultivars or locations, such as old variety collections. If a high percentage of infection is found in all areas during these initial surveys this may suffice to indicate that suppression is not feasible. If initial surveys indicate a relatively low incidence of infection or distribution limited to a certain area, more intensive sampling will be needed to either pinpoint isolated foci of infection or to more accurately determine infection levels.

If the number of trees or the area involved is limited, the most accurate method is to sample every tree. If infection rates are low, composite samples from five to 10 trees can be tested initially, and the composites that test positively retested individually. If a large area must be evaluated based on a limited number of samples, a "W" or a systematic sampling scheme is recommended to provide good coverage while using a consistent pattern that can be followed with only limited training. A new sampling protocol known as hierarchical sampling has been developed and partially validated that involves sampling at one spatial scale to predict CTV incidence at the individual tree scale (Gottwald and Hughes, unpublished). The method takes advantage of CTV infection patterns which are indistinguishable from random at low disease incidence. Samples from four adjacent trees in a square are bulked and the bulk sample assessed via ELISA for the presence or absence of CTV in the bulk. A systematic sampling pattern for these 4-tree bulks is laid out such that ca. 25% of the trees in the grove are sampled (Figure 4). One important criteria for this sampling method is that the starting positions of the sampling pattern be chosen at random. The number of individual infected trees in a grove can be predicted from the number of ELISA positive 4-tree bulks, by the use of a simple formula. Thus CTV incidence at the individual tree scale can be predicted by measuring CTV incidence at the next higher spatial scale, i.e. groups of 4-trees. The main advantage of such "hierarchical" or composite sampling is the increase in precision of disease

estimates gained due to sampling a larger number of trees (25%) maintaining the total number of samples processed via ELISA reasonably low (6.25%) (Gottwald and Hughes, unpublished).

Following the detailed survey areas or blocks of trees within areas can be ranked by percentage infection. Areas with the most infection (hot spots) are normally targeted for first attention. These are the areas where infection incidence will most quickly expand beyond the scope of reasonable tree removal, and are also the areas where infection incidence can be reduced most economically.

Suppression programs require regular periodic surveys to monitor infection incidence. The survey interval and the level of tree removal may need to be adjusted up or down periodically as data is collected and analyzed.

Strain-selective Regulation

In some areas (such as Florida), mild or decline-inducing isolates of CTV are common, but stem pitting isolates are still rare. In these areas preventative regulatory strategies may still be applied on a selective basis. Quarantines and selective registration of budwood free of stem pitting isolates can be especially important to prevent introduction and increase of especially dangerous isolates of CTV. For example, a new quality citrus tree Program for Florida requires that budwood sources be free of severe isolates of CTV, but permits propagation of budwood infected with mild isolates which are already endemic. Selective certification of budwood sources, however, requires strain-specific detection methods. Presently, indexing by graft-inoculation directly to susceptible cultivars is the most reliable method, but regular testing of many budwood source trees for stem pitting by graft-inoculation to indicators is expensive and slow. In Florida, the monoclonal antibody MCA13, which detects, but does not differentiate most stem-pitting and decline inducing isolates of CTV (35), is used as a presumptive screening tool to eliminate trees infected with stem pitting and decline from the certification program.

Information is now being aquired about the differences between CTV isolates at the molecular level. Once the regions of the viral genome which contrtol the expression of decline and stem-pitting are known, precise detection methods to distinguish these isolates can be developed. This will greatly augment efforts to survey for the

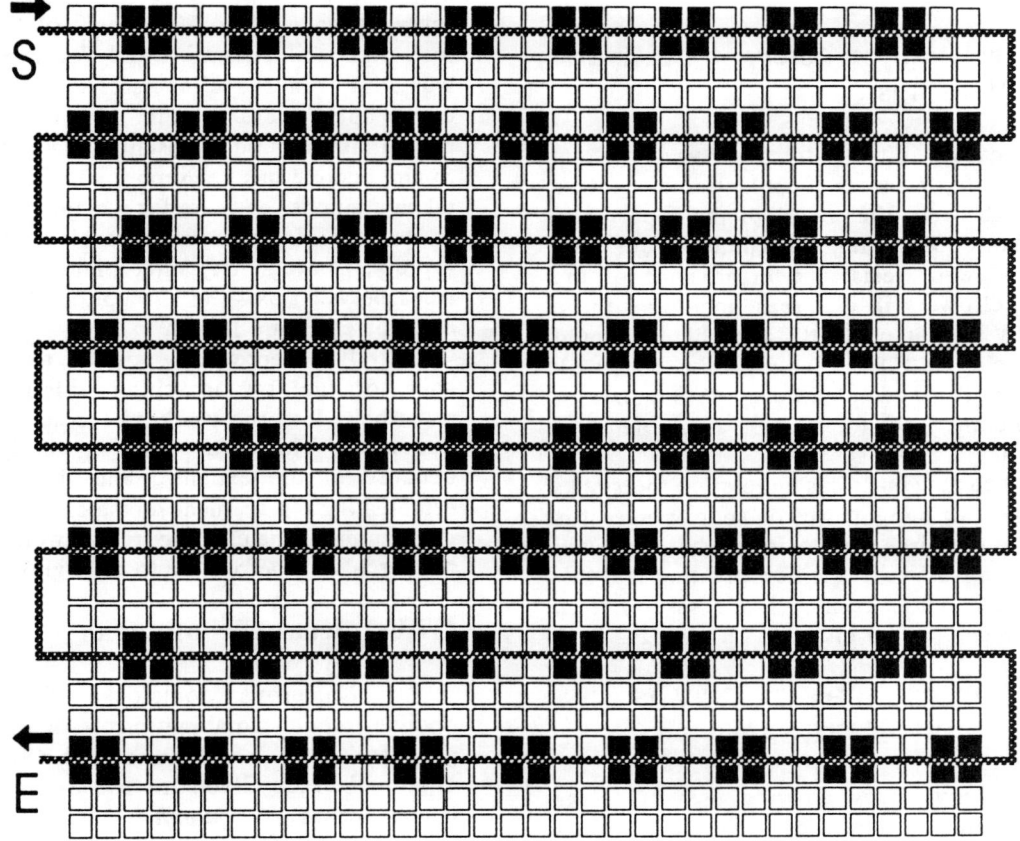

Figure 4. Diagrams for sampling patterns that can be used to do a systematic hierarchical sampling for citrus tristeza virus. Gray line indicates path traveled by survey crew member starting at position S and ending at position E. Black squares denote position of individual sampled trees. Samples from each group of four trees on bulked for processing and assay via ELISA. The disease incidence of individual trees is estimated from the incidence of ELISA positive bulk samples via a simple formula. In the above example 25% of the trees in the plot are sampled, however the number of actual biological assays is equivalent to sampling only 6.25% gaining greater precision with fewer samples.

presence of specific isolates in order to better manage CTV as well as selective certification and suppression programs.

To date, CTV eradication or suppression campaigns have been based on removal of all infected trees regardless of the severity of the strain they contain. As strain selective probes are developed for CTV, it should be possible to selectively eliminate only those trees which are infected with harmful isolates and to leave those carrying more benign isolates. This would reduce the cost of widespread tree removal, and eliminate grower objections to removal of trees which pose little harm and may contain mild isolates capable of providing crop protection. A selective eradication was successful in California when a virulent sweet orange stem pitting isolate of CTV was eliminated within a limited area where other benign isolates were endemic (41). In this case, there was a single focus of infection, natural spread was still limited, the problem was recognized promptly, and there were technical resources for extensive indexing on indicator plants.

Use of Resistant or Tolerant Cultivars in New Plantings

Because citrus trees have a long productive life, growers must anticipate future problems. Growers often overlook the opportunity to avoid future losses simply by planting cultivars with tolerance or resistance in advance of the problem. Growers in all areas, including those where CTV is not currently a problem, should consider using a CTV-tolerant scion rootstock combination for new plantings. A number of decline-tolerant rootstocks are available (see section IV).

When new plantings on tolerant rootstocks are made near existing groves on sour orange, it is important to use virus-free scions or those infected with a benign isolate. If the new planting is CTV-infected it can act as an inoculum reservoir from which CTV can spread into other plantings.

CONTROL WHERE CTV IS WIDESPREAD

In many citrus growing areas CTV is already endemic or too widespread to suppress infection. in areas experiencing a primary invasion of CTV, the immediate problem is loss of existing plantings on sour orange which must be replaced. In areas where CTV has long been present, especially where the brown citrus aphid is endemic, trees will usually already be on decline-tolerant rootstocks, and the main concern is to prevent or reduce losses from stem pitting. While regulatory approaches still may have application (see strain selective regulation in the previous section) to restrict introduction or spread of specific strains, the main emphasis will be on use of tolerant or resistant cultivars, mild strain cross protection and disease management.

Host Tolerance and Resistance

The control of tristeza-induced decline is to replace declining trees on sour orange with new trees on CTV-tolerant rootstocks (46). Incidence of decline can vary markedly and the rate of decline may affect the replacement strategy. If rates are low, individual trees can be replaced as they decline, and production of the grove can be maintained at a profitable level. However, grove care becomes difficult when trees are of many different ages, and often the entire grove is replaced once 35 to 50 percent of the trees are lost. In Brazil, Argentina and Venezuela tree loss was rapid since decline isolates and the brown citrus aphid were present. In areas where the brown citrus aphid is absent, such as Florida and Spain, decline has been more sporadic, and it has taken a number of years for a majority of the trees to decline. In areas where decline is advancing rapidly, growers may inter plant tolerant young trees between existing susceptible trees to offset anticipated future losses.

It is theoretically possible to save trees on sour orange by inarching tolerant seedling rootstocks into the trunk of the scion above the budunion. This procedure is rarely practical because of the expense, and the difficulty in establishing replacement rootstock seedlings under the canopy of the existing tree as well as the need to inarch well before the trees begins to decline.

CTV-tolerant Rootstocks

Numerous rootstocks form tristeza-decline-tolerant combinations with sweet orange, grapefruit and mandarin scions (8). Popular choices include Cleopatra and Sunki mandarins, rough lemon (*C. jambhiri*), *C. volkameriana*, rangpur lime (*C. limonia* Osb.) trifoliate orange and a number of trifoliate orange hybrids such as Troyer and Carrizo citranges and Swingle citrumelo (*C. paradisi* X *P. trifoliata*). Unfortunately, CTV is only one factor which must be considered in rootstock selection. Other diseases, such as phytophthora foot rot and citrus blight, affect many CTV-tolerant rootstocks. Citrus blight, a serious citrus decline disease of uncertain etiology, poses a dilemma for growers in Florida, South America, South Africa, and Australia who must cope with it plus tristeza decline (8). Rough lemon, *C. volkameriana*, rangpur lime, trifoliate orange and citranges are all CTV tolerant, but are susceptible to blight, while sour orange is one of the most blight-tolerant rootstocks, but is highly susceptible to CTV decline.

Other virus, viroid and viruslike pathogens that do not affect sour orange often cause serious diseases of other rootstocks. In fact, the search for alternative rootstocks inadvertently led to the discovery of many of the currently recognized virus and viroid diseases, and recognition that budwood free of graft-transmissible pathogens is often essential for successful use of other rootstocks.

While sour orange is adapted to a very wide variety of soil conditions, many other rootstocks are less tolerant to certain soil conditions, and can be used only in appropriate situations. For example, trifoliate orange and its hybrids are not well adapted to highly calcareous soils. Rootstocks also affects fruit quality, yield, cold hardiness, and other attributes of the scion. Rough lemon and rangpur lime are examples of rootstocks that produce large crops but, lack cold hardiness and do not impart high fruit quality. The disease susceptibility and horticultural features of a number of rootstocks has been recently reviewed (8).

Effort have been made to find sour orange selections with decline tolerance. A number of selections of sour orange which vary considerably in phenotype have been tested and all were susceptible to CTV-induced decline. Smooth flat seville "sour "orange and Gou Tou "sour" orange have moderate to good decline tolerance respectively, but these are natural hybrids of unknown parentage and not true sour oranges (8). It is unlikely that any true sour orange with acceptable levels of decline tolerance will be found, however, with rapid advances in genetic engineering there are new possibilities to curtail CTV-decline effects in sour orange so it can be used again in areas where CTV is endemic. For example development of CTV-immune scions could remove some of the hazards in using sour orange.

Stem Pitting Tolerant Scions

The options to prevent damage from stem pitting in scions through cultivar selection are limited. there are no truly CTV-tolerant limes, although Persian limes (*C. latifolia*) are more tolerant to CTV than small acid limes. All grapefruit cultivars are affected by stem pitting, and red grapefruit seem to be somewhat more

severely affected than white varieties. Some pummelos (*C. grandis* L.) are tolerant or resistant to some strains of CTV, and these have been successfully cultivated in some regions where some stem pitting strains of CTV are endemic. However, this resistance is not universal, and certain CTV isolates can cause severe stem pitting (43). Sweet oranges vary in susceptibility to stem pitting. Pera orange, a major variety in Brazil, is quite susceptible while Valencia is much more tolerant. However, even valencia can be severely damaged by some isolates.

Most mandarins are generally tolerant to CTV, but some mandarins, some mandarin hybrids, and some tangelos (*C. reticulata* X *C. paradisi*) can be affected by stem pitting. In some areas of the world, mandarins are the principal cultivars grown, partly because of their tolerance to stem pitting. However, reliance on mandarin cultivars severely limits market potential and growers continue to seek ways to continue growing sensitive cultivars in areas impacted by stem pitting.

Future Sources of Resistant Cultivars

The ultimate long term solution to CTV-induced diseases continues to be in the development of highly resistant or immune rootstock and scion cultivars. Extensive efforts are being made to develop citrus cultivars with improved CTV resistance by conventional breeding, somatic hybridization (27), and genetic engineering. A number of CTV tolerant or immune rootstocks have already been produced. Edible breeding lines which carry a CTV immunity gene from trifoliate orange already have been developed and are being used to develop CTV immune scions (3,22). Since citrus is heterozygous, the problem is to combine CTV immunity with the necessary horticultural characteristics. It also takes many years for hybrids to fruit and lose juvenile characteristics.

An alternative approach is to genetically engineer existing cultivars which have the desired horticultural characteristics and to add CTV resistance or immunity. Several cultivars have already been successfully transformed with a CTV coat protein gene (26). Other genes are available, and, based on precedents in other crops, these may also be useful to improve CTV resistance.

Another strategy is to use resistance genes from other hosts. For example, if the immunity gene from trifoliate orange can be identified, it could conceivably be used to transform susceptible citrus hosts if it can be vectored and expressed satisfactorily (17).

Continued rapid advances in transformation technology are anticipated and genetically engineered lines will soon be available for experimental testing. Evaluation of transformed citrus plants is still time consuming since the transformed cultivar must still be grown to fruiting age to determine yield and fruit quality even after resistance is confirmed. The durability of resistance in tranformed plants must also be determined. Since citrus is clonally propagated, however, it is not necessary to transfer the resistance to progeny plants.

Mild Strain Cross Protection (MSCP)

The basic premise of MSCP is that plants systemically infected with a mild isolate of a given virus will be protected when inoculated subsequently by more severe isolates (18). MSCP has been used as a control strategy for CTV stem pitting in areas where severe stem pitting isolates are widespread, and the brown citrus aphid is epidemic. Growers have few other options, since changing rootstocks does not prevent stem pitting in the scion, and lime, grapefruit and sweet orange cultivars tolerant or resistant to stem pitting are not available. Although MSCP is often only partially effective, and carries certain intrinsic risks, it can be useful in situations where heavy production losses are certain unless it is used.

MSCP is currently used for grapefruit in south Africa and Australia, and for Pera sweet oranges in Brazil (26). In these countries protective isolates have been selected from vigorous trees which remained in areas where most trees had been severely affected by stem pitting (9). Mildness of these isolates and their protective ability were confirmed in experimental tests (Cross protection effects have usually been evaluated on the field performance of the "protected" plant, and not on a more strict determination that infection or replication of the challenge virus is inhibited by the protecting isolate). These protective isolates have been disseminated largely by using mild strain-infected budwood to propagate new trees. In South Africa, the certification program has been used to deploy mild isolates for protective purposes (44).

MSCP also occurs naturally in citrus areas with long histories of stem pitting, and no formal cross protection programs. Each time growers select outstanding trees as budwood sources for propagation of new groves they are unknowingly either selecting trees infected with a protecting isolate, or trees that have escaped infection by severe isolates. In some cases the presumed "tolerance" of a cultivar may be associated with an unrecognized infection by a protecting isolate of CTV. In this case selections of the same cultivar freed of CTV will show more severe symptoms when inoculated with a severe isolate than the original.

The selection of protecting isolates is easiest in areas where stem pitting is common and large populations of trees have already been screened by natural challenge. The risks associated with deliberate use of infected budwood are also minimal in these situations since no new virus component is added.

Use of cross protection as a preventative strategy for stem pitting in areas where stem pitting is uncommon is not currently recommended (37). To use mscp in this situation, one must determine first that the proposed protective isolates are mild in the target cultivar, and also benign to other major cultivars present. The selective interaction of various CTV isolates with some cultivars means that isolate severity must be evaluated for all cultivars which could potentially become infected. For example, an isolate found mild in sweet orange, and chosen to provide protection to sweet orange against stem pitting, cannot automatically be assumed to be harmless to other cultivars.

Mild is a relative term and different interpretations have been given in various countries. Direct comparisons of isolates from different countries under the same conditions have shown that 'mild' protecting isolates from South Africa and Brazil are, in fact, much more severe than isolates considered mild in Florida and Spain. Results in several countries have shown that performance of 'protected' plants is inferior to that of virus-free plants until the latter become naturally infected with severe isolates.

The second requirement for successful MSCP is that there be an adequate protecting effect. Successful MSCP presumably requires a certain minimum degree of relationship between the protecting and challenge isolates at least in certain areas of the viral genome, even though the mechanism are not fully understood. Examples of protection between different isolates and examples of apparent lack of protection between others have been reported (39,40,45). If CTV is a group of related viruses rather than a single entity as current sequencing results suggest, the differences in cross protection ability observed among different isolates may be further evidence for differences in relationship between isolates. Diversity among CTV isolates in Asia, the probable origin of CTV, has been established. While CTV isolates in the secondary centers of citriculture presumably can be traced to common ancestors in Asia, isolates present in individual countries may not be the same. Different components of the total range of CTV isolates have been introduced into various countries via infected plants.

Mild isolates of CTV have been described in many citrus growing areas, but these are not necessarily protective against severe isolates in the same country, or if protective in one country against a certain challenge isolate, may not offer similar protection against a different isolate. For example, a mild protecting isolate selected under South African conditions, and found effective against severe stem pitting isolates, may not offer similar protection against a severe stem pitting isolate from another country. In some cases, isolates in secondary locations may have also evolved due to various local selection pressures. Thus to use cross protection in a preventative way one would have to know the source or sources of challenge isolates.

Field observations indicate that protection is generally not permanent, and that protective effects may break down over time, especially where challenge pressure is high. From a commercial standpoint, cross protection is a means to extend productive life of a planting faced with injury from stem pitting, but does not provide a permanent cure to stem pitting problems (37).

Some efforts have been made to develop MSCP for trees on sour orange rootstocks (2,37,48). Results to date have been less successful than with MSCP for stem pitting. Since replication of the challenge isolate anywhere in the canopy may cause a problem at the budunion, MSCP may have to be more complete than for stem pitting.

MSCP in the Future

Selection of mild protecting isolates has been largely empirical. With better knowledge of the sequence differences between CTV isolates, and how the CTV genome is organized, it should be possible to more accurately select effective and mild protecting isolates (33). It should also

become feasible to genetically engineer attenuated isolates from severe sources with protective characteristics, and to modify protective isolates so they will not be aphid transmissible. This would greatly reduce risks to non-target hosts when using cross protection in areas where diverse cultivars are grown, and it will reduce the complexity in selecting and evaluating protecting isolates. In contrast to transformation approaches which require that each cultivar be separately transformed and evaluated for resistance and horticultural characteristics, an infectious protecting isolate can be easily transferred to a number of different cultivars by grafting. Defective interfering RNAs (DIs) have recently been associated with CTV infection (29) may also have applications for cross protection.

Vector Control

Since the natural spread of CTV is attributed to aphids, aphid control is sometimes suggested as an approach to reduce CTV problems. Aphid vector suppression has not been practiced in commercial citrus, and is not considered an effective strategy except, perhaps, in a citrus nursery. Unlike direct aphid feeding damage which requires large infestations to cause damage, even a small number of winged aphids may transmit virus infections (9). However, under specific circumstances vector control may have more potential than previously recognized.

Effective aphid control can be achieved if the target pest, especially migrating (winged) forms are kept at low levels. This is difficult where the melon aphid and the spirea aphid are the principal CTV vectors because both aphids have a wide host range and migrating can originate on other crop or weed hosts. In addition, CTV acquisition and inoculation can occur in less than 30 minutes, the time required for aphids to reach vascular tissue and begin phloem ingestion.

Chemical Control

If aphid control is based on insecticides applied to protect healthy trees, these must be extremely potent and kill the aphid quickly to prevent inoculation. Long term chemical control is difficult because citrus is a perennial crop and control must continue year after year. In temperate regions, CTV spread is seasonal, and is concomitant with flushes of new growth in spring and fall. Under subtropical or tropical regions, new shoot growth is continuous as long as rainfall or irrigation is sufficient. Therefore, maintaining aphid control with insecticides is expensive, requires repeated applications, and has deleterious effects on the environment and non-target organisms. Control of the brown citrus aphid may actually be easier to achieve since the aphid reservoir is confined to citrus.

While use of insecticides for general long term control of aphids to prevent CTV spread does not appear realistic, there are situations where chemical control could be effective for the short term. For example, chemical suppression of a new and limited infestation of the brown citrus aphid may be useful to contain it until its distribution is known and the potential to eradicate or contain it are evaluated and quarantines can be imposed. While protecting healthy plants from inoculation with CTV by aphids is difficult, application of an insecticide to inoculum source trees could help prevent aphid spread of CTV to nearby healthy plantings. This strategy may be useful in conjunction with eradication or suppression programs to limit further spread until infected trees can be removed, especially if the infected trees are localized. If it can be determined that spread of CTV in some regions is sporadic and only occurs during short periods which can be defined then a single properly timed insecticide application may be useful to slow further spread.

Biocontrol

Aphids have many natural enemies including pathogens, predators, and parasitoids. In general, predators can dramatically reduce aphid populations, but don't work consistently, are not specific to specific insects, and build up too late to reduce aphid populations (12) before virus spread occurs. Entomopathogens require a build-up in aphid population to create an epizootic, are inconsistent, are dependent on high rainfall (humidity) and temperature. Because of these factors, entomopathogens frequently require augmentation to be useful commercially.

Parasitoids are better suited as natural enemies for aphid vectors because they are much more insect host specific, and their populations increase proportionally to the aphid population. Biocontrols directed at preventing aphids from reaching citrus may be feasible if parasitoids can be introduced and established in aphids on alternate crop hosts to mitigate buildup of migrant aphid populations (28). The melon aphid and the

spirea aphid have known parasitoids (42). Much less is known about natural enemies of the brown citrus aphid, but population fluctuations have been observed. Searches are underway to find efficient parasitoids of the brown citrus aphid in Asia where it originated (50).

It is not clear what level of vector control must be achieved to reduce rate of CTV spread, but pilot tests are underway to correlate effect of brown citrus aphid suppression with the rate of CTV spread. It is unlikely that movement of CTV can be prevented solely by vector management. It is likely that vector management may become important as an integrated component with other control approaches such as suppression and MSCP.

Vector Control in the Future

Entomological research currently includes development of chemicals which can deter or repel feeding by aphid vectors to reduce CTV spread. One approach can be to genetically engineer citrus plants to repel aphids or interfere with aphid feeding and reproduction. Field observations suggest that aphid spread of CTV does vary according to host and infection efficiency could be a selection factor for new cultivars.

Exotic parasitoids are being evaluated for their ability to reduce populations of the brown citrus aphid. Once their natural hosts are determined, and a release permit is issued, they will be released in several Caribbean countries where the brown citrus aphid has become established to measure their impact.

Cellular and molecular interactions of the vector, the virus, and the host are also being studied to determine the physical and genetic basis for CTV transmission. This information will be used for engineering non-transmissible mild strains of CTV for cross protection.

Integrated Disease Management

Natural spread of CTV is a complex process that involves the interaction of the plant, pathogen, and aphid vector. To date conventional approaches to controlling CTV have generally utilized only one component and that provides some effect, but is not fully successful. An integrated disease management (IDM) strategy which can utilize benefits from several components may be much more effective. To use IDM successfully will require development of a

Table 1. Application of control approaches for citrus tristeza virus (CTV) to areas with differing incidence of infection

Control Approach	CTV incidence			
	None	Rare	Decline[a] only	Decline[b] plus SP
Quarantine	X[c]	X	(X)	0
Certification	X	X	X	(X)
Eradication	0	X	0	0
Tolerant rootstock	(X)	X	X	X
Cross protection	0	0	(X)	X
Vector control	0	X	(X)	(X)
IDM[d]	(X)	(X)	X	X
Resistant scions	0	(X)	X	X

[a] Benign isolates and isolates capable of causing decline of trees in sour orange rootstock are common, but isolates that induce significant amounts of stem pitting in grapefruit or sweet orange are rare or absent (Florida and Spain are examples).
[b] Isolates capable of causing decline and isolates causing significant levels of stem pitting (SP) are common in commercial citrus.
[c] Brown citrus aphid is usually present.
X = Useful for situation; (X) = Useful under some conditions, or potential application; 0 = not used for this situation.
[d] Integrated Disease Management.

better understanding of the virus/host/vector patho-system than is currently available and some innovative thinking, but rapid progress has been made in recent years, and the specific needs have been more clearly defined. While an IDM approach may be difficult to use, it may be the only viable option to protect much of the existing citrus acreage for the foreseeable future and until improved cultivars are developed and tested. An IDM approach can incorporate horticultural, physical, regulatory, and chemical, or biological insect control aspects to manage the disease. The scope of an IDM approach must be regional in scope to be effective. Key components of IDM for CTV could include: 1) virus management components include budwood certification, (or selective propagation), eradication or suppression of severe isolates, and cross protection; 2) crop management strategies which include use of resistant or tolerant rootstocks, timing of hedging and other horticultural operations to reduce growth flushes during periods favorable for virus transmission, and planning new plantings to reduce proximity to CTV reservoirs and alternate plant hosts of the vector; and 3) vector management strategies which could include selective use of insecticides, repellents, and biological controls.

Conclusions

The control of CTV poses a continuing and diverse challenge to citriculture worldwide. A sizable portion of world citrus production area remains largely unaffected by CTV, but is under increasing threats and in need of more intensive protection efforts. Most of the estimated 200 million trees on sour orange rootstock will eventually succumb to CTV-induced decline in the absence of protective efforts. More extensive and vigorous application of quarantines, budwood certification and use of suppression programs in select situations could prevent or delay many of these losses. The lesson from areas previously devastated by CTV is that prevention is often the most effective and certainly the most economical control strategy.

In those large areas of citrus production where one or more isolates of CTV are endemic, the control approach switches to a disease management approach. Use of CTV tolerant or resistant rootstocks provides effective control for CTV-induced decline, but cultivar options for control of stem pitting are more limited. Cultivars with improved CTV tolerance or resistance will undoubtedly be developed for future plantings. However, the immediate need to protect and maintain existing plantings remains. Mild strain cross protection has been among the more prominent approaches, but broader application of it presents a variety of problems including development of more effective protective isolates. The most effective use of cross protection may be as a component in an integrated management strategy that combines management of virus, vector and host factors to control CTV.

The choice of a control approach requires some analysis of the current situation. Preventative strategies are appropriate where incidence is low but progressively less applicable as virus incidence and stain diversity increase. Deployment of mild strain cross protection is not indicated in areas where virus incidence is low and the potential source of challenge unknown. In Table 1 the relationship of control approach to infection incidence is presented in summary form. The choices and decisions are obviously more difficult in the intermediate situation. The decision to implement (or abandon) a suppression or eradication program is certainly one that is difficult to make. As infection increases, a transition in approach from prevention to management may be required and this entails major adjustments.

A thoughtful analysis of how to control CTV in different areas will usually indicate that critical information on several different factors is lacking. Development of effective research programs to address these needs often becomes an essential component of an effective control program. The action plan developed to counter future effects of CTV and the brown citrus aphid in Florida includes a list of research objectives as a component of the plan. These address both short and long term needs.

The current intensive research on the properties and characteristics of CTV and its vectors shows that this pathosystem is much more complex than we previously realized, and that the answers needed will require a concerted and coordinated international effort by many scientists.

APPENDIX-1

Pathogen Detection, Therapy and Budwood Certification Programs for Graft-transmissible Pathogens of Citrus

Many of the control strategies for control of CTV involve use of various types of detection procedures and use of virus-free budwood or budwood infected with only selected CTV isolates. Information on detection method, therapy

procedures and certification programs is covered in other chapters. The basic principals also apply to citrus. Detailed information on detection methods, therapy procedures and certification programs specific to CTV and other citrus pathogens can be found in a number of recent publications. In addition to those sources mentioned specifically here over 650 citations for CTV and other pathogens are found in a recent review by Lee et al. (26).

A Handbook on detection of graft-transmissible pathogens of citrus has been recently published by FAO (38). This provides comprehensive information on the detection of citrus pathogens by a variety of techniques, including the classic graft inoculation procedures. Detailed instructions for ELISA have been published (14). Some additional serological detection methods are described in several other publications (16,36). Detection technology is changing rapidly and use of PCR and hybridization assays with labeled pathogen specific probes is increasing (5,47). Often several procedures may be available. The choice of procedure is influenced by the speed that results are needed, as well as accuracy, sensitivity, cost, availability of special reagents, and need for specialized facilities and trained personnel. for example, accuracy and sensitivity are highly important when testing foundation sources of budwood to ensure that they are pathogen-free while low cost and rapid large scale sample processing are less critical. For large scale surveys, such as those used for eradication or suppression, low cost and rapid processing of large numbers of samples is essential.

Therapy, sanitation, quarantines and certification programs for citrus have been well reviewed by Navarro recently (32). The linkage between these approaches and their operational details are shown. A recent review of the South African budwood certification scheme (44) provides detailed descriptions of how a budwood scheme can be operated to propagate and distribute mild strain- protected budwood.

Some countries may not have technical resources or time for the development of complete certification programs, and may wish instead to import budwood from other countries which have well established programs. Guidelines for safe international movement of citrus germplasm have been developed (24) and should be consulted to minimize the risks involved.

REFERENCES

1. Bar-Joseph, M., and Lee, R. F. 1990. Citrus tristeza virus, revised description. CMI/AAB Description of plant viruses, No. 353. 7 Pages.

2. Bar-Joseph, M., Marcus, R., and Lee, R. F. 1989. The continuous challenge of citrus tristeza virus control. Ann. Rev. Phytopathol. 27:291-316.

3. Barrett, H. C. 1990. US119, an intergeneric hybrid citrus scion breeding line. Hort. Science 25:1670-1671.

4. Bennett, C. W., and Costa, A. S. 1949. Tristeza Disease of Citrus. J. Agr. Res. 78:207-237.

5. Bove, J.M, Garnier, M., Ahlawat, Y.S., Chakraborty, N. K., and Varma. A. 1993. Detection of the Asian strains of the greening BLO by DNA-DNA hybridization in Indian orchard trees and Malaysian *Diaphorina citri* psyllids. Pages 258-263 in: Proc. 12th Conf. Intern. Organ. Citrus Virol. P. Moreno, J. Da Graca and L.W. Timmer, eds. IOCV, Riverside.

6. Cambra, M., Serra, J., Villalba, D., and Moreno, P. 1988. Present situation of the citrus tristeza virus in the Valencian Community. Pages 1-7 in: Proc. 10th Conf. Intern. Organ. Citrus Virol. L.W. Timmer, S.M. Garnsey and L. Navarro, eds. IOCV, Riverside.

7   Campbell, C.L., and Madden, L.V. 1990. Introduction to Plant Disease Epidemiology. John Willey and Sons, New York. 532 Pages.

8. Castle, W.S., Tucker, D.P.H., Krezdorn, A.H., and Youtsey, C.O. 1993. Rootstocks for Florida citrus, 2nd ed. IFAS, Univ. of Florida, Gainsesville.

9. Carver, M. 1989. Biological control of aphids. Pages 141-165 in: World Crop Pests, Aphids - Their Biology, Natural Enemies and Control, Vol. 2C. A.K. Minks, and P. Harrewijn, eds. Elsevier, Amsterdam.

10. Costa, A.S., and Muller, G.W. 1980. Tristeza controlled by cross protection. A U.S.-Brazil cooperative success. Plant Dis. 64:538-541.

11. Fishman, S., Marcus, R., Talpaz, H., Bar-Joseph, M., Oren, Y., Salomon, R., and Zohar, M. 1983. Epidemiological and economic models for spread and control of citrus tristeza virus disease. Phytoparasitica 11:39-49.

12. Frazer, B.D. 1988. Predators. Pages 217-230 in: World Crop Pests, Aphids - Their Biology, Natural Enemies and Control, Vol. 2B. A.K Minks and P. Harrewijn, eds. Elsevier. Amsterdam.

13. Garnsey, S.M., Civerolo, E.L., Gumpf, D.J., Yokomi, RK., and Lee, R.F. 1991. Development of a worldwide collection of citrus tristeza virus isolates. Pages 113-120 in: Proc. 11th Conf. Intern. Organ. Citrus Virol. R.H. Brlansky, R.F. Lee and L.W. Timmer, eds. IOCV, Riverside.

14. Garnsey, S.M. and Cambra, M. 1991. Enzyme-linked immunosorbent assay (ELISA) for citrus pathogens. Pages 193-216 in: Graft-transmissible Diseases of Citrus-handbook for Detection and Diagnosis. C.N. Roistacher, ed. FAO, Rome.

15. Garnsey, S.M., Barrett, H.C., and Hutchison, D.J. 1987. Identification of citrus tristeza virus resistance in citrus relatives and its potential applications. Phytophylactica 19:187-191.

16. Garnsey, S. M., Permar, T. A., Cambra, M., and Henderson, C. T. 1993. Direct tissue blot immunoassay (DTBIA) for detection of citrus tristeza virus (CTV). Pages 39-50 in: Proc. 12th Conf. Intern. Organ. Citrus Virol. P. Moreno, J. Da Graca and L. Timmer, eds. IOCV, Riverside.

17. Gmitter, F. G., Jr., Xiao, S. Y., Huang, S., Hu, X. L., Garnsey, S. M., and Deng, Z. 1995. Localized linkage map of the citrus tristeza virus resistance gene region. Theor. Appl. Gen. (submitted for Pub.)

18. Gonsalves, D., and Garnsey, S. M. 1989. Cross-protection techniques for control of plant virus diseases in the tropics. Plant Disease 73:592-597.

19. Gottwald, T. R. 1993. CTV Epidemiology and Disease Survey Strategies. Pages 106-109 in: Citrus tristeza virus and *Toxoptera citricidus* in Central America: Development of management strategies and use of biotechnology for control. 2nd Citrus Tristeza Virus Workshop, Maracay, Venezuela, September, 1992. R. Lastra, R. Lee, M. Rocha-Pena, C. Niblett, F. Ochoa, S. Garnsey and R. Yokomi, eds. IFAS, University of Florida.

20. Gottwald. T. R., Cambra, M. and Moreno, P. 1993. the use of serological assays to monitor spatial and temporal spread of citrus tristeza virus in symptomless trees in eastern Spain. Pages 51-61 in: Proc. 12th Conf. Intern. Organ. Citrus Virol. P. Moreno, J. V. da Graca and L. W. Timmer, eds. IOCV, Riverside.

21. Gottwald, T. R., Garnsey, S. M., and Yokomi, R. K. 1993. Potential for spread of citrus tristeza virus and its vector the brown citrus aphid. Proc. Fla State Hort. Soc. 106: 85-94.

22. Hearn, C. J., Garnsey, S. M., and Barrett, H. C. 1993. Transmission of citrus tristeza virus resistance from breeding line US 119. HortScience 28:123 (abstract).

23. Hughes, G., Madden, L. V. and Mankvold, G. P. 1995. Cluster sampling for disease incidence data. Phytopathology 85: (In press)

24. International Board for Plant Genetic Resources. 1991. Technical Guidelines for the safe movement of citrus germplasm. FAO/IBPGR, Rome. 50 Pages.

25. Karasev. A. V., Boyko, V. P., Gowda, S., Nikolaeva, O. V., Hilf, M. E., Koonnin, E. V., Niblett, C. L., Cline, K., Gumpf, D. J., Lee, R. F., Garnsey, S. M., Lewandowski, D. J., and Dawson, W. O. 1995. Complete sequence of the citrus tristeza virus genome. Virology 28:511-520.

26. Lee, R. F., Baker, P. S., and Rocha-Pena, M. A. 1994. The Citrus Tristeza Virus (CTV). Intern. Inst. of Biol. Control, Ascot, Berks, UK. 197 Pages.

27. Louzada, E. S., Grosser, J. W., Gmitter, F. G., Jr., Nielsen, B., Deng, X. X., Tusa, N., and Chandler, J. L. 1992. Eight new somatic hybrid citrus rootstocks with potential for improved disease resistance. HortScience 17:1033-1036.

28. Mackauer, M.., and Way, M. J. 1976. *Myzus persicae* Sulz. An aphid of world importance, Pages 51-119 in: Studies in Biological Control. V. L. Delucchi ed. Cambridge Univ. Press. Cambridge.

29. Mawassi, M., Karasev, A. V., Mietkiewska, E., Gafny, R., Lee, R. F., Dawson, W. O., and Bar-Joseph, M. 1995. Defective RNA molecules associated with citrus tristeza virus. Virology 208: 383-387.

30. Moreno, P., Guerri, J., Ballester-Olmos, J. F., Fuertes-Polo, C., Albiach, R., and Martinez, M. 1995. Variations in pathogenicity and double stranded RNA (dsRNA) patterns of citrus tristeza virus isolate induced by host passage. Pages 8-15

in: Proc. 12th Conf. Intern. Organ. Citrus Virol. P. Moreno, J. V. da Graca, and L. W. Timmer, eds. IOCV, Riverside

31. Moreno, P., Piquer, J., Piña, J. A., Juarez, J., and Cambra, M. 1988. Spread of citrus tristeza virus in a heavily infested citrus area in Spain. Pages 71-76 in: Proc. 10th Conf. Intern. Organ. Citrus Virol. L. W. Timmer, S. M. Garnsey and L. Navarro, eds. IOCV, Riverside.

32. Navarro, L. 1993. Citrus sanitation, quarantine and certification programs. Pages 383-391 in: P. Moreno, J. V. da Graca, and L. W. Timmer, eds. Proc. 12th Conf. Intern. Organ. Citrus Virol., IOCV, Riverside

33. Niblett, C. L., Pappu, H. R., Pappu, S. S., Febres, V. J., Manjunath, K. L., Lee, R. F., Grosser, J. W., and Schell, J.S. 1993. Progress on the characterization and control of citrus tristeza virus. Proc. Fla. State Hort. Soc. 106:99-102.

34. Pappu, H. R., Karasev, A. V., Anderson, E. J., Pappu, S. S., Hilf, M. E., Febres, V. J., Eckloff, R. M. G., McCaffery, M., Boyko, V.,Gowda, S., Dolja, V. V., Koonin, E. V., Gumpf, D. J., Cline, K. C., Garnsey, S. M., Dawson, W. O., Lee, R. F., and Niblett, C. L. 1994. Nucleotide sequence and organization of eight 3' open reading frames of the citrus tristeza closterovirus genome. Virology 199:35-46.

35. Permar, T. A., Garnsey, S. M., Gumpf, D. J., and Lee, R. F. 1990. A monoclonal antibody that discriminates strains of citrus tristeza virus. Phytopathology 80:224-228.

36. Rocha-Pena, M. A., and Lee, R. F. 1991. Serological techniques for detection of citrus tristeza virus. J. Virol. Methods 34:311-331.

37. Rocha-Pena, M. A., Lee, R. F., Lastra, R., Niblett, C. L., Ochoa-Corona, F. M., Garnsey, S. M., and Yokomi, R. K. 1995. Citrus tristeza virus and its aphid vector *Toxoptera citricida*. Plant Dis. 79: 437-445.

38. Roistacher, C.N. 1991. Graft-transmissible diseases of citrus: Handbook for detection and diagnosis. FAO, Rome. 286 Pages.

39. Roistacher, C.N., and Dodds, J. A. 1993. Failure of 100 mild citrus tristeza virus isolates from California to cross protect against a challenge by severe sweet orange stem pitting isolates. Pages 100-107 in: Proc. 12th Conf. Intern. Organ. Citrus Virol. P. Moreno, J. V. da Graca and L. W. Timmer, eds. IOCV, Riverside

40. Roistacher, C. N., Dodds, J. A., and Bash, J. A. 1988. Cross protection against citrus tristeza seedling yellows and stem pitting viruses by protective isolates developed in greenhouse plants. Pages 91-100 in: Proc. 10th Conf. Intern. Organ.

Citrus Virol. L. W. timmer, S. M. Garnsey and L. Navarro, eds. IOCV, riverside.

41. Roistacher, C. N., and Moreno, P. 1991. The worldwide threat from destructive isolates of citrus tristeza virus-A review. Pages 7-19 in: Proc. 11th Conf. Intern. Organ. Citrus Virol. R. H. Brlansky, R. F. Lee and L. W. Timmer, eds. IOCV, Riverside.

42. Tang, Y. Q., Yokomi, R. K., and Gagne, R. J. 1994. Life history and description of *Endaphis maculans* (Diptera: Cecidomyiidae), an endoparasitoid of aphids in Florida and the Caribbean Basin. Ann. Entomol. Soc. Amer. 87:523-531.

43. Tsai, M. C., Su, H. J., and Garnsey, S. M. 1993. Comparative study on stem-pitting strains of CTV in the Asian countries. Pages 16-19 in: Proc. 12th Conf. Intern. Organ. Citrus Virol. P. Moreno, J. Da Graca and L. Timmer, eds. IOCV, Riverside.

44. von Broembsen, L., and Lee, A. T. C. 1988. South Africa's citrus improvement programme. Pages 407-416 in: Proc. 10th Conf. Intern. Organ. Citrus Virol. L. W. Timmer, S. M. Garnsey and L. Navarro, eds. IOCV, Riverside.

45. Wallace, J. M. 1978. Virus and viruslike Diseases. Pages 67-184 in: The Citrus Industry Vol. 4. W. Reuther, E. C. Calavan and G. E. Carman, eds. Div. of Agric. Sci., Univ. of California, Berkeley. 362 Pages.

46. Whiteside, J. O., Garnsey, S. M., and Timmer, L. W. 1988. Compendium of Citrus Diseases. APS Press, St. Paul, MN. 80 Pages.

47. Yang, X., Hadidi, A., and Garnsey, S. M. 1992. Enzymatic amplification of citrus exocortis and cachexia viroids from infected citrus hosts. Phytopathology 82:279-285.

48. Yokomi, R. K., Garnsey, S. M., Permar, T. A., Lee. R. F., and Youtsey, C. O. 1991. Natural spread of severe isolates of citrus tristeza virus isolates in citrus preinoculated with mild CTV isolates. Pages 86-92 in: Proc. 11th Conf. Intern. Organ. Citrus Virol. R. H. Brlansky, R. F. Lee and L.W. Timmer, eds. IOCV, Riverside, CA.

49. Yokomi, R. K., Lastra, R., Stoetzel, M. B., Damsteegt, V. D., Lee, R. F., Garnsey, S. M., Gottwald, T. R., Rocha-Pena, M., and Niblett, C. L. 1994. Establishment of the brown citrus aphid (Homoptera: Aphididae) in central America and the Caribbean Basin and transmission of citrus tristeza virus. J. Economic Ent. 87:1078-1085.

50. Yokomi, R. K., Tang, Y. Q., Nong, L., and Kok-Yokomi, M. L. 1993. Potential mitigation of the threat of the brown citrus aphid, *Toxoptera citricida* (Kirkaldy), by integrated pest management. Proc. Fla State Hort. Soc. 106:81-85.

CHAPTER 49

Banana Bunchy Top Disease: Current and Future Strategies for Control

J.L. Dale and R.M. Harding

Bananas and plantains (*Musa* spp) are one of the world's largest and most important agricultural commodities. This tropical and sub-tropical crop contributes significantly to many national domestic and export incomes and in some regions is the principal source of carbohydrate. However, the production of this large crop is limited by a wide range of diseases including fungal, viral, bacterial and nematode diseases. The most important of the viral diseases is banana bunchy top disease (BBTD).

BANANA BUNCHY TOP DISEASE

Symptoms of the Disease

BBTD was first recorded in Fiji in 1879 (28). It is probable that this early report was in part a result of the very distinctive symptoms of the disease and its devastating effect on production. Infected plants have a bunched appearance at the top of the plant with narrow, upright leaves which are yellowed at the margins (Figure 1 a,b). The leaves become increasingly smaller and the plant becomes dwarfed. Very characteristic of the disease are the small dark green streaks on the pseudostem, petioles and leaves (Figure 1 c) resulting in a dot-dash or "Morse code" pattern in these leaves (Figure 1 d). Depending on the time of infection, the infected plants usually do not produce fruit and therefore the disease can result in a 100% yield loss.

Geographical Distribution

The "modern" history of BBTD has been reasonably well documented (9). From the first record of BBTD in Fiji, the disease was subsequently recorded in Taiwan around 1900 (40)

and Egypt in 1901 (28). It is believed that BBTD was imported into Australia and Sri Lanka in 1913 in infected suckers from Fiji (26). The disease was probably introduced into India from Sri Lanka about 1940 (28). Interestingly, a very similar disease, abaca bunchy top disease (ABTD), in abaca (*Musa textalis*) was recorded in the Philippines in 1910 (31) but BBTD was not recorded in bananas in the Philippines until 1960 (8). It is probable, however, that BBTD and ABTD are closely related diseases.

BBTD is now present in a number of countries in Africa, Asia and the South Pacific but its geographical distribution is erratic within regions (Figure 2). The disease continues to move with recent recorded outbreaks in Pakistan in 1988 (39) and in Hawaii in 1989 (17). Importantly, the disease has not been recorded in South or Central America or in the Caribbean and is not universally distributed in Africa or Asia (Figure 2).

The history of the spread of BBTD and its present distribution suggest that BBTD is either a relatively new virus of bananas or that it had very limited distribution and/or incidence in bananas prior to the mid 1800s. The accepted center of origin and early domestication of bananas is Indochina and South East Asia (37). It has been suggested that domesticated bananas spread from South East Asia (perhaps Malaysia and Vietnam) through Oceania and India and were widespread in these regions prior to the movement of bananas from India through the Middle East to north Africa by 600AD (37). Bananas were first observed by Western explorers in West Africa in the late 15th century. Finally, the Spanish and Portuguese were responsible for the introduction of bananas to the Americas around 1516 via the Canary Islands. Since these early introductions, there has been a continual and, until recently, a largely unregulated

Figure 1. Symptoms of banana bunchy top disease: (a,b) Plants in advanced stages of the disease with narrow, upright leaves showing marginal chlorosis and bunching at the apex; (c) pseudostem showing dark green streaks, and (d) leaf showing typical "morse code " pattern of green streaks and dots.

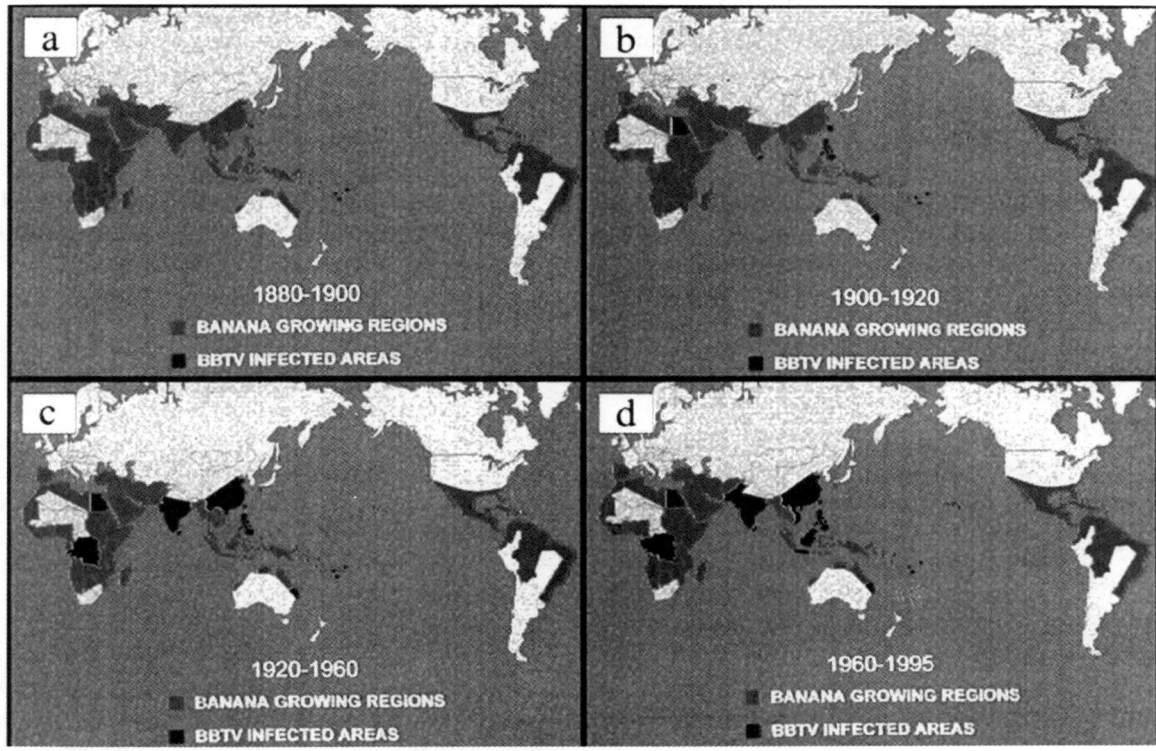

Figure 2. Distribution of banana bunchy top virus from (a) 1880-1900, (b) 1900-1920, and (d) 1960-1995

international exchange of banana planting material.

Virtually all commercial or widely grown cultivars of bananas and plantains are selections from naturally occurring hybrids within or between *M. acuminata* and *M. balbisiana*. The majority of cultivars are sterile, parthenocarpic triploids and as such are all propagated vegetatively. Suckers and divided corms have traditionally been used as planting material and it can be assumed that bananas were moved to Oceania, Africa and the Americas as suckers, corms or potted plants. BBTD is efficiently transmitted in all vegetative parts of bananas (9) and therefore if the disease had been widespread in bananas when bananas were first and subsequently moved from South East Asia, the disease should be present in the Americas.

Economic Impact

There are no accurate estimates of the international economic impact of BBTD despite the disease being widespread in Asia and the South Pacific regions. In most instances, the incidence of the disease is not well documented. However, there are two examples of outbreaks of BBTD that have been well studied.

BBTD was first recognized in Australia in 1913 in plantations on the border between the states of Queensland and New South Wales. From these plantations, the disease rapidly spread south into New South Wales and north into south east Queensland. By 1926, 90% of the original banana plantings in New South Wales were out of production; output fell from 460,000 cases of bananas in 1922 to 140,000 cases in 1925. In the Currumbin district of south east Queensland, the number of banana plantations fell from 100 in 1922 to 4 in 1925 with production falling from

4,400 tons to 110 tons in the same period (9). This drastic impact on production was almost certainly due to BBTD alone.

Similar devastation has been recorded in Pakistan. BBTD was first recorded in Pakistan in 1988 in Sindh province (23). In this province, approximately 50,000 ha were growing bananas in 1988, 23,500 ha of which produced "taxable" bananas and a further estimated 25,000 ha of "untaxable" bananas. By the end of 1992, only 8,000 ha of "taxable" bananas were in production, a reduction of about 60%. This reduction has been attributed to BBTD (38).

BBTD is widespread in Vietnam and the Philippines where the incidence in small holdings varies considerably up to about 50% infection (personal observation) and the disease has been particularly devastating in Tonga.

It appears that after an initial introduction, BBTD has the potential to spread very rapidly but with the subsequent implementation of control measures such as roguing and use of disease free planting material, the disease can be held at a level acceptable for continued production.

Transmission

BBTD is not sap transmissible but is persistently transmitted by the banana aphid (*Pentalonia nigronervosa*) (26). No other aphids are known to transmit the disease. Hafner *et al.* (14) have recently demonstrated that the virus associated with BBTD, banana bunchy top virus (BBTV) does not replicate in the aphid vector.

Magee (26) also demonstrated that the disease is transmitted through the suckers and divided corms of infected plants. Drew *et al.* (12) have demonstrated that BBTD can be transmitted through micropropagated plantlets established from infected plants. However, plantlets in tissue culture did not have the characteristic BBTD symptoms and were indistinguishable from non-infected control plants. Further, once these plantlets were established in the greenhouse, 73% of plants developed characteristic symptoms whereas the remaining 27% of plants did not develop symptoms and appeared to be not infected. These results have important ramifications for the propagation and distribution of bananas in tissue culture, a practice that is becoming increasing popular.

Host Range

The only confirmed hosts of BBTD are species within the genus Musa (*M. acuminata, M. balbisiana* and their hybrids, and Fei bananas) and *Ensete ventricosum (Musaceae)* (26). Ram and Summanwar (33) reported that *Colocasia esculenta* could be symptomlessly infected with BBTD but this report has not been confirmed. Interestingly, none of the commercial or widely grown cultivars of banana and plantain have been shown to be immune to BBTD but there does appear to be variation between cultivars in the rate of infection and the severity of symptoms (19,26,27). There are no reports on the reaction of wild diploid *Musa* spp to BBTD challenge.

If BBTD is a relatively new virus disease of bananas, a possibility suggested by Karan *et al.* (21), there must exist alternate hosts of the disease. However, it would appear that if such hosts exist, they do not play an important role in the epidemiology of BBTD in commercial agriculture.

Epidemiology

BBTD has been reported to have a single aphid vector but also moves very effectively in vegetative planting material. It appears that aphids are involved in short distance movement while infected planting material is responsible for both short and long distance movement of the disease. Wild or uncultivated bananas and neglected plantations are also an important source of the disease. There are no confirmed non-*Musa* hosts of BBTD.

The epidemiology of BBTD has only been studied to any great extent in the atypical subtropical environment of northern New South Wales in Australia (1-3). Allen (2) reported that from an initial focus of infection the mean distance of new infections, presumably resulting from aphid transmission, was 17.2 m with 70% of new infections within 20 m and 99% of new infections within 86 m. Allen and Barnier (3) found that new plantations established from disease free planting material had an 88% probability of becoming infected if they were adjacent to an infected plantation whereas this dropped to a 27% probability if the nearest infected plantation was 50 to 1000 m away. Importantly, no infections were recorded in 19 plantations which were more than 1000 m from the nearest infected plantation.

BANANA BUNCHY TOP VIRUS

The Luteovirus Hypothesis

The virus causing BBTD, banana bunchy top virus (BBTV) was originally classified as a possible luteovirus (29) as the disease had many characteristics of luteovirus infections: the disease (i) was not sap transmissible, (ii) was persistently transmitted by aphids, (iii) resulted in yellows-type symptoms and (iv) induced phloem damage in infected plants. Subsequent evidence supported this original classification. Dale et al. (11) extracted dsRNA from BBTD infected plants that was not present in non-BBTD infected plants and the electrophoretic pattern of this dsRNA was similar to the electrophoretic patterns of dsRNA extracted from plants infected with known luteoviruses. Iskra et al. (18) purified 28 nm isometric virus-like particles (VLPs) from BBTD infected plants. In contrast, Wu and Su (43) purified 20-22 nm isometric VLPs from infected plants and reported that these particles contained ssRNA of about 6.0 kb. They described these particles as those of a small luteovirus and generated monoclonal antibodies for the detection of these particles.

BBTV is a ssDNA Virus

Harding et al. (15) and Thomas and Dietzgen (41) purified 18-20 nm isometric VLPs from infected plants (Figure 3a) using modifications of the method of Wu and Su (43). These particles however contained ssDNA about 1 kb. The molecular weight of the coat protein subunit of these particles was approximately 20,000. Using either DNA probes (15) or monoclonal antibodies, it was demonstrated that these VLPs were transmitted from infected plants to non-infected plants via aphids. These VLPs are now accepted as the virions of banana bunchy top virus (BBTV).

Subsequently, Harding et al. (16) reported the sequence of the first component of the BBTV genome. This component was shown to be circular ssDNA and 1.111 kb. It was encapsidated as only one sense and contained a potential stem-loop structure, the loop sequence of which was almost identical to the invariant loop sequence of geminiviruses (24). This component also contained a single large open reading frame (ORF) in the virion sense which potentially encoded a replicase associated protein based on the presence of a nucleotide binding motif. Burns et al. (6) determined that BBTV has a multi-component genome with at least six components. Subsequent sequence analysis of these six ssDNA components, named components 1 to 6, has shown that (i) the components vary in size from 1.018 kb to 1.111 kb, (ii) each component is circular and is encapsidated in one sense only, (iii) all components share two common regions, the Major Common Region (CR-M) and the Stem-Loop Common Region (CR-SL) and (iv) five of the six components have single large ORFs in the virion sense (7) (Figure 3 b,c). Xie and Hu (45) have reported the sequence of three BBTV genomic components from a Hawaiian BBTV isolate; these components, 1, 3 and 4 are almost identical to components 1, 2 and 5 respectively of Harding et al. (16) and Burns et al. (7).

Karan et al. (21) has demonstrated that BBTV component 1 was present in all BBTV infected bananas from 11 countries using DNA probes and the polymerase chain reaction (PCR) but not in healthy bananas. Further, Karan (20) has shown that BBTV components 2 to 6 were present in the four BBTV isolates tested from four countries.

Wu et al. (44) have sequenced two additional BBTV genomic components from a Taiwanese BBTV isolate; both these components potentially encode replicase associated proteins. The two components are clearly different from BBTV component 1 and from each other. Yeh et al. (46) have sequenced a further BBTV genomic component from a Taiwanese isolate which appears to be a defective component that would encode a replicase associated protein. The significance of these additional replicase associated protein encoding components remains to be resolved.

BBTV is clearly not a member of any previously described plant virus group; it is most similar to the geminiviruses but differs from that group in that (i) its virions are isometric and not geminate, (ii) its genome consists of at least six components of circular ssDNA each about 1 kb and not one or two components of ssDNA each of about 2.7 kb, (iii) its coat protein has a molecular weight of about 20,000 and (iv) it is transmitted by aphids. Rather, BBTV is probably a member of a new plant virus group which could also contain subterranean clover stunt virus (5), faba bean necrotic yellows virus (22), coconut foliar decay virus (34) and milk vetch dwarf virus (36).

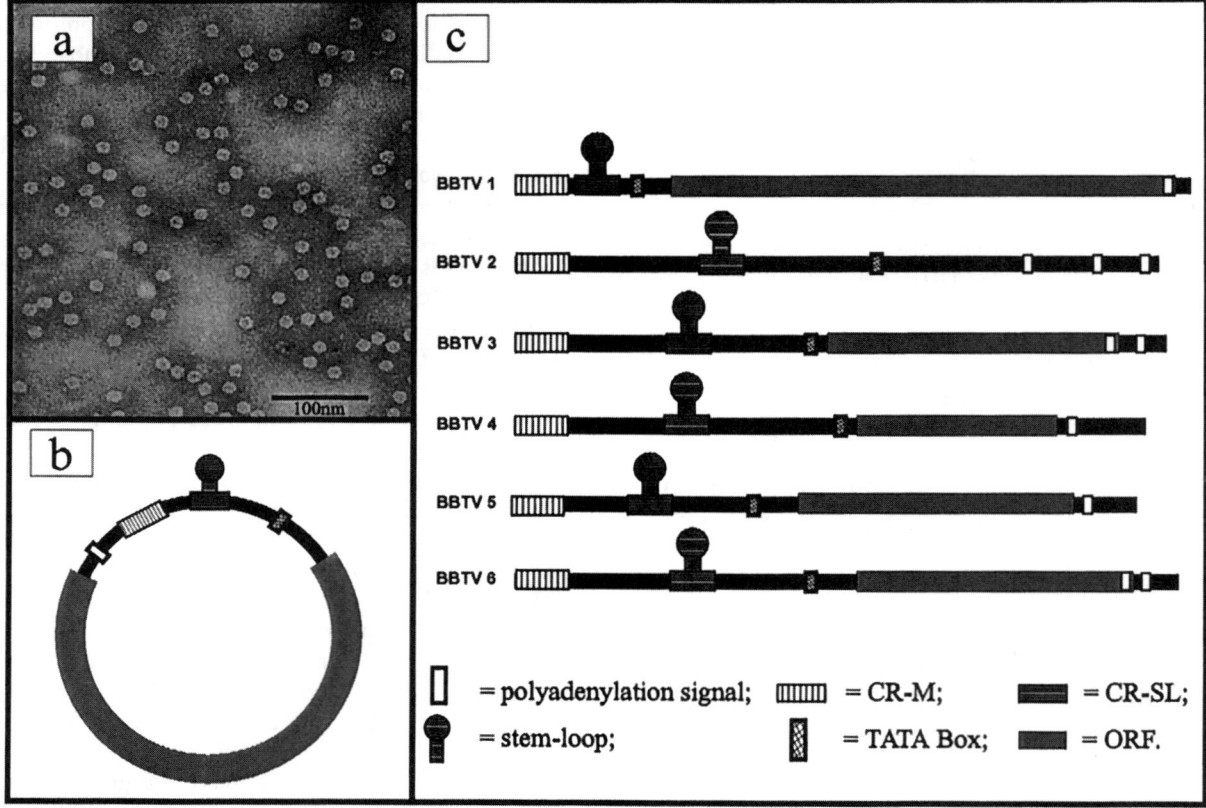

Figure 3. Morphology and genome organization of banana bunchy top virus. (a) Electron micrograph of purified banana bunchy top virus, (b) general organization of all banana bunchy top virus genome components, and (c) linear representation of each component.

Strains of BBTD and BBTV

There is strong evidence for the existence of strains of BBTD. Abaca bunchy top disease in abaca (*Musa textalis*) was recorded in the Philippines in 1910. This disease had very similar biological properties to BBTD including symptoms and transmission. BBTD was not recorded in the Philippines until 1960. Magee (26) demonstrated that BBTD could be transmitted to abaca but Ocfemia and Buhay (32) could not transmit ABTD to banana. Symptoms variation in naturally infected bananas has also been reported. Vakili (42) described symptoms in Giant Cavendish in Vietnam as either the "banana strain" or the "abaca strain". There have also been sporadic reports of mild or symptomless infections of BBTD but none of these have been confirmed or characterized.

Karan *et al.* (21) has sequenced and compared BBTV DNA component 1 from isolates from 10 countries. This comparison clearly indicated that there were two groups of isolates,

the Asian group (Taiwan, Philippines and Vietnam) and the South Pacific group (Australia, Burundi, Egypt, Fiji, India, Tonga and Western Samoa). Analysis of BBTV DNA components 2 to 6 has confirmed the division of isolates into two groups (20). It is possible that these two groups represent two strains of BBTV but, as yet, there are no other known biological properties to support this hypothesis.

BBTV as the Cause of BBTD

It has now been demonstrated that the isometric ssDNA virus, BBTV, is always associated with BBTD and that it is transmitted via aphids with the disease and almost certainly is one of the viruses or the only virus causing BBTD. However, attempts to infect bananas either via aphids previously exposed to purified BBTV virions or via microprojectiles coated with ssDNA extracted from purified BBTV virions have been unsuccessful. These failures may be the result of technical difficulties or it is possible that a second virus, possibly a luteovirus, is also involved in the disease. This is a major issue yet to be resolved.

CONTROL OF BANANA BUNCHY TOP DISEASE

BBTD has been and continues to be controlled by a range of strategies; however, it remains a major limitation to production in some regions and a continuing threat to others. Importantly, it has not been eradicated from any country into which it has been introduced. BBTD therefore has become a constant component of the cost of production either through disease losses or cost of control strategies.

Control of BBTD and International Movement of *Musa* Germplasm

For whatever reason, BBTD is not present in all banana producing countries or regions; it is not present in the Americas or the Caribbean and it has erratic distribution in Africa, Asia and the South Pacific. Even within countries, its distribution can be incomplete. For instance, in Australia, BBTD is present in south east Queensland and northern New South Wales but not present in the banana growing regions of north Queensland, the Northern Territory or northern Western Australia. The major threat of BBTD to these regions is that the disease may be introduced in infected planting material.

There is increasing movement of banana germplasm internationally. All this germplasm is moved as vegetative material except for the seeds of wild diploids used in breeding programs or for conservation. Therefore any viruses in this germplasm will be moved with it. The International Plant Genetic Resources Institute (IPGRI) and the Food and Agriculture Organization of the United Nations (FAO) have developed extensive recommendations regarding the movement of banana germplasm and planting material between regions (13). The primary recommendation is that, where possible, *Musa* germplasm should be moved as *in vitro* plantlets. This should ensure freedom from fungal, bacterial and nematode infections. However, as Drew *et al.* (12) demonstrated, BBTD can be transmitted through plantlets in vitro where those plantlets have been derived from an infected plant, symptoms of the disease may not be evident in those plantlets and a proportion of those plantlets may be "cured" of the disease while *in vitro*.

Therefore, it is essential that accurate and sensitive detection methods are used to test in vitro material prior to release in "non-infected" countries or regions. Prior to 1990, the only reliable detection method for BBTD was observation, that is visual inspection for symptoms over an extended period. This was not satisfactory because (i) not all inspectors were familiar with the symptoms of BBTD, (ii) the symptoms of BBTD are often mild or non-existent under low light conditions and (iii) there always exists the possibility of strains of BBTV that are mild or symptomless in individual cultivars.

Since the purification and characterization of BBTV, a range of sensitive and specific diagnostic techniques have become available. Wu and Su (43) generated a range of monoclonal antibodies which could be used in ELISA. Thomas and Dietzgen (41) developed both monoclonal and polyclonal antisera for the detection of BBTV using ELISA and some of these have been commercialized. Harding *et al.* (15) developed a DNA probe for the detection of BBTV DNA component 1 using either dot blot or Southern blot analysis and Hafner *et al.* (14) developed RNA probes for the detection of either virion or complementary sense BBTV DNA component 1. More recently, Karan *et al.* (21), (Shamloul and Hadidi, unpublished) developed a range of oligonucleotide primers for the amplification of all known sequence variants of BBTV DNA component 1. Other conserved primers could be designed from the sequences of other BBTV DNA components (7,20).

Thus, it is now possible to restrict the spread of BBTV internationally through (i) establishment and movement of plantlets *in vitro*, (ii) a rationally designed procedure for sampling plantlets *in vitro* and (iii) the use of one or preferably more sensitive and accurate diagnostic procedures. All these components are available for BBTV. These same guidelines can be used for the provision of BBTV-free planting material within countries or regions.

There are, however, at least three other viruses infecting bananas that should also be restricted in movement. Cucumber mosaic cucumovirus is widespread in bananas but not considered a serious threat to production. Banana streak badnavirus (25) was originally thought to have a very restricted distribution but is now believed to be widely geographically distributed. Banana bract mosaic potyvirus (4) has only been recorded in the Philippines and India. Sensitive and specific diagnostic procedures are either available or being developed for these viruses and should be incorporated into banana virus testing programs.

Current Control of BBTD Within Countries or Regions

Most of the attempts to control BBTD within infected regions or countries have centred on the destruction of infected plants, the provision of virus tested planting material and to some extent the implementation of internal quarantine zones. These strategies have had varying degrees of success. Only two attempts at eradicating BBTD have been successful.

The best example of the control of BBTD and perhaps one of the best examples of the use of phytosanitary strategies for the control of a plant virus disease is the Australian BBTV control campaign. The disease had virtually devastated the banana industry in northern New South Wales and south east Queensland by the mid to late 1920s. Subsequently, the two Australian states instituted strict government legislation to control the disease based on the recommendations of Magee (26). These recommendations were developed from a knowledge of the biological properties of the disease: the symptoms of BBTD were distinct and easily recognised by trained personnel, the disease was transmitted by aphids and was also transmitted through infected planting material. The control measures involved:

a)   the implementation of quarantine zones and buffer zones within and through which the movement of banana planting material was strictly controlled; this was particularly important in Queensland where the disease was and continues to be confined to the southern growing areas of the state. Two outbreaks of BBTD in north Queens land were quickly eradicated.

b)   the requirement for permits before plantations could be established and these plantations could only be planted with planting material derived from disease free bananas.

c)   destruction of neglected or abandoned plantations.

d)   a restriction on the number and type of bananas that could be grown non-commercially

e)   procedures for the inspection for and

eradication of infected bananas. This has involved the appointment of trained inspectors and the requirement for regular inspection of all commercial and non-commercial bananas in affected areas. Infected bananas are destroyed with the recommendation that bananas within a 20 m radius of an infected banana also be destroyed as 70% of secondary infections occur within 20 m of the initial infection (2). Thus, when the incidence of the disease has reached 5%, the plantation becomes uneconomic.

These control measures have been remarkably effective. For instance, the area planted to bananas increased in New South Wales from a low of about 1,200 ha in 1925 to about 12,000 ha in 1955 with similar increases in southern Queensland and the number of BBTD-infected bananas detected has also dramatically decreased. For example, 42,000 BBTD-infected bananas were detected in New South Wales in 1937 while over the past 30 years the number has decreased to 1,000 to 2,000 plants per year.

Thus, BBTD is no longer considered to be a limiting factor in the production of bananas in Australia. However, the disease has not been eradicated and there is an on-going cost of control.

There are probably three major reasons that these methods have been very successful in Australia where as attempts to control BBTD in other countries have not been as successful:

1.   the control measures are backed by strong government legislation which is strictly enforced

2.   there is an on-going information and education program directed towards producers and there is considerable support from producers for the control campaign

3.   the banana industry in northern New South Wales and southern Queensland is very different from the "typical" banana producing regions in that: (i) there is a definite winter period which results in a slowing of infection rates (from infection to symptoms takes about 19 days in summer and about 125 days in winter (1)) and thus provides a "break" in the infection cycle, (ii) the banana industry is well organized and administered, (iii) there are relatively few small plantations which are often geographically distant to

other plantings, and are restricted to the coastal fringes, (iv) there are relatively few non-commercial or subsistence banana plantings and (v) there are a few wild bananas to act as reservoirs for the disease.

These conditions rarely apply in other banana producing regions which are normally characterised by: (i) "wet" and "dry" seasons as opposed to summer and winter, (ii) large numbers of producers from substantial commercial plantations to numerous subsistence plantings and (iii) large numbers of bananas growing wild. However, the use of virus free planting material and the eradication of infected plants appears in many instances to slow the progression of the disease to a level where it is economically viable to produce bananas.

There is also evidence that there are at least two strains of BBTV (21) and it is possible that these strains differ in their virulence, transmissibility or other characteristics which could affect ability to spread and cause disease.

Control of BBTD: Possible Future Options

There are a range of possible future options for the control of BBTD including both conventional and non-conventional approaches.

Virus-free Planting Material

With the advent of *in vitro* propagated bananas and the availability of a range of sensitive and specific methods for BBTV diagnosis and detection, it is possible to provide large quantities of virus-tested banana germplasm and planting material. This should greatly reduce the risk of introducing BBTV into new areas, either continents, countries or regions as well as reducing the risk of establishing BBTV-infected plantations. The International Network for the Improvement of Bananas and Plantains (INIBAP) has established a "transit center" at Leuven in Belgium. This center maintains an extensive in vitro collection of *Musa* germplasm from which virus-tested banana and plantain cultivars can be obtained.

Inspection and Roguing

Regular inspection of plantings and eradication of infected bananas remains an important strategy for the control of BBTD within a planting. Early and efficient detection and eradication are key elements of this strategy and require knowledge and ability to diagnose BBTD from early symptoms. It is unlikely that the recently developed specific diagnostic methods such as ELISA, nucleic acid probes or PCR will have an impact on early detection as (i) these are expensive and relatively technically demanding techniques and therefore not generally available outside specialised virology laboratories and (ii) Hafner *et al.* (14) found that while BBTV could be detected using both nucleic acid probes and PCR at the site of inoculation, corm/meristem, roots and pseudostem prior to symptom expression, detection of the virus in new leaves coincided with symptom expression.

Resistant Cultivars

Cultivars resistant or immune to BBTV would be the most effective means for control of the disease. Unfortunately, there are no known commercial cultivars of either bananas or plantains that are immune to BBTV. Further, the vast majority of commercial cultivars are selections of what appear to be natural hybrids. However, conventional banana breeding is becoming more active with breeding programs established in Brazil, Honduras, Guadeloupe and Nigeria. Unfortunately, BBTW does not occur in any of these countries and the emphasis of these breeding programs does not currently include BBTV resistance.

There are reports of resistance in some cultivars. For instance, Jose (19) tested a range of banana cultivars and found that Kanchikela and Venattukunnan were the most resistant. Neither were immune but fewer plants became infected under the experimental conditions used. It is therefore possible to select cultivars that could potentially slow the progression or reduce the incidence of the disease. There has been no concerted effort to select such cultivars and investigate rates of infection under commercial conditions.

As it has been demonstrated that commercial cultivars of banana can be conventionally bred, it should be possible to breed a BBTV resistant cultivar if a suitable source of resistance or immunity were identified. Potential sources of such resistance could be the wild, fertile diploid *Musa* spp still existing in south east Asia and Papua New Guinea. It would be important that potential sources of resistance be screened with a range of variants or strains of BBTV (21).

Transgenic Resistance

Transgenic virus resistance, based on virus-derived transgenes, has been widely demonstrated to be an effective strategy for the control of plant viruses. This strategy could be particularly relevant to bananas as there is no known immunity to BBTV in bananas and conventional banana breeding programs have yet to be proven to be generally effective (10). The two essential components from which to develop virus resistant transgenic bananas, a transformation and regeneration system and the nucleotide sequence of most if not all of the BBTV genome, are now available. Recently, both May *et al.* (30) and Sagi *et al.* (35) reported the regeneration of genetically transformed bananas. May *et al.* (30) transformed the cultivar Grand Nain using *Agrobacterium*-mediated transformation while Sagi *et al.* (35) transformed the cultivar Bluggoe using microprojectile bombardment. Six ssDNA components of BBTV have been sequenced (7,16,45) and the viral replicase associated protein identified.

These developments should lead to the incorporation of BBTV-derived resistance genes into bananas which hopefully will result in BBTV resistant transgenic bananas.

CONCLUSIONS

Banana bunchy top disease is the most important of the virus diseases of bananas and has had a major economic impact on banana production in those countries and regions where it occurs. Banana bunchy top virus has been isolated and identified as an isometric virus with a circular ssDNA genome. This virus is always associated with the disease and it appears to occur as two strains.

In one notable instance, BBTD has been effectively controlled but not eradicated through the use of virus-free planting material and roguing; however, similar attempts in other instances have been less effective. The recent availability of specific and sensitive diagnostic methods for BBTV and the development of large scale *in vitro* micropropagation of bananas should facilitate the international movement of BBTV-free banana germplasm and the distribution of BBTV-free planting material. This should considerably reduce the risk of introducing BBTV into continents, countries or regions currently free of the virus.

However, it is unlikely that the provision of BBTV-free planting material together with roguing will be sufficient to control the disease in the future in most countries and regions where it currently occurs. The most attractive future strategy for control is cultivar resistance. It is unlikely that such cultivars will be developed through conventional breeding programs at least in the short term as sources of resistance haven't been identified, most conventional banana breeding programs are in their infancy and all of these are located in countries where BBTV does not occur. The components from which to develop transgenic bananas resistant to BBTV are available and this would appear to be the most likely source of BBTV resistant cultivars in the foreseeable future.

REFERENCES

1. Allen, R.N. 1978. Epidemiological factors influencing the success of roguing for the control of bunchy top disease of bananas in New South Wales. Aust. J. Agric. Res. 29:535-544.
2. Allen, R.N. 1978. Spread of bunchy top disease in established banana plantations. Aust. J. Agric. Res. 29:1223-1233.
3. Allen, R.N., and Barnier, N.C. 1977. The spread of banana bunchy top disease between banana plantations in the Tweed River District during 1975-76. N.S.W. Plant Dis. Surv. 46:27-28.
4. Bateson, M.F., and Dale, J.L. 1995. Banana bract mosaic virus: characterisation using potyvirus specific degenerate PCR primers. Arch. Virol. 140:515-527.
5. Boevink, P., Chu. P.W.G., and Keese, P. 1995. Sequence of subterranean clover stunt virus: affinities with geminiviruses. Virology 207:354-361.
6. Burns, T.M., Harding, R.M., and Dale, J.L. 1994. Evidence that banana bunchy top virus has a multiple component genome. Arch. Virol. 137:371-380.
7. Burns, T.M., Harding, R.M., and Dale, J.L. 1995. The genome organisation of banana bunchy top virus: analysis of six ssDNA components. J. Gen. Virol. 76: 1471-1482.
8. Castillo, B.S., and Martinez, A.L. 1961. Occurrence of bunchy top disease in bananas in the Philippines. FAO Plant Protect. Bull. 9:74-75.
9. Dale, J.L. 1987. Banana bunchy top: an economically important tropical plant virus disease. Adv. Virus Res. 33:301-325.
10. Dale, J.L. 1990. Banana and Plantain. In "Agricultural Biotechnology: Opportunities for International Development". G. J. Persley, ed. CAB International, Uk.
11. Dale, J.L., Phillips, D.A., and Parry, J.N. 1986. Double-stranded RNA in banana plants with bunchy top disease. J. Gen. Virol. 67:371-375.
12. Drew, R.A., Moisander, J.A., and Smith, M.K. 1989. The transmission of banana bunchy top virus

in micropropagated bananas. Plant Cell, Tissue Organ Cult. 16:187-193.

13. Frison, E.A., and Putter, C.A.J. 1989. FAO/IBPGR Technical guidelines for the safe movement of *Musa* germplasm. Food and Agriculture Organization of the United Nations, Rome/International Board of Plant Genetic Resources, Rome. 13.

14. Hafner, G., Harding, R.M., and Dale, J.L. 1995. Movement and transmission of banana bunchy top virus DNA component one in banana. J. Gen. Virol.76:2279-2285.

15. Harding, R.M., Burns T.M., and Dale, J.L. 1991. Virus-like particles associated with banana bunchy top disease contain small single-stranded DNA. J. Gen. Virol. 72:225-230.

16. Harding, R.M., Burns, T.M., Hafner, G., Dietzgen, R.G., and Dale, J.L. 1993. Nucleotide sequence of one component of the banana bunchy top virus genome contains a putative replicase gene. J. Gen. Virol. 74: 323-328.

17. Hu, J.S., Xu, M.Q., Wu, Z.C., and Wang, M. 1993. Detection of banana bunchytop virus in Hawaii. Plant Dis. 77:952.

18. Iskra, M.L., Garnier, M., and Bove, J.M. 1989. Purification of banana bunchy top virus (BBTV). Fruits 44:63-66.

19. Jose, P.C. 1981. Reaction of different varieties of banana against bunchy top disease. Agric. Res. J. Kerala 19:108-119.

20. Karan, M. 1995. Sequence diversity of DNA components associated with banana bunchy top virus. PhD thesis. Queensland University of Technology.

21. Karan, M., Harding, R.M., and Dale, J.L. 1994. Evidence for two groups of banana bunchy top virus isolates. J. Gen. Virol. 75:3541-3546.

22. Katul, L., Vetten, H.J., Maiss, E., Makkouk, K.M., Lesemann, D-E., and Casper, R. 1993. Characterisation and serology of virus-like particles associated with faba bean necrotic yellows. Ann. Appl. Biol. 123:629-647.

23. Khalid, S., Soomro, M.H., and Stover, R.H. 1993. First report of banana bunchy top virus in Pakistan. Plant Dis. 77:101.

24. Lazarowitz, S.G. 1992. Geminiviruses: genome structure and gene function. Crit. Rev. Plant Sci. 11: 327-349.

25. Lockhart, B.E.L. 1986. Purification and serology of a bacilliform virus associated with banana streak disease. Phytopathology 76:995-999.

26. Magee, C.J. 1927. Investigation on the bunchy top disease of the banana. Counc. Sci. Ind. Res. Bull. No. 30.

27. Magee, C.J. 1948. Transmission of bunchy top to banana cultivars. J. Aust. Inst. Agric. Sci. 14:18-24.

28. Magee, C.J. 1953. Some aspects of the bunchy top disease of banana and other *Musa* spp. J. Proc. Royal Soc. NSW 87:3-18.

29. Matthews, R.E.F. 1982. Classification and nomenclature of viruses. Intervirology 17:1-200.

30. May, G.M., Afza, R., Mason, H.A., Wiecko, A., Novak, F.J., and Arntzen, C.J. 1995. Generation of transgenic banana (*Musa acuminata*) plants via Agrobacterium-mediated transformation. Bio/Technology 13:486-492

31. Ocfemia, G.O. 1926. Progress report on bunchy top of abaca or Manilla hemp. Phytopathology 16:894.

32. Ocfemia, G.O., and Buhay, G.G. 1934. Bunchy top of abaca or Manila hemp: II Further studies on the transmission of the disease and a trial planting of abaca seedlings in a bunchy top devastated field. Philipp. Agric. 22:567-581.

33. Ram, R.D., and Summanwar, A.S. 1984. *Colocasia esculenta* (L) Schott. A reservoir of bunchy top disease of banana. Curr. Sci. 53:145-146.

34. Rohde, W., Randles, J.W., Langridge, P., and Hanold, D. 1990. Nucleotide sequence of a circular single-stranded DNA associated with coconut foliar decay virus. Virology 176:648-651.

35. Sagi, L., Panis, B., Remy, S., Schoofs, H., De Smet, K., Swennen, R., and Cammue, B.P.A. 1995. Genetic transformation of banana and plantain (*Musa* spp.) via particle bombardment. Bio/Technology 13:481-485.

36. Sano, Y., Isogai, M., Satoh, S., and Kojima, M. 1993. Small virus-like particles containing single-stranded DNAs associated with milk vetch dwarf disease in Japan. 6th International Congress of Plant Pathology, Montreal, August, 1993. Abstract 17.1.27.

37. Simmonds, N.W. 1962. The evolution of bananas. Longman, London.

38. Soomro, M.H., and Khalid, S. 1993. Banana bunchytop virus confirmed in Pakistan. Infomusa 2: 17.

39. Soomro, M.H., Khalid, S., and Aslam, M. 1992. Outbreak of banana bunchy top virus in Sindh, Pakistan. FAO Plant Protect. Bull. 40:95-99.

40. Sun, S.K. 1961. Studies on the bunchy top disease of bananas. Spec. Publ. College Agric. Taiwan Univ. 10:82-109.

41. Thomas, J.E., and Dietzgen, R.G. 1991. Purification, characterisation and serological detection of virus-like particles associated with banana bunchy top disease in Australia. J. Gen. Virol. 72:217-224.

42. Vakili, N.G. 1969. Bunchy top disease of bananas in the central highlands of South Vietnam. Plant Dis. Rep. 53:634-638.

43. Wu, R.Y., and Su, H.J. 1990. Purification and characterisation of banana bunchy top virus. J. Phytopathology 128:153-160

44. Wu, R.Y., You, L.R., and Soong, T.S. 1994. Nucleotide sequences of two circular single-stranded DNAs associated with banana bunchy top virus. Phytopathology 84:952-958.

45. Xie, W.S., and Hu, J.S. 1995. Molecular cloning, sequence analysis and detection of banana bunchy top virus in Hawaii. Phytopathology 85:339-347.

46. Yeh, H., Su, H., and Chao, Y. 1994. Genome characterisation and identification of viral-associated dsDNA component of banana bunchy top virus. Virology 198:645-652.

CHAPTER 50

The Control of African Cassava Mosaic Virus Disease: Phytosanitation and/or Resistance?

J.M. Thresh, G.W. Otim-Nape and D. Fargette

Cassava (*Manihot esculenta* Crantz) is an important staple food crop in many parts of sub-Saharan Africa. The main disease affecting the crop is African cassava mosaic disease (ACMD) which is caused by any of the African cassava mosaic geminiviruses (ACMVs) that are transmitted by the whitefly *Bemisia tabaci* (Gennadius). In previous papers it was stressed that ACMD is an important but generally underestimated problem in many countries and one that has received inadequate attention (23,25,26). It was also emphasized that substantial increases in productivity could be achieved by the widespread adoption of known methods of control.

The various possible approaches to controlling ACMD are discussed in recent reviews (9,11,27). The two most feasable and widely adopted methods are through the use of phytosanitation and resistant varieties. Additional information on these approaches is presented in this chapter which is based largely on experience gained in the very different conditions of Côte d'Ivoire and Uganda. However, the concepts developed are more generally applicable in developing appropriate control strategies for use elsewhere.

PHYTOSANITATION

This term is used here in a general way for the various means of improving the health status of cassava planting material and for decreasing the availability of sources of infection from which further spread of ACMD can occur through the activity of the whitefly vector.

There are three main features of phytosanitation for the control of ACMD:

- Crop hygiene involving the removal of all

diseased cassava from within and immediately around areas to be used for new plantings.

- The use of ACMV-free stem cuttings as planting material.

- The removal (roguing) of diseased plants from within crop stands.

Crop Hygiene

This is a basic means of facilitating the control of many pests and diseases by removing the debris and surviving plants of previous crops to decrease the risk of carry-over of inoculum to any new plantings at the site or nearby. Little attention has been given to adopting this approach with ACMD and the benefits to be gained have not been demonstrated. They could be substantial because cassava plants, including those affected by ACMD, regenerate readily from stems left in or on the ground at harvest. Moreover, farmers often harvest piecemeal and from the most vigorous unaffected plants within a stand and then establish new cuttings in the gaps created. This means that young plants often develop beneath or immediately alongside older infected ones. These are potential sources of virus inoculum and also of other pathogens and pests including the cassava green mite (*Mononychellus tanajoa* Bondar) and the cassava mealybug (*Phenacoccus manihoti* Matile-Ferrero). There is a need to quantify the risks involved, because the only relevant information on spread of ACMD from foci within plantings has been obtained in experiments in which the introduced sources of infection were the same age as the plants being assessed (6).

670

ACMV-free Planting Material

Cassava is propagated routinely from hard-wood stem cuttings. Experience in Uganda and elsewhere in Africa is that farmers usually obtain these from their own plantings or from neighbors; there is also some use of material brought in from outside the locality or provided from official sources or by Non-Governmental Organizations (NGOs). In many areas there is extensive infection of the planting material with ACMD, which may have very deleterious effects. This is because plants that develop from infected cuttings are important sources of infection and they grow and yield substantially less than uninfected plants, or those infected later by whiteflies (9). Moreover, the use of infected cuttings can lead to a high incidence of ACMD and substantial crop loss, even in areas where there is little or no spread by vectors.

There are many reasons for the widespread use of infected planting material. One is that farmers are unaware of the damage caused by ACMD, in some instances because virtually all the available material is infected and there are no uninfected plants to serve as standards and indicate the potential productivity of the varieties being grown. Even if farmers are aware of the problem they seldom have access to sources of uninfected material, either because this is not available or because they cannot afford the cost.

There are obvious advantages to be gained from the general adoption of ACMV-free cuttings as planting material, which would greatly enhance productivity and decrease the extent of infection in the locality and the opportunity for further spread by vectors. This is a basic approach to disease control and one that has been widely adopted with many vegetatively-propagated crops (12). However, little attempt has been made to promote the use of ACMV-free stocks of cassava, other than to meet the requirements of Plant Health Authorities responsible for regulating the movement of vegetative material between countries and continents.

It is sometimes argued that specialist expertise and sophisticated facilities are required to obtain ACMV-free material by meristem-tip and/or heat therapy. 'Foundation' plants obtained in this way could then be subjected to rigorous tests for ACMV and other viruses and maintained in insect-proof conditions to prevent contamination by vectors. There are also those who advocate setting up expensive schemes for the production and distribution of ACMV-free planting material of cassava similar to those used in developed countries for potato and many horticultural crops (12). These involve the periodic replacement of the stocks being grown as those being used become infected and less productive.

Such costly and elaborate schemes may eventually be adopted with cassava, but they are currently impracticable and inappropriate, given the huge quantities of planting material required, the limited budgets available for extension-type activities and the poverty of many of those growing cassava. Nevertheless, there is abundant evidence from experience in Kenya, Côte d'Ivoire and Uganda that big improvements in the health status of the planting material that is available to farmers can be achieved by simply selecting cuttings from symptomless plants and raising them at sites where there is little or no spread by vectors and where roguing is practiced routinely. This was the approach adopted in the 1980s in Kenya (3) and Côte d'Ivoire (10), where substantial quantities of ACMV-free material were raised for use in epidemiological trials. However, arrangements were not developed for more extensive propagation and distribution to farmers on a large scale.

Such schemes have since been developed in Uganda where there is now a huge demand for planting material of improved virus-resistant varieties to replace the local mainly susceptible ones that have been severely affected during the current epidemic (18,28). About 17,500 ha of virus-resistant material are now available at various propagation sites in different parts of the country and these are being used to supply ACMV-free cuttings to farmers. Three main approaches have been adopted in building up this system of multiplication, involving institutions, groups of farmers operating together and individuals. Each system has advantages and disadvantages, as discussed by Otim-Nape et al. (18). The experience is that with each system some additional re-selection and roguing is necessary, even if the initial selection of cuttings is done under optimum growing conditions. The amount of roguing required depends on the susceptibility of the variety and the inoculum pressure prevailing at the propagation site. This explains why the need for roguing has been much greater at sites in central and eastern Uganda where ACMD spreads rapidly, than in southern areas around Lake Victoria where until recently little or no spread occurred.

Current operations in Uganda require a substantial commitment of staff and resources and a large budget, some of which is being provided by outside donors and NGOs, including the UK Gatsby Charitable Foundation. One of the objectives of current work is to develop simple but effective measures that can be adopted readily by farmers and practiced routinely on a sustainable basis without the need for continued financial, technical and logistical support that will be difficult to provide. The basic concept rests on the assumption that farmers can obtain an average of about ten hard-wood cuttings from each mature source plant used to provide planting material. Thus sufficient uninfected cuttings could be obtained from existing stands in which the incidence of ACMD does not exceed 90%, provided that the uninfected plants can be identified with reasonable certainty.

This is an important proviso because there are several reasons why this is seldom feasible. Firstly, symptoms on mature plants of the type used routinely to provide cuttings tend to be inconspicuous and may be completely absent if there has been much leaf abscission during prolonged periods of drought. Moreover, symptoms may be masked or obscured by the damage due to cassava green mites, or cassava mealybug, or by zinc or other mineral deficiency symptoms (1). Another difficulty is that symptoms of ACMD may be absent or restricted to only a few shoots of plants infected at a late stage of growth by whiteflies. Nevertheless, many of the cuttings collected from such plants contain ACMV and soon develop shoots that express symptoms.

There is a particular difficulty in areas where much of the cassava is harvested and consumed during the prolonged dry season, some weeks or even months before new crops are established following the onset of the rains. This is the situation in many dry savannah areas including parts of Tanzania, Malawi and northern Ghana, where stems have to be collected at harvest and retained without leaves until required for planting. Stems are frequently collected from plants that are almost leafless or seriously affected by cassava green mite and selection of ACMV-free material is difficult or impossible. In such circumstances there may be scope for adopting the practice recommended in India, where farmers are advised to inspect plants at the time of optimum growth and symptom expression. All ACMD-affected plants are then marked with paint so that they can be avoided later when cuttings are collected for further plantings (16).

Even the most rigorous selection regimes are not completely effective and some infection is to be expected, even in cuttings collected solely from symptomless healthy-looking plants. The proportion is likely to vary depending on such factors as variety, environmental conditions, the intensity of pest attack and the overall infection pressure in the locality. However, only limited evidence is available and few attempts have been made to encourage farmers to select cuttings from symptomless plants. This is a serious omission as there are likely to be substantial benefits from adopting such practices, especially in areas where conditions for crop growth are favorable, uninfected plants are readily available and the inoculum pressure is generally low. In such circumstances farmers could soon achieve a big improvement in the health status and productivity of their plantings at little cost or inconvenience and with substantial benefits. This explains why the approach was advocated for use in Zanzibar (29) and later in the important cassava-growing areas of coastal and western Kenya once uninfected plants were selected and it was shown that the inoculum pressure in these areas was generally low (3,20). There are similar opportunities elsewhere in Africa, including those parts of Malawi, Zambia, Zimbabwe, Burundi, Rwanda, Tanzania, Cameroon and Chad, where rates of spread are low and uninfected plants can be found without difficulty and sometimes predominate (25, 26).

The scope for selection in areas of relatively high inoculum pressure is unclear and further experience is required before definitive recommendations are possible. It seems inevitable that latent infection will cause greater problems and that the overall approach will be less effective. Whether it will be rendered ineffective, except with resistant varieties or when combined with subsequent roguing, has not been determined.

There have been few attempts anywhere in Africa to exploit the benefits of selective propagation and cuttings are frequently taken from obviously infected source plants by farmers and also at official propagation sites and for use in agronomy or other experimental trials. A change from such unsatisfactory practices is long overdue and there is also a need for socio-economic assessments of current methods and farmers' attitudes to cutting selection. Preliminary studies of this type have been made already in parts of Uganda and show that some farmers select the plants from which they obtain cuttings and may do so rigorously on the basis of plant vigor, or the absence of symptoms, or some combination of the two (18). This suggests that other farmers could be persuaded to do so if they are provided with sufficient education and training, together with clear evidence of the benefits to be gained. Evidence of this type has seldom been sought, and is an obvious priority in any further research if current attitudes are to be changed and progress is to be made.

Roguing

Roguing is a well known means of disease control of wide applicability (22). It has been recommended repeatedly as a means of controlling ACMD, but there is little experimental or other evidence to indicate the effectiveness and applicability of this approach, or the most appropriate procedure to follow. This emphasizes the need for additional research and for more detailed assessments of roguing than any yet undertaken.

There is general agreement that roguing should be adopted as an essential feature of official schemes for the selection and maintenance of ACMV-free stocks for release to farmers, as discussed earlier. In these circumstances the health status of the material is of paramount importance and it is accepted that some losses are inevitable during roguing operations. Stringent roguing regimes are justified and it is appropriate to follow the recommendation that plantings should be inspected at least weekly for the first two to three months of growth, so as to find and remove any infected plants that occur (11).

The situation is very different in farmers' plantings, where other considerations apply and there is a general and understandable reluctance to remove any plants that might contribute to yield. Frequent roguing is certainly not justified where there is little spread of ACMV into or within plantings by whiteflies. Moreover, roguing is inappropriate and ineffective and leads to a progressive and unacceptable decrease in stand in areas of high inoculum pressure where much spread occurs. This was demonstrated in each of the only two experiments on roguing to have been reported. At a site in the lowland forest area of Côte d'Ivoire, the spread of ACMD was rapid and the final incidence of infection was similarly high (77-78%) in rogued and unrogued plots of the same range of varieties (4). Rapid spread also occurred in a second experiment in the same locality where the final incidence was 67% in

rogued and 87% in unrogued plots established with ACMV-free cuttings of a local Ivorian variety (6).

Another argument against roguing is that there is evidence from Côte d'Ivoire, Kenya and Uganda that the spread of ACMV is mainly between cassava plantings and not from internal foci within them (3,6,17). This suggests that roguing has little effect in reducing spread within treated fields and so is of little or no benefit to those adopting the practice, although it can be expected to decrease the risk of spread to other plantings nearby. From this it can be inferred that roguing is likely to be most effective when practiced by groups of farmers or throughout whole localities. This was the approach recommended and adopted in Uganda in the 1950s when resistant varieties and the use of roguing were introduced as the official control policy and strictly enforced by local authority statute (13).

Such drastic measures are no longer appropriate or acceptable to farmers and they cannot be enforced. Consequently, the current approach in Uganda is to develop simple, less demanding procedures that can be used by farmers to sustain the health status of the material being grown, at little inconvenience or expense. One possibility is to rogue once or twice soon after planting as the cuttings begin to sprout and infected ones usually develop conspicuous symptoms. Roguing can be done quickly and easily at this early stage of growth and there is still time to fill the gaps created by planting additional cassava or other crop plants. Any later infections that occur are allowed to remain, which also avoids gaps and the decrease in yield that is otherwise likely to occur because of the reduced stand. This is because plants infected during growth by whiteflies sustain little or no reduction in yield, although it is important to avoid propagating from such plants as the cuttings obtained would be affected much more severely. (9,24).

Another possibility being investigated in Uganda is to compare selection and roguing to determine their effectiveness in maintaining the health status of plantings when used singly and also in combination. It is already apparent that farmers will adopt these techniques if provided with sufficient training and justification (17). However, the situation in many parts of Uganda is unusual in that the current epidemic of ACMD has caused such severe damage that farmers are desperate to maintain production by whatever means available. Consequently, they are willing to make considerable effort to safeguard cassava, which is their main staple food. It remains to be established whether farmers will respond similarly in endemic areas where ACMD is a long-standing and so less obvious problem and where acceptable, albeit sub-standard, yields are obtained without any evident need to change variety or to use specific control measures.

RESISTANT VARIETIES

Although what is now known as ACMD was first reported in 1894 it did not become a serious problem until the late 1920s and early 1930s, when agriculturists in East, Central and West Africa and also in Madagascar became aware of the need for ACMV-resistant varieties. Attitudes at the time were clearly influenced by the successful deployment of resistant varieties to combat the threat posed by sugarcane mosaic virus disease (21). Various largely unsuccessful attempts were made in several countries to identify ACMV-resistant varieties from those being grown at the time in Africa, or that were imported from elsewhere. Much greater success was achieved by crossing cassava with *M. glaziovii* Muell.-Arg. to produce hybrids that were back-crossed to cassava to produce progeny having satisfactory root yield and quality. This approach was adopted independently in Tanzania and Madagascar (14) and Tanzanian seed of resistant genotypes was introduced to Nigeria where selections were made and ultimately used as sources of resistance in the IITA breeding program (2). This began in 1971 and it is now the largest and most influential in Africa as it supplies parental material, seed and breeding lines to many national programs (15).

The ACMV-resistant varieties selected in Tanzania, Madagascar, IITA and elsewhere in Africa have several important characteristics which seem to be closely associated and may be manifestations of the same basic virus resistance mechanism:

They are resistant, but not totally immune to infection with ACMV. This means that the proportion of plants infected depends on the prevailing inoculum pressure, but is consistently less than in susceptible varieties of the same age exposed to similar amounts of inoculum.

- They develop symptoms of ACMD that are less conspicuous than those of sensitive (intolerant) varieties.

- The symptoms are often restricted to certain shoots or branches and become inconspicuous or disappear as the plants age. This means that resistant varieties tend to be less severely affected by ACMD than sensitive ones, as several different studies have shown a general relationship between symptom severity and yield loss (24).

- ACMV is less completely systemic in resistant varieties than in susceptible ones. An important consequence of this effect is that a substantial proportion of the cuttings collected from infected plants are free of ACMV, even if taken from plants that were infected as cuttings or at an early stage of growth by whiteflies. This is the so-called 'reversion' phenomenon which has long been known and is likely to be of crucial epidemiological importance, as indicated by recent modelling studies (7,8). However, the phenomenon has not been adequately studied and remains poorly understood.

- Resistant varieties contain lower concentration of ACMV than susceptible ones, as demonstrated serologically by Fargette *et al.* (5). This suggests that resistant varieties are of limited potency as they are poor sources of inoculum from which vectors can acquire and transmit virus to uninfected plants. Moreover, the general adoption of such varieties is likely to lead to a decrease in the amount of inoculum present in the locality and so restrict spread of ACMV to any susceptible varieties being grown.

A likely consequence of these features is that stands of the most virus-resistant varieties sustain little damage or yield loss due to ACMD, even under conditions of very high inoculum pressure. This is to be expected because only a proportion of the plants are infected and those infected sustain little damage and lead to little further spread. Moreover, ACMV occurs in only a proportion of the cuttings collected from infected plants of resistant varieties so that the incidence of ACMD in the cuttings is usually less than in the stand from which they were collected. This 'cleansing' effect is likely to be further enhanced if infection decreases vegetative vigor, or farmers discriminate in favor of vigorous and/or symptomless plants when selecting cuttings for new plantings. For these reasons the incidence of infection is unlikely to increase progressively in successive cycles of propagation, as might otherwise be expected with a vector-borne virus of a vegetatively-propagated crop.

These suppositions are consistent with the results of modelling studies using realistic estimates of host-plant resistance, the extent of reversion, the intensity of cutting selection and rates of spread (7,8). The simulations indicate that the incidence of infection increases in successive cycles of crop production to reach asymptotes at long-term equilibrium values that can be substantially less than 100%. The actual incidences depend on the values of the parameters adopted and are influenced by seasonal and other environmental factors that determine host susceptibility, vector populations, inoculum pressure and the extent of reversion.

There is little experimental data to confirm or deny the validity of the modelling approach and the concept of equilibrium. Such evidence will be difficult to obtain as it will require long-term trials with different varieties in a wide range of agro-ecological environments. There is also a need for additional quantitative information on the reversion phenomenon and on the effects of ACMD on the yield of mixed stands of infected and uninfected plants of a wide range of varieties including the most resistant ones (24). Until such information is available there will be continuing uncertainty on the role of ACMV-resistant varieties and the most appropriate ways in which they should be deployed. For example, it is unclear whether reversion or resistance to or tolerance of infection is the main feature of ACMV-resistant varieties that is being exploited or whether it is some combination of the two. It is particularly important to determine whether phytosanitation is justified or advantageous with such resistant varieties. This is a long-standing issue yet to be resolved. One view is that selection and roguing are not required if the varieties being grown are sufficiently resistant. The other is that a yield penalty is incurred unless phytosanitation measures are adopted. The issue will only be resolved by assessing the performance of a wide range of varieties in different environments with different levels of inoculum pressure. The crucial question is whether equilibria occur at which the incidence and severity of ACMD are sufficiently high to decrease the productivity of entire stands in which

competition and compensation effects could be important. One possibility is that phytosanitation is appropriate and beneficial for the less resistant varieties, but not for the most resistant ones. This leads to a possible paradox if it is shown that phytosanitation is only effective with the most resistant varieties with which it has the least beneficial effect. Such outcomes would have a considerable impact on attitudes to control and lead to complications if the most appropriate strategy depends on the variety adopted and the circumstances under which it is grown. This would be a difficult approach to introduce to extensionists and farmers, many of whom have still to adopt even the simplest recommendations on selection and roguing.

DISCUSSION

This chapter is concerned mainly with general concepts and some of the statements and inferences are not supported by detailed references or data. This is because few relevant studies have been carried out, or the results apply only to certain areas or have not been published. Moreover, the attitude of farmers to the various control measures is crucial and yet their role has seldom been considered.

It is particularly important to determine the extent to which attitudes towards the control of ACMD are influenced by the practice of harvesting and eating leaves, which are an important part of the diet in Zaire, Cameroon and some other cassava-producing countries. It has been claimed that consumers prefer ACMD-affected leaves for their superior taste or palatability, or because they require less cooking oil when being prepared for consumption. However, no data are available from carefully controlled experiments involving representative consumers and it is important to develop protocols for preference trials that exclude the possibility of subjective bias. It will also be necessary to consider symptoms of different severity, because small severely damaged leaves which have a reduced lamina and high fiber content are unlikely to be sought, whereas those with slight symptoms may be favored on taste, sweetness or other criteria. This may explain why breeders in Zaire select genotypes that display mild symptoms of ACMD rather than those with severe symptoms or that remain completely symptomless (15).

As argued previously there is an urgent need for additional research on this and other topics which should be carried out across the whole of the very diverse range of agro-ecological environments in which cassava is grown in Africa and where ACMD is prevalent (24). It is particularly important to consider areas of both high and low inoculum pressure and to carry out experiments in different seasons and over several years to ensure the general validity of the results obtained. Moreover, the studies should be done in close collaboration with farmers and extensionists on representative holdings so that the control recommendations that emerge are relevant, practicable and appropriate in relation to current farming practices and socio-economic circumstances.

REFERENCES

1. Asher, C.J., Edwards, D.G., and Howeler, R.H. 1980. Nutritional disorders of cassava. Department of Agriculture. 48 Pages. University of Queensland, St Lucia, Queensland.

2. Beck, B.D.A. 1982. Historical perspectives of cassava breeding in Africa. Pages 13-18 in: Proceedings of a Workshop on Root Crops in Eastern Africa, Kigali, Rwanda, 23-27 November 1980. S.K. Hahn and A.D.R Ker, eds. IDRC, Canada, 177e.

3. Bock, K. 1983. Epidemiology of cassava mosaic disease in Kenya. Pages 337-347 in: Plant Virus Epidemiology. R.T. Plumb and J.M Thresh, eds. Blackwell Scientific, Oxford.

4. Colon, L. 1984. Contribution à l'étude de la résistance variétale du manioc (Manihot esculenta, Crantz). vis-à-vis de la mosaïque africaine du manioc. Etude rélisée dans le cadre du program ORSTOM: Etude de la mosaïque africaine du manioc.

5. Fargette, D., Colon, L.T., Bouveau, R., and Fauquet, C. 1996. Components of resistance of cassava to African cassava mosaic virus. Eur. J. of Plant Pathol. 102:645-654.

6. Fargette, D., Fauquet, C., Grenier, E., and Thresh, J.M. 1990. The spread of African cassava mosaic virus into and within cassava fields. J. of Phytopathol. 130:289-302.

7. Fargette, D., Thresh, J.M., and Otim-Nape, G.W. 1994. The epidemiology of African cassava mosaic geminivirus: reversion and the concept of equilibrium. Tropical Science 34:123-133.

8. Fargette, D., and Vié, K. 1995. Simulation of the effects of host resistance, reversion and cutting selection on incidence of cassava mosaic virus and yield losses in cassava. Phytopathology 85:370-375.

9. Fauquet, C., and Fargette, D. 1990. African cassava mosaic virus: Etiology, epidemiology and control. Plant Dis. 74:404-411.

10. Fauquet, C., Fargette, D., and Thouvenel, J.C. 1987. Selection of healthy cassava plants obtained by reversion in cassava fields. Pages 146-149 in: Proceedings International Seminar on African Cassava Mosaic Disease and its Control. Yamoussoukro, Côte d'Ivoire 4-8 May 1987. CTA/FAO/ORSTOM /IITA/IAPC.

11. Guthrie, J. 1987. Controlling African Cassava Mosaic Disease. Page 11. CTA, Wageningen.

12. Hollings, M. 1965. Disease control through virus-free stock. Annu. Review of Phytopathol. 3:367-396.

13. Jameson, J.D. 1964. Cassava mosaic disease in Uganda. East African Agricultural and Forestry Journal 29:208-213.

14. Jennings, D.L. 1994. Breeding for resistance to African cassava mosaic geminivirus in East Africa. Tropical Science 34:110-122.

15. Mahungu, N.M., Dixon, A.G.O., and Kumbira, J.M. 1994. Breeding for multiple pest resistance in Africa. African Crop Science Journal 2:539-552

16. Malathi, V.G., Thankappan, M., Nair N.G., Nambisan, B., and Ghosh S.P. 1987. Cassava mosaic disease in India. Pages 189-198 in: Proceedings International Seminar on African Cassava Mosaic Disease and its Control, Yamoussoukro, Côte d'Ivoire 4-8 May, 1987. CTA/FAO/ORSTOM/IITA/IAPC

17. Otim-Nape, G.W. 1993. Epidemiology of the African cassava mosaic geminivirus disease (ACMD) in Uganda. PhD. Thesis. 256 Pages. University of Reading, UK.

18. Otim-Nape, G.W., Bua, A. and Baguma, Y. 1994. Accelerating the transfer of improved production technologies: controlling African cassava mosaic virus disease in Uganda. African Crop Science Journal 2:479-496.

19. Otim-Nape, G.W., Bua, A., Thresh, J.M., Baguma, Y., Ogwal, S., Ssemakula, G.N., Acola, G., Byakakoma, B., and Martin, A. 1997. Cassava Mosaic Virus Disease in Uganda: the Current Pandemic and Approaches to Control. NARO/NRI/ODA Publication, Chatham Maritime, U.K. 65 Pages.

20. Robertson, I.A.D. 1987. The whitefly Bemisia tabaci (Gennadius) as a vector of African cassava mosaic virus at the Kenya coast and ways in which the yield losses in cassava (Manihot esculenta Crantz) caused by the virus can be reduced. Insect Science and its Application 8:797-801.

21. Storey, H.H. 1936. Virus diseases of East African plants: VI. A progress report on studies of the disease of cassava. East Africa Journal of Agricultural Science 2:34-39.

22. Thresh, J.M. 1988. Eradication as a virus disease control measure. Pages 155-194 in: Control of Plant Diseases: Costs and Benefits. B.C. Clifford and E. Lester, eds. Blackwell Scientific Publications, Oxford.

23. Thresh, J.M., Fargette, D., and Mukiibi, J. 1994. Research on African cassava mosaic virus: the need for international collaboration. Pages 271-274 in: Root Crops for Food Security in Africa. M.O. Akoroda, ed. International Society for Tropical Root Crops, CTA/IITA.

24. Thresh, J.M., Fargette, D., and Otim-Nape, G.W. 1994. Effects of African cassava mosaic geminivirus on the growth and yield of cassava. Tropical Science 34:43-54.

25. Thresh, J.M., Fargette, D. and Otim-Nape, G.W. 1994. The viruses and virus diseases of cassava in Africa. African Crop Science Journal 2:459-478.

26. Thresh, J.M., Fishpool, L.D.C., Otim-Nape, G.W., and Fargette, D. 1994. African cassava mosaic disease: an under-estimated and unsolved problem. Tropical Science 34:3-14.

27. Thresh, J.M., and Otim-Nape, G.W. 1994. Strategies for controlling African cassava mosaic geminivirus. Advances in Disease Vector Research 10:215-236.

28. Thresh, J.M., Otim-Nape, G.W., and Jennings, D.L.

1994. Exploiting resistance to African cassava mosaic virus. Aspects of Applied Biology 39:51-60.

29. Tidbury, G.E. 1937. A note on the yield of mosaic-diseased cassava. The East African Agricultural Journal 3:119.

INDEX